THE EYE IN SYSTEMIC DISEASE

THE EYE IN SYSTEMIC DISEASE

Edited by

DANIEL H. GOLD, M.D.
Associate Clinical Professor of Ophthalmology, University of Texas Medical Branch, Galveston, Texas

THOMAS A. WEINGEIST, M.D., Ph.D.
Professor and Head, Department of Ophthalmology, The University of Iowa, Iowa City, Iowa

With 320 contributors

J. B. LIPPINCOTT

PHILADELPHIA   GRAND RAPIDS   NEW YORK
ST. LOUIS   SAN FRANCISCO   LONDON   SYDNEY   TOKYO

Acquisitions Editor: Darlene Barela Cooke
Project Coordinator: Lori J. Bainbridge
Production: Spectrum Publisher Services, Inc.
Compositor: Bi-Comp, Inc.
Printer/Binder: Maple Press Company

3 5 6 4 2

Library of Congress Cataloging in Publication Data

The Eye in systemic disease / edited by Daniel H. Gold, Thomas A.
  Weingeist.
       p.   cm.
    Includes bibliographical references.
    ISBN 0-397-50722-4
    1. Ocular manifestations of general diseases.   I. Gold, Daniel H.,
    1942- . II. Weingeist, Thomas A.
    [DNLM: 1. Eye Manifestations.   WW 475 E975]
  RE65.E94   1990
  617.7′1—dc20
  DNLM/DLC   89-13618
  for Library of Congress   CIP

The authors and publisher have exerted every effort to ensure that drug selection and dosage set forth in this text are in accord with current recommendations and practice at the time of publication. However, in view of ongoing research, changes in government regulations, and the constant flow of information relating to drug therapy and drug reactions, the reader is urged to check the package insert for each drug for any change in indications and dosage and for added warnings and precautions. This is particularly important when the recommended agent is a new or infrequently employed drug.

This book is dedicated to the memory of
Samson Weingeist, M.D.: Father, Physician,
and Teacher, who brought us together

and

Paul Henkind, M.D., Ph.D.: Mentor, and
Friend, from whose search for understanding
came the genesis of this book.

Foreword

The editors have attempted to approach and discuss ocular involvement in systemic diseases from a fresh point of view, incorporating the most recent and sophisticated clinical and laboratory findings. The result has been exceedingly successful. This novel presentation and arrangement clarify many aspects of difficult problems.

The editors recruited an impressive array of experts as co-authors, and they have admirably succeeded in presenting concisely the essential features of the systemic disorder followed by a lucid discussion of the ocular involvement.

This book presents the state of the art in this field, which for the first time is being raised to the standards of the 1990s. It should be of benefit not only to ophthalmologists, but also to any other medical practitioner who wants the latest information on ocular manifestations of systemic diseases.

Frederick C. Blodi, M.D.
Professor Emeritus
The University of Iowa

Preface

Ever-increasing specialization is made necessary and inevitable by the information explosion of our times. It is, under these circumstances, easy to lose sight of the underlying interconnectedness of things. This is true not only in ophthalmology and medicine in general, but also in many other areas of life. We see it in the problems afflicting humanity on personal, local, national, and global levels.

This same information explosion has, somewhat paradoxically, also enabled us to see a more fundamental unity within the diversity. We find that medical problems that may seem different or independent when viewed at a superficial level are actually manifestations of a common underlying pathophysiologic mechanism acting simultaneously at different sites throughout the body.

The aim of this book is to examine the interrelationships between the eye and systemic disease. We wish to present the ocular complications of systemic disorders in their larger context, as local manifestations of a more widespread disease process. Toward this end, the book reviews the ocular and systemic manifestations of individual diseases with an emphasis on the interrelationships between them. A detailed discussion and description of the individual ocular and systemic complications of each disease is outside the scope of this text, as is any significant focus on their therapy. Only a multivolume encyclopedic work could possibly accomplish such a task. We wish instead to present an overview of the subject, enabling the reader to better understand *why* these complications occur, and to see underlying patterns that produce the constellation of signs and symptoms that we recognize as individual clinical disease entities.

The book is intended for our colleagues in ophthalmology and for those in the broader general medical community and allied health care fields. We hope it will be of value as a concise reference text for both the primary health care provider and the specialist in need of information about how the eye is affected by systemic diseases. There are references at the end of each chapter to provide a point of entry into the published literature for those who wish to obtain more detailed information about a particular systemic disease and its ocular manifestations.

One of the greatest challenges in creating this book was its organization. Any system of disease classification is arbitrary and changes constantly with our increasing knowledge about the diseases themselves. Many disorders can reasonably be classified in more than one category. We have attempted to adhere to fairly traditional clinical disease categories, while at the same time taking into account new information that allows the grouping of diseases previously considered to be unrelated (*e.g.*, peroxisomal disorders).

Because the scope of the book is already broad, we made a decision not to include the additional spectrum of neurologic disorders that have ocular manifestations. Of course, many systemic diseases affect the central nervous system, producing a host of neuro-ophthalmologic complications, and these are discussed under the heading of the individual disease entities.

This book is the product of the collective efforts of many hundreds of people, foremost being the contributing authors and their staffs. Our wives, Barbara and Catherine, and our children shared, in their own ways, the burden of creating the book. Special thanks go out to Dr. Bernard Milstein, Dr. Allan Fradkin, the staff of The Eye Clinic of Texas, the University of Iowa and its Ophthalmology Department and staff, and the Department of Ophthalmology of the University of Texas Medical Branch in Galveston. Without the support and encouragement of those around us, this book could never have been written. The research staff of the Moody Medical Library in Galveston provided assistance and guidance through the many long hours spent in creating the structure and outline of the book and reviewing the submitted manuscripts. The coordination and the secretarial assistance needed to create this book were provided courtesy of the talent and hard work of Susan Gray, Mary Ann Evans, Cheryl Bell, Jane Williamson, and Cindy Randolph in Galveston, and Ramona Weber and Pat Zahs in Iowa City. Ray Northway and his photographic staff at The University of Iowa provided invaluable assistance. Additional thanks go to the staff of J. B. Lippincott Publishers, and especially to our editor Darlene Cooke, whose encouragement and persistence were crucial to the completion of this project.

We are deeply indebted to our colleagues, students, and teachers, for it is through their advice, questions, presentations and patient referrals that we have come to our present state of knowledge about this field. They are part of the age-old quest for greater understanding of the diseases which affect us, and their dedication to applying their knowledge to ease the pain and suffering of their fellow human beings lies at the heart of the medical profession.

Daniel H. Gold, M.D.
Thomas A. Weingeist, M.D., Ph.D.

To Our Readers

The scope of the field covered by this text is too broad for any of us to acquire the knowledge and experience to encompass all of it. There is an ever-increasing flow of information about disease processes, their underlying mechanisms, and systemic and ocular manifestations.

We encourage our readers to become collaborators with us in this project by sharing their perspectives regarding systemic diseases with significant ocular complications that were not included in this book, as well as new information about those disorders that are included.

If this book is of sufficient value to our colleagues to warrant the publication of another edition, we will make certain the information from our readers is transmitted to those who have accepted the responsibility for creating new chapters or revising and updating existing ones.

Daniel H. Gold, M.D.
Thomas A. Weingeist, M.D., Ph.D.

Contents

Part 8 Infectious Diseases   155

Part 9   Inflammatory Diseases of Unknown Etiology   277

Part 10   Malignant Disorders   297

Part 11   Metabolic Disorders   309

A. Disorders of Amino Acid Metabolism

Part 12   Muscular Disorders   415

Part 13   Phakomatoses   437

Part 14   Physical and Chemical Injury   459

Part 19   Skin and Mucous Membrane Disorders   581

Part 20   Vascular Disorders   645

Part 21   Vitamin and Nutritional Disorders

673

Contributors

Everett Ai, M.D.
Director, Retina Unit
Department of Ophthalmology
Pacific Presbyterian Medical Center;
Assistant Clinical Professor
University of California, Berkeley
San Francisco, California

Daniel M. Albert, M.D.
Director, David G. Cogan Eye Pathology Laboratory;
David G. Cogan Eye Pathology Professor
Harvard Medical School
Massachusett's Eye and Ear Infirmary
Boston, Massachusetts

Duncan P. Anderson, M.D.C.M., F.R.C.S.(C)
Associate Professor of Ophthalmology
McGill University
Montreal, Quebec, Canada

Karl E. Anderson, M.D.
Professor of Preventive Medicine and Community
    Health, Internal Medicine, and Pharmacology and
    Toxicology
University of Texas Medical Branch
Galveston, Texas

David J. Apple, M.D.
Professor and Chairman
Department of Ophthalmology;
Vallotton Chair of Biomedical Engineering
Albert Florens Storm Eye Institute and Medical
    University of South Carolina
Charleston, South Carolina

Penny A. Asbell, M.D.
Associate Professor of Ophthalmology
Department of Ophthalmology
Mount Sinai Hospital
New York, New York

Marcos Avila, M.D.
Retina Service Director
Goiania Hospital
Goiania–Go, Brazil

Ann Sullivan Baker, M.D.
Associate Physician, Infectious Disease Unit;
Consultant, Infectious Diseases
Massachusetts Eye and Ear Infirmary;
Assistant Professor
Harvard Medical School
Boston, Massachusetts

Susan Barkay, M.D.
Head, Department of Ophthalmology
Central Emek Hospital
Afula, Israel

Norman W. Barton, M.D., Ph.D.
Development and Metabolic Neurology Branch
National Institute of Neurological Diseases and Stroke
Bethesda, Maryland

Jules Baum, M.D.
Professor of Ophthalmology
Tufts University School of Medicine;
Senior Surgeon
New England Medical Center
Boston, Massachusetts

George R. Beauchamp, M.D.
Pediatric Ophthalmology/Strabismus
The Cleveland Clinic Foundation
Cleveland, Ohio

Denys Beauvais, M.D.
Department of Ophthalmology
University of Wisconsin
Madison, Wisconsin

William E. Bell, M.D.
Departments of Pediatrics and Neurology
University of Iowa College of Medicine
Iowa City, Iowa

Jose Berrocal, M.D.
Professor of Ophthalmology
Department of Ophthalmology
University of Puerto Rico
San Juan, Puerto Rico

Robert F. Betts, M.D.
Associate Professor of Medicine, Infectious Disease
  Unit
University of Rochester
School of Medicine
Rochester, New York

Christopher F. Blodi, M.D.
Assistant Professor, Department of Ophthalmology
The University of Iowa
Iowa City, Iowa

Frederick C. Blodi, M.D.
Professor Emeritus, Department of Ophthalmology
The University of Iowa
Iowa City, Iowa

Stephen J. Bogan, M.D.
Clinical Associate, Department of Ophthalmology
Emory University
Atlanta, Georgia

William P. Boger, III, M.D.
Emerson Hospital
Concord, Massachusetts;
Children's Hospital Medical Center
Harvard Medical School
Boston, Massachusetts

James P. Bolling, M.D.
Consultant, Mayo Clinic, Jacksonville;
Assistant Professor, Mayo Graduate School
  of Medicine;
Clinical Instructor
University of Florida College of Medicine
Jacksonville, Florida

William A. Boothe, M.D.
Texas Eye Care
Plano, Texas

Dennis W. Boulware, M.D.
Associate Professor of Medicine;
Chief, Section of Rheumatology
Tulane University School of Medicine
New Orleans, Louisiana

R. Wayne Bowman, M.D.
Assistant Professor
Department of Ophthalmology
University of Texas Southwest Medical Center
  at Dallas
Dallas, Texas

Gary C. Brown, M.D.
Wills Eye Hospital
Retina Vascular Unit;
Professor of Ophthalmology
Thomas Jefferson University
Philadelphia, Pennsylvania

Seymour Brownstein M.D., C.M., F.R.C.S.(C)
Professor, Departments of Ophthalmology and
  Pathology;
Director, Ophthalmic Pathology;
Faculty of Medicine, McGill University;
Senior Ophthalmologist and Assistant Pathologist,
  Royal Victoria Hospital;
Consultant, Montreal Children's Hospital;
Consultant, Sir Mortimer Davis Jewish General
  Hospital
Montreal, Quebec, Canada

Helmut Buettner, M.D.
Associate Professor, Department of Ophthalmology
Mayo Medical School
Mayo Clinic
Rochester, Minnesota

John D. Bullock, M.D., M.S., F.A.C.S.
Professor and Chairman
Department of Ophthalmology
Wright State University School of Medicine
Dayton, Ohio

Ronald M. Burde, M.D., F.A.C.S.
Professor and Chairman
Department of Ophthalmology
Albert Einstein College of Medicine
Montefiore Medical Center
Bronx, New York

John A. Burns, M.D.
Clinical Professor, Department of Ophthalmology
Children's Hospital of Columbus
The Ohio State University
Columbus, Ohio

Robert P. Burns, M.D.
Professor and Chairman
Department of Ophthalmology
University of Missouri–Columbia
  Medical School
Columbia, Missouri

W. Laurence Burt, M.D., F.R.C.S.(C)
Associate Professor, Department of Ophthalmology
University of Western Ontario;
Chairman, Department of Ophthalmology
St. Joseph's Health Centre of London
London, Ontario, Canada

Thomas C. Burton, M.D.
Professor, Vice-Chairman
Department of Ophthalmology
Medical College of Wisconsin, Milwaukee
Milwaukee, Wisconsin

Frank K. Butler, Jr., M.D.
Chief Resident, Ophthalmology
Department of Ophthalmology
National Naval Medical Center
Bethesda, Maryland

Kenneth V. Cahill, M.D.
Clinical Assistant Professor
Department of Ophthalmology
Children's Hospital of Columbus
The Ohio State University
Columbus, Ohio

Joseph H. Calhoun, M.D., F.A.C.S.
Director, Pediatric Ophthalmology and Strabismus
Wills Eye Hospital;
Clinical Professor of Ophthalmology
Thomas Jefferson University
Philadelphia, Pennsylvania

J. Douglas Cameron, M.D.
Associate Professor
Department of Ophthalmology
Department of Laboratory Medicine and Pathology
University of Minnesota
Minneapolis, Minnesota

R. Jean Campbell, M.D., Ch.B.
Professor of Ophthalmology, Departments of
    Pathology and Ophthalmology
Mayo Graduate School of Medicine
Rochester, Minnesota

Herbert L. Cantrill, M.D.
Clinical Associate Professor
Department of Ophthalmology
University of Minnesota
Minneapolis, Minnesota

Ronald E. Carr, M.D.
Professor of Ophthalmology
New York University Medical Center
New York, New York

Robert P. Castleberry, M.D.
Professor of Pediatrics
Director, Pediatric Hematology/Oncology
University of Alabama at Birmingham
Birmingham, Alabama

Chi-Chao Chan, M.D.
Medical Officer
Laboratory of Immunology
National Eye Institute
Bethesda, Maryland

John W. Chandler, M.D.
Professor and Chairman
Department of Ophthalmology
University of Wisconsin
Madison, Wisconsin

Devron H. Char, M.D.
Director, Ocular Oncology Unit
Professor, Department of Ophthalmology, Radiation
    Oncology, and the Francis I. Proctor Foundation
University of California, San Francisco
San Francisco, California

Alan Y. Chow, M.D.
Clinical Assistant Professor
Loyola Stritch School of Medicine
Maywood, Illinois

Andrew E. Choy, M.D.
Associate Clinical Professor of Ophthalmology
University of California, Los Angeles
School of Medicine
Long Beach, California

Georgia A. Chrousos, M.D.
Associate Professor
Center for Sight
Georgetown University Medical Center
Washington, D.C.

Elaine L. Chuang, M.D.
Assistant Professor
Bascom Palmer Eye Institute
University of Miami
Miami, Florida

Gerhard W. Cibis, M.D.
Chief of Ophthalmology
Childrens Mercy Hospital;
Clinical Associate Professor
Kansas University School of Medicine
Kansas City, Missouri

Michael Cobo, M.D.
Associate Professor, Department of Ophthalmology
Duke University Medical Center
Durham, North Carolina

Steven B. Cohen, M.D.
Clinical Assistant Professor
Department of Ophthalmology
UMD—New Jersey Medical School
Vitreoretinal Associates of New Jersey
Belleville, New Jersey

Candace C. Collins, M.D.
Slidell Memorial Hospital
Slidell, Louisiana
Children's Hospital
Louisiana State University Eye Center
New Orleans, Louisiana

Patrick Coonan, M.D.
Consultant, Retina Unit
Department of Ophthalmology
Pacific Presbyterian Medical Center
San Francisco, California

James J. Corbett, M.D.
Professor, Departments of Neurology and
    Ophthalmology
University of Iowa College of Medicine
Iowa City, Iowa

Alfred J. Cossari, M.D.
Assistant Professor of Clinical Surgery
    (Ophthalmology)
State University of New York at Stony Brook
Stony Brook, New York

J. Brooks Crawford, M.D.
Clinical Professor, Department of Ophthalmology;
Director, Hogan Eye Pathology Laboratory
Beckman Vision Center
University of California
San Francisco, California

J. Oscar Croxatto, M.D.
Chairman, Department of Pathology
Fundacion Oftalmologica Argentina
    Jorge Malbran
Buenos Aires, Argentina

Matthew E. Dangel, M.D.
Assistant Professor, Department of Ophthalmology
The Ohio State University
Columbus, Ohio

Matthew D. Davis, M.D.
Professor of Ophthalmology
University of Wisconsin
Madison, Wisconsin

Chandler R. Dawson, M.D.
Professor of Ophthalmology
University of California, San Francisco;
Director, Francis I. Proctor Foundation
San Francisco, California

Gary G. Deckelboim, M.D.
Clinical Instructor, Department of Ophthalmology
Mount Sinai School of Medicine
New York, New York;
Attending Ophthalmologist
General Hospital Center
Passaic, New Jersey

Jean-Jacques De Laey, M.D., Ph.D.
Professor and Chairman
Department of Ophthalmology
University Hospital
Ghent, Belgium

J. W. Delleman
Professor, Ophthalmology and Genetics
Netherlands Ophthalmic Research Institute
University Medical Center
The Netherlands

James A. Deutsch, M.D.
Assistant Clinical Professor of Ophthalmology
Mount Sinai School of Medicine
New York, New York

Gary Diamond, M.D.
Associate Professor of Ophthalmology and Pediatrics
Hahnemann University School of Medicine
Philadelphia, Pennsylvania

Kathleen B. Digre, M.D.
Assistant Professor, Departments of Neurology and
    Ophthalmology
University of Utah
Salt Lake City, Utah

William J. Dinning, F.R.C.S., M.R.C.P., D.O.
Consultant Ophthalmogist
West Middlesex and Ealing Hospitals;
Honorary Consultant Ophthalmologist,
Moorfields Eye Hospital,
London, England

Eric D. Donnenfeld, M.D.
Clinical Associate Professor of Ophthalmology
Cornell University Medical College;
Co-Director, External Diseases Department
Manhattan Eye, Ear and Throat Hospital
New York, New York

Paul B. Donzis, M.D.
Clinical Instructor in Ophthalmology
Jules Stein Eye Institute
University of California at Los Angeles School of
   Medicine
Los Angeles, California

Donald J. Doughman, M.D., F.A.C.S.
Professor and Chairman,
Department of Ophthalmology
University of Minnesota
Minneapolis, Minnesota

Arlene V. Drack, M.D.
Resident in Ophthalmology
Center for Sight
Georgetown University Medical Center
Washington, D.C.

Neil Dreizen, M.D.
Department of Surgery, Ophthalmology Subdivision
Southern Ocean County Hospital
Manahawkin, New Jersey;
Courtesy Staff
Wills Eye Hospital
Philadelphia, Pennsylvania

Richard F. Dreyer, M.D.
Oregon Lions Sight and Hearing Institute
Portland, Oregon

F. Jane Durcan, M.D.
Assistant Professor
Department of Ophthalmology
University of Utah
Salt Lake City, Utah

George S. Ellis, Jr., M.D.
Associate Professor, Ophthalmology;
Clinical Associate Professor, Pediatrics
Tulane University Medical Center;
Director, Tulane Pediatric Ophthalmology
Children's Hospital
New Orleans, Louisiana

Frederick J. Elsas, M.D.
Associate Professor Clinical Ophthalmology
University of Alabama at Birmingham
Eye Foundation Hospital
Birmingham, Alabama

Lawrence S. Evans, M.D., Ph.D.
Loyola University Medical Center
Maywood, Illinois
Hines VA Hospital
Hines, Illinois

R. Linsy Farris, M.D.
Professor of Clinical Ophthalmology
College of Physicians and Surgeons, Columbia
   University;
Attending, Edward S. Harkness Eye Institute
Presbyterian Hospital;
Director, Department of Ophthalmology
Harlem Hospital Center
New York, New York

David S. Felder, M.D.
Chief Ophthalmic Plastic Surgeon
Humana Hospital—Cypress
Fort Lauderdale, Florida

Andrew P. Ferry, M.D.
Professor and Chairman
Department of Ophthalmology;
Professor of Pathology
Medical College of Virginia of Virginia Commonwealth
   University
Richmond, Virginia

Timothy ffytche, FRCS
Ophthalmic Surgeon
St. Thomas Hospital and Hospital for Topical Diseases
London, England

**Alistair R. Fielder, M.B.B.S., F.R.C.S.,
   E.C.Ophth.**
Professor of Ophthalmology
University of Birmingham
Birmingham, United Kingdom

Michele R. Filling-Katz, M.D.
Assistant Professor, Departments of Neurology and
   Ophthalmology
University of Louisville
Louisville, Kentucky

John W. Floberg, M.D.
Noran Neurological Clinic
Minneapolis, Minnesota

George J. Florakis, M.D.
Instructor in Clinical Ophthalmology
Edward S. Harkness Eye Institute
Columbia-Presbyterian Medical Center
New York, New York

Robert Folberg, M.D.
Associate Professor of Ophthalmology and Pathology;
Director, Eye Pathology Laboratory
The University of Iowa College of Medicine
Iowa City, Iowa

Joseph Z. Forstot, M.D., F.A.C.P., F.A.C.R.
Boca Raton Community Hospital
Boca Raton, Florida

S. Lance Forstot, M.D., F.A.C.S.
Associate Clinical Professor
Department of Ophthalmology
University of Colorado School of Medicine
Denver, Colorado

C. Stephen Foster, M.D., F.A.C.S.
Director, Immunology Service;
Director, Hilles Immunology Laboratory
Massachusetts Eye and Ear Infirmary;
Associate Professor of Ophthalmology
Harvard Medical School
Boston, Massachusetts

Thomas D. France, M.D.
Professor, Department of Ophthalmology
Director, Pediatric Ophthalmology and Strabismus
University of Wisconsin Medical School
Madison, Wisconsin

George T. Frangieh, M.D.
Chief Resident/Clinical Instructor
Department of Ophthalmology
University of Southern California
Doheny Eye Institute
Los Angeles, California

Rudolf M. Franklin, M.D.
Professor of Ophthalmology
Louisiana State University Medical Center
New Orleans, Louisiana

Alan H. Friedman, M.D.
Clinical Professor of Ophthalmology and Pathology
Mount Sinai School of Medicine
New York, New York

Peter D. Fries, M.D.
Assistant Clinical Professor
University of California;
Director, Oculoplastics and Orbital Service
Letterman Army Medical Center
San Francisco, California

Alec Garner, M.D., Ph.D., F.R.C.P., F.R.C.Path., F.C.Ophth.
Professor and Director of Pathology
Institute of Ophthalmology
University of London;
Honorary Consultant in Pathology
Moorfields Eye Hospital
London, England

James A. Garrity, M.D.
Consultant in Ophthalmology
Mayo Clinic and Mayo Foundation;
Assistant Professor in Ophthalmology
Mayo Medical School
Rochester, Minnesota

Hanna J. Garzozi, M.D.
Deputy Head, Department of Ophthalmology
Central Emek Hospital
Afula, Israel

David Gendelman
Clinical Instructor
Massachusetts Eye and Ear Infirmary
Boston, Massachusetts

Mrinmay Ghosh, M.D., F.R.C.S.C.
Associate Professor of Ophthalmology
University of Toronto;
Senior Ophthalmologist
Toronto General Hospital
Toronto, Ontario, Canada

Conrad L. Giles, M.D.
Co-Chief, Department of Ophthalmology
Children's Hospital of Michigan;
Clinical Professor of Ophthalmology
Kresge Eye Institute
Wayne State University School of Medicine
Detroit, Michigan

Daniel H. Gold, M.D.
Associate Clinical Professor of Ophthalmology
University of Texas Medical Branch
Eye Clinic of Texas
Galveston, Texas

Richard E. Goldberg, M.D., F.A.C.S.
Clinical Professor of Ophthalmology
Jefferson Medical College of Thomas Jefferson
    University;
Associated Surgeon, Retina Service and Co-Director,
    Retinal Vascular Unit Wills Eye Hospital;
Philadelphia, Pennsylvania

Rosalie B. Goldberg, M.S.
Senior Associate, Department of Pediatrics and Plastic
 Surgery
Center for Congenital Disorders
Montefiore Medical Center
Albert Einstein College of Medicine
Bronx, New York

Edward Jay Goldman, M.D.
Clinical Assistant Professor
Department of Ophthalmology
University of Maryland
Baltimore, Maryland

W. Richard Green, M.D.
Professor of Ophthalmology
Associate Professor of Pathology
The Johns Hopkins University School of Medicine
Baltimore, Maryland

Arthur S. Grove, Jr., M.D.
Assistant Professor of Ophthalmology
Harvard Medical School;
Associate Surgeon in Ophthalmology
Massachusetts Eye and Ear Infirmary
Boston, Massachusetts

Barrett G. Haik, M.D., F.A.C.S.
Professor of Ophthalmology
Tulane University Medical School
New Orleans, Louisiana

Paul W. Hardwig, M.D.
Instructor in Ophthalmology
Mayo Medical School
Mayo Clinic, Department of Ophthalmology
Rochester, Minnesota

**Sohan Singh Hayreh, M.D., Ph.D., D.Sc.,
 F.R.C.S.**
Professor of Ophthalmology
Director, Ocular Vascular Unit
Department of Ophthalmology
University of Iowa Hospitals and Clinics
Iowa City, Iowa

Thomas R. Hedges, Jr., M.D.
Chief of Ophthalmology
Pennsylvania Hospital
University of Pennsylvania
Philadelphia, Pennsylvania

Thomas R. Hedges, III, M.D.
Associate Professor
Department of Ophthalmology
Tufts New England Medical Center
Boston, Massachusetts

Robert W. Hered, M.D.
Chief, Division of Pediatric Ophthalmology
Nemours Children's Clinic
Jacksonville, Florida

Roger L. Hiatt, M.D.
Professor and Chairman
Department of Ophthalmology
University of Tennessee, Memphis
Memphis, Tennessee

David A. Hiles, M.D.
Clinical Professor of Ophthalmology
University of Pittsburgh
Children's Hospital and Eye and Ear Hospitals
 of Pittsburgh
Pittsburgh, Pennsylvania

Edward J. Holland, M.D.
Assistant Professor
Director, Cornea and External Disease
Department of Ophthalmology
University of Minnesota
Minneapolis, Minnesota

Gary N. Holland, M.D.
Assistant Professor of Ophthalmology
Jules Stein Eye Institute
University of California, Los Angeles
School of Medicine
Los Angeles, California

LTC Donald A. Hollsten, M.D., USA
Assistant Chief of Ophthalmology
Director of Ophthalmic Plastic and Reconstructive
 Surgery
Department of Ophthalmology
Brooke Army Medical Center
San Antonio, Texas

Jonathan C. Horton, M.D., Ph.D.
Department of Neurological Surgery
Neuro-Ophthalmology Unit
University of California, San Francisco
San Francisco, California

Rufus O. Howard, M.D., Ph.D.
Clinical Professor of Ophthalmology
Yale Medical School
New Britain, Connecticut

Edward L. Howes, Jr., M.D.
Professor of Pathology and Ophthalmology
Vice-Chairman of Pathology
School of Medicine,
University of California, San Francisco
Chief of Anatomic Pathology
San Francisco General Hospital
San Francisco, California

Michael S. Insler, M.D.
Associate Professor of Ophthalmology
Director, Contact Lens Service
Louisiana State University Medical Center
Louisiana State University Eye Center
New Orleans, Louisiana

W. Bruce Jackson, M.D., F.R.C.S.(C)
Ophthalmologist-in-Chief
Royal Victoria Hospital;
Professor and Chairman
Department of Ophthalmology
McGill University
Montreal, Quebec, Canada

Samuel G. Jacobsen, M.D., Ph.D.
Assistant Professor
Department of Ophthalmology
Bascom Palmer Eye Institute
University of Miami School of Medicine
Miami, Florida

Alex E. Jalkh, M.D.
Clinical Associate Scientist
Eye Research Institute of Retina Foundation;
Assistant in Ophthalmology
Massachusetts Eye and Ear Infirmary
Harvard Medical School
Boston, Massachusetts

Ira Snow Jones, M.D.
Professor Emeritus
Harkness Eye Institute
Presbyterian Hospital
Columbia University
Manhattan Eye, Ear, and Throat Hospital
New York, New York

Jeffrey Josef, M.D.
Fellow in Retina and Vitreous
Albert Einstein College of Medicine
Montefiore Medical Center
Bronx, New York

G. Frank Judisch, M.D.
Professor of Ophthalmology
Department of Ophthalmology
The University of Iowa Hospitals and Clinics
Iowa City, Iowa

Muriel I. Kaiser-Kupfer, M.D.
Chief, Ophthalmic Genetics and Clinical Services
  Branch
National Eye Institute
Bethesda, Maryland

Robert E. Kalina, M.D.
Professor and Chairman
Department of Ophthalmology
University of Washington
Seattle, Washington

Geoffrey R. Kaplan, M.D.
Attending Staff, Washington National Eye Center
Washington Hospital Center
Washington, D.C.;
Clinical Assistant Professor
Howard University College of Medicine
Washington, D.C.

Zeynel A. Karcioglu, M.D.
Professor of Ophthalmology and Pathology
Tulane University School of Medicine
New Orleans, Louisiana

James A. Katowitz, M.D.
Associate Professor
University of Pennsylvania
School of Medicine;
Director, Ophthalmic Plastic and Craniofacial
  Reconstructive Surgery
Children's Hospital of Philadelphia
Philadelphia, Pennsylvania

Norman N. K. Katz, M.D., F.R.C.S.(C.)
Professor, Department of Ophthalmology and Visual
  Sciences
University of Louisville School of Medicine
Louisville, Kentucky

Thomas P. Kearns, M.D.
Emeritus Consultant in Ophthalmology
Mayo Clinic;
Emeritus Fred C. Anderson Professor of
  Ophthalmology
Mayo Medical School
Rochester, Minnesota

Ronald V. Keech, M.D.
Assistant Professor
Department of Ophthalmology
The University of Iowa
Iowa City, Iowa

Rodney I. Kellen, M.D., F.R.C.S.(S.A.)
Assistant Professor of Ophthalmology
College of Medicine
University of Saskatchewan
Saskatoon, Saskatchewan, Canada

Nancy G. Kennaway, Ph.D.
Professor of Medical Genetics
Oregon Health Sciences University
Portland, Oregon

Kenneth Kenyon, M.D.
Cornea and External Disease
Massachusetts Eye and Ear Infirmary
Boston, Massachusetts;
Associate Professor
Harvard Medical School
Boston, Massachusetts

Mourad K. Khalil, M.D., F.R.C.S.(C)
Associate Professor of Ophthalmology
Faculty of Medicine
McGill University;
Senior Ophthalmologist
Assistant Ophthalmic Pathologist
Department of Pathology
Montreal General Hospital;
Chief, Subdepartment of Ophthalmology
Queen Elizabeth Hospital
Montreal, Quebec, Canada

Marilyn C. Kincaid, M.D.
Associate Professor
St. Louis University,
Bethesda Eye Institute
St. Louis, Missouri

Cleveland Kirkland, Jr., M.D.
Assistant Clinical Professor of Ophthalmology
Texas Tech University Health Sciences Center
San Angelo, Texas

Gordon K. Klintworth, M.D., Ph.D.
Departments of Pathology and Ophthalmology
Duke University Medical Center
Durham, North Carolina

David L. Knox, M.D.
Associate Professor
The Johns Hopkins University
School of Medicine
Baltimore, Maryland

Roger Kohn, M.D.
Professor, Department of Ophthalmology
Division of Ophthalmic Plastic and Reconstructive
  Surgery
University of California, Los Angeles
School of Medicine
Santa Barbara, California

Jay H. Krachmer, M.D.
Professor, Department of Ophthalmology
Director, Cornea/External Disease Service
The University of Iowa
Director, Iowa Lions Cornea Center
Iowa City, Iowa

Stephen P. Kraft, M.D., F.R.C.S.(C)
Assistant Professor, Department of Ophthalmology
University of Toronto;
Staff Ophthalmologist
Hospital for Sick Children
Toronto, Ontario, Canada

Frank L. Kretzer, Ph.D.
Associate Professor of Ophthalmology and Cell Biology
Cullen Eye Institute
Baylor College of Medicine
Houston, Texas

Gregory B. Krohel, M.D.
Professor of Ophthalmology and Neurology
Albany Medical College
Albany, New York

Jan W. Kronish, M.D.
Assistant Professor
Bascom Palmer Eye Institute
University of Miami School of Medicine
Miami, Florida

Mark J. Kupersmith, M.D.
Professor of Ophthalmology and Neurology
New York University;
Director, Neuro-ophthalmology
New York Eye and Ear Infirmary
New York, New York

Burton J. Kushner, M.D.
Clinical Professor
Department of Ophthalmology
University of Wisconsin
Madison, Wisconsin

Moshe Lahav, M.D.
Professor of Ophthalmology
Tufts University School of Medicine;
Chief of Ophthalmology
Boston Veterans Administration Medical Center
Boston, Massachusetts

Peter R. Laibson, M.D.
Professor of Ophthalmology
Director, Cornea and External Disease
Thomas Jefferson University;
Wills Eye Hospital
Philadelphia, Pennsylvania

H. Michael Lambert, M.D., F.A.C.S., LTC, USAF MC
Wilford Hall, USAF Medical Center;
University of Texas Health Science Center
San Antonio, Texas;
Uniformed Services University of Health Sciences
Bethesda, Maryland

Jeffrey Day Lanier, M.D.
Clinical Associate Professor of Ophthalmology
University of Texas Health Sciences Center
Houston, Texas

Mary Ann Lavery, M.D.
Assistant Professor of Ophthalmology
University of Texas Medical Branch
Galveston, Texas

David A. Lee, M.D.
Assistant Professor of Ophthalmology
Jules Stein Eye Institute
University of California, Los Angeles
School of Medicine
Los Angeles, California

Michael A. Lemp, M.D.
Professor and Chairman
Department of Ophthalmology
Director/Center for Sight
Georgetown University Medical Center
Washington, D.C.

Robert M. Lewen, M.D.
Department of Ophthalmology
Allegheny General Hospital
Pittsburgh, Pennsylvania

Richard Alan Lewis, M.D.
Associate Professor
Cullen Eye Institute
Baylor College of Medicine
Houston, Texas

Ruth M. Liberfarb, M.D., Ph.D.
Retina Associates
Eye Research Institute of Retina Foundation
Massachusetts Eye and Ear Infirmary
Massachusetts General Hospital
The Children's Hospital
Harvard Medical School
Boston, Massachusetts

Jacques Libert, M.D.
Professor of Ophthalmology
Universite Libre de Bruxelles;
Head, Department of Ophthalmology
Hospital St. Pierre
Brussels, Belgium

Thomas J. Liesegang, M.D.
Associate Professor of Ophthalmology
Mayo Clinic—Jacksonville
Jacksonville, Florida

Alan D. Listhaus, M.D.
Fellow, Retina Service
Department of Ophthalmology
University of California at San Diego
San Diego, California

Sornchai Looareesuwan, M.D., D.T.M. & H.
Associate Professor, Department of Clinical Tropical Medicine
Hospital for Tropical Disease
Faculty of Tropical Medicine
Mahidol University
Bangkok, Thailand

Irene H. Ludwig, M.D.
Staff Attending Ophthalmologist
The Mary Imogene Bassett Hospital
Cooperstown, New York;
Clinical Assistant Professor of Ophthalmology
Albany Medical College
Albany, New York
Instructor in Clinical Ophthalmology
Columbia University, New York

Bryant Lum, M.D.
Department of Ophthalmology
Alton Ochsner Medical Foundation
New Orleans, Louisiana

Larry E. Magargal, M.D., F.A.C.S., F.I.C.S.
Co-Director Retina Vasular Unit
Associate Surgeon Retina Service
Wills Eye Hospital;
Associate Professor of Ophthalmology
Thomas Jefferson University
Philadelphia, Pennsylvania

Raga Malaty, M.D., Ph.D.
Professor of Ophthalmology
Louisiana State University Medical Center
Louisiana State University Eye Center
New Orleans, Louisiana

Mark J. Mannis, M.D., F.A.C.S.
Associate Professor
Director, Cornea and External Disease Service
Department of Ophthalmology
University of California, Davis
Sacramento, California

Ahmad M. Mansour, M.D.
Assistant Professor
University of Texas Medical Branch
Galveston, Texas

Sorana Marcovitz, M.D.C.M., F.R.C.P.C.
Associate Professor Internal Medicine (Endocrinology)
McGill University
Montreal, Quebec, Canada

Curtis E. Margo, M.D.
Professor of Ophthalmology and Pathology
University of Florida;
Chief, Section of Ophthalmology
Veterans Hospital
Gainesville, Florida

A. Martin-Casals, M.D., F.A.C.S.
Associate Professor
Ponce School of Medicine
Ponce, Puerto Rico

John Matthews, M.D.
Southeastern Eye Center
Greensboro, North Carolina

Alice Y. Matoba, M.D.
Chief, Ophthalmology Service
Houston Veterans Administration Medical Center;
Assistant Professor of Ophthalmology
Cullen Eye Institute
Baylor College of Medicine
Houston, Texas

Thomas F. Mauger, M.D.
Assistant Professor, Department of Ophthalmology
Southern Illinois University
Springfield, Illinois

Craig I. McClain, M.D.
Director, Division of Digestive Diseases and Nutrition
University of Kentucky Medical Center
Lexington UAMC
Lexington, Kentucky

James P. McCulley, M.D.
Professor and Chairman
Department of Ophthalmology
University of Texas Southwestern Medical Center
   at Dallas
Dallas, Texas

Clement McCulloch, M.D.
Emeritus Professor of Ophthalmology
University of Toronto
Toronto, Canada

Peter J. McDonnell, M.D.
Assistant Professor
Department of Ophthalmology
University of Southern California
Doheny Eye Institute
Los Angeles, California

Shauna K. McKusker, M.D.
Cheyenne Eye Clinic
Cheyenne, Wyoming

William McLeish, M.D.
Bascom Palmer Eye Institute
University of Miami
Miami, Florida

David M. Meisler, M.D.
Department of Ophthalmology
Cleveland Clinic Foundation
Cleveland, Ohio

Travis A. Meredith, M.D.
Professor of Ophthalmology
Emory University School of Medicine
Atlanta, Georgia

Marilyn B. Mets, M.D.
Assistant Professor
University of Chicago;
Chief of Pediatric Ophthalmology and Ophthalmic
   Genetics
Michael Reese Hospital and University of Chicago
   Hospitals
Chicago, Illinois

Joseph B. Michelson, M.D., F.A.C.S.
Head, Division of Ophthalmology
Director, Retina-Uveitis Service
Scripps Clinic Medical Group, Inc.
La Jolla, California

Michael E. Migliori, M.D.
Clinical Assistant Professor of Ophthalmology
Brown University Program in Medicine
Providence, Rhode Island

Marilyn T. Miller, M.D.
Clinical Professor
University of Illinois at Chicago
Eye and Ear Infirmary
Chicago, Illinois

Neil R. Miller, M.D., F.A.C.S.
Frank B. Walsh Professor of Neuro-ophthalmology
Professor of Ophthalmology, Neurology and
    Neurosurgery
The Johns Hopkins Medical Institutions
Balitmore, Maryland

Joel S. Mindel, M.D., Ph.D.
Professor of Ophthalmology
Associate Professor, Pharmacology
Mount Sinai School of Medicine
New York, New York

Helen Mintz-Hittner, M.D.
Clinical Professor, Departments of Ophthalmology and
    Pediatrics
Baylor College of Medicine
Houston, Texas

Manabu Mochizuki, M.D., Ph.D.
Associate Professor
Department of Ophthalmology
Tokyo University Branch Hospital
Tokyo, Japan

Bartly J. Mondino, M.D.
Professor of Ophthalmology
Jules Stein Eye Institute
University of California, Los Angeles
School of Medicine
Los Angeles, California

Adolfo Gomez Morales, M.D.
Consultant, Department of Orbit and Neuro-
    ophthalmology
Hospital Nacional de Oftalmologia Santa Lucia
Buenos Aires, Argentina

Craig M. Morgan, M.D.
Assistant Clinical Professor, Departments of Surgery
    and Pediatrics
Marshall University School of Medicine;
Huntington Eye Associates
Huntington, West Virginia

Donald A. Morris, M.D.
Associate Clinical Professor of Ophthalmology
Albert Einstein College of Medicine;
Attending Surgeon, New York Eye and Ear Infirmary;
Associate Attending, Montefiore Hospital;
Director of Ophthalmology
North Center Bronx Hospital
Bronx, New York

Robert P. Murphy, M.D.
Associate Professor of Ophthalmology
Associate Director
The Retinal Vascular Center
Johns Hopkins University
Baltimore, Maryland

Jeffrey C. Murray, M.D.
Associate Professor in Pediatrics
Division of Medical Genetics
University of Iowa Hospitals and Clinics
Iowa City, Iowa

Maria A. Musarella, M.D. F.R.C.S.(C)
Ophthalmic Geneticist and Director
Retinitis Pigmentosa Unit;
Research Associate, Departments of Ophthalmology
    and Genetics
Hospital for Sick Children;
Assistant Professor, Department of Ophthalmology
University of Toronto
Toronto, Ontario, Canada

Leonard B. Nelson, M.D.
Co-Director Pediatric Ophthalmology
Wills Eye Hospital;
Associate Professor of Ophthalmology and Pediatrics
Thomas Jefferson University Medical College
Philadelphia, Philadelphia

Jeffrey A. Nerad, M.D., F.A.C.S.
Director of Oculoplastic and Orbital Surgery
Associate Professor, Department of Ophthalmology
The University of Iowa
Iowa City, Iowa

David A. Newsome, M.D.
Clinical Professor of Ophthalmology
Tulane University School of Medicine;
Director, Sensory and Electrophysiology Research Unit
Touro Infirmary
New Orleans, Louisiana

Erik Niebuhr, M.D., Ph.D.
Associate Professor of Medical Genetics
University of Copenhagen
Copenhagen, Denmark

Verinder S. Nirankari, M.D., F.A.C.S.
Associate Professor
University of Maryland Hospital
Homewood Hospital Center South
Mercy Medical Center
Baltimore, Maryland

Kenneth G. Noble, M.D.
Associate Professor of Clinical Ophthalmology
New York University Medical Center
New York, New York

Reijo Norio, M.D.
Director, Department of Medical Genetics
Vaestoliitto, The Finnish Population and Family
    Welfare Federation;
Assistant Professor in Clinical Genetics
University of Helsinki
Helsinki, Finland

Robert Nozik, M.D.
Research Ophthalmologist
Proctor Foundation
Clinical Professor of Ophthalmology
University of California at San Francisco
San Francisco, California

Guy E. O'Grady, M.D.
Professor of Ophthalmology
Bascom Palmer Eye Institute
University of Miami School of Medicine
Miami, Florida

John F. O'Neill, M.D.
Associate Professor of Ophthalmology and Pediatrics
Georgetown University School of Medicine
Washington, D.C.

James C. Orcutt, M.D., Ph.D.
Associate Professor, Department of Ophthalmology;
Adjunct Associate Professor, Otolaryngology, Head
    and Neck Surgery
University of Washington;
Chief of Ophthalmology Section
Seattle Veterans Administration Medical Center
Seattle, Washington

Juan Orellana, M.D.
Assistant Professor
Mount Sinai School of Medicine
New York Eye and Ear Infirmary
Beth Israel Medical Center
New York, New York

David M. Orenstein, M.D.
Associate Professor of Pediatrics
University of Pittsburgh School of Medicine;
Director, Cystic Fibrosis Clinic
Children's Hospital of Pittsburgh
Pittsburgh, Pennsylvania

Shivanand R. Patil, Ph.D.
Division of Medical Genetics
Department of Pediatrics
University of Iowa Hospitals and Clinics
Iowa City, Iowa

Robert L. Peiffer, Jr., D.V.M., Ph.D.
Professor and Director of Laboratories
Department of Ophthalmology and Pathology
University of North Carolina
Chapel Hill, North Carolina

Jay S. Pepose, M.D., Ph.D.
Associate Professor of Ophthalmology
Assistant Professor of Pathology
Washington University School of Medicine
St. Louis, Missouri

Edward S. Perkins, M.D., Ph.D.
Professor Emeritus
Department of Ophthalmology
The University of Iowa
Iowa City, Iowa

Henry D. Perry, M.D.
Associate Professor of Ophthalmology
Cornell University Medical College
New York, New York;
Chief, Cornea Service
Department of Ophthalmology
North Shore University Hospital
Manhasset, New York

Gholam A. Peyman, M.D.
Professor, Department of Ophthalmology
Louisiana State University Eye Center
Louisiana State University Medical Center
New Orleans, Louisiana

Stephen C. Pflugfelder, M.D.
Assistant Professor of Ophthalmology
Bascom Palmer Eye Institute
University of Miami School of Medicine
Miami, Florida

Warren W. Piette, M.D.
Associate Professor, Department of Dermatology
University of Iowa College of Medicine
University of Iowa Hospitals and Clinics
Iowa City, Iowa

Alan D. Proia, M.D., Ph.D.
Assistant Professor of Ophthalmology and Pathology
Duke University Medical Center
Durham, North Carolina

Jose S. Pulido, M.D., M.S.
Assistant Professor, Vitreoretinal Service
Department of Ophthalmology
University of Iowa Hospitals and Clinics
Iowa City, Iowa

John J. Purcell, Jr., M.D., F.A.C.S.
Associate Clinical Professor of Ophthalmology
St. Louis University School of Medicine;
Director, Department of Ophthalmology
St. Marys Health Center
St. Louis, Missouri

Allen M. Putterman, M.D.
Professor of Clinical Ophthalmology and Chief,
  Oculoplastic Service
University of Illinois;
Associate Professor and Director, Oculoplastic Surgery
University of Chicago and Michael Reese Hospitals
Chicago, Illinois

Graham E. Quinn, M.D.
Associate Professor of Ophthalmology
Children's Hopsital of Philadelphia;
Department of Ophthalmology
University of Pennsylvania School of Medicine;
Consultant, Pennsylvania Hospital Newborn
  Pediatrics;
Medical College of Pennsylvania;
Philadelphia, Pennsylvania

Hugo Quiroz, M.D.
Chief of Retina Service
Hospital for the Prevention of Blindness
Professor of Ophthalmology
Universidad Nacional
Autonoma de Mexico
Mexico City, Mexico

Edward Raab, M.D.
Professor of Ophthalmology
Associate Professor of Pediatrics
Director of Pediatric Ophthalmology and Strabismus
Mount Sinai School of Medicine
City University of New York
New York, New York

Sharon S. Raimer, M.D.
Professor
Departments of Dermatology and Pediatrics
University of Texas Medical Branch
Galveston, Texas

Christina Raitta, M.D.
Associate Professor
Department of Ophthalmology
University of Helsinki
Helsinki, Finland

Robert Ramsay, M.D.
Clinical Professor
University of Minnesota
Fairview Southdale Hospital
Abbott Northwestern Hospital
Metropolitan-Mount Sinai
Minneapolis Children's Medical Center
Minneapolis, Minnesota

Peter A. Rapoza, M.D.
Assistant Professor
Department of Ophthalmology
University of Wisconsin–Madison
Madison, Wisconsin

Paul David Reese, M.D.
Assistant Professor
Department of Ophthalmology
Tufts–New England Medical Center
Boston, Massachusetts

Scott C. Richards, M.D.
Resident, Department of Ophthalmology
Henry Ford Hospital
Detroit, Michigan

A. Glen Rico, M.D.
Resident, Department of Ophthalmology
University of California, Irvine
College of Medicine
Long Beach, California

Melvin I. Roat, M.D.
Assistant Professor of Ophthalmology
Corneal Surgery, External Diseases and Uveitis
University of Pittsburgh
Pittsburgh, Pennsylvania

Richard M. Robb, M.D.
Ophthalmologist-in-Chief
Children's Hospital;
Associate Professor of Ophthalmology
Harvard Medical School
Boston, Massachusetts

Merlyn M. Rodrigues, M.D., Ph.D.
Professor, Department of Ophthalmology
University of Maryland
Baltimore, Maryland

Kenneth G. Romanchuk, M.D.
Associate Professor, Department of Ophthalmology
College of Medicine
University of Saskatchewan
Saskatoon, Saskatchewan, Canada

Kenneth N. Rosenbaum, M.D.
Director, Clinical Genetics
Childrens Hospital National Medical Center
Washington, D.C.

Steven I. Rosenfeld, M.D.
Clinical Instructor
Bascom Palmer Eye Institute
University of Miami School of Medicine
Miami, Florida

Stuart T. D. Roxburgh, F.R.C.S., F.E.Opth.
Honorary Senior Lecturer in Ophthalmology
Dundee University Medical School;
Consultant Ophthalmologist
Tayside Health Board
United Kingdom

Andrew P. Schachat, M.D.
Associate Professor of Ophthalmology
Director, Ocular Oncology Service
The Johns Hopkins Hospital
Wilmer Ophthalmological Institute
Baltimore, Maryland

David J. Schanzlin, M.D.
Professor and Chairman
Department of Ophthalmology
Bethesda Eye Institute
St. Louis University
St. Louis, Missouri

Robert J. Schechter, M.D.
Associate Clinical Professor of Ophthalmology
Jules Stein Eye Institute
University of California, Los Angeles
School of Medicine
Los Angeles, California

Joel A. Schulman, M.D.
Associate Professor, Department of Ophthalmology
Louisiana State University
School of Medicine
Shreveport, Louisiana

Ivan R. Schwab, M.D.
Associate Professor in Cornea, External Disease, and
    Uveitis
Department of Ophthalmology
University of California at Davis
Davis, California

Thomas Schwartz, M.D.
Department of Ophthalmology
University of Texas Medical Branch
Galveston, Texas

Gregory H. Scimeca, M.D.
Senior Resident
Department of Ophthalmology
Temple University
Philadelphia, Pennsylvania

Robert C. Sergott, M.D.
Associate Professor
Attending Surgeon and Neuro-Ophthalmologist
Wills Eye Hospital
Philadelphia, Pennsylvania

David Sevel, M.D., Ph.D., F.A.C.S., M.C.Opht.
Scripps Clinic
La Jolla, California

Eric P. Shakin, M.D.
Instructor, Retina Service
Wills Eye Hospital
Philadelphia, Pennsylvania

Michael B. Shapiro, M.D.
Assistant Clinical Professor
University of Wisconsin
Davis–Deuhr Eye Associates
Madison, Wisconsin

John D. Sheppard, M.D.
Assistant Professor of Ophthalmology
Medical College of Hampton Roads
Director, Ocular Microbiology
Director, Uveitis Service
Norfolk, Virginia

Carol L. Shields, M.D.
Instructor, Oncology Service
Wills Eye Hospital
Philadelphia, Pennsylvania

Jerry A. Shields, M.D.
Director, Ocular Oncology Service
Wills Eye Hospital;
Professor of Ophthalmology
Thomas Jefferson University
Philadelphia, Pennsylvania

Nader Shoukrey, Ph.D.
Senior Research Associate
Research Department
King Khaled Eye Specialist Hospital
Riyadh, Saudi Arabia

Stephen H. Sinclair, M.D.
Assistant Professor of Ophthalmology
Hahnemann University School of Medicine
Philadelphia, Pennsylvania

Harold W. Skalka, M.D.
Professor and Chairman
Combined Program in Ophthalmology
Eye Foundation Hospital
University of Alabama at Birmingham
Birmingham, Alabama

Daniel N. Skorich, M.D.
The Duluth Clinic
Assistant Clinical Professor, Ophthalmology
University of Minnesota
Duluth Medical School
Duluth, Minnesota

Don B. Smith, M.D.
Clinical Assistant Professor, Neurology
University of Colorado Health Science Center
Denver, Colorado

Morton E. Smith, M.D.
Professor, Ophthalmology and Pathology
Washington University Medical School
St. Louis, Missouri

Richard S. Smith, M.D.
Professor and Chairman
Department of Ophthalmology
Albany Medical Center
Albany, New York

Scott R. Sneed, M.D.
Assistant Professor
W. K. Kellogg Eye Center
University of Michigan Medical Center
Ann Arbor, Michigan

Anna Marie Sommer, M.D.
Associate Professor and Director, Division of Genetics
Department of Pediatrics
Children's Hospital of Columbus
The Ohio State University
Columbus, Ohio

Sarkis Soukiasian, M.D.
Clinical Fellow
Massachusetts Eye and Ear Infirmary
Boston, Massachusetts
Clinical Fellow in Ophthalmology
Harvard Medical School
Boston, Massachusetts

D. J. Spalton, M.R.C.P., F.R.C.S.
Consultant Ophthalmic Surgeon and Consultant to
    Medical Eye Unit
St. Thomas' Hospital
London, England

Raymond M. Stein, M.D., F.R.C.S.(C)
Attending Ophthalmologist
Mt. Sinai Hospital;
Lecturer, University of Toronto;
Bochner Eye Institute
Toronto, Ontario, Canada

Paul G. Steinkuller, M.D.
Assistant Professor
Department of Ophthalmology
Baylor College of Medicine
Houston, Texas

Garth Stevens, Jr., M.D.
Assistant Professor
Medical College of Virginia;
Chief of Ophthalmology
Richmond Veterans Administration Medical Center
Richmond, Virginia

Edwin M. Stone, M.D., Ph.D.
Fellow Associate
Department of Ophthalmology
The University of Iowa College of Medicine
Iowa City, Iowa

Mary Seabury Stone, M.D.
Assistant Professor, Department of Dermatology
University of Iowa Hospitals and Clinics
Iowa City, Iowa

Kerstin Stromland, M.D.
Department of Ophthalmology
University of Goteborg
Sahlgren's Hospital
Goteborg, Sweden

W. P. Daniel Su, M.D.
Consultant (Staff), Dermatology Department
Mayo Clinic;
Professor in Dermatology
Mayo Medical School
Rochester, Minnesota

Alan Sugar, M.D.
Professor of Ophthalmology
University of Michigan
W. K. Kellogg Eye Center
Ann Arbor, Michigan

Joel Sugar, M.D.
Professor of Ophthalmology
University of Illinois Eye and Ear Infirmary
Chicago, Illinois

Khalid F. Tabbara, M.D.
Professor and Chairman
Department of Ophthalmology
College of Medicine
King Saud University
Riyadh, Saudi Arabia

Hugh R. Taylor, M.D., F.R.A.C.S.
Associate Professor
The Johns Hopkins University School of Medicine and
    School of Hygiene and Public Health
Baltimore, Maryland

Howard H. Tessler, M.D.
Assistant Professor, Department of Ophthalmology
University of Illinois
Eye and Ear Infirmary
Chicago, Illinois;
St. Therese Medical Center
Victory Memorial Hospital
Maywood, Illinois

Zvi Tessler, M.D.
Lecturer of Ophthalmology
Ben Gurion University
Beer-Sheba, Israel

Richard A. Thoft, M.D.
Professor and Chairman
Department of Ophthalmology
University of Pittsburgh
Pittsburgh, Pennsylvania

Andrea Cibis Tongue, M.D.
Clinical Associate Professor of Ophthalmology
Oregon Health Science University
Portland, Oregon

Elise Torczynski, M.D.
Associate Professor of Ophthalmology
The University of Chicago
Chicago, Illinois

Elias I. Traboulsi, M.D.
Center for Sight
Georgetown University Medical Center
Washington, D.C.

Clement L. Trempe, M.D., F.R.C.S.(C)
Director of Clinical Research
Eye Research Institute of Retina Foundation
Boston, Massachusetts

Brenda J. Tripathi, Ph.D.
Associate Professor
Department of Ophthalmology and Visual Science
The University of Chicago
Chicago, Illinois

Ramesh C. Tripathi, M.D., Ph.D.
Professor
Department of Ophthalmology and Visual Science
The University of Chicago
Chicago, Illinois

Jonathan D. Trobe, M.D.
Professor of Opthalmology
Assistant Professor of Neurology
University of Michigan
Division of Neurology
Ann Arbor, Michigan

David T. Tse, M.D.
Associate Professor
Department of Ophthalmology
Oculoplastic Service
Bascom Palmer Eye Institute
University of Miami
Miami, Florida

Amy V. Tso, M.D.
OptiCare Eye Health Center
Waterbury, Connecticut

David L. Valle, M.D.
Howard Hughes Medical Institute
Research Laboratories;
The Johns Hopkins University School of Medicine
Baltimore, Maryland

M. I. van Schooneveld, M.D.
Ophthalmology and Genetics
Netherlands Ophthalmic Research Institute
The Netherlands

Michael W. Varner, M.D.
Associate Professor, Department of Obstetrics/
    Gynecology
University of Utah
Salt Lake City, Utah

Michael P. Vrabec, M.D.
Assistant Professor of Ophthalmology
University of Vermont
College of Medicine
Burlington, Vermont

Michael D. Wagoner, M.D.
Assistant Clinical Professor
Harvard Medical School
Massachusetts Eye and Ear Infirmary
Boston, Massachusetts

Joseph B. Walsh, M.D.
Professor and Chairman
Department of Ophthalmology
New York Medical College
New York Eye and Ear Infirmary
New York, New York

Frederick M. Wang, M.D.
Associate Clinical Professor
Department of Ophthalmology
Montefiore Medical Center
Albert Einstein College of Medicine
Bronx, New York;
Associate Attending
New York Eye and Ear Infirmary
New York, New York

Mette Warburg, M.D.
Chief Medical Officer
Paediatric Ophthalmology and Handicaps
Gentofte Hospital
University of Copenhagen
Gentofte, Denmark

Robert C. Watzke, M.D.
Professor of Ophthalmology
Director, Retina and Vitreous Service
Oregon Health Science University
Portland, Oregon

Mary Waziri, B.S., M.D.
Department of Pediatrics (Genetics)
University of Iowa Hospitals and Clinics
Iowa City, Iowa

Mark J. Weiner, M.D.
Massachusetts Eye and Ear Infirmary
Joslin Diabetes Center
Harvard Medical School
Newton, Massachusetts

Thomas A. Weingeist, M.D., Ph.D.
Professor and Head, Department of Ophthalmology
The University of Iowa
Iowa City, Iowa

Avery H. Weiss, M.D.
Associate Professor, Department of Ophthalmology
and Pediatrics
University of South Florida
College of Medicine
Tampa, Florida

John J. Weiter, M.D., Ph.D.
Associate Professor
Department of Ophthalmology
Tufts University School of Medicine;
Harvard Medical School;
Associate Clinical Professor
Eye Research Institute of Retina Foundation
Boston, Massachusetts

Richard G. Weleber, M.D.
Professor of Ophthalmology and Medical Genetics
The Oregon Health Science University
Portland, Oregon

Duane C. Whitaker, M.D.
Associate Professor, Department of Dermatology
The University of Iowa
Iowa City, Iowa

William E. Whitson, M.D.
Corneal Consultants of Indiana
Indianapolis, Indiana

Michael Wiedman, M.D.
Harvard Medical School
Massachusetts Institute of Technology
Tufts University Medical School
Massachusetts Eye and Ear Infirmary
Boston, Massachusetts

James Wiens, M.D.
Fellow, External Disease and Cornea
McGill University
External Disease and Cornea
Montreal, Quebec, Canada

Kirk R. Wilhelmus, M.D.
Associate Professor
Department of Ophthalmology
Cullen Eye Institute
Baylor College of Medicine
Houston, Texas

George A. Williams, M.D.
Associate Clinical Professor
Kresge Eye Institute
Detroit, Michigan;
William Beaumont Hospital
Royal Oak, Michigan

John M. Williams, Sr., M.D.
Eye Care Center—Retina Center
Longmont, Colorado

W. Bruce Wilson, M.D.
Associate Clinical Professor of Ophthalmology
University of Colorado
Health Science Center
Denver, Colorado

Jonathan D. Wirtschafter, M.D., F.A.C.S.
Professor, Neuro-Ophthalmology, Orbital and
  Oculoplastic Surgery
University of Minnesota
Minneapolis, Minnesota

Robert A. Wiznia, M.D., F.A.C.S.
Assistant Clinical Professor of Ophthalmology and
  Visual Science
Yale University School of Medicine
New Haven, Connecticut

John D. Wright, Jr., M.D.
Lexington Eye Associates, Inc.
Lexington, Massachusetts

Yuval Yassur, M.D.
Professor of Ophthalmology
Ben Gurion University
Beer-Sheba, Israel

Honogliang Yin, M.D., Ph.D.
Visiting Research Fellow
Department of Surgery
University of North Carolina
Chapel Hill, North Carolina

Marco A. Zarbin, M.D., Ph.D.
The Wilmer Ophthalmologic Institute
The Johns Hopkins Hospital
Baltimore, Maryland

Hans Zellweger, M.D.
Professor Emeritus
Department of Pediatrics
University of Iowa College of Medicine
Iowa City, Iowa

PART 1

Cardiac Disorders

Chapter 1

Endocarditis

DANIEL H. GOLD

Endocarditis occurs from infancy to old age and is usually associated with an underlying defect of cardiac tissue. Normal endothelial cells are not thrombogenic, but damage to or denudation of endothelial surfaces triggers local deposition of fibrin and platelets, producing a nonbacterial thrombotic endocarditis.[1] The initial endothelial damage may result from abnormal hemodynamics of congenital or acquired cardiac lesions (eg, ventricular septal defect, patent ductus arteriosus, rheumatic valve disease, mitral valve prolapse), from injection of foreign particles by intravenous drug abusers, or from local immune complex deposition (Libman-Sacks endocarditis) in disorders like systemic lupus erythematosus.

Infective endocarditis develops when circulating microorganisms adhere to the fibrin-platelet vegetations of nonbacterial thrombotic endocarditis. The adherent organisms proliferate locally and promote deposition of additional fibrin and platelets, giving rise to the classic septic vegetations of this disorder.

Bacteria are the most common causative agents, especially α-hemolytic streptococci, which are part of the normal upper respiratory tract flora. Group D streptococci are the second most common organisms and are often associated with gastrointestinal or genitourinary disease. Staphylococci are more frequently encountered in prosthetic valve endocarditis and in intravenous drug abusers. Fungi and even chlamydia can also cause infective endocarditis.

The virulence of the infecting organism and the host's response to it determine whether the endocarditis takes a rapid acute course or a more indolent subacute or chronic course. Although most cases of infective endocarditis develop from microbial colonization of nonbacterial thrombotic endocarditis lesions, the disease can occur in the absence of any known predisposing cardiac lesion.

Systemic Manifestations

The complications of endocarditis may be primarily cardiac, with arrythmias and decreased myocardial perfusion resulting in fatigue, dyspnea, pain, palpitations, and increasing heart failure.[2] Extracardiac manifestations are most often the result of distant embolization of infected material from friable cardiac vegetations. These emboli may produce arterial obstruction and tissue infarction, localized abscesses, or mycotic aneurysms anywhere in the body. Extracardiac complications are also produced by an entirely different pathophysiologic mechanism, immune complex vasculitis. Combinations of antibody with microbial, cardiac, or other antigens have been detected as circulating immune complexes and demonstrated in blood vessels of the skin and kidneys.[2,3] Immune complex vasculitis may play a major role in the more chronic phases of subacute infective endocarditis; embolic phenomena are more important in the acute disease.

Central nervous system involvement includes strokes, brain abscesses, mycotic aneurysms, and suppurative meningitis. Musculoskeletal manifestations are common, including arthralgias, arthritis, myalgias, back pain, and osteomyelitis.[2,3] Renal embolization or immune complex deposition produces focal or diffuse glomerulonephritis, infarction, or abscess.[2,3]

Mucocutaneous lesions are characteristic of infective endocarditis, especially widespread petechial hemorrhages and linear splinter hemorrhages beneath fingernails and toenails. Osler's nodes (tender pea-sized red or purple nodules) and Janeway's lesions (nontender red macules) are classic findings. These hemorrhagic complications are the result of embolic or vasculitis-induced damage to the involved blood vessels.[2,3]

Ocular Manifestations

The ocular complications of endocarditis can readily be seen as local manifestations of pathophysiologic mechanisms that are at work in many other tissues and organs. Focal hemorrhagic phenomena throughout the body have their ocular counterparts in conjunctival and retinal hemorrhages. Petechial hemorrhages may occur in the palpebral or bulbar conjuctiva. Retinal hemorrhages are usually superficial and may be single or multiple.[4,5]

The classic ocular complication of infective endocarditis is the white-centered retinal hemorrhage known as Roth's spot (Fig. 1-1). The white center may represent a focal retinitis caused by an infected embolus, but it may also be the nonspecific fibrin-platelet clot seen with white-centered retinal hemorrhages in many other disorders (eg, anemia, leukemia). Although the ocular hemorrhages of infective endocarditis have usually been ascribed to embolic disruption of blood vessels, an immune complex vasculitis may play a role in their pathogenesis.

Branch or central retinal artery occlusion represents but a part of the widespread arterial obstructive disease associated with endocarditis, and it has been seen with both nonbacterial thrombotic endocarditis and infective

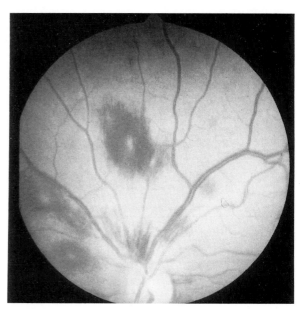

FIGURE 1-1 White-centered retinal hemorrhages in infective carditis. (Courtesy Joseph B. Walsh, M.D.)

endocarditis.[2,6] Occlusion of smaller vessels produces the cotton-wool spots seen clinically and microscopically.[4]

Inflammatory ocular complications are the result of septic emboli from infected cardiac vegetations.[7] More virulent organisms may produce an acute endophthalmitis with loss of the eye.[8] Less destructive infectious agents can present as focal retinitis, choroiditis, vitritis, or uveitis.[4,5,7] In some cases of infective endocarditis, uveitis may be the presenting feature of the disease.

Central nervous system involvement produces a host of neuro-ophthalmic manifestations.[2,3] Brain abscesses, mycotic aneurysms, and hemorrhagic or ischemic strokes may be associated with visual field defects, cranial nerve palsies with diplopia, nystagmus, or papilledema.

Treatment of the ocular complications of infective endocarditis requires identification and appropriate therapy for the underlying systemic infectious agent. Vitrectomy with intraocular administration of antimicrobial agents may be useful in both diagnosis and therapy of patients with endophthalmitis.

References

1. Sullam PM, Drake TA, Sande MA. Pathogenesis of endocarditis. *Amer J Med.* 1985;78(suppl 6B):110.
2. Hermans PE. The clinical manifestations of infective endocarditis. *Mayo Clin Proc.* 1982;57:15.

3. Krayenbuhl HP, Rickards AF. New aspects of bacterial endocarditis. *Eur Heart J.* 1984;5(suppl C):1.
4. Kennedy JE, Wise GN. Clinicopathologic corrections of retinal lesions. *Arch Ophthalmol.* 1965;74:658.
5. Dienst EC, Gartner S. Pathologic changes in the eye associated with subacute bacterial endocarditis. *Arch Ophthalmol.* 1944;31:198.
6. Brown GC, Magargal LE, Shields JA, et al. Retinal arterial obstruction in children and young adults. *Ophthalmol.* 1981;88:18.
7. Burns CL. Bilateral endophthalmitis in acute bacterial endocarditis. *Am J Ophthalmol.* 1979;88:909.
8. Treister G, Rothkoff L, Yalon M, et al. Bilateral blindness following panophthalmitis in a case of bacterial endocarditis. *Ann Ophthalmol.*1982;14:663.

Chapter 2

Mitral Valve Prolapse Syndrome

GREGORY H. SCIMECA and LARRY E. MAGARGAL

Mitral valve prolapse (MVP) is a retrograde displacement of one or more of the mitral valve leaflets into the left atrium during systole. It is generally considered to be a benign but highly variable clinical syndrome with a prevalence in the general population of 0.5% for men and up to 10% for young women.[1,2] Diagnosis is often based on the characteristic auscultatory features, which include one or more systolic clicks followed by a late systolic murmur, and it may be confirmed by echocardiography (Fig. 2-1) or angiocardiography.[3] Although various causes of MVP syndrome have been documented, including connective tissue disorders, rheumatic and congenital heart disease, cardiomyopathy, and coronary artery disease, in the majority of patients the cause is unknown.[4] There does appear to be a high incidence of MVP in patients with asthenic habitus and any of a variety of congenital thoracic deformities, which may reflect an underlying connective tissue disorder.[5,6] The predominant pathologic feature in MVP is myxomatous proliferation of the middle layer of the mitral valve leaflet, with an increase in the quantity of acid mucopolysaccharide secondary to a fundamental but as yet undefined abnormality in collagen metabolism.[7] This leads to redundancy of the mitral leaflets, prolapse during systole, and alterations in intracardiac hemodynamics.

Although MVP is generally considered to be a benign condition in most cases, a number of complications can occur.[3] These are typically cardiac in nature and include chest pain, fatigue, dyspnea, light-headedness, and palpitations. Less frequently, endocarditis, progressive mitral regurgitation, myocardial ischemia, and sudden death also occur.[3]

Extracardiac complications of MVP syndrome include cerebral and retinal ischemic events.[8-15] Previously, it was postulated that these ischemic events were related to emboli emanating from the abnormal mitral valve. More recent studies have demonstrated a higher incidence of systemic platelet coagulant hyperactivity in patients with MVP than in control subjects.[9] These findings are consistent with the view that exposure of flowing blood to the abnormal hemodynamics or valve surfaces in MVP gives rise to platelet stimulation in vivo, resulting in intravascular thromboembolism.[10]

Systemic manifestations of platelet coagulant hyperactivity in patients with MVP, which most notably involves the central nervous system, may range from transient ischemic attacks to completed stroke.[8-11] Neuro-ophthalmologic symptoms may include scintillating or fortification scotomas, transient visual obscurations, and hemifield defects.[10]

The ocular complications of MVP syndrome may occur as local manifestations of systemic platelet coagulant hyperaggregability. In one study of patients under the age of 45 with MVP, 22% had symptoms of amaurosis fugax compared to 1% of age-matched controls.[12] Central retinal, branch retinal, and choroidal arteriolar occlusion have also been seen in patients with MVP syndrome.[13,14] In a series of six patients with retinal vascular occlusive disease who underwent noninvasive cardiac and systemic testing, all were found to have prolapsed mitral valves.[13] One 14-year-old girl in

Supported in part by the Retina Research and Development Foundation of Philadelphia and the Pennsylvania Lions Sight Conservation Corporation.

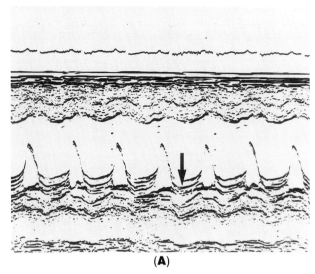

(A)

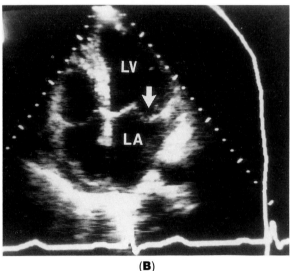

(B)

FIGURE 2-1 (A) M-mode echocardiogram reveals the hammacheal-shaped appearance of the valve leaflets (*arrow*) as they prolapse posteriorly. **(B)** Two-dimensional echocardiogram (apical four-chamber view) demonstrating mitral valve prolapse (*arrow*) into the left atrium (LA) during systole (LV, left ventricle).

this series developed extensive neovascularization of the posterior pole (atypical Eales' disease).[13] One large series of 32 patients with MVP found myopia, unusually deep anterior chamber angles with prominent iris processes, and lenticular opacities to be common ocular features.[6]

The association of MVP and platelet hyperactivity in retinal venous occlusion has also been suggested.[15,16] In

a review of 69 cases of central retinal vein occlusion (CRVO) in persons under age 50 years, a high prevalence (73%) of platelet dysfunction and MVP was found.[15] In another study of patients under age 50 with CRVO and no other predisposing ocular or systemic conditions, 64% of those who underwent echocardiography were found to have a prolapsing mitral valve.[16]

In view of the apparent association between mitral valve prolapse and platelet hyperactivity in "young healthy patients" with amaurosis fugax, retinal arterial or venous obstruction, and similar neuro-ophthalmologic sequelae, it seems appropriate to recommend to such patients a therapeutic trial of antiplatelet agents such as acetylsalicylic acid or sulfinpyrazone.[16] Although it is recommended by some,[17] anticoagulant therapy and high-dose oral steroid therapy have significant systemic side effects and therefore do not appear to be indicated, as the visual prognosis in most cases is favorable when more conservative management strategies are used.

References

1. Brown OR, Klaster FE, DeMots H. Incidence of mitral valve prolapse in the asymptomatic normal. *Circulation.* 1975; S2(suppl II):11-77.
2. Procacci PM, Savran SV, Schreiter SL, et al. Clinical frequency and implications of mitral valve prolapse in the female population. *Circulation.* 1975; S2(suppl II):11-78.
3. Braunwald E. Valvular heart disease. In: Braunwald E et al. eds. *Harrison's Principles of Internal Medicine,* 11th ed. New York, NY: McGraw-Hill; 1987;962-963.
4. O'Rourke RA, Crawford MH. The systolic click-murmur syndrome: clinical recognition and management. *Curr Probl Cardiol.* 1976;1:1.
5. Zema MJ, Chiaramida S, DeFilipp GJ, et al. Somatotype and idiopathic mitral valve prolapse. *Cathet Cardiovasc Diagn.* 1982;8:105.
6. Traboulsi EI, Aswad MI, Jalkh AE, et al. Ocular findings in mitral valve prolapse syndrome. *Ann Ophthalmol.* 1987;19:354-359.
7. Davies MJ, Moore BP, Braimbridge MV. The floppy mitral valve. Study of incidence, pathology and complications in surgical, necropsy and forensic material. *Br Heart J.* 1978; 40:368.
8. Barnett HJM, Jones MW, Boughner DR, et al. Cerebral ischemic events associated with prolapsing mitral valve. *Arch Neurol.* 1976;33:777-782.
9. Kostuk WJ, Boughner DR, Barnett HJM, et al. Strokes: a complication of mitral valve prolapse. *Lancet.* 1977;2:313-316.
10. Walsh PN, Kansu TA, Corbett JJ, et al. Platelets, thrombo-embolism and mitral valve prolapse. *Circulation.* 1981;63:552-559.
11. Barnett HJM, Boughner DR, Taylor W, et al. Further evidence relating mitral valve prolapse to cerebral ischemic events. *N Engl J Med.* 1980;302:139-144.

12. Lesser RL, Heenemann MH, Borkowski H, et al. Mitral valve prolapse and amaurosis fugax. *J Clin Neuro-ophthalmol.* 1981;1:153-160.

13. Caltrider ND, Irvine AR, Kline HJ, et al. Retinal emboli in patients with mitral valve prolapse. *Am J Ophthalmol.* 1980;90:534-539.

14. Waldoff HS, Gerber M. Desser KB, et al. Retinal vascular lesions in two patients with prolapsed mitral valve leaflets. *Am J Ophthalmol.* 1975;79:382-385.

15. Magargal LE, Gonder JR, Maher V. Central retinal vein obstruction in the young adult. *Trans Pa Acad Ophthalmol Otolaryngol.* 1985;37(2):148-153.

16. Gonder JR, Magargal LE, Walsh PN, et al. Central retinal vein obstruction associated with mitral valve prolapse. *Can J Ophthalmol.* 1983;18:220-222.

17. Hayreh SS. Optic disc vasculitis. *Br J Ophthalmol.* 1972; 56:652-670.

PART 2

Chromosomal Disorders

Section A

Deletion Syndromes

Chapter 3

Deletion of the Long Arm of Chromosome 13

13q− Syndrome

EDWIN M. STONE

Disorders associated with deletions of part of the long arm of the thirteenth chromosome (13q−) range from severe multisystem congenital syndromes to retinoblastoma, a lethal intraocular tumor of childhood that can be inherited in an autosomal-dominant fashion. The multisystem syndromes usually result from large, karyotypically obvious alterations in the chromosome, whereas retinoblastoma can be caused by small occult deletions or mutations involving a portion of a single gene.

The short arm of chromosome 13 (13p) makes up only 5% of the chromosome's total length, and to date no syndromes or diseases have been traced to this small region by gene mapping. In contrast, the long arm contains over 3% of the haploid human genome, and several important genes have been identified there, including a DNA excision repair enzyme, type IV col-

lagen, esterase D, ferritin, coagulation factors VII and X, proprionyl coenzyme A carboxylase, and a lymphocyte cytosolic protein. In addition, three heritable diseases have been mapped to the chromosome: Wilson's disease,[1] retinoblastoma,[2] and osteosarcoma.[2]

In the 1960s, a number of children were reported to have congenital anomalies associated with a deletion or a ring formation of a D group chromosome (13, 14, or 15). Lejeune et al.[3] synthesized the features previously reported in the literature into a clinical syndrome associated with deletion or ring formation of a D group chromosome. Allerdice et al.[4] used autoradiography to identify the involved chromosome as 13 in two patients and named the condition "the 13q deletion [13q−] syndrome."[4] The most striking features of the syndrome were absence of thumbs, a broad nasal bridge (giving rise to a "Greek profile"), forward-slanted ("rabbit") incisors, and microcephaly. Neibuhr[5] reviewed 72 cases of presumed 13q deletion and divided the patients into four groups based on specific clinical features. Karyotyping of subjects from each group revealed (1) the microcephaly and Greek profile features of the syndrome were associated with deletions extending approximately from 13q33 to the terminus; (2) loss of thumbs required more extensive deletion, perhaps to band q21; and (3) if the deletion involved band q14, the patient often had retinoblastoma. De Grouchy and Turleau[6] divided the 13q− patients into only two groups, those with retinoblastoma and those without. This simplification is reasonable because the type and number of clinically detectable abnormalities present in any particular patient vary according to the size and location of the deletion and, in reality, these variations form a continuous spectrum of disease rather than a few distinct entities. The presence of retinoblastoma merits special attention because without treatment this tumor is almost uniformly fatal. Another clinical feature that has prognostic importance is the absence of thumbs. In Neibuhr's series, the absence of thumbs was associated with a fivefold higher mortality in the first 6 months of life,[5] presumably because of the larger deletions present in children with this sign.

Systemic Manifestations

The catalog of systemic manifestations in gross 13q deletions is long and includes many nonspecific findings present in other chromosomal syndromes. The most characteristic features are the Greek profile, rabbit incisors, large malrotated ears, and absent thumbs. Microcephaly is almost always present and often reflects an intracranial defect such as arhinencephaly, holoprosencephaly, aplasia of the falx cerebri, or agenesis of corpus callosum. Anencephaly also occurs occasionally. In addition to the broad nasal bridge and forward-slanting incisors already mentioned, other facial dysmorphisms include asymmetry of the face, a short philtrum, and a small chin.

The most striking skeletal abnormalities associated with 13q deletions involve the hands: absent thumbs, agenesis of the first metacarpal, and fused fourth and fifth metacarpals with syndactyly. The fifth toe is occasionally absent, and supernumerary ribs, hip dislocations, and lumbosacral agenesis also occur.

The genitalia and perineum are often abnormal. The most common findings include hypospadias, epispadias, bifid scrotum, undescended testis, cryptorchidism, bifid uterus, and anal atresia. Visceral malformations include atrial and ventricular septal defects, aplasia of the gallbladder, and aplasia or hypoplasia of the kidneys.

It is difficult to find a unifying theme for all the clinical manifestations found in children with 13q deletions, but a few generalizations can be made. First, most of the structures that are affected by a 13q deletion are formed during the second and third months of embryogenesis. For example, the cartilage models of the thumbs appear about the 40th day, as do the hillocks of tissue that eventually fuse into the external ears. Some of the clinical features undoubtedly result from an imbalance of developmental regulatory elements that are encoded on chromosome 13. Unfortunately, our current understanding of developmental gene regulation in humans is very limited, but one can get a glimpse of the regulatory processes at work by comparing the effects of trisomy 13 and monosomy 13. For example, monosomy at certain loci of 13q results in absence of thumbs, while trisomy 13 (Patau's syndrome) often results in extra fingers on the ulnar side of the hand. Similarly, 13q deletions often give rise to extra bone and soft tissue in the midface (Greek profile), whereas children with trisomy 13 have a paucity of bone and soft tissue (cleft palate) in structures formed from the embryonic frontonasal prominence. These genetic "oppositions" suggest that regulatory elements on chromosome 13 are involved in positive control of finger formation and negative control of tissue proliferation in the frontonasal prominence.

Ocular Manifestations

A variety of external signs have been noted in children with 13q deletions, most notably hypertelorism, epicanthus, palpebral ptosis, and microphthalmia. The most serious ophthalmic manifestation of the deletion syndrome is retinoblastoma. In Neibuhr's series, 11 of 62 children (18%) with a clinically detectable 13q deletion syndrome had retinoblastoma.[5] Deletions are also an important cause of retinoblastomas in children without dysmorphic features. In fact, 5 to 10% of all retinoblastomas may be associated with deletion of a portion of 13q.[7] Unlike the patients in Neibuhr's series, children with small interstitial deletions usually appear completely normal except for their intraocular tumors. It is important to detect such deletions for at least two reasons. First, it establishes the heritability of the tumor; that is, although only about 40% of all retinoblastomas are heritable, 100% of those that occur on the basis of a germinal 13q deletion are heritable in an autosomal-dominant fashion with extremely high penetrance. Second, persons with germinal mutations of the retinoblastoma gene are at increased risk for other malignancies (usually osteosarcoma) in addition to their retinoblastoma. Abramson et al[8] reported a 15% incidence of second malignancies in 693 patients with heritable retinoblastoma.

Fortunately, clinically occult deletions of the retinoblastoma gene can often be detected even when they are karyotypically invisible. The gene that encodes the enzyme esterase D is extremely close to the retinoblastoma gene, so that deletions of the latter often involve the former. Thus, small deletions near 13q14 can be detected by assaying esterase D activity in red blood cells.[9] Now that the retinoblastoma gene itself has been cloned, an even more direct method of detecting deletions of the retinoblastoma gene (Southern blotting) is possible.[2] This test will soon be available commercially, at which time it will be feasible to screen all patients with retinoblastoma for tiny interstitial deletions of parts of 13q.

References

1. Frydman M, Bonne-Tamir B, Farrer L, et al. Assignment of the gene for Wilson disease to chromosome 13: linkage to esterase D locus. *Proc Natl Acad Sci USA.* 1985;82:1819.
2. Friend S, Bernards R, Rogelj S, et al. A human DNA segment with properties of the gene that predisposes to retinoblastoma and osteosarcoma. *Nature.* 1986;323:643.
3. Lejeune J, Lafourcade J, Berger R, et al. Le phenotype [Dr]. *Ann Genet.* 1968;11:79.
4. Allerdice P, Davis J, Miller O, et al. The 13q− deletion syndrome. *Am J Hum Genet.* 1969;21:499.
5. Niebuhr E. Partial trisomies and deletions of chromosome 13. In: Yunis JJ. *New Chromosomal Syndromes.* New York, NY: Academic Press; 1977;273-295.
6. De Grouchy J, Turleau C. *Clinical Atlas of Human Chromosomes.* New York, NY: John Wiley & Sons; 1984;224-245.
7. Liberfarb R, Bustos T, Miller W, et al. Incidence and significance of a deletion of chromosome band 13q14 in patients with retinoblastoma and in their families. *Ophthalmology.* 1984;91:1695.
8. Abramson D, Ellsworth R, Kitchin F, et al. Second nonocular tumors in retinoblastoma survivors. *Ophthalmology.* 1984;91:1351.
9. Cowell J, Thompson E, Rutland P. The need to screen all retinoblastoma patients for esterase D activity: detection of submicroscopic chromosome deletions. *Arch Dis Child.* 1987;62:8.

Chapter 4

Chromosome 18: Deletion of the Short Arm, Long Arm, and Ring 18 Syndromes

HANS ZELLWEGER and SHIVANAND R. PATIL

Chromosome breaks occur with great frequency, notably during meiosis. Most often they are transient, and the two chromosome segments reunite. Chiasma formation involving breaks on identical sites of the chromatids or homologous chromosomes leads to exchange (crossing over, recombination) of identical parts of sister or nonsister chromatids. Spontaneous healing of chromosome breaks and crossing over are physiologic phenomena.

However, breaks represent also the first step in the pathogenesis of diversified structural chromosome anomalies (aneusomies) such as deletions, inversions, duplications, ring formations, and translocations. If a chromosome break fails to heal, the chromosome remains split into two segments. The segment without the centromere tends to disappear, while the remaining one represents a deleted or partially deleted chromosome. If breaks occur in both arms of a chromosome, the break points may unite, forming a ring chromosome. Deletions of the long arm of chromosome 18 are designated as 18q−; deletions of the short arm of chromosome 18 are designated as 18p−; and for ring chromosome 18, the symbol r18 has been adopted.

Structural chromosome anomalies (aneusomies) are less frequent than numerical anomalies (aneuploidies). Of over 68,000 cytogenetically analyzed live newborns, 260 (0.4%) showed aneuploidy, 43 had structural abnormalities, and five of them (0.008%) had a deletion. Deletions of 18p and 18q and r18 are among the better-known deletion syndromes and will be discussed here with special emphasis on the eye findings (Table 4-1).

TABLE 4-1 Eye Findings Associated with Defects of Chromosome 18

Finding	18p−	18q−	r18
Enophthalmos		+	
Microphthalmia	+	+	+
Strabismus	++	++	+
Upper lid ptosis	++	++	+
Slanting eyes (up or down)	+		
Hypertelorism	++	+++	+++
Hypotelorism	+		
Epicanthal folds	++	++	++
Nystagmus	+	+++	+
Myopia		+	+
Hypermetropia			+
Microcornea		+	+
Corneal opacities, leukoma	+	+	+
Corneal staphyloma		+	
Keratoconus	+		
Astigmatism		+	+
Absence of anterior chamber		+	
Brushfield's spots		+	
Iris atrophy	+		
Partial aniridia			+
Coloboma	+	+	+
Cataracts	+	+	+
Intraocular cartilage and smooth muscle			+
Glaucoma	+	+	
Optic atrophy		++	+
Retinopathy	+	+	+
Retrobulbar cyst			(+)
Amblyopia	+	+	+
Cyclopia	+		
Holoprosencephaly	+		+

+++ Over 75% of cases
++ In 50% of cases or less
+ Infrequent
(+) One case
(Data from References 1,5,6,8,10-13,17, and 18.)

Deletion of the Short Arm of Chromosome 18

The 18p− syndrome was first described in 1963 by De Grouchy et al.[1] Since then, about 100 cases have been published and others have been observed that were not published. Deletion 18p is more common in females, the ratio being 3:2. The condition appears to be maternal age dependent. Thirty (53%) of 57 women were over 30 years old when their 18p− child was born, and 12 (21%) were aged 40 years or over.[2]

About two thirds of the affected persons show de novo deletion; some present de novo translocation of 18p with another chromosome or an isochromosome, 18q. Both conditions result in monosomy 18p. Some 18p− cases result from a parental translocation involving 18p or a parental mosaicism with an 18p− clone.

The clinical picture varies greatly and no specific clinical syndrome has been described to date. Prenatal and postnatal growth failure, subnormal mental development, and dysmorphic features are almost constant findings.[3] Mental retardation ranges from severe to (more often) mild; in over half the cases IQ is between 50 and 70, and a few of the patients have low normal intelligence. In rare cases, autism or schizophrenia is noted. The stature of affected adults is below the normal mean. The skull is brachycephalic; its circumference is small-normal or, rarely, microcephalic. Muscle hypotonia and seizures have been noted in some cases. Deletion 18p patients have normal sexual development. Several have had children, some with normal and some with abnormal chromosomes. Transmission of the 18p deletion from parent to child was reported by Rethore.[2] A mother with short stature, moderate mental retardation, microcephaly, and a 46,XX/46,XX,18p− mosaicism had two severely affected children with monoclonal 18p− syndrome. Dysmorphic features include broad nasal bridge, low-set, large, and floppy ears, "shark" mouth, high arched palate, receding chin, alopecia, webbed neck, and deep nuchal hairline. Teeth are late erupting and are prone to caries. Palms of the hands are large, and fingers are short. Dermatoglyphics show an abundance of whorls; the palmar triradius is often distal.

The 18p− syndrome has been associated with the autoimmune endocrinopathy complex. Diabetes mellitus, lymphocytic thyroiditis, hypothyroidism, hyperthyroidism, hypoparathyroidism, eczema, and rheumatoid arthritis have been described. Serum IgA has been deficient in a number of cases.[4]

Over 10% of patients display severe malformations of the brain, such as holoprosencephaly, cebocephaly, arhinencephaly, aplasia of the corpus callosum, and septum pellucidum; meningoceles were also reported. Other malformations include congenital heart disease, unilateral kidney aplasia, vertebral anomalies, dislocated hips, cleft palate, hypospadias, and ectopic testes.

Ocular Manifestations

Ptosis, epicanthal folds, strabismus, and hypertelorism are found in 40 to 65% of cases. Among rarer eye findings are corneal leukoma, keratoconus, microophthalmia, glaucoma, coloboma, and cyclopia.[5,6]

Prognosis

Life expectancy depends on the severity of accompanying malformations. Some patients die early, but some are known to have lived 60 years.[6] Prenatal diagnosis is possible, and amniocentesis is indicated for pregnant women with the 18p− syndrome.[7]

Deletion of the Long Arm of Chromosome 18

The 18q− syndrome is about as frequent as 18p−, with over 100 cases reported. Again, there is some female preponderance, the ratio being 3:2. Maternal age is less significant than in the 18p− syndrome; only 4% of the mothers were 40 years and over and 19 (37%) of 52 women were over 30 when they had their 18q− child.

Most of the cases are de novo deletions, usually of the distal segment of 18q (18q21.2). Some presented translocation de novo resulting in monosomy 18q. In rare instances, the 18q− syndrome resulted from a familial translocation, a familial pericentric inversion of chromosome 18,[8] or from an X/18 translocation with inactivation of the translocated autosome. About 10% of the reported cases are mosaics with an 18q− clone. Familial 18q− is rare.

The phenotypic features of the 18q− syndrome are rather characteristic, allowing clinical diagnosis of the condition. The main features are prenatal and postnatal growth failure with adult height rarely above 150 cm, delayed psychomotor development, muscle hypotonia, and rather typical dysmorphism. Mental retardation ranges between profound and mild; an IQ of about 70 is rarely encountered. Microcephaly is found in 70% of the cases. Some patients are aggressive and moved to self-mutilation. Visual impairment and hearing difficulties may contribute to the delay of development. Cerebral malformations are rarer than in 18p−, though agenesis of the corpus callosum and hydrocephalus have been reported.

Dysmorphic features include large ears with prominent antihelix and antitragus, narrow or atretic external ear canals, hypoplastic midface, shark mouth, short neck, widely spaced nipples, cutaneous dimples, notably over the acromion, large hands with proximally inserted thumbs and long tapering fingers, and proximally inserted second toes overlapping the third toes. Dermatoglyphics show an increased number of whorls and a transverse palmar crease in some cases.

Malformations such as cleft palate occur in one third of the cases; cleft lip is less frequent. Congenital heart lesions (notably pulmonic stenosis), supernumerary ribs, hypogonadism with hypoplastic scrotum, ectopia testis, hypospadias, and clubfeet have also been described. IgA deficiency is found in less than 50%, and recently, a patient has been found with peptidase A deficiency.[9] Peptidase A is coded by a gene in the 18q23 region.

Ocular Manifestations

Hypertelorism and nystagmus are found in four fifths of cases. Epicanthal folds, strabismus, ptosis, and pale optic discs are other frequent findings. Corneal opacities, corneal staphyloma, microcornea, microophthalmia, absence of the anterior chamber, glaucoma, myopia, astigmatism, coloboma of the iris, choroid, and retina have also been reported.[5,6,10-12]

Prognosis

Expectation of survival is guarded for those with severe malformations, yet the syndrome seems not to shorten life expectancy for others.

Ring Chromosome 18 Syndrome

Over 60 patients with the 18r syndrome have been reported since it was described originally by Wang et al. in 1962. The patients show symptoms of both the 18q− and the 18p− syndrome in greatly varying combinations. Severe clinical manifestations with very short life alternate with very mild cases and even phenotypic normalcy. Course and symptoms vary even in familial cases[13-15]. Faugeras and Barthe[15] described a small infant with multiple malformations compatible with the 18r syndrome. Although chromosome studies were not successful, 18r chromosomes were found in the father, the aunt, and the paternal grandmother of the proband. They were all phenotypically normal, although the two women were of somewhat short stature (150 cm). The phenotypic variability has been explained by variations in the size of the deleted segments, although other possible causes have been discussed by Hecht and Vlietinck.[16] The clinical symptoms of the 18r syndrome have been discussed extensively.[1,6,17] We limit our description to the eye findings.

Ocular Manifestations

Hypertelorism and epicanthal folds have been noted in 50% of the patients. Other eye findings occur with less frequency—nystagmus, strabismus, ptosis, astigmatism, microcornea, corneal leukoma, myopia, hypermetropia, cataracts, coloboma of the iris and choroid, and fundus anomalies.[5,12,17] A complex malformation in a cebocephalic newborn was thoroughly studied by

Yanoff et al.[18] The ocular findings included micro-ophthalmia, iris coloboma with intraocular cartilage and smooth muscle, iridic hypoplasia, and a retro-ocular cyst that communicated with the vitreous through a dehiscence of the posterior wall of the eye. Cataracts and retinal dysplasia were other findings in this interesting and thoroughly examined case.

Conclusion

Whenever one or several of the eye findings listed in Table 4-1 are found in combination with mental retardation, various dysmorphic features, or even malformations of the central nervous system, a structural abnormality of chromosome 18 should be taken into consideration and chromosomal analysis should be undertaken.

References

1. De Grouchy J, Turleau C. Clinical atlas of human chromosomes. 2nd ed. New York, NY: John Wiley & Sons; 1984.
2. Rethore MO, Pinet J. New deletion and ring chromosomes. In: Vinken P, Bruyn G, and Klawans, H, eds. *Handbook of Clinical Neurology*, Revised series 6(50). New York, NY: Elsevier Science Publishing Co., 1987.
3. Kiss P, Osztovics M, Kosztolanyi G, et al. Partial deletion of short arm of chromosome 18. *Acta Paed Hungar.* 1984;25:263-269.
4. Jones KL, Carey DE. Graves disease in a patient with the del(18p) syndrome. *Am J Med Genet.* 1982;11:449.
5. Howard RO. Classification of chromosomal eye syndromes. *Int Ophthalmol.* 1981;4:77-91.
6. Schinzel A. Catalogue of unbalanced chromosome aberrations in man. Berlin: Walter de Gruyter, 1984.
7. Sutton SD, Ridler MA. Prenatal detection of monosomy 18p and trisomy 18q mosaicism with unexpected fetal phenotype. *J Med Genet.* 1986;23:258-260.
8. Martin AD, Simpson JL, Deddish RB, et al. Clinical implications of chromosomal inversions: a pericentric inversion in no. 18 segregating in a family ascertained through an abnormal proband. *Am J Perinatol.* 1983;1:81-88.
9. Felding I, Kristoffersson V, Sjöström H, et al. Contribution to the 18q− syndrome. A patient with del (18)(q22.3qter). *Clin Genet.* 1987;31:206-210.
10. Insley J. Syndrome associated with a deficiency of part of the long arm of chromosome no 18. *Arch Dis Child.* 1967; 42:140-146.
11. Curran JP, Al-salihi FL, Allderdice PW. Partial deletion of the long arm of chromosome 18. *Pediatrics.* 1970;46:721-728.
12. Wilson WA, Alfi OS, Donnell GN. Ocular findings in cytogenetic syndromes. *Ophthalmology.* 1979;86:1184-1190.
13. Christensen KH, Friedrich U, Jacobsen P, et al. Ring chromosome 18 in mother and daughter. *J Ment Defic Res.* 1970;14:49-66.
14. Donlan MA, Donlan CH. Ring chromosome 18 in a mother and son. *Am J Med Genet.* 1986;24:171-174.
15. Faugeras C, Barthe D. Transmission d'un chromosome 18 en anneau sur deux generations, chez des sujets a phenotype normal. *J Genet Hum.* 1986;34:313-320.
16. Hecht F, Vlietinck RF. Autosomal rings and variable phenotypes. *Humangenetik.* 1973; 18:99-100.
17. Kunze J, Stephan E, Tolksdorf M. Ring chromosome 18, ein 18p−/18q− deletions syndrom. *Humangenetik.* 1972;15:289-318.
18. Yanoff M, Rorke LB, Niederer BS. Ocular and cerebral abnormalities in chromosome 18 deletion defect. *Am J Ophthalmol.* 1970;70:391-402.

Chapter 5

Deletion of the Short Arm of Chromosome 4

Wolf-Hirschhorn Syndrome, 4p− Syndrome

JEFFREY C. MURRAY

Since the establishment of the human chromosome complement as 46 in 1956, large numbers of chromosome abnormalities have been identified. These include abnormalities of chromosome number, such as Down's syndrome or trisomy 21 (an extra chromosome 21), first identified in 1959, and abnormalities of chromosome structure, such as deletion of the short arm of chromosome 4 (4p− syndrome) or Wolf-Hirschhorn syndrome, first identified in 1965. In this nomenclature "p" represents the short arm of a chromosome, "q," the long arm, and a minus sign (−) preceded by a number and a letter, absence (deletion) of material on that chromosome and arm. With the advent of chromosome banding techniques in the early 1970s it became possible to identify which specific regions of chromosomes were structurally altered in deletion syndromes. The 1980s have seen a proliferation of the techniques of molecular biology, which now make it possible to begin to ask which DNA segments in a deleted region are specifically involved in the particular expression of that condition.

Deletion syndromes are those in which a portion of a specific chromosome of affected persons is deleted, or missing. The chromosome makeup then includes one normal chromosome of that numbered pair. A person with 4p− has one normal chromosome 4 and one chromosome 4 with a portion of the short arm (p) deleted. A variety of such deletions involving other chromosomes have also been described. They are called syndromes when similar deletions as identified cytogenetically are associated with a common, recognizable set of clinical features. Other such syndromes and their associated deletions include cri-du-chat (5p−), Wilm's tumor–aniridia association (11p−), retinoblastoma (13q−), Prader-Willi syndrome (15q−), and Miller-Dieker syndrome (17p−).

Systemic Manifestations

Several texts and reviews have covered various features in the 4p− syndrome (Table 5-1).[1-4] Among the most remarkable consistencies have been the specificity of the craniofacial appearance. The long nasal root and hypertelorism provide a very characteristic appearance that is often recognizable in the newborn nursery. Preus et al.[4] have organized the recurring features into a scoring system that may be predictive of the presence of the disorder in certain persons. It is especially useful in siuations when results of an initial cytogenetic evaluation were normal. Advances in cytogenetic technology and individual sample variation make it reasonable to repeat the cytogenetic evaluation when there are indications that a specific disorder may be present in a particular clinical setting.

Many of the systemic features of 4p−, such as developmental delay, failure to thrive, congenital heart disease, and genitourinary abnormalities, are common to other chromosome disorders. Such manifestations likely represent the complex disturbances that result from the loss of multiple genes in the deleted region. Also of interest is the experience of ourselves and others that infants who survive the first year or two of life, when a variety of life-threatening events may occur, seem to do better in both physical and cognitive development than might be predicted from their initial presentation. Although the long-term prognosis is still guarded, it may improve somewhat as time passes.

TABLE 5-1 Features of 4p− Syndrome

Feature	Frequency 10-50%	Frequency >50%
Systemic		
Low birth weight		X
Failure to thrive		X
Developmental delay		X
Craniofacial		
Microcephaly		X
Prominent glabella		X
Hypertelorism		X
Ear anomalies		X
Cleft lip and/or palate	X	
Micrognathia		X
Organ system		
Arthrogryposis		X
Congenital heart disease	X	
Hypospadias	X	
Hemangiomas		X
Hypotonia		X
Seizures		X
Ocular system		
Strabismus		X
Nystagmus	X	
Ptosis	X	
Epicanthal folds		X
Iris coloboma	X	
Microophthalmia	X	
Cataracts	X	

Ocular Manifestations

As Table 5-1 shows, the ocular manifestations of 4p− are prominent and diverse. They range from craniofacial disturbances such as hypertelorism and eyebrow flaring, through nonspecific findings such as nystagmus (related to visual impairment in many cases), strabismus, and ptosis, to defined structural abnormalities, such as iris coloboma and cataracts. Cataracts have been described in about 25% of cases, but specifics of type and location have been noticeably lacking. In most cases, it appears that cataract formation was not noted at birth, although there is at least one exception to this. Coloboma, both iridic and retinal, and ectopic pupils and microophthalmia are also frequent findings. Although they are found in other chromosome disorders, they appear most frequently in 4p−, 13q−, and 22p+ (addition to 22p) syndromes. As discussed below, this particular feature may provide more insights into normal ocular development as detailed molecular knowledge of the deleted region accumulates. The ophthalmologic care of children with 4p− is predicated on the specific abnormalities present, taking into account the importance of adequate vision for the child and the likelihood that cognitive skills will improve (although not become normal) as time passes.

Discussion

The cytogenetic description of 4p− has demonstrated that deletions of various sizes in the terminal end of 4p can result in similar phenotypes. Although most such cases arise spontaneously and carry less than 1% risk for recurrence, up to 10% may result from balanced translocations or inversions in chromosome 4 found in clinically normal parents. Such cases may have a substantial recurrence risk, so all cases of 4p− require investigation of parental karyotypes as well.

Recently, the locus for Huntington's disease was localized to the region involving 4p− cases.[5] This has generated enormous interest in characterizing the molecular genetics of this portion of chromosome 4 and may eventually lead to defining specific genes or DNA sequences involved in the ocular abnormalities found in 4p−.[6] Several genes of interest are already known to lie in this region, including the *ras* pseudo-oncogene, and it is hoped that as the region becomes better defined specific correlations can be made between the consistent phenotype of 4p− and the genes whose structures have been disturbed. It will be by merging clinical descriptions with cytogenetic and molecular explanations that the fundamental questions of disturbed ocular development will be answered. The availability of a molecular characterization of the region may also lead to detecting persons with 4p− whose chromosomes appear normal but who may harbor cytogenetically "invisible" deletions.

References

1. Schinzel A. Chromosome 4 deletions. In: Schinzel A, ed. *Catalogue of Unbalanced Chromosome Aberrations in Man.* Hawthorne, NY: De Gruyter; 1983;161–164.
2. Smith DW. 4p− Syndrome (chromosome number 4 short arm deletion syndrome). In: Smith DW, ed. *Recognizable Patterns of Human Malformation.* Philadelphia: WB Saunders; 1982:36-37.
3. Wilson MG, Towner JW, Coffin GS, et al. Genetic and clinical studies in 13 patients with the Wolf-Hirschhorn syndrome [del(4p)]. *Hum Genet.* 1981;59:297-307.
4. Preus M, Ayme S, Kaplan P, et al. A taxonomic approach to the del (4p) phenotype. *Am J Med Genet.* 1985; 21:337-345.
5. Gusella JA, et al. Deletion of Huntington's disease-linked G8 (D4S10) locus in Wolf-Hirschhorn syndrome. *Nature.* 1985; 318:75-78.
6. Gusella JA, Gilliam TC, MacDonald ME, et al. Molecular genetics of human chromosome 4. *J Med Genet.* 1986;23:193-199.

Chapter 6

Deletion of the Short Arm of Chromosome 5

Cri du Chat Syndrome, 5p− Syndrome

METTE WARBURG and ERIK NIEBUHR

A deletion of the short arm of chromosome five (5p−) was first described in 1963 by Lejeune.[1] The clinical features depend on the loss of the area around the p15.2 band (Fig 6-1.)[2] When affected infants cry they sound like a cat mewing, hence the name Cri du Chat syndrome.

Systemic Manifestations

A round face is characteristic in infancy, but in later years it usually becomes elongated (Fig. 6-2). The root of the nose is almost always broad, and many authors have

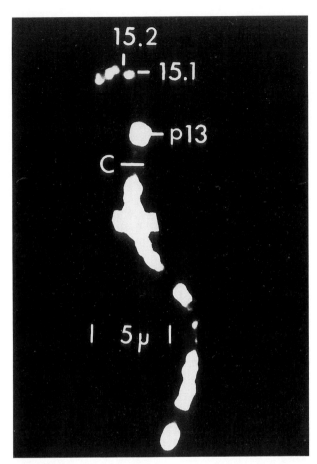

FIGURE 6-1 A normal chromosome 5 (BrdU - acridine orange staining) demonstrating the area around 5p15.2, which is the site for the deletion with the Cri-du-Chat syndrome.

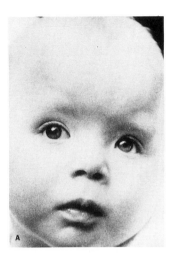

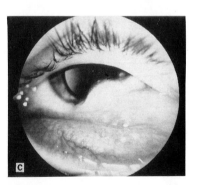

FIGURE 6-2 (**A**) Infant and (**B**) adult with 5p− syndrome. Note the round face in the infant and the coarser, more oblong features of the adult. The root of the nose is broad. The epicanthal folds are most pronounced in the infant. Both patients have flat protruding ears. (**C**) Epibulbar epidermoid in a patient with hemifacial microsomia and 5p−. This case was originally described by Ladekarl.[5]

described this as hypertelorism or telecanthus, but measurements of the distance between the inner and outer canthi as well as the canthal indexes are usually within normal values.[3,4] Facial asymmetry occurs in about one third of all patients, and preauricular tags have been reported in about 20%.[2] Two patients with typical hemifacial microsomia have been described.[5,6] Typical transverse simian creases are found in 64.5% of patients with 5p−,[7] an even higher incidence than the 45% observed in patients with Down's syndrome.[8]

Less frequent is congenital heart disease (15%), mainly persistent patent ductus arteriosus or atrial septal defects. Cutaneous hemangiomas are occasionally observed. Mental retardation is usually profound, producing slow progress in social and motor skills; a limited vocabulary may be present, and a few patients even speak in short sentences.

Ocular Manifestations

Epicanthal folds are frequent in infancy but tend to disappear later. The palpebral fissures look small but are within the limits of normal.[4] Esotropia and exotropia are very common. Among 15 patients examined ophthalmologically by one of us (MW) four had excessive myopia, two high hypermetropia, one astigmatism, and seven emmetropia. Spectacles were used by two of the myopic and one of the hypermetropic patients. Among these 15 patients, one had congenital cataracts, two showed optic nerve atrophy and one of these had optic disc hypoplasia. Tortuous retinal arterioles were observed in one of the patients.

Similar features were described in other patients with the 5p− syndrome.[3,9-11] The pupils may react with constriction to the application of 2.5% mecholyl, and tearing may be decreased.[12] It is not difficult to do refractions, and slit lamp and ophthalmoscopic examinations are easy to perform when the patients lie down. We found that spectacles should always be prescribed in high ametropias.

Vital Statistics

The incidence of the 5p− syndrome is approximately 1 in 50,000 neonates, and the prevalence among mentally retarded persons with IQ below 50 is 1:350. There is an unexplained excess of females (4:3). The infants are born at expected term, but birth weight is usually low. Survival in infancy is the rule (Fig. 6-3), but a few infants die as a result of aspiration or congenital heart disease.

Prevention

In about 10 to 15% of patients, the deletion is familial and caused by translocations or rarer cytogenetic aberrations,[13] so cytogenetic examination should always include at least the parents. Carriers of balanced chromosomal aberrations may then be offered prenatal diagnosis by chorionic villus biopsy or amniocentesis. DNA fragments localized to the region in which the deletion occurs are in the process of being defined. Such probes will improve prenatal diagnosis.[14] So far, no patient has

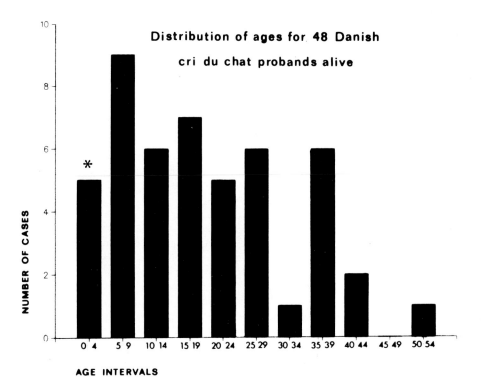

FIGURE 6-3 Distribution of age among 48 Danish patients with the 5p− syndrome. The mean observation period was 13 years. Three probands who died within the first year of life are not included.

reproduced. They all experience normal puberty; females have regular menstrual periods and males produce seminal discharge.

References

1. Lejeune J, Lafourcade J, Berger R, et al. Trois cas de délétion partielle du bras court d'un chromosome 5. *C R Acad Sci [III]*. 1963;257:3098-3102.
2. Niebuhr E. The Cri du Chat syndrome: epidemiology, cytogenetics, and clinical features. *Hum Genet*. 1978;44:227-275.
3. Schechter RJ: Ocular findings in a newborn with Cri du Chat syndrome. *Ann Ophthalmol*. 1978;10:339-344.
4. Niebuhr E. Anthropometry in the Cri du Chat syndrome. *Clin Genet*. 1979;16:82-95.
5. Ladekarl S. Combination of Goldenhar's syndrome with the Cri du Chat syndrome. *Acta Ophthalmol (Cop)*. 1968; 46:605-610.
6. Neu KW, Friedman JN, Howard-Peebles PN. Hemifacial microsomia in Cri du Chat (5p−) syndrome. *J Craniofac Genet Develop Biol*. 1982;2:295-298.
7. Holt SB, Niebuhr E. Dermatoglyphics in "Cri du Chat" syndrome (5p−). *Birth Def*. 1979;XV(6):565-589.
8. Øster J. Mongolism. *A Clinicogenealogical Investigation Comprising 526 Mongols Living on Seeland and Neighbouring Islands in Denmark*. Copenhagen: Danish Science Press;1953.
9. Kitsiou-Tzeli S, Dellagrammaticas HD, Papas CB, et al. Unusual ocular findings in an infant with Cri-du-Chat syndrome. *J Med Genet*. 1983;20:304-307.
10. Suerinck E, Noël B, Rethoré MO. Ring chromosome 5 in two malformed boys with Cri du Chat syndrome. *Clin Genet*. 1978;14:125-129.
11. Clark DI, Howard PJ, Patterson A. Ocular findings in a patient with deletion short arm chromosome 5 (Cri du Chat) and ring chromosome 14. *Trans Ophthalmol Soc UK*. 1986;105:723-725.
12. Howard R. Ocular abnormalities in the Cri du Chat syndrome. *Am J Ophthalmol*. 1972;73:949-954.
13. De Capoa A, Warburton D, Breg WR, et al. Translocation heterozygosis: a cause of five cases of the Cri du Chat syndrome and two cases with a duplication of chromosome number five in three families. *Am J Hum Genet*. 1967;19:586-603.
14. Overhauser J, Beaudet AL, Wasmuth JJ. A fine structure physical map of the short arm of chromosome 5. *Am J Hum Genet*. 1986;39:562-572.

Chapter 7

Group D Ring Chromosomes

MARIA A. MUSARELLA

A karyotype shows all the chromosomes of a single cell. By convention, they are arranged by pairs of homologues from the largest to the smallest (numbered from 1 to 22 and X and Y). The 23 pairs of chromosomes are divided according to the position of the centromere and to size into seven major groups: group A, chromosomes 1 through 3; group B, chromosomes 4 and 5; group C, chromosomes 6 through 12 and X; group D, chromosomes 13 through 15; group E, chromosomes 16 through 18; group F, chromosomes 19 and 20; and group G, chromosomes 21, 22, and Y. Although individual chromosomes are now usually discussed, the group designations still serve some purpose in describing chromosomes of similar size and structure. With the development of banding techniques, different chromosomes and their regions or bands can be distinguished by differences in staining intensity. High-resolution banding has further subdivided the individual bands.

Abnormalities of the chromosomes may be either numerical (extra or missing chromosomes) or structural (deletions, translocations, rings). Structural rearrangements are due to chromosome breakage and abnormal reunion. A ring chromosome is a type of deletion that results from breakage and loss in the terminal ends of each arm of a chromosome, followed by union of the broken ends. This mechanism presumes the loss of genetic material, particularly from the distal segments. Additionally, ring chromosomes can be unstable and can induce complex cell division events resulting in variable duplication or deficiency, mosaicism, and unpredictable phenotypic expression.[1]

This chapter is concerned with the phenotypes and ocular features that reflect ring formations of the group D chromosomes (Dr).

Phenotype Associated with the Dr Anomaly

The first observation of a ring chromosome in humans was made by Wang et al. in 1962.[2] Formation of ring chromosomes can result in considerable phenotypic variation. Although the syndromes that result from ring chromosomes are very heterogeneous, the clinical variation is least in the case of ring chromosome 13 (13r). The 13r syndrome was first reported in 1968; since then there have been several more reports and a defined syndrome has evolved.[3-7] The clinical picture is very specific, characterized by marked mental retardation, microcephaly, and craniofacial dysmorphism. Its incidence in the Caucasian population is estimated to be 1 in 58,000.[4] In all patients so far described the break points have occurred within the regions bounded by bands 13q21 to 13q34. Niebuhr and Ottosen[8,9] identified two 13r subgroups clinically—Group 1, those with normal thumbs, and Group 2, those with absent or hypoplastic thumbs. In Group 1, the chromosome break involved the loss of bands 13q33 and 13q34. This was associated clinically with microcephaly, hypertelorism, epicanthic folds, broad nasal bridge, a triangular cranium, and large ears. Anogenital malformations, protruding maxilla, and nonspecific ocular abnormalities were other features in this subgroup. Niebuhr's second group was associated with break points at 13q21 and loss of bands 13q22, q31, q32, q33, and q34. In addition to clinical features of Group 1, the thumbs of the second group were absent or hypoplastic. This classification is felt by some to reflect neither the clinical spectrum seen with this syndrome nor the great diversity of break points observed. Other anomalies of the hands include supernumerary fingers and bifid thumbs. Although the phenotype of patients with the 13r syndrome is characteristic, cytogenetic studies are essential for diagnosis.

In the literature approximately twenty 14r patients have been reported.[10-18] A number of characteristic malformations and stigmata are consistently seen in all described cases with 14r chromosome—seizures, psychomotor and growth retardation, microcephaly, dolichocephaly, high forehead, downward slanting palpebral fissures, epicanthic folds, short neck, and neurologic abnormalities, including tremor, athetosis, hypotonia, and hypertonia. All patients have had recurrent respiratory infections, and several had congenital heart defects and café au lait spots. The combination of seizures, mental retardation, and minor craniofacial dysmorphism in a child strongly warrants chromosome investigation for a 14r anomaly.

Since the introduction of the different banding techniques, at least 25 patients with 15r have been reported.[19-23] The clinical manifestations produced by 15r syndrome are different from those observed in the other Dr syndromes. While the 13r and 14r syndromes give rise to several clinical pictures, the clinical manifestations of the 15r syndrome are reduced, and there is no identifiable phenotype. Although no truly characteristic facial dysmorphism has been described, prenatal and postnatal growth retardation, slight to moderate mental retardation, and minor facial anomalies have been a constant feature present in all reported cases. Notably, the growth retardation can be extreme and has been described in adults as well as children with 15r syndrome. Microcephaly, anomalies of the radial axis, micrognathia, and a bird-headed profile are some of the other malformations that have also been reported.

Ocular Manifestations

Except possibly for 14r, the ocular findings associated with Dr abnormalities are essentially nonspecfic. Reported ocular features in patients with 13r have included hypertelorism, epicanthic folds, narrow palpebral fissures, mongoloid slants, strabismus, microphthalmia, iris colobomas, and nystagmus. Ocular anomalies in the 14r phenotype include hypertelorism, epicanthic folds, strabismus, downward slanting, small or normal palpebral fissures, and nystagmus. In one patient with 14r reported twice, Brushfield's spots, pigmentary mottling in the peripheral retina with poor ERG response, and pinpoint white opacities of the macula were noted.[18,23] Abnormal retinal pigmentation without macular or iridic lesions has been described in an additional child.[17] Ring chromosome 15 appears to have the least ocular involvement of the Dr chromosomes. Changes, when they do occur, appear to involve solely the adnexa. Hypertelorism and narrow lid fissures are the only findings reported to date.

References

1. Lejeune J. Modèle théorique de la repartitions des duplications et des deficiences dans les chromosomes en anneau. *C R Acad Sci.* 1967;264:2588.
2. Wang H-C, Melnyk J, McDonald LT, et al. Ring chromosomes in human beings. *Nature.* 1962;195:733.
3. Lejeune J, Lafourcade J, Berger R, et al. Le phenotype (Dr). Etude de trois cas de chromosomes D en anneau. *Ann Genet (Paris).* 1968;11:79.
4. Martin NJ, Harvey PJ, Pearn PJ. The ring chromosome 13 syndrome. *Hum Genet.* 1982;61:18.
5. Freid K, Rosenblatt M, Mundel G, et al. Ring chromosome 13 syndrome. *Clin Genet.* 1975;7:203.
6. Fryns JP, Deroover J, Van den Berghe H. Malformative syndrome with ring chromosome 13. *Humangenetik.* 1974;24:2335.
7. Parcheta B, Wisiewski L, Piontek E, et al. Clinical features in a case with ring chromosome 13. *Eur J Pediatr.* 1985;144:412.

8. Niebuhr E, Ottosen J. Ring chromosome D(13) associated with multiple congenital malformations. *Ann Genet.* 1973;16:157.

9. Niebuhr R. Partial trisomies and deletions of chromosome 13. In: Yunis JJ: *New Chromosomal Syndromes.* New York, NY: Academic Press; 1977.

10. Sparkes RS, Klisak I, Sparkes MC. Extended evaluation of previously reported twins with ring 14 chromosome. *Ann Genet.* 1977;20:273.

11. Amarose AP, Dorus E, Huttenlocker PR, et al. A ring 14 chromosome with deleted short arm. *Hum Genet.* 1980;54:145.

12. Iselius L, Ritzen M, Bui TH, et al. Ring chromosome 14 in a mentally retarded girl. *Acta Paediatr Scan.* 1980;69:803.

13. Bowser Riley S, Buckton KE, Ratcliffe SG, et al. Inheritance of a ring 14 chromosome. *J Med Genet.* 1981;18:109.

14. Lippe BM, Sparkes RS. Ring 14 chromosome: association with seizures. *Am J Med Genet.* 1981;9:301.

15. Schmidt R, Eviatar L, Notowsy HM, et al. Ring chromosome 14: a distinct clinical entity. *J Med Genet.* 1981;18:304.

16. Fryns JP, Kubien E, Kleczkowska A, et al. Ring chromosome 14. A distinct clinical entity. *J Genet Hum.* 1983;31:367.

17. Gilgenkrantz S, Morali A, Vidailhet M, et al. Le syndrome r(14). Trois nouvelles observations. *Ann Genet.* 1984;27:73.

18. Raoul O, Razavi F, Lescs M-C, et al. Chromosome 14 en anneau. I. Une observation de r(14) homogene. *Ann Genet.* 1984;27:88.

19. Fryns JP, Timmermans J, D'Hondt F, et al. Ring chromosome 15 syndrome. *Hum Genet.* 1979;51:43.

20. Yunis E, Leibovici M, Quintero I. Ring(15) chromosome. *Hum Genet.* 1981;51:207.

21. Fryns JP, Jaeken J, Devlieger H, et al. Ring chromosome 15 syndrome. *Acta Paediatr Belg.* 1981;34:47.

22. Fryns JP, Kleczkowska A, Buttiens, et al. Ring chromosome 15 syndrome. Further delineation of the adult phenotype. *Ann Genet.* 1986;29:45.

23. Guillot M, Duffier JL, Perignon F, et al. Anomalies ocularie du phentype 46, XYr(14). *Arch Fr Pediatr.* 1983;40:433.

Section B

Sex Chromosome Disorders

Chapter 8

Klinefelter's Syndrome

GARY G. DECKELBOIM and JOEL S. MINDEL

Klinefelter's syndrome is a sex chromosome trisomy in which the usual chromosomal complement is 47,XXY, although approximately 15% of patients are mosaic, with two or more distinct cell lines.[1] It is the most frequent major abnormality of sexual differentiation, occurring in one in 500 males. The most common mosaic pattern is 46,XY/47,XXY, but others have been described. Those affected with the syndrome are phenotypically male.

Studies of the Xg blood group have helped elucidate the mechanisms by which the abnormal sex chromosome patterns may occur.[2] The Xg blood group, described by Mann et al.[3] in 1962, is an X-linked red blood cell group, in contrast to the more familiar autosomal blood groups such as ABO and Rh. For example, if both a father and his Klinefelter's son are Xg(a+) and the mother is Xg(a−), the son had to have received the Xg^a gene, and the X chromosome that carries it, from the father. Similarly, in a family in which the father is Xg^a positive and the mother and the Klinefelter's sons are Xg^a negative, the X chromosome could not have been transferred from father to son. If both parents were Xg^a

positive and the affected son is Xg^a negative, the mother must have been a heterozygote Xg^{a+}/Xg^{a-}, the son inheriting the Xg^{a-} in duplicate.

By means of investigations such as these, it has been determined that approximately 50% of patients derive both non-identical X chromosomes from the mother, indicating that nondisjunction must have occurred during the primary division of oogenesis. Another 40% of patients received one of the X chromosomes from the mother and the other X chromosome from the father, indicating nondisjunction during spermatogenesis. Finally, 10% of Klinefelter's patients have two X chromosomes that are identical and of maternal origin, pointing to nondisjunction either during the second meiotic division of oogenesis or early after zygote formation.

Systemic Manifestations

The one or more extra X chromosomes present in some way interfere with normal development of the testes. Sperm cell and testosterone production are variably abnormal, and most of the systemic features of the syndrome are attributable to this. In general, the greater the number of X chromosomes in the karyotype, the more severe the defect in sexual development and the more abnormal the phenotype. However, unless chromosome analysis is performed for another reason, the syndrome generally is not recognizable before puberty, when secondary sex characteristics are expected to appear.

Death of the germ cell line leads to hyalinization of the seminiferous tubules; azoospermia and testicular atrophy (testicle length usually less than 2 cm) are the rule. This generally leads to sterility, although some mosaics have been reported to be fertile. The average plasma testosterone value is half normal, but there is considerable variability. Follicle-stimulating hormone (FSH) and luteinizing hormone (LH) levels are typically elevated, as are plasma estradiol levels. The androgen-to-estrogen ratio is, therefore, variable and this range of values accounts for the different degrees of androgenization and feminization noted clinically. Gynecomastia appearing at puberty is common. The risk of breast cancer may be up to twenty times that of normal males but is still significantly lower than that of normal females. Obesity and varicose veins occur in one third to one half of patients. Mental capacity varies from normal to moderate retardation. Axillary and facial hair may be sparse.

Ocular Manifestations

Ocular abnormalities are uncommon in Klinefelter's syndrome.[4] Bilateral anophthalmos associated with complete absence of the optic nerves, chiasm, and optic tracts has been reported,[5] as has optic atrophy.[6] Other abnormalities noted in the syndrome include epicanthal folds, hypertelorism, strabismus, dislocated lens, focal posterior keratoconus, unilateral corneal opacities, diffuse choroidal atrophy with midperipheral bone-corpuscular pigment clumping, retinitis pigmentosa, and uveal coloboma.[7-10] None of these reported entities can be clearly attributed to the abnormal chromosomal complement present in Klinefelter's syndrome.

Perhaps the most interesting demonstrations of ocular pathophysiology in Klinefelter's syndrome are the X-linked recessive color vision abnormalities, such as protanopia and deuteranopia. Reports of dyschromatopsia vary from none of 72 Klinefelter's patients[11] to one of six patients.[12] Francois et al.[13] reviewed the literature and found a total of ten patients out of 292 Klinefelter's were reported to have a congenital dyschromatopsia, yielding an incidence of approximately 3.5%.

While this is lower than the 8 to 10% incidence of dyschromatopsia in normal males, it is higher than the 0.5% incidence in normal females. The explanation is that 90% of Klinefelter's patients have two nonidentical X chromosomes and 10% have two identical maternally derived X chromosomes. This is in contrast to the normal population, 50% of whom (females) have two nonidentical X chromosomes and 50% of whom (males) have a single X chromosome. As a larger proportion of Klinefelter's patients have two nonidentical X chromosomes than the normal population, the incidence of dyschromatopsia would be lower in Klinefelter's syndrome. The incidence of color-defective vision in Klinefelter's patients who have two identical X chromosomes is the same as in the normal male population, but as only 10% of Klinefelter's patients have two identical X chromosomes, the overall theoretical incidence of dyschromatopsia should be in the range of 2%.

References

1. Thompson JS, Thompson MW. *Genetics in Medicine*. Philadelphia, PA: WB Saunders; 1980:175.
2. Froland A, Sanger R, Race R. Xg blood groups of 78 patients with Klinefelter's syndrome and some of their parents. *J Med Genet*. 1968;5:161.
3. Mann JD, Cahan A, Gelb A, Fisher N, et al. A sex-linked blood group. *Lancet*. 1962;1:8.
4. Duke-Elder S. *System of Ophthalmology*. Summary of Systemic Ophthalmology. St. Louis, Mo: CV Mosby; 1976; XV:37.
5. Welter DA, Lewis LW, Scharff L, et al. Klinefelter's syndrome with anophthalmos. *Am J Ophthalmol*. 1974;77:895.
6. Scouras J, Cuendet JF. Atrophie du nerf optique en le syndrome de Klinefelter. *Ann Oculist (Paris)*. 1973;206:115.
7. Gruskin, AB. In: Harley RD, ed: *Pediatric Ophthalmology*. Philadelphia, Pa: WB Saunders; 1983:973.
8. Wolkstein MA, Atkin AK, Willner JP, et al. Diffuse choroidal atrophy and Klinefelter syndrome. *Acta Ophthalmol*. 1983;61:313.
9. Hashmi MS, Karseras G. Uveal colobomata and Klinefelter syndrome. *Br J Ophthalmol*. 1976;60:661.
10. Francois J, Matton-van Leuven M, Gombault P. Uveal coloboma and true Klinefelter syndrome. *J Med Genet*. 1970;7:213.
11. Polani PI, Bishop PMF, Lennox B, et al. Color vision studies and the X chromosome constitution of patients with Klinefelter's syndrome. *Nature*. 1958;182:1092.
12. Rohde RA. Chromatin-positive Klinefelter's syndrome. Clinical and cytogenetic studies. *J Chronic Dis*. 1963;16:1139.
13. Francois J, Berger R, Saraux H. *Les Aberrations Chromosomiques en Opthalmologie*. Paris:Masson et Cie; 1972: 455.

Chapter 9

Turner's Syndrome

Gonadal Dysgenesis

ELIAS I. TRABOULSI and GEORGIA A. CHROUSOS

A chromosomal complement of 45,X is the most common cytogenetic abnormality in abortuses (18%); it is estimated to be present in 9% of spontaneous abortions and in 1.4% of all conceptuses. Only 1 in 3500 female newborns has only one X chromosome and shows the Turner's syndrome phenotype (compared to 1 in 1000 who has an extra X chromosome). The lethality of the 45,X karyotype prenatally and its relative benignity after birth is explained by the high proportion of mosaicism among survivors; 40% of patients with the Turner's phenotype are mosaics and have a variety of karyotypes with a structural alteration of the X chromosome or mosaicism involving one or more cell lines with abnormal number or structure (eg, X/XX, X/XXX, isochromosome for the long arm of X, deletions of parts of Xp or Xq, and ring X chromosomes).[1]

Turner was the first to describe the syndrome of sexual infantilism, short stature, webbing of the neck, and cubitus valgus. Ford et al.[2] demonstrated the 45,X karyotype in 1959. The phenotype is female, and results of the buccal mucosa chromatin test are usually negative. Other characteristics of this syndrome include a low nuchal hairline, webbing of the neck, characteristic facies, abnormal dermatoglyphics with a high ridge count, a shield chest with widely spaced nipples, coarctation of the aorta, and pedal lymphedema, especially in newborns. There usually is sexual infantilism, with sparse pubic hair and primary amenorrhea. The external genitalia are female, however the ovaries are most often reduced to a streak of connective tissue. Because of the high proportion of mosaics, and depending on the point during embryogenesis when the chromosomal abnormality occurs and the abnormal cell lines develop, the phenotype varies from mild to flagrant. Patients with Turner's syndrome generally have normal intelligence. They may develop psychological problems during puberty because of their appearance and their failure to develop secondary sex characteristics. Cyclic estrogen and progesterone therapy allows development of secondary sex characteristics and menstruation.

Ocular Manifestations

Strabismus is 10 times more common in patients with Turner's syndrome than in the general population. It is nonfamilial and is thought to be directly related to the phenotype. Up to one third of patients have esotropia or exotropia, the former being four times more common than the latter (Fig. 9-1).[3] External ocular abnormalities

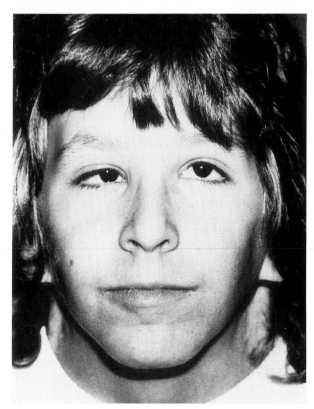

FIGURE 9-1 Facial features of a patient with Turner's syndrome. Note right ptosis and esotropia. Also present are epicanthus, low-set ears, and webbed neck. (From Chrousos.[3])

include ptosis in 16% of cases, and hypertelorism, epicanthus (-blepharalis or -inversus), and antimongoloid slanting of the palpebral fissures each in 10% of cases. The prevalence of red-green color deficiency is 10% in female Turner's patients as compared with 8% in normal males and 0.4% in normal females; this is explained by the X-linked recessive inheritance of this color vision defect. Other rarely reported ocular associations with this syndrome include periodic alternating nystagmus in one patient, microcornea in one patient, conjunctival cysts in one patient, blue sclerae, corneal nebulae, and cataracts.[3,4]

The approach to the management of strabismus in Turner's syndrome is the same as in the general pediatric population—early correction of refractive errors, prevention of amblyopia, and surgical ocular alignment. Ophthalmologists should be aware of the associated cardiac abnormalities and of the increased surgical risk for those patients. If severe ptosis threatens visual development or causes cosmetic problems, surgery can be performed.

References

1. Thompson MW. The sex chromosomes and their disorders. In: Thompson JS and Thompson MW, eds. *Genetics in Medicine*. Philadelphia, Pa: WB Saunders; 1986:144-145.
2. Ford CE, Jones KW, Polani PE, et al. A sex-chromosome anomaly in a case of gonadal dysgenesis (Turner's syndrome). *Lancet* 1959;1:711-713.
3. Chrousos GA, Ross JL, Chrousos G, et al. Ocular findings in Turner syndrome. A prospective study. *Ophthalmology*. 1984;91:926-928.
4. Lessell S, Forbes AP. Eye signs in Turner's syndrome. *Arch Ophthalmol*. 1966;76:211-213.

Section C
Trisomy Syndromes

Chapter 10
Cat Eye Syndrome

ALAN Y. CHOW

The association of iris coloboma and anal atresia was described by Haab[1] in 1878. In 1965, Schachenmann et al.[2] reported the presence of an additional small acrocentric chromosome, which was sometimes "satellited" on the short arms, in three patients with similar phenotypic findings. Other findings in their patients included hypertelorism, a slight antimongoloid slant of the palpebral fissures, microphthalmia, coloboma of the choroid, strabismus, preauricular fistulas, kidney malformations, congenital dislocation of the hips, and mild mental retardation. The mother of one patient consistently displayed an extra chromosome, and several relatives of the same patient were mosaics for the chromosomal abnormality. Schachenmann suggested a relationship between the extra chromosome and the clinical symptoms.

Since Schachenmann's report, over 50 cases have been described,[3] and the syndrome has become known as the "cat eye syndrome" (CES) for the prominent iris colobomas often present (Fig. 10-1). Some affected patients have also been shown to have congenital cardiac abnormalities,[4-8] cleft lip and palate,[6,9] or abnormalities of the lungs, pancreas, spleen,[6] intestines,[3] skeletal system,[3,4,7,9,12] and biliary tract.[5,6]

The early 1970s saw much debate in the literature over the etiology of the extra small chromosome, which was similar to but only about half as large as a G group chromosome. Pfeiffer et al.[12] in 1970 suggested that it may originate from duplication of chromosome 14 due to its characteristic of late replication. Late labeling on autoradiography of the supernumerary chromosome, however, was also consistent with chromosome 13, 14, 18, or 22.[9] With the advent of various chromosome banding techniques in the mid-1970s and of studies on family members of patients with CES, the possibilities were narrowed to partial deletions of either chromosome 13 or 22, and most studies suggested partial trisomy of chromosome 22 (22pter→q11).[3,8-10] Owing to the small size of the chromosome, G-, C-, and Q-banding techniques did not always conclusively identify the fragment.

More recently, investigators using the new radioactive chromosome probe, presented strong evidence in favor of an extra chromosome 22 fragment being respon-

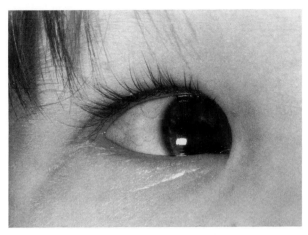

FIGURE 10-1 Inferior iris coloboma in the cat eye syndrome.

TABLE 10-1 Clinical Manifestations of Cat Eye Syndrome

Manifestations	Prevalence (%)
Systemic	
Mental retardation	80%
Microcephaly	45%
Ear anomalies	90%
Cardiovascular anomalies	55%
Renal or urinary tract anomalies	20%
Anal atresia or stenosis with or without fistula	75%
Ophthalmic	
Coloboma	65%
Hypertelorism	50%
Antimongoloid slant	40%
Microophthalmia	25%

sible for CES. McDermid et al.[11] in 1986 used a DNA probe isolated and localized to 22q11 by hybridization in situ to metaphase chromosomes and showed that multiple copies of the sequence were found in CES patients. In their two CES patients with the supernumerary chromosome, four copies were found in one patient, and three copies in the other. There was no evidence that sequences from other chromosomes were involved. From this new evidence CES therefore seems to result from the presence of either three or four copies of DNA sequences from the q11 portion of chromosome 22.

Many clinically affected persons have the supernumerary chromosome in all their cells; however, patients with mosaic patterns and even a normal chromosome constituency are common.[3,4] One explanation may be that the extra chromosome is lost in some cells in early embryogenesis after its effects have been exerted. On the other hand, phenocopies of various origins may also account for the CES reports with normal chromosomes, as the abnormalities described are seen in many other conditions. The early timing of the effects from the supernumerary chromosome during the first 2 months of gestation is supported by the types of defects seen (ie, iris and choroidal coloboma, anal atresia, and congenital heart malformations), anomalies that are generally thought to develop from events that occur during the first 2 months of embryogenesis.

Systemic Manifestations

The systemic manifestations of CES involve multiple organ systems and are variable. The most prominent systemic finding, occurring with approximately 75% prevalence, is anal atresia, often associated with a rectoperineal or rectovaginal fistula (Table 10-1). Prompt surgical correction is necessary in some cases to alleviate

bowel obstruction. Mental retardation, usually of mild to moderate severity, is seen in 80% of CES patients. Other common abnormalities are low-set, malformed ears with preauricular tags or fistulas (90%) and microcephaly (45%).

Cardiovascular anomalies occur with a prevalence of about 55%. They include tetralogy of Fallot, total anomalous pulmonary venous return, tricuspid atresia, and atrial and ventricular septal defects.[6]

Renal and urinary tract malformations, which were described in Haab's first case, are found in approximately 20% of CES patients and are variable. They include congenital absence of a kidney and ureter, obstructive uropathy, hydropelvis, bladder neck obstruction with vesicoureteral reflux, submeatal stenosis, and horseshoe kidney.

Musculoskeletal deformities are reported less commonly and may include long thumbs, hemivertebrae, congenital hip dislocation, mandibular hypoplasia, and maxillary hypoplasia.

Less common miscellaneous findings are extrahepatic biliary atresia, umbilical hernia, and on autopsy, partial anular pancreas, multiple accessory spleens, and incomplete lobulation of the lungs.

Ocular Manifestations

The ocular anomalies are often the findings that draw attention to the patient, and the most prominent of these are the iridic and choroidal colobomas (65%). These are found in the location for "typical colobomas" located inferonasally and result from improper closure of the embryonic fissure (see Fig. 10-1). They may be unilateral or bilateral, complete or incomplete. Depending on the severity of the choroidal coloboma or macular involvement, vision may be normal to poor in the affected eye. Strabismus, when present, appears to be related to poor vision. Hypertelorism is found in 50% of cases with an antimongoloid slant of the eyes in 40% of patients, producing a characteristic appearance. Microphthalmia occurs less frequently, in about 25% of patients. Congenital cataracts have also been reported occasionally.

Treatment

Treatment for CES is usually surgical and may involve correction of cardiac abnormalities, imperforate or stenotic anus or rectum, renal malformations, biliary tract atresia, strabismus, and congenital cataract. Genetic counseling and workup for parental, sibling, and relative involvement are important, since many cases of CES have familial involvement. Transmission appears to be irregularly dominant. A workup should be performed in patients born with iris coloboma and anal anomalies, especially if associated with ear deformities or congenital heart disease.

References

1. Haab O. Beitrage zu den angeborenen Fehlern des Auges. *Graefes Arch Klin Ophthalmol.* 1878;24;257.

2. Schachenmann G, Schmid W, Fraccaro M, et al. Chromosomes in coloboma and anal atresia. *Lancet.* 1965;2:290.

3. Guanti G. The aetiology of the cat eye syndrome reconsidered. *J Med Genet.* 1981;18:108.

4. Neu RL, Assemany SR, Gardner LI. "Cat eye" syndrome with normal chromosomes. *Lancet.* 1970;1:949.

5. Peterson RA. Schmid-Fraccaro syndrome ("cat eye" syndrome). *Arch Ophthalmol.* 1973;90:287.

6. Freedom RM, Park SG. Congenital cardiac disease and the "cat eye" syndrome. *Am J Dis Child.* 1973;126:16.

7. Cory CC, Jamison DL. The cat eye syndrome. *Arch Ophthalmol.* 1974;92:259.

8. Bofinger MK, Shirley WS. Cat eye syndrome. *Am J Dis Child.* 1977;131:893.

9. Weber FM, Dooley RR, Sparkes RS. Anal atresia, eye anomalies, and an additional small abnormal acrocentric chromosome (47,XX,mar+): report of a case. *J Pediatr.* 1970;76:594.

10. Schinzel A, Schmid W, Fraccaro M, et al. The "cat eye syndrome": dicentric small marker chromosome probably derived from a no. 22 (tetrasomy 22pter − q11) associated with a characteristic phenotype. *Human Genet.* 1981;57: 148.

11. McDermid HE, Duncan AM, Brasch KR, et al. Characterization of the supernumerary chromosome in cat eye syndrome. *Science.* 1986;232:646.

12. Pfeiffer RA, Heimann K, Heiming E. Extra chromosome in "cat eye " syndrome. *Lancet.* 1970;2:97.

Chapter 11

Trisomy 13

Patau's Syndrome

FREDERICK C. BLODI

Trisomy 13 (originally assumed to be trisomy 13-15, belonging to the D group) is one of the chromosome changes that most frequently and consistently causes ocular anomalies. The ocular findings are so distinctive that they may be of help in establishing the clinical diagnosis.[1-8] The frequency of the anomaly is 2 to 4.5 in 10,000 births and the incidence increases with maternal age. It occurs with equal frequency in both sexes. It is usually a lethal condition: 50% of affected infants die before the age of 1 month, and 95% die before the age of 1 year, though longer survival has been reported.

The main extraocular defects involve the central nervous system (microcephaly, holoprosencephaly, ar-

hinencephaly), the face (cleft lip and palate, malformed, low-set ears), the skeleton, and the cardiovascular, genitourinary, and gastrointestinal tracts. The patients are retarded, deaf and blind and have seizures and apneic spells.

The severe ocular anomalies are incompatible with vision and in nearly all affected children both eyes are involved (Table 11-1). The whole gamut of anomalies has been referred to as dysembryogenesis (certainly a more correct term than the old name "mesodermal dysgenesis"). Before the chromosomal anomaly was appreciated, the condition was referred to as retinal dysplasia, though more ocular structures than the retina alone are affected.

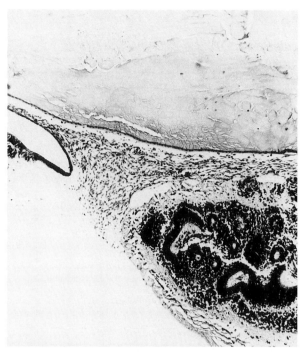

FIGURE 11-1 Vascularized connective tissue (persistence and hypertrophy of the primary vitreous) behind the lens surrounded by dysplastic retina. The lens is cataractous, and epithelial cells lie beneath the posterior capsule (hematoxylin and eosin stain, original magnification ×75).

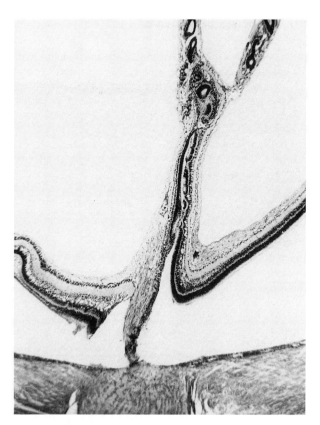

FIGURE 11-3 Dysplastic retina extends from an area near the optic nerve head, like a frontal fold toward the posterior lens pole (hematoxylin and eosin stain, ×35).

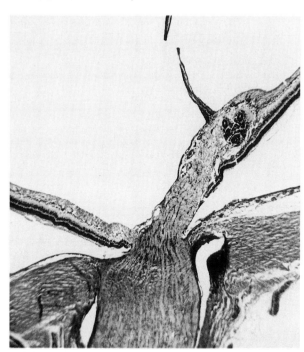

FIGURE 11-2 A persistent hyaloid artery extends from the retina near the optic nerve head into the vitreous. There is dysplastic retina beneath the origin of the hyaloid artery (hemotoxylin and eosin stain, ×45).

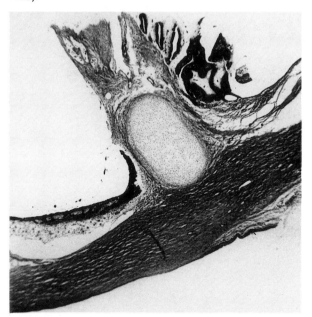

FIGURE 11-4 Choroidal coloboma filled with vascularized connective tissue and cartilage. Intraocular cartilage surrounded by dysplastic retina (hematoxylin and eosin stain, ×35).

FIGURE 11-5 Malformed iris stump, closed chamber angle (hematoxylin and eosin stain, ×45).

TABLE 11-1 Ocular Anomalies of Patau's Syndrome

Microphthalmos, even apparent anophthalmos, synophthalmos, cyclops

Colobomas

Persistence and hyperplasia of the primary vitreous (Fig. 11-1)

Cataract; retention of nuclei in the lens fibers and persistence of lens epithelium beneath the posterior half of the lens capsule

Persistence of hyaloid artery, tunica vasculosa lentis (Fig. 11-2)

Retinal dysplasia, rosettes, retinal detachment, folds (Fig. 11-3)

Cartilaginous metaplasia of fibrovascular tissue in or near coloboma (Fig. 11-4)

Corneal opacities, Peters's anomaly

Hypoplasia of optic nerve

Angle anomalies (Fig. 11-5)

Persistence of pupillary membrane

Hypertelorism, epicanthus, absent brows

References

1. Cogan DG, Kuwabara T. Ocular pathology of the 13-15 trisomy syndrome. *Arch Ophthalmol.* 1964;72:246.
2. Gieser SC, Carey JC, Apple DJ. Human chromosomal disorders and the eye. In: Renie WA, ed. *Goldberg's Genetic and Metabolic Eye Diseases.* Boston, Mass: Little, Brown; 1986:202–209.
3. Hassold TJ, Jacobs PA. Trisomy in man. *Ann Rev Genet.* 1984;18:69.
4. Hoepner J, Yanoff M. Ocular anomalies in trisomy 13-15. *Am J Ophthalmol.* 1972;74:729.
5. Jay M. *The Eye in Chromosome Duplications and Deficiencies.* New York, NY: Marcel Dekker; 1977.
6. Mets MB, Maumenee IH. The eye and the chromosome. *Survey Ophthalmol.* 1983;28:20.
7. Patau K, Smith DW, Therman E, et al. Multiple congenital anomaly caused by an extra autosome. *Lancet.* 1960;1:790.
8. Reese AB, Blodi FC. Retinal dysplasia. *Am J Ophthalmol.* 1950;33:23.

Chapter 12

Trisomy 18

Edwards' Syndrome

RUFUS O. HOWARD

Trisomy 18, initially described by Edwards in 1960, is a disease caused by an extra chromosome 18.[1,2] This error occurs more often in infants of older mothers and has multiple causes. Nondisjunction at meiosis can produce an extra chromosome 18 in germ cells. If this abnormal germ cell is fertilized, trisomy 18 will occur and will affect all fetal tissues. If non-disjunction occurs following fertilization, trisomy will be present in only a fraction of body cells, and the person will demonstrate mosaicism for trisomy 18. In trisomy 18 mosaicism clinical findings are similar to or slightly less severe than for full trisomy 18. Replication of only the distal segment of the long arm appears to be sufficient to cause the clinical and anatomic features of the syndrome. A partial trisomy 18 can arise if a balanced translocation is present in one parent and this abnormal chromosome is fertilized. The incidence is approximately 0.3 per 1000 live births, and this error in cell division is identified more frequently (3:1) in females.

In trisomy 18, all body cells and all tissues and organs have a triple, rather than the usual double, dose of the normal genes located on chromosome 18. This results in growth retardation in utero and in major organ and tissue malformations. Trisomy 18 is a recognized cause of spontaneous abortion and neonatal death. Among affected live-born infants, life expectancy is only 1 month for 70%, 1 year for 10%, and 10 years for 1%.

Systemic Manifestations

There are characteristic findings at birth for children with trisomy 18. The birth weight is low (mean 2180 g) even following a gestation of 38 weeks or more. Mothers are usually older (mean 32.5 years). Polyhydramnios, a small placenta, and a single umbilical artery are common. Resuscitation may be required in the neonatal period. A weak cry, apneic episodes, hypotonia, and poor sucking are usually recognized. The child usually dies in spite of excellent care.

A clinical diagnosis may be made on the basis of the following findings, which are present in more than 50% of newborn trisomy 18 children. The head exhibits a prominent occiput and narrow bifrontal diameter. The ears are low set and malformed; the palpebral fissures are small. A small mouth, narrow palatal arch, micrognathia, and a short neck with excess skin are additional features (Figs 12-1, 12-2). The sternum is short, and nipples may be hypoplastic. Cardiac murmurs are caused by structural defects. Inguinal or umbilical hernias or diastases recti are typical abdominal defects. The pelvis is narrow, limiting hip adduction, and hip dislocation is common. Genital abnormalities include cryptorchidism, hypoplasia of the labia majora, and clitoral prominence. The hands may be clenched, with a tendency for the index finger to overlap the third finger, and for the fifth finger to overlap the fourth. Nails are hypoplastic. Feet may exhibit rocker-bottom deformity. There is hypoplasia of skeletal muscle and adipose tissue. Children who survive exhibit severe growth and mental retardation.

The widespread and severe dysgenesis is apparent from autopsy findings. In the central nervous system, hydrocephalus, meningomyelocele, cerebellar hypoplasia, corpus callosum defects, microgyria, and deficiency of myelination have been reported.

The lungs may be abnormally segmented, or the right lobe may be absent. Structural defects of the heart have included ventricular and auricular septal defect, patent ductus arteriosus, bicuspid aortic or pulmonic valves, nodularity of valve leaflets, pulmonic stenosis, anomalous coronary artery, tetralogy of Fallot, coarctation of aorta, dextrocardia, aberrant subclavian artery, and intimal proliferation in arteries with arteriosclerotic change and medial calcification. The diaphragm shows muscle hypoplasia, with or without eventration.

Abnormalities reported in the intestinal tract include tracheoesophageal fistula, pyloric stenosis, extrahepatic biliary atresia, hypoplastic gallbladder, gallstones, Meckel's diverticulum, incomplete rotation of the colon, ectopic pancreatic or splenic tissue, omphalocele, and imperforate anus.

The renal defects of horseshoe kidney, ectopic kidney, polycystic kidney, hydronephrosis, and double ureter have been described.

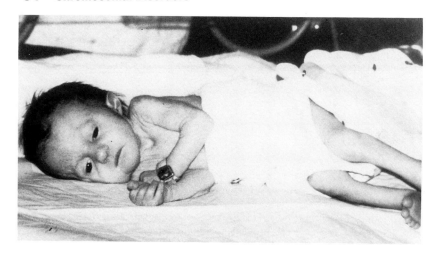

FIGURE 12-1 A clinical diagnosis of trisomy 18 can be made on the basis of neonatal abnormalities. A chromosome determination confirms the diagnosis.

Laboratory Findings

The enzyme peptidase A has been localized to band 23 of the long arm of chromosome 18 (18q23). A gene dosage effect has been demonstrated when the patient's enzyme level is compared to the median parent value (the patient has 1.5 times the parent's enzyme activity).

Ocular Manifestations

The eye findings may be normal or significantly abnormal, but abnormalities are less severe and less frequent than in trisomy 13. Short palpebral fissures are present in more than 50% of children, and hypoplasia of orbital ridges, inner epicanthal folds, ptosis of eyelids, and corneal opacities are present in 10 to 50%. Slanted palpebral fissures, hypertelorism, coloboma of iris, microphthalmos, and cataract occur in less than 10% (see Fig. 12-2).

With histopathologic examination, malformations of the central visual pathways have been identified: poorly myelinated, small optic tracts, absent corpus callosum, and absent occipital lobes. The ocular defects identified on light and electron microscopic examination include irregular thickness of Descemet's membrane, corneal endothelium folded and multilayered in patches, immature iridocorneal angle, irregular epithelial pigmentation of iris, variable length of ciliary processes, tunica vasculosa lentis, microscopic cataracts, persistent hyperplastic primary vitreous, retinal dysplasia, especially in the regions adjacent to colobomas of the optic nerve, hypopigmentation of retinal pigment epithelium, decreased choroidal vasculature, and hypoplastic optic nerves. In the posterior retina, large photoreceptor cells with large nuclei and more cytoplasm than normal, combined with prominent glial processes, are consistent with immaturity and not with abnormal development.[3]

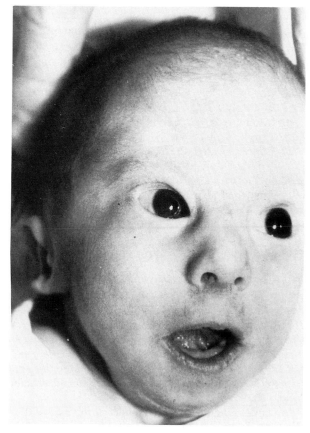

FIGURE 12-2 Trisomy 18 due to a balanced translocation in one parent.

References

1. Edwards JH, Harnden DG, Cameron AH, et al. A new trisomic syndrome. *Lancet.* 1960;1:787.

2. Smith DW. *Recognizable Patterns of Human Malformation.* 2nd ed. Philadelphia, Pa: WB Saunders; 1976.

3. Fulton AB, Craft JL, Zakov ZN, et al. Retinal anomalies in trisomy 18. *Graefes Arch Clin Exp Ophthalmol.* 1980;213:195-205.

Chapter 13

Trisomy 21

Down Syndrome

THOMAS D. FRANCE and MICHAEL B. SHAPIRO

First categorized by the English physician Langdon Down in 1866,[1] Down syndrome is a clinical entity with a wide variety of characteristic findings, many of which are apparent at birth. It is the most common type of mental retardation; an estimated 11,000 affected infants are born each year.[2]

Chromosome Composition and Systemic Manifestations

An abnormal chromosome composition was shown to be the cause of the syndrome by Lejeune, Gauthien, and Turpis in 1959.[3] The chromosomal abnormality is most commonly trisomy 21, but it may also include a translocation or mosaicism.[4] In particular, the long arm (q) of chromosome 21 is always present in excess in patients with Down syndrome.

The impaired development in utero leads to persistence of epicanthal folds, cardiac defects, unusual palm creases, and muscle flaccidity. With development, other abnormalities become evident, including short stature, mental retardation, and poor development of genitalia. The extra portion of chromosome 21 contributes an extra "dose" of genetic material, which produces these phenotypic findings.

Chromosome 21 contains about 1.5% of the total genetic material, or about 1500 genes. To date, fewer than 20 of these have been identified.[5] Several, however, seem to be related to functions that are known to be abnormal in Down syndrome. The existence of the gene loci for two enzymes of purine biosynthesis, GARS and AIRS, is particularly important, since alterations in purine metabolism have been associated with mental retardation[6] and abnormalities of purine metabolism are suggested in Down syndrome.[7]

While some of these characteristics are invariably present, such as mental retardation, others are less common. Congenital heart defects occur in only 40% of patients. Down syndrome is associated with an increased prevalence of other disease processes, including leukemia and other hematologic disorders; endocrine disease including hypothyroidism, abnormalities of the hypothalamus and the pituitary and adrenocortical glands, and a higher than usual prevalence of diabetes mellitus. These patients have an increased incidence of respiratory infections unrelated to any structural abnormalities of the respiratory tract.

Recent work to map the chromosome has resulted in a number of discoveries. The oncogene *ets*-2, present on 21q22, has been related to the onset of the previously mentioned leukemia in these patients. It appears that in at least one type of acute myelogenous leukemia, the leukemic cells have a reciprocal chromosomal translocation involving fragments of chromosomes 8 and 21.

Another enzyme, cytosol superoxide dismutase (SOD) is assigned to this portion of the chromosome and has been shown to be of importance in the aging process.[8] Premature aging has been described in Down syndrome. There appears to be an almost universal change in the central nervous system in Down patients over the age of 35, similar to that seen in Alzheimer's disease. In addition, the gene for amyloid beta protein resides here, which material has been shown to be a major component of the neurofibrillary plaques that accumulate in the brain of persons with Alzheimer's disease and identical with the protein that accumulates in the lesions in the brain of all "older" persons with Down syndrome. In spite of the aging process in these patients it has been found that almost none of them have signs of atherosclerosis.

Finally, the gene encoding α-A-crystallin protein, a structural component of the lens, has been mapped to this portion of the chromosome, which may explain the significant increase in cataract formation (see below).

Ocular Manifestations

In addition to prominent epicanthal folds, the shape of the palpebral fissure itself is a noticeable feature in patients with Down syndrome and is responsible for Down's original description of such patients as "mongols." Typically, the palpebral fissure is slanted upward temporally about 10 degrees, and is about 7 mm shorter than normal. The height is usually normal (Fig. 13-1).[9]

Blepharitis is frequently troublesome to patients with Down syndrome. Various studies report prevalences from 2 to 67%.[10,11] It is not certain whether the high incidence of eyelid infections in this group results from poor hygiene or decreased resistance to infection. Strabismus occurs in 21 to 44% of the Down syndrome population. The vast majority are esodeviations commonly accompanied by a vertical deviation as well. Standard medical and surgical techniques can produce

satisfactory results.[12] Nystagmus has been reported in 5 to 17% of patients with Down syndrome. The nystagmus is usually horizontal and not associated with other ocular abnormalities.[13]

The iridic characteristics of these patients have been well described.[14,15] Almost 90% have a blue or light gray iris, and Brushfield spots or speckles—white to light yellow spots in the periphery of the iris—are seen in more than three quarters of patients with Down syndrome (Fig. 13-2). Brushfield spots, however, are seen in 10 to 20% of normal persons, and therefore cannot be considered pathognomonic or diagnostic for Down syndrome. The cause of these iris changes is not yet known.

Keratoconus has been reported in up to 15% of patients with Down syndrome.[16] This can range from early keratoconus seen only with corneal topographic studies to large cones with grossly positive Munson sign. Acute hydrops is also seen in these patients. The cause of the increased incidence of keratoconus is also unknown. Several investigators have suggested eye rubbing as a probable cause,[17] but one cannot dismiss the possibility of an inherent defect in the structure or composition of the corneas in patients with Down syndrome. Various factors probably play a part in the development of keratoconus in these patients. More recently, epikeratophakia has become a useful surgical modality for keratoconus and oftentimes it is ideal for patients with Down syndrome.

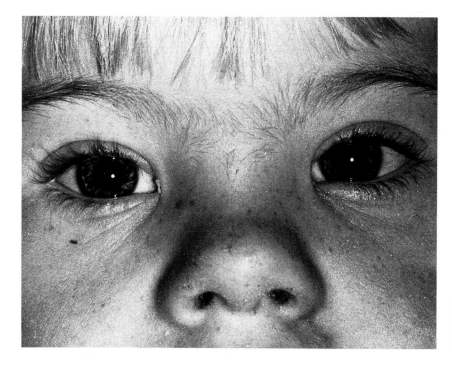

FIGURE 13-1 Five-year-old child with Down syndrome. Note upward obliquity of the palpebral fissures.

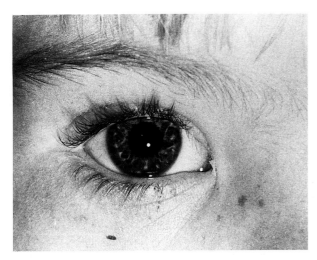

FIGURE 13-2 Iris detail of a patient with Down syndrome. The white elevated dots on the iris are typical Brushfield spots.

Significant cataracts can be observed in 10 to 50% of patients with Down syndrome. These lens changes have been described as cortical flake opacities as well as arcuate opacities.[18]

A recent study has shown that more than one quarter of patients with Down syndrome have myopia greater than 5 diopters.[9] Additionally, astigmatism of more than 3 diopters was found in 25%. The reason for this may be similar to the causes of keratoconus, that is, eye rubbing or underlying structural abnormalities in the cornea.

The current advances in chromosome mapping will undoubtedly lead to further discoveries as to the causes of the systemic and ocular defects seen in Down syndrome. Similarly, investigation of these patients will lead to a better understanding of many other conditions, such as aging, cataracts, and atherosclerosis, in patients not affected with the disorder.[19]

References

1. Down JLH. Observations of ethnic classifications of idiots. *London Hosp Rep.* 1866;3:259.
2. Benda CE. *The Child with Mongolism,* New York, NY: Grune and Stratton; 1960:4-5.
3. Lejeune J, Gauthien M, Turpis R. Les chromosomes humaines en culture des tissues. *SR Acad Sci.* 1959; 248:602.
4. Wright SW, Day RW, Muller H, et al. The frequency of trisomy and translocations in Down's syndrome. *J Pediatr.* 1967;70:420.
5. Patterson D. The causes of Down syndrome. *Sci Am.* 1987;257:52-60.
6. Lesch M, Nyhan WL. A familial disorder of uric acid metabolism and central nervous system function. *Am J Med.* 1964;36:561-567.
7. Coburn SP, Luce MW, Mertz ET. Elevated levels of several nitrogenous nonprotein metabolites in mongoloid blood. *Am J Ment Def.* 1965;69:814-817.
8. Fridovich T. The biology of oxygen radicals. *Science.* 1978;201:875-880.
9. Shapiro MB, France TD. The ocular features of Down's syndrome. *Am J Ophthalmol.* 1985;99:659-663.
10. Cullen JF, Butler HG. Mongolism (Down's syndrome) and keratoconus. *Br J Ophthalmol.* 1963;47:321-330.
11. Skeller E, Oster J. Eye symptoms in mongolism. *Acta Ophthalmol.* 1951;29:149-161.
12. Hiles DA, Hoyme SH, McFarlane F. Down's syndrome and strabismus. *Am Orthopt J.* 1974;24:63-68.
13. Eissler R, Longenecker LP. The common eye findings in mongolism. *Am J Ophthalmol.* 1962;54:398-406.
14. Brushfield T. Mongolism. *Br J Child Dis.* 1924; 21:241-258.
15. Donaldson D. The significance of spotting of the iris in mongoloids. *Arch Ophthalmol.* 1961;4:26-31.
16. Walsh SZ. Keratoconus and blindness in 469 institutionalized subjects with Down syndrome and other causes of mental retardation. *J Ment Defic Res.* 1981;25:243-251.
17. Karseras AG, Ruben M. Aetiology of keratoconus. *Br J Ophthalmol.* 1976;60:522-525.
18. Lowe RF. The eye in mongolism. *Br J Ophthalmol.* 1949; 33:131-174.
19. Scoggin CH, Patterson D. Down's syndrome as a model disease. *Arch Intern Med.* 1982;142:462-464.

Chapter 14

Trisomy 22

MARILYN B. METS

The literature on trisomy 22 is confused because of technical difficulties in karyotyping. Schinzel[1-4] addressed this issue and categorized the partial and full trisomies and their phenotypic manifestations. Cat eye syndrome (CES) is a partial trisomy or partial tetrasomy of 22pter→q11.[1] There are also cases of derivative chromosome 22.[2] Such patients are trisomic for 22pter-q11, like those with CES, but they are also trisomic for 11q23→qter and are phenotypically different from CES patients.[5] In addition, there is partial trisomy 22, which involves the distal segment of 22q.[3] Unlike these partial trisomies, full trisomy 22 is usually lethal.[4]

Cat Eye Syndrome

CES was first reported in 1878 by Haab, who noted a syndrome of uveal malformations, imperforate anus, and renal abnormalities.[6] This syndrome was found to be associated with an extra chromosome by Schachenmann et al. in 1965.[7] It is now believed that this

chromosome is an isochromosome of the short arm (p) and of the proximal portion of the long arm (q) of chromosome 22. Later, Gerald et al. coined the term "cat eye syndrome" because of the similarity of the vertical iris coloboma and the cat's vertical pupil (Fig. 14-1).[8]

The most consistent features of CES are ocular colobomas, which may involve the iris, ciliary body, and the optic nerve; anal atresia; and preauricular skin tags and fistulas. Affected persons are usually identified immediately after birth. Other systemic abnormalities include renal abnormalities, such as unilateral absence, unilateral or bilateral hypoplasia, and cystic dysplasia; umbilical hernias; cardiac defects such as tetralogy of Fallot, anomalous pulmonary drainage, and single ventricles; and low normal intelligence to moderate retardation. Less frequently seen but also characteristic are microtia with atresia of the external auditory canal, intrahepatic or extrahepatic biliary atresia, and malrotation of the gut.[9] It should be noted that there is considerable variability in phenotypic manifestations,

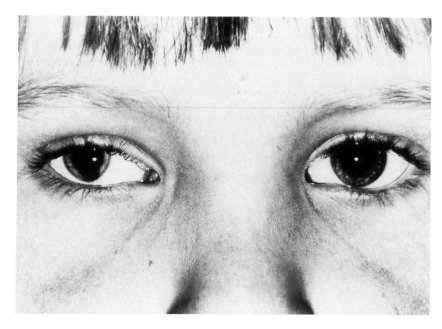

FIGURE 14-1 The right eye of a 12-year-old girl with CES shows the iris coloboma for which the syndrome was named. (Courtesy of Dr. Richard Weleber.)

even among members of one family. Schinzel refers to two families in which the mother was normal phenotypically or had only preauricular abnormalities and the child had full phenotypic manifestations.[1] The most common ocular anomalies are uveal colobomas, microophthalmia (unilateral or bilateral), hypertelorism, and antimongoloid palpebral fissures. Cataracts and strabismus are occasionally reported. There is one case report of a unilateral ocular retraction syndrome.[10] Vision is highly variable and depends primarily on the extent of the choroidal and optic nerve colobomas.

Derivative Chromosome 22

Patients with derivative chromosome 22 have a parent, most often the mother, with a balanced translocation of 11q and 22q—t(11q22q). This is now thought to be one of the most frequent sites specific for reciprocal translocations reported in man.[3,5] Their trisomy on chromosome 22 appears to involve the same region as that in CES, however these patients are also trisomic for 11q23→qter.

The most constant systemic manifestations include mental retardation, preauricular skin tags or pits, ear anomalies (low-set or large), palate anomalies (high arched or cleft), micrognathia, and congenital heart disease.[11] The heart disease is most often acyanotic and due to an atrial septal defect, ventricular septal defect, or patent ductus arteriosus.[12] Affected males may have undescended testicles, inguinal hernias, and micropenis. Less constant abnormalities include prenatal growth deficiency, microcephaly, long philtrum, excess nuchal skin, kidney, anal, and hip anomalies, and thirteen ribs.[11]

The reported eye manifestations are far less characteristic than in CES and include strabismus, epicanthus, hypertelorism, and hypotelorism, and mongoloid and antimongoloid palpebral fissures.[13] We have recently examined a patient who has a unilateral small choroidal coloboma (Fig. 14-2). This may occur more frequently than has been reported. Without the more obvious iris involvement, it could easily be missed.

Trisomy of Distal Segment of 22q

The partial trisomy of chromosome 22, which involves the distal segment of 22q has been described in only three patients.[3,14] Two brothers described by Schinzel had a familial trisomy 22q13→qter. Both died in the neonatal period. Phenotypic manifestations included intrauterine growth retardation, congenital hydrocephalus, cleft palate, genital hypoplasia with cryptorchidism,

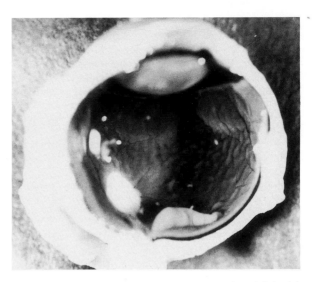

FIGURE 14-2 This is the right globe of a child with trisomy 22 who died at 12 days of age. The anterior segment is above and the sectioned optic nerve, below. Inferiorly to the right there is a retinal fold, and inferiorly to the left is a chorioretinal coloboma.

and hypospadias. The facial features showed mongoloid palpebral fissures, hypertelorism, small nose with a prominent bridge, prominent upper lip, and small mandible. The second sibling also had renal hypoplasia, arhinencephaly, and pentalogy of Fallot. All described partial trisomies of chromosome 22 have some ocular manifestations (Table 14-1). The eye abnormalities are most extensive and most characteristic in CES.

TABLE 14-1 Ocular Manifestations of Incomplete Trisomy 22

Cat eye syndrome
Uveal colobomas
Antimongoloid palpebral fissures
Hypertelorism
Cataracts
Ocular retraction syndrome
Derivative chromosome 22
Strabismus
Mongoloid and antimongoloid palpebral fissures
Hypertelorism and hypotelorism
Epicanthus
Partial trisomy 22
Mongoloid palpebral fissures
Hypertelorism

References

1. Schinzel A, Schmid W, Fraccaro M, et al. The "cat eye syndrome": dicentric small marker chromosome probably derived from a no. 22 (tetrasomy 22pter→q11) associated with a characteristic phenotype: report of 11 patients and delineation of the clinical picture. *Hum Genet*. 1981;57:148.

2. Schinzel A, Schmid W, Auf der Maur P, et al. Incomplete trisomy 22: I. Familial 11/22 translocation with 3:1 meiotic disjunction. Delineation of a common clinical picture and report of nine new cases from six families. *Hum Genet*. 1981;56:249.

3. Schinzel A. Incomplete trisomy 22. II. Familial trisomy of the distal segment of chromosome 22q in two brothers from a mother with a translocation, t(6;22)(q27;q13). *Hum Genet*. 1981;56:263.

4. Schinzel A. Incomplete trisomy 22. III. Mosaic-trisomy 22 and the problem of full trisomy 22. *Hum Genet*. 1981;56:269.

5. Phelan MC, Curtis R, Flannery DB, et al. Cytogenetics: an 11q;22q translocation in two families. *Proc Greenwood Genet Center*. 1987;6:22.

6. Haab O, et al. Chromosomes in coloboma and anal atresia. *Lancet*. 1965;ii:290.

7. Schachenmann G, Schmid W, Fraccaro M, et al. Chromosomes in coloboma and anal atresia. *Lancet*. 1985;290.

8. Gerald PS, Davis C, Say BM, et al. A novel chromosomal basis for imperforate anus (the cat's eye syndrome). *Pediatr Res*. 1968;2:297.

9. Gieser SC, Carey JC, Apple DJ. Human chromosomal disorders and the eye: In: Renie, W, ed. *Goldberg's Genetic and Metabolic Eye Disease*. 2nd ed. Boston, Mass: Little, Brown; 1985:185-240.

10. Weleber R, Walknowska J, Peakman D. Cytogenetic investigation of cat-eye syndrome. *Am J Ophthalmol*. 1977;84:477.

11. Zackai EH, Emanuel BS. Site-specific reciprocal translocation, t(11;22) (q23;q11), in several unrelated families with 3:1 meiotic disjunction. *Am J Med Genet*. 1980;7:507.

12. Lin AE, Benar J, Chin AJ, et al. Congenital heart disease in supernumerary der(22),t(11;22) syndrome. *Clin Genet*. 1986;29:269.

13. Fraccaro M, Lindsten J, Ford CE, Iselius L. The 11q;22q translocation: a European collaborative analysis of 43 cases. *Hum Genet*. 1980;56:21.

14. Jotterand-Bellomo M. Trisomic 22. *Arch Genet (Zar)*. 1976-77;49/50:134.

PART 3

Collagen Disorders

Chapter 15

Ankylosing Spondylitis

EDWARD S. PERKINS

Ankylosing spondylitis is a form of seronegative rheumatism affecting the sacroiliac and vertebral joints. It is seen mainly in Caucasian populations and is rare in blacks in the United States or in Africa. Men are affected more frequently than women by a ratio of 4:1. It is possible, however, that the disease may be present but subclinical in women, as radioisotope studies have detected evidence of sacroiliitis in women who have no clinical evidence of the disease. There is a definite familial incidence linked to the hereditary pattern of histocompatibility antigens.

The typical patient is an adult male, and the earliest symptoms are lower back stiffness and pain on waking followed by difficulty in bending down. Fortunately the severe kyphotic deformity seen previously in the later stages of the disease is much less common now, as a result of earlier recognition and treatment, but patients are still seen who have difficulty placing the chin on the chin rest of a slit lamp microscope.

On clinical examination, the patient is found to have limitation of flexion, extension, and rotation of the lower spine. There may be a reduced chest expansion, and in the later stages other joints, particularly the hips, the shoulder girdle, and the temporomandibular joint may be affected.

The most important diagnostic measure is radiography of the sacroiliac joints, and considerable experience is required to define the early changes. Radioisotope scanning is a more sensitive method of detecting inflammatory changes in the sacroiliac joints and is particularly suitable for women, as it avoids radiation to the abdomen and pelvis. The erythrocyte sedimentation rate is elevated in 80% of cases, and rheumatoid factor is rarely present. Other laboratory investigations are not helpful.

Apart from the ocular changes the most serious complications are aortic valve reflux, which may precipitate cardiac failure, fibrotic lung changes, and amyloidosis. A comprehensive collection of papers on the clinical features, radiologic findings, association with HLA antigens, pathology, complications, and treatment is contained in a volume edited by Moll.[1]

The cause of the disease is unclear. Earlier theories implicated infection from the prostate reaching the spine via Batson's veins, and it is true that many patients with ankylosing spondylitis have a history of chronic prostatitis. More recently it has been suggested that *Klebsiella pneumoniae*, an organism commonly found in the gastrointestinal tract, may be involved.[2] Ebringer et al. found the organism more frequently in ankylosing spondylitis patients with active disease than in controls or in patients with inactive disease. Positive cultures were also obtained more frequently during episodes of acute anterior uveitis than in patients without uveitis or in control subjects.[3] It was suggested that *Klebsiella pneumonia* may act as a trigger mechanism causing spinal or ocular disease in a genetically susceptible host; these results have not, however, been substantiated by other investigators.[4]

The most striking feature of ankylosing spondylitis is the very strong association with the histocompatibility antigen HLA-B27 first reported by Brewerton et al.[5] in 1973. They found that over 90% of patients with the disease had this antigen, compared with 6% of a control population. The risk of developing ankylosing spondylitis is 30 times greater for persons who carry the B27 antigen. These discoveries help to explain the racial and familial distributions of the disease.

Association Between Ankylosing Spondylitis and Uveitis

It has been recognized for many years that about one third of patients with ankylosing spondylitis have attacks of acute anterior uveitis at some time during or before the onset of the disease, and, conversely, ankylosing spondylitis and Reiter's disease were found to be associated with anterior uveitis in about 40% of men with this form of uveitis.[6] It is, therefore, the most common systemic condition known to be associated with anterior uveitis in men.

The validity of the clinical association between acute anterior uveitis and ankylosing spondylitis was strengthened by the finding that 50% of patients with this type of uveitis were positive for HLA-B27 antigen[7] and that in those cases where there was some associated rheumatic disease the incidence rose to 90%. These results have been amply confirmed by subsequent workers, and although the exact role of the HLA-B27 antigen is still not clear, genetic studies in Finland[8] have revealed several families in which more than one relative is affected by anterior uveitis or rheumatic disease, a high proportion of whom had the B27 antigen.

Although it is much less common to find ankylosing spondylitis or other rheumatic diseases associated with acute anterior uveitis in women than in men, the incidence of HLA-B27 positivity is very similar. Combining figures from Brewerton et al.[7] and Mapstone and Woodrow[9] shows that 55% of both women and men with acute anterior uveitis were positive for B27 but that the association with rheumatic disease was 30% in women compared with 68% in men. This difference is statistically significant, and, if, as seems likely, the cause of the uveitis is the same in all patients with HLA-B27, many women are in some way protected from developing the associated disease. It may be that the rheumatic disease is subclinical in women, and it should be remembered that some cases of sacroiliitis may not have been detected because of a reluctance to expose women to x rays.

There is no evidence that the B27 antigen is the cause of the joint disease or the uveitis, and at the moment it is possible to say only that the possession of B27 increases the susceptibility to such conditions. Saari et al.[8] suggest that a pleiotropic gene associated with HLA-B27 with autosomal dominant inheritance, incomplete penetrance, and variable expressivity may determine susceptibility to acute anterior uveitis in linkage disequilibrium with HLA-Cw1 antigen. Scharf and Zonis[10] wrote an excellent review of the association between histocompatibility antigens and uveitis in 1980.

Clinical Features of Uveitis Associated with Ankylosing Spondylitis

The typical features of the uveitis associated with ankylosing spondylitis are acute onset of pain, redness, photophobia, and blurred vision in one eye. There is marked ciliary injection, fine keratic precipitates, and an outpouring of protein into the anterior chamber, which may form a coagulum that can be seen by slit lamp microscopy. The pupil is constricted and ocular tension reduced. The vitreous remains clear, and apart from some hyperemia of the disc, the fundus remains normal. The first attack usually occurs in young adulthood but a similar type of uveitis may be seen in young boys with juvenile rheumatism, who are found to be B27 positive and who develop typical ankylosing spondylitis later.

A cardinal feature of the uveitis is that it is liable to recur, sometimes in the same eye but often alternating from one eye to the other, for many years. In spite of the severity of the individual attacks such complications as cataract and macular edema are rare, provided the attacks are treated early with adequate doses of corticosteroids and mydriatic eye drops.

Pathogenesis of Uveitis

The acute onset and response to corticosteroids are strongly suggestive of an allergic reaction in the eye with the release of potent vasoactive substances such as prostaglandins and vasoactive peptides. Presumably the anterior uveal tract becomes sensitized to an antigen that reaches it via the blood circulation, and the inflammatory response is due to an antigen-antibody reaction in the eye. A similar type of uveitis can be produced in animals by intravitreal injection of bovine serum albumin or other foreign protein. The antigen responsible for the human disease is unknown, but in view of the common finding of chronic prostatitis resulting from nonspecific venereal infection in many men, an infecting organism is a possible source. A similar type of uveitis can occur in association with Reiter's disease, a post-infective condition that can also follow nonspecific urethritis or a Salmonella or Yersinia bowel infection. Attempts to grow an organism from the aqueous have been unsuccessful, so a direct infective process seems unlikely and an immune reaction to some fragment of the infecting organism may be the initiating factor.

References

1. Moll JMH. *Ankylosing Spondylitis*. Edinburgh: Churchill Livingstone; 1980.

2. Ebringer RW, Cawdell DR, Cowling P, et al. Sequential studies in ankylosing spondylitis. Association of *Klebsiella pneumoniae* with active disease. *Ann Rheum Dis*. 1978;37:146-151.

3. Ebringer R, Cawdell D, Ebringer A. Klebsiella pneumoniae and acute anterior uveitis in ankylosing spondylitis. *Br Med J*. 1979;1:383.

4. Beckingsale AB, Williams D, Gibson JM, et al. Klebsiella and acute anterior uveitis. *Br J Ophthalmol*. 1984;68:866-868.

5. Brewerton DA, Caffrey M, Hart FD, et al. Ankylosing spondylitis and HLA 27. *Lancet*. 1973;1:904-907.

6. Catterall RD, Perkins ES. Uveitis and urogenital disease in the male. *Br J Ophthalmol*. 1961;45:109-116.

7. Brewerton DA, Caffrey M, Nicholls A, et al. Acute anterior uveitis and HLA 27. *Lancet*. 1973;2:994-996.

8. Saari KM, Solja J, Hakli J. et al. Genetic background of acute anterior uveitis. *Am J Ophthalmol*. 1981;91:711-720.

9. Mapstone R, Woodrow JC. HLA 27 and acute anterior uveitis. *Br J Ophthalmol*. 1975;59:270-275.

10. Scharf Y, Zonis S. Histocompatibility antigens (HLA) and uveitis. *Surv Ophthalmol*. 1980;24:220-228.

Chapter 16

Cranial Arteritis

Temporal Arteritis, Giant Cell Arteritis

J. BROOKS CRAWFORD

In 1890 Hutchinson[1] described a man with painful "red streaks" on his temples that prevented him from wearing his hat; he called the disease "thrombotic arteritis." In 1932 Horton, Magath, and Brown[2] reported the first two cases in which the temporal artery was biopsied and showed a subacute granulomatous, stenosing arterial disease which they called temporal arteritis. In 1938 Jennings[3] pointed out that blindness was a complication of the disease, and in 1958 Wagener and Hollenhorst[4] described the ocular lesions in 100 eyes of 58 patients.

"Cranial arteritis" or "giant cell arteritis" is a better term for the disease, because the inflammation is not confined to the temporal arteries but may involve any large or medium-sized artery, especially other cranial arteries.

Epidemiology

Cranial arteritis usually occurs in patients over 60 years of age. It is slightly more common in women and is rare in blacks and orientals. Epidemiologic studies of residents in Olmsted County, Minnesota, showed an incidence of 33 cases in 100,000 residents between the ages of 60 and 69 and 844 cases in 100,000 residents over the age of 80.[5]

Systemic Manifestations

Systemic symptoms, which antedate the clinical diagnosis by an average of 5 months, include headaches, arthralgia, myalgia, fever, malaise, weight loss, anemia, and depression. Involvement of carotid, vertebral, basilar, or coronary arteries may produce central nervous system or cardiac disease, which is sometimes fatal. Facial artery involvement produces jaw claudication, tongue pain, and tongue infarction. Dermatologic manifestations due to skin ischemia include vesicles, hemorrhagic bullae, focal skin crusting, necrosis, and ulceration.[6] The physician is often confronted by atypical, perplexing signs and symptoms of the disease, which can mimic so many other diseases in the elderly. Paulley and Hughes[7] stated, "When elderly people begin to fail mentally or physically, this should be one of the first disorders to be considered and not the last."

Polymyalgia rheumatica is a disease in patients over the age of 50 that produces pain and morning stiffness in the muscles of the neck, shoulders, upper arms, buttocks, and thighs. Although polymyalgia rheumatica and cranial arteritis share some signs and symptoms, they seem to be different diseases. However, approximately 20 to 40% of patients with polymyalgia rheumatica also have cranial arteritis and are therefore at risk for blindness.[8] Blindness is rare in patients with polymyalgia rheumatica who do not have the signs and symptoms of cranial arteritis.

Other vasculitides such as polyarteritis, hypersensitivity vasculitis, Wegener's granulomatosis, and systemic lupus erythematosus differ clinically and pathologically from cranial arteritis. Significant differences include the lack of pulmonary and kidney involvement, a better response to corticosteroid therapy, and a better long-term prognosis in patients with cranial arteritis.

Ocular Manifestations

The most dreaded manifestation of cranial arteritis is blindness. Severe vision loss occurs in 30 to 50% of patients.[6,8] The onset is sudden and generally irreversible. It may occur without any premonitory signs or symptoms (occult temporal arteritis). When one eye is involved, the other often becomes affected in 1 to 10 days, emphasizing the urgent need for correct diagnosis and rapid administration of corticosteroids.

Anterior ischemic optic neuropathy (AION), the most common cause of blindness, is the result of optic nerve infarction from involvement of the short posterior ciliary arteries. The disc is swollen and pale and often surrounded by small hemorrhages in the adjacent peripapillary retina (Fig. 16-1). Central retinal artery occlusion, or branch retinal artery occlusion, is a rarer cause of blindness, accounting for 1% of the vision losses in the series of patients described by Wagener and Hollenhorst.[4] However, in investigating the causes of central retinal artery occlusion, Cullen found 10% were due to cranial arteritis.[9] Cortical blindness due to occipital lobe infarction (blindness with a normal fundus and no afferent pupillary defect), ischemic retinopathy (focal retinal ischemia with cotton-wool spots), and retrobul-

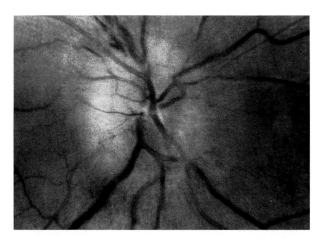

FIGURE 16-1 Anterior ischemic optic neuropathy in patient with cranial arteritis. Note the pale, swollen disc with small hemorrhages in the peripapillary retina just above the disc. (Courtesy of Ronald Burde, M.D.)

bar ischemic optic neuropathy (vision loss, an afferent pupillary defect, and a normal fundus) are relatively rare causes of vision loss in cranial arteritis.

Ophthalmoplegia, not necessarily associated with a particular cranial nerve, is a common and sometimes presenting sign, occurring both with and without diplopia and sometimes with ptosis and miosis.[10] It may be transient. Barricks et al.[11] proved that it is sometimes due to ischemic necrosis of the extraocular muscles rather than of the ocular motor nerve.

Amaurosis fugax, another important symptom, occurs in 2 to 19% of patients with cranial arteritis.[8] Unusual ocular signs and symptoms include visual hallucinations, orbital inflammatory pseudotumor (proptosis, lid edema, conjunctival inflammation and chemosis, episcleritis, and iridocyclitis), anterior segment ischemia (with corneal edema, chemosis and iritis), acute ocular hypotony, and bilateral marginal corneal ulceration.[12-15]

Although the ocular manifestations of cranial arteritis usually occur after the onset of systemic signs and symptoms, they may be the first and only manifestations of the disease. The importance of prompt recognition of so-called occult temporal arteritis cannot be overemphasized.

Laboratory Diagnosis

The erythrocyte sedimentation rate, preferably measured by the Westergren technique, is the primary laboratory measure and often exceeds 70 mm/hr. Unfortunately, this test is not specific, is often moderately elevated in otherwise normal elderly people, and is normal in approximately 2% of patients with biopsy-proven cranial arteritis.[16] To establish a diagnosis absolutely, the typical pathologic findings must be demonstrated in an involved artery.

Temporal Artery Biopsy

Although a segment of the frontal branch of the superficial temporal artery may fail to show the disease because of irregular involvement of the artery, a properly performed and analyzed biopsy will correctly establish the diagnosis and the need for corticosteroid therapy in 94% of patients.[17] The presence of "skip areas" has been clearly demonstrated. Klein et al.[18] found skip lesions in 28% of 60 patients; in one patient, only 55 of 2334 consecutive 6-μm sections from a 2.5-cm artery biopsy segment were involved. However, skip lesions in biopsies from patients seen by ophthalmologists are rare; some false-negative results stem from improper technique in obtaining or preparing the biopsy and others from failure to appreciate the nature of healing and healed lesions.[19-23]

Although opinions differ about the need for a temporal artery biopsy and its value in making the diagnosis, many internists and ophthalmologists, including this author, believe that a biopsy is generally indicated because of frequent errors in clinical diagnosis and the serious complications of prolonged corticosteroid therapy in elderly patients (25% of patients treated for giant cell arteritis experienced symptomatic vertebral fractures[24]). The branch of the temporal artery anterior to the main trunk should be selected for biopsy. If a segment of that artery is inflamed or tender, it should be chosen; otherwise, a 3- to 5-cm segment can be selected from either side. Some authorities advocate bilateral biopsy, especially if frozen sections on the first side are negative[8,25]; others have found little advantage to performing bilateral biopsy.[22] Cross-sections of the artery at 0.25-mm to 0.50-mm intervals should be stained with hematoxylin and eosin and Verhoeff-Van Gieson stains for elastic tissue. The biopsy can, and in some cases must, be done after corticosteroid therapy has been administered; to ensure accurate diagnosis, it is best to perform the biopsy within 2 weeks of the beginning of corticosteroid treatment.

Pathology

A granulomatous arteritis concentrated in the media of the artery and associated with smooth muscle and internal elastic membrane destruction is pathognomonic. Despite the designation "giant cell arteritis," the presence of giant cells is inconsistent and not necessary to make the diagnosis. Nonspecific inflammation in the wall of the artery occurs adjacent to the areas of granulomatous inflammation and should prompt the pathologist to search for more diagnostic areas. Eventually a panarteritis leads to inflammatory, thrombotic, or fibrotic occlusion of the vessel lumen.

Although there is a close correlation between the presence and amount of elastic tissue in an artery and its involvement by giant cell arteritis, ultrastructural evaluation of diseased arteries suggests that degeneration of smooth muscle cells in the media may be the initial pathologic change, followed by fragmentation of elastic fibers and granulomatous reaction.[21,26-28]

Treatment

Treatment with high doses of corticosteroid (60 to 100 mg of prednisone daily) is indicated to prevent permanent loss of vision. Some patients continue to lose vision in spite of large doses of oral prednisone. In these patients, intravenous methyl prednisolone (1000 mg every 12 hours for 5 days) may prevent blindness if started within 36 hours.[29] If vision diminishes when a patient sits or stands, a supine or shock position should be maintained until the corticosteriod therapy becomes effective.[30] After signs and symptoms subside, the corticosteroids can be decreased gradually.

Summary

According to Kearns,[31] "Cranial giant cell arteritis is the prime medical emergency in ophthalmology, there being no other disease in which the prevention of blindness depends so much on prompt recognition and early treatment." Any elderly patient with the typical or atypical ocular manifestations of cranial arteritis, especially AION, central retinal artery occlusion, amaurosis, or extraocular muscle palsies, should be thoroughly examined for the presence of associated systemic signs or symptoms, and the erythrocyte sedimentation rate should be measured. If systemic manifestations exist or if the sedimentation rate is abnormal, steroids should be started immediately and arrangements should be made for temporal artery biopsy.[6,8]

References

1. Hutchinson J. Diseases of the arteries. On a peculiar form of thrombotic arteritis of the aged which is sometimes productive of gangrene. *Arch Surg (Lond)*. 1890:1:323-329.
2. Horton BT, Magath TB, Brown GE. An undescribed form of arteritis of the temporal vessels. *Mayo Clin Proc*. 1932;7:700-701.
3. Jennings GH. Arteritis of the temporal vessels. *Lancet*. 1938;1:424-428.
4. Wagener HP, Hollenhorst RW. The ocular lesions of temporal arteritis. *Am J Ophthalmol*. 1958;45:617-630.
5. Hauser WA, Ferguson RH, Holley KE, et al. Temporal arteritis in Rochester, Minnesota, 1951-1967. *Mayo Clin Proc*. 1971;46:597-602.

6. Keltner JL. Giant-cell arteritis. Signs and symptoms. *Ophthalmology*. 1982;89:1101-1110.
7. Paulley JW, Hughes JP. Giant cell arteritis, or arteritis of the aged. *Br Med J*. 1960;2:15-62.
8. Goodman BW. Temporal arteritis. *Am J Med*. 1979;67:839-852.
9. Cullen JF. Occult temporal arteritis—a common cause of blindness in old age. *Br J Ophthalmol*. 1967;51:513-525.
10. Dimant J, Grob D, Brunner NG. Ophthalmoplegia, ptosis, and miosis in temporal arteritis. *Neurology*. 1980;30:1054-1058.
11. Barricks ME, Traviesa DB, Glaser JS, et al. Ophthalmoplegia in cranial arteritis. *Brain*. 1977;100:209-221.
12. Gibbs P. Temporal arteritis: onset with pseudotumor orbit or red eye. *Practitioner*. 1974;213:205-211.
13. Winter BJ, Cryer TH, Hameroff SB. Anterior segment ischemia in temporal arteritis. *South Med J*. 1977;70:1479.
14. Cullen JF. Ophthalmic complications of giant cell arteritis. *Surv Ophthalmol*. 1976;20:247-260.
15. Gerstle CC, Friedman AH. Marginal corneal ulceration (limbal guttering) as a presenting sign of temporal arteritis. *Ophthalmology*. 1980;87:1173-1176.
16. Healey LA, Wilske KR. *The Systemic Manifestations of Temporal Arteritis*. New York, NY: Grune and Stratton; 1978.
17. Hall S, Lie JT, Kurland LT, et al. The therapeutic impact of temporal artery biopsy. *Lancet*. 1983;2:1217-1220.
18. Klein RG, Campbell RJ, Hunder GG, Carney JA. Skip lesions in temporal arteritis. *Mayo Clin Proc*. 1976;51:504-510.
19. Cohen DN, Smith TR. Skip areas in temporal arteritis: myths versus fact. *Trans Am Acad Ophthalmol Otolaryngol*. 1974;78:772-783.
20. Hedges TR, Gieger GL, Albert DM. The clinical value of negative temporal artery biopsy specimens. *Arch Ophthalmol*. 1983;101:1251-1254.
21. Albert DM, Searl SS, Craft JL. Histologic and ultrastructural characteristics of temporal arteritis. The value of the temporal artery biopsy. *Ophthalmology*. 1982;89:1111-1126.
22. McDonnell PJ, Moore GW, Miller NR, et al. Temporal arteritis: a clinicopathologic study. *Ophthalmology*. 1986;93:518-530.
23. Brownstein S, Nicolle DA, Codiere F. Bilateral blindness and temporal arteritis with skip areas. *Arch Ophthalmol*. 1983;101:388-391.
24. Huston KA, Hunder GG, Lie JT, et al. Temporal arteritis: a 25-year epidemiologic, clinical and pathologic study. *Ann Intern Med*. 1978;88:162-167.
25. Hall S, Hunder GG. Is temporal artery biopsy prudent? *Mayo Clin Proc*. 1984;59:793-796.
26. Enzmann D, Scott WR. Intracranial involvement in giant-cell arteritis. *Neurology*. 1977;27:794-797.
27. Reinecke RD, Kuwabara T. Temporal arteritis. I. Smooth muscle involvement. *Arch Ophthalmol*. 1969;82:446-453.
28. Kuwabara T, Reinecke RD. Temporal arteritis. II. Electron microscopic study on consecutive biopsies. *Arch Ophthalmol*. 1970;83:692-697.
29. Rosenfeld SI, Kosmorsky GS, Klingele TG, et al. Treatment of temporal arteritis with ocular involvement. *Am J Med*. 1986;80:143-145.
30. Hollenhorst RW. Effect of posture on retinal ischemia from temporal arteritis. *Arch Ophthalmol*. 1967;78:569-577.
31. Kearns TP. Collagen and rheumatic disease: ophthalmic aspects. In: Mausolf FA, ed. *The Eye and Systemic Disease*. St. Louis, Mo: CV Mosby Co.; 1975:114.

Chapter 17

Discoid Lupus Erythematosus

MICHAEL S. INSLER and DENNIS W. BOULWARE

Discoid lupus erythematosus (DLE) may occur as a discrete dermatologic condition or in conjunction with systemic lupus erythematosus (SLE).[1,2] DLE is typically a benign chronic cutaneous condition; approximately 25% of patients have some laboratory abnormalities, most commonly those with widespread skin disease.[3,4] Chronic discoid skin lesions may be the initial manifestation of SLE in 10 to 20% of patients.[5]

Although the most common clinical manifestations are seen in the skin, the serosal membranes, joints, kidneys, and eyes can be affected with systemic involvement.[6-8] Most clinical manifestations of SLE can be accounted for by the inflammatory response to one of three immune-mediated mechanisms: specific antibody-mediated injury to membrane-bound antigens (eg, antierythrocyte hemolytic anemia); antigen-antibody complexes deposited in tissues (eg, glomerular immune complexes and lupus nephritis); or specific anti-phospholipid antibodies that interfere with endothelial prostacyclin production and lead to clinical vaso-occlusive disease (eg, infarction). These basic mechanisms also explain many of the ophthalmic manifestations seen in SLE and in chronic DLE.[1,2,7]

Systemic Manifestations

The cutaneous manifestations of SLE are among the most common findings in this autoimmune disease and include the butterfly rash, photosensitivity, chronic discoid lesions, alopecia, and other nondescript maculopapular eruptions.[4,5,9] Sun-exposed areas are particularly prone to develop erythematous lesions as a primary or secondary manifestation of SLE.

In cutaneous lupus erythematosus, there is a facial involvement in 80-90% of patients, which may take the form of either the malar rash, acute erythema, chronic discoid lesions, or nonscarring alopecia.[10] The erythematous lesions are usually well demarcated and may show atrophy, telangiectasia, follicular plugging, and hypopigmentation or hyperpigmentation. Lesions of the trunk and extremities occur in more widespread disease. In one series of 56 patients, 26 had localized and 30 had widespread (above and below the neck) disease.[4] Women are affected twice as often as men, and blacks are more frequently affected than Caucasians. Onset peaks during the childbearing years. The diagnosis is clinically confirmed by skin biopsy and laboratory testing. The level of antinuclear antibody (ANA) is generally elevated, as is the erythrocyte sedimentation rate (ESR).[4]

Among patients who have DLE as a manifestation of SLE a majority have the findings of classic SLE. Abnormal or positive laboratory test results include elevated ANA titers, elevated ESRs, abnormal renal function, and elevated antinegative DNA antibody titers. Associated manifestations of systemic involvement can include arthralgias, arthritis, Raynaud's phenomenon, and cardiac and renal disease.[1,2]

Ocular Manifestations

DLE involvement of the eyelids has been reported approximately 40 times in the past.[10-17] In approximately one half of these cases, eyelid involvement was the sole or presenting sign of chronic discoid lupus. The duration of symptoms generally ranges between 1 and 2 years, and the skin lesions may take on the same pattern of erythema and telangiectasia typical of the skin lesions elsewhere (Fig. 17-1). The lid margin may reveal red, scaly plaques with follicular plugging. Involvement of the conjunctiva can result in ulceration and scarring.[17]

When DLE affects the eyelids alone, the disorder is often not correctly diagnosed and may go untreated. Untreated chronic discoid lupus may lead to eyelid deformities caused by scarring, with resultant severe eyelid dysfunction, including distortion, loss of cilia, and restricted eyelid mobility.[9] Rarely, corneal infiltrates and vascularization can occur. Skin biopsy is especially helpful in the diagnosis; histopathologic findings in-

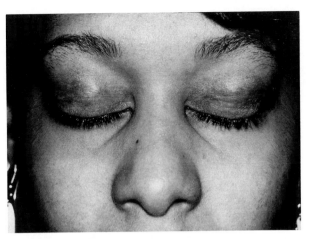

FIGURE 17-1 Thirty five-year-old black woman with eyelid involvement due to chronic DLE. The rash improved dramatically following 3 months of systemic Plaquenil therapy.

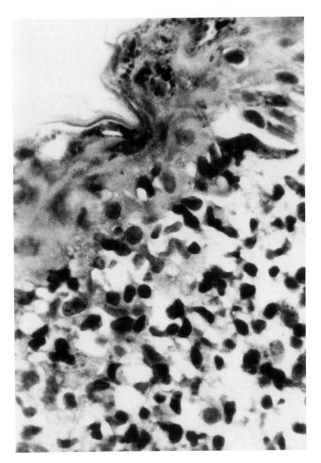

FIGURE 17-2 Dermoepidermal junction shows vacuole formation and the predominantly lymphocytic infiltrate (hematoxylin and eosin stain, ×200).

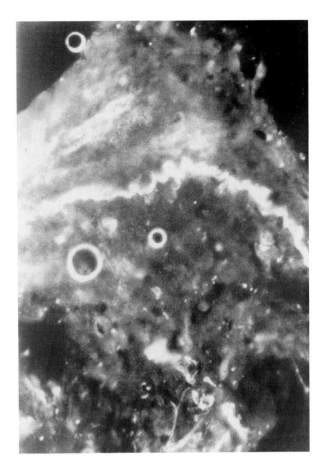

FIGURE 17-3 Immunofluorescence staining shows homogenous band of IgG at the dermoepidermal junction (original magnification ×200).

clude moderate hyperkeratosis, lymphocytic infiltration at the dermoepidermal junction, and telangiectasia (Fig. 17-2).[10] Direct immunofluorescent examination may show deposition of immunoglobulins, complement, and fibrin at the dermoepidermal junction (Fig. 17-3).

DLE involving the eyelids generally is a benign disorder and responds to the judicious use of sunscreens, topical or intralesional corticosteroids, chloroquine, or hydroxychloroquine therapy.[9,10,17-19] Eyelid involvement may be the initial presentation of systemic lupus with multisystem involvement and numerous ocular complications.

References

1. Rodnan GP, Schumacher HR, eds. *Primer on the Rheumatic Diseases.* 8th ed. Atlanta, Ga: Arthritis Foundation; 1983:49.
2. Wallace DJ, Dubois EL. Definition, classification, and epidemiology of systemic lupus erythematosus. In: Wallace DJ, Dubois EL, eds. *Dubois' Lupus Erythematosus.* 3rd ed. Philadelphia, Pa: Lea & Febiger; 1987:15-32.
3. Millard LG, and Rowell NR. Abnormal laboratory test results and their relationship to prognosis in discoid lupus erythematosus. *Arch Dermatol.* 1979;115:1055.
4. Callen JP. Chronic cutaneous lupus erythematosus: clinical, laboratory, therapeutic, and prognostic examination of 62 patients. *Arch Dermatol.* 1982;118:412.
5. Tuffanelli DL, Dubois EL. Cutaneous manifestations of systemic lupus erythematosus. *Arch Dermatol.* 1964;90:377.
6. Williams B, Hull DS. Lupus erythematosus keratoconjunctivitis. *Southern Med J.* 1986;79:631.
7. Aronson AJ, Ordonez NG, Diddie KR, et al. Immune-complex deposition in the eye in systemic lupus erythematosus. *Arch Intern Med.* 1979;139:1312.
8. Tuffanelli DL. Lupus erythematosus. *Am Acad Dermatol.* 1981;4:127.
9. Callen JP. Therapy of cutaneous lupus erythematosus. *Med Clin North Am.* 1982;66:795.
10. Donzis PB, Insler MS, Buntin DM, et al. Discoid lupus erythematosus involving the eyelids. *Am J Ophthalmol.* 1984;98:32.
11. Klauder JV, DeLong P. Lupus erythematosus of the conjunctiva, eyelids and lid margins. *Arch Ophthalmol.* 1932;7:856.
12. Kearns W, Wood W, Marchese A. Chronic cutaneous lupus involving the eyelid. *Ann Ophthalmol.* 1982;14:1009.
13. Zvi R, Schewach-Millet M, et al. Discoid lupus erythematosus of the eyelids. *J Am Acad Dermatol.* 1986;15:112.
14. Feiler-Ofry V, Isler A, Hanau D, et al. Eyelid involvement as the presenting manifestation of discoid lupus erythematosus. *J Pediatr Ophthalmol Strabismus.* 1982;16:395.
15. Huey C, Jakobiec FA, Iwamoto T, et al. Discoid lupus erythematosus of the eyelids. *Ophthalmology.* 1983;90:1289.
16. Roegnik HH, Martin JS, Eichorn P, et al. Discoid lupus erythematosus: diagnostic features and evaluation of topical corticosteroid therapy. *Cutis.* 1980;25:281.
17. Farkas TG. Discoid lupus erythematosus of the eyelids. *Ophthalmol Dig.* 1973;35:21.
18. Portnoy JZ, Callen JP. Ophthalmologic aspects of chloroquine and hydroxychloroquine therapy. *Int J Dermatol.* 1983;22:273.
19. Handa F, Chopra A, Aggarwal RR, et al. Ocular changes after treatment of chronic discoid lupus erythematosus with chloroquine. *Indian J Dermatol.* 1983;28:145.

Chapter 18

Polyarteritis Nodosa

JOHN J. PURCELL, Jr.

Polyarteritis is an immune complex vasculitis that affects small and medium-sized arteries. The cause is unknown, but the disease has been reported to follow serum sickness and drug reactions. It may run an acute, subacute, or chronic course and is more frequent in males. There is no racial or familial predisposition; it can occur at any age but is more common in the fourth or fifth decade. Since the vessels to any organ can be involved, the clinical manifestations are protean and often confusing.

Immunoglobulin deposits in blood vessels of polyarteritis patients were first demonstrated in 1975.[1] This was the first direct evidence of an immunoallergic cause of polyarteritis. Recently IgM, IgG, and complement were demonstrated within the small and medium-sized arterioles of the conjunctiva and skin lesions of a patient with polyarteritis.[2] Hepatitis antigenemia is present in 30% of patients with necrotizing vasculitis, suggesting that immune-mediated phenomena are important in the pathogenesis of hepatitis B associated with polyarteritis.[3] The immune complexes damage the vessel walls, with destruction and necrosis of all coats of the vessel wall. Rupture, hemorrhage, thrombosis, or healing follows, and the extent of damage causes the spectrum of systemic disease manifestations seen clinically.

Systemic Manifestations

Single or multiple organ systems may be involved. Abdominal pain, myalgias, peripheral neuritis, evidence of renal disease, or hypertension associated with weight loss, fever, and leukocytosis are predominant clinical features.

Cutaneous or subcutaneous nodules on any part of the body that may be painful and wax and wane are important diagnostic features. Osler's nodes (tender purple nodes) are cutaneous manifestations of vasculitis (Fig. 18-1). Other nonspecific erythematous rashes, urticaria, petechiae, purpura, or hemorrhagic bullae may occur.

Myalgias and arthralgias are present when the vessels to muscles or joints are involved. Peripheral neuritis or central nervous system symptoms may occur with involvement of these vessels. Gastrointestinal involvement may produce abdominal pain, nausea, vomiting, diarrhea, intestinal bleeding, and hepatomegaly. Pericarditis and myocarditis have been reported.

Pulmonary complications occur in about 25% of patients. Classic polyarteritis does not involve the main pulmonary artery, though arteritis of its smaller branches does occur. When lung involvement is present, it is usually seen with granulomatous reactions in a setting of asthma and severe allergic problems.[4]

Almost 80% of patients have renal involvement. Focal or diffuse glomerulonephritis, renal infarcts, and hemorrhagic cystitis have been reported. Hypertension may lead to uremia, congestive heart failure, and death.

Ocular Manifestations

The ocular manifestations of polyarteritis depend on which vessels of the eye are affected by the vasculitis. Since this is usually a disorder that segmentally affects vessels, the eye may be involved in a variety of ways.[5]

Although only 10% of cases are felt to have ocular involvement, the actual incidence may well be higher because the ocular findings may be minimal or overshadowed by systemic manifestations.

Retinopathy is one of the more commonly reported

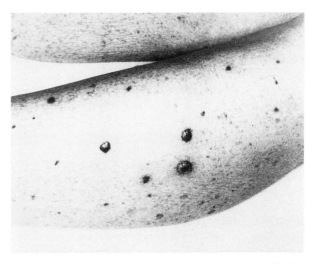

FIGURE 18-1 Painful nodular, hemorrhagic skin lesions of left leg in a patient with polyarteritis.

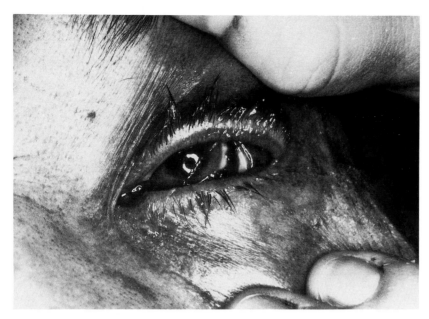

FIGURE 18-2 Necrotizing scleritis due to polyarteritis.

ocular complications of polyarteritis. This may be due to direct immune complex destruction of vessel walls or be secondary to renal disease and severe hypertension.[6] This leads to retinal ischemia, edema, cotton-wool spots, irregular caliber of retinal vessels, retinal vascular occlusion, and retinal hemorrhages.[7]

Papilledema or papillitis and subsequent optic atrophy can occur. Inflammation of orbital vessels may lead to exophthalmos.[8] Involvement of the central nervous system and vasa vasorum of the peripheral nervous system may produce extraocular muscle palsies, homonymous hemianopsia, amaurosis, Horner's syndrome, or nystagmus. Fibrinous iridocyclitis results from acute iris vessel involvement, vascular damage, and leakage of protein into the anterior chamber.[2] Necrotizing scleritis and sclerokeratitis occur with involvement of scleral vessels (Fig. 18-2). A peripheral furrowed ulceration may spread around the limbus to form a ring ulcer. Central keratitis may result in vascularization, scarring, or perforation.

Raised waxy-looking areas of conjunctival infarction and necrosis as well as subconjunctival hemorrhage and chemosis have been reported (Fig. 18-3).

Cogan's syndrome consists of interstitial keratitis and vestibuloauditory symptoms. Many patients also have associated cardiovascular disturbances. Chasson[9] reviewed 53 cases of Cogan's syndrome and found a strong association with systemic vasculitis and polyarteritis.

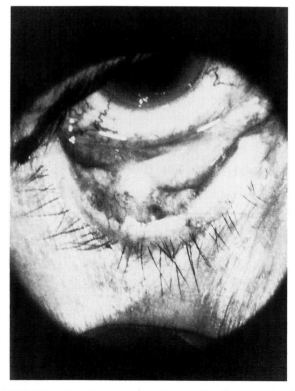

FIGURE 18-3 Areas of conjunctival edema and necrosis due to conjunctival vascular involvement in polyarteritis.

Polyarteritis is treated with systemic steroids, and localized ocular therapy with topical steroids or cycloplegia may be required for anterior segment involvement. Immunosuppressive therapy may be indicated if patients do not respond to systemic steroids.

References

1. Sams WMJ, Clayrian HN, Kohler PF. Human necrotizing vasculitis: immunoglobulins and complement in vessel walls of cutaneous lesions and normal skin. *J Invest Dermatol.* 1975;64:441.
2. Purcell JJ Jr., Birkenkamp R, Tsai CC. Conjunctival lesions in periarteritis nodosa, a clinical and immunopathologic study. *Arch Ophthalmol.* 1984;102:736.
3. Parker CW. *Clinical Immunology.* Philadelphia, Pa: WB Saunders; 1980:478.
4. Churg J, Strauss L. Allergic granulomatosis, allergic angiitis, and periarteritis nodosa. *Am J Pathol.* 1951;27:277.
5. Stillerman ML. Ocular manifestations of diffuse collagen disease. *Arch Ophthalmol.* 1951;45:239.
6. Sheehan B, Harriman DGF, Bradshaw JPP. Polyarteritis with ophthalmic and neurological complications. *Arch Ophthalmol.* 1958;60:537.
7. Wise GN. Ocular periarteritis nodosa: report of two cases. *Arch Ophthalmol.* 1952;48:1.
8. Kimbrell OCJ, Wheliss JA. Polyarteritis nodosa complicated by bilateral optic neuropathy. *JAMA.* 1967;201:139.
9. Chesson EBD, Blaning WZ, Alroy J. Cogan's syndrome: a systemic vasculitis. *Am J Med.* 1976;60:549.

Chapter 19

Polymyositis/Dermatomyositis

JOSEPH B. WALSH and JEFFREY JOSEF

Polymyositis/dermatomyositis (PM/DM) is an inflammatory disease in which skeletal muscles are diffusely damaged by a nonsuppurative inflammatory process.[1] It may occur as a solitary entity or associated with other collagen vascular diseases such as rheumatoid arthritis, systemic lupus erythematosus, scleroderma, and periarteritis nodosa. The hallmark of the disease is muscle weakness, which usually develops over a period of several weeks and may be acute in onset. Pain is more common in the acute cases, and muscle weakness out of proportion to atrophy is present in the chronic forms. The proximal limb muscles are usually more severely affected. Approximately one third of adult cases have been associated with cutaneous changes and are classified as DM.[2] There is a bimodal peak incidence, with the onset of the juvenile form between 5 and 14 years of age and of the adult form between 45 and 64 years.

Systemic Manifestations

Diagnostic criteria for PM/DM are as follows[2]:

1. Muscle weakness, usually proximal, progressing over several months
2. Biopsy evidence of muscle fiber necrosis and regeneration
3. Elevated serum levels of creatine kinase, aldolase, or myoglobin
4. Electromyography (EMG) changes compatible with multifocal myopathy.

The diagnosis is definite when all four of the above criteria are satisfied and probable when three are present. DM is diagnosed by the above criteria in addition to cutaneous changes consisting of a dusky erythematous rash in a butterfly distribution over the face with a typical heliotrope discoloration of the eyelids with telangiectasia. In the adult form, the proportion of PM is about 70% and of DM 30%, and in both cases more females than males are affected. In the juvenile form of the disease males predominate over females and the ratio of DM to PM is 2:1. In the juvenile forms calcification of subcutaneous tendons is common. Of note is the enzyme elevation, which can vary with the clinical course, and the EMG findings, which can be normal in 10% of patients, owing to the multifocal nature of the disease process. It is important when this diagnosis is suspected to investigate a large sample of proximal and distal limb muscles for both EMG and histologic confirmation.

The pathophysiology remains unknown but there is strong evidence of an immune complex–mediated vasculopathy in which the complement system is activated early. It has been noted on muscle biopsy that there is cytolytic membrane attack complex (C5b-9) in all intramuscular microvasculature, producing ischemic changes. The predominance of activated T cells, including suppressor cytotoxic and natural killer cells, suggests the presence of cell-mediated muscle damage in this disorder.[2-4] Studies have also shown an increased incidence of specific HLA types in both adult and juvenile DM/PM, suggesting a genetic predisposition.

Inflammatory trigger mechanisms such as an antecedent viral infection have also been suggested. In addition, drugs may induce a DM/PM picture, especially penicillamine, tamoxifen, and BCG vaccination.

Patients with the adult form of PM/DM are at increased risk for malignancies. In one study of 58 patients followed for 20 years, there was a 26% prevalence of malignancy: in males, cancers of the lung and gastrointestinal tract and in females, of the breast and ovary. There was a higher incidence of malignancy in DM than in PM.[5] It has been suggested that a normal (as opposed to elevated) level of creatine kinase at the time of diagnosis of DM portends a poor prognosis because of a high incidence of malignancy observed in these cases.[6]

As mentioned previously, PM/DM may occur in conjunction with other collagen diseases. One association is Sjögren's syndrome. Four such patients all had elevation of serum IgG, IgM, rheumatoid factor, and antinuclear antibodies, suggesting this may be a systemic autoimmune-related disorder with inflammation of the exocrine glands and muscles.[7] The authors suggested that this inflammation results from B-cell or humoral hyperactivity producing inflammation of both the muscles and the glands.

Active myocarditis may result in cardiac failure and conduction defects such as heart block or arrhythmia. Cardiac involvement may be seen in 70% of children, whereas it is rarer (30%) in adults.[3] Pulmonary involvement may produce a diffuse interstitial infiltration and fibrosis, which can precede the onset of myopathy.[2] In juvenile DM involvement of the intestinal tract may cause difficulty swallowing, and the vasculitis may lead to ulceration and perforation of the gastrointestinal tract.

Differential Diagnosis

Differential diagnosis includes infectious myopathies such as trichinosis, endocrinopathies such as hyperthyroidism, myasthenia gravis, muscular dystrophy, cutaneous lupus erythematosus, and other systemic diseases with myopathies, such as sarcoidosis.

Ocular Findings

In dermatomyositis, the typical heliotrope discoloration of the upper eyelids, the butterfly distribution of the erythematous facial rash, and periorbital edema, are the hallmarks of the disease. Conjunctival edema may be seen in as many as 70% of cases.[8]

Nystagmus, extraocular muscle imbalance, and exophthalmos have also been found. The reported incidence of extraocular muscle imbalance has ranged between 2 and 10% of patients with PM/DM. In several such cases reported in the literature that were reexamined several years later, however, myasthenia gravis was diagnosed in addition to the PM. The authors suggest that involvement of the extraocular muscles in DM is unusual enough that it may be said that it does not occur.[9]

Membranous conjunctivitis was reported in a 16-year-old boy with biopsy-confirmed DM.[10] A 13-year-old girl with a 9-year history of DM presented with findings of conjunctival vessel occlusion (Fig. 19-1) but no other ocular abnormalities.[11] Intraocular involvement may include iritis, but this is poorly documented.

A rare but better-documented finding is retinopathy. In 1938 Bruce was the first to report cotton-wool spots and occasional hemorrhages. He described three patients, two children and one adult.[12] Late sequelae include optic atrophy, pigmentary maculopathy, Elschnig's spots, and arteriovenous anastomoses.[13,14] These findings suggest an occlusive vasculitis affecting retina, optic nerve, and choroid. No confirmatory ocular pathology is available, however.[13,14] One 5-year-old boy with cotton-wool spots had complete resolution with treatment and did not develop optic atrophy, although no visual field results were reported.[15] The authors note that retinopathy develops more commonly in children with DM than in adults because systemic vasculitis is associated more commonly with the juvenile than the adult form.

Therapy

In PM/DM, the mainstay of therapy is systemic corticosteroids. Without treatment, the mortality rate with PM/DM is over 60% at 5 years. With systemic steroid therapy it is approximately 9 to 13% when secondary malignancies are excluded.[1] Before steroids, 32% of children with juvenile DM died and 33% were severely impaired by muscle restriction. With steroid therapy the mortality rate is approximately 7%.[3] Other treatments that have been suggested are immunosuppressive drugs such as methotrexate and azathioprine. Plasmapheresis and total-body low-dose irradiation have been used in refractory cases.

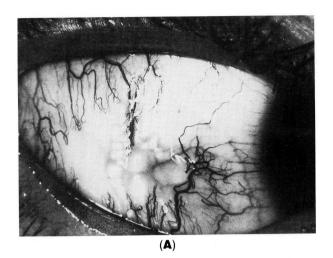

(A)

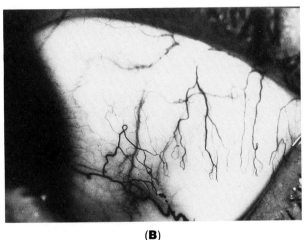

(B)

FIGURE 19-1 Avascular zone in temporal region of conjunctiva in a 13-year-old girl with long-standing dermatomyositis. **(A)** OD: avascular center has nodules. **(B)** OS: vascular segmentation is noted. (From Van Nouhuys.[11])

References

1. Bradley WL. Inflammatory disease of muscle. In: Kelly WM, Harris ED, Ruddy S, et al., eds. *Textbook of Rheumatology*. 2nd ed. Philadelphia, Pa: WB Saunders; 1985;79:1225-1245.

2. Mastaglia FL, Ojeda VJ. Inflammatory myopathies. *Ann Neurol*. 1985;17:215-227, 317-323.

3. Pachman LM. Juvenile dermatomyositis. *Pediatr Clin North Am*. 1986;33:1097-1117.

4. Kissel JT, Mendell JR, Rammogan KW. Microvascular deposition of complement membrane attack complex in dermatomyositis. *N Engl J Med*. 1986;314:329-334.

5. Callen JP, Hyla JF, Bole GG, Kay DR. The relationship of dermatomyositis and polymyositis to internal malignancy. *Arch Dermatol*. 1980;116:295-298.

6. Fudman EN, Schnitzer TJ. Dermatomyositis without creatine kinase elevation: a poor prognostic sign. *Am J Med*. 1986;80:329-332.

7. Ringel SP, Ferstet JZ, Tan EM, et al. Sjögren's syndrome and polymyositis and dermatomyositis. *Arch Neurol*. 1982;39:157-163.

8. Hollenhorst RW, Henderson JW. The ocular manifestations of the diffuse collagen diseases. *Am J Med Sci*. 1951;221:211-222.

9. Scoppetta C, Morante M, La Sali C, et al. Dermatomyositis spares extra-ocular muscles. *Neurology*. 1985;35:141.

10. Sammartino A, Lucariello A, Esposito L, et al. A rare presentation of bilateral membranous conjunctivitis in dermatomyositis. *Ophthalmologica*. 1982;184:97-102.

11. Van Nouhuys CE, Jengers RCA. Bilateral avascular zones of the conjunctivae in a patient with juvenile dermatomyositis. *Am J Ophthalmol*. 1987;104:440-442.

12. Bruce GM. Retinitis in dermatomyositis. *Trans Am Ophthalmol Soc*. 1938;36:282-297.

13. Munro S. Fundus appearances in a case of acute dermatomyositis. *Br J Ophthalmol*. 1959;43:548-558.

14. Harrison SM, Frenkel M, Grossman BJ, Matalon R. Retinopathy in childhood dermatomyositis. *Am J Ophthalmol*. 1973;76:786-790.

15. Cohen BH, Sedwick LA, Burde RM. Retinopathy of dermatomyositis. *J Clin Neuro-Ophthalmol*. 1985;5:177-179.

Chapter 20
Reiter's Syndrome

EDWARD J. HOLLAND

Reiter's syndrome (RS) is classically characterized by the triad of arthritis, conjunctivitis, and urethritis. More recently it has been recognized that a presumptive diagnosis of RS can be made on the findings of a seronegative, oligoarticular arthritis with urethritis or cervicitis.

The pathogenesis of RS is not clear; however the disease process appears to occur when patients with a genetic predisposition are exposed to certain microorganisms. The genetic marker is the HLA-B27 histocompatibility antigen, which is present in 75 to 90% of these patients. Epidemiologic studies have shown that persons with this antigen have a 20 to 35% risk of developing RS after certain bacterial infections, and the term "reactive arthritis" is sometimes used for these patients' condition.

The two types of infections that may lead to RS are genitourinary (postvenereal) and gastrointestinal (postdysenteric).[1] The pathogens that have been most commonly implicated in postvenereal RS are *Chlamydia trachomatis* and *Ureaplasma urealyticum*. *Yersinia enterocolitica*, *Salmonella* species, *Shigella* species, and *Campylobacter jejuni* have been associated with postdysenteric RS. Additional support for the association of microorganisms and RS comes from a study of HLA-B27–positive patients with RS who did not have a history of a genitourinary or gastrointestinal infection.[2] Several of the patients had histologic evidence of chronic inflammation on biopsies obtained during ileocolonoscopy, suggesting that asymptomatic infection of the ileum may be important in the pathogenesis of this disease.

The exact role of the HLA-B27 antigen in RS remains unclear, and several hypotheses have been postulated. First, the HLA-B27 gene may be only a marker for this and other diseases because of its location on the sixth chromosome adjacent to the proposed immune response (IR) genes. The HLA-B27 gene and the IR genes are in linkage disequilibrium, which would explain why 10 to 25% of patients with RS do not have the HLA-B27 antigen. Second, the HLA-B27 antigen may act as a receptor to infecting organisms that leads to an abnormal immune response. Finally, molecular mimicry may exist between the HLA-B27 antigen and microbial antigens. This can lead to disease by the host mounting an immune reaction to "self" antigens through cross-reactivity. The HLA-B27 antigen is present on all nucleated cells and may be the reason why RS and other B27–related diseases are associated with a variety of clinical features affecting many organ systems.

Systemic Manifestations

The most common manifestations of RS include arthritis, eye inflammation, urethritis, and mucocutaneous lesions.[3] The arthritis usually affects the lower extremities and is typically acute, oligoarticular, and asymptomatic. Other rheumatologic manifestations are tenosynovitis, dactylitis, sacroiliitis, plantar fasciitis, and periostitis of the calcaneus. Sausage-shaped digits often occur and are related to the enthesopathy of this disorder.

Genitourinary involvement is most commonly intermittent urethritis. The urethral discharge is usually serous and moderate in amount, and it can be asymptomatic. Females may have cervicitis, nonspecific vaginitis, or urethritis. Less common urinary disorders include prostatitis, cystitis, and urethral strictures.

Mucocutaneous lesions are seen in 30 to 40% of RS patients. The most characteristic lesion is keratoderma blennorrhagicum, which consists of papules, vesicles, or pustules, usually located on the palms and soles but occasionally seen on the limbs, trunk, and scalp (Fig. 20-1). Superficial lesions of the oral mucosa and the glans penis can be seen which are painless and may not be detected by the patient. Nail abnormalities can occur, including subungual pustules and hyperkeratosis. Pitting of the nails does not occur, which distinguishes this from the nail abnormality seen in psoriatic arthritis.

Other less common systemic manifestations include fever, pericarditis, aortic insufficiency, cardiac conduction defects, systemic amyloidosis, and neurologic problems.

Laboratory evaluation is usually unremarkable except for the presence of the HLA-B27 antigen. The sedimentation rate can be normal or elevated, rheumatoid factor and antinuclear antibodies are negative, and synovial fluid analysis is nondiagnostic.

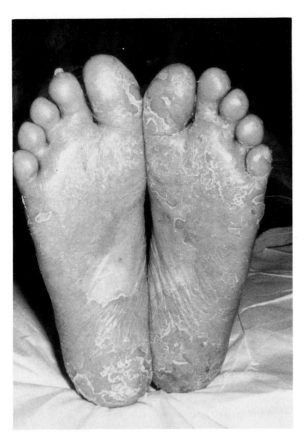

FIGURE 20-1 Keratoderma blennorrhagicum of the soles in a young male with RS.

Ocular Manifestations

Conjunctivitis is the most common eye finding in RS (58%).[4] It is often mild and associated with a mucopurulent discharge. The inflammation usually resolves without treatment after 7 to 10 days, and cultures are negative. Discomfort is minimal, and patients often do not seek ophthalmic evaluation. Rarely, a painful conjunctivitis with profusely purulent discharge can occur. A nongranulomatous iridocyclitis is the second most common ocular manifestation. The episodes are acute, with mild to moderate amounts of cells and flare, keratic precipitates, and occasional posterior synechiae. Keratitis, scleritis, episcleritis, and posterior uveitis have been reported in these patients but are less common. The keratitis has been described as a superficial punctate keratopathy with anterior stromal infiltrates and micropannus.

The pathophysiologic mechanisms underlying the ocular manifestations of this disorder are not understood. One study of HLA-B27–positive patients with anterior uveitis and RS revealed that antibodies to certain serotypes of chlamydia cross-react with HLA-B27 lymphocytes of patients with sacroiliitis.[5] These antibodies also bind to human conjunctiva, which could, in turn, produce the conjunctivitis seen clinically. Another possibility is that the conjunctivitis is caused by microorganisms transmitted from the genital tract or gut to the eye. *Chlamydia trachomatis* can often be isolated from the conjunctiva and urinary tract of patients with inclusion conjunctivitis; however, attempts to isolate organisms from patients with RS have not been successful.

Differential Diagnosis and Treatment

Differential diagnosis of RS includes gonococcal arthritis, but in this disease the urethral discharge is usually more prominent and Gram's stain of the discharge or synovial fluid should demonstrate gonococcal organisms. Psoriatic arthritis (PA) can also be confused with RS, though the skin involvement in PA is usually more extensive and the course of the disease more constant. Ankylosing spondylitis, especially in the early stages, may resemble RS, but extensive involvement of the axial skeleton occurs in this disorder. Inflammatory bowel disease (ulcerative colitis, Crohn's disease) and Whipple's disease can be associated with arthritis and uveitis. The histopathology of the bowel lesions is characteristic of these disorders. Behçet's disease can have oral and genital ulcers, skin manifestations, and arthritis in association with ocular inflammation; however the arthritis is usually self-limited and the uveitis often severe.

The onset of the individual manifestations in RS usually evolves over days to weeks. The course of this disease is characterized by recurrences in most patients, and up to 15 to 25% develop a permanent disability. Treatment is most often related to the musculoskeletal complications. Nonsteroidal anti-inflammatory medications are the mainstay of treatment. Physical therapy can be helpful. Local corticosteroid and cytotoxic agents such as azathioprine and methotrexate may be necessary for more severe cases.[6,7] The conjunctivitis usually requires no treatment. Iridocyclitis and the keratitis respond to topical steroids. The use of antibiotics has been controversial as there is no direct effect on the arthritis and a questionable effect on the urethritis.

References

1. Keat A. Reiter's syndrome and reactive arthritis in perspective. *N Engl J Med.* 1983;309:501.

2. Mielants H, Veys EM. Reiter's syndrome and reactive arthritis. *N Engl J Med.* 1984;320:1539.
3. Gerber RC. Diagnosis and management of Reiter's syndrome. *Compr Ther.* 1984;10:51.
4. Lee DA, Barker SM, Su WPD, et al. The clinical diagnosis of Reiter's syndrome: ophthalmic and nonophthalmic aspects. *Ophthalmology.* 1986;93:350.
5. Wakefield D, Penny R. Cell-mediated immune response to chlamydia in anterior uveitis: role of HLA-B27. *Clin Exp Immunol.* 1983;51:191.
6. Colin A. A placebo controlled, crossover study of azathioprine in Reiter's syndrome. *Ann Rheum Dis.* 1986;45:653.
7. Lally EV, Ho G Jr. A review of methotrexate therapy in Reiter syndrome. *Semin Arthritis Rheum.* 1985;15:139.

Chapter 21

Relapsing Polychondritis

R. WAYNE BOWMAN and JAMES P. McCULLEY

Relapsing polychondritis is an uncommon multisystem disease of unknown cause. While it is reported in all races, the disease occurs predominantly in whites during the fourth decade of life. The median age at diagnosis is 51 years (range 11 to 84 years). Sex distribution is equal. Tissues with a high glycosaminoglycan content, such as cartilage, aorta, sclera, and cornea, are primarily involved. Autoantibodies to native type II collagen (cartilage) have been demonstrated, which suggests an autoimmune cause. This may explain the clinical overlap between polychondritis and rheumatoid arthritis, systemic lupus erythematosus, ulcerative colitis, Wegener's granulomatosis, and Behçet's syndrome (the MAGIC syndrome).[1]

Relapsing polychondritis (RP) may develop from an immune response to previously sequestered or altered collagen. Histopathology of lesions in RP reveals cellular infiltration by lymphocytes, plasma cells, and polymorphonuclear leukocytes at the chondrofibrous junction. Immunofluorescence has demonstrated immunoglobulin (primarily IgG) and C3 deposition. Circulating antibodies to type II collagen have been detected by Foidart et al.[2] Some patients had antibodies to native type II collagen only; however, in other patients who had antibodies to both native and denatured type II collagen, the titers of the former were always higher. The titer correlated with disease activity, and in contrast to reports of antibodies in rheumatoid arthritis, antibodies to types I, III, and IV collagen were not found.[3] These findings suggest that the antibodies were formed before destruction of cartilage and implicate antinative type II collagen antibodies in the pathogenesis of relapsing polychondritis.[2] An animal model of auricular chondritis and arthritis has been developed in outbred Wistar rats immunized with native type II collagen.[4] Electron-dense deposits that corresponded with deposits of IgG and C3 were seen near the surface of chondrocytes; circulating IgG reactive to native type II collagen was also detected. Deposited circulating immune complexes in a systemic vasculitis could become trapped in cartilage and produce injury. Decreased T-helper or -inducer and increased T-suppressor or -cytotoxic cells have also been observed.

Systemic Manifestations

Otorhinolaryngeal disease is one of the cardinal manifestations of relapsing polychondritis. Auricular chondritis is secondary to immune destruction of cartilage collagen and may be the initial manifestation of the disease in up to 39% of patients, eventually developing in 88.6%.[3] A typical attack consists of sudden pain, swelling, and erythema of the auricular cartilage with sparing of the ear lobe. Recurrences are common and are usually less severe. With loss of cartilage support the ears become soft and pliable. Bilateral, recurrent auricular chondritis almost always signifies relapsing polychondritis. Nasal chondritis occurs in over one half of patients and can result in a saddle-nose deformity (Fig. 21-1).[3]

Audiovestibular damage occurs in approximately 50% of patients, being the presenting sign in 6%. Conductive hearing loss can be due to involvement of the eustachian tube cartilage, auricular cartilage collapse, inflammatory closure of the external auditory canal, or serous otitis media. Vasculitis of the vestibular or cochlear branch of the internal auditory artery may

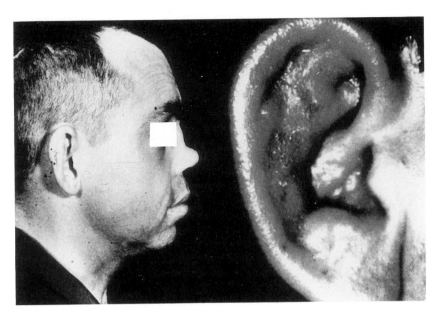

FIGURE 21-1 Auricular chondritis and saddle-nose deformity. (Courtesy of Arthritis Foundation.)

cause sensorineural hearing loss. Vascular involvement may also produce vestibular dysfunction with dizziness, vertigo, ataxia, tinnitus, nausea, and vomiting.[5]

Respiratory tract involvement is a presenting sign in 25% of patients, and over 50% develop laryngotracheal symptoms. In 10% of patients it is the cause of death. Symptoms such as hoarseness, cough, dyspnea, wheezing, choking, and tenderness over the anterior tracheal cartilage result from involvement of the laryngotracheal cartilage and airway obstruction. Either local or diffuse, fixed or dynamic airway obstruction may be seen. Inflammation of the glottis, larynx, or subglottic tissue is more common in younger patients and women, and may require tracheostomy. A dynamic airway obstruction develops when damage to bronchial and tracheal cartilage increases cartilage compliance, leading to upper airway narrowing during respiration. Inflammation and scarring of the respiratory tract cartilage leads to impairment of ciliary clearance of mucus and to decreased effectiveness of coughing. Lower airway disease can occur as well and contributes to the higher rate of pulmonary infections, respiratory failure, and subsequent morbidity due to pulmonary disease. Evaluation of the airway should include tomography, computed tomography (CT), and spirometry with a flow-volume loop. Additional airway resistance measurements may be required.[6]

Cardiovascular manifestations include valvular heart disease, arrhythmias, large-vessel vasculitis, and aneurysms. Thoracic or abdominal aortic aneurysms occur in as many as 10% of patients and a Takayasu-type aortic arch syndrome can occur. Aneurysms develop due to aortitis and may be multiple and silent until a dissection or rupture occurs. Aortitis of the ascending aorta may produce aortic regurgitation. Aortic regurgitation or mitral valve insufficiency may also develop secondary to inflammation of the valve anulus and papillary muscles and valvulitis. Arrhythmias may occur, including complete heart block and supraventricular tachycardia, probably owing to the proximity of the cardiac conduction system and the aortic root.

Arthritis may be the presenting symptom in 33% of patients and eventually develops in 50 to 80%. It is oligoarthritis or polyarthritis that is seronegative, nondeforming, and nonerosive. It tends to involve the parasternal costochondral joints as well as small and large joints and can last weeks to months.[3]

More than half of the patients with relapsing polychondritis present with inflammatory skin findings such as erythema, swelling, pain, and, less commonly, discoloration of the skin of the ears, joints, and nose. Most of the skin findings probably occur secondary to an underlying vasculitis and include palpable purpura, urticaria, angioedema, erythema multiforme, livedo reticularis, panniculitis, and erythema nodosum. The most common single lesion is leukocytoclastic vasculitis.[6] Migratory superficial thrombophlebitis may also occur.[6]

Renal involvement consists of immune glomerulonephritis, usually a focal proliferative type with crescent formation.[7] IgG and complement have been demonstrated in glomeruli. A normal urinalysis, most commonly proteinuria or microhematuria, occurs in 25% of patients. An elevated creatinine is seen in 10%.

Neurologic manifestations of relapsing polychondritis are multiple and include involvement of the second, sixth, seventh, and eighth cranial nerves, headaches, encephalopathy, seizures, hemiplegia, ataxia, mononeuritis multiplex, mixed motor sensory neuropathy, and temporal arteritis.[6] Vasculitis of the central or peripheral nervous system, with or without systemic vasculitis, is responsible.

Ocular Manifestations

Ocular findings in relapsing polychondritis are rarely specific enough to be diagnostic, although in retrospect they are the presenting symptom in up to 20% of patients and will eventually develop in approximately 60%. The lids, orbit, episclera, sclera, conjunctiva, cornea, uvea, retina, and optic nerve may be involved. Periorbital lid edema may mimic lid cellulitis and may even be bilateral. Proptosis with orbital and peribulbar chemosis can simulate an orbital pseudotumor. Extraocular muscle palsy (most commonly the lateral rectus) can occur as a direct result of vasculitis or secondarily, due to central nervous system involvement. Dacryocystitis has resulted from nasal collapse.

Episcleritis and scleritis are the most common ocular manifestations, and are as frequent in men as in women. Episcleritis is the more common of the two and may be simple or diffuse, unilateral or bilateral, and recurrent. Type II collagen is native to sclera, and the antibodies to type II collagen may become localized in the sclera, resulting in clinical disease. Other species such as fish and birds have a cartilaginous plate in the eye, and it is possible that human sclera still harbors sequestered glycoprotein antigens that are very similar to cartilage.[8] The scleritis may be anterior (diffuse or necrotizing) or posterior. There is frequently an associated uveitis. Episcleral and scleral involvement tend to occur in conjunction with other disease activity, especially in the nose and joints.

Corneal involvement resembles immune complex deposition or ischemic vasculitis seen in other autoimmune diseases. Corneal findings include small epithelial or subepithelial peripheral corneal infiltrates, peripheral corneal thinning, pannus, and melting, usually associated with scleritis (Fig. 21-2).[9] Superficial corneal opacities, peripheral limbal guttering with crystalline deposits, and punctate keratitis from keratoconjunctivitis sicca have also been noted. Keratoconjunctivitis sicca has generally been mild unless there is associated Sjögren's syndrome. Corneal perforations with or without underlying rheumatoid disease have also occurred.[10]

Iridocyclitis may occur in as many as 30% of patients and is often associated with scleritis.[6] Also associated with posterior scleritis are retinal pigment epithelial defects and sensory retinal detachments. Retinopathy, including microaneurysms, flame-shaped hemorrhages,

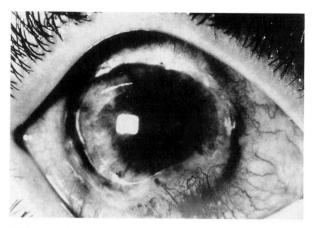

FIGURE 21-2 Marginal keratitis and thinning. (Courtesy of Alice Matoba, M.D.)

soft exudates, and cotton-wool spots, are probably due to vasculitis and immune complex deposition, as was the occlusion of a central and branch retinal vein in one patient. Chorioretinitis has also been seen.

Optic neuritis, ischemic optic neuropathy, disc edema, and field defects have all been described.[11] The ischemic optic neuropathy occurred in conjunction with a systemic vasculitis. Unilateral disc edema occurred in association with posterior scleritis. Visual field defects consistent with occipital or optic radiation infarcts have been documented in one patient with systemic vasculitis.

Treatment of the ocular complications of relapsing polychondritis involves control of the systemic immune process. Therapy with systemic corticosteroids to suppress the vasculitis remains the initial treatment of choice. Oral nonsteroidal anti-inflammatory drugs, dapsone, cytotoxic drugs, and immunosuppressants such as cyclosporine may also have a therapeutic role. Dapsone may be particularly effective since it inhibits both lysosomal enzymes and the alternate complement pathway. Cyclosporine A, with its effects on cell-mediated immunity, may prove beneficial, but experience with its use in relapsing polychondritis is limited, and potentially serious renal side effects limit its widespread use. It is possible that by using these or other as yet undiscovered drugs in addition to oral steroids the disease can be controlled with lower doses of each, thereby minimizing their side effects.

References

1. Firestein GS, Gruber HE, Weisman MH, et al. Mouth and genital ulcers with inflamed cartilage: MAGIC syndrome. *Am J Med.* 1985;79(1):65-72.

2. Foidart JM, Shigeto A, Martin G, et al. Antibodies to type II collagen in relapsing polychondritis. *N Engl J Med.* 1978;299:1203-1207.

3. White JW. Relapsing polychondritis. *Southern Med J.* 1985;78(4):448-451.

4. Cremer MA, Pitcock JA, Stuart JM, et al. Auricular chondritis in rats. An experimental model of relapsing polychondritis induced with type II collagen. *J Exp Med.* 1981;154(2):535-540.

5. Pearson CM, Kline HM, Newcomer VD. Relapsing polychondritis. *N Engl J Med.* 1960;263:51-58.

6. Isaak BL, Liesegang TJ, Michet CJ Jr. Ocular and systemic findings in relapsing polychondritis. Ophthalmology 1986:93(5):681-689.

7. Espinoza LR, Richman A, Bocanegra T, et al. Immune complex-mediated renal involvement in relapsing polychondritis. *Am J Med.* 1981;71:181-183.

8. Magargal LE, Donoso LA, Goldberg RE, et al. Ocular manifestations of relapsing polychondritis. *Retina.* 1981;1(2):96-99.

9. Michelson JB. Melting corneas with collapsing nose. *Surv Ophthalmol.* 1984;29:148-154.

10. Matoba A, Plager S, Barger J, McCulley JP. Keratitis in relapsing polychondritis. *Ann Ophthalmol.* 1984;16(4):367-370.

11. Sundaram MBM, Rajput AH. Nervous system complications of relapsing polychondritis. *Neurology.* 1983;33:513-515.

Chapter 22

Adult Rheumatoid Arthritis

WILLIAM E. WHITSON and JAY H. KRACHMER

Rheumatoid arthritis (RA) is a chronic, inflammatory, multisystem disease of unknown cause that has as its foremost finding a polyarthropathy that is most often symmetric. It has a worldwide distribution and no racial or ethnic predilection. Large studies have failed to show any consistent familial or genetic pattern. No association with HLA-A or HLA-B haplotypes has been found, however an association does exist between HLA-Dw4 and HLA-DR4 and erosive seropositive RA in Caucasian patients.[1]

Onset is in the third to the fourth decade and women are three times more likely to be affected than men. The course is variable, with a broad spectrum of clinical manifestations ranging from mild joint discomfort to severe debilitating disease. Constitutional symptoms of fever, malaise, and weight loss often herald the onset of RA. Most commonly this is followed by symmetric articular involvement that slowly progresses over years, resulting in the characteristic joint deformities that are easily recognized.

The diagnosis of RA is based on clinical criteria, since no specific biologic marker exists; however, approximately 70% of patients with RA test positive for rheumatoid factor (RF), whereas RF is found in only 1 to 5% of the general population.[2] RF is predominantly an IgM immunoglobulin with reactivity against autologous IgG; it is an autoantibody. The exact stimulus for formation of RF and its biologic role are uncertain.

The initial pathophysiologic insult in RA appears to be synovial microvascular injury with edema, vascular occlusion, and infiltration of polymorphonuclear leukocytes (PMNs). Hypertrophic and hyperplastic synovial tissue then protrudes into the joint space in slender villous projections. The normally acellular synovial stroma becomes packed with mononuclear cells, including T and B lymphocytes, which are surrounded by a mantle of plasma cells. Perivascular follicular aggregations without true germinal centers may be seen. Destructive joint changes stem from two sources—an outpouring of enzymes from PMNs causing tissue digestion, and direct attack on local structures by the pannus of granulation tissue.

Systemic Manifestations

The onset of RA may be abrupt but is usually insidious, with morning stiffness, limitation of movement, signs of inflammation, and joint pain. Any diarthrodial joint may be affected, although the thoracic and lumbar spine are generally spared.[1] Proximal interphalangeal and metacarpophalangeal joints of the hands are the joints most commonly affected and give rise to the characteristic ulnar deviation and swan-neck deformities of the fingers. Synovial swelling in the wrist may lead to carpal tunnel syndrome with compression of the median nerve. Elbow and shoulder involvement may occur; hip involvement is infrequent. Knee joint disease is common and particularly disabling, as the knee is a major weight-bearing structure. Effusions may occur along with typical enlargement of the gastrocnemius-semimembranosus bursa forming Baker's cyst. Ankle and foot involvement produce pain on walking and may lead to hallux valgus and metatarsal head subluxation. Cervical spine disease leads to limitation of neck mobility and to pain that often radiates from the occiput over the top of the head to localize behind one eye. Vertebral artery compression may cause syncope on neck flexion. Subluxation of the odontoid process vertically into the foramen magnum may cause severe nerve damage. Other joints that may be affected include the temporomandibular, sternoclavicular, and cricoarytenoid joints.

Extra-articular Manifestations

Patients with either seropositive RA or severe joint disease have a higher incidence of extra-articular manifestations than do patients without these findings. Rheumatoid nodules, which occur in 20 to 25% of all patients with RA at some time during the course of their disease, are subcutaneous granulomas commonly seen on the extensor surface of the forearms, the olecranon, and the Achilles' tendons.[2] Histologically, a central area of fibrinoid necrosis is surrounded by a palisade of connective tissue, then granulation tissue, and finally lymphocytes and plasma cells.

Vasculitis, which may be related to deposition of antigen-antibody complexes, may manifest as polyneuropathy, skin ulcerations, visceral ischemia with bowel infarction or perforation, myocardial infarction, or cerebral ischemia. Although vasculitis is usually insidious, a rare form of fulminant vasculitis clinically similar to polyarteritis may be a direct cause of death.

Pericarditis is an autopsy finding in 40% of patients with RA. Granulomatous lesions may occur in the pericardium, epicardium, and myocardium. Pulmonary findings include pleurisy, with or without effusions, intrapulmonary rheumatoid nodules (known as Caplan's syndrome when associated with pneumoconiosis), diffuse interstitial fibrosis, and vasculitis leading to pulmonary hypertension. Anemia is the most common hematologic abnormality but is usually mild. Felty's syndrome is the association of RA, splenomegaly, and leukopenia and may be complicated by recurrent infections, often by gram-positive organisms. Lymphadenopathy may be found proximal to inflamed joints or in areas not associated with articular inflammation.

Ocular Manifestations

The pathophysiologic basis of the ocular manifestations of RA is no different from that seen in the rest of the body. Localized autoimmune phenomena fall into two categories, direct inflammatory cell infiltration, with or without destruction, such as that seen in the lacrimal gland or sclera, and immune vasculitis seen in episcleritis or in cerebral vessels leading to neuro-ophthalmic disorders.

Sjögren's Syndrome

Sjögren's syndrome was originally described as the association of xerophthalmia, xerostomia, and RA. The definition has been expanded to include additional connective tissue diseases such as systemic lupus erythematosus, periarteritis nodosa, psoriatic arthritis, polymyositis, progressive systemic sclerosis, and several others. Sjögren's syndrome, with its xerophthalmia, xerostomia, and a variety of systemic complications, may also occur in the absence of any systemic disease. Lacrimal deficiency alone is called keratoconjunctivitis sicca (KCS).

Sjögren's syndrome is seen in about 25% of patients with RA, while about 60% of patients who present with dry eye and dry mouth have an associated connective tissue disease (usually RA), which may or may not yet be manifest.[3] Sjögren's syndrome is a multisystem autoimmune disorder primarily involving the lacrimal glands, salivary glands, and mucous-secreting glands of the gastrointestinal and respiratory tract. In the lacrimal gland the acini are invaded and destroyed by a mononuclear cell infiltrate and eventually are replaced by fibrous tissue and resultant loss of the aqueous portion of the tears. Systemic manifestations of Sjögren's syndrome can include salivary gland enlargement (especially the parotid gland), dysphagia, hoarseness, cough, renal tubular disease, glomerulonephritis, atrophic vaginitis, vasculitis including Raynaud's, and a spectrum of lymphoproliferative disorders.

KCS is the most common ocular manifestation of RA and is present in 10 to 25% of patients.[4] Onset is usually in the fourth to fifth decade, and the female-to-male ratio is 9:1. Symptoms of KCS include grittiness, burning, foreign body sensation, and dryness. Signs include irregular tear film, decreased tear break-up time, mucous debris on cornea and in the conjunctival sac, filamentary keratitis (mucous and epithelial cells adherent to the cornea), corneal mucous plaques, papillary conjunctivitis, and diffuse fine pitting and irregularity of the epithelium (superficial punctate keratopathy, SPK) which is usually present on the lower one third to one half of the cornea. Fluorescein often fails to stain the

SPK, but it stains vividly with 1% rose bengal, which attaches to devitalized cells.

Schirmer's tear test is an attempt to quantitate physiologic tear secretion. It is helpful as a diagnostic test when the amount of wetting on the filter paper strips placed with one end over the lid margin into the conjunctival sac is either very low (<5.5 mm) or very high (>10 to 15 mm). Intermediate measurements are difficult to evaluate, as about 15% of patients with KCS have wetting greater than 5.5 mm and many normal patients have readings of less than 10 mm. Because the ocular surface is compromised by epithelial dryness, patients with KCS frequently suffer from chronic or recurrent ocular infections, especially, but not limited to, *Staphylococcus aureus* blepharitis, conjunctivitis, and corneal ulcers.

Episcleritis

Episcleritis is morphologically divided into simple and nodular forms, both of which are characterized by vascular engorgement and edema of the episcleral tissue without involvement of the underlying sclera.[5] The inflammation is usually sectorial but may be diffuse. Pain is not a common complaint; however, the inflamed areas are often tender to the touch. Nodular episcleritis differs from simple episcleritis by forming subconjunctival nodules that are mobile over the sclera. Histologically, episcleral and scleral nodules are the ocular counterparts to subcutaneous rheumatoid nodules found elsewhere on the body. Episcleritis is usually mild and self-limited and requires no treatment unless there is associated uveitis.

Scleritis

Rheumatoid scleritis is characterized by inflammation, overlying vascular engorgement, and edema of the sclera, and is conveniently divided into anterior and posterior forms roughly delimited by the equator of the globe.[5] Anterior scleritis is subdivided into diffuse, nodular, necrotizing with inflammation, and necrotizing without inflammation (scleromalacia perforans). In contradistinction to episcleritis, pain is a common and often severe accompaniment to scleritis. Diffuse and nodular scleritis are milder and less damaging than the necrotizing varieties. Necrotizing scleritis with inflammation is the most destructive form and may be associated with overlying avascularity, necrosis, staphyloma, and rarely perforation. Necrotizing scleritis without inflammation is also called scleromalacia perforans, although this name is misleading since perforation rarely occurs. It is characterized by thinning of the sclera in a seemingly quiet eye, and progressive tissue loss can result in conjunctiva directly overlying the uvea. Staphylomas can occur if the intraocular pressure is elevated.

Posterior scleritis is usually associated with diffuse anterior scleritis but may occur in isolation. Uveal effusions, exudative retinal detachments, and disc swelling may accompany the scleral inflammation, whereas contiguous orbital inflammation and extraocular myositis may occur and result in proptosis or limitation of ocular movement.

The development of scleritis in a patient with RA is an ominous sign of a severe underlying systemic vasculitis that may be fatal if left untreated.[5]

Corneal Manifestations

Corneal changes are commonly seen in RA—the drying changes of KCS discussed above, the infrequent association of episcleritis with peripheral corneal stromal edema, opacification, and vascularization, and the corneal changes associated with scleritis.[6] Sterile corneal ulcers are seen in two categories of patients with RA: those with severe KCS and those without lacrimal deficiency. Cautious use of topical steroids is usually safe if normal lacrimal function is present, but it may predispose to corneal perforation in patients with KCS.

Significant corneal disease may occur with or without accompanying scleritis and is the leading cause of vision loss in patients with RA. These changes are primarily confined to the peripheral cornea, perhaps representing a cellular reaction to the diffusion of immunologic substances from the adjacent limbic vessels into the avascular corneal stroma. Four patterns of corneal involvement in RA have been described:

Sclerosing keratitis is an area of thickened, opacified, and vascularized peripheral cornea that develops directly adjacent to limbic scleritis.

Acute stromal keratitis may occur in diffuse or nodular scleritis as an acute inflammatory cell infiltrate in the otherwise clear peripheral corneal stroma, which may then go on to ulcerate and melt.

Peripheral corneal furrowing is a circumferential thinning of the corneal stroma, usually beginning just inside the limbus at the 3- and 9-o'clock positions. The epithelium, though generally intact, may ulcerate, with progressive thinning leading to perforation.

Keratolysis is a severe progressive melting of the entire corneal stroma that may lead to descemetocele formation or perforation. It occurs in association with necrotizing scleritis.

References

1. Krane SM. Rheumatoid arthritis. In: Rubenstein E, Federman DD, eds. *Scientific American Medicine.* New York, NY:1985:1-19.
2. Rodnan GP, Schumacher HR, Zaviﬂer NJ, eds. *Primer on the Rheumatic Diseases.* 8th ed. Atlanta, Ga: Arthritis Foundation; 1983;38-46.
3. Sjögren H, Bloch KJ. Keratoconjunctivitis sicca and the Sjögren syndrome. *Surv Ophthalmol.* 1971;16:145-159.
4. Lamberts DW. Dry eye and tear deficiency. *Int Ophthalmol Clin.* 1983;23(1):123-130.
5. Watson PG, Hayreh SS. Scleritis and episcleritis. *Br J Ophthalmol.* 1976;60:163-191.
6. Robin JB, Schanzlin DJ, Verity SM, et al. Peripheral corneal disorders. *Surv Ophthalmol.* 1986;31:1-36.

Chapter 23

Juvenile Rheumatoid Arthritis

CONRAD L. GILES

Juvenile rheumatoid arthritis (JRA) is potentially both a crippling and a blinding disease. The ocular complications of JRA primarily affect the uveal tract of the young patient, and the clinical findings may be quite subtle. This clinical subtlety places great demands on good communication between ophthalmologists and pediatric rheumatologists or immunologists who manage children with JRA.

The association of JRA and inflammation of the uveal tract in children was first observed by Ohm[1] in 1910. More recently, a better definition of the distinct clinical subgroups of JRA has made it possible to identify specific clinical manifestations that are most susceptible to ocular inflammatory disease.[2]

Classification

Table 23-1 indicates the five subgroups of JRA. Systemic-onset disease, the first subgroup, affects 20% of patients with JRA. In these children, several joints are affected, and the disease is associated with fever, hepatosplenomegaly, and anemia. Ocular complications seldom occur in this subgroup or in group 2 or 3. These two subgroups, representing patients with polyarticular disease (involvement of more than four joints) at the time of presentation, constitute 40% of patients with JRA, but it is the patients in subgroups 4 and 5 who are at risk of ocular involvement. These children present with pauciarticular rheumatoid arthritis, of early or late onset, and represent about 40% of JRA patients. About 50% of

patients with pauciarticular disease later develop polyarticular signs.

The child with early-onset JRA is usually a girl (8:1) under 10 years of age, whose antinuclear antibody (ANA) factor is positive about 50% of the time. In patients with both JRA and chronic iridocyclitis, Key and Kimura[3] reported an incidence of 91% (41 of 45) and Chylack et al.[4] an incidence of 78% (28 of 36) ANA positivity with pauciarticular onset. Patients in this subgroup have a 50% risk of exhibiting chronic iridocyclitis (Fig. 23-1).

By contrast, the child with late-onset JRA is generally a boy (10:1) in his teens who is ANA negative. Often, HLA-B27 antigens are present and the sacroiliac joint is involved. Acute iridocyclitis is the primary ocular manifestation for these patients.

Ocular Manifestations

The dramatic differences in the ocular manifestations of these two groups of patients are obvious. The young girl with chronic indolent anterior uveitis is at far greater risk than the teenage boy who presents with classic symptomatic acute iridocyclitis. The "white iritis" (iritis without a red eye) seen in the early-onset disease (group 4) occurs far more often if the patient is ANA positive. If undetected, uveal inflammation can have devastating complications, such as cataracts and glaucoma. Band keratopathy may also be associated with ocular inflammation in JRA (Fig. 23-2). Consequently, early de-

TABLE 23-1 Subgroups of JRA

Subgroup	Ratio of Girls to Boys	Age at Onset	Joints Affected	Serology Genetic	Extra-articular Manifestations
1 Systemic onset	8:10	Any	Any	ANA negative RF negative	High fever, rash, organomegaly, polyserositis, leukocytosis, growth retardation
2 Rheumatoid factor–negative, polyarticular	8:1	Any	Any	ANA 25% RF negative	Low-grade fever, mild anemia, malaise, growth retardation
3 Rheumatoid factor–positive, polyarticular	6:1	Late childhood	Any	ANA 75% RF 100%	Low-grade fever, anemia, malaise, rheumatoid nodules
4 Pauciarticular early onset	8:1	Early childhood	Few large joints (hips and sacroiliac joints spared)	ANA 50% RF negative	Few constitutional complaints, chronic iridocyclitis in 50%
5 Pauciarticular later onset	1:10	Late childhood	Few large joints (hip and sacroiliac involvement common)	ANA negative RF negative HLA-B27 75%	Few constitutional complaints, acute iridocyclitis in 5 to 10% during childhood

ANA, antinuclear antibody; RF, rheumatoid factor.
(Modified from Schaller.[2])

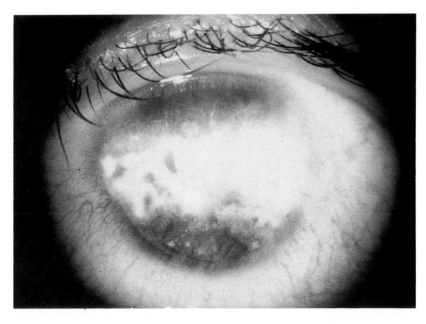

FIGURE 23-1 Advanced band keratopathy in a child with chronic anterior uveitis and pauciarticular JRA. (Courtesy of Kresge Eye Institute.)

tection of the mild inflammatory signs in these patients is important to allow the ophthalmologist to begin the treatment necessary to prevent such serious complications. Conversely, for the older child with sacroiliac involvement, early treatment is the rule rather than the exception. The definite symptoms of photophobia, tearing, redness, and pain bring this patient to the ophthalmologist at the first sign of inflammation. For these teenagers, therefore, the prognosis for preserving ocular function is excellent.

Treatment

For treatment of the primary inflammation, corticosteroids such as topical prednisone acetate 1% and atropine drops 1% are recommended. Periocular steroids are rarely necessary; however, the serious complications of chronic anterior uveitis demand more aggressive treatment. For advanced cataracts, surgery is usually required. With the advent of new vitrectomy techniques, eyes previously thought to be lost because of cataract formation and hypotony have responded to the aggressive use of lensectomy-vitrectomy procedures. Band keratopathy requires chelation with EDTA after the corneal epithelium has been removed. Medical therapy will control 50% of cases of glaucoma. Surgery may be necessary for those patients whose pressure cannot be controlled with drugs, although it is only 40 to 50% successful. Surgical procedures include trabeculectomy, cyclocryotherapy, and trabeculodialysis.

Prospective Management

In cooperation with rheumatologists and pediatricians, ophthalmologists should perform a routine examination of the susceptible groups of JRA patients on a regular basis. This examination should include the following:

Complete ophthalmologic examination of all patients with pauciarticular JRA (groups 4 and 5)
Slit-lamp examination every 6 months for patients with late-onset pauciarticular disease (group 5), with or without spondylitis, until they are 14 years old
Slit lamp examination every 3 months for patients with early-onset pauciarticular disease (group 4) until they are 14 years old.

Early identification of young patients with ocular inflammation associated with JRA permits the ophthalmologist to treat the child with uveitis successfully. With early detection, proper treatment, and careful follow-up, the most serious complications can almost be eliminated. Glaucoma following damage to the outflow channels, either by direct inflammatory reaction or secondary to peripheral anterior synechiae, presents the greatest challenge to successful treatment. Early detection of chronic iridocyclitis in patients with JRA significantly reduces the risk of blindness.

For the sake of the child with JRA, it is essential that pediatrician, rheumatologist, and ophthalmologist com-

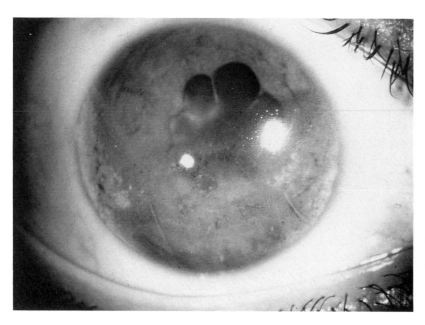

FIGURE 23-2 Early band keratopathy associated with posterior synechiae formation in a young patient with chronic uveitis and JRA. (Courtesy of Kresge Eye Institute.)

municate with one another, in order to detect the early signs of uveal inflammation. Only in this way can we hope to prevent the tragic complications of glaucoma and cataracts in these patients.

References

1. Ohm J. Bandformige Hornhauttrubung bei einem neunjahrigen Madchen und ihre Behandlung mit subkonjunktivalen jodkaliumein Spritzungen. *Klin Monatsbl Augenheilkd.* 1910;48:243-246.
2. Schaller JG. The seronegative spondyloarthropathies of childhood. *Clin Orthop.* 1979;143:76.
3. Key SN, Kimura SJ. Iridocyclitis associated with juvenile rheumatoid arthritis. *Am J Ophthalmol.* 1975;80:425-429.
4. Chylack LT, Bienfang DC, Bellows AR, Stillman JS. Ocular manifestations of juvenile rheumatoid arthritis. *Am J Ophthalmol.* 1975;79:1026-1033.

Chapter 24

Scleroderma

IVAN R. SCHWAB

Scleroderma (systemic sclerosis) was named for and is characterized by the classic hidebound stage of this multiorgan, multistage disease. Although the dermal sclerosis is of great concern to the patient and is the principal definitive diagnostic component, the skin change is not lethal. Other organ systems, such as the musculature and alimentary tract, exhibit similar pathologic changes that eventually lead to discomfort and disability but rarely to death; however, the visceral involvement that develops in the kidneys, lungs, and heart is serious, irreversible, and quite possibly fatal.

Systemic Manifestations

The term "scleroderma" is viewed by the majority of physicians as generic, representing several diseases characterized by dermal collagen alteration. The most serious is systemic sclerosis, which is characterized by a high level of organ involvement and which many authorities view as the only true scleroderma. The CREST syndrome is a variant of systemic sclerosis that includes *c*alcinosis, *R*aynaud's phenomenon, *e*sophageal hypomotility, *s*clerodactyly, and *t*elangiectasia. It carries a better prognosis but may develop into systemic sclerosis. Other related diseases include morphea (localized scleroderma), mixed connective tissue disease, and some inherited syndromes. Curiously, the histopathologic characteristics of all these diseases are quite similar and cannot be differentiated on the basis of biopsy findings.

Epidemiologically, systemic sclerosis affects women three or four times more frequently than men and typically occurs between ages 25 and 50. While no known geographic or racial predilection exists, people in certain occupations, including coal and gold mining, appear to have a higher prevalence of the disease.[1]

Although Raynaud's phenomenon, perhaps the most common manifestation of the disease, may not bring a patient to her physician, the puffy hands and face of the early phase of systemic sclerosis, a common and disconcerting symptom, *do*. This initial puffy phase has characteristic histologic findings, such as lymphoid and mononuclear cell infiltration of the subcutaneous tissue, but it is the fibrosis and the atrophy, especially of the small arteries, seen in the later stages that lead to morbidity and mortality.[2] Significantly, the progression of the disease defies prediction, and spontaneous remissions may occur, although progressive fibrosis and atrophy generally continue steadily over months to years.

Classic teaching on systemic sclerosis emphasizes the skin and esophageal disease. Certainly, the skin involvement, found in 90% of these victims, accounts for considerable morbidity.[3] Patients are very aware of the debility due to their taut, depigmented, hidebound skin. The esophagus is the second most commonly involved organ and hypomotility, esophageal reflux, esophagitis, and dysphagia are common.[2] Lung, kidney, or cardiac disease, though less common, represents a more ominous threat to the patient since involvement of any of these viscera may be life threatening. New antihypertensive medications promise to alter this grim prognosis.

Renal involvement, so common in most collagen-vascular diseases, occurs in approximately 35% of patients and represents considerable mortality, since kidney failure is the leading cause of death in scleroderma.[3]

Pulmonary involvement presents in several ways, including diffuse interstitial lung disease. Although it is the second leading cause of death in systemic sclerosis, pulmonary involvement is often underestimated. Overt shortness of breath indicates late disease, but it may also be the initial respiratory symptom.

Systemic sclerosis may involve the heart directly, although less commonly than it affects the kidneys or lungs.[3] Alternatively, cardiac involvement may be seen as a result of severe systemic or pulmonary hypertension. Inflammatory and fibrotic disease of the pericardium and myocardium and ischemic heart disease are well recognized, if poorly understood. Primary cardiac involvement in systemic sclerosis may well progress to death, since it leads to severe loss of the microcirculatory bed, precluding any benefit from bypass surgery.

Pathogenesis

Visceral dysfunction occurs with, and is probably caused by, small arterial changes consisting of myointimal cell proliferation, connective tissue deposition, and a reduction in lumen size. Although the inciting cause for these changes is obscure, it may be repeated episodes of endothelial cell injury that lead to vasospasm and local ischemic changes. Tantalizing clues suggesting immune mechanisms have been found by several investigators. Such immune-mediated phenomena as antinuclear antibodies (ANA)[4] and anti–smooth muscle antibodies[5] are prevalent in patients with systemic sclerosis. Cell-mediated immune mechanisms and cytotoxic lymphocytes have been found to be active in systemic sclerosis, especially in the latter stages of the disease. Equally powerful but simpler evidence for active immune mechanisms in systemic sclerosis can be found in the clinical company the disease keeps. Patients with systemic sclerosis have a high prevalence of Sjögren's syndrome, systemic lupus erythematosus, and dermatomyositis. It is conceivable that these diseases are all related in some basic way.

At least one other possible disease mechanism remains under active consideration, but it may be related to the overall immune process. Deposition of collagen has been found in all involved organ systems. This extensive fibrosis, unique to systemic sclerosis, suggests other pathologic mechanisms rather than immune-mediated processes, at least as we currently understand immunology.

Ocular Manifestations

Keratoconjunctivitis sicca as a component of Sjögren's syndrome is a well-recognized condition associated with systemic sclerosis.[6-10] Dry eye signs such as rose bengal staining, abnormal results on Schirmer's test, increased filaments, or mucous threads have been prevalent in as many as 25% of patients with systemic sclerosis.[11] Histopathologic examination of the salivary glands of such patients shows lymphocytic infiltration and duct cell proliferation typical of Sjögren's syndrome. Not surprisingly, collagen infiltration and deposition typical of systemic sclerosis were also found.[10]

Dermal sclerosis, best described as a woody texture to the upper eyelid skin, is probably as common as keratoconjunctivitis sicca. Similarly, lid tightness, as manifested by difficulty in upper lid eversion, is as common as dry eye.[11] Lid skin manifestations are to be expected, as the thinness of eyelid skin makes fibrosis or edema easy to identify.

Lid telangiectasia has been described[8,9] and should come as no surprise, since these changes are common on the face and anterior upper chest. Morphea (localized scleroderma) of the eyelid has been well documented but is relatively uncommon.[12]

Foreshortening of the fornix can be measured and has been documented in this chronic fibrotic disease.[8,11] Foreshortening as well as upper tarsal scarring should not be unexpected in systemic sclerosis, since subepithelial fibrosis of mucosal tissues is a principal characteristic. Small-vessel disease of the conjunctiva, documented by some observers,[9] has not been examined in an age- and sex-matched fashion.

Less common anterior segment abnormalities such as keratitis,[13] keratoconus,[14] and ocular myopathy[15] have been described but are not found frequently. Other abnormalities such as cataract, vitreous frosting, and elevated intraocular pressure have been described but may be related merely to age and not to systemic sclerosis.[8]

Funduscopic changes appear in at least 50% of patients with systemic sclerosis,[6] and as the disease progresses the prevalence probably increases. Retinal abnormalities would include relatively minor changes such as arteriovenous (A-V) nicking, small capillary hemorrhages, and cotton-wool spots. But, the spectrum of retinopathy includes the more ominous changes of retinal edema, exudates, and frank papilledema, as seen in the later stages of the disease.[16,17] Not surprisingly, patchy areas of nonperfusion can be found in the choroid, even in patients without hypertension.[6]

The retinal and choroidal microvascular bed should attract considerable attention from those interested in systemic sclerosis, which is, after all, a disease of small arteries. Indeed, some studies have evaluated the retinal and choroidal blood flow in vivo by fluorescein angiography, but the basic question remains unsettled by these studies.[6,11] Is the small choroidal or retinal arterial system involved in systemic sclerosis as the small vessels of other organs are involved? Certainly, end-stage hypertensive retinopathy and choroidopathy are well documented, but these changes may well be only secondary to hypertension and not primarily due to systemic sclerosis. What few histologic reports are available suggest that the changes in the retinal and choroidal vasculature are indistinguishable from changes seen in malignant hypertension.[16,17] Interestingly, primary cerebral scleroderma is not seen, although secondary hypertensive changes may occur. If the eye is "protected," as the brain is, then all microvascular changes seen in the eye may well be secondary. The current techniques available to study the retinal and choroidal vascular system in vivo would seem to be ideally suited to a prospective longitudinal study to answer these questions.

Treatment of the ocular signs and symptoms is limited to symptomatic management of the dry eye. Artificial tears, punctal occlusion, and other regimens should be considered when necessary. The retinopathy and choroidopathy should be managed on a systemic basis since these findings probably represent end-stage disease.

References

1. Rodnan GP, Benedek TG, Medsger TA Jr, et al. The association of progressive systemic sclerosis (scleroderma) with coal miners' pneumonoconiosis and other forms of silicosis. *Ann Intern Med.* 1967;66:323-334.
2. Fleischmajer R, Perlish JS, Reeves JRT. Cellular infiltrates in scleroderma skin. *Arthritis Rheum.* 1977;20:975-984.
3. Campbell PM, LeRoy EC. Pathogenesis of systemic sclerosis: A vascular hypothesis. *Sem Arthritis Rheum.* 1975;4:351-368.
4. Greenwald CA, Peebles CL, Nakamura RM. Laboratory tests for antinuclear antibody (ANA) in rheumatic diseases. *Lab Med.* 1978;9(4):19-27.
5. Kitridou RC, Fleischmajer R, Lagosky PA. Antismooth muscle antibody in scleroderma. *Arthritis Rheum.* 1975;18:526. Abstract.
6. Grennan DM, Forrester J. Involvement of the eye in SLE and scleroderma. *Ann Rheum Dis.* 1977;36:152-156.
7. Kirkham TH. Scleroderma and Sjögren's syndrome. *Br J Ophthalmol.* 1969;53:131-133.
8. Horan EC. Ophthalmic manifestations of progressive systemic sclerosis. *Br J Ophthalmol.* 1969;53:388-392.
9. West RH, Barnett AJ. Ocular involvement in scleroderma. *Br J Ophthalmol.* 1979;63:845-847.
10. Alarcon-Segovia D, Ibanez G, Hernandez-Ortiz J, et al. Sjögren's syndrome in progressive systemic sclerosis (scleroderma). *Am J Med.* 1974;57:78-85.
11. Schwab IR, DiBartolomeo A, Farber M. Ocular changes in scleroderma. *Invest Ophthalmol Vis Sci.* 1986;27(suppl):97.
12. El Baba F, Frangieh GT, Iliff WJ, et al. Morphea of the eyelids. *Ophthalmology.* 1982;89:1285-1288.
13. Manschot WA. Generalized scleroderma with ocular symptoms. *Ophthalmologica.* 1965;149:131-137.
14. Agatston HJ. Scleroderma with retinopathy. *Am J Ophthalmol.* 1953;36:120-121.
15. Arnett FC, Michels RG. Inflammatory ocular myopathy in systemic sclerosis (scleroderma). *Arch Intern Med.* 1973;132:740-743.
16. Ashton N, Coomes EN, Garner A, et al. Retinopathy due to progressive systemic sclerosis. *J Pathol Bacteriol.* 1968;96:259-268.
17. Farkas TG, Sylvester V, Archer D. The choroidopathy of progressive systemic sclerosis (scleroderma). *Am J Ophthalmol.* 1972;74:875-886.

Chapter 25

Sjögren's Syndrome

R. LINSY FARRIS

Although Sjögren's syndrome (SS) was for many years considered only an association of dry eye, dry mouth, and rheumatoid arthritis (RA), today we recognize it as an independent autoimmune disorder in which dry eyes and dry mouth result from lymphocytic infiltration of the lacrimal and salivary glands, which sometimes occurs in association with an array of connective tissue diseases (CTD).[1] Exocrine gland function is affected not only in the ocular and oral glands but over the entire body. Organ-directed autoimmune activity produces a wide spectrum of extraglandular features in the lung, kidney, skin, stomach, liver, muscle, thyroid, nerves, and bone marrow. As in the case of most autoimmune diseases, there is a preponderance of afflicted middle-aged and elderly women.

In the absence of an associated CTD, dry mouth and dry eye are referred to as the "sicca complex, or primary SS." Secondary SS refers to the full triad of dry eyes (xerophthalmia), dry mouth (xerostomia), and a CTD or collagen disease such as RA, scleroderma, or systemic lupus erythematosus (SLE).[2]

Systemic Manifestations

The systemic manifestations of SS are generally the result of a lymphocyte-mediated toxicity directed against tissues and organs throughout the body. They may take many forms, including chronic thyroiditis, hepatitis, biliary cirrhosis, gastric achlorhydria, adult celiac disease, pancreatitis, nephritis, superficial vasculitis, pneumonitis, peripheral neuropathy, and cerebral vasculitis. Vasculitis in the skin produces a rash of purple lesions on the lower legs or paler, pink hivelike welts.[3] Leg ulcers, polymyositis, and myopathy may also occur. The central nervous system is reported to be affected in two thirds of persons with cutaneous vasculitis, resulting in paralysis, disordered thinking, loss of speech, muscle weakness, and headaches. Lymphoproliferation, pseudolymphoma, and frank lymphoid malignancy occur in a significant percentage of patients, and there is a 40-fold greater likelihood of lymphoma in association with SS.

Blood studies reveal that approximately one third of the patients are anemic and one half have hypergamma-globulinemia with diffuse elevation of all classes. Cryoglobulinemia may occur in 20%, resulting in Raynaud's phenomenon. Autoantibodies vary in their expression in different subgroups of SS. Autoantibodies to Ro (SSA) and La (SSB) are commonly found in primary SS and in SS associated with SLE, but not in SS associated with RA. Patients with a CTD and SSA and SSB antibodies are more likely to develop sicca. B-lymphocyte hyperactivity may be due to a deficiency of suppressor hyperactivity or B lymphocyte–produced autoantibodies to T suppressor cells leading to a reduction in their number.[4] The lack of T suppressor lymphocytes appears to perpetuate B-lymphocyte hyperactivity, which leads to the production of autoantibodies, immune complexes, and hypergammaglobulinemia.[5] Circulating immune complexes are found in patients with primary and secondary SS and may be responsible for some of their systemic complications, such as hyperviscosity, vasculitis, renal disease, and some immunologic defects, such as receptor blockade[6]; however, the role of immune complexes in the pathogenesis of the fundamental lesion of SS is unclear, since there is no correlation between the levels of these complexes and disease activity. Rheumatoid factor is detected in 90%, antinuclear factor in 70%, and thyroglobulin antibody in about 35%. A decrease in T lymphocytes occurs in one third of patients.

Salivary insufficiency is responsible for the distressing symptom of dry mouth—fissures and ulcers of the tongue, buccal membranes, and lips that cause difficulty in chewing, swallowing, and talking. The incidence of dental caries increases dramatically, and hoarseness develops. Nasal dryness as well as dryness of the posterior pharynx, larynx, and tracheobronchial tree lead to hoarseness, epistaxis, recurrent otitis media, bronchitis, and pneumonia. Parotid gland enlargement occurs in approximately half of all patients.

Ocular Manifestations

The ocular manifestations of keratoconjunctivitis sicca (KCS) occur in about 90% of all SS patients. The diagnosis of KCS, or dry eye, may not be easy, because the lacrimal glands can produce reflex tearing in the early stages of the disease before complete destruction of the acini by lymphocytic infiltration has taken place. The symptoms are not specific and range from simple eye "awareness" to severe pain with inability to keep the eyes open. Although redness of the eye occurs in the more advanced stages of dryness, more subtle clinical signs such as decreased size of the wedge of tears resting on the lower lid, excess debris in the tear film, and viscous-looking tear film may be the only signs of early lacrimal insufficiency. Intermediate signs of tear deficiency are rose bengal and fluorescein staining of the cornea and conjunctival epithelium due to drying, the stains being taken up by degenerated and desquamated cells, respectively. Visualization by means of a slit lamp or ophthalmoscope without illumination, using only room illumination, will prevent reflex tearing that tends to mask the more subtle clinical signs of KCS. Advanced cases of KCS set the stage for serious ocular complications, including infection, symblepharon, pannus, marginal corneal gutter ulceration, and central corneal ulceration with perforation. The increased incidence of infection in KCS appears to be related to decreased conjunctival defense mechanisms secondary to an abnormal tear film and decreased antibacterial components of the aqueous tears.

Clinical and laboratory tests may support a diagnosis of KCS, but they are never absolutely diagnostic, since false-positive and -negative results occur to a degree determined by the selected cutoff or referent value. One may adjust this cutoff value to make the test more sensitive or more specific. Schirmer's test done without anesthesia and using a cutoff value of 3 mm of wetting in 5 minutes was found to give 100% specificity in one group of KCS patients when compared to a control group of normal patients without disease who were being fitted for contact lenses.[7] Rose bengal staining is also highly specific when a cutoff score of 3.5 is selected. Unfortunately, both Schirmer's test without anesthesia and rose bengal staining have a sensitivity of only about 50%, not much better than a toss of a coin, so many dry eyes go undiagnosed when only these tests are used. The measurement of tear osmolarity in minimally stimulated or basal tears has proven to be a highly sensitive and highly specific test for KCS.[8] Laboratory equipment and a trained technician are required to test tear osmolarity, which has been found to be significantly elevated in lacrimal insufficiency, owing to the effect of evaporation on the smaller tear volume. A referent value of 312 mOsm/L has been used to produce an overall efficiency of about 85% in the detection of KCS.

In summary, SS frequently manifests itself initially with ocular symptoms of a dry eye, but it may affect many organs, causing a wide variety of clinical abnormalities. Perhaps more than any other condition, SS presenting with ocular disease is the most likely to be associated with systemic disease.

References

1. Whaley K, Buchanan WW. Sjögren's syndrome and associated diseases. In: Parker CW, ed. *Clinical Immunology*. Philadelphia, Pa: WB Saunders; 1981:632-666.
2. Scully C. Sjögren's syndrome: clinical and laboratory features, immunopathogenesis, and management. *Oral Med*. 1986;62:510-523.
3. Alexander EL, Provost TT. Cutaneous manifestations of primary Sjögren's syndrome: a reflection of vasculitis and association with anti-Ro (SSA) antibodies. *J Invest Dermatol*. 1983;80:386-391.
4. Talal N. Sjögren's syndrome and connective tissue disease with other immunologic disorders. In: McCarty DJ, ed. *Arthritis and Allied Conditions*. 9th ed. Philadelphia, Pa: Lea & Febiger; 1979:810-824.
5. Strand V, Talal N. Advances in the diagnosis and concept of Sjögren's syndrome (autoimmune exocrinopathy). *Bull Rheum Dis*. 1980;30:1046-1052.
6. Ichikawa Y, Yoshida M, Takaya M, et al. Circulating natural killer cells in Sjögren's syndrome. *Arthritis Rheum*. 1985;28:182-187.
7. Farris RL, Gilbard JP, Stuchell RN, Mandel ID. Diagnostic tests in keratoconjunctivitis sicca. *CLAO J* 1983;9:23-28.
8. Gilbard JP, Farris RL, Santamaria J. Osmolarity of tear microvolumes in keratoconjunctivitis sicca. *Arch Ophthalmol*. 1978;96:677-681.

Chapter 26

Systemic Lupus Erythematosus

D. J. SPALTON

Systemic lupus erythematosus (SLE) is a multisystem disorder with such a wide range of manifestations that almost any organ in the body can be affected. It has an incidence of 1 in 2000 in the United States, but there are marked racial and sex differences: it is nine times more common in females, in black females the incidence has been reported to be as high as 1 in 250, and in Southeast Asia the disease may be more common than rheumatoid arthritis (RA). While the disease may occur in children or the elderly, the peak incidence is in women in the 20- to 30-year age group. Men tend to be affected more mildly than women. Milder forms of SLE are being recognized, and the typical patient has a relapsing or remitting illness rather than relentless progression. Life expectancy is good. At least 90% of patients are alive 10 years after diagnosis; renal and cerebral involvement indicate a poorer prognosis.

The basic pathology is not specific to SLE. Histopathology shows vasculitis of small vessels and capillaries with fibrinoid necrosis. This fibrinoid material, which contains fibrin, immunoglobulins, and complement, may also be found elsewhere in the body, for example in the pleura or synovium. Hematoxylin bodies of basophilic nuclear debris may be found in these lesions. They are a counterpart of LE cells, which were once an important diagnostic feature of SLE, and represent circulating neutrophils containing phagocytosed nuclear material. They are no longer diagnostically important. The clinical manifestations of SLE result from massive and complex B-cell overactivity leading to production of a diverse range of autoantibodies, immune complexes, and to activation of complement with a defect in T-cell immunoregulation. Not all autoantibodies are pathogenic; some are epiphenomena and others may cross-react with one or more antigens. During a relapse, there is a fall in T suppressor cells and an increase in T helper cells, demonstrating the defect of immunoregulation.

Diagnosis

The diagnosis of SLE is based on the clinical picture and on results of laboratory investigations. The American Rheumatological Association produced a diagnostic system in 1982 which was based on eight clinical manifestations and three laboratory investigations; presence of any four of the eleven serially or simultaneously during a period of observation clinched the diagnosis.[1] In practice, however, positive tests for antinuclear antibodies (ANAs) and double-stranded DNA antibodies are extremely helpful in making a diagnosis, and a negative ANA is extremely rare with active disease. Different types of extractable nuclear or cytoplasmic autoantibodies can be demonstrated, and some correlate with clinical subsets of the disease, such as SSA and SSB with Sjögren's syndrome or anti-RNP with Raynaud's phenomenon. Titers of ANAs or DNA antibodies can be used to monitor disease activity, although they do not always correlate well. Antiphospholipid antibodies have also attracted some attention.[2] These are responsible for the false-positive results of tests for syphilis in SLE patients and are found in high titers in a subset of patients who have increased thrombotic tendencies or repeated abortions.

Several drugs have been reported to induce an SLE-like illness; the most common are hydralazine and procainamide. Many patients taking these drugs develop positive ANAs, but not all develop clinical problems. DNA antibodies are usually absent. The drugs are thought to react with nuclear antigens to produce a drug-antigen complex that is immunogenic. There are clinical differences between idiopathic and drug-induced lupus (more elderly patients, more balanced sex distribution, fewer skin and renal manifestations), and the disease improves on withdrawal of the drug. SLE patients are very susceptible to drug allergies, and drugs such as sulfonamides and oral contraceptives may exacerbate idiopathic SLE.

A wide range of disorders can be considered in the differential diagnosis of SLE. They include subacute bacterial endocarditis, sarcoid, tuberculosis, lymphoma, septicemia, AIDS, and other collagen disorders (eg, RA, Wegener's granulomatosis).

Systemic Manifestations

The commonest systemic involvement is musculoskeletal. Myalgia is common and myositis can occur. Tenosynovitis is the commonest manifestation and may vary from mild to deforming, relapsing to relentlessly progressive. Joint involvement usually affects the proximal interphalangeal joints, metacarpophalangeal joints, wrists, and knees. Joint erosions are typically rare, even in patients with severe arthritis. The classic cutaneous manifestation of SLE is the butterfly facial rash, which is found in about 30% of patients. Other areas involved are the scalp and forehead, ears, chest, back, and upper arms. The skin lesions evolve from erythema to atrophy. Biopsy of skin lesions may show typical pathologic changes helpful in making the diagnosis. Other cutaneous manifestations are alopecia, vasculitis, and photosensitivity and susceptibility to drug allergy. Not all patients with cutaneous SLE develop systemic disease.

The incidence of renal involvement in SLE depends on the thoroughness of the investigation; only 40 to 50% of patients have clinical involvement but nearly all show changes on renal biopsy. These are not unique to SLE but can be helpful in establishing a long-term prognosis. Glomerulonephritis is produced by deposition of circulating immune complexes and activation of complement. Renal involvement is indicated by asymptomatic proteinuria or hematuria, nephrotic syndrome, or renal failure. Hypertension is a common accompaniment of renal involvement.

The commonest neurologic features of lupus are fits, psychiatric disorders, and transverse myelitis. Headaches and migraines are common. Benign intracranial hypertension is an important association for the ophthalmologist. Vasculitis is not usually found pathologically in the brain, and the neurologic signs may instead reflect thrombotic episodes, antineuronal antibodies, or disturbances of the blood-brain barrier. High doses of steroids may produce psychiatric symptoms and confuse the diagnosis.

Pleural effusion, pericarditis, and myocarditis can occur. Normochromic and normocytic anemia are extremely common, and hemolytic anemia can occur. Neutropenia, lymphocytopenia, and thrombocytopenia are also frequent findings. The sedimentation rate usually rises with relapse.

Ocular Manifestations

Orbital disease is rare, but myositis has been reported in a patient with longstanding SLE and systemic myositis.[3] Subcutaneous nodules in the lids were found in three blacks with a variant of DLE.[4]

A sicca syndrome is not uncommon. It is usually mild and best shown by staining with rose bengal dye within the palpebral aperture. Spaeth[5] identified a superficial punctate keratitis with fluorescein in 88% of patients hospitalized for SLE but noted that results of Schirmer's test were normal, leaving in doubt whether the changes were due to drying or direct epithelial damage. This is,

however, a rather higher incidence than is usually recognized, and the sicca syndrome associated with SLE does not usually require more than symptomatic treatment. Marginal corneal ulceration like that seen in Wegener's granulomatosis is not a feature of SLE.

Scleritis is a common manifestation of SLE and may be the presenting feature; any patient who presents with isolated scleritis should be screened with this in mind. It generally takes the form of a diffuse anterior or nodular scleritis and responds to systemic steroids, usually initially without much scleral necrosis. It is said that attacks may get more severe as the systemic disease deteriorates and that there may be some correlation with renal involvement, but this is unsubstantiated. Scleromalacia perforans, as seen in chronic rheumatoid arthritis, appears to be exceptionally uncommon in SLE.

The retinopathy associated with SLE is a composite of changes due to systemic hypertension, localized small-vessel vasculitis, anemia, and vascular occlusion. Retinopathy tends to occur during the active phase of the systemic disease. Sixty-one patients were followed by Gold et al.[6] for a period of 10 years; retinopathy was uncommon and was found only when the disease was active. Lanham et al.[7] evaluated 52 patients with SLE admitted for hospital treatment. They found a retinopathy in 15, but its presence was not related to neurologic or cutaneous disease activity. Six patients had cotton-wool spots, four had optic disc leakage, two of whom lost vision from ischemia. Five patients had normal fundi on ophthalmoscopy but leakage of fluorescein on angiography attributed to mildly active small-vessel disease.

Cotton-wool spots in the posterior pole are common during the active phases of disease and can be found in the absence of systemic hypertension. They represent microinfarcts produced by a vasculitis of small retinal vessels. Peripheral blotchy retinal hemorrhages and Roth's spots are frequent findings. In my own experience these tend to reflect an underlying anemia, and in particular thrombocytopenia, more than exacerbation of SLE. Subacute bacterial endocarditis or septicemia can occur in these patients, who are often on immunosuppressive drugs, and these must be excluded when Roth's spots are present. Measurement of C-reactive protein and of the blood sedimentation rate are helpful in these cases, as the former is elevated with infections but remains normal during relapse of SLE. Major vascular retinal occlusions are an uncommon feature that affects the retinal arterioles; peripheral or central retinal vein occlusion is exceptionally uncommon. In one case report histologic examination showed the retinal arteriole to be occluded without vasculitis (although this was found in the choroid).[8] This is in keeping with the vascular changes seen with neurologic involvement and indicates that local inflammatory changes are unlikely to be the cause of the retinal occlusion. Other possibilities are embolization from a Libman-Sachs endocarditis, deposition of immune complexes, and thrombosis associated with antiphospholipid antibodies. Although a choroidal vasculitis can be found histologically, anterior or posterior uveitis is not seen as a clinical entity in SLE.

Neuro-ophthalmic manifestations of SLE are common.[9] Jabs et al.[10] reviewed five patients with optic neuropathy and one with a chiasm lesion. Patients may also present with anterior or posterior ischemic optic neuropathy. Vision loss is usually acute, but one of their patients had insidious symptoms, and SLE must be excluded in any patient with an optic neuropathy of unknown cause. The pathology appears to be due to either demyelination or axonal destruction, and the results of steroid treatment are variable. Homonymous hemianopic field defects are perhaps less common than anterior visual pathway problems in SLE and indicate involvement of larger caliber vessels. In the future it will be necessary to investigate such patients for antiphospholipid antibodies which indicate an increased risk of thrombosis preventable by anticoagulation.

Ocular motor palsies are comparatively uncommon in SLE, but gaze palsies or internuclear ophthalmoplegia can occur as a result of brain stem involvement.

References

1. Tan E, Cohen A, Fries J, et al. The 1982 revised criteria for the classification of systemic lupus erythematosus. *Arch Rheum*. 1982;25:1271-1277.
2. Anticardiolipin antibodies, a risk factor for venous and arterial thrombosis. *Lancet*. Editorial. 1985;912-913.
3. Grimson BS, Simons KB. Orbital inflammation, myositis and systemic lupus erythematosus. *Arch Ophthalmol*. 1983;101:736-738.
4. Nowinski T, Bernardino V, Naidoff M, et al. Ocular involvement in lupus erythematosus profundus (panniculitis). *Ophthalmology*. 1982;89:1149-1154).
5. Spaeth GL. Corneal staining in systemic lupus erythematosus. *N Engl J Med*. 1967;276:1168-1171.
6. Gold DH, Morris DA, Henkind P. Ocular findings in systemic lupus erythematosus. *Br J Ophthalmol*. 1972;56:800-804.
7. Lanham JG, Barrie T, Kohner EM, et al. SLE retinopathy, evaluation by fluorescein angiography. *Ann Rheum Dis*. 1982;41:473-478.
8. Graham EM, Spalton DJ, Barnard RO, et al. Cerebral and retinal vascular changes in systemic lupus erythematosus. *Ophthalmology*. 1985;92:444-448.
9. Lessell S. The neuro-ophthalmology of systemic lupus erythematosus. *Document Ophthalmol*. 1979;47:13-42.
10. Jabs DA, Miller NR, Newman SA, et al. Optic neuropathy in systemic lupus erythematosus. *Arch Ophthalmol*. 1986;104:564-568.

Chapter 27

Wegener's Disease

ARTHUR S. GROVE, Jr.

Wegener's disease is characterized by necrotizing granulomatous vasculitis, which can involve multiple organ systems including the respiratory tract, the kidneys, and the eyes (Table 27-1). Although its cause is unknown, it may be a consequence of hypersensitivity to an unidentified antigen. Circulating immune complexes are present in some patients with Wegener's disease, and immune-related compounds have been found in biopsy specimens from involved tissues.[1] It is not clear whether such immune complexes have a primary role in tissue damage or whether they are secondarily involved.

Wegener's disease may present as either a dissemi-nated or a localized abnormality. The common disseminated form has necrotizing granulomas of the respiratory tract, vasculitis, and glomerulonephritis, and was described by Klinger and Wegener[2] in the 1930s. A limited form of Wegener's disease may involve only one part of the body, usually without renal abnormalities.[3,4]

Histologically, Wegener's disease involves primarily collagen and blood vessels. The prototypical lesions are necrotizing granulomas, which have frequently given the name "Wegener's granulomatosis" to this disorder. These lesions consist of a zone of fibrinoid or granular necrosis which is eventually surrounded by palisading

TABLE 27-1 Approximate Frequencies of Organ Involvement in Wegener's Disease

System	Frequency of (%) Involvement
Lungs	95
Sinuses	90
Kidneys	85
Nasopharynx	75
Eyes	60
Joints	50
Skin	40
Ears	35
Nervous system	20
Heart	15

(Data adapted from Fauci.[1])

histiocytes and frequently by multinucleated giant cells. Vascular lesions include phlebitis, granulomatous vasculitis, and necrotizing vasculitis similar to that seen in periarteritis nodosa. The renal abnormalities are typically focal necrotic lesions of the glomeruli, which can evolve into irreversible glomerular sclerosis.[4]

Systemic Manifestations

The upper and lower respiratory tracts are the most frequently involved tissues in Wegener's disease. The pulmonary lesions characteristically appear on radiographs as multiple bilateral infiltrates or thin-walled cavities, usually in the lower lobes. The nose and sinuses may be involved by nonspecific inflammation with mucosal crusting, structural deformities, and secondary bacterial infections.[5] Destructive lesions of the midface and nose can be caused by Wegener's disease as well as by malignant (polymorphic) reticulosis, malignant lymphoma, nasal carcinoma, chronic infection, and rarely by idiopathic midline destructive disease (IMDD).[5,6]

Kidney involvement can be diagnosed by the presence of proteinuria and red blood cell casts. Renal biopsy frequently demonstrates focal glomerular lesions. Renal failure resulting from glomerular sclerosis is a major cause of death in patients with Wegener's disease.

Polyarthralgias may be caused by joint involvement, and necrotizing granulomas may lead to ulceration of the skin. Lesions of the pharynx and ear may produce otitis media. Less often, the central nervous system and peripheral nerves may be involved and the heart may be affected by coronary vasculitis and pericarditis.[5]

Ocular Manifestations

The eye and ocular adnexal structures can be involved as a part of disseminated Wegener's disease or as a localized abnormality (without renal involvement). The most common eye features are progressive marginal ulcerative keratitis, scleritis, episcleritis, and conjunctival inflammation (Figs. 27-1, 27-2). These abnormalities are often bilateral and may be the initial sign of Wegener's disease. Posterior segment involvement includes retinitis with venous congestion, uveitis, and optic neuropathy.[1,7–9]

Adnexal involvement may be manifested by orbital lesions, causing exophthalmos, edema, pain, and limited eye movement. The nasolacrimal ducts are often obstructed, and the lacrimal glands may be enlarged by Wegener's disease (Fig. 27-3).[10] Eyelids may be swollen, with the appearance of nodules and occasional ptosis.[8,9] Diagnosis is established by biopsy.

Treatment of Wegener's disease usually involves the use of systemic cytotoxic agents, especially cyclophosphamide. In addition, oral corticosteroids may produce improvement in patients with fulminant pulmonary disease and in those with vasculitis of the skin, the pericardium, and the eye.[8,9]

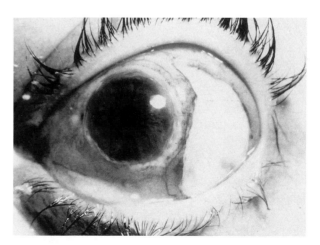

FIGURE 27-1 Necrotizing scleritis and corneal ring ulcer. (From Gold DH, Ocular manifestations of connective tissue disorders. In: Duane TD, ed., *Clinical Ophthalmology*, Philadelphia, Pa.: 1988; Vol. 5; Chapter 26:20.)

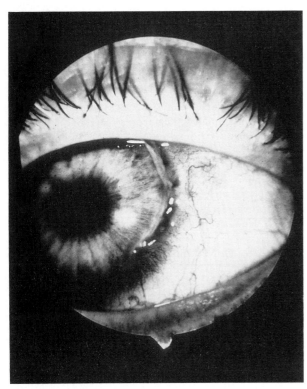

FIGURE 27-2 Corneal limbal furrow ulcer. (From Ferry AP, Leopold IH. Marginal (ring) corneal ulcer as presenting manifestation of Wegener's granuloma: a clinicopathologic study. *Trans Am Acad Ophthalmol Otolaryngol.* 1970;74:1276.)

References

1. Fauci AS, Haynes BF, Katz P. The spectrum of vasculitis: clinical, pathological, immunologic, and therapeutic considerations Part I. *Ann Intern Med.* 1978;89:660.
2. Straatsma BR. Ocular manifestations of Wegener's granulomatosis. *Am J Ophthalmol.* 1957;44:789.

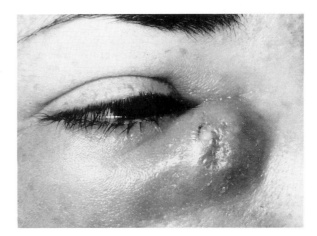

FIGURE 27-3 Dacryocystitis in a patient with Wegener's disease.

3. Cassan SM, Coles DT, Harrison EG. The concept of limited forms of Wegener's granulomatosis. *Am J Med.* 1970;49:366.
4. McCluskey RT, Fienberg R. Vasculitis in primary vasculitides, granulomatoses, and connective tissue diseases. *Hum Pathol.* 1983;14:305.
5. Batsakis JG. Wegener's granulomatosis and midline (nonhealing) "granuloma." *Head Neck Surg.* 1979;1:213.
6. Chu FC, Rodrigues MM, Cogan DG, et al. The pathology of idiopathic midline destructive disease (IMDD) in the eyelid. *Ophthalmology.* 1983;90:1385.
7. Haynes BF, Fishman ML, Fauci AS, et al. The ocular manifestations of Wegener's granulomatosis. *Am J Med.* 1977;63:131.
8. Robin JB, Schanzlin DJ, Meisler DM, et al. Ocular involvement in the respiratory vasculitides. *Surv Ophthalmol.* 1985;30:127.
9. Bullen CL, Liesegang TJ, McDonald TJ, et al. Ocular complications of Wegener's granulomatosis. *Ophthalmology.* 1983;90:279.
10. Boukes RJ, Kruit PJ, van Balen ATM, et al. Lacrimal gland enlargement in Wegener's granulomatosis. *Orbit.* 1985; 4:163.

PART 4
Endocrine Disorders

Chapter 28

Diabetes Mellitus

MATTHEW D. DAVIS

Systemic Manifestations

Diabetes mellitus is a deficiency in the production, secretion, and/or action of insulin, a small protein secreted by the beta cells of the islets of Langerhans of the pancreas and involved in many aspects of metabolism, including regulation of blood glucose concentration from moment to moment and facilitation of uptake and storage of carbohydrates, fats, and proteins following a meal. Primary diabetes is generally divided into two types, one characterized by severe beta cell loss and consequent total or nearly total lack of insulin (juvenile-onset, type I, or insulin-dependent diabetes mellitus [IDDM]) and the other by disorders in the regulation of insulin secretion and/or in its action in peripheral tissues, principally liver, fat, and muscle (adult-onset, type II, or non–insulin-dependent diabetes mellitus [NIDDM]). The hallmark of both types is intolerance to ingested carbohydrate, leading to hyperglycemia.

IDDM is characterized by abrupt clinical onset, usu-ally before the age of 20 or 30 years, with polyuria, thirst, weight loss, and excess lipolysis with hepatic conversion of plasma free fatty acids to ketones; the patient's lot is permanent dependence on injections of exogenous insulin to prevent acidosis, dehydration, and death in hyperglycemic coma. NIDDM, on the other hand, is typically discovered after age 40 years, usually on a routine examination or during investigation of an intercurrent illness, less frequently because of mild polyuria or weight loss, and occasionally because of symptoms arising from chronic diabetic complications in patients whose diabetes has presumably been unrecognized for many years. Ketoacidosis and coma do not occur. The majority of patients with newly diagnosed NIDDM are obese, and weight loss is a major therapeutic goal, since through it alone nearly normal glucose tolerance may be restored. Typical patients are easily placed in the appropriate category, but others seem intermediate and are difficult to classify. Dietary management is fundamental in both types, in IDDM always in combination with insulin and in NIDDM sometimes in combination with oral hypoglycemic agents or insulin.[1,2]

Histologic studies of patients who die within a few

Supported in part by an unrestricted grant from Research to Prevent Blindness, Inc.

months of the clinical onset of IDDM have shown round cell infiltration of the islets, suggesting viral infection or autoimmune inflammation. The former possibility has been suggested by case reports of diabetes occurring after mumps, rubella, *Coxsackie* B4, and other viral infections, by the finding of increased anti–*Coxsackie* B4 antibodies in recently diagnosed patients, and by seasonal incidence patterns of IDDM. Recent evidence that beta cell loss is probably occurring gradually over several years before nearly total loss precipitates overt diabetes suggests that viral illnesses preceding the clinical onset of IDDM may simply serve to uncover the diabetic state by adding the stress of infection, rather than playing a fundamental causal role. Suggestions that autoimmunity may be important came initially from the occasional association of IDDM with other diseases thought to have an autoimmune pathogenesis, such as thyrotoxicosis, Hashimoto's thyroiditis, and Addison's disease, and the finding of increased prevalence of organ-specific autoantibodies to endocrine tissues in the sera of persons with IDDM. More recently, islet cell antibodies and abnormalities of cellular immunity have been demonstrated in patients with newly diagnosed IDDM. In a recent report cytotoxic suppressor T lymphocytes were found to be the predominant inflammatory cells of insulitis in a child dying in ketoacidosis at the time of diagnosis of diabetes. Genetic makeup is thought to be an important determinant of autoimmune processes, and nearly all patients with IDDM have HLA haplotypes DR3 and/or DR4 (versus about 50% of the general population). Family members of IDDM patients who are HLA identical or haploidentical are at higher risk of developing IDDM, as are those who have islet cell antibodies. Thus, the current hypothesis of the cause of IDDM proposes an underlying genetic predisposition to an autoimmune disorder that is triggered by some environmental factor and characterized by a subclinical stage in which abnormalities of insulin secretion develop as beta cells are damaged, with occurrence of the typical abrupt clinical onset only after nearly all of these cells have been destroyed. This hypothesis has been strengthened by early trials of cyclosporine, which selectively blocks activation of T lymphocytes, in patients with newly diagnosed IDDM, in some of whom it appears that remission without the need for insulin can be induced for at least a year.[3-6]

The cause of NIDDM is less clear. Studies of identical twins show concordance approaching 100%, indicating a stronger genetic component than for IDDM (in which concordance is probably 50% or less). Unlike the nearly total loss of beta cells characteristic of well-established IDDM, NIDDM patients typically retain 50% or more of normal beta cell mass but demonstrate abnormalities of insulin secretion or resistance to its actions in peripheral tissues.[2,3]

Since 1921, when Banting and Best's discovery of insulin provided a solution for the acute metabolic problems of IDDM, the chronic complications of diabetes—nephropathy, retinopathy, neuropathy, and accelerated atherosclerosis—have emerged as the most important problems faced by diabetic patients and their physicians. The hypertension that is also frequently present in diabetes increases the risks attending the cardiovascular and renal complications. Atherosclerosis is of particular importance because it is the major cause of death and disability in patients with NIDDM, who constitute at least 80% of all diabetic patients, and because it occurs frequently in longstanding IDDM as well. There is an increased risk of myocardial infarction, of cerebrovascular accident, and particularly of vascular occlusion in the smaller arteries of the lower legs and feet, which tends to be bilateral and is often severe enough to lead to gangrene and amputation. No clear relationship has been demonstrated between these macrovascular complications and the level of glycemic control. Suggested as pathogenic factors are increased levels of blood lipids (especially triglycerides and low-density lipoproteins), favoring deposition in vessel walls; increased platelet aggregation, favoring thrombosis; microangiopathy of the vasa vasorum, leading to leakage of fats and proteins into large vessel walls; and premature cellular senescence of the vascular cells, a concept that has also been invoked for the beta cell as a pathogenetic factor in NIDDM.[1,7,8]

Diabetic neuropathy is also a frequent complication in both types of diabetes. Distal symmetric polyneuropathy is the most common variety, beginning in the lower extremities with impairment of touch and/or pain sensation. Autonomic neuropathy is a common cause of impotence and may also lead to orthostatic hypotension, gastrointestinal disturbances, or neurogenic bladder. The pathophysiology of diabetic neuropathy is poorly understood. Mechanisms suggested have included basement membrane thickening and increased permeability of capillaries supplying the nerves, as well as metabolic abnormalities in the Schwann cells or axons secondary to excessive activity of the polyol pathway (see below). The possible pathogenic role of glycemic control is uncertain. Cranial nerve palsies, usually in the sixth or the third (with the pupil spared) nerves, are due to vascular occlusions and typically recover spontaneously within several months.[9]

Perhaps the most extensively investigated chronic complications of diabetes are those that affect the kidney and retina; the majority of studies have been carried out in IDDM patients. Thickening of capillary basement membranes in skeletal muscle, renal glomeruli, retina, and other tissues is a characteristic finding in IDDM that begins within several years after diagnosis and progresses slowly. Degree of basement membrane thicken-

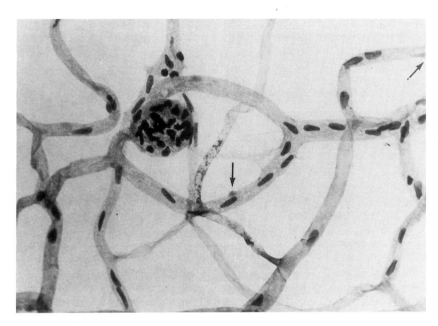

FIGURE 28-1 Trypsin digest preparation of dog retinal vessels after 60 months of alloxan diabetes shows saccular capillary microaneurysm and hypocellular and acellular capillaries. Oblique arrow indicates pericyte nucleus, vertical arrow, a pericyte ghost. Oval nuclei of endothelial cells greatly outnumber pericyte nuclei. (Courtesy of R.L. Engerman.)

ing is not closely related to severity of nephropathy or retinopathy, and neither its pathogenesis nor its possible importance in the development of these complications is clear. In the glomerulus diffuse and nodular deposits of basement membrane-like material in the mesangial region, rather than glomerular capillary basement membrane thickening, appear to be the abnormalities causing occlusion of the glomerular capillaries and renal failure. The relationship between nephropathy and glycemic control will be considered after a discussion of diabetic retinopathy.[10-12]

Ocular Manifestations
Diabetic Retinopathy*

The earliest anatomic abnormalities in diabetic retinopathy are loss of the intramural pericytes of the capillaries, development of acellular (nonperfused) capillaries, and formation of capillary microaneurysms, hypercellular saccular outpouchings from the capillary wall that typically vary from 15 to 60 μm in diameter (Fig. 28-1). The sequence of these changes is not clear; some observers suggest that pericyte loss occurs first, others think that capillary closure may be the earlier event. Angiographically, microaneurysms appear to antedate capillary nonperfusion, but the resolution of angiography is probably not sufficient to recognize capillary loss at its earliest stage. Mechanisms suggested for microaneu-

rysm formation have included vasoproliferation, weakness of the capillary wall (perhaps from loss of pericytes), abnormalities of the adjacent retina, and increased intraluminal pressure, but there is no convincing evidence for any of these alternatives. Nor is there a convincing pathogenic explanation for the closure of retinal capillaries. Thickening of their basement membranes does not appear to be sufficient to obstruct capillary lumens. The increased aggregation of red cells and adhesiveness of platelets commonly observed in diabetes have been proposed as possible factors favoring thrombosis, but there is no direct evidence to support such a mechanism. Deficient autoregulation of the retinal circulation has been found early in IDDM, before the development of retinopathy, and increases in retinal blood flow have been reported paralleling large increases in blood sugar. These hemodynamic abnormalities are reminiscent of those described in the glomerulus, suggesting a similar hypothesis (ie, that they may somehow lead to subsequent retinal capillary closure). One of the more appealing current lines of investigation proposes that the retinal capillary pericytes, which are postulated to have a vasomotor function, are damaged by the accumulation of sorbitol during periods of hyperglycemia and that this leads to the loss of autoregulation and increased blood flow. The conversion of glucose to sorbitol (and of galactose to galactitol) via the polyol pathway is catalyzed by the enzyme aldose reductase. Aldose reductase has low substrate affinity for glucose, so during normoglycemia this pathway is inactive, but during hyperglycemia it may become important in cells that do not require insulin for glucose penetration (eg.,

---

* For more detailed discussion and references see Reference 13.

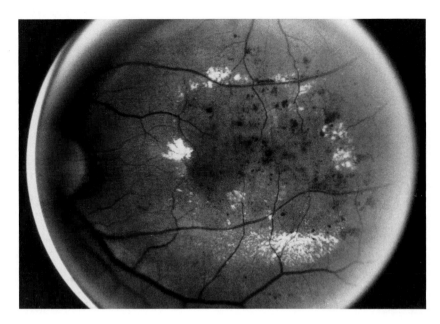

FIGURE 28-2 In this left eye a zone of thickened retina with surrounding partial hard exudate rings is centered temporal to the macula and involves the center of the macula.

retina, brain, vascular cells) if they contain aldose reductase. Sorbitol does not readily cross cell membranes, nor does fructose, to which sorbitol is metabolized to some extent in some cells, and these substances may accumulate within the cells to damaging concentrations, or possibly their excess production may lead to secondary metabolic abnormalities, even if high concentrations are not reached. This hypothesis is supported by the demonstration of aldose reductase in human retinal pericytes, by the production of retinopathy indistinguishable from that of diabetes by feeding a galactose-enriched diet to normal dogs and by the prevention of basement membrane thickening in retinal capillaries of galactosemic and diabetic rats treated with aldose reductase inhibitors. Trials to see whether aldose reductase inhibitors slow the development or progression of retinopathy are currently under way.

Although the basic pathogenesis of diabetic retinopathy remains obscure, its natural course has been described in some detail and is best understood in relation to five fundamental pathologic processes: formation of retinal capillary microaneurysms, excessive vascular permeability, vascular occlusion, proliferation of new blood vessels and accompanying fibrous tissue on the surface of the retina and optic disc, and contraction of these fibrovascular proliferations and the vitreous.

The retinal capillary microaneurysm, although seen in a variety of other conditions as well (such as branch retinal vein occlusion, radiation retinopathy, hyperviscosity syndromes, and idiopathic telangiectasis of the retinal vessels), is the hallmark of diabetic retinopathy and its earliest reliable sign. Microaneurysms may show little change over periods of several years, may become

nonperfused and clinically invisible, or may show thickening of their walls, which is sometimes sufficient to occlude their lumens.

When the number of microaneurysms in an eye exceeds ten, fluorescein angiography usually demonstrates retinal capillary abnormalities, consisting of focal fluorescein leakage from microaneurysms or more diffuse leakage from capillaries, capillary dilatation, or capillary nonperfusion (capillary dropout). By ophthalmoscopy, such eyes can usually be seen to have retinal hemorrhages, hard lipid exudates, or more advanced lesions (see below) in addition to microaneurysms and retinal edema (Fig. 28-2). Hard exudates are made up mainly of lipid, most of which has presumably leaked from the plasma across the excessively permeable walls of microaneurysms and capillaries. Hard exudates may be sprinkled across the fundus in no particular pattern, but more often are arranged in partial or complete rings, each ring marking the circumference of a roughly circular zone of thickened (edematous) retina that surrounds a cluster of microaneurysms (Fig. 28-2). The lipid appears to remain dispersed within the retina in the edematous zones and to become deposited at their edges as water and other small molecules are resorbed across the walls of more normal capillaries. Retinal edema is not easy to recognize with direct opthalmoscopy because the thickened retina maintains normal or near-normal transparency and its increased thickness is difficult to appreciate without a stereoscopic examining method. When retinal edema involves the area within a disc diameter or two of the center of the macula (macular edema), visual acuity is threatened, although it generally does not become impaired unless the center of the macula is involved.

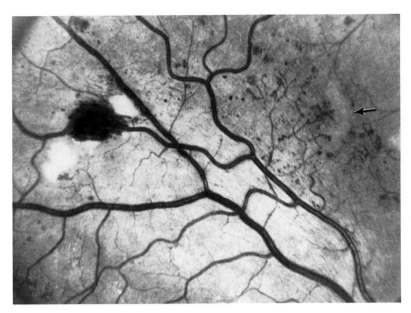

FIGURE 28-3 Severe NPDR. On the left side of the figure are two prominent soft exudates with a large blot hemorrhage between them. Venous beading is apparent where a large vein passes by the upper exudate. On the right side of the figure there is a faint soft exudate (*arrow*), as well as many prominent IRMAs. (Courtesy of the ETDRS Research Group.)

In the earliest stages of diabetic retinopathy individual closed capillaries can be seen histologically, but not ophthalmoscopically. As retinopathy becomes more severe larger patches of capillary closure appear, each typically supplied by an occluded terminal arteriole. Such patches are often marked initially by overlying cotton-wool spots (soft exudates), which tend to fade over periods of many months. Adjacent to patches of capillary closure, microaneurysms and tiny tortuous vessels are frequently seen. It is difficult to determine whether these vessels are dilated preexisting capillaries or intraretinal new vessels, and the term "intraretinal microvascular abnormalities" (IRMA) is used to include both possibilities. Histologically these vessels are hypercellular, and most observers agree that the stimulus for them, as well as for neovascularization on the surface of the retina, is probably partial ischemia of the inner retinal layers caused by capillary closure. As capillary closure becomes extensive dark red blot hemorrhages often appear, as do segmental dilatations of retinal veins (venous beading). When these lesions are prominent, nonproliferative diabetic retinopathy (NPDR) is considered *severe*, or *preproliferative* (Fig. 28-3). If capillary closure becomes very extensive these intraretinal lesions tend to disappear, and this, combined with fewer visible small venous and arteriolar branches, may produce a misleading *featureless* appearance, with little or nothing visible between large vessels except occasional white, threadlike arterioles.

When new vessels appear on the surface of the retina or optic disc, diabetic retinopathy is said to have entered the proliferative stage (PDR). New vessels arise most frequently posteriorly, within 45 degrees of the optic disc, and are particularly common on the disc itself (Fig.

28-4A). Eyes with new vessels on or near the disc are at greater risk of visual loss, and new vessels here (on or within one disc diameter of the disc or in the vitreous cavity anterior to this area) are commonly designated NVD and are considered separately from new vessels elsewhere (NVE). Initially new vessels appear bare, but later fibrous tissue appears adjacent to them, and these fibrovascular proliferations become adherent to the vitreous. The vessels may grow perceptibly in as little as a week or two or may remain unchanged for months, but *eventually they tend to regress,* sometimes almost completely (Fig. 28-4B). Hemorrhage into the vitreous from the new vessels is one of the principal mechanisms of visual loss in PDR. The other principal mechanism is detachment or distortion of the retina caused by contraction of the fibrovascular proliferations or the vitreous, contraction which pulls the proliferations themselves (and often the retina adherent to them) forward or tangentially (see Fig. 28-4).

Photocoagulation, the principal treatment for PDR, aims to induce regression of new vessels or to prevent their development, thus eliminating the source of vitreous hemorrhage and inhibiting the proliferation of fibrous tissue. After its introduction in 1960 by Meyer-Schwickerath, xenon-arc (white light) photocoagulation was used for direct destruction of patches of new vessels on the surface of the retina. By 1970 it was becoming apparent that extensive photocoagulation seemed to have a beneficial *indirect* effect, since it was sometimes followed by regression of new vessels and diminution of retinal edema and vascular congestion in areas of the retina not directly treated. Such indirect treatment, in which hundreds of usually milder, smaller laser burns are scattered throughout the fundus in a checkerboard

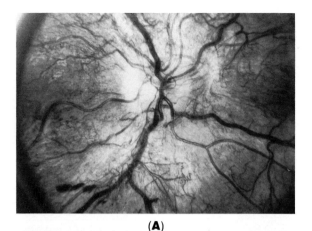

(A)

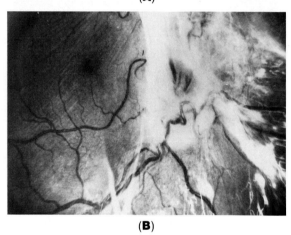

(B)

FIGURE 28-4 **(A)** In this right eye there are extensive new vessels on the surface of the disc and retina. Fibrous tissue accompanying the new vessels is faintly visible adjacent to the temporal vascular arcades and superonasal to the disc. The first major bifurcation of the inferior temporal vein occurs about 1 disc diameter below the disc. The macula is in its normal position, centered just temporal to the left edge of the figure. Visual acuity was 20/20. **(B)** Four years later, following contraction of the fibrous proliferations (which occurred several months after the preceding photograph), the center of the macula has been dragged up and nasally. The bifurcation of the inferior temporal vein has been pulled upward to the disc margin. New vessels have regressed completely; some of them now appear as networks of fine white lines. Visual acuity was 20/30. No treatment had been carried out. (Courtesy of the DRVS Research Group.)

pattern (*scatter* or *panretinal* photocoagulation), has now largely replaced direct treatment of new vessels. Although clinical trials have demonstrated that scatter photocoagulation reduces the risk of severe vision loss by 50% or more, at the price of a small risk (about 15%) of a mild permanent decrease in visual acuity (rarely more than one line on the Snellen chart), its mechanism of action is not clear.

When severe vitreous hemorrhage or retinal distortion or detachment occurs in spite of photocoagulation or when it has not been carried out, removal of the cloudy vitreous or fibrous tissue by combined suction and cutting with instruments introduced through the pars plana (pars plana vitrectomy) is often effective in restoring some vision. Photocoagulation has also been used to reduce macular edema by obliterating leaky microaneurysms with direct treatment using small (50- to 100-μm) argon laser burns. In areas of diffuse capillary leakage, "grid" treatment has been advocated—scattering small mild burns in these areas (except for the area within 500 μm of the center of the macula). The mechanism of direct treatment seems obvious, but it is not so clear how grid treatment reduces edema. Controlled trials have demonstrated a 50% reduction in moderate vision loss by photocoagulation in eyes in which edema involved or threatened the center of the macula.

The prevalence of retinopathy, and particularly of PDR, is greater in IDDM and in this group is closely related to duration of diabetes. In a recent population-based study, prevalence of PDR in persons taking insulin whose age at diagnosis of diabetes was under 30 years rose rapidly from about 2% in those with diabetes of 5 to 10 years' duration to 50% in those with more than 20 years'. Prevalence of PDR among those aged 30 years or over at diagnosis of diabetes and with duration of 20 years or more, was about 25% in those taking insulin and 5% in those not taking it. After 15 years of diabetes the prevalence of macular edema was about 12 to 20%, with less variation by diabetes type. But diabetes diagnosed after age 30 years is more common, and probably 50% or more of diabetes patients with PDR belong to this group, as do about 80% of those with macular edema.

Other Ocular Manifestations

Aside from retinopathy the most common mechanism by which diabetes leads to blindness is neovascular glaucoma, which is caused by growth of new vessels and accompanying transparent fibrous proliferations on the iris and in the anterior chamber angle, which block the trabecular meshwork. In its early stages, when a major portion of the angle remains free of anterior synechiae, regression of new vessels usually follows extensive retinal photocoagulation, and full-blown glaucoma can be avoided. When this is not the case, marked elevation of intraocular pressure occurs, often with pain, and treatment is difficult. Chronic open-angle glaucoma is more prevalent in diabetic patients than in the general population of the same age, but management does not pose any special problems. Before age 55 to 65 years cataract is also more common in diabetic patients than in the general population, but in older patients there is probably little or no difference in prevalence. Snowflake cataracts, consisting of small granular opacities beneath the anterior and posterior lens capsules, possibly related to periods of very severe hyperglycemia, are seen occasionally in young patients with IDDM, but they tend to remain stable. Autopsy studies in diabetic patients who were hyperglycemic shortly before death have shown the iris pigment epithelial cells to have a characteristic vacuolated appearance due to large accumulations of glycogen. The corneal epithelium of diabetic patients is particularly vulnerable to injury and slow to heal, factors to remember when prescribing contact lenses or carrying out surgical procedures. Diabetic patients, particularly those in ketoacidosis, are vulnerable to infection, one of the most serious of which is orbital cellulitis from *Mucor* species, fungi of the Phycomycetes class that typically cause a thrombosing arteritis and ischemic necrosis in the nose, paranasal sinusus, and orbit that spread rapidly to intracranial structures. Prompt diagnosis and antifungal treatment are essential if a fatal outcome is to be avoided.[14,15]

Glycemic Control and Complications

A question of obvious importance in the management of diabetes is the possible role of poor glycemic control in the pathogenesis of its long-term complications. Although observational studies have found that nephropathy, retinopathy, and perhaps neuropathy, tend to be more prevalent and more severe in groups of patients who have higher blood sugar levels, it is not clear whether this relationship reflects diabetes that is more severe and thus more difficult to control, or poorer control itself, or both.[16]

For patients who already have severe NPDR or PDR there is little evidence to suggest that improving glycemia control will beneficially influence retinopathy. Moreover, several case reports and small case series suggest that rapid progression of retinopathy may be precipitated by sudden improvement of longstanding poor control. Patients in whom such improvement is planned need close ophthalmologic observation.

References

1. Cahill GF. Current concepts of diabetes. In: Marble A, Krall LP, Bradley RF, et al., eds. *Joslin's Diabetes Mellitus.* Philadelphia, Pa: Lea & Febiger; 1985: 1-11.

2. Kahn CR. Pathophysiology of diabetes mellitus: an overview. In: Marble A, Krall LP, Bradley RF, et al., eds. *Joslin's Diabetes Mellitus.* Philadelphia, Pa: Lea & Febiger; 1985: 43-50.

3. Koldany A, Busick EJ, Eisenbarth GS. Diabetes mellitus and the immune system. In: Marble A, Krall LP, Bradley RF, et al., eds. *Joslin's Diabetes Mellitus.* Philadelphia, Pa: Lea & Febiger; 1985: 51-64.

4. Bottazzo GF, Dean BM, McNally JM, et al. In situ characterization of autoimmune phenomena and expression of HLA molecules in the pancreas in diabetic insulitis. *N Engl J Med.* 1985;313:353-360.

5. Feutren G, Assan R, Karsenty G, et al. Cyclosporin increases the rate and length of remissions in insulin-dependent diabetes of recent onset. Results of a multicentre double-blind trial. *Lancet.* 1986;2:119-124.

6. Bougneres PF, Carel JC, Castano L, et al. Factors associated with early remission of Type I diabetes in children treated with cyclosporine. *N Engl J Med.* 1988;318:663-670.

7. Fein FS, Scheuer J. Heart disease in diabetes. In: Ellenberg M, Rifkin H, eds. *Diabetes Mellitus, Theory and Practice.* New Hyde Park, NY: Medical Examination Publishing Co; 1983:851-861.

8. Knatterud GL, Klimt CR, Goldner MG, et al. Effects of hypoglycemic agents on vascular complications in patients with adult-onset diabetes. VIII. Evaluation of insulin therapy: final report. *Diabetes.* 1982;31:1-81.

9. Ellenberg M. Diabetic neuropathy. In: Ellenberg M, Rifkin H, eds. *Diabetes Mellitus, Theory and Practice.* New Hyde Park, NY: Medical Examination Publishing Co; 1983: 777-801.

10. Friedman EA. Diabetic renal disease. In: Ellenberg M, Rifkin H, eds. *Diabetes Mellitus, Theory and Practice.* New Hyde Park, NY: Medical Examination Publishing Co; 1983:759-776.

11. Krowlewski AS, Warram JH, Rand LI, et al. Risk of proliferative diabetic retinopathy in juvenile-onset type I diabetes: a 40-year follow-up study. *Diabetes Care.* 1986;9:443-452.

12. Williamson JR, Tilton RG, Chang K, Kilo C. Basement membrane abnormalities in diabetes mellitus: relationship to clinical microangiopathy. *Diabetes Metabol Rev.* 1988;4:339-370.

13. Davis MD. Diabetic retinopathy, a clinical overview. *Diabetes Metabol Rev.* 1988;4:291-322.

14. Aiello LM, Rand LI, Sebestyer JG, et al. The eye and diabetes. In Marble A, Krall LP, Bradley RF, et al., eds. *Joslin's Diabetes Mellitus.* Philadelphia, Pa: Lea & Febiger; 1985;600-634.

15. L'Esperance FA, James WA. The eye and diabetes mellitus. In: Ellenberg M, Rifkin H, eds. *Diabetes Mellitus, Theory and Practice.* New Hyde Park, NY: Medical Examination Publishing Co; 1983:727-757.

16. Klein R, Klein BEK, Moss SE, et al. Glycosylated hemoglobin predicts the incidence and progression of diabetic retinopathy. *JAMA.* 1988;260:2864-2871.

Chapter 29

Hyperthyroidism and Hypothyroidism

PETER D. FRIES and DEVRON H. CHAR

Thyroid dysfunction is one of the more common endocrine abnormalities; hyperthyroidism occurs more frequently than hypothyroidism and there are myriad causes of each condition. Hyperthyroidism is usually autoimmune or idiopathic in origin; occasionally it is due to overdose of a prescribed thyroid supplement. It is relatively uncommon in children. Causes of hypothyroidism include congenital defects, thyroiditis, therapeutic ablation or irradiation, iodine deficiency, maternally transmitted iodides or antithyroid agents. Early clinical and diagnostic signs in both hypothyroidism and hyperthyroidism can be subtle and insidious in onset. The ophthalmologist often is the first physician to examine these patients and has the opportunity to establish the correct systemic diagnosis.

Hyperthyroidism

Hyperthyroidism frequently presents with ocular findings before systemic disease is diagnosed. Most cases of hyperthyroidism are due to Graves' disease. Graves' disease consists of one or more of a triad of manifestations: hyperthyroidism often with diffuse goiter, ophthalmopathy, and dermopathy. Thyroid adenoma, thyroid carcinoma, hyperpituitarism, and overuse of thyroid medication can also cause hyperthyroidism, but the incidence of ocular manifestations with these entities is far lower than in Graves' disease.

Systemic Manifestations

The systemic signs and symptoms of hyperthyroidism include nervousness, irritability, fatigue, weight loss, emotional lability, heat intolerance, sweating, weakness, and palpitations. Nearly all of these signs and symptoms can be explained on the basis of increased titers of triiodothyronine (T_3) and thyroxine (T_4). The classic effect of the thyroid gland is on basal metabolic rate, but numerous molecular actions of thyroid hormones are known—specific stimulation of mitochondrial oxidative metabolism, catecholamine potentiation of various biochemical processes, nuclear regulation of protein synthesis, and specific regulatory effects on membrane physiology, among others. Increased thyroid hormone increases basal metabolic rate, which is manifested as increased energy level and calorie loss. Increased energy level produces hyperthermia with resultant sweating and heat intolerance. Calorie loss leads to weight loss. Direct cardiac effects of thyroid hormones can produce serious cardiac complications, especially in elderly patients.

Ocular Manifestations

The primary ocular signs of Graves' disease include lid retraction, lagophthalmos, periorbital and conjunctival edema, restrictive extraocular myopathy, and proptosis (Fig. 29-1). A list of thyroid-related eye signs is shown in Table 29-1. Most ocular findings are secondary to enlarged and infiltrated extraocular muscles. Enlarged episcleral vessels over the rectus muscle insertions, diplopia, and glaucoma can occur because of the enlargement, restriction, and vascular engorgement of the

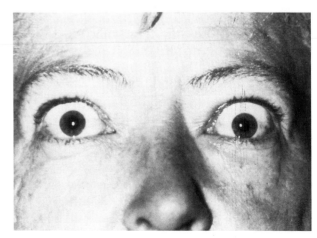

FIGURE 29-1 Exopthalmos and eyelid retraction in thyroid eye disease. (From Char.[1])

TABLE 29-1 Eye Signs in Graves' Disease

Ballet's	Paralysis of one or more extraocular muscles
Boston's	Uneven jerky motion of the upper lid on inferior movement
Cowen's	Extensive "hippus" of the consensual pupillary light reflex
Dalrymple's	Upper lid retraction
Enroth's	Edema of the lower lid
Gellinek's	Abnormal pigmentation of the upper lid
Gifford's	Difficult eversion of the upper lid
Goffroy's	Absent creases in the forehead on upward gaze
Griffith's	Lower lid lag on upward gaze
Knie's	Uneven pupil dilation in dim light
Kocher's	Spasmatic retraction of the upper lid during fixation
Loewi's	Dilation of the pupil with 1/1000 epinephrine
Mean's	Increased superior scleral show on upgaze
Möbius'	Deficient convergence
Payne/ Trousseau	Dislocation of globe
Pochin's	Reduced amplitude of blinking
Riesman's	Bruit over eyelid
Rosenbach's	Tremor of the gently closed lids
Sainton's	Frontalis contraction after cessation of levator activity
Snellen/ Donder's	Bruit over the eye
Suker's	Inability to maintain fixation on extreme lateral gaze
Stellwag's	Incomplete and infrequent blinking
Vigouroux's	Puffiness of the lids
Von Graefe's	Upper eyelid lag on downgaze
Wilder's	Jerking of eyes on movement from abduction to adduction

(From Char.[1])

rectus muscles. Initial inflammation and edema in the extraocular muscles usually progresses to muscle hypertrophy and enlargement (Fig. 29-2). Since the orbit has a fixed volume, proptosis usually results. Long-standing inflammation leads to fibrosis of the muscles, resulting in lid retraction, lagophthalmos, and restrictive extraocular myopathies. Keratoconjunctivitis sicca, corneal abrasion, and corneal ulcers occur secondary to marked proptosis (exophthalmos) and lid retraction. Vision loss can be due to corneal problems or to apical compression of the optic nerve by enlarged rectus muscles.

The pathophysiology of orbital involvement in hyperthyroidism is uncertain. Possibly antigens expressed and shared between extraocular muscle and thyroid cells result in an autoimmune disorder with extraocular muscle inflammation, edema, mucopolysaccharide and fatty infiltration, and subsequent fibrosis.[1] The role of immunogenetic factors or either suppressor cell or immunoglobulin abnormalities is unclear.[2-6] Most clinicians and investigators consider the ophthalmopathy to be a component of Graves' disease; however, the temporal relationship between systemic hyperthyroidism and thyroid ophthalmopathy can be tenuous, and some investigators believe they may be separate diseases.[1] Patients may exhibit systemic thyroid disease prior to, coincident with, or after the onset of ophthalmopathy.

In an orbital referral practice, 20% of thyroid ophthalmopathy patients examined have euthyroid ophthalmopathy, all standard thyroid laboratory tests being negative (Table 29-2). Approximately 40% of these patients eventually develop clinical or laboratory signs of Graves' disease.[3] The development of newer assays for thyroid-stimulating receptor antibodies should reduce the number of patients whose laboratory studies produce negative results.[7]

Hypothyroidism
Systemic Manifestations

Hypothyroidism occurring from birth and resulting in developmental abnormalities is termed "cretinism," whereas "myxedema" connotes a severe form of hypothyroidism with deposition of mucopolysaccharides (hyaluronic acid and chondroitin sulfate B) in the dermis and in other tissues.[8] The systemic signs of infantile hypothyroidism (temperature instability, feeding problems, prolonged physiologic jaundice, somnolence, hoarse cry, delayed developmental milestones) are progressive and generally relate to the low levels of thyroid hormone. Untreated, these children develop short stature, coarse facial features with a broad, flat nose, large tongue, sparse hair, dry skin, and a protuberant abdomen. Ocular signs of cretinism include hypertelorism, synophrys, and epicanthus.

Hypothyroidism in adulthood presents insidiously with lethargy, constipation, cold intolerance, and in women, menorrhagia. Motor weakness, inactivity, loss of appetite, paradoxical weight gain, dry skin, and brittle hair subsequently appear. In its classic state patients with myxedema have a dull expressionless face, sparse hair, pale cool skin, enlarged tongue and periorbital edema. If not treated myxedema coma may ensue, which can be fatal.

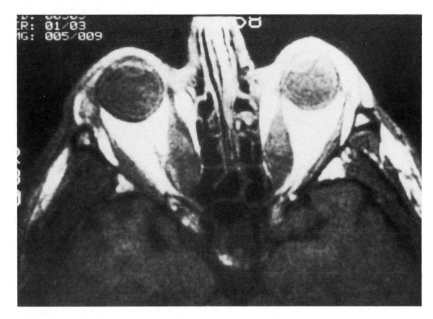

FIGURE 29-2 Axial magnetic resonance scan shows enlargement of the extraocular muscles with relative sparing of tendons, a pattern typical of thyroid dysfunction but inconsistent with "pseudotumor myositis." (From Char.[1])

Ocular Manifestations

The ocular signs associated with hypothyroidism are numerous.[9] The temporal one third to one half of the eyebrows or eyelashes is lost. Conjunctival edema and periorbital edema of a boggy, nonpitting nature frequently manifest early in the adult course. Blepharoptosis, nyctalopia, corneal edema, keratoconus, strabismus, papilledema, exophthalmos, cataract, and optic atrophy have all been reported in patients with hypothyroidism. The pathophysiologic mechanisms for these ocular findings have not been elucidated. Nyctalopia may occur secondary to decreased synthesis of retinene required for dark adaptation. Hair, eyebrow, and eyelash loss may be a manifestation of an overall nutritional deficiency. The conjunctival and periorbital edema exhibited in hypothyroidism is due to the dermal deposition of the mucopolysaccharides hyaluronic acid and chondroitin sulfate B.[8] Although the exact cause of the deposition is unknown, the hydrophilic substances produce edematous changes wherever they are deposited. This results in the characteristic doughy, nonpitting periorbital and lid edema and in edema in other involved organs and tissues. The other eye signs in hypothyroidism occur so infrequently that they may be entirely serendipitous.

References

1. Char DH. *Thyroid Eye Disease,* 2nd ed. New York: Churchill-Livingstone, 1989.
2. Kriss JP, Koniski J, Herman MM. Studies on the pathogenesis of Graves' ophthalmopathy. *Recent Prog Horm Res.* 1975;31:533-566.
3. Kohn LD, Winand RJ. Relationship of thyrotropin to exophthalmos-producing substance. Formation of an exophthalmos-producing substance by pepsin digestion in pituitary glycoproteins containing both thyrotropic and exophthalmogenic activity. *J Biol Chem.* 1971;246:6570-6575.
4. Rotella CM, Zonefrati R, Toccafondi R, et al. Ability of monoclonal antibodies to the thyrotropin receptor to increase collagen synthesis in human fibroblasts: an assay which appears to measure exophthalmogenic immunoglobulins in Graves' sera. *J Clin Endocrinol Metab.* 1986;62: 357-367.
5. Volpe R. Immunoregulation in autoimmune thyroid disease. *N Engl J Med.* 1987;316:44-45.
6. Frecker M, Stenszky V, Balazs C, et al. Genetic factors in Graves' ophthalmopathy. *Clin Endocrinol.* 1986;25:479-485.
7. Jiang N-S, Fairbanks VF, Hay ID. Assay for thyroid stimulating immunoglobulin. *Mayo Clin Proc.* 1986;61:753-755.
8. Gorman CA. The presentation and management of endocrine ophthalmopathy. *Clin Endocrinol Metab.* 1978;7:67-96.
9. Malito RS. Ocular features of hypothyroidism. *Br J Ophthalmol.* 1972;56:546-549.

TABLE 29-2 Sequential Laboratory Approach to the Diagnosis of Thyroid Ophthalmopathy

1. Serum TSH
2. Calculated free T_4 index, free T_3 index
3. Thyrotropin receptor stimulating antibodies (TRAb's)
4. Antithyroid antibodies

(From Char.[1])

Chapter 30
Parathyroid Disorders

YUVAL YASSUR and ZVI TESSLER

Parathyroid hormone (PTH), together with vitamin D and with calcitonin, are the mean agents that regulate calcium metabolism in the body. This is achieved through PTH's influence on calcium input in the renal tubules, intestine, and skeleton. PTH also participates in systemic phosphorus homeostasis by its effect on phosphorus transportation in the renal tubules.[1]

Hyperparathyroidism

Hyperparathyroidism usually occurs in adults between the 3rd and 7th decade of life and is characterized by overproduction of PTH, which results in hypercalcemia and hypophosphatemia. Primary hyperparathyroidism is the result of parathyroid adenoma in about 80% of cases, of parathyroid hyperplasia in about 15%, and of parathyroid carcinoma in about 4%.[2,3] Secondary hyperparathyroidism occurs when phosphate retention in renal failure induces compensatory hypocalcemia, which in turn stimulates overproduction of PTH by the parathyroid glands. Tertiary hyperparathyroidism occurs when after prolonged periods of such stimulation the parathyroid glands function independently of the calcium concentration in the serum and sustain their hypersecretion autonomously.

Systemic Manifestations

The specific systemic complications of hyperparathyroidism are renal stones and bone disease. The pathogenetic risk factors for stone formation are (1) acidification disturbances, which are attributed to the tubular effect of PTH[4]; (2) an unexplained tendency for urine to crystallize[5]; and (3) hypercalciuria, especially in the presence of elevated serum 1,25 dihydroxy vitamin D and resultant intestinal hyperabsorption of calcium.[5] The bone disease is either osteitis fibrosa cystica (patchy demineralization) or osteopenia (diffuse demineralization) and is attributed to the resorptive effect of PTH in bone.[6,7]

Other nonspecific features of primary hyperparathyroidism are (1) neuropathic muscle weakness, atrophy, and hyperreflexia in which PTH or phosphorus levels might have a role[8]; (2) central nervous system symptoms of lethargy, confusion, depression, personality changes, or overt psychosis, which are associated with the hypercalcemia[9]; (3) gastrointestinal symptoms, including ill-defined abdominal pain, increased gastric basal acid output, and pancreatitis, attributed to the hypercalcemia and pancreatic duct calculi[3,10-12]; (4) hypertension, probably caused by calcium-related increase in peripheral vascular resistance; and (5) ECG abnormalities, including shortened QT intervals and prolonged PR interval, which may result in first-degree heart block.[13]

Ocular Manifestations

The principal ocular signs consist of calcification in the cornea, sclera, and conjunctiva due to the hypercalcemia and to the probable contribution of physiologic local alkalosis in these tissues.[14] The local alkalosis is probably the result of diffusion into the surrounding air of the carbon dioxide produced in the corneal and conjunctival epithelium. Red, gritty eyes, in the presence of corneal calcification (band keratopathy), may be the presenting signs of hyperparathyroidism.[15] Biomicroscopic examination of the cornea within the palpebral fissure reveals a diffuse superficial opacity separated from the limbus by a clear area. The opacities consist of calcium hydroxyapatite crystals deposited intracellularly in the corneal epithelium, endothelium, and Bowman's membrane. Calcium deposits in the conjunctival epithelium which appear as white perilimbic and scleral flecks may be found as well.[16] The asymptomatic scleral calcifications may be demonstrated by computed tomography (CT).[17]

In secondary hyperparathyroidism the use of histochemical techniques demonstrates calcium in the pigment epithelium of the iris and ciliary process, suprachoroidea, and adjacent ciliary muscle.[18] An orbital rim mass which was diagnosed histologically as a brown tumor (a reaction of bone to increased PTH) was described as a rare ophthalmic manifestation of hyperparathyroidism.[19]

Hypoparathyroidism

Hypoparathyroidism may occur in children and in adults. It is due either to PTH deficiency (true hypoparathyroidism) or to PTH resistance (pseudohypoparathyroidism). Both are characterized by hypocalcemia and hyperphosphatemia. The deficiency in hormones might be the consequence of surgery or it can be idiopathic. The postoperative type usually follows inadvertent damage to the parathyroids during thyroidectomy. In most instances it consists of interruption of the blood supply to the end-arteries of the parathyroid glands rather than actual removal of parathyroid tissue.[20] The prevalence of postoperative hypoparathyroidism has been estimated to be up to 3%.[21] The idiopathic type is rare and tends to go misdiagnosed for prolonged periods.[22] Parathyroid antibodies can be demonstrated to be present in this type in about one third of cases.[23] The cause of PTH–resistant hypoparathyroidism (pseudohypoparathyroidism) is target organ resistance to the hormone. It is a rare disorder, inherited as either an X-linked or autosomal dominant trait. The pathophysiology of pseudohypoparathyroidism is not clear. Several suggested explanations are refractoriness of PTH, 1,25 dihydroxycholecalciferol deficiency,[24] and guanine nucleotide regulatory protein deficiency.[25]

Systemic Manifestations

The systemic manifestations of hypoparathyroidism are attributed primarily to hypocalcemia. Mild symptoms might be weakness, fatigue, and paresthesias. On physical examination two signs that are consequences of neuromuscular excitability of hypocalcemia may be elicited: Chvostek's sign (twitch of facial muscles following percussion on the facial nerve anterior to the ear) and Trousseau's sign (carpopedal spasm following external pressure greater than the systolic blood pressure around the upper arm by a sphygmomanometer). Severe hypocalcemia causes tetany, which might be life threatening if laryngeal spasm and generalized convulsions occur. Chronic hypoparathyroidism may present with mental retardation, hypoplasia of tooth enamel, partial loss of body hair, brittle nails, and monilial infections of nails and oral mucosa. Metastatic calcifications may be found in the basal ganglia on skull films.[26] Clinical manifestations of pseudohypoparathyroidism include atypical short stocky build, round face, shortened metacarpal and metatarsal bones, and subcutaneous calcifications and ossifications. Pseudopseudohypoparathyroidism, which is a rare genetic disorder, differs from pseudo hypoparathyroidism in that, although the phenotype is the same, serum calcium and phosphorus levels are entirely normal.

Ocular Manifestations

The ocular complications of hypoparathyroidism are local manifestations of the hypocalcemia. The most prominent ocular sign, lens opacities, is found in about 50 to 60% of patients.[27] The cataracts are frequent in patients with tetany, but visual symptoms due to the cataract may be the presenting sign of hypoparathyroidism in some patients.[27] The exact mechanism by which tetanic cataracts are produced by hypocalcemia is not fully understood. Experiments on rat lenses cultured in calcium-free medium demonstrated leakiness of the lens membrane. This permeability change, followed by localized hydration, is the likely explanation for the lens opacities.[28] Tetanic cataracts are typically bilateral. They present with small white or polychromatic crystals lying in the anterior and posterior cortex just beneath the lens capsule and separated from it by a clear zone. Lens opacities have also been described in pseudohypoparathyroidism and in pseudo-pseudohypoparathyroidism. Two less frequent ocular signs are keratoconjunctivitis and papilledema. Keratoconjunctivitis, occasionally with superficial corneal stromal vascularization, may present with photophobia and blepharospasm.[29] Papilledema and increased intracranial pressure may present with convulsions and may lead to an erroneous diagnosis of brain tumor.[30] The ocular signs improve after restoration of plasma calcium levels to normal by calcium and vitamin D supplements.[30,31]

In conclusion, it must be emphasized that ocular manifestations are only one aspect of parathyroid disorders. Ophthalmologic care and treatment are just part of the systemic treatment of hyperparathyroidism and hypoparathyroidism; however, ocular complications may persist and need further treatment long after the systemic condition is controlled.

References

1. Parfitt AM, Kleerekoper J. The divalent ion homeostatic system, physiology and metabolism of calcium, phosphorus, magnesium, and bone. In: Maxwell MH, Kleeman CR, eds. *Clinical Disorders of Fluid and Electrolyte Metabolism.* 3rd ed. pp. 269-398, New York, NY: McGraw-Hill; 1980.
2. Watson L. Primary Hyperparathyroidism. *Clin Endocrinol Metab.* 1974;3:215.
3. Habener JF, Potts JT. Parathyroid physiology and primary hyperparathyroidism. In: Avioli LV, Krane SM, eds. *Metabolic Bone Disease.* vol 2. New York, NY: Academic Press; 1978.
4. Mallette LE, Bilezikian JP, Heath DA, et al. Primary hyperparathyroidism: clinical and biochemical features. *Medicine.* 1974;53:127.
5. Broadus AE. Nephrolithiasis in primary hyperparathyroidism. In: Coe FL, Brenner BM, Stein JH, eds. *Contemporary Issues in Nephrology.* vol 5. Edinburgh: Churchill Livingstone; 1980.

6. Meunier P, Vignon G, Bernard J, et al.: Quantitative bone histology as applied to the diagnosis of hyperparathyroid states. In: Frame B, Parfitt AM, Dunmcan H, eds. *Clinical Aspects of Metabolic Bone Disease.* Amsterdam: Excerpta Medica; 1973.

7. Genant HK, Heck LL, Lanzl LH, et al. Primary hyperparathyroidism. A comprehensive study of clinical, biochemical and radiological manifestations. *Radiology.* 1973;109:513.

8. Patten BM, Bilezikian JP, Mallette LE, et al. Neuromuscular disease in primary hyperparathyroidism. *Ann Intern Med.* 1974;80:182.

9. Peterson P. Psychiatric disorders in primary hyperparathyroidism. *J Clin Endocrinol Metab.* 1968;28:1491.

10. Barreras RF. Calcium and gastric secretion. *Gastroenterology.* 1973;64:1168.

11. Wilson SD, Singh RB, Kalkhoff RK, et al. Does hyperparathyroidism cause hypergastrinemia? *Surgery.* 1976;80:231.

12. Mixter CG, Keynes WM, Chir M, et al. Further experience with pancreatitis as a diagnostic clue to hyperparathyroidism. *N Engl J Med.* 1962;266:265.

13. Van der Ark CR, Ballantyne F III, Reynolds EW Jr. Electrolytes and the electrocardiogram. *Cardiovasc Clin.* 1973;5:285.

14. David DS. Mineral and bone homeostasis in renal failure. Pathophysiology and management. In: David DS, ed. *Renal Failure and Management.* New York, NY: John Wiley; 1977;1.

15. Petrohelos M, Tricoulis D, Diamanticos P. Band keratopathy with bilateral deafness as a presenting sign of hyperparathyroidism. *Br J Ophthalmol.* 1977;61:494.

16. Jansen OA: Ocular calcifications in primary hyperparathyroidism. *Acta Ophthalmol.* 1975;53:173.

17. Kollarits CR, Moss ML, Cogan DG, et al. Scleral calcifications in hyperparathyroidism: demonstration by computed tomography. *J Comp Assist Tomogr.* 1977;1(4):500.

18. Berkow JW, Fine BS, Zimmerman LE. Unusual ocular calcification in hyperparathyroidism. *Am J Ophthalmol.* 1968;66(5):812.

19. Naiman J, Green RW, D'Heurle D, et al. Brown tumor of the orbit associated with primary hyperparathyroidism. *Am J Ophthalmol.* 1980;90:565.

20. Broadus AE. Mineral metabolism. In: Felig P, Baxter JD, Frohman LA, eds. *Endocrinology and Metabolism.* New York, NY: McGraw-Hill; 1981;1057.

21. Yendt E. The parathyroids and calcium metabolism. In: Ezrim C, Godden J, Volpe R, et al., eds. *Systematic Endocrinology.* Hagerstown, Md: Harper & Row; 1973; 119.

22. Nagant de Denxchaisnes C, Krane SM. Hypoparathyroidism. In: Avioli LV, Krane SM, eds. *Metabolic Bone Disease.* vol 2. New York, NY: Academic Press; 1978.

23. Blizzard RM, Chee D, Davis W. The incidence of parathyroid and other antibodies in the sera of patients with idiopathic hypoparathyroidism. *Clin Exp Immunol.* 1966;1:119.

24. Lawayin S, Norman DA, Zerwekh JE, et al. A patient with pseudohypoparathyroidism with increased serum calcium and 1,25 dihydroxyvitamin D after exogenous parathyroid hormone administration. *J Clin Endocrinol Metab.* 1979; 49:783.

25. Levine M, Downs R, Singh M, et al. Deficiency of guanine nucleotide regulatory protein: Postreceptor site of defect in pseudohypoparathyroidism. *Proceedings of the 7th International Conference on Calcium Regulatory Hormones.* 1980.

26. Albright F, Reifenstein EC. Parathyroid glands and metabolic bone disease. Baltimore, Md: Williams & Wilkins; 1948.

27. Blake J. Eye signs in idiopathic hypoparathyroidism. *Trans Ophthalmol Soc UK.* 1976;96:448.

28. Bunce GE. Nutrition and cataract. *Nutr Rev* 1979;37:337.

29. Walsh FB, Murray R. Ocular manifestations of disturbances in calcium metabolism. *Am J Ophthalmol.* 1953;36:1657.

30. Walsh FB. *Clinical Neuro-Ophthalmology.* Baltimore, Md: Williams and Wilkins; 1957;694.

31. Stieglitz LN, Kind HP, Kazdan JJ, et al. Keratitis with hypoparathyroidism. *Am J Ophthalmol.* 1977;84:467.

Chapter 31

Multiple Endocrine Neoplasia Syndrome

R. JEAN CAMPBELL

The syndrome of multiple endocrine neoplasia is a rare disorder that is inherited in an autosomal dominant pattern compatible with a two-mutation model. The word "multiple" refers not only to the various endocrine glands that may be affected but also to almost universal bilaterality and multicentricity of tumor growth. Because the disorder has symptoms and clinical findings that may be confusing to the physician it is convenient to classify the syndrome into groups according to the organs involved.[1,2]

The entity has two major subgroups (Table 31-1). Multiple endocrine neoplasia type I (MEN-I), also known as Wermer's syndrome, consists of neoplasia or hyperplasia of the pituitary, pancreas, and parathyroid glands. In multiple endocrine neoplasia type II (MEN-II), medullary carcinoma of the thyroid, pheochromocytoma(s), and hyperparathyroidism may occur. MEN-II has two subtypes: MEN-IIa, or Sipple's syndrome, and MEN-IIb. Patients with MEN-IIa are phenotypically normal. Those with MEN-IIb have an easily recognizable abnormal phenotype with marfanoid habitus, starry-eyed gaze, and enlargement of the tongue, lips, and eyelids due to mucosal neuromas (Fig. 31-1). Only MEN-IIb has ocular manifestations. The organs involved are of neural crest origin, but the defect(s) at the molecular level still have to be determined.[3]

Systemic Manifestations

The organs commonly involved in MEN-I are the pituitary, the pancreatic islets, and the parathyroid glands. Other organs are less frequently affected, but tumors may arise in the thyroid gland and adrenal cortex; bronchial and intestinal carcinoid, schwannoma, thymoma, gastric polyps, lipomas, and hibernomas may occur. The eyes do not have characteristic findings.

Both subtypes of MEN-II may have associated medullary carcinoma of the thyroid and pheochromocytoma(s). Hyperparathyroidism more commonly occurs in subtype IIa; it is rare in subtype IIb. Recognition of the characteristic facial appearance of patients with MEN-IIb allows early detection of these neoplastic processes in an otherwise asymptomatic patient, and timely surgery avoids the development of malignancy.

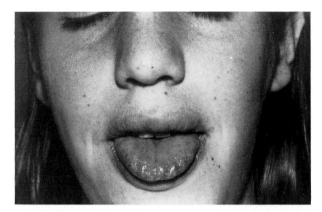

FIGURE 31-1 Irregular enlargement of tongue due to mucosal neuromas in patient with MEN IIb. (From Riley and Robertson.[5])

Medullary carcinoma arises in the parafollicular C cells of the thyroid, which are of neural crest origin. Tumor growth, which is usually bilateral and multifocal, is preceded by hyperplasia of the C cells, which secrete calcitonin. The increased secretion of plasma immunoreactive calcitonin (iCT) is determined by radioimmunoassay, and an elevation of the iCT value is the first manifestation of medullary thyroid carcinoma in an asymptomatic patient. When elevations of calcitonin are borderline, infusion with pentagastrin will increase the iCT elevation in affected patients. Investigation of family members is of the utmost importance.

If the asymptomatic state is not detected, the patient may present with a thyroid nodule, evidence of metastases, or symptoms referable to abnormalities of the adrenal or parathyroid glands. The median age of patients with such a presentation is 21 years, unlike the 51 years for patients with sporadic nonfamilial medullary thyroid carcinoma. Of all such tumors, 19% are familial and bilateral.

Histologically, the tumor is recognized by its characteristic deposition of amyloid. Its behavior is aggressive, and the prognosis is adversely affected by late clinical recognition, inadequate surgical removal, presence of metastases, multifocality, site, age, and MEN-IIb phenotype.

Neoplasia of the adrenal medulla in patients with MEN-II ranges from small, focal, nodular hyperplasia to pheochromocytoma, which may be large and is bilateral and multicentric in approximately 50% of patients with MEN-IIb. Estimation of the urinary catecholamines (epinephrine, norepinephrine, dopamine) and a high ratio of epinephrine to norepinephrine are useful screens for the asymptomatic patient at high risk. Computed tomography is helpful for localizing tumors in adrenal and extra-adrenal sites.

TABLE 31-1 Types of Multiple Endocrine Neoplasia

	Synonym	Organs Affected
Type I	Wermer's syndrome	Pituitary, pancreas, parathyroid
Type IIa (normal phenotype)	Sipple's syndrome	Thyroid, adrenal, parathyroid
Type IIb (ganglioneuroma phenotype)		Thyroid, adrenal, parathyroid (rare); mucous membranes of conjunctiva, lips, and tongue (involved by neuromas); corneal nerves prominent

Parathyroid disease is extremely rare in the MEN-IIb subtype, although minor histopathologic abnormalities may exist. This frequency is in contrast to that in MEN-IIa, in which the incidence of disease ranges from 29 to 60% of patients.

The habitus of patients with MEN-IIb is characteristically marfanoid, and the joints show increased laxity. Many patients have prognathism. The proliferation of mucosal neuromas contributes to the facial appearance, and the neuromas affect the lips, tongue, and eyelids (see Fig. 31-1).

Ocular Manifestations

The characteristic ocular findings in MEN-IIb include prominent corneal nerves, neuromas of the palpebral and bulbar conjunctiva, and keratoconjunctivitis sicca.[4] The prominent corneal nerves form a filigree pattern across the whole cornea but are visible only by slit lamp biomicroscopy (Fig. 31-2). They are not visible to the naked eye. Histologically, studies have shown that the nerve bundles at the corneoscleral limbus are composed of myelinated and unmyelinated axons associated with Schwann cells and are surrounded by a prominent neural sheath.[5,6] The associated capillary vasculature is similar to that of corneal neovascularization. At the junction of the anterior one third and the posterior two thirds of the corneal stroma, the nerve bundles, unaccompanied by vessels, enter the cornea proper, and here the axons are of varied diameter and are associated with Schwann cells, but they are unmyelinated. It has been suggested that increased numbers, as well as increased size, of axons and increased numbers of Schwann cells explain the prominent clinical appearance, but detailed morphometric studies are still needed.

The conjunctiva contains neuromas that consist of enlarged nerves arranged in bundles and that are most obvious at the corneoscleral limbus. Involvement of the palpebral conjunctiva results in thickened lids (see Fig. 31-3). The neuromas are seen immediately proximal to the eyelid margins, a site that normally is not visible to the examiner, and consist of bundles of myelinated and nonmyelinated axons. These prominent nerve bundles extend from the limbus to the equator.

Large nerve bundles have also been seen in the ciliary body and the iris, where they consist of myelinated and nonmyelinated axons.[4] Ganglion cells are present in increased numbers in the ciliary body and are present in the iris root and the uveal meshwork.

Recognition of the spectrum of physical findings in the patient with MEN-IIb not only allows early detection of neoplasia but also indicates the need to screen family members.[7]

FIGURE 31-2 Biomicroscopy of cornea showing prominent corneal nerves. (From Riley and Robertson.[5])

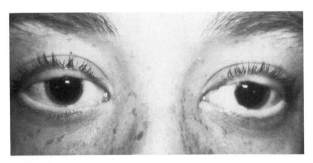

FIGURE 31-3 Patient with MEN IIb with conjunctival neuromas and thickened eyelids. (From Riley and Robertson.[5])

References

1. Montgomery TB, et al. Multiple endocrine neoplasia type IIB. A description of several patients and review of the literature. *J Clin Hypertens.* 1987;3(1):31.
2. Schimke RN. The multiple endocrine neoplasia syndromes. *Cancer Treat Res.* 1983;17:249.
3. Delellis RA, et al. Multiple endocrine neoplasia (MEN) syndrome: cellular origins and inter-relationships. *Int Rev Exp Pathol.* 1986;28:163.
4. Robertson DM, Sizemore GW, Gordon H. Thickened corneal nerves as a manifestation of multiple endocrine neoplasia. *Trans Am Acad Ophthalmol Otolaryngol.* 1975;79:722.
5. Riley FC Jr, Robertson DM. Ocular histopathology in multiple endocrine neoplasia type 2B. *Am J Ophthalmol.* 1981;91:57.
6. Spector B, et al. Histologic study of the ocular lesions in multiple endocrine neoplasia syndrome type IIB. *Am J Ophthalmol.* 1981;91(2):204.
7. Lips CJ, et al. Multiple endocrine neoplasia syndromes. *CRC Crit Rev Oncol Hematol.* 1984;2(2):117.

Chapter 32

Pituitary Disorders

DUNCAN P. ANDERSON and SORANA S. MARCOVITZ

The pituitary gland may be considered the master gland of the body; its main function is hormone secretion. The anterior pituitary (adenohypophysis) secretes prolactin (PRL), somatotropic (growth) hormone (SH or GH), adrenocorticotropic hormone (ACTH), thyroid-stimulating hormone (TSH), follicle-stimulating hormone (FSH), luteinizing hormone (LH), and melanocyte-stimulating hormone (MSH). The posterior pituitary (neurohypophysis) secretes oxytocin and vasopressin (antidiuretic hormone, ADH).

Pituitary disorders produce symptoms as a result of abnormal hormone secretion or of direct mechanical effects on adjacent structures. In general, systemic manifestations are due to hormonal hypersecretion or hyposecretion, whereas visual symptoms are related to mechanical compression of the optic nerve, chiasm, tracts, and cavernous sinus.[1,2] The commonest pituitary disorder is pituitary adenoma, which accounts for approximately 12% of symptomatic intracranial tumors and may be found in up to 30% of the population in unselected autopsy series.[3]

The pathologic classification of these adenomas has changed over the past two decades with the advent of electron microscopy and immunohistochemistry, which now permit detailed analysis of intracellular structures and positive identification of hormones contained within secretory granules. Nevertheless, these advances have had little impact on the clinical diagnosis and management of patients, partly because not all patients are treated surgically and partly because the morphologic features do not correlate closely with tumor size, degree of hypersecretion, or quality of response to various types of therapy.

From the point of view of clinical management, the classification based on measurements of circulating levels of pituitary hormones is the most useful. The availability of radioimmunoassays for the detection of anterior pituitary peptides, especially prolactin, makes it possible to diagnose hypersecreting adenomas early in their development, when they are small and before they produce any space-occupying effects.

Adenomas that are not associated with hypersecretion of any of the currently known anterior pituitary hormones are often not diagnosed until they produce either hypopituitarism or neuro-ophthalmic manifestations due to their mass effect.[4]

Systemic Manifestations

Prolactinomas are the commonest type of pituitary adenoma and prior to the prolactin radioimmunoassay were classified as nonsecreting chromophobe adenomas. Their manifestations are the result of the biologic effects of prolactin. Hyperprolactinemia can cause arrested puberty in both sexes. In adult women the most common presenting complaints are secondary amenorrhea, galactorrhea, infertility, and headaches; adult men present with sexual dysfunction, infertility, gynecomastia, or headaches.[5]

The second commonest pituitary adenoma is the somatotropic adenoma (previously called eosinophilic adenoma) which hypersecretes GH, producing acromegaly with its classical manifestation of coarsened facial features, hand and foot enlargement, changes of dental alignment, and glucose intolerance. About one third of patients with acromegaly have adenomas that hypersecrete both growth hormone and prolactin and thus have associated symptoms of hyperprolactinemia. Morphologic studies show that some of these tumors are mixed growth hormone cell and prolactin cell adenomas and that a few may be composed of a uniform population of acidophilic "stem cells" that contain both types of hormone granules in the same individual cells. This morphologic distinction does not seem to have any prognostic implications.[4]

ACTH adenomas (previously called basophilic adenomas) are the third commonest type of pituitary tumor and cause bilateral adrenal hyperplasia in response to overproduction of ACTH, producing clinical manifestations of Cushing's disease: hypertension, glucose intolerance, truncal obesity, thin skin with purple striae, muscle weakness, and psychological changes.[6,7]

TSH adenomas are very rare and most commonly cause clinical manifestations of hyperthyroidism. The true diagnosis is often overlooked initially unless serum TSH is measured, and the clinical and biochemical manifestations of hyperthyroidism may be attributed to Graves' disease.[8,9]

Gonadotropin-cell adenomas hypersecrete LH, FSH, or unpaired alpha peptide subunits. The alpha chains, which normally combine with specific beta peptides to form LH, FSH, or TSH, have no intrinsic biologic activity. Intact LH and FSH secreted in excess quantities have a paradoxical suppressive effect on gonadal function, but the resultant hypogonadism may not be diagnosed promptly in men because of gradual onset and vague symptoms, while among women most of the reported patients are postmenopausal. Thus this type of tumor tends to mimic nonsecreting adenomas in its clinical course and to be associated with large size, presence of space-occupying effects, and neuro-ophthalmic manifestations.[10,11]

In contrast with the above-mentioned secreting adenomas, the so-called nonsecreting adenomas may present either without systemic manifestations or with symptoms of hypopituitarism. The patient develops symptoms of gonadotropin deficiency (amenorrhea, loss of libido, diminution of secondary sex characteristics), hypothyroidism (cold intolerance, lethargy, constipation, dry skin), and adrenal insufficiency (hypoglycemia, weakness, hypotension).[12]

Several other types of neoplasms may be located in the region of the sella turcica and may mimic true adenomas in their clinical manifestations, radiologic appearance, and associated hormonal abnormalities and ophthalmic findings. These include craniopharyngiomas, subdiaphragmatic meningiomas, germinomas, and arachnoid cysts. The endocrine manifestations may include hypopituitarism and hyperprolactinemia; the latter is due not to hypersecretion by the neoplastic cells but to compression of the pituitary stalk, which causes the normal prolactin cells to be deprived of hypothalamic prolactin-inhibiting factor.

Ocular Manifestations

The development of sensitive and specific tests of hormone abnormalities and the resultant earlier treatment have led to a reduced incidence (about 10%) of neuro-ophthalmic abnormalities in hypersecreting pituitary adenomas.[5] More often, visual abnormalities are found in nonsecreting pituitary adenomas, craniopharyngiomas, suprasellar meningiomas, and aneurysms.[7] These abnormalities result from local compression of the optic nerves, chiasm, tracts, and occasionally other cranial nerves.

A suprasellar mass must measure at least 10 mm before it can compress the chiasm and interrupt its blood supply, producing bitemporal hemianopia. The patient complains of poor side vision if peripheral fibers are involved and if central fibers are involved will read only the left side of the eye chart with the right eye and the right side of the eye chart with the left eye.[13]

If the chiasm is post-fixed, there will be ipsilateral optic nerve compression producing mildly impaired vision and color vision with a central scotoma and afferent pupillary defect. As the mass impinges on the chiasm, a superotemporal defect in the contralateral eye (due to involvement of the anteroinferior crossing chiasmal fibers) will be added to the ipsilateral central scotoma, producing a "junctional scotoma."[13]

In a pre-fixed chiasm the ipsilateral optic tract will be compressed, producing an incongruous contralateral homonymous hemianopia with a subtle contralateral afferent pupillary defect.[14]

Optic atrophy does not develop early in compressive disease, but when it does, the pattern of atrophy may help in diagnosis. In optic nerve compression, the temporal disc will be pale due to atrophy of the papillomacular fibers. In chiasmal compression horizontal "bow-tie" pallor of each disc is seen due to atrophy of the crossing chiasmal fibers on the disc. In optic tract compression, there is the typical "bow-tie" atrophy in the contralateral eye and a more diffuse atrophy in the ipsilateral eye.[14]

Pupillary abnormalities consist of an ipsilateral afferent pupil defect (Marcus-Gunn pupil) in optic nerve compression and a bilateral hemiafferent pupil defect (Wernicke's hemianopic pupil) in chiasmal and tract compression.

A nonparetic sensory type of diplopia may occur as a result of a complete bitemporal hemianopia, owing to the fact that the two remaining nasal fields do not have enough overlap to maintain fusion. These patients complain of variable double vision and of trouble with depth perception due to "postfixational blindness" (any object distal to the fixation point falls on the sightless nasal hemiretinas).[13]

True paretic diplopia may occur if a pituitary tumor expands laterally into the cavernous sinus producing third, fourth, or sixth nerve palsy, but in this case the fields are usually normal and face pain or hypesthesia may be associated.[2]

Marked suprasellar extension may compress the hypothalamus and third ventricle, producing papilledema and diabetes insipidus as well as appetite, sleep, and personality disturbances. This mode of presentation is commonly seen in craniopharyngiomas in children, since the visual symptoms of a young child are often overlooked. In children most of these tumors are calcified, whereas in adults they are rarely calcified and often present with vision loss.[13]

Anterior growth of an intrasellar tumor can result in a mass presenting in the sphenoid sinus mimicking sinus disease. Other causes of "chiasmal syndromes" include trauma, ectopic pinealoma, posterior fossa tumor with dilated third ventricle, empty sella syndrome, and chiasmal glioma.[13]

Treatment of pituitary adenomas and other parasellar lesions that cause visual abnormalities is neurosurgical excision. This may be accomplished by the simpler and safer transsphenoidal route, but if there is excessive suprasellar extension, a transfrontal craniotomy may have to be performed. X-ray treatment is used as an adjunct both preoperatively and postoperatively in many centers and for patients who are "poor surgical risks." Prolactinomas and some GH–secreting adenomas may be treated with bromocriptine (2, bromo-α-ergocryptine; a dopamine agonist). This substance has been shown to be effective in suppressing prolactin and sometimes growth hormone secretion and in reducing tumor size in many patients with prolactinomas and in some cases of mixed prolactin- and GH–secreting tumors.[6,7] More recently a long-acting analogue of the hypothalamic hormone somatostatin (Sandoz 201-995) has been studied in patients with acromegaly and TSH–secreting tumors and has been shown to be effective in suppressing hormone hypersecretion and in some cases in reducing tumor size.[15,16] At present its use is limited by the fact that it must be administered by multiple daily subcutaneous injections, but the prospect of a somatostatin analogue that could be administered as a nasal spray is attractive. Intrasellar tumors may be treated medically with bromocriptine, which has been shown to be effective in suppressing prolactin and GH secretion and in reducing tumor size.[6,7]

In the future, improved access to endocrinologic and radiologic diagnostic techniques should lead to diagnosis of hypersecreting pituitary adenomas at a very early stage in their development, making associated ophthalmic manifestations an extreme rarity. For so-called nonsecreting adenomas, perhaps the discovery of other as yet unknown hormones that these tumors may secrete will lead to improved early diagnosis. Ultimately, it is hoped that advances in understanding the etiology of pituitary adenomas will lead to the development of specific substances capable of correcting the abnormalities and preventing the mass effects associated with tumor growth.

References

1. Duane TD. *Clinical Ophthalmology*. vol 5. Philadelphia, Pa: JB Lippincott; 1986;1-2.
2. Baker AB, Baker LH. *Clinical Neurology*. vol 4. Philadelphia, Pa: JB Lippincott; 1986;9-14.
3. Burrow GN, Wortzman G, Newcastle NB, et al. Microadenomas of the pituitary and abnormal sellar tomograms in an unselected autopsy series. *N Engl J Med*. 1981; 304:156-158.
4. Horvath E, Kovacs K. Ultrastructural classification of pituitary adenomas. *Can J Neurol Sci*. 1976;3:9-18.
5. Anderson DP, Faber P, Marcovitz S, et al. Pituitary tumors and the ophthalmologist. *Ophthalmology*. 1983;90:1265-1270.
6. Kreiger HP. Neuroendocrinology of sellar disease and pituitary hormone secretion. In: Krieger DT, Hughes JC, eds. *Neuroendocrinology*. New York, NY: HP Publishing Co; 1980.
7. Zimmerman EA. Tumours of the pituitary gland. In: Rowland LP, ed. *Merrit's Textbook of Neurology*. 7th ed. Philadelphia, Pa: Lea & Febiger; 1984;244-250.
8. Benoit R, Pearson-Murphy BE, Robert F, et al. Hyperthyroidism due to a pituitary TSH-secreting tumor with amenorrhea-galactorrhea. *Clin Endocrinol*. 1980;12:11-19.
9. Hill SA, Falko JM, Wilson CB, et al. Thyrotropin-producing pituitary adenomas. *J Neurosurg*. 1982;57:515-519.
10. Whitaker MD, Prior JC, Scheithauer D, et al. Gonadotropin-secreting pituitary tumor: report and review. *Clin Endocrinol*. 1985;22:43-48.
11. Klibanski A, Ridgeway C, Zervas NT. Pure alpha subunit-secreting pituitary tumors. *J Neurosurg*. 1983;59:585-589.
12. Daughaday WH. The anterior pituitary. In: Wilson JD, Foster DW, eds. *Textbook of Endocrinology*. Philadelphia, Pa: WB Saunders Co; 1985;568-613.
13. Duane TD. *Clinical Ophthalmology*. vol 2. Philadelphia, Pa: JB Lippincott; 1986;1-8.
14. Miller NR. *Walsh and Hoyt's Clinical Neuro-Ophthalmology*. 4th ed, vol 1. Baltimore, Md: Williams & Wilkins; 1982;127-129.
15. Comi RJ, Gorden P. The response of serum growth hormone levels to the long-acting somatostatin analog SMS 201-995 in acromegaly. *J Clin Endocrinol Metab*. 1987;64:37-42.
16. Comi RJ, Gesundheit N, Murray L, et al. Response of thyrotropin-secreting pituitary adenomas to a long-acting somatostatin analogue. *N Engl J Med*. 1987;317:12-18.

PART 5

Gastrointestinal Disorders

Chapter 33

Gardner's Syndrome

JAMES C. ORCUTT

Gardner described an inherited syndrome that included intestinal polyposis, soft tissue tumors, and benign osseous growths.[1] Gardner's syndrome is an autosomal dominant, inherited disorder. The penetrance is quite variable; however, intestinal polyposis is common and the prevalence of colon adenocarcinoma is nearly 100% by age 50 years.[2,3] Colectomy before appearance of carcinomatous changes in the intestinal polyps improves survival,[2] making early detection of persons at risk important. Awareness of the cutaneous, osseous, and ophthalmic findings in Gardner's syndrome improves the likelihood of early detection.

Systemic Manifestations

Common cutaneous findings in Gardner's syndrome are epidermoid inclusion cysts, or sebaceous cysts. Epidermoid cysts may occur anywhere on the body and are commonly multiple. Additional cutaneous findings reported in patients with Gardner's syndrome include desmoid tumors, lipomas, leiomyomas, neurofibromas, incisional fibromas, and pigmented spots on the torso and limbs. Carcinoma of the thyroid gland, fibrosarcomas, retroperitoneal mixed tumors, and mesenteric fibrous tumors also occur.

Osseous findings in Gardner's syndrome include osteomas, cortical thickening in long bones, exostoses, and dental abnormalities. Osteomas commonly occur in the skull and facial bones. Osteomas of the mandible occur in about 70% of patients with Gardner's syndrome and are rare in patients without Gardner's syndrome. The presence of a mandibular osteoma should alert the clinician to the possibility of Gardner's syndrome. Osteomas of the facial bones commonly extend into sinuses. The osteomas are benign and usually stop growing as the patient ages. Dental abnormalities include multiple carious teeth and multiple impacted, unerupted, supernumerary teeth. Abnormal dental findings occur in about 50% of patients with Gardner's syndrome.

Ocular Manifestations

Ophthalmic findings in Gardner's syndrome include retinal pigment epithelial hypertrophy,[3-5] orbital osteo-

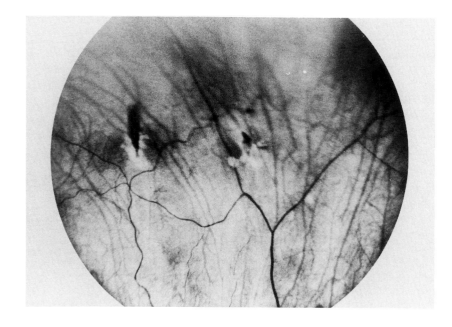

FIGURE 33-1 Congenital retinal pigment epithelial hypertrophy. Left peripheral fundus of a 44-year-old woman with stigmata of Gardner's syndrome. Similar pigment epithelial changes were seen in the right eye. The patient's father, paternal aunt, three of six paternal cousins, and the patient's only son are similarly affected.

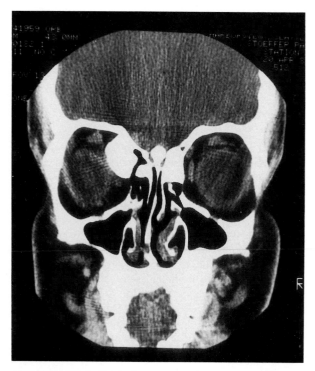

FIGURE 33-2 Orbital osteoma. CT scan of a 30-year-old female with Gardner's syndrome demonstrating an osteoma of the left orbit arising from the ethmoid bone. The patient had a mandibular osteoma removed 6 years earlier and has undergone prophylactic colectomy. Two additional osteomas project into the ethmoid sinuses.

mas,[2,6] epidermoid cysts of the eyelid,[2] and possibly angioid streaks.[7,8] Retinal pigment epithelial hypertrophy appears as brown or black lesions ranging in size from 0.1 to several disc diameters. Some are completely pigmented; others are surrounded by a depigmented halo or are almost completely depigmented. Typically, lesions of pigment epithelial hypertrophy are not associated with Gardner's syndrome and are unilateral, isolated, and asymptomatic. The fundi of patients with Gardner's syndrome demonstrate multiple lesions in both eyes in 78 to 100% of cases (Fig. 33-1).[3-5] Retinal pigment epithelial hypertrophy in patients with Gardner's syndrome is probably congenital. Bilateral multiple areas of retinal pigment epithelial hypertrophy are rarely seen in patients without Gardner's syndrome (≤5%).[4,5] Congenital hypertrophy of the pigment epithelium may therefore be an excellent marker by which to identify persons at risk of developing intestinal carcinoma.

Orbital osteomas in Gardner's syndrome generally arise from the ethmoid or frontal bone (Fig. 33-2). Proptosis is the usual presenting feature; diplopia and vision loss occur late in the clinical course. Removal of the osteoma is indicated if the mass continues to expand, threatening vision or extraocular movement or creating a facial deformity. Epidermoid cysts of the eyelid may enlarge, obstructing vision or leading to tearing abnormalities. Excision of epidermoid cysts is the treatment of choice.

Angioid streaks have been reported in two patients with familial polyposis coli. One patient had additional stigmata suggesting Gardner's syndrome,[7] whereas the other carried a diagnosis of pseudoxanthoma elasticum.[8] Patients with angioid streaks may suffer visual loss from subretinal neovascularization.

Gardner's syndrome is comprised of intestinal polyposis, soft tissue tumors, and osseous tumors. Dominant inheritance implies that 50% of the offspring of affected persons are at risk to develop Gardner's syndrome. All patients with the syndrome eventually develop intestinal carcinomas, but early prophylactic colectomy reduces the risk. Identifying persons who carry the abnormal gene at an early age would allow early prophylactic treatment, reducing their risk of later malignancy. Frequent clinical findings that may identify the 50% of offspring who carry the abnormal gene, that are detectable at an age well before malignant transformation of intestinal polyps occurs, but that are rare in the general population, include bilateral congenital hypertrophy of the retinal pigment epithelium, mandibular osteomas, and impacted, supernumerary teeth.

References

1. Gardner EJ. Follow-up study of a family group exhibiting dominant inheritance for a syndrome including polyps, osteomas, fibromas, and epidermal cysts. *Am J Hum Genet.* 1953;14:376.
2. Jones EL, Cornell WP. Gardner's syndrome: review of the literature and report on a family. *Arch Surg.* 1966;92:287-300.
3. Lewis RA, Crowder WE, Eierman LA, et al. The Gardner syndrome: significance of ocular features. *Ophthalmology.* 1984;91:916-925.
4. Blair NP, Trempe CL. Hypertrophy of the retinal pigment epithelium associated with Gardner's syndrome. *Am J Ophthalmol.* 1980;90:661-667.
5. Traboulsi EI, Krush AJ, Gardner EJ, et al. Prevalence and importance of pigmented ocular fundus lesions in Gardner's syndrome. *N Engl J Med.* 1987;316:661-667.
6. Whitson WE, Orcutt JC, Walkinshaw MD. Orbital osteoma in Gardner's syndrome. *Am J Ophthalmol.* 1986;101:236-241.
7. Awan KJ. Familial polyposis and angioid streaks in the ocular fundus. *Am J Ophthalmol.* 1977;83:12-125.
8. O'Holleran M, Merrell RC. Pseudoxanthoma elasticum and polyposis coli. *Arch Surg.* 1981;116:476-477.

Chapter 34

Hepatic Disease

KENNETH G. ROMANCHUK

Liver failure secondary to hepatic disease affects the eye. Although there are many causes of hepatic disease and the liver is involved in so many biochemical processes, fortunately complications of liver failure that affect the eye are relatively few.

Jaundice is one of the hallmarks of liver failure and often is detected by a yellowish-green discoloration of the normally white sclera of the eye. Jaundice occurs because bilirubin (derived from the normal breakdown of hemoglobin) cannot be excreted into the bile, so its blood level rises and the bilirubin pigment accumulates in tissues. It accumulates particularly in the vascularized conjunctiva and subconjunctival connective tissue (Tenon's) overlying the white sclera,[1] so that the normally white appearance of sclera as seen through the normally semitransparent conjunctiva is altered and

scleral icterus results. Superficial sclera and peripheral cornea may also be stained, as is evident when a jaundiced eye is enucleated and the conjunctiva and subconjunctival connective tissue have been removed. No visual loss occurs from the accumulation of bilirubin pigment itself.

Visual impairment in the form of night blindness (impaired dark adaptation) and xerophthalmia (xerosis of the conjunctiva and cornea) may occur in patients with hepatic disease and a marked reduction in plasma vitamin A levels.[2] In acute hepatitis, low plasma levels of vitamin A are apparently due to failure of the release of vitamin A from the liver into the plasma, since the hepatic reserves of the vitamin are apparently normal. Chronic liver disease with cirrhosis may also be associated with night blindness; poor dietary intake of vitamin

A, malabsorption of fat (vitamin A is a fat-soluble vitamin), impaired hepatic storage of retinyl esters, and decreased synthesis or hepatic release of retinal-binding protein are all thought to be possible contributing factors.[3] Vitamin A absorption is enhanced by bile salts, so depletion may occur when synthesis and secretion of bile salts are impaired in chronic liver disease and by prolonged biliary tract obstruction. Vitamin A is a necessary biochemical component of the visual pigment rhodopsin of the rod photoreceptors, and when deficient, the photochemical reaction of the rods is impaired, causing the symptom of night blindness. Dark adaptation testing shows impairment, and electroretinography shows abnormalities. With prolonged vitamin A deficiency there may be progression to actual structural change in the photoreceptors and in the retinal pigment epithelium. Xerophthalmia associated with vitamin A deficiency is caused by loss of the mucus-secreting conjunctival goblet cells with keratinization of the conjunctiva. Early changes are a superficial punctate keratopathy of the cornea and Bitot's spots (triangular patches of gray, foamy material near the limbus). There may be progression to degeneration of the cornea, with loss of its lucency, thinning, and even perforation. Vitamin A supplementation may reverse the changes early on but may not be effective when the changes have become advanced.

Patients with severe impairment of liver function due to advanced hepatic cirrhosis but who are euthyroid may exhibit eyelid retraction (Dalrymple's sign) and eyelid lag on downward gaze (Graefe's sign).[4] These two eye signs are usually associated with Graves' disease and are thought to be due to increased circulating levels of sympathetic amines in plasma. It is postulated that these increased levels of plasma sympathetic amines also occur in liver disease due to their impaired metabolism by the liver.

Some eye signs are fairly specific, so when they occur in association with hepatic disease they suggest the diagnosis of a particular genetic disease. The Kayser-Fleischer corneal ring occurs in hepatolenticular degeneration (Wilson's disease) although it is not pathognomonic, since it also occurs in other diseases such as multiple myeloma, liver diseases other than Wilson's, and copper intraocular foreign bodies.[5-7] Wilson's disease is an autosomal recessive disease in which excessive amounts of copper are deposited in the liver, basal ganglia, cerebral cortex, and corneas, which eventually leads to cirrhosis, renal tubular absorption abnormalities, and motor abnormalities. Clinical manifestations usually occur between age 6 and middle age; initial presentation is split equally between liver disease, neurologic symptoms, and both.[8] The exact biochemical defect is unknown but may be related to defective incorporation of copper into the serum enzyme ceru-

loplasmin that binds the copper or to a defect in the normal excretion of copper into the bile.[8,9] In the first years of life, increasing amounts of copper are stored in the liver. Eventually, necrosis of the liver cells occurs with release of the copper into the blood and deposition in other tissues. Copper is deposited into the peripheral cornea at the level of Descemet's membrane,[10] in a golden brown, brownish green, greenish yellow, or deep reddish ring starting at the limbus and extending from 1 to 3 mm. Copper is also present in other layers of the cornea but is not clinically visible.[7] The earliest site of visible deposition is in an arc in the superior peripheral cornea from about the 10 o'clock to the 2 o'clock position and then in the inferior peripheral cornea from 5 o'clock to 7 o'clock. Slowly it spreads circumferentially and broadens until the superior and inferior arcs meet. Visual acuity is not affected by the corneal change. The Kayser-Fleischer rings have apparently disappeared following liver transplantation[11] and also following therapy with the copper-chelating agent penicillamine.[12] Cataracts resembling sunflowers (central pigmented lens opacities with tapering extensions) may also occur in hepatolenticular degeneration and are apparently due to deposition of electron-dense granules of copper in the anterior and posterior lens capsule. Other ocular findings include nystagmus, saccadic and pursuit gaze abnormalities, and defects of convergence and accommodation, the latter apparently due to a periaqueductal lesion.[7,13,14]

Arteriohepatic dysplasia (Alagille's syndrome) is a familial disorder involving the liver, cardiovascular system, bones, eyes, and less consistently the habitus, central nervous system, kidneys, and endocrine system. Typically, affected patients present with persistent neonatal jaundice, and liver biopsy specimens show cholestasis and a paucity of intrahepatic bile ducts. Autosomal dominant transmission is suspected.[15] A rather unique combination of ocular findings occurs in arteriohepatic dysplasia, providing an expedient and inexpensive means of distinguishing this syndrome from other forms of neonatal cholestasis. Three patients with arteriohepatic dysplasia all had bilateral posterior embryotoxin, Axenfeld's anomaly, and a pigmentary retinopathy. Other ocular findings included exotropia, an ectopic pupil, band keratopathy, choroidal folds, anomalous optic discs, and infantile myopia.[16]

References

1. Yanoff M, Fine BS. *Ocular Pathology, a Text and Atlas.* 2nd ed., Hagerstown, Md: Harper & Row; 1982:284.
2. Harris AD, Moore T. Vitamin A in infective hepatitis. *Br Med J.* 1947;1:553-539.
3. Russel RM, Morrison SA, Smith FR, et al. Vitamin-A reversal of abnormal dark adaptation in cirrhosis, study of

effects on the plasma retinol transport system. *Ann Intern Med.* 1978;88:622-626.

4. Summerskill WHJ, Molnar GD. Eye signs in hepatic cirrhosis. *N Engl J Med.* 1962;266:1244-1248.

5. Ellis PP. Ocular deposition of copper in hypercupremia. *Am J Ophthalmol.* 1969;68:423-427.

6. Frommer D, Morris J, Sherlock S, et al. Kayser-Fleischer-like rings in patients without Wilson's disease. *Gastroenterology.* 1977;72;1331-1335.

7. Wiebers DO, Hollenhorst RW, Goldstein NP. The ophthalmologic manifestations of Wilson's disease. *Mayo Clin Proc.* 1977;52:409-416.

8. Danks DM. Hereditary disorders of copper metabolism in Wilson's disease and Menkes' disease. In: Stanbury JB, Wyngaarden JB, Fredrickson DS, et al. ed. *The Metabolic Basis of Inherited Disease,* 5th ed. New York, NY: McGraw-Hill; 1983:1251-1268.

9. Frommer DJ. Defective biliary excretion of copper in Wilson's disease. *Gut.* 1974;15:125-129.

10. Liebergall GS. Eye in hepatolenticular degeneration. *Am J Ophthalmol.* 1963;55:1260-1263.

11. Schoenberger M, Ellis PP. Disappearance of Kayser-Fleischer rings after liver transplantation. *Arch Ophthalmol.* 1979;97:1914-1915.

12. Sternlieb I, Scheinberg IH: Penicillamine therapy for hepatolenticular degeneration. *JAMA.* 1964;189:146-152.

13. McCrary JA. Magnetic resonance imaging diagnosis of hepatolenticular degeneration. *Arch Ophthalmol.* 1987;105:277.

14. Curran RE, Hedges TR, Boger WP. Loss of accommodation and the near response in Wilson's disease. *J Pediatr Ophthalmol Strabismus.* 1982;19:157-160.

15. LaBrecque DR, Mitros FA, Nathan RJ, et al. Four generations of arteriohepatic dysplasia. *Hepatology.* 1982;4:467-474.

16. Romanchuk KG, Judisch GF, LaBrecque DR. Ocular findings in arteriohepatic dysplasia (Alagille's syndrome). *Can J Ophthalmol.* 1981;16:94-99.

Chapter 35

Inflammatory Bowel Disease

DAVID L. KNOX

Patients with inflammatory bowel disease may have ocular complications. They occur in approximately 10% of patients with Crohn's disease and less frequently and severely in patients with ulcerative colitis. In 1932, B. B. Crohn[1] first defined the disorder which bears his name as an entity distinct from ulcerative colitis. It has taken four decades for the subtle differences to be broadly appreciated by medical practitioners.[2] Even so, there are still areas of overlap, coincidence, confusion, and subgrouping of syndromes.

Systemic Manifestations

While most patients have symptoms—diarrhea, generalized malaise, systemic symptoms—Crohn's disease can present as an extraintestinal complication in a patient whose gut is entirely asymptomatic, but who is found by endoscopy, radiology, or biopsy to have classic findings. Causal mechanisms have not been convincingly demonstrated or clarified.[2] The histopathology of Crohn's disease emphasizes the findings of granulomas containing giants cells in focal ulcers anywhere in the gut from tongue to rectum. These ulcers tend to form fistulas between loops of gut and from gut to vagina,

bladder, skin, and retroperitoneum. Ulcerative colitis is more diffuse both on endoscopic and histologic examination, which finds acute and chronic inflammatory cells but no giant cell granulomas.

Treatment is difficult, requiring for specific complications the joint efforts of internists, gastroenterologists, general surgeons, immunologist-rheumatologists, and often other subspecialists. General maintenance, diet, vitamin replacement, Asulfidine, corticosteroids, metronidazole, immunosuppressive drugs, and surgical resection of affected organs or areas of involvement are all utilized. The ophthalmologist becomes a member of this team when a patient with inflammatory bowel disease develops an ocular complication or when an ocular problem leads to the diagnosis of the intestinal disorder.

Extraintestinal complications occur in almost every organ system: skin, joints, liver, neural tissue, and bone marrow and in organs affected by fistulae.

Ocular Manifestations

Ocular complications can be divided into primary, secondary, and coincidental complications.[3] Primary com-

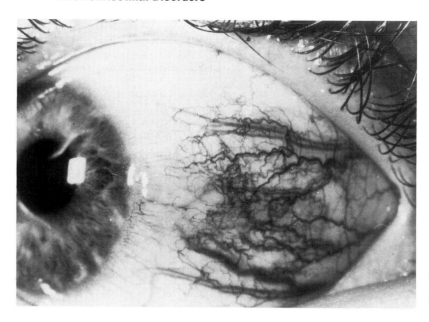

FIGURE 35-1 Large area of episcleritis appeared with reactivation of Crohn's disease.

plications are those that occur with increased activity of the gut disease and respond to treatment of the gut.

The primary complications, in order of frequency, are episcleritis, uveitis (acute iritis, chronic iridocyclitis, or panuveitis), and keratopathy (anterior stromal opacities or limbal infiltrates); macular edema, central serous retinopathy, and proptosis from orbital pseudotumor are seen less frequently. Most dramatic and urgent are optic neuropathies.[4-6] In one instance in my experience, a chiasmal syndrome responded to resumption of oral corticosteroid therapy. The development of episcleritis

in a patient thought to be in remission or in a tapering phase of oral corticosteroid therapy can be taken as a sign of increased activity of the gut disease (Figs. 35-1, 35-2). Biopsy of one of these lesions has shown giant cells.[7] It can also be used to differentiate Crohn's disease and ulcerative colitis, in which episcleritis has not been reported. Multiple gray densities of the anterior cornea is a specific morphologic entity that has been seen only in patients with Crohn's disease (Fig. 35-3).[4] Histopathologic and electron microscopic examination of a biopsied corneal lesion have provided no clues as to the

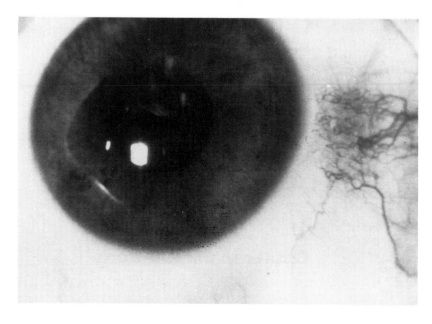

FIGURE 35-2 Small area of episcleritis was reactivated with increased activity of Crohn's disease.

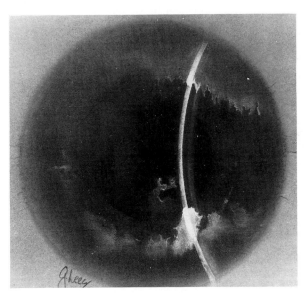

FIGURE 35-3 Anterior stromal keratopathy in active Crohn's disease.

disease mechanism.[8] In some patients chronic uveitis has responded to removal of the affected portion of the gut.

Secondary complications occur in the eye because of some primary complication such as malnutrition from poor diet or surgical loss of the absorbing section of the gut. Resulting vitamin A deficiency has produced decreased tear formation or night blindness. Retroperitoneal abscess has been responsible for exudative choroiditis.[3] Candida infection of intravenous lines for parenteral nutrition has produced endophthalmitis. Corticosteroids have produced cataracts, and scleritis has induced exudative inflammation in adjacent retina and optic nerve.

Coincidental complications are events that occur so commonly in perfectly healthy persons that they are not related to bowel disease, for example, conjunctivitis, recurrent corneal erosion or ulcer, glaucoma, and retinal artery narrowing.

Ulcerative colitis has been associated with ocular complications in a few instances. Chronic iridocyclitis was present in one young man who also had hepatitis as a complication of his ulcerative colitis. Persistent hepatitis has become an indication for total colectomy to prevent severe cirrhosis. Papillitis with edema and loss of vision occurred in another patient.

It is the author's opinion that inflammatory bowel disease, especially Crohn's disease, can join syphilis and sarcoidosis as one of the great imitators that produce a large variety of ocular and systemic manifestations.

References

1. Crohn BB, Ginzburg L, Oppenheimer GD. Regional ileitis; pathologic and clinical entity. *JAMA*. 1932;99:1323-1329.
2. Kirsner JB, Shorter RG. Recent development in "nonspecific" inflammatory bowel disease. *N Engl J Med*. 1982;306:775-785, 837-848.
3. Knox DL, Schachat AP, Mustonen E. Primary, secondary and coincidental ocular complications of Crohn's disease. *Ophthalmology*. 1984;91:163-173.
4. Knox DL, Snip RC, Stark WJ. The keratopathy of Crohn's disease. *Am J Ophthalmol*. 1980;90:862-865.
5. Weinstein JM, Koch K, Lane S. Orbital pseudotumor in Crohn's colitis. *Ann Ophthalmol*. 1984;16:275-278.
6. Sedwick LA, Klingele TG, Burde RM, et al. Optic neuritis in inflammatory bowel disease. *J Clin Neuro-ophthalmol*. 1984;4:3-6.
7. Blase WP, Knox DL, Green WR. Granulomatous conjunctivitis in a patient with Crohn's disease. *Br J Ophthalmol*. 1984;68:901-903.
8. van Vliet, AA, van Balen AT. Keratopathic dans la maladie de Crohn. *Ophthalmolgica Basel*. 1985;190:72-76.

Chapter 36

Pancreatitis

JOHN M. WILLIAMS and GUY E. O'GRADY

Pancreatitis is a condition in which autodigestion of the pancreas occurs. There are many causes; the most common include alcohol ingestion, gallstones, metabolic disorders, and certain drugs. There are acute and chronic varieties of pancreatitis, and they can be differentiated by their clinical course and by the extent of damage to the pancreas. Acute pancreatitis is a self-limited disorder, whereas chronic pancreatitis permanently damages the pancreas. In the United States, acute pancreatitis is most often caused by excessive alcohol ingestion, and gallstones are the next most frequent cause; in England the reverse is true.[1]

Normally, the pancreas produces enzyme precursors such as trypsinogen, chymotrypsinogen, and pro-elastase, which are activated after they are secreted into the duodenum. In pancreatitis, these enzymes are activated within the pancreas, resulting in edema, vascular damage, and necrosis of pancreatic and peripancreatic tissue. Proteolytic enzymes may also gain access to the circulation and the lymphatic system, causing tissue damage in organs distant from the original focus of necrosis and inflammation.[2] Other proteases such as complement, plasmin, and Hagemann's factor may be activated in pancreatitis.[3] Activation of the complement cascade results in formation of the anaphylatoxins C_{3a} and C_{5a}. These anaphylatoxins are chemotactic for leukocytes, increase capillary permeability, cause smooth muscle contraction, and provoke histamine release from mast cells.[2] C_{5a} has also been shown to cause the formation of granulocyte aggregates, which can occlude and damage small blood vessels.[4]

Systemic Manifestations

Abdominal pain is the most frequent symptom of acute pancreatitis. In addition, nausea, vomiting, and abdominal distension frequently occur. The patient may be tachycardic and hypotensive and may have a low-grade fever. Ten to 20% of patients may have pulmonary findings such as basilar rales, pleural effusion, and atelectasis.[1] The patient may also develop a pancreatic phlegmon, abscess, or pseudocyst. Diffuse intravascular coagulation, renal failure, and gastrointestinal hemor-

rhage can also occur. Central nervous system involvement can be manifested as acute psychosis.

Laboratory abnormalities in acute pancreatitis can include elevated serum amylase (>200 Somogyi units/ml) and lipase, leukocytosis, hyperglycemia, hypertriglyceridemia, hyperbilirubinemia, and hypoxemia. After 48 to 72 hours amylase values usually return to normal.[1]

Ocular Manifestations

Sudden visual loss has been reported in acute pancreatitis that is secondary to an ischemic retinopathy characterized by multiple cotton-wool spots and nerve fiber layer hemorrhages in the peripapillary retina of both eyes (Fig. 36-1, A). Fluorescein angiography reveals arteriolar and capillary nonperfusion as well as hypofluorescence in the areas of the cotton-wool spots. This fundus picture is identical to that seen in Purtscher's retinopathy, a disorder in which sudden vision loss occurs after nonocular trauma.[5] The retinopathy associated with pancreatitis is most often seen in young male patients with acute alcoholic pancreatitis and highly elevated amylase levels, and it is usually diagnosed soon after the onset of pancreatitis.[6] Typically, the patient complains of a sudden drop in visual acuity bilaterally accompanied by multiple scotomas in the central portion of the visual field. If the patient is suffering from delirium tremens or altered mental status (which can accompany acute alcoholic pancreatitis), it may be several days before complaints of visual loss are communicated to the physician. Visual loss can range from mild to profound, and the magnitude of the deficit is related to the degree of involvement of the macula and the papillomacular bundle by the ischemic process. Afferent pupillary defects can also occur. The recovery of vision is usually related to the severity of the initial visual deficit.[6] The cotton-wool spots and nerve fiber layer hemorrhages generally resolve within 2 to 8 months, leaving a fundus picture of optic atrophy, focal depressions in the retina where the cotton-wool spots were, attenuation, and sheathing of the retinal vessels, and mild mottling of the retinal pigment epithelium (Fig. 36-1, B).[7]

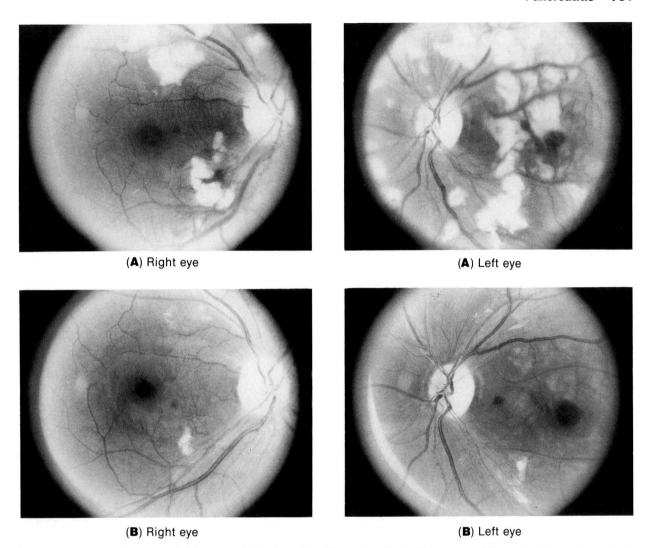

(A) Right eye   **(A)** Left eye

(B) Right eye   **(B)** Left eye

FIGURE 36-1 **(A)** Twenty-eight-year-old black male with acute alcoholic pancreatitis and retinopathy: visual acuity 20/70, OD; 20/200, OS. **(B)** Same patient 3 weeks later: visual acuity 20/20, OD; 20/70 OS.

Originally, it was thought that fat emboli were the cause of the retinopathy of pancreatitis.[8] Enzymatic destruction of omental fat was proposed as the initiating factor, which resulted in the formation of small fat emboli that lodged in the peripapillary capillaries of the retina.

Recently, activated complement, specifically the C_{5a} component, has been shown to cause granulocyte aggregation and leukoembolization in vivo.[4] Plasma samples of patients with acute pancreatitis have been shown to cause granulocyte aggregation in vitro.[9] The size of the granulocyte aggregates (up to 80 μm) is such that they can easily occlude vessels in the radial peripapillary capillary network (15–150 μm in diameter). This network is a series of end-vessels that supply the superficial retina in an area circumscribed by the optic disc and macula.[10] Presumably, blockage of these capillaries by granulocyte aggregates that degranulate and cause endothelial damage results in the ischemic retinopathy associated with pancreatitis.[9]

Histologic abnormalities have been reported in a case of retinopathy of pancreatitis.[11] Arteriolar occlusions were found in the retina and choroid; however, the long interval between the diagnosis of the disorder and the patient's death precluded identification of the material that caused the occlusions.

There is no known treatment for the retinopathy of pancreatitis. Steroids and protease inhibitors have been considered, but neither has been shown to be clinically effective.

Chronic pancreatitis can occasionally lead to diabetes mellitus by virtue of the islet cell destruction that occurs.

Rarely, diabetic retinopathy can be seen in these patients. It is followed and treated in the same manner as diabetic retinopathy in type I and type II diabetes mellitus.

Simple visual acuity testing and dilated fundus examination are important in patients with pancreatitis and vision complaints. Evaluation of the plasma of these patients with sensitive radioimmunoassay techniques may further delineate the role of C_{5a} in the retinopathy of pancreatitis.[6]

References

1. Greenberger NJ, Toskes PP, Isselbacher KJ. Diseases of the pancreas. In: Braunwald E, et al., eds. *Harrison's Principles of Internal Medicine.* 11th ed. New York, NY: McGraw-Hill; 1987:1372-1378.
2. Goldstein IM, Cala D, Radin A, et al. Evidence of complement catabolism in acute pancreatitis. *Am J Med Sci.* 1978;275:257.
3. Whicher JT, Barnes MP, Brown A, et al. Complement activation and complement control proteins in acute pancreatitis. *Gut.* 1982;23:944.
4. Hammerschmidt DE, Harris PD, Wayland JH, et al. Complement-induced granulocyte aggregation in vivo. *Am J Pathol.* 1981;102:146.
5. Purtscher O. Noch unbekannte Befunde nach Schädeltrauma. *Ber Versamm Dtsch Ophthalmol Ges.* 1910;36:294.
6. Williams JM, Gass JDM, O'Grady GE. Retinopathy of pancreatitis and the complement cascade. *Invest Ophthalmol Vis Sci.* 1986;27(suppl):105.
7. Snady-McCoy L, Morse PH. Retinopathy associated with acute pancreatitis. *Am J Ophthalmol.* 1985;100:246.
8. Inkeles DM, Walsh JB. Retinal fat emboli as a sequela to acute pancreatitis. *Am J Ophthalmol.* 1975;80:935.
9. Jacob HS, Goldstein IM, Shapiro I, et al. Sudden blindness in acute pancreatitis. *Arch Intern Med.* 1981;141:143.
10. Michaelson IC, Campbell ACP. Anatomy of the finer retinal vessels. *Trans Ophthalmol Soc UK.* 1940;60:71.
11. Kincaid MC, Green WR, Knox DL, et al. A clinicopathological case report of retinopathy of pancreatitis. *Br J Ophthalmol.* 1982;66:219.

Chapter 37

Whipple's Disease

CLEMENT L. TREMPE and MARCOS P. AVILA

Systemic Manifestations

Whipple's disease is a rare multisystem chronic granulomatous bacterial infectious disease that can be fatal if left untreated. Gastrointestinal tract, joints, heart, lung, muscles, brain, and eyes are affected to varying degrees. Steatorrhea (diarrhea associated with fat malabsorption) is often the principal feature of far advanced disease; however, early in the disease process, the gastrointestinal tract is not always significantly affected.[1-3] Other features of the disease contribute to the clinical picture in varying degrees, such as arthritis, abdominal pain, diarrhea, generalized lymphadenopathy, weight loss, fever, weakness, and skin changes (diffuse hyperpigmentation, subcutaneous nodules, erythema nodosum).[4-6] It is important for the ophthalmologist to know that vision problems are often part of the early disease process.

As a chronic multifocal granulomatous disorder, Whipple's disease can be very similar to sarcoidosis or to infection caused by *Mycobacterium avium* or *M. intracellulare*.[7,8] Para-aortic and hilar nodes are enlarged in 25% of patients with Whipple's disease.[10] On microscopy, the lymph nodes contain noncaseating granulomas that may be mistaken for a sarcoid lesion.[9,10] Sarcoid noncaseating granulomas are periodic acid–Schiff (PAS) negative, in contrast to the usual PAS–positive lesions of advanced Whipple's disease. Early in the Whipple's disease process, the noncaseating granulomas are often PAS negative, and at that stage of the disease the pathologic appearance can easily be confused with sarcoidosis.[11-14] Even electron microscopy will not demonstrate the characteristic bacilliform bodies of Whipple's disease in most early cases. A theoretical argument could be made for repeated biopsy and a thorough electron microscopic search for the classic Whipple's bacilliform bodies[15,16] in every case of biopsy-proven sarcoidosis. If this is not done, it is likely that cases of Whipple's disease will continue to be misdiagnosed as sarcoidosis. For practical reasons, it is not always possible to do extensive electron microscopy in every case of sarcoidosis, but the clinician should suspect Whipple's disease in any patient with sarcoidosis who shows deterioration, especially following treatment with prednisone and if gastrointestinal, central nervous system, or ocular complications occur. In such cases, a diagnostic therapeutic antibiotic trial can be done, because patients with Whipple's disease usually respond within a week of treatment whereas those with sarcoidosis do not.

M. avium and *M. intracellulare* infections of the gastrointestinal tract may be misdiagnosed as Whipple's disease, because they all have similar findings on duodenal biopsy. Large foamy macrophages are present in the lamina propria. These macrophages contain PAS–positive rodlike material, but by electron microscopy the bacillary bodies have a characteristic appearance different from those seen in Whipple's disease.[17] They are easily cultured, and acid-fast tissue stains (Fite acid fast and Ziehl-Neelsen stain) are very helpful in differentiating the two conditions, because they are strongly positive in cases of myobacterium infection but are negative in Whipple's disease.[7,8] On clinical grounds, atypical mycobacterial infection usually occurs in patients with immunodeficiency disorders while Whipple's disease does not.

Causes

There is imposing evidence that a single organism causes Whipple's disease because of the characteristic electron microscopic appearance of the bacillus in all cases and also because Whipple's bacilli from various patients give a characteristic indirect immunofluorescent cross-reactivity pattern with antisera to *Staphylococcus* groups B and G and to shigella B. Corynebacteria, also referred to as diphtheroids, are small, pleomorphic, gram-positive, non–spore-forming rods.[19] The *Corynebacterium anaerobium* species are the organisms that have most consistently been cultured from infected lymph nodes using proper anaerobic techniques and culture medium. *Propionibacterium acnes,* formerly called *C. parvum* or *C. acnes,* belongs to the class of anaerobic corynebacterium, and this bacteria or a variant form is most likely the causative organism for Whipple's disease.[20] This organism is a facultative intracellular parasite, and as with other enterobacteriaceae such as *Yersinia* species, it is possible that its resistance to phagocytic intracellular killing could be the result of cross infections by bacteriophages that exist among the different species of corynebacteria.[21] The indigestibility of its cell wall structure may also prevent its destruction by polymorphonuclear cells and macrophages.[22] The persistence of organisms within inflammatory cells causes macrophage activation and the granulomatous reaction seen in this type of infection.[23,24]

Another reason why the Whipple's bacillus does not seem to induce an effective immune response in the host appears to be a basic abnormality in the intracellular degradation process of monocytes and macrophages.[25] This intracellular degradation remained impaired 3 and 9 months after the start of therapy in one reported case of Whipple's disease.[25] Further investigation will be necessary to determine whether the disease process is due mainly to a resistant bacillus, or to a defect in monocytes and macrophages, or possibly to other contributing factors.

Because Whipple's disease can be caused by organisms similar to *P. acnes* and because so-called sterile endophthalmitis following intraocular surgery can be caused by the same type of organism, these two diseases may represent opposite ends of the spectrum of one disease process. Cases of "Whipple's disease" diagnosed by histologic examination of vitrectomy samples in patients with a negative jejunal biopsy could fall within this spectrum.[26,27]

Because of the increasing importance of vitrectomy in the diagnosis of intraocular inflammatory processes, it is necessary to learn the proper technique to obtain specimens for anaerobic culture and to make sure that the bacteriology laboratory obtains the specimens without delay and performs the cultures properly.[28]

The frequent recurrence of infection following vitrectomy and appropriate antibiotic treatment in cases of *P. acnes* endophthalmitis, and the recurrence of infection in Whipple's disease when antibiotics are stopped could be due to a similar pathophysiologic process. Acne of the skin is a similar chronic recurrent infection caused by the same or similar organisms. Chronic gum infection by anaerobic bacteria is another related problem that affects over 32 million American adults according to the American Academy of Periodontology.

The immune system usually tolerates the ubiquitous bacteria that are considered part of the normal flora of the skin and gastrointestinal tract. At the same time, the immune system has to recognize and destroy those organisms when they penetrate through defects in integumental or mucosal tissues. A breakdown in this delicate immunologic balance between tolerance and reaction is most likely one of the basic causes of chronic recurrent infections that can be so destructive and sometimes fatal.

Ocular Manifestations

The ocular signs and symptoms of Whipple's disease can be divided into three groups[29]:

Group 1: Ocular involvement secondary to CNS damage; ocular signs include ophthalmoplegia, nystagmus, gaze palsy, ptosis, and papilledema.

Group 2: CNS involvement concomitant with intraocular signs.

Group 3: Intraocular signs without apparent CNS involvement.

Ocular signs include keratitis, uveitis, inflammatory vitreous opacities, vitreous hemorrhage, retinal hemorrhage, and diffuse retinal and choroidal vasculitis (Fig. 37-1).[26-30]

CNS damage is the major cause of ocular manifestations. It has been reported in undiagnosed advanced cases and in diagnosed cases even after antibiotic treatment. Patients who exhibit only intraocular signs at the beginning of the disease may have fatal CNS involvement later; this progression may result from failure (1) to diagnose the disease in its early stages and (2) subsequently to institute antibiotic therapy.

Successful treatment of bacterial infections requires accurate bacteriologic diagnosis. Because the organism implicated in Whipple's disease is very difficult to culture, broad-spectrum antibiotics are used empirically (Fig. 37-2), most commonly tetracycline, erythromycin, ampicillin, or penicillin. In patients with intraocular involvement, antibiotics that achieve effective concentrations in tissue beyond the blood-ocular barrier should be used. Although chloramphenicol may have serious side effects on the hematopoietic system, it has better ocular penetration than other antibiotics commonly used in Whipple's disease. In patients with serious intra-ocular involvement, this drug should probably be used in the acute phase of the disease and another antibiotic should be used subsequently.

Because the possibility of recurrence is ever present, even after apparently successful treatment, it is mandatory to periodically evaluate the systemic and ocular condition of these patients.[29,30] Early signs of intraocular involvement should be regarded as a possible recurrence of the disease, and antibiotic therapy should be resumed.

The pathogenesis of the intraocular manifestations, as well as the systemic manifestations, is still unclear. An immune mechanism hypothesis proposed for Whipple's disease assumes that the ocular lesions are purely inflammatory and are due to a hypersensitivity phenomenon, the focus of the disease being located in the jejunum or elsewhere in the body.[31,32] A secondary hematogenous extension of the infection to the eye appears more likely, however, in view of the identification of the Whipple's bacillus from vitrectomy samples.[26,33]

References

1. Mansbach CM II, Shelburne JD, Stevens RD, et al. Lymph-node bacilliform bodies resembling those of Whipple's disease in a patient without intestinal involvement. *Ann Intern Med* 1978;89:64-66.

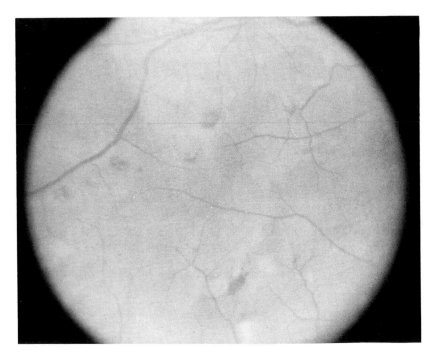

FIGURE 37-1 Left eye of a patient with Whipple's disease, before treatment, shows round, subretinal hemorrhages and small, round, grayish subretinal lesions temporal to macula.

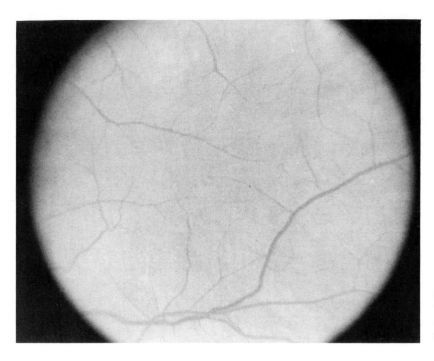

FIGURE 37-2 The same eye after treatment with antibiotic, shows resolution of hemorrhages and few residual, grayish, subretinal lesions.

2. Lopatin RN, Grossman ET, Horine J, et al. Whipple's disease in neighbors. *J Clin Gastroenterol.* 1982;4:223-226.

3. Maizel H, Ruffin JM, Dobbins WO III. Whipple's disease: a review of 19 patients from one hospital and a review of the literature since 1950. *Medicine.* 1970;49:175-205.

4. Comer G, Brandt L, Abissi, CJ. Whipple's disease: a review. *Gastroenterology.* 1983;78:107-14.

5. Hendrix JP, Black-Shaffer B, Withers RW. Whipple's intestinal lipodystrophy: report of 4 cases. *Arch Intern Med.* 1950;85:91-131.

6. Bienvenu P, Groussin P, Metman EH, et al. Maladie de Whipple avec localization cardiaques. *Ann Cardiol Angelol (Paris).* 1976;25:207-216.

7. Vincent ME, Robbins AH. Mycobacterium avium-intracellulare complex enteritis: pseudo-Whipple's disease in AIDS. *Am J Roentgenol.* 1985;144:921-922.

8. Roth R, Owen, RL, Keren DF, et al. Intestinal infection with mycobacterium avium in acquired immune deficiency syndrome (AIDS): histological and clinical comparison with Whipple's disease. *Digest Dis Sci.* 1985;30:497-504.

9. Wilcox GM, Tronic BS, Schecter DJ, et al. Periodic acid Schiff–negative granulomatous lymphadenopathy in patient with Whipple's disease. *Am J Med.* 1987;83:165-170.

10. Weiner SR, Utsinger P. Whipple's disease. *Semin Arthritis Rheum.* 1986;15:157-167.

11. Rodarte JR, Garrison CO, Holley KE, et al. Whipple's disease simulating sarcoidosis: a case with unique clinical and histologic features. *Arch Intern Med.* 1972; 129:479-482.

12. Torzillo PJ, Bignold L, Khan GA. Absence of periodic acid-Schiff-positive macrophages in hepatic and lymph node granulomata in Whipple's disease. *Aust NZ J Med* 1982;12:73-75.

13. Pequignot H, Morin Y, Grandjouan MS, et al. Sarcoidose et maladie de Whipple: association? relation? *Ann Med Interne* 1976;127:797-806.

14. Cho C, Linscheer WG, Hirschkorn MA, et al. Sarcoidlike granulomas as an early manifestation of Whipple's disease. *Gastroenterology.* 1984;87:941-947.

15. Dobbins WO, Kawiski H. Bacillary characteristics in Whipple's disease and electron microscopic study. *Gastroenterology.* 1981;80:1468-1475.

16. Chears WC, Ashworth CT. Electron microscopic study of the intestinal mucosa in Whipple's disease: demonstration of encapsulated bacilliform bodies in the lesion. *Gastroenterology.* 1961;41:129-138.

17. Gillin JS, Urmacher C, West R, et al. Disseminated mycobacterium avium-intracellulare infections in acquired immunodeficiency syndrome mimicking Whipple's disease. *Gastroenterology.* 1983;85:1187-1191.

18. Denholm RB, Mills PR, More IAR. Electron microscopy in the long-term follow-up of Whipple's disease. *Am J Surg Pathol.* 1981;5:507-516.

19. Barksdale L. *Corynebacterium diphtheriae* and its relatives. *Bacteriol Rev.* 1970;34:378.

20. Verdier M, Jaubert D, Sebald M, et al. Isolement de propionibacterium acnes a partir d'un ganglion cervical au cours d'une maladie de Whipple. *Gastroenterol Clin Biol.* 1987;11:524-526.

21. Carnes HR. Actions of bacteriophages obtained from corynebacterium diphtheriae on *C ulcerans* and *C ouis. Nature* 1968;217:1066.

22. Webster GF, Leyden JJ, Musson RA, et al. Susceptibility of proprionibacterium acnes to killing and degradation by human neutrophils and monocytes in vitro. *Infect Immun.* 1985;49:116.

23. Scott MT, Milas, L. The distribution and persistence in vivo of corynebacterium parvum in relation to its antitumor activity. *Cancer Res.* 1977;37:1673.

24. Pringle AT, Cummings CS. Relationship between cell wall synthesis in proprionibacterium acnes and ability to stimulate the reticuloendothelial system. *Infect Immun.* 1982;35:734.

25. Bjerknes R, Laerum OD, Odegaard S. Impaired bacterial degradation by monocytes and macrophages from a patient with treated Whipple's Disease. *Gastroenterology.* 1985;89:1139-1146.

26. Selsky EJ, Knox DL, Maumenee AE, et al. Ocular involvement in Whipple's Disease. *Retina.* 1984;4:103-106.

27. Margo CE, Pavan PR, Groden LR. Chronic vitritis with macrophagic inclusions. *Ophthalmology.* 1988;95:156–61.

28. Ormerod LD, Paton BG, Haaf J, et al. Anaerobic bacterial endophthalmitis. *Ophthalmology.* 1987;94:799-807.

29. Avila MP, Jalkh AE, Feldman E, et al. Manifestations of Whipple's disease in the posterior segment of the eye. *Arch Ophthalmol.* 1984;102:384-390.

30. Leland TM, Chambers JK: Ocular findings in Whipple's disease. *South Med J.* 1978;71:335-338.

31. Groll A, Valberg LS, Simon JB, et al: Immunological defect in Whipple's disease. *Gastroenterology.* 1972;63:943-950.

32. Keren DF, Weisburger WR, Yardley JH, et al. Whipple's disease: Demonstration by immunofluorescence of similar bacterial antigens in macrophages from three cases. *Johns Hopkins Med J.* 1976;139:51-59.

33. Font RL, Rao NA, Issarescu S, et al. Ocular involvement in Whipple's disease: Light and electron microscopic observations. *Arch Ophthalmol.* 1978;96:1421-1436.

PART 6

Hearing Disorders

Chapter 38

Cogan's Syndrome

L. MICHAEL COBO

Cogan's syndrome is a multisystem disease of young adults that is usually manifested as ocular inflammatory disease and vestibuloauditory dysfunction.[1,2] While ocular inflammatory disease secondary to Cogan's syndrome is generally limited and responsive to corticosteroids, permanent hearing loss often results, and early systemic immunosuppressive treatment is necessary if this complication is to be avoided.[3,4] Furthermore, a small subpopulation of patients with Cogan's syndrome develops an associated systemic necrotizing vasculitis commonly manifested as proximal aortitis and aortic insufficiency.[5,6]

The pathogenesis of Cogan's syndrome remains a matter of speculation. An infectious agent is suggested by the common temporal association of upper respiratory tract infection and onset of Cogan's syndrome. While no infectious agent has been identified, the immune-mediated events that result in Cogan's syndrome may result from sensitization during the initial infectious episode. The pathology of Cogan's syndrome demonstrates that the cornea, cochlea, and conjunctiva are infiltrated with activated T cells and macrophages, suggesting that the disease is immune mediated and responsive to immunosuppressive treatment.

Systemic Manifestations

Typical and atypical forms of Cogan's syndrome exist. The typical form consists of corneal inflammatory disease and iritis associated with Meniere's disease-like attacks of nausea, vomiting, tinnitus, vertigo, and progressive hearing loss. Vestibuloauditory symptoms may be present before or after the onset of keratitis, and on occasion the events may be separated by several months. If Cogan's syndrome is not treated, severe hearing loss occurs in 60% of patients. Prednisone treatment within 2 weeks, however, preserves hearing in 80% of patients.[4]

Atypical Cogan's syndrome consists of vestibuloauditory dysfunction associated with ocular or orbital inflammatory disease other than keratitis (eg, episcleritis, scleritis, posterior uveitis, papillitis, and orbital pseudotumor). This subcategory, comprising 30% of

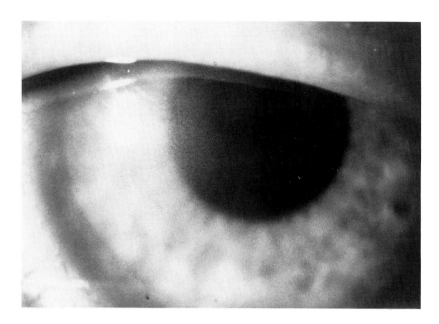

FIGURE 38-1 Patchy midstromal interstitial keratitis consistent with noncorticosteroid-treated interstitial keratitis in a patient with typical Cogan's syndrome.

reported cases, frequently overlaps with other rheumatic diseases including polyarteritis, Wegener's granulomatosis, and sarcoidosis as well as Vogt-Koyonagi-Harada disease and chlamydial infection. Patients with atypical Cogan's syndrome are more likely to develop severe visual loss (27%) and systemic vasculitis (21%).[2]

Systemic vasculitis associated with Cogan's syndrome affects larger vessels and is manifested as proximal aortitis, aortic insufficiency, and ischemic disease of other organ systems.[2,5,6] While it occurs in only 10% of Cogan's syndrome patients, this complication warrants continued observation by a physician who is cognizant

of the varied manifestations of this rare disease. Prolonged systemic immunosuppression of this life-threatening complication is generally necessary and may require prednisone, cyclophosphamide, or cyclosporin A.

Ocular Manifestations

The classic ocular manifestation of Cogan's syndrome is nonsyphylitic interstitial keratitis consisting of patchy midstromal infiltration of the corneal stroma, accompanied by corneal vascularization in the later stages

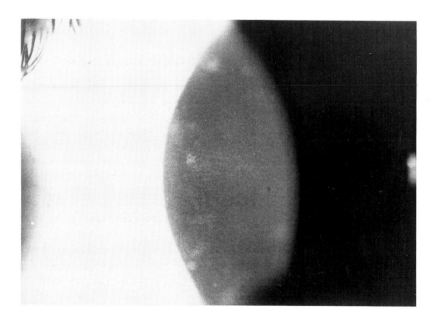

FIGURE 38-2 Faint subepithelial infiltrates in a patient treated with topical corticosteroids during the acute phase of the disease. This patient ultimately developed atypical Cogan's syndrome with posterior scleritis and systemic necrotizing vasculitis.

of the disease (Fig. 38-1).[1] In general, the keratitis is responsive to topical corticosteroids. A recent report indicates that nummular anterior stromal keratitis may be a more common corneal manifestation of Cogan's syndrome and may lead to misdiagnosis as viral keratitis (Fig. 38-2).[3] It may be that this is an early manifestation of interstitial keratitis that is suppressed by topical corticosteroids, halting progression to the classic interstitial keratitis described by Cogan in patients seen months after the onset of the disease in an era when corticosteroids were unavailable.

Other ocular inflammatory lesions seen in Cogan's syndrome include episcleritis, scleritis, uveitis, retinochoroiditis, papillitis, and orbital pseudotumor. These are generally seen in patients with atypical Cogan's syndrome, are more likely to result in visual loss, and often require chronic systemic immunosuppression.[2]

Early diagnosis of Cogan's syndrome is critical if hearing acuity is to be preserved. It is incumbent on the ophthalmologist to recognize and associate the ocular manifestations of this protean and rare disease with concurrent vestibuloauditory dysfunction, thus permitting appropriate early treatment.

References

1. Cogan DG. Syndrome of non-syphilitic interstitial keratitis with vestibuloauditory symptoms. *Arch Ophthamol.* 1945;33:144.
2. Haynes BF, Kaiser-Kupfer MI, Mason P, et al. Cogan's syndrome: studies in thirteen patients, long-term followup, and a review of the literature. *Medicine.* 1980;59:426.
3. Cobo LM, Haynes BF. Early corneal findings in Cogan's syndrome. *Ophthalmology.* 1984;91:903.
4. Haynes BF, Pikus A, Kaiser-Kupfer M, et al. Successful treatment of sudden hearing loss in Cogan's syndrome with corticosteroids. *Arthritis Rheum.* 1981;24:501.
5. Cheson BD, Bluming AZ, Alroy J. Cogan's syndrome: a systemic vasculitis. *Am J Med.* 1976;60:549.
6. Cogan DG, Dickerson GR. Non-syphilitic interstitial keratitis with vestibuloauditory syndrome: a case with fatal aortitis. *Arch Ophthalmol.* 1964;71:172.

Chapter 39

Norrie's Disease

RUTH M. LIBERFARB

Norrie's disease, congenital progressive oculoacousticocerebral dysplasia, characterized by retinal malformation, deafness, and mental retardation or deterioration, was first described by Norrie[1] in 1927 and then in more detail by Warburg.[2-5] A recent literature review revealed about 300 reported cases, initially only from Scandinavia but subsequently from various other countries and in all races.[6]

Warburg established that Norrie's disease has an X-linked recessive pattern of inheritance; it affects only males and is carried by completely unaffected females. The disease shows itself in affected males as congenital or early childhood blindness; progressive sensorineural hearing loss and mental retardation or deterioration may appear later. The gene penetrance is complete, but affected males exhibit variable expressivity. All affected males have ocular changes, which may include detached retina, vitreoretinal hemorrhage, retrolental mass, cataract, glaucoma, optic nerve atrophy, choroidal hypercellularity, and phthisis bulbi. The dramatic ocular changes occurring in infancy bring affected males to clinical attention.

Given the ocular changes, differentiating Norrie's disease from other intraocular tumorlike masses is often difficult. The differential diagnosis includes retinopathy of prematurity (retrolental fibroplasia), bilateral retinoblastoma, intrauterine infection, Lowe's oculocerebrorenal syndrome, X-linked retinoschisis, persistent hyperplastic primary vitreous, and Coats' disease. The findings of a significant family history with only affected males, progressive sensorineural hearing loss, mental retardation, and mental deterioration support the diagnosis of Norrie's disease.

Systemic Manifestations
Hearing Loss

Progressive sensorineural hearing loss has been observed in one fourth to one third of the patients studied, with the onset in the second or third decade of life.[4,5]

Electrophysiologic analysis of hearing loss indicated a purely cochlear lesion[7]; however, auditory response audiometry demonstrated a pattern consistent with both cochlear and retrocochlear disease (as seen with presbycusis) in one 18-year-old patient.[8] Temporal bone histopathology from another patient who died at age 78 years correlated with both cochlear and retrocochlear disease.[8] The cochlea is seen as the primary site of degeneration, and neuronal loss is secondary to the cochlear pathology.[8] Temporal bone histopathology showed severe degenerative changes involving the cochlea. There was nearly total loss of dendritic fibers, but axonal fibers were preserved, indicating that neuronal degeneration was secondary to neuronal degeneration of the organ of Corti. The mild-to-moderate fibrosis of the auditory nerves with many myelinated axons remaining suggested that degeneration of ganglion cells occurred later in life.

Central Nervous System Effects

Mild to severe mental retardation has been observed in two thirds of the patients, and progressive mental deterioration and psychotic symptoms appearing at various ages have been noted in some cases.[4,5]

There have been only a few reports of histopathologic changes in the brains of patients with Norrie's disease. In one case with severe mental retardation, brain examination showed atrophy of the visual pathways and lateral geniculate bodies, incomplete stratification of the cortex, and abnormalities of the mesencephalon and pons.[4] The brain of a severely retarded affected male who died at age 77 years showed densely gliotic optic nerves, small lateral geniculate bodies with anomalous lamination, thin visual cortex with atrophy of pyramidal cells, and fibrosis of the auditory nerves.[9] The histologic appearance of the lateral geniculate nucleus suggested that it was deprived of retinal afferents for a protracted period beginning before lamination was complete (i.e., before 25 weeks' gestation) and extending into the later stages of gestation. There were no specific neuropathologic changes that could explain the mental retardation.

Ocular Manifestations

The clinical course of Norrie's disease begins with multiple and recurrent vitreous hemorrhages.[10] Dilated iris vessels develop later, and some investigators have noted iris neovascularization.[11] Organization of vitreous hemorrhage leads to vitreous and retinal detachment; severe rubeosis iridis results in angle closure and glaucoma; other secondary changes in the retina, choroid, and ciliary body lead to phthisis.[12]

Histologic studies of enucleated globes from affected males with Norrie's disease showed the following findings: marked hypoplasia and dysplasia of retinal elements; the poorly differentiated dysplastic retina thrown into folds and entirely detached from the underlying pigment epithelium, and primitive neurons arranged in folds or rosettes; widespread intraretinal and preretinal gliosis; organized vitreous hemorrhage; formation of granulation tissue; persistent hyperplastic primary vitreous with fibrovascular membranes; marked hyperplasia of the ciliary and retinal pigment epithelium; intraocular osseous metaplasia; extensive iris and preretinal vascularization; cataract; glaucoma; band keratopathy; and phthisis bulbi.[11,12]

On the basis of frequently observed histologic ocular changes, Warburg[4] postulated that the basic lesion is a genetically determined biochemical defect that causes a primary arrest in retinal development with associated malformations of the primary and secondary vitreous. Neuroectodermal changes may also occur in the auditory nerve and cerebral cortex, accounting for the deafness and mental abnormalities. Apple et al.[11] postulated that retention and hyperplasia of preretinal ectodermal and mesodermal components of the primary vitreous seem to be responsible for the characteristic preretinal membranes, secondary hemorrhage, and total retinal detachment.

Management

Ocular surgery in Norrie's disease has consisted of extracapsular cataract extraction or enucleation for painful, blind, phthisical eyes. Lens and vitreoretinal surgery has been performed on three eyes in two reported cases.[6,10] Although surgery was unsuccessful in preventing blindness, two of the three eyes had not become phthisical during 2 years of follow-up of one patient[10] and 3 years in the second case.

Progressive sensorineural hearing loss with onset in the second and third decades may occur in Norrie's disease. Although initial audiometry may be normal, serial audiometric surveillance is imperative, especially since the blind patient becomes more dependent on the sense of hearing. Because hearing loss is secondary to cochlear damage, affected patients are potential implant candidates.[8]

Since there is no cure available for Norrie's disease, the best management approach is early diagnosis in a family and genetic counseling for known carriers and those at risk to be carriers.

Genetic Transmission

In the past few years, human gene probes have been generated that map close enough to several disease loci to make them useful for diagnostic purposes, carrier detection, and prenatal diagnosis. Gal et al.[13] found that the gene causing Norrie's disease is closely linked to the DXS7 locus on band 11 of the short arm of the X chromosome using the Taq I restriction fragment length polymorphism (RLFP) L1.28. They found a small deletion involving DXS7 in an affected male using L1.28/Taq I. De la Chapelle et al.[14] also found a small deletion of DXS7 in four members of a European kindred with Norrie's disease. Using probe L1.28/Taq I in the study of a chorion villus sample from a male fetus whose mother had the deletion, the fetus was found to be unaffected.[14]

That deletions are probably not a common mechanism in Norrie's disease is shown by the fact that DXS7 was not deleted in other families studied.[15,16] It is hoped that as additional RFLPs are found in the region of the locus for Norrie's disease, carrier detection and prenatal diagnosis will become available for all families of affected males.

References

1. Norrie G. Causes of blindness in children; twenty-five years' experience of Danish institutes for the blind. *Acta Ophthalmol.* 1927;5:357-386.
2. Warburg M. Norrie's disease; a new hereditary bilateral pseudotumour of the retina. *Acta Ophthalmol.* 1961;39:757-772.
3. Warburg M. Norrie's disease (atrofia bulborum hereditaria): a report of eleven cases of hereditary bilateral pseudotumour of the retina, complicated by deafness and mental deficiency. *Acta Ophthalmol.* 1963;41:134-146.
4. Warburg M. Norrie's disease; a congenital progressive oculo-acoustico-cerebral degeneration. *Acta Ophthalmol.* 1966; Supp. 89:1-147.
5. Warburg, M. Norrie's disease; differential diagnosis and treatment. *Acta Ophthalmol.* 1975;53:217-236.
6. Liberfarb RM, Eavey RD, De Long GR, et al. Norrie's disease: A study of two families. *Ophthalmology.* 1985;92:1445-1451.
7. Parving A, Elberhing C, Warburg M. Electrophysiological study of Norrie's disease. *Audiology.* 1978;17:293-298.
8. Nadol JB, Jr. Personal communication.
9. Williams R. Personal communication.
10. Jacklin HN. Falciform fold, retinal detachment, and Norrie's disease. *Am J Ophthalmol.* 1980;90:76-80.
11. Apple DJ, Fishman GA, Goldberg MF. Ocular histopathology of Norrie's disease. *Am J Ophthalmol.* 1974;78:196-203.
12. Blodi FC, Hunter WS. Norrie's disease in North America. *Doc Ophthalmol.* 1969;26:434-450.
13. Gal A, Bleeker-Wagemakers LM, Wienker TF, et al. Identification of the gene for Norrie's disease by linkage to the DXS7 locus. *Cytogenet Cell Genet.* 1985;40:633.
14. de la Chapelle A, Sankila EM, Lindlof M, et al. Norrie disease caused by a gene deletion allowing carrier detection and prenatal diagnosis. *Clin Genet.* 1985;28:317-320.
15. Bleeker-Wagemakers LM, Friedrich U, Gal A, et al. Close linkage between Norrie disease, a cloned DNA sequence from the proximal short arm, and the centromere of the X chromosome. *Hum Genet.* 1985;71:211-214.
16. Sims KB, Ozelius L, Corey T, et al. Norrie disease gene is distinct from the monoamine oxidase gene. *Am J Hum Genet.* (In press.)

Chapter 40

Usher's Syndrome

DAVID A. NEWSOME

Profound deafness present from birth and loss of vision due to a pigmentary retinal degeneration were linked clinically over a century ago by Liebreich.[1] The familial occurrence of deafness and pigmentary retinopathy had been described shortly before, in 1858, by von Graefe,[2] but the eponym was applied to this dual sensory loss after the 1914 publication of Usher's[3] report.

Minimal Diagnostic Criteria

The earliest clinical manifestation of Usher's syndrome is profound sensorineural hearing loss, either present from birth or noticed very shortly thereafter. This hear-

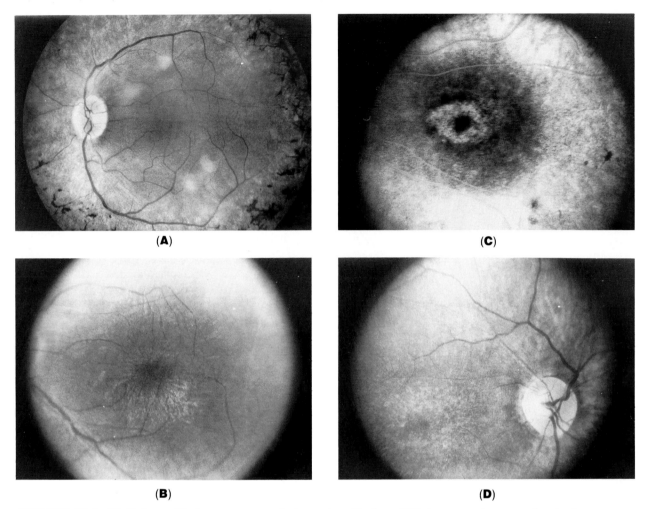

(A)

(C)

(B)

(D)

FIGURE 40-1 **(A)** Note mottled appearance of pigment epithelium 360 degrees, absence of spicules superonasally, normal optic nerve color. **(B)** Prominent epiretinal surface wrinkling changes. **(C)** Fluorescein angiography emphasizes the bull's-eye pigment epithelial window defect and widespread depigmentation peripheral to the macular area. **(D)** Both eyes of this 7-year-old male, night blind from age 2 years with 20/400 acuity OU and severe congenital sensorineural hearing loss, had similar diffuse mottling of the macular pigment epithelium. Peripheries showed widespread depigmentation.

ing loss is generally said to be nonprogressive, although progression has been described. Many experts also classify persons who have a somewhat milder but still severe sensorineural hearing loss that produces the speech pattern typical of persons with prelingual deafness as having Usher's disease.[4] This hearing loss definitely may progress. Associated with the sensorineural hearing loss is a pigmentary retinopathy that is generally indistinguishable from retinitis pigmentosa.[5] The fundus picture varies from generalized depigmentation and mottling of the pigment epithelium without bone spicule formation to heavy spiculation (Fig. 40-1, A, B). The rate of vision loss is variable but appears to be somewhat more rapid in persons with earlier onset of symptomatic night blindness.

The electroretinogram is the most useful electrodiagnostic test and is usually nonrecordable. Several other genetic syndromes, such as albinism with deaf mutism, and nonheritable conditions, such as rubella embryopathy, must be ruled out (Table 40-1). The minimal diagnostic criteria are presented in Table 40-2.

Genetics

Usher's syndrome is an autosomal recessive condition. The penetrance of the genes appears to be 100% in homozygous persons. Offspring from unions in which only one parent is affected are usually normal.[6,7]

The carrier state of Usher's has been said to be manifested in slight reductions in higher-frequency

TABLE 40-1 Differential Diagnosis of Usher's Syndrome

Syndrome	Inheritance	Comment
Usher's	Autosomal recessive	No solid clinical evidence for dominant or X-linked inheritance
Alport's[18,19]	Autosomal dominant, recessive, or X-linked	Chronic glomerulonephritis, white dots in retina, anterior lenticonus
Alström's[20,21]	Autosomal recessive	Diabetes mellitus, cataract, optic nerve drusen, optic atrophy, acanthosis nigricans, hypogonadism
Cockayne's[22,23]	Autosomal recessive	Dwarfism, "fingerprint figures" in lymphocytes by electron microscopy
Diallinas-Amalric[24]	(?) Autosomal recessive	Normal vision, stable maculopathy
Enamel dysplasia and deafness[25]	(?) Probable autosomal recessive	May have maculopathy and progressive vision loss
Refsum's[26-29]	Autosomal recessive	Elevated serum phytanic acid, polyneuropathy, ataxia
Rubella embryopathy[30,31]	—	Salt-and-pepper fundus; ERG may be normal
Waardenburg's[32,33]	Autosomal dominant	Hypertelorism, heterochromia, poliosis

hearing of a sensorineural type and in some alteration in the amplitude-latency functions of the electroretinogram.[7-10] Although the expression of the disease in terms of degree of hearing loss and natural history of the pigmentary retinopathy can be similar within a sibship, a considerable amount of phenotypic variation within a family is not rare.[11]

Demographics and Epidemiology

Usher's syndrome is rare. In the United States the incidence is estimated to be about 3 per 100,000 population.[8]

TABLE 40-2 Minimal Diagnostic Criteria in Usher's Syndrome

Criterion	Comment
Congenital sensorineural deafness	*Two principal types:* 1. Profound deafness to all frequencies, present from birth 2. Severe, high frequencies more affected, some speech discrimination occasionally, may be progressive; may become apparent after 1 to 2 years of age
Progressive pigmentary retinopathy (clinically indistinguishable from retinitis pigmentosa)	*Two modes of onset* of symptomatic night blindness: 1. About age 5 or younger 2. About age 15 to 20 *Electroretinogram:* Usually nonrecordable *Visual field:* Progressive peripheral loss with retention of central "tunnels" until late, except in atypical cases with macular involvement
Associated findings	*Neuropsychiatric disorders,* ataxia in some patients, decreased plasma polyunsaturated fatty acids, absent caloric responses in some patients

Natural History and Visual Function

The typical patient with Usher's syndrome has profound congenital sensorineural deafness and cannot discriminate audiometric tones, even at the 80- to 100-dB level. Some persons have a milder type of sensorineural hearing loss, which has been variously reported as stable or progressive.[12]

The earliest visual symptom in Usher's patients is night blindness. There appear to be two peak ages of onset of symptomatic night blindness: prior to about age 5 years, and between 15 and 20 years. In either case, the loss of vision is progressive, and mobility becomes a problem with constriction of the visual fields. Some persons with Usher's syndrome lose visual field quite rapidly and become totally blind; others appear to progress more slowly and can retain at least some vision into the 6th decade.[13]

Visual acuity in early to moderately advanced Usher's patients ranges from about 20/15 to 20/50 and may be good even with a bull's-eye maculopathy (Fig. 40-1, C). As with retinitis pigmentosa, some persons have atypical early maculopathy and consequently show much greater vision loss in the presence of relatively full peripheral fields (Fig. 40-1, D). Other causes of early visual loss include cystoid macular edema and cataract, usually of the posterior subcapsular type.

Ocular Manifestations
Anterior Segment

Findings in the anterior segment with split lamp biomicroscopy include lens changes, usually of the posterior subcapsular type. They are usually but not always bilateral and are seen up to 50% of the time in Usher's patients over about age 30. The vitreous, even in young persons with Usher's, is detached posteriorly with or without collapse and syneresis, and with some cells and condensations.

Posterior Segment

Depending on the age of the patient and the state of the disease, optic nerve color can range from normal to pale (see Fig. 40-1,A,D). Mild to exuberant optic nerve drusen can be seen in 5 to 10% of patients with Usher's syndrome. Some degree of vascular attenuation related to the degree of retinal degeneration is present. The foveal reflex is lost early, and macular epiretinal membrane changes are present in almost all eyes (see Fig. 40-1,B). There is widespread depigmentation of the retinal pigment epithelium with relative sparing early on of the macular pigment epithelium of most patients (Fig. 40-1,A). Bull's-eye lesions of varying degrees of subtlety can be present, and they become more prominent with age (see Fig. 40-1,C). They are compatible with relatively good vision. Some patients have severe atrophy of the macular pigment epithelium early on, and in such patients visual acuity is usually 20/100 or worse (see Fig. 40-1,D). Cystoid or diffuse macular edema can be present with varying degrees of reduced vision. Bone spicule formation and perivascular intraretinal pigment migration are variable and increase with time. Intraretinal or subretinal white spots and a glistening appearance at the level of the pigment epithelium are common.

Pathology

No clinicopathologic studies of ocular tissues from Usher's syndrome patients have been published.

Associated Systemic Manifestations

Published reports have linked vestibular ataxia, psychosis, often of the paranoid type, and mental retardation with Usher's syndrome.[8,14] The prevalence of these associated findings in groups of Usher's patients is highly variable, depending on the report. Cerebral atrophy has also been reported recently with Usher's syndrome.[15]

Stimulated by parallels between hair cells of the organ of Corti, which are dysfunctional in Usher's, and the photoreceptors with their connecting cilia, investigators found that males with Usher's syndrome have an increased prevalence of abnormal sperm, as judged by reduced motility and abnormal sperm tails, both grossly and electron microscopically.[16] In other studies, the plasma phospholipid content of the polyunsaturated fatty acids docosahexaenoate and arachidonate were significantly reduced in plasma from Usher's patients.[17] The clinical significance of these observations is unknown at present.

References

1. Liebreich R. Abkunft aus Ehen unter Blutsverwandten als Grund von Retinitis Pigmentosa. *Deutsche Klin.* 1861;13: 53-55.
2. Von Graefe A. Exceptionelles Verhalten des Gesichtsfeldes bei Pigmententartung der Netzhaut. *Arch Ophthalmol.* 1858;4:250-253.
3. Usher C. On the inheritance of retinitis pigmentosa, with notes of cases. *R Lond Ophthalmol Hosp Rep.* 1914;19:130-256.
4. Morgan A, Boulud B. Evolutivité de syndrome d'Usher. *J Fr Otorhinolaryngol.* 1975;24:614-620.

5. Newsome DA. Retinitis pigmentosa, Usher's syndrome, and other pigmentary retinopathies. In: Newsome DA, ed. *Retinal Dystrophies and Degenerations*. New York, NY: Raven Press; 1987;161-194.

6. Kloepfer HW, Laguite JK, McLaurin JW. The hereditary syndrome of congenital deafness and retinitis pigmentosa (Usher's Syndrome). *Laryngoscope*. 1966;76:850-862.

7. McLeod AC, McConnell FE, Sweeney A, et al. Clinical variations in Usher's syndrome. *Arch Otolaryngol*. 1971;94:321-334.

8. Vernon M. Usher's syndrome: deafness and progressive blindness. *J Chron Dis*. 1969;22:133-151.

9. Holland GM, Cambie E, Kloepfer W. An evaluation of genetic carriers of Usher's syndrome. *Am J Ophthalmol*. 1972;74:940-947.

10. Altshuler KZ. Character traits and depressive symptoms in the deaf. In: Wortis J, ed. *Recent Advances in Biological Psychiatry*. vol VI. New York, NY: Plenum Press; 1964.

11. Fishman GA. Usher's syndrome: visual loss and variations in clinical expressivity. *Perspect Ophthalmol*. 1979;3:97-103.

12. Haas EBH, Van Lith GHM, Rijnders J, et al. Usher's syndrome with special reference to heterozygous manifestations. *Doc Ophthalmol*. 1970;28:167-190.

13. Fishman G, Vasquez V, Fishman M, et al. Visual loss and foveal lesions in Usher's syndrome. *Br J Ophthalmol*. 1979;63:484-488.

14. Mangstich M, et al. Atypical psychosis in Usher's syndrome. *Psychosomatics*. 1983;24:674-675.

15. Bloom TD, Fishman MD, Mafee MF. Usher's syndrome: CNS defects determined by computed tomography. *Retina*. 1983;3:108-113.

16. Hunter DG, Fishman GA, Mehta RS, et al. Abnormal sperm and photoreceptor axonemes in Usher's syndrome. *Am J Ophthalmol*. 1986;104:385-389.

17. Bazan NG, Scott BL, Reddy TS, et al. Decreased content of docosahexaenoate and arachidonate in plasma phospholipids in Usher's syndrome. *Biochem Biophys Res Comm*. 1986;141:600-604.

18. Sohar E. Renal disease, inner ear deafness and ocular changes. *Arch Intern Med*. 1956;97:627-630.

19. Gubler M, Levy M, Broyer M, et al. Alport's syndrome: a report of 58 cases and a review of the literature. *Am J Med*. 1981;70:493-505.

20. Sebag J, Albert DM, Craft JL. The Alstrom syndrome: ophthalmic histopathology and retinal ultrastructure. *Br J Ophthalmol*. 1984;68:494-501.

21. Millay RH, Welcher RG, Heckenlively JR. Ophthalmologic and systemic manifestations of Alstrom's disease. *Am J Ophthalmol*. 1986;102:482-490.

22. Pearce WG. Ocular and genetic features of Cockayne's syndrome. *Can J Ophthalmol*. 1972;7:435-444.

23. Proops R, Taylor AMR, Insley J. A clinical study of a family with Cockayne's syndrome. *J Med Genet*. 1981;18:288-293.

24. Diallinas NP. Les altérations oculaires chez les sourds-muets. *J Genet Hum*. 1959;8:225-262.

25. Bateman JB, Riedner ED, Levin LS, et al. Heterogeneity of retinal degeneration and hearing impairment syndromes. *Am J Ophthalmol*. 1980;90:755-767.

26. Refsum S, Salomonsen L, Skatvedt M. Heredopathia atactica polyneuritiformis in children. *J Pediatr*. 1949;35:335.

27. Refsum S. Heredopathia atactica polyneuritiformis phytanic-acid storage disease, Refsum's disease: a biochemically well-defined disease with a specific dietary treatment. *Arch Neurol*. 1981;38:605-606.

28. Stokke O, Skrede S, Ed J, et al. Refsum's disease, adrenoleuco-dystrophy, and the Zellweger syndrome. *Scan J Clin Lab Invest*. 1984;44:463-464.

29. Goldman JM, Clemens ME, Gibberd FB, et al. Screening of patients with retinitis pigmentosa for heredopathia atactica polyneuritiformis (Refsum's disease). *Br Med J*. 1985; 290:1109-1110.

30. Amalric MP. Nouveau type de dégénérescence tapéto-rétinienne au cours de la surdimutité. *Bull Soc Ophthalmol Fr*. 1960;73:196-212.

31. Newsome DA. Pigment epithelial dystrophies. In: Newsome DA, ed. *Retinal Dystrophies and Degenerations*. New York, NY: Raven Press; 1987;195-231.

32. Waardenburg PJ. A new syndrome, combining development anomalies of the eyelids, eyebrows and nose root with pigmentary defects of the iris and head and with congenital defects. *Am J Hum Genet*. 1951;3:195.

33. Goldberg MF. Waardenburg's syndrome with fundus and other anomalies. *Arch Ophthalmol*. 1966;76:797-810.

PART 7

Hematologic Disorders

Chapter 41

Anemia

GARY C. BROWN and RICHARD E. GOLDBERG

Anemia can be defined as a decrease in the circulating red blood cell mass. In adult males a hematocrit value of 40% (hemoglobin, 14%) is the lower limit of normal, whereas in adult females a hematocrit of 37% (hemoglobin, 12%) indicates anemia.[1] The abnormality may also be accompanied by thrombocytopenia, a reduction in the number of circulating platelets.

The causes of anemia are legion, although several broad classes can be cited, among them hypoproliferative anemia due to deficient red blood cell production, anemia secondary to blood loss, and hemolytic anemias.[1] Encompassed within the first category are anemias secondary to iron deficiency, chronic disease, leukemia, thalassemia, infection and to numerous other causes.

Although conjunctival pallor or hemorrhage and cranial nerve palsies may be seen with varied forms of anemia,[2] the ocular signs most commonly are manifested in the fundus, particularly in the posterior pole. Major ophthalmoscopic features include streak hemorrhages in the superficial retina, dot and blot hemorrhages in the deeper retina, cotton-wool spots, and hard exudates (Figs. 41-1, 41-2).[3-5] In some instances the hemorrhages are white centered.[6] Occasionally, blood can accumulate under the internal limiting membrane of the retina or break through into the subhyaloid space and vitreous cavity.[3,6] These ocular hemorrhages have their systemic counterparts in the cutaneous and mucous membrane hemorrhages found in some anemic patients.

The retinal veins are occasionally dilated, and in some cases tortuous.[7] As an extreme, the fundus may even develop the clinical features of central retinal vein obstruction.[8-10] Retinal vascular endothelial incompetence due to anemic hypoxia is believed to be the cause of this ophthalmoscopic picture.[8] Reversal of the anemia has been demonstrated to ameliorate both visual impairment and retinopathy in severely anemic persons with the clinical appearance of central retinal vein obstruction.[8,9]

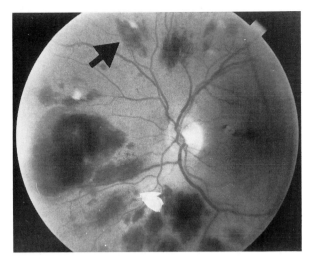

FIGURE 41-1 Anemic retinopathy. Multiple intraretinal hemorrhages and cotton-wool spots are visible. A white-centered hemorrhage (Roth's spot), bordered superiorly by a cotton-wool spot, is indicated by the arrow. A deposition of fibrin has been shown histopathologically to correlate with the white center of the hemorrhage.[6]

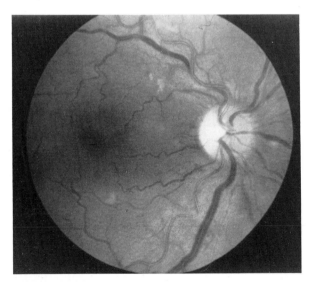

FIGURE 41-2 Right fundus of a 52-year-old woman with severe anemia due to massive gastrointestinal bleeding. The retinal veins are dilated, and several cotton-wool spots are present.

Evidence suggests that severe anemia may exacerbate diabetic retinopathy. Shorb[11] reported three patients with mild to moderate background diabetic retinopathy who developed marked iron deficiency anemia and then rapidly progressed to severe proliferative retinopathy.

Retinopathy due to anemia probably does not occur unless the hemoglobin titer is less than 8 g/100 ml.[3,4] Furthermore, either severe anemia (hemoglobin <8 g/100 ml) or thrombocytopenia (<50,000/mm^3) alone does not usually cause retinopathy. Rubenstein and associates[3] found that only 10% of persons with severe anemia alone had retinal hemorrhages, whereas 70% developed hemorrhages when both anemia and thrombocytopenia were present. None of their six thrombocytopenic patients without anemia demonstrated any retinopathy. They observed that anoxia alone (anemia) usually does not cause retinal hemorrhage as long as sufficient platelets are present to ensure the integrity of the capillary endothelium.

The incidence of retinopathy increases with the profundity of the anemia. Merin and Freund[4] found a 27% incidence of retinopathy in persons with a hemoglobin level below 8 g/100 ml, but each of their three patients with a level below 3 g/100 ml demonstrated retinal changes.

Children appear to be more resistant to the development of anemic retinopathy than adults. When Merin and Freund[4] divided their anemic patients with hemoglobin levels below 8 g/100 ml by age, they noted no cases of retinopathy among the 36 patients under 15 years of age, but a 46% incidence in the 53 who were 15 or older. It has been proposed that aging of the retinal blood vessels or prolonged exposure of the adult vessels to the deleterious effects of the anemia may account for this phenomenon.

The visual acuity may be decreased in anemic retinopathy owing to retinal and preretinal bleeding, or to macular edema in those unusual cases in which a central retinal vein picture develops. Therapy is directed toward correcting the underlying systemic problem and usually results in resolution of the retinopathy within a period of weeks.[8,9]

References

1. Spivak JL. The anemic patient. In: Harvey AM, et al. eds. *The Principles and Practice of Medicine.* Norwalk, Conn: Appleton-Century-Crofts; 1984;467-471.
2. Percival SPB. The eye and Moschcowitz's disease. *Tr Ophthal Soc UK.* 1970;90:375-382.
3. Rubenstein RA, Yanoff M, Albert DM. Thrombocytopenia, anemia, and retinal hemorrhage. *Am J Ophthalmol.* 1968;65:435-439.
4. Merin S, Freund M. Retinopathy in severe anemia. *Am J Ophthalmol.* 1968;66:1102-1106.
5. Holt JM, Gordon-Smith EC. Retinal abnormalities in diseases of the blood. *Br J Ophthalmol.* 1969;53:145-160.
6. Wong VG, Bodey GP. Hemorrhagic retinoschisis due to aplastic anemia. *Arch Ophthalmol.* 1968;80:433-435.

7. Aisen ML, Bacon BK, Goodman AM, et al. Retinal abnormalities associated with anemia. *Arch Ophthalmol.* 1983;101:1049-1052.

8. Kirkham TH, Wrigley PFM, Holt JM. Central retinal vein occlusion complicating iron deficiency anemia. *Br J Ophthalmol.* 1971;55:777-780.

9. Kurzel RB, Angerman NS. Venous stasis retinopathy after long-standing menorrhagia. *J Reprod Med.* 1978;20:239-242.

10. Mansour AM. Aplastic anemia simulating central retinal vein occlusion. *Am J Ophthalmol.* 1985;100:478-479.

11. Shorb SR. Anemia and diabetic retinopathy. *Am J Ophthalmol.* 1985;100:434-436.

Chapter 42

Coagulation Disorders

GEORGE A. WILLIAMS

The normal hemostatic mechanism is an intricately balanced system that precisely regulates formation and clearance of blood clots.[1] A wide variety of stimuli, including traumatic, inflammatory, and metabolic events, can stimulate coagulation through activation of complex enzyme cascades. The biochemical integrity of these cascades is dependent on the presence and function of many factors, some of which are proteases responsible for the conversion of zymogens into other proteases. Other factors act as catalysts or cofactors, enhancing the enzyme activity of various proteases. The end-product of these hemostatic cascades is the formation of polymerized fibrin, the major component of clotted blood. The sequential activation of these factors is separated into two distinct yet interdependent pathways; the extrinsic and intrinsic pathways. Although these pathways involve both different activating stimuli and different factors, their end-product, activated factor X, is the same. Activated factor X then stimulates the formation of thrombin, which subsequently generates fibrin monomers that are in turn polymerized into fibrin strands by factor XIII. Simultaneously with the activation of the coagulation system, the fibrinolytic system is also activated, resulting in the formation of plasmin, which clears fibrin.

Within the extrinsic, intrinsic, and fibrinolytic cascades there are complex autoregulatory mechanisms that tightly control their respective enzymatic activities. The interdependency of these separate pathways is perhaps best demonstrated in disease states in which individual factors are lacking or are dysfunctional due to hereditary or acquired disorders. The severity of hemorrhagic complications resulting from the absence of a single factor underscores the interdependence of the individual factors and pathways in the maintenance of hemostasis. These deficiencies can produce a wide spectrum of clinical manifestations, ranging from subtle bleeding events to fulminant hemorrhagic crises. This chapter will briefly review some of the coagulation disorders, both hereditary and acquired, that have been associated with ocular disease.

Systemic Manifestations

The systemic manifestations of coagulation disorders are protean. Hemorrhagic complications are typical in persons with hereditary deficiencies of clotting factors and in those taking drugs that induce anticoagulation. This type of bleeding is characterized by joint, muscle, or intraperitoneal hemorrhage, often following relatively minor trauma.

The most common inherited coagulation disorders are X-linked deficiencies of factors VIII (hemophilia A) and IX (hemophilia B) which are necessary for the generation of activated factor X by the intrinsic pathway. These disorders are clinically indistinguishable and are characterized by insidious onset of hemorrhage hours to days after injury. The severity of the bleeding correlates with the extent of the respective factor's deficiency. Patients with less than 1% of normal factor activity have severe hemorrhagic disease, usually presenting with complications at birth. Levels of factor activity from 1 to 5% of normal are considered moderate, and bleeding may not become manifest until the patient begins crawling or walking. Persons with mild hemophilia (levels of from 5 to 25% of normal) may not be diagnosed until adolescence or until they experience major trauma or surgery. The classic presentation of hemophilia is pain and swelling in a weight-bearing joint due to hemarthrosis. Chronic synovitis may result, with long-term disabling sequelae. Intramuscular hematomas are common, and large hematomas may compress vital structures. Central nervous system bleeding accounts for 25% of hemophiliacs' deaths and has a mortality rate of 34%.[2] While therapeutic replacement of the deficient factor has greatly improved the prognosis in hemophilia, it has also resulted in other severe problems, such as hepatitis and acquired immunodeficiency syndrome (AIDS) from virally contaminated blood products.

Deficiencies of other factors are much less common than the hemophilias. Factor XI deficiency is an autosomal recessive disorder common only in Ashkenazi Jews that has relatively minor clinical manifestations, such as epistaxis, hematuria, and menorrhagia. It is usually detected because of bleeding following surgery. The disparity in clinical severity between factor XI deficiency and the hemophilias is explained by the ability of the extrinsic system to bypass factors XII and XI through direct activation of factor IX. Thus, an intact extrinsic pathway can compensate for dysfunction early in the intrinsic pathway.[1]

Bleeding disorders associated with anticoagulant therapy are probably more common than inherited coagulation deficiencies.[3] Anticoagulants are used therapeutically to prevent or limit thrombosis in a wide variety of conditions such as pulmonary embolism, ischemic cerebrovascular disease, cardiac disease, and postoperative immobilization.

Heparin is used for acute anticoagulation, and the coumarin group of anticoagulants are used for long-term therapy. Heparin is a naturally occurring mucopolysaccharide that enhances the activity of antithrombin III, the major endogenous anticoagulant factor. In the presence of heparin, antithrombin III more effectively neutralizes thrombin and other activated coagulant factors, thereby slowing coagulation. The coumarin anticoagulants interfere with vitamin K metabolism, thereby affecting production of thrombin and factors VII, IX, and X and thus slowing coagulation.

The incidence of clinically significant bleeding associated with anticoagulation is dependent on the underlying primary disease process, the intensity of the anticoagulant therapy, and the duration of therapy.[3] Minor hemorrhage such as epistaxis, hematuria, ecchymosis, and hemoptysis occur in up to 25% of anticoagulated patients. The incidence of major hemorrhages requiring transfusion or hospitalization is as high as 8%, and fatal bleeding occurs in as many as 5%.

Ocular Manifestations

Considering the severity of the coagulation disorders previously discussed and the vascularity of the eye, the incidence of ocular complications associated with coagulation disorders is surprisingly small. This may reflect the fact that the eye is well-protected from the traumatic events that often precipitate bleeding in persons with dysfunctional coagulation.

Ocular manifestations of hemophilia are primarily neurologic, resulting from central nervous system hemorrhage. Pupillary abnormalities, cranial nerve palsies, blurred vision, and papilledema have been described following intracranial bleeding.[2] Repeated retinal hemorrhages were noted in one patient with factor XI deficiency.[4]

Anticoagulation with therapeutic agents has been implicated in a variety of hemorrhagic ocular complications. Isolated cases of vitreous hemorrhage have been associated with anticoagulation[5,6]; however, these reports do not provide information on the type of anticoagulation used, the duration of therapy, or parameters of blood coagulation at the time of hemorrhage. Spontaneous hyphema during coumarin-type anticoagulation in both phakic and iris-fixated pseudophakic eyes has also been described.[7,8] Massive subretinal and vitreous hemorrhage in age-related macular degeneration has been associated with anticoagulant therapy.[9] These hemorrhages, which invariably result in permanent loss of vision, can present with sudden onset of severe ocular pain and elevated intraocular pressure. Histopathologic analysis suggests the source of the hemorrhage is large choroidal vessels that vascularize disciform scars. Anticoagulation may predispose these vessels to bleed.

The use of anticoagulant medication in patients with potentially hemorrhagic ocular conditions such as proliferative diabetic retinopathy or choroidal neovascularization presents a difficult management problem for both ophthalmologist and internist. The risks of ocular hemorrhage must be weighed against the potentially life-threatening complications of the systemic disease for which anticoagulant therapy is employed. While it seems reasonable to infer an increased risk of ocular hemorrhage during anticoagulant therapy, the actual incidence of this event is unknown, and there is no evidence to support withdrawal or denial of appropriately indicated anticoagulant therapy because of fear of ocular bleeding.

References

1. Rosenberg RD. Hemorrhagic disorders I. Protein interactions in the clotting mechanism. In: Becks WS, ed. *Hematology*. Cambridge, Mass: MIT Press; 1985.

2. Eyster ME, Bill FM, Blaff PM. Central nervous system bleeding in hemophiliacs. *Blood.* 1978;51:1179-1188.
3. Levine MN, Hirsh J. Hemorrhagic complications of anticoagulant therapy. *Semin Thromb Hemost.* 1986;12: 39-57.
4. Leiba H, Ramof B, Many A. Heredity and coagulation studies in ten families with factor XI deficiency. *Br J Haematol.* 1965;11:645.
5. Butner RW, McPherson AR. Spontaneous vitreous hemorrhage. *Ann Ophthalmol.* 1982;14:268.
6. Oyakawa RT, Michels RG, Blase WP. Vitrectomy for non-diabetic vitreous hemorrhage. *Am J Ophthalmol.* 1983;96:517-525.
7. Koehler MP, Sholiton DB. Spontaneous hyphema resulting from warfarin. *Ann Ophthalmol.* 1983;15:858-859.
8. Schiff FS. Coumadin related spontaneous hyphemas in patients with iris fixated pseudophakos. *Ophthalmic Surg.* 1985;16:172-173.
9. Baba FE, Jarrett WH II, Harbin TS, et al. Massive hemorrhage complicating age-related macular degeneration. Clinicopathologic correlation and role of anticoagulants. *Ophthalmology.* 1986;93:1581-1592.

Chapter 43

Disseminated Intravascular Coagulation

HELMUT BUETTNER

Disseminated intravascular coagulation (DIC) is a complex acquired syndrome in which fibrin thrombi form in the small blood vessels throughout the body, most commonly in the skin, brain, heart, lungs, kidneys, adrenal glands, spleen, and liver. DIC always occurs in response to another disease; it is never a primary disorder. The syndrome has been described in association with numerous disorders, the most important of which are listed in Table 43-1. The syndrome can be triggered when procoagulant substances are released into the blood (e.g., amniotic fluid embolism), when blood comes into contact with tissue thromboplastin in extensively damaged tissues (e.g., severe burns), or when procoagulants are produced in the blood (e.g., massive hemolysis), activating the extrinsic and intrinsic clotting mechanisms.[1,2] As intravascular thrombi form, large numbers of platelets are consumed, resulting in thrombocytopenia. Plasminogen binds to fibrin and is converted by activators to the fibrinolytically active enzyme plasmin. When the formation of plasmin exceeds the capacity of its naturally occurring inhibitors to inactivate it, plasmin then not only consumes certain clotting factors but also degrades fibrinogen and fibrin. The resulting fibrinogen-fibrin split products act as inhibitors of fibrin thrombus formation. The consumption of fibrinogen and platelets, the anticoagulative action of the fibrinogen-fibrin split products, and the fibrinolytic action of plasmin all contribute to a hypocoagulative state with severely deranged hemostasis.[1,2] This entity is also referred to as consumptive coagulopathy, or consumptive thrombohemorrhagic disorder.[2]

Systemic Manifestations

The major clinical features of acute DIC include the sudden onset of serious bleeding; shock, frequently out of proportion with obvious blood loss; and, because of widespread intravascular clot formation, poor perfusion of various organ systems, resulting in skin hemorrhages and renal, central nervous system, pulmonary, gastrointestinal, hepatic, and adrenal dysfunction.

The most obvious changes of acute DIC are often the patches of hemorrhagic necrosis of the skin. Occlusion of renal arterioles or glomerular capillaries results in hematuria, oliguria, or even anuria. Involvement of the central nervous system in DIC is manifested by such nonspecific changes as an altered state of consciousness convulsions, or coma rather than focal lesions characteristic of major vessel occlusion. The picture of the acute respiratory distress syndrome can result from pulmonary dysfunction caused by interstitial hemorrhage. Vascular occlusion in the gastrointestinal tract may

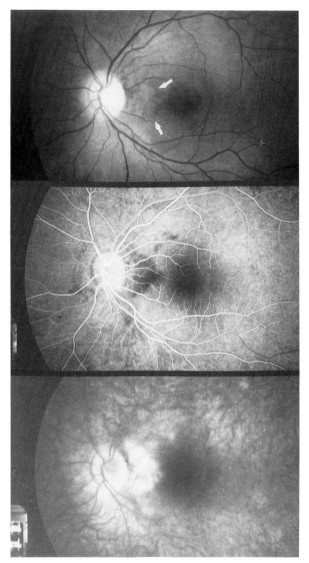

FIGURE 43-1 Gray patch of choroid temporal to disc (*white arrows*) with overlying serous retinal detachment (*top*) in an eclamptic patient with transient DIC shortly after delivery of twins by cesarean section. Early filling defect of the choroid (*middle*) and later leakage of dye into the subretinal space (*bottom*) were demonstrated by fluorescein angiography.

cause submucosal necrosis with superficial ulcerations, often leading to massive gastrointestinal bleeding. When hemorrhagic necrosis of the adrenal cortex develops in DIC, a fulminant form of sepsis (Waterhouse-Friderichsen syndrome) may ensue.[1,2] In chronic DIC, however, these clinical features are generally less obvious and may be observed with only minor bleeding or none at all. Since the pathophysiology of DIC is rather complex, in many cases it is impossible to establish

whether a given clinical or laboratory manifestation is the result of the underlying disorder or of DIC triggered by it.

The laboratory diagnosis of DIC is supported by the results of a number of tests, none of which alone is specific for the diagnosis. In general, thrombocytopenia, hypofibrinogenemia, and prolongation of the prothrombin-, thrombin-, and partial thromboplastin time are indicative of DIC in the presence of appropriate clinical findings. A reduction of the clotting factors and the presence of fibrinogen-fibrin split products indicate that widespread thrombus formation and increased fibrinolysis have developed, both of which are characteristic phenomena of DIC.[1,2]

Treatment of DIC is directed toward eliminating the underlying disease—evacuating the uterus in obstetric complications or treating infection. If the primary disease cannot be controlled promptly and if bleeding is a serious complication, clotting factors and platelets must be replaced.[1] When massive and widespread intravascular clotting dominates the picture of DIC, judicious use of heparin is indicated. Shock must be treated vigorously. In addition, general supportive treatment is given, as the particular circumstances dictate.[2]

Ocular Manifestations
Clinical Findings

Occlusion of small vessels by platelet and fibrin thrombi and hemorrhage in DIC can affect all vascularized ocular tissues, but occlusion of the vessels of the choroid is the dominant ocular abnormality. Clinical findings include patchy loss of the normal choroidal pattern with gray, yellow, or reddish brown discoloration, at times surrounded by choroidal hemorrhage, resulting in varying degrees of vision loss.[3,4] Often a serous retinal detachment is associated with these abnormalities[3,5,8-12] which are almost always located in the posterior and midperipheral portions of the fundus. In the presence of a serous retinal detachment the choroidal changes may be demonstrable only by fluorescein angiography in the form of patchy filling defects of the choriocapillaris, from which dye may later leak into the subretinal space (Fig. 43-1).[3] A serous retinal detachment may not develop, especially when DIC follows a more protracted and chronic course.[3] Subretinal, retinal, preretinal, and vitreous hemorrhage, hemorrhages into the iris and anterior chamber, and conjunctival hemorrhages are observed less frequently than choroidal changes in the adult patient with DIC. When observed, they are frequently associated with terminal uremia and septicemia; however, in neonates and infants with DIC, retinal, preretinal, and vitreous hemorrhages as well as hyphemas have been described more often.[5-7]

TABLE 43-1 The Most Common Causes of DIC

Obstetric complications: Abruptio placentae, intrauterine fetal death, abortion (septic, saline-induced), eclampsia
Infections: Bacterial (septicemia), viral, rickettsial
Hematopoetic disorders: Acute leukemia, intravascular hemolysis
Vascular disorders: Aneurysms, hemangiomas, prosthetic grafts, collagen-vascular diseases
Neoplasms: Carcinomas, other neoplasms
Massive tissue injury: Extensive burns and other tissue trauma, extracorporal circulation
Miscellaneous: Graft rejection, drug- and snake venom reactions, advanced liver disease

(Modified from Wintrobe.[1])

Histopathologic Findings

The histopathologic hallmark of ocular involvement in DIC is thrombotic occlusion of vessels in the posterior and midperipheral choroid.[3,4,8,9] The stroma surrounding the occluded vessels is massively infiltrated by hemorrhage. In more chronic cases of DIC, macrophages laden with blood breakdown products are present throughout the hemorrhage, and organization and recanalization of intravascular thrombi occur.[3,4] Scleral and episcleral vessels, posterior ciliary arteries, and vortex veins are typically unaffected by the occlusive process. In the area of thrombotic occlusion of the choroid, and particularly the choriocapillaris, the RPE exhibits varying degrees of disruption, ranging from patchy thinning to focal proliferation.[3-5,8,9] The overlying photoreceptor cells show degenerative changes ranging from disorganization of inner and outer segments to drop-out of entire photoreceptor cells, resulting in attenuation of the photoreceptor cell layer.[3,4] The inner retina, specifically the retinal vasculature, and the vitreous are normal in most cases of ocular involvement in DIC in adults (Fig. 43-2). Rarely hemorrhages into the retina, vitreous, or anterior segment are observed.[10] It is

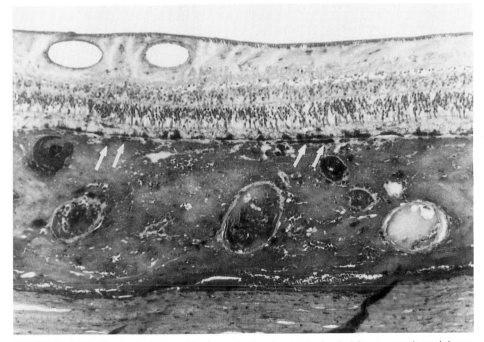

FIGURE 43-2 Thrombotic occlusion of choriocapillaris (*white arrows*) and large choroidal vessels in various stages of organization and recanalization, diffuse choroidal hemorrhage, patchy RPE changes with degeneration of overlying photoreceptor cells, normal inner retina and retinal vasculature in chronic DIC (phosphotungstic acid-hematoxylin, original magnification ×120).

not certain whether they are the result of DIC alone or of uremia or septicemia terminally complicating DIC.

The predilection of the choroidal vasculature for thrombotic occlusion has been attributed to its unique anatomic features. Unlike other vascular beds, where blood flows through a series of dichotomous branches before reaching the terminal vascular bed, the posterior ciliary arteries empty abruptly into choroidal sinusoids and veins. The resulting rapid deceleration of blood flow favors local precipitation of fibrin thrombi in the hypercoagulable state of DIC.[3]

There is no specific treatment for the ocular complications of DIC. Once the systemic disorder resolves, whether spontaneously or with treatment, the ocular complications resolve as well. The choroidal vessels become recanalized, and subretinal fluid is reabsorbed. Where the choroidal vessels had been occluded, the RPE may become rather mottled. Visual function improves significantly in most cases.

References

1. Wintrobe MM, et al, eds. *Clinical Hematology.* Philadelphia, Pa: Lea and Febiger; 1981;1213-1228.
2. Marder VJ. Consumptive thrombohemorrhagic disorders. In: Williams WJ, et al., eds. *Hematology.* New York, NY: McGraw-Hill; 1983;1433-1461.
3. Cogan DG. Ocular involvement in intravascular coagulopathy. *Arch Ophthalmol.* 1975;93:1-8.
4. Samples, JR, Buettner H. Ocular involvement in disseminated intravascular coagulation (DIC). *Ophthalmology.* 1983;90:914-916.
5. Ortiz JM, Yanoff M, Cameron JD, et al. Disseminated intravascular coagulation in infancy and in the neonate; ocular findings. *Arch Ophthalmol.* 1982;100:1413-1415.
6. Azar P, Smith RS, Greenberg MH. Ocular findings in disseminated intravascular coagulation. *Am J Ophthalmol.* 1974;78:493-496.
7. Wiznia RA, Price J. Vitreous hemorrhages and disseminated intravascular coagulation in the newborn. *Am J Ophthalmol.* 1976;82:222-226.
8. Percival SPB. Ocular findings in thrombotic thrombocytopenic purpura (Moschcowitz's disease). *Br J Ophthalmol.* 1970;54:73-78.
9. Percival SPB. The eye in Moschcowitz's disease (thrombotic thrombocytopenic purpura); a review of 182 cases. *Trans Ophthalmol Soc UK.* 1970;90:375-382.
10. Yoshioka H, Yamanouchi U. Case of retina and vitreous body hemorrhage due to thrombocytopenic purpura with special reference to pathohistological findings of the eye. *Nagasaki Med J.* 1962;37:234-237.
11. Case records of the Massachusetts General Hospital, weekly clinicopathological exercises. Case 74-1963. *N Engl J Med.* 1963;269:1195-1203.
12. Clinicopathologic conference. A patient with renal failure. *Mayo Clin Proc.* 1964;39:792-805.

Chapter 44

Dysproteinemias

JUAN ORELLANA and ALAN H. FRIEDMAN

Multiple myeloma, Waldenström's macroglobulinemia, and benign monoclonal gammopathy are three diseases in which there is an excessive monoclonal gammopathy and/or polypeptide proteins with an associated decrease in normal immunoglobulins. Characteristically, we describe five classes of immunoglobulins that are produced during the course of these diseases. Composed of light and heavy chains, each class possesses specific antigenic properties that elicit antibodies specific to that class. The immunoglobulin class is named by the heavy chain it contains (eg, alpha chains define the immunoglobulin A [IgA] class).

Multiple Myeloma

Multiple myeloma is a neoplasm of plasma cells seen in persons between the ages of 40 and 70.[1,2] It rarely occurs before the age of 30, although a disease similar to myeloma has been described in the pediatric population. Patients with multiple myeloma have punched-out lesions of the skull, vertebrae, and ribs in 70% of cases. They have a normocytic-normochromic anemia, hypercalcemia, and an increased susceptibility to infections.[2,3] The diagnosis is suggested in patients with bone pain, fatigue, anemia, and an increased erythrocyte sedimentation rate. Multiple osteolytic lesions, pathologic fractures, bone tumors, osteoporosis, hypercalcemia, uremia, cryoglobulinemia, pyroglobulinemia, hyperglobulinemia, Bence-Jones proteinuria, and protein electrophoretic abnormalities contribute to confirming the diagnosis of multiple myeloma. Seventy percent of myelomas are of the IgG variety, 29% of the IgA type, and the remaining 1% are of the IgD type.

The ocular manifestations of multiple myeloma can be divided into orbital, conjunctival, corneal, uveal, retinal, and optic nerve findings. The ocular signs of multiple myeloma are the result of myeloma infiltrates in and around the globe or of associated serum or hematologic abnormalities.[4,5] Proptosis can be an initial manifestation of myeloma due to orbital involvement. Patients can also present with diplopia and varying degrees of visual impairment. Orbital disease is not common; indeed Pasmantier and Azar[6] found no orbital involvement in 57 autopsy subjects with myeloma with extraskeletal spread. The conjunctiva may be involved in multiple myeloma by virtue of these patients' increased blood viscosity. Observing the conjunctival vessels will reveal blood sludging and deposition of crystals, the latter representing immunoglobulin deposition.

Crystals in the corneal epithelium and stroma have been described in patients with myeloma.[7] Immunoglobulin deposition in the corneal epithelium has also been documented. Pinkerton and Robertson analyzed the crystals in their patient and found them to be composed of intracytoplasmic proteinaceous material similar to that found in the patient's bone marrow.[8]

Pars plana cysts were initially described by Allen and later confirmed by others in autopsy specimens.[9-12] The cysts are located in the pars plana or pars plicata and are PAS positive, deeply eosinophilic with hematoxylin and eosin, and negative on alcian blue stain. They are located between the pigmented and nonpigmented ciliary epithelium. The cysts are clear in vivo but become opaque when exposed to formalin or other fixatives during tissue processing. Immunoelectrophoresis demonstrates that the protein content of the cysts and of serum in myeloma patients are the same. A permeable pigment epithelium may allow passage of blood proteins and so may be the origin of the cysts.

Choroidal tumors, infiltrates, and choroidal destruction have been reported in myeloma patients.[13,14]

Dilated veins, hemorrhages, cotton wool spots, and preretinal hemorrhages are retinal signs of increased blood viscosity. Spalter[15] attributed these findings to the altered serum viscosity caused by increased levels of circulating immunoglobulins.

Initially, the veins (and occasionally the arterioles) exhibit dilatation, which is followed by vessel tortuosity. Constriction at the arteriovenous crossings produces beading of the veins, which can ultimately result in frank occlusion, sometimes accompanied by vitreous hemorrhage.

The optic nerve, lacrimal gland, sclera, or iris can become infiltrated by myeloma cells. Copper deposition in Descemet's membrane of the cornea and in the anterior lens capsule can be seen.

Waldenström's Macroglobulinemia

Waldenström's macroglobulinemia is a malignant proliferation of lymphocytes that produce IgM, causing an increase in serum viscosity. Characteristically, patients are older men who experience insidious onset of weakness, fatigue, and epistaxis.[16] They are predisposed to recurrent infections, dyspnea, recurrent episodes of congestive heart failure, weight loss, and neurologic symptoms.[17] Purpura, peripheral adenopathy, pallor, and mild hepatosplenomegaly are common findings. Nearly one fourth of the patients have neurologic abnormalities and one third have renal abnormalities. The increased macroglobulin production increases the serum viscosity, thereby producing the retinopathy.[15] With a serum viscosity value above 4, patients commonly exhibit bleeding gums and epistaxis. Neurologically, patients develop dizziness, vertigo, paresthesias, headaches, nystagmus, ataxia, and seizures. Plasmapheresis or medications such as chlorambucil are therapeutic; they decrease the IgM level, which in turn decreases serum viscosity.[18] Although excessive IgM can be seen in other diseases, such as lymphoma, chronic lymphocytic leukemia, toxoplasmosis, and chronic hepatitis, the IgM is polyclonal and appears in a much lower concentration than that seen in Waldenström's macroglobulinemia.

Since these patients have such a high level of macroglobulin production, antibody synthesis is diminished. Patients have presented with orbital cellulitis secondary to sinusitis. There can be blood sludging in the conjunctival vessels, and careful observation of the blood flow at the slit lamp will demonstrate clumping and segmentation of the erythrocytes.[19] Both the conjunctiva and the cornea can demonstrate crystals. Immunofluorescence and immunoperoxidase techniques show that these crystals react positively for immunoglobulin and kappa light chains.

Retinal findings have been reported in 50% of patients with macroglobulinemia. Classically, venous changes are present, presumably due to serum hyperviscosity caused by excessive macroglobulin production.[15,20] The earliest sign is usually dilated tortuous veins (Figs. 44-1 to 44-3). Increased congestion produces flame-shaped hemorrhages as well as an increase in

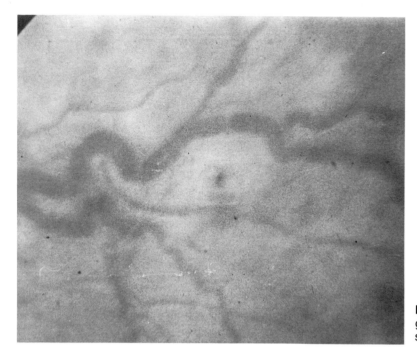

FIGURE 44-1 Waldenström's macroglobulinemia: Fundus photograph shows dilated retinal veins.

vessel diameter. Additional changes include retinal edema, vascular occlusion, and disc edema.[21]

Histopathologic studies have demonstrated PAS–positive material in the lumens of retinal vessels. The peripheral vessels have microaneurysms and decreased numbers of pericytes and endothelial cells. Immunofluorescent techniques show that eosinophilic PAS material in the retina consists of IgM deposits.[22]

Benign Monoclonal Gammopathy

Benign monoclonal gammopathy is a benign condition in which there is a monoclonal band on serum protein electrophoresis without any other sign of myelomatosis.[23] Although the condition is benign, there are in-

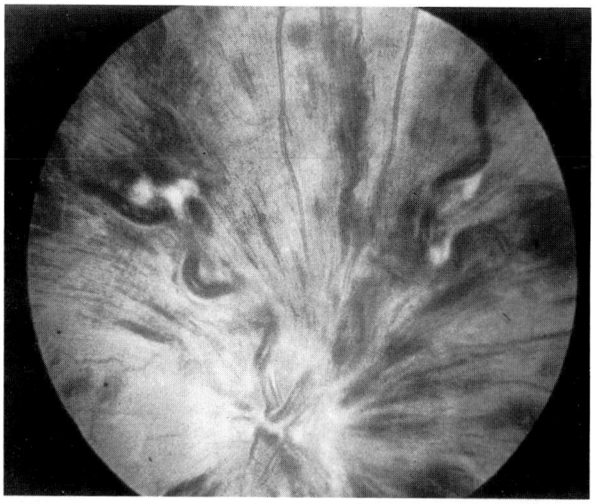

FIGURE 44-2 Waldenström's macroglobulinemia: Fundus photograph reveals central retinal vein occlusion.

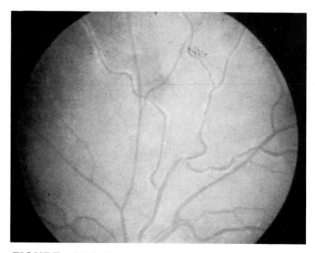

FIGURE 44-3 Waldenström's macroglobulinemia: Fundus photograph following plasmapheresis reveals amelioration of retinal picture. (Courtesy Ronald Carr, M.D.)

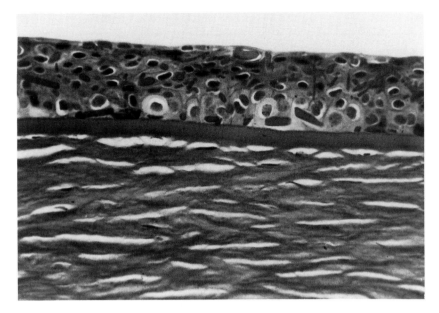

FIGURE 44-4 Benign monoclonal gammopathy: Crystalline deposits in corneal epithelium. (Masson's trichrome stain, original magnification ×900; courtesy Gordon Klintworth, M.D.)

stances in which it has progressed to myeloma. Benign monoclonal gammopathy may thus be a precursor to myeloma. Since a gammopathy may accompany many systemic conditions, such as lymphoma, leukemia, or neoplasms, these conditions must be excluded before the diagnosis of benign monoclonal gammopathy can be made.

Crystals in the conjunctiva and cornea have been described in benign monoclonal gammopathy (Fig. 44-4).[24] They can be derived from paraproteins or from cholesterol. The cholesterol hypothesis was supported by Laibson and Damiano,[25] who believed the crystals were derived from cholesterol stearate. Analysis of superficial corneal crystals in another study showed deposits derived from paraproteins from the limbal vessels.[26] Rodrigues et al.[27] found corneal deposits composed of immunoglobulins and kappa light chains. Clinically, the opacities are gray-white, yellow, or gray-brown and are arranged locally or diffusely across the cornea.

References

1. Azar HA, Potter M. *Multiple Myeloma and Related Diseases.* vol. I. Hagerstown, Md: Harper & Row; 1973;1-52.
2. Kyle RA. Multiple myeloma: review of 869 cases. *Mayo Clin Proc.* 1975;50:29-40.
3. Rosen BJ. Multiple myeloma. A clinical review. *Med Clin North Am.* 1975;59:375-386.
4. Ashton N. Ocular changes in multiple myelomatosis. *Arch Ophthalmol.* 1965;73:487-494.
5. Baker TR, Spencer WH. Ocular findings in multiple myeloma. *Arch Ophthalmol.* 1974;91:110-113.
6. Pasmantier MW, Azar HA. Extraskeletal spread in multiple plasma cell myeloma: a review of 57 autopsied cases. *Cancer.* 1969;23:167-174.
7. Aronson SB, Shar R. Corneal crystals in multiple myeloma. *Arch Ophthalmol.* 1959;61:541-546.
8. Pinkerton RMH, Robertson DM. Corneal and conjunctival changes in dysproteinemia. *Invest Ophthalmol.* 1969;8:357-364.
9. Allen RA, Miller DH, Straatsma BR. Cysts of the posterior ciliary body (pars plana). *Arch Ophthalmol.* 1961;66:302-313.
10. Sanders TE, Podos SM. Pars plana cysts in multiple myeloma. *Trans Am Acad Ophthalmol Otolaryngol.* 1966;70:951-958.
11. Sanders TE, Podos SM, Rosenbaum LJ. Intraocular manifestations of multiple myeloma. *Arch Ophthalmol.* 1967;77:789-794.
12. Johnson BL, Storey JD. Proteinaceous cysts of the ciliary epithelium. I. Their clear nature and immunoelectrophoretic analysis in a case of multiple myeloma. *Arch Ophthalmol.* 1970;84:166-170.
13. Stock W. A myeloma of the inner eye. *Klin Monatsbl Augenheilkd.* 1918;61:14-18.
14. Delaney WVJR, Liaricos SV. Chorioretinal destruction in multiple myeloma. *Am J Ophthalmol.* 1968;66:52-55.
15. Spalter HF. Abnormal serum protein and retinal vein thrombosis. *Arch Ophthalmol.* 1959;62:868-881.
16. MacKenzie MR, Fudenberg BH. Macroglobulinemia: an analysis of forty patients. *Blood* 1972;39:874-889.
17. Tubbs RR, Hoffmar GC, Chir B, et al. IgM monoclonal gammopathy. Histopathologic and clinical spectrum. *Cleveland Clin Quart.* 1976;43:217-235.
18. Schwab PJ, Okun E, Fahey JL. Reversal of retinopathy of Waldenström's macroglobulinemia by plasmapheresis. *Arch Ophthalmol.* 1960;64:515-521.
19. Ackerman AL. The ocular manifestations of Waldenström's macroglobulinemia and its treatment. *Arch Ophthalmol.* 1962;67:701-707.

20. Thomas EL, Olk RJ, Markman M, et al. Irreversible visual loss in Waldenström's macroglobulinemia. *Br J Ophthalmol.* 1983;67:102-106.

21. Orellana J, Friedman AH. Ocular manifestations of multiple myeloma, Waldenström's macroglobulinemia and benign monoclonal gammopathy. *Surv Ophthalmol.* 1981;26:157-169.

22. Friedman AH, Marchevsky A, Odel JG, et al. Immunofluorescent studies of the eye in Waldenström's macroglobulinemia. *Arch Ophthalmol.* 1980;98:743-746.

23. Waldenström J. Studies on conditions associated with disturbed gamma globulin formation (gammopathies). *Harvey Lect.* 1961;56:211-231.

24. Plam E. A case of crystal deposits in the cornea: precipitation of a spontaneously crystalizing plasma globulin. *Acta Ophthalmol.* 1947;25:165-169.

25. Laibson PR, Damiano VV. X-ray and electron diffraction of ocular and bone marrow crystals in paraproteinemia. *Science.* 1969;163:581-583.

26. Efferman RA, Rodrigues MM. Unusual superficial stromal corneal deposits in benign monoclonal gammopathy. *Arch Ophthalmol.* 1980;98:78-81.

27. Rodrigues MM, Krachmer JH, Miller SD, et al. Posterior corneal crystalline deposits in benign monoclonal gammopathy. *Arch Ophthalmol.* 1979;979:124-128.

Chapter 45

Leukemia

MARILYN C. KINCAID

"Leukemia" is a blanket term for any of a group of blood disorders in which there is an abnormal proliferation of white cells. The leukemia is named for the predominant cell, and there are several classification schemes.[1,2] There are two main types, lymphoid, including both B cell and T-cell leukemias, and myeloid, including erythroleukemia and leukemias of the other components of the myeloid cell line. Leukemias are also classified in terms of their presentation and behavior as acute or chronic.

Leukemic cells overgrow the normal cells in the bone marrow, not because of rapid cell division—if anything, the cycle is slower in the neoplastic cells—but because they remain immature and never stop dividing. Unlike normal cells, which mature, function, die, and then are removed from circulation, the leukemic cells proliferate in an uncontrolled manner. The disease progresses as the leukemic cells take over the bone marrow and invade other tissues. Patients usually die of infection when competent leukocytes are no longer produced in adequate numbers, of hemorrhage from lack of platelets, or of anemia from a lack of red cells.[1,2]

Recently, much has been learned about the pathogenesis of the leukemias.[1] It has been observed that ionizing radiation and exposure to certain toxins can induce some types of leukemia. At least one type, human T-cell leukemia, has sometimes occurred in clusters of patients and is associated with a retrovirus.[2] Several chromosomal abnormalities are now known to be typical.[1] The best known of these is the Philadelphia chromosome in chronic myelogenous leukemia; it is a translocation of a portion of chromosome 22 to chromosome 9, it is postzygotic, and occurs only in the neoplastic cell line. Certain leukemias are more frequently associated with certain types of aneuploidy, particularly acute myelogenous leukemia in Down's syndrome. Other translocations, deletions, and additions have recently been associated with other leukemias. In at least some of them, the chromosomal abnormality occurs close to the locus of one of several oncogenes.[2]

Systemic Manifestations

Symptoms of leukemia depend to a considerable extent on the type of leukemia.[2] Chronic lymphocytic leukemia, for example, may be an unsuspected finding in asymptomatic older persons. In contrast, acute leukemias present with fever, history of hemorrhage or bruising, and node enlargement. The fever is generally due to infection, often with unusual organisms.

Physical examination reveals hepatomegaly and splenomegaly, as well as lymph node enlargement. There may be ecchymoses of the skin and hemorrhages in the oral cavity and retina. The spinal fluid may contain blast cells, indicating central nervous system involvement, which occurs in up to 70% of patients with acute leukemia. Death results from the effects of tumor bulk, infection, anemia, and thrombocytopenia.

Typically the diagnosis is made by bone marrow aspiration and biopsy as well as inspection of a peripheral smear. The white cell count may actually be diminished, and platelets are always depressed in acute leukemia[2]; however, the initial white cell count in chronic myelocytic leukemia is generally over 100,000 per cubic millimeter. Immunologic surface markers of the neoplastic cells allow classification of the leukemia, from which, in turn, treatment response and prognosis can be predicted.[1,2]

Other diagnostic modalities include biopsy of other involved tissues, such as superficial lymph nodes. Radiographs may show bone involvement, particularly in acute leukemia, and computed tomographic (CT) scanning may disclose visceral lymph node enlargement.[2]

The first use of chemotherapeutic agents for acute leukemia, then a rapidly fatal disease, was in 1947.[2] Since then, advances have revolutionized the care of these patients. It is now possible to speak of cure in many instances, particularly in children with acute lymphocytic leukemia.[1] Chemotherapeutic agents are administered in specific protocols, depending on the type of leukemia. Adjuvant radiotherapy is also useful, particularly for enlarged nodes and for the central nervous system. Transfusions may be necessary to control the secondary cytopenias, and antibiotics for superinfections.

Another therapeutic advance is bone marrow transplantation, useful in patients with nonlymphocytic leukemia.[2] Patients undergo destruction of their marrow with radiation and chemotherapeutic agents, then receive marrow from an HLA tissue–matched donor, preferably a relative. A major complication is graft versus host disease, wherein the immunocompetent transfused lymphocytes attack the graft recipient.

Ocular Manifestations

Almost any ocular tissue may be affected by leukemia, by direct infiltration by leukemic cells, or by the secondary effects of the neoplasm.[3,4] The choroid is most commonly involved, and may be massively thickened by the leukemic cells, but in many cases this involvement is subclinical.[4] Secondary clinical sequelae that may occur include serous retinal detachments accompanied by changes in the retinal pigment epithelium.

Retinal changes are the most commonly seen clinical manifestation of leukemic involvement of the eye. They are nonspecific and are generally due to combinations of the accompanying anemia, thrombocytopenia, and leukocytosis.[3,4] Such manifestations include vascular sheathing and tortuosity, pallor, hemorrhages, exudates, cotton-wool spots, and neovascularization at the periphery or the disc.[4,5] At one time, tumors of leukemic cells in the retina were relatively common, but this complication is now quite rare.[4] Secondary infection of the retina and overlying vitreous by opportunistic infection may also occur. Subretinal hemorrhage in chronic leukemia leading to secondary angle closure glaucoma has been reported.[6]

The optic nerve may be directly infiltrated, leading to compression.[3] When this occurs in the optic nerve head, a picture similar to papillitis results, and there may be a modest reduction in vision; however, when the retrolaminar optic nerve is invaded by leukemic cells, visual loss can be profound, since the dural sheath makes expansion impossible. This constitutes an ocular emergency, and prompt irradiation may save vision.[3] The nerve may be indirectly involved by a rise in intracranial pressure leading to papilledema. Moreover, the different manifestations can exist concurrently, sometimes making differential diagnosis difficult.[4] Prophylactic cranial irradiation is now frequently administered to children with acute leukemia, but sometimes the eyes are not included in the irradiation field.[4]

Since the cornea is avascular, corneal involvement is indirect, secondary to compromised vascular perfusion which leads to limbic ulcers. These have been reported with acute myelogenous leukemia.[4,7] Peripheral corneal infiltrates and corneal edema have been reported as the presenting signs in a patient with chronic myelomonocytic leukemia.[8]

The orbit may be a focus of leukemic tumors, particularly granulocytic sarcoma in acute myelogenous leukemia.[3,4] In the past, these tumors have been called chloromas, because of the greenish color imparted by the enzyme myeloperoxidase. Sometimes, particularly in children, the orbital tumors are the presenting manifestation of acute mylogenous leukemia; these children seem to have a very poor prognosis despite aggressive therapy.[4] Leukemic involvement by direct infiltration has been reported in the eyelid[4,9] and the lacrimal drainage apparatus, where it masquerades as dacryocystitis.[4,10]

Thus, the eye and adnexa manifest leukemia in the same way as nonocular tissues do, since involvement reflects direct infiltration, secondary hematologic changes, or infection. Likewise, treatment is generally directed toward the underlying disease process.

References

1. Schrier SL. The leukemias and the myeloproliferative disorders. section 5, chap VIII. In: Rubenstein E, Fedelmann D, eds. *Scientific American Medicine*. New York, NY, 1987.
2. del Regato JA, Spjut HJ, Cox JD. Leukemias. In: del Regato

JA, Spjut HJ, eds. *Ackerman and del Rogato's Cancer: Diagnosis, Treatment, and Prognosis.* 6th ed. St. Louis, Mo: CV Mosby Co; 1985;1010-1035.

3. Rosenthal AR. Ocular manifestations of leukemia. A review. *Ophthalmology.* 1983;90:899-905.
4. Kincaid MC, Green WR. Ocular and orbital involvement in leukemia. *Surv Ophthalmol.* 1983;27:211-232.
5. Delaney WV Jr, Kinsella G. Optic disk neovascularization in leukemia. *Am J Ophthalmol.* 1985;99:212-213.
6. Kozlowski IMD, Hirose T, Jalkh AE. Massive subretinal hemorrhage with acute angle-closure glaucoma in chronic myelocytic leukemia. *Am J Ophthalmol.* 1987;103:837-838.
7. Font RL, Mackay B, Tang R. Acute monocytic leukemia

recurring as bilateral perilimbal infiltrates. Immunohistochemical and ultrastructural confirmation. *Ophthalmology.* 1985;92:1681-1685.
8. Wijetunga JG, Mitchell TR, Black PD. Chronic myelomonocytic leukemia presenting as a keratouveitis. *Ann Ophthalmol.* 1986;18:199-200.
9. Thall E, Grossniklaus H, Cappaert W, et al. Acute monocytic leukemia presenting in the eyelid. An immunohistochemical and electron microscopic study. *Ophthalmology.* 1986;93:1628-1631.
10. Benger RS, Frueh BR. Lacrimal drainage obstruction from lacrimal sac infiltration by lymphocytic neoplasia. *Am J Ophthalmol.* 1986;101:242-245.

Chapter 46

Multiple Myeloma and Other Plasma Cell Dyscrasias

WARREN W. PIETTE and CHRISTOPHER F. BLODI

The term "plasma cell dyscrasia" encompasses a wide variety of diseases or syndromes that have in common at least one of two processes—abnormal production of immunoglobulin and abnormal proliferation of plasma cells or plasmacytoid lymphocytes.[1] Each of these mechanisms results in separate pathophysiologic consequences and deserves separate consideration.

The abnormality in the production of immunoglobulin can be related to either the type produced or the amount synthesized. The abnormal types of immunoglobulins usually include those that precipitate in the cold (cryoglobulins), that agglutinate red blood cells in the cold (cold agglutinins), that are amyloidogenic (light-chain–type amyloid), or that induce nephropathy or neuropathy (e.g., light-chain–associated nephropathy or neuropathy). The proteins responsible for these conditions are usually monoclonal, and though the responsible clone of plasma cells is usually malignant, plasma cells that are clonal but not malignant can also secrete disease-producing monoclonal proteins.

The production of a greatly increased amount of immunoglobulin can also cause disease by inducing a platelet function defect or by producing the hyperviscosity syndrome. Patients with either monoclonal or polyclonal gammopathies frequently develop an acquired platelet function defect, which may be responsible for clinically evident problems of hemostasis. There are several mechanisms, all poorly understood, but most seem to be related to the presence of an abnormal type or quantity of serum immunoglobulin.

In contrast, though the hyperviscosity syndrome may present in a dramatic fashion, it is quite rare. The absolute amount of protein is important in producing the hyperviscosity syndrome, but the character of the protein is also important. For instance, IgM molecules, which are large and asymmetric, easily interact to impede serum flow; therefore the hyperviscosity syndrome most commonly occurs in the setting of Waldenström's macroglobulinemia, where large amounts of monoclonal IgM are produced. While the hyperviscosity syndrome is much rarer in myeloma patients, it has occurred when the monoclonal IgG or IgA produced is capable of forming aggregates or polymers in the serum. The hyperviscosity syndrome very rarely occurs in the setting of a polyclonal gammopathy, but it has been reported in rheumatoid arthritis patients with high titers of rheumatoid factor and in some patients with systemic lupus erythematosus.

The abnormal proliferation of plasma cells or plasma-cytoid lymphocytes, as seen in multiple myeloma and Waldenström's macroglobulinemia, leads to a different set of problems. Multiple myeloma cells are usually confined to the bone marrow until late in the course of disease, but proliferation in the marrow can lead to anemia, leukopenia, or thrombocytopenia. While rare cases of nonmalignant extramedullary solitary plasma-cytomas exist, the development of multiple plasmacytomas usually indicates the presence of a large total tumor mass stage of IgG or IgA multiple myeloma, or onset of IgD multiple myeloma, a rare but aggressive myeloma subset. Waldenström's macroglobulinemia results from the malignant proliferation of a slightly less well-differentiated cell, an IgM-producing plasmacytoid lymphocyte. These patients typically show early extra-medullary malignant cell growth, primarily in the liver, spleen, and lymph nodes, often resulting in dysfunction of these or other organs.

Other, less easily categorized, pathophysiologic mechanisms are also important. The lytic bone lesions typical of multiple myeloma probably result in part from a plasma cell–produced factor that stimulates osteoclast-mediated bone resorption. Suppression of normal serum immunoglobulin secretion is common in multiple myeloma and in other malignant plasma cell disorders. The severe anemia seen in some myeloma patients may be due not only to marrow infiltration by malignant cells but also to a poorly understood suppression of hemato-poiesis.

Finally, there are a few diseases that may present with prominent ocular or periocular findings and are associated with a higher than expected incidence of an associated monoclonal gammopathy or even frank multiple myeloma. These would primarily include necro-biotic xanthogranuloma with paraproteinemia, xanthoma disseminatum, normolipemic generalized plane xanthomas, and scleromyxedema.[2] The reason for this association is not known.

Systemic Manifestations

Not surprisingly, there is a wide spectrum of disease manifestations that occur in association with the plasma cell dyscrasias.[1,2] The specific disease manifestations in an individual patient are usually governed by the relative maturity of the responsible B cell (plasmacytoid lymphocyte or plasma cell), the type of cellular proliferation (benign or malignant), and the amount and characteristics of the immunoglobulin or immunoglobulins produced.

It is important to realize that 3% of people over age 70 may have a small serum M-spike, indicating the presence of a small amount of serum monoclonal protein. This monoclonal gammopathy is not ac-companied by any symptoms and its significance is unknown. In this group of patients, the incidence of progression to a malignant plasma cell dyscrasia is 10% or less.

Both benign and malignant plasma cell dyscrasias may present with the syndromes of cryoglobulinemia, cold agglutinin disease, hyperviscosity syndrome, and systemic light-chain amyloidosis. Cryoglobulinemia is defined by the presence of a cold-precipitable serum immunoglobulin. The cryoglobulin may be either a monoclonal protein or an immune complex. If it is monoclonal and is unstable at near body temperature, it may gel in vessels in acral body areas, leading to acral cyanosis, livedo reticularis, or purpura. The hyperos-motic conditions in the kidney may also induce protein precipitation, leading to renal injury. If the protein binds to another immunoglobulin in the serum, mixed cryo-globulinemia results. Though this also involves a cryo-globulin, the clinical disease pattern is more typically a nonacral immune complex disease characterized by ar-thritis, palpable purpuric lesions of necrotizing vasculi-tis, and glomerular injury.

Cold agglutinin disease results when an immuno-globulin, usually an IgM antibody, is able to agglutinate the patient's own red blood cells at the slightly lower body temperatures present in acral regions of the body. This often results in hemolysis, and acute episodes of hemolysis may be associated with paroxysms of lower back and flank pain, hematuria, and fever with rigors. Acral cyanosis is common, and in rare instances cutane-ous necrosis may occur.

The hyperviscosity syndrome may be due to an increase in the cellular components of the blood, as in polycythemia vera, or to greatly increased serum immu-noglobulin levels, as in the plasma cell dycrasias. Im-paired circulation with sludging in small blood vessels throughout the body results in purpura, hemorrhage, impaired vision, and a variety of neurologic findings. This syndrome occurs in up to 50% of Waldenström's macroglobulinemia patients but in only 2% of persons with multiple myeloma. Some of the bleeding abnormal-ities are due to acquired platelet dysfunction.

The malignant plasma cell dyscrasias most likely to present with one or more of these syndromes are multiple myeloma and Waldenström's macroglobulin-emia. These malignancies also present with or develop other more characteristic findings. Multiple myeloma typically occurs in patients 40 years of age or older, the mean age of diagnosis being around 60 years. Patients may present with pallor, weakness, fatigue, and dys-pnea on exertion due partly to anemia. Thrombocytope-nia or acquired platelet function defects may result in spontaneous hemorrhage or easily induced purpura or bleeding. The frequently associated suppression of nor-mal serum immunoglobulin production results in an increased incidence of infection, which in some patients may be complicated by acquired neutropenia. The cytopenias may be due not only to the tumor but also to cytotoxic chemotherapy. Osteolytic lesions of bone are characteristic of the disease, and are due to bone resorption, which may also lead to hypercalcemia, particularly in bedridden patients. Renal disease is quite common, often resulting in renal failure.

Waldenström's macroglobulinemia may present with findings similar to those of multiple myeloma, but it is more likely to show early evidence of extramedullary malignant cell growth and a higher incidence of the hyperviscosity syndrome, monoclonal cryoglobulinemia, and cold agglutinin disease. Osteolytic bone lesions are uncommon, and renal injury is less likely to occur.

The POEMS (*p*olyneuropathy, *o*rganomegaly, *e*ndocrinopathy, *M*-spike, *s*kin changes) syndrome is characterized by multisystem symptoms and findings. Though apparently rare in the United States, it has been reported there and may in fact be more common than published reports suggest. In a series of 102 patients from Japan with this syndrome, 62% were found to have papilledema.[3]

Ocular Manifestations

The most common ocular manifestations of multiple myeloma and other plasma cell dyscrasias are due to overproduction of monoclonal immunoglobulins (amyloidosis is discussed in Chapter 133). The most common intraocular site of accumulation of these immunoglobulins is the ciliary body, where it results in cyst formation. These collections of proteinaceous material and occasional cells, which are generally clear, can be found in 35 to 50% of all patients with multiple myeloma.[4] The cysts form between the pigmented and nonpigmented layers of the ciliary body epithelium, and because of their location in a portion of the eye that is difficult to examine, often they are not diagnosed in a clinical setting. Numerous cysts can be found in a single eye, and those of the pars plana can in time become confluent. Presumably, the noncellular portion of the cyst is made up of the same immunoglobulin molecule systemically overproduced.

While ciliary body cysts are the most common ocular manifestation of multiple myeloma, they have no pathologic sequelae. Ciliary body and pars plana cysts can also occur in other illnesses. It is important to realize that most patients who are found to have such cysts have no underlying systemic disorder whatsoever. The cyst contents may rupture into the vitreous, but they do not increase a patient's risk for rhegmatogenous retinal detachment.

Recently, iris microcysts have been a frequent finding in eyes of myeloma patients studied postmortem. However, these cysts also have been found in many other malignancies and have no clinical significance.[5]

Retinal changes are also frequently seen in patients with plasma cell dyscrasias, particularly multiple myeloma. Some of these changes can affect vision. Retinal capillary microaneurysms, flame and white-centered hemorrhages, and focal nerve fiber layer infarctions can be seen.[6]

Some patients with the hyperviscosity syndrome may develop branch or central retinal vein occlusion with associated tortuous vessels, intraretinal hemorrhages, retinal edema, and vitreous hemorrhage. Systemic treatment and plasmapheresis can have a significant beneficial effect on patients with retinal changes secondary to hyperviscosity.[7]

Anterior segment changes are generally rare. Crystalline corneal deposits in various layers have been reported in a few patients with multiple myeloma. They have been shown to consist of immunoglobulin chains overproduced systemically. Lamellar or penetrating keratoplasty may be necessary to clear the visual axis. Band keratopathy as a sequela to prolonged hypercalcemia can also be seen in patients with multiple myeloma and can be treated in a conventional manner with chelating drops.

Orbital tumors secondary to plasma cell dyscrasias are uncommon. Less than 1% of all orbital tumors are plasmacytomas, but these tumors may be the initial manifestation of the patient's systemic illness. These tumors may cause symptoms by orbital volume displacement. Proptosis, choroidal folds, and restriction of eye movement may be presenting signs. Tumor involvement of the optic nerve can cause decreased vision. These soft tissue tumors seem to respond well to either local external beam irradiation or systemic chemotherapy.

References

1. Williams JW, Beutler E, Ersley AJ, et al. *Hematology.* 3rd ed. New York, NY, McGraw-Hill; 1983.
2. Piette WW. Myeloma, paraproteinemias, and the skin. *Med Clin North Am.* 1986;70:155-176.
3. Nakanishi T, Sobue I, Toyokura Y, et al. The Crow-Fukase syndrome: a study of 102 cases in Japan. *Neurology (Cleveland).* 1984;34:712-720.
4. Baker TR, Spencer WH. Ocular findings in multiple myeloma. *Arch Ophthalmol.* 1974;91:110-113.
5. Knapp AJ, Gartner S, Henkind P. Multiple myeloma and its ocular manifestations. *Surv Ophthalmol.* 1987;31:393-351.
6. Holt JM, Gordon-Smith EC. Retinal abnormalities in diseases of the blood. *Br J Ophthalmol.* 1969;53:145-160.
7. Luxenburg M, Mausolf F. Retinal circulation in the hyperviscosity syndrome. *Am J Ophthalmol.* 1970;70:588-598.

Chapter 47

Pernicious Anemia

RODNEY I. KELLEN and RONALD M. BURDE

Pernicious anemia, one of the causes of megaloblastic anemia, is the result of vitamin B_{12} deficiency produced by decreased intrinsic factor secretion by the stomach.[1] The condition was first described by Thomas Addison of Guy's Hospital. The majority of cases occur in persons in or beyond their fifth decade, and the incidence in people of Scandinavian ancestry and in young black women is unusually high. Pernicious anemia is thought to be an autoimmune disease, because both antiparietal cell antibodies and blocking anti–intrinsic factor antibodies have been detected. Other so-called autoimmune diseases also are associated with pernicious anemia (e.g., Hashimoto's thyroiditis, vitiligo, and rheumatoid arthritis).

Systemic Manifestations

The onset of pernicious anemia is insidious, paralleling the slow depletion of vitamin B_{12} stores, resulting in megaloblastic anemia and neurologic dysfunction in the way of subacute combined degeneration. The diagnosis of pernicious anemia requires the demonstration of poor vitamin B_{12} absorption that can be reversed by the addition of exogenous intrinsic factor (i.e., Schilling's test). Vitamin B_{12} deficiency results in megaloblastosis and anemia as a result of deranged DNA synthesis. In the blood the erythrocytes are variable in shape and size but are normochromic. Macrocytes and macro-ovalocytes are prominent, mean cell volumes ranging from $100/\mu m^3$ to $140/\mu m^3$. Hypersegmented neutrophils are also a prominent feature of the peripheral blood smear (Fig. 47-1). Iron deficiency may mask the erythrocyte changes in megaloblastic anemia but not the neutrophil hypersegmentation. The bone marrow is extremely cellular, and megaloblastic changes are seen in all cell lines, but especially in the erythroid series.

The neurologic dysfunction seen in vitamin B_{12} deficiency may be due to the incorporation of abnormal fatty acids into the myelin surrounding axons.[2] The typical neurologic syndrome of vitamin B_{12} deficiency, "subacute combined system" disease of the spinal cord, is due to dysfunction and degeneration of the dorsal and lateral columns. The gradual onset of symmetric paresthesias in feet and fingers, associated with vibratory and proprioceptive sensory disturbance progressing to spastic ataxia, is characteristic but not invariable. Usually these are late features that may occur in the absence of anemia. Pathologically there is focal swelling of the myelin sheaths with vacuole formation. The nerve fibers then undergo degenerative changes involving both the myelin sheath and the axons, resulting in gliosis in chronic cases.

Ocular Manifestations

From an ophthalmologic point of view, the major impact of pernicious anemia is on the afferent visual system. In 1936, Cohen[3] was the first to describe visual dysfunction due to optic atrophy in a patient with pernicious anemia. The ocular findings predated the other clinical manifestations of pernicious anemia and occurred in the absence of any other neurologic signs. Optic neuropathy is now a well-recognized, albeit relatively rare, complication of pernicious anemia.[4-9] Vision loss is gradual, ranging from 20/40 to over 20/200, but not usually to the level of "no light perception." A central or cecocentral scotoma can be plotted both with a small white target and with a red target. Disc pallor may be present in varying degrees. Patients with optic neuropathy associated with pernicious anemia are most often men who smoke. Some authors consider the optic neuropathy to be part of the tobacco-alcohol amblyopia syndrome[6,8]; however, cases are reported in nonsmoking women, which seems to refute this hypothesis.[7] Abnormal (delayed) visual evoked responses have been demonstrated in patients with untreated pernicious anemia who had no other ocular abnormalities.[10]

Dyschromatopsia is well-documented in patients with pernicious anemia. Using the Farnsworth-Munsell 100-hue test, Chisholm[11] demonstrated color discrimination abnormalities in the yellow-blue and the violet-blue-green axes. He concluded that there are two components to the pernicious anemia–associated dyschromatopsia: one related to optic nerve dysfunction and the other to retinal visual receptor or pigment epithelial dysfunction.

Hemorrhagic retinopathy with dilated veins may

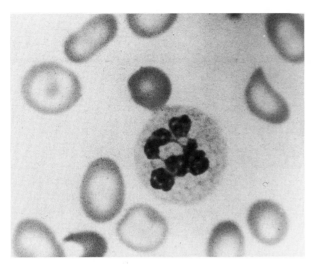

FIGURE 47-1 A peripheral blood smear demonstrating macrocytes, macro-ovalocytes, and a hypersegmented neutrophil (original magnification approximately ×3000).

occur.[12] Usually the severity of the retinopathy is a function of the degree of anemia. Although not specific, the retinal hemorrhages often have a white center, thought to be an aggregation of white cells. Extraretinal hemorrhage, pallor of the fundus and optic disc, and retinal edema with exudates are all less common features of the retinopathy of pernicious anemia. The hemorrhages probably reflect vascular damage due to hypoxia because the severity of the hemorrhagic retinopathy does not parallel the level of thrombocytopenia but rather the degree of anemia.

There is a single case report of a patient with upward gaze paralysis associated with pernicious anemia.[13] No other oculomotor disturbances have been reported.

Sjögren's syndrome, another autoimmune disease, has been described in association with pernicious anemia, but there is not an increased prevalence compared to age-matched controls.[14] Graves' disease and pernicious anemia have also been described in one patient, who showed lid lag and lid retraction but no restrictive ophthalmopathy.[15]

Treatment

Therapy consists of parenteral administration of vitamin B_{12}, preferably hydroxycobalamin. The normal daily requirement is between 2 and 5 μg daily; however, because liver stores of 2 to 5 mg must be replaced and any excess is easily excreted by the kidneys, much larger doses are given initially. A recommended treatment regimen includes intramuscular injection of 1000 μg

daily for 2 weeks, then twice weekly for an additional 4 weeks, and then monthly for the lifetime of the patient.[1] The neurologic dysfunction is reversible with therapy, provided permanent axonal damage has not occurred. Permanent changes are rare before 12 months after the onset of symptoms. Once the hemoglobin is restored to more normal levels the hemorrhagic retinopathy is reversible. The single reported case of upgaze paralysis responded to parenteral cyanocobalamin therapy.

References

1. Beck WS. Megaloblastic anemias. In: Wyngaarden JB, Smith LH Jr, eds. *Cecil Textbook of Medicine.* Part 1. Philadelphia, Pa: WB Saunders; 1982;853-860.
2. Victor M, Adams RD. Deficiency diseases of the nervous system. In: Petersdorf RG, Adams RD, Braunwald E, et al., eds. *Harrison's Principles of Internal Medicine.* 10th ed. New York, NY: McGraw-Hill; 1983;2115-2117.
3. Cohen H. Optic atrophy as the presenting sign in pernicious anæmia. *Lancet.* 1936;2:1202-1203.
4. Lerman S, Feldmahn AL. Centrocecal scotomata as the presenting sign in pernicious anemia. *Arch Ophthalmol.* 1961;65:381-385.
5. Cohen of Birkenhead, Lord. Where medicine and ophthalmology meet—some personal experiences. *Trans Ophthalmol Soc UK.* 1964;84:185-214.
6. Freeman AG, Heaton JM. The ætiology of retrobulbar neuritis in addisonian pernicious anæmia. *Lancet.* 1961;1:908-911.
7. Adams P, Chalmers TM, Foulds WS, et al. Megaloblastic anaemia and vision. *Lancet.* 1967;2:229-231.
8. Foulds WS, Chisholm IA, Steward JB, et al. The optic neuropathy of pernicious anemia. *Arch Ophthalmol.* 1969;82:427-432.
9. Juneja I. Amblyopia as the presenting symptom in pernicious anemia. *Neurol-India.* 1972;18(4):220-222.
10. Troncoso J, Mancall EL, Schatz NJ. Visual evoked responses in pernicious anemia. *Arch Neurol.* 1979;36:168-169.
11. Chisholm IA. The dyschromatopsia of pernicious anaemia. Acquired Colour Vision Deficiencies, International Symposium, Ghent 1971. *Mod Probl Ophthalmol.* 1972;11:130-135.
12. Rubin B, Munion L. Retinal findings in an adolescent with juvenile pernicious anemia. *NY State J Med.* 1982;82(8):1239-1241.
13. Sandyk R. Paralysis of upward gaze as a presenting symptom of vitamin B_{12} deficiency. *Eur Neurol.* 1984;23:198-200.
14. Williamson J, Paterson RWW, McGavin DDM, et al. Sjögren's syndrome in relation to pernicious anaemia and idiopathic Addison's disease. *Br J Ophthalmol.* 1970;54:31-36.
15. Fierro VS Jr, Freeman JS, Marazini P. Graves' disease in association with pernicious anemia: report of a case and review of the literature. *J Am Osteopath Assoc.* 1985;85:747-750.

Chapter 48

Platelet Disorders

GEORGE A. WILLIAMS

Platelets play a crucial role in normal hemostatic mechanisms, performing two major functions: they form hemostatic plugs at sites of vascular disruption and they participate in activating the coagulation system plasma proteins. Upon exposure to subendothelial extracellular matrix, platelets adhere to the wound site. This adhesion leads to platelet activation, a process involving changes in platelet shape and release of granules containing various procoagulant factors. In conjunction with fibrinogen, additional platelets aggregate, enlarging the hemostatic plug. Simultaneously with platelet activation coagulation is initiated. Platelets are crucial to the generation of thrombin and other coagulation factors. The end result is the formation of a platelet-fibrin clot, which reestablishes vascular integrity. Platelet disorders due to either abnormalities in platelet number or platelet dysfunction result in a wide variety of clinical presentations.

Systemic Manifestations

When evaluating a patient for a presumed platelet disorder, the two most helpful laboratory tests are bleeding time and platelet count. A prolonged bleeding time indicates a platelet disorder. The platelet count establishes whether the platelet disorder is due to a quantitative disturbance of platelets such as thrombocytopenia or to impaired platelet formation.[1] Most clinically significant platelet disorders are due to either thrombocytopenia or a platelet dysfunction syndrome.

Thrombocytopenia may result from (1) platelet production defects due to myeloproliferative disorders or drug-induced bone marrow toxicity, (2) platelet sequestration due to hypersplenism, or (3) increased platelet destruction due to either immune-mediated or non–immune-mediated mechanisms. The clinical hallmark of thrombocytopenia is petechiae; however, when the platelet count is very low, purpura, mucosal bleeding, or even deep tissue bleeding may be seen.

Thrombocytopenia due to platelet production defects is characterized by an abnormal bone marrow biopsy. This may demonstrate either diffuse hematopoietic disease, as occurs with marrow aplasia or infiltration, or simply abnormal megakaryopoiesis. Hypersplenism usually results in relatively mild thrombocytopenia. Clinically significant bleeding rarely occurs unless a bleeding disorder is present concurrently.

Increased platelet destruction due to either immune or nonimmune mechanisms often results in profound thrombocytopenia. Immune-mediated mechanisms include the production of antiplatelet antibodies following drug ingestion or as in idiopathic thrombocytopenic purpura (ITP).[1] ITP is an autoimmune disease typically seen in young women that may be associated with Graves' disease.[2] The clinical presentation of both ITP and drug-induced platelet destruction ranges from asymptomatic petechiae to central nervous system hemorrhage. A thorough drug history is necessary to rule out drug exposure; quinine and quinidine are the agents most studied, but others, including heparin, thiazide diuretics, phenytoin, diazepam, and acetaminophen, have been implicated. Apparently antibodies to these drugs also have an affinity to receptors on the platelet membrane, resulting in platelet destruction.[1]

The best example of non–immune-mediated acceleration of platelet destruction is thrombotic thrombocytopenic purpura (TTP). This poorly understood entity is characterized by diffuse platelet activation and aggregation resulting in five major clinical manifestations: microangiopathic hemolytic anemia, thrombocytopenia, fever, central nervous system dysfunction, and renal disease. In the past, TTP has been confused with disseminated intravascular coagulation (DIC); however, unlike in DIC, coagulation tests in TTP are normal, and TTP and DIC are now regarded as separate entities.[1]

Platelet dysfunction syndromes are characterized by a prolonged bleeding time with a normal platelet count. They may be inherited or acquired disorders. Thrombasthenia and Bernard-Soulier disease are inherited disorders of platelet membrane receptors resulting in abnormal aggregation or adhesion, respectively. Abnormal adhesion of platelets also occurs in von Willebrand's disease, owing to a deficiency of von Willebrand's factor, which mediates platelet adhesion to subendothelial surfaces. Acquired platelet dysfunction may occur with uremia, liver disease, or macroglobulinemia, but perhaps it occurs most commonly after ingestion of aspirin or other nonsteroidal anti-inflammatory agents. These drugs inhibit platelet cyclo-oxygenase and thereby diminish platelet aggregation and granule release. Aspirin irreversibly acetylates cyclo-oxygenase and thus prolongs the bleeding time until sufficient new platelets are formed, which takes about 7 days.[1]

Ocular Manifestations

Clinically significant ocular manifestations of platelet disorders are well-described. Drug-induced thrombocytopenia may cause spontaneous bleeding, such as hyphema and vitreous hemorrhage, or may result in hemorrhagic complications after ocular surgery.[3-5] Bilateral serous retinal detachments, choroidal hemorrhage, and optic neuritis have also been associated with drug-induced thrombocytopenia.[6]

Retinal hemorrhages may occur with ITP or thrombocytopenia of other causes. The strong association of ITP with Graves' disease[2] should remind the ophthalmologist to assess platelet function in patients with Graves' disease and unexplained ocular bleeding, and particularly in Graves' patients undergoing ocular or orbital surgery.

Retinal hemorrhages, extraocular muscle palsies, papilledema, and various visual field defects have been described in TTP. Serous retinal detachments may also occur in TTP, presumably due to occlusion of the choriocapillaris by microthrombi. Diffuse pigment epithelial changes may follow resolution of the serous detachments.[7]

Congenital platelet dysfunction syndromes may also have ocular manifestations. Thrombasthenia may present with spontaneous preretinal or vitreous hemorrhage.[8] Abnormalities of platelet aggregation have been associated with familial exudative vitreoretinopathy[9] and congenital miosis.[10] Both thrombocytopenia and abnormal platelet protein metabolism have been described in retinitis pigmentosa.[11]

Aspirin use creates an acquired platelet dysfunction state, which may result in ocular bleeding, particularly following trauma or surgery.[4] Aspirin has been associated with an increased rate of rebleeding following traumatic hyphema.[12] Macroglobulinemia may also cause platelet dysfunction resulting in retinal hemorrhages.[8] Presumably the elevated abnormal plasma proteins interfere with platelet membrane receptor function.

References

1. Handin RI. Hemorrhagic disorders II. Platelets and purpura. In: Beck WS, ed. *Hematology*. Cambridge, Mass: MIT Press; 1985;433-456.
2. Adrouny A, Sandler RM, Carmel R. Variable presentation of thrombocytopenia in Graves' disease. *Arch Intern Med.* 1982;142:1460-1464.
3. Ackerman J, Goldstein M, Kanarek I. Spontaneous massive vitreous hemorrhage secondary to thrombocytopenia. *Ophthalmic Surg.* 1980;11:636-637.
4. Paris GL, Waltuch GF. Salicylate-induced bleeding problems in ophthalmic plastic surgery. *Ophthalmic Surg.* 1982;13:627-629.
5. Shaw HE, Smith SW, North-Coombes JD. Quinine-induced thrombocytopenia complicating eyelid surgery. *Arch Ophthalmol.* 1987;105:1176.
6. Klepach GL, Wray SH. Bilateral serous retinal detachment with thrombocytopenia during penicillamine therapy. *Ann Ophthalmol.* 1981;13:201-203.
7. Lambert SR, High KA, Cotlier R, et al. Serous retinal detachments in thrombotic thrombocytopenic purpura. *Arch Ophthalmol.* 1985;103:1172-174.
8. Vaiser A, Hutton WL, Marengo-Rowe AJ, et al. Retinal hemorrhage associated with thrombasthenia. *Am J Ophthalmol.* 1975;80:258-262.
9. Chaudhuri PR, Rosenthal AR, Goulstine DB, et al. Familial exudative vitreoretinopathy associated with familial thrombocytopathy. *Br J Ophthalmol.* 1983;67:755-758.
10. Stormorken H, Sjaastd O, Langslet A. A new syndrome: Thrombocytopathia, muscle fatigue, asplenia, miosis, migraine, dylexia and ichthyosis. *Clin Genet.* 1985;28:367-374.
11. Vaaden MJ, Hussain AA, Chan IPR. Studies on retinitis pigmentosa in man. I. Taurine and blood platelets. *Br J Ophthalmol.* 1982;66:771-775.
12. Crawford JS, Lewandowski RL, Chan W. The effect of aspirin on rebleeding in traumatic hyphema. *Am J Ophthalmol.* 1975;80:543-545.

Chapter 49

Polycythemia

KENNETH G. NOBLE

Classification

Polycythemia ("many cells") may be conveniently divided into two types, primary and secondary. Both types are characterized by increased blood volume, increased blood viscosity, and decreased blood flow, all of which result in the characteristic signs and symptoms of the disorder. In addition, the clinical manifestations of secondary polycythemia may be related, in part, to the underlying disease.

Polycythemia Vera (Vaquez's Disease)

Polycythemia vera is an idiopathic disorder of the hematologic system characterized by a marked increase in the absolute number of red blood cells and in the total blood volume. Leukocytosis, thrombocytosis, and bone marrow hyperplasia are usually accompanying features.

Whereas the normal red cell count is 4.5 to 5.5 million/mm^3, increases to 6 to 8 million/mm^3 are not uncommon. The individual red corpuscles usually appear normal. Total red cell volume may be up to three times normal and blood viscosity up to eight times normal. The bone marrow usually shows hyperplasia of all elements, but there is nothing diagnostic to distinguish polycythemia vera from secondary polycythemia, or even from hypercellular marrow in a normal person.

The disease has an insidious onset, affects the middle-aged and elderly, and runs a chronic course. On occasion other family members may develop the disease (6% of the Polycythemia Study Group had registered patients with affected family members).[1] It is extremely uncommon in blacks and comparatively common in Jews. Males are affected slightly more often than females.

Secondary Polycythemia (Erythrocytosis)

An increase in the number of red blood cells (erythrocytosis) may be the result of a variety of nonhematologic disorders. Secondary erythrocytosis is related to an increase in the erythropoiesis-stimulating factor (erythropoietin) as a result of the stimulus of hypoxia or of aberrant production from a tumor, cyst, or vascular anomaly (Table 49-1).

Hypoxia, in a manner not yet known, induces the kidneys to secrete a hormone, erythropoietin, which in turn induces the bone marrow to proliferate and produce more red blood cells. Tissue hypoxia may result from a number of causes: (1) decrease in atmospheric oxygen (at increasingly higher altitudes there is a continued decrease in oxygen tension), (2) pulmonary disease (inadequate oxygenation of blood in the pulmonary circulation), (3) congenital heart disease (a right-to-left shunt in which unoxygenated blood bypasses the pulmonary circulation), (4) hypoventilation syndromes (primary or secondary disorder of the respiratory center in the central nervous system), or (5) abnormalities of hemoglobin, inherited or acquired (the abnormal hemoglobin has an increased affinity for oxygen and a "reluctance" to release it to the tissue).

Various tumors, cysts, and vascular anomalies are associated with the inappropriate production of erythropoietin that is identical to its normal counterpart and may be produced directly by the lesion or as a result of tissue anoxia from an enlarging mass.

Systemic Manifestations

Polycythemia may present with a multitude of nonspecific complaints (headache, weakness, dizziness, sweating, weight loss, and blurred vision) that may suggest a diagnosis of neurasthenia, or there may be no symptoms at all, in which case the disease is sometimes noted fortuitously on a routine examination. While no organ system is spared the effects, the skin, mucous membranes, and neuromuscular system are the primary sites of evident disease.

The intense red skin color (rubor) is particularly striking in the lips, cheeks, and tip of the nose (resembling the picture of chronic alcoholism). The bluish discoloration of cyanosis may be seen in the distal

TABLE 49-1 Etiologies of Secondary Polycythemia

I. Hypoxia

    A. High altitude

        1. Acute mountain sickness
        2. Chronic mountain sickness (Monge's disease)

    B. Pulmonary disease

        1. Chronic obstructive disease
        2. Diffuse pulmonary infiltrates (fibrous, granulomatous)
        3. Kyphoscoliosis
        4. Multiple pulmonary emboli
        5. Cor pulmonale (arterial desaturation, pulmonary artery hypertension)
        6. Ayerza's syndrome (asthma, bronchitis, dyspnea, cyanosis, congestive failure, right ventricular hypertrophy due to pulmonary artery narrowing as a consequence of inflammation, e.g., syphilis, arteriolar sclerosis, or congenital hypoplasia)
        7. Cavernous hemangioma
        8. Arteriovenous fistula

    C. Congenital heart disease

        1. Pulmonary stenosis (usually with patent foramen ovale, persistent patent ductus arteriosus, or defective ventricular or atrial septum)
        2. Persistent truncus arteriosus
        3. Complete transposition of the great vessels
        4. Tetralogy of Fallot

    D. Hypoventilation syndromes

        1. Ondine's curse (idiopathic disease of the medullary respiratory center)
        2. Secondary diseases of the respiratory center (bulbar poliomyelitis, vascular thrombosis, encephalitis)
        3. Pickwickian syndrome (extreme obesity, somnolence, hypercapnia, cyanosis that may be reversed by weight loss)

    E. Hemoglobin abnormalities

        1. Inherited (many varieties)
        2. Acquired
            a. Drugs (nitrites, nitrates, sulfonamides, aniline and nitrobenzene compounds)
            b. Chemicals (carboxyhemoglobin from heavy smoking, phosphorous)

II. Inappropriate erythropoietin production (in relative order of frequency)

    A. Hypernephroma

    B. Hepatocellular carcinoma

    C. Cerebellar hemangioma

    D. Cystic kidney disease

    E. Leiomyosarcoma of uterus

    F. Hydronephrosis

    G. Rarely other renal tumors, pheochromocytoma, carcinomas

extremities. Ecchymosis and purpura are common, and a wide variety of skin lesions have been seen (dry skin, eczema, acne, urticaria). The mucous membranes are deep red, and bleeding of the nose and gums may occur. An extremely common and specific complaint, related to the plethoric changes in the skin, is intense itching after a bath, which is particularly aggravated by the vasodilating effect of hot water.

Cerebral blood flow is significantly diminished with elevated levels of red blood cells, which presumably accounts for many of the previously described nonspecific symptoms and for the signs of grand mal seizure, myoclonus, paralysis, and various psychological disturbances. Limb pain may be due to bone pressure from swollen hyperplastic marrow.

Involvement of other organ systems may produce dyspnea, hoarseness, hemoptysis, hemothorax, abdominal fullness and tenderness (due to splenomegaly or to peptic ulcer), gastrointestinal and urogenital bleeding, and gout (secondary gout due to hyperuricemia).

Ocular Manifestations

The ophthalmic manifestations of polycythemia are seen in the ocular vessels and are related to the increase in blood viscosity with slowing of the circulation and resulting hypoxia. The high oxygen demands of retinal tissue may account for retinal manifestations early in the course of polycythemia and for the fact that retinal abnormalities may be worse than the systemic manifestations of the disease.[2]

It is felt that the retinopathy is more likely to occur in primary than in secondary polycythemia, because the former tends to have higher blood cell counts and more hyperviscosity.[2] Actually, the changes are identical in both and vary only in degree.

The initial change is a darkening, dilatation, and tortuosity of the retinal veins; the arterioles remain normal. The next changes are in the color of the disc, which becomes hyperemic, and of the entire fundus, which assumes a deeper, darker purple hue.

The first evidence of microvascular decompensation is the appearance of small scattered superficial and deep hemorrhages. This may be due in part to selective degeneration of the intramural pericytes.[3] This appearance resembles that of venous stasis retinopathy. Retinal edema may also result from the breakdown of the blood-retinal barrier.

Optic disc swelling (papilledema) of 3 diopters or more may occur in approximately 10% of cases.[4] The causes of the disk swelling may be multiple and may include such factors as occlusion of small vessels of the nerve heads,[5] increased intracranial pressure (which may be associated with hemangiomatous tumors of the posterior fossa, which themselves lead to a secondary erythrocytosis), or an increase in the venous pressure with passive congestion. On occasion, papilledema may be the presenting sign in this disorder.[5]

The patient becomes symptomatic (blurred vision) with the development of retinal or optic disc edema. However, the most serious vision-threatening consequence of polycythemia is central retinal vein (CRV) occlusion, which may be unilateral or bilateral.[6] This is an uncommon complication, and an additional underlying retinal vascular disease may be necessary for frank occlusion to occur.[2] The prognosis for CRV occlusion is more favorable in polycythemia than in disease of other causes[7]; however, this may reflect the fact that most cases have less arteriolar and capillary insufficiency than the typical occlusion related to arteriosclerosis and hypertension.

The conjunctival vessels are the other area where manifestations of the disease commonly occur. The vascular congestion generally parallels the skin rubor. On rare occasions the dilated tortuous vessels may simulate recalcitrant conjunctivitis while in fact they are the presenting sign of polycythemia.[8]

Visual symptoms may occur in the absence of visible eye signs—amaurosis fugax, hemianopia, and other visual field scotomas, visual hallucinations, and blurred vision. These are almost certainly a result of cerebrovascular insufficiency.[9]

Treatment

The institution of treatment in primary and secondary polycythemia is based on the nature and severity of the symptoms. In the secondary form, treatment of the underlying disease, if possible, is always considered first.

Symptomatic relief is accomplished by lowering the red cell titer and reducing the total blood volume. This can be accomplished by removing blood (phlebotomy), destroying reds cells (phenylhydrazine therapy), or suppressing red cell production (irradiation with radioactive phosphorus [^{32}P] or chemotherapy). Splenectomy should be performed only in the late stages of the disease and only because of symptoms due to its massive size. Potent antihistamines may alleviate the symptoms of pruritus, urticaria, and upper gastrointestinal distress, since these symptoms have been correlated with increased blood levels of histamine.[10]

Prognosis

The most common causes of death are intravascular thrombosis and hemorrhage. It appears that life expectancy is prolonged and the quality of life better for wisely managed patients than for untreated patients.

References

1. Hoffman R, Wasserman LR. Natural history and management of polycythemia vera. *Adv Intern Med.* 1979;24:255.
2. Wise GN, Dollery CT, Henkind P. *The Retinal Circulation.* Hagerstown, Md: Harper & Row; 1971;392,396,397.
3. de Oliveira LF. Pericytes in diabetic retinopathy. *Br J Ophthalmol.* 1960;50:134.
4. Wagener HP, Rucker CW. Lesions of the retina and optic nerve in association with blood dyscrasias. In: Sorsby A, ed. *Modern Trends in Ophthalmology.* London: Butterworth; 1948;300.
5. Kearns TP. Changes in the ocular fundus in blood diseases. *Med Clin North Am.* 1956;40:1209.
6. Duke-Elder S. *System of Ophthalmology.* Vol X, Diseases of the retina. London: Kimpton; 1967;383-385.
7. Ballantyne AJ, Michaelson IC. *The Fundus of the Eye.* Baltimore, Md: Williams & Wilkins; 1980;342-343.
8. Lindsey J, Insler MS. Polycythemia rubra vera and conjunctival vascular congestion. *Ann Ophthalmol.* 1985;17:62.
9. Thomas D, Marshall J, Ross Russell R, et al. Cerebral blood flow in polycythemia. *Lancet.* 1977;2:161.
10. Gilbert HS, Warner RRP, Wasserman LR. A study of histamine in myeloproliferative disease. *Blood.* 1966;28:795.

Chapter 50

Sickle Cell Disease

STEVEN B. COHEN and GEOFFREY R. KAPLAN

The sickle cell hemoglobinopathies are a group of inherited diseases characterized by the production of abnormal hemoglobin that, under certain conditions, can cause abnormalities in erythrocyte function. Normal adult hemoglobin, hemoglobin A, consists of two alpha and two beta peptide chains. Erythrocytes that contain hemoglobin A exclusively maintain the configuration of a biconcave disc, which ensures cell pliability and allows the cells to flow easily through capillaries. Sickle hemoglobin, hemoglobin S, occurs because of a genetic defect resulting in the substitution of a single amino acid, valine, for glutamic acid at the 6 position of the beta peptide chain. This hemoglobin maintains normal physical chemical properties in the oxy conformation, but in the deoxy conformation, the hemoglobin molecules aggregate into long polymers. This property gives erythrocytes containing hemoglobin S the tendency to assume a rigid, sickle shape, especially in the presence of hypoxia. The rigid configuration of these cells impedes blood flow by promoting impaction of the erythrocytes in small vessels, resulting in increased hypoxia. A vicious cycle occurs: the increasing hypoxia causes more erythrocyte sickling and further impedance to blood flow.[1]

While production of hemoglobin S defines the sickle hemoglobinopathies, additional genetic defects contribute to the spectrum of the disease. The most clinically significant are the production of hemoglobin C, which is due to a substitution of the amino acid lysine for glutamic acid at the 6 position of the beta chain, and beta thalassemia, which is due to a defect in the rate of production of the entire beta chain.

Inheritance of a gene for hemoglobin S from each parent results in a homozygous condition known as sickle cell anemia, or SS disease. Inheritance of a gene for hemoglobin S from one parent and for hemoglobin C from the other parent results in a doubly heterozygous condition known as sickle-C or SC disease. Inheritance of the gene for hemoglobin A from one parent and the gene for hemoglobin S from the other results in another doubly heterozygous condition, sickle trait, or AS disease.

Systemic Manifestations

In general the systemic manifestations of the sickle cell hemoglobinopathies are more pronounced in the homozygous (SS) form of the disease and fewer manifestations are exhibited in the heterozygous states (SC, S-thal, AS).

The hallmark of SS disease is a chronic compensated hemolytic anemia with a hematocrit between 18 and 30% and a hemoglobin between 6.5 and 10 g/dl. Compensation by increased erythropoiesis results in an elevated reticulocyte count, ranging from 10 to 25%. The anemia worsens during episodes of decreased erythropoiesis caused by intermittent adverse conditions, such as infection or folic acid deficiency. The anemia present in SC or S-thal disease is milder than that in SS disease.[2,3]

The primary disabilities suffered by patients with sickle cell disease are related to vaso-occlusive phenomena resulting from the previously discussed properties of S hemoglobin. Painful sickle cell crisis is characterized by excruciating pain in various parts of the body, particularly the back, chest, or extremities. The frequency is variable, and precipitating causes, while not usually identifiable, may include dehydration, cold weather, or infections.[3]

Acute vaso-occlusive complications can include cerebrovascular accidents, acute chest syndrome from involvement of the pulmonary vessels, hepatic crisis, priapism, or acute renal papillary infarction.[2,3]

Chronic organ damage may include a number of organ systems. Chronic skin ulcers may occur in the lower extremities. Bone infarcts can occur, and aseptic necrosis at the head of the femur is common. Microinfarction of the renal medulla results in the inability to concentrate urine, and papillary infarcts may result in prolonged painless hematuria. Autosplenectomy in patients with SS disease is common.[4]

Additional manifestations not related to vaso-occlusion are increased susceptibility to infections, cholelithiasis, and abnormal growth and development.[2-4]

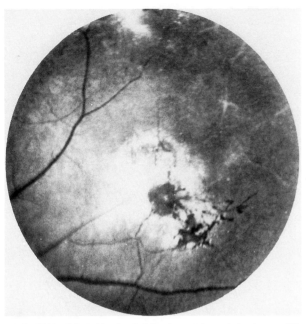

FIGURE 50-1 Typical black sunburst lesion with pigment hyperplasia and iridescent granules.

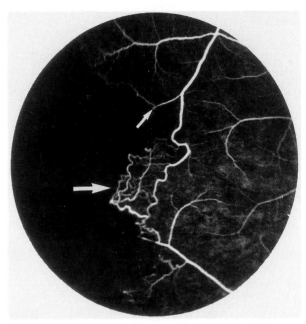

FIGURE 50-2 Fluorescein angiogram demonstrates arteriolar occlusion (*small arrow*) and arteriolar-venular anastomoses (*large arrow*).

Ocular Manifestations

The ocular complications of sickle cell disease are also due to vaso-occlusive phenomena; however, unlike the systemic manifestations, the complications are most prominent in SC and S-thal disease, less common in SS disease, and rare in AS disease. Although sickle cell disease primarily affects the retina, other ocular findings may be seen. These include lid edema secondary to sickle crisis, conjunctival sickling sign, iris atrophy, iris neovascularization, angioid streaks, and optic disc sign.[5]

Peripheral retinal vascular occlusion is the primary pathologic process in sickle cell retinopathy. Focal occlusion may result in silver wire arteriolar changes. When a medium-sized arteriole is suddenly occluded, a salmon patch hemorrhage may result. These reddish orange hemorrhages are bordered anteriorly by the internal limiting membrane and vary in size from one fourth to 1 disc diameter. As the hemorrhage resolves a small retinoschisis cavity remains, which may contain iridescent granules. Chorioretinal scars known as "black sunbursts" are also found (Fig. 50-1). These may be another end-point of salmon patch hemorrhages, presumably due to hemorrhage involving deeper layers of the retina.[6]

Permanent occlusion of the peripheral arterioles causes ischemia of the peripheral retina, resulting in the spectrum of proliferative sickle cell retinopathy which is classified by five progressive stages. Stage I is the occlusion of peripheral arterioles (Fig. 50-2). Formation of peripheral arteriolar–venular anastomoses is the hallmark of stage II (see Fig. 50-2), and progression to peripheral retinal neovascularization is seen in stage III (Figs. 50-3, 50-4). The neovascularization appears classically in a sea fan configuration. Subsequent vitreous hemorrhage characterizes stage IV disease. Progression of the processes in stage III or IV disease may result in either tractional or rhegmatogenous retinal detachment. This defines stage V disease.[7] While some patients follow the entire course of proliferative sickle retinopathy through all five stages, the condition may arrest spontaneously at any stage, without further progression.

In addition to the peripheral retinal changes described, occlusion of perimacular arterioles may result in macular ischemia. Although central vision may rarely be affected by these changes, visual acuity commonly remains normal and only subtle changes in vision function may be detected.[8,9]

Treatment for the ocular complications of sickle retinopathy is related to the stage of the disease. No prophylactic treatment has yet been shown to be effec-

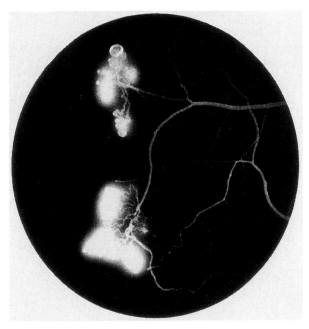

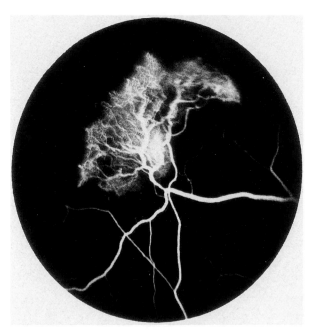

FIGURE 50-3 Fluorescein angiogram demonstrates neovascularization at two sites in sea fan configurations.

FIGURE 50-4 Fluorescein angiogram demonstrates neovascularization in a sea fan configuration.

tive in preventing neovascularization, so no treatment is presently recommended for stage I and stage II disease. However, observation for progression should continue. Once stage III develops interventional therapy may be used.

Feeder vessel photocoagulation of the retinal neovasculature has been shown to be effective in closing sea fans and reducing the incidence of visual loss.[10] However, with this modality there is a high rate of complications such as choroidal neovascularization and retinal detachment.[11] Scatter argon laser photocoagulation has therefore been used in the management of stage III disease,[12] apparently without the significant risks of feeder vessel treatment. A prospective clinical trial is currently under way to evaluate its effectiveness. Once stage IV disease with symptomatic nonclearing vitreous hemorrhage or stage IV disease with retinal detachment occurs, standard vitrectomy and retinal detachment techniques may be used for treatment.[13] Special considerations to prevent anterior segment ischemia should be taken perioperatively. These include use of supplemental intraoperative and postoperative oxygen, adequate pupillary dilation with parasympatholytic drugs, use of local anesthesia whenever possible, avoidance of sympathomimetic agents, and maintenance of lowered intraocular pressure to maximize intraocular perfusion while avoiding hemoconcentration.

During surgery, transection or traction on rectus muscles and transscleral coagulation in the horizontal meridians should be avoided. Cryotherapy should be used rather than diathermy, with care taken to avoid the long posterior ciliary arteries, and subretinal fluid should be drained whenever possible to create a high buckle while minimizing elevation of intraocular pressure. In addition, the protective effects of exchange transfusion to obtain 50 to 60% hemoglobin A and a hematocrit value under 40% should be considered to minimize intraoperative and postoperative complications. The risks of transfusion reactions, hepatitis, and acquired immunodeficiency syndrome (AIDS) must be carefully considered when the use of exchange transfusion is contemplated.

References

1. Horne MK II. Sickle cell anemia in rheologic disease. *Am J Med.* 1981;70:288-298.
2. Bonn HF. Disorders of hemoglobin. In: Braunwald E, et al., ed. *Harrison's Principles of Medicine.* 11th ed. New York, NY: McGraw-Hill; 1987;1518.
3. Mentzer WC, Wang WC. Sickle cell disease: Pathophysiology and diagnosis. *Pediatr Ann.* 1980;9:10.
4. Forget BG. Sickle cell anemia and associated hemoglobinopathies. In: Wyngaarden JB, Smith, LH, eds. *Cecil Textbook*

of Medicine. 17th ed. Philadelphia, Pa: WB Saunders; 1985;927–932.

5. Nagpal KC, Goldberg MF, Rabb MF. Ocular manifestations of sickle hemoglobinopathies. *Surv Ophthalmol.* 1977;21:391.

6. Asdourian GK, Nagpal KS, Goldbaum M, et al. Evolution of the retinal black sunburst lesion in sickle hemoglobinopathies. *Br J Ophthalmol.* 1975;59:713-716.

7. Goldberg MF. Classification and pathogenesis of proliferative sickle retinopathy. *Am J Ophthalmol.* 1971;71:649.

8. Asdourian GK, Nagpal KC, Busse B, et al. Macular and perimacular vascular remodeling in sickling hemoglobinopathies. *Br J Ophthalmol.* 1976;60:431.

9. Kaplan GR, Van Houten PA, Goldberg MF, et al. Computer assisted area analysis of macular ischemia in sickle retinopathy. *Invest Ophthalmol Vis Sci.* 1981;28(suppl):111.

10. Jampol LM, Condon P, Farber M, et al. A randomized clinical trial of feeder vessel photocoagulation of sickle cell retinopathy. I. Preliminary results. *Ophthalmology.* 1983;90:540-545.

11. Condon P, Jampol LM, Farber MD, et al. A randomized clinical trial of feeder vessel photocoagulation of sickle cell retinopathy. II. Update and analysis of risk factors. *Ophthalmology.* 1984;91:1496-1498.

12. Rednam KRV, Jampol LM, Goldberg MF. Scatter retinal photocoagulation for proliferative sickle retinopathy. *Am J Ophthalmol.* 1982;93:594.

13. Jampol LM, Green JL, Goldberg MF, et al. An update: vitrectomy surgery and retinal detachment repair in sickle cell disease. *Arch Ophthalmol.* 1982;100:541-593.

PART 8

Infectious Diseases

Section A

Bacterial Diseases

Chapter 51

Botulism

JOHN W. FLOBERG and JONATHAN D. TROBE

Botulism is a form of descending paralysis caused by an exotoxin produced by the gram-positive anaerobic bacillus *Clostridium botulinum.* Preformed toxin gains entry when the human host ingests contaminated food. Toxin may also be elaborated in a contaminated wound or in an intestinal tract overgrown with *Cl. botulinum.*

Most cases of botulism are food borne, usually from inadequately sterilized home-canned vegetables (70%), meat, or fish (30%). From 1976 to 1984, 124 food-borne outbreaks involving 308 cases were reported in the United States; the mortality rate was 7.5%.[1] Four outbreaks accounted for 40% of the cases, and fresh foods that are inadequately heated have been increasingly implicated in addition to canned foods. Wound botulism occurs much less frequently; only 16 cases (12.5% mortality) were reported in the same period. Most involved traumatic penetrating wounds and compound fractures; however, intravenous drug abusers can develop asymptomatic infected cysts.[2]

There is a third group of patients who have botulism from an unknown source. Thirty-one cases were reported from 1976 to 1984, with a 29% fatality rate. Some of these cases represent an adult form of "infant botulism," first described in 1976. Infants with this form of botulism are presumed to have ingested spores, which germinate in the gastrointestinal tract and produce toxin. Adults with infant botulism have underlying gastrointestinal disease or have been treated with antibiotics, so that the normal gut flora is replaced by toxin-producing *Cl. botulinum.*[3-5]

Three types of botulinum toxin produce human illness: A (60%), B (30%), and E (10%). Type A, the most toxic, is generally found west of the Mississippi River; type B is found to the east; type E is associated mainly with seafood. The toxin produces its effects by interfering with transmission at cholinergic nerve terminals in the peripheral and autonomic nervous system, affecting both striated and smooth muscle.[6-8] Neuromuscular transmission is normally accomplished by the calcium ion–mediated fusion of synaptic vesicles with the nerve membrane and the subsequent release of acetylcholine (ACh) into the synaptic cleft. Botulinum toxin interferes with the calcium-mediated step, either by blocking entry of calcium into the cell or by interrupting the interaction between calcium and the synaptic vesicle. With prolonged intoxication, the synapse ceases to function and there is denervation atrophy. In the recovery phase, nerve terminals sprout to form neuromuscular junctions at new locations.

Systemic Manifestations

In cases of food-borne botulism, the most common presenting symptoms reflect gastrointestinal distress (nausea, vomiting, cramping, diarrhea), cranial nerve dysfunction (diplopia, dysarthria, dysphagia), autonomic dysfunction (dry mouth and dry eyes), and limb weakness. The gastrointestinal symptoms begin within 2 hours of ingestion in half the cases. Patients usually consult a physician within 48 hours, by which time a pattern of descending bulbar, respiratory, and limb paralysis may be emerging and respiratory distress may be imminent. Nystagmus, ataxia, and extensor plantar reflexes have been described, suggesting impairment of central cholinergic transmission. In spite of frequent (15%) complaints of paresthesias, results of sensory examination are always normal. In wound botulism and in the adult form of infant botulism, symptom onset may be delayed several days, as the toxin is not ingested but is produced in situ.

Ocular Manifestations

Ocular manifestations occur frequently and early in botulism. Apart from gastrointestinal distress, the most common presenting symptoms are blurred and double vision. The blurred vision is the result of failure of accommodation, whereas double vision is due principally to weakness of the lateral recti. Other ocular motility deficits are less marked. The other prominent eye signs are ptosis (70%), dilated, sluggishly reactive pupils (50%), and reduced tear secretion. Pupillary signs may be absent at the time of presentation, becoming apparent only as the pattern of descending paralysis develops.[6-9]

Diagnosis

The differential diagnosis of ocular and bulbar muscle weakness proceeding to involve the diaphragm and extremities includes myasthenia gravis, the descending form of Guillain-Barré Syndrome (GBS; about 4% of such cases), ischemia involving the vertebrobasilar circulation, tick paralysis, and diphtheria. None of the entities on this list has the particular evolution of somatic and autonomic ocular, bulbar, and extremity weakness, together with gastrointestinal distress, that is so typical of botulism. Muted forms of botulism that lack all these elements may, however, present a diagnostic puzzle. In

this regard, a positive Tensilon (edrophonium) test by no means excludes the diagnosis of botulism. Twenty-five percent of patients with botulism show improved extremity and ocular muscle strength immediately after Tensilon injection.[4,7]

Electromyographic (EMG) and nerve conduction studies can be critical diagnostic aids but may be falsely negative in 25% of cases. Both severely and mildly involved limbs must be examined, and repetitive nerve stimulation must be performed. In mildly weak muscles, a modest incremental response in amplitude of the compound muscle action potential is noted at rapid rates of stimulation, similar to but of a lesser degree than that seen in Eaton-Lambert syndrome and unlike the decremental response seen in myasthenia gravis. Nerve conduction velocities are always normal, whereas they are frequently reduced in GBS.[10,11]

Treatment

The emergency treatment of botulism involves administration of antitoxin and intubation in anticipation of impending cardiorespiratory distress. Guanidine, an agent that increases the duration of the action potential at the nerve terminal, strengthens muscles (except very weak ones) including those involved in respiration.

Clinical Course

Patients who survive the acute phase of the illness usually recover complete ocular and motor function within months, though the complaint of dry eyes may persist as long as 5 years.[12] The overall mortality rate for botulism is about 10%, owing to early cardiorespiratory failure and to late complications of prolonged ventilatory support and nosocomial infection. Many deaths have occurred in patients who had cardiopulmonary arrest prior to elective intubation, either because the diagnosis was not suspected or because clinical deterioration was not anticipated. The prognosis is worse for type A botulism than for type B.

In summary, botulism is a disease that demands timely diagnosis. The early and prominent involvement of the ocular system frequently triggers referral to an ophthalmologist. The findings of symmetric external and internal ophthalmoplegia, together with gastrointestinal and autonomic symptoms and early signs of descending paralysis should prompt emergent disposition.

References

1. MacDonald K, Cohen M, Blake P. The changing epidemiology of adult botulism in the United States. *Am J Epidemiol.* 1986;124:794-799.

2. Rapoport S, Watkins P. Descending paralysis resulting from occult wound botulism. *Ann Neurol.* 1984;16:359-361.

3. Bartlett J. Infant botulism in adults. *N Engl J Med.* 1986;315:254-255.

4. Chia J, Clark J, Ryan C, et al. Botulism in an adult associated with food-borne intestinal infection with Clostridium botulinum. *N Engl J Med.* 1986;315:239-240.

5. Isacsohn M, Cohen A, Steiner A, et al. Botulism intoxication after surgery in the gut. *Isr J Med Sci.* 1985;21:150-153.

6. Cherington M. Botulism—ten year experience. *Arch Neurol.* 1974;30:432-437.

7. Hughes J, Blumenthal J, Merson M, et al. Clinical features of types A and B food-borne botulism. *Ann Intern Med.* 1981;95:442-445.

8. Terranova W, Palumbo J, Breman J. Ocular findings in botulism type B. *JAMA.* 1979;241:475-479.

9. Konig H, Gassman H, Jenzer G. Ocular involvement in benign botulism B. *Am J Ophthalmol.* 1975;80:430-432.

10. Cherington M. Electrophysiologic methods as an aid in diagnosis of botulism. *Muscle Nerve.* 1982;5:S28-S29.

11. Oh S. Botulism: electrophysiological studies. *Ann Neurol.* 1977;1:481-485.

12. Mann J. Prolonged recovery from type A botulism (letter). *N Engl J Med.* 1983;309:1522-1523.

Chapter 52

Brucellosis

LAWRENCE S. EVANS and HOWARD H. TESSLER

Brucellosis is a bacterial zoonosis caused by any of six species of the genus *Brucella,* of which four are pathogenic in man. It is a minor cause of eye disease today and is seldom seen in the United States. Like tuberculosis and syphilis, brucellosis exhibits mimicry and great variability in its course and organ involvement. Dalrymple-Champney's monograph summarizes knowledge of brucellosis to 1959,[1] while Duke-Elder and Perkins[2] reviewed its ophthalmic manifestations.

Mediterranean intermittent fever, also known as Malta fever and by several other names, has apparently been known since antiquity. Even in the preantibiotic era the disease was seldom fatal, sometimes passing with no sequelae but at other times leading to a chronic febrile illness. David Bruce,[3] a young British Army surgeon, reported nine deaths among British soldiers on the island of Malta in 1887. He was able to identify a bacterium from the spleens of five soldiers and cultured it on agar in four cases. He named the organism *Micrococcus melitensis;* it is a small, gram-negative, non–spore-forming, aerobic coccobacillus. Subsequently, the genus was renamed *Brucella* in honor of Bruce.

Herds of domestic animals constitute the largest potential reservoirs of brucellosis. In animals the disease manifests itself as a chronic genitourinary tract infection, leading to abortions, retained placentas, and epididymitis. Sheep, goats, swine, and cattle have in common the presence of erythritol, a growth factor for brucellae in the placentas of pregnant females and within the seminal vesicles of males. This factor is not found in human tissues.[4]

Brucellosis is a worldwide problem, not just as a disease of man but also because of the economically significant food waste that occurs when it affects domestic animals. In many nations of northern Europe and in the United States the disease has been almost eradicated through livestock vaccination and quarantine. In the United States the number of cases in humans has dropped from thousands yearly in the 1940s to only about 200 a year now,[4,5] a trend that parallels the effectiveness of control in animals. In developing countries, especially in the Middle East, parts of Latin America, and around the Mediterranean, the disease is still commonly transmitted by drinking raw goat's milk. Its incidence is believed to be underreported.[6,7] Asymptomatic infection may be ten times more common than symptomatic disease.[8]

Systemic Manifestations

The four *Brucella* species known to be pathogenic in man are *B. melitensis, -abortus, -suis,* and *-canis.* Infection may be acquired by consuming infected meat or dairy products or through contact with infected live animals, aborted fetuses, meat, or hides. Infection also occurs via the pulmonary route and through contact with abraded skin or the conjunctiva. Slaughterhouse workers, farmers, dairy workers, and veterinarians are at risk of occupational exposure. Nearly 60% of recent cases occurred in slaughterhouse workers.[8] Accidental infection has occurred from live *B. abortus* vaccine being splashed into the eyes of veterinarians. In fact, the conjunctiva is a route used for animal vaccination. Transmission of the disease from human to human is very rare, but even the unlikely instance of human-to-animal transmission has been reported.

Acute brucellosis may be a fulminant illness, especially in infection by *B. melitensis,* or it may be a self-limited, mild disease. The severity of disease in man varies not only with the species but also with the particular strain. *B. melitensis* causes the most acute, severe disease in man (Malta fever). The classic symptoms are fever averaging 103° F that tends to peak daily in the afternoon (hence the name "undulant fever"), drenching sweats, weakness, malaise, headache, anorexia, and weight loss. Frequently there is enlargement of the liver, spleen, testicles, or lymph nodes. Myalgias occur in over half of patients and arthralgias, in about one fourth. *B. abortus,* the most frequent cause of brucellosis in the United States, and *B. canis* usually cause mild, chronic, self-limited disease. *B. suis* infections tend to cause suppurative lesions with noncaseating granulomas and abscesses. Relapses sometimes occur years after an apparent cure. Subclinical brucellosis, detectable by serologic testing, is often occupation related.[1,4]

Brucellosis should be suspected in a febrile patient with myalgia and arthralgias if there has been occupational exposure or consumption of unpasteurized dairy products. Although acute brucellosis may have no localizing signs, the disease may involve any organ system. After an initial bacteremia the organisms tend to localize in the reticuloendothelial organs and form granulomas. They may be sequestered intracellularly in macrophages, which is thought to protect them from antibiotics and host immune defenses. This probably contributes to the tendency for chronicity and relapse. Bone, liver, lungs, spleen, lymph nodes, heart, nervous system (especially the leptomeninges), and kidney are commonly involved. Death may occur from endocarditis.[4,9]

Diagnosis may be difficult. Blood cultures are positive in only some 50% of acute cases, and the yield of organisms dwindles sharply as the disease progresses. Culture from a granuloma, lymph node, or bone marrow may be positive when blood culture is not.

The serum agglutination test (STA) using antigens of lipopolysaccharides from *B. abortus* is considered the most reliable serologic test. *Brucella* skin testing can give a false-positive STA result, as can cholera vaccine or infection with *Vibrio cholerae, Yersinia enterocolitica,* or

Francisella tularensis (which used to be considered a *Brucella* species).[2] A complement fixation test has been developed, and enzyme-linked immunoadsorbent assay (ELISA) and radioimmunoassay (RIA) tests show promise.[4,9] Stemshorn gives a summary of developments in laboratory diagnosis.[10]

Oral tetracycline, 500 mg four times a day for 4 to 6 weeks, given with intramuscular streptomycin, 1 gram twice a day for 2 weeks, is the current treatment of choice, as it has the lowest rate of relapse. Sulfonamides are also used, as have been other antibiotics, but none has been demonstrated to be better than tetracycline with streptomycin. A Jarisch-Herxheimer reaction with shock may occur after antibiotics are given. Surgical drainage of an abscess may be necessary as may removal of chronically infected tissue such as the spleen.[4,9]

Ocular Manifestations

Brucellosis has been reported to affect every ocular structure. Optic nerve involvement, almost always bilateral, appears to be the most common serious manifestation. It may take the form of papillary congestion, retrobulbar neuritis, papilledema, atrophy, or arachnoiditis of the chiasm. Puig Solanes et al.[11] report that 44 of 413 cases of *Brucella* infection (probably all *B. melitensis*) affected the optic nerve. As some of these patients had meningitis, this reflects a predilection of the organism to infect the leptomeninges. The visual fields show bilateral contractions of the field and enlarged blind spots. Vision loss may be profound.

Uveitis has been reported with chronic brucellosis. Non-specific chronic iritis associated with thickening of the iris, posterior synechiae, and keratic precipitates may be present. Choroiditis with multiple nodular exudates and adjacent retinal edema and hemorrhages is said to be more characteristic for brucellosis.[2] It has been suggested that brucellosis uveitis is a response to immune complex deposits.[12] Brucellae have never been cultured from the eye in uveitis.[7,12] Endophthalmitis presenting with hypopyon and progressing to phthisis has been seen in the acute bacteremic stage and is likely a metastatic infection instead of a noninfectious immune complex phenomenon.[2]

Rolando et al.[7] reported retinal detachment in a case of chronic systemic brucellosis with iritis. They were unable to culture brucellae from aqueous humor or subretinal fluid, although both gave positive agglutination test results.

Corneal inflammation may exhibit nummular (coin-shaped) infiltrates or ulcers and can be accompanied by iritis. Corneal involvement has been reported in farmers and others with serologic and skin test evidence of exposure to *Brucella*.[13] Most of these patients had subepithelial coin-shaped opacities consisting of multiple white dots. The lesions did not stain and varied in number and location over time. Some caused facets to appear as the nummular lesions evolved. *Brucella* organisms have never been cultured or demonstrated on histopathologic examination. Treatment approaches have included arsphenamine and desensitization with *Brucella* vaccine[13]; improvement has been variable. Woods[13] was able to obtain nummular corneal infiltrates in rabbits with and without systemic *Brucella* infection by scarifying the cornea with *Brucella* organisms in culture. Thus, although it is rare, *Brucella* could be a cause of nummular corneal infiltrates.

Other affections of the eye include eyelid edema, conjunctivitis (possibly phlyctenular), icterus, and cranial nerve palsies (especially of the abducens nerve) when meningitis is present.[2,11]

Making the diagnosis of brucellosis in such cases is difficult without a history of systemic disease, since most eye findings occur late, when results of diagnostic blood cultures are usually negative. A recommendation in the ophthalmology literature for skin testing is dated and was disputed when first made.[14,15] It would seem best to employ serologic tests before skin testing, which if done first can cause false-positive serologic results. There is little in the modern literature on treatment. A 3- to 4-week course of systemic antibiotics with systemic steroids has been used.[7]

In summary, brucellosis involving the eye is extremely rare, and many of the past case reports lack good photographic documentation. A diagnosis of ocular brucellosis should be considered in slaughterhouse workers and others exposed to farm animals. Optic nerve edema and atrophy and nummular keratitis have been the most characteristic findings attributed to *Brucella* organisms.

References

1. Dalrymple-Champneys W. *Brucella Infection and Undulant Fever in Man.* London: Oxford University Press; 1960.
2. Duke-Elder S, Perkins ES. *Diseases of the Uveal Tract.* In: Duke-Elder S. *System of Ophthalmology.* vol IX. London: Kimpton; 1966;236-242.
3. Bruce D. Note on the discovery of a microorganism in Malta fever. *Practitioner.* 1887;39:161.
4. Salata R, Raudin J. Brucellosis species. In: Mandell GL, Douglas RG, Bennett JE. eds. *Principles and Practice of Infectious Diseases,* New York, NY: John Wiley; 1985.
5. *Brucellosis Surveillance, Annual Summary 1978.* Atlanta, Ga: Center for Disease Control; 1979. U.S. Dept of Health and Human Services.
6. Sharda DC, Lubani M. A study of brucellosis in childhood. *Clin Pediatr.* 1986;25:492.
7. Rolando I, Carbone A, Maro D, et al. Retinal detachment in chronic brucellosis. *Am J Ophthalmol.* 1985;99:733.
8. Bennett JE. Brucellosis. In: Wyngaarden JB, Smith LH, eds. *Cecil Textbook of Medicine.* Philadelphia, Pa: WB Saunders; 1985;1614-1617.
9. Kaye D, Petersdorf RG. Brucellosis. In: Braunwald E, et al., eds. *Harrison's Principles of Internal Medicine,* 11th ed. New York, NY: McGraw-Hill; 1987.
10. Stemshorn BW. *Recent Progress in the Diagnosis of Brucellosis, 3rd International Symposium on Brucellosis.* Basel: S. Karger; 1984:325.
11. Puig-Solanes M, Heatley J, Arenas F, et al. Ocular complications in brucellosis. *Am J Ophthalmol.* 1953;36:675.
12. O'Connor GR. Endogenous uveitis. In: Kraus-Mackin E, O'Connor GR, eds. *Uveitis, Pathology and Therapy.* New York, NY: Thieme-Stratton; 1983;67.
13. Woods AC. Nummular keratitis and ocular brucellosis. *Arch Ophthalmol.* 1946;35:490-508.
14. Foggit KD. Ocular disease due to brucellosis. *Br J Ophthalmol.* 1953;38:273.
15. Dalrymple-Champneys W, Letter. *Br J Ophthalmol.* 1954;38:636.

Chapter 53

Diphtheria

JOHN W. FLOBERG and JONATHAN D. TROBE

Diphtheria is a biphasic illness produced by *Corynebacterium diphtheriae*. It consists of an acute pharyngeal or skin infection followed weeks later by myocarditis and/or peripheral neuropathy. Because of immunization, it is rarely seen in industrialized nations today. The Center for Disease Control reported five or fewer cases per year in the United States from 1980 to 1984. Three fourths of these cases occurred in patients over 19 years of age, who presumably had an inadequate titer of antibody. Outbreaks in countries with excellent immunization programs are seen largely among those living in overcrowded conditions with poor health care. From 1972 to 1975 there was an epidemic of 414 cases of cutaneous diphtheria in the Skid Road area of Seattle. In developing countries with inadequate immunization programs, diphtheria continues to be a common problem with high morbidity and mortality rates.[1-4]

C. diphtheriae, a gram-positive bacillus, produces damage by means of an exotoxin. Expressed after a bacteriophage infects the bacillus, the exotoxin causes a paranodal and segmental pattern of demyelination. Dorsal root ganglia and adjacent dorsal and ventral roots and the nodose ganglion of the vagus nerve are particularly affected. These regions appear to be vulnerable, because they lack the barriers to diffusion of toxin present in the other segments of peripheral nerve and in the central nervous system. The toxin is apparently able to enter Schwann cell cytoplasm, where it inhibits protein synthesis, and especially myelin synthesis. It may be that, as myelin turnover proceeds at the normal rate, synthesis of new myelin fails. This may explain the long delay in the appearance of neural manifestations after the initial infection.

Systemic Manifestations

The most common presentation of diphtheria is with symptoms referable to the oropharynx—sore throat and painful swallowing. The major sign is a grayish white membranous exudate that becomes densely adherent to the tonsils, oropharynx, and other faucial structures. If this membrane extends to involve the airway and bronchial tree, the patient is at high risk for respiratory compromise.

Most of the cases of diphtheria reported in tropical countries and many of those reported in the United States in the last 15 years have been of the cutaneous variety. The skin can become infected at the site of a burn, wound, or abrasion. Alternatively, a preexisting infection (impetigo, eczema, pyoderma) can become superinfected with *corynebacteria*.

The incidences of neurologic and cardiac complications depend on the location and severity of the initial infection.[5,6] Pharyngeal diphtheria appears to be associated with a far greater incidence of these complications than does skin diphtheria. In pharyngeal diphtheria, palatal paralysis occurs in 15%, accommodative weakness in 10%, extraocular muscle weakness in 3%.[5] Of those with palatal paralysis, 50% develop peripheral neuropathy.[6] However, the incidence of peripheral neuropathy varies with the intensity of the pharyngeal infection, from 75% with severe infection to 12% with moderate infection and only 2% with mild infection.[6] The overall incidence of myocarditis varies from 10 to 25%. By contrast, after skin diphtheria, peripheral neuropathy occurs in only 3 to 5% of cases, whereas myocarditis is extremely rare.[7]

The neurologic symptoms can be divided into early and late ones. In pharyngeal (faucial) and sometimes in cutaneous diphtheria, the earliest sign is palatal weakness, which appears within 4 weeks after infection. It is manifested as nasal speech and regurgitation of liquids. The late complications typically begin in the 4th week with paralysis of accommodation. In the 6th week, cranial neuropathies appear, followed by a generalized peripheral neuropathy. Cranial nerve dysfunction is demonstrated principally by weakness of the jaw, pharynx, larynx, sternocleidomastoid, and tongue muscles, and less commonly by weakness of extraocular and facial muscles. The signs of peripheral neuropathy include absence of muscle stretch reflexes; weakness of limbs (proximal more than distal, legs more than arms); weakness of diaphragm, trunk, and neck muscles; numbness, tingling, and sensory deficits; and bowel and bladder retention or incontinence. The descending pattern of paralysis seen in botulism is not characteristic of diphtheria.[5]

The most serious complication of diphtheria is myocarditis, which usually occurs within 12 days. Its pathogenesis is unclear. Electrocardiographic changes and rhythm disturbances (ventricular tachycardia, atrial fibrillation, heart block) are the main features, presenting a high risk of sudden death.

Ocular Manifestations

Accommodative paresis is often the first "late" neurologic sign, appearing as an isolated finding at about 4 weeks. An "inverse" light-near dissociation has also been described: pupillary constriction to a nearby target is lost while constriction to light is preserved. The physiologic basis of this dissociation is not known, as both the light and near response circuits form synapses in the ciliary ganglion and use the short ciliary nerves as a final common pathway. Other ocular findings—third and sixth cranial nerve palsies—occur less commonly and later (8 to 12 weeks).

Diagnosis

The differential diagnosis of diphtheria is similar to that of botulism and includes: myasthenia gravis, Guillain-Barré syndrome, Eaton-Lambert syndrome, and tick paralysis. What is distinctive about diphtheria is its biphasic presentation: initial faucial or skin infection and palatal weakness, followed weeks later by accommodative paresis and cranial and peripheral neuropathy. Diagnosis depends on identification of the organism in stained smears and in culture. In cases of polyneuropathy, nerve conduction studies show minimal involvement early in the course and then findings worsen, even as the patient is showing clinical improvement.

Treatment

Treatment depends on timely administration of equine antitoxin to prevent the complications of cardiac and neurologic involvement. Eradication of the underlying infection does not reduce the risk of complications.

Clinical Course

The acute infectious phase of the illness carries a mortality rate as high as 20%, especially if there is laryngeal involvement and airway compromise. Factors that influence outcome in both phases of the illness are age and immune status. Very young, very old, and nonimmunized persons show increased morbidity and mortality rates. The longer the delay in giving antitoxin, the higher the rate of late effects. The overall mortality rate of diphtheria is 10%. Complete recovery from the cranial and peripheral neuropathy is the rule, though it may take up to 8 months.

In summary, diphtheria is a biphasic illness with an acute infection and a subsequent myocardiopathy and neuropathy. The ophthalmologist becomes involved during the second phase, to evaluate the blurred (and, less commonly, double) vision. The finding of isolated accommodative failure or "inverse" light-near dissociation should suggest diphtheria.

References

1. Christenson B. Is diphtheria coming back? *Ann Clin Res.* 1986;18:69-70.
2. Pederson A, Spearman J, Tronca E, et al. Diphtheria on Skid Road, Seattle, 1972-75. *Public Health Rep.* 1977;92:336-342.
3. Belsey M, Sinclair M, Roder M, et al. *Corynebacterium diphtheriae* skin infections in Alabama and Louisiana. *N Engl J Med.* 1969;280:135-141.
4. Brooks G, Bennett J, Feldman R. Diphtheria in the U.S. 1959-1970. *J Infect Dis.* 1974;129:172-178.
5. Layzer R. *Neuromuscular Manifestations of Systemic Disease.* Philadelphia, Pa: FA Davis Co; 1985;152-153.
6. Lupton M, Klawans H. Neurological complications of diphtheria. In: Vinken P, Bruyn G, eds. *Handbook of Clinical Neurology,* vol 33. Amsterdam: Elsevier North Holland; 1978;481-485.
7. Swartz M, Weinberg A. Miscellaneous bacterial infections with cutaneous manifestations. In: Fitzpatrick T, Eisen A, Wolff K, eds. *Dermatology in General Medicine.* New York, NY: McGraw-Hill; 1987;2148-2149.

Chapter 54

Metastatic Bacterial Endophthalmitis

JULES BAUM

Today metastatic bacterial endophthalmitis is a rare disease. Prior to the advent of antibiotics, the incidence of the disease and death from the primary bacterial infection were, as would be expected, much higher. Two excellent reviews of this subject, one in 1976,[1] the other in 1986,[2] allow comparison of the various ocular and systemic components of the disease and contrast "present" patterns to those observed before the advent of antibiotics. The first article reviews 102 cases, the second, 72 more recent cases.

Before antibiotics, nonocular bacterial infections unchecked by natural host defense mechanisms sometimes led to septicemia, potentially allowing pathogens to enter the eye via the arteries. While current antibiotic therapy has reduced the incidence of septicemia, two new risk factors not significant 50 years ago are associated with many cases of metastatic bacterial endophthalmitis seen today—immunosuppression and intravenous drug abuse. Patients in the former category include those with diabetes, who survive much longer than they did 50 years ago, and those taking corticosteroids. Other high-risk categories are patients with certain morphologic abnormalities (e.g., abnormal heart valves) and those who have recently undergone surgery. Only 29% of patients with metastatic bacterial endophthalmitis are known to have no apparent predisposition to infection.[2]

Systemic Manifestations

Table 54-1 summarizes the sources of the primary infection in patients who developed metastatic bacterial endophthalmitis before 1976[1] and since 1976.[2] Meningitis is still by far the most common primary source of infection in both series, although recently the incidence of bacterial endophthalmitis secondary to this disease has decreased dramatically. Primary sources of infection more common today than in the past are endocarditis and infections of the urinary tract. Puerperal infection is absent in the more recent series, while *Bacillus cereus*

metastatic endophthalmitis has made a dramatic appearance in intravenous drug abusers in the last decade. In the review by Greenwald et al.,[2] positive blood cultures were found in 71% of patients who developed metastatic bacterial endophthalmitis. In that same series, 10 of the 72 reported patients died. All patients who died were debilitated.

Bacterial pathogens responsible for the induction of metastatic bacterial endophthalmitis in both review series are listed in Table 54-2. Over the years various streptococcal species have accounted for approximately 20% of ocular infections, and the sum of streptococcal infections implicates these species as the most common agents of metastatic bacterial endophthalmitis today. *Neisseria meningitidis*, the organism that most often causes bilateral infection, is still a prime pathogen, although the frequency of infection by this organism has decreased markedly in recent years.

B. cereus (primarily in intravenous drug abusers) and *Staphylococcus aureus* rank as frequent offenders. Curiously, *S. epidermidis* septicemia does not lead to metastatic ocular infection. Patients who develop metastatic endophthalmitis caused by *N. meningitidis* and *Hemophilus influenzae* seem to have no predisposition to infection.[2]

Ocular Manifestations

Metastatic endophthalmitis is twice as common in the right eye as in the left, probably because of the shorter, more direct route of arterial blood flow from the carotid artery to the eye on the right side. Of all cases, 25% are bilateral.

Ocular pain and impaired vision are the principal symptoms of all forms of bacterial endophthalmitis. The severity of the signs depends on the primary focus and virulence of the infection. Signs of infection include eyelid edema, chemosis, conjunctival injection, corneal haze and debris, and a decreased or absent red reflex.

TABLE 54-1 Primary Infection in Patients Who Develop Metastatic Bacterial Endophthalmitis

Infection	1935-1975		1975-1985	
	(No)	(%)	(No)	(%)
Meningitis	59	55	19	26
Endocarditis	1–2		10	14
Urinary tract	7	6	10	19
Abdominal (GI)			8	11
Skin	12	11	5	7
Lungs	9	8	4	6
Puerperal sepsis	6	9	—	—
Others	15	14	16	22

(Modified from Shammas[1] and Greenwald et al.[2])

TABLE 54-2 Pathogens in Metastatic Bacterial Endophthalmitis

Organism	1935-1975		1976-1985	
	(No)	(%)	(No)	(%)
Streptococcus pneumoniae	14	(13)	2	(3)
Streptococcus spp	9	(8)	10	(14)
Staphylococcus aureus	11	(10)	7	(10)
Bacillus cereus	1		11	(15)
Listeria monocytogenes	1		3	(4)
Clostridium perfringens	1		—	
Neisseria meningitidis	56	(52)	8	(11)
Hemophilus influenzae	—		7	
Actinobacillus spp	—		2	
Escherichia coli	4	(3.7)	5	(7)
Klebsiella pneumoniae	1		5	
Serratia spp	—		3	(4)
Salmonella spp	—		3	
Pseudomonas aeruginosa	2		2	
Proteus spp	2		—	
Nocardia asteroides	6	(5.5)	3	
Unidentified	—		1	

(Data from Shammas[1] and Greenwald et al.[2])

Characteristically, visualization of the fundus is precluded by vitreal inflammation.

Greenwald et al.[2] have classified metastatic bacterial endophthalmitis by anatomic location and clinical characteristics. The categories include anterior and posterior focal, anterior and posterior diffuse, and panophthalmitis.

In focal disease (less common than the other types) discrete foci of infection may be observed clinically in iris, ciliary body, choroid, or retina. Bacteria and inflammatory cells have been observed to occlude vessels in the areas of involvement. The prognosis following antibiotic therapy is excellent in anterior focal disease and slightly less so in the posterior variety.

Anterior diffuse disease is not commonly seen clinically; it has been produced experimentally by intravenous bacterial dissemination. Clinically the infection generally responds well to antibiotic therapy if treatment has commenced before posterior extension has occurred. Anterior segment infection may be seen clinically following group B streptococcal meningitis in the newborn.

The prognosis for retention of useful vision (better than finger counting) following diffuse posterior metastatic bacterial endophthalmitis is dismal, much worse than the overall prognosis of endogenous bacterial endophthalmitis. Every case reviewed in this category has ended in blindness, phthisis, or enucleation. No eye with this type of infection has been shown to recover useful vision, regardless of the type of therapy.[2] It has been postulated that in metastatic infection, a septic embolus occludes the central retinal artery and embolic fragments are then disseminated peripherally, so ischemia as well as infection may add to the poor prognosis. Furthermore, gram-negative infections and infections with B. cereus are more common than in exogenous bacterial endophthalmitis, and patients are often immunodeficient.

References

1. Shammas HF. Endogenous *E. coli* endophthalmitis. *Surv Ophthalmol.* 1977;21:429.
2. Greenwald MJ, Wohl LG, Sell CH. Metastatic bacterial endophthalmitis: a contemporary reappraisal. *Surv Ophthalmol.* 1986;31:81.

Chapter 55

Tularemia

DANIEL N. SKORICH and ROBERT NOZIK

Tularemia is the zoonotic disease caused by the bacterium *Francisella tularensis*. It is an uncommon disease: the Center for Disease Control recorded approximately 200 to 300 cases annually between 1980 and 1985. The overall incidence of tularemia has been decreasing since the 1940s, perhaps because of enhanced awareness of the disease in groups at highest risk for infection, such as meat handlers and laboratory personnel.

The organism, initially named *Bacterium tularense*, was first isolated in 1912 from ground squirrels dying from a plague-like illness in Tulare County, California.[1] The first proven human case of tularemia was described in 1914, in a meat cutter with conjunctivitis.[2] However, the basic understanding of the disease was elucidated through the dedicated efforts of Dr. Edward Francis of the United States Public Health Service. In a classic series of clinical and laboratory studies, Dr. Francis showed that wild rabbits are the major warm-blooded vector for the disease. Transmission to man is via direct contact with infected animal tissues or by biting insects.[1] Ticks were subsequently shown to be the main reservoir for the organism, but man can contract the disease from several other sources including mosquitoes, deerflies, muskrats, beavers, squirrels, woodchucks, sheep, and game birds.[3] The organism was subsequently renamed *Francisella tularensis* in honor of Dr. Francis. The eighth edition of *Bergey's Manual of Determinative Bacteriology* recognizes two biovars. Type A, found only in North America, is highly virulent for rabbits and man but can be recovered from other animals and insects. Type B is less common and produces milder disease.

F. tularensis is a small, pleomorphic, aerobic, nonmotile gram-negative (poorly staining) bacillus.[4] The organism is extremely virulent and should not be handled in a laboratory without special facilities and trained technicians. A special medium is required for culture because of its specific requirement for cysteine or other sulfhydryl compounds in amounts that exceed those normally present in nutrient media. Satisfactory growth occurs slowly on cysteine–glucose–blood agar or coagulated egg yolk medium, although growth may occur on supplemented chocolate agar or in the media used in certain automated blood culture systems (BACTEC).[5] Animal inoculation may be a more sensitive method for recovering the organism, but it is expensive and, because of the highly infectious nature of the organism, risky for laboratory personnel.

Systemic Manifestations

Human tularemia is an acute, febrile, granulomatous disease. Untreated, the mortality rate is approximately 3 to 6%.[3] The incubation period ranges from 2 to 10 days. Infection follows exposure to contaminated animal carcasses, insect bites, ingestion of undercooked meat or contaminated water, or inhalation of dust-borne organisms. Infection can occur following exposure to as few as ten organisms. Some evidence suggests that the organism can penetrate intact skin. However, the majority of cases follow introduction of the organism through small skin breaks or onto mucous membranes.

The disease is classified into four main types based on clinical presentation: (1) ulceroglandular (the primary lesion is an ulcerative skin papule that is accompanied by regional lymph node enlargement and tenderness; (2) oculoglandular (primary conjunctivitis accompanied by enlarged regional lymph nodes (Fig. 55-1); (3) glandular (no primary skin lesion but enlarged regional lymph glands; and (4) typhoidal (no primary lesion or gland enlargement).[1] Other forms of human tularemia include oropharyngeal, which may lead to a pseudomembranous tonsilitis with lymphadenopathy, and primary pneumonic tularemia, most commonly seen in lab workers. Because the above classification is burdensome, some authors advocate recognizing only ulceroglandular and typhoidal types, on the basis of clinical signs, whether or not pneumonia or pharyngitis complicates the disease.[3]

The onset of disease is sudden and is characterized by fever, chills, headache, vomiting, and body aches. In the ulceroglandular form of the disease, the most common type, patients notice pain and swelling in the region of the lymph nodes that drain the site of inoculation. The gland pain may precede the development of a painful, inflamed papule that indicates the site of infection. This papule breaks down into an ulcer with raised edges. Untreated, the ulcer can persist for 2 weeks or more.

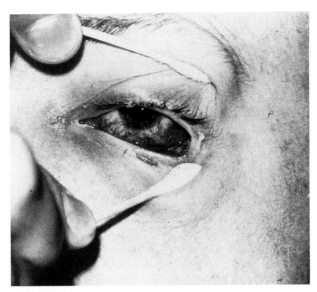

FIGURE 55-1 Oculoglandular tularemia. Conjunctivitis with swollen preauricular lymph node. (From Anderson.[8])

Suppuration of the glands frequently occurs; spontaneous drainage is infrequently noted.

Typhoidal tularemia presents as an acute septicemia. The patient is noted to have a high recurrent fever with other nonspecific systemic signs and symptoms. The fever may persist for several weeks. Other associated findings include hepatosplenomegaly, various skin rashes including erythema nodosum, diarrhea, pulmonary abnormalities, pharyngitis, and pericarditis.[3]

Pneumonia, pharyngitis, and pericarditis may complicate both ulceroglandular and typhoidal tularemia, but they are more frequently associated with the typhoidal form. The pneumonia seen in tularemia can resemble pneumonia caused by *Mycoplasma*, *Legionella*, psittacosis, Q fever, tuberculosis, histoplasmosis, blastomycosis, and viruses.[3]

Ocular Manifestations

In both pediatric and adult populations, *F. tularensis* is an uncommon cause of unilateral granulomatous conjunctivitis associated with markedly enlarged preauricular, submandibular, or cervical nodes (otherwise known as Parinaud's oculoglandular syndrome). The conjunctiva shows multiple yellow conjunctival granulomas with shallow ulcers (Fig. 55-2).[6] The differential diagnosis of oculoglandular syndrome includes cat scratch disease, tularemia, sporotrichosis, tuberculosis, syphilis, coccidioidomycosis, actinomycosis, blastomycosis, pasteurellosis, yersiniosis, listeriosis, mumps, lymphogranuloma venereum, sarcoidosis, and chancroid.[7]

Diagnosis and Management

The diagnosis of tularemia is based on historical, clinical, and laboratory data. Serum agglutinin titers are the most useful laboratory study. Titers are usually positive by 2 weeks and reach their peak by 4 to 8 weeks. A titer of

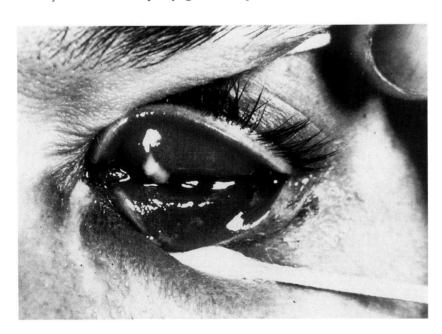

FIGURE 55-2 Characteristic yellow-based necrotic conjunctival ulcer in oculoglandular tularemia. (From Anderson.[8])

1:160 suggests a presumptive diagnosis of tularemia, but the diagnosis is confirmed by subsequent evidence of rising titers.[5]

Cultures are frequently negative, although the organism can be recovered from blood, conjunctival swabs, sputum, and gastric and pharyngeal washings.[3] The low rates of recovery reflect the organism's special growth requirements. Animal inoculation, as mentioned earlier, is a more sensitive, although hazardous, method of recovering the organism.

Streptomycin is the treatment drug of choice. The recommended dose in children is 30 to 40 mg of streptomycin per kg daily given in two divided intramuscular doses for 7 days, or alternatively, 40 mg/kg/day in two divided doses for 3 days then 15 to 20 mg/kg/day for 4 days more.[5] In adults the dose is 1 to 2 g daily in divided doses for 7 to 10 days. Other useful drugs include gentamicin and tobramycin. Tetracycline and chloramphenicol are not as useful in severe disease because of significant relapse rates.[3]

References

1. Francis E. A summary of present knowledge of tularemia. *Medicine* (Baltimore). 1928;7:411.
2. Wherry WB, Lamb BH. Infection of man with *Bacterium tularense. J Infect Dis.* 1914;15:331.
3. Evans ME, Gregory DW, Schaffner W, et al. Tularemia: a 30-year experience with 88 cases. *Medicine.* 1985;64:251.
4. Eigelsbach HT. Francisella tularensis. In: Lennette EH, et al., eds. *Manual of Clinical Microbiology.* 4th ed. Washington, DC: American Society for Microbiology; 1985;316.
5. Jacobs RF, Narain JP: Tularemia in children. *Pediatr Infect Dis.* 1983; 2:487.
6. Halperin SA, Gast T, Ferrieri P. Parinaud's oculoglandular syndrome caused by *Francisella tularensis. Clin Pediatr.* 1985;24:520.
7. Chin GN, Hyndiuk RA. Parinaud's oculoglandular conjunctivitis. In: Duane TD, ed. *Clinical Ophthalmology,* vol 4. Hagerstown, Md: Harper & Row; 1981.
8. Anderson R. Oculoglandular tularemia. *J Iowa Med Soc.* 1970;60:21.

Section B

Chlamydial Diseases

Chapter 56

Chlamydial Infections

JOHN D. SHEPPARD and CHANDLER R. DAWSON

Human chlamydial infections are widespread, including trachoma and sexually transmitted disease.[1] Trachoma still afflicts 600 million people in developing countries, of whom 6 million are blind.[2] Sexually transmited chlamydial infections are the leading sexually transmitted disease in industrialized countries. Newborns exposed at birth to genital chlamydiae may develop neonatal ophthalmia or pneumonia.

Microbiology

Chlamydiae are gram negative obligate intracellular energy parasites incapable of producing their own adenosine triphosphate (ATP). Their life cycle consists of two forms, elementary bodies (EBs) and reticulate bodies (RBs). The infectious EBs are extracellular, metabolically inactive, 300 nm particles capable of surviving in the extracellular environment. EBs are toxic to cultured epithelial and fibroblast cell lines and lethal to mice when given intravenously. RBs develop intracellularly from EBs. They are metabolically active, 1200 nm particles actively engaged in protein synthesis, ATP

transport from the host cell, and rapid multiplication. RBs are susceptible to mechanical stress, osmotic gradients, and trypsin lysis. The RBs form a microcolony in the host cell cytoplasm, which is the inclusion body seen by light microscopy. As metabolic constituents become scarce, the RBs reorganize into EBs, apparently in response to a lack of glutathione.[3]

This unique, biphasic life cycle differentiates chlamydiae from all other microorganisms, placing them in their own order, Chlamydiales. There is one genus, *Chlamydia*, and two species, *C. trachomatis* and *C. psittaci*. Almost invariably, the *C. trachomatis* serotypes A, B, Ba, and C are associated with trachoma, whereas serotypes D through K are found in genitourinary disease and in neonatal and adult inclusion conjunctivitis. Some overlap of infectivity patterns between these two serotype groups has been documented. The older term TRIC agents refers to *trachoma/inclusion conjunctivitis* serotypes A through K. *C. trachomatis* serotypes L1, L2, and L3 cause lymphogranuloma venereum (LGV). Because they invade deeper, subepithelial tissues, LGV serotypes are now regarded as a distinct biovar or species

169

subdivision, despite extensive antigenic cross-reactivity with other *C. trachomatis* serotypes.

Chlamydiae possess serotype-specific, species-specific, and genus-specific surface markers, some of which induce at least a partially protective immune response. Although most chlamydial infections are associated with increased antibody titers, reinfection is possible in both ocular and genitourinary disease.[4] Genus-specific chlamydial lipopolysaccharide (LPS) is similar to the LPS of gram negative bacteria such as *Salmonella* species.[5] Other cell wall antigens may be important in producing hypersensitivity, localized epithelial inflammation, and eventually scarring of the conjunctiva or fallopian tubes.

C. psittaci agents cause a wide variety of economically important animal diseases and occasionally produce human psittacosis. In general, *C. trachomatis* organisms are human pathogens, man being the sole natural host, while *C. psittaci* organisms are pathogens in other vertebrate species and only occasionally cause human infection. A *C. psittaci* strain found only in humans, called TWAR, has been implicated as a major cause of epidemic respiratory infections. Manifestations include follicular conjunctivitis, rhinitis, pharyngitis, laryngitis, bronchitis, and pneumonia.[6]

Genitourinary Infections

Chlamydial infections in women are an important cause of cervicitis, urethritis, salpingitis, pelvic inflammatory disease, and subsequent tubal sterility.[7] Chlamydial infection is associated with cervical erosions and cervical cellular atypia. Affected women often have high rates of concomitant herpes simplex and papilloma virus infection, so a direct association between chlamydial infection and premalignant lesions has not been made. Cervical infection, which is often asymptomatic, is felt to be the primary reservoir of human genital infection. Because of fallopian tube scarring, increased rates of ectopic pregnancies have also been attributed to chlamydial infection. Perihepatitis and peritonitis (Fitz-Hugh-Curtis syndrome) are rare sequelae of chlamydial infection in young women.

Chlamydial infections in men are a leading cause of nongonococcal and of postgonococcal urethritis. About one third of patients who have gonorrhea have concurrent chlamydial infections, and about one third of those with nongonococcal urethritis have positive chlamydial cultures.[1] At least 60% of women who have sexual contact with infected men develop chlamydial cervicitis. About two thirds of epididymitis cases are due to chlamydial infection. Sterile males with low sperm counts frequently have a history of chlamydial infection.[8]

Homosexual men have a high incidence of chlamydial proctitis and pharyngitis. There is no current evidence that human immunodeficiency virus infection is associated specifically with increased risk of chlamydial infection, although groups at risk for acquired immunodeficiency syndrome (AIDS) also are at greater risk for all sexually transmitted diseases.

LGV serotype infections cause inguinal adenopathy with late destructive sequelae of the lower gastrointestinal tract, fistulas, and anogenital erosions. Initial infection is often asymptomatic and may resolve spontaneously. One to 4 weeks following contact, shallow ulcerative lesions of the penis, vagina, or labia may develop. Several weeks after appearance of the primary ulcerative lesion, grossly visible secondary regional lymphadenopathy develops. These buboes are painful in men, but they can remain asymptomatic in women: men are 20 times more likely than women to develop symptomatic LGV. LGV infections are uncommon in the United States and rarely involve the eye. Ocular infections are seen more often in laboratory workers and physicians who have accidently contracted conjunctivitis than in patients with genital LGV.

Reiter's Syndrome

Reiter's syndrome is a postinfectious disease, presumably mediated by autoimmune mechanisms. It is strongly associated with the HLA-B27 locus and usually is found in males. Manifestations include the classic triad of polyarthritis, papillary conjunctivitis, and urethritis, in addition to keratoderma blennorrhagicum, circinate balanitis, keratitis, and recurrent unilateral acute iridocyclitis. Reiter's syndrome is usually preceded by a sexually transmitted chlamydial infection or by acute gram negative dysentery due to *Salmonella, Shigella,* or *Yersinia* species. The conjunctivitis is usually a papillary, chlamydia culture negative episode associated with initial onset of the syndrome, but the onset of Reiter's syndrome has been associated with follicular conjunctivitis due to ocular chlamydial infection.[9] The uveitis is usually self-limited, but recurrent episodes can involve either eye. Reiter's uveitis can cause permanent damage if it is not treated aggressively at the onset of each episode.

Neonatal Infections

Children born to mothers with known positive chlamydial cervical cultures have about a 50% chance of developing culture positive chlamydial infections and a 60% chance of showing serologic evidence of exposure.[10] The general estimate in the United States is that 5% of expectant mothers have chlamydial cervical infection and that rates are as high as 40% in selected high-risk populations.[11] Exposed infants have up to a

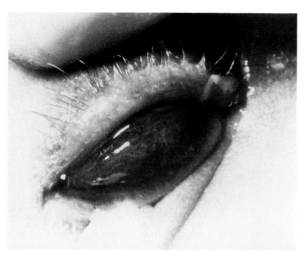

FIGURE 56-1 Infant with chlamydial ophthalmia neonatorum, serosanguinous discharge, papillary hypertrophy, and periocular edema.

50% chance of developing inclusion conjunctivitis and a 20% chance of developing chlamydial pneumonia. Other concurrent chlamydial infections may include rhinitis, pharyngitis, otitis media, bronchiolitis, and vaginal and rectal colonization.[12]

Exposed infants usually develop chlamydial ophthalmia during the first 5 to 21 days of life. Neonatal chlamydial conjunctivitis presents with a mucopurulent discharge and keratitis. It may persist for months if not treated and can cause conjunctival scarring and superficial corneal neovascularization. Follicles are not seen with chlamydial infections until infants are 6 to 12 weeks of age, because mature conjunctival lymphoid tissue is not present at birth (Fig. 56-1).

In many hospitals, topical erythromycin ointment is now used routinely to prevent chlamydial and gonococcal ophthalmia neonatorum, since Credé silver nitrate prophylaxis is ineffective against chlamydiae. Single-dose erythromycin prophylaxis, however, is not completely effective in preventing neonatal chlamydial conjunctivitis.[13] Recommended treatment is oral erythromycin, 40 to 50 mg/kg/day in four divided doses, for 3 weeks. Topical antibiotic therapy alone is not effective in controlling the conjunctivitis or in preventing infection such as pneumonia at other sites. Parents of infected neonates must be treated with oral erythromycin, tetracycline, or doxycycline at the same time to prevent the potentially damaging sequelae of genital infection. Nursing mothers should receive oral erythromycin therapy.

Adult Inclusion Conjunctivitis

Ocular infection with genital chlamydial serotypes produces acute follicular conjunctivitis that may persist for months if not treated (Fig. 56-2). Patients with adult inclusion conjunctivitis present with irritation, photophobia, redness, and a mucopurulent discharge, often without systemic or genital manifestations. The incubation period from studies performed on blind volunteers is 4 to 12 days.[9] This disease is most prevalent in young sexually active adults who have acquired a new sex partner in the 2 months preceding the onset of

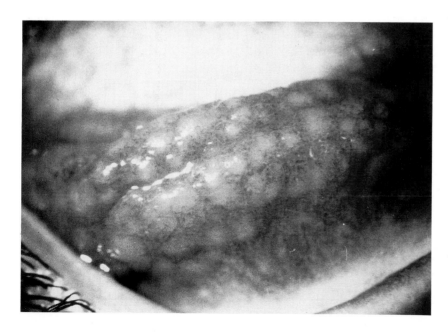

FIGURE 56-2 Adult inclusion conjunctivitis with follicular hypertrophy most pronounced in the inferior tarsal conjunctiva.

symptoms. Infection is probably acquired by direct contact with genital tract secretions. Chlamydia was thought to be the leading cause of swimming pool conjunctivitis (now usually caused by adenovirus infection) prior to the era of routine chlorination. Older children, baby sitters, and other adults exposed to infected infants occasionally develop the disease.

Epithelial keratitis develops during the second week. Most infected patients develop other forms of corneal involvement, including marginal and central infiltrates, subepithelial opacities, limbal swelling, and superficial vascularization. Although often benign and self-limited, the disease may become chronic with superficial corneal vascularization. Conjunctival scarring has been reported by some authors. Asymptomatic ocular infection has been documented in patients with known genital chlamydial infection. Recommended treatment consists of erythromycin (250 mg by mouth four times a day), tetracycline (250 mg by mouth four times a day), or doxycycline (100 mg by mouth four times a day) for 3 weeks, given simultaneously to patients and to their known sexual contacts.

Trachoma

Trachoma is a major health problem in many developing countries where sanitation and clean water supplies are limited. Endemic trachoma is a chronic follicular keratoconjunctivitis, associated with secondary bacterial conjunctivitis in many regions (Fig. 56-3). The chronic inflammation leads to varying degrees of conjunctival scarring and corneal vascularization. The conjunctival

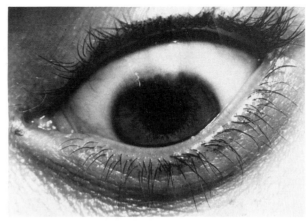

FIGURE 56-4 Herbert's pits, excavated sites of quiescent superior limbal follicles, are diagnostic of trachomatous inflammation.

scars gradually contract, causing trichiasis and entropion, which in turn produce constant ocular surface abrasion, and eventually, corneal ulceration and opacification (Figs. 56-4, 56-5). Tear film abnormalities due to compromised surfacing, goblet cell destruction, and poor wetting following lacrimal ductule obstruction may also contribute to corneal scarring.

In hyperendemic communities, active infectious trachoma has an insidious onset in children, and usually subsides by adolescence. Post-inflammatory scarring persists throughout life. The severity of infectious trachoma may vary from one household to another, even in communities with a high rate of disease. After the infectious phase has subsided, patients with cicatricial

FIGURE 56-3 Active trachoma in a 4-year-old Egyptian boy; there is intense superior tarsal follicular and papillary hypertrophy.

FIGURE 56-5 End-stage trachoma with corneal opacification, entropion, and trichiasis causing constant corneal abrasion.

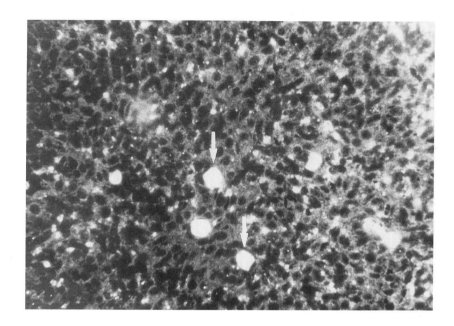

FIGURE 56-6 Positive McCoy cell tissue culture fibroblast monolayer. Arrows indicate fluorescent antibody–stained inclusions (original magnification ×400).

trachomatous disease continue to present for the care of late complications such as entropion, tear deficiencies, and corneal blindness.

Present World Health Organization recommendations for trachoma control include mass treatment campaigns with topical tetracycline ointment to both eyes, twice daily for 5 consecutive days or once daily for 10 days each month for 6 months each year, repeated as necessary.[3] Results are significantly improved if patients

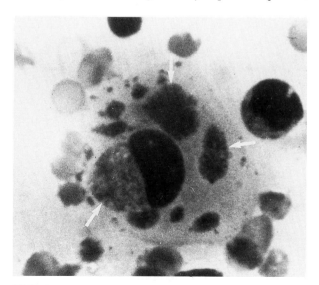

FIGURE 56-7 Giemsa-stained conjunctival scraping. Arrows indicate cytoplasmic basophilic Halbastaedter-Prowaszek inclusions (original magnification ×1000).

with severe inflammation are also treated on the same schedule with oral doxycycline. Treatment benefits are only temporary, however, unless they are accompanied by improvements in hygiene and economic development. While previous human vaccine trials with purified whole EBs have failed, advances in molecular biology have led to the prospect of potentially successful elementary body (EB) cell surface subunit vaccines. Trachoma is no longer a problem in the United States and other industrial nations.

Laboratory Diagnosis

Isolation of *C. trachomatis* in McCoy cell culture (Fig. 56-6) has become the diagnostic standard for chlamydial infection. Cell culture takes 3 to 6 days and requires an expensive tissue culture facility. For the diagnosis of ocular infection, Giemsa staining of conjunctival smears is highly specific but insensitive, except for chlamydial ophthalmia neonatorum (Fig. 56-7). More rapid and less expensive tests are now widely available to both laboratories and practicing physicians.[14] These tests include enzyme immunoassays utilizing polyclonal rabbit serum (Chlamydiazyme, Abbott), and direct immunofluorescence tests utilizing monoclonal species-specific antibody (MicroTrak, Syva; Fig. 56-8).[15] Measurement of serum antibody is not useful in adult infections because a large portion of the population have antibody to either *C. trachomatis* or cross-reactive TWAR antigens. Elevated IgM antibody is highly specific for neonatal chlamydial pneumonia and other systemic chlamydial infections in newborns.

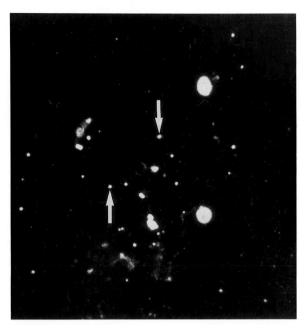

FIGURE 56-8 Positive conjunctival scraping by direct fluorescent antibody test. Arrows show uniform apple green EBs among dull orange–staining epithelial cells (original magnification ×600).

References

1. Schachter J. Chlamydial infections. *N Engl J Med.* 1978;298:428-435, 490-495, 540-549.
2. Schachter J, Dawson CR. Human Chlamydial Infections. Littleton, Mass: PSG Publishing, 1978.
3. Bavoil P, Ohlin A, Schachter J. Role of disulfide bonding in outer membrane structure and permeability in *Chlamydia trachomatis. Infect Immun.* 1984;44:479.
4. Zhang YX, Stewart S, Joseph T, et al. Protective monoclonal antibodies recognize epitopes located on the major outer membrane protein of *Chlamydia trachomatis. J Immunol.* 1987;138:575-581.
5. Caldwell HD, Hitchcock PJ. Monoclonal antibody against a genus specific antigen of *Chlamydia* species: location of the epitope on chlamydial lipopolysaccharide. *Infect Immun.* 1984;44:306.
6. Kuo C-C, Chen H-H, Wang S-P, Grayston JT. Identification of a new group of *Chlamydia psittaci* strains called TWAR. *J Clin Microbiol* 1986;24:1034-1037.
7. Schachter J. *Sexually Transmitted Chlamydial Infections. Transitions,* Palo Alto, Calif: Syva Syntex Laboratories, Inc; 1981;4:5.
8. Alexander ER, Harrison HR. Chlamydial infections. In: Evans AS, Feldman HA, eds. *Bacterial Infections of Humans.* New York: Plenum Medical Book Co; 1982;159-185.
9. Dawson CR, Schachter J. TRIC agent infections of the eye and genital tract. *Am J Ophthalmol.* 1967;63:1288.
10. Schachter J, Grossman M, Sweet RL, et al. Prospective study of perinatal transmission of *Chlamydia trachomatis. JAMA.* 1986;255:3374-3377.
11. Hardy PH, Hardy JB, Nell EE, et al. Prevalence of six sexually transmitted disease agents among pregnant inner-city adolescents and pregnancy outcome. *Lancet.* 1984;2:333-337.
12. Sheppard JD, Chandler JW. Prevention of *Ophthalmia neonatorum* and oculogenital diseases. In: Friedlander MH, ed. *Prevention of Eye Disease.* New York: Liebert, 1988: 3-18.
13. Oriel JD: Ophthalmia neonatorum: relative efficacy of current prophylactic practices and treatment. *J Antimicrobial Chemo.* 1984;14:209.
14. Sheppard JD, Kowalski, Meyer MP, et al. Immunodiagnosis of adult chlamydial conjunctivitis. *Ophthalmol.* 1988;95:434-443.
15. Stevens RS, Tam MR, Kuo C-C, Nowinski RC. Monoclonal antibodies to *Chlamydia trachomatis:* antibody specificities and antigen characterization. *J Immunol.* 1982;128:1083-1089.

Section C

Helminthic Diseases

Chapter 57

Cysticercosis

ALEX E. JALKH and HUGO QUIROZ

Cysticercosis is a parasitic infestation of different body organs by *Cysticercus cellulosae*, the larval stage of *Taenia solitum*. Persons between 10 and 30 years of age, males and females equally, are most commonly affected. The usual sites of infestation are the eye, the subcutaneous tissue, and the brain, ocular involvement being most common.

Taenia Solium Life Cycle

Humans are the usual definitive host. The intestinal tract harbors a single adult bladeworm, the last segments of which contain a large number of eggs and are eliminated in the stool. Common intermediary hosts are pigs, and less frequently other animals that ingest food or water contaminated by the eggs. When gastric or duodenal juices dissolve the shells of the eggs, the embryos are released. Aided by their hooks, they perforate the intestinal mucosa, enter the blood circulation, and reach the organs of the host, usually the muscles, heart, and brain. As the embryos reach the mature larval stage, they lose the hooks, acquire a head or scolex, and become larger and vesicular, a process that usually takes about 3 months. When humans eat contaminated pork that is undercooked, the cysticercus is freed in the gastrointestinal tract, where its suckers attach to the mucosa. Within a few weeks, it develops into an adult tapeworm and the regular life cycle is complete.

Systemic Manifestations

Human systemic cysticercosis occurs when man is the intermediary rather than the definitive host. This happens when contaminated food or water is ingested. Less frequently, autoinfestation can occur through reflex peristalsis of eggs from a harboring adult parasite. The released embryo perforates the gastrointestinal mucosae, reaches the circulation, and then localizes, primarily in the eye, the subcutaneous tissue, and the brain.

The systemic manifestations include subcutaneous cysts and inflammatory nodules, which may show typical calcifications on soft tissue x-ray studies later in the

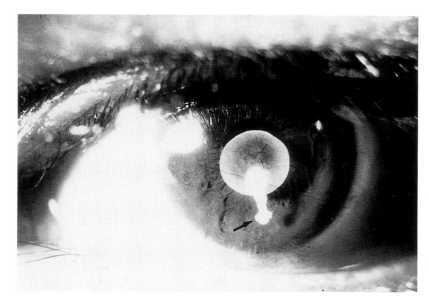

FIGURE 57-1 Anterior segment photograph shows cysticercus cyst floating freely in the anterior chamber. Note scolex of cysticercus (*arrow*). (Courtesy of Archives of Photography, Association para Evitar la Ceguera Hospital, Mexico City.)

course of the disease. Central nervous system manifestations consist of epilepsy, neurologic deficits, or psychiatric disorders. Signs of meningoencephalitis have been seen in cases with multiple cysts.

Ocular Manifestations

Ocular cysticercosis, first described by Sommerring in 1930,[1] affects the posterior segment of the eye three times more frequently than the anterior segment.[2,3]

The parts of the eye affected by *C. cellulosae* are, in order of frequency, the subretinal space, the vitreous cavity, the conjunctiva, the anterior segment (ciliary body, iris, and anterior chamber), and the orbit. The parasite reaches the posterior segment through the posterior ciliary arteries, enters the choroid, and localizes in the subretinal space (often in the macula). Occasionally, it perforates the retina and invades the vitreous cavity, where it can float freely.

The intraocular parasite causes an inflammatory reaction induced by toxins released from the cysticercus.[4] Frequently, the reaction is mild to moderate, producing iritis, vitritis, or both, with redness and photophobia. The parasite seems to be tolerated better in the vitreous than in the subretinal space, where the cysticercus often induces focal chorioretinitis. Occasionally, it produces a severe and fulminating intraocular inflammation leading to endophthalmitis. This usually occurs with the death of the parasite and subsequent release of large amounts of toxins. Recent studies have shown that the inflammatory reaction accompanying intraocular cysticercosis is more closely related to the host immune response than to the cysticercus itself.

The passage of the cysticercus from the subretinal space into the vitreous cavity produces a retinal break that can lead to rhegmatogenous retinal detachment, particularly when it is associated with vitreous traction; however, the focal chorioretinitis caused by the presence of the parasite in the subretinal space can seal this break, and subsequently, the chorioretinal adhesion produced by subretinal scarring can prevent retinal detachment.[5]

The vision in eyes with ocular cysticercosis can be well preserved unless the parasite is localized in the macular area, an associated retinal detachment is present, or the intraocular inflammation is severe. Biomicroscopy of the anterior chamber and anterior vitreous cavity reveals varying degrees of flare and cells and may show the presence of the parasite. When the ocular media are clear, the intraocular cysticercus with its translucent, round, cystic shape and a head or scolex can be detected easily by biomicroscopy (Fig. 57-1) or indirect ophthalmoscopy. When the cysticercus floats freely in the anterior chamber or the vitreous cavity, it shows characteristic movements of undulation, contraction, and expansion that give it the appearance of a "living, mobile pearl." When the scolex protrudes from the cyst, it shows pendulous movements; but when it is invaginated, it appears to be a dense white spot within the cyst (Fig. 57-2). The latter is seen when the parasite is trapped under the retina (Fig. 57-3) or conjunctiva.

In cases of opaque media secondary to cataract, corneal opacities, severe inflammation, or intraocular hemorrhage, ultrasonography is valuable to detect the

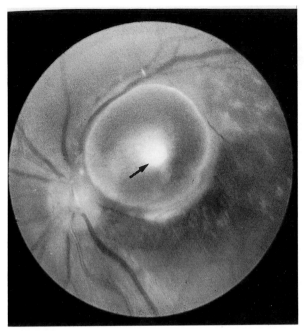

FIGURE 57-2 Fundus photograph of left eye shows cysticercus cyst floating freely in the vitreous cavity in front of the posterior pole. Dense white dot in the center of the cyst represents the invaginated scolex (*arrow*).

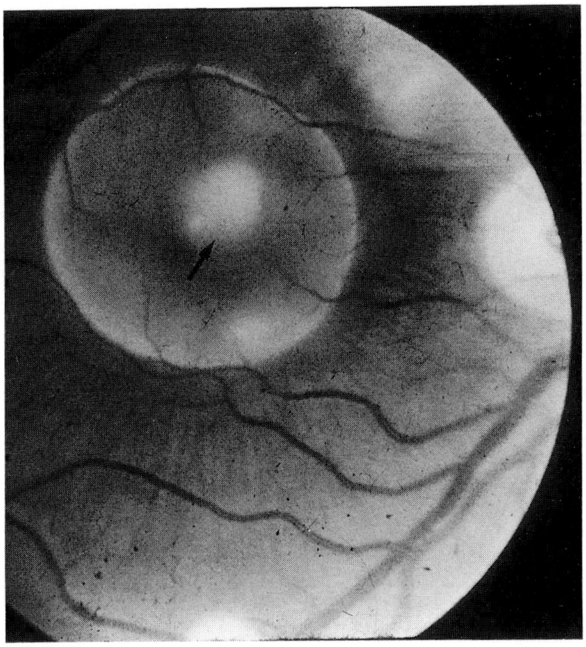

FIGURE 57-3 Fundus photograph of right eye shows subretinal cysticercus in the macular area. Whitish portion in the center represents the invaginated scolex (*arrow*).

cyst,[5] which shows characteristic ameboid movements on B-scan. Echoes within the cystic cavity originating from the scolex are often detected.

Diagnostic laboratory studies include blood eosinophilia, serum ELISA for cysticercosis, and positive stool parasitology for *Taenia* eggs. Anterior chamber tap is helpful in showing eosinophils in the aspirate. A positive ELISA reaction by the anterior chamber aspirate is a highly specific diagnostic test for intraocular cysticercosis. A CT scan should be performed to rule out brain involvement, even when central nervous system manifestations are not present.

Treatment

Left untreated, intraocular cysticercosis can lead to severe ocular damage. This has been reported in about 80% of cases when cysts are not removed.[4] Antihelminthic drugs have been used mainly to treat central nervous system and skin cysticercosis, but they have been ineffective against intraocular cysticercosis.[6] The treatment of choice is local surgical removal of the cysticercus cyst.

Diathermy, cryotherapy, and photocoagulation, the conventional methods of destroying the cysts, have had limited success because of the risk of severe inflammatory reaction when toxins are released from necrotic larval tissues. Currently, pars plana closed vitrectomy, using suction and gentle aspiration with a vitrectomy probe, is the best method to remove intravitreal cysticercus cysts. The portion of the vitreous adjacent to the cyst then is removed to eliminate released toxins. Similarly, anterior chamber cysts are aspirated by a vitrectomy probe inserted into the anterior chamber through a corneal incision.

In cases of subretinal cysticercus, a sclerotomy and choroidal incision is made over the cyst, which is removed through the incision by applying gentle pressure on the globe.[7] When a rhegmatogenous retinal detachment is present it is repaired by conventional scleral buckling procedures.

References

1. Duke-Elder S. *System of Ophthalmology*. vol 9. *Diseases of the Uveal Tract*. St. Louis, Mo: CV Mosby; 1966;478-488.
2. Malik SRK, Gupta AK, Choudhry S. Ocular cysticercosis. *Am J Ophthalmol*. 1968;66:1168-1171.
3. Segal P, Mrzyglod S, Smolarz-Dudarewicz J: Subretinal

cysticercosis in the macular region. *Am J Ophthalmol.* 1964;57:655-664.

4. Lech JR. Ocular cysticercosis. *Am J Ophthalmol.* 1949;32:523-548.

5. Kruger-Leite E, Jalkh AE, Quiroz H, et al. Intraocular cysticercosis. *Am J Ophthalmol.* 1985;99:252-257.

6. Santos R, Chavarria M, Aguierre AE. Failure of medical treatment in two cases of intraocular cysticercosis. *Am J Ophthalmol.* 1984;97:249-250.

7. Topilow HW, Yimoyines DJ, Freeman HM, et al. Bilateral multifocal intraocular cysticercosis. *Ophthalmology.* 1981; 88:1166-1172.

Chapter 58

Echinococcosis

Hydatid Disease

ADOLFO GOMEZ MORALES and J. OSCAR CROXATTO

Echinococcosis is an infection caused by cestodes of the genus *Echinococcus*.[1] Four species are defined: *E. granulosus, E. multilocularis, E. oligoarthus,* and *E. vogeli.* They differ in morphology, biology, host, and endemic areas. The metacestode forms are called hydatid cysts, and the disease they cause, hydatid disease. Although *E. multilocularis* and *E. vogeli* are known to cause human infections, classic cystic hydatid disease is referred to as *E. granulosus* infection.[2]

E. granulosus infection occurs throughout the world. The disease is more prevalent in rural areas of South America, the Middle East, North Africa, Central Europe, and Mediterranean countries.

The life cycle involves two mammalian hosts. *E. granulosus* adults live in the small intestine of many carnivores, the dog being the most important. Eggs discharged in the feces are ingested by the intermediate host, most often sheep. Humans act as intermediate hosts and become infected when they ingest ova accidentally. Eggs hatch in the intestine, releasing the oncosphere. The passage of the motile oncosphere through the intestinal mucosa and into the host's venous (portal circulation) and lymphatic system may be facilitated by lytic secretions present in the gut. Factors that determine the final location are not well-defined, but they include anatomic and physiologic factors of the host, and the strain of *E. granulosus.* Larvae of the northern sylvatic strain limited to higher latitudes of North America and Eurasia are located predominantly in the lungs. The process of cyst development involves degeneration of the oncospheral stage and emergence of the metacestode stage. Growth ranges from 1 to 5 cm each year. Cysts may remain viable for long periods or may undergo degeneration and calcification. The life cycle is completed when tissues containing viable cysts are ingested by the definitive host.

The hydatid cyst consists of an inner germinative layer and an outer acellular laminated membrane (cuticular layer). The cyst is surrounded by an adventitial layer that represents a granulomatous reaction produced by the host. Internally, secondary cysts (brood capsules) bud from the germinative layer, in which protoscolices are produced by polyembryony. A viable protoscolex is ovoid, measures about 100 μm, and shows four suckers, calcareous bodies, and a rostellum with hooklets withdrawn in the region behind the sucker.

Ten or more different antigens have been demonstrated in the hydatid fluid and tissues of the metacestode. The antibody response elicited by these antigens is apparently limited by the integrity of the cyst and varies with location. Circulating IgG, IgM, IgA, and IgE have been detected in the serum of affected individuals. The germinative layer prevents or regulates their passage into the cyst. Lysis of protoscolices is mediated by the alternate pathway of complement activation. Anticomplementary substances and cytotoxic factors present in the hydatid fluid interfere with immunologic mechanisms, accounting for the long survival of the metacestode in tissues. Rupture of the cyst is followed by sudden stimulation of antibodies, giving rise to anaphylactic reactions. IgM and IgE levels decrease soon after removal of the cysts, while raised titers of IgG antibodies persist for several years. Healthy metacestodes in locations other than the liver may be seronegative.

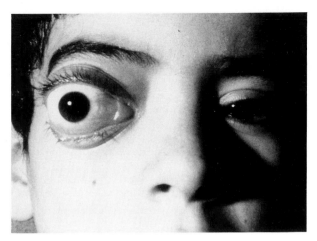

FIGURE 58-1 This 15-year-old boy has progressive exophthalmos of the right eye and mild inflammatory signs.

Systemic Manifestations

Many human infections remain silent and asymptomatic and are found on routine x-ray examination or at autopsy. Clinical manifestations are determined by the location of the cysts and their size. Children or adults may be infected.

Hydatid cysts are most frequently located in the liver (70%) and the lungs (up to 30%) and less frequently, the spleen, kidney, heart, central nervous system, and elsewhere. The incidence of multiple cysts varies. Forty percent of patients with pulmonary cysts can have liver involvement.

Large cysts of the liver produce abdominal pain and palpable masses, associated with jaundice if bile ducts are obstructed. When pulmonary cysts become symptomatic, they cause cough, thoracic pain, and hemoptysis. Rupture of the cyst in the bronchial tree causes a hydatid vomica, which contains portions of the wall and contents.

Plain radiographic studies are useful to reveal lung lesions; only calcified cysts are visible elsewhere with this technique. Computed tomography (CT) and ultrasonography are of diagnostic value in these patients.

The diagnosis should be suspected in patients from areas where the organism is endemic who are occupationally exposed to sheep or dogs. Aspiration of the cyst should not be performed because of the risk of dissemination or anaphylaxis. Several immunodiagnostic techniques that vary in sensitivity and sensibility are available. Recently, the arc 5 double-diffusion (DD5) test has been found useful in patients with compatible clinical signs and symptoms.[2]

Symptomatic cysts should be removed surgically without rupturing them, to avoid seeding.

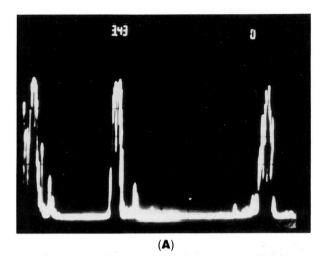

(A)

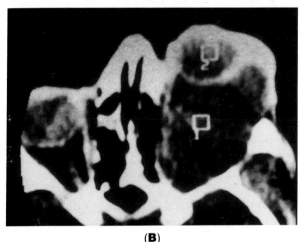

(B)

FIGURE 58-2 **(A)** A-scan ultrasonography of the patient depicted in Figure 58-1 shows an echo-free zone between two high-reflectivity echoes corresponding to the orbital cyst. **(B)** CT discloses right exophthalmos and a large retrobulbar cystic mass with well-defined borders. Note the displacement of the inner orbital wall.

Ocular Manifestations

Motile oncospheres that escape the portal and pulmonary vascular systems may reach the carotid artery and be distributed to the brain, eye, and orbit. Subretinal, vitreous, and anterior chamber cysts have been reported occasionally.[3,4] The orbit represents 1% of all systemic locations. Orbital hydatid cysts accounted for 19.8% of orbital tumors in Iraq and 5% in our series.[5]

Patients with orbital involvement are usually children or young adults who complain of slowly progressive exophthalmos, with or without pain (Fig. 58-1). Impaired extraocular motility and visual disturbances may be associated. Some patients may present with chemosis, lid edema, and orbital cellulitis because of rupture or secondary infection. Plain orbital radiographs may show increased soft tissue density and orbital enlargement. Ultrasonography discloses a cystic mass without internal reflectivity (Fig. 58-2,A). CT reveals a well-circumscribed mass with sharply defined borders (Fig. 58-2,B).[6] Internal density ranges from 9 to 38 Hounsfield units (Hu). Because of the integrity of the cysts, results of serologic tests are usually negative. Orbital hydatid cysts are rarely associated with involvement of other organs.[5]

Treatment is surgical–excision en bloc of the cuticular membrane and contents through an orbital approach.

References

1. Schantz PM. Echinococcosis. In: Steele JH, ed. *Zoonoses*. vol 1. Boca Raton, Fla: CRC Press; 1983;231-277.
2. Monzón CM, Coltorti EA, Varela-Diaz VM. Application of antigens from *Taenia hydatigena* cyst fluid for the immunodiagnosis of human hydatidosis. *Z Parasitenkd*. 1985;71:533-537.
3. Duke-Elder S, Perkins ES. Diseases of the uveal tract. In: Duke-Elder S, ed. *System of Ophthalmology*. London: Henry Kimpton; 1966;IX:473-478.
4. Scherz W, Meyer Schwickerath G, Piekarsky G, et al. Echinococcus in der Vorderkammer. *Klin Mbl Augenheilk*. 1973;163:66-70.
5. Talib H. Orbital hydatid disease in Iraq. *Br J Surg*. 1972;59:391-394.
6. Aouchiche M, Benrabah R, Abanou A, et al. Aspects tomodensitométriques du kyste hydatique intra-orbitaire. A propos de 10 observations. *J Fr Ophtalmol*. 1983;6:901-916.

Chapter 59

Loa Loa

DAVID GENDELMAN

Loa loa is an uncommon disease outside Africa. It is caused by a parasitic nematode that in its adult stages occupies the subcutaneous tissues. This roundworm is found in West and Central Africa in rain forests and swamp lands.

In the subcutaneous tissue the adult releases sheathed microfilariae into the host's blood following a diurnal pattern. The horsefly of the genus *Chrysops* ingests the larvae from the host's blood, acting as the intermediate vector. The microfilariae are in turn injected into the skin of man during the fly's next feeding.[1]

Systemic Manifestations

In humans, the worm matures over the 1st year and may live as long as 15 years. It has been found migrating in the subcutaneous space of virtually every part of the body. The most common symptom of systemic loa loa is the Calabar swelling, a hot pruritic swelling ranging from 2 to 3 cm to many times this size.[2] The Calabar swelling usually resolves in a few days and is associated with eosinophilia equal to 60 to 90% of the white blood cell count. In addition to subcutaneous involvement, polyarticular arthritis, endomyocardial fibrosis, or meningoencephalitis may occur; the latter is associated with parasitemia in excess of 50,000 microfilariae per cm^3.

Eosinophilia and the presence of microfilariae are diagnostic of loa loa (Fig. 59-1). Microfilariae are about 8 μm across and 250 μm long. The anterior end is blunt and devoid of cellular nuclei. An excretory pore is situated one third of the way from the caudad end of the organism.

Treatment consists of diethylcarbamezine (DEC), which in low concentrations slows the microfilariae and allows them to be destroyed in the liver. Higher concen-

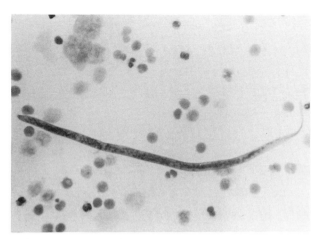

FIGURE 59-1 Peripheral blood smear showing microfilariae and eosinophils (hematoxylin-eosin, original magnification ×40).

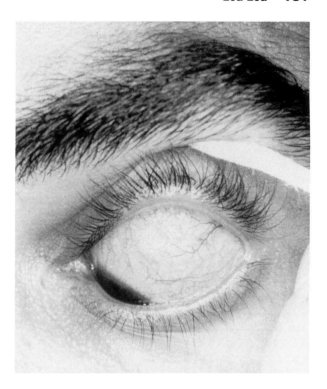

FIGURE 59-2 Subconjunctival worm in the superotemporal quadrant of the right eye.

trations of DEC are lethal to the microfilariae. DEC causes significant release of antigen when the organism dies, resulting in fever, malaise, arthralgia, meningoencephalitis, purpura, or coma. Careful titration of DEC combined with steroid and antihistamine use is recommended. An infectious disease consultation is helpful in the management of this problem.

Ocular Manifestations

The ocular complications of loa loa are variable, depending on the site that the parasite infests. Most typically the worm is noted on the surface of the eye beneath the conjunctiva (Fig. 59-2). Duke-Elder has suggested that temperature changes may bring the parasite to the surface of the eye.[3] When it is in the subconjunctival space a combination of itching, pain, and irritation may arise. In addition, the unnerving complaint of a moving foreign body sensation is described. A number of local remedies have been suggested to drive the worm from the eye, including applications of cold compresses and of onions. Most often it is useful to surgically remove the subconjunctival worm. If systemic medications are allowed to kill the filariae in the subconjunctival space an intense inflammatory reaction may result from the necrotic worms. This in turn may produce not only conjunctivitis but keratitis or iridocyclitis.[4]

Several suggestions have been made for the surgical removal of the parasite from the eye. One technique involves grasping the worm with forceps, then passing a silk suture under the worm and tying it tightly immediately after application of anesthesia. The worm can then be removed by dissecting beneath the conjunctiva.[4] An alternative approach suggested by Ghartey utilizes a cryoprobe to immobilize the worm. In this technique forceps are used to first grasp the worm in the subconjunctival space while, simultaneously, a topical anesthetic is applied. The cryoprobe is then applied immediately over the middle of the worm. At this time the conjunctiva is incised, and the motionless parasite is removed.[1] Any technique should be applied rapidly, as the worm travels quickly (approximately 1 cm per minute).

The loa loa worm may be present in other ocular tissues. Toussaint[5] reported a case of loa loa retinopathy that was characterized clinically by superficial hemorrhagic sheaths and yellow exudates on the retina. Ocular histopathologic examination revealed microfilariae in all retinal and choroidal blood vessels. Subcutaneous migration of filariae has also been described in the eyelid, and the worm has been found in the vitreous. Parasites in the anterior chamber have caused uveitis. Treatment of ocular loa loa requires identification of the organism and appropriate therapy for the underlying parasite. Surgical removal of parasites may be useful. A full understanding of the complications of medical management is important in treating this disease.

References

1. Gendelman D, Blumberg R, Sadun A. Ocular pathology for clinicians. Ocular loa-loa with cryoprobe extraction of subconjunctival worm. *Ophthalmology.* 1984;91:300-303.
2. Manson-Bahr PEC, ed. *Manson's Tropical Diseases.* 18th ed. London: Balliere Tindall, 1982;161-174, 727-729.
3. Disease of the outer eye: conjunctiva. In: Duke-Elder W, ed. *System of Ophthalmology.* vol. VIII. St. Louis, Mo: CV Mosby; 1965;403-405.
4. Rogell G. Infectious and inflammatory diseases. In: Duane TD, ed. *Clinical Ophthalmology.* vol. 5. Philadelphia, Pa: Harper & Row; 1982;19.
5. Toussaint D, Danis P. Retinopathy in generalized loa-loa filariasis; a clinicopathological study. *Arch Ophthalmol.* 1965;74:470-476.

Chapter 60

Onchocerciasis

River Blindness

ROBERT P. MURPHY and HUGH R. TAYLOR

Onchocerca volvulus is a filarial parasite that infects humans to cause the disease onchocerciasis, or river blindness. Onchocerciasis is a leading cause of blindness in the world and is the leading cause of blindness in the endemic areas of equatorial Africa and Central America.[1] In hyperendemic areas everyone is infected, and half the people are blinded by onchocerciasis before they die.[2] No other disease has an impact of this magnitude on a community. The vectors, blackflies of the family Simuliidae, breed in rapidly flowing water in the endemic areas and feed on the blood of humans and cattle.

The infected adult female blackfly can transmit up to six infective larvae to a human during a blood meal. The larvae migrate freely through subcutaneous tissues of the host. After going through several intermediary stages, they grow to their adult size (the female worms may grow to a length of 1 m) in firm, rubbery nodules that contain both adult male and female worms. Although some of these nontender nodules are palpable beneath the skin, many more are distributed throughout the body, where they are not accessible to ordinary examination. Mature adult worms, which can live over a decade, are coiled and confined in the nodules. During their lifetime, they produce millions of microfilariae,

which escape from the nodules to migrate throughout the host. The microfilariae can live for up to 18 months. The microfilariae are transmitted from humans back to the vector when another female blackfly takes a blood meal. Microfilariae in the skin enter the blackfly along with blood during feeding. There they migrate from the gut to the flight muscles and after several molts become third-stage infective larvae ready for transmission back to the host during the next blood meal.

Live microfilariae are motile and pass easily through body and ocular tissue without either exciting an inflammatory response or activating the host's immune system.[3] The pathologic changes in the eye and the rest of the body are related either directly or indirectly to microfilarial death. Death of individual microfilariae causes a limited, localized inflammatory response, particularly in untreated disease, when relatively small numbers of parasites are dying at any time. The accumulation over the years of numerous microscopic areas of inflammation and scarring throughout the body results in the systemic and ocular damage characteristic of onchocerciasis. Because the skin and eye are favored sites for the microfilariae, these tissues are most severely damaged over time. In the eye even small amounts of tissue damage can impair vision.

FIGURE 60-1 Chorioretinal scarring in onchocerciasis. Note the optic nerve pallor and the numerous areas of pigmented chorioretinal scars, especially temporal to the macula and inferior and nasal to the optic nerve.

Systemic Manifestations

In areas where the disease is endemic, infection with microfilariae is evident within a few years, and the intensity of infection increases with exposure. Systemic manifestations increase in extent with time. Pruritus is one of the commonest and earliest symptoms of infection. Histopathologically, perivasculitis is one of the earliest skin changes. With heavy infection, microfilariae can be seen, usually at the junction of dermis and epidermis. Late changes include hyperkeratosis, acanthosis, parakeratosis, inflammation, and collagen disruption.[4] With advanced involvement there is fibrosis of the dermis with atrophy of the overlying epithelium ("lizard skin") with hyperpigmentation and hypopigmentation ("leopard skin"). Pruritus is common in all stages. Although microfilariae have been found in almost every organ of the body and in many bodily secretions, damage to the remainder of the body is usually minimal and not clinically significant.

Ocular Manifestations

Living and dead microfilariae can be seen in all parts of the eye, as can inflammatory damage. The earliest evidence of ocular involvement is usually seen in the anterior chamber, where live organisms can be identified at the biomicroscope. Since they float freely in the anterior and posterior chamber, positioning the patient in a head-down position for 2 minutes prior to examination can facilitate their identification by allowing them to

aggregate near the concave endothelial surface of the cornea. They appear as small, slender, motile, wriggling white organisms approximately 300 μm long. Live microfilariae can also be seen in the cornea as coiled threads. Dead microfilariae are easier to detect because they are straighter and more opaque, often surrounded by a fluffy white inflammatory infiltrate. More advanced corneal involvement includes punctate keratitis and sclerosing keratitis, which can lead to blindness. The presence of anterior uveitis is variable, ranging from mild nongranulomatous inflammation to a more severe uveitis with posterior synechiae.

Blindness from anterior segment disease is more common in the dry plains of West Africa, where sclerosing keratitis predominates. Blindness in Central Africa often results from anterior uveitis. In the rain forests of West Africa, blindness usually results not from anterior segment disease but from chorioretinitis.

The spectrum of posterior segment involvement ranges from active motile microfilariae in the vitreous and retina without visible scarring to large geographic areas of chorioretinal scarring and atrophy with severe vision loss.[5] In the intermediate stages, transient small hemorrhages, cotton-wool spots, and small white retinal opacities may be seen with or without therapy. Retinal pigment epithelial atrophy and subretinal fibrosis are late findings associated with the inflammatory stage of the disease and may result from either local inflammation from dying microfilariae or from circulating antiretinal antibodies or immune complex deposition.[6] Extensive chorioretinal scarring or macular scarring with total destruction of the retina and choroid is responsible for severe visual loss in posterior segment disease (Fig. 60-1).

The associations between onchocerciasis and optic neuritis, optic atrophy, and glaucoma are not clear. Inflammation from microfilarial activity can cause optic neuritis, but many cases of optic nerve damage appear to result from the inflammation associated with treatment with diethylcarbamazine or suramin. Optic atrophy can also result from extensive peripheral visual field loss. Open-angle glaucoma is seen with onchocerciasis, especially in Africa, but it is probably a separate disease not pathogenically related to onchocerciasis. Secondary glaucoma can occur as a result of onchocerciasis-induced uveitis.

Treatment

The newly developed drug, ivermectin, promises to revolutionize the treatment of onchocerciasis.[7] A single oral dose of 6 to 12 mg repeated on an annual basis has a dramatic effect on reducing the microfilaria levels and ocular disease.[8] Ivermectin kills microfilariae but not adult worms, so treatment should be continued for about 10 years, until the adults die.

References

1. WHO Expert Committee on Onchocerciasis. Third report. Geneva: World Health Organization; 1987;1-167. WHO technical report series 752.
2. Prost A. The burden of blindness in adult males in the savanna villages of West Africa exposed to onchocerciasis. *Trans R Soc Trop Med Hyg.* 1986;80:525-527.
3. Greene BM, Gbakima AA, Albiez EJ, et al. Humoral and cellular immune responses to *Onchocerca volvulus* infection in humans. *Rev Infect Dis.* 1985;7:789-795.
4. Connor DH, George GH, Gibson DW. Pathologic changes of human onchocerciasis: implications for future research. *Rev Infect Dis.* 1985;7:809-819.
5. Bird AC, Anderson J, Fuglsang H. Morphology of posterior segment lesions of the eye in patients with onchocerciasis. *Br J Ophthalmol.* 1976;60:2-20.
6. Donnelly JJ, Semba RD, Xi M-S, et al. Experimental ocular onchocerciasis in cynomolgus monkeys. III. Roles of IgG and IgE antibody and autoantibody and cell-mediated immunity in the chorioretinitis elicited by intravitreal *Onchocerca lienalis* microfilariae. *Trop Med Parasitol.* 1988;39:111-116.
7. Taylor HR. Onchocerciasis—a potential revolution for its treatment. *Internat Ophthalmol.* 1987;11:83-85.
8. Taylor HR, Murphy RP, Newland HS, et al. Comparison of the treatment of ocular onchocerciasis with ivermectin and diethylcarbamazine. *Arch Ophthalmol.* 1986;104:863-870.

Chapter 61

Schistosomiasis

KHALID F. TABBARA and NADER SHOUKREY

Schistosomiasis is a chronic infection caused by diecious trematodes of the genus *Schistosoma.* There are three major species that commonly infect humans: *S. haematobium*, found throughout much of Africa and parts of the Middle East, causes urinary schistosomiasis; *S. mansoni* overlaps with *S. haematobium* in Africa and the Middle East and causes intestinal schistosomiasis; *S. japonicum*, confined to the Far East, causes Asiatic intestinal schistosomiasis.

Being digenetic, schistosomes have two phases of multiplication in their life cycle, one sexual in the adult stages parasitizing the definitive human host, the other asexual, in the larval stages developing in the interme-diate snail host. Infected snails shed fork-tailed larval schistosomes, the cercariae, which are infective to humans in fresh water. The swimming, nonfeeding cercariae must reach their definitive host within 2 days or die. If successful, they can penetrate the skin or mucous membranes, dropping off their tails and transforming into schistosomules. These enter lymphatics and venules and hence are carried through the venous circulation to the right side of the heart and the lungs. After that they proceed to the liver, but the route is not clearly known. In the portal circulation, schistosomules grow rapidly and copulate. Males carry fertilized females of *S. haematobium* to the vesical veins, of *S. mansoni* to the inferior mesenteric veins, and of *S. japonicum* to the superior mesenteric veins, where egg laying takes place. The destination of adult schistosomes is not always the vesical and mesenteric plexuses, however, as they are occasionally found in other anatomic sites in the definitive host. Eggs firmly wedged into vessel walls burrow their way out of the venules into tissues of the bladder wall (*S. haematobium*) or intestinal wall (*S. mansoni, S. japonicum*) and eventually into the lumens of these organs to be excreted with the urine and feces. Many eggs are either dislodged from the vessel walls and carried through the bloodstream to the liver, lungs, or other ectopic foci or remain trapped in tissues of the bladder and intestinal walls. Defecation or urination in fresh water causes eggs to hatch, liberating ciliated larvae, the miracidia, which must find their intermediate mollusk host within 24 hours or die. Once inside the snail host, miracidia undergo cyclopropagative development culminating in the formation of the cercariae.

Primary cercarial invasion of the skin evokes both humoral and cellular immune responses which cause mild cercarial dermatitis. During subsequent skin invasions, the dermatitis is acute, possibly owing to immobilization of reaginic antibodies which bring about signs of immediate hypersensitivity at the sites of cercarial invasion. Schistosomules derived from a primary cercarial invasion are protected from immune-mediated attack by sustaining continuous turnover of the membranocalyx, leading to loss of antigenic components that can be identified by host antibodies, and by incorporating host antigens on their surface.[1] Similarly, adult schistosomes in veins evade the host immune response by acquiring host antigen on their surface; however, circulating adult worm antigen induces immunity to reinfection by killing schistosome developmental stages by an antibody-dependent, cell-mediated response in which eosinophils play a major role.[2]

Unlike live adult worms, which produce protective immunity but are themselves shielded from immune attack, dead worms and eggs cause an active response. Dead adult schistosomes are carried through the bloodstream to the liver or lungs, where they elicit immune reactions and may become seeds for thrombus formation. Schistosome egg antigens induce delayed, cell-mediated granulomatous hypersensitivity reactions involving mainly eosinophils, mononuclear phagocytes, and lymphocytes. This immune reaction includes modulating mechanisms that influence granuloma size and the resulting pathologic changes in the chronic disease. Moreover, egg count seems to be inversely proportional to the immune response, and a number of humoral and cellular mechanisms have been implicated in suppressing lymphocyte reactivity.

Both acute schistosomiasis (Katayama fever), associated with primary heavy infections, and glomeru-lonephritis, frequently associated with schistosomal disease, are believed to be forms of immune complex disease. Katayama fever may be the outcome of cross-reaction between egg and worm antigens that causes a rapid increase in antibody levels and the formation of antigen-antibody complexes and acute serum sickness. Renal injury, on the other hand, may be initiated by the retention of circulating schistosome immune complexes in renal glomeruli.

Systemic Manifestations

Skin invasion by cercariae of human schistosomes in a primary exposure causes mild transitory inflammation with itching at the site of penetration. The cercarial dermatitis is often marked, with maculopapular pruritic rash after repeated exposures or when elicited by non-human parasites (swimmer's itch), but it is usually absent in inhabitants of areas of endemic disease. Passage of schistosomules through the lungs may cause cough and hemoptysis, and their maturation in the liver occasionally causes mild to severe hepatitis. In a primary exposure, manifestations of acute schistosomiasis (the Katayama syndrome) are prominent about the time egg laying begins, particularly in *S. japonica* and heavy infections of *S. mansoni*, but seldom in *S. haemotobia* or in inhabitants of endemic areas. The patient becomes febrile and commonly has generalized lymphadenopathy, hepatosplenomegaly, eosinophilia, and elevated immunoglobulin levels. Following the acute phase, the established infection, evolving usually in sites where a significant proportion of eggs is deposited, is often silent, and the occurrence of symptoms commonly correlates with the intensity of infection.

The principal organs involved in schistosomiasis haematobia are the urinary bladder and ureters. Cystitis from eggs deposited in the bladder wall causes the early predominant manifestations of *S. haematobium* infections; namely, terminal or total hematuria, dysuria, and urinary frequency. The deposited eggs give rise to granulomatous polypoid patches, which become fibrotic. As the eggs calcify, the fibrous patches evolve into sandy patches, which, in heavy infections, become confluent, producing a calcified bladder. Nonhealing ulcers may form, usually overlying sandy patches, and cause severe pain, particularly on urination. Egg deposition in the ureter provokes similar pathologic changes, with the occurrence of obstructive uropathy manifesting as hydroureters, hydronephrosis, and py-elonephritis. Patients with urinary schistosomiasis have a high incidence of carcinoma of the bladder, predominantly of the squamous cell type and less commonly of the transitional cell type.

In infections of *S. mansoni* and *S. japonicum*, the intestine and liver are the organs principally affected.

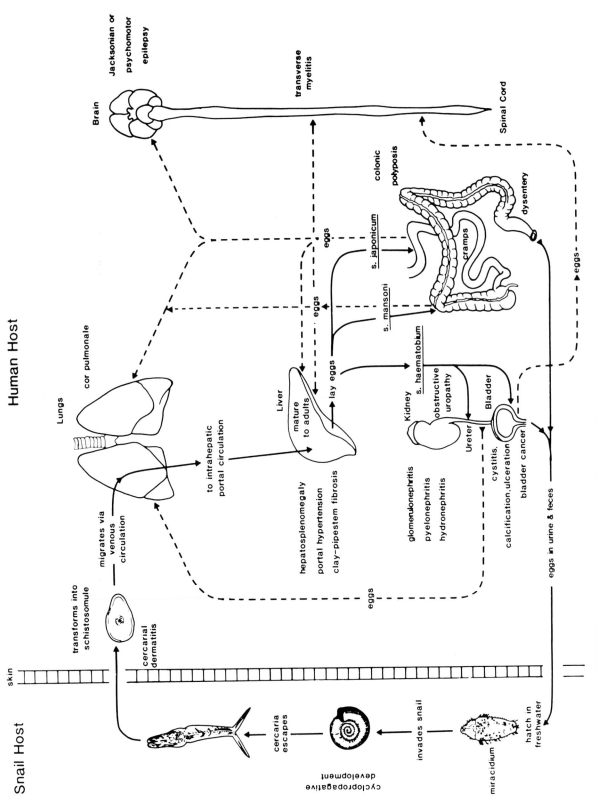

FIGURE 61-1 Life cycles and pathology of the three schistosome species that infect humans.

The granulomatous, hemorrhagic, and ulcerative mucous membrane from eggs deposited in the intestinal wall causes crampy abdominal pains and mild to severe schistosomal dysentery, depending on the severity of infection. Nonadenomatous inflammatory polyps of the colon are frequently associated with heavy infections in Egypt but rarely in other areas of endemic infection. Eggs carried back to the liver through portal blood cause granulomas and deposition of fibrous tissue culminating in Symmers' "clay pipe stem" fibrosis around branches of the portal vein and presinusoidal block of portal blood flow. The subsequent portal hypertension in the grossly enlarged liver and the development of portosystemic collaterals result in marked splenomegaly, esophageal varices (hematemesis), and, less commonly, ascites. Glomerulonephritis is not uncommonly associated with S. mansoni infection.

Schistosomal cor pulmonale may be seen with infections of all three species in humans owing to egg deposition in the lungs and subsequent pulmonary hypertension. Cerebral schistosomiasis is a discrete syndrome occurring in areas where S. japonicum is endemic and manifesting as convulsive attacks, Jacksonian epilepsy, and psychomotor epilepsy. S. mansoni and S. haematobium more frequently involve the spinal cord than the brain, the main clinical feature being transverse myelitis. A schematic representation of the evolution of the three main forms of human schistosomiasis is shown in Figure 61-1.

Ocular Manifestations

Ocular involvement in schistosomiasis is particularly manifested in the acute phase of the disease (Kayatama's syndrome), with urticaria, swelling, and edema of the eyelids. Anterior nongranulomatous uveitis, retinal vasculitis, orbital pseudotumor, and cataract are occasionally observed in patients with known schistosomiasis.[3] These ocular lesions, in which neither eggs nor worms are detected, may be due to immune complex deposits similar to those found in schistosomal glomerulonephritis.

Ocular schistosomiasis due to hematogenous spread of schistosome eggs or worms from the vesical and mesenteric veins to the eye is either quite unusual or possibly subclinical. Extensive review of the literature reveals only 14 cases of histologically proven ocular schistosomiasis.[3,4] Patients in most of these cases are children between the ages of 5 and 12 years who inhabit endemic areas. The ocular lesions are confined to the conjunctiva, retina, and choroid, and manifest as irregularly scattered yellowish white nodules of various sizes with some concentration close to blood vessels. Lesions in the choroid are attributable to S. mansoni, whereas S. haematobium seems to have a predilection for localization in the conjunctiva. Microscopic examination reveals typical eggs and typical granuloma formation with infiltrating polymorphonuclear leukocytes (PMNs), plasma cells, lymphocytes, and eosinophils. In a case of dacryoadenitis caused by S. haematobium, there were egg granulomas, excessive fibrous deposition, and scar tissue formation which led to widespread destruction of the lacrimal gland. Schistosome worms have also been reported in ocular tissue: paired male and female S. haematobium worms were detected in a branch of the superior orbital vein during sectioning of a granulomatous lesion removed from the palpebral conjunctiva in the region of the caruncle, while, on another occasion, an immature S. mansoni male worm was removed from the anterior chamber of a 19-year-old patient, where it caused hyphema and blurred vision.

There are two possible pathways by which schistosome eggs may reach tissues of the eye: embolism of eggs through the arterial system and local deposit of eggs following anomalous parasite migration. Arterial embolism seems to be the more likely route of invasion, which is facilitated by pulmonary arteriovenous fistulas caused by eggs deposited in the lungs and by the portopulmonary azygous anastomosis favored by portal hypertension, particularly in patients with hepatosplenic schistosomiasis. Another possibility by which ocular schistosomiasis may occur is the direct entry of cercariae through the conjunctiva and the maturing of the parasite in one of the conjunctival veins.[5] However, it has been shown that ocular entry of the parasite in experimental models is not responsible for unusual eye lesions, resulting only in hepatointestinal schistosomiasis. The possibility of direct contamination of the conjunctiva by eggs on the fingers should also be considered in a patient with urinary schistosomiasis.

References

1. Mahmoud AAF. Schistosomiasis. In: Warren KS, Mahmoud AAF, eds. *Tropical and Geographical Medicine*. New York, NY: McGraw-Hill; 1984;443-457.
2. Caulfield JP, Lenzi HL, Elsas P, et al. Ultrastructure of the attack of eosinophils stimulated by blood mononuclear cell products on schistosomula of *Schistosoma mansoni*. *Am J Pathol*. 1985;120:380.
3. El-Antably SA, El-Hoshi MA. Cataract and parasitic infestation in Egypt. *Bull Ophthalmol Soc Egypt*. 1975;68:275.
4. Pittela JEH, Orefice F. Schistosomatic choroiditis. II. Report of first case. *Br J Ophthalmol*. 1985;69:300.
5. Welsh HN. Bilharzial conjunctivitis. *Am J Ophthalmol*. 1968;66:933.

Chapter 62

Toxocariasis

Visceral Larva Migrans

GEORGE S. ELLIS, Jr., and CANDACE C. COLLINS

Toxocara canis (ascarid) is a parasite acquired by human ingestion of embryonated *T. canis* eggs and a very common species of intestinal worms found in dogs. Human infection is not necessary to complete the life cycle. The eggs do not mature into adult reproductive forms in the human or adult dog. Evidence exists for the presence of immune mechanisms in the adult dog that inhibit ovum maturation.

Life Cycles

Puppies are the vectors for human infection. They are infected by various means, including ingesting infective eggs in soil, eating small mammals that are themselves infected by eggs ingested from soil (paratenic hosts), prenatal migration of larvae from tissues of the mother to the fetal pups, and ingestion of mother's milk by the newborn pup. The mother may be infected by ingesting larvae passed in the feces of hyperinfected pups. After the puppies become infected, the life cycle is complete. The puppies then excrete ova in feces, through which humans become infected.[1]

Pathophysiology of Human Disease

Human disease occurs when the person eats dirt infested with the ova or ingests ova on the puppy's coat by licking. In the human intestine, the ova mature into second-stage larvae, which penetrate the gut wall. Tissue migration from the person's gut through the portal circulation, mesenteric vessels, and intestinal lymphatic channels allows the larvae to reach the liver, lungs, brain, and eyes.

As the larvae migrate through the tissue, a focal eosinophilic reaction occurs, invariably around degenerating larvae or leaking cysts. Eosinophils have a direct cytotoxic effect on these parasites. A dead larva elicits a stronger immune response than a live one, a fact

that has implications for ocular treatment. Initially, there is a marked IgE production with resultant eosinophilia and local eosinophilic inflammatory infiltration. Subsequently, the inflammatory reaction develops into a granulomatous lesion. The eosinophilic granuloma has a characteristic morphology and may appear as the Splendore-Hoeppli phenomenon (central parasite with necrotic changes and surrounding antigen-antibody complexes).[2]

Protective immunity does not seem to be significant. The fact that the full life cycle occurs only in puppies suggests that adult dogs have some immunity. A weak immune response may also account for the failure of larvae to mature in humans. In addition to antibody production, cell-mediated events play a significant role.

Larval dissemination may cause systemic symptoms such as fever and malaise and signs of hepatosplenomegaly, leukocytosis, and eosinophilia. In the eye, the patient notes decreased vision and has findings of posterior uveitis. All of these are manifestations of the body's immune response.

T. canis eye infection causes posterior uveitis. Since it may present as a large vitreoretinal mass, leukocoria is a prominent sign and *Toxocara* is quite important in the differential diagnosis. Floaters, vitreous cell, cystoid macular edema, and retinal detachment are often found. "Spillover" anterior uveitis is common and produces pain, redness, photophobia, anterior chamber cell and flare, and keratic precipitates.

Systemic Manifestations

Children aged 1 to 4 years are most often affected. The widespread presence of the larvae in various tissues is responsible for the systemic manifestations of visceral larva migrans (VLM). Findings are related to antigenic response and are characterized by fever, eosinophilia, hepatosplenomegaly, pulmonary infiltrates, seizures, and recurrent upper respiratory infections. If all are

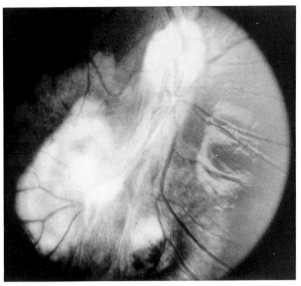

FIGURE 62-1 Toxocara chorioretinal lesion involving inferior portion of ocular fundus. (Courtesy of Barrett G. Haik, M.D.)

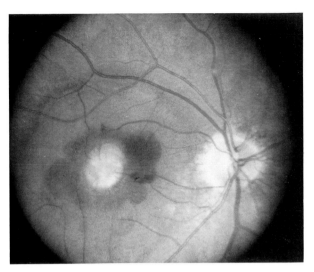

FIGURE 62-2 Toxocara granulomatous lesion with subretinal hemorrhage. (Courtesy of Rudolph M. Franklin, M.D.)

evident, the child may be extremely ill. Frank anaphylaxis may occur if a large amount of antigen is released at one time. Deaths have been recorded, but complete recovery is usual in 6 to 12 months. Fortunately, most cases are subclinical, without evidence of generalized infection.[3]

Ocular Manifestations

Usually a child about 7 years old presents with unilateral ocular disease without a history of accompanying or previous systemic symptoms of VLM. Signs include leukocoria, decreased vision, unilateral uveitis, retinal detachment, and dragged disc (distortion of the normal retinal vascular pattern, with the retinal vessels drawn toward a peripheral cicatricial focus). The differential diagnosis of such a presentation includes retinoblastoma, unilateral pars planitis, endophthalmitis (bacterial or fungal), exudative retinitis, and Coats' disease.

The larvae reach the eye through hematogenous spread; ocular infection is almost always a unilateral event. The larvae enter the eye through either the retinal or ciliary circulation via the ophthalmic artery. The presence of the larvae can cause mechanical retinal dysfunction secondary to the presence of a worm in the subretinal space. Larvae can also be found free in the vitreous. Because they contain antigenic foreign protein, they elicit an immune response.

In the human eye, this immune response can have

drastic consequences. The uveitis can cause hypotony secondary to ciliary body shutdown, which may lead to phthisis bulbi. Cataracts are a common consequence of intraocular inflammation. A granuloma may be located in the posterior pole and can decrease central vision. If the granuloma is peripheral, the resulting vitreous traction can cause retinal detachment or the appearance of a dragged disc (Figs. 62-1, 62-2).

Because the eye is a small, contained space, infection within can lead to total destruction. Infection with *T. canis* has four recognizable ocular presentations, all related to the immune response. Fourteen different types of ocular toxocariasis have been described.[4]

The most common presentation is *chronic destructive endophthalmitis*, which accounts for two thirds of all cases. A focal retinal eosinophilic granuloma with the larvae in the center of necrosis is usually seen and is associated with a localized retinal detachment secondary to serous exudation. Occasionally a cyclitic membrane with posterior synechiae is present.

Localized posterior granuloma is another common form of ocular involvement. The presentation is similar to that of chronic endophthalmitis, except that there is no retinal detachment and the lesion is well-defined. It is usually hemispheric and retinal in location. Tension lines radiate from the granuloma, and fibrous bands may extend to the pars plana and into the vitreous. As a circumscribed lesion, it may be confused with retinoblastoma. If the nematode enters the eye via the ciliary circulation, a *peripheral granuloma* may develop. The widespread vitreous reaction and endophthalmitis are usually absent in these forms, but traction bands are

TABLE 62-1 *Toxocara canis* Serum ELISA Titer in 333 5- to 6-year-old Children Without Physical Evidence of VLM*

	ELISA titer							
	<1/8	1/8	1/16	1/32	1/64	1/128	1/256	>1/256
Percentage of children whose titer was equal to or higher than that listed (cumulative)	100%	65%	32%	23%	13%	7%	4%	1%

* No child had ocular toxocariasis.

common. Traction on the retina may cause a detachment or give the appearance of a dragged disk. Finally, the infection may cause *vitreous or subretinal abscess.*[5]

Laboratory Evaluation

The serum ELISA test is positive in infected patients. It is thought to be positive if found in a 1:8 dilution, though detection in undiluted solution may be significant.[6] Seropositivity with no evidence of disease is common. In one study more than 30% of 333 kindergarten children were seropositive (ELISA ≥ 1/16; Table 62-1) but not one had evidence of ocular disease.[7] Because of the prevalence of ELISA positivity, the ELISA test cannot be used to differentiate *Toxocara* infection from retinoblastoma in a patient with leukocoria.[7] Blood serotype anti-A and anti-B titers may be elevated; however the patient's blood type must be considered in this interpretation.

It is not common to make a definitive diagnosis in VLM. The only way to make the diagnosis is a liver biopsy that shows the presence of larvae. No hematologic or liver enzyme abnormalities are pathognomonic for VLM or ocular *Toxocara* infection.

Management and Treatment

The treatment of VLM is usually supportive, since the disease is self-limited. If symptoms are severe, steroids may be used to control inflammation. Thiabendazole (25 mg/kg/day) has also been used.

Antihelminthic drugs are not used in ocular disease. The presence of the worm in the eye elicits an inflammatory reaction that is exacerbated if the worm dies. The inflammation is the enemy. Photocoagulation and cryotherapy may kill the worm, so they are to be avoided.

Ocular treatment consists of topical steroids to control anterior inflammation as needed. Periocular methylprednisolone acetate (40 to 60 mg) or systemic prednisone can be used for posterior disease. Vitrectomy and retinal detachment repair are done when indicated. If the diagnosis of retinoblastoma cannot be ruled out by the characteristic calcifications noted on ultrasound, enucleation may be necessary.[4]

Summary

Ocular *toxocariasis* is a visually devastating disorder without a good treatment. Prevention remains important. Pica should be discouraged, and all puppies should be dewormed with piperazine. When leukocoria occurs, retinoblastoma must be excluded. Early enthusiasm for the ELISA test in confirming the diagnosis may have been unfounded and premature.

References

1. Faust EC, Beaver PC, Jung RC. Animal agents and vectors of human disease. 4th ed. Philadelphia, Pa: Lea and Febiger, 1975.
2. Garner A, Klintworth GK, eds. *Pathobiology of Ocular Disease.* New York, NY: Marcel Dekker; 1982.
3. Beaver PC. Visceral larva migrans. In: Vaughan VC, et al., eds., *Textbook of Pediatrics.* 11th ed. Chap. 11.115. Philadelphia, Pa: WB Saunders Co; 1979.
4. Pavan-Langston D, ed. *Manual of Ocular Diagnosis and Therapy,* 2nd ed. Boston, Mass: Little, Brown; 1985.
5. Kanski JJ. *Clinical Ophthalmology.* St. Louis, Mo: CV Mosby Co; 1984.
6. Pollard ZF. ELISA for diagnosis of ocular toxocariasis. *Ophthalmology.* 1979;86:743.
7. Ellis GS, Pakalnis VA, Worley G, et al. *Toxocara canis* infestation. *Ophthalmology.* 1986;93:1032.

Chapter 63

Trichinosis

KHALID F. TABBARA and NADER SHOUKREY

Trichinosis is a disease caused by roundworms of the species *Trichinella spiralis*. A zoonosis, the disease is particularly common in rodents, bush pigs, bears, and hogs that are fed unsterilized garbage. Man is an incidental host, and trichinosis is especially prevalent in pork-eating cultures. Humans become infected with trichina worms after eating raw or undercooked pork or other meat that contains viable encysted larvae. The cyst wall is digested in the stomach and the infective larvae are released. They burrow into the duodenal and jejunal mucosa and develop into adult worms that emerge into the intestinal lumen and copulate. Fertilized females reburrow into the mucosa of the upper intestine, larviposit, and are eventually rejected in the feces. Newborn larvae migrate principally to striated muscle by both hematogenous and lymphatic spread. Skeletal muscles of the limbs and diaphragm are most often parasitized, followed in order by the tongue, masseters, extraocular muscles, and laryngeal muscles. The larvae coil up between muscle fibers, become encapsulated, and develop into infective third-stage larvae as they grow to about 10 times hatching size within 3 weeks. Encysted larvae may remain viable for many years, even after the capsule wall has calcified.

Cell-mediated immunity plays a significant role in expelling adult trichina worms from the mucosa of the upper intestine into the feces.[1] Administration of corticosteroids suppresses this local form of tissue immunity and lengthens the intestinal phase of infection. Serum antibodies are aimed specifically against the migrating larvae; they belong to the classes IgM, IgG, and IgA and are not protective. Effective mechanisms for destroying the migrating larvae incorporate primarily granulocytes. T lymphocytes, on the other hand, are believed to mediate the hypereosinophilia that characterizes the larval migration phase of trichinosis.[2]

Systemic Manifestations

Cases of *T. spiralis* infection are predominantly subclinical and can be detected only by autopsy surveys. Eosinophilia may serve, however, as an indicator for the possibility of asymptomatic trichinosis. Whether infection will be symptomatic or subclinical is most likely determined by the number of viable cysts ingested, and later by the number of larvae per gram of muscle. Clinical manifestations of the intestinal beginning phase of the disease are characterized by nonspecific gastroenteritis with diarrhea or constipation, nausea, vomiting, anorexia, abdominal cramps, and weakness. Many infected persons report no gastrointestinal distress. The larval migration phase that follows is associated with hypereosinophilia, irregular high fever, myositis, periorbital edema, fatigue, and headache. Eosinophilia, ranging from 10 to 80%, is a major laboratory finding in trichinosis, and its absence indicates a poor prognosis.

Trichinosis is anomalous among helminthic diseases in that it causes fever that may reach 40°C. Muscle pains, particularly in the masseters, may be quite severe, and there may be some interference with muscle function, causing difficulty in mastication, breathing, and swallowing. Serum muscle enzymes such as creatine phosphokinase and lactate dehydrogenase are elevated in most patients.[2,3] The terminal phase of the disease, that of encystation, is marked by generalized toxemia due to destruction and absorption of host tissue and of many of the encysting larvae. In fatal cases of heavy *Trichinella* infections, death is generally caused by invasion of the central nervous system (meningoencephalitis), heart (myocarditis), lungs (bronchopneumonia), and less frequently kidney (nephritis), and peritoneum (peritonitis). The life cycle of *T. spiralis* in humans, and the pathology it causes at the various stages of infection, are diagrammatically illustrated in Figure 63-1.

Ocular Manifestations

Periorbital edema coupled with fever and large-scale eosinophilia provide presumptive evidence of trichinosis. This important ocular sign appears as early as the 7th day of infection and is rarely absent in patients who develop clinical symptoms of the disease. Edema of the eyelids, and the less commonly occurring retinal hemorrhages, conjunctival edema, and subconjunctival hemorrhages and petechiae,[2-4] are probably related to the vasculitis developing during the larval migration phase

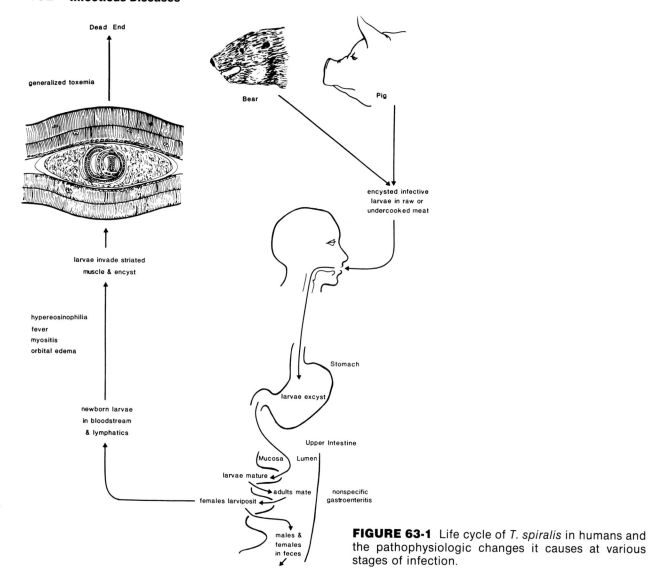

FIGURE 63-1 Life cycle of *T. spiralis* in humans and the pathophysiologic changes it causes at various stages of infection.

and are responsible for eliciting splinter hemorrhages.[5] In some patients, conjunctivitis, photophobia, diplopia, visual field changes, or oculomotor dysfunction may occur.[6] Muscle pain, particularly associated with eye movements and mastication, may be quite severe and is usually one of the symptoms that prompts the patient to seek medical advice. In such cases, the parasite is encysting in extraocular muscles and masseters, as shown on autopsy.[7]

References

1. Harari Y, et al. Anaphylaxis-mediated epithelial C1-secretion and parasite rejection in rat intestine. *J Immunol.* 1987; 138:1250.

2. Kazura JW. Trichinosis. In: Warren KS, Mahmoud AAF, eds. *Tropical and Geographical Medicine.* New York, NY: McGraw-Hill; 1984;427-430.

3. Simon JW, et al. Further observations on the first documented outbreak of trichinosis in Hong Kong. *Trans R Soc Trop Med Hyg.* 1986;60:313.

4. Barrett-Connor E, Davis CF, Hamburger RN, et al. An epidemic of trichinosis after ingestion of wild pig in Hawaii. *J Infect Dis.* 1976;133:437.

5. Markell EK, Voge M. *Medical Parasitology.* Philadelphia, Pa: WB Saunders Co; 1981;258-262.

6. Ancelle T, et al. Horsemeat-associated trichinosis—France. *JAMA.* 1986;256:177.

7. Givner I. Ocular signposts to systemic diagnosis. The Mark J. Schoenberg Memorial Lecture. *Am J Ophthalmol.* 1965;60:792.

Section D

Mycobacterial Diseases

Chapter 64

Leprosy

Hansen's Disease

T. J. FFYTCHE

Leprosy is still regarded by many people as a disease of medieval or biblical times that no longer exists in the developed world but survives only in deprived areas of India, Africa, Asia, and South America. The crippling deformities of the face and limbs and the social stigma that accompanies the disease are, of course, well-known, and the term "leper" continues to be used derogatively in contemporary writing and speech. The reality is quite different: leprosy, far from being an ancient and dead disease, is still on the increase and is endemic in almost every underdeveloped country in the world. Estimates of its numbers extend to 15 million.

New methods of treatment and the development of an effective vaccine may eventually produce a significant reduction in the incidence of the disease, but the residue of neural, skeletal, dermatologic, and ocular complications will need active care and therapy for many years.

Ophthalmologists should remember that leprosy carries the highest incidence of ocular complications of any

systemic disease and that much of the damage to the eye caused by leprosy is amenable to therapy and in many instances is preventable. Blindness, when added to loss of sensation and to deformity, destroys the ability for self-care and independence that is such an essential factor in the life of the patient, and the onset of visual failure remains for many the most terrifying aspect of the disease. Yet many of the conditions that lead to blindness arise from simple neglect or ignorance, and the ophthalmologist must play an active role not only in managing existing complications but also in educating patients and leprosy workers in methods of preventing them.[1-3]

The Disease

Leprosy is caused by infection by *Mycobacterium leprae,* a gram-positive intracellular bacillus initially observed by Hansen in 1873 (making it the first reported bacterial

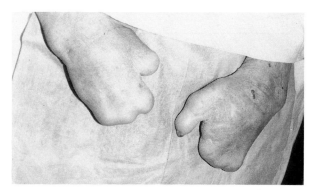

FIGURE 64-1 Advanced deformity of the hands in chronic lepromatous leprosy. This sort of disability is compounded by co-existing visual impairment.

pathogen in humans). Man is the only significant natural reservoir of the disease, although it can be induced in the nine-banded armadillo and in the thymectomized mouse.

The disease is a complex one and still poorly understood: its mode of transmission remains undetermined. In endemic areas a high proportion of the population are exposed to the bacilli, but over 90% are immune and do not develop the disease. Susceptible persons react to the organism in a variety of ways, depending on their immune response, which may be genetically and racially determined. Some have a high degree of resistance, and a few organisms provoke a strong cell-mediated response (*tuberculoid leprosy, TL*); others have no immunity at all, with billions of organisms present in the body causing damage by tissue destruction and infiltration (*lepromatous leprosy, LL*). These are the two polar forms of the disease, but a full classification based on immune status includes intermediate forms: borderline lepromatous (BL), borderline (BB), and borderline tuberculoid (BT).[4] LL is more frequently seen in Asiatic and European races, whereas the tuberculoid form occurs more in African and Indian races.

Although the disease is primarily one of slow progression, certain patients develop acute reactions related to alterations in immune status.[5] BB patients may develop *reversal reactions* (type 1), and lepromatous patients may be subject to *erythema nodosum leprosum* (ENL), or type 2, reactions. Both reactions can present some of the most difficult aspects of the disease to manage and both have important implications from an ophthalmologic point of view.

Leprosy is therefore a chronic disease of the skin, nerves, and eyes that progresses slowly, with occasional acute reactions, through dermatologic presentation to neurologic damage characterized by sensory and motor impairment resulting in contractures and deformities.

Tissue and skeletal destruction ensue, affecting the face and extremities (Fig. 64-1), and blindness may occur in a significant proportion of cases (7 to 10%). Yet leprosy does not kill; patients have a normal life span and, when faced with the relentless mutilation and incapacity that used to be an inevitable consequence of infection, it is easy to understand why the disease was and still is so feared, being regarded in many cultures as a manifestation of divine punishment.

The Organism

M. leprae has certain unique properties that are responsible for the varied clinical manifestations that culminate in the crippling deformities that characterize the chronic condition and also render the eye particularly susceptible to infection and infiltration. The organism is slow growing, it has a particular affinity for neural tissue (Schwann cells), and it flourishes best in the cooler areas of the body. Thus the disease evolves slowly, with the clinical manifestations confined to the superficial cooler tissues of skin, muscle, and nerves, particularly those that are near the surface (small sensory and autonomic fibers and certain large nerves such as the ulnar, posterior tibial, and facial). The anterior segment of the eye is also affected, since it contains large numbers of small unmyelinated nerves in the cornea, iris, and ciliary body and has a temperature several degrees below body temperature.

Ocular Manifestations

With such a varied pathophysiologic background to the disease, it is not surprising that the ocular manifestations are diverse and subject to many outside influences. Factors such as the age, race, and sex of the patient, the type of leprosy, its status and duration, and above all the nature and success of therapy and preventive measures, all influence the severity of eye complications and affect the incidence of blindness.[6]

Six basic mechanisms of ocular involvement that can be identified are summarized below.

Direct Invasion of the Anterior Segment

Anterior segment invasion occurs exclusively in LL. Organisms reach the eye through the bloodstream, although some arrive by direct spread from neighboring structures. The bacilli probably lodge and multiply initially in the ciliary body and iris. Their slow growth delays the manifestations of intraocular damage, making it rare for the eye to show signs of ocular involvement in the first few years of the disease.

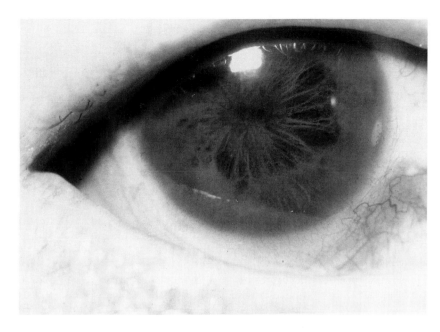

FIGURE 64-2 Chronic iridocyclitis in lepromatous leprosy showing a miotic pupil associated with atrophy of the dilator muscle of the iris.

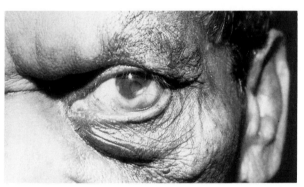

FIGURE 64-3 Limbal nodule probably arising from the ciliary body and infiltrating the sclera. The iris is being pulled into the lesion and a staphyloma is likely to develop.

Corneal involvement includes thickening and destruction of the corneal nerves with resultant hypesthesia. A superficial stromal keratitis, commencing in the superior temporal quadrant, may cause opacification, although this seldom affects vision.[7]

The ciliary body and iris become invaded at an early stage of the disease. It is not usually possible to detect ciliary body involvement clinically except through gradual failure of accommodation, and in the later stages, ocular hypotension. The iris, however, may demonstrate progressive deterioration in function, with dilator atrophy and increasing miosis, and this chronic form of iridocyclitis with pinpoint nonreacting pupils (often combined with cataract) is a common cause of visual impairment (Fig. 64-2). Clumps of active and degenerate

bacilli may form white bodies known as "iris pearls," and occasionally an organized leproma may develop in the iris, ciliary body, or cornea.[8]

The sclera may be involved by direct infiltration of organisms from the ciliary body with the formation of nodules and eventually a staphyloma (Fig. 64-3).

Direct involvement of the peripheral choroid can occur but rarely extends behind the equator, so that its effect on vision is negligible. The posterior parts of the eye, which are relatively warmer, are not significantly involved.

Acute Inflammations: Type 1 and Type 2 Reactions

Type 1 reactions can involve the facial nerve, particularly its zygomatic branches to the orbicularis oculi, which control eye closure and blinking. This reaction results in acute lagophthalmos and can rapidly lead to exposure keratopathy if untreated.

Type 2 reactions (ENL) in lepromatous cases are immune-complex disturbances that provoke episcleritis, scleritis, and iridocyclitis. All of these may recur intermittently, and iritis and scleritis require energetic therapy, as they can rapidly cause blindness. Many eyes are anesthetic and therefore neglected in this situation, so the painless red eye should be regarded with great suspicion by leprologist and ophthalmologist.

Impaired Lid Closure

In all forms of the disease, paralysis of branches of the facial nerve may lead to loss of blinking, inadequate lid

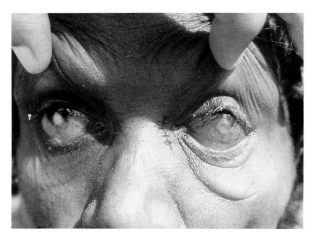

FIGURE 64-4 End-stage exposure keratopathy caused by a combination of impaired corneal sensation and defective closure of the lids is a preventable situation.

closure, and lagophthalmos. This may occur acutely as part of a type 1 reaction or chronically. Urgent medical or surgical therapy is required to protect the eye, since untreated exposure keratopathy can cause corneal opacification, ulceration, and perforation.

Impaired Corneal Sensation

Diminished or absent corneal sensation can occur in all forms of leprosy. It affects transparency of the cornea through loss of its normal protection and also by disturbing corneal metabolism. The combination of impaired lid closure and reduced corneal sensation is a major cause of blindness in the disease (Fig. 64-4).

Damage to Adnexae

Structural damage to the lids, lacrimal passages, and nose may all affect the eye adversely, either directly or through their influence on tear drainage. Madarosis, blepharochalasis, dystichiasis, entropion, and trichiasis occur in every form of the disease as a result of local infiltration and neural damage. Their effect on sight can be considerable, particularly through their influence on the health and clarity of the cornea.

Secondary Infection

Ocular and adnexal damage from all forms of leprosy renders the eye susceptible to secondary infection, often exacerbated by defective metabolism and poor healing.

Summary

The prevalence of ocular complications in leprosy patients varies according to the populations studied (6 to 90%). Some complications are of academic interest only, but the majority affect vision either directly or indirectly.[9]

Preservation of sight by preventive and medical measures as well as by active surgical intervention remains one of the most important aspects of this complex disease. Ophthalmologists and leprologists must realize that leprosy is an important cause of blindness worldwide and that the leprosy patient, already handicapped by loss of sensation and deformity, needs vision even more than others in order to remain independent. Visual impairment in leprosy too often is caused by ignorance and failure to give practical advice to patients on matters of eye protection and hygiene. The responsibility for this lies firmly with the medical services, and neglect of this will sustain the high incidence of blindness in this distressing disease.

References

1. Brand M. The care of the eye. *The Star.* Carville, La; 1980.
2. Brand M, ffytche TJ. Ocular complications of leprosy. In: Hastings RC, ed. *Leprosy.* Edinburgh: Churchill Livingstone; 1985.
3. Joffrion VC, Brand M. Leprosy of the eye—a general outline. *Lepr Rev.* 1984;55:105-114.
4. Ridley DS, Jopling WH. Classification of leprosy according to immunity, a five group system. *Int J Lepr Other Mycobact Dis.* 1966;34:255-273.
5. Turk TL. Leprosy as a model of subacute and chronic immunologic disease. *Invest Dermatol.* 1976;67:457-463.
6. ffytche TJ. Ocular leprosy. *Trop Doct* 1985;15:118-125.
7. Allen JH, Byers JL. The pathology of ocular leprosy. I. Cornea. *Arch Ophthalmol.* 1960;64:216-220.
8. Allen JH. The pathology of ocular leprosy. II. Miliary lepromas of the iris. *Am J Ophthal.* 1966;61:987-992.
9. Lamba PA, Santoshkumar D, Arthanariswaran R. Ocular leprosy. A new perspective. *Lepr India.* 1983;55:490-494.

Chapter 65

Tuberculosis

ROBERT L. PEIFFER, Jr., ROBERT M. LEWEN, and HONOGLIANG YIN

Tuberculosis is an acute or chronic communicable disease caused by the acid-fast bacterium *Mycobacterium tuberculosis*, which most commonly involves the lungs but may affect virtually any organ or tissue in the body. The principal focus of pulmonary infection usually produces no symptoms. Secondary tuberculosis, which is commonly associated with systemic signs and symptoms of disease, may result from reactivation of a primary lesion or by reinfection from exogenous sources. Although morbidity and mortality rates from tuberculosis have declined steadily for several decades, the disease persists and is an important health problem in the United States, where in recent years approximately 25,000 to 30,000 new cases of clinical tuberculosis have been reported annually.[1]

Systemic Manifestations

Tubercle bacilli can gain entrance to the body by several routes. The only one of practical importance in the United States is the respiratory tract. Tuberculosis is transmitted by airborne particles 1 to 5 μm in diameter; the organisms spread from the lungs to regional lymph nodes, producing lymphadenitis. Subsequently, lymphatic drainage delivers the tubercle bacilli to the systemic circulation, whence it has the potential to spread to all organs of the body. During this primary phase of tuberculosis the infection is usually subclinical; the likelihood that the disease is radiographically or clinically apparent is approximately 5%.

The most common outcome for the initial infection with *M. tuberculosis* is healing with granuloma formation; this occurs over a period of months and is for most persons accompanied by the development of tuberculin skin test reactivity. Generally, the granulomas remain stable, and frequently they calcify. In a minority of cases (5 to 15%), occasionally but not always associated with stress or other disease, the granulomas break down; tubercle bacilli multiply and disperse, producing systemic disease.

Ocular Manifestations

The incidence of ophthalmic manifestations is approximately 1 to 2%.[2] Ocular and periocular involvement may occur as a consequence of active infection from hematogenous or contiguous spread of viable bacilli or as a local manifestation of an allergic or hypersensitivity reaction to circulating tuberculoproteins. Variations in the size and virulence of the infecting or inciting inoculum and in the nature and degree of the host's immune responses, eventuate in remarkably pleomorphic presentations of ocular tuberculosis. With the exception of the crystalline lens, any other tissues of the eye and its adnexa may be affected. Tubercle formation of the lids, conjunctiva, cornea, sclera, and uveal tract are all histologically confirmed examples of ocular tuberculosis, which are usually but not invariably associated with obvious systemic disease[3-7]; the choroid appears to be exceptionally predisposed to infection. Multiple uveal granulomas are not uncommonly associated with miliary tuberculosis. Solitary granulomas of the uveal tract may mimic other proliferative ocular diseases, such as malignant melanoma. Histologically, the classic pattern of tuberculous granulomatous inflammation is caseation necrosis in the centers of the granulomas in which smooth rods can be demonstrated by the Ziehl-Neelsen's or the auramine-rhodamine[5] method. The areas of caseation are surrounded by a zonal type of reactive granulomatous infiltration with epithelioid cells, Langhans' giant cells, plasma cells, and lymphocytes (Figs. 65-1 to 65-3). Tuberculosis remains the classic example of a disease that is controlled almost entirely by cell-mediated immunity involving the macrophage as the effector cell and the lymphocyte (especially the T cell) as the immunoresponsive cell.[8]

Latent interstitial keratitis and phlyctenular keratoconjunctivitis are believed to be allergic sequelae of systemic tuberculosis, and granulomatous uveitis and retinal vasculitis associated with vitreous hemorrhage (Eale's disease) have long been presumed to be causally

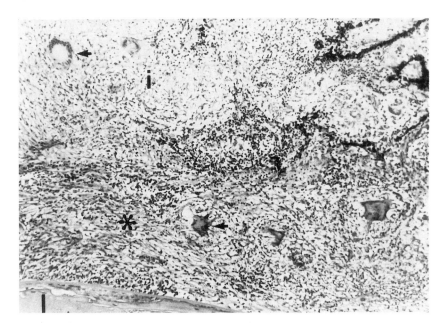

FIGURE 65-1 Diffuse granulomatous anterior uveitis associated with tuberculosis panophthalmitis. Langhan's giant cells (*arrows*) are seen within the iris (i) and an inflammatory retroiridal membrane (*) that has joined iris to lens (l); (hematoxylin-eosin, original magnification ×125).

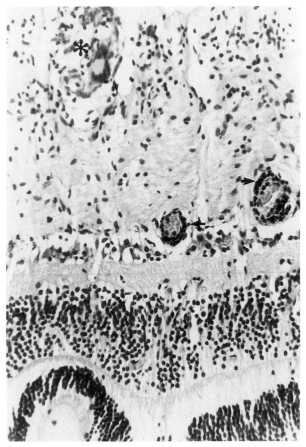

FIGURE 65-2 Retinal lymphocytic perivascular infiltrate (*arrows*) and small granulomas within the nerve fiber layer (*) are observed in a patient with systemic tuberculosis (hematoxylin-eosin, original magnification ×310).

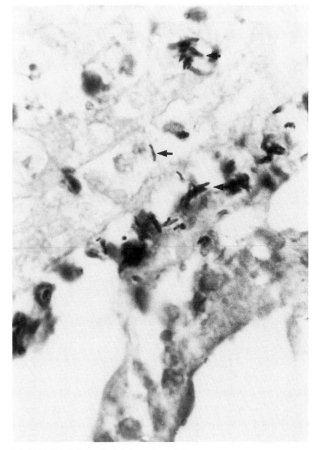

FIGURE 65-3 Acid-fast bacilli (*arrow*) are identified in the anterior uvea of the patient depicted in Figure 65-1 (Ziehl-Neelsen, original magnification ×1250).

associated with the disease. These conditions are more frequently identified in patients with primary tuberculosis.[9]

Paralleling the decrease in the reported incidence of pulmonary tuberculosis, there has been a dramatic decline in the number of diagnosed cases of tuberculous uveitis. In 1960 Woods reported 22% of a series of uveitis cases to be causally related to tuberculosis[10]; 13 years later, Schlaegel reported a 0.3% incidence.[11] The rare presentation of tuberculous uveitis in the 1980s was recently confirmed by Henderly et al.,[12] who found an incidence of 0.2% among more than 400 cases of uveitis in which it was possible to make a specific diagnosis based on history, physical findings, and laboratory studies. Improved detection of previously unrecognized disease entities and changing epidemiologic trends in the population appear to have played major roles in the remarkable reduction of the often presumptive diagnosis of ocular tuberculous uveitis. However, as it is potentially curable, tuberculosis should always be considered in the diagnostic work-up of patients with granulomatous uveitis.

A decade ago it was believed that of the five known species of *Mycobacterium* only human and bovine were pathogenic to man. Recently patients with the acquired immunodeficiency syndrome (AIDS) have been found to be infected with other types of mycobacteria. Evolving trends in epidemiology and infectious disease, including the emergence of AIDS, may elevate the significance of mycobacteria in terms of ocular involvement in systemic disease.

References

1. Glassroth J, Robins AG, Snider DE Jr. Tuberculosis in the 1980's. *N Engl J Med*. 1980;302:1441-1450.
2. Donahue HC. Ophthalmologic experience in a tuberculosis sanatorium. *Am J Ophthalmol*. 1967;64:742-748.
3. Ellingsworth RS, Wright T. Tubercles of the choroid. *Br Med J*. 1948;2:365-368.
4. Darrell RW. Acute tuberculosis panophthalmitis. *Arch Ophthalmol*. 1967;78:51-54.
5. Cangemi FE, Friedman AH, Josephberg R. Tuberculoma of the choroid. *Ophthalmology*. 1980;87:252-258.
6. Jabbour NM, Faris B, Trempe CL. A case of pulmonary tuberculosis presenting with a choroidal tuberculoma. *Ophthalmology*. 1985;92:834-837.
7. Lyon CE, Grimson BS, Peiffer RL, et al. Clinicopathologic correlation of a solitary choroidal tuberculoma. *Ophthalmology*. 1985;92:845-850.
8. Collins FM. The immunology of tuberculosis. *Am Rev Respir Dis*. 1982;125:42-49.
9. Shumomura Y, Tader R, Yuam T. Ocular disorders in pulmonary tuberculosis. *Folia Ophthalmol Jpn*. 1979;30:1973-1978.
10. Woods AC, Abraham IW. Uveitis survey. *Am J Ophthalmol*. 1961;51:761-780.
11. Schlaegel TF Jr. Differential diagnosis of uveitis. *Ophthalmol Digest*. 1973;
12. Henderly DE, Gentsler AJ, Smith RE, et al. Changing patterns of uveitis. *Am J Ophthalmol*. 1987;103:131-136.

Section E

Mycotic Diseases

Chapter 66

Histoplasmosis

ROBERT C. WATZKE

Histoplasmosis is a systemic disease caused by the inhalation and lymphogenous spread of *Histoplasma capsulatum*, and characterized by involvement of various organs in an intracellular granulomatous inflammation. There is epidemiologic evidence that the choroid and retina can be involved during the primary infection and by significant late complications. *H. capsulatum* is a fungus that grows in the soil in certain areas of the world. The reason for such geographic variability is unknown, but all populations surveyed confirm this.[1] In the United States, there is tremendous geographic variation of histoplasmosis, as indicated by surveys of skin test positivity.[2]

Human infection is acquired by inhalation of the mycelial form of *H. capsulatum* in air contaminated by soil particles and not by transmission from any animal species. While soil contaminated by chicken and pigeon droppings has been associated with human histoplasmosis epidemics, the organism does not cause avian infection in nature. In fact, the only nonhuman species that are spontaneously infected with histoplasmosis are bats.[3] The high concentration of *H. capsulatum* in soil contaminated with chicken and pigeon droppings is probably due to its high nitrogen content.[4]

Systemic Manifestations

After inhalation, the mycelial phase is transformed within a few hours into the yeast phase. Depending on the number of organisms inhaled and the resistance of the host, a small area of bronchopneumonia may occur. Macrophages are soon involved and become filled with the yeast forms of *H. capsulatum*. A granulomatous inflammation occurs in the lung, with encapsulation and caseation. The hilar nodes become enlarged and conglomerated, producing a typical x-ray pattern.

In most persons the primary lung infection heals and the patient becomes asymptomatic but demonstrates a positive complex when examined by chest radiography. Hematogenous spread of yeast forms of *H. capsulatum* during the primary infection can occur, so almost any organ can be involved. In fact, there is a very high incidence of small, calcified foci in the liver and spleen of

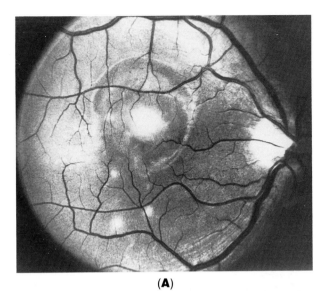

(A)

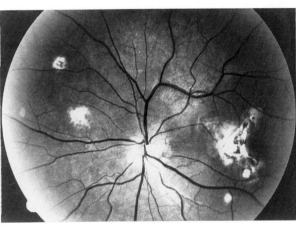

(B)

FIGURE 66-1 The right and left fundi of a patient with the presumed ocular histoplasmosis triad. **(A)** The right eye has a few typical choroidal lesions in the posterior pole and a subretinal neovascular membrane in the fovea. **(B)** The left eye has several old "histo spots" or scars with a peripapillary scar and a hyperplastic scar in the fovea.

persons studied at autopsy in areas where histoplasmosis is endemic. *H. capsulatum* can be found in as many as 87% of these foci.[5]

Whether the patient recovers or experiences progression of systemic histoplasmosis depends on the balance of hypersensitivity and immunity in that person. Patients taking immunosuppressive drugs and those who have immune system depression are especially susceptible to disseminated histoplasmosis. The skin test invariably becomes positive a few weeks after exposure, but a positive skin reaction must be interpreted by a skilled observer and according to strict criteria. Cross-reaction to blastomycosis, coccidioidomycosis, and adiospiromycosis occurs.[6] Skin test positivity decreases with age. Nevertheless, most of the epidemiology of histoplasmosis rests on skin test surveys, a relatively unreliable epidemiologic method.

Ocular Manifestations

Ocular histoplasmosis comprises two types. First is panophthalmitis or granulomatous uveitis, which is seen in persons with systemic histoplasmosis. The uvea and retina are involved in a granulomatous infection in which *H. capsulatum* can be demonstrated in inflammatory cells. The ocular tissues are necrotic, and the eye is usually lost.[7-11] Occasionally, *H. capsulatum* causes a single solitary granuloma.[12] These types of ocular infections usually occur in an immunocompromised host.

The "presumed ocular histoplasmosis syndrome" is a construct used to explain the occurrence of inflammatory choroidal foci in persons living in areas where histoplasmosis is endemic. The ocular picture is typical. There are few too many choroidal spots, which are approximately 100 to 500 μm in diameter without overlying retinal or vitreous reaction. These spots are asymptomatic when they first appear and are thought to represent an inflammatory reaction to *H. capsulatum* organisms brought to the choroid through the bloodstream during the initial pulmonary infection.

These choroidal spots are the hallmark of the presumed ocular histoplasmosis syndrome, giving it its alternative name "multifocal inner choroiditis."[13] The lesions initially consist of an aggregate of lymphocytes in the choroid beneath an intact Bruch's membrane.[14] *H. capsulatum* organisms have not been demonstrated in these lesions except for two reports, one controversial and one unconfirmed.[15,16] The choroidal lesions gradually evolve into atrophic scars with breaks in Bruch's membrane and growth of new blood vessels from the choroid into the subretinal space.[17,18] At this stage, no organisms can be found. One of the puzzling features of the disease is the appearance of new foci in the fundi of infected persons over many years, even after they have moved out of an endemic area.[19,20]

The choroidal spots evolve into atrophic chorioretinal scars with a characteristic central or eccentric fleck of pigment. At this stage, they are called "histo spots." A population survey of persons living in an area where the disease is endemic showed a prevalence of 4.6%.[21] No comparable population survey has been done of people from an area where *H. capsulatum* is not endemic.

When a choroidal scar is present near the fovea the patient is forever at risk of loss of vision from the development of a subretinal neovascular membrane. Because of continued atrophy and proliferation of the retinal pigment epithelium, new vessels from the chorio-capillaris proliferate into the subretinal space. This is a common event in many chorioretinal scars wherever Bruch's membrane is destroyed. Once the new vessels invade the subretinal space, fluid leaking through the imperfect endothelial cell junctions causes a serous detachment, hemorrhage, and progressive scar formation.[22] The causative factors involved in the evolution of chorioretinal lesions into the hyperplastic fibrovascular disciform scars are not known.

Eyes of patients from endemic areas also show scarring around the optic disc. Choroidal neovascular membranes can evolve from such peripapillary scars and can threaten central vision. If an eye has "histo spots," a peripapillary scar, and a subretinal fibrovascular membrane in the macula, it is said to have the histoplasmosis triad (Fig. 66-1). When subretinal neovascularization near the center of the fovea occurs, laser photocoagulation is indicated to protect the patient from loss of central vision by destroying the new vessel membrane before it reaches the central fovea.[23]

In spite of extensive epidemiologic and experimental studies, the organisms of *H. capsulatum* have not yet been conclusively and unequivocally found in eyes that manifest the histoplasmosis triad nor has any explanation been substantiated for the occurrence of the triad in patients living in areas where histoplasmosis is not endemic.[24,25] After decades of study, the role of histoplasmosis as a cause of ocular infection is still unclear. Until more material for histologic study is obtained early in the disease, the diagnosis of ocular histoplasmosis must remain provisional.

References

1. Edwards PQ, Klaer JH. Worldwide geographic distribution of histoplasmosis and histoplasmin sensitivity. *Am J Trop Med Hyg.* 1956;5:233-257.
2. Edwards LB, Acquaviva FA, Livesay VT, et al. An atlas of sensitivity of tuberculin, PPD-B, and histoplasmin in the United States. *Am Rev Respir Dis.* 1969;99(2):1-132.
3. Schwarz J. *Histoplasmosis.* New York, NY: Praeger Publications; 1981;105.
4. Zeidberg LD, Ajello L, Dillon A, et al. Isolation of *H. capsulatum* from soil. *Am J Pub Health.* 1952;42:930-935.
5. Schwarz J. *Histoplasmosis.* New York, NY: Praeger Publications; 1981;299.
6. Schwarz J. *Histoplasmosis.* New York, NY: Praeger Publications; 1981;151.
7. Craig EL, Suite T. *Histoplasma capsulatum* in human ocular tissue. *Arch Ophthalmol.* 1974;91:285.
8. Hoefnagels KLJ, Pijpers PM. *Histoplasma capsulatum* in a human eye. *Am J Ophthalmol.* 1967;63:715.
9. Klintworth GK, Hollingsworth AS, Lusman PA, et al. Granulomatous choroiditis in a case of disseminated histoplasmosis. *Arch Ophthalmol.* 1973;90:45.
10. Schwarz J, Salfelder K, Viloria HJE. *Histoplasma capsulatum* in vessels of the choroid. *Ann Ophthalmol.* 1977;9:633.
11. Scholz R, Green WR, Kutys R, et al. *Histoplasma capsulatum* in the eye. *Ophthalmology.* 1984;91:1100-1104.
12. Zimmerman LE. Discussion. In: Krill AE, Chisti MI, Klien BA, et al. Multifocal inner choroiditis. *Trans Am Acad Ophthalmol Otolaryngol.* 1969;73:222.
13. Krill AE, Chisti MI, Klein BA, et al. Multifocal inner choroiditis. *Trans Am Acad Ophthalmol Otolaryngol.* 1969;73:222-242.
14. Makley TA, Craig EL, Werling K. Histopathology of ocular histoplasmosis, In: TF Schlaegel, ed., Update on ocular histoplasmoses, *Int Ophthalmol Clin.* 1983; 23(2):1-18.
15. Roth AM. *Histoplasma capsulatum* in the presumed ocular histoplasmosis syndrome. *Am J Ophthalmol.* 1977;84:293.
16. Khalil M. Histopathology of presumed ocular histoplasmosis. *Am J Ophthalmol.* 1982;94:369.
17. Weingeist TA, Watzke RC. Ocular involvement by histoplasma capsulatum, *Int Ophthalmol Clin.* 1983;23:33-47.
18. Meredith TA, Green WR, Key SN, et al. Ocular histoplasmosis: clinicopathologic correlation of 3 cases, clinical pathological review. *Survey Ophthalmol.* 1977;22:189-205.
19. Lewis ML, Van Newkirk MR, Gass JDM. Follow-up study of presumed ocular histoplasmosis syndrome. *Ophthalmology.* 1980;87:390.
20. Watzke RC, Claussen RW. The long-term course of multifocal choroiditis (presumed ocular histoplasmosis). *Am J Ophthalmol.* 1981;91:750-760.
21. Ganley JP. Epidemiological characteristics of presumed ocular histoplasmosis. *Acta Ophthalmol.* 1973;119 (suppl): 32, Copenhagen.
22. Miller H, Miller B, Ryan SJ. Correlation of choroidal subretinal neovascularization with fluorescein angiography. *Am J Ophthalmol.* 1985;99:263.
23. Macular Photocoagulation Study Group: Argon laser photocoagulation for ocular histoplasmosis. *Arch Ophthalmol.* 1983;101:1347-1357.
24. Braunstein RA, Rosan DA, Bird AE. Ocular histoplasmosis syndrome in the United Kingdom. *Br J Ophthalmol.* 1974;58:893.
25. Craandijk A. *Focal Macular Choroidopathy.* The Hague: Dr. W. Junk, B.V., Publishers; 1979.

Chapter 67

Metastatic Fungal Endophthalmitis

JOEL A. SCHULMAN and GHOLAM A. PEYMAN

The prevalence of disseminated fungal infections has increased significantly in recent years. Advances in therapy of chronic or debilitating disease have placed more patients at risk of developing such hematogenous infections. Fungal septicemia is also being diagnosed and reported with increased frequency, owing in part to increased awareness of the disease and improvement in techniques used to diagnose fungal infections. Fungemia is primarily a nosocomial infection. Hart and associates[1] noticed a fourfold increase in fungal infections in compromised hosts, comparing the periods from 1960 to 1963 and from 1964 to 1967; a different group of investigators[2] reviewing records at Memorial-Sloan Kettering Cancer Center reported that fungal infections increased 15-fold between 1950 and 1959.

Multiple factors influence the development of disseminated fungal infections. Prolonged chemotherapeutic, immunosuppressive, or steroid therapy may interfere with host defense mechanisms, enhancing susceptibility to mycotic infection. Therapeutic interventions, such as hyperalimentation, surgery, prolonged antibiotic administration, the use of indwelling catheters, and prolonged irradiation are at times complicated by fungal infection. The prolonged survival of patients with malignancies has resulted in a population predisposed to develop disseminated fungal infections. Fungemia has also been recognized with increased frequency in intravenous drug abusers who use unsterile injection paraphernalia.[1,2]

A complication of fungemia is endophthalmitis. Intraocular fungal infections occur from either endogenous or exogenous sources. Endogenous fungal endophthalmitis is a result of hematogenous spread and begins as a retinitis or chorioretinitis with secondary vitreous involvement. Fungi may also be introduced into the eye after ophthalmic surgery, penetrating injuries, or intraocular extension of a mycotic corneal ulcer. The resulting exogenous infection initially involves the vitreous only and later spreads to the retina and choroid.

Fungi usually considered to be saprophytic can cause endophthalmitis. Endogenous mycotic endophthalmitis has become increasingly important to the ophthalmologist. A wide variety of organisms are responsible for this infection. The most common cause is *Candida albicans*, but the less virulent species *C. tropicalis, C. stellatoidea,* and *C. parapsilosis* have also been implicated. Additional fungal organisms less frequently reported to cause this bloodborne infection include *Aspergillus fumigatus, Blastomyces dermatitidis, Histoplasma capsulatum, Sporothrix schenckii, Cryptococcus neoformans, Coccidiodes immitis,* and organisms of the order *Mucorales* and of the genuses *Fusarium* and *Penicillium*.[3-5]

The importance of eye lesions in the diagnosis of systemic candidiasis has only recently been recognized. In one report, endophthalmitis was present in 10 (37%) of 27 patients with *C. albicans* fungemia.[6] A second study demonstrated a 30% incidence of candidal endophthalmitis in 82 patients with candidemia.[7] Both studies noted the risk of developing candidal endophthalmitis was greatest for patients receiving hemodialysis or parenteral hyperalimentation. Both studies concluded that periodic eye examinations are necessary, even in the absence of symptoms, for all patients with *C. albicans* sepsis.

Henderson and associates[8] prospectively followed ocular findings in 131 surgical intensive care patients who were fed by hyperalimentation. Fourteen of the 131 (10.6%) patients developed candidal endophthalmitis. Seven patients in this group had negative results on serial blood cultures. These results strongly suggest that careful serial examination by an ophthalmologist using indirect ophthalmoscopy is also warranted in seriously ill patients receiving hyperalimentation, regardless of blood culture status.

Clarkson and Green,[4] commenting on predisposing factors in candidal endophthalmitis, noted 100 patients with this ocular infection. Eighty-five patients received antibiotics, seventeen, systemic steroids, and eight, combination therapy with both agents. Major surgery was performed on 53 persons, 25% of which were procedures for malignancy. Six patients had cirrhosis of the liver attributable to alcohol abuse, and nine had diabetes.

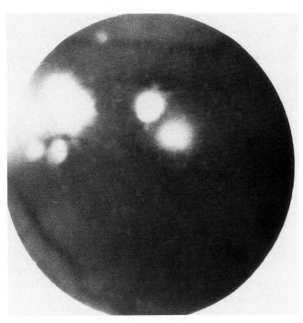

FIGURE 67-1 Fundus photograph demonstrates localized abscess formation in the posterior pole of a young drug-addicted patient.

Unlike bacterial endophthalmitis, candidal endophthalmitis usually has an insidious onset with a slow, indolent course. The organism enters the eye through the chorioretinal circulation. The eye may be asymptomatic in the earliest stages of this mycotic intraocular infection. Impaired vision, which may be accompanied by photophobia, is usually the first symptom of ocular infection. The eye may be painful or may appear red. Early in the course of the disease, iritis may be present, which characteristically is followed by hypopyon. Focal small white fluffy lesions develop in the retina and choroid, followed by an overlying vitreous inflammatory reaction. Vitreous extension of these lesions may produce microabscesses (Fig. 67-1), which enlarge and coalesce, appearing at this stage as solitary or multiple large white masses of varying size. Perivascular sheathing or infiltrates and retinal hemorrhages may be present. Eventually, the vitreous may become opaque, obscuring underlying retinal detail. Late vitreous organization may lead to traction retinal detachment. The differential diagnosis of endogenous candidial endophthalmitis includes such entities as toxoplasmic retinochoroiditis, familial amyloidosis, viral retinitis, nematode endophthalmitis, phacoanaphylactic reactions, primary reticulum cell sarcoma of the eye, uveitis due to sarcoidosis, and sterile inflammation.

The offending organism in bloodborne fungal endophthalmitis may be identified by culturing various anatomic fluids, including blood, urine, and sputum, or smears from indwelling catheters when appropriate. When cultures are negative, diagnostic vitreous tap and anterior chamber paracentesis may be necessary. Culture of the vitreous offers greater diagnostic accuracy then aqueous paracentesis.[9] In some cases, culture of the vitreous removed at surgery may be necessary to confirm the diagnosis.

The virulence of the offending organism influences the visual prognosis in endogenous fungal endophthalmitis. Virulent organisms elaborate enzymes and proteolytic byproducts that are toxic to the retina. A second factor influencing permanent visual acuity is the interval between diagnosis and treatment. Delay in treatment, especially in the presence of virulent organisms, adversely affects visual results.[9]

The management of endogenous fungal endophthalmitis is controversial. Animal studies have demonstrated that intraocular involvement begins in blood vessels of the choroid and retina. The blood-retina barrier is disrupted; then foci of infection extend to areas in proximity to the chorioretinal blood vessels and spread to the vitreous.[10] Because of damage to the blood-retina barrier, parenterally administered antimicrobial agents may achieve therapeutic concentrations in the retina and vitreous at the site of infection.

Optimal treatment regimens for endogenous fungal endophthalmitis have not been established. Intravitreal injection of antimicrobials, alone or combined with vitrectomy, may be required in more advanced cases of endophthalmitis when the infection significantly involves the vitreous. Intravitreal injection usually results in higher vitreous levels of antifungal agents than can be achieved via more conventional routes of administration (parenteral or subconjunctival). Vitrectomy improves diffusion of antifungal drugs into the vitreous while removing fungal organisms and their inflammatory exotoxins and proteolytic enzymes, which may damage the eye. Organization of vitreous may result in cyclitic membranes and traction bands, which when left unchecked can cause hypotony or retinal detachment. Vitrectomy removes this organized vitreous and provides a clearer optical axis.[9]

Only a few drugs are available for the treatment of fungal endophthalmitis. Amphotericin B, introduced in 1954, has been the mainstay in the treatment of intraocular mycotic infections, but its parenteral use is limited by systemic toxicity. Only a small margin of safety has been described following intravitreal administration of amphotericin B.[9] Liposomes are membranelike vesicles used to encapsulate drugs. Liposomal binding of antifungal agents such as amphotericin B prior to intravitreal injection may reduce intraocular drug toxicity and prolong clearance by controlling the rate of drug release.[11]

The use of ketaconazole to treat intraocular fungal infections has been limited.[12] Favorable results have been reported in two persons with endogenous fungal endophthalmitis treated with oral ketaconazole in combination with another antifungal.[13,14] Ketaconazole is an imidazole derivative that demonstrates good ocular penetration following oral administration[13]; however, because of the potential for inducing liver toxicity, the British Society for the Safety of Medicine advises against using this drug in most instances.[14]

Flucytosine (5-fluorocytosine) is an antifungal agent with little toxicity but a markedly limited spectrum of activity. Synergy of amphotericin B and flucytosine has been demonstrated for pathogenic yeasts, but despite initial susceptibility, secondary resistance, which may develop in strains of candida organisms, limits use of this agent.[9] Combined therapy with flucytosine and amphotericin B or ketaconazole usually prevents emergence of resistant organisms.[9]

The efficacy of treatment with other antifungal agents, including clotrimazole, nystatin, and miconazole, has been difficult to evaluate from clinical and experimental reports, since a variety of factors determine the outcome of mycotic endophthalmitis (the infecting organism, variations in the pathogenicity of organisms, host immune mechanisms, and so on).[3] A promising new group of antifungals are the triazole derivatives. Several agents in this class of antimicrobials have shown significant activity in different animal models of fungal infection and may prove efficacious in treating endogenous fungal endophthalmitis.[15]

References

1. Hart PD, Russell E Jr, Remington JS. The compromised host and infection II. Deep fungal infection. *J Infect Dis.* 1969;120:169-191.
2. Hutter RVP, Collins HS. The occurrence of opportunistic fungus infection in a cancer hospital. *Lab Invest.* 1962;11:1035-1048.
3. Jones DB. Chemotherapy of experimental endogenous *Candida albicans* endophthalmitis. *Trans Am Ophthalmol Soc.* 1980;78:846-895.
4. Clarkson JG, Green WR. Endogenous fungal endophthalmitis. In: Duane TD, ed. *Clinical Ophthalmology.* vol 3. Philadelphia, Pa: Harper & Row; 1981;1-44.
5. Swan SK, Wagner RA, Myers JP, et al. Mycotic endophthalmitis caused by *Penicillium* after parenteral drug abuse. *Am J Ophthalmol.* 1985;100:408-410.
6. Parke DW II, Jones DB, Gentry LO. Endogenous endophthalmitis among patients with candidemia. *Ophthalmology.* 1982;89:789-796.
7. Griffin JR, Foos RY, Pettit TH. Relationship between candida endophthalmitis, candidemia, and disseminated candidiasis. *Concilium Ophthalmologicum, 22nd, 1974.* Paris 1976;2:661-664.
8. Henderson DK, Edwards JE, Montgomerie JZ. Hematogenous Candida endophthalmitis (HCE) in surgical patients receiving parenteral hyperalimentation fluids (PHG). *Clin Res.* 1979;27:41A.
9. Peyman GA, Schulman JA. *Intravitreal surgery.* Norwalk, Conn: Appleton-Century-Crofts; 1986;407-455.
10. Cohen M, Edwards JE Jr, Hensley TJ, et al. Experimental hematogenous *Candida albicans* endophthalmitis: electron microscopy. *Invest Ophthalmol Vis Sci.* 1977;16:498-511.
11. Tremblay C, Barza M, Szaka F, et al. Reduced toxicity of liposome associated amphotericin B injected intravitreally in rabbits. *Invest Ophthalmol Vis Sci.* 1985;26:711-718.
12. Barrie T. The place of elective vitrectomy in the management of patients with *Candida* endophthalmitis. *Graefes Arch Clin Exp Ophthalmol.* 1987;225:107-113.
13. Goodman DF, Stern WH. Oral ketoconazole and intraocular amphotericin B for treatment of postoperative *Candida parapsilosis* endophthalmitis. *Arch Ophthalmol.* 1987;105:172.
14. Weiss JL, Parker WT. *Candida albicans* endophthalmitis following penetrating keratoplasty. *Arch Ophthalmol.* 1987;105:173.
15. Richardson K, Brammer KW, Marriott MS, et al. Activity of UK-49, a bis-triazole derivative, against experimental infections with *Candida albicans* and *Trichophyton mentagrophytes*. *Antimicrob Agents Chemother.* 1985;27:832-835.

Chapter 68

Mucormycosis

ROGER KOHN

Rhino-orbital mucormycosis is a fungal infection of class Phycomycetes and order Mucorales that is notable for its high morbidity and mortality rates.[1-11] Early recognition of this disorder is essential, particularly in medically compromised patients. Early medical and surgical management improve the prognosis.[1,3,6,8,10] Historically, few patients have survived without blinding, mutilating surgery to eradicate the infection. In a previous study we reported eight cases of rhino-orbital mucormycosis managed at various U.C.L.A. hospitals over a 12-year period. This represented the largest series of survivors who did not require exenteration and who had unaltered visual acuity.[10]

Our cases included five males and three females, whose mean age was 38 years. Two had juvenile-onset diabetes (one of whom was in ketoacidosis). Five had adult-onset diabetes (four of whom were in ketoacidosis). An eighth patient had no apparent predisposing condition. Only two of the patients were found to have the classic black eschar on the palate or within the nose. Following treatment, all patients survived with no resultant loss of vision.

Treatment included (1) early definitive diagnosis, (2) correction of diabetic ketoacidosis or other concomitant metabolic derangement, (3) wide local excision and debridement of all involved and devitalized oral, nasal, sinus, and orbital tissue, (4) establishing adequate sinus and orbital drainage, (5) daily irrigation and packing of the involved orbital and paranasal areas with amphotericin B, and (6) intravenous amphotericin B. Follow-up periods ranged from 6 months to 9 years (mean 4 years).

Phycomycetes are ubiquitous fungi occurring in soil, air, skin, and food.[1-3,5,8,10] Inoculation is by inhalation into the nasopharynx and oropharynx.[1-3,8,10] At this stage most patients generate phagocytic containment of the organisms. Those whose cellular or humoral defense mechanisms have been compromised by disease or immunosuppressive treatment may generate an inadequate response.[6,10] The fungus may then spread to the paranasal sinuses and subsequently to the orbits, meninges, and brain by direct extension.[3,6,8,10]

Mucormycosis preferentially invades the walls of blood vessels, resulting in vascular occlusion, thrombosis, and infarction.[3,6-8,10] This frequently affects the ophthalmic artery and in more serious cases may involve the internal carotid artery and cavernous sinus.

Mucormycosis infections are opportunistic and rarely occur in healthy persons. They are most commonly seen in diabetics, particularly those in ketoacidosis.[1-3,6,7,10] Other predisposing conditions include various malignancies (leukemia, lymphoma, multiple myeloma), anemia, generalized infections (septicemia, tuberculosis), fluid-electrolyte imbalance (thermal burns, extensive wounds, malnutrition, dehydration, severe diarrhea), and disorders of the gastrointestinal (gastroenteritis, hepatitis, cirrhosis), renal (glomerulonephritis, acute tubular necrosis, uremia), pulmonary (alveolar proteinosis), and cardiovascular systems (congenital heart disease). Patients treated with antibiotics, folic acid antagonists, chemotherapeutic agents, corticosteroids, or ionizing radiation may also be predisposed.[2,3,7,10]

Symptoms that may suggest mucormycosis in susceptible persons include multiple cranial nerve palsies, unilateral periorbital or facial pain, orbital inflammation, eyelid edema, acquired blepharoptosis, proptosis, acute

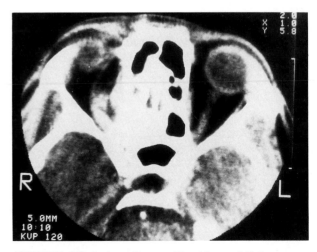

FIGURE 68-1 A patient with rhino-orbital mucormycosis demonstrates on CT scan a right ethmoidal sinusitis and subperiosteal abscess extending into the right medial orbit.

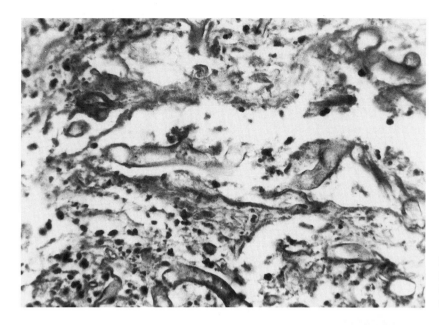

FIGURE 68-2 Large, branching, nonseptate hyphae characteristic of mucormycosis (hematoxylin-eosin, ×40).

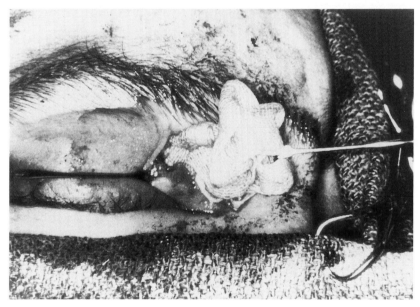

FIGURE 68-3 A patient with rhino-orbital mucormycosis demonstrates an amphotericin B packing placed in the ethmoidal sinus and superior medial orbit.

motility changes, internal or external ophthalmoplegia, headache, and acute loss of vision.[3,6,10]

Evaluation should include radiologic studies of the paranasal sinuses and either a computed tomographic (CT) or magnetic resonance (MR) study of the orbit, sinuses, and anterior cranial areas (Fig. 68-1). Tissue should be obtained for fungal culture, microscopic examination, and histopathologic examination.[6] The fungus may be seen with routine hematoxylin and eosin stains,[6] but special fungal stains (KOH, methenamine silver) may also prove helpful. Although the characteris-

tic large, branching, nonseptate hyphae of mucormycosis (Fig. 68-2) will often be recovered on paraffin section studies, absolute reliance on this modality may result in delayed diagnosis. Therefore, frozen section studies at the time of original biopsy are also recommended, to afford earlier diagnosis and institution of appropriate management.

Early diagnosis, while the disease is still somewhat anatomically confined, is essential to a more favorable outcome.[1,3,6,8,10] This allows prompt initiation of appropriate treatment and in some cases may obviate mutilating surgery, including exenteration. Our 12-year retrospective series at U.C.L.A. represents the largest reported series with a favorable outcome.

The adjunctive use of daily irrigation and packing of the involved orbital and sinus areas with amphotericin B (1 mg/cm^3) should not be overlooked. Mucormycosis has a vaso-occlusive characteristic, which impedes effective delivery of intravenous amphotericin B. Irrigation and packing may augment local delivery of this fungistatic agent (Fig. 68-3).

The extent of surgical excision should balance the degree of morbidity and mutilation against the life-threatening risk this organism may represent. In limited cases surgical excision may be confined to tissues that are clearly infarcted. Should infection be extensive, as demonstrated by widespread necrosis, aggressive surgery, including exenteration of the orbit and of any involved paranasal sinuses may prove necessary and life saving.

It is important to recognize that the prolonged survival of debilitated patients, along with the increased therapeutic use of antibiotics and immunosuppressive agents, has increased the frequency of this devastating condition. Our heightened awareness is the first step toward a more favorable outcome.

References

1. Baum J. Rhino-orbital mucormycosis occurring in an otherwise healthy individual. *Am J Ophthalmol.* 1967;63:335.
2. Blodi F, Hannah F, Wadsworth J. Lethal orbitocerebral phycomycosis in otherwise healthy children. *Am J Ophthalmol.* 1969;67:698.
3. Bullock J, Jampol L, Fezza A. Two cases of orbital phycomycosis with recovery. *Am J Ophthalmol.* 1974;78:811.
4. Burns R. Mucormycosis of the sinuses, orbit, and central nervous system. *Trans Pac Coast Oto-ophthalmol Soc.* 1959;40:83.
5. Ferry A. Cerebral mucormycosis. *Surv Ophthalmol.* 1961;6:1.
6. Ferry A, Abedi S. Diagnosis and management of rhino-orbitocerebral mucormycosis (phycomycosis). *Ophthalmology.* 1983;90:1096.
7. Fleckner R, Goldstein J. Mucormycosis. *Br J Ophthalmol.* 1969;53:542.
8. Gass J. Acute orbital mucormycosis. Report of two cases. *Arch Ophthalmol.* 1961;65:214.
9. Gass J. Ocular manifestations of acute mucormycosis. *Arch Ophthalmol.* 1961;65:226.
10. Kohn R, Hepler R. Management of limited rhino-orbital mucormycosis without exenteration. *Ophthalmology.* 1985;92:1440.
11. Straatsma B, Zimmerman L, Gass J. Phycomycosis. A clinopathologic study of 51 cases. *Lab Invest.* 1962;11:963.

Section F
Protozoan Diseases

Chapter 69
Amebiasis

IRENE H. LUDWIG and DAVID M. MEISLER

Amebas are protozoans of the class Rhizopodea, order Amoebidae, which move by pseudopodal extensions and reproduce mitotically. Most exist in both trophozoite and cyst forms, the cyst allowing the organism to survive unfavorable environmental conditions.

Parasitic Amebas

Amebas of the family Entamoebidae are obligate parasites, reproducing only in the digestive tract of invertebrates and vertebrates.[1] *Entamoeba histolytica* is a significant cause of morbidity and mortality worldwide. *Dientamoeba fragilis* (the only one without a cyst form) is thought to be pathogenic in some cases. Organisms such as *Entamoeba coli, E. gingivalis, E. hartmanni, Endolimax nana,* and *Iodamoeba bütschlii* are commensal in man, but their presence in stool indicates ingestion of contaminated material. Although they are asymptomatic in healthy persons, their nonpathogenic designation appears to be changing with the increasing number of immunocompromised persons, in whom these organ-isms may produce disease (G Turek, personal communication).

E. histolytica is infective in the cyst form, which, when ingested, survives gastric acid to pass into the small intestine, where the cyst wall dissolves and the quadrinucleate metacyst divides into eight small trophozoites, which may then colonize and invade the intestinal wall at the cecal and sigmoidorectal levels. Trophozoites that pass into the intestinal lumen encyst themselves and pass with the feces out of the host, where they can survive several weeks. Cysts survive freezing, but survival decreases at elevated temperatures. Humans are the primary hosts, and an infected person may shed millions of cysts daily, transmitting them to new hosts by direct contact or indirectly through contaminated food and water. Infection rates correlate with poor sanitation and socioeconomic conditions and crowding.[2]

Diagnosis is made by microscopic examination of stool; morphologic differences identify the species of trophozoites and cysts. The *E. histolytica* trophozoite is 10 to 60 μm in diameter, has inclusions that contain red blood cells, few vacuoles, and fingerlike pseudopodia.

The cyst is 5 to 20 μm and contains four nuclei.[1] Strains of *E. histolytica* of differing virulence have been shown to exist, which are distinguished by isoenzyme markers.[2]

Free-Living Amebas

The free-living amebas complete their life cycle without requiring a host but occasionally produce disease opportunistically. *Naegleria fowleri,* an ameboflagellate of the family Vahlkampfidae, and three species of the family Acanthamoebidae are the only known pathogens, although other species do exist. Free-living amebas are ubiquitous in nature. They are found in fresh water and seawater, air, and soil. Large numbers exist in warm, polluted waters that contain organic material,[3] particularly hot tubs and swimming pools.[4]

When local factors such as pH, oxygen, food supply, and cell crowding become unfavorable, the trophozoite rounds up, secretes a protective cyst wall, and reduces its metabolic activity. *Acanthamoeba* cysts are 10 to 25 μm in diameter and double walled; they survive desiccation and subzero temperatures.[3-5] *N. fowleri* cysts are smaller, single walled, and poorly resistant to dryness and freezing, and they multiply readily at temperatures too high for reproduction of *Acanthamoeba* species (40 to 46°C).[6] The *Acanthamoeba* trophozoite is characterized by its size (15 to 45 μm), locomotor pseudopodia, and spinelike acanthapodia, whereas *Naegleria* is 8 to 30 μm in diameter, has blunt pseudopodia, and is capable of transformation into a temporary flagellate stage. Both *Acanthamoeba* and *Naegleria* possess a large nucleus with a dense nucleolus surrounded by a clear zone, food vacuoles, an osmoregulatory contractile vacuole, and abundant cytoplasm.[7]

Systemic Manifestations
Entamoeba Histolytica

The presence in the bowel of *E. histolytica* (as diagnosed by fecal examination or serologic testing) is asymptomatic in most persons. In a minority (1 to 20%), a probable combination of organism virulence factors plus poor diet and sanitation contribute to the development of symptomatic amebic dysentery with invasion of the bowel wall by the organisms. The gradual onset (1 week to 6 months) is manifested by increasing abdominal pain and frequent watery stools containing mucus and blood. Left untreated, fever, dehydration, nausea, vomiting, and anemia may develop. At this stage, antiamebic drugs and correction of fluid and electrolyte imbalances usually cure the problem, although relapse is not uncommon.[8]

The most frequent complication of amebic dysentery (3% in adults, 9% in children) is peritonitis, which carries a 40% mortality risk. Acute bowel perforation is the cause in some patients, but more commonly, a slow leak from the diffusely diseased bowel seeds the peritoneal cavity with bacteria and amebas. Other complications seen less commonly are amebic bowel strictures, ameboma (granulation tissue) of the large bowel, postdysenteric ulcerative colitis that persists for several weeks following elimination of *E. histolytica* from the bowel, hemorrhage due to erosion of a vessel in the bowel wall, perianal cutaneous amebiasis, and intussusception.[8]

Hepatic amebic abscess caused by extraintestinal extension presents with localized abdominal pain (usually right-sided upper quadrant), weight loss, malaise, fever, and cough. In many cases there is no history of past or current amebic dysentery. The liver is enlarged, with localized tenderness. The hepatic abscess may extend into the thorax, with resultant hepatobronchial fistula, pleural effusion, or pulmonary abscess. A hazardous complication is rupture of a left lobe abscess into the pericardium. Rupture into the peritoneum or bowel may occur, and amebic brain abscesses, though rare, have been described.[9]

Free-Living Amebas

Primary amebic meningoencephalitis is a rare, fulminant disease produced by *N. fowleri,* which is usually fatal within 1 week of onset. One cure has been reported.[10] The patients are young and previously were healthy. Usually they have a history of summertime swimming in pools or man-made lakes 2 to 3 days before the onset of symptoms. Headache, fever, and vomiting begin and progress to drowsiness, confusion, neck stiffness, and finally coma and death. The organisms invade through the nasal mucosa and cribiform plate, producing diffuse hemorrhagic necrosis of gray and white matter with marked acute inflammatory changes.[3]

Granulomatous amebic encephalitis is generally a disease of the chronically immunosuppressed caused by *Acanthamoeba* species. *A. castellani* and *A. culbertsoni* have been implicated. The CNS is invaded by hematogenous spread from a respiratory or skin ulcer focus. Focal neurologic deficits of any type, along with symptoms of increased intracranial pressure, gradually progress to coma and death. The diagnosis is usually made postmortem.

Acanthamoeba organisms have been isolated from skin lesions,[11] from a purulent ear discharge,[12] and from a patient with osteomyelitis.[13] An acquired immunodeficiency syndrome (AIDS) patient developed *A. castellani* sinusitis.[14] The organisms have also been recovered from the throats of normal persons.[15]

Ocular Manifestations

Entamoeba Histolytica

Ocular involvement by intestinal amebiasis is uncommon but has been described. The reports are circumstantial and anecdotal, relating improvement in the ocular conditions to successful medical treatment of amebiasis. No pathologic confirmation of intraocular infection by E. histolytica has been reported. Alternate hypotheses to that of direct ocular infection by E. histolytica include associated immunologic mechanisms, infestation by multiple parasites (with different organisms affecting the eye but responding to the same medications), and mere coincidence.

Iritis was reported to resolve with antiamebic therapy in several cases.[16-18] A case of bilateral anterior and posterior uveitis with posterior pole chorioretinitis improved when the patient's amebic dysentery was treated, leaving pigmentary changes.[19]

Several series of amebic dysentery patients with macular lesions that improved coincident with antiamebic therapy have been described. The lesions were subretinal "cysts" at or near the fovea, with central scotomas, subretinal hemorrhage, surrounding pigmentation, and in some, an opaque yellow mass within the retinal cyst. Final acuities were variable.[20,21]

In one patient orbital proptosis attributed to E. histolytica resolved with systemic therapy.[22] A case of severe orbital infection with proven E. histolytica by spread from an eyelid ulcer led to orbital exenteration and eventually death, in an infant.[23]

Free-Living Amebas

Ocular involvement in amebic meningoencephalitis due to Naegleria or Acanthamoeba species has not been described beyond mention of occasional visual field deficits or papilledema.[3] This may not, however, reflect the true frequency of these findings, as it is doubtful that ophthalmologic evaluations are performed on many of these critically ill patients.

One unusual case of Acanthamoeba meningoencephalitis presented with unilateral nongranulomatous uveitis in a previously healthy 7-year-old boy who had a history of likely amebic exposure by immersion in stagnant water. Postmortem examination revealed trophozoites in the ciliary body.[24]

A. castellani and A. polyphaga are known to produce a chronic, recalcitrant keratitis caused by direct corneal invasion without systemic disease. A history of antecedent corneal trauma or contact lens wear is common, and exposure to stagnant water, soil, or contaminated contact lens paraphernalia is sometimes demonstrable. The amebic corneal ulcer may mimic herpes simplex or fungal keratitis. Intense conjunctival reaction, scleritis, epithelial stippling, and subepithelial opacities are present initially; they progress to stromal infiltration (often ring shaped), recurrent breakdown and healing of the epithelium, and left untreated, to descemetocele formation or perforation. Corneal nerves may appear inflamed. Pain is marked throughout. Early treatment with topical and oral antiamebic medications or corneal epithelial debridement may result in cure, but penetrating keratoplasty has often been required. Recurrence in the graft is not uncommon. Advanced disease has led to enucleation. Prevention—educating the public about adequate contact lens disinfection and avoidance of amebic exposure while wearing contact lenses—may reduce the frequency of this serious infection.[4,25,26]

References

1. Brown HW, Neva FA. *Basic Clinical Parasitology*. 5th ed. Norwalk, Conn: Appleton-Century-Crofts; 1983.
2. Guerrant RL. Amebiasis: introduction, current status, and research questions. *Rev Infect Dis*. 1986;8(2):218-227.
3. Martinez AJ. *Free-living amebas: natural history, prevention, diagnosis, pathology, and treatment of disease*. Boca Raton, Fla: CRC Press; 1985.
4. Ludwig IH, Meisler DM. Acanthamoeba keratitis. In: Tabbara K, Hyndiuk R, eds. *Infections of the Eye*. Boston, Mass: Little, Brown; 1986;665-678.
5. Meisler DM, Ludwig IH, Rutherford I, et al. Susceptibility of *Acanthamoeba* to cryotherapeutic method. *Arch Ophthalmol*. 1986;104:130-131.
6. Griffin JL. Temperature tolerance of pathogenic and nonpathogenic free-living amoebas. *Science*. 1972;178:869.
7. Schuster FL. Small amebas and ameboflagellates. In: Levandowsky M, Hutner SH, eds. *Biochemistry and Physiology of Protozoa*. 2nd ed. New York, NY: Academic Press; 1979;216–285.
8. Adams EB, MacLeod IN. Invasive amebiasis. I. Amebic dysentery and its complications. *Medicine*. 1977;56:315-23.
9. Adams EB, MacLeod IN. Invasive amebiasis. II. Amebic liver abscess and its complications. *Medicine*. 1977;56:325-334.
10. Seidel JS, et al. Successful treatment of primary amebic meningoencephalitis. *N Engl J Med*. 1982;306:346.
11. Gullett J, et al. Disseminated granulomatous *Acanthamoeba* infection presenting as an unusual skin lesion. *Am J Med*. 1979;67:891.
12. Lengy J, Jakovijevich R, Talis B. Recovery of a hartmanelloid amoeba from a purulent ear discharge. *Trop Dis Bull*. 1971;68:818.
13. Barochovitz D, et al. Osteomyelitis of a bone graft of the mandible with *Acanthamoeba castellani* infection. *Human Pathology*. 1981;12:573-576.
14. Gonzalez MM, et al. Acquired immunodeficiency syndrome associated with *Acanthamoeba* infection and other opportunistic organisms. *Arch Pathol Lab Med*. 1986;110:749-751.
15. Wang SS, Feldman HA. Isolation of *Hartmannella* species from human throats. *N Engl J Med*. 1967;277:1174.

16. Mills L. Amoebic iritis occurring in the course of nondysenteric amoebiasis. *Arch Ophthalmol*. 1923;52:525-545.
17. Cunningham J. Ocular findings in amoebic dysentery. *Trans Far East Assoc Trop Med*. 1927;1:303-306.
18. Samaniego JMV. Ocular manifestations of some tropical diseases. *Am J Ophthalmol*. 1951;34:1574-1578.
19. Harris D, Birch CL. Bilateral uveitis associated with gastrointestinal *Entamoeba histolytica* infection: a case report. *Am J Ophthalmol*. 1960;50:496-500.
20. Braley AE, Hamilton HE. Central serous choroidosis associated with amebiasis. A record of nine cases. *Arch Ophthalmol*. 1957;58:1-14.
21. King RE, Praeger DL, Hallet JW. Amebic choriodosis. *Arch Ophthalmol*. 1964;72:16-22.
22. Mortada A. Orbital pseudo-tumors and parasitic infections. *Bull Ophthalmol Soc Egypt*. 1968;61:393-399.
23. Beaver PC, et al. Cutaneous amebiasis of the eyelid with extension into the orbit. *Am J Trop Med Hyg*. 1978; 27(6):1133-1136.
24. Jones DB, Visvesvara GS, Robinson NM. *Acanthamoeba polyphaga* keratitis and *Acanthamoeba uveitis* associated with fatal meningoencephalitis. *Trans Ophthalmol Soc UK*. 1975;95:221.
25. Avran JD, Starr MB, Jakobiec FA. Acanthamoeba keratitis. A review of the literature. *Cornea*. 1987;6(1):2-26.
26. Ludwig IH, Meisler DM, Rutherford I, et al. Susceptibility of *Acanthamoeba* to soft contact lens disinfection systems. *Invest Ophthalmol Vis Sci*. 1986;27:626-628.

Chapter 70

Leishmaniasis

MERLYN M. RODRIGUES

Human leishmaniasis, a zoonosis transmitted by a blood-sucking vector sandfly of the genus *Lutzomyia* in the New World and *Phlebotomus* in the Old World is caused by parasitic protozoa. Reservoir hosts include dogs and cats and other animals. The disease can be produced by at least 14 different species and subspecies of the genus *Leishmania*.[1] Human infection is either cutaneous or visceral (kala-azar).[2] Visceral disease caused by *L. donovani*, is characterized by hepatosplenomegaly, generalized fever, pancytopenia, and cachexia and has a high mortality rate. Cutaneous leishmaniasis, caused by *L. tropica* and *L. aethiopica* in the Old World and by *L. mexicana* and *L. braziliensis*, in the New World, manifests clinically as single or multiple ulcers affecting the face or extremities and is rarely fatal.[1] The sores develop at the site of the sandfly bite, usually the extremities, face, and ear. After inoculation into the skin, leishmanias multiply by binary fission within the histiocytes that rupture and release the organisms into the bloodstream, infecting histiocytes elsewhere. The incubation period usually ranges from 2 to 8 weeks but may be as long as a few years. Initial erythematous papules become indurated and frequently ulcerated. Satellite lesions may develop along the course of draining lymphatics.

Ophthalmic leishmaniasis includes lesions of the conjunctiva[5] and the cornea,[6] but the most common form of ocular involvement is eyelid lesions, which occur as ulcers (Fig. 70-1) or nodules.[3,4,7] Scarring and cicatrization may be followed by eyelid abnormalities. Conjunctival (limbal) nodules have been described.[5] Direct corneal involvement is rare, although keratitis and uveitis have been reported in association with cutaneous leishmaniasis.[7] Bilateral anterior uveitis has also been reported following apparently successful treatment of visceral leishmaniasis.[8]

Diagnosis

The following diagnostic methods are used:

1. Needle aspiration of the edges of cutaneous and mucocutaneous lesions.[9]
2. Impression smear, or "tissue imprints," of the cut surface of a biopsy specimen stained with either Wright's or Giemsa stain. The latter is preferred, since it stains the kinetoplast and nucleus more intensely (Fig. 70-2).
3. Staining of paraffin sections. Organisms are usually detected with routine hematoxylin and eosin stains, but they can be more clearly visualized with special stains, such as Wilder's reticulin. The amastigotes are small oval cells, measuring 2.5 to 3.0 μm in diameter.
4. Transmission electron microscopy. In an eyelid lesion

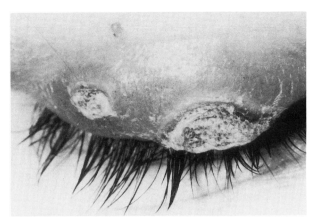

FIGURE 70-1 Upper lid ulcers—larger primary lesion and smaller satellite lesion. (From Chu et al.[7])

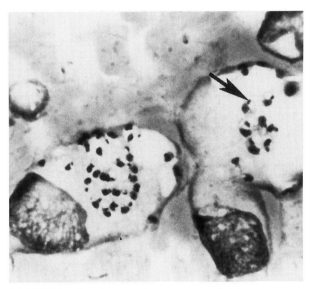

FIGURE 70-2 Impression smear of biopsy tissue shows intracellular organisms with a prominent nucleus and dotlike kinetoplast (*arrow*), (Giemsa, ×1300).

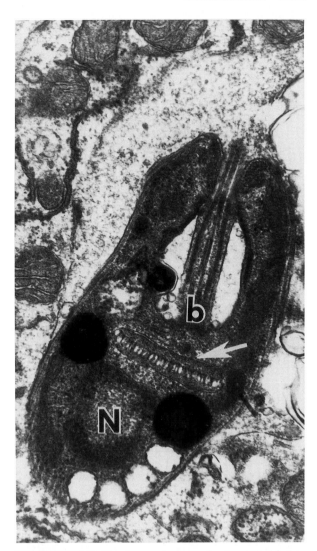

FIGURE 70-3 Electron micrograph shows an intracellular organism containing a nucleus (N), kinetoplast-containing mitochondrion (*arrow*), and electron-dense bodies. The basal body (b) and flagellum are indicated (×35,100).

the intracellular organisms were often within the cytoplasm of histiocytes (Fig. 70-3). The protozoa were lined by a double-unit membrane with prominent subcutaneous microtubules. Each organism had a nucleus, kinetoplast, and an intracellular flagellum with 9 + 2 axoneme structures.[10] Research into species differentiation of amastigotes has shown that differences in microtubule number and diameter may be used as a method of distinguishing different species.[10]

5. Recent advances in the diagnosis of leishmaniasis:

a. Monoclonal antibodies.
b. Recombinant DNA techniques. This methodology permits direct diagnosis from patients' lesions without the need to isolate the parasite.[1]

References

1. Wirth DF, Rogers WO, Barker R. et al. Leishmaniasis and malaria: new tools for epidemiologic analysis. *Science* 1986;234:975-979.

2. Neva FA. Diagnosis and treatment of cutaneous leishmaniasis. In: Remington JS, Swartz MN, eds. *Current Clinical Topics in Infectious Diseases.* New York, NY: McGraw-Hill; 1982.
3. Ferry AP. Cutaneous leishmaniasis (oriental sore) of the eyelid. *Am J Ophthalmol.* 1977;84:349-354.
4. Rentsch FJ. Klinik and Morphologie der Leishmaniose der Lider. *Klin Mbl Augenheilk.* 1980;177:75-79.
5. Sodaify M, Aminlari A, Resaer H. Ophthalmic leishmaniasis. *Clin Exp Dermatol.* 1981;6:485-488.
6. Cairns JE. Cutaneous leishmaniasis (oriental sore). A case with corneal involvement. *Br J Ophthalmol.* 1968;52:481-483.
7. Chu FC, Rodrigues MM, Cogan DG, et al. Leishmaniasis affecting the eyelids. *Arch Ophthalmol.* 1983;101:84-91.
8. Dechant W, Rees PH, Kager PA, et al. Post kala-azar uveitis. *Br J Ophthalmol.* 1980;64:680-683.
9. Neva F. Recent advances in the diagnosis and management of leishmaniasis and American trypanosomiasis. In: Leech JH, Sande MA, Root RK, eds. *Contemporary Issues in Infectious Disease,* vol 7. New York, NY: Churchill Livingstone, 1988:243-258.
10. Gardener PJ, Shchory L, Chance ML. Species differentiation in the genus *Leishmania* by morphometric studies with the electron microscope. *Ann Trop Med Parasitol.* 1977;71:147-155.

Chapter 71

Malaria

SORNCHAI LOOAREESUWAN

Malaria is the most important parasitic infection in the world; at any one time approximately 300 million people are infected and it is estimated that there are at least a million malaria deaths each year. In parts of the tropics where transmission of the parasite (by female mosquitoes of the genus *Anopheles*) is intense, malaria is principally a disease of children. Where transmission is less intense or less geographically uniform, symptomatic malaria occurs in both adults and children. Four species of malaria parasites infect man, *Plasmodium falciparum, P. vivax, P. malariae,* and *P. ovale,* of which *P. falciparum* is the most important, being responsible for almost all deaths.

After inoculation by the biting mosquito, there is a period of development within hepatocytes which usually lasts between 1 and 3 weeks. This is followed by invasion of erythrocytes, and all pathologic events result from this stage of infection. The parasite grows within the red cell while consuming hemoglobin. Eventually after 2 days (or in the case of *P. malariae,* 3 days) the growing trophozoite has developed into a large ball of daughter parasites, the schizont, which then bursts the

red cell host to liberate the young merozoites. These in turn infect more red cells and the cycle continues.[1] There is progressive hemolytic anaemia with fever. In falciparum malaria the mature trophozoites and schizonts are not seen in peripheral blood smears because of sequestration in vital organs such as the brain, heart, and liver.[2,3] The infected red cells stick to capillary and venular endothelium ("cytoadherence") until they are destroyed at schizogony (bursting of the schizont). Sequestration is thought to account for severe and sometimes lethal organ dysfunction in falciparum malaria.[4]

Systemic Manifestations

Malaria is characterized by fever with sweats, myalgia, headaches, malaise, and sometimes rigors. There is hemolytic anaemia and splenic enlargement. The importance of the fever pattern has been overemphasized in the past. The classic tertian (every second day) and quartan (every third day) fever spikes are rarely seen in the modern era of antimalarial drugs. Vital organ dysfunction is rare in the benign malarias (*P. vivax, P. malariae, P. ovale*), although nephrotic syndrome resulting from an immune complex glomerulonephritis is a peculiar complication of chronic *P. malariae* infection. In these infections the proportion of red cells parasitized

I am grateful to Dr. N. J. White, Professor D. A. Warrell, Professor D. Bunnag, and Professor T. Harinasuta for their help and the support from the Wellcome Trust of Great Britain as part of the Wellcome-Mahidol University, Oxford Tropical Medicine Programme.

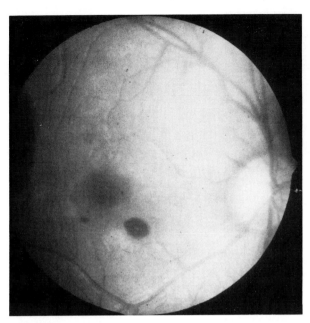

FIGURE 71-1 Large and small hemorrhages near the macula in a patient with cerebral malaria. (From Looareesuwan S, et al.[10])

seldom exceeds 2%, whereas in the potentially lethal falciparum malaria the degree of parasitemia often exceeds this figure. Cerebral malaria or coma in falciparum malaria is a characteristic feature of severe infections. The mortality rate of cerebral malaria is approximately 20%.[5] Cerebral malaria commonly coexists with other severe manifestations, such as acute renal failure, lactic acidosis, hypoglycemia, acute pulmonary edema, profound anemia, and complications such as aspiration pneumonia and bacterial infections.[5,6]

Ocular Manifestations

A wide variety of ophthalmic manifestations have been attributed to malaria.[7-9] In many cases the evidence for cause and effect is unconvincing. Certainly jaundice is common in severe falciparum malaria, and several authors have noted that the conjunctiva commonly shows a yellow tinge. Subconjunctival hemorrhage may occur, particularly in cases of severe falciparum malaria with disseminated intravascular coagulation (approximately 5% of patients with cerebral malaria overall).[5] Contrary to early reports, malaria is not associated with keratitis, although coincident herpes simplex infection may occur. The cornea may be injured if the eyes of cerebral malaria patients are not taped or covered. The corneal reflexes of

comatose patients are usually preserved. There is no evidence that uveal tract involvement occurs in malaria.

Retinal hemorrhages occur in severe falciparum malaria, although their frequency varies geographically. In Thailand we reported retinal hemorrhages in 14.6% of patients with cerebral malaria. In 17 of the 21 cases the hemorrhages were multiple, and in 14 cases bilateral. Retinal hemorrhages were associated with several indices of severity, including anemia, and were not observed in patients with uncomplicated infections.[10] In Goroka Hospital, New Guinea, Davis et al.[11] observed retinal hemorrhages in 14 of 50 consecutive malaria patients admitted; eight of these patients were fully conscious, and two had *P. vivax* infection.[11] There was a strong correlation with the degree of anemia. In Zaire, Kayembe et al.[12] reported retinal hemorrhages in 31% of their patients. Many other authors have reported retinal hemorrhages, usually in severely ill anemic patients. The hemorrhages (Fig. 71-1) may occur at any of the retinal layers. They commonly have pale centers and may resemble Roth's spots. Histologic studies of retinal vessels have suggested that stasis of parasitized red cells may be responsible for the lesions.[13]

Interference with vision, or the light reflex in unconscious patients, is most unusual, and complete resolution is the rule. Occasional exudates have been observed in severe malaria, and papilledema is a very unusual finding. We have observed two transient sixth nerve palsies in patients recovering from cerebral malaria, and this has also been noted by others, although it seems to be a rare finding overall. Transient palsies of the third and fourth nerves and occasional pupillary abnormalities have been noted in the literature but are probably also very rare. Commonly there is divergent or disjugate gaze (Fig. 71-2) in cerebral malaria, with preservation of the oculocephalic and oculovestibular reflexes. We have seen nystagmus in 21 of 175 cerebral malaria patients while they were comatose, and 62% of the nystagmus was the pendular type. It was also reported in uncomplicated falciparum malaria.[14]

Diagnosis and Management

The diagnosis of malaria depends entirely on demonstration of the parasite by microscopic examination of a suitably stained thick and thin blood smear. There is no specific treatment for the ocular manifestations of malaria. All patients with severe malaria should be treated promptly with intravenous quinine, or with chloroquine in areas where the parasite remains sensitive to this drug.[15]

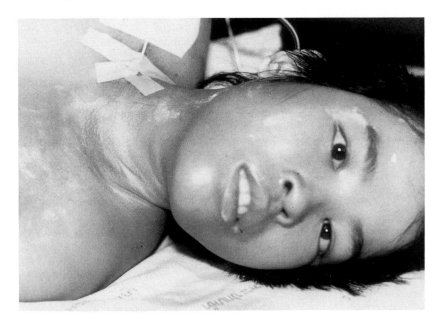

FIGURE 71-2 Disjugate position of the eyes in a Thai woman with cerebral malaria.

References

1. Bradley DJ, Newbold CT, Warrell, DA. Malaria. In: Weatherall DJ, Ledingham JGG, Warrell DA, eds. *Oxford Textbook of Medicine.* 2nd ed. vol 1. Oxford: Oxford Medical Publications; 1987;5.474-5.502.
2. Spitz S. The pathology of acute falciparum malaria. *Milit Surg.* 1946;99:555-572.
3. MacPherson GG, Warrell MJ, White NJ, et al. Human cerebral malaria: a quantitative ultrastructural analysis of parasitized erythrocyte sequestration. *Am J Pathol.* 1985;119:385-401.
4. David PH, Hommel M, Miller LH, et al. Parasite sequestration in *Plasmodium falciparum* malaria: spleen and antibody modulation of cytoadherence of infected erythrocytes. *Proc Nat Acad Sci (USA).* 1983;80:5076-5079.
5. Warrell DA, Looareesuwan S, Warrell MJ. Dexamethasone proves deleterious in cerebral malaria: a double-blind trial in 100 comatose patients. *N Engl J Med.* 1982;306(6):313-319.
6. White NJ, Looareesuwan S. Cerebral malaria. In: Kennedy PGE, Johnson RT, eds. *Infections of the Nervous System.* London: Butterworths; 1987:118-144.
7. Bywater HH. Notes on malarial conditions of the eyes. *Trans Ophthalmol Soc UK.* 1922;42:359-365.
8. Bell RW. Ophthalmologic findings in malaria. *Ann Ophthalmol.* 1975;7:1439-1442.
9. Duke-Elder S. *System of Ophthalmology.* vol XV. 1976;94-95.
10. Looareesuwan S, Warrell DA, White NJ, et al. Retinal hemorrhages, a common physical sign of prognostic significance in cerebral malaria. *Am J Trop Med Hyg.* 1983;32(5):911-915.
11. Davis MW, Vaterlaws HL, Simes J, et al. Retinopathy in malaria. *Papua New Guinea Med J.* 1982;25:19-22.
12. Kayembe D, Maertens K, De Laey JJ. Complications oculaires de la malaria cérébrale. *Bull Soc Belge Ophthalmol.* 1980;190:53-60.
13. Dudgeon LS. A case of malignant malaria. *Trans Ophthalmol Soc UK.* 1921;41:236-238.
14. Senanayake N. Delayed cerebellar ataxia: a new complication of falciparum malaria. *Br Med J.* 1987;294:1253-1254.
15. WHO Malaria Action Programme. Severe and complicated malaria. *Trans R Soc Trop Med Hyg.* 1986;80(suppl):1-50.

Chapter 72

Toxoplasmosis

JOSE BERROCAL

Toxoplasmosis is the most common cause of inflammatory disease of the posterior segment of the eye, accounting for 40% of all posterior uveitis. The organism is *Toxoplasma gondii*, an intracellular protozoan organism that measures about 2 μm by 7 μm and contains a well-defined nucleous. It is neurotrophic, having a particular affinity for invading neural tissue.[1]

Toxoplasma organisms can infect man and much of the animal kingdom. The cat is its definitive host, but it has been found in all orders of mammals and in some birds and reptiles. It is a worldwide infection, though warm and moist areas have a higher incidence of disease than dry, cold ones.

Systemic toxoplasmosis may be acquired or congenital. The most common route of infection is via ingestion of the organism, usually in undercooked meat. Infection from contaminated soil may occur in gardens or sandboxes, where oocysts from cat feces can be found. Other modes of transmission include blood transfusion, laboratory accidents, human milk, and ingestion of raw eggs.

Congenital Systemic Toxoplasmosis

Congenital toxoplasmosis develops after transplacental transmission of the parasite from a recently infected pregnant mother to her fetus.[2] A high antibody titer is found in the mother who has just delivered a diseased child. Once a mother has had toxoplasmosis and has become immune, she generally does not infect subsequent children. Toxoplasmosis in pregnancy is difficult to diagnose and is usually missed, since most infections are subclinical. The diagnosis of toxoplasmosis acquired during pregnancy does not indicate a hopeless prognosis for the newborn, since half of the infants are born healthy. However, 14% of fetuses develop severe congenital toxoplasmosis if the mother is infected in the first trimester of pregnancy. The chances of getting infected in the last trimester are quite small. Congenital toxoplasmosis may be characterized by hydrocephalus and necrotizing encephalitis leading to any

of the following: scattered cerebral calcifications in healing granulomas, neonatal jaundice, hepatosplenomegaly, mental retardation, convulsions, microcephaly, and microophthalmos. However much milder forms may occur, affecting only a few tissues or just one. The retina may be one such tissue, exhibiting a locally restricted, limited form of congenital toxoplasmosis in 1.2% of cases. The organism may destroy the fetus, resulting either in stillbirth or a severe form of congenital toxoplasmosis.

The triad of retinochoroiditis, hydrocephalus, and calcifications in the brain have been recognized as the classic findings of congenital systemic toxoplasmosis.[1] Congenital toxoplasmosis in infants of mothers infected during pregnancy can be divided into two groups: subclinical infection (90%) and active disease (10%). Among symptomatic infants two forms have been described. The more common form is the neurologic type (69%) and the other is generalized (31%).

The clinician must have a high index of suspicion in order to diagnose the disease. Diagnosis is especially important in pregnant women. If the pregnant mother is found to be seronegative early in her pregnancy, she must be considered at risk of infection. As far as the infant is concerned, an IgM antibody test is useful, but it must be done as soon as possible after delivery.

Both cells and antibodies take part in immunity to toxoplasmosis. Cellular immunity protects; antibody or B-cell immunity merely delays death. Cellular transfer of immunity from the mother does not occur. Congenital toxoplasmosis is a protracted infection that leaves numerous encysted parasites in the tissues. Deficiency in cellular immunity in the infant explains the greater severity and more prolonged course in the newborn. Of the 3 million babies born annually in the United States 3000 have congenital toxoplasmosis.[3]

Pregnant women can contract toxoplasmosis by eating uncooked meat with toxoplasma cysts or through contact with cats or soil contaminated with their oocysts. After cats eat the cyst in infected animal tissues they shed oocysts in their feces after an incubation period of 4 days.[4]

Acquired Systemic Toxoplasmosis

There is an extremely high incidence of systemic toxoplasmosis throughout the world. A person's chances of having had toxoplasmosis are approximately the same as his or her age. Therefore most cases of systemic toxoplasmosis are subclinical in nature. The lymphadenopathic form is by far the most common clinical type. There are no symptoms, but there is localized lymphadenopathy. It is responsible for approximately 15% of otherwise unexplained lymphadenopathies. Toxoplasmic mononucleosis gives rise to ocular toxoplasmosis in about 1% of cases. In more than 90% of immunocompromised toxoplasma patients, encephalitis is the cause of death. Toxoplasma has been associated with malignancies and with simultaneous intracellular infections with cytomegaloviruses and herpes virus.

Other clinical manifestations of acquired toxoplasmosis include polymyositis, encephalitis, pneumonitis, exanthems, and psychiatric disturbances. Polymyositis presents with fatigue, followed by weakness of the arms and legs. There is usually a generalized tender lymphadenopathy. Encephalitis shows the usual triad of fever, clouding of consciousness, and meningismus. Lymphadenopathy is usually present.

Ocular Manifestations

The major ocular complication of toxoplasmosis is retinochoroiditis. This infection of the neural retina is the ocular counterpart of central nervous system involvement by this neurotrophic parasite. The encysted parasite may remain dormant in the retina for years. Recurrent active retinitis is thought to be due to rupture of cysts, which produces acute inflammation in the area.

Ocular toxoplasmosis is no longer divided into congenital and acquired forms. We believe at present that most cases are congenital in origin.[5] Involvement of the macular area is typical, though peripheral lesions may be seen. There may be smaller "satellite" lesions near a larger "parent" scar. Bilateral involvement is common, though active disease is usually monocular.

Most cases of active retinochoroiditis occur at the edge of a previous scar.[6] It is strongly necrotizing and usually occurs between the ages of 11 and 40 years. Active lesions are generally single and are located posterior to the equator. There is exudation of leukocytes into the vitreous, with marked vitritis. Juxtapapillary toxoplasmosis destroys the retinal nerve fiber layer, with a resulting nerve fiber visual field defect extending from the normal blind spot.

Other ocular complications include papillitis from optic nerve involvement, papilledema from ruptured cerebral cysts producing increased intracranial pressure, and vitritis. Granulomatous iridocyclitis, a common finding, is probably produced by immune-mediated mechanisms, since no toxoplasma organisms have ever been found in the anterior part of the eye.[7] Glaucoma may result from iritis-induced damage to the aqueous drainage system, or it may be secondary to the steroids used to treat the disease. Steroid therapy or chronic inflammation may promote secondary "complicated" cataract formation. Cystoid macular edema can occur due to the iridocyclitis or a juxtafoveal retinitis.

Rhegmatogenous retinal detachments have been described. The responsible retinal breaks are usually found adjacent to a toxoplasmosis scar. Microophthalmos, nystagmus, and strabismus have also been reported. Patients with retinochoroiditis could develop more generalized systemic toxoplasmosis if they acquire diseases that suppress their immune response.

Exudates located along the retinal arteriolar tree in a boxcar fashion have been designated "segmental periarteritis." Usually they are found near an area of fresh retinochoroiditis. They may be due to an immune complex reaction in vessel walls, when antigens located in or near the vessel wall make contact with circulating antibodies.

Histopathologic examination reveals varying degrees of coagulative necrosis and granulomatous inflammatory reaction in the choroid, sclera, and episclera—so-called segmental panophthalmitis.[8] Melanin granules from the pigment epithelium are found dispersed in the necrotic retina. Encysted colonies of *T. gondii* are observed, as well as groups of crescent-shaped organisms in necrotic retina. The retina may show perivascular lymphocytic infiltration, edema, and gliosis.[9] The iris and ciliary body usually exhibit chronic granulomatous or nongranulomatous inflammation. Encysted colonies may be formed in the optic nerve and papillae.

The necrotizing granulomatous retinochoroiditis is due to multiplicative activity of the organism itself. Toxoplasma divide in the cytoplasm of the parasitized cell until the cell ruptures. *T. gondii* may remain viable in the eye for years in the form of thick-walled cysts.[9]

The treatment depends on the location and severity of the active retinitis. Any retinitis near the macula or optic nerve threatens central vision and should be treated. Also any fresh, large, exudative lesion should be treated.

The triple therapy of pyrimethamine-sulfonamide-steroids is the accepted drug regimen. One complication of pyrimethamine is a depletion of the bone marrow. Folinic acid prevents the bone marrow depression associated with pyrimethamine. Oral steroids should never be given without concomitant antimicrobial therapy.

Clindamycin has been proven to be effective in the

treatment of animals with toxoplasmic retinochoroiditis when used subconjunctivally or orally (300 mg given orally every 6 hours). The use of subconjunctivally administered clindamycin is currently being studied. Since the antitoxoplasma medication kills only extracellular parasites and not those that remain encysted, it follows that effective antitoxoplasma therapy of an acute recurrent attack does not protect the patient against future recurrences.

References

1. Alford CA, et al. Congenital toxoplasmosis: clinical, laboratory and therapeutic considerations, with special references to subclinical disease. *Bull NY Acad Med.* 1974;50:160.
2. Desmot G, Couvreur J. Toxoplasmosis in pregnancy and its transmission to the fetus. *Bull NY Acad Med.* 1974;50:146.
3. Feldman HA. Toxoplasmosis: an overview. *Bull NY Acad Med.* 1974;50:110-127.
4. Hutchinson W, et al. The life cycle of the coccidian parasite, *Toxoplasma gondii*, in the domestic cat. *Trans R Soc Trop Med Hyg.* 1971;65:380-399.
5. Saari M, et al. Acquired toxoplasmic chorioretinitis. *Arch Ophthalmol.* 1976;94:1485.
6. Frenkel JK. Chorioretinitis associated with positive test for toxoplasmosis. Acta XVII International Congress of Ophthalmology, Vol. 3. Toronto, Canada: University of Toronto Press; 1955; 1965.
7. Smith RE, Nozik RM. *Uveitis, A Clinical Approach to Diagnosis and Management.* Baltimore, Md: Williams & Wilkins; 1983.
8. Zimmerman LE. Ocular pathology of toxoplasmosis. *Surv Ophthalmol.* 1961;6:832.
9. Remington JS, Little HL. Attempts at isolation of *Toxoplasma* from the human retina. *Am J Ophthalmol.* 1973;76:566.

Chapter 73

Trypanosomiasis

KHALID F. TABBARA and NADER SHOUKREY

Trypanosomiasis is a disease caused by hemoflagellate protozoa of the genus *Trypanosoma*. African trypanosomiasis (African sleeping sickness) is associated with *T. brucei* throughout tropical Africa, and American trypanosomiasis (Chagas' disease) with *T. cruzi* in Central and South America. African sleeping sickness exists in two geographically and clinically distinguishable forms caused by two morphologically indistinguishable subspecies of *T. brucei*. West African or Gambian sleeping sickness is caused by *T. b. gambiense*, which is transmitted from man to man by riverine tsetse flies. The more virulent East African or Rhodesian sleeping sickness is caused by *T. b. rhodesiense*, which is transmitted from sylvatic animals to man by savanna-woodland tsetse flies. *T. cruzi*, the causative agent of Chagas' disease, is transmitted from domestic and paradomestic animals to man by reduviid bugs. The disease can also be acquired by blood transfusions and by transplacental passage of the parasite.

Insect vectors ingest the bloodstream pleomorphic trypomastigotes, commonly characterized by an undulating membrane and flagellum, when taking blood meals from infected animals or man. After passing through several developmental changes in the body of the insect, the parasites transform finally into metacyclic trypomastigotes, which are infectious to humans. The metacyclic forms of *T. brucei* are located in the salivary glands of tsetse flies and enter the bite wound with the saliva during biting. In reduviid bugs, the metacyclics of *T. cruzi* are found in the rectum, pass out in the feces during feeding, and enter the skin when the intensely pruritic lesions are scratched.

Once injected into the bite wound, the nonmultiplicative metacyclics of *T. brucei* transform into the long, slender multiplicative bloodstream trypomastigotes, which proliferate by longitudinal binary fission. They invade the bloodstream and lymphatic tissue and remain confined to extracellular spaces, namely, bloodstream, tissue fluids, and, in the terminal stages of infection, cerebrospinal fluid (CSF). Interestingly, the rounded, immobile, and mainly intracellular amastigotes have been detected in the choroid plexus in experimental animals. Later in the course of infection, the multiplicative bloodstream trypomastigotes of *T.*

brucei differentiate into short stumpy nonmultiplicative forms, which are thought to be preadapted to life in the tsetse fly.

On the other hand, the metacyclics of *T. cruzi,* once introduced into the bite wound, are phagocytosed by macrophages and other cells, locally and in lymph nodes to which they are conveyed by lymphatic spread. Phagocytosed metacyclics transform into amastigotes, which multiply by binary fission, producing amastigote pseudocysts. Amastigotes that are released into interstitial spaces by rupture of the pseudocyst actively invade other cells or soon die. Some amastigotes differentiate intracellularly into trypomastigotes, which are released into the vasculature. The circulating trypomastigotes are incapable of multiplication and are relatively resistant to interiorization by macrophages in the nonimmune host. They soon enter other cells, preferentially cardiac muscle, smooth muscles of the digestive tract, and autonomic nerve ganglia, where they assume the amastigote form and proliferate or are withdrawn by another insect vector.

Metacyclics of *T. brucei* injected into the bite wound are predisposed in more than one way to shield themselves against attack by host antibodies. Not only do they comprise populations of heterogenous antigenic types, but also both metacyclics and their descendants, the bloodstream trypomastigotes, are prone to rapid antigenic variation.[1] Presumably this is expressed through replacing the surface coat of antigenic glycoprotein synthesized by the metacyclic *T. brucei* in the fly salivary gland with a new and different glycoprotein. Antigenic variation of the parasite is reflected in one of the most specific features of African trypanosomiasis, namely, fluctuating parasitemia; each rise in the number of trypomastigotes in the blood and lymphatic fluids marks the appearance of new generations of immunologically distinct trypomastigotes subsequent to destruction of most of the old generations by host variant-specific antibodies. This is in conformity with the substantial rise in serum immunoglobulins (specific and unrelated antibodies, autoantibodies, immune complexes) in patients with African trypanosomiasis. This generalized condition of B-lymphocyte proliferation is thought to be responsible for the characteristic marked elevation of IgM levels in peripheral blood and in CSF, the latter being diagnostic of African trypanosomiasis.

Although *T. cruzi,* the causative agent of American trypanosomiasis, carries at least two major surface antigens,[2] the parasite does not exhibit any antigenic variation in the host. *T. cruzi* evades the host immune response by maintaining itself in the intracellular amastigote form within the cytoplasm of nonimmune cells, thereby sheltered from any possible serum antibodies. Protective host antibodies belong to the class

IgG, and their effects are expressed through immunophagocytosis of the parasite by macrophages. The elevated parasitemia observed during the acute phase of Chagas' disease may be related to suppression of both humoral and cell-mediated immune mechanisms and the great resistance of bloodstream trypomastigotes to phagocytosis by nonactivated macrophages. The parasitemia is, however, very low during the chronic phase of the disease, owing to the establishment of strong acquired immunity.

Systemic Manifestations

The earliest sign of African trypanosomiasis is an inflammatory reaction localized at the bite wound. This may develop into a diagnostic, 3- to 4-cm, round swelling, the trypanosomal chancre, which resolves spontaneously. Systemic illness follows a basically subclinical incubation period of 1 to 3 weeks (occasionally 2 to 5 years) and is characterized by irregular fever with lymphadenopathy, and sometimes with splenomegaly. Blood concentrations of parasites are high during fever and decline sharply in afebrile periods. Enlargement of lymph nodes in the supraclavicular and posterior cervical groups (Winterbottom's sign) is diagnostic of Gambian trypanosomiasis, whereas anemia and thrombocytopenia are more severe in Rhodesian trypanosomiasis. The evolving disease is characterized by headache, malaise, anorexia, lethargy, and weight loss.

Onset of central nervous system (CNS) involvement is delineated by symptoms of meningoencephalomyelitis. Trypomastigotes migrate from the bloodstream into the brain, where they occur mainly in the frontal lobe, pons, and medulla and also invade the CSF from the choroidal plexus. There is perivascular round cell infiltration of the Virchow-Robin spaces, and the CSF reveals increased amounts of proteins, an elevated level of IgM that is independent of its elevated level in blood, lymphocytosis, mononucleosis, and morular cells. Objective signs of CNS involvement include tremor, athetosis, cerebellar ataxia, mental deterioration, daytime somnolence, and cachexia. Eventually the typical sleeping sickness stage is reached: somnolence becomes almost continuous, and ultimately true coma develops, and death. Although CNS invasion occurs early in the more acute Rhodesian trypanosomiasis, the rapidly deteriorating condition of the patient does not usually permit development of the typical sleeping sickness, cardiac failure being the most common cause of death in these patients.

Chagas' disease is seen in its most acute form in children under 5 years of age, particularly infants. The

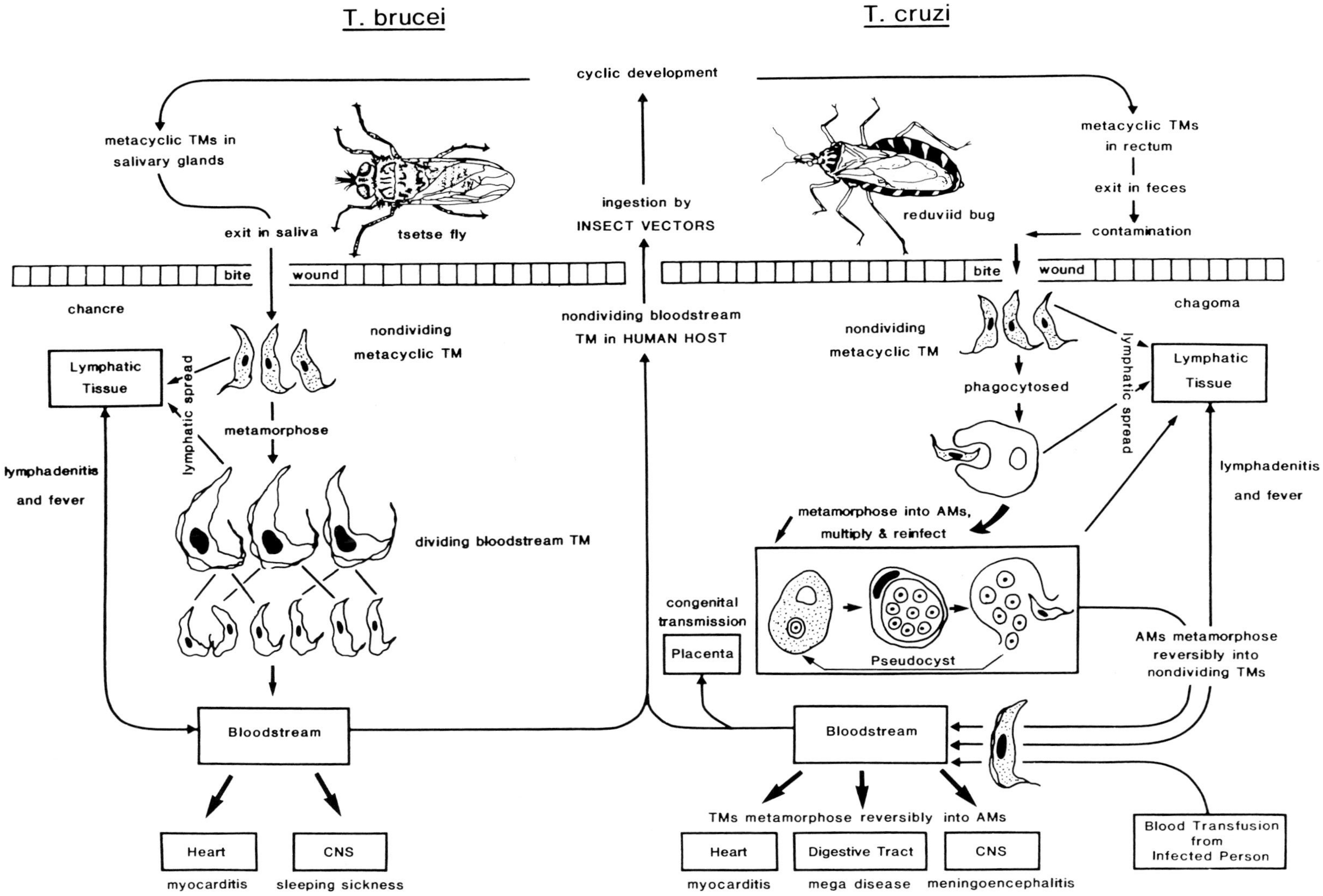

FIGURE 73-1 Life cycles and pathology of *T. brucei* and *T. cruzi.*

chronic form of Chagas' disease is, however, the most frequent and serious consequence of *T. cruzi* infection, and occurs in older children and adults, usually following an acute attack. The disease may begin with a 1- to 3-cm inflammatory nodule at the site of inoculation, the primary chagoma, which subsides gradually. Similar nodules (lipochagomas) may appear throughout the body, apparently by hematogenous spread of the parasite and preferential invasion of fat cells in the infected areas. When the chagoma is around the eye, there is a diagnostic unilateral palpebral edema and conjunctivitis (Romana's sign). The classical manifestations of the disease are signs of cardiomyopathy associated with irregular high fever, parasitemia, localized or generalized lymphadenitis, anasarca, hepatosplenomegaly, and possibly meningoencephalitis, particularly in young children. The principal gross changes are in the heart: cardiomegaly, interstitial myocarditis, and hypotension; patients may die of irreversible cardiac failure or ventricular fibrillation.

Acute Chagas' disease may terminate in death or recovery, or the patient may enter the chronic phase of the infection. Congenital transmission can take place in both the acute and the chronic stage of the disease. The majority of chronic infections are asymptomatic and can be discovered only by xenodiagnosis or serologic tests. Classic features of the chronic disease are cardiomyopathy and, less commonly, megaviscera. Chagasic cardiomyopathy exhibits signs of progressive congestive, mainly right-sided, cardiac failure. Histologic examination reveals the presence of fibrosis (cardiomegaly), degenerating myocardial fibers, infiltration of mononuclear cells, and presence of pseudocysts. Characteristic alterations in the electrocardiogram show partial or complete atrioventricular block, complete right-sided bundle branch block, or premature ventricular contractions, along with abnormalities of the QRS complexes and of the P and T waves. Emboli from the ventricular apices or atria may cause pulmonary or cerebral infarcts. In megaviscera there is extensive dilatation of various tubular organs, particularly of esophagus and colon and less frequently of stomach, duodenum, and ureter. Denervation due to profound reduction in neurons of the myenteric plexus is the basis for Chagas' megadisease. Megaesophagus is usually characterized by dysphagia, and megacolon by symptoms of prolonged constipation, meteorism, and volvulus. An outline of the life cycles and pathology of *T. brucei* and *T. cruzi* is illustrated in Figure 73-1.

Ocular Manifestations

Although typical oculoglandular complex (Romana's sign) is relatively rare in Chagas' disease, when present it is an important diagnostic sign because of its prominent location.[3] This clinical sign may appear about 1 or 2 weeks after infection, when the portal of entry of the parasite is the orbicular region. It manifests as marked edema of the periorbital region, often with total closure of the eye, reddish violet discoloration of the skin, conjunctivitis, and dacryocystitis.[4] There is ipsilateral swelling of the submaxillary lymph nodes, and the edema usually expands to involve the cheek, and occasionally the neck on the same side. Sometimes the contralateral eyelid is involved with bilateral dacryocystitis and bilateral facial edema. These manifestations are not always caused by the presence of *T. cruzi* but may be induced by allergic responses elicited by repeated bites of uninfected reduviid bugs. On the other hand, ocular involvement in African trypanosomiasis is characterized by urticarial swelling of the eyelids with enlargement of the preauricular glands.[5]

Interstitial keratitis with neovascularization is not uncommon, and is occasionally associated with iridocyclitis. Experimental *T. brucei* infection in cats and dogs induces intense protein leakage in the anterior chamber with cyclitis, lacrimation, and facial edema, and the parasite can be recovered from the anterior chamber.[6,7]

References

1. Borst P, Cross GAM. Molecular basis for trypanosome antigenic variation. *Cell.* 1982;29:291.
2. Nogueira N, Unkeless J, Cohn Z. Specific glycoprotein antigens on the surface of insect and mammalian stages of *Trypanosoma cruzi. Proc Natl Acad Sci USA.*
3. Romana C. Acerca de un sintoma inicial de valor para el diagnostico de forma aguda de la enfermedad de Chagas. La conjuntivitis esquizotripanosica unilateral (hipotesis sobre puerta de entrada conjuntival de la enfermedad). *Publ MEPRA.* 1935;22:16.
4. Ackerman WA, et al. Chagas' disease. *Int J Dermatol.* 1986;5:320.
5. Rodger FC. Ophthalmology in the tropics. In: Manson-Bahr PEC, Apted FIC, eds. *Manson's Tropical Diseases.* London: Bailliere Tindall; 1982;604-605.
6. Mortelmans J, Neetans A. Ocular lesions in experimental *Trypanosoma brucei* infection in cats. *Acta Zool Pathol Antverp.* 1975;62:149.
7. Ikede BO. Ocular lesions in sheep infected with *Trypanosoma brucei. J Comp Pathol (Engl).* 1974;84:203.

Section G
Rickettsial Diseases

Chapter 74
Rickettsial Infections

DENYS BEAUVAIS and JOSEPH B. MICHELSON

Rocky Mountain spotted fever is the most common of the acute febrile exanthematous illnesses caused by *Rickettsia rickettsii.* These gram-negative coccobacilli possess the characteristics of both viruses and bacteria. Like viruses, they are obligate intracellular parasites. Like bacteria, however, they possess both DNA and RNA and are susceptible to antibiotics, which makes them amenable to treatment with oral chloramphenicol and tetracyclines.

Infection occurs in a wide variety of hosts, and humans are affected by the bite of two major species of infected ticks. Dogs have been implicated in the transmission of infected ticks to humans, and further, have been shown to be a reservoir of rickettsiae, infecting ticks themselves. *Dermacentor andersoni,* the wood tick, is distributed in the area of the Rocky Mountain states, and is most active during spring and summer. *D. variabilis,* the dog tick, is found extensively in the Eastern United States, especially in the South, which now accounts for more than 50% of reported cases. Rickettsial infection appears to be almost ubiquitous in the continental United States, with the exception of only Maine and

Vermont. Dr. Howard Ricketts is credited with isolating and identifying *R. rickettsii* as the etiologic agent of Rocky Mountain spotted fever.

Young adults and children are most often affected, probably because they spend more time in outdoor activities and recreation and with pets. Rocky Mountain spotted fever, a potentially lethal febrile illness, may result from infection with only a few organisms. They are deposited on the skin when ticks defecate or bite and feed. Then the rickettsiae enter endothelial cells of capillaries and venules, where multiplication occurs, new organisms are released, and infection ensues. The mortality rate has been reported to be between 3 and 8%.[1-3]

Systemic Manifestations

A focal or proliferative vasculitis is the hallmark of these disorders, which may involve all organ systems, because the organisms parasitize and multiply in vascular endothelial cells. The intima and the media become necrotic,

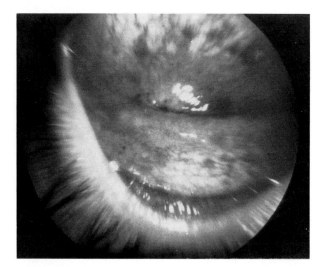

FIGURE 74-1 A 31-year-old white man had severe myalgia, arthralgia, headache, chills, and fever, and a sore, red eye, with conjunctival papillae, chemosis, and petechiae on both the bulbar and palpebral conjunctiva. Note prominent petechiae on the bulbar conjunctiva. Visual acuity was 20/20. The patient's history revealed tick exposure 7 days earlier. (Courtesy of David A. Snyder, M.D.)

After the first few days of illness, edema usually develops and is a most important diagnostic clue. Initially, the edema may be localized in the periorbital region and subsequently spreads and becomes generalized, involving the extremities. Splenomegaly is present in about one third of the patients. After the disease progresses, hepatomegaly can often be observed. Respiratory symptoms are infrequent, although rickettsial pneumonitis may occur. Myocardial involvement may occur in advanced stages of the disease. Neurologic findings are seen in the later stages, although neurologic symptoms have been reported as the initial manifestations of the disease. Headaches and lethargy are common; confusion, delirium, seizures, cranial nerve deficits, and pathologic reflexes may also occur in seriously ill patients.[2-4]

Ocular Manifestations

The ocular manifestations of rickettsial infection are usually limited to petechial lesions on the bulbar conjunctiva with conjunctivitis (Fig. 74-1). Other ocular involvement has been noted but not commonly reported. Conjunctival vasculitis may result in subconjunctival hemorrhages (Fig. 74-2). Iritis, endophthalmitis, neuroretinitis, optic nerve engorgement, and optic disc pallor, as well as corneal ulcers have all been seen in typhus fever. The vasculitis that is common to all the rickettsial infections is probably the cause of the retinal and optic disc edema that has been reported, and of the retinal vein engorgement, retinal hemorrhages, retinal exudates, panuveitis, and vitreous opacities.

Several groups of authors have reported a retinopathy of Rocky Mountain spotted fever. In a study of six patients, Presley[5] described venous engorgement, retinal edema, papilledema, cytoid bodies (nerve fiber layer infarcts), retinal hemorrhages, and an arteriole occlusion. Raab, Leopold, and Hodes[6] described a 7-year-old boy with papilledema, venous distension and occlusions, white exudates, hemorrhages, and arteriole branch occlusions. Cherubini and Spaeth described a 54-year-old woman with a miotic nonreactive right pupil, a normal-sized, sluggish left pupil, and small white keratic precipitates in both eyes with anterior segment inflammation. The fundi had congested retinal veins inferiorly.[7]

Similarly, Smith and Burton[8] reported a case of an 11-year-old girl demonstrating disc edema, dilated tortuous retinal veins, and cotton-wool spots, reflecting nerve fiber layer infarcts in the retina. Fluorescein angiography demonstrated nonperfusion secondary to small vessel obstruction, coexistent with vasculitic patterns.

causing thrombosis with microinfarction, and these lesions extend into arterioles, and occasionally into venules. In the more severe cases of rickettsial infection, angiitis results in focal ischemic damage to the liver, heart, and central nervous system (CNS). Death may occur, from peripheral vascular collapse or from a supervening bacterial infection. The individual rickettsial diseases are distinguished by their clinical course and are classified in three groups: spotted fever group, typhus group, and the Q- and trench fever group.

A history of tick bite can be obtained in 70 to 80% of patients with Rocky Mountain spotted fever. Early in the course of the disease, symptoms are nonspecific: fever, headache, generalized myalgia. A high level of suspicion of rickettsial infection must be maintained for febrile illness in young adults and children, since fatal cases are most often associated with delay in diagnosis rather than delay in obtaining medical attention. Fever, appropriate rash, and travel or habitation in areas where the disease is endemic are the most helpful criteria.

The rash that accompanies Rocky Mountain spotted fever characteristically appears about the 3rd to the 5th day of illness and may be macular, maculopapular, or occasionally petechial at onset. Classically, the rash affects the extremities first, then spreads centripetally.

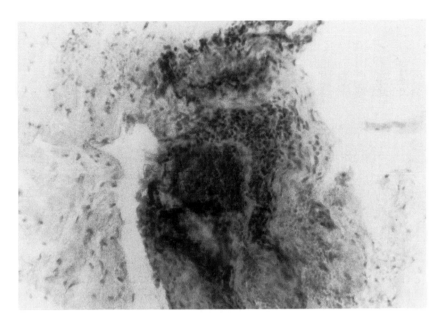

FIGURE 74-2 Histopathology of conjunctival biopsy demonstrates a mononuclear cell infiltration with perivasculitis (Giemsa stain ×150; courtesy of David A. Snyder, M.D.)

The patient demonstrated leakage of fluorescein from the optic nerve head indicative of the vasculitis of the papilla as well. Duffey and Hammer[9] reported a case with uveitis, retinal vasculitis, and an iris nodule similar to a typhus nodule of the CNS reported in typhus rickettsial disease. An extensive review of the ocular manifestations of scrub typhus, in which the retinal findings are similar to those described in Rocky Mountain spotted fever, was reported by Scheie[10] in 1948.

The histopathologic mechanism of necrotic vascular walls causing thrombosis with microinfarction can explain the ophthalmoscopic and fluorescein angiographic abnormalities which have been observed by the authors mentioned above. Thrombosis of retinal precapillary arterioles results in focal areas of infarction with capillary nonperfusion, especially in the nerve fiber layer of the retina, resulting in cotton-wool spots.

Perivascular staining is usually adjacent to the areas of capillary nonperfusion, which is a direct consequence of the intense angiitis. Similarly, occlusion of the small vessels in the substance of the optic nerve head produces endothelial incompetence, leakage, and subsequent optic disc edema. The disc edema may be responsible for partial obstruction of the central retinal vein. It is also likely that occlusion of these small vessels can produce infarction of the optic nerve head, with or without swelling of the papilla.

References

1. Sulenski ME, Green WR. Ocular histopathologic features of a presumed case of Rocky Mountain spotted fever. *Retina*. 1986;6:125-130.
2. Woodward TE. The rickettsioses. In: Adams RD, Martin JB, Braunwald E, et al, eds. *Harrison's Principles of Internal Medicine*. 10th ed. New York, NY: McGraw-Hill; 1983;1069-1072.
3. Woodward TE, Hornick RB. *Rickettsia rickettsii* (Rocky Mountain spotted fever). In: Mandell GL, Douglas RG Jr, Bennett JE, eds. *Principles of Practice of Infectious Disease*. New York, NY: John Wiley & Sons; 1985;1082-1087.
4. Kelsey DS, Rocky Mountain spotted fever. *Pediatr Clin North Am*. 1979;26:367-376.
5. Presley GD. Fundus changes in Rocky Mountain spotted fever. *Am J Ophthalmol*. 1969;67:263-264.
6. Raab EL, Leopold IH, Hodes HL. Retinopathy and Rocky Mountain spotted fever. *Am J Ophthalmol*. 1969;68:42-46.
7. Cherubini TD, Spaeth GL. Anterior nongranulomatous uveitis associated with Rocky Mountain spotted fever. *Arch Ophthalmol*. 1969;81:363-365.
8. Smith TW, Burton TC. The retinal manifestations of Rocky Mountain spotted fever. *Am J Ophthalmol*. 1977;84:259-262.
9. Duffey RJ, Hammer ME. The ocular manifestations of Rocky Mountain spotted fever. *Ann Ophthalmol*. 1987;19:301-306.
10. Scheie HG. Ocular changes associated with scrub typhus. *Arch Ophthalmol*. 1948;40:245-267.

Section H

Spirochetal Diseases

Chapter 75

Leptospirosis

SUSAN BARKAY and HANA GARZOZI

Leptospirosis is a spirochetal epidemic or endemic disease. Leptospiras are harbored by animal hosts, mostly lower mammals such as rats, mice, and raccoons, but domestic animals including cattle, swine, and dogs also can be affected. Infections in natural hosts are usually not significant, although in some countries they cause veterinary problems. In carrier animals an asymptomatic infection of renal tubules occurs, with long-lasting leptospiruria. Leptospiras can survive for weeks in moist soil or alkaline water in warm seasons.[1-3]

Humans can acquire the disease through contact with urine of affected animals or through direct contact with tissues of dead animals via abraded skin or mucous membranes.

Leptospirosis is an occupational hazard to farmers, dairymen, veterinarians, rice and sugar cane field workers, trappers, and others, but it is also a recreational hazard to hunters, fishermen, and swimmers in infected areas. Direct transmission from person to person is possible but rare. It is primarily a disease of young adults, chiefly men, contracted during the summer and autumn months.[4]

Pathologic leptospiras belong to the species *Leptospira interrogans*. Within the one species, on the basis of distinct, different agglutinogenic properties, some 170 serovars (serotypes) have been differentiated, which are indistinguishable by morphologic, cultural, and physiologic properties. These 170 serotypes fall into about 20 serogroups on the basis of common overlapping antigenic components.[1,2]

Leptospiras penetrate tissues mechanically. Animal experiments show rapid invasion of the leptospiras into the bloodstream after peritoneal inoculation. They are present in virtually all organs after 24 hours and can be recovered from cerebrospinal fluid (CSF), the brain, and the anterior chamber of the eyes without hemorrhage or signs of irritation. Similar observations were made in several cases of human disease.[5] The tissue damage caused by leptospira is probably of a toxic nature; virulence seems to depend on toxin production.[5]

Systemic Manifestations

Leptospirosis is a biphasic illness. The first, the lepto-spiremic phase, begins after an incubation period of 3 to 26 days and lasts about 4 to 9 days; then, after an asymptomatic interval of 1 to 3 days, the second phase ("immune phase") follows. Leptospiras invade the bloodstream and CSF and cause an acute febrile illness with abrupt onset, severe frontal, bitemporal or retro-orbital headaches, muscle pains, and great sensitivity to palpation. Sensory disturbances, nausea, vomiting, diarrhea, occasional pulmonary manifestations, pharyngeal injection, skin rash, and hemorrhages occur. Jaundice in the first phase is not a usual finding. A very characteristic sign is an almost asymptomatic conjunctival effusion, which although important is often overlooked.[1,5,6] At this stage of the disease, leptospiras can be cultured from blood or CSF with semisolid (Fletcher's or Stuart's) medium; darkfield microscopy or stained preparations are unreliable, although they were once widely used.

The first phase ends with the disappearance of leptospiras from blood and CSF on the 4th to 9th day. After the 10th day the leptospiras are excreted in the urine for 1 to 11 months, but negative urine cultures do not exclude leptospirosis. The first phase is characterized by a homogeneous clinical picture, whereas the second, the immune phase shows individual variations, some patients being almost without symptoms while others are severely ill. About 50% of the patients have high fevers with meningismus and 25% have meningitis with elevated protein values and pleocytosis in CSF.[1,5]

Encephalitis, Guillain-Barré syndrome, involvement of the optic, abducens, facial, and auditory nerves, radiculitis, and peripheral nerve lesions are further complications. In some cases the disease takes a very severe course. The fever is persistent and extremely high; jaundice appears after the first days, with hepatomegaly. Renal manifestations occur—pyuria, hematuria, and peak elevations of blood urea nitrogen as a sign of acute tubular necrosis. Hemorrhagic manifestations, epistaxis, hemoptysis, gastrointestinal bleeding, and subarachnoidal hemorrhage occur as a result of diffuse vasculitis and capillary injury.[1,5] The picture is known as Weil's syndrome.

The clinical signs of the immune phase parallel the appearance of leptospira antibodies in the blood at the end of the 1st week, reaching their peak in the 3rd or 4th week.[1,5] During this period of the disease serologic tests should be performed. They are theoretically complicated, owing to the large number of antigenically distinct leptospiral serotypes, but often it is sufficient to test only for the organisms known to be present in the endemic area. The microscopic agglutination-lysis test against antigens of living leptospira is the most sensitive test. It is often used to confirm the result of other tests because of its high specificity.[1,3,5] The simple agglutination test using suspensions of killed leptospiras is effective but less accurate, since nonspecific reactions occur.

From the 10th day of illness until the 3rd month, a complement fixation test is also an excellent diagnostic aid. At this time, complement-fixing antibodies are usually present in the serum. Blood for serologic tests should be examined during the acute illness and also during the convalescence period. A fourfold rise in titer is considered diagnostic. If only a single specimen is available, a titer of 1:1600 gives strong presumptive evidence of the diagnosis.[5]

In the anicteric form of leptospirosis the prognosis is mostly good, but in cases with jaundice, the mortality rate ranges from 25 to 40% in untreated cases.

Postmortem findings show extensive ecchymoses and petechiae in striated muscle, kidneys, adrenals, liver, lungs, spleen, and stomach. The tubular epithelium of the kidneys shows changes ranging from swelling to desquamation and necrosis. Liver findings are less specific, with focal areas of necrosis around the portal veins. Extensive changes in skeletal muscles, from vacuoles to necrosis, are usual findings.

Ocular Manifestations

The incidence of ocular findings in leptospirosis has been reported to be from 3 to 92%. This wide range probably depends on the enthusiasm and knowledge of the clinical observers.[4]

During the first days of the disease a conjunctival injection occurs, involving the anterior part of the globe and the lower lid. The clinical picture is characteristic: engorged conjunctival and episcleral vessels anastomose with dilated pericorneal vessels, giving the impression of a reticulum. In jaundiced patients there is a yellow shade, otherwise the conjunctiva is pink. Usually there are no subjective complaints, no lacrimation, and no discharge; however, discharge, if present, contains leptospiras and may be the source of further infection.[6] This peculiar picture tends to disappear during convalescence.

In the immune phase, as part of the central nervous system involvement, palpebral herpes and optic neuritis may develop, with or without iridocyclitis.[7] This picture is usually self-limited and subsides after a short time.

Of special interest for ophthalmologists are the late uveitis cases. These occur long after the general features of the disease have subsided, in an apparently healthy person. It may appear after the illness itself has been forgotten; indeed, cases after 5 years have been reported.[3] There is usually an acute, moderate, bilateral iridocyclitis with hyperemia of the iris, but cases with severe exudative inflammation with mutton fat precipitates, dense posterior synechiae between iris and lens, and even hypopyon and secondary glaucoma, are quite common findings. Choroiditis, exudates and hemorrhages in the retina, and peculiar vitreous membranes running from the optic disc toward the anterior segment have been described, with a long-lasting course of absorption. Blindness as a final outcome has been reported.[3,4]

The pathogenesis of this uveitis as a late complication is not clear. Leptospira organisms have been isolated from the anterior chamber, but antibodies have also been found there. It has been suggested that the leptospiras can survive for long periods of time in the eye, as they do in the pelvis of the kidney, in spite of the high levels of antibodies in the general circulation. It is possible that the organisms that survive can excite inflammation in the form of late uveitis only after systemic immunity has faded.[3]

From the etiologic point of view, diagnosis of the uveitis is not simple. In endemic areas and in cases of short intervals between the general disease and uveitis, it is not difficult to find the connection; however, in many cases of severe uveitis the forerunner was, months or even years before, a febrile disease of several days' duration without significant complications. In such cases diagnosis is difficult and often presumptive, as it must rest on the history of systemic illness with characteristic anamnestic data and clinical signs of leptospirosis that sometimes occurred long before, on exclusion of other causes, and on positive serologic tests.

Treatment

Systemic treatment should be administered at the onset of the disease, during the leptospiremia. Penicillin G is preferred, but tetracyclines are also effective. There is general agreement that antibiotics are useless after the 5th day. In severe cases of tubular necrosis, hemodialysis or peritoneal dialysis is indicated.

Uveitis should be treated with locally applied mydriatics and steroids. Systemic steroid treatment is indicated in severe cases.[4]

Human leptospirosis is a disease that can be avoided with adequate prophylactic methods. Until safe polyvalent vaccines are developed, hygienic precautions should be taken in cooperation with veterinary services and health care authorities.

References

1. Braunwald E, Isselbacher KJ, Petersdorf RG, et al., eds. *Harrison's Principles of Internal Medicine.* 11th ed. New York, NY: McGraw-Hill; 1987;652-655.
2. Javetz E, Melnick JL, Adelberg EA. *Review of Medical Microbiology.* 14th ed. Los Altos, Calif: Lange Medical Publications; 1980;259-260.
3. Duke-Elder S. *System of Ophthalmology.* vol IX. *Diseases of the Uveal Tract.* London: Henry Kimpton; 1966;322-325.
4. Barkay S, Garzozi H. Leptospirosis and uveitis. *Ann Ophthalmol.* 1984;16:164-168.
5. Braude AI. *Medical Microbiology and Infectious Diseases.* Philadelphia, Pa: WB Saunders Co; 1981;437-441, 1143-1146, 1839-1847.
6. Duke-Elder S. *System of Ophthalmology.* vol VIII. *Diseases of the Outer Eye.* London: Henry Kimpton; 1965;201-203.
7. Walsh F, Hoyt W. *Clinical Neuro-Ophthalmology.* 3rd ed. vol II. Baltimore, Md: Williams & Wilkins; 1969;1548-1551.

Chapter 76

Lyme Disease

GEORGE J. FLORAKIS, MICHAEL P. VRABEC, and JAY H. KRACHMER

Lyme disease was first recognized by Steere et al.[1] in 1975 following an unusual outbreak of arthritis in Old Lyme, Connecticut. Lyme disease is a systemic disease caused by the spirochete *Borrelia burgdorferi*. The vector is the deer tick (*Ixodes dammini*). It has been reported in many areas throughout the United States, most commonly in the Northeast and in Minnesota and Wisconsin.[2] It has also been reported in Europe and Australia.[3]

Systemic Manifestations

The disease can be difficult to diagnose, as it presents with a number of different signs and symptoms which are often typically intermittent and changing over a period of several weeks. Four to 20 days after a tick bite a characteristic skin lesion, erythema chronicum migrans (ECM), usually develops. The ECM rash begins as a red macule or papule that expands to form a large annular lesion with a bright red outer border and a partially clear center (Fig. 76-1). The rash is usually preceded by or accompanied by a flulike syndrome, often consisting of a sore throat, chills, fever, headache, myalgias, malaise, regional lymphadenopathy, and arthralgias, especially in the knees. Weeks to months after the initial rash, neurologic or cardiac manifestations may develop. The neurologic signs include meningeal irritation with neck pain and stiffness or mild encephalopathy, somnolence, insomnia, memory difficulties, emotional lability, dizziness, poor balance, clumsiness, or dysesthesias. Radiculopathies and cranial neuropathies such as 6th or 7th nerve palsy may also occur. The cardiac manifestations include atrioventricular block with rapid fluctuations in the degree of the block or ST-segment and T-wave abnormalities.[2] Acute arthritis often develops within the first 6 months. Typically it localizes to the knee, although multiple other joints may be affected simultaneously. A chronic arthritis, especially of the knee, may also occur.[2]

The most common nonspecific laboratory abnormalities include increased erythrocyte sedimentation rate and aspartate transaminase level.[3] Positive antispirochetal antibody titers by ELISA or other methods are diagnostic.

The pathogenesis of the signs and symptoms of Lyme disease are still being investigated, and it is not clear whether direct spirochete infection or self-propagating inflammatory and immune mechanisms are responsible, as there is evidence of both. It is important to note, however, that the disease may be self-limited and may not progress to its later stages.

Recommended treatment includes penicillin, tetracycline, or erythromycin orally for 2 to 3 weeks. For the later stages of Lyme disease, including those with cardiac or neurologic manifestations, high-dose intravenous penicillin or intramuscular or intravenous ceftriaxone should be used.

Ocular Manifestations

Little is known about the natural history of the ocular manifestations of Lyme disease. Multiple eye findings have been described in a small number of case reports (Table 76-1.). In 1983, Steere et al.[3] found conjunctivitis in 11% of patients with Lyme disease and noted that periorbital edema was rare. Since then, however, other ocular findings have been reported.

Bertuch et al. described patchy snowball-like intrastromal corneal infiltrates at random depths in two patients.[4] The response of these infiltrates to steroids suggest an immune mechanism rather than a direct corneal infection with the spirochete.

Gallin treated two children, aged 4 and 6, both with a characteristic history of Lyme disease proven by positive titers. Both patients presented with a unilateral mild,

TABLE 76-1 Reported Eye Findings in Lyme Disease

Conjunctivitis
Periorbital edema
Corneal infiltrates
Iritis
Panophthalmitis
Anterior ischemic optic neuropathy
Optic disc edema
Cranial neuropathies

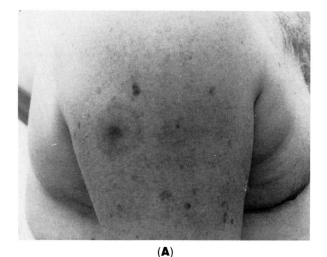

(A)

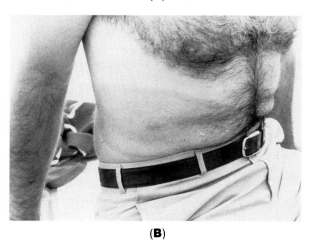

(B)

FIGURE 76-1 Rash of erythema chronicum migrans. (Courtesy Richard Strongwater, M.D.)

acute, nongranulomatous anterior uveitis. They responded well to steroids (personal communication, 1987).

The most serious ocular complication in Lyme disease, however, was reported by Steere et al.[5] in 1985, who described a patient with a painful red eye which developed 4 weeks after the onset of ECM. Initially the patient had only an acute iritis with posterior synechiae. There was no sign of vitreal or retinal involvement. This subsequently progressed to panophthalmitis with a purulent vitreous cavity. Cultures of vitreous aspirates were negative, and no spirochetes were seen on darkfield microscopy. The eye lost light perception and was enucleated. Histopathologic examination revealed advanced cell necrosis as well as occasional intact spirochetes resembling *B. burgdorferi* in the vitreous debris. This suggested direct infection of the spirochete resulting in loss of the eye.

In another report, Schechter demonstrated ischemic optic neuropathy as part of the clinical syndrome of Lyme disease.[6] His patient presented with fever, malaise, myalgias, and ECM. He also had decreased vision, papillitis, and an inferior altitudinal defect on visual field examination. Erythrocyte sedimentation rate was elevated. Results of temporal artery biopsy were negative.

Wu et al.[7] reported a 7-year-old child with Lyme disease who developed bilateral optic disc edema. Lumbar puncture revealed a normal opening pressure and 14 leukocytes. Results of computed tomography were normal. Lyme titers were positive. The disc edema resolved with intravenous penicillin therapy; however, the patient did have a residual amount of retinal pigment epithelial mottling of his maculae which persisted.

Lyme disease is a relatively recently described systemic entity whose ophthalmic manifestations are slowly being characterized. Like other spirochete infections such as leptospirosis, syphilis, and relapsing fever, Lyme disease involves systemic infection with a variety of clinical stages with remissions and exacerbations as well as eye involvement. In addition, Lyme disease can be likened to other syndromes that involve inflammation of the uveal tract as well as arthritis, such as ankylosing spondylitis, Reiter's syndrome, psoriasis, and juvenile rheumatoid arthritis.

References

1. Steere AC, Malawista SE, Snydman DR, et al. Lyme arthritis: an epidemic of oligoarticular arthritis in children and adults in three Connecticut communities. *Arthritis Rheum.* 1977;20:7.
2. Meyerhoff J. Lyme disease. *Am J Med.* 1983;75:663.
3. Steere AC, Bartenhagen NH, Craft JE, et al. The early clinical manifestations of Lyme disease. *Ann Intern Med.* 1983;99:76.
4. Bertuch AW, Rocco E, Schwartz EG. Eye findings in Lyme disease. *Connecticut Med.* 1987;51:151.
5. Steere AC, Duray PH, Kauffmann DJ, et al. Unilateral blindness caused by infection with the Lyme disease spirochete, *Borrelia burgdorferi. Ann Intern Med.* 1985;103:382.
6. Schechter SL. Lyme disease associated with optic neuropathy. *Am J Med.* 1986;81:143.
7. Wu G, Lincoff H, Ellsworth RM, et al. Optic disc edema and Lyme disease. *Ann Ophthalmol.* 1986;18:252.

Chapter 77

Relapsing Fever

RAGA MALATY and RUDOLPH M. FRANKLIN

Relapsing fever is the name given to a group of closely associated acute infectious disorders caused by blood spirochetes of the genus *Borrelia*. It occurs throughout most of the world either in the epidemic (mainly louse-borne) or endemic and sporadic (tick-borne) fashion. The main focus of louse-borne relapsing fever is now the Ethiopian high plateau, where more than 10,000 people may be infected each year.[1] As long as this and other small foci exist, another pandemic is possible. In the United States, relapsing fever is limited to the tick-borne variety and occurs mainly in the mountainous regions of the West. The relapsing pattern of the fever is associated with the phase variation in the *Borrelia* organism that infects a patient; thus, the spirochetes develop antigenic variations during subsequent attacks. The organism seems capable of adapting itself to the antibody defenses of the host that it infects; however, the number of readaptations is limited. After each relapse the organisms retreat to central nervous system (CNS), spleen, liver, and bone marrow.

Causative Organisms

The *Borrelia* organisms are transmitted to vertebrates by hematophagous arthropods. The louse-borne spirochete is known as *B. recurrentis*, and the tick-borne spirochetes are named according to their associated *Ornithodorus* species tick vectors. For most species of *Borrelia*, the usual vertebrate reservoir is a rodent. Some species, however, utilize only humans as a host, and others are primarily associated with large animals such as cattle or deer. One of the modes of transmission from the tick to the vertebrate host is by regurgitation of the gut contents (where the *Borrelia* organisms are primarily distributed) during feeding.[2] In louse-borne relapsing fever, the bite itself is not the source of the organisms, nor are the feces. Transmission occurs when the infected person crushes the irritating louse and rubs the spirochete-rich hemolymph into the bite wound.[3]

The *Borrelia* organisms then circulate and multiply in the blood until specific antibodies appear. When the antibody level is high enough, the organisms rapidly disappear from the blood, but they are still found in

organs, mainly the brain. The clearance from the blood is associated with binding of T cell-independent antibodies. This has been demonstrated experimentally in mice that are deficient in or deprived of thymus-derived lymphocytes. The animals eliminated blood-borne *Borrelia* organisms as efficiently as their normal littermates, though B cell-deprived mice could not.[4,5] This suggests that a B cell-dependent mechanism is responsible for the elimination of the spirochetes. A morphologic characteristic of clinical significance is the lipopolysaccharide fraction of the organism's gram-negative cell wall. This is a nonendotoxin pyrogen that has been implicated in the pathogenesis of the Jarisch-Herxheimer reaction often seen after antimicrobial therapy.[4]

Clinical Manifestations

The incubation period is difficult to establish and is usually associated with insect vector exposure. In cases in which it is possible to date the time of infection, it has ranged from 7 to 18 days.[6] The onset is abrupt, with sudden appearance of fever and violent headache. Eye pain is a frequent complaint and may be associated with photophobia and a burning sensation. Neurologic symptoms are common and may include touch and taste hyperesthesia, facial paralysis, hemiplegia, aphasia, and psychotic phenomena. The first attack usually lasts 1 week. Other signs include rash, jaundice, hepatosplenomegaly, and hemorrhage. The fever then drops abruptly, with hypotension, sweating, and weakness, and there is an interval of 1 to 2 weeks before the next relapse. Relapses have a tendency to become shorter and milder as the disease progresses.

Laboratory Diagnosis

Borrelia organisms can be demonstrated in thick or thin blood smears during the attack. Seventy percent of cases can be diagnosed using Giemsa or Wright's stain (Fig. 77-1) with a higher yield using acridine orange fluorescent stain.[6,7] Serologic tests are less reliable.

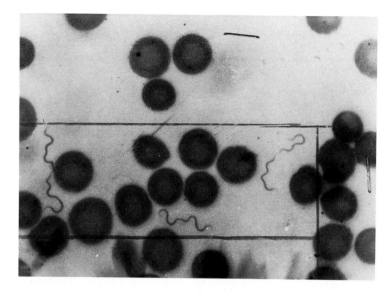

FIGURE 77-1 *B. recurrentis* spirochetes in a blood smear. (Giemsa stain, original magnification ×1650; from Gillum.[10])

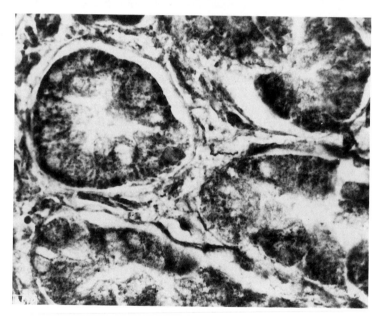

FIGURE 77-2 Spirochetes in the spleen of a patient with relapsing fever. (Warthin-Starry stain, original magnification ×820; from Gillum.[10])

Pathology

Patients with relapsing fever have abnormalities of blood coagulation, including thrombocytopenia and a prolonged thromboplastin time. The disease is accompanied by bacteremia with, typically, 10^5 to 10^6 spirochetes per μL of blood. Although these organisms are predominantly located free in the plasma space, spirochetes have also been observed in blood polymorphonuclear leukocytes. The nonendotoxin pyrogen contained in the spirochetes may act in a manner similar to endotoxin by stimulating leukocytes to produce leukocytic pyrogen, thromboplastin, and other mediators of inflammation which contribute to fever production and the coagulation disorders seen in relapsing fever.[8,9]

The most characteristic histopathologic feature of the disease is found in the spleen, where *Borrelia* organisms collect around miliary abscesses (Fig. 77-2). In other organs the changes are nonspecific, and a presumptive diagnosis is confirmed by demonstration of spirochetes, which are commonly seen in the lumens of blood vessels.[10]

Treatment

Penicillin and tetracycline are most often used to treat relapsing fever. These drugs can eliminate the *Borrelia* organisms but may induce a severe Jarisch-Herxheimer–like reaction in most patients. Tetracycline is the agent of choice in such cases, since it clears spirochetes from the blood rapidly and since the reactions are of shorter duration than those in patients treated with penicillin. The recommended dosage of tetracycline for adults is 0.5 g every 6 hours for 5 to 10 days.[1]

Ocular Manifestations

No reports include a systematic evaluation of the ocular problems associated with the disease. Conjunctivitis is sometimes present, but uveal inflammation is the commonest complication. Iritis has been reported in 15% of cases in some series.[6] This usually occurs after the second or third febrile attack. The course and severity of the iritis seems to be unrelated to the occurrence of further relapses. The intraocular inflammation may be further complicated by posterior synechiae, vitreous opacities, and exudate.[11] In one case, retinal venous occlusion with retinal hemorrhage and exudate were observed.[12] Ptosis may occur as a manifestation of central nervous system involvement. Many patients who have ocular complications suffer vision impairment.

References

1. Warrell DA, Perine PL, et al. Pathophysiology and immunology of the Jarisch-Herxheimer-like reaction in louse-borne relapsing fever: comparison of tetracycline and slow-release penicillin. *J Infect Dis.* 1983;147:908.
2. Burgdorfer W. Discovery of the Lyme disease spirochete and its relation to tick vectors. *Yale J Biol Med.* 1984;57:515.
3. Chung HL, Wei YL. Studies on the transmission of relapsing fever in North China. II Observations on the mechanism of transmission of relapsing fever in man. *Am J Trop Med Hyg.* 1938;18:661.
4. Barbour AG, Hayes SF. Biology of *Borrelia* species. *Microb Rev.* 1986;50:381.
5. Newman K Jr, Johnson RC. T-cell–independent elimination of *Borrelia turicatae. Infect Immunol.* 1984;45:572.
6. Southern PM Jr, Sanford JP. Relapsing fever: a clinical and microbiological review. *Medicine.* 1969;48:129.
7. Sciotto CG, Lauer BA, White WL, et al. Detection of *Borrelia* in acridine orange–stained blood smears by fluorescence microscopy. *Arch Pathol Lab Med.* 1982;107:384.
8. Butler T, Spagnuolo PJ, et al. Interaction of *Borrelia* spirochetes with human mononuclear leukocytes causes production of leukocytic pyrogen and thromboplastin. *J Lab Clin Med.* 1982;99:709.
9. Galloway RE, Levin J, et al. Activation of protein mediators of inflammation and evidence for endotoxemia in *Borrelia recurrentis* infection. *Am J Med.* 1977;63:933.
10. Gillum RI. Relapsing fever. In: Binford CH, Conor DH, eds. *Pathology of Tropical and Extraordinary Diseases.* Washington, DC: Armed Forces Institute of Pathology; 1976.
11. Hamilton JB. Ocular complication in relapsing fever. *Br J Ophthalmol.* 1943;27:68.
12. Bryceson ADM, Parry EHO, et al. Louse-borne relapsing fever. *Quart J Med.* 1970;39:129.

Chapter 78

Syphilis

JOSE S. PULIDO, JAMES J. CORBETT, and WILLIAM M. McLEISH

Since the introduction of penicillin, syphilis has gradually lost its place as the "great imitator" in medicine; however, a 32% increase in the incidence of primary and secondary syphilis in 1987 and a gradual increase in the incidence of congenital syphilis are ominous developments.[1] The incidence of syphilis has increased in both the homosexual and heterosexual populations, and acquired immunodeficiency syndrome (AIDS) has provided a couple of new twists to this otherwise well-described disease process. The increasing incidence of syphilis is of major concern, because syphilis increases the risk of developing a human immunodeficiency virus

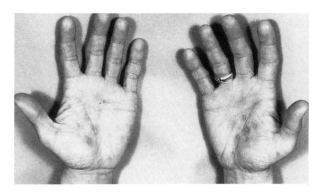

FIGURE 78-1 Classic maculopapular rash in an HIV–infected patient with syphilitic retinitis.

(HIV) infection. The genital ulceration of syphilis provides a portal of entry for HIV and also increases the number of T lymphocytes at the ulcer site.[2] This increased incidence of HIV infection is not seen with gonococcal or chlamydial infections.

Syphilis is caused by *Treponema pallidum*, a very fastidious but highly infectious spirochete.[3] There is both a humoral and a cellular response to *T. pallidum*. The humoral response consists of nonspecific antibodies that react to cardiolipins (Wassermann or Reagin) and of specific antibodies (immobilizing, agglutinating, and fluorescent treponemal) that react to specific components on treponemes. At most, these antibodies provide partial protection against syphilitic infection. Cell-mediated immunity is the dominant mechanism in controlling treponemal infections. This is significant because in HIV-infected patients, it is the cell-mediated arm of the immune response that has been amputated.

Clinical Features

Classically, syphilis has been divided into three clinical stages. Primary syphilis is usually characterized by a chancre on mucous membranes that usually appears within 3 weeks of inoculation. Females and patients who have had anal or oral intercourse may not be aware of the lesions, which may be painless and small. The chancre resolves within 3 weeks. Ocular findings in this stage are rare. Rarely a chancre may be seen on the lids or conjunctiva. The protean ocular and ophthalmic manifestations of syphilis occur in the secondary and tertiary stages. The secondary stage occurs anywhere from 2 months to 3 years after the inoculation. The classic findings are iridocyclitis, retinochoroiditis, neuroretinitis, and optic neuropathy. Many patients pass through this stage without experiencing noticeable symptoms.[4]

In an immunocompetent host, the humoral and cellular systems can control the spirochetes, and a subsequent latent stage ensues. Tertiary syphilis occurs when the immune regulation breaks down and treponemes are detectable in certain tissues. The classic findings at this stage are gummas ("benign" tertiary syphilis), syphilitic aortitis, and neurosyphilis in the forms of general paresis and tabes dorsalis. Ocular findings include iridocyclitis, retinochoroiditis, optic atrophy, and Argyll-Robertson (AR) pupils.[4]

Pathologically, the hallmark of syphilis is an obliterative endarteritis with perivascular plasma cell cuffing. In tertiary and late congenital syphilis, gummas form; histologically they consist of coagulated necrotic centers surrounded by macrophages and plasma cells.

Ocular Manifestations

Multiple and diverse manifestations are the hallmark of the disease both systemically and ophthalmologically (Table 78-1).[1–6]

Conjunctiva A chancre may present as an indolent ulcer with indurated margins. Characteristically it is unilateral, is associated with regional nodes and usually results in scarring. During secondary syphilis, mucous patches, the mucous membrane equivalent of the papillomacular rash, can be found (Fig. 78-1). The mucous patches appear as phlyctenules or gray-white nodules.

Sclera and Episclera Anterior, diffuse, nodular, or posterior scleritis may occur in the secondary, tertiary, or congenital stages.

Iris The iris may be involved during secondary syphilis, where the iris capillaries become engorged, and small dilated tufts may be noted (iris roseata). Larger, yellow-red papules (papulata) or even larger nodules have been seen in secondary and tertiary stages.

Retinal Vessels Retinal vessels may be affected independently or in association with syphilitic retinitis. Perivasculitis of both arteries and veins, vascular occlusions, and neovascularization have been seen.

Retina and Choroid In secondary lues, a gray-white patchy or confluent retinitis with overlying vitritis may be seen, which is bilateral in about half of all cases. The lesions may occur around the optic nerve and in the posterior pole with an accompanying papillitis (neuroretinitis), or they may be found only in the periphery (Fig. 78-2). Hemorrhages may occur, and syphilitic retinitis may be so necrotizing that it is in the differential diagnosis of acute retinal necrosis (ARN) syndrome

TABLE 78-1 Ophthalmic Manifestations of Syphilis

	Primary	Secondary	Tertiary	Congenital
Pathology	Chancre	Perivasculitis, meningitis, meningovasculitis, fibrosis and scarring, hydrocephalus	"Benign" gumma, fibrosis and scarring with basal pachymeningitis, hydrocephalus	As in secondary and tertiary; scarring
Lashes/lids	Chancre	Loss of lashes, papulomacular rash, ulcerative blepharitis	Gumma	As in secondary and tertiary; also rhagades-perianal fissures due to scarring
Conjunctiva	Chancre	Mucous patches, papillary conjunctivitis, phlyctenules, tarsitis	Tarsitis, gummas	As in secondary and tertiary
Lacrimal system		Dacryoadenitis	Dacryoadenitis, gummas of the lacrimal sac	As in secondary and tertiary; epiphora, punctal stenosis
Orbit			Gummas, orbital apex syndrome, periostitis—especially of sphenoid wing	As in tertiary
Cornea		Interstitial keratitis	Interstitial keratitis, keratitis, punctata profunda	Interstitial keratitis
Sclera/episclera		Episcleritis, scleritis	Episcleritis, scleritis, gummas	As in tertiary
Lens			Subluxation?	Cataract, subluxation?
Glaucoma		Secondary to iritis	During or after interstitial keratitis	As in secondary and tertiary
Anterior chamber		Flare, cell	Flare, cell	Flare, cell
Iris		Iritis	Gummas, iritis	As in secondary and tertiary
Vitreous		Vitreitis	Vitreitis	As in secondary and tertiary
Retinal vessels		Perivasculitis, neovascularization	Perivasculitis, neovascularization	As in secondary and tertiary
Retina/choroid		Retinitis, neuroretinitis, chorioretinitis, choroiditis, cystoid macular edema, posterior scleritis	As in secondary, also gummas of retina or choroid, pseudoretinitis pigmentosa, subretinal neovascular membrane, cystoid macular edema, posterior scleritis	As in secondary and tertiary; also salt-and-pepper retinopathy
Pupils			Argyll-Robertson pupil	Argyll-Robertson pupil
Optic nerve		Papillitis	Papillitis, gummas, optic atrophy	As in secondary and tertiary
Chiasm and retrochiasmal			Chiasmatic arachnoiditis and vasculopathy affecting white matter and cortex, bitemporal and homonymous field loss	Optic atrophy secondary to chiasmatic arachnoiditis or hydrocephalus
Ocular motility		Third, fourth and sixth nerve palsy, INO, gaze palsy, dorsal midbrain syndrome, other brain stem syndromes		

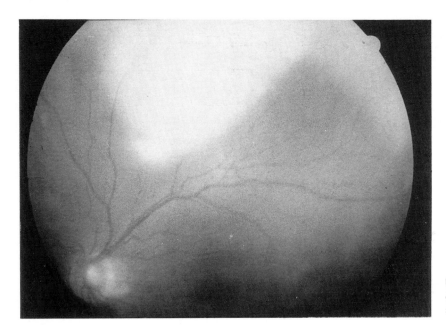

FIGURE 78-2 Fundus photograph of a patient showng a pie-shaped area of retinitis.

along with retinitis caused by cytomegalovirus (CMV), other herpesvirus diseases, toxoplasmosis, and Behçet's syndrome. When the retinitis resolves, these inflamed areas show a mottled retinal pigment epithelium.

Retinal pigment epithelium mottling, bone spiculing, and optic atrophy may be seen. This later pigmentary change is included in the differential diagnosis of pseudoretinitis pigmentosa and occurs in both the congenital and tertiary forms. Cystoid macular edema with vitritis may be seen in secondary or tertiary syphilis.

A chorioretinitis may occur in secondary and tertiary disease, and the pattern may be focal or disseminated. In these cases, choroidal involvement with outer retinal whitening is noted. In the posterior pole, it may be confused with central serous retinopathy or outer layer toxoplasmosis. The choroid may also be involved without overlying retinal pathology, although this is rare. When chorioretinal syphilis resolves, areas of chorioretinal atrophy with retinal pigment epithelium hypoplasia and hyperplasia may be seen. An exudative retinal detachment may accompany the chorioretinitis or may occur in cases of posterior scleritis. Gummas of the choroid, retina, or optic nerve may also be seen.

Neurosyphilis
Neuro-ophthalmic complications of syphilis occur as a result of (1) inflammatory reaction in the subarachnoid space during the meningitic phase, (2) vasculopathy in the meningovascular phase, (3) "benign" gumma formation, (4) fibrosis, or (5) a combination of all three. Clearly, inflammation, vascular occlusion, and fibrosis will occur at different stages of the disease, but the secondary effects of these pathologic processes may be identical.

Cranial Mononeuropathies. Cranial mononeuropathies may develop if nerves are entrapped in proliferating connective tissue in tertiary syphilis, or they may be caused by acute vascular events with occlusion of vasa nervorum in meningovascular syphilis. Oculomotor, abducens, and even trochlear nerve palsies have been reported, singly and in combinations. A combination of ocular motility disturbance and pain in the orbit, reminiscent of the Tolosa-Hunt syndrome, has been reported with periostitis of the sphenoid wing. Diplopia can also be caused by increased intracranial pressure with hydrocephalus or by compression from a gumma or syphilitic aneurysm.

Optic Nerve Involvement. The effect of syphilis on the optic disc is variable, but presumably the typical hyperemic disc swelling seen with syphilitic papillitis or papilledema is caused by altered axoplasmic flow due to inflammation, edema, and ischemia. The disc swelling has been given various titles, such as syphilitic optic neuritis, syphilitic papillitis, syphilitic papilledema, and perioptic neuritis or optic perineuritis. Presumably most of these conditions are related to syphilitic basal meningitis or are caused by increased intracranial pressure produced by either hydrocephalus or, rarely, by a gumma (Fig. 78-3).

SYPHILITIC OPTIC NEURITIS. Syphilitic optic neuritis

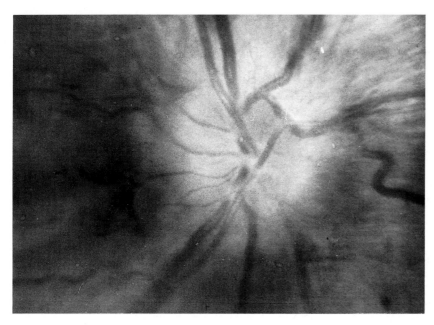

FIGURE 78-3 Optic disc swelling in a patient with syphilitic optic neuritis whose VDRL titer was 1:256.

characteristically consists of hyperemic optic discs, vitreous cells, and decreased visual acuity. Resolution of syphilitic optic neuritis may leave normal vision and normal-looking optic nerve heads, or vision may be defective and the optic discs pale. It is likely that the extent of the vascular pathology has much to do with the outcome.

SYPHILITIC PERIOPTIC NEURITIS OR OPTIC PERINEURITIS. Syphilitic perioptic neuritis or optic perineuritis is more difficult to define. It is said to be a unilateral or bilateral "inflammatory optic disc swelling without visual loss"[7]; however, cases have been reported in which there was decreased visual acuity and both anterior chamber and vitreous reaction. It is likely that this entity represents a form of ischemic optic neuropathy without loss of vision or visual field. Disc swelling is due to focal inflammatory axoplasmic stasis.

SYPHILITIC PAPILLEDEMA. Syphilitic papilledema has been reported in patients who have disc swelling with transient visual obscuration but normal visual acuity and no cellular response in the vitreous. Unfortunately, this category of luetic optic neuropathy has been muddied by reports of patients with normal cerebrospinal fluid (CSF) pressure, and it is not clear how it differs from the perineuritis.

Strictly speaking, syphilitic papilledema should be disc swelling caused by elevated intracranial pressure, as with hydrocephalus in secondary and tertiary lues and in patients with intracranial gummas.

The optic disc and the entire length of the optic nerve can be affected by syphilitic meningitis and meningovascular inflammation. Depending on the primary location of these processes and on the exact mix of inflammatory and vasculitic components, patients may have unilateral or bilateral optic disc edema, with or without vitreous or anterior chamber cells and with or without loss of visual acuity or visual field.

Swollen optic discs, whether caused by inflammation, ischemia, or increased intracranial pressure, are the precursors to vision loss and what is known as syphilitic optic atrophy. In addition to these more anterior forms of damage to the optic nerve, ischemic damage within the orbit and at the chiasm as a result of chiasmatic arachnoiditis also has the potential to cause vision loss and optic atrophy.

Pupils. The hallmark pupillary abnormality seen in syphilis is the AR pupil. First described in 1869, the AR pupil is characterized as small, irregular, poorly reactive or nonreactive to light but briskly reactive to an accommodative near effort. This form of "light-near dissociation" has been reported in patients with lesions to the brachium of the superior colliculus, such as those seen with Parinaud's (dorsal midbrain) syndrome; however, the pupils in this condition are not small. The smallness of the AR pupil presumably has to do with a disinhibited sphincter, and the causal lesion is presumed to be in the periaqueductal gray matter involving the pretectal oculomotor fibers, though there is little histopathologic confirmation. AR pupils occur in late secondary syphilis and are a characteristic but not invariable component of the two serious forms of tertiary neurosyphilis, tabes dorsalis and general paresis, where they occur in about 70% of patients.

Congenital Syphilis

Congenital syphilis, though rare, certainly has not disappeared, and with the resurgence of syphilis in heterosexuals, its prevalence is increasing. The women at greatest risk of syphilis in pregnancy are those who receive poor prenatal care. Most infections occur in utero and can occur in any trimester. During the first year of maternal infection, there is an 80 to 90% chance for the fetus to be infected. The fetal infection rate decreases after the first year of the maternal infection, and by the fourth year of maternal infection, fetal infection is rare. It is rare for the fetus to be infected during the primary stage of the disease in the mother, but a newborn may be infected while traversing the birth canal.

Fetuses with congenital syphilis are more likely to be stillborn or premature. Systemic signs of early congenital syphilis develop 2 weeks or more after delivery. Failure to thrive, hepatosplenomegaly, and anemia are seen not only in congenital syphilis but also in congenital toxoplasmosis, rubella, CMV, and herpes (TORCH). A papillomacular rash, mucous patches, and rhagades may be seen. Periostitis is common, and "snuffles" are seen in 50% of the cases. Frontal bossing, saddle nose, and sabre shins are also caused by chronic periostitis. Periostitis of the gums causes mulberry molars and barrel-shaped central incisors of the permanent dentition (Hutchinson's teeth).

Ocular findings at this time include iritis, neuroretinitis, chorioretinitis, and necrotizing retinitis. Late congenital syphilis occurs if no treatment has been given by the age of 2 years. Neurosyphilis, either symptomatic or asymptomatic, is common, particularly with sensorineural hearing loss. Syphilis should be considered in the differential diagnosis for a young person with hearing loss, with or without ocular findings. Hutchinson's triad of Hutchinson's teeth, deafness and interstitial keratitis is a classic finding in late congenital syphilis. Interstitial keratitis in congenital syphilis is bilateral, and the second eye is involved within weeks to months of the first, whereas interstitial keratitis in acquired syphilis is unilateral. Congenital interstitial keratitis usually appears between the ages of 5 and 18 years. During the initial stage of interstitial keratitis, intense lacrimation, photophobia and a "ciliary flush" from pericorneal inflammation occur. Subsequently, vascularization of the cornea occurs, with ensuing clouding of the cornea accompanied by iridocyclitis. During the second stage of the interstitial keratitis, vascularization of the cornea is at its maximum and the cornea appears pink (salmon patches of Hutchinson). In the third stage of interstitial keratitis, inflammation regresses. Scarring of the cornea occurs, and as blood flow through the corneal vessels diminishes, "ghost vessels" are formed. The scarring can depress visual acuity, and with time, corneal edema may occur secondary to corneal endothelial compromise. The pathogenesis appears to be an immune complex phenomenon.

A classic retinal finding seen in congenital syphilis is salt-and-pepper retinopathy, which is similar to rubella retinopathy. Optic atrophy may also be seen, along with retinal vessel narrowing. Areas of active gray-white chorioretinitis or inactive areas of chorioretinitis with hypopigmentation and hyperpigmentation of the retinal pigment epithelium along with patches of choroidal atrophy may occur.[4,5]

Diagnosis of Classic Syphilis

The diagnosis depends on both clinical signs and symptoms and on laboratory tests. Darkfield and direct immunofluorescence of scrapings of the skin lesions are helpful in primary and secondary syphilis. The nonspecific tests, including Venereal Disease Research Laboratory (VDRL), rapid plasma reagin (RPR), and radiosensitivity test (RST), are helpful in following response to therapy. These tests should not be used alone to confirm the disease, since there may be false-negative results in primary, latent, and tertiary disease, as well as in late congenital stages. Specific tests (fluorescent treponemal antibody absorption [FTA-Abs] and microhemagglutination-*Treponema* pallidum [MHA-TP]) are usually positive in the presence of disease, except in early primary syphilis. CSF examination is very useful in suspected secondary and tertiary cases. Every patient suspected of having secondary or tertiary ocular manifestations of syphilis should have a lumbar puncture. HIV testing, as well as examination for other sexually transmitted diseases, must be performed in all patients with syphilis.

Concurrent HIV Infections

There appears to be a symbiosis between HIV and syphilis. Just as syphilis may predispose to HIV infections, HIV may change the natural course of syphilis infections because of immunosuppression. It possibly accelerates and may also mask the syphilis infection. Very long courses of penicillin treatment are needed in these cases, and in patients who are very T lymphocyte depleted, seronegativity to syphilis may occur.

Treatment

Treatment recommendations in syphilis are evolving and continue to be very controversial. The latest recommendations of the Centers for Disease Control should be checked prior to initiating treatment.

References

1. Centers for Disease Control. Continuing increase in infectious syphilis—United States. *Arch Dermatol.* 1988;124:509-510.

2. Stamm WE, Handsfield HH, Rompalo AM, et al. The association between genital ulcer disease and acquisition of HIV infection in homosexual men. *JAMA.* 1988;260:1429-1433.

3. Freeman BA. *Burrows Textbook of Microbiology.* 22nd ed. Philadelphia, Pa: WB Saunders Co; 1985:637-644.

4. Wilhelmus KR. Syphilis. In: Insler MS, ed. *AIDS and Other Sexually Transmitted Diseases and the Eye.* Orlando, Fla: Grune & Stratton; 1987:73-104.

5. Duke-Elder S. Diseases of the outer eye, Parts I and II. In: Duke-Elder S, ed. *System of Ophthalmology.* St. Louis, Mo: CV Mosby Co; 1965:236-241, 815-831.

6. Miller NR. *Walsh and Hoyt's Clinical Neuro-Ophthalmology,* 4th ed. vol. 1. Baltimore, Md: Williams & Wilkins; 1982:239-243.

Section I

Viral Diseases

Chapter 79

Acquired Immunodeficiency Syndrome

GARY N. HOLLAND

Infection with human immunodeficiency virus (HIV), a recently discovered retrovirus, leads to an array of clinical disorders, the most serious of which is the acquired immunodeficiency syndrome (AIDS).[1] In patients with AIDS, HIV produces severe immunosuppression by infecting and destroying cells of the body's immune system. AIDS is identified by the appearance of secondary life-threatening opportunistic infections and neoplasms. The percentage of asymptomatic HIV–infected persons who will eventually develop AIDS is unknown. The incidence of AIDS and associated disorders has risen dramatically since they were first described in 1981. The mortality rate among patients who have had AIDS for 3 years approaches 100%. It is believed that AIDS is a new disorder. The first cases probably developed no earlier than 1978.

HIV is transmitted through intimate sexual contact, through receipt of contaminated blood or blood products, and transplacentally from mother to unborn child.

Groups at greatest risk for HIV infection are sexually active homosexual and bisexual males, intravenous drug abusers, hemophiliacs and other recipients of whole blood transfusions, heterosexual partners of these persons, and children of mothers with HIV infection.

A variety of ophthalmic disorders are associated with AIDS.[2-7] The majority of patients with AIDS develop one or more ocular problems during the course of their illness. These ophthalmic disorders may be early signs of disseminated or multifocal disease processes. Ophthalmic diseases fall into four major categories: vasculopathies of uncertain cause, opportunistic infections, neoplasms, and neuro-ophthalmic abnormalities.

Most, if not all, patients with AIDS have a retinal microvasculopathy characterized by swollen endothelial cells, damage to and loss of pericytes, thickened basal laminae, and narrowed capillary lumens.[6,7] These ultrastructural findings are similar to those seen in diabetic retinopathy. The microvasculopathy results in retinal

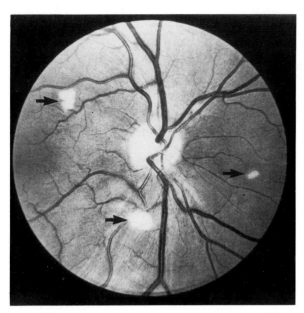

FIGURE 79-1 Cotton-wool spots, which result from retinal ischemia, appear as fluffy white spots adjacent to the major vascular arcades (*arrows*).

ischemia and stasis of axoplasmic flow in the nerve fiber layer. Resultant focal swelling of the nerve fibers with aggregation of organelles ("cytoid bodies") is seen ophthalmoscopically as cotton-wool spots, the single most common ocular manifestation of AIDS (Fig. 79-1). These fluffy, white retinal opacities are found in approximately two thirds of patients with AIDS and are occasionally seen in HIV–infected persons without full-blown AIDS. Cotton-wool spots develop and regress over a 4- to 6-week period without visible sequelae in most cases.

Other, less frequent, signs of retinal microvasculopathy include retinal hemorrhages, microaneurysms, and microvascular anomalies on fluorescein angiography.[3,6,7] The diagnostic and prognostic significance of these vascular lesions has not been established. Vascular damage by deposition of circulating immune complexes has been hypothesized as a cause.[6] Cotton-wool spots appear to be more frequent in patients with multiple opportunistic infections, but no definite association has been established with any specific infectious disease. Microvascular anomalies have also been seen in other tissues, including conjunctiva, brain, and skin, which suggests that the microvasculopathy of AIDS is a widespread process.

A number of severe, blinding ocular infections have been associated with AIDS, the most common being cytomegalovirus (CMV) retinopathy, which occurs in approximately 30% of patients.[3,6] CMV infection of the

retina leads to full-thickness retinal necrosis. The retinopathy is characterized by multiple patches of granular, yellow-white retinal opacification with indistinct borders, usually arising adjacent to retinal vessels (Fig. 79-2). It may be associated with retinal hemorrhage and vascular sheathing. The overlying vitreous remains clear. Once established, the infection progresses relentlessly to destroy the entire retina over a period of several months.[4] Other ocular tissues are rarely infected with CMV.

Ganciclovir, a new antiviral drug, appears to be effective in controlling the spread of CMV infection.[8] It does not eradicate virus from the eye, however, and prolonged low-dose maintenance therapy is necessary for continued control of the infection. Even with continued therapy, however, eventual reactivation and slow progression of disease occurs in at least 50% of patients.

Other opportunistic pathogens infect the eye less frequently than they infect nonocular sites.[6] Herpes simplex virus can cause a necrotizing retinopathy, but it is rare. Toxoplasmosis is the most common nonviral intracranial opportunistic infection associated with AIDS. In contrast, ocular toxoplasmosis with focal retinochoroidal inflammatory lesions occurs only occasionally. Although mucocutaneous candidiasis is common in patients with AIDS, only patients who continue to abuse intravenous drugs or have indwelling catheters for long periods appear to be at risk for candidemia and endogenous candidal chorioretinitis. This intraocular infection has been reported only rarely in patients with AIDS.[5] Several other pathogens have been identified in ocular tissue at autopsy, including *Cryptococcus neoformans*, *Histoplasma capsulatum*, *Mycobacterium avium-intracellulare*, and *Pneumocystis carinii*. Because these pathogens infect the eye late in the course of the AIDS illness, little is known about the clinical diseases they cause. When intraocular infections do occur, they appear to be reliable indicators of tissue-invasive infections elsewhere in the body.[6]

Patients with HIV infection, whether or not they have AIDS, appear to be at increased risk for the development of herpes zoster ophthalmicus. The vesicular lesions of the skin, as well as the associated keratitis and iridocyclitis, can be severe and prolonged.

Kaposi's sarcoma, a multifocal vascular neoplasm, may affect the conjunctiva, eyelid margins, and rarely the orbit. Eighteen percent of patients with Kaposi's sarcoma have ocular or periocular lesions.[3] The conjunctiva and eyelids may occasionally be early sites of tumor development. Conjunctival lesions, usually seen in the inferior cul de sac, are slowly spreading, bright red subconjunctival masses that resemble subconjunctival hemorrhages (Fig. 79-3). They rarely interfere with vision or eyelid function. Lesions on the eyelid margin, however, may result in entropion formation with trichi-

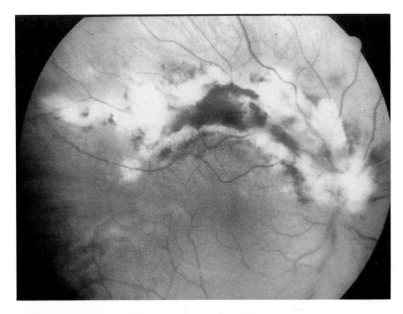

FIGURE 79-2 The characteristic signs of cytomegalovirus retinopathy, including granular-appearing patches of retinal whitening and hemorrhage, are seen adjacent to retinal vessels. The optic nerve head is swollen.

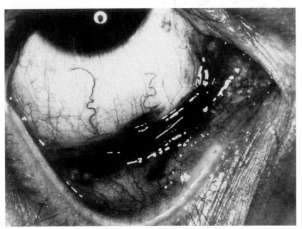

FIGURE 79-3 A slowly expanding, bright red subconjunctival lesion of Kaposi's sarcoma is seen in the inferior cul de sac.

asis or may lead to blepharoconjunctivitis when lesions ulcerate and become superinfected. Eyelid margin tumors may be treated with local irradiation. Because it is a multifocal neoplasm there is rarely a medical indication for local treatment of conjunctival tumors. Burkitt's lymphoma is another AIDS–related neoplasm that can develop in the orbit, resulting in proptosis and damage to cranial nerves and extraocular muscles.

Neuro-ophthalmic abnormalities, including cranial nerve palsies, papilledema, optic atrophy, and visual field defects, may develop in patients with AIDS. They are signs of intracranial neoplasms and infectious diseases.

HIV has been isolated from corneal tissue and from tears. There is no evidence that HIV infection of ocular tissues produces clinical disease directly. The presence of the virus, however, has led to concern that viral transmission could occur by contact with tears or by corneal transplantation. The Centers for Disease Control, therefore, have established precautions against contamination of ophthalmic instruments and contact lens fitting sets by virus-infected tears.[9] These recommendations supplement guidelines established for all health care workers to prevent accidental needle-stick injuries and direct exposure to contaminated specimens.[10] The Eye Bank Association of America now requires members to perform enzyme-linked immunosorbent assay (ELISA) tests on all cornea donors to screen for HIV antibodies. While these precautions are warranted in view of the incomplete understanding of HIV transmission, there is no evidence to date that it has occurred by corneal transplantation, by use of contaminated ophthalmic instruments, or by normal patient contact.

References

1. Spira TJ. The acquired immunodeficiency syndrome. In: Insler MS, ed. *AIDS and Other Sexually Transmitted Diseases and the Eye.* Orlando, Fla: Grune & Stratton; 1986;119-144.
2. Holland GN. Ophthalmic disorders associated with the acquired immunodeficiency syndrome. In: Insler MS, ed. *AIDS and Other Sexually Transmitted Diseases and the Eye.* Orlando, Fla: Grune & Stratton; 1986;145-172.

3. Holland GN, Pepose JS, Pettit TH, et al. Acquired immune deficiency syndrome: Ocular manifestations. *Ophthalmology.* 1983;90:859-872.

4. Palestine AG, Rodrigues MM, Macher AM, et al. Ophthalmic involvement in acquired immune deficiency syndrome. *Ophthalmology.* 1984;91:1092-1099.

5. Friedman AH. The retinal lesions of the acquired immune deficiency syndrome. *Trans Am Ophthalmol Soc.* 1984; 82:447-491.

6. Pepose JS, Holland GN, Nestor MS, et al. Acquired immunodeficiency syndrome: Pathogenic mechanisms of ocular disease. *Ophthalmology.* 1985;92:472-484.

7. Newsome DA, Green WR, Miller ED, et al. Microvascular aspects of acquired immune deficiency syndrome retinopathy. *Am J Ophthalmol.* 1986;98:590-601.

8. Holland GN, Sidikaro Y, Kreiger AE, et al. Treatment of cytomegalovirus retinopathy with ganciclovir. *Ophthalmology* 1987;94:815-823.

9. Centers for Disease Control. Recommendations for preventing possible transmission of human T-lymphotrophic virus type III/lymphadenopathy-associated virus from tears. *MMWR.* 1985;34:533-534.

10. Centers for Disease Control. Recommendations for preventing transmission of infection with human T-lymphotrophic virus type III/lymphadenopathy-associated virus in the workplace. *MMWR.* 1985;34:681-695.

Chapter 80

Cytomegalovirus Infection

HERBERT L. CANTRILL

Cytomegalovirus (CMV) is a common cause of asymptomatic infection in humans. Up to 80% of adults have serologic evidence of previous exposure to CMV. Primary infection presumably occurs as a result of respiratory droplet spread. In adults asymptomatic infection is the rule, although a heterophil antibody–negative mononucleosis syndrome may result. Following infection, a prolonged period of viral shedding occurs in urine and other body secretions, lasting 2 to 3 years. Rates of infection are higher in lower socioeconomic groups and in persons whose living conditions are crowded. Young children are a frequent reservoir of infection and show a high rate of viral shedding.[1] CMV may also be acquired by transplacental spread, blood transfusion or organ transplantation. CMV is probably also transmitted by sexual contact.[2]

Congenital infection is a significant risk for seronegative females in the child-bearing years.[3] One to two percent of newborn infants in the United States have evidence of infection at birth. About 10% of infected infants show signs of cytomegalic disease, including microcephaly, intracranial calcification, chorioretinitis, hepatosplenomegaly, and thrombocytopenia. CMV is also the cause of mild mental retardation, seizure disorders, sensorineural hearing loss, and psychomotor retardation in older children. Infection may also be acquired at the time of birth, usually in the form of pneumonitis.

Immunosuppressed persons are prone to develop serious illness. Systemic manifestations include fever, myalgias, anemia, pneumonitis, hepatitis, gastritis, enterocolitis, and chorioretinitis.[4] In the transplant population, infection of the transplanted organ is a frequent cause of graft rejection.[5] Disseminated disease is a common cause of death in immunosuppressed persons. Patients with AIDS show a high rate of CMV infection and frequently develop symptomatic disease.

Cytomegalovirus Retinitis

CMV retinitis was first described by Foerster in the newborn in 1959.[6] CMV retinitis was later described in adults, occurring as a terminal event in patients debilitated by malignancy or lymphoma.[7] By 1977, a significant number of cases had been reported in the transplant recipient population.[8] CMV retinitis is estimated to occur in 2 to 5% of heart and kidney transplant recipients. For reasons that are unclear, the bone marrow transplant population is less prone to develop this complication. Both groups show a high rate of viremia and viruria. It appears that local factors may be impor-

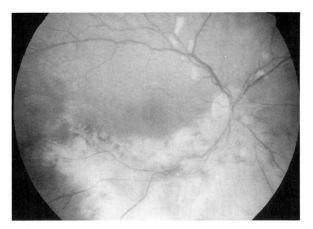

FIGURE 80-1 Fundus photograph shows a typical zone of necrotizing retinitis with associated hemorrhages and retinal vasculitis.

tant in the development of CMV retinitis in these groups. Beginning in 1980, large numbers of cases have been reported in the AIDS population. At the present time, CMV retinitis is the most common ocular opportunistic pathogen in patients with AIDS.[9]

The reason for the high incidence of infection in the AIDS population is unclear. Clinical and autopsy studies indicate a prevalence of between 17 and 40%. Some geographic differences exist, since the incidence of CMV retinitis is lower in the African population with AIDS.[10] It was originally felt that the irreversible immunosuppression of the AIDS population accounted for the high incidence of CMV retinitis. One possible explanation is interaction with other viruses, particularly human immunodeficiency virus (HIV). Experimental studies demonstrate transactivation of HIV and CMV in vitro.[11] In the retina, HIV appears to infect the retinal vascular endothelium and is probably the cause of AIDS retinopathy.[12] CMV, on the other hand, is a neurotropic virus and is less likely to infect retinal vascular endothelium.[13] In the presence of both viruses, HIV may cause endothelial damage, affording CMV access to retinal tissue.

Pathophysiology

Infection may result from an exogenous source or as a result of activation of latent virus. Mononuclear leukocytes are probably the reservoir of latent infection.[14] Systemic disease may result from either mechanism, although the most serious consequences result from exogenous infection. Regardless of the mechanism, CMV disease results from blood-borne dissemination of the virus. Tissues most frequently infected include lung, liver, spleen, intestine, adrenal gland, kidney, and the central nervous system (including the eye).

CMV has been isolated from retinal tissue in both active and inactive lesions. The virus can be detected by electron microscopy, immunohistochemistry, and by in vitro hybridization techniques. Necrosis of the entire retina is seen in zones of involvement, with preservation of normal retinal architecture in adjacent uninvolved retina. Mild mononuclear infiltration may be seen in the choroid and in the overlying vitreous. The histopathologic features include the presence of large intracytoplasmic and intranuclear inclusions.[15] The characteristic finding is the owl's eye cell, which has a large basophilic nuclear inclusion adjacent to a clear rim of nucleoplasm.

Clinical Findings

CMV retinitis may present with asymptomatic involvement of the retinal periphery. More frequently, patients complain of floaters, blurred vision, or loss of visual field. Signs of inflammation are conspicuously absent, in contrast to other forms of viral retinitis. CMV retinitis presents as a necrotizing retinitis with associated retinal hemorrhages and vasculitis (Fig. 80-1).[4] The typical lesion is a zone of white infiltration with translucent slightly irregular margins. Multicentric origin and bilateral disease are frequently seen. As the lesions mature they become more granular and ultimately transparent. Fine pigment stippling is seen at the level of the pigment epithelium. Focal deposits of lipid, calcium, and glial tissue may be seen within inactive lesions. The retinal vessels become markedly attenuated, and secondary optic atrophy develops. A zone of active necrotizing retinitis remains at the margin of the lesions.

The early lesions may be hard to differentiate from other causes of retinal vascular disease. Anemic retinopathy with scattered hemorrhages, cotton-wool spots, and white-centered hemorrhages may be confused with CMV retinitis. The lesions of AIDS retinopathy are particularly hard to distinguish from those of early CMV. CMV has also been confused with other retinal vascular diseases, including branch vein occlusion and diabetic retinopathy.

The differential diagnosis of CMV retinitis includes other causes of necrotizing retinitis such as toxoplasmosis and cryptococcal, disseminated candidal, and endogenous bacterial infections. In healthy patients, CMV retinitis may be confused with the acute retinal necrosis syndrome. Toxoplasmosis is usually associated with significant inflammation in the anterior segment and vitreous, but it may present in an atypical fashion in the immunocompromised host. Cryptococcal disease usually affects the central nervous system, with secondary

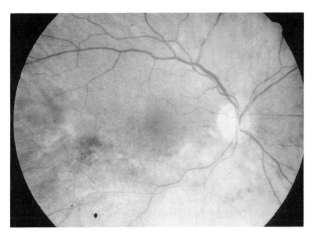

FIGURE 80-2 Same eye seen in Fig. 80-1 demonstrates results of treatment with ganciclovir. The area of previous involvement shows atrophy of the optic nerve, narrowing of retinal vessels, and a granular appearance to the retina and pigment epithelium.

involvement of the eye. Candida infection begins as a multifocal choroiditis and extends into the retina and vitreous. Endogenous bacterial infection usually results in a vitreous abscess. The acute retinal necrosis syndrome most closely resembles CMV retinitis but occurs in nonimmunosuppressed patients. In fact, CMV is one cause of the acute retinal necrosis syndrome. Other viral infections of the retina occur less frequently but may mimic CMV retinitis. These include infection with either herpes simplex or herpes zoster virus. Culture of the virus from other sites may help in the differential diagnosis.

The diagnosis of CMV retinitis is based primarily on ophthalmoscopic findings in the proper clinical setting. Laboratory tests are of limited benefit. Serologic tests for CMV antibodies are of no benefit and may even be misleading. Positive blood and urine cultures do not necessarily indicate the presence of tissue infection but do reveal the presence of systemic infection. Most patients with positive culture results do not have evidence of CMV retinitis.

Clinical Course

Without treatment, CMV retinitis progresses relentlessly to loss of vision from involvement of the optic nerve or macula or to retinal detachment. In the immunosuppressed population, withdrawal of immunosuppression may lead to resolution, but in many patients systemic CMV infection has an immunosuppressive effect itself,

preventing the return of normal immune function.[16] In the AIDS population, immunosuppression is irreversible. The rate of progression is variable and may be influenced by the type of immunosuppression. CMV retinitis typically progresses over a period of weeks with the expansion of old lesions and the development of new ones. Rarely, it progresses rapidly. Clinical observation may be unreliable in documenting disease progression. Other tests of vision function such as acuity and fields may be affected by factors other than CMV retinitis. Wide-field fundus photography is probably the most accurate means of assessing progression and regression.

There are several mechanisms of vision loss in CMV retinitis. An absolute scotoma corresponds to areas of inactive retinitis. Ischemia and edema adjacent to active zones may lead to vision loss, particularly when they are present in the macula. CMV may affect the optic nerve, causing loss of vision either by direct infection or by secondary ischemic damage to the nerve. Finally, CMV retinitis may result in retinal detachment in 15 to 30% of cases. Retinal holes typically occur near the posterior border of inactive zones of necrotizing retinitis. A significant number of patients develop proliferative vitreoretinopathy. Repair of these detachments frequently requires vitrectomy techniques, though the surgery is often unsuccessful in reattaching the retina.

Management

Before 1984 treatment of CMV retinitis was usually ineffective. Discontinuation of immunosuppressive therapy was an option in the transplant recipient population. Specific antiviral therapy has been investigated with such agents as vidarabine, idoxuridine, trifluoridine, alpha interferon, and acyclovir. These agents were either ineffective or had significant systemic toxicity. Beginning in 1983, ganciclovir a derivative of acyclovir, became available. Ganciclovir has 10 to 100 times as much activity against CMV as acyclovir. CMV lacks the viral thymidine kinase required to convert acyclovir to acyclovir triphosphate, the active form of the drug. Ganciclovir is selectively phosphorylated in CMV–infected cells by cellular enzymes. Ganciclovir triphosphate competitively inhibits viral DNA polymerase, preventing viral replication.

There is overwhelming clinical evidence that ganciclovir is effective in the management of CMV retinitis and other significant systemic CMV infections (Figs. 80-1, 80-2).[17-20] Unfortunately, the drug is virostatic and must be taken indefinitely. Maintenance treatment is complicated by significant myelosuppression, requiring discontinuation of the drug in 30 to 50% of patients. Intravitreal treatment with ganciclovir has been reported and appears to be an effective alternative in patients who cannot be treated systemically.[21,22]

Foscarnet is another antiviral agent that is effective against CMV via a slightly different form of toxicity.[23] It also inhibits viral DNA polymerase, but by a different mechanism than ganciclovir. Nephrotoxicity is the most significant adverse effect of foscarnet. Because of the different toxic effects of the two drugs, they may be used sequentially to control CMV retinitis.

Finally, zidovudine has been shown, at least anecdotally, to have some effect on CMV retinitis in the AIDS population. Zidovudine reduces the amount of HIV present, and may effect transactivation of CMV locally in tissue.

Conclusion

With the epidemic growth of the worldwide AIDS population, we can expect a corresponding increase in the number of cases of CMV retinitis. Since there is no effective means of reversing the immunosuppression of AIDS, there is a need for more effective antiviral therapy. As effective antiviral therapy becomes available and our clinical skill in the diagnosis and management of its complications improves, the ocular morbidity of this devastating opportunistic infection may be reduced.

References

1. Pass RF, Hutto C, Ricks R, et al. Increased rate of cytomegalovirus infection among parents of children attending day-care centers. *N Engl J Med*. 1986;314:1414-1418.
2. Handsfield HH, Chandler SH, Caine VA, et al. Cytomegalovirus infection in sex partners: evidence for sexual transmission. *J Infect Dis*. 1985;151:344-348.
3. Stagno S, Pass RF, Dworsky ME, et al. Congenital cytomegalovirus infection: the relative importance of primary and recurrent maternal infection. *N Engl J Med*. 1982; 306:945-949.
4. Egbert PR, Pollard RB, Gallagher JG, et al. Cytomegalovirus retinitis in immunosuppressed hosts. II. Ocular manifestations. *Ann Intern Med*. 1980;93:664-670.
5. Peterson PK, Balfour HH Jr, Marker SC, et al. Cytomegalovirus disease in renal allograft recipients: a prospective study of the clinical features, risk factors and impact on renal transplantation. *Medicine (Baltimore)*. 1980;59:283-300.
6. Foerster HW. Pathology of granulomatous uveitis. *Surv Ophthalmol*. 1959;4:296.
7. Smith ME. Retinal involvement in adult cytomegalic inclusion disease. *Arch Ophthalmol*. 1964;72:44-49.
8. Murray HW, Knox DL, Green WR, et al. Cytomegalovirus retinitis in adults: a manifestation of disseminated viral infection. *Am J Med*. 1977;63:574-584.
9. Friedman AH. The retinal lesions of the acquired immune deficiency syndrome. *Trans Am Ophthal Soc*. 1984;82:447-491.
10. Kestelyn P, Van De Perre P, Rouvroy D, et al. A prospective study of the ophthalmologic finds in the acquired immune deficiency syndrome in Africa. *Am J Ophthalmol*. 1985;100:230-238.
11. Skolnik PR, Kosloff BR, Hirsch MS. Bidirectional interactions between human immunodeficiency virus Type I and cytomegalovirus. *J Infect Dis*. 1988;157:508-513.
12. Pomerantz RJ, Kuritzkes DR, De La Monte SM, et al. Infection of the retina by human immunodeficiency virus type I. *N Engl J Med*. 1987;317:1643-1647.
13. Friedman HM, Macarak EJ, MacGregor RR, et al. Virus infection of endothelial cells. *J Infect Dis*. 1981;143(2):266-273.
14. Jordan MC, Jordan GW, Stevens JG, et al. Latent herpesviruses of humans. *Ann Intern Med*. 1984;100:866-880.
15. De Venecia G, Zu Rhein GM, Pratt MV, et al. Cytomegalic inclusion retinitis: a clinical, histopathologic and ultrastructural study. *Arch Ophthalmol*. 1971;86:44-57.
16. Rook AH. Interactions of cytomegalovirus with the human immune system. *Rev Infect Dis*. 1988;10(3):460-467.
17. Palestine AG, Stevens G, Lane HC, et al. Treatment of cytomegalovirus retinitis with dihydroxy propoxymethyl guanine. *Am J Ophthalmol*. 1986;101:95-101.
18. Henderly DE, Freeman WR, Causey DM, et al. Cytomegalovirus retinitis and response to therapy with ganciclovir. *Ophthalmology*. 1987;94:425-434.
19. Holland GN, Sidikaro Y, Kreiger AE, et al. Treatment of cytomegalovirus retinopathy with ganciclovir. *Ophthalmology*. 1987;94:815-823.
20. Collaborative DHPG Treatment Study Group. Treatment of serious cytomegalovirus infections with 9-(1,3- dihydroxy-2-propoxymethyl) guanine in patients with AIDS and other immunodeficiencies. *N Engl J Med*. 1986;314(13):801-805.
21. Henry K, Cantrill H, Fletcher C, et al. Use of intravitreal ganciclovir (dihydroxy propoxymethyl guanine) for cytomegalovirus retinitis in a patient with AIDS. *Am J Ophthalmol*. 1987;103:17-23.
22. Ussery FM, Gibson SR, Conklin RH, et al. Intravitreal ganciclovir in the treatment of AIDS-associated cytomegalovirus retinitis. *Ophthalmology*. 1988;95:640-648.
23. Walmsley SL, Chew E, Read SE, et al. Treatment of cytomegalovirus retinitis with trisodium phosphonoformate (foscarnet). *J Infect Dis*. 1988;157:569-572.

Chapter 81

Epstein-Barr Viral Infection

ALICE MATOBA

Systemic Manifestations

Epstein-Barr virus (EBV) is a DNA virus of the *Herpesvirus* genus. It is the most common cause of infectious mononucleosis (IM) syndrome, which is characterized clinically by fever, sore throat, and generalized lymphadenopathy, frequently with enlargement of the spleen and liver. Primary infection, whether asymptomatic or associated with IM, leads to a virus carrier state that persists for life.[1] Productive infection of some permissive cell in the oropharynx and nonproductive infection of circulating lymphocytes are both detectable in healthy carriers. EBV has also been strongly linked to endemic Burkitt's lymphoma and nasopharyngeal carcinoma. Recently associations with rheumatoid arthritis and Sjögren's syndrome have been suggested.[2,3]

The diagnosis of IM is based on the presence of clinical, hematologic, and serologic findings. The hematologic picture consists of leukocytosis, usually with more than 50% mononuclear cells and at least 10% atypical lymphocytes, that reaches a peak during the second or third week of illness.[4]

The serologic diagnosis of acute EBV infection is based on determination of antibody titers to EBV–specific antigens. The most readily available tests measure antibody levels against viral capsid antigen (VCA) and Epstein-Barr nuclear antigen (EBNA). Patients with symptoms of IM manifest elevated IgM and IgG levels against VCA, but EBNA antibodies are absent until several weeks or months after the onset of clinical disease. Both VCA and EBNA IgG antibodies remain detectable throughout life. The presence of elevated VCA antibody titers in association with absent or rising (fourfold increase) EBNA antibodies is diagnostic of recent EBV infection.[4] Other viral antigens that elicit a serologic response during the first few weeks of infection are early antigen (EA) and membrane antigen (MA). Since VCA, FA, and MA are expressed in lytically infected cells, a decrease in the antibodies directed against them is believed to be correlated with control of the infection by the host immune system. Recently, atypical immune responses with persistence of serologic profiles indicative of early EBV infection have been noted in patients with chronic systemic symptoms of infectious mononucleosis syndrome. The pathogenesis of this disease is not well-known, but at least in some cases maintenance of an immune profile characteristic of recent primary infection is associated with persistence of suppressor T cell activity in the circulating blood.[5,6] Patients with "chronic" EBV infection manifest persistently elevated VCA and EA antibodies with low titers of EBNA antibodies.

Ocular Manifestations

The reported ocular manifestations of EBV infection affect all parts of the eye. Entities described prior to the availability of virus-specific serologic tests were identified by their association with the clinical picture of IM and usually by one of the hemagglutination tests. Differential absorption, which renders the hemagglutination test more specific, was not always performed. Follicular conjunctivitis is commonly reported in conjunction with IM, but the incidence and course cannot be estimated with the currently available information.[7,8] Other reported manifestations include iritis, dacryoadenitis, and episcleritis.[7,9,10] Neurologic complications reported prior to the availability of EBV–specific tests include papilledema and optic neuritis.[9,11,12] More recently documented complications include ophthalmoplegia associated with Guillain-Barre syndrome and Fisher syndrome, as well as Bell's palsy.[13,14]

External and anterior segment disease reported in association with EBV infection include oculoglandular syndrome and epithelial and stromal keratitis. In 1981, Meisler and associates described an 11-year-old boy with oculoglandular syndrome that was manifested as an inflamed nodular upper tarsal conjunctival mass with preauricular and cervical lymphadenopathy.[15] The patient had no systemic symptoms but had serologic evidence of early EBV infection. In the same year Wilhelmus described a 16-year-old girl with mild fever and pharyngitis with small stellate "microdendrites" involving the central and peripheral cornea in all quadrants. Epstein-Barr virus was cultured from the conjunctiva and from tear samples.[8]

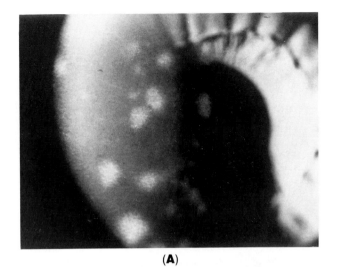

(A)

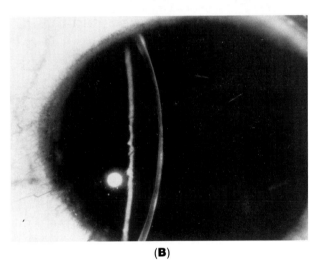

(B)

FIGURE 81-1 Subepithelial infiltration of the cornea resembling adenovirus-associated keratitis.

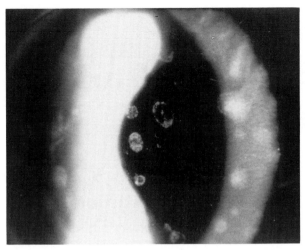

FIGURE 81-2 Multifocal, discrete, ring-shaped anterior and mid-stromal opacities.

EBV–associated stromal keratitis involves all levels of the corneal stroma. Subepithelial infiltration resembling adenovirus-associated keratitis has been reported in two patients with serologic evidence of systemic EBV infection and negative complement fixation tests for adenoviral antibodies (Fig. 81-1).[16]

We have identified seven patients with EBV–associated stromal keratitis.[17] All had serologically documented recent infection with EBV or documented prior infection with EBV virus in association with nonreactive tests for other causes of stromal keratitis. Two forms of EBV–associated keratitis were noted: (1) multifocal, discrete, granular, ring-shaped opacities scattered throughout the anterior and mid-stroma (Fig. 81-2) and (2) multifocal, nonsuppurative, full-thickness or deep

stromal infiltrative keratitis involving the peripheral cornea (Fig. 81-3). One patient experienced recurrent bouts of keratitis in both eyes. Four years after the onset of corneal inflammation, nodular scleritis developed in one eye. In 1987, Wong et al.[18] described a third form of stromal keratitis: peripheral corneal stromal edema overlying "confluent, geographic patches of white precipitates." The patient had clinical and serologic evidence of chronic EBV infection.

Other ocular manifestations of chronic EBV infection include anterior and posterior uveitis, papillitis, and choroiditis.[18,19] The choroidal lesions were described as ranging in size from 50 to 500 μm, involving the periphery or posterior pole. Active lesions were "gray to yellow infiltrates," whereas inactive lesions appeared as "punched-out areas of pigment epithelial scarring." Disciform scars involving the macula were noted in several patients. All patients had elevated VCA and/or EA antibodies with low titers of EBNA antibodies.

The treatment of ocular manifestations of EBV infection is not well-defined, owing to the relatively small numbers of reported cases. The superior tarsal conjunctival lesion of the patient with oculoglandular syndrome did not recur following excisional biopsy. The preauricular and cervical lymphadenopathy diminished without specific therapy, suggesting that this entity may be a self-limited disease.[15]

Stromal keratitis associated with EBV infection, whether subepithelial or deep stromal in location, may be treated with a topical corticosteroid such as prednisolone acetate 1%, 1 drop every 2 hours initially. Concomitant ocular antiviral therapy is not necessary if other potential causal agents, such as herpes simplex virus, have been excluded. The one reported case of microden-

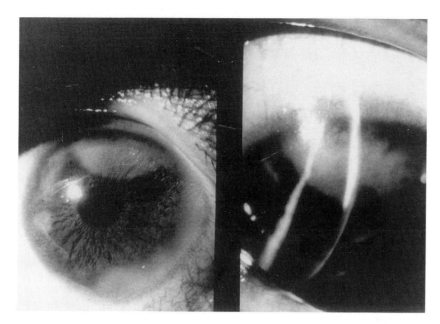

FIGURE 81-3 Multifocal, non-suppurative full-thickness or deep stromal infiltrative keratitis of the peripheral cornea. (Matoba AY, et al.[17])

dritic keratitis resolved after 2 days' treatment with topical acyclovir. The natural course of this entity is unknown, and no conclusions can be drawn about the role of topical antiviral therapy for epithelial keratitis. Topical corticosteroid therapy does not seem to be warranted.

Posterior disease, such as panuveitis and choroiditis, has been documented in patients with "chronic" EBV infection and may manifest a chronic or recurrent course. Topical or systemic corticosteroid does not appear to reliably induce a remission, but corticosteroid given in conjunction with topical and systemic acyclovir led to improvement of ocular signs in one patient with uveitis.[18]

Acyclovir [9-(2-hydroxyethoxy-methyl) guanine] is a synthetic acyclic nucleoside that is phosphorylated by virus-encoded thymidine kinase to form a potent antiviral agent with activity against herpes simplex virus types 1 and 2 and varicella zoster virus. Although EBV lacks a virus-specific thymidine kinase, acyclovir has been demonstrated to inhibit EBV DNA synthesis in vitro.[20]

Ocular formulations of acyclovir are not commercially available at this time, and penetration of topical forms of acyclovir in rabbits has been variable.[21,22] Subconjunctival injection of acyclovir solution has been reported to lead to high levels of the drug in the aqueous and vitreous of rabbits,[22] but the clinical utility of this therapeutic modality in EBV–associated ocular disease remains unknown.

EBV systemic infection has been linked to a wide variety of disorders involving all segments of the eye. While the stromal keratitis may manifest in a distinctive form, many of the entities are relatively nonspecific in appearance. Therefore, in patients with atypical or chronic inflammatory corneal or choroidal disorders EBV infection should be considered in the differential diagnosis. Since many patients have no antecedent symptoms of mononucleosis syndrome, the diagnosis often depends on serologic testing. The ever-broadening spectrum of EBV–associated ocular disease reported since the development of virus-specific serologic tests suggests that ocular manifestations of systemic EBV infection may be more common than was previously recognized.

References

1. Rickinson AB. Cellular immunological responses to the virus infection. In: Epstein MA, Achong BG, eds. *The Epstein-Barr Virus: Recent Advances.* New York, NY: John Wiley & Sons; 1986;77-125.
2. Alspaugh MA, Jensen FC, Rabin H, et al. Lymphocytes transformed by Epstein-Barr virus. Induction of nuclear antigen reactive with antibody in rheumatoid arthritis. *J Exp Med.* 1987;147:1018-1027.
3. Pflugfelder SC, Roussel TJ, Culbertson WW. Primary Sjogren's syndrome after infectious mononucleosis. *JAMA.* 1987;257:1049-1050.
4. Henle W, Henle G. Serodiagnosis of infectious mononucleosis. *Resident Staff Physician.* 1981;27:37-43.
5. Tosato G, Blaese RM. Epstein-Barr virus infection and immunoregulation in man. *Adv Immunol.* 1985;37:99-149.
6. Borysiewicz LK, Haworth SJ, Cohen J, et al. Persistence of symptoms following primary Epstein-Barr infection associated with impaired virus specific immune responses. *Quart J Med.* 1986;58(226):112-121.

7. Jones BR, Howie JB, Wilson RP. Ocular aspects of an epidemic of infectious mononucleosis. *Proc Univ Otago Med Schl.* 1952;30:1-4.

8. Wilhelmus KR. Ocular involvement in infectious mononucleosis. *Am J Ophthalmol.* 1981;89:117-118.

9. Tanner OR. Ocular manifestations of infectious mononucleosis. *Arch Ophthalmol.* 1952;51:229-241.

10. Librach IM. Ocular symptoms in glandular fever. *Br J Ophthalmol.* 1956;40:619-621.

11. Shaw EB. Infectious mononucleosis of the central nervous system with bilateral papilledema. *J Pediatr.* 1950;37:661-665.

12. Karpe G, Wising P. Retinal changes with acute reduction of vision as initial symptoms of infectious mononucleosis. *Acta Ophthalmol.* 1948;26:19-24.

13. Grose C, Henle W, Henle G, et al. Primary Epstein-Barr virus infections in acute neurologic diseases. *N Engl J Med.* 1957;292:392-395.

14. Slavick HE, Shapiro RA. Fisher's syndrome associated with Epstein-Barr virus. *Arch Neurol.* 1981;38:134-135.

15. Meisler DM, Bosworth DE, Krachner JH. Ocular infectious mononucleosis manifested as Parinaud's oculoglandular syndrome. *Am J Ophthalmol.* 1981;92:722-726.

16. Matoba AY, Jones DB. Corneal subepithelial infiltrates associated with systemic Epstein-Barr viral infection. *Ophthalmology.* 1987;94:1669-1671.

17. Matoba AY, Wilhelmus KR, Jones DB. Epstein-Barr viral stromal keratitis. *Ophthalmology.* 1986;93:746-751.

18. Wong KW, D'Amico DJ, Hedges TR, et al. Ocular involvement associated with chronic Epstein-Barr virus disease. *Arch Ophthalmol.* 1987;105:788-792.

19. Tiedeman JS. Epstein-Barr viral antibodies in multifocal choroiditis and panuveitis. *Am J Ophthalmol.* 1987;103:659-663.

20. Colby BM, Shaw JE, Elion GB, et al. Effect of acyclovir [9-(2-hydroxy ethoxymethyl) guanine] on Epstein-Barr virus DNA replication. *J Virol.* 1980;34:560-568.

21. Poirier RH, Kingham JD, de Miranda P, et al. Intraocular antiviral penetration. *Arch Ophthalmol.* 1982;100:1964-1967.

22. Schulman J, Peyman GA, Fiscella R, et al. Intraocular acyclovir levels after subconjunctival and topical administration. *Br J Ophthalmol.* 1986;70:138-140.

Chapter 82

Herpes Simplex

MARK J. MANNIS

Systemic Manifestations

Herpes simplex virus (HSV) is a ubiquitous DNA virus of which man is a natural host. There are two subtypes of the virus, HSV-1 and HSV-2, which can be distinguished virologically and by the clinical disease entities they produce.[1]

Epidemiology

The human encounter with HSV occurs very early in life. Although the infant acquires antibodies through passive placental transfer, the levels decline over the first 6 months of life. Subsequent natural environmental exposure to the virus results in a primary infection that usually occurs between 1 and 5 years of age, so the prevalence of herpes-neutralizing antibodies in the adult population ranges between 30 and 97%. The transmission of herpes simplex occurs primarily through direct contact with virus-containing secretions. Exposure to the virus at mucosal surfaces or skin abrasions permits access to the host.[1-3]

Pathogenesis

The variability of the clinical disease produced by HSV infection can be explained in part by the natural history of the virus and its relationship to the host. *Primary infection* is subclinical in over 85% of cases. Therefore, despite the widespread exposure to the virus, primary human infection, though it occurs early in life, is manifested only occasionally as a gingivostomatitis, blepharoconjunctivitis, fever of undetermined origin, upper respiratory infection, or rarely, meningoencephalitis or disseminated visceral infection. Within 3 weeks after the primary infection, the virus establishes itself in a state of *latency* in the sacral or trigeminal ganglia, during which time it is not recoverable from the host. A variety of trigger mechanisms may subsequently reactivate the virus. Depending on the host's immune status, this reactivation may produce manifest *recurrent disease* with lesions from which the virus can be recovered. Most clinically evident herpetic disease represents recurrences.

The specific manifestations and clinical course of disease produced by HSV are determined by the site of infection, the antigenic type of the virus, and the age and immune status of the host. The site of primary infection determines whether the patient develops clinical primary or recurrent orofacial lesions, genital lesions, skin lesions, herpetic whitlow, ocular disease, central or peripheral nervous system infections, or visceral disease. The viral subtype is also important. HSV-1 usually affects areas supplied by the trigeminal ganglia and is therefore associated with herpes labialis, gingivostomatitis, and blepharokeratoconjunctivitis, whereas HSV-2 is most commonly implicated in herpes progenitalis, herpes vulvovaginitis, and in neonatal herpes infections. Changes in host immune competence from exogenous immunosuppression or systemic atopy may predispose to disseminated forms of disease.[1,2]

Ocular Manifestations

HSV infection is a frequent cause of recurrent corneal disease and is one of the two commonest causes of corneal blindness in the United States, producing over 500,000 new cases of ocular disease each year.[2-4]

Primary Herpes

Since most primary herpetic infections are subclinical or involve the oral cavity, primary ocular infection is rarely encountered and many children harbor latent virus in the trigeminal ganglia by five years of age. In primary ocular infection, a child most commonly develops acute follicular conjunctivitis, often with preauricular adenopathy, watery discharge, lid edema, and in most patients, a vesicular eruption of the lids and periocular skin (Fig. 82-1a). In 50% of such cases of primary blepharoconjunctivitis, an accompanying keratitis develops within 2 weeks. It begins as a punctate epithelial keratitis and may develop into characteristic epithelial dendrites. A mild reactive iritis may accompany the epithelial disease. Primary herpetic blepharokeratoconjunctivitis usually resolves within 2 to 3 weeks. Scarring generally does not ensue from the corneal, conjunctival, or skin lesions. The diagnosis of primary ocular HSV is usually made on the basis of typical periocular vesicular skin lesions or corneal dendrites. In allergic or immunosuppressed persons, more severe forms of primary infection can occur, producing widespread skin eruption (Kaposi's varicelliform eruption), bilateral disease, and stromal inflammation with scarring.[3,5,6]

A special case of primary HSV infection is neonatal herpes, most commonly caused by HSV-2 and acquired by the infant during passage through the birth canal. Ocular disease may include conjunctivitis, keratitis, cataract, or chorioretinitis and may be part of a severe systemic syndrome including herpes encephalitis.[3,7]

Recurrent Herpes

Most herpetic eye disease encountered clinically is recurrent disease and is manifest as either epithelial keratitis, stromal keratitis, or herpetic uveitis. The pathogenesis of this recurrent disease is either infectious, immunologic, or related to structural damage from previous disease.

Epithelial Keratitis (Dendritic, Geographic, Metaherpetic)   Infectious herpes keratitis most commonly presents as a *dendritic corneal ulcer*. The patient has symptoms—foreign body sensation, irritation, tearing, and photophobia. Physical findings include ciliary flush and corneal lesions, which may range from punctate epitheliopathy to fully developed branching filigree ulcerations that stain brightly with fluorescein or rose bengal. The edges of these dendritic ulcers are raised because of the swollen virus-bearing epithelial cells, and the ulcers usually demonstrate characteristic end-bulbs (Fig. 82-1b). Dendrites usually occupy the axial cornea; they may be single or multiple and are less commonly seen at the limbus or on the conjunctiva. With recurrent attacks, corneal sensation is diminished. The majority of dendritic ulcers resolve within 2 weeks; resolution may be hastened by mechanical débridement of the infected cells or by the use of topical antiviral agents. Resolution of the active dendrite typically leaves a subepithelial ghost opacity with a ground glass appearance at the site of the ulcer.

A *geographic ulcer* is an active infectious form of epithelial HSV that represents the expansion of a dendrite. The edges of this ameboid ulcer are characteristically ragged, and marginal cells contain replicating virus (Fig. 82-1c). Such an ulcer frequently takes longer to heal and may be associated with stromal disease.

A *metaherpetic ulcer* (also called postinfectious, indolent, or trophic ulcer) is a noninfectious form of chronic epithelial defect. It is characteristically round or oval and

INFECTIOUS   IMMUNOLOGIC   STRUCTURAL

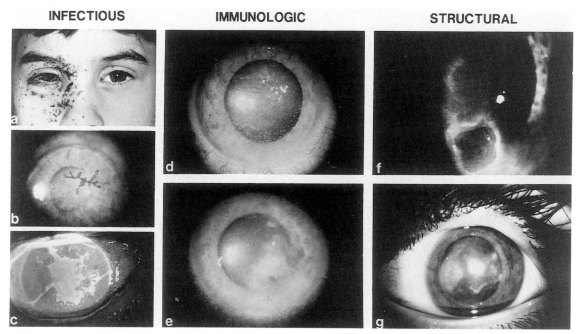

FIGURE 82-1 Principal manifestations of ocular herpes simplex. (**a**) Primary herpes simplex blepharoconjunctivitis; (**b**) HSV dendritic keratitis; (**c**) Geographic ulcer of HSV infection; (**d**) Metaherpetic ulcer; (**e**) HSV disciform keratouveitis; (**f**) Stromal necrotizing keratitis with hypopyon uveitis; (**g**) Cornea is quite scarred secondary to HSV infection.

has thickened edges. It can be distinguished from the geographic ulcer by its smooth margins (Fig. 82-1d). Such ulcers are most often seen after significant prior corneal disease, often in markedly hypesthetic corneas and in association with stromal inflammatory disease. These defects may persist for long periods of time and can be associated with stromal melting or bacterial superinfection.[3,5,6]

Stromal Keratitis (Disciform, Stromal Necrotizing)
While stromal keratitis can appear in a variety of manifestations, the two most characteristic clinical forms are disciform and stromal necrotizing keratitis.

Disciform keratitis is a noninfectious immune manifestation of HSV in which there is an inflammatory reaction in response to previously deposited viral antigen. It presents with symptoms of decreased vision, tearing, photophobia, and dull aching pain in the eye. Examination of the cornea usually reveals a discoid area of central or paracentral corneal stromal edema with overlying epithelial edema. Almost uniformly there are keratic precipitates underlying the area of stromal edema and Descemet's folds (Fig. 82-1e). An accompanying anterior uveitis may also be present. Disciform keratouveitis may last for months and leaves variable amounts of scarring

in its wake, depending on the degree of associated inflammation. Rarely it can be complicated by stromal dissolution or by scarring with vascularization.

Stromal necrotizing (interstitial) *keratitis* is another variation of noninfectious immunologic inflammatory reaction to herpetic disease of the cornea. It most commonly occurs in patients who have a significant history of previous herpetic corneal disease. The patient develops a focal or diffuse stromal infiltrate, often with associated stromal edema (Fig. 82-1f). The overlying epithelium may be intact or an ulcer may form. In the severest cases, the patient may develop dense stromal suppuration with necrosis or corneal abscess formation. These may progress to stromal melting, descemetocele formation, and perforation. The stromal inflammatory disease is often complicated by dense vascularization. Varying degrees of uveitis accompany the interstitial keratitis. This suppurative form of stromal keratitis is most often confused clinically with bacterial or fungal keratitis. Resolution leaves significant scarring and varying degrees of vascularization (Fig. 82-1g).

Other less common forms of stromal keratitis include Wesseley immune rings and limbal vasculitis.[3,5,6]

Uveitis
A mild reactive uveitis can be seen accompanying epithelial keratitis. More significant uveitis with

keratic precipitates is seen in conjunction with disciform keratitis. Stromal necrotizing keratitis is associated with varying degrees of iritis, ranging from scattered anterior chamber cells to hypopyon. Recurrent herpetic uveitis may be seen even in the absence of significant corneal disease. Such uveitis may be severe enough to produce hypopyon, anterior chamber hemorrhage, rubeosis, synechia formation, and secondary glaucoma.[3,7]

Management Principles

The appropriate management of HSV ocular disease hinges on an understanding of the pathogenesis of the clinical manifestations. The success of a course of treatment depends on whether the clinician correctly identifies the clinical manifestation as active infection as opposed to immune-mediated inflammatory disease or structural damage to the eye.

References

1. Corey L, Spear PG. Infections with herpes simplex viruses. Part 1. *N Engl J Med*. 1986;314(11):686-691.
2. Corey L, Spear PG. Infections with herpes simplex viruses. Part 2. *New Engl J Med*. 1986;314(12):749-757.
3. Hyndiuk RA, Glasser DB. Herpes simplex keratitis. In: Tabbara KF, Hyndiuk RA, eds. *Infections of the Eye*. Boston, Mass: Little, Brown and Co; 1986;343-368.
4. Pavan-Langston DR. Ocular viral diagnosis. In: Galasso GJ, Merigan TC, Buchanen RA. *Antiviral Agents and Viral Diseases of Man*. 2nd ed. New York, NY: Raven Press; 1984;207-246.
5. Easty DL. *Virus Disease of the Eye*. Lloyd-Duke Medical Books Ltd, Chicago, Il: Year Book Medical Publishers; 1985;135-178.
6. Binder PS. Herpes simplex keratitis. *Surv Ophthalmol*. 1977;21(4):313-327.
7. Oh JO, ed. Symposium of herpesvirus infections. *Surv Ophthalmol*. 1976;21:2.

Chapter 83

Herpes Zoster

THOMAS J. LIESEGANG

The varicella zoster virus (VZV) is responsible for the clinical diseases of chickenpox and zoster seen at the opposite ends of the age spectrum.[1-3] Varicella is a response of a person with no immunity; zoster is the response of a partially immune person. Man is the only natural host for the virus, a fact that has hampered research in the treatment of the disease.

Varicella is highly contagious as it is spread by airborne droplets through the respiratory tract as well as through vesicles. In temperate climates it is endemic with regular recurring seasonal prevalence. Periodic epidemics occur every 2 to 3 years. It may be clinically inapparent but can be severe and fatal in immunosuppressed patients or in adults. During the course of varicella, the virus enters the sensory nerve endings and spreads to the ganglia, where the DNA viral genome resides in a dormant state. Later the virus is reactivated owing to suppression of natural immunity with age, disease, or other unknown factors. The reactivated virus may be contained locally and may have no clinical signs, it may be suppressed at an early stage, or it may spread to the skin or eyes or enter the bloodstream. Zoster is the localized disease in which there is unilateral radicular pain and vesicular eruption limited to the dermatome of a single sensory nerve. Disease that involves any of the three branches of the ophthalmic nerve is referred to as herpes zoster ophthalmicus (HZO). Zoster occurs sporadically and independently of varicella and is not likely transmitted by contact.

In varicella and zoster, the vesicular eruption of the skin, and occasionally of the visceral organs, is almost identical. The skin lesions of zoster may involve deeper layers of the dermis with resultant scarring. The major pathologic features in nerve tissue are evident 10 to 12 days after the onset of symptoms, with necrosis of all or part of the ganglion surrounded by a lymphocytic reaction. There is an intense accompanying arteritis. After inflammation resolves, fibrous tissue develops, overlying sheaths thicken, and ganglion cells disappear.

Histopathologic studies of ocular tissue after HZO

suggest early changes related to vasculitis or direct inflammation. With chronicity, the disease progresses to a granulomatous reaction within tissue or arteries. The role of the virus in the eye and nerve tissue remains unknown; many of the ocular and neurologic features of the disease may be caused by an ischemic vasculitis, with or without direct inflammatory reaction or viral invasion.

Systemic Manifestations

Herpes zoster is typically ushered in by malaise, headache, fever, and nausea, which are followed by burning, itching, and tingling. Later there is a hot, flushed hyperesthesia with edema in the dermatome where the lesion will erupt. An astute clinician can make the diagnosis before the eruption takes place. Some patients have the neuralgia without ever developing a cutaneous eruption (zoster sine herpete). Once the skin eruption occurs, the constitutional symptoms frequently abate.

The distinctive rash of zoster is unilateral and limited to the area of skin innervated by a single sensory nerve. The dermatomes most frequently involved are the thoracic, followed by the cranial, cervical, lumbar, and sacral. The rash is rarely bilateral but a generalized eruption may occur in 2 to 5% of cases, especially in association with immunosuppression. If examined carefully, many patients demonstrate isolated vesicles in distant areas of the body, an indication of viremia. The rash progresses within 36 hours from discrete areas of erythema to macules, papules, and vesicles. The vesicles may become grouped or confluent. Lesions usually appear by the 4th day. Virus can be cultured for up to 4 days, or even longer in immunosuppressed patients. Over the next 3 to 4 days vesicles become turbid and yellow as pustules form. Because of the deeper involvement of the dermis (unlike herpes simplex), the vesicles form eschars, which may leave behind pigmented or pitted scars.

Most patients are healthy, without evident precipitating factors, although zoster is a more severe disease in advanced age or in the presence of malignancy or immunosuppression.[4] Herpes zoster is especially frequent and severe in Hodgkin's disease. Cutaneous dissemination by itself is generally a benign process, but it may continue for several months in immunosuppressed patients. Morbidity is a more significant problem with visceral dissemination to the lungs, liver, or brain.

The most frequent significant complication, seen especially in HZO, is postherpetic neuralgia, pain that persists in the original dermatome for more than 2 months. The pain and paresthesias vary among patients and at different times. They are more common and severe in the elderly. The cause of the pain is not clear; it is a delayed development that may involve anatomic and psychological restructuring of pain pathways.

Ocular Manifestations (Herpes Zoster Ophthalmicus)

The most common single nerve involved in zoster is the trigeminal nerve, especially the ophthalmic division.[3] The ophthalmic nerve is a purely sensory nerve that passes into the cavernous sinus along the lateral wall, sending branches to the tentorium and to the third, fourth, and sixth cranial nerves. As it enters the orbit through the superior orbital fissure, it divides into the frontal, lacrimal, and nasociliary nerves. The most serious consequences are evident when the nasociliary nerve is involved. Sometimes all three nerves of the ophthalmic division are affected simultaneously; even rarer is the simultaneous involvement of the ophthalmic, maxillary, and mandibular divisions of the trigeminal nerve.

Extension of the replicating virus or inflammation within the cavernous sinus may account for the involvement of the cranial nerves associated with ocular motility. There is a distinct syndrome of HZO and granulomatous arteritis of the central nervous system. Patients present at various ages with fever, headache, mental confusion, and a history of HZO within the previous 2 months. Patients may later develop contralateral hemiplegia, seventh nerve palsy, hemianopia, agraphia, or aphasia. Radiographic and histopathologic studies have demonstrated segmental angiitis affecting the carotid artery or the branches.

There are multiple and diffuse ocular complications of HZO related to multiple mechanisms, including ischemia, viral spread, and an inflammatory granulomatous reaction.

The lid vesicles of zoster evolve with pitting and pigmentation of the skin, lid ulcers, cicatricial lid retraction, and damage to the lash roots and meibomian glands. The lids may become necrotic and slough from ischemic vasculitis. Oculoplastic surgical procedures may be necessary to protect the integrity of the globe.

Orbital complications include proptosis, chemosis, ptosis, and ocular motor palsies resulting from spread of the infection within the orbit or from ischemic vasculitis. Third, fourth, and sixth nerve palsies may occur singly or in combination and are frequently asymptomatic and transient.

Conjunctival findings include hyperemia, petechial hemorrhages, follicular reaction, and occasionally conjunctival vesicles or membranous conjunctivitis. Inflammation of the episclera or sclera may occur acutely or may be delayed for 2 to 3 months. During resolution, scleral thinning may occur. Corneal complications can occur from a combination of factors, including replicating virus, limbal vasculitis, abnormal tear film, corneal exposure, nerve damage, and the host's inflammatory response.[5] Punctate epithelial keratitis and pseudodendrites may develop within the first few days of disease onset due to viral replication in the corneal epithelium. There are usually multiple, small, peripheral blotchy, swollen, epithelial cells. Later corneal complications associated with vasculitis, immunologic mechanisms, or host inflammatory reactions include anterior stromal infiltrates, sclerokeratitis, keratouveitis and endotheliitis, serpiginous ulceration, and disciform keratitis. Corneal sensation is often greatly diminished after HZO and when desensitization persists it leads to neurotrophic keratitis characterized by an unstable tear film, lack of corneal luster, or intraepithelial vesicles. It may progress to epithelial or stromal ulceration. If lid damage has also occurred, exposure keratitis may be an additional insult. The course of neurotropic keratitis is variable, depending on the return of corneal sensation. The end result of the corneal complications of HZO may be permanent endothelial damage with corneal edema or chronic interstitial keratitis accompanied by vascularization, crystalline lipid deposits, stromal scarring, thinning, or perforation.

Uveitis is a frequent finding, resulting from ischemic vasculitis and lymphocytic infiltration of the iris stroma. Eruptive lesions and localized dilated vessels may be seen on the iris, with resulting sectoral patches of iris atrophy. If severe ciliary body ischemia occurs, hyphema, hypopyon, synechia, and anterior segment ischemia and phthisis may result. Elevated intraocular pressure accompanies chronic uveitis due to trabecular inflammation, debris, and synechiae. Posterior subcapsular cataracts are commonly associated with the chronic uveitis or the steroid use in HZO.

Vascular insults may produce an ischemic optic neuropathy, central retinal vein occlusion, central retinal artery occlusion, or retinal vasculitis. The newly described acute retinal necrosis syndrome has been associated with the VZV.

Early therapy of HZO with oral acyclovir has shown promise in reducing the incidence and severity of the most common ocular complications of HZO, but it has no apparent effect on the incidence, severity, or duration of postherpetic neuralgia.[6-8] Before this agent was used, the therapy for ocular complications was palliative and not demonstrably effective. The treatment of the host of other systemic and ocular complications of herpes zoster awaits further definition of the role of several factors involved in the disease process (ie, viral spread, inflammatory reaction, and ischemia).

References

1. Weller TH. Varicella and herpes zoster. Part I. Changing concepts of the natural history, control, and importance of a not-so-benign virus. *N Engl J Med.* 1983;309:1362-1368.
2. Weller TH. Varicella and herpes zoster. Part II. Changing concepts of the natural history, control, and importance of a not-so-benign virus. *N Engl J Med.* 1983;309:1434-1440.
3. Liesegang TJ. The varicella-zoster virus: systemic and ocular features. *Am Acad Dermatol.* 1984;11:165-192.
4. Sandor EV, Millman A, Croxson C, et al. Herpes zoster ophthalmicus in patients at risk for the acquired immune deficiency syndrome (AIDS). *Am J Ophthalmol.* 1986;101:153-155.
5. Liesegang TJ. Corneal complications from herpes zoster ophthalmicus. *Ophthalmology.* 1985;92:316-324.
6. Balfour HH Jr, Bean B, Laskin OL, et al. Acyclovir halts progression of herpes zoster in immunocompromised patients. *N Engl J Med.* 1983;308:1448-1452.
7. Cobo LM, Foulks GN, Liesegang TJ, et al. Oral acyclovir in the treatment of acute herpes zoster ophthalmicus. *Ophthalmology.* 1986;93:763-770.
8. Balfour HH Jr. Acyclovir therapy for herpes zoster: advantages and adverse effects. *JAMA.* 1986;255:387-388.

Chapter 84

Influenza

ANN S. BAKER and ROBERT BETTS

The Agent

The influenza viruses (orthomyxoviruses) are RNA viruses. They are 80- to 100-nm spherical or ovoid particles with a lipoprotein envelope. The surface is covered by evenly spaced radially oriented spikes (peplomers), which are glycoprotein molecules of two types, hemagglutinin and neuraminidase. It is the antibody to these two surface components that is responsible for immunity to influenza. The virus is heat labile; its infectivity is destroyed by lipid solvents. Immediately inside the envelope is the matrix (M) protein. The inner core of the viron consists of nuclear protein (NP), three polymerase proteins, and eight strands of genomic RNA; these eight segments of the single-stranded RNA code for the structural proteins. The influenza viruses are divided into three types (A, B, and C) on the basis of antigenic differences in the NP and M proteins.

Systemic Manifestations
Pathophysiology

The influenza virus selectively injures the ciliated epithelial lining of the respiratory tract of various animals. It also agglutinates erythrocytes.

Clinical Picture

Influenza is an acute respiratory infection associated with antigenic type A and B viruses. Influenza A viruses cause outbreaks every year and are responsible for epidemics every 2 to 4 years and for pandemics approximately every 10 to 40 years. Influenza epidemics start abruptly, peak in 2 to 3 months, and subside.

Following a 1- to 2-day incubation period, there is abrupt onset of chills, fatigue, headache, malaise, and muscle pain; the temperature rises rapidly to 101 to 104°F. Constitutional symptoms and a nonproductive cough are usually more prominent than evidence for rhinitis. Conjunctival suffusion and posterior pharyngitis without exudate are also seen. Physical findings in the chest are minimal; scattered moist rales are absent.

Influenza may also present as a milder illness with an upper respiratory tract infection without fever.

Ocular Manifestations

Ocular manifestations include acute conjunctivitis, superficial punctate or interstitial keratitis, and marginal corneal ulcers; subconjunctival hemorrhages may be present. Intraocular complications are less common; the most typical is uveitis. Vascular complications in the retina include hemorrhages, retinal thrombosis, retinal edema, and stellate retinopathy. An acute optic or retrobulbar neuritis may occur as part of an encephalitic syndrome, sometimes associated with ophthalmoplegia. Finally, paralysis of convergence or divergence may occur, as well as ptosis. An irritative spasm of the intrinsic ocular muscles may give rise to transient myopia, mydriasis, or palsy. Palpebral edema is common, as are dacryoadenitis and dacryocystitis.[1] Diagnosis can be made by conjunctival, corneal or lacrimal culture.

Conjunctivitis

Catarrhal or hyperemic conjunctivitis is associated with the onset of influenza; this may be painful with periorbital swelling. The conjunctivitis lasts 4 to 5 days and then subsides. Occasionally conjunctival hemorrhages may occur. Conjunctivitis in marine biologists during the 1981 seal influenza epidemic also provided evidence for conjunctivitis associated with influenza.[2,3] A superficial punctate keratitis has also been recorded.[4]

Uveitis

The anterior uveitis associated with influenza is an acute, nongranulomatous process that may be unilateral or bilateral and that occurs during the period of convalescence; choroiditis is rare. The prognosis is generally good; persistent synechiae are not commonly seen.[5]

Ocular Muscle Palsy

Hay[6] described a watchmaker who noted diplopia 14 days after the onset of influenza. He had failure of convergence with crossed diplopia of the concomitant type when looking at near objects. Brown, et al.[7] and Wood[8] described a young woman who had poor accommodation following an illness that was presumed to be influenza.

Dacryoadenitis

Dacryoadenitis occurs during convalescence with a subacute course.

Retinal

Fraenkel[9] described retinal hemorrhages and exudative or edematous retinitis, sometimes with a star figure at the macula, in a 1920 study of 103 patients who died in the influenza epidemic of 1918 and 1919. Thirteen patients had hemorrhages in the retina, two in the macula, and three in the vitreous. Mathur also described "shining vesicular dots, three in the right eye and four in the left, round about the macula with a capillary ending at each one of them." These lesions occurred in a Hindu woman, 25 years old at the time of an influenza epidemic in 1957. The foveal reflex was absent. The macular region was irregular, and the whole area appeared somewhat darker than normal. The macular lesions seemed to be of vascular origin and were associated with erythematous spots on the limbus during the attack.[10]

More recently, Weinberg[11] described bilateral submacular hemorrhages in a nurse with presumptive influenza. The patient had a fourfold rise in titers for influenza A, a clinical course compatible with influenza, and was exposed to patients in an influenza outbreak in Pittsburgh in 1975. Fundus examination showed the presence of submacular hemorrhages in both eyes. Fluorescein angiography showed blocked choroidal fluorescein with leakage. The hemorrhages resolved completely during the ensuing weeks.

Rabon also described an acute bilateral posterior angiopathy with a secondary maculopathy and papillitis.[12] An intravenous fluorescein angiogram revealed subtle blocking defects of the perifoveal capillary network in the early phase. By the time the full venous phase developed, there were many focal areas of leakage from the perifoveal capillary network. These coalesced in later phases to produce a diffuse posterior pole hyperfluorescence and papillitis. Complement fixation testing for influenza A antibodies showed a fourfold rise from $1:4$ or less to $1:16$ over a period of 3 weeks. The authors postulated that the white retinal lesions seen in this patient may have been the result of temporary hypoxic damage with resultant intracellar edema and loss of transparency in the inner retinal layers.

References

1. Schlaegel TG, Grayson MC. Orthomyxoviridae. In: Darrell RW, ed. *Viral Diseases of the Eye*. Philadelphia, Pa: Lea & Febiger; 1985;211-218.
2. Webster RG, Geraci J, Petursson G. Conjunctivitis in human beings caused by influenza A virus of seals. *N Engl J Med*. 1981;304:911.
3. Geraci JR, St Aubin DJ, Barker IK. Mass mortality of harbor seals—pneumonia associated with influenza A virus. *Science*. 1982;215:1129-1131.
4. Doggert JH. Superficial punctate keratitis. *Br J Ophthalmol*. 1933;17:65-70.
5. Cavarra V. 1954, Acta XVII Concilium Ophthalmologicum-Canada & USA. 2:1269-1282.
6. Hay PJ. *Trans Ophthalmol Soc U.K.* 1924;44:408-410.
7. Brown E, et al. On paralysis of accommodation after influenza. *Br J Ophthalmol*. 1898;2:485.
8. Wood DJ. Accommodative failure in malaria and influenza. *Br J Ophthalmol*. 1898;2:485.
9. Fraenkel E. Ueber Augenerkrankungen bei Grippe. *Deutsch Med Wschr*. 1920;46:673.
10. Mathur SP. Macular lesion after influenza. *Br J Ophthalmol*. 1958;42:702.
11. Weinberg RJ, Nerney JJ. Bilateral submacular hemorrhages associated with an influenza syndrome. *Ann Ophthalmol*. 1983;15:710-712.
12. Rabon RJ, Louis GJ, Zegarra H, et al. Acute bilateral posterior angiopathy with influenza A viral infection. *Am J Ophthalmol*. 1987;103:289-293.

Chapter 85

Measles

Rubeola

WILLIAM E. BELL and CHRISTOPHER F. BLODI

Measles (rubeola) had been recognized for centuries but was clearly separated from other exanthematous conditions in the 17th century by Thomas Sydenham. The measles virus is an RNA–containing myxovirus in the paramyxovirus group and, along with the distemper and rinderpest viruses, is placed in the medipest subgroup.

Systemic Manifestations

An incubation period of 10 to 14 days precedes the initial symptoms of measles, which include fever, coryza, and conjunctivitis. The first distinctive features of the illness are Koplik's spots, which appear on the buccal mucosa 2 to 3 days after onset of symptoms and 2 to 3 days before onset of the cutaneous rash. The second distinctive feature of measles is the erythematous maculopapular rash, which makes its debut behind the ears and along the forehead before spreading over the face and then downward. With spread of the rash, the constitutional symptoms worsen and fever is increased. After 2 to 4 days, the rash begins to fade, and does so in the same order that it had appeared. Antibodies appear in the serum soon after onset of the rash and persist for many years, providing a reliable indication of immunity. Measles virus is present in blood, urine, and respiratory secretions during the prodromal stage of the illness but is isolated by laboratory means with considerable difficulty.

Measles is associated with a number of neurologic complications which have been separated into three forms, each with rather distinctive features. Acute measles encephalomyelitis has been estimated to occur in approximately one in 1000 cases of measles and is generally felt to be an immune-mediated disorder with or without viral invasion of brain tissue. Immunosuppressive measles encephalitis is a subacute illness with a latency period of several weeks to 6 months after exposure to the virus and occurs in persons with an underlying immunocompromising disorder. Subacute sclerosing panencephalitis occurs in previously normal children and is a progressive, deteriorative neurologic disease with onset several years after natural measles.

Unusual neurologic disturbances associated with measles include acute blindness, probably due to bilateral retrobulbar neuritis, acute transient cerebellar ataxia, and peripheral polyneuropathy, sometimes with characteristics of the Guillain-Barré syndrome.[1-4] An additional nonencephalitic complication of measles is abrupt onset of hemiplegia, usually associated with repetitive focal convulsions at the outset. This condition is most likely the result of cerebral arterial occlusion secondary to an inflammatory vasculitis. An acute encephalopathy manifested by symptoms and signs of increased intracranial pressure that rapidly leads to coma, decerebrate rigidity, and death has also been observed during the course of measles.[3] The disorder, referred to as "toxic" encephalopathy or "acute encephalopathy of obscure origin," has been described following a variety of childhood infections and has many similarities to Reye's syndrome.[5]

The neurologic signs in acute measles encephalomyelitis usually reflect both diffuse and multifocal abnormalities of the cerebrum, brain stem, and spinal cord, and thus are complex and variable from case to case. Extraocular muscle dysfunction secondary to brain stem involvement is usually either unilateral or markedly asymmetric, and it does not always resolve in those who survive the disease. Sixth nerve dysfunction can also be caused by intracranial hypertension provoked by cerebral swelling.[3,6]

A subacute and moderately rapidly progressive form of encephalitis in the immunosuppressed patient is the most recently identified form of neurologic disease caused by the measles virus.[7-10] Most cases described thus far have been in children with leukemia in remission, although the illness has occurred in an adult with Hodgkin's disease and in a 21-year-old man following renal transplantation.[10,11] The primary measles infection has usually occurred 2 to 6 months before onset of the neurologic disorder, but in rare instances no definite preceding infection with the measles virus had been recognized.[11] The child with subacute immunosuppressive measles encephalitis described by Haltia et al.[8] was found during life to have a retinopathy similar to that described with subacute sclerosing panencephalitis. In most cases, gradual deterioration is followed by death a few weeks after onset of the neurologic manifestations.

Ocular Manifestations

Most patients with measles develop conjunctivitis, often a few days before the outbreak of the rash, that affects the palpebral conjunctiva more often than the bulbar conjunctiva. A watery discharge is often associated with it. Some patients may develop subconjunctival hemorrhages; virtually all develop bilateral epithelial keratitis, which tends to affect the cornea within the palpebral fissures. These round or punctate defects can sometimes become confluent and regress rather rapidly without sequelae.[12]

Most patients suffer no symptoms with these changes and treatment is not usually necessary. Prophylactic antibiotics and antiviral medicines are not indicated in most cases.

Rare cases of uveitis, chorioretinitis, optic neuritis, and vein occlusion associated with measles infections have been reported.[13] The neurologic complications of measles give rise to a number of neuro-ophthalmic problems, as noted above.

Postmeasles blindness (PMB) is a severe problem in parts of Africa. In some areas 70% of blind children have corneal scarring from PMB, a complication whose cause is not understood. Most probably, a number of additional factors put children at higher risk of becoming blind following measles infection—malnutrition, coexisting herpes simplex infection, vitamin A deficiency, and toxic effects from home remedies. The cofactors create a situation such that, in some areas of Africa, about 1% of all children with measles develop the severe ocular sequelae of corneal ulceration.[14]

References

1. Schlossberg FR, Prizer M. Retinal changes with marked impairment of vision in measles. *Am J Ophthalmol.* 1940;23:998-1000.
2. Strom T. Acute blindness as post-measles complication. *Acta Paediatr.* 1953;42:60-65.
3. Tyler HR. Neurological complications of rubeola (measles). *Medicine.* 1957;36:147-167.
4. Berkovich S, Schneck L. Ascending paralysis associated with measles. *J Pediatr.* 1964;64:88-93.
5. Lyon G, Dodge PR, Adams RD. The acute encephalopathies of obscure origin in infants and children. *Brain.* 1961;83:680-706.
6. Hoyne AL, Slotkowski EL. Frequency of encephalitis as a complication of measles. Report of twenty cases. *Am J Dis Child.* 1947;73:554-558.
7. Aicardi J, Goutieres F, Arsenio-Nunes M-L, et al. Acute measles encephalitis in children with immunosuppression. *Pediatrics.* 1977;59:232-239.
8. Haltia M, Paetau A, Vaheri A, et al. Fatal measles encephalopathy with retinopathy during cytotoxic chemotherapy. *J Neurol Sci.* 1977;32:323-330.
9. Murphy JV, Yunis EJ. Encephalopathy following measles infection in children with chronic illness. *J Pediatr* 1976;88:937-942.
10. Wolinsky JS, Swoveland P, Johnson KP, et al. Subacute measles encephalitis complicating Hodgkin's disease in an adult. *Ann Neurol.* 1977;1:452-457.
11. Agamanolis DP, Tan JS, Parker DL. Immunosuppressive measles encephalitis in a patient with a renal transplant. *Arch Neurol.* 1979;36:686-690.
12. Dekkers NWHM. The cornea in measles. In: Darrell RW, ed. *Viral Diseases of the Eye.* Philadelphia, Pa: Lea & Febiger; 1985;239-250.
13. Bergstrom TJ. Measles infection of the eye. In: Darrell RW, ed. *Viral Diseases of the Eye.* Philadelphia, Pa: Lea & Febiger; 1985;233-238.
14. Foster A, Sommer A. Corneal ulceration, measles, and childhood blindness in Tanzania. *Br J Ophthalmol.* 1987;71:331-343.

Chapter 86

Molluscum Contagiosum

CURTIS E. MARGO

Molluscum contagiosum (MC) is a common self-limited viral infection of the skin that is usually of minor importance clinically, except when it is located around the eye, where it can incite toxic keratoconjunctivitis. MC occurs throughout the world but is more prevalent in economically disadvantaged countries. Most cases occur sporadically. Direct virus transmission is an important means of spread and is responsible for outbreaks traced to community swimming pools, Turkish baths, and gymnasiums.[1] While spread within families does occur, it is not common. Although most often an infection of children, MC can be transmitted sexually and has been transmitted by tattooing.[2,3]

While as many as 4.5% of children under the age of 10 years in some third world countries show evidence of infection, the prevalence is lower in industrialized nations.[1] The Committee on Infectious Diseases of the American Academy of Pediatrics does not recommend isolating infected children.[4]

Severe infection with MC has been reported in patients who are immunocompromised by T-lymphocyte deficiency from non-Hodgkin's lymphoma, idiopathic selective IgM deficiency, and sarcoidosis, and in a patient receiving systemic prednisone and methotrexate.[5-8] Severe MC has occurred in patients with acquired immunodeficiency syndrome (AIDS).[9-11]

Although the causative virus of MC can be obtained in abundance, it cannot be grown outside the human host. Based on its size, shape, ultrastructural characteristics, and intracellular site of replication, the MC virus is a member of the poxvirus family, which contains the largest of all vertebrate viruses.[1] The molluscum virus measures approximately 300 by 250 μm.

A defined gene library of MC–virus DNA sequences was established using restriction endonuclease techniques.[12] Cloning was achieved with bacterial plasmid vector pAT153. All DNA fragments except two terminal fragments were cloned. Fourteen independently isolated virus samples revealed two types of DNA cleavage patterns: Thirteen of 14 DNA cleavage patterns were similar and thus were termed MCV type I. One isolate from the vagina with a completely different cleavage pattern was named MCV type II. The biologic and clinical significance of these findings is not known.

Viral antigens can be demonstrated in the stratum spinosum, stratum granulosum, and stratum corneum of the skin. Virus-specific antibodies to MC are present in more than 65% of patients with skin lesions.[13] While virus-specific antibody usually is IgG, some patients produce IgM and IgA virus-specific antibodies. Treatment can induce a virus-specific antibody response.

Histopathologic features of MC are diagnostic. There is marked epithelial hyperplasia with enlargement of infected cells. Intracytoplasmic inclusions begin to develop in the lower stratum spinosum and enlarge as they migrate to the surface (Fig. 86-1, *inset*). Inclusion bodies are strongly eosinophilic and Feulgen-positive (stain for DNA). The nucleus is pushed to the side of the cell as inclusions enlarge. As infected cells move toward the surface their cell walls disintegrate, liberating inclusion bodies. Coalescence of inclusions centrally causes umbilication of the surface epithelium. The base of the lesion may show a mild degree of chronic inflammation.

Clinical Manifestations

Lesions occur on all parts of the body but are most common on the head, trunk, and genitalia. MC begins as a minute papule, enlarging to 3 to 6 mm. Lesions rarely attain a diameter of more than 3 cm. They are discrete flesh-colored to white, smooth, dome-shaped papules that may be surrounded by a mildly erythematous base. In time, the papule becomes umbilicated and a white curdlike material can be expressed from its core. Most patients have multiple lesions occurring in crops.

The vast majority of patients have no symptoms. Some complain of mild itching, and occasionally lesions become secondarily infected. MC usually involutes spontaneously in 2 to 8 months without causing scarring.

When located around the eye, MC can induce a toxic follicular keratoconjunctivitis.[14,15] Viral replication in the conjunctival and corneal epithelium probably does not occur. Not all cases of periocular MC, however, are associated with conjunctivitis (see Fig. 92-1).[15] Factors that may be related to the development of conjunctivitis include proximity of lesions to the eyelid margin and their stage of development.

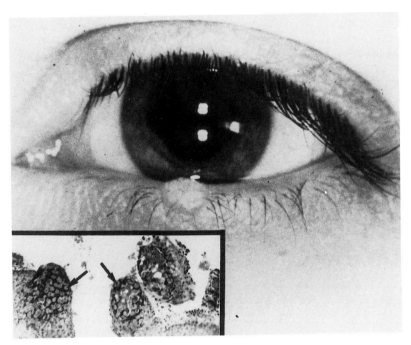

FIGURE 86-1 A solitary nodule of MC in the middle of the lower eyelid is not associated with any signs of conjunctival inflammation. (**Inset**) Biopsy of the lesion shows the classic features of molluscum, including large eosinophilic inclusion bodies (*arrows*) (hematoxylin-eosin, original magnification ×40).

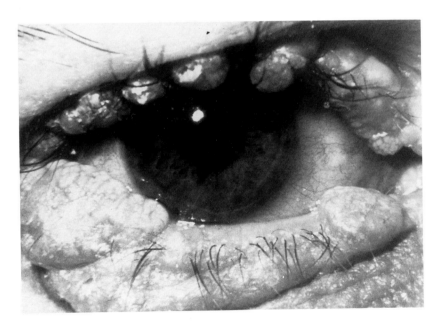

FIGURE 86-2 Severe MC in a patient with AIDS. While some of the eyelid nodules are umbilicated, the diagnosis may be obscure clinically because the lesions are confluent, larger than average, and so numerous. (Courtesy of Henry Perry, M.D.)

Once the periocular skin lesion is removed or obliterated, conjunctivitis resolves rapidly. Chronic sequelae are exceptionally rare, but secondary corneal ulceration has been described.[16] In immunocompromised patients MC may be more difficult to diagnose because lesions can be present in great numbers, be larger than usual, or become confluent (Fig. 86-2).

Since MC is a benign, self-limited disease, some physicians believe that it should not be treated.[17] Lesions located around the eye, however, pose a specific problem because they can induce conjunctival inflammation. If periocular MC is not associated with conjunctivitis, it is reasonable to observe the lesion and wait for spontaneous regression. Cases associated with conjunctivitis should be treated.

Dozens of methods have been described to treat MC.

Almost any type of noxious injury will cause involution. Simple curettage or cryotherapy is quite effective. Treatment of children can be problematic because they may be unable to tolerate local anesthesia or to hold still enough to allow safe performance of a procedure near the eye.[15]

References

1. Postlethwaite R. Molluscum contagiosum. A review. *Arch Environ Health*. 1970;21:432-452.
2. Wilken JK. Molluscum contagiosum venereum in a women's outpatient clinic: a venerally transmitted disease. *Am J Obstet Gynecol*. 1977;128:531-535.
3. Low RC. Molluscum contagiosum. *Edinburgh Med J*. 1946;53:657-670.
4. Report of the committee on infectious disease. Evanston, Il: American Academy of Pediatics; 1977:146-147.
5. Peachey RDG. Severe molluscum contagiosum infections with T cell deficiency. *Br J Dermatol*. 1977;97:49-50.
6. Mayumi M, Yamaoka K, Tsutsui T, et al. Selective immunoglobulin M deficiency associated with disseminated molluscum contagiosum. *Eur J Pediatr*. 1986;145:99-103.
7. Ganpule M, Garretts M. Molluscum contagiosum and sarcoidosis: report of a case. *Br J Dermatol*. 1971;85:587-589.
8. Rosenberg EW, Yusk JW. Molluscum contagiosum: eruption following treatment with prednisone and methotrexate. *Arch Dermatol*. 1970;101:439-441.
9. Katzman M, Elmets CA, Lederman MM. Molluscum contagiosum and the acquired immunodeficiency syndrome. *Ann Intern Med*. 1985;102:413-414.
10. Redfield RR, James WD, Wright DC, et al. Severe molluscum contagiosum infection in a patient with human T-cell lymphotrophic (HTLV III) disease. *J Am Acad Dermatol*. 1985;13:821-823.
11. Sarma DP, Weilbaecher TG. Molluscum contagiosum in the acquired immunodeficiency syndrome. *J Am Acad Dermatol*. 1985;13:682-683.
12. Darai G, Reisner H, Scholz J, et al. Analysis of the genome of molluscum contagiosum by restriction endonuclease analysis and molecular cloning. *J Med Virol*. 1986;18:29-39.
13. Shirodaria PV, Matthews RS. Observations on the antibody responses in molluscum contagiosum. *Br J Dermatol*. 1977;96:29-34.
14. Duke-Elder S. *System of Ophthalmology*. vol 7. part 1. St Louis, Mo: CV Mosby; 1965;377-379.
15. Curtin BJ, Theodore FH. Ocular molluscum contagiosum. *Am J Ophthalmol*. 1955;39:302-307.
16. Margo C, Katz NNK. Management of periocular molluscum contagiosum in children. *J Pediatr Ophthamol Strabismus*. 1983;20:19-21.
17. Weston WL, Lane AT. Should molluscum be treated? *Pediatrics*. 1981;65:865.

Chapter 87

Mumps

KIRK R. WILHELMUS

Mumps typically calls to mind the disconsolate facial appearance of epidemic parotitis, yet it is more accurately described as a generalized viral infection. Like other paramyxoviruses, mumps virus consists of a coiled nucleocapsid containing single-stranded RNA with an RNA–dependent RNA polymerase. An outer lipid envelope contains a hemolysin that mediates viral fusion to the host cell membrane, and a hemagglutinin-neuraminidase protein spike. Except for some epitope differences, there is only one distinct antigenic type. Some cross-reactivity exists with other paramyxoviruses such as parainfluenza type 1 (Sendai) and Newcastle's disease viruses. Man is the only natural host for mumps virus, although domestic dogs have developed parotitis.

Mumps is generally spread by airborne respiratory droplets. Clinically infected persons are contagious from approximately 6 days before the appearance of sialadenitis to 9 days afterward. Epidemic spread may be difficult to follow because approximately 30% of all mumps infections are subclinical or asymptomatic. During an incubation period of 14 to 18 days, viral replication occurs in the susceptible host's upper respiratory tract, regional lymph nodes, and possibly conjunctiva, leading to transient viremia. The virus is thereby spread hematogenously, giving rise to viruria and multiple organ involvement, particularly the parotid gland and the central nervous system. Viral spread is terminated by the humoral and secretory antibody responses, although the ability to infect T lymphocytes could lead to continued viral dissemination.

Systemic Manifestations

Most patients with clinical signs of mumps develop bilateral parotid gland edema and lymphocytic infiltration. This parotitis can be distinguished from lymphadenopathy by finding that it obscures palpation of the angle of the jaw. The parotid gland swelling can lead to lymphatic obstruction with cervical and presternal edema and even facial nerve palsy. Other causes of acute parotitis include parainfluenza virus types 1 and 3, influenza A, and coxsackievirus as well as microbial infections. The submaxillary (and rarely the sublingual) salivary glands may also be involved.

About 15% of patients have signs or symptoms of meningitis, such as headache, fever, loss of appetite, neck stiffness, positive Kernig's sign, photophobia, and dilated, unequal pupils, sometimes with accommodative paresis. Cerebrospinal fluid (CSF) changes are present in about 50% of patients with mumps, sometimes in the absence of meningeal signs. CSF abnormalities include pleocytosis, usually lymphocytosis, normal or elevated protein, and normal or decreased glucose titers. Some patients may develop postinfectious encephalomyelitis, characterized by lymphocytic perivasculitis and neuronal degeneration, possibly related to an autoimmune reaction or to viral neurotropism. Optic neuritis, third and sixth cranial nerve palsies, and trigeminal or facial neuritis may also occur. Eighth nerve involvement may produce nystagmus, and deafness may follow endolymphatic labyrinthitis. Brain stem involvement can produce conjugate gaze paresis, and periacqueductal inflammation can lead to hydrocephalus.

Epididymo-orchitis, generally unilateral, occurs in 20 to 30% of postpubertal males with mumps, generally beginning about 7 days after the onset of illness. Partial testicular atrophy ensues in 30 to 50%, but infertility is uncommon.

Other complications include pancreatitis and diabetes mellitus. A chronic mumps myositis of both young adults and elderly persons has recently been described that is pathologically related to oculophalangeal muscular dystrophy. Involvements of other organs are listed in Table 87-1.

Ocular Manifestations

Several eye inflammatory changes have occurred in patients with mumps.[1-3] During mumps viremia, several organs may be infected; whether the ocular manifestations are due to viral localization and multiplication has not been resolved. Parotitis may produce marked lymphedema with eyelid swelling and conjunctival chemosis. Papillary and follicular conjunctivitis can be associated with petechial and intraconjunctival hemor-

TABLE 87-1 Multisystem Involvement in Mumps Infection

System	Common	Uncommon (<5%)*
Salivary	Parotitis	Submaxillary adenitis, sublingual adenitis
Genitourinary	Orchitis†	Epididymitis, prostatitis, nephritis, oophoritis, pregnancy complications
Neurologic	Meningitis	Labyrinthitis, encephalomyelitis
Ocular	Conjunctivitis, dacryoadenitis	Stromal keratitis, episcleritis, scleritis, iridocyclitis, optic neuritis
Cardiopulmonary	Bronchitis	Myocarditis
Abdominal	Splenomegaly	Pancreatitis, hepatitis
Other	Mastitis†	Thyroiditis, arthritis, myositis, thrombocytopenia

* Mumps virus infection of various sites may be subclinical. For example, viruria and a transient decrease in glomerular filtration rate is very common, despite the absence of symptomatic renal involvement.
†More common in adults.

rhages. Rarely, the granulomatous conjunctivitis of Parinaud's oculoglandular syndrome has been reported.

Acute, usually bilateral, dacryoadenitis may accompany or precede mumps parotitis. "Lacrimal mumps" in the absence of parotid swelling has been suspected in some epidemics. Mumps dacryoadenitis resembles other causes of acute lacrimal gland inflammation.[4] Typical changes include localized pain and tenderness with swelling of the outer half of the upper eyelid, forming a characteristic S curve of the eyelid margin. A mucoid discharge is generally present, with conjunctival and episcleral hyperemia; adenitis of the accessory lacrimal glands may contribute to the chemosis. The enlarged preauricular lymph nodes may be obscured by the parotitis. Spontaneous resolution occurs over a few days to weeks without scarring, although a secondary dry eye syndrome has infrequently been noted.

Episcleritis, sometimes with peripheral corneal stromal infiltration, and scleritis have occurred. Mumps keratitis is an infrequent complication. It is not clear whether this corneal inflammation represents infection or a sterile immunogenic reaction. A fine punctate epithelial keratitis has been noted, but the most distinctive finding is a disciform stromal keratitis, usually unilateral, appearing 5 to 7 days after the parotitis. The focal stromal inflammation tends to evolve into diffuse corneal edema with striae, rarely developing marked cellular infiltration. When associated with sudden hearing loss, the clinical picture may resemble Cogan's syndrome. Unlike herpes simplex keratitis, spontaneous recovery within 2 to 3 weeks produces minimal residual scarring and no vascularization.

Iritis may occur in association with keratitis, or acute iridocyclitis may be the principal ocular manifestation. Experimentally, mumps virus can survive in the eye for several days after direct intraocular inoculation, but it has not been recovered following intravenous administration or conjunctival application. Anterior uveitis usually begins 4 to 14 days after the onset of sialadenitis and tends to occur more in adults and in persons with orchitis or other complications. Although it is usually unilateral, severe bilateral uveitis can occur in immunosuppressed patients. Secondary ocular hypertension, and rarely hypotony, can occur, possibly related to trabeculitis. Focal choroiditis and neuroretinitis have been described, sometimes leading to pigmentary chorioretinal changes.

Optic neuritis, usually unilateral, may be a manifestation of mumps meningoencephalitis, appearing 2 to 4 weeks after the onset of mumps and lasting for 10 to 20 days. It generally resolves completely, although permanent vision loss from optic atrophy has occurred. Other neuro-ophthalmic findings have been reported, such as cranial nerve palsy, central retinal vein occlusion, Parinaud's syndrome, and cortical blindness. Congenital malformations from maternal mumps are rare, although cataracts have been produced experimentally.

Diagnosis

The diagnosis of mumps is based on the clinical findings. With sialadenitis, serum amylase is usually elevated. In the absence of parotid gland swelling, diagnostic evaluation could include viral culture and rapid detection methods along with serology. Virus has been isolated from mouth and throat swabbings, urine, blood, CSF, milk, placenta, inner ear fluid, and conjunctiva. The isolation rate can be increased by centrifugation in the laboratory. For ocular involvement, conjunctival swabbings may be obtained within 5 to 7 days after onset, for inoculation onto monkey kidney, human embryonic kidney, HeLa, or other tissue culture cells. If immediate culture inoculation is not performed, the specimen should be kept on dry ice during transport to the laboratory. Cytopathic effects are confirmed by hemadsorption or immunofluorescence (IF) techniques. IF and enzyme immunoassay (EIA) can also be adapted for rapid detection of mumps virus in clinical specimens.

Several antibodies are elicited during mumps infection. Shortcomings of serologic tests are that some assays, such as hemagglutination inhibition and neutralization, are too cumbersome for routine clinical use, and other antibodies may cross-react with other paramyxoviruses. Complement fixation (CF) is used to detect S antibody directed against a soluble antigen of the viral protein capsid and V antibody directed against the hemagglutinin of the viral envelope. Because the former rises then falls within a few weeks, a titer of 1:64 or more of S antibody or a high S/V ratio provides presumptive evidence of recent mumps. IF and EIA techniques are also available to quantify virus-specific IgG and IgM levels. A high IgM or a rising IgG or CF antibody titer can be helpful in diagnosing recent mumps. Assay for virus-specific secretory IgA has been performed with saliva and may be applicable to tear film. The mumps skin test is not useful and is no longer available.

Treatment

The treatment of mumps is bed rest with respiratory precautions. Isolation is not necessary, but contact instruments should be appropriately disinfected. Pain relief includes application of warm or cold compresses, oral analgesics, a soft diet free of acidic foods, and diverting activities. Antiviral and antimicrobial agents are not indicated. Cycloplegia is useful for iritis and keratouveitis. The course of stromal keratitis, and possibly of iridocyclitis, may be shortened by a topical corticosteroid, even though mumps keratitis is generally self-limiting and heals with minimal scarring. An oral steroid usually is not recommended. Mumps immune globulin is not cost effective for treating and preventing disease.

Live attenuated mumps vaccine, also available in a single preparation with measles and rubella virus vaccines (MMR), should be administered to all children at 15 months of age. Only about one case of neurologic complications, including ophthalmoplegia and Bell's palsy, has occurred for every million doses of vaccine that have been administered. Since licensure of the vaccine in 1967, widespread immunization in the United States has contributed to the relatively low incidence of mumps, which has declined from 15 to 250 per 100,000 population to 1 to 3 per 100,000.

References

1. Woodward JH. The ocular complications of mumps. *Ann Ophthalmol.* 1907;16:7.

2. Riffenburgh RS. Ocular manifestations of mumps. *Arch Ophthalmol.* 1961;66:739.
3. Darrell RW. Mumps virus ocular disease. In: Darrell RW, ed. *Viral Diseases of the Eye.* Philadelphia, Pa: Lea & Febiger; 1985;225.
4. Jones BR. The clinical features and aetiology of dacryoadenitis. *Trans Ophthalmol Soc UK.* 1955;75:435.

Chapter 88

Pharyngoconjunctival Fever

RAYMOND M. STEIN and PETER R. LAIBSON

Pharyngoconjunctival fever (PCF) is an acute and highly infectious illness that was first described by Beal in 1907.[1] It is characterized by a triad of symptoms—fever, pharyngitis, and nonpurulent follicular conjunctivitis. These manifestations may occur singly or together and in any degree of severity. The disease is seen in all age groups but predominantly in children under 10 years of age. It often occurs in epidemic outbreaks, for example in families and schools and among service personnel.[2] The causative agent is usually adenovirus type 3, and less frequently type 4 or 7, but it can be seen with any of the 41 serotypes.[3] PCF can be contrasted with another adenovirus infection, epidemic kertoconjunctivitis (EKC; Table 88-1). The latter disease is usually caused by adenovirus types 8 and 19. It is seen most often between ages 20 and 40 and is characterized by a follicular conjunctivitis, keratitis, and preauricular adenopathy. Unlike PCF, in EKC there are usually few systemic manifestations.

Causal Agent

Although PCF has been recognized clinically for over 80 years, its association with the adenoviruses was first noted in 1954.[4] This was one year after the initial identification of the adenoviruses, which were found in adenoid tissue taken from children undergoing tonsilectomy.[5] The adenovirus is composed of a simple DNA core surrounded by a protein shell or capsid. It is highly stable, measuring 80 to 120 μm in tissue culture fluid and 50 to 65 μm in the nuclei of cells, where the viruses arrange themselves in crystal-like patterns.[2]

Replication of the virus takes place mainly in the epithelial cells of the conjunctiva and in the respiratory or intestinal tract, which constitute the main portals of entry. The exact mechanisms underlying clinical manifestations of adenoviral infection are poorly understood.[6]

Transmission

Adenovirus infection is a communicable disease that is easily spread by the patient to family members and close contacts. It is also an occupational hazard, particularly to ophthalmologists and other eye care personnel who work in situations where hand-to-eye contact is common. Every patient with a red eye who is seen in an

TABLE 88-1 Comparison of PCF and EKC

Parameter	PCF	EKC
Age group	<10 years	20-40 years
Incubation period	3-12 days	3-12 days
Adenoviral serotypes	3, 4, 7	8, 19
Follicular conjunctivitis	+++	+++
Preauricular adenopathy	+	++
Keratitis	+/−	+++
Pharyngitis	+++	−
Fever	+++	−
Pseudomembranes	+/−	+

ophthalmologist's office should be considered potentially to be infected with adenovirus and should be treated as such to avoid spread of this contagious disease. In addition to spread by hand-to-eye contact, PCF is more likely than EKC to be spread by droplet infection from the upper respiratory tract and via fecal excretion. While the virus disappears from the conjunctiva and throat after 2 weeks, fecal excretion can continue for weeks to months, accounting for the fact that many cases of PCF are contracted at public pools in the summer months.[3] Serotypes 3 and 7 have been associated with "swimming pool conjunctivitis."[7]

Incubation Period

The incubation period of adenovirus may vary from 3 to 10 or 12 days and is variable. The length of the incubation period depends on the virulence and the inoculating dose of virus at the time of exposure. The patient is asymptomatic during the incubation period but can probably spread virus toward the end of this period, before follicular hypertrophy occurs. It is at this particular time that patients are most dangerous to examine, as the unsuspecting ophthalmologist, nurse, or technician who does not wash his or her hands between patient examinations may spread the disease to other patients or to him- or herself.

Systemic Manifestations

Pharyngitis is characterized by discomfort rather than pain.[8] Examination of the throat generally shows injection of the posterior oral pharynx, which is frequently studded with lymphoid follicles. Nontender submaxillary lymphadenopathy is common even in the absence of the sore throat complaint.

Fever appears after an incubation period of 5 to 9 days and lasts from 1 to 10 days, with an average of 5 days. The onset of the fever may be gradual or sudden. The fever can rise rapidly to as high as 103 or 104°F, particularly in children.[8] The elevated temperature may be associated with malaise, myalgias, headache, abdominal discomfort, and diarrhea. The pulse and respiration rates tend to follow the temperature curve.

Ocular Manifestations

The onset of symptoms may be gradual or sudden. They range from mild itching and burning to moderately severe ocular irritation. Tearing is common, but photophobia is not marked. The hallmark of adenoviral infection is follicular hypertrophy (Fig. 88-1). This may affect one or both eyes, initially or in sequence. The lymphoid hyperplasia is usually most marked on the lower palpebral conjunctiva, where the conjunctiva has more subconjunctival tissue to form larger follicles, whereas the conjunctiva over the tarsal plate does not have much subconjunctival tissue. The follicular hypertrophy in the upper lid is usually seen along the upper border of the tarsal plate nasally and temporally. Follicular hypertrophy across the tarsal plate may be seen as small, pale pink or whitish, slightly elevated patches, sometimes with petechial hemorrhages between (Fig. 88-2). The vascular dilatation may lead to subconjunctival hemorrhages involving the entire subconjunctival area of the lids and bulbar surface, masking follicular hypertrophy. This is more common with EKC than with PCF. There

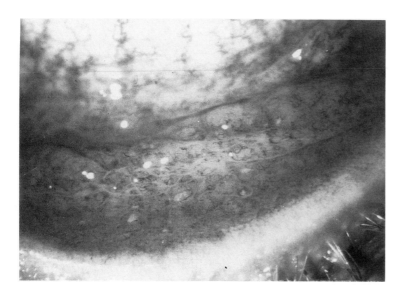

FIGURE 88-1 Follicular hypertrophy of the conjunctiva is the hallmark of adenovirus infections.

FIGURE 88-2 Petechial hemorrhages and follicular hypertrophy of the superior tarsal conjunctiva in a patient with adenovirus infection.

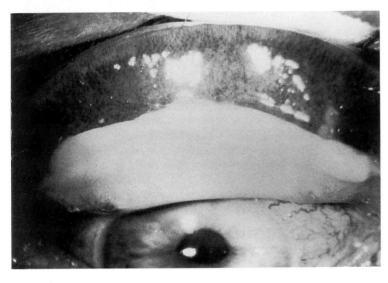

FIGURE 88-3 Pseudomembrane of the conjunctiva in a patient with adenovirus infection.

may be bleeding in the skin of the lids as well. To protect the examiner, eyelid hemorrhages without a history of trauma should be considered to be adenoviral infection until proven otherwise.

The formation of pseudomembranes (Fig. 88-3) is more common with EKC than with PCF and generally depends on the virulence of the organism and on the size of the inoculum. The pseudomembranes may mask the appearance of follicles, thereby preventing the proper diagnosis initially. Patients with pseudomembranous conjunctivitis should be considered to have adenovirus infection until it is proven otherwise. The only other disease in which pseudomembranes are common is streptococcal conjunctivitis, which is very unusual but can occur in children.

In addition to conjunctival hyperemia, chemosis and small, nontender preauricular nodes are usually present with both PCF and EKC. If an exudate develops it is almost always clear fluid or liquid (serous) and not purulent. The conjunctivitis persists from 3 to 4 days to as long as 3 weeks.[8]

Patients with PCF do not, as a rule, develop the mild to severe corneal changes typical of EKC. The keratitis in patients with PCF is usually mild and diffuse; only a small percentage develop focal keratitis and subepithelial infiltrates (Fig. 88-4). The infiltrates are usually smaller and less intense than those seen in the typical EKC syndrome produced by adenovirus types 8 and 19.[9] A fine focal or diffuse epithelial keratitis, similar to that seen in EKC, may be observed early in cases of PCF.

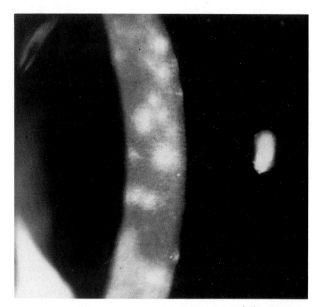

FIGURE 88-4 Subepithelial infiltrates of the cornea are uncommon in pharyngoconjunctival fever but are characteristic of epidemic keratoconjunctivitis.

Individual epithelial cells are involved with virus infection but they do not stain with fluorescein or rose bengal. In retroillumination the cornea may have myriads of pinpoint spots which appear best in retroillumination at the slit lamp. Visual acuity is usually unaffected, as the corneal surface is smooth and only individual epithelial cells are involved. By the 5th to the 7th day many of these pinpoint dots have disappeared. In areas where they have coalesced there may be a small epithelial break surrounded by a pinpoint-sized gray area. This involves the epithelium and superficial stroma just beneath the epithelium. These spots do stain with fluorescein or rose bengal and may last for a week. The subepithelial corneal infiltrate may be evident 12 to 16 days after the onset of the follicular conjunctivitis. Where the focal infiltrates have formed, antigen from the virus infection may enter the cornea through Bowman's layer. At about 2 weeks sensitized T lymphocytes react to the antigen localized in these areas and produce subepithelial infiltrates. Usually the eye that has follicular conjunctivitis first is the eye with the worse infiltrates. Whether the virus is spread systemically or by contact from one eye to the other is unknown. The subepithelial infiltrates gradually disappear over a period of weeks to months in most cases. Probably fewer than 1% of patients with subepithelial infiltrates have permanent changes in their corneas.

Laboratory Testing

The diagnosis of PCF is usually based on clinical findings of fever, pharyngitis, and follicular conjunctivitis. The diagnosis may be confirmed by viral cultures, but these are expensive and take several days before the results are known. The use of fluorescent antibody techniques allows more rapid detection, but these tests are not readily available and are expensive. Neutralizing antibody and complement fixation blood tests require prepared specimens, which must show a fourfold rise or fall in titer before a presumed diagnosis of adenovirus can be made. A Giemsa stain done on a conjunctival scraping will show a predominant lymphocytic response unless there is a pseudomembrane, in which case a predominance of polymorphonuclear cells will be present.

Treatment

Currently no effective or specific treatment for PCF is available. The disease is entirely self-limited, and management is supportive and symptomatic. Therapy should be directed toward reduction and relief of symptoms, treatment of secondary bacterial infection (which is very unusual), and prevention of spread. The patient may be more comfortable using artificial tears, lubricating ointment, cold compresses, or decongestants such as naphazoline drops. The use of systemic steroids, systemic antibiotics, or antiviral drugs is contraindicated in adenovirus infections.[10] Analgesics can be used for pain relief and antipyretics for fever control.

Complications

There are usually no sequelae in patients who have had PCF. Rarely, a pseudomembrane may leave a fine linear scar, especially in the inferior tarsal conjunctiva. In general, the disease subsides after a 3-week course. There may occasionally be persistence of subepithelial infiltrates, although not for longer than a few weeks to months.[11]

References

1. Beal R. Sur une forme particuliere de conjonctivite aigue avec follicules. *Ann Oculist.* 1907;87:1.
2. Duke-Elder S, Leigh A. *System of Ophthalmology. Diseases of the Outer Eye.* vol 8. part 2. London: Henry Kimpton; 1965;348.
3. Cooney MK. Adenoviruses. In: *Manual of Clinical Microbiology,* 4th ed. Washington, DC: American Society for Microbiology; 1985;701-704.
4. Parrott RH, Rowe WP, Heubner RJ, et al. Outbreak of

febrile pharyngitis and conjunctivitis associated with type 3 adenoidal-pharyngeal-conjunctival virus infection. *N Engl J Med.* 1954;251:1087-1090.

5. Rowe WP, Huebner RJ, Gilmore LK, et al. Isolation of a cytopathogenic agent from human adenoids undergoing spontaneous degeneration in tissue culture. *Proc Soc Exp Biol Med.* 1953;84:570-573.

6. Howe C, Coward JE. Adenoviruses and other respiratory viruses. In: *Microbiology: Basic Principles and Clinical Applications.* New York, NY: Macmillan Publishing Co; 1983;439-445.

7. Foy HM, Cooney MK, Hatlen JB. Adenovirus type 3 associated with irregular chlorination of a swimming pool. *Arch Environ Health.* 1968;17:795-802.

8. Bell JA. Clinical manifestations of pharyngoconjunctival fever. *Am J Ophthalmol.* 1957;43:11-13.

9. Kimura SJ, Hanna L, Nicholas A, et al. Sporadic cases of pharyngoconjunctival fever in northern California. *Am J Ophthalmol.* 1957;43:14-16.

10. Hecht SD, et al. Treatment of epidemic keratoconjunctivitis with idoxuridine (IDU). *Arch Ophthalmol.* 1965;73:49.

11. Deluise VP. Viral conjunctivitis. In: Tabbara KF, Hyndiuk RA, eds. *Infections of the Eye.* Boston, Mass: Little, Brown and Co; 1986;436-438.

Chapter 89

Rubella

German Measles

JOHN F. O'NEILL and ARLENE V. DRACK

Primary rubella infection is a benign, self-limited exanthematous infection that can be contracted at any age but is primarily a disease of children and young adults. It is caused by the rubella virus, an RNA togavirus with a single antigenic strain. The severe consequences of rubella occur when a mother's system is exposed to acute primary rubella infection early in gestation. This mechanism of maternal viremia as a cause of placental and fetal infection has become the model for our understanding of the pathophysiology of intrauterine infection and of the devastating effects that maternally transmitted infection may have on the developing fetus.

Our extensive knowledge of rubella virus infection in humans has evolved in three major increments over the past five decades. Sir Norman Gregg, an Australian ophthalmologist, is credited with linking *congenital malformations* in a group of infants to an earlier epidemic of German measles in 1941.[1] Those malformations included an unusual type of congenital cataract with micro-ophthalmia, congenital heart defects, deafness, and mental retardation.

Two decades elapsed before the rubella virus was positively isolated, identified, and propagated in tissue culture in 1962. Numerous clinicians and investigators were then able to document that those infants born with malformations following the severe rubella epidemic of 1963 and 1964 not only had static lesions present at birth but also suffered from a *chronic contagious infection* that began in fetal life and continued into infancy. These infants were not only "riddled with virus" but also had active ongoing infection of multiple organ systems and were shedding virus from essentially every body fluid and cavity. This combination of early malformation and chronic fetal infection has become known as the congenital rubella syndrome (CRS).

Additional insight into the nature of the condition has been gained in the past two decades, revealing a spectrum of *late onset disease* or *delayed manifestations* of congenital rubella. These disorders, which first make their appearance years after birth, include endocrinopathies (particularly diabetes mellitus), deafness, progressive rubella panencephalitis, and several ocular processes.[2-6]

Systemic Manifestations

The extent of fetal malformation and chronic infection depends first on the level of immunity of the maternal system and second on the point in gestation at which viral exposure took place. The time of greatest fetal vulnerability is the brief period of organogenesis between the 20th and 40th day. Although the overall incidence of rubella-associated defects is approximately 28% of infants exposed to maternal rubella throughout the 9-month gestation, the prevalence of major defects in the first month is 70% or higher. Major defects only rarely result from exposure in the third trimester of gestation.[3]

Rubella virus infection of the developing fetus causes a general retardation of cell growth and an actual diminution of the number of cells in various organs. This accounts for the low birth weight at full term, averaging less than 2500 g in two thirds of one large study group.[3] The associated failure to thrive may be attributed both to specific developmental defects (eg, cardiac malformations) and to the continuous destructive activity of the persisting viral infection.

There are four major areas of rubella-associated defects, and the majority of infants affected in the first trimester of gestation have multiple handicaps. (Ocular defects will be addressed separately.)

Hearing Loss

Deafness, the most common of all rubella-associated defects, is a permanent sensorineural hearing loss, most often bilateral. It may be an isolated defect if it occurs later in the second trimester, but it is present in all multi-handicapped children. Hearing deficits may become progressively worse after the first year of life, and initial onset of severe hearing loss was documented in one CRS child as late as 10 years of age, without any other explanation for such a deficit.[2]

Mental Retardation

The term "mental retardation" is utilized when IQ scores are below 70. Although brain malformations rarely occur in CRS, mental retardation is the second most common defect in the syndrome. It generally results from extensive meningitis and encephalitis associated with cerebral vasculitis. The encephalitis may continue for years, in which case it produces cumulative damage. Although this is not well-documented, late gestational rubella is thought to be associated with such "soft" neurologic signs as specific learning disabilities, diminished intellectual capacity, and minor motor disorders.

A delayed manifestation of chronic infection of the brain is *progressive rubella panencephalitis (PRP)*.[2] It is a slowly progressive and ultimately fatal disease of the central nervous system analogous to the chronic cerebral infection that follows measles, subacute sclerosing panencephalitis (SSPE). PRP begins insidiously 10 years or more after primary infection, with progressive deterioration of intellectual and motor function, and eventu-

ally, dementia and death. Rubella viral antigens have been identified in brain tissue and spinal fluid.

Cardiac Anomalies

Heart defects are the third most common malformation in CRS and generally occur following rubella exposure in the first 2 months of gestation. The most common defects are persistent patent ductus arteriosus and peripheral pulmonary artery stenosis. Other congenital malformations include aortic stenosis, coarctation of the aorta, and ventricular septal defects. These are associated with high rates of morbidity and mortality, particularly in the 2nd and 3rd months of life, accompanied by congestive failure and pneumonia.

Other Disorders

Diffuse fetal rubella infection may cause additional clinical disorders that are evident in neonatal life, including thrombocytopenia with petechiae or purpuric lesions, hemolytic anemia, hepatitis and hepatomegaly, bone defects, and pneumonitis. Postnatal viral shedding also constitutes a public health hazard of contagious spread and requires isolation techniques in the nursery and the clinical setting. CRS infants may lose their immunity and be rendered susceptible to recurrent infections later.

Delayed Manifestations

In addition to late-onset hearing defects and PRP several endocrinopathies have been identified with increasing frequency as delayed manifestations of CRS, the most common of which is diabetes mellitus. In one study, as many as 40% of long-term survivors developed overt or latent diabetes, the majority before age 35, which usually was associated with deafness or other defects.[6] Three of these patients were reported to have diabetic retinopathy. Other endocrine disorders seen less commonly include hypothyroidism, hyperthyroidism, thyroiditis, growth hormone deficiency, and hypoadrenalism. The endocrinopathies are therefore emerging as common late disease processes in CRS.

Ocular Manifestations

Primary acute rubella infection may be preceded or accompanied by a mild nonspecific conjunctivitis, or more rarely by a punctate epithelial keratitis without anterior chamber reaction. Mild keratitis may more frequently occur after onset of the rash and may resolve without sequelae. Rare cases of acute rubella retinopathy and optic neuritis have also been reported associated with primary acute rubella infection and as postvaccination complications.

Infants with congenital rubella are said to be riddled with virus and thus suffer diffuse tissue damage and extensive chronic infection; similarly every part of the eye may be involved by direct viral insult or by the effects of chronic panophthalmitis.

Microphthalmia

Microphthalmia is the ocular counterpart of generalized somatic growth retardation seen in CRS, and such "small eyes" are present in the majority of infants affected in the first 2 months of gestation. In one study of 20 culture-proven cases, every eye that was cataractous was also microphthalmic, although all microphthalmic eyes did not develop cataracts. The average transverse corneal diameter was 9.0 to 10.0 mm, as compared with 10.5 to 11.5 mm in normal infants of comparable age.[7]

Cataracts

Rubella cataracts are distinctly different in structure and appearance from other types of congenital cataracts, and although they are more often bilateral they may frequently be present in only one eye. Clinically there is a characteristic dense white nuclear sclerosis with variable surrounding cortical liquefaction. Gradual absorption of the soft necrotic cortex may occur over long periods. It is thought that viral penetration of the hyaline capsule does not occur after its formation between the 8- and 13-mm stage (4th to 6th weeks of gestation). Live virus has been recovered from the infant lens as late as 36 months after birth, and cataract surgery may be complicated both by the presence of live virus and the intense postoperative reaction related to chronic iridocyclitis.[9]

Retinopathy

Rubella retinopathy is a diffuse pigmentary disturbance that presents histopathologically as an irregular degeneration of the retinal pigment epithelium with pigment migration and clumping.[8] Clinically it appears as a salt-and-pepper type of mottling with fine areas of scattered depigmentation and pigment accumulation, mainly in the posterior pole. Although this retinal condition was thought not to affect visual acuity, subretinal neovascularization has been reported as a delayed manifestation between the ages of 8 and 17 years, which, when accompanied by subretinal hemorrhage, may severely reduce vision.[5] Pigmentary retinopathy is probably the most common ocular manifestation of congenital rubella.

Iris and Ciliary Body

Iris hypoplasia may be mild to severe, but frequently the intrinsic iris musculature is underdeveloped, which lim-

its the ability of the pupil to dilate. Chronic nongranulomatous uveitis, which primarily involves the iris and the ciliary body, may account for the severity of postoperative anterior chamber reactions in CRS and for the late manifestations of keratic precipitates and corneal hydrops, and may contribute to the late-onset type of glaucoma.[4]

Glaucoma

Glaucoma is an uncommon defect in CRS and is unusual in that it more commonly presents in microphthalmic rather than buphthalmic or enlarged eyes. Incomplete cleavage of the filtering angle of the anterior chamber has been offered as the probable mechanism,[8] and traditional methods of therapy (goniotomy, trabeculotomy, and trabeculectomy) have been only minimally effective. Late-onset glaucoma has been reported in a series of 13 persons ranging in age from 3 to 22 years. All of these persons either had previous cataract surgery or spontaneous lens absorption, and the majority of the eyes were microphthalmic.[4]

Corneal Transparency

Alterations of corneal transparency may vary considerably. Opacification has been noted to be transient or permanent, to be associated with glaucoma or not have any measured intraocular pressure increase, to be mild or sufficiently severe and progressively ectatic to require urgent keratoplasty. Obvious breaks in Descemet's membrane have followed episodes of increased intraocular pressure, and absence of Descemet's membrane with diffuse hydropic degeneration and acute bullous keratopathy has been noted.[8] Transient or fluctuating keratic precipitates have also been noted in a series of patients who previously had cataract surgery or whose cataractous lenses were spontaneously absorbed.[4]

Other ocular manifestations such as strabismus and nystagmus often accompany poor vision in CRS patients, and optic atrophy and retinal detachment have occasionally been reported.

Management

There is no definitive treatment for primary or congenitally acquired rubella infection. Appropriate treatment modalities must be utilized for the specific symptoms or systemic disorders. Current immunization programs utilize the Wistar RA 27/3 vaccine, which induces 95% immunity in vaccinees; however, the levels of rubella susceptibility in young adults still remains between 10 and 20%.[3] These facts, plus the knowledge that persons with CRS have been demonstrated to shed live virus well into adult life, suggest that in spite of near eradication of the disease, rubella infection will continue to be with us in the foreseeable future.

References

1. Gregg NM. Congenital cataract following German measles in the mother. *Trans Ophthalmol Soc Aust.* 1941;3:35-45.
2. Sever J, South MA, Shaver K. Delayed manifestations of congenital rubella. *Rev Inf Dis.* 1985;7(suppl):S164-S169.
3. South MA, Sever J. Teratogen update: the congenital rubella syndrome. *Teratology.* 1985;31:297-307.
4. Boger WP III. Late ocular complications in congenital rubella syndrome. *Ophthalmology.* 1980;87(12):1244-1252.
5. Collis WJ, Cohen DN. Rubella retinopathy, a progressive disorder. *Arch Ophthalmol.* 1970;84(1):33-35.
6. Menser MA, Forest JM, Honeyman MC, et al. Diabetes HLA antigens, and congenital rubella. *Lancet.* 1974;2:1058-1059.
7. O'Neill JF. Strabismus in congenital rubella. Management in the presence of brain damage. *Arch Ophthalmol.* 1967;77(4):450-454.
8. Zimmerman LE. Histopathological basis for ocular manifestations of congenital rubella syndrome. *Am J Ophthalmol.* 1968;65(6):837-862.
9. Wolff SM. The ocular manifestations of congenital rubella. *Trans Am Ophthalmol Soc.* 1972;70:577-614.

Chapter 90

Subacute Sclerosing Panencephalitis

JEAN-JACQUES De LAEY

Subacute sclerosing panencephalitis (SSPE) is a degenerative disease of the central nervous system that occurs in children and adolescents. It was first described by Dawson[1] in 1933 and later by van Bogaert and De Busschere[2] and by Pette and Doring.[3] Dawson[1] observed inclusion bodies in the nerve cells and neuroglial cells of a 16-year-old patient who died from SSPE and considered the disease to be of viral origin. A possible relationship with measles was first suspected by Bouteille et al.,[4] who demonstrated by electron microscopy the presence in the brain of affected patients of the nucleocapsid of a virus similar to measles virus. Connolly et al.[5] noted high titers of measles antibody in the blood and cerebrospinal fluid (CSF) and, using immunofluorescence, found measles antigens in the brain. A virus closely related to measles virus was isolated from brain tissue samples by Horta-Barbosa et al.[6] and by Payne et al.[7] Still, some studies suggest that there are antigenic differences between the slow virus that causes SSPE and the measles virus.[8]

Epidemiology

Measles virus plays an important role in the pathogenesis of SSPE. Epidemiologic studies indicate a history of measles in 93% of confirmed SSPE patients.[9,10] Almost half of them contracted measles before their second birthday, even though in industrialized countries measles is rare in this age group. About 85% of the patients show the first neurologic signs between age 5 and 14 years. The disease is two to four times more common in boys than in girls. The male-to-female ratio is inversely proportional to age at onset.[10] Another remarkable feature is that the incidence of SSPE is almost six times higher in rural (and especially farm) areas than in large cities. A number of studies underline the fact that most SSPE patients come from large families and that they contract measles at home from older siblings. Usually it is the youngest child who develops SSPE. From these different facts Aaby et al.[9] suggest that SSPE results from intensive exposure to measles.

Systemic Manifestations

According to Jabbour et al.[11] the clinical course of SSPE can be divided into four stages:

Stage 1: Cerebral (mental-behavioral) signs
Stage 2: Convulsion, motor signs

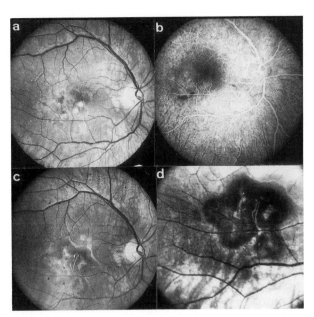

FIGURE 90-1 Macular lesion in a patient with SSPE: **(a)** appearance in June 1980; **(b)** fluorescein angiogram in June 1980; **(c)** appearance in September 1980; **(d)** appearance in November 1981.

Stage 3: Coma, opisthotonos
Stage 4: Mutism, loss of cerebral cortex function, myoclonus.

The first stage is quite insidious. Parents observe poorer school performance, some mental regression, and increased irritability. Even in this first stage, neurologic signs such as intention tremor or involuntary movements may appear. Ocular complications, particularly retinitis, may be the first reason patients seek medical advice. Stage 2, which may follow stage 1 after a few weeks or months, is characterized mainly by myoclonic jerks, consisting of a rapid flexion movement followed by a slow relaxing component. The clinical diagnosis at this stage is confirmed by typical EEG changes, namely the presence of paroxysmal bursts of high-voltage discharge occurring simultaneously with the myoclonus. The mental deterioration further progresses, although remissions may sometimes be observed.

In stage 3 spasticity increases. Decerebrate rigidity and coma follow. The body temperature increases sometimes, probably as a consequence of the dysregulation of autonomic functions. In stage 4 the severe hypertonia may diminish and myoclonus may become less frequent. This often gives rise to false hopes, but all central functions are lost. The disease is fatal. Most patients die within 5 to 12 months.

The clinical diagnosis is confirmed by the typical EEG changes, the finding of increased gamma globulins in the CSF, and high titers of measles antibodies in serum and CSF. Brain biopsy reveals perivascular lymphocytic infiltrates in neurons and glial cells.

Ocular Manifestations

Ocular complications are found in almost 50% of SSPE patients.[12] They may be divided into three groups[13]:

1. Disturbances of higher visual functions such as visual hallucinations and cortical blindness[14]
2. Motility problems: ocular muscle palsies, supranuclear palsies, ptosis, nystagmus
3. Ocular fundus changes: papilledema, optic neuritis, optic atrophy, (macular) retinitis.

Exceptionally, exophthalmos has been described in the course of SSPE.[15] It could be related either to increased intracranial pressure or to bilateral cavernous sinus thrombosis. Involvement of the optic disc is related to the cerebral manifestations of the disease. Papilledema is an essential symptom of the pseudotumor cerebri form of the disease.[16] Intranuclear inclusions may be found in the optic nerve, which would explain the optic neuritis.[17]

The most striking ocular manifestation is necrotizing retinitis, first described in 1963 by Otradovec.[18] The lesion is characterized by edema, retinal folds, and sometimes superficial hemorrhages, and by the absence of inflammatory signs in the vitreous (Fig. 90-1). Usually the lesion is solitary, but both eyes may be affected and more widespread involvement is possible outside the posterior pole.[19] The retinitis may be associated with a localized serous detachment of the neuroepithelium or of the retinal pigment epithelium (RPE). Fluorescein angiography in the acute stage may reveal some staining of the lesion. More widespread dye leakage has been described in a pattern suggestive of Harada's disease.[19] In the cicatricial stage, the fundus lesions are more or less atrophic, with irregular pigmentation. Modifications of the macular reflexes are sometimes especially striking. Preretinal membranes have been described, as have macular holes. As the fundus manifestations may antedate the neurologic signs by several months, they are often misdiagnosed. In bilateral cases they may be considered to be juvenile macular degeneration. Not uncommonly, the lesion is mistaken for *Toxoplasma* chorioretinitis.[20]

On histopathologic examination the inner nuclear layer and the ganglion cells appear to be affected more severely than the outer retinal layers.[21] These changes may be associated with disorganization of the RPE and even fragmentation of Bruch's membrane. Retinal folds and detachment of the inner limiting membrane are observed.[22] The affected retina may be modified into fibroglial scar tissue. The rarity of inflammatory cells is striking, although, depending on the intensity of the response, marked macrophagic infiltrates are sometimes seen in the retina and in the choroid.[19]

Inclusion bodies may be found in the retina.[23] With electron microscopy the presence in the retina of tubular filaments of a paramyxovirus has been demonstrated, so the fundus lesions in SSPE are to be regarded as a direct consequence of the viral infection.[24]

Prevention and Treatment

Although SSPE has been described in children who were vaccinated against measles, the incidence of SSPE has dropped dramatically in populations that received mass antimeasles vaccination.[25] The occurrence of SSPE in vaccinated children can be explained either by incomplete vaccine efficacy or by older age at vaccination (ie, these children were already exposed to measles prior to receiving the vaccination).

Most antiviral agents have little beneficial effect on the course of SSPE.[26] Amantadine hydrochloride, an anti-DNA agent that blocks the penetration of virus into host cells, has proved to have limited value in the treatment of SSPE. Inosiplex (Isoprinosine) has variable antiviral activity but may also have an immunomodulating effect. Inosiplex was found to improve or stabilize SSPE patients. The recommended dosage of Inosiplex is 100 mg/kg per day in divided doses. Treatment should be initiated as early as possible. As recurrences have been reported in patients in remission when treatment was discontinued, Inosiplex therapy should be continued even after remission is apparent.[26] Panitch et al.[27] recently reported the beneficial effect of repeated intraventricular injection of interferon.

References

1. Dawson JR Jr. Cellular inclusions in cerebral lesions of lethargic encephalitis. *Am J Pathol.* 1933;9:7-15.
2. Van Bogaert L, De Busschere J. Sur la sclérose inflammatoire de la substance blanche des hémisphères (Spielmeyer). *Rev Neurol.* 1939;71:679-701.
3. Pette H, Doring G. Uber einheimische Panencephalomyelitis von Charakter der Encephalitis japonica. *Dtsch Ztschr Nervenheilk.* 1939;149:7-44.
4. Bouteille M, Fontaine C, Verdenne C, et al. Sur un cas d'encéphalite subaigue à inclusions: étude anatomoclinique et ultrastructurale. *Rev Neurol.* 1965;113:454-458.
5. Connolly JH, Allen I, Hurwitz LJ, et al. Measles virus antibody and antigen in subacute sclerosing panencephalitis. *Lancet.* 1967;1:542-544.

6. Horta-Barbosa L, Fucillo DA, Sever JL. Subacute sclerosing panencephalitis: isolation of measles virus from a brain biopsy. *Nature.* 1969;221:974.
7. Payne FE, Baublis JV, Itabashi HH. Isolation of measles virus from cell cultures of brain from a patient with subacute sclerosing panencephalitis. *N Engl J Med.* 1969;281:585-589.
8. Steele RW, Fuccillo DA, Hensen SA, et al. Specific inhibitory factors of cellular immunity in children with subacute sclerosing panencephalitis. *J Pediatrics.* 1976;88:56-62.
9. Aaby P, Bukh J, Lisse IM, et al. Risk factors in subacute sclerosing panencephalitis: age- and sex-dependent host reactions or intensive exposure. *Rev Infect Dis.* 1984;6:239-250.
10. Modlin JF, Halsey NA, Eddins DL, et al. Epidemiology of subacute sclerosing panencephalitis. *J Pediatrics.* 1979;94:231-236.
11. Jabbour JT, Garcia JH, Lemmi H, et al. Subacute sclerosing panencephalitis. A multidisciplinary study of eight cases. *JAMA.* 1969;207:2248-2254.
12. Hiatt RL, Grizzard HT, McNeer P, et al. Ophthalmologic manifestations of subacute sclerosing panencephalitis (Dawson's encephalitis). *Trans Am Acad Ophthalmol Otolaryngol.* 1971;75:344-350.
13. Sebestyen J, Strenger J. Die ophthalmologischen Beziehungen bei der subakuten progressiven Panenzephalitis. *Klin Mbl Augenheilk.* 1964;145:202-212.
14. Lund OE, Fortster C, Bise K. Zerebral bedingte Sehstorungen als Erstsymptom bei subakuter sklerosierender Panenzephalitis (SSPE). *Klin Mbl Augenheilk.* 1983;182:290-293.
15. Cherry PMH, Faulkner JD. A case of subacute sclerosing panencephalitis with exophthalmos. *Ann Ophthalmol.* 1975;7:1579-1586.
16. Glowacki J, Guazzi GC, Van Bogaert L. Pseudo-tumoural presentation of subacute sclerosing panencephalitis. *J Neurol Sci.* 1967;4:199-215.
17. Gass JDM. Stereoscopic atlas of macular diseases. St. Louis, Mo: CV Mosby; 1977:302-303.
18. Otradovec J. Chorioretinitis centralis bei Leuco-encephalitis subacuta sclerotisans. *Van Bogaert. Ophthalmologica.* 1963;146:65-73.
19. Brudet-Wickel CLM, Hogeweg M, De Wolff-Rouendaal D. Subacute sclerosing panencephalitis (SSPE). A case report. *Docum Ophthalmol.* 1982;52:241-250.
20. Koniszewski G, Ruprecht KW, Flugel KA. Nekrotisierende retinitis bei subakuter sklerosierender Panencephalitis (SSPE). *Klin Mbl Augenheilk.* 1984;184:99-103.
21. La Piana FG, Tso MOM, Jenis EH. The retinal lesions of subacute sclerosing panencephalitis. *Ann Ophthalmol.* 1974;6:603-610.
22. De Laey JJ, Hanssens M, Colette P, et al. Subacute sclerosing panencephalitis: fundus changes and histopathologic correlations. *Doc Ophthalmol.* 1983;56:11-21.
23. Nelson DA, Weiner A, Yanoff M, et al. Retinal lesions in subacute sclerosing panencephalitis. *Arch Ophthalmol.* 1970;84:613-621.
24. Font RL, Jenis EH, Tuck KD. Measles maculopathy associated with subacute sclerosing panencephalitis. *Arch Pathol.* 1973;96:168.

25. Zilber N, Rannon L, Alter MA, et al. Measles, measle vaccination and risk of subacute sclerosing panencephalitis (SSPE). *Neurology.* 1983;33:1558-1564.
26. Taylor WJ, Du Rant RH, Dyken PR. Treatment of subacute sclerosing panencephalitis. An overview. *Drug Intell Clin Pharmacol.* 1984;18:375-381.
27. Panitch HS, Gomez-Plascencia J, Norris FH, et al. Subacute sclerosing panencephalitis remission after treatment with intraventricular interferon. *Neurology.* 1986;36:562-566.

Chapter 91

Varicella

Chickenpox

STEPHEN C. PFLUGFELDER

Varicella zoster is a DNA virus of the Herpesviridae family.[1] It is morphologically identical to the other herpes viruses, herpes simplex virus types 1 and 2, human cytomegalovirus, and Epstein-Barr virus. Like other herpes viruses, varicella zoster is capable of establishing latency after primary infection, and disease may occur from reactivation of latent virus.[2]

Systemic Manifestations

Chickenpox (varicella), the primary human infection of varicella zoster virus, spreads through airborne droplets from cutaneous lesions or respiratory secretions, and it is highly contagious to susceptible individuals. Varicella typically develops during childhood and is usually a mild, self-limited disease. The infection manifests with fever, malaise, and a cutaneous exanthem that typically lasts 7 to 10 days. The rash of chickenpox begins as macules, progresses to papules, vesicles, and finally pustules that dry and crust over. The disease can be severe or fatal in immunocompromised hosts.[2] These patients may develop extensive cutaneous eruptions, pneumonia, visceral involvement, and encephalitis.

Ocular Manifestations

Mild ocular involvement may develop with varicella.[3,4] A papillary conjunctivitis, occasionally with membrane formation, is the most common ocular manifestation. Small vesicles may appear on the lid margins or the bulbar conjunctiva. Punctate or dendritic epithelial keratitis may occur concurrently with the skin lesions. Disciform stromal keratitis, uveitis, and elevated intraocular pressure may develop following epithelial keratitis. Varicella keratitis is self-limited and does not recur. Corneal scarring is a rare complication.

Less common ocular manifestations of varicella include eyelid necrosis, permanent loss of lashes, interstitial keratitis, cranial nerve palsies, focal chorioretinitis, optic neuritis, and cataract if there is severe intraocular inflammation. Maternal varicella infection during pregnancy can result in congenital varicella syndrome, the systemic manifestations of which include microcephaly, limb deformities, deafness, and cardiac abnormalities.[2,5] Ocular anomalies such as microphthalmia, chorioretinitis, and cataracts may also occur.[2,5,6]

Varicella zoster virus establishes latency in the sensorineural ganglia after primary infection.[2,7] In most cases, herpes zoster (shingles) represents endogenous reactivation of latent varicella zoster virus. Rarely, cutaneous zoster lesions may develop in persons who harbor latent virus after they are exposed to exogenous varicella zoster virus by contact with persons who have active varicella or zoster infection.[8]

References

1. Darrell RW. Introduction (Taxonomy of Viruses) In: Darrell RW, ed. *Viral Diseases of the Eye.* Philadelphia, Pa: Lea & Febiger; 1985;1-5.
2. Weller TH. Varicella and herpes zoster. Changing concepts

of the natural history, control, and importance of a not-so-benign virus. *N Engl J Med.* 1983;309:1362-1368, 1434-1440.

3. Wilson FM. Varicella and herpes zoster ophthalmicus. In: Tabarra KF, Hyndiuk RA, eds. *Infections of the Eye.* Boston, Mass: Little, Brown and Co; 1986;369-386.

4. Liesgang TJ. The varicella-zoster virus: systemic and ocular features. *J Am Acad Dermatol.* 1984;11:165-191.

5. Taranger J, Blomberg J, Strannegard. Interuterine varicella: a report of two cases associated with hyper-A-immuno-globulinemia. *Scand J Infect Dis.* 1981;13:297-300.

6. Charles NC, Bennett TW, Margolis S. Ocular pathology of congenital varicella syndrome. *Arch Ophthalmol.* 1977;95:2034-2037.

7. Strauss SE, Reinhold W, Smith HA, et al. Endonuclease analysis of viral DNA from varicella and subsequent zoster infections in the same patient. *N Engl J Med.* 1984;311:1362-1364.

8. Arwin AM, Koropchak, Wittek AE. Immunologic evidence of reinfection with varicella-zoster virus. *J Infect Dis.* 1983;148:200-205.

PART 9

Inflammatory Diseases of Unknown Etiology

Chapter 92

Behçet's Syndrome

WILLIAM J. DINNING

Behçet's syndrome is rare in Western Europe and the United States but is found with increasing frequency from Eastern Europe and the Mediterranean across Asia, the incidence being highest in Japan (7 to 8 per 100,000 in the north).[1] There is a strong association with the possession of the HLA-B51 haplotype in Japanese, Turkish, and British patients, especially those with ocular disease.[2,3] The Japanese statistics suggest that the disease is increasing in incidence but that the male predominance may be decreasing. Findings from the much smaller series of patients seen in other parts of the world are particularly difficult to interpret. All reports have bias imposed upon them by the clinical interest of the centers from which they emanate.

Pathologic Features

The unifying pathologic process in this disease is vasculitis of small blood vessels. It first appears as infiltration around the postcapillary venules and eventually involves the capillaries and the arterioles. It consists of lymphocytes, plasma cells, polymorphonuclear leuko-

cytes, and macrophages. The small vessels become thrombosed and microinfarcts are produced. These changes underlie the ulcerative lesions of the skin and oral and intestinal mucosa and can be seen in the eye. They are also found around small areas of cerebral softening. Thrombosis of larger vessels such as peripheral veins and larger cerebral veins is thought to begin as inflammation of the vasa vasorum of these vessels.

Recent studies of the early stages of the lesions of the oral mucosa have shown that the predominant cell type in the infiltrate is the T lymphocyte. Helper and suppressor cells are equally evident and all show signs of activation. Macrophages form a significant proportion of the infiltrating cells and express HLA class II antigens. So, too, do the prickle cells of the epithelium. These observations may be interpreted to indicate a cell-mediated immune reaction.[4] It is to be noted that the same findings were made in recurrent oral ulceration, whether or not it was part of Behçet's syndrome.

The cause remains a mystery. The current concept is that an external agent such as a virus interacts with patients who possess the diathesis to produce recurrent episodes of focal vascular inflammation mediated by intermittent disturbances of the immune system. There is some evidence from DNA hybridization studies that mononuclear cells in patients with the syndrome may be persistently infected with herpes simplex 1 virus,[5] but this is very indirect, and viral particles have not been identified. A proposition to be seriously considered is that Behçet's syndrome represents one end of a clinical spectrum, the other end of which is the condition of recurrent oral aphthous ulceration that is so common in the population at large.[3] This is an attractive idea, considering the nonspecific nature of recurrent oral ulceration and the apparent differences in the spectrum of Behçet's disease itself in different populations.

Systemic Manifestations

The diagnosis is a clinical one, based on the presence of two or more of the so-called major findings, namely buccal ulceration, genital ulceration, eye lesions, and nonulcerative skin lesions, in combination with any of the "minor" findings, namely arthritis, thrombophlebitis, cardiovascular, gastrointestinal, or central nervous system disease, epididymitis, or a family history.[6,7]

Recurrent Oral Ulceration

In theory it would be possible to make the diagnosis of Behçet's syndrome without ulcers; as this is not done in practice, all patients have oral ulceration. The recurrent oral ulcers in patients with Behçet's syndrome are clinically and histologically indistinguishable from common aphthous ulcers of other persons. They may, however, be larger, may heal more slowly, and may generally be more troublesome to the patient.

Genital Ulceration

Two thirds of patients have ulcers of the penis, scrotum, vulva, vagina, or perianal region, which run a course similar to that of the buccal ulcers but tend to leave more scarring when they heal.

Nonulcerative Skin Lesions

Acneiform lesions, folliculitis, erythema nodosum, and superficial thrombophlebitis occur at some time in two thirds of patients.

Disease in Other Parts of the Body

It is becoming increasingly apparent that lesions may occur in any system of the body, although renal and hepatic disease seem rare. Acute arthritis occurs at some time in more than 50% of patients, most commonly in the knees. Thrombophlebitis of superficial veins occurs in 10 to 20% of patients, but occlusion of larger veins is uncommon. Ulcerative lesions in the gut, particularly the cecum and terminal ileum, are now being reported.

The need to revise our concepts of the true spectrum of this disease is highlighted by recent findings from magnetic resonance scans of the central nervous system.[11] Lesions difficult to distinguish from those of multiple sclerosis are common in Behçet's syndrome, even in patients without symptoms or signs of neurologic disease. They are mainly in the brain stem, with less periventricular involvement than in multiple sclerosis. Episodes of clinical progression tend to be fewer but more extensive than in multiple sclerosis. The lesions are also probably the result of small areas of vasculitis. Any type of motor, sensory, or neuropsychiatric symptom may occur.

Ocular Manifestations

About two thirds of patients have ocular disease at some time in their clinical course.[8] In Europe and America the patient's complaint is usually first diagnosed as a case of uveitis until the retinal disease, which is the most characteristic finding in the eye, is noticed. In Japan a large proportion of patients have only anterior uveitis.[9] Anterior segment inflammation may be very severe, but attacks tend to resolve rapidly. Hypopyon occurs in fewer than one third of patients.

The posterior segment disease is the cause of blindness.[10] It is a retinal vasculitis that varies greatly in severity from patient to patient. It may progress slowly or explosively. The earliest changes are seen in the retinal periphery and are often asymptomatic. Small areas of retinal edema and punctate hemorrhage may be seen, and careful examination reveals loss of the capillary beds in these areas. Sometimes the eye disease presents with occlusion of larger vessels than these, with profound visual disturbance. Branch retinal vein occlusion is common. Arteriolar occlusion leads to infarction of large areas of retina and is accompanied by intense retinal edema and hemorrhage and vitreous opacification. As the retina is progressively destroyed optic atrophy becomes evident, but vasculitis may occur in the optic nerve itself. An intriguing feature of the eye disease is its wide range of severity and asymmetry in the two eyes.

Investigation

There are no specific investigations and no characteristic serologic changes. In acute disease the serum levels of all immunoglobulin classes are elevated. Circulating immune complexes are found, suggesting that this may be an immune complex–mediated disease, but they may represent an epiphenomenon. Some workers have measured altered platelet function and decreased fibrinolytic activity in the acute phase of disease. Neutrophil chemotaxis appears to be increased, but a primary abnormality in the neutrophil has not been established.

Course and Prognosis

Attacks of systemic disease tend to subside over a period of 10 years or so. Cardiovascular or central nervous system disease poses a threat to life, but ocular disease causes by far the greatest morbidity. Up to 50% of patients with ocular disease are blind within 5 years of the onset of symptoms.[12]

Treatment

The symptoms of oral ulceration may be alleviated by local application of steroid ointments. Joint disease may be relieved by nonsteroidal anti-inflammatory agents. Patients with severe and disabling ulceration may benefit from low-dose systemic steroid therapy. Inflammation confined to the anterior segment of the eye may be managed satisfactorily by local steroid therapy, but retinal vasculitis requires systemic therapy. It responds only temporarily to high doses of systemic steroids, and steroids appear to have no effect on reducing the rate of exacerbations in patients with severe disease, for whom the ideal treatment is uncertain.[13] Recourse is usually to immunosuppressive drugs (usually one of the nitrogen mustard derivatives) or to cyclosporine A, but the toxicity of all these agents is a formidable problem and the search for safer and more effective therapy continues. It is impractical to expect a "cure" from current therapeutic regimens, and even when attacks of acute retinal inflammation appear to have been arrested, progressive narrowing of the retinal vascular tree is often observed.

References

1. Maeda K, Agata T, Nakae K. Recent trends of Behçet's disease in Japan and some of its epidemiological features. In: Inaba G, ed. *Behçet's Disease*. Tokyo: University of Tokyo Press; 1982;15-24.
2. Ohno S, Asanuma T, Sugiura S, et al. HLA-Bw51 and Behçet's disease. *JAMA.* 1978;240:529.
3. Lehner T, Batchelor JR. Classification and an immunogenetic basis of Behçet's syndrome. In: Lehner T, Barnes CG, eds. *Behçet's Syndrome*. London: Academic Press; 1979; 13-32.
4. Poulter LW, Lehner T, Duke O. Immunohistological investigation of recurrent oral ulcers and Behçet's disease. In: Lehner T, Barnes CG, eds. *Recent Advances in Behçet's Disease*. London: Royal Society of Medicine; 1986;123-128.
5. Eglin RP, Lehner T, Subak-Sharpe JH. Detection of RNA complementary to herpes simplex virus in mononuclear cells from patients with Behçet's syndrome and recurrent oral ulcers. *Lancet.* 1982;ii:1356-1361.
6. Mason RM, Barnes CG. Behçets syndrome with arthritis. *Ann Rheum Dis.* 1969;28:95-103.
7. Behçet's Disease Research Committee of Japan. Behçet's disease. Guide to diagnosis of Behçet's disease. *Jpn J Ophthalmol.* 1974;18:291-294.
8. Urayama A, Takahashi N, Sakai F. The position of ocular symptoms in Behçet's disease. In: Inaba G, ed. *Behçet's Disease*. Tokyo: University of Tokyo Press; 1982;153-160.
9. Mimura Y, Miyaura T, Mizuno K. Indication of corticosteroid, cyclophosphamide and colchicine therapies in ocular lesions of Behçet's disease. In: Henkind P, ed. *Acta XXIV International Congress of Ophthalmology*. Philadelphia, Pa: JB Lippincott; 1983;826-829.
10. Dinning WJ. An overview of ocular manifestations. In: Lehner T, Barnes CG, eds. *Recent Advances in Behçet's Disease*. London: Royal Society of Medicine; 1986;227-233.
11. Miller DH, Ormerod IEC, Gibson A, et al. MR brain scanning in patients with vasculitis: differentiation from multiple sclerosis. *Neuroradiology.* 1987;29:226-231.
12. Dinning WJ. Behçet's disease and the eye; epidemiological considerations. In: Lehner T, Barnes CG, eds. *Behçet's Syndrome*. London: Academic Press; 1979;177-181.
13. Dinning WJ. Therapy—selected topics. In: Kraus-Mackiw E, O'Connor GR, eds. *Uveitis, Pathophysiology and Therapy*. Stuttgart/New York: Georg Thieme Verlag, 1986;204-226.

Chapter 93

Chronic Granulomatous Disease

JAY S. PEPOSE

The phagocytic system of man, comprised of neutrophils, monocytes, and eosinophils, is an early-acting, nonspecific immunologic defense mechanism responsible for the ingestion and killing of pyogenic bacteria and fungi. To accomplish this function, phagocytes are capable of many sequential functions, including chemotaxis, adhesion, ingestion, and intracellular microbial lysis.

Chronic granulomatous disease (CGD) is an inherited disorder characterized biochemically by the inability of phagocytes to utilize molecular oxygen to generate important microbicidal byproducts (eg, hydrogen peroxide, hydroxyl radical, or superoxide anion) and clinically by recurrent, life-threatening infections of lung, lymph nodes, liver, and skin with catalase-positive microorganisms.[1] While CGD is a rare disease, affecting only 1 in 1 million persons, an understanding of its pathogenesis may yield important insights into an essential aspect of host immune defenses and may shed light on the mechanisms by which abnormal, chronic inflammation can lead to granuloma formation.

In a majority of families, CGD is transmitted as an X chromosome–linked, recessive trait, although both autosomal recessive and autosomal dominant inheritance with variable penetrance have been reported.[1,2] The genetic heterogeneity of the disease probably reflects different point mutations in the complex oxidative metabolic pathways of the neutrophil. Phagocytosis is normally accompanied by a respiratory burst, with oxidation of reduced nicotinamide adenine dinucleotide phosphate (NADPH) by a membrane-bound oxidase. Electron transport and reduction of oxygen at the plasma membrane may also involve a group b cytochrome, acting alone or in concert with the oxidase or other proteins. Whereas cytochrome b, flavoproteins, quinones, and several enzymes have each been implicated in various forms of the disease, the definitive "lesion(s)" in the membrane-associated NADPH-oxidase systems that cause CGD remain unknown. In addition to altered oxidative metabolism, abnormalities in neutrophil complement (C3b)–receptor expression, antibody-dependent cellular cytotoxicity (ADCC), and microtubule metabolism have also been demonstrated.[1]

Abnormal phagocyte oxidative metabolism in CGD can be diagnosed in the laboratory by abnormal chemiluminescence, deficient superoxide and hydrogen peroxide production by stimulated granulocytes, and failure of stimulated CGD leukocytes to drive the NADPH–dependent reduction of the yellow dye nitroblue tetrazolium to the blue-purple formazan (the nitroblue tetrazolium dye reduction test). In families of patients with the X-linked form of the disorder, approximately half of the females are CGD carriers as identified through these tests, harboring variable proportions of normal and abnormal cells (probably depending on the extent of X-chromosome inactivation or lyonization). A novel and exciting approach to CGD applied "reverse genetics," an experimental method in which the involved gene is characterized before the abnormal gene product is known. The gene that is abnormal in the X-linked form of CGD was cloned without reference to any specific protein by relying on its chromosome map position. Transcripts of this gene were found to be absent or abnormal in the CGD patients tested.[3] The predicted X-linked CGD protein derived from the complementary DNA sequence is a 468–amino acid polypeptide with no homology to proteins previously described in the plasma membrane NADPH-oxidase system. The X-linked CGD gene has yet to be introduced into CGD phagocytes, and the protein product of the abnormal X-linked CGD gene remains to be functionally characterized.

Systemic Manifestations

CGD is characterized by recurrent and often chronic infections involving, in decreasing order of frequency, lungs, lymph nodes, skin and soft tissues, liver, gastrointestinal tract, bone, genitourinary tract, eyes and ocular adnexa, and brain. CGD often presents as a low-grade fever associated with an elevated erythrocyte sedimentation rate, and no organisms are recovered from multiple cultures in over half of these febrile episodes. The most frequent pathogens involved are *Staphylococcus aureus*, *Aspergillus* species, *Chromobacterium violaceum*, *Pseudomonas cepacia*, and *Nocardia* species.[1] Most of the pathogens are catalase-positive microorganisms. Since catalase-positive microbes destroy the

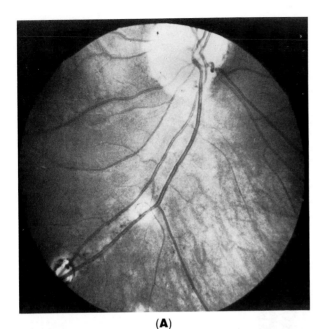

(A)

(B)

FIGURE 93-1 **(A)** This fundus photograph demonstrates atrophic peripapillary lesions and focal chorioretinal scars in chronic granulomatous disease. **(B)** Multiple focal chorioretinal lesions are seen in a perivascular distribution. (Courtesy of Lois J. Martyn, M.D.)

hydrogen peroxide they generate, the CGD phagocytes, themselves defective in hydrogen peroxide generation, are thereby deprived of an alternative source of hydrogen peroxide by which their microbicidal function could

be restored. Infections with catalase-negative microbes, such as *Streptococcus pneumoniae*, are rare in CGD.

Chronic inflammation of the gingiva and aphthous ulcers are common findings in CGD, as are granulomas and draining abscesses with chronic suppuration in lymph nodes. Prolonged antibiotic therapy and aggressive surgical intervention of localized infections are frequently required, with thorough debridement and prolonged drainage. Prophylactic antibiotics and leukocyte- or monocyte-enriched transfusions have been utilized with some clinical efficacy, and bone marrow transplantation has been performed with varying success.

Ocular Manifestations

The ocular signs of CGD fall into two groups: recurrent blepharokeratoconjunctivitis and chorioretinal lesions.[4–6] The focal, pigmented chorioretinal lesions appear to be a consistent finding in CGD (Fig. 93-1). They are usually perivascular and peripapillary in distribution and can progress to large areas of chorioretinal atrophy.[5,6] Macular lesions have not been reported, and visual acuity has not been decreased, although visual field defects have been reported corresponding to the location of chorioretinal scars.[6] Histopathologic studies of the chorioretinal lesions revealed almost total atrophy of choroid and retina and irregular proliferation of the retinal pigment epithelium at the scar margin. No inflammatory cells, microorganisms, or pigmented lipid histiocytes were observed in these lesions, and microbial cultures from fresh autopsy specimen eyes were negative. It has been postulated that the chronic chorioretinal scars may reflect prior active chorioretinal infection, similar to fundus lesions seen in patients with septicemia who may release septic emboli.[8]

Chronic blepharoconjunctivitis and marginal or punctate keratitis have been reported in patients with CGD, frequently accompanied by both pannus formation and perilimbal infiltrates.[5,7] These are characteristic findings of staphylococcal blepharitis and are thought to represent an immune reaction to staphylococcal antigens or toxins rather than a direct invasion of the cornea by *Staphylococcus aureus*.[9,10] Because of the specific defect in killing catalase-positive microbes, staphylococcal blepharitis is often a chronic condition in patients with CGD that requires rigorous lid hygiene and selective use of topical antibiotics and corticosteroids to prevent corneal neovascularization and scarring.[7]

References

1. Gallin JI, Buescher ES, Seligmann BE, et al. Recent advances in chronic granulomatous disease. *Ann Intern Med.* 1983;99:657.

2. Tauber AI, Borregaard N, Simons E, et al. Chronic granulomatous disease: a syndrome of phagocyte oxidase deficiencies. *Medicine.* 1983;62:286.

3. Royer-Pokora B, Kunkel LM, Monaco AP, et al. Cloning the gene for an inherited human disorder—chronic granulomatous disease—on the basis of its chromosomal location. *Nature.* 1986;322:32.

4. Carson MJ, Chadwick DL, Brubaker CA, et al. Thirteen boys with progressive septic granulomatosis. *Pediatrics.* 1965;35:405.

5. Martyn LJ, Lischner HW, Pileggi AJ, et al. Chorioretinal lesions in familial chronic granulomatous disease of childhood. *Am J Ophthalmol.* 1972;73:403.

6. Palestine AG, Meyers SM, Fauci AS, et al. Ocular findings in patients with neutrophil dysfunction. *Am J Ophthalmol.* 1983;95:598.

7. Rodrigues MM, Palestine AG, Macher AM, et al. Histopathology of ocular changes in chronic granulomatous disease. *Am J Ophthalmol.* 1983;96:810.

8. Myers SM. The incidence of fundus lesions in septicemia. *Am J Ophthalmol.* 1979;88:661.

9. Mondino BJ, Kowalski R, Ratajczak HV, et al. Rabbit model of phlyctenulosis and catarrhal infiltrates. *Arch Ophthalmol.* 1981;99:891.

10. Smolin G, Okumoto M. Staphylococcal blepharitis. *Arch Ophthalmol.* 1977;95:812.

Chapter 94

Kawasaki's Disease

Mucocutaneous Lymph Node Syndrome

SCOTT C. RICHARDS and DAVID J. APPLE

Kawasaki's disease (KD), or the mucocutaneous lymph node syndrome, is an acute exanthematous illness with multisystem involvement that occurs almost exclusively in infants and children.[1–4] The diagnosis depends on the presence of at least five of six commonly observed criteria: fever, bilateral conjunctival injection, changes in the mucous membranes of the upper respiratory tract, skin and nail changes of the extremities, a papulomacular rash, and cervical lymphadenopathy (Table 94-1). Many investigators have noted the similarities between KD and infantile periarteritis nodosa (IPN).[5] The pathologic entity of IPN, in fact, is virtually indistinguishable from KD both clinically and pathologically.

First described in Japan in 1967 by Tomisaku Kawasaki,[6,7] KD has now been reported in nearly all countries and races. Children of Japanese ancestry show significantly higher incidence rates than children of other races. Fifty percent of cases occur in children less than 2 years of age, and few cases have been reported in children over the age of 10. The disease is more common in boys than in girls, with a 1.6:1 male-to-female ratio. KD occurs both endemically and epidemically, with epidemics occurring at 2- to 3-year intervals in some areas of Japan and the United States. Variables of diet,

exposure, travel, or medical history have not been shown to consistently predispose to the disease.

Initially KD was believed to be a benign febrile disease that was acute and severe but entirely self-limiting. However, Japanese physicians soon recognized that approximately 2% of their KD patients died suddenly during the subacute phase of the illness. The cause of death was found to be massive myocardial infarction due to acute thrombosis of aneurysmally dilated coronary arteries. Nearly 90% of these fatalities occur in boys.

The cause of KD is unknown. The acute onset, fever, aseptic meningitis, and the occurrence of community-wide epidemics strongly suggest an infectious agent; on the other hand, the arthralgia and arthritis, vasculitis, elevated sedimentation rate, and elevated IgE levels suggest immune mediation. Epidemiologic studies suggest that KD is caused by an easily transmissible pathogen of low virulence that is spread by respiratory secretions. This proposed "Kawasaki agent" is believed to infect nearly all young children in a community but to cause recognizable disease only in children with a genetic predisposition. The Kawasaki agent must be considered an unknown until the presence of an organism can be demonstrated consistently in patients with KD.

TABLE 94-1 Principal Diagnostic Criteria of Kawasaki's Disease

1. Fever persisting for more than 5 days that is unresponsive to antibiotics
2. Bilateral nonexudative conjunctival injection
3. Oropharyngeal changes:
   a. Erythema, fissuring, and crusting of the lips
   b. Diffuse erythema of the oropharynx
   c. "Strawberry" tongue
4. Peripheral extremity changes:
   a. Induration of hands and feet
   b. Erythema of palms and soles
   c. Desquamation of finger- and toe tips
   d. Transverse grooves across finger- and toenails (Beau's lines)
5. Erythematous truncal rash
6. Cervical lymphadenopathy with lymph node diameter greater than 1.5 cm

(Modified from Kawasaki et al.,[7] Melish,[1] and Cheatham et al.[3])

Systemic Manifestations

Many of the diverse ocular and systemic findings of KD can be attributed to a widespread vasculitis, particularly of musculoelastic arteries. Kawasaki originally described six specific clinical criteria for this syndrome (see Table 94-1).[6,7] Currently, the diagnosis of KD depends on the recognition of five of these six criteria and the elimination of any other likely diseases.

The clinical course of the illness is best described as triphasic: an acute febrile phase, a subacute phase, and a convalescent phase. The first phase begins abruptly with the onset of fever, followed within 1 to 3 days by most of the other principal diagnostic criteria. The associated features of aseptic meningitis, diarrhea, and hepatic dysfunction may occur during this period, which lasts from 7 to 14 days. The subacute phase lasts from approximately the 10th to the 25th day after the onset of fever. Thrombocytosis, desquamation, arthritis, and myocardial dysfunction are common features of this phase. The convalescent phase lasts from the period when all signs of illness have disappeared until the sedimentation rate has returned to normal, usually 6 to 8 weeks from onset.

Fever, usually the first sign of the illness, occurs in all patients. It often has a remitting pattern, with multiple spikes as high as 40 °C (104 °F). The fever lasts at least 5 days; 11 days is average. A rash occurs with or soon after the onset of the fever. It can take multiple forms but is usually a generalized urticarial truncal rash with large, raised, erythematous plaques. Conjunctival involvement consists of bilateral injection of the bulbar conjunctiva (see below). The oropharyngeal manifestations include erythema and fissuring of the lips, oropharyngeal erythema, and a "strawberry" appearance of the tongue caused by hypertrophic papillae.

Changes in the hands and feet are the most distinctive features of KD. During the acute phase, marked induration of the hands and feet occurs. Desquamation of the skin of the finger- and toe tips occurs in the subacute phase and may involve the entire surface of the palms and soles. In the convalescent phase, deep transverse grooves across the fingernails and toenails may become apparent.

Cervical lymphadenopathy is the least commonly seen of the six principal criteria, occurring in 50 to 80% of patients. When present, the lymphadenopathy is generally unilateral, begins with the onset of the disease, and gradually disappears at the time of defervescence.

In addition to the six principal diagnostic criteria listed above, several associated features or complications are occasionally seen. These include pyuria and urethritis, arthritis or arthralgia, diarrhea, abdominal pain, aseptic meningitis, and hydrops of the gallbladder. Irritability and mood lability are seen in virtually all patients. Clinical cardiac disease occurs in at least 20% of patients with KD. The most common manifestations of cardiac involvement include tachycardia, arrhythmias, congestive heart failure, pericardial effusions, and mitral valve insufficiency.

The most catastrophic complication of KD is sudden death due to acute myocardial infarction, rupture of aneurysms, or arrhythmia. The most impressive pathologic changes in these fatal cases are found in the heart, and include pancarditis, acute inflammation, and aneurysms of the coronary arteries, and coronary obstruction and thrombosis. The large musculoelastic arteries of noncardiac organs are also frequently involved. Common sites of arteritis include the kidneys, adrenal glands, testes, lungs, pancreas, spleen, mesentery, and gastrointestinal tract. True aneurysms often occur secondary to inflammation and necrosis of the media.

The laboratory abnormalities in KD are nonspecific. Nearly all patients show an elevated erythrocyte sedimentation rate. Leukocytosis and sterile pyuria are often seen during the acute febrile phase. Thrombocytosis is present in essentially all patients, with evidence of a hypercoagulable state. Moderate elevation in IgE has been noted, but this has no known diagnostic or prognostic significance. Biopsy of accessible tissue, including skin, conjunctiva, and lymph nodes, has been of minimal diagnostic value to date.

Ocular Manifestations

Since ocular manifestations play a primary role in the recognition of KD, several articles have addressed the

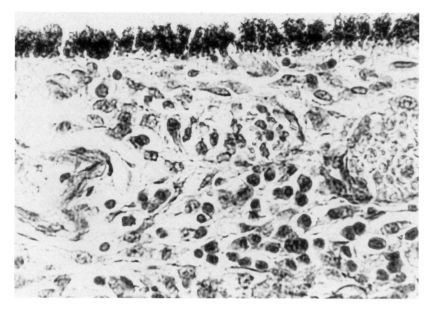

FIGURE 94-1 High-power photomicrograph of choroid and retinal pigment epithelium showing infiltrate of chronic inflammatory cells within the choroid (hematoxylin-eosin, original magnification × 400)

range of ophthalmic findings seen in patients with this syndrome.[8-12] Ocular findings associated with KD include bilateral conjunctival injection, an acute self-limiting iridocyclitis, superficial punctate keratitis, vitreous opacities, papilledema, and subconjunctival hemorrhage.

More than 90% of children with KD display a mild to moderate nonexudative vasodilatation of the conjunctival vessels that is bilateral and characteristically affects only the bulbar conjunctiva. No chemosis, follicles, or papillae generally are present. This is not a true conjunctivitis, since the conjunctival tissue itself seems to have little or no inflammation. The conjunctival injection seen in KD generally resolves without sequelae, but Ryan and Walton[10] reported one case of a 10-month-old male infant in whom bilateral scarring of the superior and inferior conjunctival fornices apparently developed in association with KD.

The presence of an anterior uveitis is common in KD patients. The iridocyclitis is mild and bilateral, with fellow eyes demonstrating equal degrees of inflammation. The acute iridocyclitis and inflammatory cells in the anterior chamber are usually noted a little later than the conjunctival injection, reaching a maximum 5 to 8 days after onset of the fever. Many of these patients with iridocyclitis demonstrate keratin precipitates. The anterior uveitis resolves completely within 2 to 8 weeks after onset of the disease without the development of any of the sequelae commonly seen with uveitis. Googe et al.[5] described a case of posterior uveitis (choroiditis) in a patient with presumed KD, but this is a rare finding (Fig. 94-1).

Several reports have been published in which a series of patients with KD were examined for ocular manifestations. Ohno and associates,[9] in a prospective study of 18 patients with KD, found bilateral injection of the bulbar conjunctiva in 16, acute nongranulomatous iridocyclitis in 14, superficial punctate keratitis in 4, vitreous opacities in 2, papilledema in 2, and subconjunctival hemorrhage in 1. Although the incidence of superficial punctate keratitis, vitreous opacities, papilledema, and subconjunctival hemorrhage were low, these ocular symptoms are important clinical manifestations of this disease.

Other ocular findings have been reported in association with KD, but they are rare. One case of focal white exudative retinal lesions has been seen,[8] with associated macular and disc edema, inflammatory cells in the vitreous, and development of a preretinal membrane. One year after the initial occurrence, this posterior segment inflammation recurred and was associated with anterior uveitis. In 1983, Font and colleagues[12] reported inner retinal ischemia secondary to the systemic vasculitis with multiple thrombotic occlusions in a patient with KD who died of cardiac complications. Retinal vascular involvement by KD is surprisingly rare, especially in light of the vasculitic process responsible for so many of the clinical manifestations of the syndrome.

Although the ocular manifestations of KD are generally mild and self-limiting, the ophthalmic examination plays an important part in diagnosis and evaluation. The principal ophthalmic manifestations of KD must be differentiated from those of juvenile rheumatoid arthritis and Stevens-Johnson syndrome, two illnesses that can mimic KD. Patients with Kawasaki's disease may occasionally present with a red eye as their primary complaint. KD should be suspected in all children who have acute bilateral iridocyclitis with bulbar conjunctival injection.

In summary, Kawasaki's disease is an acute febrile illness of young children with multisystem involvement. The most common systemic manifestations include fever, changes in the oropharynx and extremities, rash, lymphadenopathy, and cardiac abnormalities. The ocular findings in KD include bilateral conjunctival injection, uveitis, superficial punctate keratitis, and more unusual manifestations such as vitreous opacities, papilledema, and retinal vascular involvement. Many of the systemic and ocular findings in KD are attributable to the presumed infectious process or to the diffuse arteritis seen in this disease. KD is generally acute and self-limited, but is fatal in a small percentage of cases. The ocular manifestations tend to be bilateral and mild, and usually resolve completely without sequelae.

References

1. Melish ME. Kawasaki syndrome (the mucocutaneous lymph node syndrome). *Ann Rev Med.* 1982;33:569-585.
2. Hicks RV, Melish ME. Kawasaki syndrome. *Pediatr Clin North Am.* 1986;33(5):1151-1175.
3. Cheatham JP, Kugler JD, Pinskey WW, et al. Kawasaki disease in Nebraska—a review of the literature. *Nebr Med J.* 1983;68(9):282-293.
4. Kato H, Koike S, Yokoyama T. Kawasaki disease. *Pediatrics.* 1979;63:175-179.
5. Googe JM, Brady SE, Argyle JC, et al. Choroiditis in infantile peri-arteritis nodosa. *Arch Ophthalmol.* 1985; 103:81-83.
6. Kawasaki T. Acute febrile mucocutaneous syndrome with lymphoid involvement with specific desquamation of the fingers and toes in children. *Arerugi.* 1967;16:178-222.
7. Kawasaki T, Kosaki F, Okawa S, et al. A new infantile febrile mucocutaneous lymph node syndrome (MLNS) prevailing in Japan. *Pediatrics.* 1974;54:271-276.
8. Jacob JL, Polomeno RC, Chad Z, et al. Ocular manifestations of Kawasaki disease (mucocutaneous lymph node syndrome). *Can J Ophthalmol.* 1982;17:199-202.
9. Ohno S, Miyajima, T, Higuchi M, et al. Ocular manifestations of Kawasaki's disease (mucocutaneous lymph node syndrome). *Am J Ophthalmol.* 1982;93:713-717.
10. Ryan EH, Walton DS. Conjunctival scarring in Kawasaki disease: a new finding? *J Pediatr Ophthalmol Strabismus.* 1983;20:106-108.
11. Germain BF, Moroney JD, Gugino GS, et al. Anterior uveitis in Kawasaki disease. *J Pediatr.* 1980;97:780-781.
12. Font RL, Mehta RS, Streusand SB, et al. Bilateral retinal ischemia in Kawasaki's disease. *Ophthalmology.* 1983; 90:569-577.

Chapter 95

Reye's Syndrome

STEPHEN P. KRAFT

Reye's syndrome is an acute and sometimes fatal illness of unknown cause in which encephalopathy and fatty infiltration of multiple organs develop within a few days after a seemingly unremarkable viral illness.[1,2] It is most commonly seen in children between the ages of 4 and 12, but cases have been reported in infants as young as a few weeks old[3-5] and in adults as old as 60.[6] There is a slight preponderance of females in reported series.[3]

Reye's syndrome has been reported on all continents.[2,7] Its incidence in North America is approximately 1 case per year per 100,000 children under 18 years of age.[2] It affects 1 out of 2000 children infected with influenza B and 1 out of 4000 who contract varicella.[2] The disease rarely affects more than one person in a household.[8]

The disease is felt to be a self-limited derangement of mitochondria in liver, muscle, and brain.[2] It may be caused by a virus-host interaction in which the genetic makeup of the host renders the patient more susceptible to modification by exogenous factors.[2,4] Such factors may include the toxins of several viruses, including influenza A and B, varicella, and adenoviruses.[2,3,6,8,9] Epidemiologic studies have also implicated salicylates and salicylate-containing medications as causative factors when they are given during the acute phase of chickenpox or upper respiratory illnesses.[6,10] This has led to an American Academy of Pediatrics recommendation that salicylates not be given to children with influenza or varicella.[10] However, the specific pathogenic relationship between salicylate ingestion and Reye's syndrome still is not known.[2,3,6]

Systemic Manifestations

The syndrome presents as a biphasic illness. There is a prodrome consisting of a seemingly uncomplicated upper respiratory infection, most commonly influenza, or of varicella. The symptoms can include sore throat, rhinorrhea, and otitis, and the patient may be febrile.[1,2] In rare circumstances there may be primarily gastrointestinal symptoms such as diarrhea.[3]

The second stage is an acute encephalopathy characterized by nausea and vomiting, lethargy, and sudden personality change. There is a rapid deterioration in neurologic status, with progression within 4 to 60 hours to delirium, convulsions, and then coma. The mortality rate in untreated cases is over 75%, and death usually results from medullary coning or brain death.[1,2,11] Thus, it is important that the diagnosis be made early in the clinical course. It should be suspected when any child develops profuse vomiting after an upper respiratory infection. Early diagnosis in infants is more difficult as they typically present with respiratory distress, hyperventilation, seizures, and apnea, and they are rarely febrile.[2]

The criteria for the diagnosis of Reye's syndrome include the following:[3,6,7,11,12]

1. A prodromal illness with upper respiratory or gastrointestinal symptoms that may or may not include a fever, or varicella
2. Biochemical evidence of liver dysfunction including any or all of the following: three times normal elevation of serum ammonia level; elevated serum aspartate aminotransferase or alanine aminotransferase levels; prothrombin time 40% or more above normal
3. Abrupt onset of central nervous system disorder with progressive loss of consciousness
4. Exclusion of other disease states that cause encephalopathy and hepatic dysfunction
5. Negative toxic screen for salicylates.

There are other biochemical features that are typical of Reye's syndrome. Hypoglycemia is common, and it can be profound in infants under 2 years of age and in severely ill patients.[2,4,11,12] It is felt to be caused by defective gluconeogenesis.[3] The cerebrospinal fluid glucose levels are frequently below normal, although they can be in the normal range.[2,3,7] There can be elevated serum amino acids, elevated total free fatty acids, and metabolic acidosis.[5,7]

The severity of the disease at presentation correlates with the levels of serum ammonia and the prothrombin time. Hepatomegaly is not found consistently,[2,7] and the bilirubin titer is rarely significantly elevated except in the terminal stages.[4,6]

The differential diagnosis of Reye's syndrome[2,4,12] includes inherited metabolic disorders, especially urea cycle defects, "near-miss" sudden infant death syndrome (SIDS), hypoxic liver and brain disease, shock associated with severe bacterial infections, generalized viral infections, insulin-producing tumors, and drug overdoses.

Ocular Manifestations

Few reports address the ocular findings in patients with Reye's syndrome. There are no pathognomonic ocular signs.

During the acute encephalopathy phase of the disease the most consistent finding is pupil dilatation with poor or absent reaction to light.[12,13] Patients may develop papilledema.[13] There are reports of transient cortical blindness during the acute phase of Reye's syndrome.[12] One patient developed nystagmus and a unilateral facial palsy; only the palsy resolved after the patient recovered.[14] Finally, there is one case of bilateral central retinal vein occlusion, which was noted during the early stages of the encephalopathy; presumably it was due to either acute cerebral edema or acute hypoxia.[13]

Patients who recover from their encephalopathy rarely show any residual ocular abnormalities. There are reports of cortical blindness that persisted after recovery.[13] Optic atrophy has been noted in rare circumstances in patients who had severe papilledema during the acute phase.[13] There is one case report of pendular nystagmus and a comitant exotropia that developed subsequent to a severe encephalopathy.[14]

Pathology and Treatment

The consistent pathology in Reye's syndrome is that of acute hepatocellular failure. Percutaneous liver biopsy is considered essential to confirm the diagnosis and to rule out other inborn errors of metabolism in the differential diagnosis.[1-4,6,12] Fat stains, such as oil-red-O, are needed. Light microscopy typically shows a microvesicular panlobular accumulation of fat in liver cells. There is no hepatic necrosis or inflammation.[2-4] These changes are reversible if the patient survives, and the liver architecture returns to normal within 4 weeks.[2] There can be similar fatty infiltration of epithelial cells in the loop of Henle and the proximal convoluted tubules of the kidney, and in myocardium, pancreas, brain endothelial cells, and lymph nodes.[3,12] Examination of brain tissue shows cerebral edema and microscopic features of hepatic encephalopathy.[3,5,7,12] Findings on electron microscopy of liver cells are typical and include loss of glycogen, accumulation of lipid, increase in peroxisomes, and a characteristic swelling of mitochondria, which become pleomorphic and show fragmentation of cristae and protein denaturation.[2]

Treatment for Reye's syndrome has reduced the mortality rate in North America from 80% to about 25% over the past 20 years; however, up to one half of survivors have residual psychological problems, such as memory loss and mental retardation, or neurologic impairment, such as paraplegia.[2,5] The mortality and morbidity rates are highest among children in the first year of life.[2]

Therapy is directed at the two most threatening problems—hypoglycemia and elevated intracranial pressure (ICP). Blood sugar is replenished with intravenous hypertonic dextrose solution and insulin infusion. The raised ICP is treated with assisted ventilation with muscle paralysis, hyperosmotic agents such as mannitol, and careful ICP monitoring. The use of dexamethasone infusions is controversial. Other supportive measures include correction of fluid and electrolyte imbalances, cooling of the patient, neomycin enemas to reduce the ammonia level, chest physiotherapy, serial measurements of blood osmolality, central venous pressure and arterial pressure monitoring, vitamin K and fresh frozen plasma to correct coagulation abnormalities, and anticonvulsants.[2,3,5,7,11]

References

1. Reye RDK, Morgan G, Baral J. Encephalopathy and fatty degeneration of the viscera: a disease entity in childhood. *Lancet.* 1963;2:749-752.
2. Mowat AP. Reye's syndrome: 20 years on. *Br Med J.* 1983;286:1999-2001.
3. Glasgow JFT. Clinical features and prognosis of Reye's syndrome. *Arch Dis Child.* 1984;59:230-235.
4. Huttenlocher PR, Trauner DA. Reye's syndrome in infancy. *Pediatrics.* 1978;62:84-90.
5. Trauner DA. Treatment of Reye syndrome. *Ann Neurol.* 1980;7:2-4.
6. Peters LJ, Wiener GJ, Gilliam J, et al. Reye's syndrome in adults: a case report and review of the literature. *Arch Intern Med.* 1986;146:2401-2403.
7. Laxdal OE, Sinha RP, Merida J, et al. Reye's syndrome: encephalopathy in children associated with fatty changes in the viscera. *Am J Dis Child.* 1969;117:717-721.
8. Wilson R, Miller J, Greene H, et al. Reye's syndrome in three siblings: association with type A influenza infection. *Am J Dis Child.* 1980;134:1032-1034.
9. Edwards KM, Bennett SR, Garner WL, et al. Reye's syndrome associated with adenovirus infections in infants. *Am J Dis Child.* 1985;139:343-346.
10. Hurwitz ES, Barrett MJ, Bregman D, et al. Public health service study on Reye's syndrome and medications. *N Engl J Med.* 1985;313:849-857.
11. Venes JL, Shaywitz BA, Spencer DD. Management of severe cerebral edema in the metabolic encephalopathy of Reye-Johnson syndrome. *J Neurosurg.* 1978;48:903-915.
12. Massey JY, Roy FH, Bornhofen JH. Ocular manifestations of Reye syndrome. *Arch Ophthalmol.* 1974;91:441-444.
13. Smith P, Green WR, Miller NR, et al. Central retinal vein occlusion in Reye's syndrome. *Arch Ophthalmol.* 1980;98:1256-1260.
14. Wolter JR, Kindt GW, Waldman J. Concomitant exotropia: following Reye's syndrome. *J Pediatr Ophthalmol.* 1975;12:162-164.

Chapter 96

Sarcoidosis

GORDON K. KLINTWORTH

Sarcoidosis, an idiopathic systemic granulomatous disorder, affects young adults predominantly, and seldom persons under the age of 16 years.[1-3] Granulomatous disease is mediated by antigen-stimulated activated T-helper cells, which secrete lymphokines that attract, arrest, and activate monocytes and cause them to differentiate into epithelioid and multinucleated giant cells. Sarcoidosis seems to reflect an immune response to a still unidentified antigen (or antigens), probably airborne because of the high frequency of hilar lymphadenopathy and pulmonary involvement at the onset.

In sarcoidosis the immunoglobulin levels and the

TABLE 96-1 Manifestation of Sarcoidosis at First Examination (From Obenauf et al.[5])

Manifestation*	No.	%
Asymptomatic:		
Abnormal chest x-ray study	192	36.1
Symptomatic:		
Pulmonary	105	19.7
Ocular	101	19.0
Constitutional (weakness, malaise, weight loss)	86	16.2
Peripheral lymphadenopathy	48	9.0
Cutaneous	32	6.2
Arthralgias, arthritis	12	2.3
Central nervous system (seizures, hypopituitarism, facial nerve palsies, meningitis)	10	1.9
Cardiac	2	0.4

* More than one manifestation present in some patients at first examination.

absolute and relative amounts of T cells and activated T cells are increased at the site of active lesions, as in bronchopulmonary lavage fluid. In active sarcoidosis humoral immunity is heightened, causing increased levels of all major classes of immunoglobulins in the plasma, and common antigens evoke an exaggerated antibody response. Some of the antibodies in the serum react with the patient's own tissues. Such autoantibodies include rheumatoid factor and antibodies against nuclei and T lymphocytes. T-cell activation is manifested by the presence of an increased number of circulating activated lymphocytes and a spontaneous increase in DNA synthesis and lymphokine production by these cells in vitro. Paradoxically, T cells have a diminished capacity to proliferate when stimulated with phytohemagglutinin and concanavalin A or to ubiquitous antigens to which they have previously been exposed. This T-cell suppression affects mainly helper T cells, hence the ratio of suppressor to helper T cells becomes elevated. T-cell suppression is due partly to inhibitory factors, which are also lymphocytotoxic in the presence of complement. Most persons with sarcoidosis are unable to be sensitized to dinitrochlorobenzene or to develop delayed skin reactions to tuberculin and other ubiquitous antigens, or they lose their positive reactivity if they develop sarcoidosis. This anergy is restored after the sarcoidosis resolves. However, persons with sarcoidosis do not manifest impaired rejection of homografts or increased susceptibility to malignant neoplasms or infection (provided that they are not receiving long-term high-dose corticosteroid therapy). A peripheral blood T cell lymphopenia occurs in sarcoidosis, apparently due to the extravascular pooling of T lymphocytes in tissues rather than because of an immune deficiency.

Systemic Manifestations

Sarcoidosis presents with an acute or insidious onset. Sarcoid granulomas may form in almost any organ, producing varying degrees of inflammation and dysfunction (Tables 96-1, 96-2). The lungs, eyes, and skin are most often involved, but the disease is frequently subclinical, the initial abnormality being detected on a routine chest radiograph. There is a striking predilection for an intrathoracic hilar lymphadenopathy, with or without pulmonary infiltrates or fibrosis. Respiratory symptoms, when present, include cough, dyspnea, hemoptysis, and wheezing. Acute arthritis is often accompanied by lymphadenopathy and erythema nodosum (vasculitis with immunoglobulin and complement deposition). Chronic arthritis in sarcoidosis may mimic rheumatoid arthritis. Other sites of sarcoid granuloma formation include the liver, spleen, kidneys, muscles, bones, and the central and peripheral nervous system.

Ocular Manifestations

The visual system is affected in at least one fourth of cases, and such involvement may be the first clinical evidence of sarcoidosis (Table 96-3).[5] Ocular involvement is usually accompanied by systemic disease, which may be asymptomatic. The commonest ocular manifestation is anterior granulomatous uveitis, which is generally bilateral and recurrent and may become chronic.[6] It usually begins insidiously with mutton-fat keratic precipitates. The uveitis is often severe and sometimes is followed by peripheral anterior synechiae, glaucoma, and cataract. Occasionally one eye abruptly becomes red and painful, with photophobia and decreased vision.

TABLE 96-2 Systemic Involvement in Sarcoidosis

Involved Organ	No.	%
Intrathoracic:		
Hilar lymphadenopathy	371	70.0
Lung parenchyma	271	53.0
Extrathoracic:		
Ocular	202	38.0
Peripheral lymphadenopathy	148	27.8
Cutaneous (including erythema nodosum)	122	22.9
Hepatomegaly	118	22.0
Splenomegaly	69	13.0
Central nervous system	46	8.7
Musculoskeletal	38	7.2
Parotid	31	5.8
Cardiac	17	3.2
Other	27	5.1

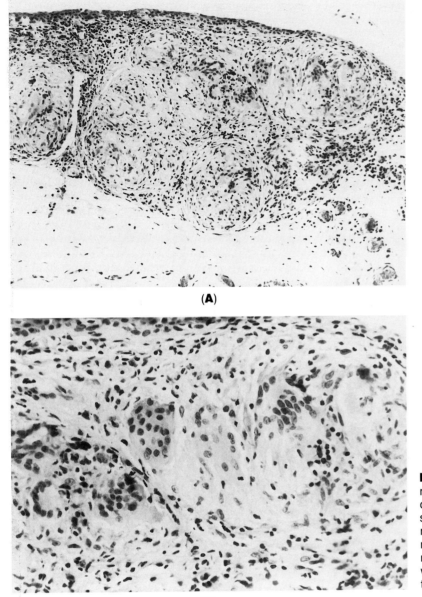

(A)

(B)

FIGURE 96-1 **(A)** Granulomatous inflammation of the conjunctiva from a patient with sarcoidosis (hematoxylin-eosin, magnification ×67). **(B)** Higher magnification of part **A** illustrates numerous multinucleated cells within sarcoid granuloma (hematoxylin-eosin, magnification ×235).

Nodules due to granulomas commonly develop on the iris at the pupillary margin (Koeppe's nodules) or in the superficial stroma (Busacca's nodules). The uveitis of sarcoidosis may be associated with a facial nerve palsy and fever (Heerfordt's syndrome). Soluble immune complexes form in acute sarcoidosis owing to the excess antigen, and their circulation is usually associated with active symptomatic disease, such as acute iritis with erythema nodosum, malaise, arthralgias, and bilateral hilar lymphadenopathy. This so-called Lofgren's syndrome has a good prognosis and usually resolves without therapy. Calcific band keratopathy may complicate

chronic iridocyclitis. This is due partly to the hypercalcemia that frequently accompanies sarcoidosis because of the conversion of 25-hydroxycholecalciferol (25-hydroxyvitamin D_3) to 1,25-dihydroxycholecalciferol (calcitriol) by macrophages in areas of granulomatous inflammation. Calcitriol increases calcium absorption by the gut. Opaque, gray-yellow, slightly elevated nodules form in the conjunctiva in about one third of patients.

The posterior segment is involved in about 25% of patients with ocular sarcoidosis, and such involvement correlates with a high incidence of cystoid macular edema and central nervous system involvement.[5] The

TABLE 96-3 Abnormalities in 202 Patients with Ophthalmic Sarcoidosis (From Obenauf et al.[5])

Abnormality	No.	%
Anterior segment disease	171	84.7
Chronic granulomatous uveitis	106	52.5
Iris nodules	23	11.4
Acute iritis	30	14.9
Cataracts	17	8.4
Conjunctival lesion	14	6.9
Band keratopathy	9	4.5
Interstitial keratitis	2	1.0
Posterior segment disease	51	25.3
Chorioretinitis	22	10.9
Periphlebitis	21	10.4
Chorioretinal nodules	11	5.5
Vitreous cells, opacities, or both	6	3.0
Vitreous hemorrhage	3	1.5
Retinal neovascularization	3	1.5
Orbital and other disease	53	26.2
Lacrimal gland	32	15.8
Optic nerve	15	7.4
Motility	4	2.0
Orbital granuloma	2	1.0

vitreous occasionally contains isolated cells or snow-white clumps of lymphocytes, epithelioid cells, and mononuclear cells arranged like a string of pearls. Aggregates of these cells may look like drippings of candle-wax around retinal veins, which may become thrombosed. Hemorrhages into the retina and vitreous sometimes follow retinal neovascularization. Multiple round, yellow-gray granulomas may form in the choroid and sometimes evolve into chorioretinal scars.

The optic nerve is often affected, together with the central nervous system. Edema or neovascularization of the optic disc, retrobulbar or optic neuritis, optic atrophy, papillitis, and granulomas of the optic nerve may occur.[7]

Granulomatous nodules of the eyelids may be the only ophthalmic abnormality. A granulomatous dacryoadenitis is common and presents clinically as painless enlargement of the lacrimal gland. When bilateral it may be associated with enlarged parotid glands (Mikulicz's syndrome). Involvement of other parts of the orbit occasionally causes unilateral proptosis because of the increase in orbital mass.[8] Sarcoidosis of the lacrimal sac is rare. Central nervous system involvement may produce neuro-ophthalmic signs and symptoms.

Diagnostic Procedures

A histopathologic diagnosis is needed when therapy with corticosteroids is planned and when the clinical diagnosis is uncertain. The granuloma of sarcoidosis contains macrophages, epithelioid cells, T and B lymphocytes, and multinucleate giant cells. Some central necrosis may be present, but caseation necrosis is not a feature. Biopsy of conjunctival granulomas is often a good way of confirming the diagnosis, but light microscopy of clinically normal conjunctiva is positive in less than 10% of cases of sarcoidosis.[9] Positive conjunctival biopsies are obtained in 30 to 50% of cases where conjunctival follicles are visualized (Fig. 96-1).[10] A transconjunctival biopsy of the lacrimal gland yields positive results in 25% of patients with sarcoidosis with clinically normal glands and in 75% of instances when the glands are enlarged. A gallium-67 scan, which allows the visualization of infiltrates in the lacrimal glands, may indicate appropriate sites for biopsy.[9] A transbronchial lung biopsy using a flexible fiberoptic bronchoscope may be necessary. Sarcoid-like granulomas occur in numerous conditions and need to be distinguished from those of sarcoidosis.

Epithelioid cells and other cells derived from monocytes are sources of angiotensin-converting enzyme (ACE). While not diagnostic of sarcoidosis, the serum ACE level may be useful in monitoring the clinical course of the patient, as it correlates with the severity of the disease.[11] The serum ACE becomes normal during remissions or when corticosteroids suppress activity, and rising levels accompany relapses.

Results of the Kveim-Siltzbach skin test, which is infrequently used in practice, are positive in most established cases of sarcoidosis. Most patients do not require therapy, and about two thirds recover completely with little or no residua. The overall mortality rate for sarcoidosis is 1 to 5%; pulmonary insufficiency is the commonest cause of death. Sudden death due to involvement of the cardiac conducting system may be the initial manifestation of sarcoidosis.

Treatment

Objectives in the therapy of ocular sarcoidosis are to alleviate symptoms and to prevent complications, which can usually be achieved by administering topical corticosteroids in conjunction with long-acting cycloplegics. Because posterior segment involvement usually accompanies serious systemic disease, its presence customarily necessitates systemic corticosteroid therapy.

References

1. James DG, Williams WJ. Sarcoidosis and other granulomatous disorders. Philadelphia, Pa: WB Saunders Co; 1985.
2. Johns CJ, ed. *Proceedings of the Tenth International Conference*

on Sarcoidosis and Other Granulomatous Disorders. New York, NY: New York Academy of Sciences;1986.

3. Scadding JG, Mitchell DN. Sarcoidosis. 2nd ed. London: Chapman and Hall; 1985.
4. Daniele RP, Dauber JH, Rossman MD. Immunologic abnormalities in sarcoidosis. *Ann Intern Med.* 1980;92:406-416.
5. Obenauf CD, Shaw HE, Sydnor CF, Klintworth GK. Sarcoidosis and its ophthalmic manifestations. *Am J Ophthalmol.* 1978;86:648-655.
6. Jabs DA, Johns CJ. Ocular involvement in chronic sarcoidosis. *Am J Ophthalmol.* 1986;102:302-307.
7. Beardsley TL, Brown SWL, Sydnor CF, et al.: Eleven cases

of sarcoidosis of the optic nerve. *Am J Ophthalmol.* 1984;97:62-77.

8. Collison JMT, Miller NR, Green WR. Involvement of orbital tissues by sarcoid. *Am J Ophthalmol.* 1986;102:297-301.
9. Weinreb RN, Tessler H. Laboratory diagnosis of ophthalmic sarcoidosis. *Surv Ophthalmol.* 1984;28:653-664.
10. Merritt JC, Ross G, Avery A. Conjunctival biopsy in sarcoidosis: 4 year NCMH experience. *North Carolina Med J.* 1983;44:636-637.
11. Thomas PD, Hunninghake GW. Current concepts of the pathogenesis of sarcoidosis. *Amer Rev Respir Dis.* 1987; 135:747-760.

Chapter 97

Vogt-Koyanagi-Harada Syndrome

MANABU MOCHIZUKI

Vogt-Koyanagi-Harada (VKH) disease is a systemic disease typically affecting the eyes, ears, meninges, hair, and skin. The tissues affected in the disease contain melanocytes, and the clinical symptoms manifested are bilateral uveitis, dysacousia, meningeal irritation, poliosis, and vitiligo. The disease affects adults primarily; most patients are between ages 20 and 50. Distinct sex differences are not found in the disease incidence. VKH disease is observed more frequently in Orientals and in people of pigmented races than in Caucasians. In Japan, the disease is found in about 8% of endogenous uveitis patients (compared to 2.5% in Brazil and only about 1% in the United States).[1-3] The predilection of VKH disease for the Japanese appears to be related to their immunogenetic background. Ohno et al.[4] have reported a close association of the disease with HLA-Bw54, -DQWa, -DR4, and -DRw53. All patients examined so far are positive for HLA-DRw53. In addition, HLA-Bw54, -DQWa, -DR4, and -DRw53 are in strong linkage disequilibrium in the Japanese, whereas such linkage disequilibrium is not seen in Caucasians.[4] Therefore, this HLA haplotype peculiar to Japanese may be responsible for their high susceptibility.

The cause and pathogenesis of VKH disease are still controversial. Viral infection was previously suggested

as a cause because of the acute onset of headache, the fatigue, and the self-limiting nature of the disease, although no specific virus has been detected to date. More recently, immune or autoimmune mechanisms have been emphasized. A number of reports, mainly by Japanese investigators, have demonstrated the involvement of autoimmunity or cytotoxic lymphocytes against melanocytes in the inflammatory processes of VKH disease.[5,6] Some studies suggest that uveitogenic retinal antigens (eg, retinal soluble antigen and interphotoreceptor retinoid-binding protein) may play a role in initiating autoimmune inflammation in the eye.[7]

Systemic Manifestations

The evolution of VKH disease can be divided into three stages, prodromal, ophthalmic, and convalescent.[1] In the prodromal stage, the majority of patients complain of severe headache, pain deep in the orbit, slight fever, vertigo, and nausea. These episodes usually last a few days and are followed by the ophthalmic stage. In the prodromal and ophthalmic stages, meningeal signs and ear symptoms are quite common. Pleocytosis of the

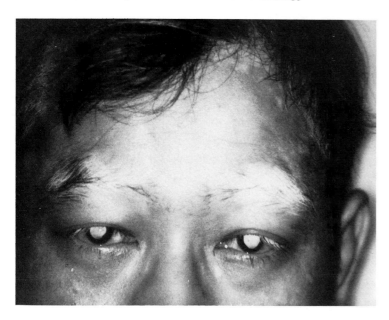

FIGURE 97-1 Poliosis of the eyelashes and the eyebrows and vitiligo of the head in a 32-year-old man. (Courtesy Dr. S Ohno.)

cerebrospinal fluid and hearing loss definitely demonstrated by audiogram are observed in most patients, but they disappear gradually in 2 to 3 months. The type of cells in the cerebrospinal fluid are mainly lymphocytes, of which 85% are OKT11 positive and 65% are OKT4 positive.[8] However, no patients show localizing neurologic signs such as cranial neuropathy or myopathy. These symptoms are thought to be evoked by inflammation due to an immune response to melanocytes in the leptomeninges and in the membranous labyrinth.[1]

About 2 to 3 months after the onset, acute uveitis begins to subside and the convalescent stage follows. Depigmentation appears in the eye, skin, and hair (Fig. 97-1). The earliest sign of depigmentation occurs in the corneal limbus about 1 month after the onset of the disease (Sugiura's sign). Systemic administration of corticosteroid appears to reduce the incidence and intensity of depigmentation. According to Ohno et al.,[4] about 85% of patients have Sugiura's sign, 64% have a blond-appearing fundus (sunset-glow fundi), 40% have poliosis, 34% have vitiligo, and 33% have alopecia.

Ocular Manifestations

The ophthalmic symptoms begin with sudden onset of blurred vision. Though it is a bilateral condition, in about 30% of patients there is a few days' delay before the second eye becomes affected. Uveitis begins primarily in the posterior segment of the eye with papilledema and circumscribed retinal edema, mostly located in the posterior pole (Fig. 84-2). As the inflammation becomes severe, an exudative retinal detachment occurs, usually in the inferior quadrants of the eye. Fluorescein angiography shows characteristic multiple areas of leakage of fluorescein from the choroid into the subretinal space, but not from the retinal vessels (Fig. 97-2). Therefore, the inflammatory changes in the fundus are not due to retinal vasculitis but to inflammation in the choroid and the retinal pigment epithelium. Inflammation in the anterior uvea is also commonly observed, with cells and flare in the anterior chamber, mutton fat–like keratic precipitates, and iris nodules. In the convalescent stage, evidence of depigmentation can be seen in the eye: at the corneal limbus about 1 month after the onset and in the ocular fundus in 2 to 3 months. The fundus shows a sunset-glow appearance combined with a mottled appearance typified by scattered white spots and pigment clumps.

Histologic changes of the eye in VKH disease are characterized by the presence of granulomatous changes consisting of giant cells and epithelioid cells surrounded by lymphocytes and plasma cells. An electron microscopic study by Matsuda et al.[5] demonstrated that lymphocytes and melanocytes were closely opposed in the uveal tissues of VKH patients, and they suggested that immune reactions directed toward the melanocyte play a significant role in the inflammatory process of the disease.

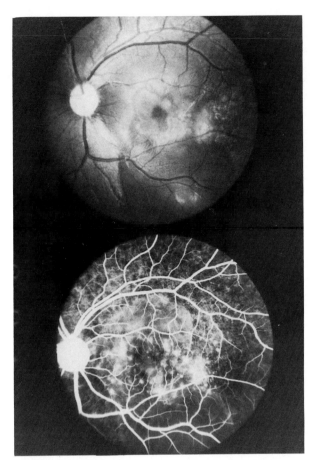

FIGURE 97-2 Fundus pictures of the eye of a 35-year-old woman with VKH disease. **(Top)** Subretinal edema at the posterior pole. **(Bottom)** Fluorescein angiogram of the same patient, showing multiple leakage of the fluorescein dye at the posterior pole from the choroid.

The treatment of VKH disease is primarily local and systemic corticosteroids. Masuda et al.[9] reported using a combination of an osmotic agent and a large dose of corticosteroid (200 mg of prednisolone), with the dose reduced by one half or two thirds every 3 days until the dose reached 40 mg of prednisolone, when slow tapering was initiated. This method is successful in most cases and is widely used for the treatment of VKH disease in Japan.

References

1. Sugiura S. Vogt-Koyanagi-Harada disease. *Jpn J Ophthalmol.* 1978;22:9.
2. de Abreu MT, Hirata PS, Belfort R Jr, et al. Uveites en Sao Paulo. Estudo epidemiologico, clinico e terapeutuco. *Arq Bras Oftal.* 1980;43:10.
3. Ohno S, Kumura SJ, O'Connor GR. Vogt-Koyanagi-Harada syndrome. *Am J Ophthalmol.* 1977;83:735.
4. Ohno S. Vogt-Koyanagi-Harada's disease. In: Saari KM, ed. *Uveitis Update.* Amsterdam: Excerpta Medica; 1984;401-405.
5. Matsuda H, Sugiura S. Ultrastructural changes of the melanocyte in Vogt-Koyanagi-Harada syndrome and sympathetic ophthalmia. *Jpn J Ophthalmol.* 1971;15:69.
6. Maezawa N, Yano A. Two distinct cytotoxic T lymphocyte subpopulations in patients with Vogt-Koyanagi-Harada disease that recognize human melanoma cells. *Microbiol Immunol.* 1984;28:219.
7. Chan CC, Palestine AG, Nussenblatt RB, et al. Anti-retinal auto-antibodies in Vogt-Koyanagi-Harada syndrome, Behçet's disease, and sympathetic ophthalmia. *Ophthalmology.* 1985;8:1025.
8. Okubo K, Kurimoto S, Okubo K, et al. Surface marker studies of cerebrospinal fluid lymphocytes in Vogt-Koyanagi-Harada's disease. *Acta Soc Ophthalmol Jpn.* 1984;89:726.
9. Masuda K. Harada's disease. A new therapeutic approach. *Jpn J Clin Ophthalmol.* 1969;23:553.

PART 10

Malignant Disorders

Chapter 98

Metastatic Malignant Tumors

JERRY A. SHIELDS, ERIC P. SHAKIN, and CAROL L. SHIELDS

General Considerations

Neoplasms metastatic to the eye and adnexa are a relatively small proportion of the tumors treated by ophthalmologists or oncologists. The vast majority of tumors metastatic to the ocular structures are carcinomas, whereas sarcomas metastatic to the eye and adnexa are quite rare. Since the eye and most of the orbit are devoid of lymph channels, metastases generally reach these structures by hematogenous spread.

Carcinoma of the breast is the primary lesion that most often metastasizes to ocular tissues.[1-4] Lung cancer accounts for the majority in men. Cancer of the prostate has a tendency to metastasize to the orbital bones, but it rarely spreads to the intraocular structures. Other tumors that occasionally metastasize to the eye and adnexa include carcinoma of the gastrointestinal tract, renal cell carcinoma, thyroid carcinoma, carcinoid tumor, and cutaneous malignant melanoma. Lymphoid tumors and leukemias, which are not discussed here, can also involve the ocular tissues as part of the systemic disease.

The vast majority of ocular metastases occur in adults. The exceptions are the orbital metastases that sometimes occur from neuroblastoma or Ewing's tumor in children.

Tumors metastatic to the ocular structures have rather typical clinical features, and the diagnosis can often be made on the basis of history and ophthalmologic examination. Many patients have a history of prior treatment for a malignancy, such as mastectomy, lung resection, gastric or intestinal resection, or nephrectomy. If such a history is not elicited, a thorough systemic evaluation may help to detect a primary tumor or other evidence of metastasis. In many cases, ancillary diagnostic studies such as ultrasonography, computed tomography (CT), and magnetic resonance imaging (MRI) may be necessary to substantiate the ocular diagnosis. Biopsy is not always necessary for establishing the diagnosis, particularly when the lesion is in a relatively inaccessible location. About one fourth of patients who present to the ophthalmologist with a metastatic tumor

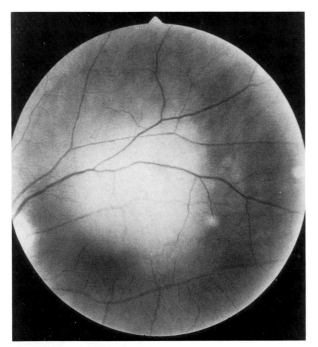

FIGURE 98-1 Fundus photograph of metastatic carcinoma to the choroid.

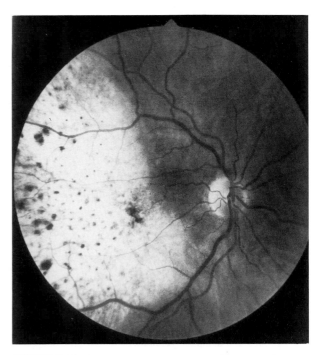

FIGURE 98-2 Fundus photograph of metastatic carcinoma to the choroid involving much of the posterior pole with overlying pigment disturbance.

do not have a history of cancer. In such cases making the diagnosis can be more challenging.

In most instances the systemic prognosis is quite poor once the patient develops metastasis to the ocular structures. However in some cases, particularly with metastasis from carcinoid tumors, the patient may survive for several years after the development of ocular metastasis.

Intraocular Metastasis

Tumors that metastasize to the intraocular structures usually involve the uveal tract and most commonly are located in the posterior choroid. Metastatic tumors to the iris and ciliary body are less common. Tumors metastatic to the retina, optic disc, and vitreous are comparatively rare.

Clinical Features

The clinical features vary with the site of the intraocular metastasis.[1] A choroidal metastasis usually appears as a creamy yellow, slightly elevated, sessile or ovoid mass in the posterior portion of the choroid (Fig. 98-1). Choroidal metastases are usually solitary but may be multifocal. An associated serous nonrhegmatogenous

retinal detachment is present in about 70% of cases.[2,3,5] Long-standing or regressed lesions may show clumps of golden brown pigment on the tumor surface (Fig. 98-2). A metastatic lesion should be differentiated from amelanotic choroidal melanoma, choroidal hemangioma, posterior scleritis, and other posterior segment conditions.

An iris metastasis appears as one or more yellow, friable, gelatinous, nodular lesions within the iris stroma (Fig. 98-3).[3] Such tumors tend to be loosely cohesive and may seed tumor cells throughout the anterior chamber, simulating intraocular inflammation. A ciliary body metastasis may be more difficult to visualize and may masquerade as an unexplained uveitis and glaucoma for a while before the diagnosis is suspected.

A metastatic tumor to the optic disc appears as a yellowish white swelling of the optic nerve head. The white infiltrative material is rather typical and should be differentiated from papilledema or optic neuritis. In many cases there is also some peripapillary choroidal involvement.

Diagnostic Approaches

Most tumors metastatic to the uveal tract can be diagnosed on the basis of a general physical and ocular evaluation; ancillary studies only supplement the diagnosis. Choroidal metastases have typical, but not patho-

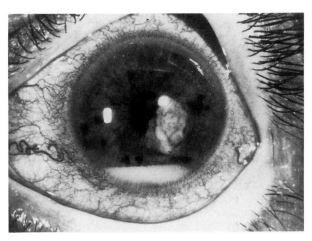

FIGURE 98-3 Metastatic carcinoma to iris, presumably of gastric origin.

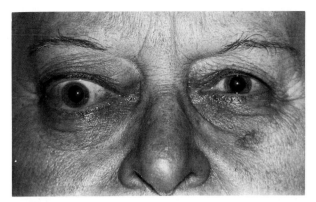

FIGURE 98-4 Proptosis of the right eye secondary to metastatic tumor to the orbit.

gnomonic, characteristics with fluorescein angiography and ocular ultrasonography. The value of CT and MRI in the diagnosis is not clearly established. In very difficult cases, transocular fine needle aspiration biopsy has been advocated.[7]

Management

The management of tumors metastatic to the uvea is generally chemotherapy or radiotherapy. If the uveal metastasis is controlled with chemotherapy alone and if visual acuity is not impaired, radiotherapy can be withheld. If there is associated vision loss, intraocular inflammation, or glaucoma that is not responding to chemotherapy, radiotherapy should be employed. In general, about 3500 cGy is delivered to the entire eye and orbit in fractionated doses over a 3- to 4-week period.

Orbital Metastasis

The orbit is a fertile ground for the deposition of metastatic tumors from distant sites.[6] The majority of orbital metastases come through hematogenous routes and settle in the soft tissues. Many have a tendency to infiltrate the extraocular muscles. Metastasis from prostatic carcinoma has a greater tendency to affect the orbital bones as well.

Clinical Features

A metastatic tumor to the orbit is usually characterized by rather rapid onset and progression of proptosis and displacement of the globe (Fig. 98-4).[4] Some patients with sclerosing metastatic tumors, such as scirrhous

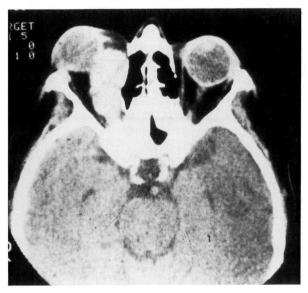

FIGURE 98-5 Computed tomogram of patient shown in Figure 98-4 shows irregular mass in superomedial portion of the orbit.

infiltrating ductal carcinoma of the breast, may develop paradoxical enophthalmos on the affected side, owing to contraction of the fibrotic orbital tumor.

Diagnosis

In recent years orbital CT has become the most important diagnostic modality for metastatic tumors to the orbit. It has largely replaced standard radiography, arteriography, venography, and ultrasonography as the method for establishing the diagnosis. The classic CT appearance is that of an irregular or diffuse homogeneous orbital mass (Fig. 98-5). With CT, the tumor may

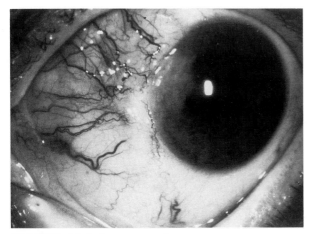

FIGURE 98-6 Breast carcinoma metastatic to conjunctiva and peripheral cornea. Note the fleshy vascular nature of the lesion and the involvement of the peripheral cornea.

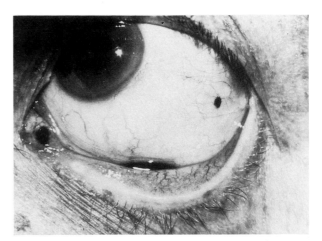

FIGURE 98-7 Metastatic melanoma to the conjunctiva from a cutaneous melanoma in the skin of the buttocks. Note the three separate foci. (Courtesy of Chester Pryor, M.D.)

closely resemble a lymphoid tumor or inflammatory pseudotumor.[4]

The diagnosis of metastatic tumor to the orbit is usually established by some form of biopsy. In the case of small circumscribed anterior orbital tumors, excisional biopsy may be possible. Large, poorly defined tumors may be diagnosed by incisional biopsy. In cases of posterior orbital tumors that are less surgically accessible or when there is evidence of widespread metastasis, fine-needle aspiration biopsy for cytologic studies may be adequate to substantiate the diagnosis.[4]

Management

Orbital metastases are managed by chemotherapy and radiotherapy in a manner similar to that for uveal metastasis.

Conjunctival Metastasis

Tumors metastatic to the conjunctiva are relatively rare and generally occur in patients who have known metastatic disease elsewhere. They generally appear as one or more fleshy stromal masses (Fig. 98-6). Metastatic melanoma to the conjunctiva is generally black (Fig. 98-7). Treatment is usually local resection or radiotherapy.

Palpebral Metastasis

Tumors metastatic to the eyelids account for less than 1% of all eyelid malignancies.[8] Eyelid metastases are usually deep to the surface epithelium and may initially

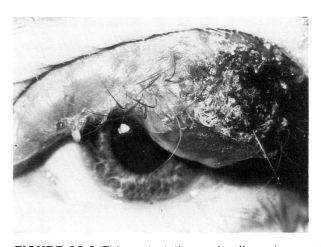

FIGURE 98-8 This metastatic renal cell carcinoma lesion on the eyelid was initially believed to be a chalazion.

simulate chalazion (Fig. 98-8). Metastatic melanoma to the eyelid may be dark blue or black. The treatment of eyelid metastasis is usually local resection or radiotherapy.

Summary

Tumors metastatic to the eye and adnexa represent a unique group of neoplasms. Most develop in adults and occur secondary to carcinoma of the breast or lung. Intraocular metastasis usually involves the uveal tract and generally appears in the posterior choroid. Retinal metastasis is extremely rare. Similar metastases can appear in the soft tissues of the orbit, the eyelids, and the conjunctiva. Management usually consists of biopsy followed by chemotherapy or radiotherapy. Long-term prognosis is generally poor.

References

1. Ferry AP, Font RL. Carcinoma metastatic to the eye and orbit. I. A clinicopathologic study of 227 cases. *Arch Ophthalmol.* 1974;92:276-286.
2. Stephens RF, Shields JA. Diagnosis and management of cancer metastatic to the uvea: a study of 70 cases. *Ophthalmology.* 1979;86:1336-1349.
3. Shields JA. Metastatic tumors to the uvea and retina. In: *Diagnosis and Management of Intraocular Tumors.* St. Louis, Mo: CV Mosby; 1983.
4. Shields JA. Metastatic tumors to the orbit. *Diagnosis and Management of Orbital Tumors.* Philadelphia, Pa: WB Saunders Co.; 1988.
5. Shakin ES, Shields JA, Augsburger JJ. Metastatic tumors to the uvea and optic disc. A study of 300 cases. (Unpublished data.)
6. Shields CL, Shields JA, Peggs M. Tumors metastatic to the orbit. *J Ophthalmol Plast Reconstr Surg.* 1988;42:78-80.
7. Augsburger JJ, Shields JA. Fine needle aspiration biopsy of solid intraocular tumors. Indications, instrumentation and techniques. *Ophthalmic Surg.* 1984;15:34-40.
8. Mansour AM, Hidayat AA. Metastatic eyelid disease. *Ophthalmology.* 1987;94:667-670.

Chapter 99

Non-Hodgkin's Lymphoma
Reticulum Cell Sarcoma

ANDREW P. SCHACHAT

In most cases the lymphomas can be classified according to the cell of origin.[1] Reticulum cell sarcoma has also been referred to as diffuse histiocytic lymphoma, large cell lymphoma, malignant lymphoma of the uveal tract, and intraocular lymphoma. Both B- and T-cell lineage has been claimed.[2-7] When a clearer picture of the immunologic derivation of this condition is available, it may be appropriate to change the name of the disease. For now, it seems most reasonable simply to classify the disease as a non-Hodgkin's lymphoma.

Clinical Features

Recognition of reticulum cell sarcoma is important because it is one of the few life-threatening diseases that an ophthalmologist is likely to diagnose. Typically there is a delay of 1 to 2 years between onset of symptoms and diagnosis.[8] The condition should be suspected in older patients with "idiopathic uveitis."

The disease usually presents in patients over age 50 years, although it has been reported in a 27-year-old.[6] In our series, the mean age at the time of diagnosis of ocular reticulum cell sarcoma was 61 years and the age range was 36 to 82 years.[8] There is no sexual or racial predisposition. Although it is often asymmetric, reticulum cell sarcoma involves both eyes in roughly 80% of patients.

The ocular symptoms are usually those of posterior uveitis: loss of vision and floaters. Pain is not prominent, and externally the eyes are usually quiet. The clinical findings may include decreased acuity, anterior segment inflammation (keratic precipitates, aqueous cells, and flare), vitreous cells, and subretinal infiltrates.[8] In a few cases there are characteristic large multifocal yellowish sub–pigment epithelial tumors (Fig. 99-1).[9,10]

Both true inflammatory signs (circulating inflammatory cells) and pseudoinflammatory signs (circulating tumor cells) may be seen. Histopathologic study shows that inflammatory cells are often admixed with tumor cells, and it is possible that in a given patient the cells may initially be inflammatory rather than neoplastic.[11] Alternatively, the tumor itself may produce an inflammatory reaction.

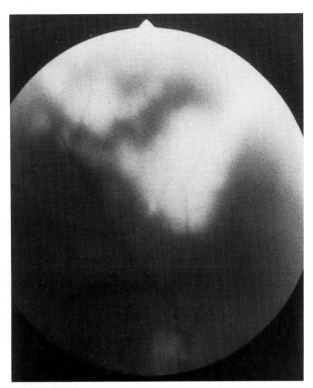

FIGURE 99-1 Yellowish subpigment epithelial infiltrate in a patient with reticulum cell sarcoma. The photograph is hazy because of the associated vitritis. (Courtesy Hilel Lewis, M.D. and William Mieler, M.D.)

Ocular reticulum cell sarcoma is associated with systemic lymphoma. In our experience, associated central nervous system (CNS) disease was present in 60% of patients. Ocular and visceral lymphoma was present in 14% of patients. Six percent of patients had ocular, CNS, and visceral disease, whereas 20% had what appeared to be isolated ocular disease. Ophthalmologists feel that ocular signs and symptoms tend to predate CNS and visceral symptoms, but this may represent ascertainment bias.[8,12] In the setting of CNS signs and symptoms or visceral findings consistent with lymphoma the diagnosis should be strongly suspected.

Differential Diagnosis

When anterior segment findings predominate, anterior uveitis of any cause as well as iris tumors (chiefly metastases) and diffuse iris melanomas should be considered. Simulating posterior segment conditions include viral retinitis and toxoplasmosis. When the choroidal features predominate, amelanotic melanoma, choroidal metastases, choroidal hemangioma, benign

reactive lymphoid hyperplasia, and all causes of disseminated choroiditis should be considered.[13] Unfortunately, the majority of patients with reticulum cell sarcoma have been followed—and often treated—for a year or more for a diagnosis of idiopathic uveitis or vitritis before the diagnosis of reticulum cell sarcoma was established.[8]

Diagnostic Evaluation

In the past, the diagnosis of reticulum cell sarcoma usually was not made until autopsy or enucleation. With the advent of diagnostic vitrectomy, ocular cytologic diagnosis is now possible. Most eye pathology laboratories process the specimen with a millipore filter and employ Papanicolaou's stain.[14] In a recent review the diagnosis was made by vitreous biopsy in 54% of patients, by enucleation in 14%, and not until postmortem examination in 31% of cases.[8]

Because systemic findings are common, the ophthalmologist should enlist the aid of an internist or oncologist in the systemic evaluation. Lumbar puncture and CSF cytopathology may yield positive findings (29% of cases in our experience) and thus obviate diagnostic vitrectomy.[8] Computed tomography (CT) and magnetic resonance imaging (MRI) may demonstrate focal CNS lesions.

Treatment

Radiation therapy is generally recommended for patients with ocular reticulum cell sarcoma (Figs. 99-2, 99-3). Four thousand rad given in divided doses can improve the ocular disease, although acuity may not normalize.[15] Focal CNS radiation therapy often ameliorates focal CNS symptoms, but it may not prolong survival. The role of prophylactic radiation therapy is controversial. Intrathecal chemotherapy may be of value when there is associated CNS disease. Promising results of treatment of intraocular lymphoma with high doses of systemic cytosine arabinoside have been summarized by Baumann, et al.[16] New developments in the treatment of primary CNS lymphoma, chiefly multiagent chemotherapy in conjunction with osmotic blood-brain modification, may prove to be beneficial for patients with intraocular lymphoma.[17]

Currently we recommend bilateral ocular radiation in cases of proven ocular reticulum cell sarcoma and CNS radiation if focal CNS disease is present. Prophylactic CNS radiation and intrathecal chemotherapy may have a role. Systemic treatment recommendations should be made only after consultation with oncologists, radiation therapists, and other physicians involved in the patient's care.

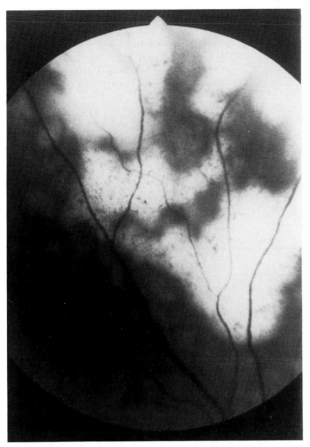

FIGURE 99-2 Three months following external beam radiation, the sub-pigment epithelial infiltrate (Fig. 99-1) has been resorbed, leaving areas of pigment atrophy. (Courtesy Hilel Lewis, M.D. and William Mieler, M.D.)

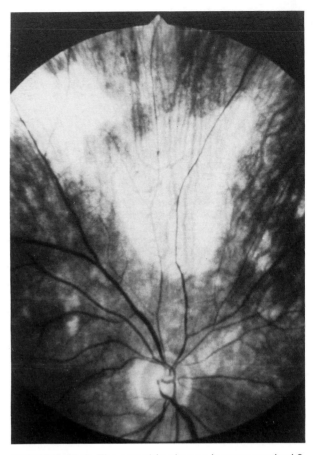

FIGURE 99-3 The atrophic change is more marked 2 years later. (Courtesy Hilel Lewis, M.D. and William Mieler, M.D.)

Prognosis

Past reports indicate that most patients die within 2 years of the ocular diagnosis, usually as a result of progressive CNS disease.[15,18,19] In our series the median survival time was 39 months. The mean interval from diagnosis of reticulum cell sarcoma at any site to death was 20 months.

References

1. DeVita VT, Jaffe ES, Hellman S. Hodgkin's disease and non-Hodgkin's lymphomas. In: DeVita VT, Hellman S, Rosenberg SA, eds. *Cancer, Principles and Practice of Oncology*. 2nd ed. Philadelphia, Pa: JB Lippincott; 1985.
2. Jones SE, Fuks Z, Bull M, et al. Non-Hodgkin's lymphomas IV. Clinicopathologic correlation in 405 cases. *Cancer.* 1973;31:806-823.
3. Kaplan HJ, Meredith TA, Aaberg TM, et al. Reclassification of intraocular reticulum cell sarcoma (histiocytic lymphoma): immunologic characterization of vitreous cells. *Arch Ophthalmol.* 1980;98:707-710.
4. Saga T, Ohno S, Matsuda H, et al. Ocular involvement by a peripheral T-cell lymphoma. *Arch Ophthalmol.* 1984;102:399-402.
5. Michelson JB, Michelson PE, Bordin GM, et al. Ocular reticulum cell sarcoma: presentation as retinal detachment with demonstration of monoclonal immunoglobulin light chains on the vitreous cells. *Arch Ophthalmol.* 1981;99:1409-1411.
6. Qualman SJ, Mendelsohn G, Mann RB, et al. Intraocular lymphomas: natural history based on a clinicopathologic study of eight cases and review of the literature. *Cancer.* 1983;52:878-886.
7. Elner VM, Hidayat AA, Charles NC, et al. Neoplastic angioendotheliomatosis. *Ophthalmology.* 1986;93:1237-1247.

8. Freeman LN, Schachat AP, Knox DL, et al. Clinical features, laboratory investigations and survival in ocular reticulum cell sarcoma. *Ophthalmology.* 1987;94:1631-1639.

9. Barr CC, Green WR, Payne JW, et al. Intraocular reticulum-cell sarcoma: Clinicopathologic study of four cases and review of the literature. *Surv Ophthalmol.* 1975;19:224-239.

10. Gass JDM, Sever RJ, Grizzard WS, et al. Multifocal pigment epithelial detachments by reticulum cell sarcoma. *Retina.* 1984;4:135-143.

11. Kennerdell JS, Johnson BL, Wisotzkey HM. Vitreous cellular reaction: association with reticulum cell sarcoma of the brain. *Arch Ophthalmol.* 1975;93:1341-1345.

12. Rockwood EJ, Zakov ZN, Bay JW. Combined malignant lymphoma of the eye and CNS (reticulum-cell sarcoma). *J Neurosurg.* 1984;61:369-374.

13. Shields JA. Intraocular lymphoid tumors and leukemias. In: *Diagnosis and Management of Intraocular Tumors.* St. Louis, Mo: CV Mosby; 1983;638.

14. Engel HM, Green WR, Michels RG, et al. Diagnostic vitrectomy. *Retina.* 1981;1:121-149.

15. Margolis L, Fraser R, Lichter A, et al. The role of radiation therapy in the management of ocular reticulum cell sarcoma. *Cancer.* 1980;45:688-692.

16. Baumann MA, Ritch PS, Hande KR, et al. Treatment of intraocular lymphoma with high-dose Ara-C. *Cancer.* 1986;57:1273-1275.

17. Neuwelt EA, Frenkel EP, Gumerlock MK, et al. Developments in the diagnosis and treatment of primary CNS lymphoma, a prospective series. *Cancer.* 1986;58:1609-1620.

18. Shields JA. *Diagnosis and Management of Intraocular Tumors.* St. Louis, Mo: CV Mosby; 1983;634-640.

19. Char DH, Margolis L, Newman AB. Ocular reticulum cell sarcoma. *Am J Ophthalmol.* 1981;99:1048-1052.

Chapter 100

Remote Effects of Cancer

RICHARD F. DREYER

Paraneoplastic syndromes result from nonmetastatic effects of a malignancy on remote organs. The central nervous system (CNS) can be involved with symptoms of memory loss, disorientation, and seizures. In other tissues, benign proliferative changes occur, as in the malignant form of acanthosis nigricans, where symmetric pigmented skin lesions develop, usually in intertriginous spaces, in patients who have a systemic malignancy.[1] Another example of benign hypertrophic lesions occurring in association with systemic malignancy is sebaceous adenoma of the eyelid, reported in association with gastrointestinal carcinomas.[2] In the CNS, histopathologic study has identified astrocytic proliferation and fibrous gliosis in the amygdaloid nucleus, hippocampus, fornix, and mamillary bodies.[3] Several possible mediators of this remote effect have been suggested, including toxins secreted by the tumor, antibodies produced in response to the systemic malignancy, which then attack normal tissue, and reduced immunocompetence, which allows viral infection to occur.

In the eye two paraneoplastic syndromes have been described. The first has been called cancer-associated retinopathy[4] and was initially described in three patients with oat cell carcinoma.[5] It has now been reported in breast, uterine, and adrenal carcinomas. Early transient vision loss progresses to severe visual loss, usually resulting in bare light perception. Visual field testing often reveals ring scotomas. The ocular fundus may be normal when symptoms first occur, but arteriolar narrowing and periarteriolar sheathing often appear. Disc pallor occasionally follows. The pigment epithelium is normal in some patients but atrophic in others.[5,6]

Light microscopy reveals rod and cone outer segment degeneration, with pigmented macrophages appearing in areas of photoreceptor damage.[5] Electron microscopic studies reveal a decrease in the rod and cone populations near the fovea and massive photoreceptor loss outside the macula.[6] Outer segment disc membranes are disrupted, and the retinal pigment epithelium contains increased amounts of melanin; normal phagocytosis of outer segments is apparently absent (Fig. 100-1).[6]

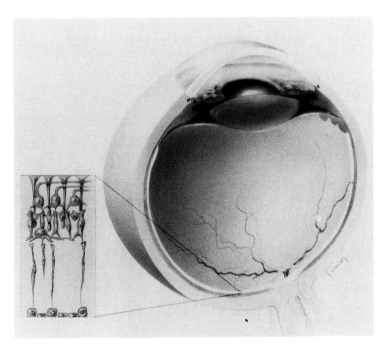

FIGURE 100-1 Schematic diagram of changes seen in cancer-associated retinopathy. There is prominent thinning of the retinal arterioles; the population of photoreceptors is thin and atrophic.

It is recognized that oat cell, pancreatic, and other carcinomas produce physiologically active hormones, including adrenocorticotropic hormone, melanocyte-stimulating hormone, parathyroid hormone, and oxytocin.[7] Melanocyte-stimulating hormone is also a cleavage product of adrenocorticotropic hormone. Since systemic malignancies secrete melanocyte-stimulating hormone, Buchanan et al.[6] postulated that there was a derangement of normal melanin synthesis in pigment epithelial cells. They identified ultrastructural evidence of abnormally increased production and degradation of melanin and speculated that accelerated melanin synthesis interfered with the normal phagocytic activity of the retinal pigment epithelium. Such a loss of phagocytosis of outer segments would result in photoreceptor derangement.

An alternative mechanism of photoreceptor damage in cancer-associated retinopathy could be autoimmune. Antibodies produced against a malignancy might attack normal retinal tissue. Two patients are reported whose vision loss responded to corticosteroid treatment.[8,9] In a later study, Thirkill et al.[4] identified retinal autoantibodies in the serum of four patients with cancer-associated retinopathy. Antiretinal ganglion cell antibodies have been identified in two patients with oat cell carcinoma.[10] It seems likely that an autoimmune mechanism is present; whether or not abnormal melanin synthesis also plays a role remains to be determined.

A second ocular paraneoplastic syndrome has been described that consists of bilateral, diffuse, melanocytic pigmented tumors involving the entire uveal tract. This has been recognized in patients with adenocarcinomas, pancreatic carcinomas, and squamous and undifferentiated pulmonary carcinomas. The ocular changes include rapid loss of vision, symptoms of uveitis, conjunctival injection, anterior chamber cell and flare, pigmented or nonpigmented iris nodules, vitreous cells, cataracts, and serous retinal detachment overlying pigmented choroidal tumors (Fig. 100-2). As in cancer-associated retinopathy, visual loss may precede a diagnosis of systemic carcinoma. This syndrome has been reported in association with pancreatic carcinoma, colon adenocarcinoma, squamous cell carcinoma of the lung, and uterine carcinoma.[11-13] Although scleral invasion by these melanocytic cells has been identified, metastasis has not occurred.

In this syndrome, the uveal tract is diffusely involved with pigmented tumors appearing in the iris, choroid, and ciliary body. Histopathologic study has revealed them to be benign tumors composed mostly of spindle-shaped cells. Mitotic figures and other malignant cytologic features are rare; however, prominent nucleoli and epithelioid cells have been observed. As Barr et al.[12] pointed out, because the tumors appear to be benign histologically and metastases have not occurred and because death from the systemic malignancy is likely, enucleation or other treatment is not recommended.

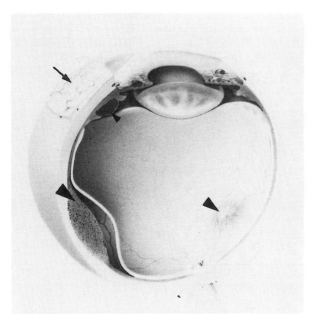

FIGURE 100-2 This composite schematic drawing summarizes the features of the diffuse melanocytic uveal tumors seen in association with systemic malignancies. Conjunctival congestion (*arrow*), hyperpigmented iris nodules, and infiltration of the ciliary body (*small arrowhead*) can occur. A cataract is present. Posterior choroidal tumors are seen (*large arrowhead*) with overlying retinal detachment.

The cause of these tumors remains uncertain. As discussed above, many tumors produce physiologically active hormones, including melanocyte-stimulating hormone. Such substances may in some way cause the appearance of these benign melanocytic tumors.

An alternative explanation is that some derangement of inhibition of cell growth has occurred that allows growth of both the systemic malignancy and coincidentally of the benign melanocytic choroidal tumors.

References

1. Brown J, Winkelmann RK. Acanthosis nigricans: a study of 90 cases. *Medicine.* 1968;47:33.
2. Jakobiec FA. Sebaceous adenoma of the eyelid and visceral malignancy. *Am J Ophthalmol.* 1974;78:952.
3. Corsellis JAN, Goldberg J, Norton JR. Limbic encephalitis and its association with carcinoma. *Brain.* 1968;91:481.
4. Thirkill CE, Roth AM, Keltner JL. Cancer-associated retinopathy. *Arch Ophthalmol.* 1987;105:372.
5. Sawyer RA, Selhorst JB, Zimmerman LE. Blindness caused by photoreceptor degeneration as a remote effect of cancer. *Am J Ophthalmol.* 1976;81:606.
6. Buchanan TAS, Gardiner TA, Archer DB. An ultrastructural study of retinal photoreceptor degeneration associated with bronchial carcinoma. *Am J Ophthalmol.* 1984;97:277.
7. Rees LH, Bloomfield GA, Rees GM, et al. Multiple hormones in a bronchial tumor. *J Clin Endocrinol Metabol.* 1974;38:1090.
8. Keltner JL, Roth AM, Chang S. Photoreceptor degeneration; possible autoimmune disorder. *Arch Ophthalmol.* 1983;101:564.
9. Klingele TG, Burde RM, Rappazzo JA, et al. Paraneoplastic retinopathy. *J Clin Neuroophthalmol.* 1984;4:239.
10. Kornguth SE, Klein R, Appen R, et al. Occurrence of anti-retinal ganglion cell antibodies in patients with small cell carcinoma of the lung. *Cancer.* 1982;50:1289.
11. Ryll DL, Jean-Campbell R, Robertson DM, et al. Pseudometastatic lesions of the choroid. 1980;87:1181.
12. Barr CC, Zimmerman LE, Curtin VT, et al. Bilateral diffuse melanocytic uveal tumors associated with systemic malignant neoplasms, a recently recognized syndrome. 1982; 100:249.
13. Gass JDM. *Stereoscopic Atlas of Macular Diseases.* St. Louis, Mo: CV Mosby; 1987.

PART 11

Metabolic Disorders

Section A

Disorders of Amino Acid Metabolism

Chapter 101

Albinism

EVERETT AI and PATRICK COONAN

Albinism refers to a group of clinical syndromes characterized by hypopigmentation due to heritable metabolic defects in the pigment cell (melanocyte) system of the eye and skin (Fig. 101-1). In general, albinism is due to a nonlethal mutation with a manifest incidence of 1 in 5,000 to 1 in 100,000 population, depending on the area studied.[1,2] Most affected persons lead relatively normal lives, limited only by their poor vision, light-sensitive skin, and the social stress generated by their unusual appearance.

Pathophysiology

Melanin is the dark pigment of the skin, hair, eyes, and brain. In most forms of albinism, the melanocytes are distributed normally throughout the eye and skin but fail to synthesize adequate amounts of melanin. Such is the case in oculocutaneous albinism; in ocular albinism, the number of melanosomes is decreased. The pathway for melanin synthesis begins with the conversion of tyrosine to dihydroxyphenylalanine (DOPA), and then to DOPA quinone. Both steps are catalyzed by the enzyme tyrosinase. Albinism is caused by the absence, deficiency, or ineffectiveness of this enzyme.[3]

Systemic Manifestations

Certain forms of albinism may demonstrate systemic manifestations that have important clinical implications and which may be helpful in identifying a particular syndrome. While the findings seen in ocular albinism are usually limited to the eye, there may be a brief period during infancy and early childhood when pale skin and snow-white hair are noted.[4,5]

Oculocutaneous (generalized, universal) albinism involves the skin and eye and may demonstrate important nonophthalmic associations. The examiner should pay particular attention to the possible presence of deafness, mental retardation, progressive cerebral degeneration, coagulation defects, and evidence of reticuloendothelial incompetence. In addition, the lack of skin pigmentation may cause easy sunburning, and in some patients, a tendency to develop skin carcinomas on exposed body surfaces.

311

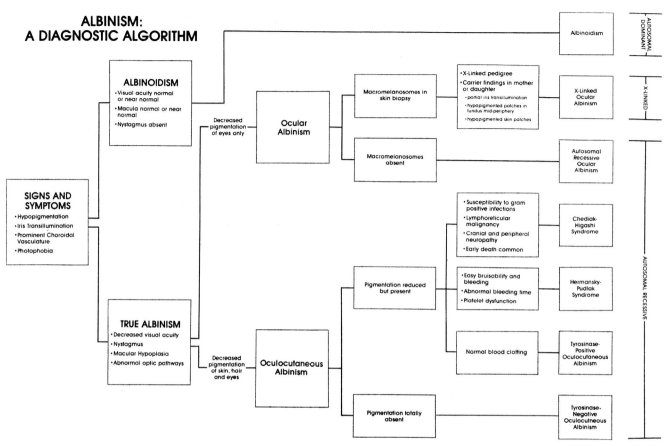

FIGURE 101-1 Classification of albinism.

Ocular Manifestations

Oculocutaneous Albinism

Albinism has long been divided into subgroups on the basis of clinical descriptions and pedigree studies. The introduction of the hair bulb incubation test for tyrosinase activity now permits a more orderly classification. This test is performed by incubating the lower portion of plucked hair shafts in solutions containing tyrosine. The hair bulbs are then fixed and examined microscopically for melanin content. Those from normally pigmented subjects produce intense pigmentation. Hair bulbs from albinos react in one of three ways: (1) no pigment production (tyrosinase-negative albinism), (2) dense pigment production (tyrosinase-positive albinism), and (3) partial and variable pigment production (yellow mutant albinism). Based on the results of the hair bulb incubation test and clinical presentation, the following classification may be used:

1. Tyrosinase test negative
   a. Tyrosinase-negative (Ty-neg) oculocutaneous albinism (complete perfect albinism), autosomal recessive
2. Tyrosinase test positive
   a. Tyrosinase-positive (Ty-pos) oculocutaneous albinism (complete imperfect albinism), autosomal recessive
   b. Hermansky-Pudlak syndrome (HPS, albinism with hemorrhagic diathesis), autosomal recessive
   c. Chediak-Higashi syndrome (CHS), autosomal recessive
   d. Cross syndrome (gingival fibromatosis, hypopigmentation, microphthalmia, oligophrenia, athetosis), autosomal recessive
3. Tyrosinase test variable
   a. Yellow mutant oculocutaneous albinism (Amish albinism, xanthous albinism), autosomal recessive.

Tyrosinase-Negative Oculocutaneous Albinism Affected patients typically have pink skin and white hair and suffer from skin photosensitivity. These patients have severe photophobia and demonstrate pendular nystagmus and a visual deficit to the 20/80 level or worse. Their blue-gray irides transilluminate, and the ocular fundus appears to be essentially devoid of pigmentation. Visual loss may result from any combination resulting from the dazzling effect of poor light filtration, high refractive error, hypoplasia of the fovea, and strabismus. Furthermore, albinos have abnormal optic nerve pathways, with most fibers from the temporal retina decussating to the opposite side at the optic chiasm, thereby preventing binocular vision. In addition, there is disorganization and defects in the lateral geniculate body and anomalous development of the geniculocortical tracts.

Tyrosinase-Positive Oculocutaneous Albinism While the presence of hair bulb tyrosinase suggests somewhat milder symptoms, there is great individual variation. Affected patients have flaxen yellow hair and lightly pigmented skin and suffer from only moderate degrees of photophobia and nystagmus. There is some color to the iris, and the fundus is lightly pigmented. Although patients may lack pigment at birth, they gradually acquire it with age and may even experience an increase in visual acuity over time.

Hermansky-Pudlak Syndrome This syndrome consists of oculocutaneous albinism with a clotting disorder due to platelet defects and is accompanied by the accumulation of a ceroidlike material in the reticuloendothelial system and elsewhere throughout the body.[6] Restrictive lung disease and ulcerative colitis develop in the 3rd and 4th decades of life, and the life span is shortened.

Chediak-Higashi Syndrome This syndrome is a rare, usually fatal disorder in which reticuloendothelial incompetence leads to susceptibility to infection and death. Patients are often mentally retarded, and there may be generalized lymphadenopathy. The principal laboratory finding is the presence of giant abnormal cytoplasmic granules that occur within leukocytes, especially neutrophils.[7] Chediak-Higashi syndrome should be ruled out whenever a child presents with oculocutaneous albinism.

Ocular Albinism

Ocular albinism has minor accompanying skin changes, if any. It occurs in X-linked recessive (Nettleship-Falls) and autosomal-recessive forms. Macromelanosomes are found on light microscopy of skin biopsy specimens from patients with the X-linked recessive form and are not seen in the autosomal recessive variety. This finding, when verified by electron microscopy, can be used to differentiate the two forms.

X-linked ocular albinos have essentially normally pigmented skin and light-colored hair. Their albinism is mostly restricted to the eye, where they may demonstrate all the ocular stigmata of oculocutaneous albinism. Female carriers may be detected by the presence of translucent irises and a mosaic pattern of pigmentation in the fundus. Autosomal-recessive ocular albinism (AROA) has all the features of X-linked ocular albinism, except that females are affected as severely as males, indicating incompatibility with an X-linked recessive pattern of inheritance.

Albinoidism

Oculocutaneous and ocular forms of albinoidism have an autosomal dominant pattern of inheritance and are differentiated from the various types of albinism by the absence of photophobia, nystagmus, and foveal hypoplasia. With rare exceptions, patients have normal vision, and the most important aspect of this entity may be its possible confusion with the various types of albinism.[8]

Treatment

The major difficulties encountered by patients with albinism are their sensitivity to sunlight, their impaired vision, and their psychosocial problems. Albinos should be instructed to avoid direct sunlight whenever possible and to use sunscreen lotions whenever necessary. The skin of older patients should be routinely evaluated for the presence of premalignant and malignant lesions.

The use of tinted glasses may help to reduce photophobia in albino patients. Finally, the use of high doses of ascorbic acid may be of benefit to patients with Chediak-Higashi syndrome. Patients affected by the Hermansky-Pudlak syndrome should avoid aspirin and other drugs that could interfere with platelet function. Platelet transfusions may be necessary in the event of major surgery or massive bleeding.

Conclusion

Albinism is comprised of a diverse group of clinical syndromes characterized by a congenital deficiency of pigmentation. The approach to affected persons should be multidisciplinary, with special attention being paid to the possible presence of accompanying systemic disorders as seen with Chediak-Higashi and Hermansky-Pudlak syndromes. The classification of albinism into the various subgroups on the basis of clinical signs and laboratory tests allows a determination of visual prognosis and provides the baseline for genetic counseling.

References

1. Froggat P. Albinism in Northern Ireland. *Ann Human Genet (London).* 1960;24:213-238.
2. Renie WA, ed. *Goldberg's Genetic and Metabolic Eye Disease.* Boston, Mass: Little, Brown and Co. 1986;490-498.
3. Witkop CJ Jr, Quevedo WC Jr, Fitzpatrick TB. Albinism and other disorders of pigment metabolism. In: Stanbury JB, Wyngaarden JB, Fredrickson DS, Goldstein JL, Brown MS, eds. *The Metabolic Basis of Inherited Disease.* 5th ed. New York, NY: McGraw-Hill; 1983;301-346.
4. Falls HF. Albinism. *Trans Am Acad Ophthalmol Otolaryngol.* 1953;57:324.
5. Ohrt V. Ocular albinism with changes typical of carriers. *Br J Ophthalmol.* 1956;40:721.
6. Hermansky F, Pudlak P. Albinism associated with hemorrhagic diathesis and unusual pigment reticular cells in the bone marrow: report of two cases with histochemical studies. *Blood.* 1959;14:162.
7. White JG, Clawson CC. The Chediak-Higashi syndrome: the nature of the giant neutrophil granules and their interactions with cytoplasm and foreign particulates. *Am J Pathol.* 1980;98:151.
8. Witkop CJ Jr. Abnormalities of pigmentation. In: Emery AEH, Rimoin DL, eds. *The Principles and Practice of Medical Genetics.* Edinburgh: Churchill Livingstone; 1982.

Chapter 102

Cystinosis

MURIEL I. KAISER-KUPFER and CHI-CHAO CHAN

In nephropathic cystinosis, a rare autosomal recessively inherited storage disease, nonprotein cystine accumulates in cells at levels that are 50 to 100 times normal.[1] The primary defect in this disorder is impairment of the carrier-mediated transport of cystine across the lysosomal membrane into the cytoplasm.[2] When cystine reaches particularly high concentrations, crystal formation occurs within the lysosomes of many body tissues, including kidneys, bone marrow, intestine, liver, spleen, pancreas, thyroid, lymph nodes, and the ocular tissues of cornea, conjunctiva, iris, and retinal pigment epithelium (RPE).[3] The diagnosis of cystinosis may be suspected by the clinical course and confirmed by demonstrating the presence of crystals in the corneal stroma on slit lamp biomicroscopy and by elevated cystine levels (5 to 10 nmole ½ cystine/mg/cell protein) in cultured fibroblasts or polymorphonuclear leukocytes (PMNs) from patients. Obligate heterozygotes have half the normal amount of lysosomal cystine transport[4] and mildly elevated cystine levels in fibroblasts or PMNs.

Prenatal diagnosis is possible by amniocentesis or chorionic villus biopsy.[1]

Late-onset, intermediate, or adolescent cystinosis presents later in childhood with the same sequence of events, but the severity of the condition appears milder.[5]

Benign or adult cystinosis represents a rare variant. Renal function remains normal, there is no retinopathy, and discovery of crystals in the cornea is an incidental finding in otherwise asymptomatic patients.[6]

Systemic Manifestations

Children with nephropathic cystinosis appear clinically normal at birth. By 6 to 12 months of age, they usually manifest dehydration, electrolyte imbalances, and failure to thrive secondary to renal tubular Fanconi's syndrome. Phosphaturia leads to hypophosphatemic rickets. Progressive renal failure ensues, necessitating renal

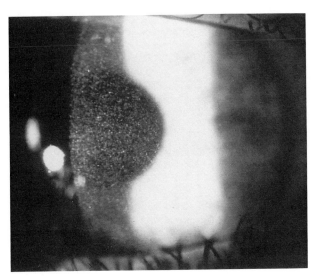

FIGURE 102-1 Slit-lamp photograph of densely packed corneal crystals in a 9-year-old patient.

transplantation by age 9 or 10 years. However, the continued accumulation of cystine within other tissues has resulted in serious extrarenal complications such as hypothyroidism,[7] hypohidrosis,[8] diabetes mellitus, and neurologic defects.[9] An encouraging advance has been the introduction of cysteamine, a cystine-depleting agent that has stabilized renal function and improved growth of young patients.[10]

Ocular Manifestations

Refractile cystine crystals seen by slit lamp biomicroscopy generally appear in the cornea by 1 year of age and are pathognomonic for cystinosis.[3] The crystals appear patchy at first in the anterior one third of the cornea, but with time, they become densely packed, occupying the full thickness of the cornea (Fig. 102-1).[11] Crystals have also been described in conjunctiva and iris.[3]

Histologic studies demonstrated the corneal crystals to be needle-shaped, whereas those in conjunctiva were rectangular or hexagonal.[12,13] The morphologic differences are attributable to the close packing of stromal cells between collagen bundles and the loose structure of conjunctival cells.

Photophobia generally develops within the first few years of life. Although there appears to be some variability in its severity in ambient light, all patients eventually experience discomfort with bright illumination. Bowman's membrane has shown thinning, irregularity, and even the development of breaks.[14,15] These changes, combined with the deposition of cystine crystals, may

contribute to recurrent corneal erosions and pain. In addition, the marked accumulation of iris crystals may induce photophobia.[15]

A generalized patchy depigmentation of the retinal periphery between the ora serrata and the equator has been noted as early as 5 to 10 weeks of life. The clinical appearance suggests involvement of the RPE. This is supported by evidence of "window defects" seen on fluorescein angiographic studies and by histopathologic findings of cystine in RPE and focal degeneration of the RPE.[12,16]

With patients surviving into the second decade as the result of renal transplantation, new findings have demonstrated serious ocular complications, including blindness, posterior synechiae, abnormalities in cone thresholds, dark adaptation, reduced or extinguished responses on electroretinograms, color vision deficits, glaucoma, and intractable photophobia and blepharospasm.[17]

The continuous accumulation of cystine crystals in ocular tissues may correspond to the expression of major histocompatibility class II antigens on certain ocular resident cells, such as keratocytes, fibroblasts, and vascular endothelial cells. The expression of class II antigens by these cells may stimulate them to phagocytose more cystine crystals and may possibly contribute to the local amplification of pathologic responses.[14]

Treatment of the ocular complications of nephropathic cystinosis includes local symptomatic therapy, such as artificial tears, topical lubricants, and thin bandage soft contact lenses. When topical therapy has failed and the symptoms of incapacitating photophobia result from recurrent erosions, rarely one may elect to perform penetrating keratoplasty.[18]

The experience of the National Collaborative Cysteamine Study has confirmed the effectiveness of long-term oral administration of cysteamine in improving renal function and growth of young patients.[10] Because of the clinical evidence of multisystem complications in older renal transplant recipients, cysteamine may be important in all patients to prevent continued deposits of cystine in tissues other than the kidney (eg, RPE). If this is so, the retina might be protected against photoreceptor damage. Cysteamine, a free thiol, crosses the plasma and lysosomal membrane, participating in a disulfide interchange reaction, forming cysteine and cysteine-cysteamine mixed disulfide, both of which exit cystinotic lysosomes, bypassing the cystine transport mechanism, which is blocked in cystinosis.

Oral cysteamine does not appear to prevent corneal crystal deposition. Preliminary results in a randomized clinial trial have shown that frequent instillation of topical cysteamine (0.1%) can reverse the deposition of crystals in central cornea.[11] Further experience with topical cysteamine is necessary.

References

1. Gahl WA. Cystinosis coming of age. *Adv Pediatr.* 1986;33:95-126.

2. Gahl WA, Bashan N, Tietze F, et al. Cystine transport is defective in isolated leukocyte lysosomes from patients with cystinosis. *Science.* 1982;217: 1263-1265.

3. Schulman JD, ed. Cystinosis. (DHEW Publication No. (NIH) 72-249). Washington, DC: US Government Printing Office; 1973.

4. Gahl WA, Bashan N, Tietze F, et al. Lysosomal cystine counter transport in heterozygotes for cystinosis. *Am J Hum Genet.* 1984;36:277-282.

5. Goldman H, Scriver CR, Aron K. Adolescent cystinosis: comparisons with infantile and adult forms. *Pediatrics.* 1971;47:979-988.

6. Lietman PS, Frazier PD, Wong VG, et al. Adult cystinosis: a benign disorder. *Am J Med.* 1966;40:511-517.

7. Chan AM, Lynch MJG, Bauley JD, et al. Hypothyroidism in cystinosis: a clinical, endocrinologic and histologic study involving sixteen patients with cystinosis. *Am J Med.* 1970;48:678-692.

8. Gahl WA, Hubbard VS, Orloff S. Decreased sweat production in cystinosis. *J Pediatr.* 1984;104:904-905.

9. Gahl WA, Kaiser-Kupfer MI. Complications of nephropathic cystinosis after renal failure. *Pediatr Nephrol.* 1987;1:260-262.

10. Gahl WA, Reed GF, Thoene JG, et al. Cysteamine therapy for children with nephropathic cystinosis. *N Engl J Med.* 1987;316:971-977.

11. Kaiser-Kupfer MI, Fujikawa L, Kuwabara T, et al. Removal of corneal crystals by topical cysteamine in nephropathic cystinosis. *N Engl J Med.* 1987;316:775-779.

12. Sanderson PO, Kuwabara T, Stark W. Cystinosis: a clinical, histopathologic and ultrastructural study. *Arch Ophthalmol.* 1974;19:270-274.

13. Cogan DG, Kuwabara T. Ocular pathology of cystinosis with particular reference to the elusiveness of corneal crystals. *Arch Ophthalmol.* 1960;63:51-57.

14. Kaiser-Kupfer MI, Chan C-C, Rodrigues M, et al. Nephropathic cystinosis: immunohistochemical and histologic studies of cornea, conjunctiva and iris. *Curr Eye Res.* 1987;6:617-622.

15. Kenyon KR, Sensenbrenner JA. Electron microscopy of cornea and conjunctiva in childhood cystinosis. *Am J Ophthalmol.* 1974;270-274.

16. Wong VG, Leitman PS, Seegoiller JE. Alterations of pigment epithelium in cystinosis. *Arch Ophthalmol.* 1967;77:361-369.

17. Kaiser-Kupfer MI, Caruso RC, Minkler DS, et al. Long-term ocular manifestations in nephropathic cystinosis. *Arch Ophthalmol.* 1986;104:706-711.

18. Kaiser-Kupfer MI, Datiles MB, Gahl WA. Corneal transplant in a boy with nephropathic cystinosis. *Lancet.* 1987;1:331.

Chapter 103

Hartnup Disease

DAVID A. HILES and ROBERT W. HERED

Hartnup disease is a rare familial metabolic disorder characterized by a defect in renal transport of a specific subgroup of amino acids, resulting in a diagnostic pattern of hyperaminoaciduria. In 1956 Baron et al.[1] first described the disorder as it affected two members of the Hartnup family. A history of consanguinity is typical of affected families and autosomal recessive inheritance is suspected.[2]

The underlying biochemical defect in Hartnup disease is a malfunction in the active transport system for monoamino-monocarboxylic amino acids.[2] This is one of four specific high-capacity amino acid transport systems that function at the level of the renal tubules and the columnar epithelial cells of the jejunum. In this disease, certain amino acids are subject to renal hyperexcretion and intestinal malabsorption. Tryptophan has been the amino acid most studied, but others affected by Hartnup disease are alanine, serine, threonine, asparagine, glutamine, valine, leucine, isoleucine, phenylalanine, tyrosine, histidine, and citrulline.[2] The pattern of aminoaciduria is constant and unchanged by nicotinamide or antibiotic administration. The diagnosis is made by urine chromatography. All other renal tubular function and clearance tests are normal; no reducing sugars are noted.

Hartnup patients excrete elevated amounts of the affected amino acids in the feces, mirroring the urinary pattern of excretion. Although the intestinal transport of affected monoamino acids is reduced, separate unaffected transport systems exist for the absorption of these same amino acids as peptides.[3] As a result, only a relative amino acid deficiency exists, which becomes clinically important in times of inadequate dietary intake. The threshold below which diets are inadequate in nicotinamide is raised. Although the key to diagnosis is the specific pattern of aminoaciduria, the cause of the clinical signs of the disease is felt to be related more closely to the results of intestinal malabsorption.[2]

Tryptophan is an essential nutrient that is metabolized to nicotinic acid and serotonin and that is not synthesized by man. In addition to active transport deficiencies, a defect in the metabolism of tryptophan to nicotinic acid is suspected in Hartnup disease. A deficiency of nicotinic acid is also found in pellagra, another disease whose clinical findings include photosensitivity skin rash and neurologic disorders.

By a separate pathway tryptophan is normally converted to certain indolic acids by intestinal flora and mammalian tissues. When intestinal malabsorption of tryptophan is reduced, as in Hartnup disease, colonic bacteria can produce indolylpropionic acid, which is absorbed and metabolized to indolylacrylic acid and other compounds that are then excreted in the urine. Urinary excretion of indoles is variable in Hartnup disease, but almost all patients have shown abnormally high urinary excretion at some time. Certain of the indoles, such as indolylacrylic acid, have been suggested as being responsible for the toxic clinical episodes seen in the disease. Levels of urinary indoles are particularly high during the acute bouts of ataxia and skin rash, and may remain elevated after clinical recovery although great variability is seen.[2]

Owing to the role of bacterial flora in producing indoles, oral neomycin and chlortetracycline have been shown to temporarily reduce urinary indole excretion. These antibiotics do not, however, reduce the aminoaciduria.[3]

Systemic Manifestations

The clinical features of Hartnup disease vary in severity between patients and over time, as the disease features exacerbations and remissions. The prognosis is generally good; however, nine reported patients have succumbed to complications of the disease.[2,4] With routine screening for aminoaciduria, the biochemical abnor-

mality of Hartnup disease has been detected in patients with normal physical and mental health.[2,5]

Minor generalized nutritional defects occur in Hartnup patients owing to the poor absorption and increased loss of essential amino acids. Most Hartnup patients are short in stature; more severe manifestations of the disease are probably avoided because of normal intestinal absorption of peptides by the other unimpaired transport systems.

Patients intermittently develop a dry, red scaly rash on sun-exposed regions of the body. The appearance and location of the rash are similar to those seen in pellagra.

Acute severe cerebellar ataxia most frequently occurs at times when the rash is most severe. Nystagmus and diplopia may accompany ataxia. Complete recovery generally occurs over several weeks. Headaches and syncope have been reported. Mental disturbances ranging from emotional lability to overt psychosis may also occur.[2] Although complete resolution of disease exacerbations is the rule, cases of progressive neurologic deterioration have been reported.[4] Dietary inadequacies of tryptophan have been linked to exacerbations.

Other precipitating factors include fever, sunlight exposure, stress, and sulfonamide administration. The role of certain indoles in producing acute episodes is not fully understood.

Ocular Manifestations

The ocular findings of Hartnup disease include nystagmus, which typically occurs during exacerbations of cerebellar ataxia. Both horizontal and vertical nystagmus have been reported.[6] "Divergent squint," diplopia, tics, and blepharoptosis may also occur during these episodes.[1,6] Photosensitivity and skin rash may involve the eyelids.[7] Conjunctival xerosis may accompany the skin rash.[8]

Tahmoush et al.[4] have described a case of a severely affected patient with bilateral optic atrophy and loss of visual fixation. The histopathologic evaluation of this patient revealed diffuse cerebral atrophy, which was most apparent in the occipital lobes. Loss of axons and myelin with intense gliosis occurred in the geniculocalcarine tracts, and the lateral geniculate nuclei were small. The optic nerves and tracts showed mild loss of axons and myelin. The loss of cells in the visual cortex was implicated in causing retrograde degeneration of the lateral geniculate body. The cerebellum showed marked loss of Purkinje cells. Cerebral white matter does not appear to be primarily affected in Hartnup patients, in contrast to other inborn errors of amino acid metabolism.[4]

Treatment

Oral nicotinamide therapy has caused improvement in both the dermatitis and neurologic disorders, although treatment success is difficult to document because of the frequency of spontaneous recovery. The threshold below which diets are considered inadequate in nicotinamide has been elevated for patients with this disease.[2]

Because of the possible toxic effect of certain indoles, oral neomycin may be helpful in treating a severe episode. Monoamine oxidase inhibitors are contraindicated.

References

1. Baron DN, Dent CE, Harris H, et al. Hereditary pellagra-like skin rash with temporary cerebellar ataxia, constant renal amino-aciduria, and other bizarre biochemical features. *Lancet*. 1956;2:421-428.
2. Jepson JB. Hartnup disease. In: Stanbury JB, Wyngarden JB, Fredrickson DS, eds. *The Metabolic Basis of Inherited Disease*. 3rd ed. New York, NY: McGraw-Hill; 1972;1486-1503.
3. Leonard IV, Marrs TC, Addison JM, et al. Absorption of amino acids and peptides in Hartnup disease. *Clin Sci Molec Med*. 1974;46:15.
4. Tahmoush AJ, Alpers DH, Feigen RD, et al. Hartnup disease. Clinical pathological and biochemical observations. *Arch Neurol*. 1976;33:797-807.
5. Wilcken B, Yu JS, Brown DA. Natural history of Hartnup disease. *Arch Dis Child*. 1977;52:38-40.
6. Halvorsen K, Halvorsen S. Hartnup disease. *Pediatrics*. 1963;31:29-38.
7. Francois J. Ocular manifestations in aminoacidopathies. *Adv Ophthalmol*. 1972;25:28-103.
8. Srikantia SG, Venkatachalam PS, Reddy V. Clinical and biochemical features of a case of Hartnup disease. *Br Med J*. 1964;1:282-285.

Chapter 104

Homocystinuria

ROBERT C. RAMSAY

Homocystinuria was initially described as a clinical entity in 1962, while a group of children were being tested to determine the cause of their mental retardation. Homocystinuria is an inborn error of metabolism caused by a deficiency of cystathionine B-synthase (CS).[1] It is inherited as an autosomal-recessive trait and has a worldwide distribution. CS catalyzes the conversion of homocystine to cystathionine, and the major clinical manifestations result from elevated plasma homocystine levels. The disorder is associated with considerable heterogeneity in the mutations that produce deficiency of CS activity, so that the age of onset and severity of clinical manifestations vary widely among affected persons.[2] Newborn screening tests suggest that the incidence of a homozygous deficiency of CS is approximately 1 : 200,000. The incidence of heterozygosity for CS deficiency in the normal population is estimated to be between 1 : 70 and 1 : 200.

Systemic Manifestations
Mental Retardation

Mental retardation is a major systemic manifestation of homocystinuria. Recent studies have shown that approximately 50% of affected persons have at least average intelligence. When present, the mental retardation tends to be slowly progressive. Cerebral cellular dysfunction is secondary to reduced levels of cystathionine as well as toxicity from homocystic acid (following conversion from homocystine).

Major Seizure Disorder

As homocystic acid is a potent excitoxic amino acid, major motor seizures are frequent in patients with homocystinuria.

Skeletal Deformities

Homocystine blocks cross-linking reactions in the formation of collagen, resulting in a structural defect. Vertebral osteoporosis, the most common skeletal abnormality, is often complicated by scoliosis, kyphosis, and chest wall deformities. Lengthening of the long bones may occur, producing an appearance similar to that of Marfan's syndrome patients. The hair of affected patients is often blonde owing to elevated serum methionine levels.

Occlusive Vascular Disease

The principal cause of death in patients with homocystinuria is thrombosis or thromboembolism.[2] The increased frequency of arterial and venous thromboses results in a greatly reduced life expectancy: expected mortality rate is 50% before age 20 and 75% by age 30. Although spontaneous thromboses are usual, surgery and general anesthesia significantly increase the risk. Angiography is particularly associated with thrombosis. Chronic chemical injury of the vascular endothelium leading to microemboli and healing by atherosclerosis is the major mechanism responsible for the increased rate of thrombosis. Heterozygosity for homocystinuria may be an important factor in at least some patients who present with premature (before 50 years of age) peripheral and cerebral vascular occlusive disease.[3] Treatment with therapeutic doses of vitamin B_6, as a cofactor for CS, may reduce the incidence of thromboembolic events. Thrombosis may also be reduced by treating with aspirin and dipyridamole, to reverse the decreased platelet survival time found in patients with homocystinuria. Measures to prevent thrombosis during or following anesthesia include treatment with vitamin B_6, aspirin, and dipyridamole, avoidance of dehydration, rapid mobilization, and use of intravenous dextran to reduce platelet adhesion.[4]

Ocular Manifestations

Bilateral subluxation of the crystalline lens (ectopia lentis) is the major ocular finding in homocystinuria.[5,6] The dislocation is bilateral and symmetric. Approximately 30% of affected patients have subluxated lenses in infancy. The incidence approaches 80% by age 15 years. Typically the direction of subluxation is inferior and nasally, whereas that associated with Marfan's syndrome is more commonly superior and temporal. Complete dislocation of the lens into the anterior chamber or vitreous is more common in homocystinuria than in Marfan's syndrome. Subluxation occurs secondary to collagen cross-linkage abnormalities that affect the lens zonules.

Acute glaucoma may result from subluxation of the lens into the anterior chamber. Retinal detachment may occur spontaneously, but more commonly it follows surgical intervention to remove a subluxed lens. Retinal arterial occlusions have been reported.[7] Optic nerve atrophy may follow either acute glaucoma or retinal vascular occlusions.

References

1. Skovby F. Homocysteinuria. *Acta Paêdiatr Scand.* (Suppl.) 1985;321:1-2.
2. Mudd HJ, Skovby F, Levy HL, et al. The natural history of homocysteinuria due to cystathionine B-synthase deficiency. *Am J Hum Genet.* 1985;37:1-31.
3. Boers GH, Smals AB, Trijbels FJ, et al. Heterozygosity for homocystinuria in premature peripheral and cerebral occlusive arterial diseases. *N Engl J Med.* 1985;313:709-715.
4. Parris WC, Quimby CW. Anesthetic considerations for the patient with homocysteinuria. *Anesth Analg.* 1982;61:708-710.
5. Cross HE, Jensen AD. Ocular manifestations in the Marfan syndrome and homocysteinuria. *Am J Ophthalmol.* 1973;75:405-420.
6. Nelson KB, Maumenee IH. Ectopia lentis. *Surv Ophthalmol.* 1982;27:143-160.
7. Wilson RS, Ruiz RS. Bilateral central retinal artery occlusion in homocysteinuria. *Arch Ophthalmol.* 1969;82:267-268.

Chapter 105

Hyperornithinemia

Gyrate Atrophy

MURIEL I. KAISER-KUPFER and DAVID L. VALLE

Gyrate atrophy (GA) of the choroid and retina, a rare autosomal recessive inherited chorioretinal dystrophy, is characterized by hyperornithinemia (500 to 1200 μm; normal is 75 ± 25 μm) and deficiency of activity of the mitochondrial matrix enzyme, ornithine δ-amino transferase (OAT). Hyperornithinemia in GA was first identified by Simell and Takki[1] in 1973 in the Finnish population; however, since then, it has become recognized worldwide in patients of different ethnic and racial backgrounds.[2] The salient clinical findings are confined to the eye and include myopia, cataracts, night-blindness, and a characteristic midperipheral to peripheral chorioretinal atrophy that causes progressive loss of visual function leading to vision of hand movements or less in most cases.

Systemic Manifestations

Although the major pathology and disability relates to the ocular manifestations, there are systemic changes, which for the most part are of minor clinical importance. Some patients have fine sparse scalp hair with areas of alopecia. Microscopic examination of the hair shaft showed intermittent dark cores within the medullary zone, which were not a deposit but appeared to be due to the unusual refractive properties of loosely formed macrofilaments distributed within wide spaces containing a structureless electron-lucent but compact substance that was insoluble in both water and organic solvents. These abnormalities were thought to be due to a defect in maturation of the hair.[3]

A number of reports have documented the presence of ultrastructural abnormalities in skeletal muscle fibers of GA patients.[3-6] Type 2 skeletal muscle fibers were reduced in diameter and contained tubular aggregates, deposits of abnormally staining material demonstrated by hematoxylin-eosin, Gomori ATPase, or NAD$^+$-tetrazolium reductase techniques. The significance of these findings is unclear, since they have been considered to be nonspecific signs in other diseases that

affect muscle, such as myotonic dystrophy and myasthenia gravis. Although the majority of patients do not exhibit signs of muscle weakness, a few of our patients have been found to have abnormalities on electromyography and demonstrate clinical evidence of muscle weakness.

It is noteworthy that cultured muscle from patients with GA, although it develops normally, has collections of tubular and mitochondrial abnormalities as well as a deficiency in OAT activity. Also, the addition of 20 μm of ornithine to muscle cell culture medium was toxic to GA cells within 48 to 72 hours, whereas normal cells remained unaffected for as long as 14 days in this medium.[7]

No alteration in liver function was noted, but liver mitochondria have shown nonspecific abnormalities such as elongation, branching, and segmentation.[4] The majority of patients are of normal intelligence and did not exhibit seizures despite mild to moderate diffuse slow wave recording on electroencephalography.[3]

Ocular Manifestations

Myopia was a constant finding and was present during the first decade. Nightblindness was a common complaint, although not universal, since a few patients neither reported it nor had abnormal dark adaptometry. Nearly all patients developed posterior sutural cataracts by the 2nd decade. The incidence of these lens opacities is not only higher than that seen in retinitis pigmentosa but it occurs at an earlier age and often requires surgery by the 2nd or 3rd decade.

The chorioretinal atrophy appears to begin with selective loss of the retinal pigment epithelium (RPE) peripherally. In our two youngest patients, less than 3 and 1 year of age, small, discrete depigmented areas within areas of diffuse pigment mottling were noted, primarily in the mid- and far periphery. Although the ophthalmoscopic changes were subtle, substantial evidence of both rod and cone involvement was demon-

FIGURE 105-1 A photomontage of the fundus of a 40-year-old patient with GA. Note atrophy with scalloped borders in midperiphery.

strated by a significant depression of the recorded electroretinogram.

Typically, psychophysical testing showed the development of constricted visual fields by Goldmann perimetry, color vision defects of a tritan pattern, and markedly depressed or extinguished electroretinogram and electro-oculogram recordings.

Patients between age 6 and 10 years developed discrete, defined circular areas of chorioretinal atrophy one quarter to one disc diameter in the peripheral retina. These discrete atrophic areas coalesced and in time eventually extended to encircle the retina completely (Fig. 105-1). There was an increase in the pigment clumping surrounding the atrophic areas. In the most advanced stages, extensive involvement of the posterior pole was noted, resulting in total atrophy with narrowed retinal vessels.

Although an association has been documented between age and progression of lesions, our experience points to considerable variation between patients, suggesting clinical evidence of genetic heterogeneity.[1] This evidence is further substantiated by biochemical heterogeneity as manifested by the few patients who have shown a significant reduction of plasma ornithine when given therapeutic doses of pyridoxine hydrochloride (500 mg per day).[6,8]

Despite intensive research efforts during the last decade, the pathophysiologic mechanisms relating deficiency of OAT and subsequent ornithine accumulation to chorioretinal degeneration and cataract formation remain known. No histopathologic description of the ocular pathology in GA is available; however several observations point to the RPE as the site of primary insult. Clinical observations in very young GA patients show multiple window defects of the RPE. The observations by Takki[10] on fluorescein angiography demonstrated damaged RPE with window defects and underlying intact choriocapillaris, and the animal experiments of Kuwabara et al.[11] clearly demonstrated that in both rats and monkeys intravitreal injection of ornithine causes edema of RPE cells followed by degeneration with a secondary loss of photoreceptors.

Any hypothesis about mechanism needs to explain the specific sensitivity of the ocular tissues as well as the slow rate of progression. Of particular importance is the need to explain the lack of ocular involvement in the hyperornithinemia, hyperammonemia, homocitrullinuria (HHH) syndrome. The HHH syndrome is a recessively inherited disorder wherein OAT activity is normal and there is no ocular involvement. Ornithine becomes elevated (range 360 to 630 μm), presumably owing to a defect in the transport protein that mediates ornithine entry into the mitochondrial matrix.[14]

Theoretical possibilities for the pathophysiologic mechanisms of the chorioretinal degeneration in GA include a toxic effect of the accumulated precursor (ornithine) or one of its metabolites; or a deficiency of the reaction product $\triangle'$-pyrroline-5-carboxylate (P5C) or one of its metabolites. The presence of substantial OAT activity in both normal RPE and neural retina suggests that OAT function is important in these tissues, and its absence may play a major role in the pathogenesis of GA.

An attractive hypothesis for the cause of the retinal

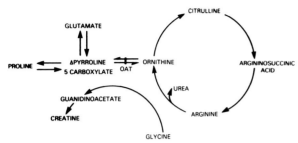

FIGURE 105-2 Diagram of pathways of ornithine metabolism. The dark vertical bar denotes the site of the metabolic block in GA.

pathology in GA implicates a deficiency of P5C, the product of the OAT reaction (Fig. 105-2). P5C production may be inadequate owing to a deficiency in the activity of OAT, and there is also an inhibitory effect of the elevated ornithine on P5C synthase, the enzyme that catalyzes the alternate limb of P5C and proline biosynthesis.[12] Decreased P5C production may result in reduced proline biosynthesis or a disruption in the regulatory role that P5C and its metabolic interconversions may exert on the activity of the hexose monophosphate shunt and the intracellular redox levels.[13]

There have been three therapeutic approaches to GA: stimulation of residual OAT activity with pharmacologic doses of pyridoxine; reduction of elevated plasma ornithine concentrations by restricting the intake of its precursor, arginine; and administration of creatine.

Although very few patients[6,8] have shown a significant reduction in plasma concentrations of ornithine by the administration of pyridoxine, it is recommended as the first step in newly diagnosed patients, because it is a benign regimen and easy to implement.[1]

On the basis of reduced levels of guanidinoacetate and creatine in their GA patients, Sipila et al.[14,15] treated seven patients with creatine for 1 year. Despite this treatment, the chorioretinal lesions continued to progress.

Since no other treatment modalities have been of benefit, we have advocated the use of an arginine-restricted diet in selected patients. This is based on the known biochemical pathway in which arginine is the precursor of ornithine. By reducing dietary arginine, one can effectively reduce plasma ornithine, which theoretically should have a salutary effect on the course of the disease. Evaluation of therapy in GA is complicated by the slow and variable progression of the chorioretinal changes, the lack of natural history data, and the evidence for clinical heterogeneity. Our preliminary results

in patients with long-term (6 to 9 years) excellent compliance are encouraging. We must await further experience with early intervention.

References

1. Simell O, Takki, K. Raised plasma ornithine and gyrate atrophy of the choroid and retina. *Lancet*. 1973;1:1031-1033.
2. Kaiser-Kupfer MI, Valle DL. Clinical, biochemical and therapeutic aspects of gyrate atrophy. In: Osborne N, Chader J, eds. *Progress in Retinal Research*. Oxford: Pergamon Press; 1986;179-206.
3. Kaiser-Kupfer MI, Kuwabara T, Askanas V, et al. Systemic manifestations of gyrate atrophy of the choroid and retina. *Ophthalmology*. 1981;88:302-306.
4. McCulloch JC, Marliss EB. Gyrate atrophy of the choroid and retina with hyperornithinemia. *Am J Ophthalmol*. 1975;80:1047-1057.
5. Sipila I, Simmel O, Rapola J, et al. Gyrate atrophy of the choroid and retina with hyperornithinemia: tubular aggregates and type 2 fiber atrophy in muscle. *Neurology*. 1979;29:996-1005.
6. Kennaway NG, Weleber RG, Buist NRM. Gyrate atrophy of the choroid and retina with hyperornithinemia: biochemical and histologic studies and response to vitamin B_6. *Am J Hum Genet*. 1980;32:529-541.
7. Askanas V, Valle D, Kaiser-Kupfer MI, et al. Cultured muscle fibers of gyrate atrophy (GA) patients: Tubules, ornithine toxicity and 1-ornithine-2-oxacid (OAT) deficiency. *Neurology*. 1980;30:368.
8. Berson E, Shih V, Sullivan PL. Ocular findings in patients with gyrate atrophy on pyridoxine and low protein, low arginine diets. *Ophthalmology*. 1981;88:311-315.
9. Kaiser-Kupfer MI, Ludwig IH, DeMonasterio FM, et al. Gyrate atrophy of the choroid and retina: early findings. *Ophthalmology*. 1985;92:394-401.
10. Takki K. *Gyrate Atrophy of the Choroid and Retina with Hyperornithinemia*. Helsinki: University of Helsinki; 1974. Doctoral Thesis.
11. Kuwabara T, Ishikawa Y, Kaiser-Kupfer MI. Experimental model of gyrate atrophy in animals. *Ophthalmology*. 1981;33:331-334.
12. Lodata RF, Smith JR, Valle D, et al. Regulation of proline biosynthesis: the inhibition of pyrroline-5-carboxylate synthase activity by ornithine. *Metabolism*. 1981;30:908-913.
13. Phang JM. The regulator functions of proline and pyrroline-5-carboxylic acid. In: Korecker BL, Stadtman ER, eds. *Current Topics in Cellular Regulation*. Orlando, Fla: Academic Press; 1985;91-132.
14. Sipila I, Simell O, Arjomaa P. Gyrate atrophy of the choroid and retina with hyperornithinemia. Deficient formation of guanidinoacetic acid from arginine. *J Clin Invest*. 1980;66:684-687.
15. Sipila I, Rapola J, Simell O, et al. Supplemental creatine as a treatment for gyrate atrophy of the choroid and retina. *N Engl J Med*. 1981;304:867-870.

Chapter 106

Maple Syrup Urine Disease

GRAHAM E. QUINN

Patients with branched-chain ketoaciduria, or maple syrup urine disease (MSUD), have a spectrum of clinical presentations with at least three clinical types: classic, intermediate, and a thiamine-responsive form. The incidence of the disease is between 1 in 120,000 and 1 in 290,000 persons, and its inheritance pattern is autosomal recessive.[1]

Systemic Manifestations

In the classic form of MSUD, the infant appears normal for the first several days after birth, but at 3 to 5 days of age the baby becomes listless and has feeding difficulties. Within a short period, neurologic abnormalities are noted, such as a high-pitched cry, loss of tendon and Moro reflexes, and convulsions. Stupor, hypotonia, and respiratory disturbances are noted and are followed by coma and death in untreated patients. At about the time of onset of neurologic abnormalities, the body fluids of affected persons give off a maple syrup-like odor, which gives the disease its common name. The odor is noticed in urine and sweat particularly and is quite distinct. The clinical course corresponds closely to that of neonatal sepsis, and metabolic abnormalities should be considered as part of the work-up for sepsis in the neonate.

The intermediate types of MSUD are characterized by episodic and milder clinical manifestations. Infancy is unremarkable, but during the second year, usually after an episode of illness or increased intake of dietary protein, the infant becomes irritable, progressively more lethargic, and ataxic. With good supportive care, the child usually recovers and may have several more episodes before a correct diagnosis is made. The odor of maple syrup is noted in urine and sweat during these episodes. Increased urinary excretion and plasma levels of branched-chain amino acids and branch-chain ketoacids can be detected during the episodes and may lead to severe acidosis that can be fatal.[1] Another late-onset form of MSUD is characterized by persistent elevation of branched-chain amino acid and branched-chain ketoacid levels and is associated with mental retardation.[2]

A thiamine-responsive form of MSUD has also been described, associated with developmental retardation but without seizures, ataxia, or lethargy.[3] The plasma and urine concentrations of branched-chain amino acids and branched-chain ketoacids are lowered in these patients when low-protein diets and thiamine supplementation are instituted.

Biochemical Abnormality

The catabolism of the branched-chain amino acids (leucine, isoleucine, and valine) is abnormal in MSUD. The initial transamination from amino acid to ketoacid proceeds normally, but the next step, oxidative decarboxylation, is defective.[4] A defect in the activity of the enzyme alpha ketoacid dehydrogenase results in the accumulation of the ketoacid derivatives of leucine, isoleucine, and valine, and of the amino acids themselves. The enzyme complex is located in the outer face of the inner membrane of mitochondria, and it is currently unclear whether the decarboxylation of the three branched-chain acids is accomplished by a single enzyme or by three distinct enzymes. Measurement of the enzymatic activity of the branched-chain decarboxylase shows no activity in classic MSUD and a markedly reduced level (15 to 25 percent of normal) in intermediate and late-onset forms.[1]

As a consequence of this enzyme defect, there is an accumulation of the branched-chain amino acids and their ketoacid derivatives in plasma, urine, red blood cells, and cerebrospinal fluid. The plasma levels of leucine are usually higher than those of either isoleucine or valine.[4]

Ocular Manifestations

The ocular manifestations of MSUD may be the presenting signs of the disease in the neonatal period. The diagnosis of ophthalmoplegia was first reported by Zee et al.[5] in a 10-day-old infant with feeding difficulties and restricted eye movements. The child had "bilateral paresis of adduction and absence of upward gaze." Mild bilateral ptosis and poor pupillary reaction were also noted. With supportive care, the baby regained a full range of eye movements and the diagnosis of MSUD was established at 6 weeks of age.

MacDonald and Sher[6] reported an infant with MSUD who had a normal ocular examination early in his course but gradually manifested poor adduction and limited upward gaze. The "apparent" bilateral intranuclear ophthalmoplegia resolved quickly when dietary treatment was begun at 6 weeks of age. Ocular fluttering manifested by episodic bursts of conjugate vertical eye movements lasting several seconds and occurring several times per hour was documented after several weeks of dietary therapy. The authors correlated these episodes with increased serum leucine levels at times when the levels of valine and isoleucine had returned to normal.

In 1979, Chhabria, Tomasi, and Wong[7] reported a case where ophthalmoplegia and ptosis were noted in association with several cranial nerve palsies. They documented decreased corneal sensation, facial diplegia, and absence of gag reflex in a 16-day-old girl. These authors suggested that these reversible cranial nerve pareses may be most consistent with a MSUD variant and not the classic form. Mantovani et al.[8] described a patient with decreased corneal reflexes and episodic ocular flutter at 15 days of age and documented possible cortical atrophy on follow-up computed tomography (CT).

Diagnosis

The diagnosis of MSUD must be made by measuring branched-chain amino acids and branched-chain keto-acids in blood and urine using chromatography or chemical techniques. MSUD should be considered when a neonate presents with lethargy, overwhelming illness, or marked central nervous system decompensation.[1] These life-threatening pediatric illnesses clearly are not likely to be handled by an ophthalmologist. The ophthalmologist's role is to document ophthalmoplegia, ocular flutter, or other eye movement abnormalities and to suggest MSUD in the differential diagnosis of those disorders.

Treatment

No specific ocular therapy has been reported for the reversible, transient ophthalmoplegia and other cranial nerve abnormalities associated with the episodes of acute decompensation in MSUD. Acutely, exchange transfusion or peritoneal dialysis may be lifesaving for the comatose infant. Long-term therapy is directed largely at restricting the intake of the branched-chain amino acids to the amounts needed for growth. Intercurrent illness may cause acute decompensation in a usually stable patient and must be carefully attended to. With adequate dietary management, the outlook for mental development is reasonably good for all variants of this disease.[9]

References

1. Stanbury JB, Wyngaarden JB, Fredrickson DS, et al. eds. *The Metabolic Basis of Inherited Disease*. New York, NY: McGraw-Hill; 1983;451-457.
2. Schulman JD, Lustberg TJ, Kennedy JL, et al. A new variant of maple syrup urine disease (branched-chain ketoaciduria). *Am J Med*. 1970;49:188.
3. Scriver CR, Mackenzie S, Clow CL, Devlin E. Thiamine-responsive maple syrup urine disease. *Lancet*. 1971;1:310.
4. Nyhan WL. *Abnormalities in Amino Acid Metabolism in Clinical Medicine*. Norwalk, Conn: Appleton-Century-Crofts; 1984; 21-35.
5. Zee DS, Freeman JM, Holtzman NA. Ophthalmoplegia in maple syrup urine disease. *J Pediatr*. 1974;84:113.
6. MacDonald JT, Sher PK. Ophthalmoplegia as a sign of metabolic disease in the newborn. *Neurology*. 1977;27:971.
7. Chhabria S, Tomasi LG, Wong PWK. Ophthalmoplegia and bulbar palsy in variant form of maple syrup urine disease. *Ann Neurol*. 1979;6:71.
8. Mantovani JF, Naidich TP, Prensky AL, et al. MSUD: presentation with pseudotumor cerebri and CT abnormalities. *J Pediatr*. 1980;96:179.
9. Clarke JTR, Low C, et al. Committee for improvement of hereditary disease management: management of maple syrup urine disease in Canada. *Can Med Assoc J*. 1976;115:1005.

Chapter 107

Ochronosis

MARY SEABURY STONE

Ochronosis refers to yellow-brown pigmentation in the sclera, skin, cartilage, tendons or heart valves due to the deposition of phenylalanine and tyrosine degradation products. Ochronosis is primarily associated with the hereditary disease alkaptonuria, a rare, autosomal recessive disorder that occurs in 1 in 250,000 live births.[1] Alkaptonuria, which is due to the absence of the enzyme homogentisic acid oxidase in the liver and kidneys, principally affects the skin, eyes, and joints. It was the first hereditary metabolic disorder to be described. The term "alkapton" was originally used by Boedeker in 1859 to describe a substance in the urine of a patient with this disease that had great avidity for oxygen at alkaline pH. The term "ochronosis" was coined by Virchow in 1866 because of the ochre color of the pigmented deposits seen microscopically in this disease.

Ochronosis may also be acquired from ingested or topically applied substances. Exogenous ochronosis may vary from a strictly cutaneous disease to one that closely resembles alkaptonuria.

Biochemistry

Phenylalanine and tyrosine normally undergo degradation to fumaric acid and acetoacetic acid. Homogentisic acid (HGA), an intermediate in this pathway, is metabolized by the enzyme homogentisic acid oxidase. In the absence of this enzyme, HGA accumulates and is excreted unchanged in the urine. In connective tissue, excess HGA is metabolized by the enzyme homogentisic acid polyphenol oxidase to ochronotic pigment. Ochronotic pigment consists of benzoquinone acetic acid and its polymers. Excess HGA may result in dark urine, since normally colorless HGA is gradually nonenzymatically oxidized to melanin-like compounds in the urine. This oxidation of HGA is markedly hastened by alkalinization. Thus the presence of dark urine in patients with alkaptonuria is dependent on the pH of the urine.[2]

Systemic Manifestations

The systemic manifestations of alkaptonuria consist of ochronotic pigmentation of the skin and eye, arthropathy, prostatic and renal stones, and cardiovascular changes. Although the disease may be disabling, it does not usually shorten life. Alkaptonuria generally does not result in significant clinical changes until the 3rd or 4th decade, though dark urine or discoloration of diapers after cleaning in alkaline soap may lead to the diagnosis in infancy.

The visible changes of alkaptonuria are those of the skin and the eye. The skin changes result from ochronotic pigment deposition in the dermis and sweat follicles, and more prominently due to cartilage and tendon pigmentation visible through areas of thin skin. Dark brown or black cerumen may develop in childhood, and the tympanic membranes may be blue. Greenish or brown axillary pigmentation may be present by late in the 1st decade. Later on, the ears, nose tip, extensor tendons of the hands, and costochondral junctions become pigmented. This pigmentation is seen through the skin as a bluish gray color. Intrinsic skin pigmentation is most prominent in areas with increased sweat glands such as the malar area, axillae, and anogenital area. Brown perspiration in these areas may stain clothing. Bluish fingernails and very dark nevi may also be present.[1]

Ochronotic spondyloarthropathy is the most disabling manifestation of the disease. It tends to be more severe and to appear earlier in males. Low back pain is usually the first complaint; pain and stiffness of the thoracic and lumbar spine typically appear in the 4th decade. As the disease progresses, pain and stiffness also develop in the hips, knees, and shoulders. The hands and feet are generally spared. Radiographic studies demonstrate characteristic narrowing of the intervertebral spaces, early calcification of the intervertebral discs, and eventually disc collapse, which may cause considerable loss of height. In addition to joint involvement, tendinitis, tendon rupture, and tendon calcification may complicate the disease. With time the patient may become completely incapacitated.

Other organ systems also are often involved. Soft pigmented stones occur in the prostate gland of older males, and black renal calculi have been reported in patients of both sexes. Ochronotic pigment is also deposited in the heart valves and in atherosclerotic plaques. Symptomatic aortic valvular disease may result. Some authors believe accelerated atherosclerosis is also present in this entity, but others feel that adequate documentation is lacking.[1]

Ocular Manifestations

Ocular ochronosis is present in the majority of patients with alcaptonuria and usually begins in the 3rd or 4th decade with scleral pigmentation.[3] The pigmentation is usually bilateral and may vary from a tiny spot to pigmentation of the entire sclera. Typically, the pigmentation is most pronounced near the insertion of the rectus muscles, particularly the lateral rectus muscle, and is present as triangular patches. Pigmented pinguecula-like changes may be seen in the episclera, usually between the limbus and muscle insertions. Corneal involvement is less common; changes usually appear as pigmented globules of varying size in the peripheral stroma within a few millimeters of the limbus. These globules resemble oil droplets when examined by retroillumination.[3,4]

Exogenous Ochronosis

Gross and microscopic changes that may be indistinguishable from those of alcaptonuria despite normal HGA oxidase levels are found in exogenous, or acquired, ochronosis. Exogenous ochronosis, which may result from several different types of exposure, was first reported to develop in persons who used carbolic acid (phenol) dressings on a long-term basis for treatment of leg ulcers. They developed ochronosis of the skin, sclera, cartilage, and tendons, and many even developed dark urine. Ochronosis of the skin has been reported in black women who use topical hydroquinone bleaching creams. Ingestion of antimalarial drugs occasionally causes ochronotic pigmentation of the conjunctiva, sclera, face, mucosa, shins, and subungual area. Antimalarial drugs have not been reported to cause dark urine or arthritis.[5]

The mechanism of exogenous ochronosis is poorly understood. Phenol compounds, hydroquinones, and HGA have structurally related metabolic products, which may partially explain the similar pigmentary deposits. Antimalarial drugs, which are a group of acridines and 4-aminoquinolones, may cause ochronosis-like pigmentation by chemically mediating the accumulation of melanin in affected tissues.[5]

Histopathology

Under the microscope, ochronotic pigment is amber. It may be seen as extracellular granules in connective tissue and as thick curled or jagged yellow-brown clumps caused by the accumulation of ochronotic pigment in collagen bundles. Foreign body giant cells may be present near these clumps. The amber-colored granules may also be seen in endothelial cells, macrophages, and in basement membranes and secretory cells of sweat glands.[6]

Diagnosis

In patients with dark urine, arthropathy, and cutaneous ochronosis, alkaptonuria is an easy diagnosis to make. Histopathologic study of pigmented tissue will show the characteristic amber granules and curled fibers. Spine films may be diagnostic in some persons. Since normal persons do not excrete HGA in the urine, its presence in urine is absolutely diagnostic of alkaptonuria. HGA does not fluoresce. A nonspecific but useful test is the observation of darkening of urine after the addition of sodium hydroxide. Other tests for HGA excretion are based on its reducing properties and include blackening of photographic emulsion paper after application of urine, blackening of urine after treatment with $FeCl_3$ (ferric chloride), and the development of an orange precipitate when urine is mixed with Benedicts' sugar reagent. Detection and quantification of urinary HGA is possible by a direct spectrophotometric method using HGA oxidase.[1,4]

Treatment

Unfortunately, at this time there is nothing but symptomatic treatment to offer patients with alkaptonuria. As arthropathy is the primary cause of morbidity, rest, analgesia, and physical therapy are important. For advanced disease, joint replacement may be very helpful. In patients with exogenous ochronosis, withdrawal of the offending substance generally results in improvement with time, though the pigmentation may persist indefinitely.

References

1. Goldsmith LA. Cutaneous changes in errors of amino acid metabolism: alkaptonuria. In: Fitzpatrick TB, et al., eds. *Dermatology in General Medicine*. 3rd ed. New York, NY: McGraw-Hill; 1987;1642-1646.
2. LaDu BN. Alcaptonuria. In: Stanbury JB, Wyngaarden JB, Fredrickson DS, eds. *The Metabolic Basis of Inherited Disease*. 4th ed. New York, NY: McGraw-Hill; 1978;268-282.
3. Allen RA, O'Malley C, Straatsma BR. Ocular findings in hereditary ochronosis. *Arch Ophthalmol*. 1961;65:657-668.
4. Wirtschafter JD. The eye in alkaptonuria. In: Bergsma D, Bron AJ, Cotlier E, eds. *Birth Defects*. New York, NY: Alan R. Liss; 1976;279-289.
5. Cullison D, Abele DC, O'Quinn JL. Localized exogenous ochronosis. *J Am Acad Dermatol*. 1983;8:882-889.
6. Lever WF, Schaumberg-Lever G. *Histopathology of the Skin*. 6th ed. Philadelphia, Pa: JB Lippincott, 1983;423-424.

Chapter 108

Tyrosinemia

ROBERT P. BURNS

Abnormalities of tyrosine metabolism in man include neonatal tyrosinemia, the tyrosinosis of Medes, hereditary tyrosinemia (tyrosinemia I), and Richner-Hanhart syndrome (tyrosinemia II).[1] In the rare Richner-Hanhart syndrome, deficiency of hepatic soluble tyrosine aminotransferase (TAT) causes elevated blood and urine levels of tyrosine.[2] Ocular complications are found only in this syndrome; the classic clinical triad is pseudodendritic keratitis, hyperkeratosis of the palms and soles, and mental retardation.

The syndrome typically begins in the first 3 months of life with ocular symptoms that include tearing, redness, and photophobia. Ophthalmologic examination most commonly demonstrates a bilateral pseudodendritic form of keratitis that varies from mild epithelial branching to geographic corneal ulcers associated with stromal opacity and vascularization (Figs. 108-1 to 108-3). Other ocular findings may include nystagmus, strabismus, cataract, and thickening of the conjunctiva.[3-5]

Histopathologic examinations of eye tissues from Richner-Hanhart patients have been limited to conjunctival biopsy and corneal scrapings.[1,2,6] The conjunctival biopsy revealed thickening of the epithelium,

keratofibrils in the cytoplasm of the basal epithelium, plasma cell infiltration of the subepithelial connective tissue, and inclusion bodies. Some of the inclusion bodies found in fibrocytes contain fine, needlelike crystalline material. Corneal scrapings showed hyperplastic, stratified squamous epithelium.

The skin lesions in the Richner-Hanhart syndrome are usually seen within 1 year of birth, with hyperkeratotic papules and lamellar peeling limited to the palms and soles. These areas are quite painful and occasionally prevent walking. Histopathologic examination of the affected skin demonstrates nonspecific findings of acanthosis, parakeratosis, a thickened granular layer, and hyperkeratosis.[7]

Mental retardation varies from mild to severe and has been found in most of the reported cases. The complete syndrome is not seen in all patients with tyrosinemia II. Persons without eye involvement and skin findings and with normal intelligence have been reported.[5]

Other reported physical findings include seizures, vertical striations of the distal femur, and multiple congenital anomalies (microcephaly, cleft lip and palate, hernia, talipes equinovarus, and absence of one kidney).[5]

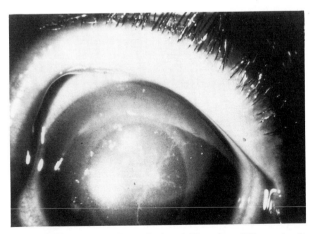

FIGURE 108-1 Pseudodendritic keratitis at 3 months of age, 6 weeks after onset. Note small central ulcer. (From Charlton et al.[3])

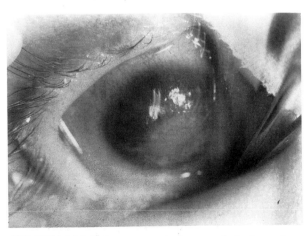

FIGURE 108-2 Larger central corneal shallow ulcer with sharp pointed margins at 1 month of age. (From Burns.[2])

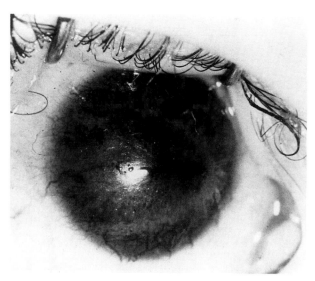

FIGURE 108-3 Irregular corneal ulceration with vascularization at age 14 months. (From Goldsmith.[4])

Tyrosinemia without hepatorenal disease is necessary for the diagnosis of the Richner-Hanhart syndrome; serum tyrosine titer ranges from 2.5 to 25 times normal. Urinary tyrosine is markedly increased, and tyrosine metabolites, including parahydroxyphenylpyruvic acid (pHPPA), parahydroxylphenyllactic acid (pHPLA), parahydroxyphenylacetic acid (pHPAA), acetyltyrosine, and paratyramine are also found in increased concentrations in the urine.[8]

Tyrosine aminotransferase (TAT) EC2.6.1.5, is the enzyme that catalyzes the first step of tyrosine metabo-

lism. It is an inducible cytoplasmic enzyme found to be most concentrated in the liver. TAT converts tyrosine to pHPPA, which is probably the rate-limiting step in the metabolism of tyrosine. TAT has been analyzed in liver biopsy specimens from three patients. In two, no soluble TAT activity was detected; the third patient had lower than normal TAT activity.[8]

The treatment of the Richner-Hanhart syndrome consists of dietary restriction of tyrosine and phenylalanine. Serum tyrosine levels fall rapidly, and the skin and corneal lesions resolve in a few days to several months.[3]

Tyrosinemia II is thought to be an autosomal-recessive condition because several affected families had histories of consanguinity.[5] Additionally, most families have had only one child with the disease. Attempts to identify heterozygotes have been unsuccessful. Tyrosine levels in the serum of parents of affected patients are normal even after dietary loading with tyrosine.[8]

It is not known why the findings in tyrosinemia II are localized in the eye, the skin of the extremities, and the central nervous system. An animal model has been found in rats fed a diet high in tyrosine. They developed keratoconjunctivitis, hyperkeratosis of the feet, alopecia, and they lost weight.[9] Histopathologic studies of the cornea in rats fed tyrosine have shown that corneal epithelial edema develops within 24 hours after the high tyrosine diet is started and progresses to form "snowflake" opacities. Epithelial lesions examined with an electron microscope demonstrate needle-shaped birefringent crystals, which disrupt cellular and nuclear membranes (Fig. 108-4) and which are thought, but not proven, to be tyrosine.[10] Tyrosine crystals are known lysosome labilizers and can mediate inflammation by

FIGURE 108-4 Electron micrograph of corneal epithelial cell of tyrosine-fed rat. Negative images of a bundle of needle-shaped crystals pierce cytoplasm and nucleus. (From Gipson, et al.[10])

this mechanism. In the rat, polymorphonuclear leukocytes invade the cornea, and blood vessels grow in from the limbus; however, despite the continued dietary tyrosine loading, corneas generally clear spontaneously within a month, probably because of the induction of hepatic TAT.

The ophthalmologist who sees a newborn or very young infant with an atypical dendritic corneal ulcer should think of herpes type II, but this should have an earlier onset than tyrosinemia, which causes symptoms at 4 to 6 weeks. Viral cultures and serum tyrosine levels should help in making the diagnosis.

References

1. Zaleski WA, Hill A. Tyrosinosis: a new variant. *Can Med Assoc J*. 1973;108:477-484.
2. Burns RP. Soluble tyrosine aminotransferase deficiency: an unusual cause of corneal ulcers. *Am J Ophthalmol*. 1972;73:400-402.
3. Charlton KH, Binder PS, Wozniak L, et al. Pseudodendritic keratitis and tyrosinemia. *Ophthalmology*. 1981;88:355-360.
4. Goldsmith LH, Reed J. Tyrosine induced skin and eye lesions: a treatable genetic disease. *JAMA*. 1976;236:382-384.
5. Fuerst D. Tyrosinemia in mink. In: Tabbara KF, Cello RM, eds. *Animal Models of Ocular Disease*. Springfield, Il: Charles C. Thomas; 1984;207-213.
6. Bienifang DC, Kuwabara T, Pueschel SM. The Richner-Hanhart syndrome. *Arch Ophthalmol*. 1976;94:1133-1137.
7. Rehak A, Sebim MM, Yadov G. Richner-Hanhart syndrome (tyrosinemia II). *Br J Dermatol*. 1981;104:469-475.
8. Buist NRM, Kennaway N, Fellman JG. Tyrosinemia type II: hepatic cytosol tyrosine amino-transferase deficiency. In: Bickel H, Wachtel U, eds. *Inherited Diseases of Amino Acid Metabolism*. Stuttgart: George Thieme Verlag; 1985;203-235.
9. Beard ME, Burns RP, Rich LF, et al. Histopathology of keratopathy in tyrosine fed rat. *Invest Ophthalmol*. 1974;13:1037-1041.
10. Gipson IK, Burns RP, Wolfe-Lande JD. Crystals in corneal epithelial lesions of tyrosine-fed rats. *Invest Ophthalmol*. 1975;14:937-941.

Section B

Disorders of Carbohydrate Metabolism

Chapter 109

Galactosemia

HAROLD SKALKA

Galactose, the sugar in milk and other dairy products, is metabolized as glucose after the following conversion:

$$\text{Galactose ATP} \xrightarrow{\text{Galactokinase (Gk)}} \text{Galactose-1-phosphate} + \text{ADP} \quad (1)$$

$$\text{Galactose-1-phosphate} + \text{UDP Glucose}$$
$$\xrightleftharpoons{\text{Galactose-1-phosphate uridyl transferase (GPUT)}} \quad (2)$$
$$\text{Glucose-1-phosphate} + \text{UDP Galactose}$$

$$\text{UDP Galactose} \xrightleftharpoons{\text{UDP Galactose-4-epimerase}} \text{UDP Glucose} \quad (3)$$

Inability to convert galactose to glucose produces galactosemia, with excessive galactose accumulation and the subsequent conversion of galactose to its sugar alcohol dulcitol (galactitol) by aldose reductase:

$$\text{Galactose} + \text{NADPH} \xrightleftharpoons{\text{Aldose reductase}} \text{Dulcitol} + \text{NADP} \quad (4)$$

Galactosemia is an autosomal recessive, inherited disease. Classic galactosemia is due to absence of GPUT.[1] The incidence of homozygous GPUT deficiency is probably about 1 in 62,000 births.[2,3] Reduced GPUT levels may be due to heterozygous GPUT deficiency or to variants of this enzyme, such as the Duarte (relatively

common), Indiana, Los Angeles, Rennes, and Berne variants. Each shows a characteristic degree of reduction in enzyme activity (eg, a heterozygote for the mild Duarte variant would be expected to show about 75% of normal activity—50% attributable to the normal allele and 25% to the Duarte allele). Electrophoretic studies are necessary to distinguish these variants.

Classic (GPUT) galactosemia produces systemic disease in infancy or early childhood and is fatal early in life unless galactose is removed from the diet. In addition to cataracts, manifestations include renal disease, hepatosplenomegaly, cirrhosis, anemia, deafness, nutritional failure and mental retardation. Galactose-1-phosphate is believed to be responsible for these systemic complications. Infantile galactosemic cataracts may be reversed or stabilized if they are detected in the first few months of life and if appropriate dietary changes are initiated.

Homozygous GK deficiency was first described as a cause of galactosemia in 1965.[4] This disorder also manifests cataract formation in the first few years of life but does not produce the fatal systemic complications seen in GPUT galactosemia; GK galactosemic babies appearing well except for cataract formation. The occurrence of GK galactosemia may be as rare as 1 in 1 million births.[2]

UDP galactose-4-epimerase deficiency has also been described,[5] but except for two cases in which GPUT-like galactosemic symptoms appeared, it has not been associated with cataract formation or other clinically evident disease.

The lens has an active hexose monophosphate shunt, providing NADPH, which drives reaction 4 (see equations above) to the right. While galactose is not toxic to the lens, the dulcitol generated by aldose reductase does not cross cell membranes and accumulates within metabolizing lens cells. (Dulcitol, unlike some other sugar alcohols, is resistant to oxidation by lens sorbitol dehydrogenase and therefore is not metabolized further.) This increases intracellular osmotic pressure, producing fluid imbibition with consequent lens fiber swelling, opacification, and finally disruption. In addition, galactose reduction within lens cells may be partially responsible for accelerated oxidation of lens protein sulfhydryl groups. This would enhance protein cross-linking, increasing insoluble lens protein (and therefore cataract formation).

Galactosemic enzyme levels may be ascertained by red blood cell assays.[6,7] Erythrocytes have many structural and biochemical similarities to lens cells and provide an easily accessible source for enzyme activity assays. Heterozygosity for GK deficiency may be expected to occur in approximately 0.2% of the population, and heterozygosity for GPUT deficiency in about 0.8%. Approximately 1% of the population, therefore, are heterozygous for the absence of one of these enzymes. GK and GPUT heterozygotes have red blood cell GK or GPUT enzyme levels that are approximately half those of normal persons. Such heterozygosity has not been associated with any systemic abnormalities, but GK heterozygosity has been suspected since the early 1970s of being a cause of early cataract formation.[8-10]

In 1980, Skalka and Prchal[11] demonstrated GK heterozygosity—and to a lesser extent GPUT heterozygosity—to be associated with an increased incidence of idiopathic bilateral presenile (age 50 or less) cataract formation. Of their 38 patients with presenile bilateral idiopathic cataracts, 47.4% had GK (14) or GPUT (5) enzyme activity levels consistent with heterozygosity (one patient in their study had reduced levels of both enzymes). They also found secondary cataracts (cataracts in patients subjected to cataractogenic insults such as trauma, intraocular inflammation, diabetes, etc.) in this age group to be associated with an increased incidence of GK heterozygosity (4 of 56 patients, or 7.1%). As noted above, only about 0.2% of the overall population are heterozygous for GK deficiency, and 0.8% for GPUT deficiency. The "presenile" cataracts in both of

these groups were almost invariably posterior subcapsular (PSC) in origin (infantile galactosemic cataracts commonly begin with an oil-droplet appearance and progress to essentially total lens opacification). In a subsequent study evaluating GK heterozygosity only, the association with presenile cataracts was corroborated by other investigators.[12]

It appears that galactosemic (GK or GPUT) heterozygosity, in conjunction with modern dietary practice, may be cataractogenic over a span of years to decades, and that such heterozygosity also increases lens vulnerability to potentially cataractogenic insults. GK heterozygosity appears to be the more potent facilitator of cataract formation, which perhaps is not surprising in view of the higher levels of galactose available for conversion to dulcitol in GK deficiency as compared with GPUT deficiency (in which galactose-1-phosphate accumulates).

Modern cataract surgery with lens implantation is a successful surgical procedure with a very high rate of visual restoration. Milk products are widely distributed in foods, processed and "natural," and lifelong compliance with dietary regimens designed to exclude or severely limit galactose exposure in the hope of possibly preventing or retarding cataract formation in heterozygotic persons at risk may be difficult to achieve. In point of fact, nothing is known of dose-response relationships in the period of years or decades over which such cataractogenesis may be operative. In the absence of additional therapeutic goals, chronic aldose reductase therapy of galactosemic heterozygotes also presents cost-benefit questions that are problematic at best.

Normal dietary habits coupled with impaired galactose utilization probably increase susceptibility to cataract formation, both primarily (PSC via dulcitol) and secondarily (presumably involving dulcitol). The cataractogenesis of heterozygous enzyme levels appears to be operative over a prolonged period of years, may not become clinically evident in many persons, and results in a treatable disease. These factors, combined with the absence of other detectable morbidity attributable to such heterozygosity, suggest that dietary preventive health measures or long-term aldose reductase therapy may not be acceptable to most patients. Identified heterozygotes (especially for GK) might prudently be informed that limiting dietary galactose may prove salubrious to their vision.

References

1. Isselbacher KJ, Anderson EP, Kurahashi K, et al. Congenital galactosemia, a single enzymatic block in galactose metabolism. *Science*. 1956;123:635-636.
2. Levy HL, Hammerson G. Newborn screening for galacto-

semia and other galactose metabolic defects. *J Pediatr.* 1978;92:871-877.

3. Beutler E. Screening for galactosemia: studies of the gene frequencies for galactosemia and the Duarte variant. *Isr J Med Sci.* 1973;9:1323-1329.

4. Gitzelmann R. Deficiency of erythrocyte galactokinase in a patient with galactose diabetes. *Lancet.* 1965;2:670-671.

5. Gitzelmann R. Deficiency of uridine diphosphate galactose-4-epimerase in blood cells of an apparently healthy infant. *Helv Paediatr Acta.* 1972;27:125-130.

6. Beutler E. Red cell metabolism. New York, NY: Grune & Stratton; 1971;79-88.

7. Beutler E. *Red Cell Metabolism.* 2nd ed. New York, NY: Grune & Stratton; 1975;91-97.

8. Beutler E, Krill A, Comings D, et al. Galactokinase deficiency: an important cause of familial cataracts in children and young adults. *J Lab Clin Med.* 1970;76:1006.

9. Harley JD, Irvine S, Gupta JD, et al. Galactokinase deficiency in cataract formation. *Med J Aust.* 1972;1:1326-1327.

10. Beutler E, Matsumoto F, Kuh W, et al. Galactokinase deficiency as a cause of cataracts. *N Engl J Med.* 1973;288:1203-1206.

11. Skalka HW, Prchal JT. Presenile cataract formation and decreased activity of galactosemic enzymes. *Arch Ophthalmol.* 1980;98:269-273.

12. Elman NJ, Miller MT, Matalon R. Galactokinase activity in patients with idiopathic cataracts. *Ophthalmology.* 1986;93:210-215.

Chapter 110

Glycogenoses

RICHARD S. SMITH

In the presence of ATP and hexokinase, glucose is phosphorylated and then transformed through a series of reactions to form glycogen. A homopolymer is created which consists of chains of glucose molecules formed through a 1–4 glycosidic linkage. A branch is formed every 12 to 18 glucose units, connected by a 1–6 glycosidic linkage. Numerous enzymes are involved in the overall process as well as in the degradation of glycogen. Absence of an enzyme may lead to accumulation of glycogen in various body tissues and may in turn result in abnormal function. In some syndromes, abnormal glycogen or partially degraded glycogen products may accumulate. A total of nine different abnormalities of glycogen storage have been identified, although only two are known to have ocular manifestations.[1]

Glycogen is normally found in skeletal muscle and liver. In muscle tissue it is located in the sarcoplasmic reticulum and around the mitochondria. The glycogenoses show abnormal amounts of glycogen in these and other locations. In skeletal muscle, these abnormal deposits are often associated with muscle fiber degeneration and swelling of mitochondria. The glycogen deposits may be free in tissue or enclosed in lysosomes. Their distribution varies with the specific syndrome.

Systemic Manifestations

The first glycogenosis was described by von Gierke[2] and bears his name. It is characterized by a deficiency of glucose-6-phosphatase and results in massive accumulations of glycogen in the liver and kidneys. Patients also have short stature, hypoglycemia, acidosis, and elevated serum lipids. Xanthomas are often found on the extensor surfaces of the extremities. Excessive bleeding from the nose and following surgery may be a major problem and appears related to defective platelet function. Despite the hepatomegaly, liver function studies are normal. Hyperuricemia is commonly seen and may produce symptoms. There is suggestive evidence that von Gierke's disease has an autosomal recessive type of inheritance.

In 1933, Pompe[3] described a disease in which there were large accumulations of glycogen in the cardiac muscle of an infant who had died of idiopathic cardiac hypertrophy. In the infantile form of the disease, early death from cardiac involvement is a hallmark. Glycogen is also increased in liver, spleen, kidney, and skeletal muscle. The enzyme defect in this instance is due to a lack of amylo-1,4-glucosidase. This enzyme is needed

FIGURE 110-1 Inferior oblique muscle from a patient with documented Pompe's disease. Large pools of glycogen (*arrow*) replace portions of the muscle (M). There is moderate swelling of the mitochondria (*X*) (original magnification ×15,000).

FIGURE 110-2 A pericyte (*arrow*) next to affected muscle contains glycogen, as does the adjacent capillary endothelial cell (*arrowheads;* original magnification ×15,000).

for the normal breakdown of glycogen by hydrolysis of the 1–4 glucosidic linkage. Alternate pathways for glycogen catabolism are available and may account for the variable manifestations of this disease. While most patients die by the age of 2 years, those with milder manifestations and typical enzyme deficiency may survive into the second decade.[4] These patients have no clinical signs of heart disease and present with severe muscle weakness. Death usually is a consequence of pneumonia and respiratory failure. A much smaller

group of patients do not manifest disease until adult life and have respiratory failure without cardiac signs. Despite the variation in these three clinical groups, all have the same enzyme deficiency.

Amylo-1,6-glucosidase (debrancher) deficiency is the defect responsible for limit dextrinosis or Cori's syndrome (type III). This defect stops the breakdown of the glycogen molecule at the site of branching and results in the accumulation of partially degraded glycogen in liver, heart, and skeletal muscle. In addition to accumulation of glycogen degradation products, patients with this syndrome are particularly prone to show hypoglycemia, since they are unable to fully mobilize stored glycogen.

A few patients have been described with deficiency of amylo-1,4-1,6-transglucosidase (brancher), which results in accumulation of an abnormal glycogen that resembles amylopectin. This is known as Anderson's disease (type IV). There are diffuse deposits of this abnormal material. Lack of muscle phosphorylase gives rise to McArdle syndrome (type V), which has late onset characterized by muscle pain and weakness. The remaining glycogenoses affect the liver and, except for type VII (lack of phosphoglucomutase), are due to defects in liver phosphorylase.

Ocular Manifestations

Glycogenoses III to IX show no clinical ocular signs, and no ocular histopathologic changes have been reported. The only eye findings reported in von Gierke's syndrome have been marginal corneal clouding and symmetric yellow paramacular retinal lesions.[5] The former may represent glycogen deposition in the cornea similar to that seen in the mucopolysaccharidoses. The retinal changes may be secondary, as all involved patients had elevated serum lipids.

The most thoroughly studied of the glycogenoses is type II, or Pompe's disease.[6-8] Using light microscopic techniques, Touissant[6] found glycogen deposits in retinal ganglion cells, in the mural cells of retinal vessels, and in ocular smooth and striated muscle. Ultrastructural studies have confirmed these findings and show abnormal deposits in the extraocular muscles and in all ocular tissues except the retinal pigment epithelium. In some instances, the ocular glycogen deposits were free and in others they were incorporated in lysosomes. Areas of muscle fiber degeneration and mitochondrial swelling were prominent features. Figures 110-1 and 110-2 show the typical findings, with large deposits of glycogen in both the muscle tissue and a nearby blood vessel. This patient, reported elsewhere,[7] had clinical strabismus. While the figures suggest muscle damage, it cannot be said with certainty that the strabismus in this patient was directly related to the glycogenosis.

References

1. Salter RH. The muscle glycogenoses. *Lancet.* 1968;i,1301.
2. Von Gierke E. Hepato-nephromegalia glykogenica. *Beitr Pathol Anat.* 1929;82:497.
3. Pompe JC. Hypertrophie idiopatique de couer. *Ann Anat. Pathol.* 1933;10:23.
4. Hug G, Garancis JC, Schubert WK, et al. Glycogen storage disease, types II, III, VIII, and IX. *Am J Dis Child.* 1966;111:457.
5. Fine RN, Wilson WA, Donnell GN. Retinal changes in glycogen storage disease type 1. *Am J Dis Child.* 1968;115:328.
6. Touissant D, Danis P. Ocular histopathology in generalized glycogenosis (Pompe's disease). *Arch Ophthalmol.* 1965;73:342.
7. Smith RS, Reinecke RD. Electron microscopy of ocular muscle in type II glycogenosis (Pompe's disease). *Am J Ophthalmol.* 1972;73:965.
8. Libert J, Martin J, Ceuterick C, et al. Ocular ultrastructural study in a fetus with type II glycogenosis. *Br J Ophthalmol.* 1977;61:476.

Chapter 111

Oxalosis

TRAVIS A. MEREDITH

Hyperoxaluria is characterized by excessive serum and urine levels of oxalic acid.[1] The two primary forms of the disease are caused by two separate enzyme deficits. In type I primary hyperoxaluria (glycolic aciduria) a block in the metabolism of glyoxalate results in the synthesis of excessive amounts of oxalate and glycolate. A deficiency of soluble α-ketoglutarate: glyoxylate carboligase has been identified. Type II disease (L-glyceric aciduria) is more rare; a defect in hydroxypyruvate metabolism is caused by a deficiency of leukocyte D-glyceric dehydrogenase. The inheritance pattern of both disorders is thought to be autosomal recessive.

Systemic Manifestations

Oxalic acid is a common organic acid. When its concentrations are excessive it combines with calcium ion to form calcium oxalate. Deposition of these oxalate crystals in tissues is termed oxalosis. Clinically recurrent calcium oxalate nephrolithiasis and nephrocalcinosis cause progressive renal failure. In inherited disease, symptoms of renal failure typically begin before the age of 5 years, although the onset and severity of symptoms may vary. The diagnosis is based on measurement of urinary oxalate excretion, but this may be difficult with advanced renal insufficiency or renal failure.

Hyperoxaluria may be acquired or secondary. Rare cases of oxalate poisoning occur in humans. Ingestion of ethylene glycol may lead to excessive crystallization of calcium oxalate in renal tubules and within the renal parenchyma. The anesthetic agent methoxyflurane has also produced hyperoxaluria with crystalline deposits in the kidneys and eyes. A hyperoxaluric syndrome has also been observed in malabsorption states. Patients with a variety of chronic gastrointestinal disorders, including those who have undergone removal of a long portion of the ileum, hyperabsorb dietary oxalate. Large doses of intravenous xylitol have also resulted in hyperoxaluria.

Treatment of these disorders attempts to decrease oxalate excretion by inhibiting oxalate synthesis and to increase calcium oxalate solubility in the urine. Large doses of pyridoxine have reduced oxalate synthesis in some patients. New stone formation may be reduced by use of phosphate or magnesium or both to increase calcium oxalate solubility in vivo. In renal transplant recipients, stones tend to recur in the donor kidney.

Ocular Manifestations

Eye findings have been described in type I hyperoxaluria and in cases secondary to effects of methoxyflurane anesthesia. The clinical picture is similar in both. The eye finding in the early stage of hyperoxaluria consists of deposition of fine yellow crystalline spots that are widespread throughout the posterior pole as far anteriorly as the equator. In observations of an acute case secondary to methoxyflurane, Bullock and Albert[2] noted multiple yellowish white punctate lesions diffusely scattered throughout the posterior pole. In more advanced cases these crystals may have a predilection to accumulate around retinal vessels (Fig. 111-1). This was dramatically demonstrated in a 33-year-old woman who was observed to have progression of crystals around the retinal vessels. Cotton-wool spots developed, and fluorescein angiography documented extensive small vessel occlusion. Disc neovascularization developed and progressed despite photocoagulation therapy.

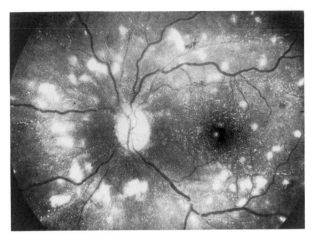

FIGURE 111-1 Marked arterial deposition of crystals with multiple intraretinal crystals and cotton-wool spots. (Courtesy Richard Lewis, M.D.)

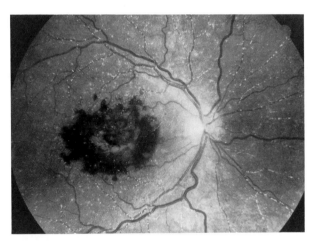

FIGURE 111-2 Chinese-pattern black macular lesion and widespread refractile crystal deposition with a perivascular deposition.

In several cases areas of deposition of black pigment have been reported in the posterior pole.[4,5] In the case illustrated in Fig. 111-2 this assumed a dramatic pattern that resembled a Chinese decorative figure. Whitish material at the center appears to be fibrous tissue, and this area stained on fluorescein angiography, suggesting that subretinal neovascularization was present.[3] The black clumping of pigment epithelium may be hyperplasia secondary to mechanical irritation by the crystalline deposits. Angiography in this case also demonstrated multiple small rings of hyperfluorescence surrounding hypofluorescent centers. These may be rings of atrophic pigment epithelium surrounding an oxalate crystal, producing the doughnutlike appearance on the angiogram.

Early descriptions of the pathology of primary hyperoxaluria noted that birefringent crystals of variable size accumulated between the pigment epithelium and the choriocapillaris. Special fixation techniques and staining in another case demonstrated more widespread deposition of crystals in vascularized tissues: conjunctiva, iris, ciliary body, inner layers of the retina to the extent of the retinal vascularization, choroid, episclera, and to a lesser degree, sclera.[3] These findings are consistent with pathologic changes elsewhere in the body that demonstrate a predilection for vascular and perivascular deposition of these crystals.[6] In a secondary case, crystals were noted in the retinal pigment epithelium, neural retina, and ciliary body.[2]

In summary, ocular hyperoxaluria is a rare condition that may be due to either of two different sets of enzyme defects or may be secondary to other disorders, such as methoxyflurane anesthesia, ethylene glycol poisoning, or malabsorption syndromes. Throughout the body there is a predilection for crystalline deposits in and around vascular structures, often leading to destruction of the kidney and subsequent renal failure. In the eye the most visible retinal crystals are within the retina and beneath the pigment epithelium. This may cause a dramatic picture of retinal crystallization, and black changes in the macula may also occur in more chronic cases, presumably owing to mechanical stimulation of the retinal pigment epithelium.

References

1. Williams HE, Smith H Jr. Primary hyperoxaluria. In: Stanbury JB, Wyngaarden JB, Fredrickson DS, eds. *The Metabolic Basis of Inherited Disease.* New York, NY: McGraw-Hill; 1978;182-204.
2. Bullock JD, Albert DM. Flecked retina. *Arch Ophthalmol.* 1975;93:26-31.
3. Meredith TA, Wright JD, Gammon JA, et al. Ocular involvement in primary hyperoxaluria. *Arch Ophthalmol.* 1984;102:584-587.
4. Gottlieb RP, Ritter JA. Flecked retina: an association with primary hyperoxaluria. *J Pediatr.* 1977;90:939-942.
5. Zak TA, Buncic R. Primary hereditary oxalosis retinopathy. *Arch Ophthalmol.* 1983;101:78-80.
6. Scowen EF, Stansfeld AG, Watts RWE. Oxalosis and primary hyperoxaluria. *J Pathol.* 1959;77:195-205.

Section C

Disorders of Lipoprotein and Lipid Metabolism

Chapter 112

Cerebrotendinous Xanthomatosis and Other Xanthomas

AHMAD M. MANSOUR and SHARON S. RAIMER

Xanthomas are localized infiltrates of lipid-containing histiocytic foam cells that are usually found within the dermis or tendons. The five general clinical types of xanthomas include: tendinous, planar, tuberous, and eruptive xanthomas, and xanthoma disseminatum. Recognition of these clinical types is important in the detection of the various associated underlying systemic and ocular diseases (Tables 112-1 and 112-2).

Tendinous Xanthomas

Clinical Features

Tendinous xanthomas arise in extensor tendons of the hands, knees, and elbows as well as in the Achilles tendons.[1] They appear as freely movable small, smooth, firm nodules of various sizes (Fig. 112-1). Tendinous xanthomas are associated with xanthelasmas, tuberous xanthomas, and coronary atherosclerosis. A frequent accompanying finding in patients with tendinous xanthomas, xanthelasmas, and high blood cholesterol is corneal arcus. Like xanthelasmas, corneal arcus often occurs in the absence of high blood cholesterol. Patients under 50 years of age with corneal arcus have a higher

TABLE 112-1 Eye Findings Associated with Xanthomas

Xanthelasma palpebrum	Peribulbar xanthomas
Corneal arcus	Lipemic diabetic retinopathy
Lipemia retinalis	Cataract

TABLE 112-2 Clinical Profile in Xanthomas

Xanthoma Type	Eye Findings*	Lipid Profile
IA (tendinous)	Xanthelasma	Beta-lipoprotein disturbance
IB (cerebrotendinous xanthomatosis)	Cataract, xanthelasma	Normal
II (planar)	Xanthelasma	Abnormal
III (tuberous)		Broad beta disease
IV (eruptive)	Lipemia retinalis	Very elevated triglycerides
V (disseminated)	Corneal and scleral xanthomas	Normal

* Additional findings not listed are secondary to associated hyperlipidemic states.[1,6]

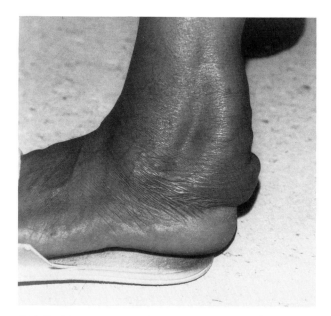

FIGURE 112-1 Tendinous xanthoma of Achilles tendon.

incidence of coronary atherosclerotic heart disease.[2] The presence of tendinous xanthomas indicates an underlying disturbance of cholesterol metabolism, except in the context of cerebrotendinous xanthomatosis.

Cerebrotendinous Xanthomatosis (CTX)

CTX is a rare autosomal-recessive lipid storage disease characterized by bilateral juvenile cataracts, tendon xanthomas, dementia, pyramidal paresis, and cerebellar ataxia.[1-3] The first specific manifestations appear in late childhood as zonular or radial cataracts, palpebral xanthelasmas, and enlargement of the Achilles tendons.[3,4] The condition is slowly progressive, and death usually occurs during the 6th decade, secondary to gradual pseudobulbar paralysis.[2] Biochemically, CTX is a lipid storage disease in which serum cholesterol levels are normal and a serum level of cholesterol is high.[1] The neurologic dysfunction results from deposition of cholestanol and replacement of cholesterol by cholestanol in the central nervous system and peripheral nerve myelin. Long-term treatment with chemodeoxycholic acid may correct the biochemical abnormalities—and may possibly reverse the progression of CTX.[5] Ophthalmoscopic findings have included juvenile cataracts and palpebral xanthelasmas. Visual loss with pallor of the optic discs has been noted late in life.[4] Cataractous lenses of CTX patients were found to have high levels of cholestanol.[4]

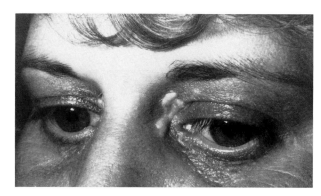

FIGURE 112-2 Planar xanthoma (xanthelasma palpebrum).

Planar Xanthomas

Planar xanthomas are yellow, soft, slightly elevated plaques. The most common type, xanthelasma, occurs on the eyelids (Fig. 112-2). Xanthelasmas are suggestive of underlying hypercholesterolemia when seen in subjects under 50 years of age.[1] A second type, xanthoma striatum palmare, appear as linear yellow lesions in the creases of the palms and fingers. These lesions usually indicate a disturbance in lipid metabolism.

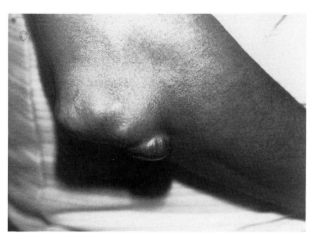

FIGURE 112-3 Tuberous xanthoma of elbow.

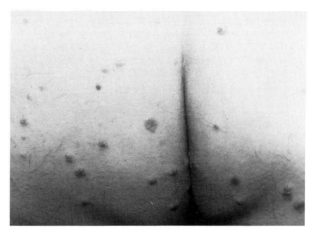

FIGURE 112-4 Eruptive xanthomas involving the buttocks.

Tuberous Xanthomas

Tuberous xanthomas appear as large yellow nodules on the extensor surfaces, mainly elbows, knuckles, and knees (Fig. 112-3). With increasing fibrosis they enlarge and become firm. Tuberous xanthomas indicate an alteration of cholesterol or triglyceride metabolism.

Eruptive Xanthomas

Eruptive xanthomas are 1- to 4-mm yellow papules surrounded by an erythematous halo, commonly distributed over the extremities and buttocks (Fig. 112-4). Eruptive xanthomas are seen exclusively with lactescent plasma secondary to high plasma triglycerides. Lipemia retinalis is a common finding.

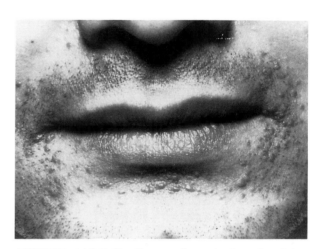

FIGURE 112-5 Xanthoma disseminatum involving the perioral region.

Xanthoma Disseminatum

Lesions of xanthoma disseminatum are rare. They appear as dark mahogany-brown papules (Fig. 112-5) over flexural creases, mucous membranes, cornea, and sclera.[1] Serum lipids are normal.

Various causes of secondary hyperlipoproteinemias are often associated with xanthomas and include diabetes mellitus, obesity, pancreatitis, nephrotic syndrome, uremia, hypothyroidism, cholestatic liver disease, dysglobulinemia, and ingestion of estrogens, corticosteroids, and Accutane.[1,6] Lipemic diabetic retinopathy is characterized by marked accumulation of hard exudates in the fundi, mimicking Coats' disease and leading to marked visual impairment.[7] This fundus disease is related to elevated levels of triglycerides and can be associated with various xanthomas, such as eruptive xanthomas and tuberous xanthomas.

References

1. Parker F. Xanthomas and hyperlipidemias. *J Am Acad Dermatol.* 1985;13:1-30.
2. Rosenman RH, Brand RJ, Shultz RI, et al. Relation of corneal arcus to cardiovascular risk factors and the incidence of coronary disease. *N Engl J Med.* 1924;291:1322-1324.
3. Berginer VM, Abeliovich D. Genetics of cerebrotendinous

xanthomatosis (CTX)—an autosomal recessive trait with high gene frequency in Sephardim of Moroccan origin. *Am J Med Genet*. 1981;10:151-157.

4. Kearns WP, Wood WS. Cerebrotendinous xanthomatosis. *Arch Ophthalmol*. 1976;94:148-150.

5. Swartz M, Burman ID, Salen G. Case report. Cerebrotendinous xanthomatosis: a cause of cataracts and tendon xanthoma. *Am J Med Sci*. 1982;283:147-152.

6. Seland JH, Slagsvold JE. The ultrastructure of lens and iris in cerebrotendinous xanthomatosis. *Acta Ophthalmologica*. 1977;55:201-207.

7. Berginer VM, Salen G, Shefer S. Long-term treatment of cerebrotendinous xanthomatosis with chemodeoxycholic acid. *N Engl J Med*. 1984;311:1649-1652.

8. Vinger PF, Sachs BA. Ocular manifestations of hyperlipoproteinemia. *Am J Ophthalmol*. 1970;70:563-573.

9. Brown GC, Ridley M, Haas D, et al. Lipemic diabetic retinopathy. *Ophthalmology*. 1984;91:1490-1495.

Chapter 113

Histiocytosis X

ROBERT FOLBERG

The term "histiocytosis X" has been applied to a collection of histiocytic proliferations that includes unifocal bone lesions (monostotic eosinophilic granuloma); multifocal bone lesions with adjacent soft tissue involvement (known as Hand-Schuller-Christian disease when it presents as lytic lesions of the calvarium with diabetes insipidus and exophthalmos); and disseminated visceral disease, frequently with skin involvement (known as Letterer-Siwe disease).[1] All of these conditions involve proliferation of a subset of histiocytes, Langerhans cells.[2] Langerhans cells reside normally in surface epithelia (including the skin, conjunctiva, and cornea).[3] They can be identified by the presence of a characteristic inclusion demonstrable by electron microscopy, Birbeck's granule (also known as the Langerhans granule) and by a characteristic immunohistochemical profile.[4] Recognizing the central role of the Langerhans cell in these proliferations, it has been suggested that these entities be renamed "Langerhans cell histiocytosis."[5]

Each of the entities is characterized by a granulomatous reaction in which a variable number of Langerhans cells is present. The morphology of an individual lesion depends on the stage of evolution. Early lesions may be characterized by a predominance of histiocytes (including Langerhans cells); later lesions may reveal a component of necrosis with a prominent eosinophilic infiltrate. Late lesions may demonstrate fibrosis with a scant cellular infiltrate.[2]

The causes and pathogenesis of these conditions are unknown. It is not possible to separate the localized forms of the disease from the disseminated conditions by pathologic examination alone.

Systemic Manifestations

In its disseminated form, Letter-Siwe disease, multiple organ systems may be involved. The skin lesions resemble those of seborrheic dermatitis. Involvement of the marrow may lead to thrombocytopenia. A chronic, draining otitis media may be a prominent feature. Enlargement of lymph nodes, liver, and spleen are not uncommon, and the lung and thymus may also be involved. The disseminated form is most commonly seen in infants.[2]

The triad of exophthalmos with multiple lytic lesions of the skull and diabetes insipidus (Hand-Schuller-Christian disease) is infrequently seen.[6] It is uncommon for multifocal osseous lesions with adjacent soft tissue involvement to progress to wider dissemination, and it is usual for monostotic lesions (solitary eosinophilic granuloma) to remain localized.[2,5,6]

The prognosis for these conditions is related to the age of the patient at presentation and to the extent of disease. Children below the age of 2 years are more likely to have disseminated disease and their prognosis is poorer.[2]

Ocular Manifestations

The lesion most likely to be encountered by ophthalmologists is the solitary eosinophilic granuloma, which characteristically affects the frontal bone, but involvement of the lateral orbital wall has also been reported.[5,7]

Eosinophilic granuloma of bone tends to be painful,

and the skin overlying the involved bone may show signs of inflammation. Because of the tendency of eosinophilic granuloma to involve the lateral aspect of the frontal bone, the clinical presentation of this condition may be mistaken for dacryoadenitis or preseptal cellulitis.[5-7]

The radiographic findings of eosinophilic granuloma are distinctive. The lesions are radiolucent, with scalloped borders and varying degrees of adjacent sclerotic reaction.[5] These findings, together with the characteristic location, should prompt a biopsy to establish the diagnosis. It is not necessary to surgically extirpate the entire lesion. Lesions that are incompletely removed surgically may regress spontaneously[6]; adjuvant radiation therapy (900 rad for children, 1500 rad for adults, fractionated over three to five sessions) may be delivered. Aggressive surgical treatment of isolated eosinophilic granuloma may lead to persistent osseous defects.[5]

Patients with histiocytosis X may present with proptosis. Orbital soft tissue involvement is usually accompanied by adjacent bone involvement.[6]

Intraocular involvement has been reported in disseminated disease. Lesions may be demonstrated in the choroid. Anterior segment involvement has been associated with glaucoma.[8] Iritis, posterior scleritis, and bilateral perforating corneal ulcers have been cited as intraocular complications of histiocytosis X.[6]

Extraconal orbital involvement in histiocytosis X may lead to compression of the optic nerve with optic disc edema and optic atrophy. The optic chiasm may be involved by an adjacent lesion in the pituitary-hypothalamic zone.[6]

Ophthalmologists encountering new patients with histiocytosis X should consult with pediatric oncologists to determine the extent of systemic involvement, to manage systemic complications, and to administer systemic therapy if indicated. A workup to exclude systemic disease may include a radiographic bone survey, chest film, bone marrow examination, lung and liver function tests, and, if appropriate, tests to exclude diabetes insipidus. Children with disseminated histiocytosis X may be treated with systemic chemotherapy, including prednisone, vincristine, and vinblastine.[6]

References

1. Lichtenstein L. Histiocytosis X: integration of eosinophilic granuloma of bone, Letterer-Siwe disease and Schuller-Christian disease as related manifestations of a single nosologic entity. *Arch Pathol.* 1953;56:84.
2. Favara BE, McCarthy RC, Mierau GW. Histiocytosis X. *Hum Pathol.* 1985;14:663.
3. Rodrigues MM, Rowden G, Hackett J, Bakos I. Langerhans cells in the normal conjunctiva and peripheral cornea of selected species. *Invest Ophthalmol Vis Sci.* 1981;21:759.
4. Birbeck MD, Breathnach AJ, Everall JD. An electron microscopic study of basal melanocytes and high level clear cells (Langerhans cells) in vitiligo. *J Invest Dermatol.* 1961;37:51.
5. Jakobiec FA, Trokel SL, Aron-Rosa D, et al. Localized eosinophilic granuloma (Langerhans' cell histiocytosis) of the orbital frontal bone. *Arch Ophthalmol.* 1980;98:1814.
6. Moore AT, Pritchard J, Taylor DSI. Histiocytosis X: an ophthalmological review. *Br J Ophthalmol.* 1985;69:7.
7. Feldman RB, Moore DM, Hood CI. Solitary eosinophilic granuloma of the lateral orbital wall. *Am J Ophthalmol.* 1985;100:318.
8. Epstein DL, Grant WM. Secondary open-angle glaucoma in histiocytosis X. *Am J Ophthalmol.* 1977;84:332.

Chapter 114

Hyperlipoproteinemia

ROBERT J. SCHECHTER

Fat Absorption and Metabolism

Plasma lipoproteins transport cholesterol, triglycerides, phospholipids, and proteins. Lipoproteins consist of chylomicrons and, according to their separation characteristics on ultracentrifugation, very low–density lipoproteins (VLDL), intermediate-density lipoproteins (IDL), low-density lipoproteins (LDL), and high-density lipoproteins (HDL). Associated with these chylomicrons and lipoproteins is a spectrum of protein types referred to as "apoproteins."

Triglyceride in the diet is packaged into chylomicrons ("exogenous" lipoprotein) by the intestinal epithelial cells. They are then transported into the lymph and enter the blood. The circulating chylomicrons attach to the capillary endothelium of certain peripheral tissues, including adipose, muscle, mammary, and liver, where they come into contact with lipoprotein lipase (LPL), which catalyzes their hydrolysis and the subsequent release of fatty acids and monoglycerides. The chylomicron remnants then detach from the endothelial surface, reenter the circulation, and are eventually cleared by processes involving the liver.

"Endogenous" triglycerides, mostly synthesized in the liver, are released into the blood as VLDL. VLDL is catabolized into transient IDL, also known as LDL_1, and then into LDL (LDL_2). Finally, remnants of the catabolism of both VLDL and chylomicrons are transferred to the HDL, which carries cholesterol and other components back to the liver (reverse cholesterol transport).

Systemic Manifestations

Hyperlipoproteinemia is referred to as primary if it results from a (possibly genetic) defect in lipoprotein metabolism. It is referred to as secondary if it is related to another underlying disease.

Types I and V: Hyperchylomicronemia without and with Elevated VLDL

Because chylomicrons are mostly triglyceride, patients with hyperchylomicronemia may have dramatically elevated triglyceride levels with normal or only moderately elevated levels of cholesterol. Familial hyperchylomicronemia results from a deficiency in LPL (which is required to metabolize chylomicrons) or from a deficiency in apoprotein C-II, a cofactor needed for LPL function. Patients with type V disease may have other apoprotein abnormalities.[1]

The accumulation of triglyceride-rich chylomicron particles in the blood may result in hepatosplenomegaly, abdominal pain and pancreatitis, central nervous system (CNS) dysfunction, eruptive xanthomas, and lipemia retinalis. Early atherosclerosis is probably not associated.[2]

Hyperchylomicronemia may occur secondary to diabetes, since LPL requires insulin for synthesis. Types I and V may also be exacerbated by obesity, alcohol intake, exogenous estrogens, renal insufficiency, dysgammaglobulinemia, and systemic lupus erythematosus.

Type IV: VLDL Elevation

Since VLDL is synthesized endogenously, its level may be associated with a large number of genetic or environmental factors. Obesity and mild diabetes are related, and type IV patients may consume too much refined carbohydrate, saturated fat, and alcohol. VLDL has about three to four times as much triglyceride as cholesterol. Clinically, triglyceride levels over 1000 mg% are rarely seen in type IV, and cholesterol levels are more moderately elevated. Type IV does not seem to be associated with an increased prevalence of xanthelasma or corneal arcus.[3]

Type III: IDL Elevation

Some patients have defects involving the complete metabolism of chylomicrons and VLDL. Cholesterol-rich remnant particles accumulate as a somewhat heterogeneous mixture of the products of incomplete chylomicron and VLDL breakdown. This disease is best thought of as a disorder of remnant removal. It is probably due to absence of certain apoproteins needed for chylomicron/ VLDL breakdown. Clinically, patients may present with characteristic palmar or tuboeruptive xanthomas, and occasionally xanthelasma. Peripheral vascular disease and accelerated coronary atherosclerosis are seen.

Although some secondary cases have been reported, most type III cases are considered primary.

Type II: LDL Elevation

This type has been subdivided into type IIa, in which LDL is elevated but VLDL is normal, and type IIb, in which both LDL and VLDL are elevated.

Two primary genetic disorders are known. Familial hypercholesterolemia (FH) is due to a defect in certain apoprotein cell surface receptors that are an essential part of the balance between intracellular cholesterol synthesis and extracellular cholesterol levels. In familial hypercholesterolemia, the cellular surface receptors are absent or deficient, resulting in high levels of plasma LDL and cholesterol. The pathophysiology of the other primary disorder, familial combined hyperlipidemia, is uncertain though its clinical picture appears to be the result of VLDL and LDL overproduction by the liver.

Clinically, patients with FH present with xanthomas of tendons, tuberous xanthomas, corneal arcus, xanthelasma, and accelerated atherosclerosis. In general, type II patterns may be seen associated with hypothyroidism and a variety of renal, hepatic, and dysgammaglobulinemic diseases.

Ocular Manifestations

Tuberous (and tendon) xanthomas may be seen in types II, IV, and V, and may involve the eyelids. Tuberous xanthomas are histologically more deeply situated than xanthelasmas. They display foamy histiocytes intermixed with multinucleated Touton-type giant cells, as well as extracellular cholesterol deposits, moderate fibrosis, and inflammation.[4]

Xanthelasmas, small, flat, orange-yellow elevations on the eyelids, may be seen in familial hypercholesterolemia (type II), and rarely in type III. Microscopically, these superficial lesions consist of focal collections of lipid-laden histiocytes.[4]

Corneal arcus is caused by deposition of cholesterol and phospholipid at the periphery of the stroma. There may be a similar infiltration of the paralimbal sclera, with sparing of portions of the limbus. The arcus is separated from the limbus by a narrow band of relatively clear cornea. Arcus will be formed preferentially just central to an area of pathologically increased limbal vascularization.[5] In general, the clinical findings of xanthelasma or corneal arcus, especially in young people, is suggestive (though by no means diagnostic) of a lipoprotein abnormality.[3] The presence of corneal arcus in men under age 50 is a distinct risk factor for coronary heart disease.[6]

Lipemia retinalis usually does not appear until triglyceride levels exceed 2000 to 2500 mg%.[7] In lipemia retinalis, the first change in the color of the retinal vessels is seen in the periphery, where they become slightly milky or salmon colored. The change then spreads toward the optic disc. The vascular light reflex becomes diffuse or is lost. As the condition clears, the vessels return to normal in the reverse sequence. Vitreous hemorrhage occurring during an episode of lipemia retinalis results in a creamy opacity in the posterior portion of the eye; such an event has been termed "lipidosis vitrealis." (The terminology was suggested at the 1984 American Academy of Ophthalmology meeting, at which the first such case was presented as a poster exhibit by Robert J. Schechter, M.D., Ronald N. Gaster, M.D., and Eric Del Piero, M.D.)

Particles with diameters less than one fourth the wavelength of visible light are not seen, and the serum containing these particles appears clear.[8] Elevation of cholesterol per se does not cause the serum to become turbid and is not a cause of lipemia retinalis; this explains why lipemia retinalis is not seen in type II disease, though it can occur in all the other types.

The presence of extensive, confluent hard exudates in persons with diabetes has been correlated with hyperlipidemia.[9] Retinitis pigmentosa patients may have an increased incidence of lipid abnormalities.[10] Hyperlipidemia types IV and V have been associated with retinal abnormalities and vein occlusion.[11]

References

1. Schaefer EJ, Levy RI. Pathogenesis and management of lipoprotein disorders. *N Engl J Med* 1985;312:1300-1310.
2. Havel RJ. Approach to the patient with hyperlipidemia. *Med Clin North Am.* 1982;66:319.
3. Segal P, Insull W Jr, Chambless LE, et al. The association of dyslipoproteinemia with corneal arcus and xanthelasma. *Circulation.* 1986;73(suppl I):I108-I118.
4. Font RL. Eyelids and lacrimal drainage system. In: Spencer

WH, ed. *Ophthalmic Pathology. An Atlas and Textbook.* Philadelphia, Pa: W.B. Saunders Co.; 1986;2248-2250.

5. Cogan DG, Kuwabara T. Arcus senilis: its pathology and histochemistry. *Arch Ophthalmol.* 1959;61:553-560.

6. Rosenman RH, Brand RJ, Sholtz RI, et al. Relation of corneal arcus to cardiovascular risk factors and the incidence of coronary disease. *N Engl J Med.* 1974;291:1322-1324.

7. Vinger PF, Sachs BA. Ocular manifestations of hyperlipoproteinemia. *Am J Ophthalmol.* 1970;70:563-573.

8. Ahrens EH Jr, Kunkel HG. The stabilization of serum lipid

emulsions by serum phospholipids. *J Exp Med.* 1949;90:409-424.

9. Brown C, Ridley M, Haas D. Lipemic diabetic retinopathy. *Ophthalmology.* 1984;91:1490-1494.

10. Converse CA, Hammer HM, Packard CJ, et al. Plasma lipid abnormalities in retinitis pigmenosa and related conditions. *Trans Ophthalmol Soc UK.* 1983;103:508-512.

11. Dodson PM, Galton DJ, Winder AF. Retinal vascular abnormalities in the hyperlipidaemias. *Trans Ophthalmol Soc UK.* 1981;101:17-21.

Chapter 115

Hypolipoproteinemias

JOSE S. PULIDO, G. FRANK JUDISCH, and MICHAEL P. VRABEC

Lipoproteins are the body's transport vehicles for lipids. They can be grouped by centrifugation and electrophoretically into essentially four different classes: chylomicrons, very low–density lipoproteins (VLDLs), low-density lipoproteins (LDLs), and high-density lipoproteins (HDLs). Chylomicrons are made in the intestinal villi and carry the absorbed lipids to the liver. VLDLs are made by the liver and transport triglycerides. LDLs and HDLs are involved in cholesterol transport and degradation. Most LDLs are derived from catabolism of VLDLs. HDLs are made by the liver and intestine as well as from catabolism of chylomicrons and VLDLs.[1]

Lipids and the fat-soluble vitamins A, D, E, and K are absorbed in the intestinal villi by the action of lipases and bile salts to form micelles, which are then transported across intestinal cell membranes. Lipids and vitamins A, E, and K are then packaged in association with proteins to make chylomicrons, which are transported via the lymphatics to the liver. Vitamin D is transported by a separate mechanism. The liver uses the vitamins and lipids or transports them. Vitamin A is transported by its own retinol-binding protein. Vitamin E appears to be transported by lipoproteins.

The core of lipoproteins is made up of proteins that have both lipophilic and hydrophilic components (amphipathic). There are eight main apoproteins (apo):

apoA-I, apoA-II, apoA-IV, apoB, apoC-I, apoC-II, apoC-III, and apoE. ApoA-I, apoA-II, and apoA-IV are found on HDLs. ApoA-I, apoC-I, apoC-II, and apoC-III are constituents of both VLDLs and HDLs. ApoE is an important component of VLDLs and chylomicrons and binds lipoprotein receptor molecules on certain cell surfaces similar to apoB.[2,3]

ApoB exists primarily in two forms: ApoB-100, which is created primarily by the liver and is found on VLDLs and LDLs, and apoB-48, which is created by cells of the intestinal villi and is found on chylomicrons. ApoB serves as the recognition site that binds lipoproteins to the lipoprotein receptor molecules on cell surfaces.[2,3] The hypolipoproteinemias can be divided into two categories of primary ocular manifestations, retinal and corneal.

Hypolipoproteinemias Associated with Retinal Abnormalities

Bassen-Kornzweig Disease

Bassen-Kornzweig disease (abetalipoproteinemia) is characterized by steatorrhea, acanthocytosis, pigmentary retinopathy, and progressive neurologic disease. It is considered to be an autosomal recessive disorder that is more common in Jewish populations but can affect members of any race and is caused by either decreased production or decreased secretion of apoB. Heterozygotic persons have normal levels of apoB and lipoproteins.[4] In patients with Bassen-Kornzweig disease, because chylomicrons cannot be formed, the intestinal villi fill with lipids. Steatorrhea ensues. Because vitamins A, E, and K cannot be transported, vitamin deficiency states result. In addition to suffering from steatorrhea, homozygotic patients are subject to a progressive neurologic disorder characterized by large-fiber neuropathy producing areflexia followed by ataxia and proprioceptive sensory loss. Mental retardation is seen in one third of the patients.

Laboratory findings in homozygotic patients include low plasma concentrations of cholesterol and triglycerides as well as low levels of VLDLs, chylomicrons, and LDLs. Because of low vitamin K levels, blood clotting abnormalities have also been noted. Acanthocytes presumed to be present from birth represent at least 50% of the red blood cells (RBCs) in the peripheral circulation. Acanthocytes are mature RBCs with multiple irregularly arranged spiny or blunt projections. The cholesterol content of the RBCs is normal or slightly increased, whereas the phospholipid and fatty acid content of the red cell membrane are usually decreased. The ratio of phosphotidyl choline to sphingomyelin is decreased. Normal red cells transfused into patients with abetalipoproteinemia become abnormal. Acanthocytosis has also been seen in patients with severe hepatocellular disease, anorexia nervosa, and severe malnutrition. Peroxidative hemolysis of RBCs in vitro is increased and can be corrected by adding vitamin E to the medium.[4,5]

The pigmentary retinopathy of abetalipoproteinemia usually is recognized by the second decade; however, electroretinographic (ERG) findings often are subnormal or nonrecordable much earlier. The fundus changes are those of either typical or atypical retinitis pigmentosa and include pigment clumping, pigment spiculing, white dots, narrowing of the retinal vessels, and optic atrophy.[6,7] The posterior fundus is more severely affected than the periphery, and sharply demarcated areas of chorioretinal atrophy may also be seen. The fundus findings are not pathognomonic, and the diagnosis can be made with certainty only by demonstrating the characteristic deficiency of lipoproteins. Visual fields show ring scotomas; dark adaptation studies are abnormal; and patients complain of progressive night blindness. Other ocular findings include angioid streaks of the fundus, exotropia, and dissociated nystagmus in lateral gaze.[8,9] Specimens examined histologically showed loss of photoreceptors, invasion of pigmented cells into the retina, and intracellular collection of lipofuscin in the areas where the RPE was still present.[6]

Treatment has been controversial. One hypothesis postulates that the low serum triglyceride levels do not allow for formation of myelin in membranes of neural tissues. Though serum triglyceride levels may be important in acanthocyte formation, there are no data to support its importance in neuronal function.[4] The other theory states that there are deficiencies of vitamins A and E. Vitamin A is important for the production of rhodopsin and the other visual pigments, and vitamin E probably serves as an antioxidant. Originally, administration of vitamin A was associated with some initial improvement; however, with time, the retinopathy progressed.[10] Recently, selected vitamin E deficiency states have been recognized in some patients. These patients tend to have areflexia and loss of proprioception, neurologic signs that are very similar to those of abetalipoproteinemia.[11] Other patients with cholestatic disease, cystic fibrosis, and short bowel syndrome also have a deficiency of vitamin E and have shown histopathologic findings similar to those of patients with abetalipoproteinemia.[4,11] In rats, vitamin E prevents both accumulation of lipofuscin and oxidation of photoreceptor membranes and of vitamin A stores. Rats made deficient in vitamins A and E have a more severe loss of photoreceptor nuclei than rats made deficient of either vitamin alone.[12]

Judisch et al.[7] reported a case of a 13-month-old child with abetalipoproteinemia and a nonrecordable ERG (Fig. 115-1). After treatment with water-soluble, oral vitamins A and E, an ERG of approximately 30% of normal could be recorded. After 7 years of follow-up, the ERG remains stable, and several subjective visual acuities of 20/40 have been recorded in both eyes, suggesting the importance of both vitamin A and E supplements for these patients.[7] Other investigators have noted similar findings.[13]

Familial Hypobetalipoproteinemia

A related disorder, familial hypobetalipoproteinemia, has similar clinical findings, but in this disease heterozygotic patients have half the normal plasma levels of LDL and VLDL. Aside from the lipoprotein abnormalities, heterozygotic persons usually show no clinical manifestations of disease. Homozygotes have either very low levels of LDL, VLDL, and chylomicrons or none at all. The apoB is indetectable in these patients. The homozygotes appear to have a disease similar to Bassen-Kornzweig, but the neurologic manifestations appear to be milder and to occur later in the disease. Supplementation of vitamins A and E is important in the homozygotic state.[4,14]

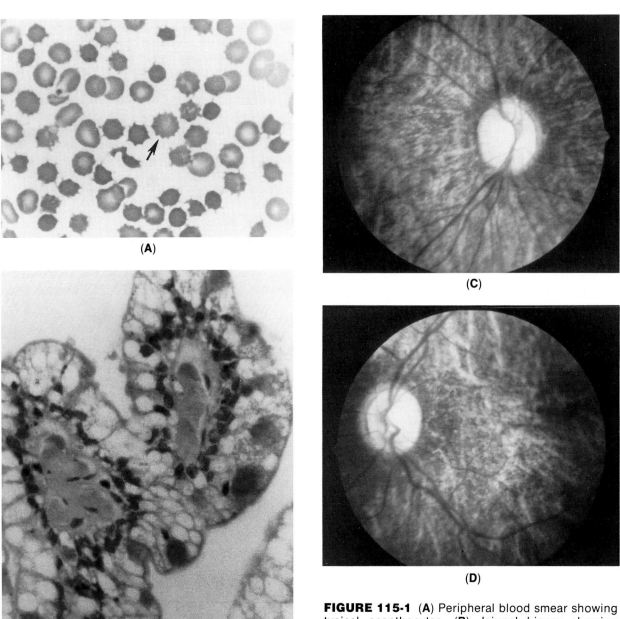

FIGURE 115-1 (A) Peripheral blood smear showing typical acanthocytes. (B) Jejunal biopsy showing empty vacuoles in the intestinal epithelium which represent loci of lipid deposition. (C) Right eye of a patient with abetalipoproteinemia at 13 months of age. Note the granular pigmentary retinopathy with patchy pigment loss. (D) Left eye of same patient at 13 months of age. (From Judisch, Rhead, Miller.[7])

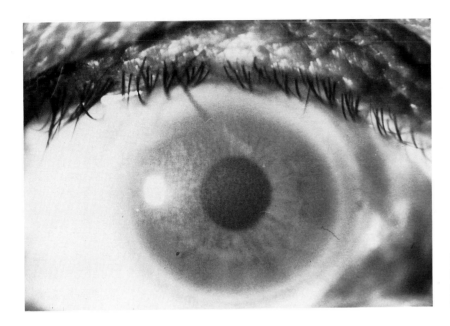

FIGURE 115-2 Corneal changes in LCAT deficiency. Note peripheral arcus with central corneal haze.

Hypolipoproteinemias Associated with Corneal Abnormalities

Tangier Disease

Tangier disease is a rare autosomal recessive disease characterized by a deficiency or an absence of HDLs in the plasma accompanied by accumulation of cholesteryl esters in many tissues throughout the body. Plasma triglyceride levels tend to be normal or high and cholesterol levels tend to be low or normal. Lipoprotein electrophoresis is pathognomonic in that low HDLs or none are noted. The VLDLs have normal amounts of triglycerides, but the cholesteryl ester content is one third that of control VLDLs. In Tangier disease, LDLs are also markedly different from normal in that they have a greater proportion of triglyceride and less cholesterol than normal LDLs. The apoA-I and -II constituents of HDLs are markedly reduced, while the ApoC is only moderately reduced. Studies have not confirmed whether the metabolic defect is a problem of decreased HDL synthesis or of increased HDL metabolism.[4,15]

Clinically, these patients show hyperplastic yellow-orange tonsils, hepatosplenomegaly, and lymphadenopathy. Biopsy of these tissues has revealed histiocytes, fibroblasts, and Schwann cells that have intracellular cholesteryl ester droplets. Because of the Schwann cell involvement, these patients develop a subtle peripheral neuropathy that appears to progress with age. Though HDLs are decreased in this disease, it is interesting to note there does not appear to be an increased incidence of atherosclerotic disease in homozygotic patients.

The ocular findings in these patients tend to reflect the continuing cholesteryl ester deposition with aging. Corneal clouding is noted in 50% of them and appears to increase with age. Conjunctival biopsies revealed intracellular lipid particles in pericytes and fibrocytes. Other ocular findings have included orbicularis oculi weakness and frank ectropion.[4,15,16]

Obligate heterozygotic patients usually appear clinically normal, though some reports have noted coronary heart disease in patients over 45 years. HDL concentrations in heterozygotic patients tend to be below the normal range. ApoA-I and apoA-II also appear to be decreased.[4]

Fish Eye Disease

Fish eye disease is an autosomal recessive disease whose main clinical manifestation is severe corneal clouding with vision loss. Affected persons develop bilateral dotlike, grey-white opacities that form a mosaic pattern. The cloudiness is in all layers of the cornea except the epithelium. HDL levels are 10% of normal, and there are low levels of apoA-I and -II. LDLs are normal, also distinguishing it from Tangier disease. Patients have high levels of VLDLs, whereas in Tangier disease the VLDLs tend to be in the normal range. Serum cholesterol is normal; in Tangier disease it is low. The exact biochemical defect is unknown. Cholesterol has been noted histopathologically in the stroma and in Bowman's membrane. No premature coronary artery disease has been reported.[4,17]

Lecithin Cholesterol Acyl Transferase Deficiency

Lecithin cholesterol acyl transferase (LCAT) deficiency is thought to be an autosomal recessive disease. LCAT esterifies free cholesterol, and without it large concentrations of unesterified cholesterol accumulate in the plasma. Homozygotic patients' HDL levels are 30% of normal. VLDLs and LDLs are elevated. ApoA-I and -II are about 50% of normal. Serum cholesterol and triglyceride levels are high or normal. Clinically these patients may develop anemia, proteinuria, uremia, and premature atherosclerosis. On ocular examination, they have corneal deposition of multiple small grey opacities, the exact composition of which has not been determined (Fig. 115-2). They are seen as both an arcus and as small isolated flecks in the corneal stroma which appear grossly as a diffuse haze. Histopathology of the cornea shows multiple tiny vacuoles in Bowman's membrane filled with electron-dense particles.[17,18]

References

1. Schaefer EJ, Levy RI. Pathogenesis and management of lipoprotein disorders. *N Engl J Med.* 1985;1300-1310.
2. Schonfeld G. Disorders of lipid transport: relationship to abnormalities of apoproteins, enzymes and cellular receptors. In: Rao DC, et al., eds. *Genetic Epidemiology of Coronary Heart Disease: Past, Present, and Future.* New York, NY: Alan R. Liss; 1984; 375-402.
3. Breslow JL. Apolipoprotein defects. *Hosp Pract.* 1985;43-49.
4. Herbert PN, Assmann G, Gotto A, et al. Familial lipoprotein deficiency: abetalipoproteinemia, hypobetalipoproteinemia, and Tangier disease. In: Stanbury JB, Wyngaarden JB, Fredrickson DS, et al. eds. *The Metabolic Basis of Inherited Disease.* New York, NY: McGraw-Hill; 1983:589-621.
5. Lange Y, Steck TL. Mechanism of red blood cell acanthocytosis and echinocytosis *in vivo. J Membr Biol.* 1984; 77:153-159.
6. Cogan DG, Rodrigues M, Chu FC, et al. Ocular abnormalities in abetalipoproteinemia. *Ophthalmology.* 1984;91:991-998.
7. Judisch GF, Rhead WJ, Miller DK. Abetalipoproteinemia. *Ophthalmologica, Basel.* 1984;189:73-79.
8. Duker JS, Belmont J, Bosley TM. Angioid streaks associated with abetalipoproteinemia. *Arch Ophthalmol.* 1987; 105:1173-1174.
9. Yee RD, Cogan DG, Zee DS. Ophthalmoplegia and dissociated nystagmus in abetalipoproteinemia. *Arch Ophthalmol.* 1976;94:571-575.
10. Gouras RE, Carr RE, Gunkel RD. Retinitis pigmentosa in abetalipoproteinemia: effects of vitamin A. *Invest Ophthalmol.* 1971;10:784-793.
11. Harding AE, Matthews S, Jones S, et al. Spinocerebellar degeneration associated with a selective defect of vitamin E absorption. *N Engl J Med.* 1985;313:32-35.
12. Robison WG, Kuwabara T, Bieri JG. Deficiencies of vitamins E and A in the rat: retinal damage and lipofuscin accumulation. *Invest Ophthalmol.* 1980;19:1030-1037.
13. Runge P, Muller DPR, McAllister J, et al. Oral vitamin E supplements can prevent the retinopathy of abetalipoproteinaemia. *Br J Ophthalmol.* 1986;70:166-173.
14. Yee RD, Herbert PN, Bergsma DR, et al. Atypical retinitis pigmentosa in familial hypobetalipoproteinemia. *Am J Ophthalmol.* 1976;82:64-71.
15. Pressly TA, Scott WJ, Ide CH, et al. Ocular complications of Tangier disease. *Am J Med.* 1987;83:991-994.
16. Chu FC, Kuwabara T, Cogan DG, et al. Ocular manifestations of familial high-density lipoprotein deficiency (Tangier disease). *Arch Ophthalmol.* 1979;97:1926-1928.
17. Carlson LA, Philipson B. Fish-eye disease. *Lancet.* 1979;2:921-923.
18. Vrabec MP, Shapiro MB, Koller E, et al. Ophthalmic observations in lecithin cholesterol acyltransferase deficiency. *Arch Ophthalmol.* 1988;106:225-229.

Chapter 116

Lipoid Proteinosis

Urbach-Wiethe Disease

JAMES A. DEUTSCH and PENNY A. ASBELL

Lipoid proteinosis (also known as Urbach-Wiethe disease and hyalinosis cutis et mucosae) is a rare autosomal recessive disorder characterized by deposition of a hyalinelike material, primarily in the skin and mucous membranes. The material, which has not yet been definitely characterized, can also be found in cells of every organ of the body.[1] The classic clinical features of the disease include hoarseness dating from early infancy due to deposits in the larynx and the occurrence of yellowish plaques, papules, and verrucous nodules in the skin and mucous membranes. These nodules are especially common on the face and extensor surfaces of the limbs. The classic ophthalmic manifestation is pearly yellow nodules along the free margins of the eyelids.[2]

The hyaline deposits are seen first in the walls of small blood vessels, the sweat glands, and the arrector pili muscles. Later they are found in the superficial dermis.[3] The deposits are positive for periodic acid-Schiff reagent and Sudan compound. Recent research suggests that the hyaline material in lipoid proteinosis may result from overproduction of noncollagenous proteins, some of which are found as normal constituents of human skin.[4] However, other researchers have proposed a defect in collagen metabolism, a disorder in basal lamina synthesis, and that the disorder is a lysosomal storage disease.[5-7]

Slightly more than 280 cases of lipoid proteinosis have been reported, primarily in patients of European extraction.[3] Of the over 50 cases documented in South Africa, all have been traced back to one German settler who migrated to the Cape in the latter half of the 17th century.[8]

Systemic Manifestations

In addition to skin and mucous membranes, infiltration of the tongue, parotid glands, teeth, respiratory and digestive tracts, and central nervous system have been reported.[2] The laryngeal involvement, which causes hoarseness dating from infancy, has been reported to cause laryngeal obstruction that necessitated tracheos-tomy.[9] Patients whose true and false vocal cords are both involved may be at increased risk of aspiration pneumonia.[2]

Central nervous system involvement includes bilateral calcific densities in the region of the hippocampi.[10] Both petit mal and grand mal seizures have been reported, as have visual and olfactory hallucinations and disturbances of short-term memory.[2]

Ocular Manifestations

The development of small yellowish papules along the free margin of all four eyelids is practically pathognomonic for the disease. Generally, these are of little clinical significance, although there is one report of resultant trichiasis causing corneal ulceration.[11] Occasionally, eyelashes or eyebrows are lost.

Drusen of the macula and retinal degeneration were reported in slightly less than half of one group studied.[2] Rarely reported changes include cystic changes of the various ocular glands, small yellow plaques of the conjunctiva, corneal opacities, and glaucoma (thought to be secondary to infiltrates in the trabecular meshwork). Additional reports include the occurrence of slow-reacting pupils, eccentric pupils, increased engorgement and tortuosity of the retinal vessels, iridocyclitis, and chorioretinitis.[2] One case of pseudomembranous conjunctivitis is reported.[3]

Although lipoid proteinosis does not usually shorten life, the disorder may cause extensive disfigurement and impairment of function. Still, it is rare that any therapeutic intervention is necessary.

References

1. Ramsey ML, Tschen JA, Wolf JE Jr. Lipoid proteinosis. *Int J Dermatol.* 1985;24(4):230.
2. Hofer PA. Urbach-Wiethe disease (lipoglycoproteinosis; lipoid proteinosis; hyalinosis cutis et mucosae). A review. *Acta Derm Venereol* [Suppl] (Stockh). 1973;53:1

3. Barthelemy H, Mauduit G, Kanitakis J, et al. Lipoid pro-teinosis with pseudomembranous conjunctivitis. *J Am Acad Dermatol.* 1986;2:367.

4. Fleischmajer R, Krieg T, Dziadek M, et al. Ultrastructure and composition of connective tissue in hyalinosis cutis et mucosae skin. *J Invest Dermatol.* 1984;82(3):252.

5. Harper JI, Duance VC, Sims TJ, et al. Lipoid proteinosis: an inherited disorder of collagen metabolism? *Br J Dermatol.* 1985;113(2):145.

6. Brocheriou C, Kuffer R, Laufer J, et al. Cutaneous-mucous hyalinosis (Urbach-Wiethe disease). Histologic and ultra-structural study of a case. *Ann Pathol.* 1984;4(4):297.

7. Bauer EA, Santa-Cruz DJ, Eisen AZ. Lipoid proteinosis: *in vivo* and *in vitro* evidence for a lysosomal storage disease. *J Invest Dermatol.* 1981;76:119.

8. Heyl T. Lipoid proteinosis in South Africa. *Dermatologica.* 1971;142(3):129.

9. Finkelstein MW, Hammond HL, Jones RB. Hyalosis cutis et mucosae. *Oral Surg Oral Med Oral Pathol.* 1982;54(1):49.

10. Friedman L, Mathews RD, Swanepoel PD. Radiographic and computed tomographic findings in lipoid proteinosis. A case report. *S Afr Med J.* 1984;65:734-735.

11. Hewson GE. Lipid proteinosis (Urbach-Wiethe syndrome). *Br J Ophthalmol.* 1963;47:242.

Chapter 117

Batten's Disease

Neuronal Ceroid Lipofuscinoses

MARY ANN LAVERY

In 1903, Batten described a family with cerebral degener-ation and macular changes.[1] This disease and its group of related cerebromacular degenerations involve storage of autofluorescent lipopigments. Initially, they were classified under "familial amaurotic idiocy," along with Tay-Sachs disease, until the hexosaminidase A enzyme defect and G_{m2} ganglioside storage of Tay-Sachs and the pathologic lipopigment storage in Batten's disease were noted to clearly differentiate these groups. Batten's storage diseases, so-called neuronal ceroid lipofuscin-oses, exhibit autosomal recessive inheritance. There are four main clinicopathologic types of Batten's lipopig-ment storage diseases (infantile, or Haltia-Santavuori; late infantile, or Jansky-Bielschowsky; juvenile, or Spielmeyer-Vogt, and adult, or Kufs') that have varying components and associated degrees of psychomotor degeneration, myoclonic epilepsy, visual loss, and reti-nal pigmentary degeneration. A review of pediatric neurology admissions in two American Medical Centers found "neuronal ceroid lipofuscinosis" to be the second most common neurodegenerative disease of infancy and childhood.[2]

Since no specific biochemical defect has been iden-tified, the diagnosis of a lipopigment storage disease depends entirely on confirmation of clinical suspicions by histopathologic demonstration of abnormal storage of lipopigments that are autofluorescent and stain with lipid stains but are insoluble in lipid solvents. Zeman applied the term "neuronal ceroid lipofuscinosis" to emphasize the pathologic storage of ceroid in the first three types (sometimes collectively termed Batten's dis-ease) and of lipofuscin (age pigment) in the adult type (Kufs' disease).[3] The abnormal lipopigment accumu-lated in Batten's and Kufs' diseases has not been defini-tively characterized. There are qualitative and quantita-tive differences between these lipopigments and normal lipofuscin, which may be related to the neuronal dam-age that occurs.[4] The lipopigment storage material, like normal lipofuscin, is found in secondary lysosomes and is thought to be derived largely from components of intracellular membranes. The abnormal lipopigment consists primarily of insoluble proteins with 20 to 40% lipid. Dolichols account for up to 5%. Some affected patients have increased dolichol levels in urine, tissues, and fractions of the storage granules.[5]

Diagnostic ultrastructural characteristics of the stored lipopigments are abundant membrane-bound cytosomal inclusions that vary from granular dense osmiophilic deposits (predominant in the infantile type but also present in the adult type), to curvilinear profiles (predominant in the late infantile type), to fingerprint stacked lamellar profiles (predominant in juvenile and adult types). Although clinical manifestations of Batten's and of Kufs' disease emphasize the cerebroretinal degeneration that occurs with accumulation of abnormal lipopigments in neurons and retina, multiple systemic sites of lipopigment accumulation can be found: skin and conjunctiva, sweat glands, rectal myenteric plexus, sural nerve, Schwann cells, striated and smooth muscle, appendix, lymphocytes, urinary sediment, vascular endothelial cells, bone marrow, viscera, and endocrine glands.[6]

In one report of the late infantile type, prenatal diagnosis by curvilinear cytosome inclusions of amniotic fluid cells was confirmed.[7] Attempts to detect heterozygotic patients have utilized leukocyte hypergranulation, skin biopsy electron microscopy, and urinary dolichol levels. No therapy has yet proven effective in stopping the progressive neurodegeneration of these diseases.

Systemic Manifestations

The infantile type (Haltia-Santavuori) begins by 24 months of age. In most of the children retardation of mental development begins by 12 months of age with loss of the ability to play and speak. Motor development ceases, and muscle hypotonia and ataxia follow. Microcephaly becomes apparent around 2 years of age. Myoclonic jerks and generalized convulsions may start at any time, averaging around 2 years of age. At age 3 years voluntary movements are lost and flexion contractures begin. Most children die in the first decade (mean 8 years). Computed tomography (CT) reveals diffuse supra- and infratentorial atrophy that increases with age. The electroencephalogram (EEG) is pathologically low in amplitude at first, decreasing to isoelectric by age 3 years. There are no vacuolated lymphocytes.

The late infantile type (Jansky-Bielschowsky) begins between 15 and 33 months of age with deterioration of speech, gait, and intellectual functions. Ataxia with grand mal and myoclonic epilepsy follows. The terminal state is spastic tetraplegia and bulbar paralysis. Lymphocyte vacuoles are not present. Early on, the electroencephalogram (EEG) may show large amplitudes with characteristic poor rhythmic activity, irregular slow waves, and polyphasic spikes to photic stimulation. Pulmonary aspiration is frequently the immediate cause of death at about 10 years of age.

The juvenile type (Spielmeyer-Vogt) commonly begins with central vision impairment at age 3 to 4 years followed by seizures at age 6 to 16 and deterioration of speech and gait at age 7 to 13 years. Dementia ensues (8 to 12 years) and spastic rigid tetraplegia occurs. Vacuoles are present in lymphocytes, showing lipopigment inclu-

sions on electron microscopy. The EEG has pathologic spike-wave activity, especially to photic stimulation. Death occurs between 12 and 25 years of age.

Age of onset of adult type (Kufs') ranges from 11 to 50 years. Myoclonic epilepsy typically begins around age 30 years. Seizures often become intractable, with signs of progressive dementia and ataxia developing. The EEG shows intense photoparoxysmal response. Giant somatosensory evoked potentials may be seen. Generalized cerebral atrophy becomes apparent on CT. Severe dementia, major seizures, and myoclonus continue until death occurs, between 17 and 60 years of age. Lymphocytes are not vacuolated.

Ocular Manifestations

The infantile type (Haltia-Santavuori) shows diffuse visual deterioration by 12 months of age, with pupil reaction slow or absent after the age of 2 years. Progressive hypopigmentation of the fundus with brownish discoloration of the macula and optic atrophy with attenuated retinal vessels occur. The ERG reading is subnormal before 20 months of age and is extinguished by 30 months of age, even if the fundus still appears normal. The visual evoked response (VER) is subnormal at age 20 months and is abolished by age 3. Stellate posterior polar cataracts and retinal degeneration with mottled hyperpigmentation have been described in long-term survivors.[8] Autopsy findings demonstrate loss of almost all photoreceptor and ganglion cells with glial proliferation and optic atrophy.

The late infantile type (Jansky-Bielschowsky) has visual loss as a later symptom, with reddish-brown macular pigment mottling in a bull's eye pattern followed by optic atrophy with attenuation of retinal arterioles from 3½ to 5 years of age. Classic retinitis pigmentosa may be observed. The ERG signal is small at first, and then unrecordable. The VER is grossly enlarged. Blindness comes between 28 months and 5 years of age. Ocular pathologic processes include loss of cones and rods, with lipopigment inclusions in retinal pigment epithelial cells. There are also loss of ganglion cells and optic atrophy, and lipopigment inclusions in the remaining ganglion cells. In addition, abnormal lipopigment inclusions are found in corneal keratocytes and endothelium as well as in choroidal and retinal vascular endothelium and pericytes.[9]

The juvenile type (Spielmeyer-Vogt) has early reduction of central vision with red-green color blindness and relatively better night vision than day vision beginning around age 5 to 7 years. A fine mottling of the macular pigment (occasionally arranged in a bull's-eye fashion) may be seen even before visual impairment is demonstrable. Retinal pigment epithelial defects on fluorescein angiography may confirm clinical suspicions. The foveolar reflex diminishes. The earliest ERG findings show B-wave reductions and abolished oscillatory potentials. The ERG becomes unrecordable by the time vision difficulties are evident, and blindness follows. A slowly progressive optic atrophy with attenuation of retinal vessels occurs. The visual evoked potential (VEP) has normal early components that become progressively more poorly defined and then disappear. Later in the course of the disease, peripheral clumped retinal aggregates are seen. Ocular pathology shows severe involvement of the deep retinal layers with loss of rods and cones resulting in blindness, with relative sparing of the ganglion cell layer. There is retinal pigment epithelial atrophy and hyperplasia with perivascular migration. Optic atrophy is present. The abnormal lipopigment inclusions are found in the retinal pigment epithelium and the ganglion cells, along with storage in conjunctiva, cornea, iris, ciliary body, and extraocular muscle cells.

The adult type (Kufs') displays storage of abnormal lipopigments in retinal ganglion cells and little or no storage in deeper retinal layers. The conjunctiva has abnormal lipopigment inclusions in fibrocytes, Schwann cells, and vascular endothelial cells. The ERG and the VER findings are normal. Loss of vision and retinal pigmentary degeneration are not seen in Kufs' disease.[10]

References

1. Batten FE. Cerebral degeneration with symmetrical changes in the macula in two members of a family. *Trans Ophthalmol Soc UK.* 1903;23:386.

2. Dyken P, Krawiecki N. Neurodegenerative diseases of infancy and childhood. *Ann Neurol.* 1983;13:351.

3. Zeman W, Donahue S, Dyken P, et al. The neuronal ceroid-lipofuscinoses (Batten-Vogt syndrome). In: Vinken PJ, Bryn GV, eds. *Handbook of Clinical Neurology.* vol 10. Amsterdam: North-Holland; 1970;588.

4. Sohal RS, Wolfe LS. Lipofuscin: characteristics and significance. *Prog Brain Res.* 1986;70:171.

5. Armstrong D, Koppang N, Rider JA, eds. *Ceroid-Lipofuscinosis (Batten's Disease).* Amsterdam: Elsevier Biomedical Press; 1982.

6. Carpenter S, Karpeti G, Andermann F, et al. The ultrastructural characteristics of the abnormal cytosomes in Batten-Kufs' disease. *Brain.* 1977;100:137.

7. Macleod PM, Dolman CL, Nickel RE, et al. Prenatal diagnosis of neuronal ceroid-lipofuscinoses. *Am J Med Genet.* 1985;22:781.

8. Bateman JB, Philippart M. Ocular features of Hagberg-Santavuori syndrome. *Am J Ophthalmol.* 1986;102:262.

9. Traboulsi EI, Green WR, Luckenbach MW, et al. Neuronal ceroid-lipofuscinosis ocular histopathologic and electron microscopy studies in the late infantile, juvenile, and adult forms. *Graefe's Arch Clin Exp Ophthalmol.* 1987;225:391.

10. Berkovic SF, Carpenter S, Andermann F, et al. Kufs' disease: a critical reappraisal. *Brain.* 1988;111:27.

Section D

Disorders of Lysosomal Enzymes

Chapter 118

Farber's Disease

Disseminated Lipogranulomatosis Ceramidase Deficiency

MARCO ZARBIN and W. RICHARD GREEN

Systemic and Ocular Manifestations

Farber's disease (disseminated lipogranulomatosis) is an inherited autosomal recessive metabolic disorder clinically characterized by hoarseness, multiple painful subcutaneous periarticular nodules, and progressive arthropathy. Other features include poor growth and development, mild lymphadenopathy, variable visceral (lungs, liver, heart, blood vessels) and neural involvement, and occasionally fever.[1-3] The general nature and biochemical aspects of the disease were recently reviewed by Moser and Chen.[2]

Although the early onset of signs and symptoms, the presence of subcutaneous nodules (near the dorsal surface of the joints and elsewhere), joint swelling, and vocal cord thickening are more or less constant features of the disease, different phenotypes have been identified on the basis of longevity.[2] The severe form encountered in two thirds of the patients has a rapidly progressive course, resulting in death by age 4 years. Visceral and neurologic findings (approximately one half of patients are mentally retarded) are present.

Five patients have been described with the intermediate phenotype. Visceral lesions have not been demonstrated in this type, although mental retardation was present in two of the patients.

The mild phenotype has been described in five patients. They had no clinically apparent visceral involvement, although four were mentally retarded. Patients with the latter two phenotypes often live longer, some through the second decade.

353

There may be a correlation between the age at which dermal nodules first appear and life expectancy.[2] Children who developed dermal nodules by the age of 4 months died by age 14 months at the latest. Although the relationship between severity of the disease and organ system involvement is not entirely clear, in the five cases in which eye lesions were clinically documented, the child died before the age of 3 years, suggesting that they all have a more severe phenotype. The eye lesions have been quite varied clinically and include blindness, xanthomatoid growths of the conjunctiva, grayish parafoveal retinal opacity, mild cherry-red macula, and nodular corneal opacities.[3]

Biochemical Features

The accumulation of ceramide, the *N*-acyl fatty acid derivative of sphingosine, is characteristic of Farber's disease, and in 9 of 27 patients described in the literature a deficiency of ceramidase (a lysosomal enzyme that catalyzes the hydrolysis of the amide linkage between the sphingosine and fatty-acid components) has been proven.[2] Family pedigree analysis and ceramidase assay in clinically unaffected obligate heterozygotes, has provided evidence for autosomal recessive inheritance of the disease. Thus, Farber's disease appears to be a lysosomal storage disease associated with an inborn error of ceramide metabolism.

Systemic Histopathology

The characteristic pathologic lesion in Farber's disease is a granuloma whose histologic and histochemical properties change with time.[4] In the early stages the lesions contain periodic acid-Schiff positive fibroblasts and fibrocytes. These cells contain dilated, rough endoplasmic reticulum and membrane-bound vacuoles, both of which enclose reticular granular material. Later, the granuloma forms. It has increased collagen levels and is populated by foam cells (which might be derived from fibroblasts or transformed histiocytes). The foam cells are replete with lysosomal structures containing numerous pleomorphic profiles of moderate electron density, termed curvilinear tubular bodies (CTBs) or Farber bodies. Granulomas have been observed in the skin, subcutaneous tissues, periarticular and synovial tissues, lymph nodes, and thymus, and to a lesser extent in other solid viscera, such as the liver, lungs, and heart. Biochemical, and ultrastructural data suggest that CTBs represent accumulated ceramide and that lamellar inclusions represent a combination of ganglioside and phospholipid.

Ocular Histopathology

Cogan et al.[5] provided the first light microscopic evidence that in Farber's disease glycolipid accumulates in retinal ganglion cells. We[3] have confirmed the presence of accumulated lipid-like material and have described the ultrastructural characteristics of the inclusion bodies in a patient with Farber's disease and macular cherry-red spots. Such changes are most apparent in the retina, particularly in the macula and midperiphery, but they are also present in other ocular tissues such as the uveal tract, lens, cornea, sclera, and conjunctiva.

The principal pathologic feature in the tissue that we examined was the presence of intracellular inclusions of variable morphology and density. Retinal ganglion cells, neurons and glia, and phagocytic cells showed the greatest accumulation of inclusions. Using high-pressure liquid chromatography, we have shown that the retina in the case we reported contained increased levels of ceramide, suggesting that some or all of the inclusions may represent stored ceramide.[6] These biochemical data suggest that the retina, like other organs of persons with this disease, exhibits decreased ceramidase activity.

Lamellar inclusions have been reported in alpha motor neurons in another case of Farber's disease. In this respect, retinal ganglion cells show the same pathologic changes as other neurons in the central nervous system.

Clinicopathologic Correlation

The abundance of lipid inclusions in retinal ganglion cells may very well account for the decreased vision and macular cherry-red spots observed in cases of Farber's disease. Nodular corneal opacities may be a manifestation of the lipid inclusions present in corneal epithelial cells, and xanthoma-like conjunctival growths may be due to lipid-laden fibrocytes in the substantia propria. Some tissues that contain lipid inclusions (eg, the uveal tract) have not demonstrated clinical involvement.

Diagnosis

As is typical of many of the sphingolipidoses,[7] in Farber's disease, retinal ganglion cells and processes in the nerve fiber and inner nuclear layers seem to have the greatest accumulation of lamellar inclusions in the posterior pole of the eye. The relative sparing of the retinal pigment epithelium is one distinction between Farber's disease and Niemann-Pick disease, or even Batten's disease (in which CTBs are seen). Although intraendothelial lamellar inclusions are seen in both Fabry's and Farber's disease, they seem to be much more numerous in the former. Also, lamellar inclusions are relatively rare in the conjunctival and corneal epithelium in Farber's disease, a finding that distinguishes this sphingolipidosis from Fabry's disease and from GM_1 gangliosidosis. In metachromatic leukodystrophy and Krabbe's disease, lamellar inclusions are confined primarily to optic nerve glia.[8] Lamellar inclusions of the GM_2 gangliosidoses seem to be confined to neuronal elements. Thus, as far as can be judged from the material examined, the ultrastructural ocular abnormalities of Farber's disease may be distinctive.

Biopsy of the peripheral nodules in Farber's disease can provide pathologic material confirming the clinical diagnosis. Our findings suggest that conjunctival biopsy probably would not have provided affirmative evidence for the diagnosis, at least in the case we have reported. Because conjunctival involvement in Farber's disease may be variable,[9] it is difficult to assess the role of conjunctival biopsy as a diagnostic aid. The frequency of abnormal biopsy findings could be a function of sampling error, disease duration, or phenotype.[10]

References

1. Farber S. A lipid metabolic disorder—disseminated lipogranulomatosis: a syndrome with similarity to, and important difference from, Niemann-Pick and Hand-Schuller-Christian disease. *Am J Dis Child*. 1952;84:499-500.
2. Moser HW, Chen WW. Ceramidase deficiency: Farber's lipogranulomatosis. In: Stanbury JB, Wyngaarden JB, Frederickson DS, et al., eds. *The Metabolic Basis of Inherited Disease*. 5th ed. New York, NY: McGraw-Hill; 1983;820-830.
3. Zarbin MA, Green WR, Moser HW, et al. Farber's disease. Light and electron microscopic study of the eye. *Arch Ophthalmol*. 1985;103:73-80.
4. Tanaka T, Takahashi K, Hakozaki H, et al. Farber's disease (disseminated lipogranulomatosis): a pathological, histochemical, and ultrastructural study. *Acta Pathol Jpn*. 1979;29:135-155.
5. Cogan DG, Kuwabara T, Moser H, et al. Retinopathy in a case of Farber's lipogranulomatosis. *Arch Ophthalmol*. 1966;75:752-757.
6. Zarbin MA, Green WA, Moser H, et al. Increased levels of ceramide in the retina of a patient with Farber's disease. *Arch Ophthalmol*. 1988;106:1163.
7. Berman ER. Sphingolipidoses and neuronal ceroid-lipofuscinosis. In: Garner A, Klintworth GK, eds. *Pathology of Ocular Disease*. part B. New York, NY: Marcel Dekker; 1982;897-930.
8. Quigley HA, Green WR. Clinical and ultrastructural ocular histopathologic studies of adult-onset metachromatic leukodystrophy. *Am J Ophthalmol*. 1976;82:472-479.
9. Zetterstrom R. Disseminated lipogranulomatosis (Farber's disease). *Acta Paediatr*. 1958;47:501-510.
10. Libert J, Danis P. Differential diagnosis of type A, B, and C Niemann-Pick disease by conjunctival biopsy. *J Submicroscop Cytol*. 1979;11:143-157.

Chapter 119

Fabry's Disease

Angiokeratoma Corporis Diffusum Alpha Galactosidase Deficiency

CLEMENT McCULLOCH and MRINMAY GHOSH

Anderson described a patient with proteinuria, deformity of the fingers, varicose veins, and edema of the legs in 1898. In the same year, Fabry described a patient with purpura nodularis. From these two early cases a disease that is protean in manifestations, often incapacitating or lethal, and hereditary in origin has been revealed. It occurs in whites, blacks, and orientals. It has come to be known as Fabry's disease, Anderson-Fabry disease, angiokeratoma corporis diffusum (universale), or alpha galactosidase deficiency, but it is most easily indexed as Fabry's disease.

Systemic Manifestations

The allele for Fabry's disease is located on the X chromosome. A defect in the hemizygous male results in the full-blown picture of the disease, while defects in the heterozygous female result in less severe manifestations. There is significant interfamilial and intrafamilial variability among both males and females.[1]

In the male the outstanding symptom is attacks of pain, usually located in the fingers, toes, palms, or soles and the large joints and often accompanied by fever and increased sedimentation rate.[1,2] Pain may also be abdominal or in the flank; diarrhea and satiety commonly occur.

Skin lesions take the form of clusters of dark red angiectases, located particularly on hips, buttocks, thighs, and genitals. Aneurysmal lesions may be present in the mouth. The renal changes include proteinuria and polyuria with fixed specific gravity. Casts, red cells, and birefringent lipid globules may be present in the urine. Azotemia develops in middle age. Hypertension is a common sequela of renal involvement.

The cardiovascular complications are many. Cardiac arrhythmias, myocardial ischemia, cardiomegaly, and cardiac infarction have all been reported. Cerebrovascular manifestations may take the form of vertigo, tinnitus, seizures, hemiplegia, hemianesthesia, aphasia, and cerebral hemorrhage.[3] Respiratory difficulty may arise due to chronic air flow obstruction.

The gonads, the adrenals, and the pituitary gland may be involved. Sterility is common. Life expectancy is reduced to 40 to 50 years.

Ocular Manifestations

The most prominent ocular finding is corneal. A fine grayish yellow sediment appears at the level of the epithelium at age 6 years. This slowly increases and takes on the typical whorled appearance of cornea verticillata[4] (Fig. 119-1). While it is not pathognomonic of Fabry's disease, its presence limits the diagnosis. A similar finding may occur with the use of chloroquine, amiodarone, chlorpromazine, prepacrine, aurodiaquin, and indomethacin.[5] The lens may show opacities, which may be fine, anterior subcapsular, cream-colored, and feathery, or posterior and star-or cross shaped.[4]

The retinal vessels may show aneurysmal dilatations, occlusions,[6] angulations, and sausagelike dilatations of the veins.[6] There may be hemorrhages into the retina or preretinal blood. If kidney disease has resulted in vascular hypertension, a hypertensive retinopathy may be present.[7]

The small vessels of the conjunctiva often show aneurysmal dilatations, tortuosity, and kinking.

Ocular abnormalities may also occur secondary to neurologic disease; hemianopia, muscle palsies, and pupillary abnormalities have been described. A common

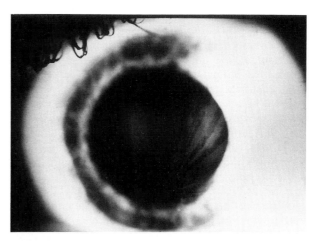

FIGURE 119-1 Clinical photograph shows typical cornea verticillatas.

finding that might lead to the correct systemic diagnosis is a continuing unexplained edema of the eyelids, particularly the upper lids.

Females can exhibit the same systemic findings as males, however they are usually milder.[3] Women can have acute attacks of pain and angiectases of the skin. Kidney, heart, and vascular involvement occur. Cerebral complications may be present. Renal and cardiovascular insufficiency can be terminal. In affected women the most helpful ocular diagnostic sign is the corneal whorl pattern, which is present in most heterozygotic persons; it is often more marked than in males.[4] Conjunctival and retinal lesions can occur, and lens opacities may be present.[8] Ocular findings secondary to cerebral or vascular complications of the disease can be expected.

Pathology

The protean manifestations are explained by the widespread intracellular deposition of glycosphingolipids, hexosylceramides, in blood vessels and in tissues all across the body.[9] This occurs in vascular endothelium and mural cells, in muscle cells, in myocardium, endocardium, in the glomerular tufts and tubules of the kidney, in ganglion cells, in neuronal sheaths, in autonomic nerves, in the cells of the sweat glands, and in the glands of internal secretion. The resultant weakening of vessel walls and infarcts leads to loss of glomerular function, myocardial weakness, and cardiac conduction abnormalities, and to hormone deficiency and sterility. The inclusions occur in the blood vessels of the central nervous system, also in the cells of the thalamus, hypothalamus, amygdala, substantia nigra, and intermediolateral cell columns of the spinal cord. Involvement of the nervous system results in attacks of pain and defective temperature control, the two early findings in the disease. Deposition in the lungs and bronchi leads to the air flow obstruction.

In the eye, hexosylceramides are deposited in the corneal epithelium, in conjunctival epithelium and goblet cells, and in the endothelium and mural cells of the blood vessels (Figs. 119-2 and 119-3).[10] The accumulation in goblet cells suggests that there is a significant discharge of the sphingolipid into the tears. It has been claimed that conjunctival change precedes corneal change.

The glycosphingolipids that accumulate are chiefly galactosyl-galactosyl-glucosyl ceramide, and glactosyl-galactosyl ceramide, although other variations occur. Variations also occur between the different body structures. The lipids accumulate because of a deficiency of the enzyme α-galactosidase A. The missing enzyme normally breaks down hexosylceramides by splitting off the sugar molecules. The complex sphingolipid molecules are normal components of cell membranes, includ-ing the walls of red blood cells. Disposal of these sphingolipids is retarded in Fabry's disease.

Heredity

The hereditary deficiency in Fabry's disease results in a loss of α-galactosidase A. The diagnostic value of the ocular abnormalities can be significant. The corneal findings are very indicative of the disease. A case of periodic pain, fever, and increased erythrocyte sedimentation rate may suddenly be explained by the discovery of cornea verticillata. The retinal and conjunctival vascular abnormalities would be confirmatory. Once a case has been identified, the family should be explored fully. Both males and females are at risk and should be sought out. Considering the severity of the disease, genetic counseling may become a key prophylactic tool.

Diagnostic Confirmation

A number of laboratory tests are available to confirm the clinical findings, and chemical confirmation of the disease should be obtained. Histologic examination of a specimen from skin, corneal epithelium, conjunctiva, kidney, or any tissue containing blood vessels will show birefringent inclusions.[11] On electron microscopy typical lamellar intracellular bodies are found. In urine and tears, birefringent bodies can be seen, and in urine, free and intracellular lamellar inclusions can be identified in concentrated sediment.[11,12] Glactosyl-ceramide levels in urine, tears, and plasma are increased. In males, the α-galactosidase level in plasma or in white blood cells

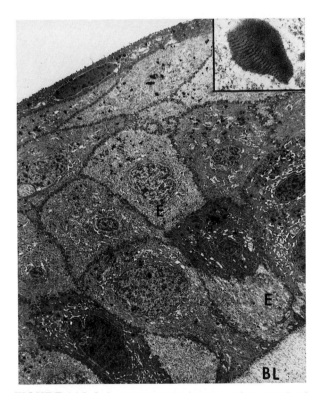

FIGURE 119-2 Lower-power electron micrograph of corneal epithelium (*E*) and Bowmans layer (*BL*), showing myriads of electron-dense deposits in epithelium. (Original magnification ×3,600; inset ×82,000)

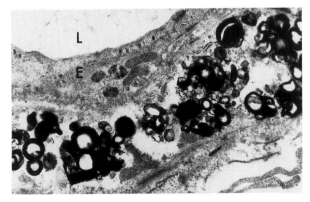

FIGURE 119-3 Electron micrograph of conjunctival vessel showing lumen (*L*), endothelium (*E*), and massive deposits in muscle cells. (Original magnification ×23,500)

will be very low or zero.[13] In the heterozygous female, α-galactosidase levels are about 50% of normal values.

Treatment

Several forms of treatment for Fabry's disease are available, although none has yet been proven to be effective. Plasma transfusion supplies alpha galactosidase and results in a lowering of blood levels of trihexosylceramide.[14] It will take years of clinical trials to determine the therapeutic value of repeated transfusions. Infusions of purified α-galactosidase have been used, and further development and trials of this form of treatment can be expected.[15] Isolation of a DNA clone encoding human α-galactosidase A lysosomal enzyme has been achieved, and synthesis of the human enzyme may be possible.[16] Transplantation, particularly kidney transplant, not only replaces a malfunctioning organ but introduces a source of the missing enzyme.[17] Further experience with this procedure will be necessary to determine its lasting value.[3,17,18] Since this is one of the few examples of an enzyme deficiency in which there are modes of treatment that offer a significant chance of clinical improvement, the ophthalmologist has an important role to play in early diagnosis of the disease.

References

1. Spence MW, Clarke JTR, D'Entremont DM, et al. Angiokeratoma corporis diffusum (Anderson-Fabry disease) in a single large family in Nova Scotia. *J Med Genet.* 1978;15(6):428-434.
2. Gadoth N, Sandbank U. Involvement of dorsal root ganglia in Fabry's disease. *J Med Genet.* 1983;20:309-312.
3. Bird TD, Lagunoff D. Neurological manifestations of Fabry disease in female carriers. *Ann Neurol.* 1978;4:537-540.
4. Sher NA, Letson RD, Desnick RJ. The ocular manifestation of Fabry's disease. *Arch Ophthalmol.* 1979;97:671-676.
5. Bron AJ. Vortex patterns of the corneal epithelium. *Ophthalmol Soc UK.* 1973;93:455-472.
6. Sher NA, Reiff W, Letson RD, et al. Central retinal artery occlusion complicating Fabry's disease. *Arch Ophthalmol.* 1978;96:815-817.
7. Bloomfield SF, David WS, Rubin AL. Eye findings in the diagnosis of Fabry's disease. *JAMA.* 1978;240:647-649.
8. Weingeist TA, Blodi FC. Fabry's disease: ocular findings in a female carrier. *Arch Ophthalmol.* 1971;85:169-176.
9. Faraggiona T, Churg J, Grishman E, et al. Light and electron-microscopic histochemistry of Fabry's disease. *Am J Pathol.* 1981;103:247-262.
10. Riegel EM, Pokorny KS, Friedman AH, et al. Ocular pathology of Fabry's disease in a hemizygous male following renal transplantation. *Survey Ophthalmol.* 1982;26:247-252.
11. Libert J, Tondeur M, Van Hoof F. The use of conjunctival biopsy and enzyme analysis of tears for diagnosis of homozygotes and heterozygotes with Fabry's disease. *Birth Defects.* 1976;12(3):221-239.
12. Johnson DL, Del Monte MA, Cotlier E, et al. Fabry's disease. Diagnosis by α-galactosidase activity in tears. *Clin Chim Acta.* 1975;63:81-90.
13. Sher NA, Reiff W, Letson RD, et al. Central retinal artery occlusion complicating Fabry's disease. *Arch Ophthalmol.* 1978;96:815-817.
14. Desnick RJ, Dean KJ, Grabowski G, et al. Enzyme therapy in Fabry's disease. *Proc Natl Acad Sci USA.* 1979;76:5326-5330.
15. Brame HG, Pyeritz RE, Folstein MF, et al. A prospective double-blind study of plasma exchange therapy for the arcoparesthesia of Fabry's disease. *Transfusion.* 1981;6:686-689.
16. Calhoun DM, Bishop DF, Bernstein HS, et al. Fabry disease: isolation of a CDNA clone encoding human α-galactosidase A. lyzosomal hydrolase. *Proc Natl Acad Sci USA.* 1985;82:7365-7368.
17. Maizel SE, Simmons RL, Kjellstrand C, et al. Ten year experience in renal transplantation for Fabry's disease. *Transport Proc.* 1981;13:57-59.
18. Sheth KJ, Roth DA, Adams MB. Early renal failure in Fabry's disease. *Am J Kidney Dis.* 1983;2:651-654.

Chapter 120

Fucosidosis

JACQUES LIBERT

Fucosidosis is a lysosomal storage disorder with autosomal recessive transmission related to a deficiency in α-L-fucosidase.[1] Intralysosomal accumulation of fucose-rich polysaccharides, glycolipids, and glycoproteins in tissues has been demonstrated biochemically and histologically. High levels of fucose-containing derivatives in urine may be the basis for the screening of the disease, whereas enzyme assays on blood, cultured cells, or tears, together with ultrastructural analysis of skin or conjunctival biopsy, confirms the diagnosis.[2,3]

About 80 patients affected with fucosidosis are known around the world. Several mutations seem to be involved in the disease and might explain clinical and biochemical heterogeneity.

Systemic Manifestations

The disease starts in all subtypes with muscle weakness and hypotonia, changing to hypertonia and spasticity. Hepatosplenomegaly is not constant, and recurrent infections are often reported.

The severe phenotype is furthermore characterized by early and rapidly progressive psychomotor retardation and variable dysmorphy. The facies may be coarse and similar to that in Hurler's syndrome or rather normal, becoming inexpressive with the progression of mental retardation. Macrocephaly, thickness of the skull, and changes in the vertebral bodies represent the most prominent x-ray abnormalities. Vacuolated blood lymphocytes and foam cells in the bone marrow are often the first signs suggesting a lysosomal storage disease. Some patients have been reported to have discrete palmar telangiectasia. Rapid neurologic decline, spastic decortication or decerebration, and profound cachexia lead to death before the age of 6 years.[1,4]

Milder symptoms and longer survival characterize the mild disease. Mental retardation appears later, and dysmorphy is usually slighter, although patients have been described who have prominent forehead, hypertelorism, broad, flattened nose, heavy eyebrows, thick lips and tongue, growth retardation, broad thorax, and lumbar hyperlordosis. The rate of psychomotor and neurologic deterioration is slower, and patients may survive into the 3rd decade. In all cases, a papular rash is described that closely resembles the angiokeratoma corporis diffusum of Fabry's disease but appears earlier and is not related to α-galactosidase deficiency.[5]

Ocular Manifestations

The only striking ocular features in fucosidosis are vascular tortuosities. They are constant in the mild phenotype and may be present in the severe one. Conjunctival vessels are tortuous, with multiple areas of saccular and fusiform microaneurysms. Retinal veins are dilated and tortuous, contrasting with normal retinal arteries (Fig. 120-1).[6,7] Neither macular cherry-red spot nor pigmentary retinopathy has been described. Corneal

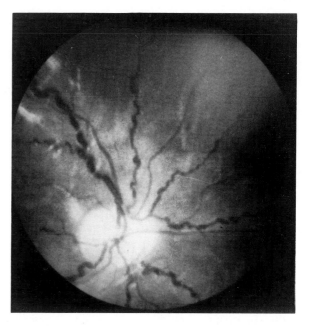

FIGURE 120-1 Dilated and tortuous retinal veins with beading were evident at age 4½ in a girl affected with the severe type of fucosidosis.

opacities have been reported occasionally but are not believed to be consistently related to the disease.

Ultrastructural studies of the eyes show massive lysosomal overloading of the endothelial cells of capillaries and veins, whereas the artery walls remain intact (Fig. 120-2). Similar vascular abnormalities are also found all over the body at autopsy.[8] These lesions explain the development of retinal vein tortuosities, conjunctival aneurysms, and cutaneous angiomatosis, since they induce a loss of mechanical resistance against blood pressure.[9]

A massive lysosomal storage process has also been described within retinal ganglion cells, particularly in the macular area. This observation contrasts with the absence of a macular cherry-red spot in fucosidosis. The nature of the stored material and its optical characteristics are probably responsible for this discrepancy.[9]

By electron microscopy, conjunctival and corneal epithelial cells, keratocytes, sclerocytes, fibroblasts, corneal endothelial cells, photoreceptors, glial cells of the optic nerve, and extrinsic muscle fibers are also involved, but the lysosomal swelling is never severe enough to produce clinical symptoms.

The presence of important and specific alterations in the conjunctiva allows the use of ultrastructural study of biopsies for an accurate diagnosis. Tear enzyme analysis also offers the ophthalmologist an easy way to demonstrate the defect in α-L-fucosidase.[3]

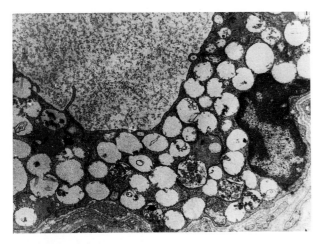

FIGURE 120-2 A severe disorganization of the cytoarchitecture of endothelial cells is evident by electron microscopy with disappearance of fibrillary elements and endoplasmic reticulum, whereas the lysosomal system is overloaded with oligosaccharides. (Original magnification ×5,000)

References

1. Durand P, Borrone C, Della Cella G, et al. Fucosidosis. *Lancet.* 1968;I:1198.
2. Durand P, Borrone C. Fucosidosis and mannosidosis; glycoprotein and glycosylceramide storage diseases. *Helv Pediatr Acta.* 1971;26:19.
3. Libert J, Van Hoof F, Tondeur M. Fucosidosis: ultrastructural study of conjunctiva and skin, and enzyme analysis of tears. *Invest Ophthalmol.* 1976;15:626.
4. Loeb H, Tondeur M, Jonniaux G, et al. Biochemical and ultrastructural studies in a case of mucopolysaccharidosis "F" (fucosidosis). *Helv Pediatr Acta.* 1969;5:519.
5. Koussef BG, Beratis NG, Strauss L, et al. Fucosidosis type 2. *Pediatrics.* 1976;57:205.
6. Snyder RD, Carlow TJ, Ledman J, et al. Ocular findings in fucosidosis. *Birth Defects.* 1976;12:241.
7. Snodgrass MB. Ocular findings in a case of fucosidosis. *Br J Ophthalmol.* 1976;60:508.
8. Libert J. La fucosidose. Ultrastructure oculaire. *J Fr Ophtalmol.* 1984;7:519.
9. Libert J, Toussaint D. Tortuosities of retinal and conjunctival vessels in lysosomal storage diseases. *Birth Defects.* 1982;18:347.

Chapter 121

Krabbe's Disease

Globoid Cell Leukodystrophy

MICHELE R. FILLING-KATZ, NORMAN W. BARTON, and NORMAN N. K. KATZ

Krabbe's disease (KD), or globoid cell leukodystrophy, is a rare autosomal recessive disorder of glycolipid metabolism caused by diminished activity of the lysosomal enzyme galactocerebroside β-galactosidase.[1] The enzyme defect is generalized and has been detected in the brain, kidney, liver, spleen, skin fibroblasts, and leukocytes from the peripheral blood. The pathologic hallmark of KD is the globoid cell, a multinucleated giant cell of nonneuronal, mesodermal origin present in brain tissue obtained from KD patients. Galactocerebroside is known to lead to globoid cell reaction in the brain.[2,3]

The major substrate for galactocerebroside β-galactosidase is the myelin sheath lipid galactocerebroside (Fig. 121-1). The enzyme, however, can also hydro-

lyze psychosine, though at a much slower rate than it does galactocerebroside. Thus, diminished activity of the enzyme leads to accumulation of psychosine in the brain. Psychosine is postulated to be specifically toxic to oligodendroglia. Its concentration is increased 10- to 100-fold over normal levels in brains of KD patients. Since the oligodendroglia, which synthesize and degrade the galactocerebroside, degenerate and die, galactocerebroside never accumulates in excessive quantities. Myelin sheaths are formed by oligodendroglia. Myelin turnover by oligodendroglia is very active during the first 18 months of life but continues at a slower rate until early adulthood. KD is pathologically characterized by severe myelin deficiency and paucity of oligodendroglia.

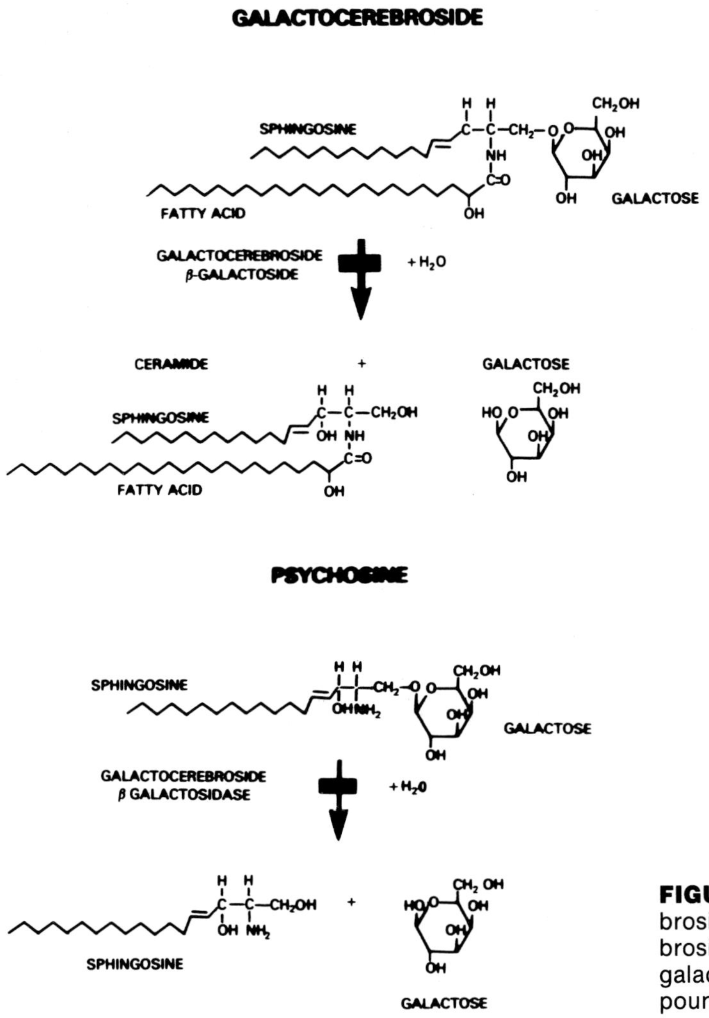

GALACTOCEREBROSIDE

PSYCHOSINE

FIGURE 121-1 Metabolic pathways of galactocerebroside (*top*) and psychosine (*bottom*). Galactocerebroside β-galactosidase cleaves the bond between the galactose and sphingosine moieties in both compounds.

Systemic Manifestations

KD presents in two major forms, infantile and adult.

Infantile KD

Signs and symptoms are present within the first 6 months of life. Tonic seizures, extensor rigidity, spasticity, unexplained hyperthermia, deafness, and progressive psychomotor retardation are prominent features. Peripheral nerve disease of varying severity is present. Head size is usually but not always decreased.[4] Other clinical manifestations may vary and include ichthyosis, infantile obesity, weight loss, and ammoniacal breath. The viscera are not enlarged. Patients with the infantile form usually do not survive beyond the first few years of life. Rarely, the disease presents in late infancy. These infants become symptomatic in the toddler stage.

Adult KD

Signs and symptoms of this phenotype do not manifest until the 2nd decade of life. Spasticity, progressive weakness, and gait difficulty are common initial symptoms. Peripheral nerve disease is rare. A 27-year-old male with KD currently being managed by the authors is believed to be the oldest currently living survivor of this disease.

Laboratory analysis usually reveals an increased level of cerebrospinal fluid protein with elevation of the albumin and the α_1 fractions and diminution in the β_1

and λ fractions. Computed tomography (CT) reveals generalized atrophy of the brain and may reveal significant white matter disease. Magnetic resonance imaging (MRI) is superior to CT for detection of the white matter abnormalities associated with this disease.

Gross examination of the brain of patients with KD reveals marked atrophy. Histopathology demonstrates characteristic nests of globoid cells in the central nervous system (CNS). Globoid cells contain twisted tubules of galactocerebroside on ultrastructural examination.[5,6] Extensive and severe demyelination is present.[6]

Ocular Manifestations

No adnexal, lid, or orbital abnormalities have been reported in KD. Abnormal ophthalmic features include optic nerve atrophy and blindness.

Visual loss may be a presenting feature, and deterioration of vision may be rapid and profound. Various pupillary abnormalities may be present at the onset of symptoms. Blindness may precede the loss of pupillary light reflexes and may be central in origin. The pupils may also be fixed at initial presentation.[7] Late in the course of the disease the pupils invariably become fixed. The appearance of the optic nerves may vary initially, but optic nerve atrophy always supervenes in later stages. The loss of the normal foveal light reflex may be the only other fundus abnormality.[6] Esotropia, nystagmus, and random searching eye movements occur and evolve concurrently with vision loss.

Microscopic examination of the globe in KD reveals abnormalities within the retina and optic nerve.[4,7-9] There is extensive loss of retinal ganglion cells and nerve fibers. Electron microscopy of the retina in one case revealed no abnormal inclusions within the remaining ganglion cells.[9]

Examination of the optic nerve in a case of KD revealed gliosis, focal atrophy, demyelination, and markedly abnormal myelin sheath structure. Numerous globoid cells were visible within the nerve. Electron microscopy of the optic nerve in several cases revealed twisted tubules consistent with galactocerebroside storage, with extensive disruption and loss of myelin.[4,7,8] In addition to the changes occurring within the optic nerve, gross examination of brains from KD patients reveals a striking degree of involvement of the myelin of the optic radiations. Thus, thinning of the retinal nerve fibers and ganglion cell layer appeared to be due to retrograde degeneration. Myelin destruction in the optic tracts is also noted and may contribute to vision loss.

Treatment

There is no known treatment for KD. Enzyme replacement is not effective for central nervous system involvement in the presence of an intact blood-brain barrier.[10] Moreover, blood-brain barrier modification techniques available currently are experimental and of short duration. Thus, in KD enzyme replacement has serious theoretical constraints. Bone marrow transplantation prolongs life but does not affect the neurologic outcome in the twitcher mouse model of KD. Currently, use of bone marrow transplantation in this disorder should be considered highly experimental. Study of the animal model of KD, the twitcher mouse, may eventually help delineate therapeutic strategies for this disease.

References

1. Suzuki Y, Suzuki K. Krabbe's globoid cell leukodystrophy; deficiency of galactocerebroside in serum, leukocytes and fibroblasts. *Science*. 1971;171:73.
2. Malone MJ, Szoke MC, Looney GL. Globoid leukodystrophy I. Clinical and enzymatic studies. *Arch Neurol*. 1975;32:606-612.
3. Malone MJ, Szoke MC, Davis DA. Globoid leukodystrophy II. Ultrastructural and chemical pathology. *Arch Neurol*. 1975;32:613-617.
4. Laxdal T, Hallgrimsson J. Krabbe's globoid cell leucodystrophy with hydrocephalus. *Arch Dis Child*. 1974;232-234.
5. Shaw CM, Carlson CB. Crystalline structures in globoid-epithelioid cells: an electron microscopic study of globoid leukodystrophy (Krabbe's disease). *J Neuropathol Exp Neurol*. 1970;29:306-309.
6. Suzuki K, Grover WD. Krabbe's leukodystrophy (globoid leukodystrophy) an ultrastructural study. *Arch Neurol*. 1970;22:385-396.
7. Brownstein S, Meagher-Villemure K, Polomeno RC, et al. Optic nerve in globoid leukodystrophy (Krabbe's disease) ultrastructural changes. *Arch Ophthalmol*. 1978;96:864-870.
8. Emery JM, Green WR, Huff DS. Krabbe's disease: histopathology and ultrastructure of the eye. *Am J Ophthalmol*. 1972;74:400-406.
9. Harcourt B, Ashton N. Ultrastructure of the optic nerve in Krabbe's leukodystrophy. *Br J Ophthalmol*. 1973;57:885-891.
10. Brady RO. Enzyme replacement therapy. In: Boyer PD, ed. *The Enzymes*. vol XVI. New York, NY: Academic Press; 1983;697-691.

Chapter 122

Gangliosidoses

EDWARD L. RAAB

The gangliosidoses are inherited conditions whose basic defect in each instance is insufficient activity of a lysosomal enzyme that results in accumulation of abnormal substances in various tissues (Table 122-1).[1] Within the GM$_1$ class, type 1 results from deficiency of isoenzymes A, B, and C of β-galactosidase, while in type 2 only B and C are involved. In the GM$_2$ group, Sandhoff's disease exhibits deficiency of both β-hexosaminidases A and B; in the other members, only the former isoenzyme is deficient.[2] Several techniques are available for investigating these enzyme deficiencies, often involving prenatal monitoring.[3,4] In a variant of classic Tay-Sachs disease both hexosaminidases A and B are normal; brain biopsy or enzyme assay of the cerebrospinal fluid is necessary for diagnosis.[5]

As in other "inborn errors of metabolism," the enzyme deficiencies that give rise to these diseases are inherited as autosomal-recessive traits. The Tay-Sachs and chronic (type 4) varieties of GM$_2$ gangliosidosis have been observed just about exclusively in persons of Ashkenazi Jewish ancestry.

Systemic Manifestations

The abnormal substances resulting from these enzyme deficiencies tend to accumulate prominently in the central nervous system, leading to progressive neurologic dysfunction. Dementia, cerebellar dysfunction (ataxia, dysarthria, dysmetria), upper and lower motor neuron involvement, and seizures are prominent systemic features, occurring with different emphasis in the several members of this group of diseases. In the generalized gangliosidoses, the liver, spleen, and kidney are

TABLE 122-1 Classification of the Gangliosidoses*

Class	Deficient Enzymes	Abnormal Substance
GM$_1$:		
Type 1 (generalized)	β-galactosidase A,B,C	GM$_1$ ganglioside Asialo GM$_1$ ganglioside Keratan sulfate
Type 2 (juvenile generalized)	β-galactosidase B,C	Sialomucopolysaccharide
GM$_2$:		
Type 1 (Tay-Sachs)	Hexosaminidase A	GM$_2$ ganglioside Asialo GM$_2$ ganglioside
Type 2 (Sandhoff)	Hexosaminidase A,B	GM$_2$ ganglioside Asialo GM$_2$ ganglioside Globoside GA$_2$
Type 3 (juvenile)	Hexosaminidase A (partial)	GM$_2$ ganglioside Asialo GM$_2$ ganglioside
Type 4 (chronic)	Hexosaminidase A (partial)	GM$_2$ ganglioside Asialo GM$_2$ ganglioside

(Modified from Musarella et al.[1])

TABLE 122-2 Reported Ocular Findings in the Gangliosidoses

Class	Cherry-Red Spot	Pigmentary Retinopathy	Optic Atrophy	Visual Deficit	Oculomotor Disturbances
GM$_1$:					
Type 1	+	−	+	+	?
Type 2	−	+	+	+	?
GM$_2$:					
Type 1	+	−	+	+	+
Type 2	+	−	+	+	?
Type 3	−	+	+	+	?
Type 4	−	−	−	−	+

involved as well by the abnormal mucopolysaccharides; in this respect the generalized forms are somewhat similar to the more familiar mucopolysaccharidoses.

There is clinical as well as biochemical heterogeneity in these conditions. Within the GM$_2$ class, Tay-Sachs disease and the Sandhoff type are characterized by rapid mental deterioration, seizures, and early death. GM$_1$ gangliosidosis gives a similar clinical picture.[6] The GM$_2$ juvenile form exhibits a slower psychomotor deterioration, while the chronic variety is marked by progressive spinocerebellar involvement.[1]

Ocular Manifestations

Variations in clinical expression apply also to the ophthalmic manifestations (Table 122-2). Type 4 is unique among the GM$_2$ variants for the absence of any ophthalmoscopic findings; it affects only ocular movements. Perimacular glycolipid deposition in the retinal ganglion cells (cherry-red spot or gray macula), pigmentary retinal degeneration, and/or optic atrophy characterize the remaining types. Accordingly, Type 4 alone is not associated with severe vision loss. This unique feature has been confirmed by visual evoked cortical response and electroretinographic studies.[1]

The ocular movement abnormalities seen in Tay-Sachs disease are thought to be secondary to generalized central nervous system deterioration, unlike the deficit of the chronic variety, which is attributable to cerebellar dysfunction. The latter is evidenced by abnormal horizontal smooth pursuit movements and inability to suppress the vestibulo-ocular reflex by fixation.[1] The pathologic counterpart of these findings is the postmortem observation, in one affected person, of diffuse subcortical and cerebellar neuronal storage. By contrast, in Tay-Sachs disease cortical involvement predominates markedly.

Other observations in type 4 GM$_2$ gangliosidosis include saccadic dysmetria without impaired velocity and defects in voluntary upward gaze.[1,7] The latter have been observed in other storage diseases and suggest involvement of the rostrad midbrain.

The ophthalmologic importance of these conditions lies mainly in the evaluation of the eyes in cases of severe neurologic deterioration in infants and children, as part of a full systemic and biochemical investigation. In the chronic form of GM$_2$ gangliosidosis, which is associated with a somewhat more favorable prognosis, impaired vision is not a feature and the ocular motility defects usually do not preclude reasonably satisfactory function.

References

1. Musarella MA, Raab EL, Rudolph SH, et al. Oculomotor abnormalities in chronic GM$_2$ gangliosidosis. *J Pediatr Ophthalmol Strab.* 1982;19:80.
2. O'Brien JS. The gangliosidoses. In: Stanbury JB, Wyngaarden JG, Fredrickson DS, et al., eds. *The Metabolic Basis of Inherited Disease.* 4th ed. New York, NY: McGraw-Hill; 1978;841.
3. Willner JP, Grabowski GA, Gordon RE, et al. Chronic GM$_2$ gangliosidosis masquerading as Friedreich's ataxia. Clinical, morphologic, and biochemical studies of nine cases. *Neurology.* 1981;31:787.
4. Berman ER. Biochemical diagnostic tests in genetic and metabolic eye disease. In: Goldberg MF, ed. *Genetic and Metabolic Eye Disease.* Boston, Mass: Little, Brown and Co; 1974;73.
5. Pullarkat RK, Reha H, Beratis NG. Accumulation of ganglioside GM$_2$ in cerebrospinal fluid of a patient with the variant AB of infantile GM$_2$ gangliosidosis. *Pediatrics.* 1981;68:106.
6. Landing BH, Silverman FN, Craig MM. Familial neurovisceral lipidosis. *Am J Dis Child.* 1965;108:503.
7. Musarella-Kelly MA, Rudolph S, Pokorny KS, et al. Defective vertical eye movements in lysosomal storage disease. *Invest Ophthalmol.* 1980;19(suppl):270.

Chapter 123

Gaucher's Disease

MICHELE R. FILLING-KATZ, NORMAN W. BARTON, and
NORMAN N. K. KATZ

Gaucher's disease (GD), the most common inherited disorder of glycolipid metabolism, is caused by diminished activity of the lysosomal enzyme β-glucocerebrosidase.[1,2] Excessive amounts of glucocerebroside accumulate in multiple organ systems, notably within the reticuloendothelial cells of the spleen, liver, and bone marrow. GD is an autosomal recessive disorder. Approximately two thirds of all patients are Jewish. Gene frequency in the Jewish population has been estimated to be 4.5%.[3] The gene for glucocerebrosidase is located on the short arm of chromosome 1 and has been cloned by several groups. Several phenotypic forms of GD exist.

Type I Gaucher's Disease

The most common form of GD is type 1, or the chronic nonneuronopathic type. There is great variation in age of onset and severity as well as in type of end-organ damage. Clinical manifestations of the disease are related largely to the deposition of glycolipid within the reticuloendothelial system. Most patients with GD develop significant hepatosplenomegaly early in life (Fig. 123-1). Hypersplenism may necessitate partial or total splenectomy. Evidence exists, though it is controversial, that suggests that splenectomy accelerates bone and liver disease. The presumptive mechanism is considered to be loss of splenic glycolipid storage capacity leading to excessive accumulation of glucocerebroside in the liver and bone marrow. Therefore, most authorities advise that splenectomy be limited to patients with severe complications related to hypersplenism or splenomegaly. Finally, there are many patients with mild disease who suffer visceromegaly only late in life.

Bone disease is a very serious problem, involving 50 to 70% of patients with GD. Acute severe attacks of pain, erythema, and soft tissue swelling (bony crises), usually associated with fever, may occur. Aseptic necrosis and subsequent collapse of the femoral head also occur and may require hip replacement surgery. Vertebral body collapse, pathologic fractures, and progressive Erlenmeyer flask deformity of the distal femurs are also common.

Pulmonary involvement is uncommon but leads to

The authors would like to thank Dr. Roscoe O. Brady for his critical review of this chapter as well as his many contributions to Gaucher's disease research over the years.

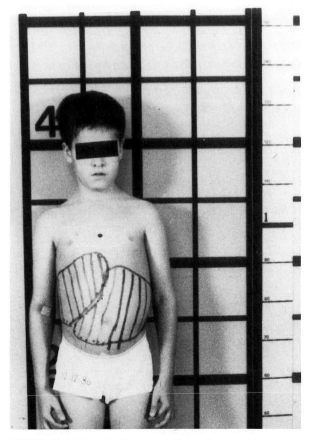

FIGURE 123-1 A 10-year-old boy diagnosed at age 3 to have GD. Severe hepatosplenomegaly is present.

grave complications such as pulmonary hypertension, interstitial lung changes, and cor pulmonale. Mild hepatic dysfunction is common in GD but may occasionally result in cirrhosis, portal hypertension, and variceal bleeding.

The diagnosis of GD should be considered in any patient with hepatosplenomegaly. Bone marrow biopsy reveals lipid-laden macrophages (foam cells). The Gaucher cell, or "tissue paper macrophage" (Fig. 123-2) is unique to GD and pathognomonic. Diagnosis should be confirmed by assay of glucocerebrosidase activity in skin fibroblasts or leukocytes from the peripheral blood.

Ocular Manifestations

To our knowledge, there have been no reports of orbital, lid, or adnexal involvement in type 1 GD. Pingueculae are common, age related, and appear initially on the nasal conjunctiva and later on the temporal side. Histologic descriptions of these lesions have been limited and describe lipid-laden histiocytes.[4] Light- and electron microscopy of the conjunctiva (in the absence of pingeculae) did not reveal any Gaucher cells.[5]

White deposits have been described in the corneal epithelium, drainage angle, and pupillary margin in two siblings.[6] Lipid-laden macrophages have been reported in the ciliary body and have been correlated to the white deposits seen clinically in the anterior segment in one case.[6] Vitreous opacities and cells have been reported in four patients with GD. Reported fundus abnormalities include perifoveal ring-shaped granular opacities, "perimacular grayness" (Fig. 123-3), and cherry-red spots

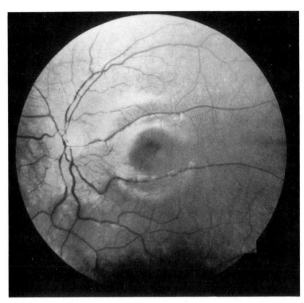

FIGURE 123-3 Fundus photo of a 15-year-old girl with type 1 GD. A perimacular gray ring is seen.

(whose existence has been questioned by Cogan).[6-13] These changes may constitute a continuum of the abnormal perimacular lipid storage process. Scattered white spots have also been reported in the fundus (Fig. 123-4). Electron microscopy of postmortem material identified these spots as clusters of lipid-laden macrophages on the inner surface and within the inner layers of the retina.[5] These retinal spots are usually seen in type 3 disease but may rarely occur in severe forms of type 1 disease. Long-term follow-up of a patient with such spots indicates that they may be causally related to white-centered retinal hemorrhages.[12] Retinal edema and hemorrhages have also been reported but may be related to the underlying systemic disease, pulmonary hypertension, and cardiac failure, which are usually end-stage complications of GD.

Type 2 Gaucher's Disease

Type 2, or infantile neuronopathic GD, is a very rare form. Even though the age at presentation may vary from birth to 18 months, the disease usually manifests within the first few months of life. Severe hepatosplenomegaly, anemia, microcephaly, and occasionally pulmonary interstital disease occur. In conjunction with developmental delay, these patients also develop spasticity, trismus, hypertonicity, hyperreflexia, and tonic neck hyperextension. Hypotonia and apathy supervene

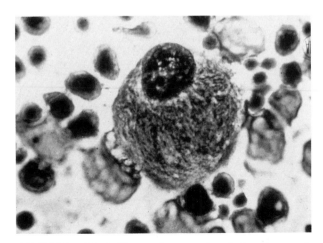

FIGURE 123-2 Photomicrograph of Gaucher cell in bone marrow aspirate demonstrates characteristic tissue paper appearance (original magnification ×380).

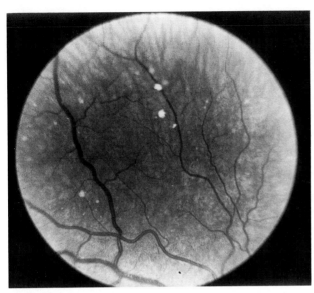

FIGURE 123-4 Fundus photo of an adult male with severe type 1 disease shows the characteristic white spots in the retina. (From Cogan, Chu, Gittinger.[9])

in the later stages of the disease. Pulmonary infection is frequently a terminal event.

Ocular Manifestations

No lid, adnexal, orbital, corneal, fundus, or optic nerve findings have been described. Neuro-ophthalmic complications develop within the 1st year of life and include cranial nerve palsies, strabismus, and cortical blindness.[14] In addition, one author (N.B.) has observed supranuclear gaze paresis in four patients with this disease.

Type 3 Gaucher's Disease

Type 3, or subacute neuronopathic, GD shares many systemic features with type 1 and type 2 disease. An isolate exists in Sweden, the Norrbottnian variant, which is clinically identical to type 3 disease. In infants and young children with GD, predicting the subsequent development of neurologic disease may be difficult. Visceral involvement in type 3 GD usually precedes neurologic involvement. Patients commonly have severe systemic disease.

Some patients with type 3 GD develop progressive dementia and myoclonic seizures in addition to the horizontal eye movement abnormalities described be-

low. In other patients the neurologic findings are restricted to eye movement abnormalities alone.

Ocular Manifestations

Lid, adnexal, and orbital abnormalities have not been reported. Retinal white spots are more common, perhaps because of the severity of the systemic disease. Splenectomy has been reported to exacerbate the fundus abnormalities, suggesting that abnormal lipid deposition in the eye increases after splenectomy.[15]

Horizontal eye movement disturbances are a feature of all type 3 GD patients. In some patients upward-looking eye movements with slow saccades may be the only oculomotor abnormality. In others, the ability to generate horizontal saccadic eye movements may be entirely absent. The most severely impaired patients have a profound supranuclear disruption of all horizontal (and, rarely, vertical) eye movements. The eye movement abnormalities occasionally resemble oculomotor apraxia. Their natural history remains unknown.

Therapy

Therapy for GD has previously been limited to the treatment of attendant complications. Enzyme replacement therapy has shown promising results in a phase 2 clinical trial under way at the National Institutes of Health. Gene therapy is undergoing extensive laboratory testing but is not yet feasible in humans.

References

1. Brady RO, Kanfer J, Shapiro D. Demonstration of a deficiency of glucocerebrocidase cleaning-enzyme in Gaucher's disease. *J Clin Invest.* 1966;45:1112-1115.
2. Carbone A, Petrozzi C. Gaucher's disease. *Henry Ford Hospital Med J.* 1968;16:55-60.
3. Matoth Y, et al. Frequency of carriers of chronic (type 1) Gaucher's disease in Ashkenazi Jews. *Am J Med Genetics.* 1987;27:561-565.
4. East T, Savin L. A case of Gaucher's disease with biopsy of the typical pingueculae. *Br J Ophthalmol.* 1940;24:611-613.
5. Ueno H, Ueno S, Matsuo N, et al. Electron microscopic study of Gaucher cells in the eye. *Jpn J Ophthalmol.* 1980;24:75-81.
6. Sasaki T, Tsukahara S. A new ocular finding in Gaucher's disease: a report of two brothers. *Ophthalmolgicia.* 1985;191:206-209.
7. Cogan D, Federman D. Retinal involvement with reticuloendotheliosis of unclassified type. *Arch Ophthalmol.* 1964;71:489-491.
8. Cogan D, Kuwabara T. The Sphingolipidoses and the Eye. *Arch Ophthalmol.* 1968;79:437-452.

9. Cogan D, Chu F, Gittinger J, et al. Fundus abnormalities in Gaucher's disease. *Arch Ophthalmol.* 1980;98:2202-2203.
10. Eyb C. Augenintergrund bei der kindlichen Gaucherschen Erkrankung. *Wien Klin Wochenschr.* 1952;64:38.
11. Glascow G. A case of amaurotic family idiocy with lipid storage disease of bone. *Aust Ann Med.* 1957;6:295-299.
12. Stark H. Uber Augenveranderungen beim Morbus Gaucher. *Klin Mbl Augenhielk.* 1983;183:216-220.
13. Yanagida N. On peculiar retinal changes in Gaucher's disease. *Acta Soc Ophthalmol Jpn.* 1950;54(suppl):432-442.
14. Sanders M, Lake B. Ocular movements in lipid storage disease. Reports of juvenile Gaucher disease and the ophthalmolplegic lipidoses. *Birth Defects: Original Article Series.* 1976;12:535-542.
15. Erikson A, Wahlberg I. Gaucher disease Norrbotian type ocular abnormalities. *Acta Ophthalmol.* 1984;63:211-225.

Chapter 124

Mannosidosis

AVERY H. WEISS

Mannosidosis is one of the lysosomal disorders in which there is a defect in the catabolism of glycoprotein. The enzyme mannosidase, which cleaves mannose from the oligosaccharide portion of the glycoprotein, has been found to be deficient. This leads to accumulation of mannose-rich oligosaccharide chains and the appearance of clinical manifestations of a multisystem condition. The entity is a rare autosomal-recessive trait; less than 100 cases are reported to date.[1]

In order to understand the basis of mannosidosis it is necessary to review the structure and function of glycoproteins. They are polypeptides with covalently attached oligosaccharide chains. The latter are coupled through hydroxyl groups of serine and threonine residues in O-linked oligosaccharides and through the free amino group of asparagine in N-linked oligosaccharides. Once synthesized and processed by the addition and removal of specific sugars, glycoproteins are either transported to cellular membranes or secreted. Membrane-bound glycoproteins seem to play a role in determining the architecture and function of the membrane. The presence of oligosaccharide chains on secreted proteins is believed to stabilize them and to play a role in targeting them to specific sites.

Catabolism of glycoproteins is accomplished by the lysosomal enzymes. A deficiency of mannosidase leads to the storage of predominantly N-linked oligosaccharides in tissues and their excessive excretion in the urine. All mannose residues except the innermost one are coupled in α-linkage and removed by α-mannosidase. Multiple forms of the enzyme have been separated by chromatographic methods, and recently the encoding

gene has been mapped to chromosome 19. All but one patient with mannosidosis reported to date have a deficiency of this enzyme. The innermost residue is in β-linkage and removed by β-mannosidase. So far, only one patient with mannosidosis has been found to have a deficiency of β-mannosidase.[2]

Systemic Manifestations

Accumulation of mannose-rich oligosaccharides in soft tissues leads to coarse features, macroglossia, hypertrophic gums, and hernias. Increased storage in the central nervous system (CNS) results in mental retardation, neurologic deficits, and sensorineural hearing loss. Hepatosplenomegaly is common. The predominant skeletal abnormalities are dysostosis multiplex and thickened calvarium. Vacuolated lymphocytes are found in the peripheral blood and foam cells in the bone marrow. The manifestations resemble those of a number of lysosomal storage diseases, and particularly the mucopolysaccharidoses. According to age of onset and severity, mannosidosis has been classified as infantile (type I) and juvenile (type II). In type I, clinical signs are evident in infancy and they are more severe. In type II, the signs and symptoms develop later.

Ocular Manifestations

The ocular manifestations are thought to be related to a local deficiency of α-mannosidase. This is most apparent

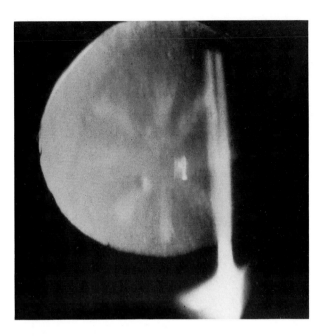

FIGURE 124-1 Radial opacification within the posterior cortex of the lens in a patient with mannosidosis. (Courtesy of Linn Murphree, M.D.)

mechanism, opacities sometimes appear in the superficial layers of the cornea.

Two additional features of ocular involvement may prove helpful in establishing the diagnosis. Electron microscopy of conjunctiva usually shows fibroblasts and endothelial cells that contain membrane-bound vacuoles suggestive of lysosomal storage.[1] Tears are an accessible fluid in which levels of mannosidase activity can be measured.[1] Other ocular manifestations have been reported, but their relationship to the underlying metabolic defect is uncertain. The higher incidence of strabismus may be related to CNS involvement. On the other hand, the reported increased pectinate ligaments in the iridocorneal angle, pallor and blurring of the optic disc, and tortuosity of retinal vessels are difficult to explain.[5,6]

References

1. Beaudet AL. Disorders of glycoprotein degradation: mannosidosis, fucosidosis, sialidosis, and aspartylglycosaminuria. In: Stanbury J, et al, eds. *The Metabolic Basis of Inherited Disease.* New York, NY: McGraw-Hill;1986:788-792.
2. Wenger DA, Sujansky E, Fennessey PV, et al. Human β-mannosidase deficiency. *New Engl J Med.* 1986;315:1201-1205.
3. Harding JJ, Crabbe MJC. The lens: development, proteins, metabolism and cataract. In: Davson H, ed. *The Eye.* Orlando, Fla: Academic Press;1984;291.
4. Carlin R, Cotlier E. Glycosidases of the crystalline lens 1. Effect of pH, inhibitors and distribution in various areas of lens and in subcellular fractions. *Invest Ophthalmol.* 1971;10:887-897.
5. Arbisser AI, Murphree AL, Garcia CA, et al. Ocular findings in mannosidosis. *Am J Ophthalmol.* 1976;82:465-471.
6. Letson RD, Desnick RJ. Punctate lenticular opacities in type II mannosidosis. *Am J Ophthalmol.* 1978;85:218-223.

in the lens, where mannose-rich glycoproteins and α-mannosidase activity have been found.[3,4] At least one third of the reported patients have lens opacities, but their location and appearance varies. In type I, the opacities are confluent in the posterior cortex and have a radial configuration; some authors consider this type of cataract pathognomonic (Fig. 124-1).[5] In type II, punctate opacities are scattered throughout the lens.[6] On occasion, anteriorly located plaquelike opacities have been noted in both types. By the same pathogenic

Chapter 125

Mucolipidoses

MATTHEW E. DANGEL and THOMAS MAUGER

The mucolipidoses are four (ML I-IV) of the more than 100 inherited metabolic disorders that have ocular manifestations. Spranger and Weidemann[1] initially defined

and classified the mucolipidoses as a group of storage diseases that exhibit signs and symptoms of both the mucopolysaccharidoses and the lipidoses. The mucolipidoses are inborn lysosomal storage disorders. The deficiency of one or more lysosomal hydrolytic enzymes prevents the physiologic degradation of specific macromolecules, which accumulate in increased quantities within the lysosomes. This abnormal storage of material within lysosomes and the resultant effect on the involved cells and tissues is the pathophysiologic mechanism that produces the characteristic findings of the mucolipidoses (Table 125-1).

These disorders can occur sporadically or by autosomal recessive inheritance and have characteristic inclusions in cultured cells. Exception in ML IV, a Hurler's-like phenotype is seen, but none of these disorders demonstrates mucopolysacchariduria.

The disorder originally classified as ML I by Spranger and Wiedemann[1] has since been redesignated the sialidoses. The lysosomal enzyme defect is a deficiency of α-N-acetylneuraminidase (sialidase).[2] In this way ML I differs from ML II and III, which result from a defect in lysosomal enzyme localization (ie, enzyme deficiency versus non-incorporation of abnormal enzyme[s] into the lysosome).

There are two phenotypic variants of sialidosis, type I (normosomatic) and type II (dysmorphic). Patients with type I disease do not have a Hurler's facies and present between 8 and 15 years of age with decreased visual acuity, myoclonus, and gait abnormalities (cherry-red spot–myoclonus syndrome). Characteristically they have a history of normal health and intellect. They may survive until their 3rd or 4th decade. The ocular findings are notable for macular graying or a cherry-red spot and occasional fine punctate corneal opacities without significant corneal clouding.

Type II sialidoses (dysmorphic form) has been further

TABLE 125-1 The Mucolipidoses

Mucolipidosis	Enzyme Abnormality	Systemic Manifestation	Ocular Manifestation
I (sialidoses)	Deficiency of α-N-acetyl-neuraminidase (sialidase)	Type I: Myoclonus and gait abnormalities	Cherry-red spot and fine punctate corneal opacities without significant clouding (type II presenting earlier than type I)
		Type II: Hurler's-like phenotype with organomegaly, renal involvement, dyostosis multiplex, and mental retardation	
II (I cell disease)	Defect in incorporation of acid hydrolases into the lysosome	Hurler's-like phenotype including growth failure, joint restriction, delayed development, upper respiratory infection, and death	Corneal stromal granularity and mild haziness reported in some cases
III (pseudo-Hurler's polydystrophy)	Defect in incorporation of acid hydrolases into the lysosome (less severe allelic variation of ML II)	Joint stiffness with delayed somatic and intellectual development	Fine corneal stromal opacities not visually significant and not involving the epithelium; hyperopic astigmatism, optic disc edema, and surface wrinkling retinopathy
IV	Deficiency of ganglioside neuraminidase	Mental retardation (notably, no physical features resembling Hurler's syndrome)	Bilateral corneal epithelial opacity is most striking. Tapetoretinal appearance with extinguished ERG in some cases

divided into a juvenile and an infantile form, which show considerable overlap. These patients progressively develop a Hurler's syndrome–like phenotype with organomegaly, renal involvement, dysostosis multiplex, and mental retardation. A cherry-red spot and myoclonus also occur, as do fine corneal opacities without significant corneal clouding. Sialidoses type I has been demonstrated to have 10 times the residual enzyme activity of type II, which correlates with the more severe involvement and earlier age of onset of sialidosis type II (dysmorphic).

Mucolipidosis type II is also known as I cell (inclusion cell) disease, owing to the presence of inclusions within fibroblasts.[3] The etiology of this disease (and also of ML III) appears to be the failure to phosphorylate the mannose residues in the oligosaccharide portion of the lysosomal acid hydrolases.[4] Mannose-6-phosphate residues on acid hydrolases are necessary to interact with receptor sites on the lysosomal membrane to facilitate incorporation of the acid hydrolase into the lysosome. This enzymatic abnormality results in exclusion of certain acid hydrolases from lysosomes (with diminished degradation and increased lysosomal storage of specific macromolecules) and the extracellular secretion of these enzymes. As a result, elevated serum levels of these abnormal lysosomal enzymes are seen in ML II and III. ML II is associated with less residual enzyme activity of *N*-acetylglucosamine phosphotransferase than is ML III and therefore has a more pronounced phenotype.

ML II is usually diagnosed in a newborn or young infant with signs such as growth failure, joint restriction, organomegaly, gargoyle facies with gingival hyperplasia, and nasal congestion. This is followed by delayed development, hypotonia, recurrent upper respiratory infections, and death, usually between ages 2 and 5 years secondary to pneumonia or congestive heart failure.[4] The cornea is clear early in life, but cases of stromal granularity and mild haziness have been reported.[3] Conjunctival subepithelial histiocytes, fibroblasts, and keratocytes contain membrane-bound vacuoles with fibrillogranular material and membranous lamellar inclusions.

Mucolipidosis type III (pseudo-Hurler's polydystrophy) is an allelic variation of the same molecular defect as ML II. It manifests less severe phenotypic expression as a result of greater residual enzyme activity.[4] Affected children present between the ages of 2 and 4 years with joint stiffness followed by delayed somatic and intellectual development. Radiographic findings are those of dysostosis multiplex. Slit lamp examination shows fine, peripheral corneal opacities that are best seen with scleral scatter illumination. It may begin in either the anterior or posterior stroma then proceeds to involve the full thickness but does not significantly affect vision. The

corneal epithelium is spared. Hyperopic astigmatism, optic disc edema, and surface wrinkling maculopathy have also occasionally been described in ML III.[5] The pathology from conjunctival biopsy is similar to that seen in ML II.

ML IV was first described by Berman et al.[6] in 1974 as a disorder presenting with severe corneal clouding and mental retardation with no physical features of a Hurler's-like syndrome. ML IV is most probably caused by a deficiency of ganglioside neuraminidase. The disorder is autosomal recessive and the initial cases were found in children of Ashkenazi Jewish origin. Organomegaly, joint restrictions, skeletal abnormalities, and abnormalities in the usual screening tests for storage diseases such as mucopolysacchariduria are notably uncommon in ML IV. Corneal clouding is present at birth or shortly thereafter and is usually the most striking clinical finding. The opacities are typically bilateral and clinically involve only the epithelium while the stroma remains clear. The corneal epithelium demonstrates cytoplasmic inclusions—both single membrane-bound vesicles filled with fibrillogranular material consistent with mucopolysaccharides and membranous lamellar bodies consistent with phospholipids.[7] Simple débridment of the epithelium results in recurrent opacity when the affected peripheral corneal or conjunctival epithelial cells migrate centrally to reepithelialize the corneal defect. However, epithelium removal combined with transplantation of healthy conjunctiva to the peripheral cornea has produced long-term clarity in one case.[8] Retinal findings include optic nerve pallor, narrow vasculature, and extinguished electroretinogram signal with fundus appearance similar to that in tapetoretinal dystrophy in some cases.[9,10]

At this time there are no specific enzyme replacements for the treatment of the mucolipidoses; so diagnosis is primarily to eliminate other treatable disorders and to provide prognostic and genetic counseling.

References

1. Spranger JW, Wiedemann HR. The genetic mucolipidoses, diagnosis and differential diagnosis. *Humangenetik.* 1970;9:113-139.
2. Beaudet AL. Disorders of glycoprotein degradation mannosidoses, fucosidosis, sialidosis, and aspartylglycosaminura. In: Stanbury JB, Wyngaarden JB, Fredrickson DS, et al. eds. *The Metabolic Basis of Inherited Disease.* 5th ed. New York, NY: McGraw-Hill; 1983;788-802.
3. Limbert J, Van Hoof F, Farriqux JP. Ocular findings in I-cell disease (mucolipidosis type II). *Am J Ophthalmol.* 1977;83(5):617-628.
4. Neufeld EF, McKusick VA. Disorders of lysosomal enzyme synthesis and localization: I-cell disease and pseudo-Hurler

polydystrophy. In: Stanbury JB, Wyngaarden JB, Fredrickson DS, et al. eds. *The Metabolic Basis of Inherited Disease.* 5th ed. New York, NY: McGraw-Hill; 1983:778-787.

5. Traboulsi EI, Maumenee IH. Ophthalmologic findings in mucolipidosis III (pseudo-Hurler polydystrophy). *Am J Ophthalmol.* 1986;102:592-597.

6. Berman ER, Livni N, Shapira E. Congenital corneal clouding with abnormal systemic storage bodies: a new variant of mucolipidosis. *J Pediatrics.* 1974;84(4):519-526.

7. Kenyon KR, Maumenee IH, Green WR. Mucolipidosis IV: histopathology of conjunctiva, cornea, and skin. *Arch Ophthalmol.* 1979;97(6):1106-1111.

8. Dangel ME, Bremer DL, Rogers GL. Treatment of corneal opacification mucolipidosis IV with conjunctival transplantation. *Am J Ophthalmol.* 1985;99:137-141.

9. Abraham FA, Brand N, Blumenthal M, et al. Retinal function in mucolipidosis IV. *Ophthalmologica Basel.* 1985;191:210-214.

10. Riedel KG, Zwaan J, Kenyon KR. Ocular abnormalities in mucolipidosis IV. *Am J Ophthalmol.* 1985;99:125-136.

Chapter 126

Mucopolysaccharidoses

GEORGE T. FRANGIEH, ELIAS I. TRABOULSI, and KENNETH R. KENYON

The systemic mucopolysaccharidoses (MPSs) are lysosomal storage diseases that result from deficiency of enzymes involved in the degradation of the glycosaminoglycans dermatan sulfate, heparan sulfate, and keratan sulfate. As a result, incompletely degraded mucopolysaccharides are excreted in the urine and accumulate in most organs. Connective tissues are primarily involved. The MPSs are inherited as autosomal recessive traits, with the exception of Hunter syndrome, an X-linked disease.[1] Characteristic clinical features occur with varying degrees of severity and overlap; these include skeletal abnormalities, gargoyle facies, hepatosplenomegaly, cardiac disease, mental deficiency, deafness, and ocular involvement.[2,3] Variability in clinical manifestations may be due to the tissue-specific differences in the structure of mucopolysaccharides and to the predominant types accumulated in various organs.

At present, seven types of MPS are recognized on clinical grounds alone. For some types, subtypes are recognized.

Corneal clouding, pigmentary retinal dystrophy, and optic atrophy are the main ocular features of the MPSs. Retinal pigmentary degeneration with bone-spicule appearance and nightblindness is seen only in MPS types that involve storage of heparan sulfate. Patients with MPS, especially types IH, IS, IH-S, IV, and VI, complain of moderately severe photophobia early in the course of their disease. Papilledema, a frequent finding, has been attributed to the hydrocephalus that occurs secondary to meningeal thickening with the storage material;

however, local factors at the nerve head, such as narrowing of the scleral canal due to posterior scleral thickening with mucopolysaccharide accumulation, may play a role in the optic nerve head swelling. Glaucoma (acute and chronic) may be seen in MPS I.

The main ocular features of the various types of MPS are summarized in Table 126-1.[4] The ocular histopathologic features have been reviewed by Spencer.[5] The following paragraphs briefly discuss the systemic and ocular findings in the various MPS types.

Hurler Syndrome (MPS IH)

In this prototype of all MPS storage diseases, activity of the lysosomal hydrolase α-L-iduronidase is absent. Intracellular degradation of heparan sulfate and dermatan sulfate is hindered, resulting in their excessive urinary excretion and their intralysosomal storage in almost every system in the body. Affected persons are clinically normal in infancy. Dwarfism and typical features progressively become manifest by age 2 or 3 years. The facies are coarse, with prominent eyebrows, synophrys and lid fullness, saddle nose, hypertrichosis, and nasal obstruction with chronic rhinorrhea. Other findings include mental retardation, deafness, stiff joints, hepatosplenomegaly, and umbilical hernias. Death occurs before age 10 years secondary to respiratory infection or heart failure.

TABLE 126-1 Inheritance, Ocular Findings, and Enzyme Defects in the Mucopolysaccharide Storage Disorders

Type	Inheritance	Corneal Clouding	Retinal Pigmentary Degeneration	Optic Atrophy	Papille-dema	Glaucoma	Enzyme Defect
Hurler, MPS IH	Autosomal recessive	+	+	Late	Frequent	Some	α-L-iduronidase
Scheie, MPS IS	Autosomal recessive	+	+	Late	Some	Frequent	α-L-iduronidase
Hurler-Scheie, MPS IH-S	Autosomal recessive	+	+	Late	Some	Some	α-L-iduronidase
Hunter, MPS II, types A,B	X-linked recessive	Late	+	+	Chronic	–	A & B: Sulfoiduronate sulfatase
Sanfilippo, MPS III, types A,B,C,D	Autosomal recessive	–	+	Rare	–	–	A: Heparan-N-sulfate-sulfatase B: N-acetyl-α-D-glucosaminidase C: Acetyltransferase D: N-acetylglu-cosamine-6-sulfate sulfatase
Morquio, MPS IV, types A,B	Autosomal recessive	+	–	–	+	–	A: Galactosamine-6-sulfatase B: β-Galactosidase
Maroteaux-Lamy, MPS VI, types A,B	Autosomal recessive	+	–	–	Rarely	–	A & B: Arylsulfatase B
Sly, MPS VII	Autosomal recessive	Mild	+	Late	+	–	β-Glucuronidase

(From Lange G, Maumenee IH.[4])

Corneal clouding is typically diffuse, with fine punctate opacities throughout the entire stroma; in some patients, slit lamp biomicroscopy may be necessary to detect the mild stromal haze. Ophthalmoscopy reveals a retinitis pigmentosa–like picture that is indistinguishable from other forms of heredofamilial retinal pigmentary dystrophies. Narrowing of retinal vessels, optic atrophy, and/or optic nerve head swelling are commonly seen. Glaucoma is a relatively rare occurrence.

Diagnosis is made by direct assay for α-L-iduronidase activity in white blood cells or cultured skin fibroblasts. A similar assay is performed on amniocytes for prenatal diagnosis.

It was recently demonstrated that residual α-L-iduronidase activity in Hurler fibroblasts is heat stable, whereas that in Scheie fibroblasts is heat labile; the enzyme activity in MPS IH-S is intermediate between the two.

No specific therapy is available. Plasma infusion treatment to correct the enzyme deficiency results in transient improvement. Bone marrow transplantation has yielded inconsistent results. Corneal transplantation may be required for severe corneal disease, but in such cases vision remains poor because of retinal degeneration or optic atrophy.

Scheie Syndrome (MPS IS)

Previously characterized as MPS V, Scheie syndrome involves the same enzyme defect as Hurler syndrome.[6] Allelism of the Hurler, Scheie, and Hurler-Scheie syndromes due to different amino acid substitutions at the same site in the cistron and resulting in different protein product properties is a likely pathogenetic mechanism in this group of MPSs. Patients with MPS IS have normal intelligence and stature and a relatively normal life span. Facial features are coarse but not grotesque. Claw-hands, stiff joints, aortic insufficiency, deafness, and hernia are common. Peripheral corneal clouding, progressing centrally with age, is the predominant ocular feature (Fig. 126-1). Pigmentary retinal degeneration occurs in the 1st decade and night blindness and visual field constriction in the 2nd and 3rd decades. Acute glaucoma may be seen. Corneal transplantation may be required for corneal clouding.

Hurler-Scheie Syndrome (MPS IH-S)

The clinical features of MPS IH-S are intermediate between those of IH and IS syndrome. Different alleles for the defect are presumably inherited from each parent; some patients, however, may have new mutations. There is defective α-L-iduronidase activity, with excessive tissue deposition and urinary excretion of heparan- and dermatan sulfate.

Dysostosis multiplex with mild dwarfism, moderate mental retardation, hirsutism, stiff joints, heart disease, deafness, and hernia occur between 2 and 4 years of age. Patients survive into their late twenties. Micrognathia (receding chin) is a distinctive facial feature. Arachnoid cysts with spinal fluid rhinorrhea are characteristic.

These cysts are also seen in Hurler syndrome, where they may lead to enlargement of the sella turcica.

Progressive corneal clouding requiring corneal transplantation by the end of the 1st decade is generally the rule. Glaucoma and retinal degeneration with night blindness are common. Increased intracranial pressure and arachnoid cysts result in papilledema and optic atrophy. Diagnosis can be made on the basis of the distinctive clinical features. Biochemical and enzymatic assays are identical to those in MPS IH and IS.

Hunter Syndrome (MPS II)

Hunter syndrome is an X-linked recessive disorder that presents in two allelic forms, A (severe) and B (mild). Sulfoiduronate sulfatase is absent in both forms.

Patients with MPS II appear normal in early infancy and childhood and may even be tall for their age. Later, they develop typical clinical features, with coarse facies, stiff joints, hoarseness, deafness, short stature, heart disease, hepatosplenomegaly, and typical nodular skin lesions. Patients with type A Hunter syndrome are mentally retarded and die in their mid-teens. Patients with type B disease have normal mental function and survive to adulthood; some have reproduced. Ocular features include pigmentary retinal degeneration with

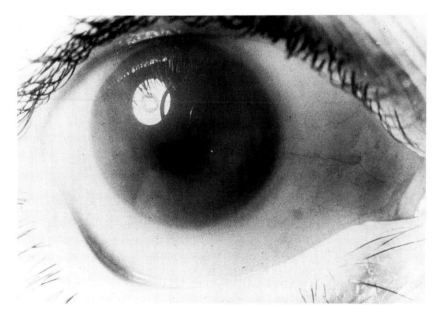

FIGURE 126-1 Clinical photograph of the right cornea of a patient with Scheie syndrome shows ground-glass appearance and diffuse stromal haze.

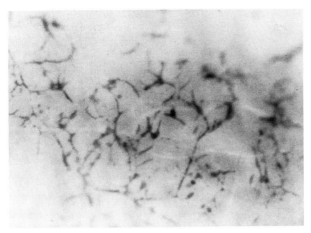

FIGURE 126-2 Gross photograph of the retina of a patient with San Filippo syndrome shows retinal pigmentary degeneration in the midperiphery. Note the bone corpuscle configuration of the migrating retinal pigment epithelium.

narrowing of retinal arterioles and nightblindness; papilledema is seen in more than half of these patients and may be associated with visual field defects. Total blindness may ensue in the 2nd decade. Corneal clouding is not a common finding, although deposits of acid MPS are found histologically in clinically clear corneas.

Diagnosis is made by assaying for sulfoiduronate sulfatase in fibrocytes or, for prenatal diagnosis, in amniocytes. There is excessive urinary excretion of chondroitin B sulfate and heparan sulfate. No specific treatment is available.

Sanfilippo Syndrome (MPS III, A, B, C, and D)

The four types of this clinically distinct MPS are due to different nonallelic mutations. Type A is most severe, with earlier onset, greater severity, and earlier death than types B, C, or D. Excessive heparan sulfate, but not dermatan sulfate, is excreted in the urine. In type A there is a deficiency of the enzyme heparan-N-sulfate sulfatase; in type B, of N-acetyl-α-D-glucosaminidase (NAG); in type C, of N-acetyltransferase; and in type D, of N-acetylglucosamine-6-sulfate sulfatase.

Clinical features include severe mental retardation and relatively mild somatic features, with slight hepatosplenomegaly, moderately shortened stature, coarse facies, hirsutism, and joint stiffness. Age of onset is 2 to 6 years; death ensues before age 20. Pigmentary retinal

degeneration with arteriolar narrowing and optic atrophy are typically seen (Fig. 126-2). Corneal clouding is absent. Diagnosis is based on specific enzyme assays.

Morquio Syndrome (MPS IV)

Clinically distinct from the other MPSs, two types of MPS IV are now recognized. Type A is due to the absence of galactosamine-6-sulfatase, and type B to β-galactosidase deficiency. In type A the activity of the enzyme neuraminidase was also found to be markedly decreased.

Characteristic systemic manifestations include marked dwarfism and dysostosis multiplex, with barrel-chest deformity, short neck, and knock-knee. Aortic valvular disease is common. Intelligence is normal. Hypoplasia of the odontoid process may lead to atlantoaxial dislocation and spinal cord compression later in the course of the disease. Death occurs late in childhood from respiratory paralysis secondary to spinal cord compression and recurrent pneumonia. Corneal clouding is the most common ocular feature, with diffuse ground-glass stromal opacification. No retinal dystrophy is seen.

Diagnosis is made on the basis of typical clinical features and excessive excretion of keratan sulfate (mostly) and chondroitin sulfate in the urine. Assay for the specific enzymes is also possible.

Corneal transplantation may be necessary in children with normal intelligence.

Maroteaux-Lamy Syndrome (MPS VI)

Severe systemic defects characterize this autosomal-recessive disease, which is due to absence of arylsulfatase B. Heparan sulfate (85%) and dermatan sulfate (15%) are excreted excessively in the urine. Mild and severe forms are recognized. Of all MPSs, the Maroteaux-Lamy syndrome shows the most striking inclusions in circulating white blood cells.

Marked polydystrophic dwarfism, lumbar kyphosis, stiff joints, genu valgum, anterior sternal protrusion, aortic stenosis, and umbilical hernia are the characteristic clinical features. Hypoplasia of the odontoid process may cause spinal cord compression and spastic paraplegia. Patients with the severe form die in their teens from hydrocephalus secondary to meningeal involvement.

Ocular findings include mild diffuse corneal opacities, papilledema, and optic atrophy.

Sly Syndrome (MPS VII)

Sly syndrome, the most recently described MPS, results from absence of β-glucuronidase. The gene coding for this enzyme has been assigned to the long arm of chromosome 7 at 7q11.23-7q21. Genetic heterogeneity probably exists in this syndrome.

The main features are psychomotor retardation, mild dysostosis multiplex, hepatosplenomegaly, and umbilical hernias. Ocular features include minimal corneal clouding, papilledema, and retinal dystrophy. Metachromatic "Alder" granules are seen in leukocytes, and absence of β-glucuronidase is demonstrated in cultured fibroblasts.

References

1. McKusick VA. *Mendelian Inheritance in Man.* 7th ed. Baltimore, Md: Johns Hopkins University Press; 1986;1125-1134.
2. McKusick VA, Neufeld EF. The mucopolysaccharide storage diseases. In: Stanbury JB, Wyngaarden JB, Fredrickson DS, et al. eds. *The Metabolic Basis of Inherited Disease.* 5th ed. New York, NY: McGraw-Hill; 1983;751-777.
3. Kenyon KR. Ocular manifestations and pathology of the systemic mucopolysaccharidoses. *Birth Defects.* 1976; 12(3): 133-153.
4. Lange G, Maumenee IH: Retinal dystrophies associated with storage diseases. In: Newsome DA, ed. *Retinal Dystrophies and Degenerations.* New York, NY: Raven Press; 1987.
5. Spencer WH. *Ophthalmic Pathology. An Atlas and Textbook.* 3rd ed. vol 1, Philadelphia, Pa: WB Saunders Co; 1985;346-352.
6. Roubiceck M, Gehler J, Spranger J. The clinical spectrum of alpha-L-iduronidase deficiency. *Am J Med Genet.* 1985; 20:471-481.

Chapter 127

Sialidosis and Galactosialidosis

JAMES A. DEUTSCH and PENNY A. ASBELL

Sialidosis and galactosialidosis are rare autosomal recessive metabolic disorders associated with deficiency of the intracellular enzyme α-neuraminidase. Several clinical variants of these lysosomal storage diseases have been observed. Whereas earlier classifications emphasized morphologic characteristics, more recent delineations of these disorders have been based on current biochemical and gene mapping research.[1-3] In this discussion sialidosis denotes clinical presentations in which only α-neuraminidase is deficient and galactosialidosis has a coexisting deficiency of β-galactosidase (Table 127-1).

Recent research in gene mapping has suggested that two genes are necessary for the expression of α-neuraminidase. Sialidosis results from a mutation on chromosome 10, presumably the one that encodes the neuraminidase structural gene. Galactosialidosis results from a mutation in a second gene needed for neuraminidase expression, located on chromosome 20. This mutation also causes increased susceptibility of β-galactosidase to proteolytic degradation.[3]

Clinical Presentations

Sialidosis (originally lipomucopolysaccharidosis) presents in two major clinical variants. The first type, known as mucolipidosis I, or sialidosis type II—infantile onset, presents between birth and 10 months of age.[1,2] Systemic findings include the development of Hurler's facies (prominent brows with hypertrichosis, frontal bossing, and saddle nose), skeletal dysplasia, visceromegaly, inguinal and umbilical hernias, and joint stiffness. Sensorineural hearing loss, mental retardation, and cardiomegaly leading to heart failure are frequently seen.[2,4,5] Reported ophthalmic manifestations include decreased visual acuity associated with the development of a cherry-red spot at the macula, corneal clouding, and lens opacities. Epicanthus, tortuosity and saccular

TABLE 127-1 Classification Systems of Sialidosis

Mueller et al.[3]	Lowden and O'Brien[1]	Kelley et al.[2]	Others[4-6,8-10]
Sialidosis	Sialidosis		Lipomucopolysaccharidosis (for all entities)
Infantile onset (dysmorphic)	A Type II, infantile onset	A Mucolipidosis I	
Juvenile onset (nondysmorphic)	B Type I	B Sialidosis	B Cherry-red spot–myoclonus syndrome
Galactosialidosis (dysmorphic)	C Type II, juvenile onset	C Mucolipidosis I	C Goldberg's syndrome GM$_1$ gangliosidosis–Type 4 The cherry-red spot–myoclonus syndrome with dementia

aneurysms of the conjunctival and retinal vessels, and esotropia have also been noted.[2,4,5] The disease usually leads to death during childhood.

The second clinical variant of sialidosis, known as sialidosis—type I, or the cherry-red macula–myoclonus syndrome, presents between the ages of 8 and 18 years in a normosomatic person with complaints of decreasing visual acuity or myoclonus.[6-8] The major systemic manifestation, myoclonus, presents initially in the limbs and can progress to massive spasms; generalized grand mal seizures are noted. Affected persons develop progressive difficulty in walking and standing and eventually become unable to perform any voluntary task. Mental retardation is absent, as are dysmorphic features.[6,7] The major ophthalmic manifestation is progressive blindness associated with the development of a cherry-red spot at the macula. The cornea and lens are usually clear, although punctate opacities of the lens have been reported. Death prior to age 30 is common, although several patients have lived into the 4th decade. A large percentage of affected persons are of Italian descent.

Galactosialidosis has been previously termed sialidosis-type II juvenile onset, Goldberg's syndrome, and cherry-red spot with dementia syndrome.[1,9,10] Affected persons present between the ages of 8 and 15 years with the progressive development of Hurler's facies, joint stiffness, and decreased visual acuity. Ataxia, myoclonus, and a cherry-red spot at the macula are seen. Hepatomegaly is not found, nor is mental retardation.[9,10] Most reported patients are still living, the oldest being 25 years old. A high proportion of those affected are of Japanese descent.

References

1. Lowden JA, O'Brien JS. Sialidosis: a review of human neuraminidase deficiency. *Am J Hum Genet.* 1979; 31:1.
2. Kelley TE, Bartoshsky L, Harris DJ, et al. Mucolipidosis I (acid neuraminidase deficiency). *Am J Dis Child.* 1981;135:703.
3. Mueller OT, Henry WM, Haley LL, et al. Sialidosis and galactosialidosis: chromosomal assignment of two genes associated with neuraminidase deficiency disorders. *Proc Natl Acad Sci USA.* 1986;83:1817.
4. Cibis GW, Tripathi RC, Harris DJ. Mucolipidosis I. *Birth Defects.* 1982;18(6):359.
5. Spranger JW, Wiedemann HR, Tolksdorf M, et al. Lipomucopolysaccharidosis. *Eur J Pediatr.* 1968;103:285.
6. Rapin I, Goldfischer S, Katzman R, et al. The cherry red spot—myoclonus syndrome. *Am J Neurol.* 1978;3:234.
7. O'Brien JS. Neuraminidase deficiency in the cherry red spot—myoclonus syndrome. *Biochem Res Commun.* 1977;79:1136.
8. Thomas GH, Tipton RE, Ch'ien IE, et al. Sialidosis (a-N-acetyl neuraminidase deficiency: the enzyme defect in an adult with macular cherry red spots and myoclonus without dementia): a new autosomal recessive disorder. *Clin Genet.* 1978;13:369.
9. Goldberg MF, Cotlier E, Fischenscher LG, et al. Macular cherry red spot, corneal clouding and beta-galactosidase deficiency. *Arch Intern Med.* 1971;128:387.
10. Wenger DA, Tarly TJ, Wharton C. Macular cherry red spots and myoclonus with dementia: coexistent neuraminidase and beta-galactosidase deficiencies. *Biochem Biophys Res Commun.* 1978;82:589.

Chapter 128

Niemann-Pick Disease

Sphingomyelin Lipidosis

JOHN J. WEITER and JOHN D. MATTHEWS

Niemann-Pick disease (NPD), or sphingomyelin lipidosis, was first described in 1914 by the German pediatrician Niemann and was characterized as a distinct clinical entity in 1927 by Pick. NPD is one of the inborn lysosomal diseases, a term utilized by Hers[1] in 1965 to call attention to a unifying theory for a series of complex storage diseases whose pathophysiology remained obscure for a long time. These storage diseases resulted from an inherited deficiency in one of the lysosomal hydrolases. A common feature of the lysosomal enzyme disorders is the requirement for the patient to be homozygous for the abnormal allele in order for the disease to be apparent. The lysosomal enzyme deficiency states that affect the eye fall into three main classes: sphingolipidoses, mucopolysaccharidoses, and mucolipidoses. NPD is one of the sphingolipidoses in which sphingomyelin has been identified as the storage product in the various organs (Fig. 128-1). Brady in 1966 identified sphingomyelinase deficiency as the responsible biochemical defect.

Crocker[2] classified four different clinical types based on age of onset, severity of neurologic involvement, and evolution of the disease. In a review by Frederickson[3] five subtypes (A-E) were distinguished. Further biochemical and genetic studies are required to improve the classification of the various clinical subtypes. The necessity of this is apparent from the various reports on patients whose disease cannot be classified into one of the existing subtypes and, so, is designated as a "new variant."[4]

These subtypes may be explained by the presence of several sphingomyelinase isoenzymes. Callahan[5] showed the deficiency of all isoenzymes in types A and B but a deficiency of only one of these isoenzymes in two other variants (types C and E).

Systemic Manifestations

NPD is a group of diseases characterized by the abnormal accumulation of sphingomyelin, a phospholipid that is a normal component of all body tissues. The rate of accumulation and the degree to which the material interferes with normal body function determine the clinical course of the disease. NPD can be subdivided into five types.[3] Type A, the acute infantile form, is the most common and shows marked hepatosplenomegaly, severe neurologic involvement, and a rapidly fatal progression by age 3 to 4 years. A cherry-red spot of the macula is frequently present.[6]

In type B, visceral involvement may become apparent in infancy or childhood. Although a macular halo may be present in the retina, function of the nervous system is normal.[7] Except for organomegaly, the patients may attain adulthood in reasonably good health. Some patients with the so-called sea-blue histiocyte syndrome may have this form of NPD.

Type C has both visceral and nervous system involvement, but in contrast to type A, its course is subacute or chronic, with death usually occurring by age 15. Type D is clinically similar to type C, except that there is no evidence of sphingomyelinase deficiency. These patients are descendants of an Acadian couple born in Yarmouth, Nova Scotia in the early 1600s.[3] Adults found incidentally to have moderate sphingomyelin excess in one or more organs but no evidence of familial involvement have been considered to have type E NPD.[3]

Ocular Manifestations

Type A (acute neuronopathic form) shows a cherry-red spot in the macula in about 50% of patients (Fig. 128-2). Corneal opacification and brownish discoloration of the anterior lens capsule are frequently seen.[6]

Type B (chronic form without nervous system involvement) may have a macular halo (Fig. 128-3) in the retina without any other ocular or visual findings.[7]

Type C (chronic neuronopathic form) may have vertical ophthalmoplegia. A slightly opaque appearance of

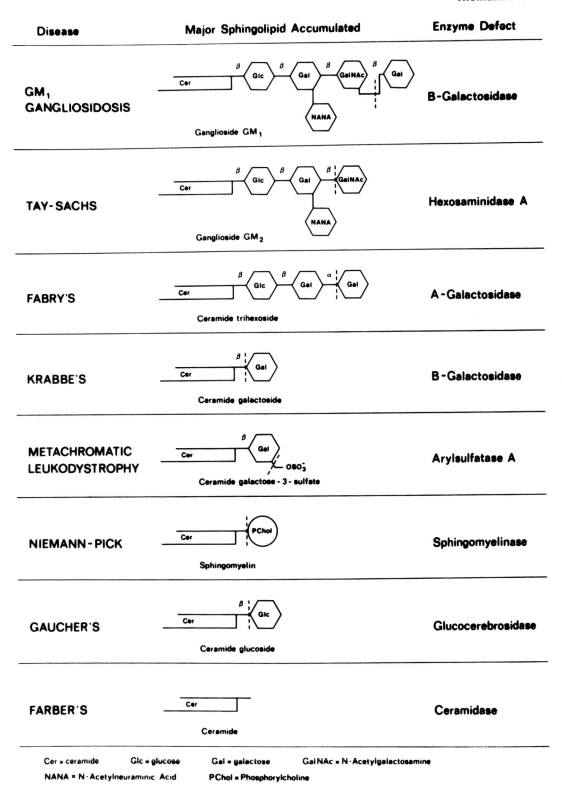

Disease	Major Sphingolipid Accumulated	Enzyme Defect

GM₁ GANGLIOSIDOSIS — Ganglioside GM₁ — **B-Galactosidase**

TAY-SACHS — Ganglioside GM₂ — **Hexosaminidase A**

FABRY'S — Ceramide trihexoside — **A-Galactosidase**

KRABBE'S — Ceramide galactoside — **B-Galactosidase**

METACHROMATIC LEUKODYSTROPHY — Ceramide galactose - 3 - sulfate — **Arylsulfatase A**

NIEMANN-PICK — Sphingomyelin — **Sphingomyelinase**

GAUCHER'S — Ceramide glucoside — **Glucocerebrosidase**

FARBER'S — Ceramide — **Ceramidase**

Cer = ceramide   Glc = glucose   Gal = galactose   GalNAc = N-Acetylgalactosamine

NANA = N-Acetylneuraminic Acid   PChol = Phosphorylcholine

FIGURE 128-1   Outline of biochemical defect in sphingolipidoses. (From Matthews et al.[7])

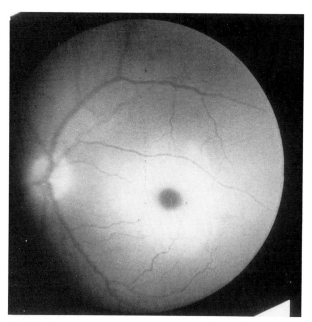

FIGURE 128-2 The fundus appearance of a 28-month-old patient with type A NPD. Note the cherry-red spot and diffuse area of macular opacity. (Courtesy of David S. Walton, M.D.)

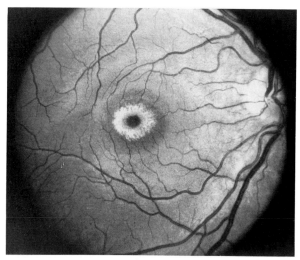

FIGURE 128-3 The fundus appearance of a 19-year-old patient with type B NPD. Note the more subtle type of cherry-red spot (termed macular halo), which appears to be pathognomonic for type B disease. (From Matthews et al.[7])

the perimacular area and minimal optic nerve pallor have been found in one case.[8] A previous report of pale grayish colored maculas and optic atrophy most likely also represents type C NPD.[4,9]

Type D (Nova Scotia variant) and type E (adult, nonneuronopathic form) have no recognized ocular abnormalities.

Histopathologic studies of NPD have shown conjunctival involvement in types A, B, and C and widespread ocular involvement in types A and C.[8] The abnormal intracellular inclusion material in NPD types A and C appeared to be similar, implying similar catabolic molecular mechanisms. The differences in clinical manifestations of this disease most likely lie in losses of specific sphingomyelinase isoenzymes combined with local cellular environmental factors that control the expression of certain isoenzymes.

Unfortunately, to date there is no specific therapy for NPD. Advances in biogenetics offer the possibility of correcting the metabolic defect through enzyme replacement in the future.

References

1. Hers HG. Inborn lysosomal diseases. *Gastroenterology.* 1965;48:625-633.
2. Crocker AC. The cerebral defect in Tay-Sachs disease and Niemann-Pick disease. *J Neurochem.* 1961;7:69-80.
3. Frederickson DS, Sloan HR. Sphingomyelin lipidosis: Niemann-Pick disease. In: Stanbury JB, Wyngaarden JB, Frederickson DS, et al. eds. *The Metabolic Basis of Inherited Disease.* New York, NY: McGraw-Hill; 1972;783-807.
4. Weiter JJ, Kolody EH. A variant of Niemann-Pick disease. *Am J Ophthalmol.* 1976;82:946-947.
5. Callahan JW, Khalil M, Gerrie J. Isoenzymes of sphingomyelinase and the genetic defect in Niemann-Pick disease, type C. *Biochem Biophys Res Comm.* 1974;58:384-390.
6. Walton DS, Robb RM, Crocker AC. Ocular manifestations of group A Niemann-Pick disease. *Am J Ophthalmol.* 1978;85:174-180.
7. Matthews JD, Weiter JJ, Kolodny EH. Macular halos associated with Niemann-Pick type B disease. *Ophthalmology.* 1986;93:933-937.
8. Palmer M, Green WR, Maumenee IH, et al. Niemann-Pick disease—type C. Ocular histopathologic and electron microscopic studies. *Arch Ophthalmol.* 1985;103:817-822.
9. Weiter JJ, Farkas TG. Retinal abnormalities in lactosyl ceramidosis. *Am J Ophthalmol.* 1973;76:804-810.

Chapter 129

Metachromatic Leukodystrophies

Sulfatide Lipidoses

MICHELE R. FILLING-KATZ, NORMAN W. BARTON, and
NORMAN N. K. KATZ

Metachromatic leukodystrophy (MLD) is an autosomal recessive disorder involving cerebroside sulfate metabolism. Many of the features of MLD are shared by another disorder, Austin's disease, also known as multiple sulfatase deficiency (MSD).[1] MSD, however, also involves a broader deficiency of activity of lysosomal and microsomal sulfatases (nine in all), thus sharing many features of mucopolysaccharidosis (MPS). The two syndromes will be considered separately.

Metachromatic Leukodystrophy

Cerebroside sulfate is catabolized by aryl sulfatase A, a lysosomal hydrolase. Deficiency in activity of this enzyme results in accumulation of metachromatically staining glycolipids, the major components of which are sulfatides. The gene encoding aryl sulfatase A is located on chromosome 22 distal to q13.[2] Allelic forms of aryl sulfatase A exist, not all of which are associated with clinical illness. "Pseudodeficiency" of aryl sulfatase A activity has been reported in clinically normal members of some MLD kindreds. Complementation studies of "pseudodeficient" fibroblasts with MLD–affected fibroblasts do not demonstrate augmentation of aryl sulfatase A activity in the fusion product, suggesting that "pseudodeficiency" is a separate allele of aryl sulfatase A.[3] In total, using complementation techniques, five allelic mutations of aryl sulfatase A are now described.[2-4] Two nonallelic forms of MLD also exist. MSD will be described in detail later in this chapter. The other nonallelic form of MLD is an activator protein deficiency. An activator protein that interacts with the sulfatide substrate is required for full expression of the activity of aryl sulfatase A. Two sibs were described with MLD whose aryl sulfatase A activity appeared normal when assayed with an artificial substrate. They were subsequently found to be missing an activator protein localized to chromosome 10, designated sphingolipid activator protein 1.[4]

The clinical features of MLD are age dependent. Two neonates have been discovered with MLD but their cases have not been well characterized. Patients with late infantile MLD usually manifest symptoms prior to the 3rd year of life, most commonly between 18 and 24 months. Prominent clinical features at presentation include weakness, diminished to absent deep tendon reflexes, and diminished tone. Developmental delay is variably present. Dementia, ataxia, dysarthria, and quadriparesis ensue swiftly, and death frequently occurs before the age of 5 years. In the late stages, decorticate to decerebrate posturing is present along with quadriplegia. Tonic seizures and total environmental disconnection are also terminally present.

The juvenile form of MLD begins prior to the 2nd decade of life, and produces death prior to the 3rd decade. Clinical presentation and disease evolution are similar to, although they occur much sooner than in, the late infantile form.

An adult-onset form presents after puberty with usual initial symptoms of personality change, psychosis, and dementia.[5] Reflexes may be well preserved or brisk, in contrast to the infantile and juvenile forms. Corticospinal tract dysfunction evolves later in the illness. Life expectancy is prolonged and death may occur several decades after the onset of initial symptoms.

Cerebrospinal fluid protein level is elevated. Metachromatic granules are present in the urine. Urine sulfatide excretion is markedly elevated. Enzyme activities may be quantitated in urine, plasma, leukocytes from the peripheral blood, cultured skin fibroblasts, and tears. Sulfatides accumulate in the central and peripheral nervous sytem as well as the gallbladder, liver, kidney, and pancreas.

Computed tomography (CT) of the brain in MLD may reveal atrophy or attenuation of the white matter. This may be marked in the periventricular white matter, particularly near the frontal horns. Magnetic resonance imaging (MRI) findings may also be striking and helpful in suggesting leukodystrophy as a general diagnosis.

Metachromasia is present on pathologic sections from peripheral nerve biopsy as well as on liver and kidney biopsy.

Ocular Manifestations

Ocular features are variable. Myelin degeneration with whorl formation and granular inclusions in a herringbone pattern are seen in multiple nerves, including those of the orbit.[6] No other lid, adnexal, or orbital abnormalities are noted. Conjunctival biopsy reveals the occasional presence of mucopolysaccharidelike inclusions within epithelial cells.[6] Electron microscopy of such specimens reveals membrane-bound osmophilic inclusions containing a fibrillogranular material. Tears can be assayed for enzyme activity of aryl sulfatase A and B but not for aryl sulfatase C.[6] Aryl sulfatase B and C levels are normal in MLD but diminished in MSD.

There are no known lens or vitreous abnormalities. The retinal surface has been repeatedly noted to have a grayish color, particularly in the perimacular region.[6-8] Progression to a cherry-red spot has also been noted on two separate occasions, but this should be considered rare in this disorder.[9] Optic nerve atrophy from retrograde degeneration is frequent, may be severe, and is usually a preterminal complication.[8] Blindness and loss of pupillary reflexes have also been noted as late and preterminal sequelae.

Retinal pathology is limited to the ganglion and glial cell layers. Osmophilic lamellar inclusions are present in the retinal layers.[7,8,10,11] Autopsy specimens from three patients with infantile MLD revealed inclusions in the nonpigmented apical ciliary epithelium.[6] Retinal vessel attenuation and narrowing may also occur late in the illness. Nystagmus and strabismus have been described, as have skew and tonic eye deviations.

Multiple Sulfatase Deficiency

In 1965, Austin et al.[1] noted a new form of autosomal-recessive metachromatic leukodystrophy that was associated with some of the findings of MPS. In addition to a

FIGURE 129-1 Composite photographs of a child with MSD. Comparison of the photographs taken at ages 2, 4, 5, and 9 years demonstrate progressive facial coarsening. Chronic skin changes are visible in the latest photograph.

deficiency of aryl sulfatase A, deficiency of other enzymes, including aryl sulfatases B and C, steroid sulfatase, and several enzymes involved in degradation of mucopolysaccharides, were also noted. Thus, the clinical features of multiple sulfatase deficiency (MSD) include all the features of MLD in conjunction with the features of MPS.

The clinical features of MSD are less age dependent than those of MLD. Patients present within the 1st decade of life, usually within the first 2 years of life. Rare patients may learn to walk and talk, but psychomotor retardation is usually present in infancy and may be profound. Microcephaly may be present as well in MSD but not in MLD. Progressive coarsening of the features may occur (Fig. 129-1). Hepatosplenomegaly is present but is not troublesome. Ichthyosis of the type seen in the X-linked steroid sulfatase deficiency can be

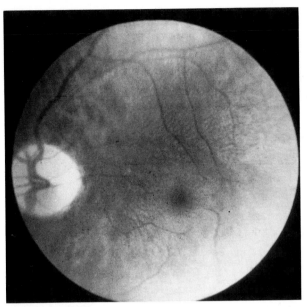

FIGURE 129-2 Fundus photograph taken at age 9 shows the characteristic gray-white macular ring. Optic nerve atrophy and retinal vessel attenuation are also visible.

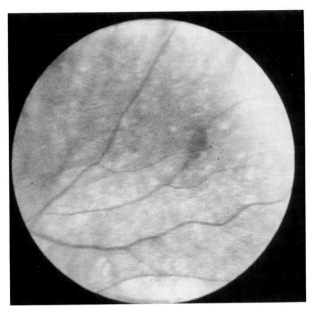

FIGURE 129-3 Peripheral fundus photograph demonstrates abnormal retinal pigment epithelium (atypical retinitis pigmentosa).

troublesome and relates to the diminished activity of steroid sulfatase.[1,12,13]

Biochemical abnormalities that overlap both MPS and MLD are present. Excretion of urinary sulfatides and both dermatan and heparan sulfate is increased.[1]

Radiographic findings in MSD are similar to those seen in mucopolysaccharide storage disease. Skeletal abnormalities seen in MPS, including broadening of the clavicular ends, broad phalanges, dysostosis multiplex, vertebral body beaking, and lumbar kyphosis are all present in MSD, albeit to a lesser degree than in some MPS syndromes. CT and MRI may both be informative. Atrophy and diffuse disturbance in white matter in MSD may lead to identification of the appropriate category of metabolic disorder and may suggest the differential diagnosis.

Ocular Manifestations

Ocular features of MSD overlap features seen in the MPS syndromes and MLD.[6,7,12] A gray-looking macula (Fig. 129-2) and retinal pigmentary change (Fig. 129-3), both inconstant features of MLD, seem to be constant features of MSD. Cherry-red spots are more common in MSD than in MLD.[9,12] Corneal clouding consistent with MPS storage may occasionally be present. Circumferential, gray lens opacities were noted in one well-documented patient.[7] Optic atrophy, retinal pigmentary

degeneration with atypical retinitis pigmentosa, decrease or absence of the electroretinogram response, and loss of pupillary light reflexes may also be present.

Treatment

No treatment is currently available for MLD or MSD. Enzyme replacement and bone marrow transplantation are subject to inherent limitations of treatment of any CNS disorder with an intact blood-brain barrier. Specific pharmacologic intervention aimed at stabilization of the mutant enzyme or at diminishing synthesis of sulfatides are of great theoretical interest but have not been clinically promising. Similarly, gene replacement or modification may be a mode of therapy in the future.

References

1. Austin JH, McAfee D, Armstrong D, et al. Abnormal sulphatase activities in two human diseases (metachromatic leukodystrophy and gargoylism). *Biochem J.* 1964;93:15C-17C.
2. Schaap T, Zlotogara J, Elian E, et al. The genetics of the aryl sulfatase locus. *Am J Hum Genet.* 1981;33:531-539.
3. Kihara H. Genetic heterogeneity in metachromatic leukodystrophy. *Am J Hum Genet.* 1982;34:171-181.
4. Stevens RL, Fluhary AL, Kihara H, et al. Cerebroside

sulfatase activator deficiency induced metachromatic leukodystrophy. *Am J Hum Genet.* 1981;33:900-906.

5. Waltz G, Harik S, Kaufmann B. Adult metachromatic leukodystrophy: value of computed tomographic screening and magnetic resonance imaging of the brain. *Arch Neurol.* 1987;44:225-227.

6. Libert J, Van Hoof F, Toussaint D, et al. Ocular findings in metachromatic leukodystrophy: an electron microscopic and enzyme study in different clinical and genetic variants. *Arch Ophthalmol.* 1979;97:1495-1504.

7. Cogan DG, Kuwabara T, Moser H. Metachromatic leukodystrophy. *Ophthalmologica.* 1970;160:2-17.

8. Quigley HA, Green R. Clinical and ultrastructural ocular histopathologic studies of adult-onset metachromatic leukodystrophy. *Am J Ophthalmol.* 1976;82:472-479.

9. Kivlin J, Sanborn G, Myers G. The cherry red spot in Tay Sachs and other storage disorders. *Ann Neurol.* 1985;17:356-360.

10. Goebel HH, Shimokawa K, Argyrakis A, et al. The ultrastructure of the retina in adult metachromatic leukodystrophy. *Am J Ophthalmol.* 1978;85:841-849.

11. Weiter JJ, Feingold M, Kolodny EH, et al. Retinal pigment epithelial degeneration associated with leukocytic arylsulfatase A deficiency. *Am J Ophthalmol.* 1978;90:768-849.

12. Bateman JB, Philippart M, Isenberg S. Ocular features of multiple sulfatase deficiencies and a new variant of metachromatic leukodystrophy. *J Pediatr Ophthalmol Strabismus.* 1984;21(4):133-139.

13. Philippart M, Isenberg SJ, Pineda GS. Metachromatic leukodystrophy: early onset, cherry red spot and ichthyosis—a new variant. *Ann Neurol.* 1980;8:218-219.

Section E

Disorders of Mineral Metabolism

Chapter 130

Hemochromatosis

ROBERT A. WIZNIA

Hemochromatosis is a disorder characterized by an abnormally high rate of intestinal iron absorption. This leads to a gradual increase in parenchymal iron deposits, largely in the form of the iron storage pigment hemosiderin. Tissue damage develops in a number of organs, including the liver, pancreas, heart, and pituitary. Genetic or primary hemochromatosis is due to an autosomal-recessive gene on chromosome 6 that is closely linked to the A locus of the HLA complex. The homozygote frequency is 0.3% in Anglo-Saxon populations.[1,2] In the primary form of this disease, iron deposition occurs only in parenchymal cells. Systemic hemosiderosis is a term used by some investigators to describe iron deposition in tissues without cirrhosis and its complicating features. In this condition, excess body iron is acquired from parenteral administration or autohemolysis (eg, in thalassemia or sideroblastic anemia). In such cases, iron is taken up by the reticuloendothelial system, including the bone marrow, spleen, and Kupffer cells of the liver. With increasing amounts of systemic iron stores, deposition begins to occur in parenchymal cells also. This leads to clinical and patho-logic features identical to those of primary hemochromatosis. The combination of parenchymal and reticuloendothelial cell iron deposits is termed acquired hemochromatosis.[3]

Ordinarily, intestinal mucosal iron absorption of 1 to 1.5 mg per day is in balance with loss of body iron. In primary hemochromatosis, the rate of iron absorption may be several times normal. This results in the deposition of excess iron in the parenchymal cells of a number of organs, including liver, pancreas, heart, pituitary, spleen, kidney, and skin. The eye similarly is subject to abnormal iron deposits in a number of tissues.

The normal body iron content is 3 to 4 g. It takes many years for persons with primary hemochromatosis to reach the level of 20 g or more generally required for clinical manifestations to develop. Thus, the condition is rarely diagnosed before age 40. There is a marked male predominance, reportedly as high as 10:1, owing to women's greater iron losses from menstruation, pregnancy, and lactation. In acquired hemochromatosis, usually owing to disorders of erythropoiesis, similar tissue and organ pathology develops because of increased mucosal iron absorption from treatment with iron preparations or from blood transfusions.

Systemic Manifestations

The liver is the organ most commonly affected, hepatomegally being present in more than 90% of cases. In primary hemochromatosis the iron is in the form of ferritin and hemosiderin in parenchymal cells. Slow progression to a characteristic perilobular fibrosis and in advanced stages to a macronodular or mixed macro- and micronodular cirrhosis is typical. Surprisingly, results of liver function tests may be normal until the condition is relatively advanced. Splenomegaly is common, although portal hypertension and esophageal varices are unusual. Hepatocellular carcinoma, the most common cause of death in patients with hemochromatosis, develops in approximately one third of the patients with cirrhosis.[3,4]

Increased skin pigmentation is evident in 90% of symptomatic patients and is due to increased melanin and iron in the cells of the basal layer (dermis) as well as in the epidermis. There is a generally diffuse metallic gray hue, which may include the eyelids (see below). Increased pigmentation of the oral mucosa is unusual.[3,4]

Diabetes mellitus is found in approximately two thirds of patients with primary hemochromatosis. Iron deposition in the parenchymal cells of the pancreas is a major factor in the development of diabetes mellitus. A family history of diabetes mellitus increases the predisposition for diabetes in those afflicted with primary hemochromatosis.[3,4]

Deposits of iron are reported surrounding the cells of the synovium in many joints. Calcium pyrophosphate crystals and calcium deposits form in these tissues as a progressive polyarthritis develops in approximately one third to one half of patients with hemochromatosis.[3,4]

Progressive deposition of iron in the myocardium may lead to diffuse cardiac enlargement and congestive heart failure in 10 to 15% of patients. A variety of cardiac arrhythmias have been reported, including supraventricular beats, paroxysmal tachycardia, atrial flutter, atrial fibrillation, and all degrees of atrioventricular block. Involvement of the pituitary gland with iron deposition may lead secondarily to testicular atrophy and decreased libido. Less commonly, adrenal insufficiency, hypothyroidism, and hypoparathyroidism may result.[3,4]

Ocular Manifestations

Ocular involvement in hemochromatosis is generally not associated with visual symptoms or with iron toxicity. Iron is deposited predominantly in two sites: the nonpigmented epithelium of the ciliary body and the sclera. The iron in the sclera is present both free within the sclera and within scleral fibroblasts. The sclera demonstrates no reaction to the iron. Increased melanin deposits may be seen in the perilimbal bulbar conjunctiva. Minimal amounts of free ferric iron are found in the conjunctival epithelium and in the perilimbal corneal epithelium. The eyelid margins, particularly on the cutaneous side, demonstrate increased pigmentation. This is greatest surrounding the follicles and is a localized example of skin involvement in hemochromatosis. The increased melanin deposition in the conjunctiva and skin is seen in about one third of cases.[5,6] Slate blue pigmentation around the optic discs (parapapillary) has been reported, although it is found in less than 20% of cases. It correlates with small iron deposits that may be detected in parapapillary retinal pigment epithelium.[7,8] Diabetic ocular involvement, especially diabetic retinopathy, may be superimposed on these findings.

There is much more extensive pathology in the case of *ocular hemosiderosis* due to a retained intraocular iron-containing foreign body or to intraocular hemorrhage (Table 130-1). Vision loss may be severe, in contrast to the general lack of visual consequences noted in hemochromatosis. In the anterior segment iron is deposited in the trabecular meshwork, cornea, iris dilator and sphincter muscles, iris epithelium, iris stroma, and lens epithelium. The ciliary epithelium also is involved. Greater consequences in terms of ocular and visual toxicity are noted in the posterior segment, where iron deposition leads to degeneration of the hyaluronic acid of the vitreous body, the acid mucopolysaccharides of the retinal perivascular tissue, and the retinal pigment epithelium. These changes lead to contraction bands in the vitreous and to proliferation and obliteration of retinal blood vessels and retinal degeneration.[5,6]

Diagnosis

Diagnosis of hemochromatosis should be considered when the combination of clinical findings includes hepatomegaly, increased skin pigmentation, diabetes mellitus, heart disease, arthritis, loss of libido, and testicular atrophy. Ocular findings of increased perilimbal conjunctival pigmentation, increased eyelid pigmentation, and parapapillary slate-blue pigmentation may support this clinical diagnosis; however, it would be unusual to make the diagnosis of primary hemochromatosis on the basis of the eye findings alone. Laboratory results which may point to the diagnosis include increased levels of serum iron and serum ferritin and degree of transferrin saturation. If any of these test results is abnormal, a liver biopsy should be performed.[3,9]

TABLE 130-1 Iron Deposition in Ocular Hemosiderosis and in Primary and Acquired Hemochromatosis

Structure	Ocular Hemosiderosis	Primary and Acquired Hemochromatosis
Conjunctival epithelium	+	+
Corneal epithelium	+	+
Corneal stroma	++	---
Trabecular meshwork	++	---
Iris epithelium	++	---
Iris dilator and sphincter muscles	++	---
Iris stroma	++	---
Lens epithelium	++	---
Nonpigmented ciliary epithelium	++	++
Sclera	+	++
Vitreous	+++	---
Retina	+++	---
Retinal pigment epithelium	++	+

Key: Iron deposits are: + minimal; ++ moderate; +++ extensive; --- not present.

Treatment

Therapy is based on removal of body iron. Weekly or twice-weekly phlebotomy is more effective than chelating agents. To restore body iron levels to normal requires 2 to 3 years of weekly phlebotomy. Chelating agents are indicated when anemia or hypoproteinemia precludes phlebotomy.[3,4]

References

1. Edwards CQ, Carroll M, et al. Hereditary hemochromatosis—diagnosis in siblings and children. *N Engl J Med.* 1977;297:7-13.
2. Simon M, Bourd M, et al. Idiopathic hemochromatosis—demonstration of recessive transmission and early detection by family HLA typing. *N Engl J Med.* 1977;297:1017-1021.
3. Powell W, Isselbacher J. Hemochromatosis. In: Braunwald E, et al., eds. *Harrison's Principles of Internal Medicine.* 11th ed. New York, NY: McGraw-Hill; 1987.
4. Milder MS, Cook JD, et al. Idiopathic hemochromatosis, an interim report. *Medicine.* 1980;59:34-49.
5. Davies G, Dymock I, et al. Deposition of melanin and iron in ocular structures in hemochromatosis. *Br J Ophthalmol.* 1972;56:338-342.
6. Roth AM, Foos RY. Ocular pathologic changes in primary hemochromatosis. *Arch Ophthalmol.* 1972;87:507-514.
7. Maddox K. The retina in haemochromatosis. *Br J Ophthalmol.* 1933;17:393-394.
8. Hudson JR. Ocular findings in haemochromatosis. *Br J Ophthalmol.* 1953;37:242-246.
9. Bassett ML, Halliday JW, et al. Diagnosis of hemochromatosis in young subjects: predictive accuracy of biochemical screening tests. *Gastroenterology.* 1984;87:628-633.

Chapter 131

Menkes' Kinky Hair Syndrome

ROGER L. HIATT

The full range of clinical findings in Menkes' kinky hair syndrome includes abnormal hair, progressive cerebral degeneration, hypopigmentation, bone changes, arterial rupture or thrombosis, and hypothermia. Premature delivery, jaundice, and hypothermia are common neonatal problems. Hair abnormalities may or may not be present at this stage, but they become more obvious by 3 months. A characteristic facial appearance, with pudgy cheeks and drooping jowls, is recognizable at a very early age. Developmental delay is generally obvious by 6 to 8 weeks, and progressive loss of skills continues until death occurs between the age of 6 months and 3 years, often as a result of intracrainal hemorrhage.[1]

The hair is tangled, lustreless, and grayish, with broken stubble in areas where the scalp is rubbed. Pili torti is found microscopically. Skeletal x-ray studies show osteoporosis and flared metaphyses, which may fracture. Wormian bones may be seen in the skull. The combination of rib fractures and subdural hematoma may lead to an erroneous diagnosis of child abuse. CT scan may show microscopic areas of brain destruction. Arteriograms show elongation, tortuosity, and variable caliber of major arteries throughout the body. Emphysema, bladder diverticulae, and retinal tears have also been described.[1]

Genetics

Menkes' kinky hair syndrome is an X-linked genetic disease characterized by disordered copper metabolism and an invariably fatal outcome in infancy or early childhood.[2] Danks et al.[3] first suggested the association with copper, based on the clinical similarity of Menkes' syndrome and copper deficiency in sheep. This disease provides scientists with a detailed look at the clinical manifestations of copper deficiency.

One very mildly affected patient has been described, who presented at the age of 2 years with mild developmental delay and marked cerebellar ataxia. Other features were present to a mild degree.[1]

Numerous pedigrees show X-linked recessive inheritance. Further support comes from the mosaic skin depigmentation seen in a heterozygous Negro girl and the pili torti seen in some heterozygotes. Mental retardation has been described in one Japanese girl, the sister of a severely affected boy, and in one French girl. This matches the experience with most other X-linked diseases and is generally attributed to chance inactivation in most cells of the embryo of the X chromosome carrying the normal allele. The incidence appears to be of the order of 1 in 100,000 births. The mildly affected boy presumably represents an allelic variant, although this has not been proven conclusively. Some heterozygotes can be recognized by the finding of pili torti in some hairs, by mosaic skin pigmentation following sun exposure, or by studies of cultured fibroblast cells.[1]

Pathophysiology

The extraordinary correspondence between physiologic defects and reduced activity of copper metalloenzymes is unique in clinical genetics. The pathognomonic feature—kinky hair or pili torti—can be ascribed to a defect in the formation of disulfide bonds through a deficiency of amine oxidase activity. The pale skin and depigmented hair are attributed to a deficiency of tyrosinase. Vascular tortuosity and diffuse aneurysms are due to defective elastin formation. The cuperoenzyme lysyl oxidase is required to cross-link the precursors of elastin and collagen. The hypothermia has been related to deficient cytochrome c oxidase activity in mitochondrial oxidative metabolism. The neurologic disorders—mental retardation, seizures, hypotonia—are due in part to cytochrome c oxidase deficiency, and perhaps also to dopamine-β-hydroxylase deficiency and altered catecholamine metabolism. Skeletal demineralization resembling scorbutic bone disease has been attributed to deficient ascorbate oxidase capacity, suspected to be copper dependent in mammals as well as plants.[4] Circulating levels of copper and ceruloplasmin also are markedly depressed in Menkes' syndrome. Several publications provide important insights into the pathogenesis of the copper deficiency state and into the range of phenotypic expressions possible in this disease.[2,4–6]

Increased accumulation of radioactive copper has

been demonstrated in cultured fibroblasts from Menkes' patients.[7-10] A team of molecular geneticists[6] who studied copper binding in protein fractions from the cultured fibroblasts of three patients with Menkes' disease characterized a cysteine-rich, aromatic amino acid–poor protein in the 10,000-dalton fraction of the cytoplasm. Using differential incorporation of tritiated amino acids, sulfur-35-cystein, and copper 64 from the media, they established that the amount of copper bound to this fraction was related constantly to its sulfur content, but that the Menkes' cells had a twofold enrichment of this sulfur-rich 10,000-dalton protein. The ready elution of the protein from a DEAE-cellulose, ion-exchange chromatograph distinguished this protein from the tenaciously bound copper-chelating proteins previously isolated from rat liver. The protein was thereby identified as metallothionein. The authors speculate that the increased accumulation or reduced efflux of copper in cultured fibroblasts, and elevated concentrations of copper in certain tissues such as the kidney and gut, are the result of a greater amount, but not a greater affinity, of the copper-binding protein.[2]

These results suggest that the basic defect must lie in some component of the intracellular copper transport system, which allows the copper to be diverted from the sites of copper enzyme synthesis and to be incorporated into metallothionein, thereby giving these cells their excessively high level of copper.[1,11]

Copper has a role in the cytochrome oxidase system.[12] Copper oxidase has been found to be reduced in a number of persons with this syndrome.[5,13-16] An inadequacy of cytochrome oxidase may result in a lack of high-energy phosphate required for nerve cell maintenance and function.[16] Copper has a role in myelin formation.[17] It is required for the phosphotidic acid synthesis needed for myelin production. The extent to which each of these biochemical pathways contributes to the electrophysiologic abnormalities noted in Menkes' kinky-hair syndrome must await further clarification of the metabolic role of copper in visual function.[16]

Ocular Manifestations

Pathologic changes occur in the brains of children with Menkes' syndrome, with diffuse neuronal degeneration and depopulation of neurons.[15,17,18] The amplitude of visual evoked response can reflect the abnormal retinal function of this syndrome or reflect additional damage to the ganglion cells in the central pathways.[16] There is a decrease in visual function with progression of the disease or with copper accumulation of the retina. Thus, the visual evoked response is abnormal. The pupillary response remains normal.[19]

In the optic nerve, there is a lack of myelin with glial replacement of atropic nerve fibers.[19]

The role of copper in visual functions is unknown. The A wave of the electroretinogram does not appear to be significantly affected by the disease process. The B wave was decreased for a child of this age.[16]

It has been suggested that copper replacement therapy is not successful because the copper is unavailable to the retinal cells because of a transport defect into the cells, or that once retinal cell damage has occurred in Menkes' syndrome it is not reversible by short-term administration of copper. The retinal pathology may involve the Muller cells or intermediate layers as well as the ganglion cell layers.[1,2,16,18,20,21]

References

1. Danks DM. Inborn errors of trace element metabolism. *Clin Endocrinol Metab.* 1985;14:591-615.
2. On the pathogenesis and clinical expression of Menkes' kinky hair syndrome. *Nutr Rev. Clinical nutrition.* 1981;39:391-393.
3. Danks DM, Campbell J, Walker-Smith BJ, et al. Menkes' kinky-hair syndrome. *Lancet.* 1972;1:1100-1102.
4. Hsiek S, Hsu JM. Zinc and copper. In: Karcioglu ZA, Sarper R, eds. *Medicine.* Springfield, Il: CC Thomas; 1980; 94-125.
5. Danks DM, Campbell PE, Stevens V, et al. Menkes' kinky hair syndrome: an inherited defect in copper absorption with widespread effects. *Pediatrics.* 1972;50:188-201.
6. Holtzman NA. Menkes' kinky hair syndrome: a genetic disease involving copper. *Fed Proc.* 1976;35:2276-2280.
7. Chan WY, Garnica AD, Rennert OM. Cell culture studies of Menkes' kinky hair disease. *Clin Chim Acta.* 1978;88:495-507.
8. Beratis NG, Price P, LaBadie G, et al. [64]Cu metabolism in Menkes and normal cultured skin fibroblasts. *Pediatr Res.* 1978;12:699-702.
9. Camakaris J, Danks DM, Ackland E, et al. Altered copper metabolism in cultured cells from human Menkes' syndrome and mottled mouse mutants. *Biochem Genet.* 1980;18:117.
10. LaBadie GU, Hirschhorn K, Katz S. et al. Increased copper metallothionein in Menkes' cultured skin fibroblasts. *Pediatr Res.* 1981;15:257-261.
11. Watanabe I, Watanabe Y, Motomura E, et al. Menkes' kinky hair disease: clinical and experimental study. *Doc Ophthalmol.* 1985;60:173-181.
12. Dowdy RP. Copper metabolism. *Am J Clin Nutr.* 1969; 22:887.
13. Danks DM, Stevens BJ, Townley RW. Menkes' kinky hair disease. Further definition of the defect in copper transport. *Science.* 1973;179:1140.
14. Danks DM, Stevens BJ, Campbell PE, et al. Menkes' kinky hair syndrome. *Lancet.* 1973;1:1100
15. Garnica A, Frias J, Easley J, et al. Menkes' kinky-hair disease. A defect in metallothionein metabolism? In: Bergsma D, ed. *Clinical Cytogenetics and Genetics.* Birth Defects: Original Article Series. Miami, Fla: Symposia Specialists; 1974.
16. Levy NS, Dawson WW, Rhodes BJ, et al. Ocular abnormali-

ties in Menkes' kinky-hair syndrome. *Am J Ophthalmol.* 1974;77:319-325.

17. Eversen CJ, Schroder RE, Wang TI. Chemical and morphological changes in brains of copper deficient guinea pigs. *J Nutr.* 1968;96:115.

18. Seelenfreund MH, Gartner S, Vinger F. The ocular pathology of Menkes' disease. *Arch Ophthalmol.* 1968;80:718.

19. Geeraets WJ. *Ocular Syndromes.* 3rd ed. Philadelphia, Pa: Lea & Febiger; 1976;292.

20. Danks DM. Of mice and men, metals and mutations. *J Med Genet.* 1986;23:99-106.

21. Leone A, Pavlakis N, Hamer DH. Menkes' disease: abnormal metallothionein gene regulation in response to copper. *Cell.* 1985;40:301-309.

Chapter 132

Wilson's Disease

Hepatolenticular Degeneration

THOMAS R. HEDGES, Jr., and THOMAS R. HEDGES III

Systemic Manifestations

Wilson's disease is a condition in which the ophthalmologist plays an important role in both diagnosis and management. Otherwise known as hepatolenticular degeneration, this condition occurs as a fatal defect in copper metabolism. It is an autosomal recessive disorder that usually presents between the ages of 8 and 16 years but may be seen as early as 6 years and as late as 50. It is due to defective excretion of copper from hepatic lysosomes and it is also associated with, but not directly due to, a decrease in the amount of circulating ceruloplasmin (<200 mg/L in 90% of affected patients). Copper deposition in the liver causes damage to hepatocytes, with eventual fibrosis and cirrhosis. Liver failure is the usual cause of death, although deposition of copper in the kidney may lead to renal failure.[1,2] Much of the morbidity in Wilson's disease is due to the neurologic dysfunction from copper toxicity in the basal ganglia, particularly the putamen and caudate nuclei (Fig. 132-1), as well as in frontal, cerebral, and cerebellar white matter. A significant amount of copper also accumulates in the thalamus and the red nuclei. This leads to incoordination, dysarthria, and tremor, athetosis, or chorea. Personality changes are also part of the spectrum of neurologic dysfunction in affected patients.[3,4]

Ocular Manifestations

The cornea is a diagnostically important site of copper deposition. Kayser first observed the corneal rings in 1902, and this finding was associated with Wilson's disease by Fleischer in 1909. Kayser-Fleischer rings are virtually pathognomonic of Wilson's disease; however, they have also been reported in other liver conditions wherein copper can be released from liver tissues and a similar ring occurs with topical application of copper to the eyes and with intraocular copper foreign bodies.[5] In Wilson's disease copper can be demonstrated throughout the cornea, but it is more clearly visible in the peripheral portions of Descemet's membrane. It develops first and is most easily seen in the superior and inferior cornea, with a clearer zone in the extreme periphery of the cornea (Fig. 132-2). The ring is usually gold or gray-brown peripherally with a more green or bluish tint centrally and is composed of granules containing copper and sulfur.[6] It is detectable in 97 to 100% of patients with neurologic evidence of Wilson's disease when properly examined with a slit lamp.[1,5] Gonioscopy may be useful in detecting the earliest changes, since they occur in the deepest levels of the cornea and may be obscured by overlying opacities. It may be seen in the absence of neurologic or hepatic disease, and it is, therefore, useful in the diagnosis of patients suspected of having Wilson's disease.

Another ocular manifestation of Wilson's disease is cataract.[5] This usually appears as a disc containing colorful spokes resembling a sunflower. It is due to accumulation of copper inside the anterior and posterior lens capsule and is usually visible with the slit lamp but not the ophthalmoscope. The cataract remains asymptomatic in the 15 to 20% of patients with Wilson's disease who develop it.[5]

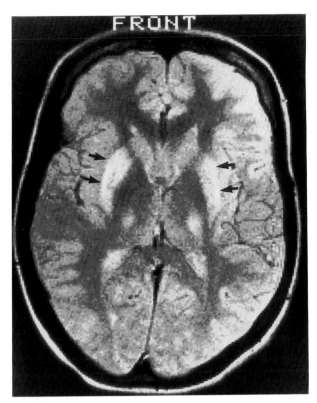

FIGURE 132-1 Magnetic resonance image (spin-density) showing increased intensity corresponding to areas of copper deposition in basal ganglia (*arrows*). (Courtesy J.A. McCrary III, M.D.)

The neurologic effects of Wilson's disease also involve the ocular motor system to a limited extent. Electro-oculographic studies have shown saccadic slowing[7] and dysmetria as well as saccadic pursuit.[8] Convergence and accommodative insufficiency occur fairly commonly in our experience and may be symptomatic.[9,10] Magnetic resonance abnormalities seen in the periaqueductal area indicate involvement of accommodative and convergence centers in the midbrain.[11]

Penicillamine has been the main drug of choice for the treatment of Wilson's disease. Kayser-Fleischer rings and sunflower cataracts commonly resolve with treatment. One of the unusual side effects of penicillamine is a form of ocular myasthenia that is Tensilon (edrophonium chloride) responsive and resolves when penicillamine therapy is discontinued.[12]

The Kayser-Fleischer ring is one of the few specific, pathognomonic signs that remain useful in clinical medicine. It is virtually diagnostic of Wilson's disease, especially in patients with tremor, dystonia, rigidity, dysarthria, or hepatic decompensation of uncertain origin. The absence of Kayser-Fleischer rings in a symptomatic patient should raise serious doubt about the diagnosis of Wilson's disease. Corneal examination may be helpful in screening family members, although the absence of copper deposition does not rule out Wilson's disease in asymptomatic relatives. Observing the appearance of Kayser-Fleischer rings and sunflower cataracts may be useful in assessing the effectiveness of penicillamine therapy for lowering tissue levels of copper.

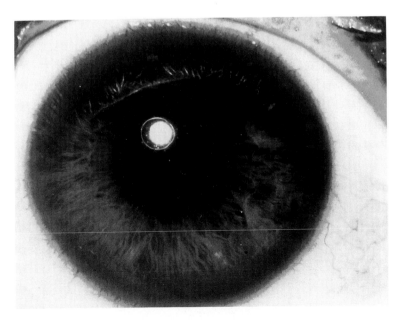

FIGURE 132-2 Kayser-Fleischer ring obscures the peripheral iris detail in a patient with Wilson's disease.

References

1. Scheinberg IH, Sternlieb I. Wilson's disease. Philadelphia, Pa: WB Saunders Co; 1984.
2. Strickland GT, Leu ML. Wilson's disease: clinical and laboratory manifestations in 40 patients. *Medicine (Baltimore)*. 1975;54:113-137.
3. Walshe JM. Wilson's disease (hepatolenticular degeneration). In: Vinken PJ, Bruyn GW, eds. *Handbook of Clinical Neurology*. New York, NY:North Holland Publ., vol 27. 1976;379-414.
4. Starosta-Rubinstein S, Young AB, Kluin K, et al. Clinical assessment of 31 patients with Wilson's disease. Correlations with structural changes on magnetic resonance imaging. *Arch Neurol*. 1987;44:365.
5. Wiebers DO, Hollenhorst RW, Goldstein NP. The ophthalmologic manifestations of Wilson's disease. *Mayo Clinic Proc*. 1977;52:409-416.
6. Johnson RE, Campbell RJ. Wilson's disease. Electron microscopic, x-ray energy spectroscopic, and atomic absorption spectroscopic studies of corneal copper desposition and distribution. *Lab Invest*. 1982;46:564-569.
7. Kirkham TH, Kamin DF. Slow saccadic eye movements in Wilson's disease. *J Neurol, Neurosurg, Psychiatr*. 1974;37:191-194.
8. Goldberg MF, von Noorden GK. Ophthalmologic findings in Wilson's hepatolenticular degeneration. *Arch Ophthalmol*. 1966;75:162-170.
9. Curran RE, Hedges TR III, Boger WP III. Loss of accommodation and the near response in Wilson's disease. *J Pediatr Ophthalmol Strab*. 1982;19:157-160.
10. Klingele TG, Newman SA, Burde RM. Accommodation defect in Wilson's disease. *Am J Ophthalmol*. 1980;90:20-24.
11. McCrary JA III. Magnetic resonance imaging diagnosis of hepatolenticular degeneration. *Arch Ophthalmol*. 1987;105:277.
12. Albers JW, Hodach RJ, Kimmel DW, et al. Penicillamine-associated myasthenia gravis. *Neurology*. 1980;30:1246-1250.

Section F

Miscellaneous Metabolic Disorders

Chapter 133

Amyloidosis

JOEL SUGAR

Amyloid is an insoluble extracellular material made up of sheets of fibrous protein. Histologically, it stains a characteristic pink with Congo red and with the same stain in polarized light exhibits a birefringent apple-green color. This staining characteristic is dependent on the physical conformation of the protein aggregate into "beta-pleated" sheets and is not a function of the specific proteins involved. Thus, amyloid can be formed from a number of different proteins, and the amyloidoses are a varied group of diseases all of which lead to accumulation of similar substances.

In primary amyloidosis, the amyloid is referred to as type AL, because it is made up of immunoglobulin light chains. In secondary amyloidosis and familial Mediterranean fever, the amyloid is type AA, because it is made up of protein A, a specific protein derived from a serum protein, SAA. In familial (neuropathic) amyloidosis, the principal protein component, prealbumin, is designated AF. Another component of amyloid, AP, is a glycoprotein common to all forms of amyloid.[1,2] This chapter will review primary systemic amyloidosis, secondary amyloidosis, familial amyloidosis, and isolated ocular forms of amyloidosis.

Primary Systemic Amyloidosis

Systemic Manifestations

Primary systemic amyloidosis is an acquired disorder, with deposition of amyloid in multiple organs. Multiple myeloma is a common associated disorder. These patients frequently present with weakness, fatigue, abdominal pain, peripheral neuropathy, and purpura. On clinical examination, they may have hepatomegaly, macroglossia, and edema. About one third have congestive heart failure because of infiltration of the heart muscle with amyloid, and one third have nephrotic syndrome because of glomerular deposition of amyloid. Almost all other organ systems may be involved as well, with pulmonary involvement leading to dyspnea, gastrointestinal infiltration leading to motility

FIGURE 133-1 Artists representation of vitreous amyloid deposits. Arrow points to anterior surface of lens. (Hitchings RA, Tripathi RC.[6])

disturbances, nerve involvement leading to numbness and autonomic dysfunction with orthostatic hypotension, and skin involvement leading to skin fragility. Tendon infiltration leads to carpal tunnel syndrome. Bleeding occurs because of an associated clotting factor deficiency, fibrinolysis, and intravascular coagulation; vascular wall integrity may be lost because of amyloid infiltration.

Laboratory studies demonstrate proteinuria, frequently abnormal serum protein electrophoresis, and monoclonal protein on serum immunoelectrophoresis. The diagnosis can be confirmed by demonstration of amyloid on rectal biopsy, bone marrow aspiration, or biopsy of involved organs. The prognosis is poor: only 25% of patients are alive 3 years after diagnosis.[3] No specific therapy exists although colchicine, corticosteroids, and antimetabolites have been recommended.

Ocular Manifestations

The most frequent ocular findings are eyelid ecchymoses and nodules. Purpura secondary to perivascular deposition of amyloid and increased vascular wall fragility is common, and the eyelids are the most common site of involvement.[4] Smooth, discrete or often confluent yellow to hemorrhagic waxy papules are common on the eyelids. Conjunctival involvement, tarsal nodules, and extraocular muscle infiltration may occur, but they are

more frequent in the localized forms of amyloidosis.[5] Vitreous involvement is much less common than in some of the familial forms of amyloidosis. When it occurs, there are glass-wool or sheetlike veils or string-of-pearls white opacities, often resembling old vitreous hemorrhage or inflammation (Fig. 133-1).[6] The opacities appear to begin near the retina and extend into the vitreous.[7] Histopathologic examination of involved eyes shows amyloid in the vitreous, retina, choroid, sclera, trabecular meshwork, and on the surface of the iris and lens.[8] Vitrectomy is effective in improving vision when the opacities are dense, but reaccumulation of opacities can occur subsequently.

Secondary Amyloidosis

Secondary amyloidosis is due to the accumulation of amyloid in organs of patients with underlying inflammatory disorders such as rheumatoid arthritis, tuberculosis, and Hansen's disease. Ocular involvement is produced by the underlying disorder; there are no ocular changes peculiar to amyloidosis.

Familial Amyloidosis
Systemic Manifestations

Familial amyloidosis with polyneuropathy, also called primary neuropathic amyloidosis, is a group of dominantly inherited disorders that have been subdivided into four types. Type I, or Portuguese type, is characterized by sensory and motor dysfunction of the lower limbs and autonomic dysfunction. In addition to persons from Portugal, families from Japan and Sweden have been described who have this type of amyloidosis. Type II, the Indiana type, has less severe limb involvement, cardiac failure, and carpal tunnel syndrome. Type III, or Iowa type, has generalized neuropathy and severe renal involvement. In type IV, the Finnish type or Meretoja's syndrome, progressive cranial neuropathy occurs, typically with facial nerve involvement and skin changes (Fig. 133-2). Cardiac involvement may occur, and at autopsy diffuse deposition of amyloid is seen.[9] Familial Mediterranean fever, an autosomal-recessive disorder, is also associated with systemic amyloid deposition.

Ocular Manifestations

Ocular involvement in types I and II familial amyloidoses consists of vitreous opacification, as described above. This responds well to vitrectomy. Scalloped pupils have also been described.[10] Type III does not cause ocular involvement. Persons with type IV do not

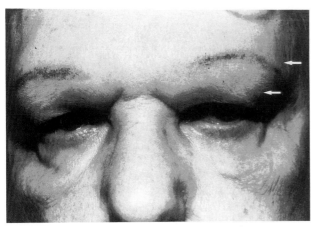

FIGURE 133-2 Patient with type IV familial amyloidosis. Note the flat facial features and extreme brow ptosis. Upper arrow denotes eyebrow pencil to simulate eyebrows. Actual eyebrow position is denoted by lower arrow.

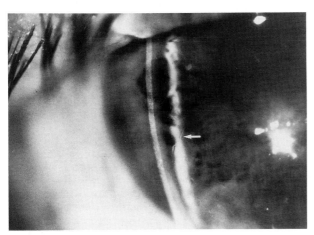

FIGURE 133-3 Cornea of patient in Figure 133-2 shows central corneal haze and peripheral lattice lines (*arrow*).

develop vitreous opacification but characteristically develop corneal amyloid deposition that looks like lattice corneal dystrophy (see below). The corneal changes differ only in that the opacification extends to the limbus and may be coarser than that of lattice dystrophy (Fig. 133-3). Some authors have referred to the corneal changes as "lattice dystrophy type II." Ptosis and cor-

neal exposure from the facial neuropathy are common, and glaucoma, presumably from trabecular meshwork accumulation of amyloid, has also been described.

Isolated Ocular Amyloidosis

Isolated ocular forms of amyloidosis are numerous. In addition to nonspecific localized accumulation of amyloid in the orbit, lacrimal gland, conjunctiva, eyelids, and sclera, some more specific forms have been described in the cornea. These include secondary amyloidosis after inflammation, as is seen in interstitial keratitis. Another common form of amyloidosis, which often is not recognized, is polymorphic amyloid stromal degeneration. This is presumably an aging change in which rodlike opacities are seen in the deep corneal stroma of elderly persons. Lattice corneal dystrophy is a localized form of amyloidosis without systemic abnormalities. Primary familial corneal amyloidosis, also called dropletlike dystrophy, is another localized form of ocular amyloidosis, in which bumpy elevations and opacities occur just beneath the corneal epithelium and in the anterior corneal stroma. None of these forms is associated with systemic amyloid deposition.

References

1. Glenner GG. Amyloid deposits and amyloidosis: I. the β-fibrilloses. *N Engl J Med.* 1980;302:1283-1291.
2. Glenner GG. Amyloid deposits and amyloidosis: II. the β-fibrilloses. *N Engl J Med.* 1980;302:1333-1343.
3. Kyle RA, Griepp PR. Amyloidosis (AL): clinical and laboratory features in 229 cases. *Mayo Clin Proc.* 1983;58:665-683.
4. Brownstein MH, Elliott R, Helwig EB: Ophthalmologic aspects of amyloidosis. *Am J Ophthalmol.* 1970;69:423-430.
5. Purcell JJ, Birkenkamp R, Tsai CC, et al. Conjunctival involvement in primary systemic nonfamilial amyloidosis. *Am J Ophthalmol.* 1983;95:845-847.
6. Hitchings RA, Tripathi RC. Vitreous opacities is primary amyloid disease: a clinical, histochemical and ultrastructural report. *Br J Ophthalmol.* 1976;60:41-54.
7. Savage DJ, Mango CA, Streeten BW. Amyloidosis of the vitreous. *Arch Ophthalmol.* 1982;100:1776-1779.
8. Schwartz MF, Green WR, Michels RG, et al. An unusual case of ocular involvement in primary systemic nonfamilial amyloidosis. *Ophthalmology.* 1982;89:394-401.
9. Purcell JJ, Rodrigues M, Chisti MI, et al. Lattice corneal dystrophy associated with familial systemic amyloidosis (Meretoja's syndrome). *Ophthalmology.* 1983;90:1512-1517.
10. Lessell S, Wolf PA, Benson MD, et al. Scalloped pupils in familial amyloidosis. 1975;293:914-915.

Chapter 134

Gout

ANDREW P. FERRY

Gout formerly was incriminated as the cause of a wide variety of ocular disorders, particularly iritis. In more recent decades this assertion has been questioned by experienced clinicians, but there had been few firm data upon which to base conclusions in this regard. A study of the eyes of 69 patients with particularly severe gout has helped fill the information gap.[1]

Systemic Manifestations

Gout is one of many arthritic disorders that affect the eyes.[2] The term "gout" denotes a heterogeneous group of diseases found exclusively in man that in their full development are manifested by (1) an increase in serum urate concentration; (2) recurrent attacks of a characteristic type of acute arthritis, in which crystals of monosodium urate monohydrate are demonstrable in leukocytes of synovial fluid; (3) aggregated deposits of monosodium urate monohydrate (tophi) occurring chiefly in and around joints of the extremities and sometimes leading to severe crippling and deformity; (4) renal disease involving glomerular, tubular, and interstitial tissues and blood vessels; and (5) uric acid urolithiasis.[3] These manifestations can occur in different combinations.

Gout occurs in both primary and secondary forms; the numerous secondary types of hyperuricemia and gout are attributed to a decrease in the renal excretion of uric acid. Diuretic therapy is currently one of the most important causes of secondary hyperuricemia in man. A true serum urate value above 7.0 mg/dl is abnormal. This value corresponds to a concentration of 7.5 to 8.0 mg/dl as determined by chemical methods and automatic analyzers.[3] In the course of its natural history, gout has four stages—asymptomatic hyperuricemia, acute gouty arthritis, intercritical gout, and chronic tophaceous gout.[3]

Ocular Manifestations

Uveitis

How extensively the gouty diathesis figures in the development of uveitis has always been a matter of dispute.[4] The older physicians, who found their expression in the forceful teaching of Jonathan Hutchinson (particularly as summarized in the first Bowman Lecture),[5] were liberal in the share they allotted to it. But even before Hutchinson's time other authorities (eg, Mackenzie in 1830) felt that the term "gouty iritis" was used too loosely.[4] Beaumont, an ophthalmologist in Bath, reported that not one among 2159 gout patients who had come to the Royal Mineral Water Hospital for treatment of gout had iritis.[6] Viewing the reverse side of the coin, Schlaegel and Coles reported, "We have not seen an instance of gout in more than 3000 patients with uveitis."[7]

Writing in the 1960s, Duke-Elder[4] summarized his own view on the subject as follows:

> Even although it seems certain that gout is now much less common than it used to be in the days when overindulgence in food and drink was the necessary attribute of a gentleman, it is quite obvious today that the terms 'gout' and 'gouty rheumatism' used to be employed in far too facile a manner, and that many cases of ocular inflammation were so labeled on quite unconvincing evidence; but at the same time it is as obvious that some of the cases described were true examples of a specific clinical entity and that similar instances are observed at the present time. The condition is, however, rare . . . most writers on the statistical incidence of uveitis nowadays make no mention of it. . . .

The experience mentioned earlier in this chapter supports the view that uveitis occurs in patients with gout only in vanishingly small numbers.[1] None of 69 patients, all of whom were afflicted with particularly severe gout, had uveitis, a history of uveitis, or objective evidence of past uveitis on detailed ocular examination (Table 134-1).[1]

Corneal Crystals and Scleral Tophi

These are extraordinarily rare in patients with gout, but corneal crystals of equivocal nature have been reported on several occasions.[1,8,9]

TABLE 134-1 Ocular Abnormalities in 69 Patients with Severe Gout

Abnormality	Patients (No.)
Uveitis or its sequelae	0
Scleral tophi	0
Corneal crystals (nature undetermined)	1
Elevated intraocular pressure	10
Asteroid hyalosis	3
Pingueculas	17
Red eyes	43

Elevated Intraocular Pressure

Beaumont cites several authorities (including Hutchinson and Nettleship) who claim that glaucoma was more prevalent in gout patients than in the remainder of the population but expresses his view that no convincing evidence to support this claim had been presented.[6] Ten of the 69 patients in the study mentioned earlier in this chapter had elevated intraocular pressures (see Table 134-1).[1] Five of them had open-angle glaucoma. The other five had ocular hypertension without evidence of glaucoma in the form of abnormal cupping of the optic nerve head or visual field damage.

Asteroid Hyalosis

Three of the 69 patients had asteroid hyalosis (see Table 134-1). Three instances of this uncommon disorder occurring in a cohort of 69 gouty subjects invites speculation about the possible increased prevalence of asteroid hyalosis in gout patients. Three cases out of 69 is obviously far more frequent than one would expect to find in virtually any population, no matter how selected.

Since the study was published, I have encountered 14 patients with asteroid hyalosis. Five of them had gout and three others had elevated serum uric acid levels, lending further support to the proposal that asteroid hyalosis occurs far more often in persons with gout than it does in the general population.[1]

Red Eyes

At least 43 of the 69 patients in this study had chronically red eyes (Tables 134-1 and 134-2).[1] The redness was bilateral and persistent. The color was more dusky red than bright red. Both the bulbar and the palpebral conjunctivae were involved. There was a tendency for the eyes to become much redder following manipulation during the course of the ocular examination. Many of the other 26 patients, whose eyes were not particularly

TABLE 134-2 Ocular Redness in 69 Patients with Severe Gout

Present	43
Absent	18
No record	8

injected at the beginning of the examination, developed severe conjunctival flushing after instillation of topical medications. We were unable to relate the degree of conjunctival or episcleral hyperemia to certain levels of blood urate. Nor did the redness recede in those patients whose blood urate levels fell in response to uricosuric therapy.

The ocular redness associated with gout has received little clinical attention. Although it was well-described by Duke-Elder[10] and was regarded by him as being relatively common, most practitioners do not consider the possibility of gout when confronted with a patient suffering from chronically red eyes. Although many of our 43 patients with red eyes expressed concern about their appearance, no ophthalmologist or other physician had ever suggested to them the possibility that the conjunctival or episcleral congestion had its basis in gout. The implications for clinical practice are clear: When evaluating a patient who is troubled by bilateral chronic conjunctival redness, gout should be considered in the differential diagnosis. Appropriate inquiries regarding the patient's general medical status should be made, and the necessary laboratory studies should be undertaken.

Throughout the centuries gout has enjoyed a royal patronage, and victims of gout have been favored subjects of caricature, satire, novels, and biographies. It is clear that the special popularity of gout stemmed from the centuries-old belief that the disease resulted from overindulgence in food and drink and from luxurious living in general. I believe that chronic ocular redness was probably present in the majority of those gouty persons, adding to their unfortunate appearance and thereby rendering them particularly easy targets for satirists.

References

1. Ferry AP, Safir A, Melikian HE. Ocular abnormalities in patients with gout. *Ann Ophthalmol.* 1985;17:632-635.
2. Ferry AP. The eye and rheumatic diseases. In: Kelley WN, Harris ED, Ruddy S, et al., eds. *Textbook of Rheumatology,* 3rd ed. Philadelphia, Pa: WB Saunders Co; 1989;579-596.
3. Kelley WN, Fox IH, Palella TD. Gout and related disorders of purine metabolism. In: Kelley WN, Harris ED, Ruddy S, et al., eds. *Textbook of Rheumatology.* 3rd ed. Philadelphia, Pa: WB Saunders Co; 1989;1395-1448.

4. Duke-Elder S, Perkins ES. Diseases of the uveal tract. In: Duke-Elder S, ed. *System of Ophthalmology.* vol IX. St. Louis, Mo: CV Mosby; 1966;642-645.

5. Hutchinson J. The Bowman Lecture on the relation of certain diseases of the eye to gout. *Br Med J.* 1884;2:995-1000.

6. Beaumont WM. Ocular disease in the gout. In: Llewellyn LJ, ed. *Gout.* St. Louis, Mo: CV Mosby; 1921;308-326.

7. Schlaegel TF, Coles RS. Uveitis and miscellaneous general diseases. In: Duane TD, ed. *Clinical Ophthalmology.* Hagerstown, Md: Harper & Row; 1978.

8. Slansky HH, Kuwabara T. Intranuclear urate crystals in corneal epithelium. *Arch Ophthalmol.* 1968;80:338-344.

9. Fishman RS, Sunderman FW. Band keratopathy in gout. *Arch Ophthalmol.* 1966;75:367-369.

10. Duke-Elder S, Leigh AG. Diseases of the outer eye. In: Duke-Elder S, ed. *System of Ophthalmology.* vol. VIII. St. Louis, Mo: CV Mosby, 1965;1060-1061.

Chapter 135

The Porphyrias

ALAN D. PROIA and KARL E. ANDERSON

The porphyrias are primarily inherited disorders characterized by deficiencies of enzymes in the heme biosynthetic pathway and, when clinically expressed, by specific patterns of overproduction and excretion of porphyrins and porphyrin precursors.[1-3] Seven types of inherited porphyria have been identified, and each is due to a deficiency of a different enzyme (Table 135-1). Abnormalities in metabolism or excretion of porphyrin precursors or porphyrins may occur in a variety of other disorders, the most striking of which are lead poisoning and hereditary tyrosinemia. In these conditions, reversible inhibition of δ-aminolevulinic acid dehydratase and neurologic symptoms reminiscent of some of the porphyrias are found.[1,2] Inherited deficiency of δ-aminolevulinic acid synthase has not been described. Because this inducible enzyme is rate limiting for heme biosynthesis (at least in the liver), it is presumed that such a mutation might be lethal.

The porphyrias can be classified as erythropoietic or hepatic, depending on whether the primary site of overproduction of porphyrins and porphyrin precursors is the bone marrow or the liver. With one exception all porphyrias are due to inherited enzyme deficiencies. Both inherited and acquired forms of porphyria cutanea tarda are described.[1] Environmental factors, including exposure to sunlight and certain drugs, hormones, or nutritional alterations, are essential in determining the clinical expression of all of the porphyrias.

Each type of porphyria is associated with accumulation and excess excretion of the substrate of the deficient enzyme. Substrates proximal to the deficient enzyme may also accumulate. For example, in acute intermittent porphyria there is excess excretion of both δ-aminolevulinic acid and porphobilinogen. These porphyrin precursors are also increased in hereditary coproporphyria and variegate porphyria when there are neurovisceral symptoms. Intermediates seemingly distal to the deficient enzyme may also accumulate. For example, uroporphyrin can form spontaneously (by autooxidation) from porphobilinogen in bladder urine when it is present in high concentrations. In porphyria cutanea tarda one or more intermediates prior to hepatic uroporphyrinogen decarboxylase, which is deficient in that disorder, can be metabolized to a series of "isocoproporphyrins" by coproporphyrinogen oxidase, the next enzyme in the pathway.

The patterns of accumulation and excretion of porphyrin precursors and porphyrins in erythrocytes, plasma, urine, and stool are variable and complex but are important for establishing a diagnosis of porphyria. Because treatment varies with the type of porphyria, it is important to establish clearly which type of porphyria is present in a given patient or family. Porphobilinogen deaminase activity can be measured in erythrocytes and other cell types. This assay is particularly useful for screening family members of patients known to have acute intermittent porphyria but is less useful than quantitation of urinary porphobilinogen for screening acutely ill patients with suspected porphyria. A full description of methods of diagnosis, clinical manifestations, and treatment of the porphyrias is beyond the scope of this review.

TABLE 135-1 Sequence of Intermediates and Enzymes in the Heme Biosynthetic Pathway, Types of Porphyria Associated with Specific Enzyme Deficiencies, Modes of Inheritance, and Chromosomal Location of the Normal or Mutant Gene (If Known)

Intermediates	Enzymes	Disease	Autosomal Inheritance	Gene Location
Glycine, succinyl CoA				
	δ-Aminolevulinic acid synthase	——	——	——
δ-Aminolevulinic acid				
	δ-Aminolevulinic acid dehydratase	ALAD porphyria	Recessive	Chromosome 9
Porphobilinogen				
	Porphobilinogen deaminase†	Acute intermittent porphyria	Dominant	Chromosome 11
Hydroxymethylbetane*				
	Uroporphyrinogen III cosynthase	Congenital erythropoietic porphyria	Recessive	——
Uroporphyrinogen III				
	Uroporphyrinogen decarboxylase	Porphyria cutanea tarda	Dominant§	Chromosome 1
Coproporphyrinogen III				
	Coproporphyrinogen oxidase	Hereditary coproporphyria	Dominant§	Chromosome 9
Protoporphyrinogen IX				
	Protoporphyrinogen oxidase	Variegate porphyria	Dominant§	——
Protoporphyrin IX; Fe^{2+}				
	Ferrochelatase‡	Erythropoietic protoporphyria	Dominant	——
Fe-protoporphyria IX (heme)				

*This intermediate spontaneously cyclizes to form uroporphyrinogen I in the absence of the cosynthase.
†Also known as uroporphyrinogen I synthase
‡Also known as heme synthetase.
§Homozygous forms of these disorders have also been described.

Systemic Manifestations

The porphyrias can also be classified according to whether they produce acute neurologic symptoms (acute porphyrias) or dermatologic manifestations (cutaneous porphyrias). The acute porphyrias include δ-aminolevulinic acid dehydratase (ALAD) porphyria (which has only recently been described), acute intermittent porphyria, hereditary coproporphyria, and variegate porphyria. The latter two diseases may also produce skin lesions. The other types of porphyria are associated with skin photosensitivity but not neurologic manifestations.

The most common initial symptom of an attack of acute porphyria is abdominal pain, often accompanied by constipation, nausea, and vomiting. Pain in the extremities, back, chest, or other areas may also occur. More severe or prolonged attacks are often associated with a peripheral neuropathy, which is primarily motor. A variety of other central and peripheral neurologic manifestations may occur, including inappropriate antidiuretic hormone secretion and seizures. Mental disturbances are common and may include insomnia, anxiety, depression, disorientation, hallucinations, and paranoia. Common physical findings include tachycardia, hypertension, and signs of ileus and neurologic involvement. With earlier diagnosis and better treatment, death from progressive neuropathy and respiratory paralysis is now rare. It is important to note that about 90% of people who inherit porphobilinogen deaminase deficiency do not experience symptomatic episodes. Factors that are known to predispose to clinical expression include sex steroid hormones and certain of their metabolites, administration of drugs, low calorie intake, and infections or other illnesses.[1] Similar considerations apply in variegate prophyria and hereditary coproporphyria. Most of the factors that cause clinical expression of these disorders induce hepatic cytochrome P_{450} and δ-aminolevulinic acid synthase and increase the metabolic consequences of partial enzyme deficiency.

The mechanism of the acute neurovisceral symptoms that occur in the acute porphyrias is not established.[1,4]

TABLE 135-2 Ocular Manifestations of the Porphyrias

Disorder	Ocular Manifestations	References
Congenital erythropoietic porphyria	Scarring of eyelids, conjunctiva, cornea	1,6-9
	Scleral necrosis with or without perforation	7-9
	Photophobia	5
Erythropoietic protoporphyria	Burning sensation of eyelids	5
	Eyelid edema	5
Porphyria cutanea tarda	Scarring of eyelids, conjunctiva, cornea	10-13
	Scleral necrosis with or without perforation	10-12
	Hypertrichosis of eyebrows	10,14
	Choroidal scars	10
	Deuteranomaly	15
	Sjögren's syndrome	16
Acute intermittent porphyria	Retinal edema	10
	Retinal hemorrhages	10,17,18
	Choroidal scars	10
	Atrophy of optic nerve	18,19
	Retinal branch vein occlusion	20
	Temporary visual loss	19
	Blindness secondary to infarction of occipital lobes	21
	Oculomotor nerve palsy	10

Porphyrin precursors and their byproducts have not been proven to be neurotoxic, but further work is needed. Deficiency of heme within neural tissues has also been postulated.[4]

Porphyria cutanea tarda is the commonest type of porphyria. The cutaneous manifestations occur on sun-exposed areas and include blisters, milia, scarring, pig-mentary changes, and facial hypertrichosis.[5] Increased skin fragility with minor trauma is common. Pseudo-sclerodermatous changes of exposed skin areas sometimes occur. Liver dysfunction is common and is usually mild, although cirrhosis and hepatocellular carcinoma can develop in advanced cases. Barbiturates and other drugs that induce δ-aminolevulinic acid synthase and can exacerbate the acute porphyrias have not been clearly associated with worsening of porphyria cutanea tarda. Exogenous factors that clearly exacerbate this disease are alcoholic beverages, estrogens, and iron. Mild or moderate hepatocellular iron overload is common, and a course of phlebotomies usually produces complete remission.

Variegate porphyria and hereditary coproporphyria may produce skin lesions that are indistinguishable from those of porphyria cutanea tarda. The most prominent manifestations of erythropoietic protoporphyria are cutaneous burning pain, itching, erythema, and edema that immediately follow sun exposure. Blisters are uncommon. Occasionally, marked protoporphyrin accumulation in the liver and liver failure occur. Congenital erythropoietic porphyria, a rare disorder, is usually associated with severe blistering, scarring, secondary infection, and mutilation of facial features and digits.[1,5] Hepatoerythropoietic porphyria is clinically a similar disorder, but it is now recognized to be the homozygous form of porphyria cutanea tarda.

Dermatologic manifestations in cutaneous porphyrias are believed to be due to porphyrins that reach the skin via the plasma. Abnormal skin responses occur on exposure to long-wave ultraviolet light, especially near 400 nm, which coincides with the absorption maximum (Soret band) for porphyrins. Light absorption excites the porphyrin molecule, which can then transfer energy via singlet oxygen and other active species of oxygen and cause tissue damage.[5] Porphyrins are reddish and fluoresce on exposure to 400-nm light. By contrast, the porphyrin precursors are colorless, do not fluoresce, and do not produce photosensitivity.

Ocular Manifestations

Ocular involvement in the porphyrias is uncommon. In the cutaneous porphyrias, portions of the eye that are most likely to be exposed to sunlight are most commonly involved, and the mechanism of damage is probably the same as for sun-exposed skin. Neurologic damage in the acute porphyrias can produce abnormalities in the optic nerve, extraocular movements, and central areas, including the visual cortex. None of the ocular or neurologic manifestations is unique to the porphyrias, and histologic examination of tissue does not reveal findings that can be considered specific. It should be kept in mind that in published reports the nomenclature and diagnostic criteria that are applied to the porphyrias vary considerably. Our understanding of the various types of porphyria and of the methods for distinguishing them has recently improved greatly. Most of the ophthalmic reports are not recent, so it is not always clear which type of porphyria is present in these cases. The reports are still instructive because all clinical manifestations of the porphyrias are essentially due either to photosensitivity or to neurologic involvement, and ocular damage due to either of these mechanisms can occur in several of the porphyrias. However, cases in which only slight abnormalities in levels of porphyrins or porphyrin precursors were noted should be interpreted with caution because such abnormalities are not specific or diagnostic. In future reports, greater emphasis on clearly documenting the type of porphyria by quantitative methods would be useful. A list of the ocular manifestations that have been reported is provided in Table 135-2.

Patients with congenital erythropoietic porphyria and marked photosensitivity may develop scarring of the eyelids, conjunctiva, and cornea (see Table 135-2). These patients are prone to episodes of scleral necrosis at the corneoscleral limbus in the interpalpebral fissure. The areas of necrosis may perforate in severe cases.

Ocular photosensitivity reactions in erythropoietic protoporphyria and porphyria cutanea tarda are less frequent and less severe than in congenital erythropoietic porphyria. Barnes and Boshoff[10] reported that choroidal scars had an increased prevalence in patients with "delayed cutaneous porphyria" (presumably porphyria cutanea tarda). An increased prevalence of deuteranomaly has also been reported.[15]

Ophthalmic abnormalities in patients with acute intermittent porphyria (see Table 135-2) are infrequent, and most probably result from neuronal damage, or perhaps vasospasm.[18-21] Neuropathy may rarely involve the oculomotor nerve, other cranial nerves, and areas of the central nervous system involved in vision, including the visual cortex.

References

1. Kappas A, Sassa S, Anderson KE. The porphyrias. In: Stanbury JB, Wyngaarden JB, Fredrickson DS, et al. eds. *The Metabolic Basis of Inherited Disease.* 5th ed. New York, NY: McGraw-Hill; 1983;1301-1384.
2. Labbe RF, Lamon JM. Porphyrins and disorders of porphyrin metabolism. In: Tietz NW, ed. *Textbook of Clinical Chemistry.* 3rd ed. Philadelphia, Pa: WB Saunders Co; 1986; 1589-1614.
3. Elder GH. Enzymatic defects in porphyria: an overview. *Semin Liver Dis.* 1982; 2:87-99.
4. Yeung Laiwah AC, Moore MR, Goldberg A. Pathogenesis of acute porphyria. *Quart J Med.* 1987;63:377-392.
5. Magnus IA. Cutaneous porphyria. *Clin Haematol.* 1980;9(2):273-302.
6. Garrod A. Congenital porphyria: a postscript. *Quart J Med.* 1936;5:473-480.
7. Girod P. Les signes ophtalmologiques de la porphyrie congénitale. *Ann Oculist (Paris).* 1969;202:937-951.
8. Douglas WHG. Congenital porphyria: general and ocular manifestations. *Trans Ophthalmol Soc UK.* 1972;92:541-553.
9. Hamard H, Guillerm AD, Onfray B, Guillaumat L. Complications oculaires des porphyries. A propos d'un cas de maladie de Günther. *J Fr Ophtalmol.* 1982;5:771-777.
10. Barnes HD, Boshoff PH. Ocular lesions in patients with porphyria. *Arch Ophthalmol.* 1952;48:567-580.
11. Sevel D, Burger D. Ocular involvement in cutaneous porphyria: a clinical and histological report. *Arch Ophthalmol.* 1971;85:580-585.
12. Chumbley LC. Scleral involvement in symptomatic porphyria. *Am J Ophthalmol.* 1977;84:729-733.
13. Sober AJ, Grove AS Jr, Muhlbauer JE. Cicatricial ectropion and lacrimal obstruction associated with the sclerodermoid variant of porphyria cutanea tarda. *Am J Ophthalmol.* 1981; 91:396-400.
14. Kingery FAJ. Eyebrows, plus or minus. *JAMA.* 1966;195:163.
15. Cullmann B, Denk R, Holzmann H. Zur Häufung von Farbensinnstörungen bei der Poprhyria cutanea tarda (Waldenström). *Graefes Arch Klin Exp Ophthalmol.* 1966;170: 201-208.
16. Ramasamy R, Kubik MM. Porphyria cutanea tarda in association with Sjögren's syndrome. *Practitioner.* 1982; 226:1297-1298.
17. Jaffe NS. Acute porphyria associated with retinal hemorrhages and bilateral oculomotor nerve palsy. *Am J Ophthalmol.* 1950;33:470-472.
18. Wolter JR, Clark RL, Kallet HA. Ocular involvement in acute intermittent porphyria. *Am J Ophthalmol.* 1972;74: 666-674.
19. DeFrancisco M, Savino PJ, Schatz NJ. Optic atrophy in acute intermittent porphyria. *Am J Ophthalmol.* 1979;87: 221-224.
20. Miller SA, Bresnick GH. Retinal branch vessel occlusion in acute intermittent porphyria. *Ann Ophthalmol.* 1979;11:1379-1383.
21. Lai C-W, Hung T-p, Lin WSJ. Blindness of cerebral origin in acute intermittent porphyria. Report of a case and postmortem examination. *Arch Neurol.* 1977;34:310-312.

Section G

Peroxisomal Disorders

Chapter 136

Adrenoleukodystrophy

W. BRUCE WILSON

Adrenoleukodystrophy (ALD) is characterized by mental dysfunction (dementia and behavior change), motor degeneration (spastic gait and paresis), and sensory loss (blindness and sensorineural hearing loss). It may be a spectrum of neurodegenerative hereditary disease but is usually separated into several closely related subtypes. The best understood form is childhood adrenoleukodystrophy (CALD). The others are neonatal adrenoleukodystrophy (NALD), adrenomyeloneuropathy (AMN), adrenoleukomyeloneuropathy (ALMN), and the heterozygote carriers of childhood adrenoleukodystrophy (CCALD). CALD and AMN are sex-linked and recessive, and the neonatal form is autosomal recessive. The childhood form starts most commonly between ages 4 and 8 years, AMN between ages 20 and 30 years, and the neonatal form is present at birth. It is now thought that the CALD gene locus is probably on the long arm of the X chromosome.[1]

The basic pathophysiologic problem seems to be a lack of oxidative enzymes to degrade or oxidize saturated unbranched very long–chain fatty acids (VLCFA), a type being C 26:0. In ALD the enzyme or enzymes seem to be carried in the tissue peroxisome, a subcellular organelle. In CALD the primary enzyme is probably a mixed-function oxidase. In the other forms of ALD the abnormal or deficient enzyme or enzymes are probably similar.[2-5] The defective metabolism of these fatty acids then leads them to collect in multiple tissues and the collections produce various pathologic expressions of these disorders. Whereas they commonly make up 1 to 2% of lipids in tissue, in ALD they may account for 40 to 50% of all lipids. Elevated levels of VLCFA probably occur in all forms of ALD. In neonatal ALD the number of liver peroxisomes is reduced, but in the childhood form it seems to be normal; so a qualitative or quantitative enzyme lack is probably the problem. A group of related peroxisomal diseases, Zellweger's and Refsum's, also involve enzyme deficiencies. The latter group seem to share a lack of phytanic acid oxidase (Table 136-1).[6-8]

The accumulation of fatty acids of a carbon chain length of 24 to 30 in the white matter of the brain probably accounts for the central nervous system (CNS) symptoms and signs in CALD. In fact, the curvilinear cytoplasmic inclusions seen in the perivascular macro-

TABLE 136-1 Adrenoleukodystrophy and Related Syndromes

	Childhood adrenoleuko-dystrophy (CALD)	Neonatal ALD (NALD)	Adrenomyelo-neuropathy (AMN)	Carriers of CALD (CCALD)	Zellweger's Syndrome (ZS)	Pseudo-Zellweger's Syndrome	Refsum's Disease	Infantile Refsum's Disease
Inheritance	Sex-linked recessive	Autosomal-recessive	Sex-linked recessive		Autosomal recessive			
Enzyme deficiency	Mixed-function oxidase	Mixed-function oxidase	Mixed-function oxidase		Phytanic-acid oxidase	Phytanic-acid oxidase	Phytanic-acid oxidase	Phytanic-acid oxidase
↑ Very long–chain fatty acids	Yes	Yes	Yes	Yes	Yes	Yes	Yes	Yes
Cytoplasmic inclusions	Yes	Yes						
↓ Peroxisomes	No	Yes, liver	No	No	Yes, liver, kidney		No	Yes, liver
Pipecolic acidemia	No	Yes	No	No	Yes	Yes		
Skin or adrenal changes	Bronzing, atrophy	Atrophy	Atrophy	No	±			
CNS changes	Demyelination, dementia, limb paresis, bladder abnormalities, blindness, hearing loss	Dementia, seizures, hypotonia, Blindness, retinitis pigmentosa, polymicrogyria, hearing loss	Limb paresis, bladder abnormalities, peripheral neuropathy	Spastic paraparesis	Dementia, seizures, hypotonia Retinitis pigmentosa, polymicrogyria		Peripheral neuropathy, retinitis pigmentosa, hearing loss	Retinitis pigmentosa
Other		Dysmorphia, hepatomegaly			Dysmorphia, hepatomegaly, renal cysts			

phages in areas of demyelination on EM study probably represent these deposits.[2,3,9-11] These same inclusions are seen in degenerating retinal ganglion cells in CALD, and in NALD they are found in macrophages of the retina and nerve and in the photoreceptors, retinal pigment epithelial cells, and neurons of the inner nuclear layer.[12-14] The cerebrum is usually affected earlier and more severely than other parts of the CNS. The primary change is demyelination that characteristically spreads from the posterior pole forward. Eventually the visual system in its entirety is demyelinated (Figs. 136-1, 136-2). Less commonly, demyelination of the stem and cord occurs and, rarely, even the cerebellum is involved. Although demyelination is the primary problem, there is also neuronal loss and multinucleated astrocytes are found. These inclusions likely lead to the finding of pigmentary retinopathy in NALD.

Although identical inclusions are found in the balloon cells of the zona fasciculata and -reticularis of the adrenal and in Schwann cells and Leydig cells of the testis, and although fatty acid deposits have been isolated from adrenal cortex, cultured skin fibroblasts, and other tissues, clinical signs from other organ systems seem isolated to the adrenal atrophy seen in CALD, NALD, and AMN.[2,3,9,15]

There are four other diseases that now seem to be either loosely or closely tied to ALD on the basis of similar peroxisomal enzyme abnormalities and overlapping clinical signs. These are Zellweger's syndrome, pseudo-Zellweger's or hyperpipecolic acidemia, and Refsum's and infantile Refsum's disease (see Table 136-1). Polymicrogyria is an example of this association and is found in neonatal ALD and in Zellweger's syndrome. Reduced numbers of peroxisomes in the liver

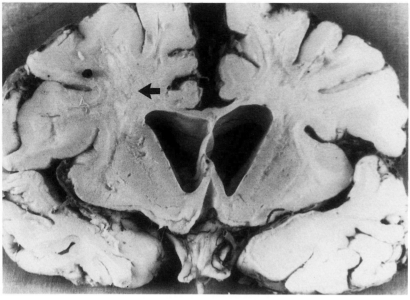

(A)

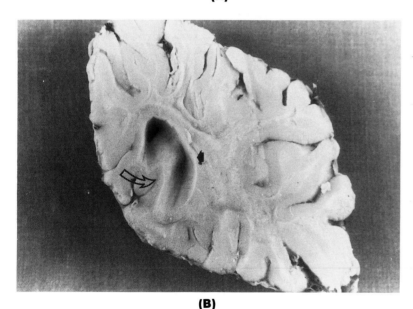

(B)

FIGURE 136-1 Coronal gross brain sections of a victim of childhood adrenoleukodystrophy. Section **(A)** is cut through the chiasm and shows diffusely softened, even somewhat necrotic-looking white matter (*arrow*). Compare with Fig. 136-2A and note that the gray and white matter are easily separated visually. The chiasm and optic nerves are shrunken. Section **(B)** is through the posterior horn of the lateral ventricle. The closed arrow is on the visual radiation, which also appears necrotic. The open arrow is on the calcar avis (compare with Fig. 136-2B).

and kidney are seen in Zellweger's and infantile Refsum's disease.[3,4] High blood levels of fatty acids are found in all four of these problems, but they may be of a somewhat different composition than those found in ALD.[3,6,12,16-18]

Systemic Manifestations

The aforementioned pathophysiologic and pathologic findings lead to the systemic findings in ALD. The adrenal glands and the skin seem to be the two organs that show clinical abnormalities outside the eye and central nervous system. Adrenal atrophy occurs in almost all cases of CALD, NALD, and AMN, along with bronzing of the skin. However, poor ACTH stimulation of cortisol occurs in fewer cases and clinical signs of adrenal insufficiency in fewer yet.[3,9,11,17] AMN probably has the most severe clinical expression of adrenal deficiency. In addition, the neonatal form has hepatic abnormalities and dysmorphia, findings shared with Zellweger's syndrome which are further evidence of the spectrum of this disease.

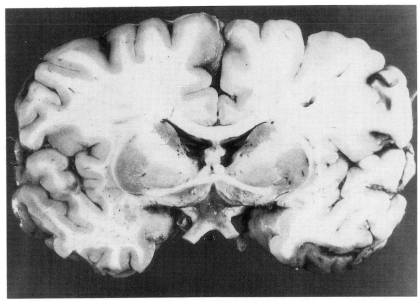

(A)

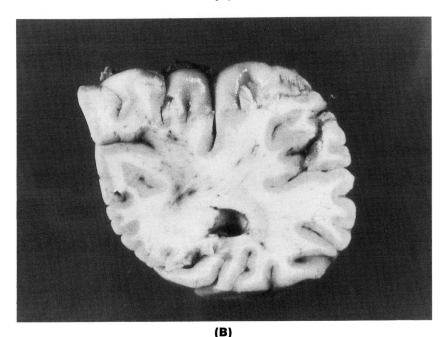

(B)

FIGURE 136-2 Sections **(A)** and **(B)** are taken from a normal age-matched brain and are meant to compare with Figs. 136-1A and B. The shape of the brain is slightly different owing to fixation flattening. Note the thick healthy white matter, including optic nerves. The ventricles are small, since no brain substance has been lost.

CNS and Ocular Manifestations

The demyelinative changes in the central nervous system that occur secondary to lipid storage lead to the classic childhood ALD picture of dementia, behavior changes, blindness, sensorineural hearing loss, occasional aphasia and visual agnosia, spastic gait and hemiparesis, occasional dysarthria, dysphagia, and incontinence.[1,3,9-11,17] Dementia and behavioral change (eg, aggression) occur early in the course of the disease. The neonatal form does not share the behavioral changes as commonly but does share the dementia. The visual changes are similar, but seizure and hypotonia are more common.[19] AMN rarely if ever shares in the dementia and behavior changes, but it has peripheral neuropathy and brain stem and spinal cord signs (eg, bladder problems and spastic paraparesis). Carriers of childhood ALD have spastic paraparesis but very little else.[16,17]

The specific sensory changes as seen in the optic nerves and visual system, and accompanied by hearing loss, are best understood in childhood ALD. Optic atrophy is usually severe but rarely early. Since cerebral demyelination usually progresses from posterior to anterior and is usually much more severe than in the stem or cord, cortical blindness may be a presenting sign.[3,9,11,17,20] Eventually the whole visual system, including chiasm, tracts, and radiations, are extensively demyelinated.[11] Vision may vary between 20/20 and no light perception and seems to be a function of the duration of the disease. Occasionally cataracts, optic nerve hypoplasia, and horizontal gaze problems are seen. Centrocecal scotomas, homonymous hemianopia, and nerve fiber defects are noted. There are electrophysiologic changes in the visual evoked response (VER) but not in the electroretinogram (ERG) and electrooculogram (EOG).[11,17] Neonatal ALD shares almost all of the optic nerve abnormalities that are seen in the childhood form and also includes hearing loss. AMN and CCALD rarely have visual symptoms. ALMN is closely related to AMN clinically but may have some intracranial signs.

The other four syndromes that are closely related to ALD share some symptoms and signs, such as dementia, seizures, and hypotonia in Zellweger's syndrome and abnormal gait, peripheral neuropathy, and sensorineural hearing loss in Refsum's disease.[3,7] Retinal pigment degeneration, which does not occur in CALD, occurs in the neonatal ALD and in Zellweger's, Refsum's, and infantile Refsum's disease, perhaps reflecting the asymmetric occurrence of cytoplasmic inclusions.[3,7,12]

Therapeutic response is generally very poor whether the modality is immunosuppressive or steroid drugs, plasma exchange, or diet.[17,21]

References

1. Moser HW, Moser AB, Kawamura N, et al. Adrenoleukodystrophy: studies of the phenotype, genetics and biochemistry. *Johns Hopkins Med J.* 1980;147:217-224.
2. Igarashi M, Schaumburg HH, Powers J, et al. Fatty acid abnormality in adrenoleukodystrophy. *J Neurochem.* 1976;26:851-860.
3. Moser HW, Moser AB, Singh I, et al. Adrenoleukodystrophy: survey of 303 cases. Biochemistry, diagnosis, and therapy. *Ann Neurol.* 1984;16:628-641.
4. Roels F, Corneilis F, Poll-The BT. Hepatic peroxisomes are deficient in infantile Refsum disease. *Am J Med Genet.* 1986;25:257-271.
5. Arias JA, Moser AB, Goldfischer SL. Ultrastructural and cytochemical demonstration of peroxisomes in cultured fibroblasts from patients with peroxisomal deficiency disorders. *J Cell Biol.* 1985;100:1789-1792.
6. Poulos A, Singh H, Paton B, et al. Accumulation and defective beta-oxidation of very long chain fatty acids in Zellweger's syndrome, adrenoleukodystrophy and Refsum's disease variants. *Clin Genet.* 1986;29:397-408.
7. Budden SS, Kennaway NG, Buist NRM, et al. Dysmorphic syndrome with phytanic acid oxidase deficiency, abnormal very long-chain fatty acids, and pipecolic acidemia: studies in 4 children. *J Pediatr.* 1986;108:33-39.
8. Moser AE, Singh I, Brown FR III, et al. The cerebro-hepatorenal (Zellweger) syndrome: increased levels and impaired degradation of very long chain fatty acids and their use in diagnosis. *N Engl J Med.* 1984;310:1141-1146.
9. Schaumburg HH, Powers JH, Raine CS, et al. Adrenoleukodystrophy. A clinical and pathological study of 17 cases. *Arch Neurol.* 1975;32:577-591.
10. Blaw ME. Melanodermic type leukodystrophy (adrenoleukodystrophy). In: Vinken PJ, Bruyn GW, eds. *Handbook of Clinical Neurology.* vol 10. New York, NY: American Elsevier; 1978;128-133.
11. Wilson WB. The visual system manifestations of adrenoleukodystrophy. *Neuro-Ophthalmology.* 1981;1:175-183.
12. Cohen SMZ, Brown FR, Martyn L, et al. Ocular histopathologic and biochemical studies of the cerebrohepatorenal syndrome (Zellweger's syndrome) and its relationship to neonatal adrenoleukodystrophy. *Am J Ophthalmol.* 1983;96:488-501.
13. Cohen SMZ, Green WR, de la Cruz ZC, et al. Ocular histopathologic studies of neonatal and childhood adrenoleukodystrophy. *Am J Ophthalmol.* 1983;95:82-96.
14. Wray SH, Cogan DG, Kuwabara Y, et al. Adrenoleukodystrophy with disease of the eye and optic nerve. *Am J Ophthalmol.* 1976;82:480-485.
15. Moser HW, Moser AB, Kawamura N, et al. Adrenoleukodystrophy: elevated C26 fatty acid in cultured skin fibroblasts. *Ann Neurol.* 1980;7:542-549.
16. Moser HW, Moser AE, Trojak JE, et al. Identification of female carriers for adrenoleukodystrophy. *J Pediatr.* 1983;103:54-59.
17. Traboulsi EI, Maumenee IH. Ophthalmologic manifestations of X-linked childhood adrenoleukodystrophy. *Ophthalmology.* 1987;94:47-52.
18. Arneson DW, Tipton RE, Ward JC. Hyperpipecolic acidemia: occurrence in an infant with clinical findings of the cerebrohepatorenal (Zellweger) syndrome. *Arch Neurol.* 1982;39:713-716.
19. Kelley RI, Datta NS, Dobyns WB, et al. Neonatal adrenoleukodystrophy: new cases, biochemical studies, and differentiation from Zellweger and related peroxisomal polydystrophy syndrome. *Am J Med Genet.* 1986;23:869-901.
20. Cogan DG, Chu FC, Reingold D, et al. Ocular motor signs in some metabolic diseases. *Arch Ophthalmol.* 1981;99:1802-1808.
21. Murphy JV, Marquardt KM, Moser HW, et al. Treatment of adrenoleukodystrophy by diet and plasmapheresis. *Ann Neurol.* 1982;12:220. *Abstract.*

Chapter 137

Adult Refsum's Disease

RICHARD G. WELEBER and NANCY G. KENNAWAY

Refsum's disease (also called adult Refsum's disease to distinguish it from infantile Refsum's disease) is an autosomal recessive disorder that was first described in 1946 under the name heredopathia atactica polyneuritiformis.[1] The disorder is a constellation of progressive multisystem symptoms and findings that include retinitis pigmentosa, peripheral polyneuropathy, cerebellar ataxia, nerve deafness, cardiac conduction defects leading to arrhythmias, ichthyosiform skin changes, and epiphyseal dysplasia.[2,3] In 1963, Klenk and Kahlke detected massive accumulation of a neutral lipid in serum, liver, and kidneys on autopsy material from a 7-year-old girl.[4] The abnormal lipid was characterized as phytanic acid, a fully saturated, branched-chain 20-carbon fatty acid that is normally present only in trace amounts in serum or tissues (less than 0.5% of total fatty acids). In Refsum's disease phytanic acid may account for 5 to 30% of total fatty acids in serum and up to 50% of fatty acids in tissues such as liver. Phytanic acid is not synthesized by humans but occurs in the diet in animal fats, dairy products, and in plants as its precursor phytol.

Phytanic acid is normally broken down into shorter-chain fatty acids before further oxidation for the generation of energy or heat. Most fatty acids are readily broken down by beta oxidation into shorter-chain units with the production of acetylcoenzyme A (acetyl-CoA) for generation of adenosine triphosphate (ATP). Both mitochondria and peroxisomes have complete enzyme systems for beta oxidation. Mitochondrial beta oxidation is the most active pathway for breakdown of fatty acids, especially those of 18 carbons or less; however, the presence of a methyl group at the third carbon position of phytanic acid blocks beta oxidation. Phytanic acid appears to be broken down exclusively by alpha oxidation of the second carbon to form α-hydroxyphytanic acid, followed by oxidative decarboxylation to form the n-1 carbon fatty acid, pristanic acid. The first step of this reaction appears to be deficient in cells from patients with Refsum's disease.[3,5]

Infantile phytanic acid storage disease, also called infantile Refsum's disease, is an entirely separate genetic disorder from classic Refsum's disease. Infantile Refsum's disease has many features in common with Zellweger's syndrome and neonatal adrenoleukodystrophy and is associated with defective biogenesis of peroxisomes, multiple enzyme deficiencies, mental retardation, deafness, and retinitis pigmentosa.[6] Phytanic acid is elevated in the serum of patients with infantile Refsum's disease and Zellweger's syndrome, disorders where peroxisomes are known to be absent, suggesting that although the initial steps of phytanic acid oxidation appear to occur in mitochondria, subsequent steps may take place in peroxisomes.

Systemic Manifestations

The mechanism by which elevation of phytanic acid levels produces the features of the disease is not known, but several theories of pathogenesis have been proposed.[3] The simplest is that the rigid, fully saturated phytanic acid molecule becomes incorporated into membranes in retina, brain, and nerve tissue, creating instability, altering normal function, and eventually leading to cell death. This theory is supported by the fact that dietary restriction of phytanic acid, if begun early, can prevent development of certain features and treatment at a later stage appears to arrest further progress and even allows recovery of some function. Another, less supported theory is that, because of its similarity to the isoprenoid side chains of certain fat-soluble vitamins, phytanic acid may interfere with the function of vitamins E and K. Certain clinical features of Refsum's disease in common with abetalipoproteinemia would tend to support this theory. Other possible pathophysiologic mechanisms include interference with function of coenzyme Q and the possibility that the defect in Refsum's disease is a more widespread, generalized defect in alpha oxidation. The pathophysiology of the ichthyosiform skin lesions is unknown. Suggested possibilities include reduction of available free cholesterol because of reaction with phytanate in skin to form cholesterol phytanate, and induced relative linoleate deficiency, since ichthyosis is seen also in essential fatty acid deficiency.

The four cardinal features—retinitis pigmentosa, peripheral polyneuropathy, cerebellar ataxia, and elevated cerebrospinal fluid protein—are present in all cases of

Refsum's disease. Other features appear to be variably present. Although the disease may present with symptoms and physical findings anywhere from the 1st to 5th decade of life, most patients have symptoms by 20 years of age. Almost all features of the disease appear to progress slowly, with notable exacerbations during febrile illness, surgery, or pregnancy. Nightblindness, weakness in extremities, and ataxia are the most common early symptoms. Variable features include progressive nerve deafness, anosmia, ichthyosis, and epiphyseal dysplasia. Syndactyly, short fourth metatarsal, hammer toe, pes cavus, and osteochondritis dissecans have all been reported. Many patients are initially thought to have a hereditary spinocerebellar ataxia, such as Friedreich's ataxia. The cerebellar dysfunction, which presents as unsteady gait, positive Romberg sign, intention tremor, and nystagmus, is usually out of proportion to the degree of peripheral neuropathy, which presents as symmetric motor and sensory losses with absent or decreased deep tendon reflexes. The elevated cerebrospinal fluid protein is not associated with pleocytosis. Sudden death may result from cardiac conduction defects with arrhythmias, especially if dieting or starvation mobilizes fatty stores, or the intake of phytanic acid in the diet increases dramatically. Plasmapheresis is life saving when severely elevated phytanic acid levels become associated with cardiac arrhythmias. Unless on a phytanic acid–restricted diet, most patients will have serum phytanic acid levels greater than 10 mg/dl.

Ocular Manifestations

All patients have retinitis pigmentosa with nightblindness that usually dates to early childhood but often requires specific questioning to elicit. Midperipheral scotomas in the visual field enlarge, coalesce, form ring scotomas, and eventually result in profound constriction of side vision (tunnel vision). Nystagmus, which is not often present with more common forms of uncomplicated retinitis pigmentosa, often occurs in Refsum's disease. Pupillary abnormalities (miosis and poor pupillary dilation) have been reported.[7] Central visual acuity can fail from early macular involvement in the degenerative process. Older patients often develop marked macular degenerative changes that further reduce vision. Cataracts, typically posterior subcapsular, often occur in midlife. The retinal vessels are attenuated early in the disease, and in later stages secondary optic nerve atrophy may ensue. Retinal pigmentary changes are more often described as fine to medium, granular brown clumps ("salt and pepper") rather than typical bone spicule formation, as seen with classic retinitis pigmentosa.

The electroretinogram is abnormal or unrecordable in all patients at an early stage of the disease.[7,8] Long-term treatment of the disease by dietary restriction of phytanic acid appears to modify the appearance and severity of retinal findings and the responses on the electroretinogram.[9]

References

1. Refsum S. Heredopathia atactica polyneuritiformis. *Acta Psychiatr Scand*. 1946;38(suppl):9.
2. Steinberg D. Phytanic acid storage disease: Refsum's disease. In: Standbury JB, Wyngaarden JB, Fredrickson DS, et al., eds. *The Metabolic Basis of Inherited Disease*. 4th ed. New York, NY: McGraw-Hill; 1978,688-706.
3. Steinberg D. Phytanic acid storage disease (Refsum's disease). In: Standbury JB, Wyngaarden JB, Fredrickson DS, et al., eds. *The Metabolic Basis of Inherited Disease*. 5th ed. New York, NY: McGraw-Hill; 1983;731-747.
4. Klenk E, Kahlke W. Über das Vorkommen der 3,7,11,15-Tetramethyl-hexadecansäure (Phytansäure) in den Cholesterinestern und anderen Lipoidfraktionen der Organe bei einem Krankheitsfall unbekannter Genese (Verdacht auf Heredopathia actactica polyneuritiformis, Refsum's syndrome). *Hoppe Seyler Z Physiol Chem*. 1963;333:133.
5. Steinberg D, Avigan J, Mize C, et al. Conversion of U-C[14]-phytol to phytanic acid and its oxidation in hederopathia atactica polyneuritiformis. *Biochem Biophys Res Commun*. 1965;19:783-789.
6. Schutgens RBH, Heymans HSA, Wanders RJA, et al. Peroxisomal disorders: a newly recognised group of genetic disease. *Eur J Pediatr*. 1986;144:430-440.
7. Refsum S. Heredopathia atactica polyneuritiformis. Phytanic acid storage disease (Refsum's disease) with particular reference to ophthalmological disturbances. *Metabol Ophthalmol*. 1977;1:73-79.
8. Berson EL. Retinitis pigmentosa and allied diseases: application of electroretinographic testing. *Int Ophthalmol*. 1981;4:7-22.
9. Hansen E, Bachen NI, Flage T. Refsum's disease: eye manifestations in a patient treated with low phytol low phytanic acid diet. *Acta Ophthalmol*. 1979;57:899-913.

Chapter 138

Infantile Refsum's Disease

RICHARD G. WELEBER AND NANCY G. KENNAWAY

Infantile Refsum's disease (also called infantile phytanic acid storage disease) was first described by Kahlke[1] as a variant of Refsum's disease in 1974 and was brought to international attention by Scotto[2] in 1982. This is an autosomal recessive disorder with certain features similar to those of Zellweger's syndrome, including mild craniofacial dysmorphism, hypotonia, hepatomegaly, moderate to severe psychomotor retardation, neurosensory deafness, retinal degeneration and, in late stages of the disease, optic atrophy.[3-5] Serum phytanic acid levels are only mildly elevated (1 to 5 mg/dl) and may occasionally be normal. Although both have deficient oxidation of phytanic acid, infantile Refsum's disease is quite distinct clinically and biochemically from the classic type (often called adult-onset Refsum's disease). Infantile Refsum's is most probably related to a generalized deficiency of peroxisomal functions, possibly produced by a defect in biogenesis of peroxisomes.[6,7] Adult-onset Refsum's disease is an autosomal-recessive disorder associated with deficient activity of phytanic acid α-oxidase, marked accumulation of phytanic acid in tissues and serum (>10 mg/dl), pigmentary reginopathy, deafness, peripheral neuropathy, cerebellar ataxia, cardiac conduction defects leading to arrhythmias, and ichthyosis.

Peroxisomes are single-membrane subcellular organelles that house a variety of metabolic reactions that use or consume molecular oxygen.[8] Peroxisomal proteins are produced on free polysomes, enter the cytosol, and are imported into pre-existing peroxisomes, which then bud or divide. They cannot be synthesized de novo. They were first described in 1954 by J. Rohdin as small organelles bounded by a single membrane and filled with homogeneous, moderately electron-dense matrix. Rohdin called the organelles "microbodies" because of their nondescript appearance. In the 1960s de Duve biochemically characterized microbodies as having catalase activity and containing hydrogen peroxide–producing oxidases. The name "peroxisome" was chosen because hydrogen peroxide was both used and destroyed in the particles.

Much has been learned recently about the functions of peroxisomes.[8-10] At present they are considered to play significant roles in the following processes: catabolism of very long–chain fatty acids (VCLFAs), biosynthesis of ether phospholipids (plasmalogens), biosynthesis of bile acids, catabolism of oxalic acid and pipecolic acid, catabolism of dicarboxylic acids (omega oxidation products), catabolism of phytanic acid, and oxidation of polyamines. In plants, similar organelles contain respiratory enzymes and the glyoxylate cycle enzymes. Rather than generate ATP, the reactions of peroxisomes generate heat and so are believed to play a role in thermogenesis.

The peroxisomal catabolism of VLCFAs is not a duplication of mitochondrial beta oxidation. The mitochondrial system handles chains of up to 18 carbons. Peroxisomes have little or no activity for fatty acid chains shorter than 8 carbons. Oxidation of VLCFAs (C24 to C26) appears to take place exclusively in peroxisomes. Unlike in the mitochondrial system, peroxisomal beta oxidation is inducible by chemicals, drugs (such as clofibrate), increased dietary intake of fats, and cold.

The first steps of biosynthesis of plasmalogens takes place in peroxisomes. Plasmalogens are extremely important lipids in membranes and account for about one third of phospholipids in the brain. The major enzymes involved in the synthesis of plasmalogens are the peroxisomal enzymes dihydroxyacetone phosphate acyltransferase (DHAP-AT) and alkyldihydroxyacetonephosphate synthase (alkyl-DHAP synthase). Subsequent reactions take place in microsomes.

The conversion of trihydroxycoprostanoic acid into cholic acid, one of the reactions involved in biosynthesis of bile acids, occurs in peroxisomes, as does catabolism of pipecolic acid, which is an intermediate in the degradation of L-lysine.

The mild elevation of phytanic acid levels in infantile Refsum's disease is poorly understood. Although alpha oxidation of phytanic acid is thought to be mitochondrial, the possibility of a peroxisomal phytanic acid oxidase has not been rigorously excluded. More likely, phytamic acid accumulation occurs secondary to failure of peroxisomal beta oxidation of the n-1 carbon fatty acid, pristanic acid, which results from mitochondrial alpha oxidation of phytanic acid.[11]

Disorders of peroxisomal function present either as disorders of single peroxisomal function or disorders of multiple peroxisomal functions.[6,10] Disorders of multiple peroxisomal functions include cerebrohepatorenal (Zellweger's) syndrome, neonatal adrenoleukodystrophy (NALD), hyperpipecolic acidemia, rhizomelic chondrodysplasia punctata (RCDP), and infantile phytanic acid storage disease (infantile Refsum's disease, or IRD). The disorders of single peroxisomal function that affect the visual system include pseudo-Zellweger's syndrome, pseudo-NALD and X-linked adrenoleukodystrophy (ALD). The currently held classification of adult Refsum's disease as a disorder of single peroxisomal function is uncertain at best.

It appears that there are considerable differences between authors as to what features are necessary to diagnose a specific peroxisomal disorder such as infantile Refsum's disease. Greater variability of features appears to characterize infantile Refsum's, and there most assuredly exists unappreciated genetic heterogeneity within this group. Kelly et al[12] have proposed the name "peroxisomal polydystrophies" to include all presently recognized syndromes associated with multiple or generalized peroxisomal dysfunction. All have craniofacial or skeletal dysmorphism, psychomotor retardation, seizures, failure to thrive, ocular abnormalities (RCDP has cataracts, but not retinopathy), impaired hearing, and multiple biochemical impairments of peroxisome function.

Systemic Manifestations

The systemic manifestations of infantile Refsum's disease are similar to those found in Zellweger's syndrome and NALD but less severe. Dysmorphic features consist of epicanthal folds, low-set ears, and single palmar creases. Patients have less severe hypotonia and dysmorphic features than those with Zellweger's syndrome and live much longer than either Zellweger's or NALD patients. Birth weight is normal, but neonatal jaundice, hemolytic anemia, bleeding diathesis, abnormal results of liver function tests, and hepatomegaly have been reported in young infants. Psychomotor retardation is not as severe as in Zellweger's or NALD. Deafness and retinal degeneration become evident within the first year of life and appear to be prominent features in infantile Refsum's. Peroxisomes are deficient or absent in fibroblasts and in liver. Serum plasmalogens are severely deficient, as are the enzymes of plasmalogen synthesis, DHAP-AT and alkyl-DHAP synthase. Phytanic acid oxidation is also deficient, and phytanic acid levels are mildly elevated (1 to 5 mg/dl), as are levels of pipecolic acid, bile intermediates, and VLCFAs. Several peroxisomal beta oxidation enzyme proteins are deficient.

Neuropathologic studies have not yet been reported on patients with infantile Refsum's disease, but demyelination was reported in hyperpipecolic acidemia. Moderate to extensive central nervous system demyelination is evident in both NALD and ALD, an X-linked disorder associated with a specific deficiency of VLCFA oxidation. Since retinal degeneration and sensorineural deafness are not seen in ALD, the deficiency of plasmalogens seen in Zellweger's, NALD, and IRD children may play a role in the retinal degeneration and sensorineural deafness of those disorders.

All of the more generalized disorders of peroxisomal function carry an associated high burden for patient and family. All result in mental retardation, which can be moderate to profound. Blindness from the retinal degeneration or cataracts and deafness further impair the ability of these unfortunate children to communicate with their parents or caretakers.

Ocular Manifestations

The ocular manifestations of infantile Refsum's disease are similar to those found in other generalized peroxisomal disorders. Patients may present with features suggesting either Leber's congenital amaurosis, because of significant visual impairment in infancy, or Usher's syndrome, because of the association of deafness with retinitis pigmentosa.[4,5] Nystagmus, strabismus, and micro-ophthalmia have been reported. The retina may show only attenuation of the retinal vessels and mild mottling of the pigment epithelium as the first fundus findings. Macular degeneration is prominent, and retinal pigmentation, when it occurs, consists of fine to coarse clumps of brown pigment rather than the black bone spicules seen in typical retinitis pigmentosa. The electroretinogram is profoundly abnormal. Initially, visual function may be sufficient for the limited needs of these children; however, if not noted earlier, by 5 to 6 years of age most appear visually impaired when they are in unfamiliar surroundings or in dim illumination. Optic atrophy, presumably secondary to the progressive severe retinal degeneration, is associated with further reduction of vision as the children approach the second decade of life. Little is known about the eventual outcome for these children's vision.

References

1. Kahlke W, Goerlich R, Feist D. Erhöhte Phytansäurespiegel in Plasma und Leber bei einem Kleinkind mit unklarem Hirnschaden. *Klin Wochenschr*. 1974;52:651-653.
2. Scotto JM, Hadchouel M, Odievre M, et al. Infantile phy-

tanic acid storage disease, a possible variant of Refsum's disease: three cases, including ultrastructural studies of the liver. *J Inherited Metab Dis.* 1982;5:83-90.

3. Poulos A, Sharp P, Whiting M. Infantile Refsum's disease (phytanic acid storage disease): a variant of Zellweger's syndrome? *Clin Genet.* 1984;26:579-586.

4. Weleber RG, Tongue AC, Kennaway NG, et al. Ophthalmic manifestations of infantile phytanic acid storage disease. *Arch Ophthalmol.* 1984;102:1317-1321.

5. Budden SS, Kennaway NG, Buist NRM, et al. Dysmorphic syndrome with phytanic acid oxidase deficiency abnormal very long chain fatty acids, and pipecolic acidemia: studies in four children. *J Pediatr.* 1986;108:33-39.

6. Schutgens RBH, Heymans HSA, Wanders RJA, et al. Peroxisomal disorders: a newly recognized group of genetic diseases. (Review.) *Eur J Pediatr.* 1986;144:430-440.

7. Poll-Thé BT, Saudubray JM, Ogier H, et al. Infantile Refsum's disease: biochemical findings suggesting multiple peroxisomal dysfunction. *J Inherit Metab Dis.* 1986;9:169-174.

8. Goldfischer S. Peroxisomes (microbodies) in cell pathology. *Int Rev Exp Pathol.* 1984;26:45-84.

9. Osumi T, Hashimoto T. The inducible fatty acid oxidation system in mammalian peroxisomes. *Trends Biochem Sci.* 1984;9:317-319.

10. Moser HW, Goldfischer SL. The peroxisomal disorders. *Hosp Pract.* 1985;20:61-70.

11. Poulos A, Sharp P, Fellenberg AJ, et al. Accumulation of pristanic acid (2, 6, 10, 14 tetra methylpentadecanoic acid) in the plasma of patients with generalized peroxisomal dysfunction. *Eur J Pediatr.* 1988;147:143-147.

12. Kelly RI, Datta NS, Dobyns WB, et al. Neonatal adrenoleukodystrophy: new cases, bichemical studies, and differentiation from Zellweger and related peroxisomal syndromes. *Am J Med Genet.* 1986;23:869-901.

Chapter 139

Zellweger's Syndrome

Cerebrohepatorenal Syndrome

ALEC GARNER and ALISTAIR R. FIELDER

The cerebrohepatorenal syndrome associated with the name of Zellweger was first described in 1964.[1] Although initially considered very rare, it has now been reported on over 100 occasions, so most pediatric ophthalmologists can expect to encounter a case sooner or later. Features include a characteristic facial appearance and multisystem abnormalities of the central nervous system, liver, kidneys, joint cartilages, and eyes.[2,3] It is inherited as an autosomal recessive trait. Clinically, Zellweger's syndrome presents in the neonatal period, and although death usually occurs within the 1st year, some children have survived into the 2nd decade.[4]

Underlying Defect

There is now good reason, despite some earlier confusion, to regard Zellweger's syndrome as being a consequence of a fundamental defect in peroxisomal function.[2,3,5] Peroxisomes are small round organelles within the cytoplasm of most cell types that are responsible for a range of oxidative functions. The variety of oxidative processes attributable to peroxisomes is reflected in diverse metabolic and morphologic defects, but it is significant that it is organs that have an abundance of these organelles that are principally affected. Among the more important metabolic abnormalities reported in Zellweger's syndrome are accumulations in the serum of very long–chain (>C22) fatty acids (VLCFAs), 5,β-prostanoic acids (the immediate precursors of the bile acids), and pipecolic acid through a minor pathway of lysine metabolism that is particularly relevant to neuronal tissues.[2] Serum transaminase levels may also be raised as a result of liver dysfunction, whereas the amount of phosphatidylethanolamine, a plasmalogen present in all cell membranes including those of erythrocytes, is reduced.[6]

Zellweger's syndrome may be seen as the prototype of genetic disorders related to peroxisomal dysfunction; other diseases with a comparable pathogenesis are hy-

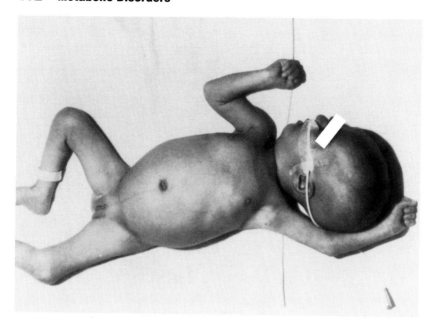

FIGURE 139-1 A newborn infant with Zellweger's syndrome: a high forehead is associated with flat supraorbital ridges and bridge of the nose.

perpipecolic acidemia, neonatal adrenoleukodystrophy (NALD), infantile Refsum's disease, and rhizomelic chondrodysplasia punctata.[3] In addition to these generalized disorders of peroxisome function there are those in which just one oxidative enzyme appears to be affected, namely, X-linked ALD and classic Refsum's disease. As might be anticipated, these partial peroxisome disorders tend to present at a later age and progress less rapidly.

The precise nature of the peroxisomal defect has yet to be defined, but it appears to concern the formation of an essential membrane protein or, alternatively, of a protein involved in the uptake by the peroxisomes of specified oxidative enzymes from the cytosol.

Systemic Manifestations

Because the metabolic derangement exists from very early fetal life, a malformation results that, in contrast to a congenital malformation due to an isolated insult, is also associated with degenerative features.

Clinically, Zellweger syndrome is apparent at birth with a characteristic facial appearance (Fig. 139-1) caused by a high forehead and a pear-shaped skull, flat supraorbital ridges, micrognathia, redundant neck skin folds, epicanthus, and upward-slanting palpebral fissures. Neurologic features include profound hypotonia, absent reflexes, fits, and severe developmental delay related to defective migration of neurones within the fetal central nervous system and rapidly supervening degenerative changes in the axons and myelin sheaths.[7,8] Hepatic

changes in the form of interstitial and portal fibrosis with diffuse parenchymal cell degeneration are associated with jaundice, and, sometimes, hypoprothrombinemia. The kidneys commonly contain multiple cortical cysts, and moderate proteinuria and aminoaciduria may develop. Other findings include cardiac defects and skeletal abnormalities with calcific stippling of the epiphyses. Death is almost invariable within a few months from birth because of a deteriorating neurologic state and infective respiratory problems.

Ocular Manifestations

Eventual blindness is almost inevitable in Zellweger's syndrome and ophthalmic involvement in one form or another is probably an integral component of the disease: it comprises both dysgenetic and degenerative disturbances.

The eyes are unusually prominent because of the facial appearance, and nystagmus is common, probably due to neurologic dysfunction, since it is present from the neonatal period. The eyeball itself is usually normal in size, but micro-ophthalmos has been reported. Should the globe become enlarged, that is an indication of glaucoma caused by a malformation of the anterior chamber that interferes with aqueous outflow.[9] Anterior insertion of the iris has been observed to be one form of abnormal development in this region, but other causes include posterior embryotoxon and iridocorneal adhesion. Corneal clouding varying between faint and dense is a frequent finding, and although the cause is not always apparent, raised intraocular pressure or deficiencies in the endothelium and Descemet's membrane may be responsible. Iris abnormalities, such as Brushfield's spots and a persistent pupillary membrane, have also been noted. Cataracts have been reported on a number of occasions, but the pathogenesis is obscure: Hittner et al.[10] have described curvilinear condensations at the corticonuclear interface in electron microscopic examination of the lens of a heterozygotic patient.

Posterior segment involvement is the principal cause of blindness and seems to be a constant feature of Zellweger's syndrome. The available evidence points to a rapidly progressive retinal dystrophy as opposed to a primary malformation. Thus, while the electroretinogram is usually absent, a normal response has been reported,[8] as has another case in which a phase of greatly reduced activity was noted before total extinction occurred.[11] The number of eyes examined histopathologically is small, but loss of photoreceptor cell outer segments with atrophy of the pigment epithelium and migration of the dispersed melanin into the neuroretina has been described.[9,11] Predictably the visual evoked potential is abnormal, but ophthalmoscopic evidence of "tapetoretinal" dystrophy is minimal during infancy, findings being limited to attenuation of the retinal arterioles and macular abnormalities variously described as retinal hole, red lesion, or pigmentary clumping. Nevertheless, if survival is long enough more obvious evidence of a pigmentary retinopathy develops, sometimes with large areas of peripheral pigmentary clumping to form so-called leopard spots. Reduced numbers of ganglion cells have also been described, possibly linked with demyelination of the optic nerve.[9,11] Deposits of a bizarre laminated structure in the subretinal space have been seen at an ultrastructural level.[9,11] The retinal findings in Zellweger's syndrome are unlikely to be specific and invite comparison with Leber's amaurosis, but a suggestion that some forms of the latter rather heterogeneous condition may also be peroxisomal disorders is as yet unsubstantiated.[12]

Conclusion

Ocular involvement in Zellweger's syndrome is usual, the most serious manifestations occurring within the retina, where the changes are probably a part of the overall neurologic disturbance. There is no effective treatment in prospect, but studies involving the measurement of peroxisome-based enzyme activity in cultured amniotic or chorionic villus cells have opened up the possibility of prenatal diagnosis.[13,14]

References

1. Bowen P, Lee CSN, Zellweger H, et al. A familial syndrome of multiple congenital defects. *Bull Johns Hopkins Hosp.* 1964;114:402-414.
2. Kelley RI. Review: the cerebro-hepato-renal syndrome of Zellweger, morphologic and metabolic aspects. *Am J Med Genet.* 1983;16:503-517.
3. Schutgens RBH, Heymans HSA, Wanders RJA, et al. Review: peroxisomal disorders, a newly recognized group of genetic diseases. *Eur J Pediatr.* 1986;144:430-440.
4. Bleeker-Wagenmakers EM, Oorthuys JWE, Wanders RJA, et al. Long-term survival of a patient with the cerebro-hepato-renal (Zellweger) syndrome. *Clin Genet.* 1986; 29:160-164.
5. Goldfischer S, Reddy JK. Peroxisomes (microbodies) in cell pathology. *Int Rev Exp Pathol.* 1984;26:45-84.
6. Heymans HSA, Bosch Hvd, Schutgens RBH, et al. Deficiency of plasmalogens in the cerebro-hepato-renal (Zellweger) syndrome. *Eur J Pediatr.* 1984;142:10-15.
7. Mei Liu H, Bangaru BS, Kidd J, et al. Neuropathological considerations in cerebro-hepato-renal syndrome (Zellweger's syndrome). *Acta Neuropathol (Berlin).* 1976;34:117-123.
8. Volpe JJ, Adams RS. Cerebro-hepato-renal syndrome of Zellweger: an inherited disorder of neuronal migration. *Acta Neuropathol (Berlin).* 1972;20:175-198.
9. Cohen SM, Brown FR, Martyn L, et al. Ocular, histopathologic and biochemical studies of the cerebro-hepato-renal syndrome (Zellweger's syndrome) and its relationship to neonatal adrenoleukodystrophy. *Am J Ophthalmol.* 1983;96:488-501.
10. Hittner HM, Kretzer FL, Mehta RS. Zellweger syndrome: lenticular opacities indicating carrier status and lens abnormalities characteristic of heterozygotes. *Arch Ophthalmol.* 1981;99:1977-1982.
11. Garner A, Fielder AR, Primavesi R, et al. Tapetoretinal degeneration in the cerebro-hepato-renal (Zellweger's) syndrome. *Br J Ophthalmol.* 1982;66:422-431.
12. Ek J, Kase BF, Reith A, et al. Peroxisomal dysfunction in a boy with neurologic symptoms and amaurosis (Leber disease): clinical and biochemical findings similar to those observed in Zellweger syndrome. *J Pediatr.* 1986;108:19-24.
13. Schutgens RBH, Heymans HSA, Wanders RJA, et al. Prenatal detection of Zellweger syndrome. *Lancet.* 1984; 2:1339-1340.
14. Hajra AK, Datta NS, Jackson LG, et al. Prenatal diagnosis of Zellweger cerebro-hepato-renal syndrome. *N Engl J Med.* 1985;312:445-446.

PART 12

Muscular Disorders

Chapter 140

Congenital Myopathies

JONATHAN D. WIRTSCHAFTER

This chapter concerns a heterogeneous group of disorders that are more or less congenital, usually mildly progressive, and are characterized by morphologic abnormalities in muscles that may be epiphenomena associated with a systemic problem.[1,2] Inheritance of these conditions is most frequently autosomal dominant, but some are autosomal recessive. Most of the disorders have a molecular genetic basis that is as yet unexplained but that should relate to the specific messenger RNAs and polypeptides elaborated by specific genes. When activated, these genes control differentiation during embryonic myogenesis and maturation of postembryonic muscle. The genes are members of multigene families, so the proper gene must be activated at the proper time to form, for example, the correct myosin heavy chain, or as another example, to replace the production of a fetal isoprotein of creatine kinase with an adult isoprotein of creatine kinase.

Considering myogenesis at a cellular level, neural crest and mesenchymal tissues give rise to myoblasts, which then divide and differentiate (Table 140-1). Myodifferentiation involves more than the cytoplasmic as-sembly of myofibrils and the development of cross-striations, because each of the cellular organelles also undergoes development and changes in function. During normal myogenesis the nucleus moves from a central to a peripheral position near the plasmalemma. Several congenital centronuclear myopathies are characterized by failure of nuclear margination. Later aspects of myodifferentiation are partially controlled by the arrival of nerves within the muscle. In addition to causing the acetylcholine receptors to concentrate in the region of the neuromuscular junction, the arrival of nerves controls whether each skeletal muscle cell differentiates into a type I, red, slow-contracting (tonic) fiber or a type II, white, fast-contracting (twitch) fiber. Abnormal differentiation is usually characterized by an abnormal predominance of type I fibers and by various cytoplasmic changes with descriptive names.[3] Central core disease refers to a condition of fibers in which central regions are unreactive with oxidative enzyme histochemistry (nicotinamide adenine dinucleotide-tetrazolium reductase, or NADH-TR) and have an unstructured core with disruption of the A, I, and Z bands.

TABLE 140-1 Some Aspects of Myodifferentiation

Development of flattened, pleomorphic mesenchymal cells

Cytoskeletal reorganization to form spindle-shaped
  myoblasts

Cessation of DNA production blocking myocyte at G1 stage
  of mitosis

Cytoplasmic fusion to form multinucleated cells becoming
  myotubules

Nuclear margination

Arrival of nerves at target tissue

Differentiation controlled by neural trophism

Growth controlled by neural trophism

TABLE 140-2 Some Nonmyopathic Causes of
Congenital Hypotonia

Suprasegmental:
  Perinatal hypoxia (type II fiber atrophy)
  Mental retardation syndromes
Spinal muscle atrophies:
  Werdnig-Hoffmann disease
Peripheral neuropathies
Neuromuscular junction disorders:
  Myasthenia (neonatal and congenital types)
  Botulism (infantile type)
Unknown:
  Prader-Willi syndrome
  Benign congenital hypotonia

Nemaline myopathy has been characterized by electron microscopy as having abnormal Z discs and other changes that give the appearance of small rods within the muscle.

The failure of the muscle fibers to grow and maintain their size results in muscle fiber hypotrophy, usually involving the type I fibers. Type I fibers are differentiated histochemically by light staining with ATPase at pH 9.4; type II fibers are characterized by dark staining with the same stain. In most human muscles type II fibers outnumber type I fibers by a ratio of 2:1.

Systemic Manifestations

The floppy infant may have problems that are not related to any myopathy. Some of these are listed in Table 140-2.

Most of the disorders in this chapter have type I fiber predominance as their common feature, and indeed there is a congenital hypotonia in which that is the only feature. The other congenital myopathies and congenital muscular dystrophies share a pattern of clinical presentation. Hypotonia is noted soon after birth and may be associated with dislocation of the hips or muscle contractures. A long, thin face and a high-arched palate are also frequent features. Motor milestones are attained late. In most cases the disorders are nonprogressive or slowly progressive, although a few patients may die of respiratory failure. As the child grows, kyphoscoliosis may develop in addition to deformities of the feet. Proximal muscle weakness may lead to a waddling gait and winging of the scapulas. Although selective proximal muscle weakness is the hallmark of inherited, toxic, and metabolic muscle disease, it is important to remem-

---

TABLE 140-3 Ocular Manifestations of Some Congenital Myopathies

Myopathy	Ptosis/EOM weakness	Face weak	Other
Without CNS Disease:			
Centronuclear (myotubular)	+	+	Cataracts
Multicore	+	±	
Congenital fiber type disproportion	Rare	Rare	
Central core	Rare	+	Malignant hyperthermia
Nemaline	Very rare	Rare	
Reducing body	+	+	
With CNS Disease:			
Fukuyama type	+	+	High myopia, optic atrophy, choroidal atrophy
Walker-Warburg syndrome	+	+	Retinal abnormality, microphthalmia, coloboma

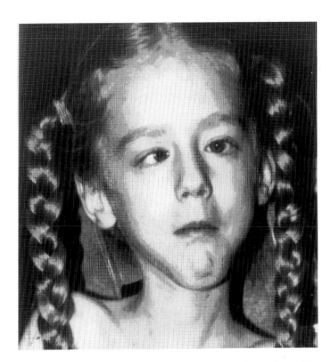

FIGURE 140-1 Congenital muscular dystrophy with central nervous system disease in a girl demonstrating ocular muscle and facial weakness. (From Brooke.[3])

ber that there are a number of neurogenic conditions that can lead to proximal weakness. These include Guillain-Barre syndrome and porphyric, diabetic, toxic, and paraneoplastic neuropathies.

Ocular Manifestations

The ocular manifestations of this varied group of disorders are selectively listed in Table 140-3. One disorder with autosomal recessive inheritance, congenital muscular dystrophy with central nervous system disease (Fukuyama type), combines myopathic features with widespread lucencies in the white matter seen on computed tomography (CT) of the brain. This may be associated with optic and choroidal atrophy. Although the disease is more prevalent in Japan it has been recognized in white Americans (Fig. 140-1) who have oculomotor palsies, ptosis, and other dysmorphic features. The pathogenesis is not known.[3,4]

The Walker-Warburg Syndrome (WWS) is an autosomal recessive disorder with a median survival of 9 months that combines congenital myopathy with congenital malformations of the brain and eye. The cerebral cortex of WWS has diffuse agyria (a smooth surface: Lissencephaly type II) with only scattered areas of polymicrogyria or macrogyria. This is reverse of the configuration of the cerebral surface in Fukuyama congenital muscular dystrophy which also has less severe retinal and cerebellar abnormalities. In WWS there are always retinal abnormalities including abnormal retinal differentiation (100%), persistent hyperplastic primary vitreous, retinal folds or detachment, optic nervehead hypoplasia, atrophy of the inner retinal ganglion cell and nerve fiber layers, poorly formed photoreceptors, and wavy patterns of retinal pigmentation that may give rise to a "leopard spot" appearance in the peripheral retina. The brain abnormalities include a dilated fourth ventricle, Dandy-Walker malformations, hypoplasia of the cerebellar vermis and hypoplasia or absence of the corpus callosum and septum pellucidum. Microphthalmia is present in about half of the cases and the anterior chamber may be absent or demonstrate the Peter anomaly. The cornea may be cloudy. The WWS is now thought to include cases previously described as cerebro-oculo-muscular syndrome.[5]

References

1. Engle AG, Banker BQ. *Myology: Basic and Clinical.* New York, NY: McGraw-Hill; 1986.
2. Schochet SS. *Diagnostic Pathology of Skeletal Muscle and Nerve.* Norwalk, Conn: Appleton-Century-Crofts; 1986.
3. Brooke MH. *A Clinicians View of Neuromuscular Diseases.* 2nd ed. Baltimore, Md: Williams & Wilkins; 1986.
4. Miller NR. *Walsh and Hoyt's Clinical Neuro-Ophthalmology.* 4th ed. vol 2. Baltimore, Md: Williams & Wilkins; 1985:787-794.
5. Dobyns WB, Pagone RA, Armstrond D, et al. Diagnostic criteria for Walker-Warburg Syndrome. *Am J Med Genet.* 1989;32:195-210.

Chapter 141

Mitochondrial Cytopathies

CURTIS E. MARGO, MICHELE R. FILLING-KATZ, and
ZEYNEL A. KARCIOGLU

By definition, mitochondrial disease is due to primary defects in mitochondrial metabolism and encompasses a heterogeneous group of syndromes. The concept of disease caused by mitochondrial dysfunction was proposed in 1962 by Luft et al.,[1] who reported the morphologic, biochemical, and clinical findings in a patient with nonthyroidal hypermetabolism.[1] Defective coupling of oxidation and phosphorylation in muscle mitochondria led to a condition in which the rate of cellular respiration was disproportionate to energy requirements and excess energy was dissipated as heat. Ultrastructural study of skeletal muscle revealed enlarged mitochondria containing crystal-like inclusions.

Since 1962, a variety of clinical conditions associated with morphologically and chemically abnormal mitochondria have been reported.[2] In 1963 Engel and associates demonstrated that abnormal mitochondria of patients with external ophthalmoplegia could be seen by light microscopy using a modified Gomori trichrome stain.[3] Fibers stained red or purple, imparting a ragged-red look to the muscle.

Biochemical studies have been limited by the large amounts of tissue that were required to isolate and purify mitochondria for polargraphic and cytochromatic spectral analysis. Diagnoses, therefore, usually have been arrived at through careful clinical evaluation and supported by adjunctive studies.

Physiology and Biochemistry

Mitochondria, present in all animal cells, provide the major source of energy mediating the aerobic oxidation of pyruvate, fatty acids, and amino acids to form carbon dioxide and water. Intracellular distribution and number of mitochondria vary with the functional characteristics of the cell; however, structure remains relatively uniform.[4] A double membrane surrounds a protein-rich matrix containing enzymes of the tricarboxylic acid, or Krebs, cycle, pools of nucleotides, and various coenzymes. The outer membrane contains fatty acid–activating enzymes and monoamine oxidase. The inner membrane, containing enzymes of respiration and oxidative phosphorylation and adenosine triphosphate (ATP), is more flexible and forms infolds (cristae) that expand and contract. Mitochondria are permeable to water, ions, urea, and short-chain fatty acids. The intracristal space contains channels for access to the inner membrane.

During the oxidation of carbohydrates, fats, and proteins, ATP is synthesized from adenosine diphosphate (ADP) and inorganic phosphate. Water-soluble enzymes of the Krebs cycle generate 14 molecules of ATP and reduced NADH, which is oxidized by the inner membrane–bound, nonaqueous soluble enzymes of the respiratory chain. Electrons are transferred along this chain of enzymes to ultimately reduce molecular oxygen. Conservation of free energy in the form of phosphate bonds is the process known as oxidative phosphorylation.

Genetics

The inheritance of nuclear DNA can be predicted by the theory of mendelian genetics. Mitochondrial DNA (Mt-DNA) replicates autonomously throughout the cell cycle and is inherited in a fashion not predicted by mendelian theory.[4] Mt-DNA is circular, double stranded, and encodes for 13 polypeptides. Nuclear DNA, however, encodes 90% of all mitochondrial protein. The complex interaction between nuclear and Mt-DNA is not clearly understood. Mt-DNA is inherited maternally, which explains why the transmission of some mitochondrial diseases cannot be predicted by mendelian genetics.[5]

Disease Classification

Classification of mitochondrial disorders based solely on clinical or morphologic criteria has proved impractical. There is no totally satisfactory term used to describe disorders of mitochondrial function. The term "mitochondrial cytopathy" has been proposed, because it is less restrictive than "mitochondrial myopathy," although it is also less specifically applicable to individual syndromes.[2]

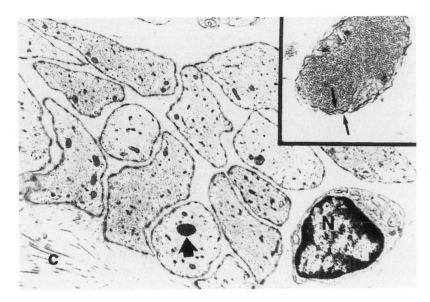

FIGURE 141-1 Skin biopsy shows arrector pili muscle from a child with pigmentary retinopathy and mental retardation who was suspected of having a primary mitochondrial disorder. Smooth muscle fibers (*N*, nucleus; *C*, collagen) contain enlarged mitochondria (*arrow*). Inset, the internal structure of the mitochondria is replaced with an orderly array of electron-dense deposits. The inner and outer mitochondrial membranes (*arrows*) are still visible. (Original magnifications: photo, ×12,500; inset, ×87,000.)

Even though there are syndromes that are highly suggestive of mitochondrial diseases, there is no single clinical phenotype that characterizes the entire group. Patients with different biochemical defects may express a similar disease pattern. Since structurally abnormal mitochondria may occur secondarily in other disorders, the diagnosis of mitochondrial cytopathy should not be based solely on morphologic abnormalities. Biochemical classification theoretically offers many advantages, but no classification has won universal approval.[6]

Diagnosis and Laboratory Evaluation

The diagnosis of a mitochondrial disorder is confirmed by the identification of a specific enzyme defect. Currently, most diagnoses are arrived at by a process of exclusion supported by morphologic and biochemical evidence of mitochondrial dysfunction.

Putative mitochondrial disorders should be systematically evaluated by excluding diseases of glycolipid metabolism, lysosomal storage, perioxisome dysfunction, and glycogen storage, as well as the acquired viral, toxic, and endocrine myopathies and encephalopathies.[2]

Certain substrates serve centrally in the catabolism of glucose and the production of energy via mitochondrially mediated aerobic oxidative phosphorylation. Abnormalities in this system cause predictable disturbances in the balance between certain key substrates. The ratio of NADH to NAD+ reflects the redox state of cytosol, which is determined primarily by the functional integrity of the oxidative phosphorylation system, the Krebs cycle, and the electron transport chain, as well as by mitochondrial transfer of reducing equivalents.

A general approach to putative mitochondrial disorders includes measurement of energy substrates, glucose, lactate, and pyruvate and determination of venous pH in a fasting state. Several strategies can be used to assess mitochondrial function under stress, including substrate depletion induced by fasting or exercise and substrate loading.[7]

Inferences about brain metabolism and mitochondrial function can be made from similar measurements of energy substrates obtained from cerebrospinal fluid.[8]

Pathology

The evaluation of mitochondrial anatomy has been greatly enhanced by electron microscopy.[9] Biopsies often reveal increased numbers and size of mitochondria. Cristae often lose their radial orientation to the outer wall and can form whorls as well as crystalloid inclusions having a variety of patterns (Fig. 141-1). Not all mitochondrial cytopathies, however, demonstrate ultrastructural abnormalities.

Clinical Syndromes

Ocular signs and symptoms are important, early, and striking features of many mitochondrial cytopathies (Table 141-1). Biochemical analyses of presumptive

TABLE 141-1 Disorders of Possible Primary Mitochondrial Origin

Disease	Ophthalmologic findings
Chronic progressive external ophthalmoplegia	Ptosis, ophthalmoplegia
Kearns-Sayre syndrome	Ptosis, ophthalmoplegia, pigmentary retinopathy
Mitochondrial myopathy, encephalopathy, lactic acidosis, and stroke-like episodes (MELAS)	Cortical blindness, hemianopic field loss
Myoclonic epilespy with ragged-red fibers (MERRF)	Optic atrophy
Cerebrohepatorenal syndrome (Zellweger's disease)	Corneal opacities, angle abnormalities, retinal degeneration, cataract, strabismus
Subacute necrotizing encephalomyelopathy (Leigh's disease)	Nystagmus, optic atrophy, strabismus

mitochondrial syndromes yield variable results, even within a specific clinical entity. Chronic progressive external ophthalmoplegia and the Kearns-Sayre syndrome both reveal morphologic abnormalities of mitochondria from skeletal muscle.[10]

Leber's optic atrophy, one of the few well-described disorders with maternal inheritance, has been suspected of being a mitochondrial cytopathy.[5,11] Characterized pathologically by morphologically normal mitochondria and clinically by progressive optic atrophy, Leber's optic atrophy has been associated with a deficiency of at least one mitochondrial enzyme.[11] Ophthalmoscopy during the acute phase may reveal hyperemia and telangiectasia of peripapillary retinal capillaries and swelling of the nerve fiber layer.

Ocular abnormalities are also prominent in the syndrome of myoclonus, encephalopathy, lactic acidosis, and strokelike syndrome (MELAS), probably a mitochondrially based disorder.[2] Visual deficits from infarction along the entire visual pathway frequently lead to unilateral hemianopic deficits and cortical blindness.

Metabolic changes occurring in subacute necrotizing encephalomyelopathy (Leigh's disease) suggest it may represent a defect in mitochondrial oxidation.[7,12] A deficiency in cytochrome *c* oxidase has been demonstrated in at least one case confirmed at autopsy.[7] Although the neuropathologic findings are characteristic of the disease, there is great variability in its clinical

manifestations. Optic atrophy, nystagmus, and strabismus have been reported in some cases.[13,14]

Genetic transmission of myoclonic epilepsy associated with ragged-red fibers (also referred to by the acronym MERRF) appears to be nonmendelian.[2] The presence of ragged-red fibers on muscle biospy and possible maternal inheritance support the possibility of a primary defect in mitochondrial metabolism. Ocular manifestations of the disease are usually overshadowed by other neurologic problems, such as ataxia, dementia, weakness, and myoclonic seizures. Optic atrophy is probably the most common ocular abnormality noted.

Cerebrohepatorenal syndrome (Zellweger's disease) may be due to a defect in the initial portion of the electron transport system.[15] There are a large number of reported ocular problems associated with this syndrome, including corneal opacities, retinal degeneration, cataract, optic atrophy, and abnormalities in the angle.[16]

Therapeutic success in the treatment of mitochondrial cytopathies has been limited. Thiamine may improve lactic acidosis and exercise tolerance. Other treatment strategies are currently limited to isolated case reports.

Until more precise methods are found to confirm the diagnosis of primary mitochondrial disease, diagnoses may be inadvertently classified according to nonspecific findings, such as lactic acidosis.[17]

Studies in patients with the Kearns-Sayre syndrome and Leber's hereditary optic neuropathy have localized the primary metabolic defect to the mitochondrial genome.[18,19]

References

1. Luft R, Ikkos D, Palmieri G, et al. A case of severe hypermetabolism of nonthyroidal origin with a defect in the maintenance of mitochondrial respiratory control: a correlated clinical, biochemical, and morphologic study. *J Clin Invest.* 1962;41:1776-1804.
2. DiMauro S, Bonilla E, Massimo Z, et al. Mitochondrial myopathies. *Ann Neurol.* 1985;17:521-538.
3. Engel WK, Cunningham GG. Rapid examination of muscle tissue: an improved trichrome stain method for fresh biopsy sections. *Neurology (Minneap).* 1963;13:919-923.
4. Green DE. Mitochondria—structure, function and replication. *N Engl J Med.* 1983;309:182-183.
5. Egger J, Wilson J. Mitochondrial inheritance in a mitochondrially mediated disease. *N Engl J Med.* 1983;309:142-146.
6. Morgan-Hughes JA, Hayes DJ, Clark JB, et al. Mitochondrial encephalmyopathies: biochemical studies in two cases revealing defects in respiratory chain. *Brain.* 1982;105:553-583.
7. Willems JL, Monnens LAH, Trijbels JMF, et al. Leigh's encephalomyelopathy in a patient with cytochrome C oxidase deficiency in muscle tissue. *Pediatrics.* 1977;60:850-857.

8. Siesjo B. *Brain Energy Metabolism.* New York, NY: John Wiley & Sons; 1978;607.

9. Kamieniecka Z, Schmalbruch H. Neuromuscular disorder with abnormal muscle mitochondria. *Int Rev Cytol.* 1980;65:321-357.

10. Karpati G, Carpenter S, Larbrisseau A, et al. The Kearns-Sky syndrome. A multisystem disease with mitochondrial abnormality demonstrated in skeletal muscle and skin. *J Neurol Sci.* 1973;19:133-151.

11. Cagianut B, Rhyner K, Furrer W, Schnebli HP. Thiosulfate-sulfur transferase (Rhodanese) in Leber's hereditary optic atrophy. *Lancet.* 1981;2:981-982.

12. Miyabayashi S, Ito T, Narisawa K, et al. Biochemical study in 28 children with lactic acidosis in relation of Leigh's encephalomyelopathy. *Eur J Pediatr.* 1985;143:278-283.

13. Howard RO, Albert DM. Ocular manifestations of subacute necrotizing encephalomyelopathy (Leigh's disease). *Am J Ophthalmol.* 1972;74:386-393.

14. Sedwick LA, Burde RM, Hodges FJ. Leigh's subacute necrotizing encephalomyelopathy manifesting as spasmus nutans. *Arch Ophthalmol.* 1984;102:1046-1048.

15. Goldfishcher S, Moore CL, Johnson AB, et al. Peroxisomal and mitochondrial defects in the cerebro-hepato-renal syndrome. *Science.* 1973;182:62-63.

16. Haddad R, Font RL, Friendly DS. Cerebro-hepato-renal syndome of Zellweger. *Arch Ophthalmol.* 1976;94:1927-1930.

17. Hayasaka S, Yamaguchi K, Mizuno K. Ocular findings in childhood lactic acidosis. *Arch Ophthalmol.* 1986;104:1656-1658.

18. Moraes CT, Salvatore D, Zeviani M, et al. Mitochondrial DNA deletions in progressive external ophthalmoplegia and Kearns-Sayre syndrome. *N Engl J Med.* 1989;320:1293-1299.

19. Singh G, Lott MT, Wallace DC. A mitochondrial DNA mutation as a cause of Leber's hereditary optic neuropathy. *N Engl J Med.* 1989;320:1300-1305.

Chapter 142

Kearns-Sayre Syndrome

JAMES A. GARRITY and THOMAS P. KEARNS

The Kearns-Sayre syndrome is a constellation of signs, the most prominent being a triad of external ophthalmoplegia, pigmentary retinopathy, and heart block.[1-3] Chronic progressive external ophthalmoplegia, first described by von Gräfe[4] in 1868, is a bilateral insidiously progressive inability to move the eyes, along with an associated ptosis. The disorder was incorrectly assumed to be of nuclear origin until 1951, when muscle biopsy showed that it was a myopathy.[5] Current evidence suggests that the syndrome should be classified as a mitochondrial cytopathy.[3,6]

Ocular Manifestations

The onset of clinical symptoms in the Kearns-Sayre syndrome usually occurs before the age of 20 years, and ptosis is most often the first manifestation. The ptosis is characteristically bilateral (Fig. 142-1, A), although asymmetric onset has been reported, and tends to become increasingly severe with time. The patient develops a compensatory head tilt. The ptosis often becomes complete and a major handicap.

The onset of limitation of ocular movements is usually symmetric and insidious to the point where the patient usually does not complain of diplopia. It is the ptosis that is usually the major ocular problem. Ocular rotation may be well preserved in downward gaze until late in the course of the disorder.

Although the pigmentary retinopathy is similar to true retinitis pigmentosa, there are important differences, and it is often referred to as "atypical retinitis pigmentosa." While retinitis pigmentosa tends to involve the periphery, the posterior pole of the eye is generally affected in the Kearns-Sayre syndrome. Severe peripapillary involvement is typical and has been called "choroidal sclerosis" because of the increased visibility of choroidal vessels owing to overlying retinal pigment epithelial atrophy. Formation of bone spicule–shaped pigment clumps is common with retinitis pigmentosa but does not usually occur in the Kearns-Sayre syndrome. The salt-and-pepper appearance of the posterior pole of the eye in the Kearns-Sayre syndrome is somewhat suggestive of the fundus seen in persons with congenital rubella.

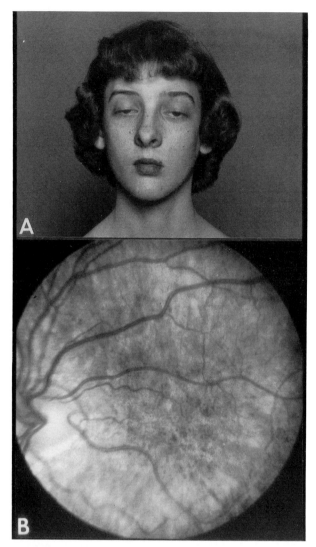

FIGURE 142-1 **(A)** Typical appearance of patient with Kearns-Sayre syndrome, a 15-year-old girl who experienced initial onset of asymmetric ptosis at the age of 13 years. Note the ptosis with posterior head tilt. **(B)** Left eye of same patient. Retinal pigment epithelium had a granular appearance that was most severe in the macula. Note atrophy of peripapillary retinal pigment epithelium and normal vascular caliber. (From Kearns.[2] By permission of the American Ophthalmological Society.)

Patients with Kearns-Sayre syndrome do not develop the posterior subcapsular cataracts, disc pallor, or the retinal vascular attenuation seen in patients with retinitis pigmentosa.

Visual symptoms, including decreased acuity and nightblindness, tend to be mild and are certainly less severe than in true retinitis pigmentosa. The clinical course of the visual involvement is generally benign. The patient is often unaware of any vision problem other than that associated with ptosis and decreased ocular motility. Vision problems related to the ptosis are best handled by a surgical approach designed for visual, not cosmetic, correction. Weak orbicularis oculi muscles and a poor Bell's reflex predispose the patient to postoperative exposure keratitis.

Systemic Manifestations

The last feature of the triad, and the principal reason for establishing the diagnosis of the Kearns-Sayre syndrome, is cardiac involvement. Patients develop cardiomyopathy, although the exact frequency and time of onset are unknown. Most reports note that the heart block has its onset approximately 10 years after the ptosis appears. The conduction system is classically involved, although idiopathic hypertrophic subaortic stenosis and mitral valve prolapse have also been reported.[7] Various degrees of heart block may be present, even a complete atrioventricular conduction defect. Pacemakers for treating symptomatic cardiac disease and prophylactic cardiac pacing are essential to prevent death from the heart block.

Other associated systemic features of the Kearns-Sayre syndrome include short stature, somatic muscle weakness, cerebellar ataxia, pendular nystagmus, vestibular dysfunction, sensorineural hearing loss, impaired mental function, elevated cerebrospinal fluid protein level, spongiform degeneration of the cerebral cortex or brain stem, basal ganglia calcification, and hypoparathyroidism.[3]

Pathophysiologic Mechanisms

Histopathologically, Kearns and Sayre described widespread loss of the retinal pigment epithelium and the photoreceptor layer. A more recent histopathologic evaluation showed atrophy of the retinal pigment epithelium, most severe in the posterior pole (Fig. 142-1, B).[8] The preservation of peripheral rods and cones seemed to mirror the health of the underlying peripheral retinal pigment epithelium. A primary defect resulting in the retinopathy may reside in the retinal pigment epithelium. Another study confirmed the presence of abnormal mitochondria in the retinal pigment epithelium.[6,7] Macrophages have also been noted in the subretinal space.[6] This implies impaired phagocytosis by the retinal epithelial cells, which may form the basis of a secondary inflammatory response that could also damage the outer retina.

Histologically, ragged-red fibers are seen in the extraocular muscles and in biopsy specimens of skeletal muscle. These fibers represent accumulations of mitochondria beneath the plasma membrane and between myofibrils. Many of the mitochondria are abnormal and are characterized by increased numbers, increased volume, deformed shape, and the presence of inclusions. Decreased mitochondrial respiration has been associated with structural mitochondrial abnormalities. The finding of increased blood lactate levels at rest or during exercise and of impaired lactate tolerance is suggestive of a mitochondrial defect in pyruvate metabolism. In one patient, the administration of coenzyme Q10 (serum levels of this were low while serum lactate and pyruvate levels were high) normalized the levels of coenzyme Q10, lactate, and pyruvate.[9]

Pharmacologic intervention with prednisone in an attempt to increase muscle strength led to hyperglycemia and to the deaths of two patients from metabolic acidosis.[10] The authors cautioned against treating patients with the Kearns-Sayre syndrome with steroids.

Cardiac involvement is probably on neurogenic and myogenic bases, because enlarged, abnormal mitochondria have been found in both the myocardial cells and the bundle of His conduction system.[11] The cardiac conduction defect may improve with metabolic therapy.[9]

The genetics of the mitochondrial cytopathies are not clear. Previously the syndrome was believed to arise on a sporadic basis; however, autosomal-dominant transmission has been documented.[8] A careful study of all family members is recommended to search for subtle evidence of involvement. Another interesting possibility relates to mitochondrial transmission. Mitochondrial DNA is believed to be maternally transmitted.[3] In this instance, nonmendelian inheritance would be expected.

Summary

The Kearns-Sayre syndrome is a rare syndrome consisting of external ophthalmoplegia, pigmentary degeneration of the retina, and cardiomyopathy. When first described in 1958, it was merely an academic curiosity, as there was no effective treatment for the heart block. Now that cardiac pacing is available, the recognition of this syndrome is more important because the associated potentially fatal heart block can be treated. Patients with this constellation of symptoms should be seen by a cardiologist. When indicated, the implantation of a pacemaker can be lifesaving.

References

1. Kearns TP, Sayre GP. Retinitis pigmentosa, external ophthalmoplegia, and complete heart block: unusual syndrome with histologic study in one of two cases. *Arch Ophthalmol.* 1958;60:280.
2. Kearns TP. External ophthalmoplegia, pigmentary degeneration of the retina, and cardiomyopathy: a newly recognized syndrome. *Trans Am Ophthalmol Soc.* 1965;63:559.
3. Miller NR. *Walsh and Hoyt's Clinical Neuro-Ophthalmology.* 4th ed. vol. 2. Baltimore, Md: Williams & Wilkins; 1985;811-823.
4. von Gräfe A. Verhandlungen ärztlicher Gesellschaften. *Berl Klin Wochenschr.* 1868;5:125.
5. Kiloh LG, Nevin S. Progressive dystrophy of the external ocular muscles (ocular myopathy). *Brain.* 1951;74:115.
6. McKechnie NM, King M, Lee WR. Retinal pathology in the Kearns-Sayre syndrome. *Br J Ophthalmol.* 1985;69:63.
7. Leveille AS, Newell FW. Autosomal dominant Kearns-Sayre syndrome. *Ophthalmology.* 1980;87:99.
8. Eagle RC Jr, Hedges TR, Yanoff M. The atypical pigmentary retinopathy of Kearns-Sayre syndrome. A light and electron microscopic study. *Ophthalmology.* 1982;89:1433.
9. Ogasahara S, Yorifuji S, Nishikawa Y, et al. Improvement of abnormal pyruvate metabolism and cardiac conduction defect with coenzyme Q10 in Kearns-Sayre syndrome. *Neurology.* 1985;35:372.
10. Flynn JT, Bachynski BN, Rodrigues MM, et al. Hyperglycemic acidotic coma and death in Kearns-Sayre syndrome. *Trans Am Ophthalmol Soc.* 1985;83:131.
11. Stefani FH. The Kearns-Sayre syndrome. Presented at the Joint Meeting of the Verhoeff Society and the European Ophthalmic Pathology Society; April 20–24, 1986; Philadelphia, Pa.

Chapter 143

Muscular Dystrophies and Ocular Myopathies

JONATHAN D. WIRTSCHAFTER

The term "muscular dystrophy" implies that genetically determined postnatal changes weaken normally developed muscles. Thus muscular dystrophy is contrasted with congenital myopathy and with muscle atrophy following disuse or denervation. These clinical distinctions may prove less important because of rapid advances in molecular genetics. Duchenne's (pseudohypertrophic) muscular dystrophy (DMD) is the most common, most severe, and best understood of these disorders, and has been associated with abnormalities in the Xp21 chromosome leading to the deficient production of a messenger RNA and its protein product, dystrophin. The exact function of dystrophin is uncertain, but it is probably a cytoskeletal protein underlying the sarcolemma. The rapid advances in the molecular biology of DMD including the cloning of DMD cDNA should serve to increase our understanding of other muscular dystrophies. For example, it will be necessary to determine if the absent protein affects membrane channels or mechanical strength of the membrane.

This X-linked recessive disorder affects about 1 out of 5000 boys with progressive weakness after age 1 year. The average patient is confined to a wheelchair by age 12, and many die of cardiorespiratory problems.[1,2] The frequency of the disease in the population suggests that one third of the cases arise from new mutations that involve the short arm of the X chromosome.[3] The mutation rate for DMD is among the highest observed for any clinical disorder. Chromosomal deletions occur in 50% of affected males. Morphologic studies of chromosomes have shown translocations involving this region. Recent studies using restriction fragment length polymorphisms (RPLFs) have zoomed in on the DMD locus at Xp21. The sum of the sizes of the restriction fragments detected by these probes is greater than 4000 kilobases (kb). In 1987 the gene for another muscular dystrophy, myotonic muscular dystrophy, was localized on chromosome 19 with RPLFs using a genomic clone LDR152.[4] The goal of current research is to increase diagnostic accuracy (presently 95% for DMD) by decreasing the distance (measured in kilobases) between the gene and the region detected by the probe. Current studies using pulse gel electrophoresis have shown that other translocations and deletions in the (DMD) region result in glycerol kinase deficiency and adrenal hypoplasia.[3]

The genetic abnormalities result in the production of abnormal and short dystrophin polypeptides that contribute to the ultimate breakdown of the muscle in DMD. A specific defect in the surface membrane of DMD patients has also been identified. In addition to the reduced amounts of normal dystrophin, there is no shortage of other phenotypic biochemical abnormalities that have been detected both in patients and in obligate carriers of DMD. These abnormalities, which may reflect the breakdown of muscle, include elevation of the serum myoglobin level. The most frequently measured abnormality is an elevation of the serum creatine kinase (CK). Unfortunately it is not sufficiently elevated to allow intrauterine diagnosis, although a normal CK titer during the first year of life is assumed to preclude the development of DMD. Elevation of CK has been used to detect carriers, but the test is best used during the first 10 years of life in sisters of affected males. The Lyon hypothesis holds that half of an obligate carrier female's cells contain an abnormal X, but there may be a lifelong decrease in the proportion of active muscle cells that contain the abnormal X and that may explain the normal CK level in older obligate carrier females. Only about half of obligate carriers can be detected by testing for CK.[2] Thus RPLF gene probes promise practical detection of the carrier state and may provide the only practical means for intrauterine diagnosis for the one third of cases that result from new mutations. More recently an immunohistochemical stain raised against a synthetic peptide fraction of dystrophin was demonstrated to stain the muscles of obligate female carriers in the mosaic pattern predicted by the Lyon hypothesis.

The pathogenic connections between the abnormal gene product and the histopathologic changes may soon be explained, so only the morphology is described here. The striking features of the muscle are fibrosis, necrosis, and phagocytosis. The outline of the fibers undergoes a metamorphosis from polygonal to circular. Frequently some basophilic fibers are noted.[5] The heart is also involved, and most patients have abnormal electrocardiograms. The final stages of the disease are characterized by extreme weakness and ventilatory failure. While DMD is presently incurable, treatment is directed toward respiratory therapy, physical medicine, and surgical relief of contractures.[2]

Systemic Muscular Dystrophies

DMD has no important ocular findings, nor does a milder, possibly allelic, disorder, Becker's muscular dystrophy.[6] Facioscapulohumeral dystrophy may be complicated by impaired eyelid closure during sleep, but the extraocular muscles are not clinically abnormal. Inheritance is autosomal dominant. Brooke[2] cites three cases of Coats' syndrome (exudative telangiectasia of the retina) concurrent with facioscapulohumeral dystrophy. Both conditions are associated with sensorineural hearing loss, and failure to recognize Coats' syndrome, and failure to treat it with retinal photocoagulation could leave the child with touch as the major remaining sensory modality. Ophthalmoplegia has been reported in a case of congenital muscular dystrophy with central nervous system involvement.[7] (See Chapter 140.)

Ocular Muscle Dystrophies (Ocular Myopathies)

Oculopharyngeal muscular dystrophy and ocular muscle dystrophy (ocular myopathy of Kiloh and Nevin) are two disorders whose clinical presentation is limited to the extraocular and pharyngeal muscles.[1,2,5] The former illness always involves the pharyngeal muscles, but most frequently the ocular muscles are involved first or simultaneously with the pharyngeal muscles. Patients with the latter condition may complain of dysphagia, but it is not severe. Although most patients survive oculopharyngeal muscular dystrophy, weakness of the proximal limb muscles may occur in the final, emaciating phases.

Both conditions have autosomal-dominant inheritance, except in a few families with ocular muscle dystrophy of autosomal recessive inheritance. There are two geographic clusters of oculopharyngeal muscular dystrophy in the United States; one is along the Canadian border, where there are descendants of a Frenchman who came to Quebec in 1634, and the other is a cluster of Hispanic families living near the Colorado–New Mexico border and in Arizona.

The onset of ocular muscle dystrophy usually occurs within the first 3 decades of life, whereas oculopharyngeal muscular dystrophy generally begins later. Diagnosis is often difficult in the early stages, particularly if the patient presents with asymmetric ptosis. Bilateral ptosis may cause the patient to tilt the head backward. Limitation of extraocular movements may be detected on examination, but diplopia is a rare complaint. More frequently the patient complains because the eyes cannot be lowered to use the reading portion of bifocal lenses. The presence of facial weakness, dysphagia, and proximal muscle weakness should cause the examiner to consider a myopathy. Laryngoscopic examination of the oropharynx and cineradiographic studies may show hypotonia and incoordination of swallowing. Cricopharyngeal myotomy is frequently of value.[9]

Biopsy of oculopharyngeal muscular dystrophy reveals muscle fibers with autophagic rimmed vacuoles and intranuclear aggregates of tubular filamentous inclusions. These inclusions are present in only 3 to 5% of the muscle nuclei.[1,5] Metallic mitochondrial inclusions have been reported in two cases.[9] Both the intranuclear inclusions and the mitochondrial inclusions may be specific for this condition. There are also nonspecific changes, such as inflammatory cells. Muscle enzymes,

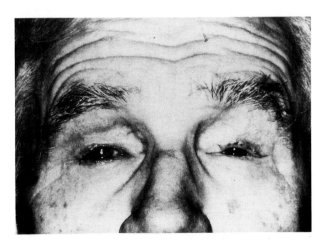

FIGURE 143-1 This 68-year-old man with oculopharyngeal muscular dystrophy demonstrates bilateral ptosis and left corneal scarring 5 years after a left ptosis procedure. A ptosis crutch is used on his right lid, and both eyes require frequent lubrication.

such as CK, may be normal or only elevated to four times normal. Results of the electrocardiographic examination are also normal.

Ophthalmic therapy consists of lid crutches or minimal ptosis surgery. It is not uncommon for such patients to experience corneal problems, including ulceration and perforation, when ptosis surgery is performed in an eye without the protection of Bell's phenomenon (Fig. 143-1).

Other myopathic conditions are considered elsewhere in this volume (see Congenital Myopathies, Kearns-Sayre Syndrome, Myotonic Dystrophy, Metabolic, Nutritional, and Endocrine Myopathies). Myasthenia gravis can usually be ruled out if the Tensilon test and acetylcholine receptor–blocking antibody tests are negative and if the muscle weakness is not intermittent. Diagnosis of the various disorders of the oculomotor nuclei and nerves is beyond the scope of this chapter. Progressive supranuclear ophthalmoplegia usually first involves the vertically acting extraocular muscles and may show disparity between full eye movements evoked by the doll's-head maneuver and limited eye movements evoked by voluntary gaze.

References

1. Engle AG, Banker BQ. *Myology: Basic and Clinical.* New York, NY: McGraw-Hill; 1986.
2. Brooke MH. *A Clinician's View of Neuromuscular Diseases.* 2nd ed. Baltimore, Md: Williams & Wilkins; 1986.
3. Sicinski P, Geng Y, Ryder-Cook AS, et al. The molecular basis of muscular dystrophy in the *mdx* mouse: a point mutation. *Science.* 1989;244:1578-1579.
4. Bartlett RJ, Pericak-Vance MA, Yamaoka L, et al. A new probe for the diagnosis of myotonic muscular dystrophy. *Science.* 1987;235:1648-1650.
5. Schochet SS. *Diagnostic Pathology of Skeletal Muscle and Nerve.* Norwalk, Conn: Appleton-Century-Crofts; 1986.
6. Wilichowski E, Krawczak M, Seemanova E, et al. Genetic linkage study between the loci for Duchenne and Becker muscular dystrophy and nine X-chromosomal DNA markers. *Hum Genet.* 1987;75:32-40.
7. Martinelli P, Gabellini AS, Ciucci G, et al. Congenital muscular dystrophy with central nervous system involvement. *Eur Neurol.* 1987;26:17-22.
8. Miller NR. *Walsh and Hoyt's Clinical Neuro-Ophthalmology.* 4th ed. vol 2. Baltimore, Md: Williams & Wilkins; 1985;807-811.
9. Pratt MF, Myers PK. Oculopharyngeal muscular dystrophy: recent ultrastructural evidence for mitochondrial abnormalities. *Laryngoscope.* 1986;96:368-373.

Chapter 144

Metabolic, Nutritional, and Endocrine Myopathies

JONATHAN D. WIRTSCHAFTER

Metabolic Myopathies

Skeletal muscle is unique in a metabolic sense because it must accommodate extraordinary variation between rest and exercise in the demand for adenosine triphosphate (ATP). Thus many muscle disorders are caused by enzyme and mitochondrial defects revealed by exercise; have symptoms of cramps or weakness; and may lead to muscle breakdown, myoglobinemia, and renal failure. The enzyme abnormalities can affect carbohydrate and lipid metabolism. The mitochondrial abnormalities are often characterized by ragged-red fibers seen on the modified Gomori trichrome stain and can lead to a muscle disorder with ptosis in conditions such as NADH-CoQ reductase deficiency. The mitochondrial abnormalities can also be part of a multisystem disorder, such as oculocraniosomatic neuromuscular disease (Kearnes-Sayre syndrome; see Chap. 142). Ocular muscles may harbor glycogen vacuoles in glycogenesis type II (Pompe's disease), but clinical weakness has not been documented.[1,2]

Malignant hyperpyrexia (malignant hyperthermia) is also a metabolic myopathy, with exhuberant coupling of muscle excitation and muscle contraction involving the regulation by adenyl cyclase of the transport of calcium and the sarcolemmal membranes. This potentially fatal

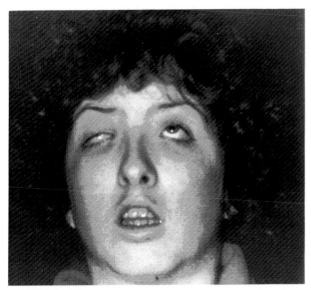

FIGURE 144-1 A patient with King-Denborough syndrome shows reverse slant of the eyelid fissure and ptosis. (From Brooke.[1])

complication of general anesthesia is mentioned here because it can complicate other myopathies or present as a feature of the King-Denborough syndrome, with weakness, short stature, laterally downward-slanting palpebral fissures, ptosis (Fig. 144-1), and other myopathic features.[3]

Periodic Paralyses

Potassium is involved in many aspects of muscle contraction, including the generation of the muscle action potential at the muscle cell membrane and the coupling of excitation and contraction involving structures such as the sarcoplasmic reticulum and the transverse tubules. The periodic paralyses constitute a diverse group of conditions often characterized by abnormalities in the extracellular concentration of potassium; however, the pathophysiology of the abnormal fluxes of potassium and other electrolytes are not yet well-explained for any of the variants, hyperkalemic, hypokalemic, or normokalemic. Even these distinctions are simplistic.[1] The attacks usually present before age 20 years. Provocative factors include rest after exercise, oversleeping, cold exposure, and dietary factors such as high carbohydrate intake, fasting, alcohol intoxication, and potassium supplementation. There are also secondary forms of hyper- and hypokalemic periodic paralyses. Licorice contains a potent mineralocorticoid that causes potassium de-

pletion, so chronic ingestion may produce a condition that resembles hyperaldosteronism. Ophthalmologists may be interested to know that acetazolamide is frequently used in the treatment of the hypokalemic form of the disease. The periodic paralyses differ from many neuromuscular disorders by the absence of autonomic nervous system dysfunction and relative lack of cranial nerve dysfunction. Table 144-1 presents the ocular features of these disorders.[1,4] Cardiology and myopathology are beyond the scope of this chapter.

Nutritional Myopathies

Malabsorption with resultant myopathy and neuropathy underlies the two treatable conditions presented here. Celiac disease (nontropical sprue) has produced oculomotor disturbances in one child whose symptoms disappeared when he ate a gluten-free diet and took vitamin therapy.[4] Impaired absorption of lipids and lipid-soluble vitamins, specifically vitamin E, now explains what was formerly described as the Bassen-Kornzweig syndrome. The failure to absorb lipids results in an absence of β-lipoproteins in the serum, and this in turn contributes to deficient development of cell membranes. The disorder now has several designations: vitamin E deficiency syndrome, chronic cholestasis, and vitamin E deficiency with neuromuscular dysfunction (encephalomyopathy).[5] Findings include acanthosis, pigmentary retinopathy, and neurologic abnormalities. Oculomotor abnormalities include nystagmus, pseudointernuclear ophthalmoplegia, abnormal saccades, and a peculiar alternating ptosis with lowering of the lid of the abducting eye. While most of these signs are neuropathic, myopathic changes have also been reported.

Endocrine Myopathies

Apart from Graves' disease and its variants such as triiodothyronine-toxicosis,[6] thyroid abnormalities cause few problems for the ocular muscles. Thyrotoxicosis may be complicated by hypokalemic periodic paralysis, whereas dysthyroidism of both the hyperthyroid and hypothyroid varieties compounds the weakness of myasthenia gravis.[2] No causal relationship between the thyroid abnormalities and myasthenia has been established, but immune-mediated and genetic factors may be involved.[4] Dysinsertion of the levator aponeurosis from the tarsal plate may explain ptosis following the congestive phase of Graves' ophthalmopathy.[4]

Cushing's syndrome may mimic early or mild

TABLE 144-1 Ocular Manifestations of Some Periodic Paralyses

Type	Inheritance or Cause	Eyelid Signs	EOM	Face	Other
Hypokalemic	Autosomal dominant (male penetrance)	Ptosis ++ Myotonia (MT)	Weak +	Weak +++	Ascending weakness
Secondary hypokalemic	Oriental ethnicity with thyroxicosis; K+ loss; licorice toxicity	Weak +	Weak +	Weak +	Potassium depletion by urinary or gastro-intestinal tracts
Hyperkalemic	Autosomal dominant	Stare, lid lag, retraction (induced by ice on lids), myotonia	Gaze MT		Myotonia (MT) may precede weakness
Secondary hyperkalemic	Renal or adrenal failure, diuretics with K+ retention	Weak +	Weak +	Myo-kymia	Cardiac problems worse than weakness
Normokalemic	Autosomal dominant				Limb muscles only

Graves' disease and is characterized by the accumulation of lipid droplets in type 1 muscle fibers. The mechanism for the accumulation is only partially understood. Cortisol releases free fatty acids from adipose tissue. Type 2 fiber atrophy of skeletal muscles may follow the administration of corticosteroids, especially the halogenated drugs.[1,2] The resultant weakness may be devastating to patients placed on high-dose steroid therapy for ocular conditions such as giant cell arteritis. Conversely, systemic and even local administration of corticosteroids may result in ptosis and other signs of weakness of the ocular muscles.

Tetany is the only important muscle complication of hypoparathyroidism, so the ophthalmologist might encounter a patient with bulbar symptoms and cataracts.

References

1. Brooke MH. *A Clinicians View of Neuromuscular Diseases.* 2nd ed. Baltimore, Md: Williams & Wilkins; 1986.
2. Engle AG, Banker BQ. *Myology: Basic and Clinical.* New York, NY: McGraw-Hill, 1986.
3. Mcpherson EW, Taylor CA. The King syndrome: malignant hyperthermia myopathy and multiple anomalies. *Am J Med Genet.* 1981;8:159-165.
4. Miller NR. *Walsh and Hoyt's Clinical Neuro-Ophthalmology.* 4th ed. vol 2. Baltimore, Md: Williams & Wilkins; 1985; 803-807, 862, 943.
5. Muller DPR, Lloyd JK, Wolff OH. Vitamin E and neurological function. *Lancet.* 1983;1:225-228.
6. Hornblass A, Yagoda A. Thyroid ophthalmopathy related to T3-toxicosis: a variant of classical Graves' disease. *Ann Ophthalmol.* 1980;12:49-52.

Chapter 145

Myasthenia Gravis

ROBERT C. SERGOTT

Myasthenia gravis (MG) is an autoimmune disease of the neuromuscular junction characterized by easy fatigability and weakness of the skeletal muscles and the muscles innervated by the cranial nerves. Except when the weakness has become profound, recovery of strength and muscle function after rest is the hallmark of both generalized MG and the ocular myasthenia syndrome that is limited to the levator palpebrae, orbicularis oculi, and extraocular muscles. MG has a 2:1 female preponderance. Persons of all ages may be affected, but the incidence is highest in the 3rd decade of life.

Pathophysiology

The weakness and fatigue produced by MG result from impaired neuromuscular transmission because of a reduced number of acetylcholine receptors (AChR) at the neuromuscular junction.[1] MG is the neurologic disease that has been demonstrated most conclusively to have an immune basis. Abnormalities of the thymus gland in 75% of MG patients and the common association of other diseases with a probable immunopathogenesis, such as Graves' disease, Hashimoto's thyroiditis, rheumatoid arthritis, pernicious anemia, and systemic lupus erythematosus, provided the first indications that MG also had an immune basis.

The possible role of antibodies (anti–AChR) directed against the AChR in the induction of a myasthenic syndrome evolved from the production of an MG–like state in animals immunized with AChR from the electric organ of eels.[2] The occurrence of transient neonatal MG in some infants of mothers with MG, presumably by transplacental passage of antibody, and the transfer of MG by injection of serum from patients into animals, further incriminated antibodies to AChR in the pathogenesis of the disorder. Eventually IgG was found to be the active fraction in the transfer experiments, and anti–AChR antibodies have been detected in the sera of most MG patients, to the point that radioimmunoassay for such antibodies is now an accepted diagnostic procedure.[3]

Three different mechanisms for antibody-mediated damage and the eventual reduction of AChR have been postulated: (1) blockage of the receptors' active sites; (2) destruction of the AChRs in a complement-dependent reaction; and (3) enhancement of the degradation rate of AChRs by cross-linking of the receptors with antibodies.[4] The relative contribution of each of these mechanisms in individual patients is not known. What factors initiate the aberrant immune response remains uncertain. Electron microscopic analysis has revealed that the myasthenic neuromuscular junction not only has fewer AChR receptors but that these structures have flattened postsynaptic folds and that the distance from presynaptic to postsynpatic membrane is increased.[5]

Systemic Manifestations

Patients with MG may have diffuse or exceedingly focal weakness of facial, extraocular, pharyngeal, respiratory, and extremity muscles. The pattern and degree of weakness vary greatly from patient to patient, but the weakness, regardless of its location and extent, recovers after rest or the administration of anticholinesterase medication. Patients exhibit a myopathic, immobile facies with ptotic upper eyelids, producing elevated eyebrows, flattened nasolabial folds, a horizontal smile, and contracted frontalis musculature.

Involvement of the tongue, laryngeal musculature, and face produces a nasal voice and a tendency to aspirate food and saliva. Impaired respiration constitutes the primary threat to life in MG. Although it was once a disease with a 30% mortality rate, MG is now very controllable with a combination of anticholinesterase medication to enhance neuromuscular transmission; corticosteroids, cytotoxic agents and thymectomy to induce immunosuppression; and plasmapheresis to reduce circulating antibody levels.[6]

Ocular Manifestations

Involvement of the eyelids and extraocular muscles occurs in over 90% of patients with generalized MG. In contrast, impaired neuromuscular transmission may affect only the eyes, and other muscles may never become weak. The pupil is never involved clinically in ocular or systemic MG.

FIGURE 145-1 This patient with ptosis and limited extraocular movement in all directions of gaze. She experienced double vision before the ptosis became complete.

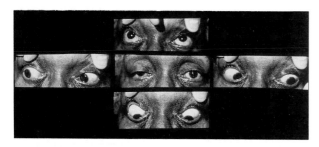

FIGURE 145-2 Both ptosis and ophthalmoplegia improved dramatically after intravenous Tensilon therapy.

Ptosis of both upper lids in combination with double vision is the classic constellation of neuro-ophthalmic symptoms. The ptosis invariably worsens with prolonged upward gaze, and this "fatigable" ptosis is usually associated with weakness of the orbicularis oculi as demonstrated by impaired forced lid closure. Isolated ptosis without diplopia is only rarely neuromuscular in origin, unless bilateral ophthalmoplegia is present. The ptosis and diplopia are more symptomatic in the evenings than in the morning hours. The double vision is always binocular and may be horizontal, vertical, or oblique. Double vision that can be localized to a single cranial nerve or to a single neuro-ophthalmic site must raise the index of suspicion that neuromuscular junction disease is not responsible for the patient's symptoms. However, myasthenia is legendary for producing "pseudointernuclear ophthalmoplegia" and "pseudo–

gaze palsies." A meticulous history usually reveals the variable, fatigable character of the double vision or the association of fluctuating ptosis, thereby suggesting myasthenia in the differential diagnosis.

The diagnosis of ocular MG is established by observing improvement of the ptosis and ophthalmoplegia after the administration of anticholinesterase medications such as intravenous edrophonium chloride (Tensilon; Fig. 145-1) and intramuscular neostigmine. While these anticholinesterase medications effectively diagnose ocular MG, usually neither these agents nor other therapeutic options for systemic MG completely resolve the ptosis and double vision. Occluding one eye with a patch is often the most effective treatment. Prism lenses are seldom effective, and most importantly, extraocular muscle and eyelid surgery are contraindicated because of the unpredictable, fluctuating course of the diplopia and ptosis.

Why certain patients present with and continue to have only purely isolated ocular MG is unknown. In a retrospective study of 108 patients who presented with ocular MG, Bever et al.[7] found that in 40% the disease remained ocular, 11% enjoyed spontaneous remission, and in 40% it progressed to generalized disease.[7] Patients who developed generalized MG usually did so within 2 years of the onset of the neuro-ophthalmic symptoms.

References

1. Fambrough DM, Drachman DB, Satyamurti S. Neuromuscular junction in myasthenia gravis: decreased acetylcholine receptors. *Science.* 1973;182:293-295.
2. Patrick J, Lindstrom J. Autoimmune response to acetylcholine receptor. *Science.* 1973;180:871-872.
3. Toyka KV, Drachman DB, Griffen DE, et al: Myasthenia gravis: study of humoral immune mechanisms by passive transfer to mice. *N Engl J Med.* 1987;316:743-745.
4. Drachman DB. Present and future treatment of myasthenia gravis. *N Engl J Med.* 1987;316:743-745.
5. Engel AG, Santa T. Histometric analysis of the ultrastructure of the neuromuscular junction in myasthenia gravis and in the myasthenic syndrome. *Ann NY Acad Sci.* 1981;183:46-63.
6. Grob D, Brunner NG, Namba T. The natural course of myasthenia gravis and effect of therapeutic measures. *Ann NY Acad Sci* 1981;377:652-669.
7. Bever CT, Aqurino AV, Penn HS, et al. Prognosis of ocular myasthenia. *Ann Neurol.* 1983;14:576-579.

Chapter 146

Myotonic Dystrophy

DON B. SMITH and CURTIS E. MARGO

Myotonic dystrophy (MD) is an inherited multisystem disease whose signs may include muscle weakness, frontal baldness, testicular atrophy, and a distinctive type of lenticular degeneration. Myotonia, the phenomenon of delayed muscle relaxation, is the clinical hallmark of the disorder. Once the muscle membrane is activated it tends to discharge repeatedly, giving rise to a characteristic "dive-bomber" pattern of waxing and waning activity on the audio component of electromyography (EMG).[1] Myotonia is not unique to MD. It is noted in several rare diseases that involve defects in the muscle membrane or metabolic environment—myotonia congenita, paramyotonia congenita, and hyperkalemic periodic paralysis.[2]

MD is an autosomal dominant disorder with variable phenotypic expression. It may go unnoticed by some patients, whereas for others it can be debilitating. Symptoms of weakness often begin in the first 2 decades of life and may be initially preceived as stiffness. A classic complaint is difficulty in relaxing one's grip. Because symptoms can be subtle and are not perceived as problems by patients, diagnosis may be delayed.

Signs of myotonia may be found in slowness of grip relaxation or in a myotonic response to percussion of affected muscles. In symptomatic adults characteristic EMG findings are invariably present, but in infants and young children both the EMG findings and clinical myotonia may be absent. When the illness begins in infancy, transmission is nearly always from the mother, and the course is more severe. As yet unknown intrauterine factors from an affected mother seem to influence disease activity in the newborn.

The pathogenesis of MD is unknown, but there are clues that it may involve abnormalities of an intrinsic membrane protein, perhaps one related to an ion channel or adenosine triphosphatase. The gene for MD has been localized to the short arm of chromosome 19. It is anticipated that its exact location and its base sequence will be known in the near future.[3] Apparent spontaneous mutations have been reported, but because of variability of disease expression the incidence of these events may be overestimated unless one carefully evaluates asymptomatic relatives.

Specific treatment awaits fundamental advances in understanding the pathophysiology. No current treatment is known to prevent progression of the weakness and muscle wasting. Symptomatic treatment of the myotonia may sometimes be accomplished by the use of phenytoin, quinine, quinidine, acetazolamide, or procainamide.[4] Imipramine has been reported to be useful not only for associated depression but also for some of the somatic aspects of the illness.[5]

Systemic Manifestations

MD involves excessive muscle contraction, but the affected muscles are weak. This apparent paradox is explained by the fact that affected muscles have reduced numbers of functioning motor units. Minor peripheral nerve abnormalities such as mildly slowed motor nerve conduction velocity—and more rarely nerve hypertrophy—are sometimes noted, but these neuropathic features are thought to have little clinical importance. The weakness appears to be due primarily to intrinsic muscle disease.

Weakness may ultimately involve all skeletal muscles, but certain patterns of weakness give characteristic findings, such as "hatchet facies," with hollow cheeks and a sagging jaw due to atrophy of the temporal and masseter muscles. Jaw weakness may predispose to temporomandibular joint dislocation. Associated abnormalities of the skull are uncommon but have been reported to include hyperostosis cranii, a small sella turcica, enlarged paranasal sinuses, and a narrow high-arched palate. In congenital MD club feet are common.

Selective degeneration of the cardiac conducting system has been noted, with electrocardiographically detectable conduction abnormalities occurring in a majority of patients.[6] Most often the problem is simple first-degree heart block. Fatal arrhythmias and heart failure are rare. The incidence of emboli from a cardiac source is low.

Primary atrophy of the testes is present in approximately 80% of men with MD.[7] While there is no consistent evidence of ovarian dysfunction in women, there may be higher than normal rates of infertility and miscarriage. Labor may be difficult because of poor uterine contractions. Frontal balding occurs in most affected men but rarely in women. Basal metabolic rates seem abnormally low in most patients, despite normal thyroid function. Hyperinsulinemia is usually present, but clinically significant diabetes mellitus is uncommon.[8]

Hypoventilation is thought to be due to diaphragmatic weakness, but abnormalities in central ventilatory regulation may also contribute to respiratory insufficiency. In some patients alveolar hypoventilation may lead to somnolence, mental dullness, cyanosis, and pulmonary hypertension.[9] Sleep apnea may occur. Because patients with MD are unduly susceptible to medications that depress ventilatory drive, they are at increased risk for surgery and anesthesia. An additional anesthesia caveat is that myotonia (but not necessarily MD) may be a risk factor for malignant hyperthermia.

Disturbances in gastrointestinal function may lead to considerable morbidity. Difficulty swallowing reflects weakness of striated muscles in the oropharynx. Abnormal esophageal peristalsis and reduced motility of the small bowel or colon are attributed to smooth muscle involvement. There have been rare case reports of smooth muscle involvement in other organs, such as the gallbladder or the urinary bladder.

Serum IgG levels are reduced in many cases, possibly due to accelerated catabolism. Minor abnormalities in platelets and red blood cell membranes have also been noted.

There are no constant morphologic abnormalities in the central nervous system, but behavioral changes may occur. Lethargy, apathy, and depression are not always explainable by the peripheral manifestations of the disease. Mental retardation is not uncommon, especially in early-onset cases. Computed tomography has shown cortical atrophy and ventricular dilatation in some patients.

Ocular Manifestations

Mild to moderate ptosis is common, but only occasionally do ptotic eyelids interfere with vision. Blepharitis is common and may be associated with keratitis or reduced lacrimal secretion. Ocular motility may be normal or severely limited, but results of forced ductions tests are typically normal.

The lens changes in MD are highly specific for the disease. Some authorities believe that virtually all patients with MD have lenticular degeneration.[10] Lens changes do not commonly interfere with vision, however, and they are often detectable only with a slit lamp. Classic lens changes consist of iridescent flecks within a thin band of anterior and posterior cortex beneath the lens capsule along with stellate opacifi-cations radiating from the posterior suture. The iridescent "crystals" appear red, blue, green, and opaque white. Ultrastructurally, they consist of whorls of plasmalemma from lens fibers.

Ocular hypotony is present in nearly one third of patients.[11,12] Low intraocular pressure is due to diminished production of aqueous, but the mechanism for this is not clear. A variety of pupillary abnormalities have been described in MD, but in most patients the pupils react normally to light. Ciliary processes are shorter than normal, and by light microscopy the ciliary epithelium is vacuolated and has a thickened basement membrane.

The macula and peripheral retina often develop pigmentary changes, but these usually do not significantly affect visual acuity.[13,14] Macular lesions are at the level of the retinal pigment epithelium and have a clinical appearance similar to that of the so-called patterned dystrophies. Pigmentary changes in the peripheral fundus are often more striking than those in the posterior pole. The electroretinogram is abnormal in almost all patients, showing low amplitudes of the b wave and reduced light sensitivity in the rod system.[15]

With a classic constellation of findings and a positive family history, the diagnosis of MD is easy to establish, but incomplete disease expression in the absence of a family history may make the diagnosis less obvious. In this situation, electromyography, careful slit lamp examination, and examination of asymptomatic first-degree relatives are useful means to verify the diagnosis. DNA probes may be the diagnostic tests of choice in the future.

References

1. Reinhardt R. The pathophysiologic basis of the myotonia and the periodic paralyses. In: Engel AG, Banker BQ, eds. *Myology*. New York, NY: McGraw-Hill; 1986.
2. Harper PS. Myotonic disorders. In: Engle AS, Banker BQ, eds. *Myology*. New York, NY: McGraw-Hill; 1986.
3. Pericak-Vance MA, Yamaoka LH, Assinder RIF, et al. Tight linkage of apolipoprotein C2 to myotonic dystrophy on chromosome 19. *Neurology*. 1986;36:1418-1423.
4. Brooke MH. *A Clinician's View of Neuromuscular Diseases*. Baltimore, Md: Williams & Wilkins Co; 1977;133.
5. Brumback RA, Carlson KM, Wilson H, et al. Myotonic dystrophy as a disease of abnormal membrane receptors: an hypothesis of pathophysiology and a new approach to treatment. *Med Hypotheses*. 1981;7:1059-1066.
6. Griggs RC, Davies RJ, Anderson DC, et al. Cardiac conduction in myotonic dystrophy. *Am J Med*. 1975;59:37-42.
7. Harper P, Penny R, Foley TP Jr, et al. Gonadal function in males with myotonic dystrophy. *J Clin Endocrinol Metab*. 1972;35:852-856.
8. Barbosa J, Nuttall FQ, Kennedy W, et al. Plasma insulin in patients with myotonia dystrophy and their relatives. *Medicine*. 1974;53:307-323.

9. Carroll JE, Zwillich CW, Weil JV. Ventilatory response in myotonic dystrophy. *Neurology.* 1977;27:1125-1128.

10. Miller NR. *Walsh and Hoyt's Clinical Neuro-ophthalmology.* 4th ed. vol 2. Baltimore, Md: Williams & Wilkins; 1985;794-803.

11. Hayasaka S, Kiyosawa M, Katsumata S, et al. Ciliary and retinal changes in myotonic dystrophy. *Arch Ophthalmol.* 1984;102:88-93.

12. Walker SD, Brubaker RF, Nagataki S. Hypotony and aqueous humor dynamics in myotonic dystrophy. *Invest Ophthalmol Vis Sci.* 1982;22:744-751.

13. Raitta C, Karli P. Ocular findings in myotonic dystrophy. *Ann Ophthalmol.* 1982;14:647-650.

14. Betten MG, Bilchik RC, Smith ME. Pigmentary retinopathy of myotonic dystrophy. *Am J Ophthalmol.* 1971;72:720-723.

15. Stanescu B, Michiels J. Temporal aspects of electroretinography in patients with myotonic dystrophy. *Am J Ophthalmol.* 1975;80:224-226.

PART 13

Phakomatoses

Chapter 147

Ataxia-Telangiectasia

Louis-Bar Syndrome

NEIL R. MILLER

Ataxia-telangiectasia is characterized by progressive cerebellar ataxia, oculocutaneous telangiectasia, and recurrent sinopulmonary infections.[1-6] It is a hereditary condition, but unlike neurofibromatosis, tuberous sclerosis, and von Hippel-Lindau disease, it appears to be inherited in an autosomal *recessive* fashion. It has been reported in identical twins.[7]

The ataxia becomes apparent shortly after the child begins to walk, whereas the telangiectasia of the bulbar conjunctiva is usually noted between 4 and 7 years of age.[2-4] The conjunctival telangiectasia is similar to that observed in patients with Sturge-Weber syndrome, but it is *bilateral* and not associated with lesions of the ocular fundi or with cutaneous angiomas (Fig. 147-1). The conjunctival changes become more pronounced with age (Fig. 147-2). Telangiectasia eventually develops on the ears, face, hard and soft palate, and the extensor surfaces of the extremities (Figs. 147-3, 147-4).[8] Other skin lesions include follicular keratosis, seborrheic der-

matitis, pigmentary disturbances, and secondary infections. Additional general features of the syndrome are mental retardation, which is present from birth but which does not usually become clinically apparent until the child reaches adolescence, growth retardation, choreoathetosis, dysarthria, drooling, and dry, coarse hair and skin.

Patients with ataxia-telangiectasia frequently develop recurrent respiratory infections, including bronchitis, bronchiectasis, and pneumonia. This increased susceptibility to infections of the respiratory tract is the most serious effect of the disease and is responsible for the majority of disease-related deaths. It appears to result from failure of the thymus to develop, with consequent deficiency or absence of immunoglobulins A (IgA), E (IgE), and G (IgG), as well as deficiency of T-cell function.[9-11] It is thus not surprising that these patients may develop lymphoid hyperplasia and that they also have an increased risk of developing leukemia,

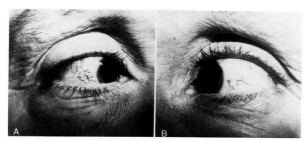

FIGURE 147-1 Conjunctival telangiectasia in a patient with ataxia-telangiectasia. These lesions are similar to those seen in patients with Sturge-Weber syndrome, but they are bilateral: **(A)** right eye; **(B)** left eye. (From Miller.[29])

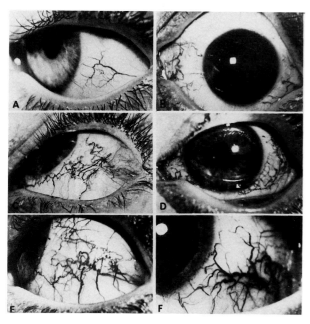

FIGURE 147-2 Conjunctival vascular changes in six ataxia-telangiectasia patients. The photographs show the increase in the severity of telangiectasia with age. **(A)** A 6-year-old patient; **(B)** 10-year-old patient; **(C, D)** two 12-year-old patients; **(E)** 16-year-old patient; **(F)** 19-year-old patient. (From Miller.[29])

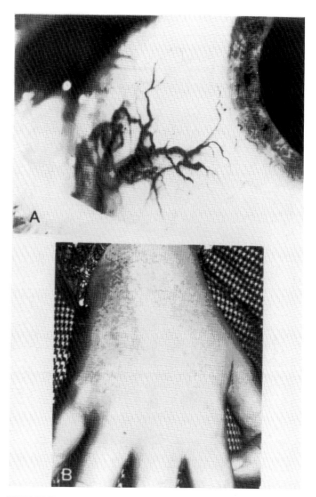

FIGURE 147-3 Ataxia-telangiectasia in a 6-year-old girl with progressive cerebellar dysfunction. **(A)** Dilated, engorged, bulbar conjunctival vessels. **(B)** Telangiectasia in the skin of the dorsum of the hand. (From Spencer.[30])

malignant lymphoma, and other systemic malignancies.[6,12-16]

Patients with ataxia-telangiectasia have a variety of abnormalities of ocular motility in addition to the almost constant finding of conjunctival telangiectasia. The characteristic initial ocular motor defects are inability to initiate voluntary saccades and abnormalities in the fast phase of vestibular nystagmus. Later, pursuit movements are impaired, and eventually, there is total oph-

thalmoplegia.[10,17-20] The ophthalmoplegia is *supranuclear*, however, as oculocephalic testing (doll's-head maneuver) reveals intact contraversive conjugate eye movements when the head or body is rotated horizontally or vertically. As difficulty performing eye movements increases, patients with ataxia-telangiectasia develop a head thrust similar to that observed in patients with congenital ocular motor apraxia; however, unlike patients with the latter disorder, those with ataxia-telangiectasia have abnormalities of both horizontal and vertical eye movements and also do not even make random saccades.

The neuropathology of ataxia-telangiectasia consists primarily of atrophy of the cerebellar cortex, particularly

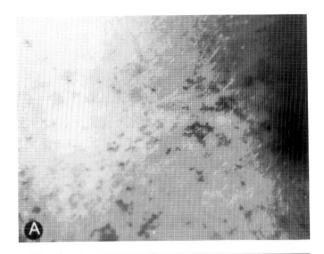

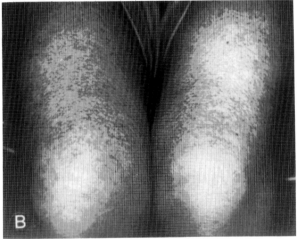

FIGURE 147-4 Skin changes in advanced ataxia-telangiectasia. **(A)** Telangiectasia in the skin of the anticubital fossa. **(B)** Telangiectasia on the thighs. (From Miller.[29])

the Purkinje and granular cells (Fig. 147-5).[3,21-25] This is associated with reduction in various neurotransmitters that normally reside in these cells, including glutamate, taurine, and γ-aminobutyric acid (GABA).[26] Eventually, degenerative changes also occur in the dentate and olivary nuclei, the spinal cord, and, ultimately, throughout most of the central nervous system.[27,28]

The treatment of patients with ataxia-telangiectasia is difficult. The prognosis for both quality and length life is extremely poor. Such patients and their families need to be counseled about the genetics of the disorder. Since it is transmitted as an autosomal-recessive trait, parents with one affected child have a 25% chance of having another child with the same condition. Patients who live long enough to marry and have children will pass the gene to all of their offspring. If the spouse does not have the gene, the children will only carry the disorder; however, if the spouse has the gene, there is a 50% chance that the children will have the disease. Obviously, if two patients with ataxia-telangiectasia marry (an unlikely occurrence), all of their children will have the disorder.

Affected patients are at increased risk of developing sinopulmonary and other infections and lymphoreticular and other malignancies,[6,14-16] so they must be made aware of the need for careful periodic examinations and of the potential consequences of even minor infectious and inflammatory illnesses. Such problems must be treated aggressively at the first sign of infection or tumor.

References

1. Louis-Bar D. Sur un syndrome progressif comprenant des telangiectasies capillaires cutanees et conjunctivales symmetriques a disposition nevoide des troubles cerebelleux. *Confin Neurol.* 1941;4:32.

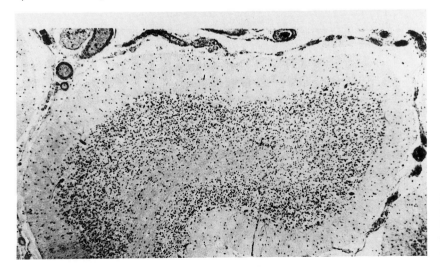

FIGURE 147-5 Neuropathology of ataxia-telangiectasia: cerebellar cortex shows severe loss of both Purkinje and granular cells. (From Stritch.[22])

2. Boder E, Sedgwick RP. Ataxia telangiectasia: a familial syndrome of progressive cerebellar ataxia, oculocutaneous telangiectasia and frequent pulmonary infection. A preliminary report on seven children, an autopsy and a case history. *Univ South Calif Med Bull.* 1957;9:15.

3. Boder E, Sedgwick RP. Ataxia-telangiectasia: a familial syndrome of progressive cerebellar ataxia, oculocutaneous telangiectasia and frequent pulmonary infection. *Pediatrics.* 1958;21:526.

4. Sedgwick RP, Boder E. Progressive ataxia in childhood with particular reference to ataxia-telangiectasia. *Neurology.* 1960;10:705.

5. Halasz P. Ataxia-telangiectasia (Louis-Bar syndrome). *Confin Neurol.* 1966;28:50.

6. Waldmann TA, Misiti J, Nelson DL, et al. Ataxia-telangiectasia: A multisystem hereditary disease with immunodeficiency, impaired organ maturation, x-ray hypersensitivity, and a high incidence of neoplasia. *Ann Intern Med.* 1983;99:367.

7. Meshram CM, Sawhney IMS, Prabhakar S, et al. Ataxia telangiectasia in identical twins: unusual features. *J Neurol.* 1986;233:304.

8. Grutzner P. Augensymptome bei Ataxia telangiectatica. *Klin Monatsbl Augenheilkd.* 1959;135:712.

9. Eisen AH, Karpati G, Laszlo T, et al. Immunologic deficiency in ataxia telangiectasia. *N Engl J Med.* 1965;272:18.

10. Cogan DG, Chu FC, Reingold D, et al. Ocular motor signs in some metabolic diseases. *Arch Ophthalmol.* 1981;99:1802.

11. Berkel AI. Studies of IgG subclasses in ataxia-telangiectasia patients. *Monogr Allergy.* 1986;20:100.

12. Harley RD, Baird HW, Craven EM. Ataxia-telangiectasia: report of seven cases. *Arch Ophthalmol.* 1967;77:582.

13. Font RL, Ferry AP. The phakomatoses. *Int Ophthalmol Clin.* 1972;12:1.

14. Becker Y. Cancer in ataxia-telangiectasia patients: analysis of factors leading to radiation-induced and *spontaneous* tumors. *Anticancer Res.* 1986;6:1021.

15. Morrell D, Cromartie E, Swift M. Mortality and cancer incidence in 263 patients with ataxia-telangiectasia. *J.N.C.I.* 1986;77:89.

16. Swift M, Morrell D, Cromartie E, et al. The incidence and gene frequency of ataxia-telangiectasia in the United States. *Am J Hum Genet.* 1986;39:573.

17. Smith JL, Cogan DG. Ataxia-telangiectasia. *Arch Ophthalmol.* 1959;62:364.

18. Thieffry S, Arthius M, Aicardi J, et al. L'ataxie-telangiectasie. *Rev Neurol.* 1961;105:390.

19. Hyams SW, Reisner SH, Neumann E. The eye signs in ataxia-telangiectasia. *Am J Ophthalmol.* 1966;62:1118.

20. Baloh RW, Yee RD, Boder E. Eye movements in ataxia-telangiectasia. *Neurology.* 1978;28:1099.

21. Centerwall WR, Miller MM. Ataxia, telangiectasia and sinopulmonary infection: a syndrome of slowly progressive deterioration in childhood. *Am J Dis Child.* 1958;95:385.

22. Stritch SJ. Pathological findings in three cases of ataxia-telangiectasia. *J Neurol Neurosurg Psychiatr.* 1966;29:489.

23. Aguilar MJ, Kamoshita S, Landing BH, et al. Pathological observations in ataxia-telangiectasia: a report of five cases. *J Neuropathol Exp Neurol.* 1968;27:659.

24. Terplan KL, Krauss RF. Histopathological brain changes in association with ataxia-telangiectasia. *Neurology.* 1969;19:446.

25. Paula-Barbosa MM, Ruela C, Tavares MA, et al. Cerebellar cortex ultrastructure in ataxia-telangiectasia. *Ann Neurol.* 1983;13:297.

26. Perry TL, Kish SJ, Hinton D, et al. Neurochemical abnormalities in a patient with ataxia-telangiectasia. *Neurology.* 1984;34:187.

27. Leon GA, Grover WD, Huff DS. Neuropathologic changes in ataxia-telangiectasia. *Neurology.* 1976;26:947.

28. Amromin GD, Boder E, Teplitz R. Ataxia-telangiectasia with a 32 year survival: a clinicopathological report. *J Neuropathol Exp Neurol.* 1979;38:621.

29. Miller NR. *Walsh and Hoyt's Clinical Neuro-ophthalmology.* 4th ed. vol 3. Baltimore, Md: Williams & Wilkins; 1983.

30. Spencer WH, Zimmerman LE. *Conjunctiva.* In: Spencer WH, ed. *Ophthalmic Pathology: An Atlas and Textbook.* 3rd ed. vol 1. Philadelphia, Pa: WB Saunders Co; 1988;109-228.

Chapter 148

Sturge-Weber Syndrome
Encephalotrigeminal Angiomatosis

BRENDA J. TRIPATHI, RAMESH C. TRIPATHI, and GERHARD W. CIBIS

Sturge-Weber syndrome is a rare congenital vascular phakomatous condition that shows no established genetic pattern or sexual or racial predilection. It is also called encephalotrigeminal (facial) angiomatosis, meningiofacial angiomatosis and its variants, Jahnke's syndrome (or nevus flammeus), Schirmer's syndrome, Lawford's syndrome, Parks-Weber-Oslow-Dimitri syndrome, Milles syndrome, and Klippel-Trenaunay syndrome; all of these entities are variations of one disorder. Clinically, Sturge-Weber syndrome is characterized by the classic triad of skin, nervous system, and ocular disorders, and it is probably the most familiar of all the phakomatoses. Although the nevus flammeus of the face is the most obvious sign, the major manifestations are due to central nervous system involvement. However, the facial port-wine stain may occur with or without one or both of the other major manifestations, leptomeningeal angiomatosis followed by cerebral gyriform calcifications, or glaucoma (with or without buphthalmos).

Because the lesions are vascular, Sturge-Weber syndrome was originally regarded as a mesodermal "phakomatosis." Abnormalities of Streeter's primordial vascular plexus were thought to occur early during embryonic development (11.5-mm stage), when the ectoderm that is destined to form the skin of the forehead lies directly over the neural tube. With the subsequent growth of the cerebral tissues, morphogenetic movements cause the abnormal vascular supply of the pia mater to be separated from that of the facial skin. However, a recent topographic analysis indicates that there is a morphogenetic link between the three cutaneous sensitive areas of the trigeminal nerve and the site of the facial stain.[1] Based on this study and on more recent knowledge about the contribution of the neural crest to the mesenchyme, Enjolras et al.[1] suggested that a morphogenetic error arising in a limited region of the cephalic neural crest could explain the abnormal vasculature in the mesectodermal derivatives of the supraocular dermis, choroid, and pia mater. Several studies now have provided evidence that the neural crest cells form the facial dermal connective tissue, the choroid of the eye, the pia mater, and the pericytes and smooth muscle cells of the vasculature. However, the endothelial lining of the vessels is still believed to be of mesodermal origin. For a full explanation of the clinical findings in Sturge-Weber syndrome, therefore, it is necessary to postulate a defective interaction during embryogenesis between the neural crest–derived elements and the vascular endothelium.[2]

Systemic Manifestations
Nevus Flammeus

The location of the nevus generally corresponds to at least a part of the distribution of the ophthalmic, maxillary, and (rarely) mandibular divisions of the trigeminal nerve. It may extend into the region of the upper cervical nerves. In Klippel-Trenaunay syndrome, the nevi are also present on other parts of the body. The lesion is unilateral and usually stops close to the midline, but not infrequently it crosses the midline. The nevus may be flat or nodular or even verrucous. It is present at birth and, with age, generally darkens and becomes nodular.[3] Histopathologically, the vessels appear normal until late childhood, at which time widely dilated, ectatic, blood-filled capillaries become apparent in the papillary dermis. The deep dermal and subcutaneous vessels are affected similarly in nodular lesions. The nevi do not regress and usually do not respond to sclerosing agents or topical steroids.

Central Nervous System Manifestations

The pathognomonic feature of Sturge-Weber syndrome is the angioma of the pia and arachnoid mater. As a result of the leptomeningeal angioma, the vascular cir-

Acknowledgement: This work was supported in part by a USPHS award EY-03747 from the National Eye Institute.

culation is altered in the underlying cerebral cortex, and this leads to secondary degeneration and subsequent calcification. The central nervous system involvement manifests itself as focal or generalized convulsions in 80% of the patients (in 50% before the age of 1 year), mental retardation in 54% of the patients, and hemiplegia contralateral to the angiomas in 31% of the cases. The majority of patients have intracranial calcifications, and on radiography they have the characteristic appearance of double-contoured, curvilinear densities that resemble "tram lines." These lesions can be detected within the 1st or 2nd decade of life, increase in size up to the 2nd decade, and then remain stable.

On cerebral arteriography, a lack of superficial veins, tortuosity and enlargement of the deep subependymal and medullary veins, and nonfilling of the superior sagittal sinus have been demonstrated.[3] This pattern may be due to either absence or closure of the cortical veins beneath the angiomatous leptomeninges, with the formation of collaterals that drain directly into the subependymal veins. Radionuclide angiography further demonstrates abnormal capillary permeability, which results in reduced perfusion throughout the involved hemisphere.[4] As seen on isotopic cisternography, there is an increased rate of absorption of cerebrospinal fluid in the area that corresponds to the abnormal venous hemangiomatosis of the leptomeninges.[5]

Patients with unilateral involvement have an ipsilateral signal attenuation on electroencephalography and, at some time, may also show a significant slow-wave disturbance on the contralateral side. On brain scan, the affected hemisphere is usually smaller in size, which can be correlated with hemiplegia but not with mental retardation.[4]

The primary pathogenic mechanism in Sturge-Weber syndrome is defective structural differentiation of the vascular wall and persistence of the sinusoid embryonal character of the vascular bed in the region of angiomatosis.[6,7] The functional abnormality of the vessels leads to secondary changes, which include fibrosis, hyaline degeneration, dilatation, and calcification of the walls, that ultimately produce venous stasis, obliteration of the lumen, and even formation of psammoma bodies. The tissues that are supplied by the abnormal vessels then manifest degenerative changes (cell loss and gliosis) and calcareous deposits, particularly in the cerebral cortex. Venous stasis can result in recurrent thrombotic episodes, with transient ischemic attacks being responsible for the gradual neurologic deterioration that occurs in patients with Sturge-Weber syndrome.[8]

Ocular Manifestations

Thirty percent of Sturge-Weber patients develop glaucoma.[9] Of these patients, 60% manifest it early enough in life (prior to age 2 years) to develop buphthalmos; the remaining 40% do not develop glaucoma until late childhood or young adulthood. Congenital glaucoma is often associated with involvement of the upper lid. Most commonly, the glaucoma is unilateral. The intraocular pressure is generally thought to be elevated only in patients whose facial nevus flammeus involves the eyelid or conjunctiva; however, exceptions do occur. Bilateral nevus flammeus with bilateral buphthalmos is a rare congenital disorder, but it has been reported, especially when the nevus is widespread, as in Klippel-Trenauney syndrome. Glaucoma also may be bilateral if the nevus flammeus crosses the midline.

The cause of the glaucoma remains enigmatic, although a number of pathogenic mechanisms appear to play a role[9,10]: (1) outflow obstruction by a chamber angle malformation which may be associated with increased vascularity of the iris; (2) occlusion of the angle with anterior synechiae, seen mainly as a secondary phenomenon in adults with choroidal hemangioma, retinal detachment, and neovascular glaucoma; (3) elevation of episcleral pressure, which reduces outflow facility; (4) hypersecretion glaucoma correlated with the neurogenic theory of trigeminal nerve distribution and heterochromia iridis of the sympathetic system; (5) elevation of intraocular pressure that results from increased permeability of the thin-walled blood vessels of the choroidal hemangioma; and (6) an adult form of glaucoma similar to primary open-angle glaucoma that may be associated with dilated conjunctival vessels, heterochromia iridis, choroidal hemangioma, and abnormal retinal vessels.

Gonioscopically, patients with Sturge-Weber syndrome may have prominent iris processes that adhere to the trabecular meshwork; this finding indicates a primary congenital defect in the development of the angle. A recent study of trabeculectomy specimens from patients with Sturge-Weber syndrome revealed not only a compact trabecular meshwork with hyalinization of the trabeculae but also the presence of amorphous material in the intertrabecular spaces of the deeper uveal and corneoscleral regions.[9] Individual trabecular beams were thickened and had a prominent cortical zone that contained "curly" collagen (Fig. 148-1). Many trabecular cells showed degenerative changes. The juxtacanalicular region contained an excess of extracellular elements (granuloamorphous material, basal lamina material, banded and nonbanded structures), and the cellular component had undergone degenerative changes. The endothelial lining of Schlemm's canal was attenuated and had very few vacuolar configurations. These alterations in patients with Sturge-Weber syndrome are similar to those that occur in the elderly and in patients with primary open-angle glaucoma, and they appear to represent premature aging of the trabecular meshwork and

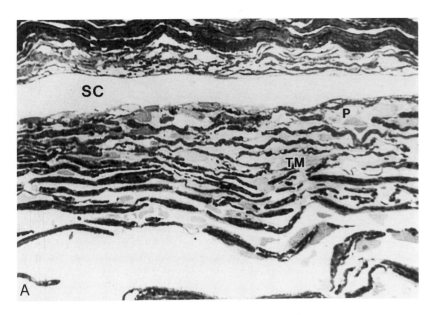

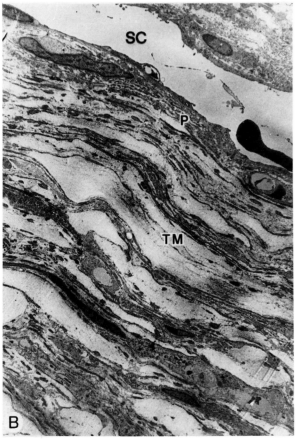

FIGURE 148-1 **(A)** Light photomicrograph of trabeculectomy specimen from a patient aged 11 years with Sturge-Weber syndrome (Klippel-Trenauney type). The patient had bilateral nevus flammeus and open-angle glaucoma. Except for some prominence of the iris processes, the angle had an unremarkable gonioscopic appearance. The intraocular pressures were uncontrolled with maximum medical therapy and the patient had a cup-to-disc ratio of 0.9. The retinal vessels were tortuous. The trabecular meshwork (*TM*) and pericanalicular region (*P*) of Schlemm's canal (*SC*) are compact. Note the extracellular material in the intertrabecular spaces (epoxy-resin section, toluidine blue, original magnification ×1200). **(B)** Survey electron micrograph of the inner wall of Schlemm's canal (*SC*), showing the compactness of the trabecular meshwork (*TM*) and the pericanalicular region (*P*) and the presence of granuloamorphous and fibrillar materials in the extracellular spaces (original magnification ×6340).

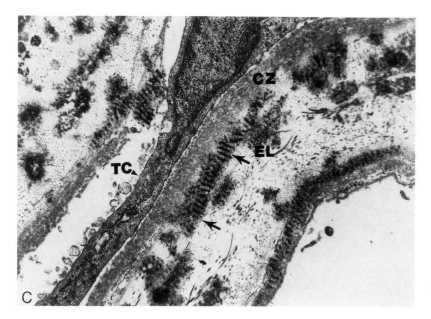

FIGURE 148-1 (*C*) Electron micrograph of uveal trabeculae. The trabecular cells (*TC*) show degenerative changes and the cortical zone (*CZ*) is thickened owing to the abundant basement membrane material and "curly" collagen (*arrows*). Elastic tissue (*EL*) is also associated with the curly collagen. These changes are similar to those seen in aging eyes and in advanced primary open-angle glaucoma (original magnification ×26,500).

Schlemm's canal system. Evidence is accumulating that during human embryonic development the trabecular meshwork, a major part of the corneoscleral structures, and the uveal tissue are derived from cells of the neural crest; thus, Sturge-Weber patients who have glaucoma may be regarded as manifesting neurocristopathy of the eye.[2,11] It is also apparent that the defect in the aqueous outflow pathway can arise early in the development of the anterior chamber, because many of these patients develop glaucoma, and even buphthalmos, early in life. Trabeculotomy in infants and trabeculectomy in older patients appear to be effective surgical procedures for the management of the glaucoma.[9]

Choroidal hemangioma is the most common ocular abnormality in Sturge-Weber syndrome, occurring in approximately 40% of cases. These hemangiomas may be relatively flat, isolated lesions that measure up to several millimeters in diameter and are located in the posterior fundus. Alternatively, they may be diffuse and involve a large area of the choroid, in which case the fundus has the appearance of red velvet or tomato catsup. The choroidal angioma is often situated temporal to the optic disc, with its greatest height in the macular region, and is surrounded by dilated choroidal and retinal vessels. Typically, congenital vascular hamartomas, which are of the cavernous type, cause no visual disturbance until the patient reaches early adulthood. During their slow growth they induce prominent cystoid degeneration of the overlying sensory retina, which eventually results in exudative retinal detachment and intractable secondary glaucoma. Calcification rarely occurs in the angiomatous choroid.

Ocular involvement in Sturge-Weber syndrome may include not only buphthalmos and glaucoma but also heterochromia iridis, conjunctival angioma, dilatation of episcleral vessels, and retinal aneurysm which is associated with arteriovenous angioma of the thalamus and midbrain. Heterochromia iridis is seen only occasionally. When present, it occurs on the same side as the facial nevus flammeus and is thought to be related to a disturbance in the sympathetic nervous system. Histopathologically, the iris melanocytes are increased in number and are deeply pigmented. They aggregate as small melanocytic hamartomas that are scattered on the anterior surface of the iris. Other ocular abnormalities, such as tortuosity of retinal vessels, retinitis pigmentosa, neuroblastoma of the retina, bilateral lens subluxation, anisocoria, and coloboma of the iris and optic nerve head, may also occur in patients with Sturge-Weber syndrome.

References

1. Enjolras O, Riche MC, Merland JJ. Facial port-wine stains and Sturge-Weber Syndrome. *Pediatrics.* 1985;76:48.
2. Tripathi BJ, Tripathi RC. Embryology of the anterior segment of the human eye. In: Ritch R, Shields B, Krupin T, eds. *The Glaucomas,* vol 1. St. Louis, Mo: CV Mosby; 1981;3-40.
3. Person JR, Perry HO. Recent advances in the phakomatoses. *Int J Dermatol.* 1978;17:1.
4. Kuhl DE, Bevilacqua JE, Miskin MM, et al. The brain scan in Sturge-Weber syndrome. *Radiology.* 1972;103:621.
5. Chang JC, Jackson GI, Baltz R. Isotopic cisternography in Sturge-Weber syndrome. *J Nucl Med.* 1970;11:551.

6. Wohlwill FJ, Yakovlev PL: Histopathology of meningo-facial angiomatosis (Sturge-Weber's disease). *J Neuropathol Exp Neurol.* 1957;16:341.
7. Di Trapani G, Di Rocco C, Abbamondi AL, et al. Light microscopy and ultrastructural studies of Sturge-Weber disease. *Child's Brain.* 1982;9:23.
8. Garcia JC, Roach ES, McLean WT. Recurrent thrombotic deterioration in the Sturge-Weber syndrome. *Child's Brain.* 1981;8:427.
9. Cibis GW, Tripathi RC, Tripathi BJ. Glaucoma in Sturge-Weber syndrome. *Opthalmology.* 1984;91:1061.
10. Weiss DI. Dual origin of glaucoma in encephalotrigeminal haemangiomatosis. *Trans Ophthalmol Soc UK.* 1973;93:477.
11. Tripathi BJ, Tripathi AC. Neural crest origin of human trabecular meshwork and its implication in the pathogenisis of glaucoma. *Amer J Ophthalmol.* 1989;107:583.

Chapter 149

Neurofibromatosis

SEYMOUR BROWNSTEIN

Von Recklinghausen's neurofibromatosis (NF), which is categorized with the hereditary disseminated hamartomatous conditions (phakomatoses), frequently affects the globe and ocular adnexa. Because it may produce severe disfigurement, NF has come to be known popularly as "elephant man's disease," although recently there has been some controversy as to whether John Merrick, the celebrated "elephant man" of 19th century London, had NF or another disease. NF occurs in about 1 in 3000 live births without any sexual or racial predilection. It is transmitted as an autosomal dominant trait with almost complete penetrance but with variable expressivity.[1] The spontaneous mutation rate is substantial in that only about one half of the patients have a positive family history for the condition. Recently, the NF gene has been linked by gene marker techniques to chromosome 17; this finding is expected to aid in prenatal and postnatal diagnosis of the disorder.[2]

Systemic Manifestations

Classically, NF has cutaneous, neurologic, skeletal, visceral, and ocular manifestations.[1] The cutaneous lesions consist predominantly of café-au-lait patches and superficial neurofibromas. Café-au-lait patches (Fig. 149-1) are pigmented macules, and the demonstration of six or more measuring over 1.5 cm in diameter is diagnostic of NF in the presence of any other designated criterion for the disease[1]: more than one neurofibroma of any type, sphenoid wing dysplasia, thinning of the long bone cortices, bilateral optic nerve gliomas, bilateral acoustic neuromas, more than one iris Lisch nodule, or a first-degree relative (parent, sibling, offspring) with NF.

There are three major types of neurofibromas.[3,4] The most common, the fibroma molluscum, is a small, localized skin tumor consisting of enlarged cutaneous nerves with a proliferation of Schwann cells and connective tissue elements. The plexiform neurofibroma is a diffuse proliferation within the nerve sheath resulting in thickened, tortuous nerves resembling a "bag of worms" (see Fig. 149-1). Elephantiasis neuromatosa is the result of diffuse schwannian proliferation in the dermis outside the nerve sheath resulting in marked thickening and folding of the skin.

In addition to the neurofibroma, other neurologic findings in NF include multiple tumors of the brain, spinal cord, meninges, cranial nerves, peripheral nerves, and autonomic nervous system. Among the skeletal manifestations are defects of the bones of the skull, vertebral anomalies, and pseudarthrosis of long bones. The visceral abnormalities include pheochromocytoma and multiple neurofibromas of the intrathoracic and intra-abdominal organs. Malignant neural and, more rarely, melanocytic tumors, including neurofibrosarcoma, malignant schwannoma, and malignant melanoma, develop in approximately 3% of patients.[1,3] The symptoms and signs in patients with NF develop progressively and frequently are not evident until late childhood, although they may be congenital or may appear early in infancy.[1,3]

With regard to pathogenesis, NF has been considered a neurocristopathy because of the neural crest origin of the cellular components of the characteristic lesions.[1,5] These include the café-au-lait spots, neurofibromas, and other tumors of the nervous system. The primary defect may be intrinsic to neural crest–derived cells, or it may

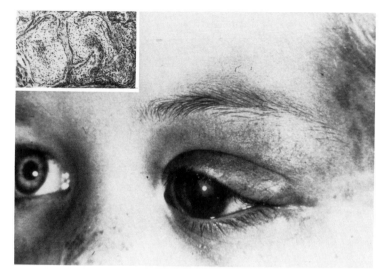

FIGURE 149-1 Main figure shows plexiform neuroma, mainly laterally in left eyelids with characteristic S-shaped contour of left upper lid, café-au-lait spots in left temple, and buph-thalmic left globe displaying an enlarged cornea, dilated pupil, ectropion iridis, and cataract. Inset discloses plexiform neuroma of left upper eyelid with enlarged nerves surrounded by thickened perineural sheaths (he-matoxylin-eosin stain, original magnification ×100). (From Brownstein and Little.[7])

require an interaction with the cellular or extracellular environment, such as with the glycosaminoglycans of the extracellular ground substance, in order to manifest the lesions of NF.

Ocular Manifestations

The ocular findings of NF include café-au-lait spots of the eyelid; neurofibromas and neurilemmomas of the lids, conjunctiva, cornea, and orbit; thickened corneal, conjunctival and ciliary nerves; melanocytic and neuronal hamartomas in the trabecular meshwork, uvea, retina, and optic nerve; glaucoma; sectoral retinal pigmentation; and absence of the greater wing of the sphenoid bone, which may lead to pulsating exophthalmos (see Fig. 149-1).[3,4] Furthermore, the intracranial tumors of NF may produce visual field defects, oculomotor and sensory nerve deficits, papilledema, and optic atrophy.

Hamartomas of the iris have been reported by Lewis and Riccardi[6] in 92% of neurofibromatosis patients aged 6 years and older. These lesions, termed Lisch nodules, have been distinguished clinically from common iris nevi by their elevation, smooth contour, and soft translucency on biomicroscopic examination and are thus a useful aid for the diagnosis of NF in problematic cases (Fig. 149-2, inset). They increase in number with age and are usually bilateral. The histogenesis of these iris hamartomas has been controversial with regard to their possible derivation from schwannian elements, but most studies have demonstrated that they are composed of melanin-containing cells, indicating that they are melanocytic hamartomas (see Fig. 149-2).[1,6,7]

In a prospective survey of 77 patients with NF, 51%

were found to have between 1 and 18 choroidal hamartomas, ill-defined, yellow-white to light brown lesions scattered throughout the posterior pole.[6] They may be diffuse or localized to the posterior pole of the eye. Histopathologic examination of the choroidal hamartoma has disclosed a variety of neuronal, melanocytic, and fibrocytic elements, including onionlike formations that contained fine neuronal fibers and that superficially resemble tactile nerve endings (Fig. 149-3).[7] These were shown to be composed of lamellae of Schwann cell processes arranged around axons; they are called "ovoid bodies."[4,7]

According to Grant and Walton,[8] the major cause of glaucoma in NF is infiltration of the angle by neurofibroma, which obstructs the aqueous outflow channels. Other causes include closure of the angle due to neurofibromatous thickening of the ciliary body, secondary fibrovascular and synechial closure of the angle, and possibly a congenital malformation (failure of normal development) of the chamber angle. Most patients with congenital glaucoma in neurofibromatosis have plexiform neuromas of the ipsilateral (usually upper) eyelid.[3,7,8] When plexiform neuromas of the eyelid are present, they are accompanied by ipsilateral congenital glaucoma in about 50% of cases.[3,4,7,8]

Approximately 25% of patients with optic glioma have NF.[9] Among patients with NF, 15% have these tumors of the anterior visual pathways.[9] Two thirds of the tumors produce no symptoms and no clinical signs on ophthalmologic examination. High-resolution computed tomography (CT) is responsible for detecting them, in contrast to the relatively lower yield from ophthalmoscopy and conventional radiography of the skull, orbits, and optic foramina. Furthermore, in one

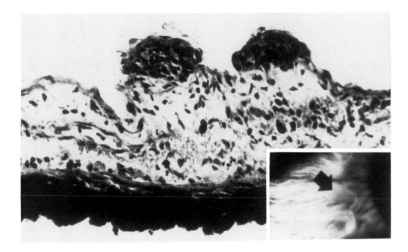

FIGURE 149-2 Inset shows characteristic Lisch nodule (*arrow*) of iris characterized by its elevation, smooth contour, and soft translucency. (Courtesy of Trevor Kirkham, M.D.) Main figure discloses melanocytic nodules covered by endothelium projecting into anterior chamber from anterior border layer of iris (hematoxylin-eosin stain, original magnification ×400). (From Brownstein and Little.[7])

study, a circumferential perineural architectural pattern featuring tumor eruption and proliferation in the subarachnoid space correlated with the presence of NF, whereas an expansile intraneural pattern correlated with the absence of NF.[10]

The management of ocular NF includes surgery for large neurofibromas of the eyelid and orbit and medical or surgical therapy of glaucoma. Usually the glaucoma is congenital and may occur before there are other obvious symptoms or signs of NF.[7] Children with congenital glaucoma should thus be examined thoroughly for associated disorders such as NF. The glaucoma frequently is intractable, resulting in buphthalmos and a blind painful globe that requires enucleation.[7] However, the diagnosis of the associated NF is important for the management of subsequent internal manifestations of the disease, for prognosticating to the parents the possible outcome of the ocular and general condition, and for genetic counseling.

References

1. Riccardi VM. Von Recklinghausen neurofibromatosis. *N Engl J Med.* 1981;305:1617-1626.
2. Barker D, Wright E, Nguyen K, et al. Gene for von Recklinghausen neurofibromatosis is in the pericentromeric region of chromosome 17. *Science.* 1987;236:1100-1102.
3. Font RL, Ferry AP. The phakomatoses. *Int Ophthalmol Clin.* 1972;12(1):1-50.

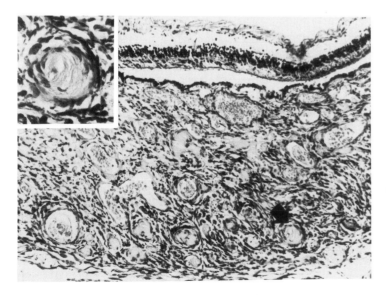

FIGURE 149-3 Main figure shows neurofibroma filling thickened choroid with loss of photoreceptor cells in overlying retina (hematoxylin-eosin stain, original magnification ×100). Inset shows ovoid body in region of arrow in main figure (hematoxylin-eosin, original magnification ×400). (From Brownstein and Little.[7])

4. Yanoff M, Fine BS. *Ocular Pathology: A Text and Atlas.* 2nd ed. Philadelphia, Pa: Harper & Row; 1982;33,34,37-39.
5. Bolande RP. Neurofibromatosis—the quintessential neurocristopathy: pathogenetic concepts and relationships. *Adv Neurol.* 1981;29:67-75.
6. Lewis RA, Riccardi VM. Neurofibromatosis. Incidence of iris hamartomata. *Ophthalmology.* 1981;88:348-354.
7. Brownstein S, Little JM. Ocular neurofibromatosis. *Ophthalmology.* 1983;90:1595-1599.
8. Grant WM, Walton DS. Distinctive gonioscopic findings in glaucoma due to neurofibromatosis. *Arch Ophthalmol.* 1968;79:127-134.
9. Lewis RA, Gerson LP, Axelson KA, et al. Von Recklinghausen neurofibromatosis. II. Incidence of optic gliomata. *Ophthalmology.* 1984;91:929-935.
10. Stern J, Jakobiec FA, Housepian EM. The architecture of optic nerve gliomas with and without neurofibromatosis. *Arch Ophthalmol.* 1980;98:505-511.

Chapter 150

Tuberous Sclerosis

Bourneville's Disease

MARY WAZIRI

Tuberous sclerosis (TS) is a multisystem, hamartomatous, genetic disorder. It is a dominantly inherited condition with a high rate of new mutations. The classic symptoms include seizures, mental retardation, and facial angiofibromas. While these continue to be the most frequently devastating manifestations of TS, a much broader spectrum of clinical expression of the gene has become apparent.

As for other disorders, the tools for diagnosis and treatment of TS have been improving. With the availability of remarkable new imaging modalities we are finding patients with TS who are more mildly affected than those who were previously recognized. Many have been found to have asymptomatic tumors of the brain, heart, kidneys, and eyes. Advances such as laser treatment, newer medications, and improved surgical techniques have improved the quality of life for many. The recent discovery of linkage of the TS gene to other genes on the long arm of chromosome 9 may prove to be one of the most significant findings to date.[1] In addition to providing a means for earlier and more definitive diagnosis, molecular genetic study may eventually lead to a better understanding of the gene's function, and possibly to a preventive approach.

Ocular Manifestations

The most characteristic eye finding suggestive of TS is the retinal astrocytic hamartoma (phakoma) found in 50 to 87% of examined patients.[2,3] Accurate identification of such a lesion may be accomplished in early infancy or in old age. Kiribuchi et al.[3] relate the higher incidence found in their study to somewhat darker fundi in their population and to the protocol used. They sedated each patient and frequently supplemented examinations with fluorescein angiography. The tumors may occur anywhere in the fundus, although typically they are located at the posterior pole adjacent to the optic disc.[4] One type, possibly more often peripheral, are lesions that are flat, smooth, and translucent, with poorly defined margins. They may be very difficult to detect except for an abnormal light reflex or blurring of the image of an underlying blood vessel. Another type, possibly more central or at a later stage, is the retinal tumor that has become calcified, is multinodular, measures from $\frac{1}{2}$ to 4 disc diameters, and may be significantly elevated. When located at the optic disc, they may resemble hyaline bodies or giant drusen, except that they tend to obscure the underlying retinal vessels.[5] Rarely, vitreous hemorrhage or glaucoma results from a TS lesion. The tumor is never malignant.[2]

Histologically, retinal hamartomas are composed of astrocytes with long processes and small oval nuclei. They contain large blood vessels and calcareous deposits. Initially they arise from the retinal ganglion cell layer and later involve all layers.[4]

In addition to the fundus hamartomas of TS, other internal eye findings may include peripheral, punched-out, depigmented areas of the retina, atypical colobomas, and optic atrophy.[2] Papilledema and sixth cranial nerve paresis may be seen in cases accompanied by increased intracranial pressure secondary to obstruction of cerebrospinal fluid drainage by brain "tubers." External eye findings may include eyelid angiofibromas, white patches of the iris or eyelashes, and strabismus.[2] The pigmentary anomalies are thought to resemble those seen in the skin.

Central Nervous System Manifestations

Seizures and mental retardation are the clinical evidence of altered central nervous system (CNS) function in tuberous sclerosis. Among patients with TS, 82 to 90% have a seizure disorder, and 41 to 47% are mentally subnormal.[2] The onset of seizures most often occurs in infancy or childhood. Myoclonic seizures that occur several to many times a day prior to 6 months of age and are often associated with a bizarre (hypsarrhythmia) electroencephalogram pattern carry the worst prognosis for future intellectual functioning. This type of seizure is supplanted later by other types and is rare after 4 years of age. Tonic clonic seizures are common only after the 1st year of life and are often associated with other types. Atypical absences and partial seizures with complex symptoms are common.[2]

Imaging techniques for detection of the CNS hamartomas (tubers) continue to become more specific and useful for correlation of clinical findings with pathology and for prediction of future intellectual functioning. Occasional low-density areas of the cerebral hemispheres are noted on CT, which have been found recently to correlate with high-signal magnetic resonance imaging (MRI) abnormalities.[6] No correlation between the computed tomographic (CT) abnormalities in TS and the severity of the neurologic disease had previously been found.[7] The typical MRI abnormality, however, is a high-signal lesion involving the cerebral cortex, probably corresponding to the cortical hamartomas noted pathologically.[6] Remarkable preliminary data suggest that lesions noted on MRI may correlate with site of seizure foci and clinical seizure pattern and that some TS patients with a severe language deficit are found on MRI to have multiple left temporal lesions.[8] Preliminary studies of positron-emission tomography suggest that this technique may be useful for studying seizure activity in TS.[2]

Other CNS abnormalities occur. Giant cell astrocytomas have been seen in 2% of TS patients, growing particularly when patients with TS are between 5 and 18 years old.[2] They rarely, if ever, become malignant and rarely bleed spontaneously, though they are well-vascularized. They may grow sufficiently to obstruct the foramina of Monro.[2,7] Ventricular dilatation and cortical

atrophy may be seen, even in the absence of obstruction to CSF flow, in patients who have had frequent and severe seizures and are mentally subnormal. Also noted has been the presence of cerebellar lesions in some patients who have no recognized clinical signs of cerebellar dysfunction.[6]

Cutaneous Manifestations

The skin manifestations of TS occur in 96% of affected patients.[2] The hypomelanotic macules found in 86% of patients with TS number from one to many. These white patches are usually 1 to 2 cm in diameter, but they vary in size. Use of an ultraviolet light (Wood's lamp) in a darkened room is helpful in locating them. Facial angiofibromas are present in 47% of TS patients, are typically first noted when the patient is between 3 and 5 years of age, and may be progressive until young adulthood. These lesions are small but often confluent, red, raised

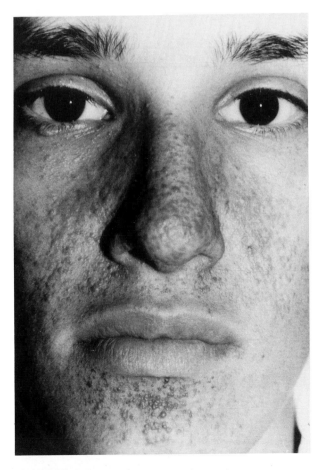

FIGURE 150-1 Angiofibroma of face in young man with tuberous sclerosis.

nodules, typically noted in the nasolabial folds, over the malar areas, and on the chin (Fig. 150-1). They are composed of hyperplastic connective and vascular tissue and secondarily displace and alter the sebaceous glands. Many have a prominent telangiectatic component and may be associated with recurrent bleeding.[2] The facial angiofibroma problem is the most emotionally devastating for patients. Problems with treatments have included regrowth of tumor and scarring; however, improvement in appearance following treatment can be expected.[9]

Other ectodermal problems occur in TS. Fibrous plaques or Shagreen's patches are confluent areas of skin tumor with a waxy, yellowish brown or flesh-colored appearance, most typically seen on the forehead, back, or legs. Ungual fibromas of the fingernails or toenails may cause abnormal nails with longitudinal grooves (Fig. 150-2). The clinician should also examine the teeth for possible pitted enamel hypoplasia. Such pits are rare in the average population, whereas 71% of persons with typical TS were found to have these clearly distinguishable lesions. The pits are usually located on the outer surfaces of the teeth, away from the gingiva.[10] Poliosis is noted on occasion.

Kidney and Heart Alterations

Angiomyolipomas and renal cysts may be found together or alone in one or both kidneys of the TS patient. The angiomyolipoma is found in 50 to 80% of affected patients.[11] The kidney tumors and cysts are usually asymptomatic, though symptoms which become apparent at an early age may include gross hematuria, pain, uremia and other signs of renal failure, and hypertension.[12]

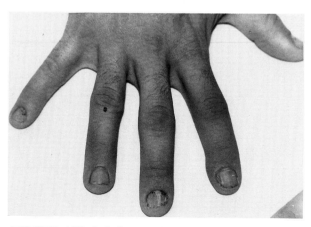

FIGURE 150-2 Subungual fibroma in a patient with tuberous sclerosis.

The cardiac rhabdomyomas associated with TS are most frequently asymptomatic. Echocardiograms of TS patients have revealed a tumor prevalence of 43%, most of which occur in the interventricular septum or in the ventricular wall.[2] Cardiac arrhythmia or obstruction of blood flow may occur.

Conclusion

A thorough evaluation is recommended. Patients and relatives at risk of having TS require a careful ophthalmologic examination with dilated pupils, a head MRI or CT study, an echocardiogram, an ultrasound or CT study of the kidneys, and a thorough physical examination that gives special attention to the skin, nails, and teeth. Prenatal diagnosis by fetal echocardiogram and ultrasound studies is possible in some cases.[13] Symptomatic treatment should be carried out as needed. Genetic counseling is indicated for patient and family.

References

1. Conner JM, Pirrit LA, Yates JRW, et al. Linkage of the tuberous sclerosis locus to a DNA polymorphism detected by v-abl. *J Med Genet*. 1987;24:544.
2. Gomez MR, ed. *Neurocutaneous Diseases*. London: 1987; Butterworths.
3. Kiribuchi K, Uchida Y, Fukuyama Y, et al. High incidence of fundus hamartomas and clinical significance of a fundus score in tuberous sclerosis. *Brain Dev* 1986;8:509.
4. Williams R, Taylor D. Tuberous sclerosis. *Surv Ophthalmol*. 1985;30:143.
5. Robertson DM. Ophthalmic findings. In: Gomez MR, ed. *Tuberous Sclerosis*. New York, NY: Raven Press; 1979.
6. Roach ES, Williams DP, Laster DW. Magnetic resonance imaging in tuberous sclerosis. *Arch Neurol*. 1987;44:301.
7. Kingsley DPE, Kendall BD, Fritz CR. Tuberous sclerosis: a clinicoradiological evaluation of 110 cases with particular reference to atypical presentation. *Neuroradiology*. 1986;28:38.
8. Curatolo P, Cusmai R. The value of MRI in tuberous sclerosis. *Neuroped*. 1987;18:184.
9. Weston J, Apfelberg DB, Maser MR, et al. Carbon dioxide laserbrasion for treatment of adenoma sebaceum in tuberous sclerosis. *Ann Plastic Surg*. 1985;15:132.
10. Lygidakis NA, Lindenbau RH. Pitted enamel hypoplasia in tuberous sclerosis patients and first-degree relatives. *Clin Genet*. 1987;32:216.
11. Monaghan HP, Krafchik BR, MacGregor DL, et al. Tuberous sclerosis complex in children. *Am J Dis Child*. 1981;135:912.
12. Hendren WG, Monfort GJ. Symptomatic bilateral renal angiomyolipomas in a child. *J Urol*. 1986;137:256.
13. Journel H, Roussez M, Plais MH, et al. Prenatal diagnosis of familial tuberous sclerosis following detection of cardiac rhabdomyoma by ultrasound. *Prenat Diagn*. 1986;6:283.

Chapter 151

Von Hippel-Lindau Disease

PAUL W. HARDWIG

Von Hippel-Lindau (VHL) disease is a hereditary disorder with protean clinical manifestations. It is a generalized developmental dysgenesis of neuroectoderm and mesoderm, but the basic defect is not known. The main features of the disease include angiomatosis of the retina, capillary hemangioblastomas of the central nervous system (CNS), and cysts or angiomatous tumors of the viscera (Table 151-1). Six lesions commonly produce significant morbidity: angiomatosis retinae; cerebellar, medullary, and spinal hemangioblastomas; renal cell

TABLE 151-1 Pathologic Lesions in Von Hippel-Lindau Disease

Site	Pathologic Lesion
Retina	Angiomatosis retinae*
Central nervous system†	
Cerebellum	Hemangioblastoma*
Medulla oblongata	Hemangioblastoma,* syringobulbia
Spinal cord	Hemangioblastoma,* syringomyelia
Important visceral structures‡	
Pancreas	Hemangioblastoma
Kidney	Clear cell carcinoma,* hemangioblastoma
Adrenal medulla and sympathetic chain	Pheochromocytoma* paraganglioma
Miscellaneous	
Cerebrum	
Meninges	
Lung	
Liver	
Spleen	
Omentum, mesocolon	
Ovary	
Bladder	
Bones	
Skin	

* Significant morbidity
† Almost always below the tentorium
‡ Lesions common to all include cyst and adenoma.

carcinoma; and pheochromocytoma.[1-3] The disease is often lethal, cerebellar hemangioblastoma and renal cell carcinoma being the most common causes of death. In contrast, the more common cysts and angiomatous tumors of several visceral organs are usually asymptomatic.

The typical hemangioblastoma consists of a small highly vascular mural nodule within a much larger fluid-filled cyst. There may be striking hypertrophy of afferent and efferent vessels. The retinal angioma is similar, but usually lacks a cyst. Histologically, the vascular tumor is a fine capillary network within a matrix of lipid-laden foam cells. There are no mitotic figures.

VHL disease has an autosomal dominant inheritance pattern with variable expression. Although a child born to an affected person has a 50% chance of being affected, it is distinctly unusual for one person to manifest the entire syndrome. Furthermore, the diagnosis of affected persons is impeded by the silence of many diagnostic lesions. The management and evaluation of persons suspected to have VHL requires periodic presymptomatic screening, as lifesaving surgical intervention can be more prudently planned with the early detection of potentially lethal lesions.[2] At present, asymptomatic lesions are most likely to be detected in the retina. Careful ophthalmoscopy is therefore essential to any screening program. The detection of early central nervous system (CNS) and visceral lesions calls for enhanced computed tomography (CT) of the head, upper cervical spinal cord, and abdomen.

Systemic Manifestations

Cerebellar hemangioblastoma is the most common CNS manifestation; its reported incidence ranges from 35 to 75%.[2,3] The tumor is usually symptomatic at the time of diagnosis and typically presents in the 2nd to 4th decade of life with signs of increased intracranial pressure or cerebellar dysfunction. Although it is eminently resectable, it is the most common cause of death.

Medullary and spinal hemangioblastomas are comparatively uncommon. The medullary lesion may be lethal if it compresses vital brain stem centers. The spinal hemangioblastoma, most often upper cervical in location, is often silent.

Renal cell carcinoma is recognized clinically in roughly one third of patients and most often presents as hematuria, obstructive nephropathy, or an abdominal mass. It tends to be the last of the major manifestations to appear. The tumor is a severe life-threatening manifestation of the disease that may produce death from metastasis or uremia.

Pheochromocytoma is usually symptomatic and is said to occur in less than 10% of affected persons. Studies have suggested that the reported frequency of pheochromocytoma in VHL varies because the tumors tend to cluster in certain predisposed families.[1]

Polycythemia has been reported in 10 to 25% of VHL patients, predominantly in those harboring cerebellar hemangioblastomas.[1,2] This polycythemia has been presumed to be related to erythropoietic activity of the CNS cyst fluid; however, polycythemia has also been reported to occur in association with occasional cases of renal cell carcinoma and pheochromocytoma.

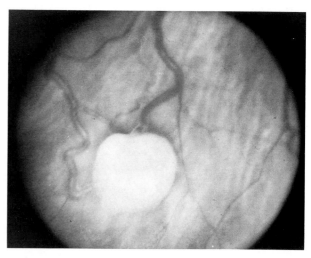

FIGURE 151-1 A typical von Hippel retinal angioma, found in the fundus midperiphery.

Ocular Manifestations

Angiomatosis retinae is found in the majority of VHL patients. It is usually the first observed manifestation of VHL disease and is frequently asymptomatic when diagnosed in the 2nd or 3rd decade of life.[1,2,4] Tumors are most often midperipheral in location. Occasionally they may occur in a juxtapapillary location, where they may be mistaken for papilledema, papillitis, or another tumor. Rarely, tumors occur in the intraorbital portion of the optic nerve. Tumors are bilateral in 30 to 50% of patients and are multiple in one eye in 33%.

The earliest lesion is a small capillary cluster similar in size and configuration to a diabetic microaneurysm.[5] The fully developed classic lesion consists of a globular, slightly raised, pink retinal tumor that is fed by a dilated, tortuous retinal artery and vein pair (Fig. 151-1). On occasion, abnormal vessels at the optic disc point to the anteriorly located tumor. As the tumor enlarges, lipid exudation into the retinal and subretinal space begins. Such extravasation may occur around the tumor, or distantly at the optic disc or within the macula. Exudative retinal detachment, neovascularization, rubeotic glaucoma, and phthisis bulbi may ensue. Although the progression of a VHL retinal angioma varies, an untreated lesion can cause blindness and enucleation.[5] In one series involving 36 eyes, recorded visual acuity ratings at last examination were about equally divided between normal and severely compromised.[2] Roughly 40% of eyes had a visual acuity of 20/20, whereas another 40% had acuities of less than 20/200. Four eyes had to be enucleated for pain and blindness. This was true whether or not the eyes had been treated.

Therapy includes photocoagulation, cryotherapy, and diathermy and is successful in obliterating preclassical and small VHL angiomas.[5-7] Injudicious treatment, particularly of larger tumors, may be followed by massive retinal detachment, retinal hole formation, vitreous hemorrhage, and retinitis proliferans. Argon laser photocoagulation is applied directly to the surface of small (<2.5 disc diameters) angiomas. Cryotherapy is useful in slightly larger tumors, in anteriorly located tumors, and in eyes with semiopaque media. Diathermy, with or without scleral buckle, may still be useful in managing large tumors that are difficult to treat in any other way.

References

1. Melmon KL, Rosen SW. Lindau's disease; review of the literature and study of a large kindred. *Am J Med.* 1964; 36:595-617.
2. Hardwig PW, Robertson DM. Von Hippel-Lindau disease: a familial, often lethal, multi-system phakomatosis. *Ophthalmology.* 1984; 91:263-270.
3. Horton WA, Wong V, Eldridge R. Von Hippel-Lindau

disease: clinical and pathological manifestations in nine families with 50 affected members. *Arch Intern Med.* 1976; 136:769-777.

4. Grossman M, Melmon KL. Von Hippel-Lindau disease. In: Vinken PJ, Bruyn GW, eds. *Handbook of Clinical Neurology.* vol. 14. *The Phakomatoses.* Amsterdam: Elsevier North-Holland; 1972;241-259.

5. Welch RB. Von Hippel-Lindau disease: the recognition and

treatment of early angiomatosis retinae and the use of cryosurgery as an adjunct to therapy. *Trans Am Ophthalmol Soc.* 1970; 68:367-424.

6. Gass JD. Treatment of retinal vascular anomalies. *Trans Am Acad Ophthalmol Otol.* 1977; 83:432-442.

7. Annesley WH, Leonard BC. Fifteen year review of treated cases of retinal angiomatosis. *Trans Am Acad Ophthalmol Otol.* 1977; 83:446-453.

Chapter 152

Wyburn-Mason Syndrome

AHMAD M. MANSOUR and THOMAS L. SCHWARTZ

Wyburn-Mason syndrome, also called Bonnet-Dechaume-Blanc syndrome, is a combination of retinal and systemic arteriovenous malformations (AVMs).[1,2] AVMs consist of direct communications between the arterial and the venous circulation. Several designations have been given to these malformations, such as cirsoid aneurysms, arteriovenous varix, racemose aneurysm, and plexiform angioma.[3]

Systemic Manifestations

AVMs in Wyburn-Mason syndrome tend to involve predominantly the central nervous system (CNS), the ocular adnexa, and the oronasopharynx (Table 152-1). AVMs are the most common form of intracranial vascular hamartoma. CNS AVMs have a predilection for the midbrain. In a literature review of 800 cases, Krayenbuhl and Yasargil[4] found 86% of AVMs to be supratentorial, with the majority supplied by the middle cerebral artery. AVMs may involve various sites of the visual pathways, resulting in papilledema, optic atrophy, or hemianopia. Other manifestations of CNS AVMs include headache, seizures, mental retardation, cranial nerve palsies, and hemiparesis. Cerebral AVMs are associated with intracerebral or subarachnoid hemorrhages, and it seems that there is a direct relationship between the tendency of AVMs to hemorrhage and their size.[5] Epistaxis and oral hemorrhage (especially during or following dental surgery) are the presenting signs of nasopharyngeal AVMs.

Ocular Manifestations

AVMs involving the orbit and eyelids produce ptosis, proptosis, dilated conjunctival vessels, and bruits. Retinal AVMs are congenital stable unilateral vascular malformations that have a predilection for the posterior pole and the superotemporal quadrant.[3] Retinal AVMs are divided into three grades: grade I, anastomosis between an arteriole and a venule; grade II, anastomosis between a branch artery and a branch vein; and grade III, diffuse marked dilatation and tortuosity of the whole vascular system. The mean age at presentation for subjects with

TABLE 152-1 Location of Systemic Arteriovenous Communications in Wyburn-Mason Syndrome*

Site	Cases	
	No.	%
Central nervous system	22	59
Orbit	7	19
Oronasopharynx	6	16
Eyelid	2	5
Skin	1	3
Lung	1	3
Spine	1	3

* A few cases have several systemic malformations. (Adapted from Mansour, Walsh, and Henkind.[3])

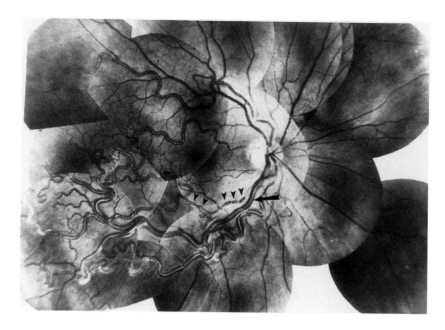

FIGURE 152-1 Grade II arteriovenous malformation of the retina involves the posterior pole of the left eye. Note the dilatation and tortuosity of the inferotemporal arcade vessels (*arrows*) and beading of the corresponding arterial tree (*arrowheads*).

retinal AVMs of grades I and II is about 42 years; the majority are detected on routine ophthalmoscopic examination.[3] Visual acuity in this category is usually normal, and a small percentage of patients have associated systemic AVMs. Grade I retinal AVMs can easily be missed. Grade I and II AVMs manifest as arterial and venous dilatation and tortuosity of a sector of the retinal circulation, along with difficulty in differentiating arteries from veins in terms of vessel caliber and color (Fig. 152-1). The mean age at presentation for subjects with grade III retinal AVMs is around 25 years. Subjects in this category present early because of vision loss secondary to a large retinal AVM or because of systemic or ocular symptoms related to AVMs of the CNS or elsewhere. Grade III retinal AVMs have the appearance of "intestinelike vessels," "snakelike convoluted vessels," or "a bag of worms." Poor visual acuity is related to the retinal ischemia (steal phenomenon) surrounding the area of AVM (Fig. 152-2); also, associated AVMs of the visual pathways lead to optic atrophy and visual field loss.

Although most retinal AVMs are stable lesions, several ocular complications have been reported: intraretinal macular hemorrhage, central retinal vein occlusion, branch retinal vein occlusion, neovascular glaucoma, and vitreous hemorrhage.[6] Retinal AVMs may undergo remodeling or spontaneous regression following retinal vein occlusion.[7,8] Venous occlusive disease in AVMs is related to "arterialization" of the venous system.[6] In one histopathologic study of retinal AVM in a human eye it was not possible to differentiate arteries

from veins, as all the vessels of the AVM demonstrated widened fibromuscular media.[9]

Small to medium-sized retinal AVMs (grades I and II) are usually isolated retinal lesions, and there is no need for thorough workup for systemic AVMs if such patients remain asymptomatic. However, large retinal AVMs are frequently associated with systemic AVMs in the context of the Wyburn-Mason syndrome, so systemic workup is warranted.

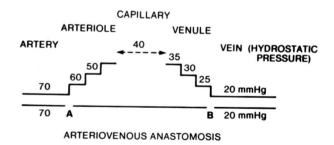

$$FLOW = \frac{GRADIENT}{RESISTANCE}$$

FIGURE 152-2 Physiologic model to explain the steal phenomenon in the area surrounding a retinal AVM. The values represent the hydrostatic pressures (in millimeters of mercury) of the retinal circulation. A direct communication between the arterial and venous system has very low resistance. The blood flow tends to be high in the AVM and low in the surrounding area (flow = gradient/resistance).

References

1. Wyburn-Mason R. Arteriovenous aneurysm of mid-brain and retina, facial naevi and mental changes. *Brain*. 1943;66:163-203.

2. Bonnet P, Dechaume J, Blanc E. L'aneurysme cirsoide de la retine (aneurysme vasemeaux). Ses relations avec l'aneurysme cirsoide de la face et avec l'aneurysme cirsoide du cerveau. *J Med Lyon*. 1937;18:165-178.

3. Mansour AM, Walsh JB, Henkind P. Arteriovenous anastomoses of the retina. *Ophthalmology*. 1987;94:35-40.

4. Krayenbuhl H, Yasargil MG. *Das Hirnaneurysma*. Basel: JR Geigy; 1958.

5. Fults D, Kelly DL, Jr. Natural history of arteriovenous malformations of the brain. A clinical study. *Neurosurgery*. 1984;15:658-662.

6. Mansour AM, Wells CG, Jampol LM, et al. Ocular complications of arteriovenous communications of the retina. *Arch Ophthalmol*. 1989;107:232-236.

7. Augsburger JJ, Goldberg RE, Shields JA, et al. Changing appearances of retinal arteriovenous malformation. *Graefes Arch Klin Exp Ophthalmol*. 1980;215:65-70.

8. Unger H-H, Umbach W. Kongenitales okulozerebrales Rankenangiom. *Klin Monatsbl Augenheilk*. 1966;148:672-682.

9. Cameron ME, Greer CH. Congenital arterio-venous aneurysm of the retina. A post-mortem report. *Br J Ophthalmol*. 1968;52:768-772.

PART 14

Physical and Chemical Injury

Chapter 153

Alcoholism

ALAN D. LISTHAUS and DONALD A. MORRIS

Nearly 10% of the adult population of the United States abuse alcohol.[1] The cost to society in terms of loss of productivity and family disruption is immeasurable. Ethyl alcohol, the active ingredient in most alcoholic beverages, is absorbed unaltered from both the stomach and the small intestine. After entering the bloodstream, it enters the organ systems of the body and is present in the cerebrospinal fluid, urine, and pulmonary alveoli. It is metabolized primarily in the liver to acetaldehyde via the enzyme alcohol dehydrogenase.[1] Alcohol seems to reduce the intestinal absorption of numerous nutrients, including glucose, amino acids, calcium, folate, and vitamin B_{12}. This impaired absorption may be a primary contributor to the malnutrition frequently encountered in the alcoholic population. Alcohol may cause hypertriglyceridemia by affecting lipid metabolism. It has also been found to interfere with carbohydrate metabolism.

Alcoholics are frequently anemic and thrombocytopenic. It is unclear whether these hematologic manifestations are secondary to malnutrition,[2] to direct depression of the bone marrow, or to a combination of the two.

Acute and chronic pancreatitis are often found in the alcoholic population. Gastritis and peptic ulcers, which often bleed, are also common. One serious cause of hematemesis is the Mallory-Weiss syndrome, which is characterized by lacerations of the mucosa at the gastroesophageal junction due to the alcohol-induced gastritis. The most ominous form of gastrointestinal involvement results from cirrhosis. Approximately 8% of chronic alcoholics develop liver disease. Sequelae of liver disease include esophageal varices, which may result in intractable bleeding that leads to death.

A variety of neurologic disorders are associated with both acute and chronic alcoholism and may be attributed to the direct toxic effects of the alcohol or to the nutritional deficiencies mentioned above.

The signs and symptoms of the fetal alcohol syndrome have been described by many authors. They include intrauterine growth retardation, mental retardation, microcephaly, cardiovascular defects (persistent patent ductus arteriosus, ventral septal defects), and decreased joint mobility. The ophthalmic manifestations of fetal alcohol syndrome include short palpebral fissures, divergent strabismus, ptosis, pale optic discs, and telecanthus.[3]

Chronic alcoholics are at risk for developing the so-called toxic amblyopia syndrome.[2,4-6] Its characteristic visual field defect is a bilateral cecocentral scotoma. It is not clear whether the alcohol acts directly on nerve or retinal tissue or whether the effect is again secondary to nutritional deprivation. There have been many studies involving alcoholics with and without nutritional deficiencies; the results remain inconclusive. There is also a synergistic effect when both alcohol and tobacco are used by the patient, the reason for which is unknown.

Methyl alcohol (wood alcohol) is sometimes used by alcoholics when they are unable to obtain ethyl alcohol. After ingestion and absorption, methyl alcohol is oxidized to formaldehyde and formic acid, producing metabolic acidosis.[7] This may result in degeneration of the retinal photoreceptors with associated degeneration of the optic nerve. It is hypothesized that the formic acid might inhibit hexokinase activity in the retinal cells.

A number of studies have found that alcoholics suffer from dyschromatopsia.[8-11] Over 80% of patients with documented Laennec's cirrhosis demonstrate a definite decrease in color perception. Rothstein[8] also found a relationship between decreased color perception and decreased folic acid levels.

Alcoholics may develop an excess of blood lipids, which in its extreme form may present as lipemia retinalis.[11,12] Markedly elevated serum triglyceride levels impart a milky color to the blood column visible in the retinal vessels. They may vary in color from pink to ivory, and the arteries and veins become indistinguishable from each other. They may appear larger and flatter than normal, and a yellowish white cuff may be seen adjacent to them. The appearance of the vessels in the periphery usually changes first, and as the triglyceride level increases the color of the vessels in the posterior pole approaches that of those in the periphery.

A commonly overlooked feature of alcohol abuse is the presence of a cataract.[14,15] It is usually of the posterior subcapsular variety and is present in people under 60 years of age. It is usually rapidly progressive and may occur in association with Dupuytren's contracture.[14,15] Corneal arcus also develops in alcoholics at an earlier age than it does in the general population. It may be a result of hypertriglyceridemia.

It is interesting to note that one of the treatments of chronic alcoholism may have an ocular side effect.

Disulfiram (Antabuse) is a relatively nontoxic substance that can markedly alter the metabolism of alcohol.[16,17] When a patient medicated with disulfiram drinks alcohol, the blood acetaldehyde concentration may increase by a factor of five to ten. This increase causes a serious systemic reaction (nausea and vomiting), which is intended to discourage further drinking. Optic neuritis may be another unwanted side effect of the disulfiram treatment.

References

1. Isselbacher KJ, Adams RD, Braunwald E, et al., eds. *Harrison's Principles of Internal Medicine.* 10th ed. New York, NY: McGraw-Hill; 1984.
2. Carroll FD. Nutritional amblyopia. *Arch Ophthalmol.* 1980;90:476-480.
3. Altman B. Fetal alcohol syndrome. *J Pediatr Ophthalmol Strab.* 1978;13(5):255-258.
4. Dunphy EB. Alcohol and tobacco amblyopia: a historical survey—XXXI Deschweinitz lecture. *Am J Ophthalmol.* 1969;68(4):568-578.
5. Victor M, Dreyfus PM. Tobacco and alcohol amblyopia. *Arch Ophthalmol.* 1965;74(11):649-657.
6. Frenkel REP, Spoor TC. Visual loss and intoxication. *Surv Ophthalmol.* 1986;30(6):391-396.
7. Robbins SL, Cotran RS. *The Pathologic Basis of Disease.* 2nd ed. Philadelphia, Pa: WB Saunders; 1979.
8. Rothstein TB, Shapiro MW, Sacks JG, et al. Dyschromatopsia with hepatic cirrhosis: relation to serum B12 and folic acid. *Am J Ophthalmol.* 1973;75(5):889-895.
9. Sakuma Y. Studies on color vision anomalies in subjects with alcoholism. *Ann Ophthalmol.* 1973;5:1277-1292.
10. Sandberg MA, Rosen JB, Berson EL. Cone and rod function in vitamin A deficiency with chronic alcoholism in retinitis pigmentosa. *Am J Ophthalmol.* 1977;84(5):658-665.
11. Verriest G, Franq P, Pierart P. Results of colour vision tests in alcoholic and in mentally disordered subjects. *Ophthalmologica.* 1980;180:247-256.
12. O'Connor PR, Donaldson DD. Lipemia retinalis. *Arch Ophthalmol.* 1972;87:230-231.
13. Grosberg SJ. Retinal manifestations in hyperlipemia of alcohol origin. *Arch Ophthalmol.* 1986;75:750-751.
14. Drews RC. Cataracts, Dupuytren's contracture and alcohol addiction. *Am J Ophthalmol.* 1974;77:418.
15. Sabiston DW. Cataracts, Dupuytren's contracture, and alcohol addiction. *Am J Ophthalmol.* 1973;76:1005-1007.
16. Gilman AG, Goodman LS, Gilman A. *The Pharmacological Basis of Therapeutics.* 6th ed. 1980.
17. Norton A, Walsh F. Disulfiram-induced optic neuritis. *Trans Am Acad Ophthalmol Otolaryngol.* 1972;76:1263-1265.

Chapter 154

Fetal Alcohol Syndrome

MARILYN T. MILLER and KERSTIN STROMLAND

A pattern of malformations has been observed in children born to women who have a history of alcohol abuse during pregnancy. The presence of certain dysmorphic features along with other symptoms and signs has been designated the fetal alcohol syndrome (FAS) by Jones et al.[1,2] These adverse effects of alcohol have been described in patients of many racial groups by authors in many countries. The frequency and spectrum of features of this syndrome have been recognized primarily during the last few decades; however, the observation that ingestion of large amounts of alcohol by women during their pregnancy may harm the fetus has been known for centuries. Aristotle observed, "Foolish drunken or harebrain women must part bring forth children like unto themselves, *'morosos et languidos.'*" Cultural taboos in ancient Carthage, based on Greek and Roman mythology, forbade drinking by a bridal couple so as not to produce defective children.[3]

During the last 20 years interest in FAS has increased. It has been estimated that at least one third to one half of children born to chronic alcoholic mothers show some of the stigmata of FAS.[2,4] Although ascertainment is difficult, its estimated incidence is 1 or 2 per 1000 live births, with partial expression in up to 3 to 5 per 1000.[4,5] While the critical amount of alcohol required to produce the syndrome or any malfunction of the fetus has not been established, a definite risk of developing the syndrome is felt to exist when intake of absolute alcohol is 89 ml per day (six average drinks).[5] Lesser amounts carry no established risks but are believed to increase the incidence of fetal anomalies.

Systemic Manifestations

The more consistent characteristics of the FAS involve (1) facial abnormalities with short palpebral fissures, thin vermilion border of upper lip, and epicanthal folds; (2) mental retardation varying in degree from mild to severe; (3) low weight and height at birth that persist in the postnatal period; and (4) abnormalities of cardiovascular and skeletal systems (Fig. 154-1).[1,2,5] Many teratogens, including alcohol, exert a wide range of effects on the developing fetus. This causes transitional stages

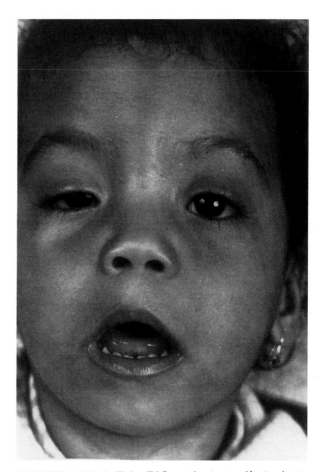

FIGURE 154-1 This FAS patient manifests bow-shaped mouth with flat, thin upper lip. Ocular findings include right ptosis and esotropia, telecanthus, and moderate myopia. (From Miller.[12])

between the complete syndromes and minor anomalies that are difficult to prove to be alcohol effects unless the offspring of large numbers of pregnant women are studied carefully.

Low birth weight and short body in infants with nearly normal gestation periods are typical of FAS. Diminished growth rate continues into the postnatal period and becomes more pronounced. Studies of growth hormone, gonadotropins, and cortisol levels have not shown any abnormality.[6] One explanation is that alcohol in some way diminishes the number of cells that develop, which in turn results in decreased size. The midface hypoplasia, small upper lip with thin vermilion border, undersized or absent philtrum, and small palpebral fissures may reflect this disturbed growth pattern. While no single feature is pathognomonic, the facies of patients with FAS is as "distinctive as that of the patient with Down syndrome" and can be readily recognized by a trained observer.[7]

Mild to severe mental retardation is present in most children who manifest the stigmata of FAS, and the level of retardation appears to be directly related to the degree of dysmorphogenesis.[7] There is evidence that the offspring of chronic alcoholics may be at risk for some decreased intellectual functioning, even when only a few physical signs of the syndrome are evident. The average IQ of children with the diagnosed syndrome in the study was 65; this depressed level remained stable in 77% of the children, even in the presence of apparently acceptable home environments.[8]

Behavioral characteristics that are reported in more than 50% of cases include irritability in infancy and hyperactivity in childhood. Affected children are described as "jumpy." More severe antisocial behavior is rare, however.[8] There may be other signs of neurologic dysfunction, such as poor sucking and altered reflexes.

Ocular Manifestations

The ophthalmic findings in the early reports were usually confined to soft tissue or motility disturbances, such as small palpebral fissures, epicanthus, ptosis, or strabismus, although myopia, microophthalmia, and blepharophimosis were reported occasionally.[2,5]

The extent and severity of the ocular manifestations have become more obvious with the emergence of fairly complete case descriptions in the pediatric literature and increased interest in the subject by ophthalmologists. The incidence of ocular pathology is difficult to establish, as few prospective studies include routine eye evaluations performed by an ophthalmologist, but evidence is accumulating that these children are at risk for many types of ocular pathology that often cause severe visual impairment.

Lids and Palpebral Fissures

The most frequently reported ocular finding of FAS in the nonophthalmic literature is an abnormality of the palpebral fissure, usually a short fissure from telecanthus, although "blepharophimosis" is commonly cited. One confusing aspect of this observation is the variety of meanings implied by the term. Does "blepharophimosis" describe fissures that are diminished in all dimensions, or is it used to mean only a short fissure? The most characteristic finding in FAS appears to be a marked increase in the distances between the medial canthi (ie, telecanthus; see Fig. 154-1).[9-16] This measurement is usually increased on an absolute scale when compared with normal, age-matched control subjects, but it is even more striking in view of the fact that these children often have delayed growth of head circumference. If the distance between the lateral canthi is not abnormal, the marked increase in the medial intercanthal distance alone results in a short horizontal fissure. In some patients the position of the lateral canthus is also mildly disturbed.[12]

The increased medial canthal distance is a soft tissue disturbance (primary telecanthus) and does not reflect an increase in interorbital distance (hypertelorism with secondary telecanthus). Normal interpupillary distances support this observation, although the diagnostic differentiation is best made by a radiologic measurement of the interorbital distance. Epicanthus is quite frequently reported in the literature; however it is difficult to draw any conclusions from this finding as it is so often a normal feature in some populations.

A high incidence of ptosis has been noted.[7,12,14,15] It is interesting that monocular or asymmetric ptosis is more frequent than balanced bilateral ptosis in many cases described in the literature (see Fig. 154-1).

Visual Acuity and Refraction

Visual acuity of children suffering from FAS is often reduced. Data about the vision of 24 FAS children compared to two control groups showed severely reduced vision in 20% of the children's eyes and moderately reduced vision in another 45% (Fig. 154-2).[14]

Refraction varies to a large extent in FAS children, and high refractive errors, mostly myopia, have been present in some patients.[12,14-16] In a series of eight patients with anterior segment anomalies, all eyes that could be refracted (or the uninvolved eyes) had a significant myopic refractive error, suggesting that the myopia may be in some way a mild teratogen-induced disturbance of the developing anterior segment of the eye.[15]

Anterior Segment

Anterior segment anomalies, including microophthalmia and forms of mesenchymal dysgenesis, have been reported in a number of patients, although they cannot be classified as a characteristic finding.[14,15] Peters' anomaly has been the developmental anomaly most

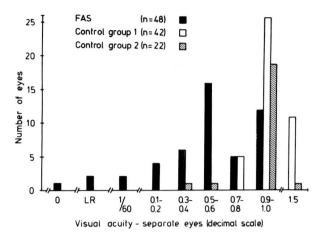

FIGURE 154-2 Corrected visual acuity in 24 children with FAS and in two control groups.

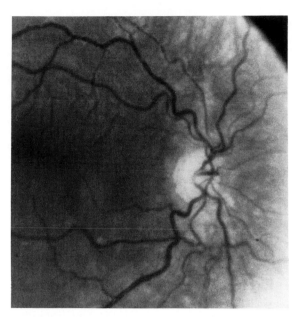

FIGURE 154-3 Optic nerve hypoplasia and anomalous retinal vessels in a child with FAS.

frequently observed, but Axenfeld's anomaly and anteriorly displaced Schwalbe's line have also been noted.[14,15] It was proposed that the anterior segment anomalies may represent a teratogenic action of alcohol during a very brief but critical period of development.[15]

Sulik, working with a C57 mouse strain, demonstrated a marked increase in ocular anomalies (microophthalmia, corneal pathology, etc.) following high doses of alcohol administered on the 7th day of gestation.[17] This experimental mouse model simulates the binge drinking effect.

Whereas cataracts are reported occasionally, they appear to be an infrequent manifestation. Iris anomalies and persistent hyaloid are occasional findings.[16]

Strabismus

Strabismus is a frequent finding, occurring in about 50% of patients with complete FAS. It appears to be nonspecific and is most commonly observed as a small variable esotropia.[12,14,16]

Fundus

A wide spectrum of anomalies of the ocular fundi are found in FAS children, ranging from discrete anomalies of the optic nerve head and the retinal vasculature to extensive malformations of several fundus structures.

Optic disc anomalies appearing as hypoplasia of the optic nerve head are the most common malformations of the fundus (up to 48%).[14] The disc is small, with sharp and often irregular margins either unilaterally or bilaterally (Fig. 154-3). A double ring surrounds all or part of the disc.

By definition, hypoplasia of the optic nerve implies a reduction in the number of axons. That does not necessarily require that the disc be small. A reduced number of axons might occur in a disc of normal or even large size. Tests of the function of the visual system are helpful in making the diagnosis of optic nerve hypoplasia. The uncooperativeness of young or retarded children makes application of such methods difficult in FAS children, where the diagnosis of optic nerve hypoplasia often must be made by clinically identifying a small disc.

Anomalies of the retinal vessels occur in the eyes of children with FAS. The vessels are tortuous and exhibit abnormal width and course over the retinal surface (see Fig. 154-3). The tortuosity occurs mainly in arteries, but it can be observed also in veins. Earlier than normal branching of the vessels and a greater than normal number of small, unspecified vessels can be observed. A combination of anomalies of the optic disc and the retinal vessels is a typical finding in the fundus of an FAS child (see Fig. 154-3).[14]

In summary, ophthalmic malformations resulting in severe visual impairment are common results of heavy alcohol abuse during pregnancy. Some characteristic anomalies could be speculated to result from drinking smaller amounts of alcohol over a longer critical period, whereas malformations noted only in a small percentage of cases may represent ingestion of more alcohol in a briefer, more critical period.

References

1. Jones KL, Smith DW. Recognition of the fetal alcohol syndrome in early infancy. *Lancet.* 1973;1:999.
2. Jones KL, Smith DW, Ullehand CN, et al. Pattern of malformation in offspring of chronic alcoholic mothers. *Lancet.* 1973;1:1267.
3. Warner RH, Rossett HL. The effects of drinking on off-spring. *J Stud Alcohol.* 1975;36:1395.
4. Olegard R, Sabel KG, Aronsson M, et al. Effects on the child of alcohol abuse during pregnancy—retrospective and prospective studies. *Acta Paediatr Scand (Suppl).* 1979;275:112.
5. Clarren SK, Smith DW. The fetal alcohol syndrome. *N Engl J Med.* 1978;298:1063.
6. Root AW, Reiter EO, Andriola M, et al. Hypothalamic pituitary function in the fetal alcohol syndrome. *J Pediatr.* 1975;87:585.
7. Smith DW. The fetal alcohol syndrome. *Hosp Pract.* 1979;14:121.
8. Streissguth AP, Herman CS, Smith DW. Intelligence, behavior and dysmorphogenesis in the fetal alcohol syndrome: A report on 20 patients. *J Pediatr.* 1978;92:363.
9. Altman B. Fetal alcohol syndrome. *J Pediatr Ophthalmol.* 1978;13:255.
10. Rabinowicz IM. Examining eye abnormalities in children with the fetal alcohol syndrome. *Ophthalmology.* 1980;87(suppl):93.
11. Stromland K. Eye ground malformations in the fetal alcohol syndrome *Neuropediatrics.* 1981;12:97.
12. Miller M, Israel J, Cuttone J. Fetal alcohol syndrome. *J Pediatr Ophthalmol Strab.* 1981;18(4):6.
13. Stromland K. Eyeground malformation in the fetal alcohol syndrome. *Birth Defects: Original Article Series.* 1982;18:651.
14. Stromland K. Ocular abnormalities in the fetal alcohol syndrome. *Acta Ophthalmol.* 1985;63(suppl):171.
15. Stromland K. Ocular involvement in the fetal alcohol syndrome. *Surv Ophthalmol.* 1986;(31)3:277.
16. Miller MT, Epstein RJ, Sugar J, et al. Anterior segment anomalies associated with the fetal alcohol syndrome. *J Pediatr Ophthalmol Strab.* 1984;21(1):8.
17. Sulik KK, Johnston MC, Webb MA. Fetal alcohol syndrome: embryogenesis in a mouse model. *Science.* 1981;214:936.

Chapter 155

Child Abuse

ANDREA CIBIS TONGUE

Child abuse has occurred throughout recorded civilization, and it continues today in spite of protective measures and attitudes adopted by society. The medical profession was generally ignorant of the extent and manifestations of physical child abuse until the 1960s. In 1961 Kempe organized a multidisciplinary symposium on "the battered child syndrome."[1] This symposium was responsible for professional and public recognition of the problem of child abuse and for the subsequent laws in all states that make it mandatory for professionals dealing with children to report any suspected cases of child abuse. In 1974 The Federal Child Abuse Prevention and Treatment Act was passed, and in 1975 Title XX of the Social Security Act made protective child services mandatory in each state.[2] Children's Protective Services (CPS) agencies are responsible for investigating all reported cases of suspected abuse and for the child's immediate and long-term welfare and protection. In 1976, 417,033 cases of child abuse were reported nationally, in 1984, 1.7 million.[2,3] About 42% of the reported cases are substantiated, but there are some estimates that the actual incidence of severe physical abuse is seven times as great as the reported rate.[2] Major physical injuries secondary to abuse occur in about 3.3% of reported cases. CPS reported 500 deaths secondary to abuse in 1984. The actual death rate is not known and is probably higher than that reported because of physicians' and medical examiners' reluctance to make the diagnosis.[4]

It is critical for physicians to recognize the signs and symptoms of abuse. Delayed or missed diagnosis may subject the child to continued abuse, which may result in major physical and mental impairment, and in some cases death. CPS agencies also depend on the physician's statement and testimony concerning the diagnosis. Without it they are almost inevitably unable to pursue substantiation and appropriate disposition of a suspected child abuse case.

Systemic Manifestations

Nonaccidental trauma should be suspected and considered to be the primary diagnosis if (1) the injury is not explained by the alleged cause, (2) the injury is unlikely to occur accidentally in the child's age group, (3) multiple injuries (acute or chronic) are present, (4) there are recurrent injuries not readily explained by alleged cause, and (5) there is significant delay in obtaining medical care for the injury. Although physical manifestations of child abuse are protean, the three systemic signs of physical abuse most likely to be encountered by the ophthalmologist are bruises, skeletal fractures, and subarachnoid or subdural hemorrhages.

Bruises most likely to be secondary to nonaccidental trauma are any bruises in infants not yet crawling or walking and bruises in varying states of healing on body parts not exposed to repeated trauma (ie, buttocks, lower back, genitals, inner thighs, cheek, earlobe, upper lip, frenulum, neck). Young children rarely get bruises from falling down stairs except over bony prominences. The color of bruises indicates the approximate age of the injury. A bruise may remain red for several hours, blue or purple for 1 to 3 days, green for 4 to 7 days, yellow 1 to 2 weeks, and it disappears by 2 to 4 weeks.[5] Bruises with sharp outlines or patterns of an object are almost inevitably the result of intentional trauma. Adult bite marks are sharply demarcated, hyperpigmented, double crescent–shaped, with a dental arch of greater than 4 cm.[6]

Fractures in children under 1 year of age are highly suggestive of nonaccidental trauma unless they are secondary to birth injury. Most fractures in child abuse occur in children under 18 months of age. Multiple fractures at any age without a well-established cause are most compatible with nonaccidental trauma. Although epiphyseal or metaphyseal fractures in children under age 5 years (excluding newborns), spiral fractures of the humeral shaft, symmetric clavicular fractures, and capital femoral epiphyseal fractures in children before age 2 years are thought to be highly suggestive of nonaccidental trauma. The most common fracture seen in child abuse cases is the nonspecific diaphyseal fracture.[7,8]

If fractures are present, multiple fractures are the rule. Rib fractures are the single most common fracture, are usually asymptomatic, and the breaks are along the costovertebral junction, probably as a result of lateral chest compression. Skull fractures are rarely the result of falling from a height less than 1 meter and in one series were associated with other fractures in about 50% of abuse patients.[7]

Radiologic examination of chest, abdomen, skull, spine, and extremities should be performed in all children under 1 to 2 years of age who have signs of probable abuse. Older children should have radiologic examinations based on clinical symptoms of skeletal injury. Unsuspected fractures were found in 22% of infants under 1 year of age but in only 9% of children over 2 years.[7] Acute fractures of cartilaginous bone may not be detectable until new bone formation occurs 12 to 14 days after injury, a fact to keep in mind, since infants have predominantly cartilaginous bone.[8]

Central nervous system (CNS) trauma with resultant intracranial hemorrhage occurs primarily in infants under 2 years of age, secondary to either physical battering or to vigorously shaking the infant without supporting its head (shaken baby syndrome[9]). The intracranial hemorrhages in the shaken baby syndrome are felt to be primarily secondary to the to and fro motion of the infant's soft brain inside the skull, which causes stretching and shearing of the less elastic blood vessels, particularly the cerebral bridging vessels. Contusion injuries of the frontal and occipital lobes may also result from the relatively excessive movement of the brain inside the skull when it is subjected to whiplash forces. It is, however, impossible to say how frequently CNS injuries in the abused child are the result of shaking rather than of direct trauma.

Cranial computed tomography (CT) is an exceedingly valuable diagnostic tool for detecting cerebral injuries. The most common abnormalities are subarachnoid hemorrhage with a predilection for the parietooccipital interhemispheric fissure and cerebral edema. Cerebral hemorrhage and subdural hematoma occur less frequently.[10] Cerebral atrophy can often be demonstrated on follow-up CT scans and is often already present within a week after hospitalization for the acute CNS manifestations.

Ocular Manifestations

Intraocular hemorrhages are present in the majority of infants and young children with CNS trauma from physical abuse. Ophthalmoscopy is therefore an important part of the examination of any child who has signs of CNS injury, particularly when no external signs of trauma exist, as in the shaken baby syndrome. Misdiagnosis and delayed diagnosis are common, particularly in the shaken baby syndrome, where the infants most frequently present with apnea, respiratory arrest, seizures, coma, or obtundation. A scenario frequently reported by the family is that the infant was fine until he or she choked while bottle feeding and stopped breathing. Resuscitative efforts are then instituted. Infectious diseases such as meningitis, metabolic disorders, seizure disorders, and sudden infant death syndrome may be primarily suspected unless careful ophthalmoscopy reveals retinal hemorrhages. The presence of retinal hemorrhages in an afebrile infant with signs or symptoms of CNS injury and no clear-cut history of CNS trauma (eg, car accident, fall from a height of greater than 1 meter, being run over by a car) makes child abuse the primary diagnosis.

The appearance of retinal hemorrhages in children suffering from CNS trauma is not pathognomonic. The hemorrhages may be unilateral or bilateral and may involve all layers of the retina as well as the choroid and vitreous, as has been demonstrated by histologic examination.[11,12] The hemorrhages may be discrete or extensive. Small blot hemorrhages with pale centers are common. Preretinal or subhyaloid hemorrhages, particularly in the posterior pole, are also frequently seen. Their appearance clinically is one of a dome-shaped, sharply circumscribed mound of blood. Deeper hemorrhages may also be sharply circumscribed and dome shaped, giving rise to the appearance of retinoschisis.[13] Disc edema and vascular congestion are commonly seen in the acute phase of the CNS injury, highly suggestive of elevated central retinal venous pressure. In infants the retinal vascular system may be particularly susceptible to abrupt changes in pressure. The vessels may be more fragile than in the adult, the normally lower systemic blood pressure, higher cardiac rate, as well as rapid onset of metabolic acidosis with hypoxia may affect the cerebrovascular hemodynamic forces more readily than in the adult. Since cerebral anoxia and ischemia occur secondary to microvascular changes in experimental intracranial hypertension, it is not unreasonable to postulate concomitant retinal anoxia and ischemia.[14] It is most likely that the retinal hemorrhages are a result of increased intracranial pressure with resultant venous stasis and rupture of retinal blood vessels rather than of acceleration-deceleration forces on the retina and retinal vessels. Probably the extent and location of the retinal ischemia is the most critical factor in producing loss of vision, just as the extent and location of cerebral ischemia is the most critical factor in the resultant neurologic dysfunction. Cortical atrophy and infarction, which can be massive in spite of relatively minor acute intracranial hemorrhages in these infants, may play a major role in the permanent visual deficit or may simply reflect the extent of cerebral—and possible concomitant retinal—anoxia.

Vitreous hemorrhages are the result of leakage of blood from beneath the relatively thin internal limiting membrane of the infant's eye. They may be present within 1 or 2 days after a child's presentation with CNS symptoms and initial ophthalmoscopic examination, but they can occur later. The interval required for vitreous hemorrhages to clear is predominantly related to their size; some hemorrhages may take 6 to 12 months to clear completely. Small superficial retinal hemorrhages may clear within 1 to 3 weeks, deeper and more massive retinal hemorrhages may take 4 to 5 months to clear. The

appearance of the fundus, however, changes within 2 to 4 days after the initial presentation of the child to the hospital. Initially, in the CNS–injured and symptomatic child, venous congestion is often present, the disc may be mildly to markedly edematous, and the retinal hemorrhages appear fresh and bright red. Choroidal hemorrhages may give rise to a dark or almost black appearance, however. Within a few days of treatment, if CNS edema and pressure are reduced, venous congestion decreases, disc edema lessens or disappears, and the hemorrhages begin to fade if they are small and superficial. Nonhemorrhagic and possibly ischemic retina appears almost white in contrast to the red and darker hemorrhagic areas that give rise to a reticular and necrotic appearance, particularly in the more darkly pigmented fundi (Fig. 155-1). Some of the larger and more massive hemorrhages may be surrounded by intraretinal yellowish or whitish demarcation lines. True exudates are rarely seen, although deposits suggestive of cholesterol crystals, as seen in Coats' disease, may also develop several months after the initial injury.

Chronic changes include chorioretinal or retinal scars, which may be permanent, intraretinal and preret-

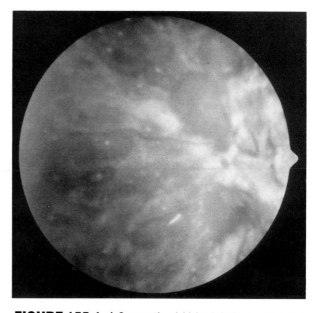

FIGURE 155-1 A 3-month-old black infant with bilateral massive intraretinal hemorrhages associated with CNS hemorrhage and periocular bruising. Note pale grayish nonhemorrhagic retina giving the fundus appearance of acute hemorrhagic necrosis. Breakthrough of blood into the vitreous obscured view of fundus 1 week later.

inal gliosis and fibrosis, loss of pigment epithelium, macular pigmentary disturbances, and distortion with absence of foveal reflexes and normal macular anatomy. Some of these changes, particularly in the macula, may eventually disappear. Intraretinal demarcation lines and apparent cysts may be present for years. Optic atrophy is frequently seen within a month or two after injury, usually in children who have concomitant significant CNS injury.

The visual prognosis is best for children with relatively superficial retinal hemorrhages and minor CNS injury. Blindness may result from cortical atrophy, optic atrophy, retinal ischemia, or necrosis. Clinically the retina may appear normal, yet the child may be totally blind secondary to cortical injury. Hemianopia is not an uncommon finding and in a young child may be missed unless it is specifically sought. Amblyopia can occur in cases of unilateral hemorrhage, particularly if it involves the macula. Patching treatment is recommended for infants whose macular hemorrhages or macular distortion persist for more than 1 month after injury.

In summary, the most common ophthalmic sign of child abuse is retinal and vitreous hemorrhage secondary to CNS injury. CT scans and lumbar punctures are important in determining whether CNS hemorrhage is present; however, results of these studies may be negative when CNS hemorrhage is actually present. Ophthalmic examination is extremely important in establishing the probable diagnosis of child abuse in children with CNS injuries whose history is unremarkable for trauma, and it should be performed in any child with symptoms or signs of unexplained CNS injury. Misdiagnosis and missed diagnosis of the CNS–injured physically abused child is still prevalent today. The incidence of retinal hemorrhages in CNS injuries secondary to shaken baby or battered child syndrome is not definitely established but is most likely greater than 90%.

References

1. Kempe CH, Silverman FN, Steele BF, et al. The battered child syndrome. *JAMA.* 1962;181:17.
2. The American Humane Association. *Highlights of Official Child Neglect and Abuse Reporting, 1982.* Denver, Colo: American Humane Association; 1984.
3. American Association for Protecting Children. *Highlights of Official Child Neglect and Abuse Reporting, 1984,* Denver, Colo: American Humane Association; 1986.
4. Helfer RE. Where to now, Henry: a commentary on the battered child syndrome. *Pediatrics.* 1985;76:993.
5. Wilson EF. Estimation of the age of cutaneous contusions in child abuse. *Pediatrics.* 1977;60:748.
6. Heins M. The ''battered child'' revisited. *JAMA.* 1984; 251:3295.
7. Merten DF, Radkowski MA, Leonidas JC. The abused child: a radiological reappraisal. *Radiology.* 1983;146:378.
8. Worlock P, Stower M, Barbor P. Patterns of fractures in accidental and non-accidental injury in children: a comparative study. *Br Med J.* 1986;293:100.
9. Caffey J. The whiplash shaken infant syndrome: manual shaking by the extremities with whiplash-induced intracranial and intraocular bleeding, linked with residual permanent brain damage and mental retardation. *Pediatrics.* 1974;54:396.
10. Zimmerman RA, Bilaniuk T, Bruce D, et al. Computed tomography of craniocerebral injury in the abused child. *Radiology.* 1979;130:687.
11. Ober RR. Hemorrhagic retinopathy in infancy: a clinicopathologic report. *J Pediatr Ophthalmol Strabismus.* 1980; 17:17.
12. Lambert SR, Johnson TE, Hoyt CS. Optic nerve sheath and retinal hemorrhages associated with the shaken baby syndrome. *Arch Ophthalmol.* 1986;104:1509.
13. Greenwald MJ, Weiss A, Oesterle CS, et al. Traumatic retinoschisis in battered babies. *Ophthalmology.* 1986;93:618.
14. Shigeno T, Artigas J, Katoh G, et al. Cerebral edema following experimental subarachnoid hemorrhage. In: Cervos-Navarro J, Fritschka E, eds. *Cerebral Microcirculation and Metabolism.* New York, NY; Raven Press; 1981;427-431.

Chapter 156

Decompression Sickness

FRANK K. BUTLER, Jr.

Decompression sickness refers to any systemic dysfunction produced by the formation of bubbles of inert gas in body tissues as a result of a decrease in the ambient pressure. Gas in solution obeys Henry's law, which states that the amount of gas in solution is a function of the ambient absolute pressure. When the ambient pres-

The opinions and assertions expressed are the private ones of the author and do not necessarily reflect the views of the Navy Department, the Naval Service in general, or the Department of Defense.

sure is reduced slowly, the excess inert gas (usually nitrogen) in solution is carried by the venous blood to the lungs, where it diffuses into the alveolar air. If the pressure is decreased too rapidly, gas comes out of solution and forms bubbles in the venous blood, periarticular tissues, or other parts of the body. Ocular involvement in this condition was first described in 1670 by Robert Boyle, who observed gas bubbles in the anterior chamber of the eye of a viper that had been experimentally exposed to increased pressure.

The presence of gas bubbles may or may not result in the clinical manifestations of decompression sickness. Doppler ultrasonography of asymptomatic divers following long, deep dives often detects the presence of venous gas emboli in the subclavian veins and heart.[1] These bubbles usually cause no clinical disease and are removed by the lungs unless they are present in such large amounts that pulmonary blood flow is impaired.[1] Divers who develop decompression sickness may have symptoms develop through a number of different mechanisms. Bubble accumulation in the epidural venous plexus can produce venous stasis infarction of the spinal cord.[2] Localized bubble formation in other parts of the body such as the inner ear and periarticular tissues may produce signs or symptoms referable to the involved region. Gas bubbles in the venous blood may also gain access to the arterial circulation through a functionally patent foramen ovale or pulmonary arteriovenous shunts. These arterial gas emboli travel to the distal circulation and produce tissue ischemia in the area served by the blocked vessel. Secondary inflammatory and thrombotic changes in the microvasculature may cause continued ischemia even after resolution of the gas embolus. Arterial gas emboli generally result in damage to the cerebral cortex.

Decompression sickness has been most commonly associated with diving, but it may also strike aviation crews, tunnel workers, and other persons who are exposed to sudden reductions in pressure. Prevention of decompression sickness has been a major factor in the selection of cabin atmospheres and space suit design for space flight.

Systemic Manifestations

Decompression sickness encompasses a spectrum of severity which ranges from minimally symptomatic to very rapidly fatal. Severity correlates roughly with the magnitude of the pressure change, the time spent at the increased pressure, and the time over which the reduction in pressure takes place, but there is wide individual variation in susceptibility to decompression sickness. Adherence to published decompression tables greatly reduces the incidence of this disease but does not

eliminate it entirely. The U.S. Navy classification system recognizes two categories of decompression sickness. Type 1 is less serious and includes cases limited to joint pain, skin changes, and lymphatic pain. Neurologic or cardiorespiratory symptoms are classified as Type 2 decompression sickness. Neurologic manifestations may include numbness, weakness, paresthesias, loss of consciousness, cranial nerve palsies, and peripheral neuropathies. Cardiorespiratory effects may include massive pulmonary gas embolism or myocardial infarction. This classification of decompression sickness provides a basis for therapy; both Type 1 and Type 2 are treated with recompression and hyperbaric oxygen, but Type 2 cases are treated with a longer and sometimes deeper recompression schedule. Fortunately, Type 1 decompression sickness is the most commonly encountered, with 89% of cases presenting as joint pain.[3] Onset generally occurs shortly after the reduction in pressure, with 55% of patients developing symptoms within 1 hour and 93% within 12 hours.[4] Recompression therapy for decompression sickness must be initiated as soon as possible after the onset of symptoms, since delay may result in rapid deterioration of the patient's condition leading to death or permanent neurologic damage.

Ocular Manifestations

Although there have been only isolated reports of ocular decompression sickness in the ophthalmic literature,[5] this entity is well-reported in the diving medicine literature.[1,4,6-9] Ocular manifestations include nystagmus, diplopia, tunnel vision, scotomas, homonymous hemianopia, orbicularis pain, cortical blindness, convergence insufficiency, and central retinal artery occlusion. The incidence of visual symptoms in patients with decompression sickness was found to be 7% in one large series[4], but incidences as high as 12% have been reported.[6] One study of altitude decompression sickness found that visual symptoms were the most common neurologic presentation observed.[8] Ophthalmologists seldom encounter the disease in an acute setting because diving medicine physicians are trained to treat suspected decompression sickness, including visual symptoms, with immediate recompression and hyperbaric oxygen therapy. Delays for referral to an eye physician in the acute setting are contraindicated. Standard treatment tables are very effective, with reported cure rates of 85 to 100%. Since complete resolution of symptoms is the usual outcome of treatment, most patients with visual symptoms prior to therapy are asymptomatic after treatment and are not referred to ophthalmologists for follow-up.

Failure to recognize decompression sickness, incomplete response to treatment, or a recurrence of symptoms following treatment may bring the patient with ocular decompression sickness to the eye physician. If an ophthalmologist encounters a patient with visual symptoms presenting shortly after diving or other exposure to increased or decreased ambient pressure, the patient should be referred to a diving medicine consultant on an emergency basis. Delays for diagnostic tests such as CT scans are contraindicated, especially when other signs or symptoms of decompression sickness are present. Patients who have had a significant interval since the onset of symptoms may be treated on a less emergent basis. Recompression therapy and hyperbaric oxygen therapy should be administered even when presentation is delayed, since such treatment may be effective despite delays of up to several weeks.[10] Although other entities may cause ocular dysfunction after diving,[11,12] the potential for rapid deterioration of the patient's condition requires that acute visual symptoms following exposure to increased or decreased ambient pressure be regarded as decompression sickness until proven otherwise.

References

1. Edmonds C, Lowry C, Pennefather J. *Diving and Subaquatic Medicine*. 2nd ed. Morman, Australia: Diving Medical Centre; 1981.

2. Hallenbeck J, Bove A, Elliott D. Mechanisms underlying spinal cord damage in decompression sickness. *Neurology*. 1975;25:308-316.

3. *U.S. Navy Diving Manual*. vol 1. rev 1. NAVSEA Publication 0994-lp-001-9010; Washington, DC: U.S. Department of the Navy; 1975.

4. Rivera JC. Decompression sickness among divers—an analysis of 935 cases. *Military Med*. 1964; 129:314.

5. Liepmann M. Accommodative and convergence insufficiency after decompression sickness. *Arch Ophthalmol*. 1981;99:453.

6. Summitt JK, Berghage TE. *Review of diving accident reports 1968*. Navy Experimental Diving Unit Research Report 11-70.

7. Van Der Aue O, Dufner G, Behnke A. The treatment of decompression sickness—an analysis of one hundred and thirteen cases. *J Industr Hyg Toxicol*. 1947;29:359.

8. Davis J, Sheffield P, Shuknecht L, et al. Altitude decompression sickness: hyperbaric therapy results in 145 cases. *Aviation Space Environ Med*. 1977;48:722.

9. Hart BL, Dutka AJ, Flynn ET. Pain-only decompression sickness affecting the orbicularis oculi. *Undersea Biomed Res*. 1986;13:461.

10. Butler F, Pinto C. Progressive ulnar palsy as a late complication of decompression sickness. *Ann Emerg Med*. 1986;15:736.

11. Simon D, Bradley M. Corneal edema in divers wearing hard contact lenses. *Am J Ophthalmol*. 1978;85:462.

12. Wright W. Scuba divers delayed toxic epithelial keratopathy from commerical mask defogging agents. *Am J Ophthalmol*. 1982;93:470.

Chapter 157

Altitude Illness

MICHAEL WIEDMAN

Altitude illness has become increasingly important as a syndrome encountered by large numbers of recreational athletes and tourists.[1] In the United States each year, 1 million hikers, skiers, and climbers rapidly ascend to altitudes over 3000 m, risking potentially hazardous medical consequences. Throughout the world, large population groups live above 3,000 m.[2] With increasing industrialization and access to these areas, there has been an influx of people who are not genetically conditioned to such high altitude.

High-altitude retinopathy (HAR) has a dual significance. Foremost are the ophthalmic consequences of retinal hemorrhages. Of further importance, HAR is an early warning signal of potentially lethal systemic altitude illness. Thus it is advantageous for ophthalmologists to appreciate the multiple, interrelated systemic components of altitude illness.

Altitude illness is composed of four clinical entities, any one or more of which may be present in varying degrees.[3] The subdivisions are acute mountain sickness (AMS), HAR, high-altitude cerebral edema (HACE), and high-altitude pulmonary edema (HAPE).

AMS is characterized by headache, insomnia, anorexia, lethargy, nausea, impotence, and disorientation. Predisposing factors include rapidity of ascent, high altitude, youth, and inadequate acclimatization. Treat-

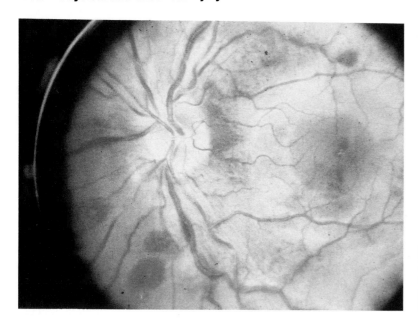

FIGURE 157-1 Diffuse intraretinal hemorrhages—grade IV A,B (1)—4 disc areas in size are located in the peripapillary paramacular and more peripheral retina.

ment involves hydration, high-carbohydrate diet, gradual acclimatization, oxygen or carbon dioxide, and acetazolamide.

HAR is characterized by increasingly dilated retinal veins and arteries, diffuse or punctate preretinal hemorrhages, usually located peripherally but occasionally in the macula, vitreous hemorrhage, papillary hemorrhage, peripapillary hyperemia, and papilledema.[1,4-6]. Predisposing factors include previous HAR, rapidity of ascent, and the presence of other components of altitude illness. Treatment is that of the altitude illness entities noted, including descent if the hemorrhages are diffuse or if the macula is threatened. Also, caution is advised regarding any subsequent reascent.

HACE is characterized by progressive severe headache, impaired cortical function and judgment, irrationality, projectile vomiting, diplopia, ataxia, depressed sensorium, and coma. Predisposing factors are AMS, rapid ascent, and high altitude. The treatment involves the use of osmotic agents, oxygen, and dexamethasone, and the patient must be transported to a lower altitude. The prognosis is guarded.

HAPE is characterized by dyspnea, tachypnea, dry cough, blood-tinged sputum, tachycardia, cyanosis, moist rales, respiratory failure, and death. Early treatment with oxygen, furosemide, and rapid return to lower altitude may be lifesaving. A portable hyperbaric bag may be useful to mimic descent.

Moderate to severe AMS is experienced by approximately one sixth of unacclimatized hikers, skiers, and climbers, who ascend to 3000 m and by one fourth of those who ascend to 5000 m. In contrast, HACE or HAPE occurs in 5 to 10% of persons at 5000 m, and the percentage rises steeply above this altitude. But the incidence of HAR is the most striking—50% in persons who ascend to 5000 m and approaching 100% in those who ascend above 6500 m.

The relationship of HAR to HACE is characterized by contiguous pathophysiology. If the fundi are examined carefully and with dilation, retinal hemorrhages or vascular enlargement is seen in all cases of clinically evident cerebral edema. This progresses to papilledema in severe HACE. A process similar to the observed altitude retinopathy is occurring simultaneously in the brain. This concurrent cerebral edema and cerebral hemorrhage develops because of hypoxia. Such edematous and hemorrhagic changes were noted by Singh and colleagues during skull trephinations.[7] Therefore, the fundi of patients with suspected or frank cerebral edema should be searched carefully for HAR as corroborating evidence and for guidance in treatment.

Proper medical care dictates that sojourners ascending to high altitude be examined funduscopically and monitored for HAR as an early warning sign of the impending, more serious aspects of altitude illness.

Classification of High-Altitude Retinopathy

A systematic method of reporting the incidence and evaluating the severity of HAR is useful for both visual

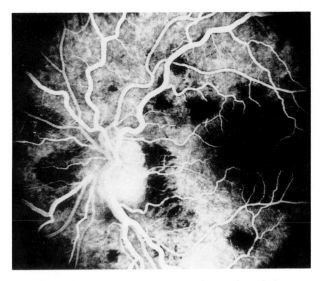

FIGURE 157-2 Fluorescein angiography of the patient in Figure 157-1 shows widely dilated arterioles and venules, no leakage, and complete hemorrhagic block of underlying fluorescence.

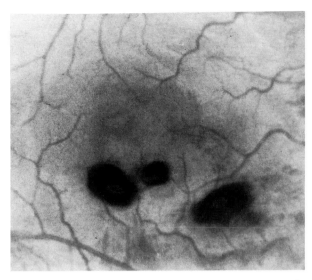

FIGURE 157-3 Paramacular hemorrhages—grade IV A,B (1,2)—with encroaching intraretinal hemorrhages lie nasal to fovea.

and systemic prognoses.[5] Since such hemorrhages have a significant relationship to AMS and to cerebral edema, the classification of HAR is of prognostic value in systemic altitude illness. A classification system is presented here:

Grade I
  A. In grade I HAR dilated retinal veins are present. These are seen in increasing degrees through all subsequent classified grades. Although retinal arterioles also dilate with hypoxia, this is not readily evident. A venule-to-arteriole ratio of approximately 3:2 persists, with a darker venous hue.
  B. The hemorrhage extends up to 1 disc area. Disc area is a more appropriate measure than disc diameter, since the latter is more variable with regard to disc height and width. Retinal hemorrhages may be intraretinal and preretinal. Intraretinal hemorrhages are characterized as being under or adjacent to retinal vessels, so that a clear view of such vessels is retained. Preretinal hemorrhages occur anteriorly, obscuring underlying vessels.

Grade II
  A. In grade II HAR, there are moderately dilated retinal veins, with a venule-to-arteriole ratio of 3.5:2.
  B. The hemorrhages extend up to 2 disc areas.

Grade III
  A. Grade III HAR has greatly dilated retinal veins, with a venule-to-arteriole ratio of 4:2.
  B. 1. Hemorrhages may extend up to 3 disc areas.
     2. Hemorrhages may be paramacular.
     3. Hemorrhages may be vitreous (diffuse or dense, minor, extending less than 3 disc areas).
Grade IV (Figs. 157-1 to 157-3)
  A. Retinal veins in a grade IV HAR are engorged, with a venule-to-arteriole ratio of 4.5:2. Venous hue is blue-purple.
  B. 1. Hemorrhages may cover more than 3 disc areas.
     2. Hemorrhages may be macular (distinctly within the macular zone).
     3. Hemorrhages may be vitreous (diffuse or dense, major, extending more than 3 disc areas).
     4. Papilledema may be apparent, or measured, greater than 2 diopters.

Additional Classification Data

Grading should be specific (eg, OD: grade III A,B (1,2,3); OS: grade IV A,B (1,2). In addition, the physician should document baseline visual acuity and any subsequent vision loss. The presence of concurrent HACE or HAPE should be recorded.

It is important to report the altitude at which HAR and any systemic complications were first noted. Finally, treatment and resolution of the case also warrant reporting.

Altitude sickness occurs at the moderate altitudes encountered in skiing at popular resorts or hiking in the Sierras, the Alps, or the Rocky Mountains. As one ascends to altitudes higher than 3000 m, altitude illness of increasing severity occurs, occasionally in life-threatening proportion.

HAR may be present in association with AMS and without clinical evidence of HACE or HAPE. However, and very importantly, if a patient has the symptoms or signs of cerebral edema, careful fundus examination will show a corresponding grade of HAR. Furthermore, if the fundi of a patient with HACE were monitored during the days or weeks prior to the development of clinical cerebral changes, a finding of HAR would indeed be an important early warning to the physician of impending more serious cerebral complications.

It is incumbent on physicians in high-altitude ski, touring, and climbing areas to recognize this relationship, perform funduscopic examinations, grade any existing HAR, and be alert to the development of further systemic altitude illness.

References

1. Wiedman M. High altitude retinal hemorrhage. *Arch Ophthalmol.* 1975;93:401.
2. West JB, Sukhamay L. Preface. In: West JB, Sukhamay L, eds. *High Altitude and Man.* Bethesda, Md: American Physiologic Society; 1984;v.
3. Houston CS. High altitude illness. *JAMA.* 1976;236:2193.
4. Wiedman M. Les hémorrhagies retiniennes des sommets. *Concilium Ophthalmologicum 1974 (Paris).* 1975;22(2):544.
5. Wiedman M. High altitude retinal hemorrhages: a classification. In: Henkind P, ed. *Acta XXIV International Congress of Ophthalmology.* Philadelphia, Pa: JB Lippincott; 1980; 421-424.
6. Lubin JR, Rennie D, Hackett P, et al. High altitude retinal hemorrhages: a clinical and pathological case report. *Ann Ophthalmol.* 1982;14:1071.
7. Singh I, Khana PK, Srivastava MC. Acute mountain sickness. *N Engl J Med.* 1968;280:175.

Chapter 158

Purtscher's Retinopathy

ELAINE L. CHUANG and ROBERT E. KALINA

Ocular Manifestations of Distant Trauma

Ocular involvement following injuries that do not involve direct trauma to the eye has been classified under a variety of syndromes, all involving the posterior fundus. Most notable among them are posttraumatic fat embolism syndrome, traumatic asphyxia (cyanosis), Valsalva retinopathy, hydrostatic pressure syndrome, and traumatic retinal angiopathy (Purtscher's retinopathy).[1-5] The fundus may have a similar appearance in each of these conditions, and they may be distinguishable only on the basis of history and extraocular signs and symptoms.

Purtscher's Retinopathy

The most widely recognized entity following remote injury is traumatic retinal angiopathy (lymphorrhagia retinae, traumatic liporrhagia retinalis, retinal teletraumatism), first described by Purtscher in 1910.[6] The five patients included in his original and subsequent publication had cotton-wool spots and intraretinal hemorrhages involving the posterior fundus, as well as edema of the macula and peripapillary retina. Purtscher believed the ocular involvement to be related to head trauma and postulated a hydrostatic mechanism of retinal vascular injury via lymphatic and cerebrospinal fluid channels transmitted through the optic nerve sheath and the perivascular sheaths of retinal vessels.

Though the mechanism proposed by Purtscher has been discarded on anatomic grounds, there is still controversy over the pathogenesis of the retinal vasculopathy. The issue is further clouded by extension of the designation "Purtscher's retinopathy" to include sequelae of nontraumatic disorders such as acute pancreatitis, but such associations may provide clues to common pathophysiology.

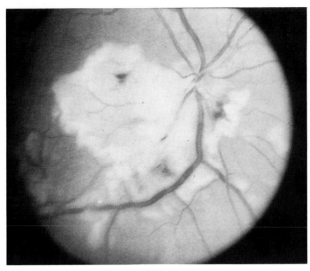

(A)

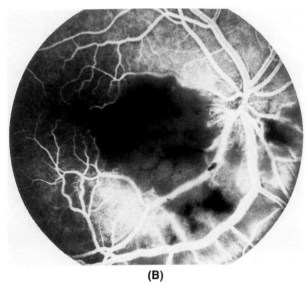

(B)

FIGURE 158-1 **(A)** Right fundus of a 39-year-old man with bilateral Purtscher's retinopathy. White area of retinal infarction extends from disc margin into fovea. Small retinal hemorrhages are also present. **(B)** Early phase fluorescein angiography demonstrates vascular occlusion and severe nonperfusion in the macular area.

Ocular Manifestations and Clinical Course

The usual mechanism of injury is chest compression that may have seemed otherwise quite innocuous. The typical fundus picture appears soon after injury and consists of cloudlike opacities located superficially in the retina.[2,5-7] These cotton-wool spots are ischemic infarcts of the nerve fiber layer and are located primarily in the posterior fundus. The retinal veins frequently are engorged. Hemorrhages are generally superficial but may be preretinal. Optic nerve swelling is frequently associated, as is macular thickening. Alert patients invariably report blurred vision, perhaps moments after the injury. Initial visual loss may be mild or severe.

Cases of unilateral trauma-associated Purtscher's retinopathy have been reported. Most reports documented unilateral orbital or skull fractures or other evidence of direct trauma to the region of the globe,[8] but truly unilateral cases following distant trauma have been described. Venous dilatation confined to the side of local injury suggests a distal site of vascular compromise in these patients.

Fluorescein angiography has been carried out within hours following injury in patients with Purtscher's retinopathy.[9] The earliest studies documented retinal capillary nonperfusion and arteriolar and venular closure matching the regions involved by ophthalmoscopy (Fig. 158-1). Late staining of vessel walls is characteristic, as is optic disc edema. Choroidal perfusion has been noted to be angiographically normal.

New retinal hemorrhages and cotton-wool spots may appear, or lesions may extend over succeeding days. Such a phenomenon remains compatible with the concept of a single acute vascular insult as signs of evolving ischemia become more clinically evident. With resolution of the acute picture, retinal atrophy and optic disc pallor develop in the more severe cases. Alterations remain confined to the inner retina, with the exception of central retinal pigment epithelial granularity in some.

Pathogenesis

Historically, fat embolism had been postulated most frequently to be responsible for retinal vascular occlusion in Purtscher's retinopathy. It certainly may coexist in cases associated with fractures, specifically those involving long bones. In the absence of fractures, however, there is no clinical or histologic evidence to support fat embolism as the cause.

Intra-arterial embolism of air into the retina also has been proposed, paralleling widespread organ involvement seen after experimental explosive chest compression. However, Purtscher's retinopathy is characteristically seen following much milder injuries and does not include evidence of branch retinal artery occlusion.

More acceptable has been the concept of intravascular transmission of a hydrostatic pressure wave. When it originates from chest compression, as exemplified by the development of Purtscher's retinopathy following motor vehicle accident seat belt injury,[9] such an impulse can pass unimpeded through the valveless venous drainage system of the head and neck to reach the retinal veins. Thereafter, either vascular endothelial damage may occur directly or retinal vascular autoregulation may lead to active constriction and subsequent damage. The latter sequence of events is supported by the apparent absence of involvement of the choroidal circulation, a nonautoregulated system. Similar mechanisms of increased intracranial venous pressure may explain the picture following head trauma alone. Trauma such as whiplash injury, birth asphyxia, and acceleration or deceleration injuries may also lead to the same intraluminal phenomena.

The particular distribution of lesions in Purtscher's retinopathy also can be explained on the basis of vascular anatomy.[10] In the peripapillary region where the retina is thickest, the capillary network is particularly extensive and includes a radial peripapillary system. This arrangement—plus proximity to the central retinal vessels—may place this area at highest risk for damage. Involvement of one eye alone may reflect factors such as individual variations in large vessel structure or relative protection of one side of the circulation by position of the head or neck at the time of injury. Other persons may suffer circulatory shock or more proximal vascular injury that adds further insult to perfusion.

It is not surprising that a uniform interpretation does not exist for the cause of the retinal vascular damage in Purtscher's retinopathy. The nature of trauma provides ample opportunity for the interaction of various pathophysiologic mechanisms, even in instances in which direct ocular trauma can be excluded.

Different mechanisms probably lead to the retinal manifestations in pancreatic disease. This uncommon association has been described solely in the setting of long-standing alcohol abuse complicated by acute pancreatitis.[11] Fat emboli have been demonstrated in extraocular sites in fatal cases; one case of histologic examination of the eyes performed 3 weeks after the onset of the retinopathy documented occlusion of retinal and choroidal vessels with fibrin, but lipid was not sought by histochemical techniques at this late stage.[12]

An alternative pathogenic mechanism in acute pancreatitis is suggested by the finding of increased complement levels in such patients.[13] Granulocyte aggregation occurs in the presence of complement component C5a, thought to be generated by activated pancreatic proteases. It is postulated that this leads to leukoembolization and microinfarction in various locations, including the retina. The simultaneous development of acute renal failure in such patients has been described and may provide supporting evidence for this mechanism. Available histologic material has not yet demonstrated the presence or absence of such microemboli.[14]

These patients show identical ophthalmoscopic findings, except for the lack of venous dilatation. In addition, all cases of retinopathy associated with acute pancreatitis published in the English literature have been bilateral. Such features support a different pathophysiology than that seen in posttraumatic retinal angiopathy.

Differential Diagnosis

Because the retina has only a limited repertoire of responses, many other conditions may produce a fundus picture identical or similar to that of Purtscher's retinopathy. Details of history and associated systemic symptoms and signs should be sufficient in most cases to differentiate traumatic retinal angiopathy from other entities characterized by nerve fiber infarcts and retinal hemorrhages (Table 158-1).

Prognosis and Treatment

The natural history of Purtscher's retinopathy tends to be one of functional and vascular recovery. Though most patients regain nearly normal central visual acuity within weeks to months, angiographic, visual field, and other psychophysical evidence of prior compromise may be detectable permanently.[5] Visual outcome is determined by the initial region and severity of involvement but is usually 20/40 or better in the absence of severe retinal atrophy or direct optic nerve injury. No late complications have been described.

No effective therapy for Purtscher's retinopathy has yet been demonstrated or systematically evaluated. There may be a theoretical rationale for the use of corticosteroids or other agents such as vasodilators in severe cases if sequelae to the initial injury include

TABLE 158-1 Systemic Disorders that Mimic Purtscher's Retinopathy

Diabetes mellitus
Systemic hypertension
Severe anemia
Leukemia
Collagen vascular diseases
Systemic lupus erythematosus
Dermatomyositis
Scleroderma
Acquired immunodeficiency syndrome
Amniotic fluid embolism
Disseminated intravascular coagulation
Fat embolism syndrome
Other endogenous embolization
Atrial myxoma
Bacterial endocarditis

inflammatory or other responses of a cellular nature. For example, the possibility of mechanisms such as leukocytic aggregation, as postulated in acute pancreatitis, has not been assessed in posttraumatic retinal angiopathy. In pancreatitis iself, inhibition of postulated protease activity is not recommended because hypercoagulability may result.[13]

References

1. Chuang EL, Miller FS, Kalina RE. Retinal lesions following long bone fractures. *Ophthalmology.* 1985;92:370-374.
2. Duke-Elder S, MacFaul PA. *System of Ophthalmology,* vol. xiv. part 1. St. Louis, Mo: CV Mosby; 1972;693-733.
3. Duane TD. Valsalva hemorrhagic retinopathy. *Am J Ophthalmol.* 1973;75:637-642.
4. Lyle DJ, Stapp JP, Button RR. Ophthalmologic hydrostatic pressure syndrome. *Am J Ophthalmol.* 1957;44:652-657.
5. Archer DB. Richardson Cross lecture: traumatic retinal vasculopathy. *Trans Ophthalmol Soc UK.* 1986;105:361-384.
6. Purtscher O. Angiopathia retinae traumatica. Lymphorhagien des augengrundes. *Graefes Arch Ophthalmol.* 1912;82:347-371.
7. Marr WG, Marr EG. Some observations on Purtscher's disease: traumatic retinal angiopathy. *Am J Ophthalmol.* 1962;54:693-705.
8. Burton TC. Unilateral Purtscher's retinopathy. *Ophthalmology.* 1980;87:1096-1105.
9. Kelley JS. Purtscher's retinopathy related to chest compression by safety belts: fluorescein angiographic findings. *Am J Ophthalmol.* 1972;74:278-283.
10. Wise GN, Dollery CT, Henkind P. The retinal circulation. New York, NY: Harper & Row; 1971;22-29.
11. Inkeles DM, Walsh JB. Retinal fat emboli as a sequela to acute pancreatitis. *Am J Ophthalmol.* 1975;80:935-938.
12. Kincaid MC, Green WR, Knox DL, et al. A clinicopathological case report of retinopathy of pancreatitis. *Br J Ophthalmol.* 1982;66:219-226.
13. Jacob HS, Craddock PR, Hammerschmidt DE et al. Complement-induced granulocyte aggregation. *N Eng J Med.* 1980;302:789-794.
14. Pratt MV, de Venecia G. Purtscher's retinopathy: a clinicohistopathological correlation. *Surv Ophthalmol.* 1970;14:417-423.

Chapter 159

Tobacco-Alcohol Amblyopia

MARK J. KUPERSMITH

The optic neuropathy associated with tobacco and alcohol abuse is bilateral, painless, and progressive if left untreated.[1] Most of the cases occur in men over age 30. Acuity loss, dyschromatopsia, and central or centrocecal visual field defects occur early in the illness without significant abnormalities of the fundus. The acuity rarely falls below 20/400. The centrocecal or central field defects are usually symmetric but may be asymmetric, usually sparing the periphery. The pattern and degree of vision loss and optic disc pallor can be indistinguishable from optic neuropathy of various nutritional causes or from "Jamaican" optic neuropathy.[2] Usually a central area of greater density of field loss exists, while the remainder of the scotoma does not become more uniformly dense until late in the course of untreated cases. In early cases, a scotoma for perception of a red test object may be the only field defect determinable. Though early in the disease the optic disc appears normal, after several months temporal pallor and thinning of the papillomacular nerve fiber layer develop. Eventually in an untreated patient, temporal disc pallor becomes more extensive, with severe loss of the papillomacular bundle. With treatment of tobacco-alcohol optic neuropathy, the scotoma retracts from the temporal area closest to fixation first and acuity improves, unless the case is of long standing associated with a permanent defect and severe atrophy of the disc.

The combination of alcohol, tobacco, and poor nutrition probably have a synergistic effect on the optic nerve. Severe malnutrition alone causes optic neuropathy as well as other neurologic defects.[3] Vitamin B_{12} and vitamin B_1 deficiencies are also associated with optic neuropathy.[4,5] The toxic effect of alcohol on the central nervous system when nutrition is maintained is less severe than in cases when malnutrition is present. Some authors believe alcohol abuse or tobacco, in particular pipe or cigar smoking,[6] can cause optic neuropathy. In contrast, other authors report recovery of vision with improved nutrition even when tobacco use continues.[7,8] The biochemical hypotheses of the toxic role of tobacco on vitamin B_{12} metabolism has been described in many reviews.[1] The elevated serum thiocyanate level found in normal smokers has led to the suggestion that cyanide released into the blood by tobacco smoke may accumulate in the body tissues and serve as a toxin. Cyanide can be detoxified by conversion to thiocyanate via conjugation with sulfur, and patients with optic neuropathy may not have the expected rise in thiocyanate, suggesting defective metabolism of cyanide. A diet low in protein could result in low blood levels of sulfur-containing amino acids such as cystine, cysteine, and methionine, which can lead to inadequate thiocyanate production. The hydroxyl group of naturally occurring vitamin B_{12} (hydroxycobalamin) exchanges with excessive tissue levels of cyanide.[9] Alcoholics may have a diminished ability to detoxify cyanide because of altered vitamin B_{12} metabolism. Alcoholics can have low vitamin B_{12} levels because of poor diet, poor absorption of B_{12} from the ileum because of diminished intrinsic factor levels,[10] or severe liver disease, which impairs B_{12} storage. Cyanocobalamin, the vitamin B_{12} in most multivitamins, will not exchange with the cyanide because of the presence of a cyano group. Increased serum levels of cyanocobalamin, presumably an inactive form of vitamin B_{12} in cyanide detoxification, have been described in patients with tobacco-alcohol optic neuropathy.[11]

The role of cyanide toxicity as the cause of tobacco-alcohol optic neuropathy is far from clear. Large doses of cyanide were required to cause demyelinating lesions of the optic nerves in only 25% of the rats treated. These animals had additional significant lesions in multiple areas of the brain leading to seizures and coma,[12] which are not seen in uncomplicated cases of tobacco-alcohol optic neuropathy. Additionally, nonhuman primates on a vitamin B_{12}–free diet that were given repeated small doses of cyanide typical of tobacco consumption failed to develop optic neuropathy.[13] However, this negative finding may be an artifact related to the long time needed to cause a true vitamin B_{12} deficiency in primates[14] or to develop cyanide toxicity. Cyanide-induced demyelinating lesions may not exactly parallel the lesions, including axonal loss, in the optic nerves of patients with tobacco-alcohol optic neuropathy.[15,16]

There is also little evidence that an alteration in vitamin B_{12} metabolism causes tobacco-alcohol optic neuropathy. Serum vitamin B_{12} levels are usually normal in patients with tobacco-alcohol optic neuropathy. Also, patients with tobacco-alcohol optic neuropathy do not develop the myelopathy or severe megaloblastic anemia that can result from overt B_{12} deficiency in pernicious anemia[17] or from nitrous oxide–induced dysfunction of B_{12} metabolism at a biochemical level.[18] Patients with tobacco-alcohol optic neuropathy do not

have a delayed visual evoked potential, which finding is typically seen in patients with pernicious anemia or multiple sclerosis.[19,20]

Since controlled studies of the multiple factors that could possibly cause tobacco-alcohol optic neuropathy have not been done, it is impossible to identify which of the above-mentioned mechanisms might be the cause of the optic neuropathy in these patients. It is likely that each factor plays a variable role, depending on the individual patient. There must be some metabolic susceptibility to the effects of tobacco and alcohol in these patients, because most persons who use ethanol and tobacco do not develop optic neuropathy. There is little disagreement that these patients should stop using alcohol and tobacco and consume a well-balanced diet with adequate amounts of complete protein and vitamins. Hydroxycobalamin is more effective than cyanocobalamin for improving vision.[1] Parenteral hydroxycobalamin, 1000 μg, should be administered several times per week for at least 1 month and then weekly for several months. A full 6 months' therapy should be given before it is concluded that there has been no benefit. The main difficulty in treating these patients lies in the basic problem of the addiction to alcohol and tobacco. Many patients have continued poor nutrition and continue to use alcohol, tobacco, or both despite the direst predictions of visual loss.

References

1. Brockhurst RJ, Boruchoff SA, Hutchinson BT, et al. *Controversy in Ophthalmology.* Philadelphia, Pa: WB Saunders; 1977;843-874.
2. Carroll FD. Jamaican optic neuropathy in immigrants to the United States. *Am J Ophthalmol.* 1971;71:261-265.
3. Obal A. Malnutrition amblyopia. *Am J Ophthalmol.* 1951;34:857-865.
4. Foulds WS, Chisholm IA, Stewart JB, et al. The optic neuropathy of pernicious anemia. *Arch Ophthalmol.* 1969;82:427-432.
5. Hoyt CS, Billson FA. Optic neuropathy in ketagenic diet. *Br J Ophthalmol.* 1979;63:191-194.
6. Harrington DO. Amblyopia due to tobacco, alcohol, and nutritional deficiency. *Am J Ophthalmol.* 1962;53:967-972.
7. Carroll FD. The etiology and treatment of tobacco-alcohol amblyopia. Part I. *Am J Ophthalmol.* 1944;27:713-715.
8. Potts AM. Tobacco amblyopia. *Surv Ophthalmol.* 1973;17:313-339.
9. Boxer GE, Rickards JC. Studies on metabolism of carbon of cyanide and thiocyanate. *Arch Biochem.* 1952;39:7-26.
10. Lindenbaum J, Lieber CS. Alcohol-induced malabsorption of vitamin B_{12} in man. *Nature.* 1969;224:806.
11. Wilson J, Linnell JC, Matthews DM. Plasma-cobalamin in neuro-ophthalmological diseases. *Lancet.* 1971;I:259-261.
12. Lessell S. Experimental cyanide optic neuropathy. *Arch Ophthalmol.* 1971;86:194-204.

13. Bronte-Stewart J, Pettigrew AR, Foulds WS. Toxic optic neuropathy and its experimental production. *Trans Ophthalmol Soc UK*. 1976;96:355-358.

14. Agamanolis DP, Chester EM, Victor M, et al. Neuropathology of experimental vitamin B_{12} deficiency in monkeys. *Neurology*. 1976;26:905-914.

15. Victor M, Mancall EM, Dreyfus PM. Deficiency amblyopia in the alcoholic patient. *Arch Ophthalmol*. 1960;64:1-33.

16. Victor N, Dreyfus PM. Tobacco-alcohol amblyopia. *Arch Ophthalmol*. 1965;74:649-657.

17. Adams RD, Kubik CS. Subacute combined degeneration of the brain in pernicious anemia. *N Engl J Med*. 1934;231:1-9.

18. Layzer RB. Myeloneuropathy after prolonged exposure to nitrous oxide. *Lancet*. 1978;2:1227-1230.

19. Kupersmith MJ, Weiss PA, Carr RE. The visual-evoked potential in tobacco-alcohol and nutritional amblyopia. *Am J Ophthalmol*. 1983;95:307-314.

20. Halliday AM, McDonald WI, Mushin J. Delayed visual evoked response in optic neuropathy. *Lancet*. 1972;1:982-985.

PART 15

Pregnancy

Chapter 160

Pregnancy

KATHLEEN B. DIGRE and MICHAEL W. VARNER

Pregnancy is associated with complex metabolic, hormonal, and vascular changes. Many are mediated via the 30 to 50% increase in maternal blood volume and cardiac output during normal pregnancy or the progesterone-mediated retention of extracellular fluid. Although normal pregnancy should not produce permanent or serious changes in the eye, it may affect eye diseases, and an abnormal pregnancy may have serious ocular consequences.

Normal Pregnancy

Many women complain of some visual difficulties, especially late in pregnancy. These most frequently include minor refractive changes (especially hyperopia) and difficulty with accommodation. The changes may be partly related to the progesterone-mediated increase in fluid content of the cornea or lens, respectively. They resolve by 6 weeks postpartum. New refractions or fitting for contact lenses, especially in the 3rd trimes-ter, should be deferred until at least several weeks postpartum, to allow these physiologic changes to regress.[1]

In normal pregnancies, intraocular pressure usually falls in the second half of pregnancy, returning to prepregnancy pressures by 2 months postpartum. Although the precise mechanism for the drop is unknown, some have postulated reduced venous pressure in the head, which occurs because of increased venous capacitance in the lower extremities.

Occasionally, women experience mild blepharoptosis in pregnancy. Although this may be cosmetically bothersome, women can be assured that there is little long-lasting effect and that the condition will probably improve after pregnancy.[2] Pregnancy may also be associated with eyelid hyperpigmentation, owing to a progressive increase in the titer of melanocyte-stimulating hormone.

In normal pregnancy the retina does not change. Retinal sheen, often considered by obstetricians to be a sign of preeclampsia, is a normal finding in the retinas of young males and females and has no diagnostic significance.

There are no known changes in the optic nerves and chiasm in pregnancy. While there have been reports in the older literature of constriction of visual fields and visual field defects, these changes have not been substantiated by recent studies.

Effects of Pregnancy on Eye Disease

Diabetes is a relatively common disease in women of reproductive age, and pregnancy has many effects on diabetes, such as fasting hypoglycemia, postprandial hyperinsulinemia, and elevated levels of free fatty acids and triglycerides. As a result, there now exists a sizable body of information about diabetic retinopathy in pregnancy. In any diabetic patient, pregnant or not, retinopathy is related primarily to the duration of the disease.[3] Women with gestational diabetes are not at great risk of developing retinopathy in pregnancy, but those who have had the disease for more than 10 to 15 years have a 63 to 82% chance of having retinopathy at the beginning of pregnancy. In a controlled study,[4] 15% of a population of pregnant women with diabetes developed benign retinopathy and 29% showed progression of benign retinopathy; however, at 6 months postpartum, the amount of benign retinopathy in the group regressed to control levels. Therefore, benign diabetic retinopathy may worsen during pregnancy but may also regress postpartum. During pregnancy, the presence of untreated proliferative retinopathy appears to increase the risk of loss of vision. Photocoagulation treatment of the retinopathy before pregnancy may prevent worsening[3]; however, laser photocoagulation may be performed during pregnancy if indicated.

Many of the vascular eye tumors, such as hemangiomas—whether they are in choroid, retina, orbit, or brain—first become symptomatic in pregnancy because of the increased maternal blood volume and cardiac output.[1] Experience with these tumors in pregnancy is limited.

Uveal melanomas may present more commonly than expected in pregnancy. Although no estrogen receptors were encountered on the tumors, the investigators proposed that other unknown hormone alterations that take place in pregnancy may be responsible.[5]

Pituitary tumors, especially prolactin-secreting microadenomas, can become symptomatic in pregnancy. With the advent of bromocriptine therapy, many women with pituitary tumors have not required surgery. Women with microadenomas may experience not only the normal modest increase in size of the pituitary, but some (up to 35%) may acquire a symptomatic

pituitary enlargement that causes headache or bitemporal field deficits. Treatment is necessary when vision is threatened.[1] Meningiomas have been shown to be clearly influenced by pregnancy, since many of them contain estrogen receptors and may rapidly increase in size with the normal rise of the blood estrogen level in pregnancy.

Pseudotumor cerebri, or benign intracranial hypertension, is a syndrome of papilledema with increased intracranial pressure without localizing signs. Its diagnostic clinical signs are normal computed tomographic (CT) scan, elevated opening pressure on lumbar puncture, and normal cerebrospinal fluid indices. It may cause symptoms during pregnancy. It is not known whether pregnancy actually exacerbates the condition or whether pregnancy is an unrelated association, since benign intracranial hypertension is most frequently seen in women of child-bearing age. In either case, pregnancy per se, does not affect eventual visual outcome.[6] These women require close follow-up with frequent visual field examinations. If vision is threatened, surgical treatment—usually optic nerve sheath decompression—may be indicated.

Changes in the Eye Associated with Pathologic States of Pregnancy

Severe preeclampsia and eclampsia are pregnancy-specific conditions that can affect the visual pathway in many ways. The syndrome of preeclampsia consists of hypertension, proteinuria, and edema without evidence of underlying microvascular disease. The clinical manifestations are the results of generalized arteriolar vasospasm and loss of capillary integrity. Severe preeclampsia is a more advanced state of preeclampsia, and the diagnosis of eclampsia is made when the woman develops seizures or coma in addition to the above findings. These pathophysiologic changes may affect any maternal organ system, including the eye and the visual system. The visual system may be affected anywhere from the retina, choroid, and optic nerve to the occipital pole. However, the effects of preeclampsia or eclampsia on the eye are primarily in the retina, choroid, and optic nerve. The retinal vasculature is often involved. Frequently seen changes include segmental retinal arteriolar spasm, retinal ischemia, cotton-wool spots, retinal hemorrhage, and retinal edema.[7] The finding of "retinal sheen," which many obstetric textbooks report as a common abnormality, is normal in young, healthy retinas.

Since the pathophysiology of preeclampsia and

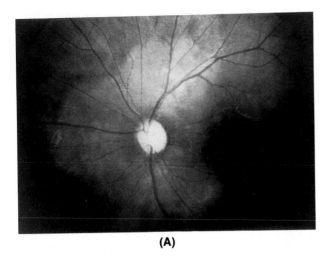

(A)

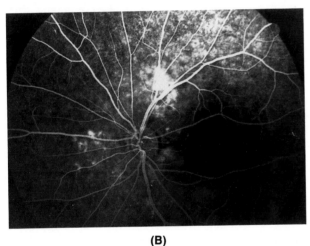

(B)

FIGURE 160-1 **(A)** Fundus photo of a G1, P1 woman with severe preeclampsia at 35 weeks gestation. There is a typical peripapillary serous retinal detachment present. **(B)** Fluorescein angiogram on the same eye shows patchy fluorescence consistent with choroidal infarction.

eclampsia involves primarily the microvasculature, it is not surprising that the choroid, an extremely vascular structure, is also involved in the disease. In fact, choroidal ischemia and infarction are probably the most common findings in toxemia.[8] Ischemic injury to the choroid may cause Elschnig's spots, focal infarcts of the choroid and retinal pigment epithelium. They appear acutely as yellow-white focal lesions deep in the retina and chronically as small yellowish scars with central pigment deposits. With fluorescein angiography, leakage from the choriocapillaris can also be seen. These changes in the choroid resemble those found in experi-

mental models of malignant hypertension.[9] Damage to the choroid may also cause a collection of fluid, which can produce serous retinal detachments. If the macula is not involved, these women may be asymptomatic. Frank retinal detachments occur in approximately 1% of patients with severe preeclampsia and eclampsia (Fig. 160-1).

Since the disc's blood supply is from the same origin as the choroid's, disc edema can occur as well; the swelling may represent focal ischemia to the optic nerve.[10] Postpartum fluorescein angiography often shows abnormalities of the choroid and evidence of papilledema. For unknown reasons, delivery of the baby usually causes improvement in all signs of preeclampsia. Fortunately, the eye abnormalities—even retinal detachment—frequently resolve spontaneously after delivery.

Cortical blindness due to occipital lobe ischemia has been reported. Although it is not a common occurrence, profound vision loss can develop. Recovery usually occurs unless there is significant infarction. Presumably, the vision loss is due to microvascular infarcts in the brain, often at the gray-white junction.

Postpartum and Lactation Effects on the Eye

It is not known whether there are any puerperal or lactational effects on the eye beyond those associated with the resolution of pregnancy-associated changes. There are reports in the early literature of lactation neuritis—unilateral or bilateral optic neuritis in the postpartum period. Lactation neuritis may be an incidental occurrence.

Pregnancy has multiple effects on every organ of the body; it is to be expected that the eye may reflect these changes and may be influenced by conditions that are peculiar to pregnancy.

References

1. Teich SA. Common disturbances of visual and ocular movement and surgery of the eye in the pregnant patient. In: Gleicher N, ed. *Principles of Medical Therapy in Pregnancy.* New York, NY: Plenum Medical Book Company; 1985; 861-874.
2. Sanke RF. Blepharoptosis as a complication of pregnancy. *Ann Ophthalmol.* 1984;16:720-722.
3. Dibble CM, Kochenour NK, Worley RJ, et al. Effect of pregnancy on diabetic retinopathy. *Obstet Gynecol.* 1981;59:699-704.
4. Moloney JBM, Drury MI. The effect of pregnancy on the natural course of diabetic retinopathy. *Am J Ophthalmol.* 1982;93:745-756.
5. Seddon JM, MacLaughlin DT, Albert DM, et al. Uveal

melanomas presenting during pregnancy and the investigation of oestrogen receptors in melanomas. *Br J Ophthalmol.* 1982;66:695-704.

6. Digre KB, Varner MW, Corbett JJ. Pseudotumor cerebri in pregnancy. *Neurology.* 1984;34:721-729.

7. Jaffe G, Schatz H. Ocular manifestations of preeclampsia. *Am J Ophthalmol.* 1987;103:309-315.

8. Fastenberg DM, Fetkenhour CL, Choromokos E, et al. Choroidal vascular changes in toxemia of pregnancy. *Am J Ophthalmol.* 1980;89:362.

9. Hayreh SS, Servais EG, Virdi PS. Fundus lesions in malignant hypertension: VI Hypertensive Choroidopathy. *Ophthalmology.* 1986;93:1383-1400.

10. Beck R, Gamel JW, Willcourt RJ, et al. Acute ischemic optic neuropathy in severe preeclampsia. *Am J Ophthalmol.* 1980;90:342.

PART 16

Pulmonary Disorders

Chapter 161

Cystic Fibrosis

JOHN D. SHEPPARD and DAVID M. ORENSTEIN

Cystic fibrosis (CF) is the most common cause of chronic obstructive pulmonary disease and pancreatic insufficiency during the first 3 decades of life. It is the most common lethal inherited disease of Caucasians, displaying autosomal recessive inheritance with nearly complete penetrance. About 5% of white Americans carry the CF gene. Assuming an incidence of 1 in every 2000 live births, over 1500 white infants are born with CF every year in the United States. In contrast, 1 in 17,000 black Americans is born with CF. In Hawaii, only 1 in 90,000 live-born infants of Oriental ancestry is homozygous for CF.[1]

Genetics

Affected persons have a normal karyotype. There appear to be no strong associations between CF and other genetic diseases. Homozygote detection is readily determined by pilocarpine iontophoresis quantitative sweat analysis; chloride and sodium concentrations are invariably elevated due to impermeability of various CF epithelia to the chloride ion, resulting in generalized eccrine gland dysfunction.[2] Clinical identification of heterozygotes currently is not possible, but advances in molecular genetics have localized the CF gene to chromosome 7.[3] This has enabled prenatal diagnosis and carrier detection in most families with an affected child.[4]

Pathophysiology

The elusive basic defect involves both eccrine and exocrine glands. Many of the clinical manifestations are a result of organ passage obstruction by mucous secretions that display abnormal physiochemical behavior and markedly increased viscosity. In Europe CF is usually referred to as "mucoviscidosis." Ion transport defects that prohibit normal passage of chloride ions across epithelial cell membranes in glandular structures may explain the pathophysiology of CF. The most characteristic abnormalities in CF diagnosed by pathologists include focal biliary cirrhosis, distinctive obstructive lesions of the male genital tract, meconium ileus, *Staphylococcus* and *Pseudomonas* colonization of respiratory secretions with bronchial and bronchiolar inflammation, amyloidosis, and pancreatic fibrosis with exocrine atrophy and duct obliteration. The severity of these findings is variable and generally age related.

Systemic Manifestations

The classic clinical features of CF include chronic pulmonary obstruction and infection, malnutrition, growth failure, developmental delay, and, frequently, premature death. Clinical presentation and prognosis may vary considerably, even among siblings.[5]

When oral vitamin A is given to patients with CF in doses adequate to maintain normal hepatic stores, plasma vitamin A levels remain low. Plasma retinol-binding protein concentration is also significantly lower in CF patients than in control subjects. Vitamin A supplementation and pancreatic enzyme administration, resulting in improved digestion and absorption of fat and the fat-soluble vitamins, have been partially responsible for the markedly improved prognosis for CF in the last 3 decades.[1]

Ninety percent of CF patients die of progressive pulmonary involvement complicated by *P. aeruginosa* or *S. aureus* infection. These two organisms have a striking association with the disease.[6] *P. cepacia* colonization or infection portends a particularly poor overall prognosis. Most CF patients eventually become colonized with mucoid strains of *P. aeruginosa,* which may contribute to destructive pulmonary immune complex formation. Infection with these organisms is characteristically refractory to both systemic and aerosol antibiotic therapy.

Although no primary endocrine abnormality is associated with CF, diabetes mellitus is 25 times more common in CF patients than in the general population. It occurs in 2% of pediatric patients and in 13% of CF patients over 25 years old. Overall, about 43% of CF patients have abnormal oral glucose tolerance tests.[1] Pancreatic beta cells responsible for insulin production as well as glucagon-producing pancreatic alpha cells fall victim to progressive exocrine cicatricial ablation. This may explain the relative infrequency of ketosis in CF diabetics, since glucagon is considered essential for the development of ketosis.

Ocular Manifestations

Ocular surface pathology in CF is generally subclinical, although tear function may be abnormal. Tear lysozyme and Schirmer's basic test have been shown to be depressed in CF, while the prevalence of biomicroscopic corneal fluorescein staining and blepharitis are increased.[7] Thus, eccrine tear aqueous deficiency as well

as exocrine ocular sebaceous secretory abnormalities parallel the systemic glandular manifestations. Alterations of normal tear electrolyte concentrations, particularly calcium, have been reported. There is no evidence for suppressed ocular surface immunity. Patients with well-controlled CF can successfully wear hard or soft contact lenses.

Xerophthalmia and nyctalopia have been reported occasionally in CF, particularly in circumstances of vitamin therapy noncompliance or social deprivation. Qualitative and quantitative conjunctival biopsy goblet cell findings resemble those of normal control subjects and show no signs of stagnation.[8]

Corneal endothelial permeability is increased in CF, with a compensatory increased mean endothelial cell pump rate and a decreased mean endothelial cell area. This endothelial dysfunction is aggravated by hyperglycemia.[9]

CF patients may exhibit preganglionic oculosympathetic paresis when pupillary responses are tested pharmacologically. Decreased dilation following instillation of 5% topical cocaine corresponds to increased disease severity.[10]

Funduscopic changes in advanced CF are not pathognomonic for the disease but represent complications arising from a combination of hypercapnia, chronic ischemia, and often diabetes mellitus. Retinal hemorrhage, venous tortuosity and enlargement, cystoid macular edema, and papilledema have all been reported.[11] Retinal vascular abnormalities generally parallel pulmonary deterioration. Vitreous fluorophotometry has demonstrated increased retinal endothelial permeability and a significant breakdown in the blood-retinal barrier preceding clinically apparent retinopathy in CF patients with fasting hyperglycemia. This occurs with equal frequency and severity in CF patients and in matched control subjects with insulin-dependent diabetes and thus does not appear to be attributable to CF itself.[12]

Retrobulbar neuritis due to an idiosyncratic response to chloramphenicol by CF patients has been reported.[13] Symptoms include acute vision loss, color vision abnormalities, orbital pain, and further decrease in acuity when reading (Uhthoff's symptom). There can be associated systemic symptoms, including painful paresthesias in the fingers and toes, and generalized peripheral neuritis. Signs include sluggishly reactive pupils, enlarged blind spots, central scotomas, and occasionally papillitis. The neuropathy is thought to be reversible with discontinuation of chloramphenicol, although recovery may take months and permanent deficits may remain. Visually asymptomatic patients with CF who have used chloramphenicol can have abnormalities of color vision and of visual evoked potentials. Optic neuropathy can recur with reinstitution of chloramphenicol therapy and can lead to optic atrophy, although many patients who resume chloramphenicol therapy do not suffer further clinical insult. Most CF patients with optic neuropathy have advanced pulmonary disease and have been taking chloramphenicol for many months at daily doses ranging from 30 to 60 mg/kg. CF patients who have not taken chloramphenicol have contrast sensitivity–testing defects, which may be exacerbated by subsequent chloramphenicol use.[7] Optic nerve compromise due to vitamin deficiency, hypoxia, other antibiotics, or a primary effect of the disease itself, may render CF patients more susceptible to the effects of chloramphenicol. Low serum levels of vitamin E may be synergistic with chloramphenicol in producing both chronic and acute optic neuropathy. Some patients with acute neuropathy have responded to theraputic doses of vitamin E or B, allowing continuation of chloramphenicol therapy.[1]

As survival times continue to increase, patients with CF will be seeking the advice of nonpulmonary specialists more frequently. Understanding this unique disease and awareness of its ophthalmic manifestations will allow physicians to deliver optimal care to these patients.

References

1. Lloyd-Still JD. *Textbook of Cystic Fibrosis.* Boston, Mass: John Wright PSG; 1983.
2. Arehart-Treichel J. Cystic fibrosis linked to chloride ions' inability to cross certain cells. *JAMA.* 1984;252:2519-2527.
3. Knowlton RG, Cohen-Haguenauer O, Van Cong N, et al. A polymorphic DNA marker linked to cystic fibrosis is located on chromosome 7. *Nature.* 1985;318:380-382.
4. Tsui LC, Barker D, Braman JC. Cystic fibrosis locus defined by a genetically linked polymorphic DNA marker. *Science.* 1985;230:1054-1055.
5. di Sant'Agnese PA, Davis PB. Cystic fibrosis in adults. *Am J Med.* 1979;66:121-132.
6. May JR, Herrick NC, Thompson D. Bacterial infection in cystic fibrosis. *Arch Dis Child.* 1972;11:108-114.
7. Sheppard JD, Orenstein DM, Chao C-C, et al. The ocular surface in cystic fibrosis. *Ophthalmology.* 1989;96:(in press).
8. Holm K, Kessing SV. Conjunctival goblet cells in patients with cystic fibrosis. *Acta Ophthalmol.* 1975;53:167-172.
9. Lass JH, Spurney RV, Dutt RM, et al. A morphologic and fluorophotometric analysis of the corneal endothelium in Type I diabetes mellitus and cystic fibrosis. *Am J Ophthalmol.* 1985;100:783-788.
10. Spaide RF, Diamond G, D'Amico RA, et al. Ocular findings in cystic fibrosis. *Am J Ophthalmol.* 1987;103:204-210.
11. Bruce GM, Denning CR, Spalter HF. Ocular findings in cystic fibrosis of the pancreas. *Arch Ophthalmol.* 1960;63:391-401.
12. Rodman HM, Waltman SR, Krupin T, et al. Quantitative vitreous fluorophotometry in insulin treated cystic fibrosis patients. *Diabetes.* 1983;32:505-511.
13. Harley RD. Optic neuritis and optic atrophy following chloramphenicol use in cystic fibrosis patients. *Trans Am Acad Ophthalmol Otolaryngol.* 1970;74:1011-1031.

Chapter 162

Respiratory Insufficiency

AMY Y. TSO and STEPHEN H. SINCLAIR

Abnormalities in pulmonary gas exchange (or in the hematogenous transport of exchanged gases) induce alterations in hemodynamics and in tissue function in all systemic organs. Typically, pulmonary insufficiency causes systemic arterial hypoxemia, hypercapnia, or both. In this chapter we review the effects of altered respiratory gases on retinal hemodynamics and then discuss clinical manifestations associated with respiratory failure. The lung also serves an additional function as a sieve for abnormal particulate and cellular matter arising on the venous side of the systemic circulation. Its failure may allow embolic material to gain access to the arterial circulation. This chapter concludes with a discussion of Purtscher's retinopathy, which may share a common pathogensis with the adult respiratory distress syndrome.

Mechanisms of Hypoxemia and Hypercapnia with Pulmonary Insufficiency

While both hypoxemia (reduced oxygen content) and hypercapnia (elevated carbon dioxide level) may result from pulmonary dysfunction, hypoxemia is more common. Clinical causes of hypoxemia are numerous but may include central nervous system depression of respiratory muscles (drug-induced, spontaneous, or paralytic), injury to the chest wall, increased airway resistance in asthma or emphysema, pulmonary edema or infection, congenital heart disease with right-to-left shunting, carbon monoxide poisoning, and anemia.

Hypercapnia, on the other hand, usually does not occur alone but typically is observed with hypoxemia when due to ventilation-perfusion abnormalities or to hypoventilation. Patients wih mild chronic obstructive pulmonary disease (emphysema) and mild degrees of ventilation-perfusion mismatch many present with hypoxemia but without hypercapnia if ventilation is increased; however, carbon dioxide retention always develops as the mismatch worsens. Hypoventilation, on the other hand, always causes hypercapnia and may be exacerbated by administering enriched oxygen mixtures in the setting of chronic obstructive lung disease. Hypercapnia does not occur with hypoxemia when the latter is due to impaired diffusion, because carbon dioxide diffuses many times more rapidly than oxygen and its diffusion is seldom limited. Hypercapnia is also usually not associated with hypoxemia when the latter is due to abnormalities of oxygen transport, such as carbon monoxide poisoning, anemia, or circulatory deficiency.

Hypoxemia, Hypercapnia, and the Retinal Circulation

The retinal vasculature has no autonomic neural control but is fastidiously autoregulated by the "local milieu" of the tissue surrounding the small (15- to 30-μm) arterioles. These vessels are exquisitely sensitive to changes in arteriolar gas content, although changes in their caliber cannot be appreciated by ophthalmoscopy; hence retinal flow changes may occur without appreciable diameter changes in the observable major vessels. Commonly, however, the large (80- to 120-μm) retinal vessels constrict or dilate in concert with the resistance vessels. The clinical technique of fluorescein angiography provides a qualitative analysis of the distribution of blood flow, while the research techniques of laser Doppler velocimetry and the blue field entopic simulation technique provide quantitative measurements of large vessel and macular capillary retinal blood flow.

Hypoxemia

Small alterations in blood oxygen content cause changes in blood flow (and therefore in the arteriolar resistance vessels) as well as in the diameter of the major retinal arteries and veins. A change from breathing room air (20% oxygen) to breathing 100% oxygen is associated

with a 12 to 16% decrease in large artery and vein diameters and with a 61% decrease in flow, as well as a 30 to 40% decrease in macular capillary flow.[1-3] With a decrease in inspired oxygen to 10.5% there is an observed increase in macular capillary flow of 39% and an increase in large artery and vein diameter of 10%.[4,5]

Hypercapnia

Acute hypercapnia (breathing 7% carbon dioxide) has been demonstrated to produce a 24% increase in macular capillary flow, a similar increase in large vessel flow, but no observable increase in large vessel diameter.[6] Therefore, its effect must be limited to small arteriolar resistance vessels.

Clinical Alterations with Respiratory Insufficiency

Clinical manifestations of hypoxemia occur at arterial oxygen pressures of 75 mm Hg or less. A characteristic skin cyanosis usually is manifested when the absolute concentration of desaturated hemoglobin exceeds 5 g/ml. This is associated with duskiness of the conjunctival and retinal vessels. The associated finger and toe clubbing and plethoric facies are observed with secondary polycythemia. In the more advanced states with increased pulmonary vascular resistance, right heart failure occurs with increased systemic venous pressure, peripheral edema, and hepatomegaly. Carbon dioxide retention in far advanced cases results in mental status changes and tremors.[7]

Changes in retinal blood flow with chronic hypoxemia have not been examined; however, chronic hypoxemia produces dilation of the major retinal arteries and veins, which is worse when the viscosity is increased because of secondary polycythemia. In the aged person, the arterial and venous dilatation is segmental because of replacement of the normally elastic smooth muscle in the vessel wall by segmental fibrosis.[8] Segmental constriction also occurs at arteriovenous crossings by sclerosis of the venous adventitia where the artery and vein share a common adventitial sheath.[8] With severe pulmonary decompensation, superficial or deep retinal hemorrhages may occur, with macular and optic disc edema.[9]

Chronic Obstructive Lung Disease and Systemic Exocrine Cystic Fibrosis

Chronic obstructive pulmonary disease (COPD) is characterized by the limitation of expiratory air flow secondary to chronic bronchitis or emphysema, or usually to a combination of both. The ocular findings are related to alterations in the blood gases; the funduscopic changes typically depend on the degree of hypoxemia and hypercapnia and upon the secondary polycythemia.

Systemic cystic fibrosis is an autosomal recessive disorder that causes dysfunction of almost all exocrine and endocrine glands by the abnormal obstruction of mucin ducts (see Chapter 161). Pancreatic insufficiency occurs in children and adolescents with biliary obstruction and cirrhosis. Pulmonary dysfunction occurs because of the accumulation of viscous bronchiolar mucin plugs. Some of the associated ocular abnormalities are similar to those observed in adults with COPD and are related to the severity of hypoxemia and carbon dioxide retention.

Congenital Heart Disease and Cyanosis

An increase in diameter of the large retinal vessels (both arteries and veins) with tortuosity has been reported in infants and in children with congenital cyanosis secondary to intracardiac or pulmonary right-to-left shunts (Fig. 162-1). These vascular changes are differentiated from those of congenital retinal vessel tortuosity, which causes tortuosity of the veins or arteries, but not both, and which produces minimal vascular dilatation.

High-Altitude Retinopathy

Among unacclimated climbers at altitudes above 15,000 to 17,000 feet (arterial oxygen pressure of 75 to 80 mm Hg), dilatation and tortuosity of the major veins and arteries is observed, with intraretinal capillary hemorrhages in approximately 20 to 35% and occasional optic disc edema or rare vitreous hemorrhage (see Chapter 157).[10-17] The cause of the retinal hemorrhages is controversial, but they may be produced by endothelial fragility caused by the hypoxia, accompanied by an increase in capillary hydrostatic pressure due to increased ophthalmic artery pressure, increased cerebrospinal fluid pressure, or Valsalva maneuvers occurring with the exercise.[10,16,18]

Reversible functional abnormalities are also reported at high altitudes, including selective red color vision loss and increased light adaptation recovery time.[19] While most vision abnormalities are reversed with descent to normal altitudes, persistent scotomas have been reported with permanent impairment of central vision.[16]

Carbon Monoxide Poisoning

Carbon monoxide poisoning produces conjunctival and retinal large vessel changes similar to those that occur with hypoxemia, except that the vessels are not cyanotic but have a cherry-red color contributed by the carboxyhemoglobin. Exposure for as little as 1 hour to concentrations of 0.1% inspired carbon monoxide, results in a blood saturation of 50 to 80% carboxyhemoglobin and can result in coma, respiratory depression, and death.

Purtscher's Retinopathy

Central fundus changes of multiple cotton-wool spots (nerve fiber–layer infarcts) occurring with an occasional striate intraretinal hemorrhage and preretinal hemorrhage (Fig. 162-2) were originally reported by Purtscher in 1910 in a patient with head trauma (see Chapter 158).[20] The findings are now recognized as one manifestation of a systemic vascular insult occurring not only with major trauma from skull or long bone fracture but also with minor blunt chest trauma without fractures, with alcoholic pancreatitis, and post-partum.[21-25] Following trauma, there may be a latency period of a few hours to several days before symptoms develop and the retinopathy appears. The retinopathy, which may be unilateral or bilateral, evolves over several days and in most cases resolves over several months, leaving good vision; however, in many reported cases, residual scotomas or poor central vision have persisted. With the onset of the retinopathy, other systemic manifestations often occur, such as petechial hemorrhages of the conjunctiva, upper chest, neck, and face, and cerebral symptoms of restlessness, confusion, and coma. These systemic signs often occur together with the acute onset of respiratory failure characterized by hypoxia, dyspnea, tachypnea, and diffuse pulmonary edematous infiltrates (the adult respiratory distress syndrome, ARDS).

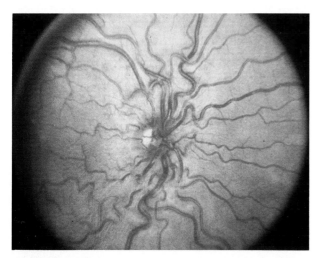

FIGURE 162-1 Funduscopic findings in a patient with congenital heart disease and cyanosis. Note dilated and tortuous retinal vessels.

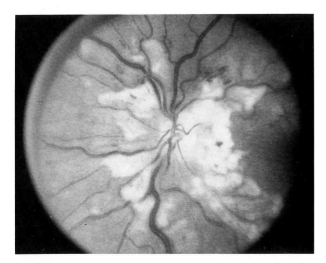

FIGURE 162-2 Purtscher's retinopathy in a patient with head trauma. Multiple nerve-fiber layer infarcts are observed over and surrounding the optic nerve head.

Many pathologic mechanisms have been proposed for Purtscher's retinopathy, but it is now recognized that the nerve fiber–layer lesions represent small precapillary or capillary infarcts caused by systemic arterial emboli. Furthermore, it is thought that the pathogenic mechanisms of Purtscher's retinopathy may be similar to those of ARDS because of their contemporaneous occurrence in traumatized patients and because of the similar vascular pathology. Mechanisms that have been proposed include fat microemboli and leukoaggregates caused by complement activation.[24,26-31]

Whatever the cause for the microemboli, they most likely arise on the venous side of the systemic circulation and have been assumed to gain access to the systemic arterial circulation by escaping the normal trapping of the lung microvasculature. The simple theory of the overwhelmed lung trapping mechanism, however, does not explain several observations. Many causes of severe ARDS (eg, systemic sepsis, multiple blood transfusions, thermal burns) have not been reported to produce Purtscher's-like retinopathy, and almost no cases of Purtscher's retinopathy reported in the literature have been observed to occur with the most severe cases of ARDS. On the contrary, Purtscher's retinopathy most often occurs in patients who suffer relatively mild trauma and have no severe signs of ARDS. These observations suggest three alternative mechanisms for the generation of the systemic anteriolar emboli (Fig. 162-3): (1) an occult right-to-left shunt (such as an atrial septal defect), (2) an intrapulmonary shunt due to the trauma, or (3) activation of a ''second limb'' of sequences

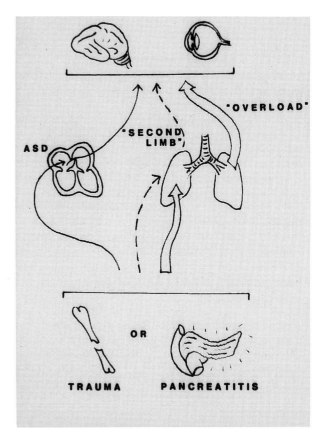

FIGURE 162-3 Possible pathways by which venous microemboli might produce systemic arterial manifestations.

by pathologic mechanisms that arise within the pulmonary venous circulation or on the systemic arterial side, which then produce the systemic arterial microemboli. Such alternative mechanisms need to be explored.

References

1. Grunwald JE, Riva CE, Petrig BL, et al. Effect of pure O_2 breathing on retinal blood flow in normal and in patients with background diabetic retinopathy. *Curr Eye Res.* 1984;3:239-241.

2. Frayser R, Hickman JB. Retinal vascular response to breathing increased carbon dioxide and oxygen concentrations. *Invest Ophthalmol Vis Sci.* 1964;3:427.

3. Petrig BL, Riva CE, Grunwald JE, et al. Effect of graded oxygen breathing on macular capillary leukocyte velocity. *Invest Ophthalmol Vis Sci.* 1986;27:221.

4. Fallon TJ, Maxwell D, Kohner EM. Autoregulation of retinal blood flow in diabetic retinopathy under conditions of hypoxia and hyperoxia. *Invest Ophthalmol Vis Sci.* 1983;27:244.

5. Hickman JB, Frayser R, Ross JC. A study of retinal venous

blood oxygen saturation in human subjects by photographic means. *Circulation*. 1963;27:375.

6. Petrig BL, Grunwald JE, Baine JC, et al. Changes in macular capillary leukocyte velocity and segmental retinal blood flow during normoxic hypercapnia. *Invest Ophthalmol Vis Sci*. 1986;27:245.

7. Austen FK, Carmichael MW, Adams RD. Neurologic manifestations of chronic pulmonary insufficiency. *N Engl J Med*. 1957;257:579.

8. Yanoff M, Fine BS. *Ocular Pathology*. Philadelphia, Pa: Harper & Row; 1982.

9. Spalter HF, Bruce GM. Ocular changes in pulmonary insufficiency. *Trans Am Acad Ophthalmol Otolaryngol*. 1964;68:661.

10. Duguet J, Dumont P, Bailliart JP. The effects of anoxia on retinal vessels and retinal arterial pressure. *Aviation Med*. 1947;18:516.

11. Frayser R, Houston CS, Gray G, et al. The response of the retinal circulation to altitude. *Arch Intern Med*. 1971;127:708.

12. Frayser R, Houston CS, Bryan AC, et al. Retinal hemorrhage at high altitude. *N Engl J Med*. 1970;282:184.

13. Wiedman M. High altitude retinal hemorrhage. *Arch Ophthalmol*. 1975;93:401.

14. Singh I, Khanna PK, Srivastava MC, et al. Acute mountain sickness. *N Engl J Med*. 1969;280:175.

15. Hickman JB, Frayser R. Studies of the retinal circulation in man: observations on vessel diameter, arteriovenous oxygen difference and mean circulation time. *Circulation*. 1966;33:302.

16. Shults WT, Swan KC. High altitude retinopathy in mountain climbers. *Arch Ophthalmol*. 1975;93:404.

17. Wilson R. Acute high altitude illness in mountaineers and problems of rescue. *Ann Intern Med*. 1973;78:421.

18. Hansen JE, Evans WO. A hypothesis regarding the patho-physiology of acute mountain sickness. *Arch Environ Health*. 1970;21:666.

19. Kobrick JL. Effects of hypoxia and acetazolamide on color sensitivity zones in the visual field. *J Appl Physiol*. 1970;2:741.

20. Purtscher O. Noch unbekannte Befunde nach Schadel-trauma. *Berl Deutsch Ophth Ges*. 1910;36:294.

21. Marr WG, Marr EG. Some observations on Purtscher's disease: traumatic retinal angiopathy. *Am J Ophthalmol*. 1962;54:693.

22. Kelley JS. Purtscher's retinopathy related to chest compression by safety belts. *Am J Ophthalmol*. 1972;74:278.

23. Inkeles DM, Walsh JB. Retinal fat emboli as a sequela to acute pancreatitis. *Am J Ophthalmol*. 1975;80:935.

24. Jacob HS, Goldstein IM, Shapiro I, et al. Sudden blindness in acute pancreatitis: possible role of complement-induced retinal leukoembolization. *Arch Intern Med*. 1981;141:134.

25. Kincaid MC, Green WR, Knox DL, et al. Clinicopathological case report of retinopathy of pancreatitis. *Br J Ophthalmol*. 1982;66:219.

26. DeVoe AG. Ocular fat embolism. *Arch Ophthalmol*. 1950;43:857.

27. Chuang EL, Miller FS, Kalina RE. Retinal lesions following long bone fractures. *Ophthalmology*. 1985;92:370.

28. McCarthy B, Mammen E, LeBlanc LP, et al. Subclinical fat embolism: a prospective study of 50 patients with extremity fractures. *J Trauma*. 1973;13:8.

29. Craddock PR, Hammerschmidt D, White JG, et al. Complement (C5a)-induced granulocyte aggregation in vitro. *J Clin Invest*. 1977;60:260.

30. Ernst E, Hammerschmidt DE, Bagge U, et al. Leukocytes and the risk of ischemic diseases. *JAMA*. 1987;257;2318.

31. Jacob HS, Craddock PR, Hammerschmidt DE, et al. Complement-induced granulocyte aggregation. *N Engl J Med*. 1980;302:789.

PART 17

Renal Disorders

Chapter 163

Alport's Syndrome

PETER J. McDONNELL, W. RICHARD GREEN, and
DAVID J. SCHANZLIN

Alport's syndrome is one of several heredofamilial renal disorders associated primarily with glomerular injury.[1] The nephritis is accompanied by high-frequency sensorineural deafness and ocular abnormalities.

Genetic Aspects

Males tend to be affected more frequently and more severely than females, and their disease is more likely to progress to renal failure. Females are not completely spared. The most common presenting sign is gross or microscopic hematuria, frequently accompanied by erythrocyte casts. Proteinuria may occur, and, rarely, the nephrotic syndrome develops. Symptoms typically appear between the ages of 5 and 20 years but have been noted as early as 5 months. The onset of overt renal failure occurs between ages 20 and 50.[2] The mode of inheritance in most kindreds is autosomal dominant. Feingold et al.[3] reported four families with autosomal recessive inheritance. In some families a sex-linked dominant mode of genetic transmission is seen. Penetrance and expressivity are variable. There is less likelihood that boys who receive the gene from their fathers will develop the disease (13%) compared with sons of affected females, whose children have an equal chance of inheriting the disorder regardless of their sex (Table 163-1).[4,5] The disease has a wide geographic distribution and has been reported in patients of different ethnic and racial backgrounds. It is the most common of the heritable renal diseases.

TABLE 163-1 Risk of Developing Kidney Disease Among Offspring of Parents with Alport's Syndrome

Affected Parent	Sons (%)	Daughters (%)
Mother	42	45
Father	13	53

(From Preus and Fraser.[4])

Auditory Defects

The auditory defects may be subtle, requiring extensive testing. Hearing loss is present in about half the patients; its onset is usually in the first decade and it is more pronounced in males. The hearing loss is usually progressive; it may be asymmetric or even unilateral. This dysfunction is thought to be due to degeneration of the striae vascularis and the hair cells of the organ of Corti.[6]

Renal Disease

Renal biopsy demonstrates glomerulosclerosis, vascular narrowing, tubular atrophy, and interstitial fibrosis. Electron microscopic examination of the basement membrane of glomeruli and tubules shows irregular thickening, lamination of the lamina densa, and foci of rarefaction. The pathogenesis of this disorder is unknown but it may be defective synthesis of glycopeptide (noncollagenous) components of basement membrane. As the lesions progress, the kidney shrinks and an end-stage chronic glomerulonephritis results.

There is no specific therapy for the nephritis; steroids and cytotoxic agents are ineffective. Standard therapeutic measures for renal insufficiency and its complications are used when specific problems arise. Recurrence of the disease after renal transplantation has not been described.

Ocular Manifestations

Ocular abnormalities are estimated to occur in about 10% of patients with Alport's syndrome.[7,8] Ocular structures that may be affected include the conjunctiva, cornea, lens, retina, and optic nerve head (Table 163-2).

Buchbinder et al.[9] reported finding calcium crystals in the conjunctiva of brothers with Alport's syndrome. The crystals were surrounded by foreign body type giant cells. The presence of the crystals was attributed to the hypercalcemia associated with the renal failure.

Chavis and Groshong[10] and Freidburg[11] each reported corneal arcus occurring in a young patient with Alport's syndrome. Davies[12] reported a patient with corneal endothelial pigmentation. This was in the setting of pigment dispersion syndrome, with pigment granules deposited in the anterior chamber angle, on the lens zonules, and on the peripheral retina. Huck et al.[13] reported a patient with bilateral, nonsimultaneous corneal ring abscesses.

The most typical ocular abnormalities to occur with Alport's are lenticular lesions: anterior lenticonus (Fig. 163-1) and anterior polar cataracts. Histopathologic

TABLE 163-2 Ocular Abnormalities Reported in Alport's Syndrome

Site	Finding	Reference
Conjunctiva	Calcium crystals with granulomatous response	Buchbinder[9]
Cornea	Juvenile arcus	Chavis[10] Friedburg[11]
	Ring abscesses	Huck[13]
	Endothelial pigmentation	Davies[12]
Anterior chamber angle		
	Pigment dispersion	Davies[12]
	Angle-closure glaucoma	Fiore[16]
Lens	Anterior lenticonus	Brownell[14] Huck[13]
Retina	Retinitis pigmentosa	Fiore[16]
	Fundus albipunctatus	Perrin[8] Davies[12]
	Abnormal electroretinogram	Davies[12] Huck[13]
Optic nerve head	Drusen	Friedburg[11]

examination of anterior lenticonus in this setting is remarkable for a thinned anterior lens capsule, a decreased number of lens epithelial cells, and bulging of the lens substance.[14] Electron microscopy reveals capsular dehiscences, reflecting fragility of the anterior capsule.[15] This thinning of the anterior lens capsule, the basement membrane of the lens epithelial cells, echoes the basement membrane abnormalities noted in the renal glomerulus. Thus, the glomerular lesions and anterior lenticonus may reflect an underlying disorder of basement membrane synthesis. Posterior cortical opacities may also occur.

The retina may contain numerous discrete white intraretinal dots, resembling fundus albipunctatus. Perrin et al.[8] observed symmetrical bilateral changes resembling fundus albipunctatus surrounding the foveal area in 29 of 79 patients with Alport's syndrome (Fig. 163-2). Patients with these retinal changes experienced onset of renal failure at a significantly earlier age than patients with normal fundi. The electroretinogram may be abnormal, with a reduced b wave, while the electrooculogram findings are normal.[12,13] Fiore et al.[16] described a kindred with familial nephropathy with retinitis pigmentosa and closed-angle glaucoma that was transmitted in an autosomal dominant pattern. The syndrome was thought to represent a variant of Alport's syndrome. As a consequence of the severe renal failure that may occur, patients may develop marked hypertension and hypertensive retinopathy. In addition, uremia may contribute to retinal and optic disc edema.[17] Drusen of the optic nerve head have also been described in patients with Alport's syndrome.[11]

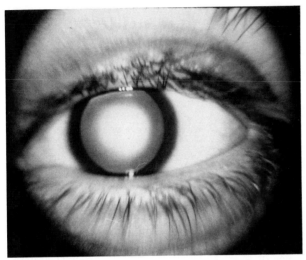

FIGURE 163-1 Retroillumination of anterior lenticonus in a patient with Alport's syndrome demonstrates central optical changes related to anterior bulging of thinned anterior lens capsule.

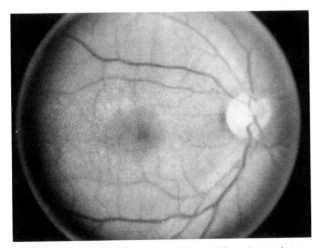

FIGURE 163-2 Discrete yellow-white deep intraretinal dots in the posterior pole of a patient with Alport's syndrome and renal failure present in about one third of patients. These retinal changes are consistent with poor prognosis for renal function.

A Disorder of Basement Membrane Production

In summary, Alport's syndrome is a hereditary disorder characterized by hematuria, hearing loss, and ocular abnormalities. The ophthalmologist may help determine the patient's prognosis, because the presence of retinal changes correlates with the early onset of renal failure. The abnormalities of basement membrane of the renal glomerulus and anterior lens capsule suggest an underlying disorder of synthesis of at least one component of basement membrane. Butkowski et al.[18] reported that the basic defect in this syndrome lies in the noncollagenous portion of the type IV collagen molecule. Until clinicopathologic correlation is performed, it is not possible to state whether the other ocular changes, such as the retinal changes that resemble those of fundus albipunctatus, also reflect a disorder of basement membrane.

References

1. Alport AC. Hereditary familial congenital hemorrhagic nephritis. *Br Med J.* 1927;1:504.
2. O'Neill WM, et al. Hereditary nephritis: a re-examination of its clinical and genetic features. *Ann Intern Med.* 1978;88:176.
3. Feingold J, Bois E, Chompret A, et al. Genetic heterogeneity of Alport syndrome. *Kidney Int.* 1985;27:672.
4. Preus M, Fraser FC. Genetics of hereditary nephropathy with deafness (Alport's disease). *Clin Genet.* 1971;2:331.
5. Hasstedt SJ, Atkin CL. X-linked inheritance of Alport syndrome: family P revisited. *Am J Hum Genet.* 1983; 35:1241.
6. Gregg JV, Becker SF. Concomitant progressive deafness, chronic nephritis, and ocular lens disease. *Ann Ophthalmol.* 1963;69:293.
7. Drummond KN. Hereditary or familial diseases. In: Vaughan VC, McKay RJ, Behrman RE, eds. *Nelson Textbook of Pediatrics,* 11th ed. Philadelphia, Pa: WB Saunders; 1979;1523.
8. Perrin D, Jungers P, Grunfeld JP. Perimacular changes in Alport's syndrome. *Clin Nephrol.* 1980;13:163-167.
9. Buchbinder MC, Gindi JJ, Schanzlin DJ, et al. Conjunctival crystals in Alport's syndrome. *Ophthalmology.* 91 (suppl):123,1984.
10. Chavis RM, Groshong T. Corneal arcus in Alport's syndrome. *Am J Ophthalmol.* 1973;75:793.
11. Friedburg D. Pseudoneuritis and drusenpapillae beim Alport-Syndrom. *Klin Monatsbl Augenheilk.* 1968;152: 379.
12. Davies PD. Pigment dispersion in a case of Alport's syndrome. *Br J Ophthalmol.* 1970;54:557.
13. Huck D, Meythaler H, Rix R. Alport-Syndrom mit Hornhautbeteilung und Veranderungen des ERG. *Klin Monatsbl Augenheilk.* 1976;168:553.

14. Brownell RD, Wolter JR. Anterior lenticonus in familial hemorrhagic nephritis. *Ann Ophthalmol.* 1974;71:481.

15. Streeten BW, Robinson MR, Wallace R, et al. Lens capsule abnormalities in Alport's syndrome. *Arch Ophthalmol.* 1987;105:1693.

16. Fiore C, Santoni G, Reggiani FM, et al. Familial nephropathy with retinitis pigmentosa and closed angle glaucoma. *Ophthal Paediatr Genet.* 1985;5:39.

17. Leishman R. The cardiovascular system. In: Sorsby A, ed. *Modern Ophthalmology.* 2nd ed. vol 2. London: Butterworths, 1972;511-516.

18. Butkowski RJ, Wieslander J, Wisdom BJ, et al. Properties of the globular domain of type IV collagen and its relationship to Goodpasture antigen. *J Biol Chem.* 1985;260:8564-8570.

Chapter 164

Familial Juvenile Nephronophthisis

RONALD E. CARR

Systemic Manifestations

In 1951 Fanconi et al.[1] described a familial renal disease that they called "familial juvenile nephronophthisis" (FJN). In the two families described, the initial symptoms occurred between ages 2 and 3 years and consisted of polydipsia, polyuria, and nycturia. In spite of this, no proteinuria was seen and signs of renal insufficiency became apparent only in the late stages of the disorder.

Some years prior to the report by Fanconi, Smith and Graham[2] described a disorder that they called "medullary cystic disease of the kidney." The disorder presented in young adulthood with symptoms similar to those of FJN. Subsequent studies showed the majority of cases to have an autosomal dominant inheritance pattern, and because of this FJN and medullary cystic disease were regarded as separate entities. However, the findings that the clinical course and the pathologic processes were identical in both disorders led to the recognition that they were one disease with two modes of inheritance.[3] Prior to that, the literature was somewhat confusing, and a multitude of names were given to the diseases (Table 164-1). In this paper the term "familial juvenile nephronophthisis" (FJN) is used.

As the condition progresses, the short stature associated with many renal diseases is noted. Ultimately, anemia and progressive renal failure ensue, with low creatine clearance rate, azotemia, and inability to concentrate urine. Despite these changes, hypertension is very uncommon. Survival has been prolonged as renal transplants and dialysis have improved.

The affected kidneys are small with multiple intrarenal cysts in both cortex and medulla. Histologically there is a diffuse, chronic interstitial fibrosis with variable areas of tubular atrophy and dilatation. Renal tubule sheaths stain positive with the periodic acid-Schiff stain and hyalinized glomeruli and periglomerular fibrosis are seen. One study of renal biopsy specimens demonstrated large interstitial masses of a glycoprotein that stained with monospecific antibody to human Tamm-Horsfall (T-H) protein.[4] It may be that the T-H protein is involved in the pathogenesis of FJN, but the specific biochemical or enzyme abnormality has yet to be clarified.

In addition to the systemic changes induced by chronic renal disease, a number of skeletal, hepatic, and neuromuscular disorders have been reported, either

TABLE 164-1 Other Names for Familial Juvenile Nephronophthisis

Familial juvenile nephronophthisis
Medullary cystic disease
Childhood type
Adult type
Cystic disease of the renal medulla
Progressive hereditary nephropathy
Familial disease of the renal medulla
Chronic idiopathic tubulointerstitial nephropathy
Familial uremic medullary cystic disease

TABLE 164-2 Systemic Disorders Associated with
Renal-Retinal Dysplasia

Congenital hepatic fibrosis (Boichis' syndrome)[12]
Cerebellar ataxia[13]
Mental retardation[1]
Mitochondrial cytopathy[14]
Asphyxiating thoracic dystrophy (Jeune's syndrome)[15]
Cone-shaped epiphyses of the hands*,[16]

* Cone-shaped epiphyses of the hands, cerebellar ataxia, skeletal anomalies, retinitis pigmentosa, FJN (Saldino-Mainzer syndrome).

singly with FJN or in various combinations (Table 164-2). All have been reported in conjunction with renal-retinal dysplasia. (FJN and its ocular manifestations are discussed below.)

Ocular Manifestations

In 1960 Contreras and Espinosa[5] described the association of a generalized degeneration of the retina with FJN. Similar findings were reported 1 year later by Senior et al.[6] and Loken et al.[7] In all of the cases the affected children presented with poor vision, nystagmus, and fundus changes indicative of a generalized abnormality of the retina, a clinical picture resembling Leber's congenital amaurosis. Histologic studies of the eyes showed a generalized degeneration of the retina that was also similar to that found in Leber's congenital amaurosis. The spectrum of retinal changes in this disorder was further enlarged by Meier and Hass,[8] who described six siblings in a family, four of whom had fundus changes typical of retinitis pigmentosa, with narrowed arterioles and multiple clumps or strands of pigment in the retina. While Leber's congenital amaurosis may be considered a type of retinitis pigmentosa, under the broader framework of the "generalized heredoretinal degenerations" it is usually classified separately because of the congenital nature of the disorder, its relative lack of progression, and the presence of nystagmus. The cases described by Meier and Hass were more typical of retinitis pigmentosa with later onset of symptoms of poor night vision and poor peripheral vision, no nystagmus, and good central vision in the early stages.

Since these initial reports these associations have been noted in numerous publications, and the term "renal-retinal dysplasia" has been used to describe these syndromes. In all instances, the association of a generalized retinal disease with FJN, whether it be Leber's or retinitis pigmentosa, has shown an autosomal-recessive

mode of inheritance. It is interesting that in several of the families with typical retinitis pigmentosa and FJN, certain members may have either the renal disorder or the retinal disorder, while others have both diseases.

Reports of "atypical" retinitis pigmentosa in association with FJN include retinitis punctata albescens, retinitis pigmentosa sine pigmento, and central retinitis pigmentosa. All can be considered ophthalmoscopic variants of generalized heredoretinal degeneration.

The sectorial retinitis pigmentosa associated with FJN in one report was probably a mild cone-rod form of retinitis pigmentosa.[9] Another family was described in which some members had renal-retinal dysplasia and others had renal disease and congenital stationary nightblindness. These findings were probably the result of two separate genes from this consanguineous marriage and not of directly associated abnormalities.[9]

Several reports have noted mild electroretinographic (ERG) or electro-oculographic (EOG) abnormalities in heterozygotic patients.[10,11] These findings are not consistent, however, and cannot yet be used to determine the carrier of this autosomal recessive gene.

References

1. Fanconi G, et al. Die familiare juvenile Nephronophthise. *Helv Paediatr Acta.* 1951;6:1.
2. Smith CH, Graham JB. Congenital medullary cysts of the kidneys with severe refractory anemia. *Am J Dis Child.* 1945;69:369.
3. Strauss MB, Sommers SC. Medullary cystic disease and familial juvenile nephronophthisis. *N Engl J Med.* 1967;227:863.
4. Vernier RL, Resnick J. Medullary cystic disease. The possible role of Tamm-Horsfall protein. *Kidney Int.* 1976;9:450.
5. Contreras BC, Espinosa SJ. Discussión clinica y anatomopatológica de enfermos que presentaron un problema diagnóstico. *Pediatr. (Santiago).* 1960;3:271.
6. Senior B, Friedman AF, Braudo JL. Juvenile familial nephropathy and tapetoretinal degeneration. A new oculorenal dystrophy. *Am J Ophthalmol.* 1961;52:625.
7. Loken AC, et al. Hereditary renal dysplasia and blindness. *Acta Paediatr Scand.* 1961;50:177.
8. Meier DA, Hass JW. Familial nephropathy with retinitis pigmentosa. *Am J Med.* 1965;39:58.
9. Godel V, et al. Hereditary renal-retinal dysplasia. *Doc Ophthalmol.* 1980;49:347.
10. Abraham FA, et al. Electrophysiologic study of the visual system in familial juvenile nephronophthisis and tapetoretinal dystrophy. *Am J Ophthalmol.* 1974;78:591.
11. Polak BCP, et al. Carrier detection in tapetoretinal degeneration in association with medullary cystic disease. *Am J Ophthalmol.* 1983;95:487.
12. Boichis H, et al. Congenital hepatic fibrosis and nephronophthisis. *Quart J Med.* 1973;42:221.

13. Mainzer R, et al. Familial nephropathy associated with retinitis pigmentosa, cerebellar ataxia and skeletal abnormalities. *Am J Med.* 1970;49:556.

14. Egger J, Lake BD, Wilson J. Mitochondrial cytopathy. A multisystem disorder with ragged red fibers on muscle biopsy. *Arch Dis Child.* 1981;56:741.

15. Donaldson MDC, et al. Familial juvenile nephronophthisis, Jeune's syndrome, and associated disorders. *Arch Dis Child.* 1985;60:426.

16. Saldino RM, Mainzer F. Cone-shaped epiphyses (CSE) in siblings with hereditary renal disease and retinitis pigmentosa. *Radiology.* 1971;98:39.

Chapter 165

Lowe's Syndrome

Oculocerebrorenal Syndrome

GERHARD W. CIBIS, RAMESH C. TRIPATHI, and BRENDA J. TRIPATHI

Lowe's syndrome, first described in 1952 by Lowe, Terrey, and MacLachlan,[1] is an X-linked hereditary syndrome whose manifestations include congenital cataracts, miotic pupils, mental retardation, and defective renal tubular reabsorption of bicarbonate, phosphate, and amino acids. Half of all patients develop congenital glaucoma.

Systemic Manifestations

Systemic findings include decreased muscle mass and hypotonia with absent or decreased deep tendon reflexes. There is a characteristic hypotonic facial appearance and a prominent forehead (Fig. 165-1). Affected persons from unrelated families look very much alike, as if they were siblings. Members of all races have been reported to be affected. Muscle biopsies show small muscle fibers without major structural abnormalities. Such a pattern is known to occur with poor development of the anterior fetal brain. The electroencephalogram usually shows nonspecific abnormalities. Anticonvulsant drug therapy may be needed. The hypotonia leads to joint dislocations and scoliosis. Patients tend to be short; the cause may be intrinsic to the syndrome or related to the renal abnormalities.

Renal dysfunction begins as tubular dysfunction but may progress to renal failure in late childhood or adolescence. Fanconi's syndrome of multiple defects in proximal tubular function develop. Fanconi's syndrome may range from mild and incomplete to severe. The proximal renal tubular dysfunction usually is not present at birth but develops by 3 months to 1 year of age. Clinical assessment for Lowe's syndrome in the newborn period therefore depends on the systemic appearance and eye findings. Generalized aminoaciduria and tubular proteinuria tend to be the first evidence of renal involvement. Bicarbonate loss leads to acidosis, which, along with phosphate loss, may require vigorous replacement therapy. It also leads to osteoporosis and rickets. Phosphate and vitamin D metabolite supplements are needed. The severe metabolic problems lessen after age 5 to 8 years, although death from progressive renal deterioration has been reported. Survival into the 20s and 30s is occurring with improved management.

The degree of mental retardation is showing itself to be less severe as better and more aggressive correction of systemic, developmental, and visual problems allows maximization of patient potential. Physical rehabilitation, speech, vision, and ultimately occupational therapy, should all be aggressively pursued from early infancy onward.

The exact biochemical defect in Lowe's syndrome is not known. Abnormalities in both mucopolysaccharide and mitochondrial metabolism have been reported.[2,3] Lowe's syndrome patients excrete abnormally large amounts of low-sulfated chondroitin-4-sulfate in their urine.[2] Depressed sulfation rather than excessive removal of sulfation is the cause.[4] There is an elevation of nucleotide pyrophosphatase activity in the cells of Lowe's patients. Carriers have activity levels intermediate between those of patients and of normal individuals. Such elevation of activity may lower intracellular levels of various nucleotides of metabolic importance, accounting for biochemical defects such as depressed oxidative phosphorylation in mitochondria.[3,4] A general

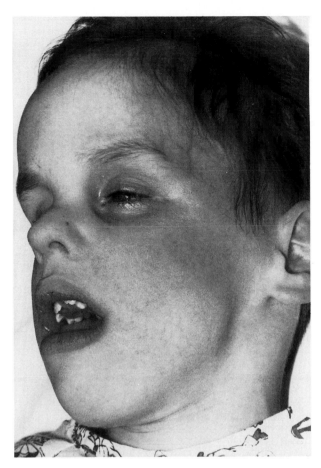

FIGURE 165-1 Typical facies with frontal bossing of a Lowe's patient. Each eye has a corneal keloid.

Ocular Manifestations

A small discoid cataractous lens with peculiar capsular and epithelial changes is present at birth in nearly 100% of affected males. There is an absence of demarcation of nucleus and cortex, which suggests a defect early in embryonic differentiation between young and older lens cells. There is a small nuclear bow, and retention of nuclei by many deeper cells indicates retarded maturation. If the posterior lens cells fail to form primary lens fibers, secondary lens fibers move toward the central region of the lens and fill the defect. This would account for the lack of demarcation between the nucleus and cortex and for the discoid shape of the lens. If the primary lens fibers degenerated rapidly after their initial development and the fibers that originated from the equatorial region remained in place or migrated slowly, a ring-shaped (Soemmering's) lens would result. Posterior proliferation of lens epithelium causes posterior lenticonus adherent to condensed anterior vitreous. The lens capsule is a product of growth and remodeling of lens cells. Irregularities in thickness of the lens capsule in Lowe's patients signify defective metabolism of the subcapsular epithelium.

An anatomically anomalous chamber angle typical of congenital glaucoma is present in Lowe's syndrome. The iris root inserts high on the trabecular meshwork. Ciliary processes are rudimentary and anteriorly displaced. Iris blood vessels have abnormal fenestrations. These anomalies may lead to a disturbance in the blood-aqueous barrier that could be active in the eye during embryogenesis or could contribute later to corneal keloid formation. There is segmental hypoplasia of the pupillary dilator muscle with adhesions to the anterior lens surface. Glaucoma develops in 50% of patients, usually by 6 years of age.

Corneal keloids are characteristic of Lowe's syndrome. They are the major cause of visual disability after age 6 or 7 years in patients whose cataracts, glaucoma, and refractive error have been successfully managed. The keloids extend through the full corneal thickness, and are thus not amenable to lamellar keratoplasty. Why they form is not known, but the abnormal blood-aqueous barrier (based on anatomic anomalies previously cited) associated with trauma, inflammation, phenytoin (Dilantin) therapy, and congenital predisposition may all play a role. How the tendency for keloid formation should influence therapy, for example use of contact lenses or epikeratophakia, in these patients, is speculative. We believe minor surface trauma associated with the other factors cited plays a role in keloid formation. Since cataract extraction is bilateral, glasses are a good form of vision rehabilitation. We discourage cosmetic contact lens use. How these patients would react to epikeratophakia is unknown. One could argue

defect in glycosaminoglycan (GAG) metabolism leading to excretion of low-sulfated GAGs may account for the renal problems of Lowe's patients, since sulfated GAGs form the selective permeability barrier of the kidney glomerulus.

How these metabolic defects may relate to the ophthalmic findings is a matter of speculation. A defect in sulfate metabolism of an enzyme involved in connective tissue formation could account for abnormal corneal findings.[5] The Lyon hypothesis[6] would explain how some cells in the lenses of carrier females, where the affected X chromosome was active in that cell, would produce clinical manifestation of the metabolic defect as progressive punctate and placoid cataracts.[7] Genetic expression of the abnormal metabolic gene in all the lens cells of affected males can account for the cataract development during embryogenesis.[8] Development of a congenital glaucomatous anomalous chamber angle might be secondary.

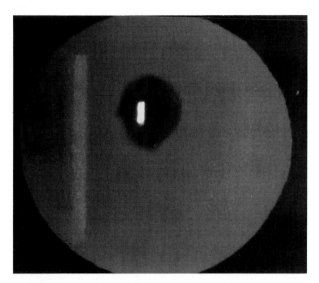

FIGURE 165-2 Obligate carrier for Lowe's with innumerable flecks, seen in direct and indirect illumination, and a posterior subcapsular opacity. (From Cibis.[7])

that, acting as a pressure bandage, epikeratophakia might have a protective effect against keloids.

Strabismus and amblyopia occur in Lowe's syndrome, either from poor vision or from poor fusion control secondary to the bilateral aphakia. Their pathogenesis is therefore no different from other low vision or aphakic patients, where poor vision or defective binocular fusion can lead to strabismus and amblyopia. Therapy is similar.

Female carriers in families with Lowe's syndrome gene can be diagnosed in the 2nd decade of life, when genetic counseling is of most interest to them, on the basis of typical punctate and plaquelike cataracts (Fig. 165-2). The presence of typical cataracts is highly diagnostic, but their absence does not rule out the carrier state. The typical white punctate opacities are often in wedge-shaped aggregates deep in the cortex just outside the nucleus, indicating that they began to form in infancy. More superficial opacities indicate that they continue to form well into adult life, as do the subcapsular plaquelike cataracts seen predominantly in older

carriers. Molecular linkage analysis has localized the Lowe's gene to the distal long arm of the X chromosome at Xq24–q26.[9] This assignment of the gene has the potential of improving carrier detection and prenatal diagnosis.

Prompt attention to visual problems is necessary to maximize the learning and developmental potential of these patients. The cataracts are dense and should be removed immediately on diagnosis. There is an anomalous chamber angle formation in these patients, and a pars plicata approach is indicated to minimize synechiae formation, further disruption of the chamber angle, and endothelial corneal damage. Glaucoma when it occurs requires surgery—either goniotomy or, as is our preference, trabeculotomy. Pupilloplasty may be needed if there is iris dilator muscle hypoplasia, to ensure adequate postsurgical management of refraction and fundus examination. Congenital nystagmus commonly develops even with optimal ophthalmic care.

References

1. Lowe CU, Terrey M, MacLachlan EA. Organic-aciduria, decreased renal ammonia production, hydrophthalmos and mental retardation. AMA *J Dis Child*. 1952;83:164-184.
2. Wisniewski K, Kieras FJ, French JH, et al. Ultrastructural, neurological and glycosaminoglycan abnormalities in Lowe's syndrome. *Ann Neurol*. 1984;16:40-49.
3. Gobernado JM, Lousa M, Gimeno A, et al. Mitochondrial defects in Lowe's oculocerebrorenal syndrome. *Arch Neurol*. 1984;41:208-209.
4. Yamashina I. Biochemical studies on Lowe syndrome. *Proc 1st International Conference on Lowe's Syndrome*. 1986;19-20.
5. Cibis GW, Tripathi RC, Tripathi BJ, et al. Corneal keloid in Lowe's syndrome. *Arch Ophthalmol*. 1982;100:1795-1799.
6. Lyon MF. Sex chromatin and gene action in the mammalian X chromosome. *Am J Hum Genet*. 1962;14:135-148.
7. Cibis GW, Waeltermann JM, Whitcraft CT, et al. Lenticular opacities in carriers of Lowe's syndrome. *Ophthalmology*. 1986;93:1041-1045.
8. Tripathi RC, Cibis GW, Tripathi BJ. Pathogenesis of cataracts in patients with Lowe's syndrome. *Ophthalmology*. 1986;93:1046-1051.
9. Silver DN, Lewis RA, Nussbaum RL. Mapping the Lowe oculocerebrorenal syndrome to Xq24-q26 by use of restriction fragment length polymorphisms. *J Clin Invest*. 1987;79:282-285.

Chapter 166

Renal Failure, Dialysis, and Transplantation

EDWARD L. HOWES, Jr.

Renal failure may be acute or chronic.[1-3] Acute renal failure is of rapid onset and is often reversible. It is characterized by acute loss of renal function manifested by oliguria or anuria. This sudden change may be due to vascular disease—malignant hypertension or a vasculitis such as periarteritis nodosa; rapidly progressive glomerulonephritis; acute interstitial nephritis, drug-related or infectious; acute tubular necrosis; or acute urinary obstruction. These alterations lead to varying combinations of intrarenal hemodynamic changes, tubular obstruction, backward leak of urine through damaged tubules, and a decrease in glomerular filtration rate. It is primarily the vascular causes of acute renal failure that may produce concurrent ocular manifestations. Malignant hypertension may occur alone or may be superimposed on longstanding primary or secondary benign hypertension, and it can lead to acute renal failure as well as retinopathic manifestations of severe hypertension. A number of inflammatory diseases of blood vessels may have ocular manifestations or produce acute renal shutdown. These include various forms of vasculitis or vaso-obstructive disease such as periarteritis nodosa, Wegener's granulomatosis, and thrombotic thrombocytopenic purpura.

Chronic renal failure occurs whenever there is a permanent reduction in glomerular filtration rate. This reduction may be slowly progressive, leading to chronic renal insufficiency and, over a period of years, end-stage renal disease. Hypertension may complicate the picture; it is found in 80% of patients with end-stage renal disease regardless of cause. The clinical syndrome of uremia is an indication of far advanced destruction of renal tissue and resultant loss of renal excretory function. It is characterized biochemically by retention of nitrogenous products such as urea and by a particular set of signs and symptoms that affect multiple organ systems. Many diverse types of renal disease may have a similar picture of chronic renal insufficiency. These include different forms of glomerulonephritis, interstitial renal disease, nephrosclerosis resulting from primary hypertension, diabetic glomerulosclerosis, anal-gesic nephropathy, and congenital polycystic kidneys.

Systemic Manifestations

Chronic renal failure develops when loss of functioning nephrons reaches a critical point.[1-3] As nephrons are destroyed, remaining ones hypertrophy, maintaining a similar level of function at the expense of reserve. Ultimately the ability of the kidney to excrete or retain certain solutes is lost. These changes result in fluid and electrolyte imbalance, with retention of sodium and extracellular fluid, potassium and phosphate. The results are edema and ascites. Buffering capacity is lost because of retention of hydrogen ion and loss of bicarbonate, and metabolic acidosis results. Activation of the renin-angiotensin system produces hyperaldosteronism. The retention of phosphate is associated with hypocalcemia and secondary hyperparathyroidism. Hypocalcemia may be due in part to a direct effect of phosphate on bone, promoting calcium storage resulting in a decrease in the amount of circulating calcium. Hypocalcemia is also enhanced because less calcium is absorbed from the gut. Vitamin D is of major importance in intestinal absorption of calcium, and the metabolism of vitamin D is deranged because of renal tubular damage, for one of the most active forms of vitamin D, 1,25-dihydroxy-cholecalciferol is produced in tubules. Hyperparathyroidism may have associated changes in the skeletal system that can be complex. Renal osteodystrophy is manifested by osteomalacia due to vitamin D deficiency; osteitis fibrosa cystica secondary to hyperparathyroidism; and at times osteosclerosis due to phosphate retention.

The uremic syndrome probably is not due to retention of any one substance.[1-3] Blood urea is a good chemical indicator of this condition, but urea itself is not toxic. Multiple substances are retained, including creatinine, urates, sulfates, phosphates, phenols, guanidine, low-molecular-weight polypeptides, and electrolytes; enhanced levels of parathormone and aldosterone are

507

produced. A normochromic normocytic anemia occurs. In uremia, coagulation is altered because of a qualitative effect of the uremic state on platelet function. The cardiovascular system is affected because of more severe arteriosclerosis and hypertension. Because of hypertension, anemia, fluid overload, and acidosis, congestive heart failure develops. Fibrinous pericarditis, multiple ulcerations of the gastrointestinal tract, dermatitis, and pneumonitis may occur. Uremic encephalopathy is accompanied by a slow-wave pattern on electroencephalograms and is characterized by loss of alertness and memory, confusion, and hallucinations. Many of these changes of uremia are reversed by renal dialysis.

Ocular Findings in Renal Failure

Hypertension may result from many renal diseases, or it may be the primary cause of renal failure. Malignant or accelerated hypertension can result in both a severe hypertensive retinopathy and acute renal failure and may be superimposed on chronic vascular disease in both kidneys and eyes. Often the retinopathic changes are particularly severe in renal failure, more so than blood pressure changes alone can account for. This has led to the somewhat controversial concept of a renal retinopathy, in which particularly severe retinal and disc edema are superimposed on vascular changes and are associated with prominent cotton-wool spots.[4,5] Diastolic blood pressure measurements in the face of these severe changes can be modest. This striking retinopathic picture has been attributed to the tissue effects of retained nitrogen products.[5]

Retinopathy is always present in advanced diabetic nephropathy and tends to be more severe as renal failure develops. In fact most patients with renal failure resulting from diabetic nephropathy have proliferative retinopathy (reported incidence 70%).[6] These changes have been attributed to the underlying vascular disease in both tissues, to hypertensive vascular disease, or to a superimposed effect of retained nitrogen products.

A nonrhegmatogenous retinal detachment (ie, not caused by a retinal tear or hole) may occur in patients in renal failure.[4,7] This event has been attributed both to metabolic changes resulting from renal failure and to vascular changes. Nitrogen and water retention and dilutional hyponatremia have been suggested as factors that predispose to detachment. Hypertension may lead to obstruction of the choriocapillaris with leakage of fluid into the subretinal space and consequent retinal detachment.[7]

Visual acuity may be altered in renal failure because of ocular and cerebral effects. Optic nerve and disc edema may occur secondary to renal failure, and cere-

bral edema or vascular spasm may alter occipital or parietotemporal lobe function.[4] Cataracts may develop in chronic renal failure, occurring as lens stippling. At least some of these are probably associated with secondary hyperparathyroidism.[4]

Derangements of calcium and phosphate metabolism resulting from renal dysfunction may affect the eye in a number of ways. Conjunctival and corneal changes are extremely common. They are due to deposition of calcium phosphate and apatite crystals in the cornea at the limbus and conjunctiva in the interpalpebral fissure. The exposure to air in these locations presumably causes a decrease in carbon dioxide concentration, resulting in a more alkaline medium and decreased solubility of the calcium phosphate product.[4] The crystals may be large enough to cause irritation, acute inflammation, and a red eye.[4] A band keratopathy may also occur.[4] The histopathologic changes of pinguecula and pannus formation may also be found.[8] The degree of conjunctival and corneal involvement does not always correlate with the calcium phosphate product, the length of the renal failure, or the patient's age.[8] With severe derangement in calcium and phosphate metabolism, secondary and tertiary hyperparathyroidism develop. The bone changes associated with hyperparathyroidism, osteitis fibrosa cystica, may rarely be manifested in the orbital bone as so-called brown tumors.[9]

Ocular Findings in Hemodialysis and Renal Transplantation

During the early days of hemodialysis, deterioration of vison and blindness were common. These changes were usually secondary to worsening of hypertensive and diabetic retinopathies. With increased control of hypertension and advances in photocoagulation for diabetic retinopathy, progression of visual and retinal changes have been sharply curtailed. Currently less than 5% of patients on dialysis develop blindness.[6] Continuous ambulatory hemodialysis may allow for still greater improvement in visual status.[10]

Acute ocular changes can occur during or immediately following dialysis. These changes are usually vascular in etiology: central retinal vein occlusion, cortical blindness, or anterior ischemic optic neuropathy.[11-13]

Long-term changes in patients on renal dialysis alone include: posterior subcapsular cataracts, calcifications of cornea and conjunctiva, and pigmentary changes in the posterior segment.[14] Following successful renal transplantation visual acuity often remains stable but retinopathy can progress and proliferative retinopathies do develop in these patients.[6] These changes can be anticipated for the most part, are treatable by photo-coagulation, and do not contraindicate transplantation.

Immunosuppressive therapy following renal transplantation introduces the possibility of enhanced cataract formation, glaucoma, ocular infections, and ocular and periocular neoplasms. Cataract is a common complication and most are attributable to corticosteriod (usually prednisone) therapy and most often are posterior subcapsular in location. The incidence can be as high as 58%.[15] Steroid-induced glaucoma is another potential difficulty for these patients.[4]

Although cytomegalovirus retinopathy is the most common ocular infection following immunosuppression for renal transplant, a number of different infectious processes have been described.[16] Immunosuppression is usually sustained by a combination of prednisone and azathioprine, and infections are attributed to altered immune function. Infections include bacterial and fungal endophthalmitis caused by *Staphylococcus*, *Pseudomonas*, *Candida*, and *Nocardia* species and coccidoidomycosis and aspergillosis. Herpes simplex keratitis and presumed retinitis, toxoplasmosis, and pneumocystis infection have also been described.[17,18]

In addition to an increased incidence of ocular infections, immunosuppressed renal transplant recipients are also apt to develop particular types of cancers—squamous cell carcinoma and lymphoma. The emergence of these neoplasms has been attributed to alteration of immunologic surveillance, although more direct oncogenic effects of immunosuppressive agents remain a possibility. Squamous cell carcinoma of the lids, ocular lymphoma, and activation of a uveal malignant melanoma, have been described.[19-21]

References

1. Robbins SL, Cotran RS, Kumar V. *The Pathologic Basis of Disease*, 3rd ed. Philadelphia, Pa: WB Saunders; 1984; 21:991-1006.
2. Stein JH, Fried TA. Acute renal failure. In: Stein JH, ed. *Internal Medicine*. 3rd ed. Boston, Mass: Little, Brown; 1987;755-762.
3. Luke RG, Strom T. Chronic renal failure. In: Stein JH, ed. *Internal Medicine*, 3rd ed. Boston, Mass: Little, Brown; 1987;762-782.
4. LaPiana FG. Renal disease. In: Duane TD, ed. *Clinical Ophthalmology*. vol 5. Cambridge, Mass: Harper & Row; 1984.
5. Bar S, Savir H. Renal retinopathy—the renewed entity. *Metab Pediatr Syst Ophthalmol*. 1982;1:33.
6. Grenfell A, Watkins PJ. Clinical diabetic nephropathy: natural history and complications. *Clin Endocrinol Metab*. 1986;15:783.
7. Wagdi S, Dumas J, Labelle P. Retinal detachment in renal insufficiency. A report of 3 cases. *Can J Ophthalmol*. 1978;13:157.
8. Demco TA, McCormick AQ, Richards JSF. Conjunctival and corneal changes in chronic renal failure. *Can J Ophthalmol*. 1979;9:208.
9. Parrish CM, O'Day DM. Brown tumor of the orbit. Case report and review of the literature. *Arch Ophthalmol*. 1986;104:1199.
10. Rottembourg J, et al. Continuous ambulatory peritoneal dialysis in diabetic patients. The relationship of hypertension to retinopathy and cardiovascular complications. *Hypertension*. 1985;7:125.
11. Barton CH, Vaziri ND. Central retinal vein occlusion associated with hemodialysis. *Am J Med Sci*. 1979;277:39.
12. Moel DI, Kwyn YA. Cortical blindness as a complication of hemodialysis. *J Pediatr*. 1978;93:890.
13. Servilla KS, Groggel GC. Anterior ischemic optic neuropathy as a complication of hemodialysis. *Am J Kidney Dis*. 1986;8:61.
14. Hilton AF, et al. Ocular complications in haemodialysis and renal transplant patients. *Austral J Ophthalmol*. 1982;10:247.
15. Ramsay RC, et al. The visual status of diabetic patients after renal transplantation. *Am J Ophthalmol*. 1979;87:305.
16. Egbert P, et al. Cytomegalovirus retinitis in immunosuppressed hosts. *Ann Intern Med*. 1980;93:664.
17. Yovinsky E, et al. Disseminated herpes simplex infection with retinitis in a renal allograft recipient. *Ophthalmology*. 1983;90:175.
18. Knox DL. Ocular complications of chronic renal disease. *Transplant Proc*. 1987;19(2 suppl):73.
19. Stewart WB, et al. Eyelid tumors and renal transplantation. *Arch Ophthalmol*. 1980;98:1771.
20. Ziemanski MC, Godfrey WA, Lee KY, et al. Lymphoma of the vitreous associated with renal transplantation and immunosuppressive therapy. *Ophthalmology*. 1980; 87:596.
21. Spees EK Jr, et al. Reactivation of ocular malignant melanoma after renal transplantation. *Transplantation*. 1980; 29:421.

Chapter 167

Wilms' Tumor—Aniridia Syndrome

Miller's Syndrome

LEONARD B. NELSON

A palpable mass in the abdomen of a child is an ominous clinical finding. The two most common malignant tumors involving the upper abdomen in childhood are neuroblastoma arising in the adrenal gland or in retroperitoneal tissues and Wilms' tumor.[1]

Wilms' tumor is an embryonal malignancy of mixed origin that arises in the kidney. It accounts for about 20% of malignant tumors in children; its incidence in the general population is between 1 in 10,000 and 1 in 50,000 live births.[2] Approximately 40% of Wilms' tumors are hereditary; mode of transmission is autosomal dominant.[3] The diagnosis is made at a median age of 3 years, although for hereditary cases it is approximately 2 years.[3] Approximately 80% of children with Wilms' tumor are diagnosed by the age of 5 years.

The presenting manifestation of Wilms' tumor in childhood is an abdominal swelling or mass in 50 to 60% of patients, pain in 20 to 30%, and hematuria in 5 to 10%.[4] Hypertension may occur from increased renin levels caused by renal ischemia, probably secondary to compression of the renal artery by the tumor.

Systemic Manifestations

In 1964, Miller et al.[5] reviewed 440 cases of Wilms' tumor and found that six children had sporadic aniridia. In another large group of patients with Wilms' tumor, aniridia was found in 1% of the patients; this contrasts significantly with the frequency of aniridia in the general population (1 in 50,000 to 1 in 100,000).[6,7]

There are now over 60 reported cases of associated Wilms' tumor and aniridia.[7] Affected children typically have sporadic aniridia, Wilms' tumor that develops before age 2 to 3 years, severe mental retardation (75%), genitourinary abnormalities (66%), craniofacial dysmorphism (75%), occasional microcephaly, and frequent growth retardation.[7] The dysmorphic child typically has a long narrow face, prominent nose, and low-set ears with poor lobulation. As Fraumeni and Glass pointed out, aniridic children with Wilms' tumor differ in two

important respects from aniridia patients generally: (1) the aniridia is severe and usually associated with other systemic congenital anomalies and (2) familial aniridia is rare.[8] The only patient in whom Wilms' tumor occurred with a positive parental history of aniridia was also unique in having no other nonocular associated anomalies.[8]

In 1978, Francke et al.[9] found a deletion of the short arm of chromosome 11 (11p−) in the Wilms' tumor–aniridia syndrome. Numerous reports subsequently confirmed the location of the critical deleted segment of 11p13 for the Wilms' tumor–aniridia association. Although the Wilms' tumor–aniridia association is usually accompanied by 11p deletion, some aniridia patients have been described who have the 11p deletion but have not developed Wilms' tumor. One report describes a set of monozygotic twins with aniridia and the 11p deletion, of whom only one developed the Wilms' tumor.[9]

The gene for red blood cell catalase has recently been mapped to 11p13 using tissues derived from persons with deletion or triplication of different segments of 11p.[10] In subsequent studies involving 11p deletion, reduction of catalase levels has been confirmed.[11] Recognition that the Wilms' tumor–aniridia association may occur with normal chromosomes or with an 11p deletion suggests that a subtle submicroscopic deletion of 11p13 may result in aniridia or Wilms' tumor.[7] Abnormal catalase levels in these patients may serve as a genetic marker for detecting subtle deletions of chromosome 11 that are not detectable with present methods.

Ocular Manifestations

Aniridia is a bilateral uncommon panocular disorder affecting not only the iris but also the cornea, anterior chamber angle, lens, retina, and optic nerve.[7] Although iris hypoplasia is the most apparent clinical manifestation of the aniridic eye, it is not the major determinant of visual function. Visual impairment is better correlated with absence of the macular reflex, optic nerve hypoplasia, and the development of cataracts, glaucoma, and corneal opacification.[7]

Aniridia is inherited in an autosomal dominant manner in approximately two thirds of affected children. Of the sporadic cases of aniridia, two thirds represent a new autosomal dominant condition.

Although aniridia is a rare malformation, its recognition should immediately alert ophthalmologists, pediatricians, and other physicians to the possible presence of other ocular and systemic abnormalities. Between 25 and 33% of patients who exhibit sporadic aniridia develop Wilms' tumor prior to age 3 years.

Because of the increased risk of developing Wilms' tumor in children with sporadic aniridia, a thorough physical examination, especially abdominal palpation, is necessary semiannually during the first 5 years of life.[7] Careful documentation should also be made to detect other congenital systemic abnormalities, especially of external genitalia, head, and external ears.[7]

Besides physical examination, radiographic evaluation should be performed on patients with aniridia in an effort to detect Wilms' tumor early, while it is still curable. The evaluation should be performed semiannually for patients with sporadic aniridia and annually for patients with familial aniridia. The radiographic evaluation is necessary for approximately the first 5 years of life. Although the intravenous pyelogram was the radiographic method of choice in the past to document Wilms' tumor, renal ultrasound, which is less invasive and quicker and requires no contrast material, should be considered when it is available.[12]

References

1. Campbell PE. Abdominal masses: tumors of the kidney. In: Jones PG, Campbell PE, eds. *Tumors of Infancy and Childhood.* Melbourne: Blackwell Scientific; 1976.
2. Glenn JF, Rhame RC. Wilms' tumor: epidemiological experience. *J Urol.* 1961;85:911.
3. Knudson AG, Strong LC. Mutation and cancer: a model for Wilms' tumor of the kidney. *J Natl Cancer Inst.* 1972;48:313.
4. Rubin P. Cancer of the urogenital tract: Wilms' tumor and neuroblastoma. *JAMA.* 1968;204:981.
5. Miller RS, Fraumeni JF Jr, Manning MD. Association of Wilms' tumor with aniridia, hemihypertrophy, and other congenital malformations. *N Engl J Med.* 1964;270:922.
6. Pendergrass TW. Congenital anomalies in children with Wilms' tumor: a new survey. *Cancer.* 1976;37:403.
7. Nelson LB, Spaeth GL, Nowinski TS, et al. Aniridia. A review. *Surv Ophthalmol.* 1984;28:621.
8. Fraumeni JF Jr, Glass AG. Wilms' tumor and congenital aniridia. *JAMA.* 1968;206:825.
9. Francke V, Riccardi VM, Hittner HM, et al. Interstitial del (11p) as a cause of the aniridia–Wilms' tumor association: band localization and a heritable basis. *Am J Hum Genet.* 1978;30:81A.
10. Humen C, Turleau C, deGrouchy J, et al. Regional assignment of catalase (CAT) gene to band 11p13. Association with the aniridia–Wilms' tumor–gonadoblastoma (WAGR) complex. *Ann Genet.* 1980;23:165.
11. Narahara K, Kikkawa K, Kimira S, et al. Regional mapping of catalase and Wilms' tumor–aniridia, genitourinary abnormalities, and mental retardation triad loci to the chromosome segment 11p1305-p1306. *Hum Genet.* 1984;66:181.
12. Friedman AL. Wilms' tumor detection in patients with sporadic aniridia. Successful use of ultrasound. *Am J Dis Child.* 1986;140:173.

PART 18

Skeletal Disorders

Section A

Cranial Deformity Syndromes

Chapter 168

Craniosynostosis

JOHN D. WRIGHT, Jr., and WILLIAM P. BOGER III

Three factors have contributed to increased attention to the ocular manifestations of the craniosynostosis syndromes. First, more patients with rare syndromes are concentrated in major institutions under the care of multidisciplinary craniofacial treatment teams. Second, the importance of chronic hydrocephalus as a cause of optic atrophy and mental retardation complicating craniosynostosis has been increasingly well recognized over the past few decades.[1-3] As a result of vigorous early neurosurgical intervention many of these patients are growing up to become intelligent and gregarious persons for whom strabismus surgery can have significant benefits. The third factor converging in this story is our evolving understanding of vertically acting extraocular muscles and of the abnormalities that give rise to A-pattern and V-pattern strabismus.

Craniosynostosis without Facial Involvement

"Craniostenosis" and "craniosynostosis" refer to conditions in which at least one suture line of the skull or face fuses prematurely. Tessier has differentiated these two terms, suggesting that the term "craniostenosis" be used for the malformation and that "craniosynostosis" be used for the cause of the malformation.[4] Most authors, however, use the terms interchangeably, and this has been our custom as well.[5] The existing literature on these entities can be understood only by using these terms synonymously; series of patients reported by one author to have craniostenosis come from the same population of patients that are reported by other authors to have craniosynostosis. The fusion of the cranial bones across a suture line has sometimes been seen on prenatal radiographs of the mother's abdomen. The reason for the premature fusion is unknown. Fusion along one suture line appears to limit bone growth perpendicular to that suture. The brain continues to grow, however, so excessive expansion takes place where open sutures allow it.[3,6,7] Craniostenosis involving only the cranial bones may manifest a variety of calvarial shapes, depending on the suture line that is fused:

Sagittal suture (long, narrow skull; boatlike skull), scaphocephaly, dolichocephaly
Coronal suture (wide skull), brachycephaly

Metopic suture, trigonocephaly

Unilateral coronal suture (asymmetric skull), plagio-cephaly

Compound synostosis of several sutures (as a result of premature closure of both coronal and sagittal sutures, skull expands vertically), acrobrachycephaly (peaked, wide skull), oxycephaly (peaked or tower skull; synonyms: turricephaly, acrocephaly).

Ocular Anomalies Associated with Craniosynostosis

Optic Atrophy

Intracranial pressure is frequently elevated in children who have closure of multiple cranial sutures, for example in oxycephaly.[1,3,8-10] It can also be elevated when only one suture is involved, but this occurs less often.[2,3] The mechanism of increased intracranial pressure usually appears to be restriction of normal brain growth created by the confined intracranial volume. This form of elevated intracranial pressure has been approached neurosurgically by strip craniectomy with polyethylene film applied to the bone edges to prevent rapid reclosure. Hydrocephalus is an additional mechanism of elevated intracranial pressure that appears to coexist with craniosynostosis more often than would be expected by chance.[2,3]

In the craniosynostosis syndromes, neurosurgical intervention may be crucial to avoid the ravages of increased intracranial pressure. Although the older literature talks about primary optic atrophy from direct local effects on the nerve, this mechanism of blindness must be extremely rare, if indeed it occurs at all.[7-9,11] Optic atrophy in craniosynostosis syndromes almost always is due to the secondary effects of increased intracranial pressure and chronic papilledema.[2,3] Craniosynostosis syndromes must be differentiated in this regard from other skeletal disorders that can involve the skull and facial structures, such as craniometaphyseal dysplasia, which is more commonly associated with direct effects on the optic nerve when vision is affected.[12] Narrowing of the optic canal from bony overgrowth is *not* a characteristic of the craniosynostosis syndromes.

Characteristic Vertical Strabismus Associated with Plagiocephaly

The strabismus associated with plagiocephaly is perhaps the clearest example of a unique mechanism for strabismus caused by a structural abnormality due to craniostenosis.[13] Unilateral coronal synostosis (plagioce-phaly) occurs sporadically as a limited form of cranial synostosis. There is distinctive, asymmetric growth of the frontoorbital region, with elevation of the brow and flattening of the forehead on the involved side. Orbital radiographs show a "harlequin" deformity of the affected orbit, an exaggerated obliquity of the orbital roof, elevated greater and lesser wings of the sphenoid bone, and absence of the ipsilateral coronal suture.

Patients with unilateral coronal synostosis quite often have hypertropia of the eye on the side of the synostosis.[13] The hypertropia increases with gaze away from the involved side and lessens with gaze toward the involved side, but results of Bielschowsky's head tilt test are only occasionally positive in these patients. The roof of the orbit is elevated and foreshortened, and the trochlea is displaced posteriorly, thus weakening the depressive action of the reflected superior oblique tendon. The inferior oblique muscle is not similarly weakened, probably because the orbital floor is less affected by the cranial abnormality than is the orbital roof.[13]

Early neurosurgical intervention for the plagiocephaly may cure the vertical strabismus in some patients. The procedure involves moving the superior orbital rim forward, pulling the galea and orbital periosteum along with the advanced orbital rim. It is noteworthy, however, that even though early surgical correction allows the fronto-orbital cranium to assume a more symmetric contour, some orbital asymmetry usually remains, and most patients who exhibit vertical strabismus in the context of plagiocephaly continue to demonstrate vertical strabismus after cranial surgery.[13]

Craniosynostosis Involving Both the Cranial and Facial Structures (Crouzon's and Apert's Syndromes)

Deformities become more complicated when multiple sutures are fused, and eponyms are commonly appended to syndromes in which facial structures are involved. Craniostenosis has been estimated to occur in 1 in 1900 live births, the facial skeleton being affected in approximately 10%.[14,15] Crouzon[16] reported that craniofacial dysostosis may have a familial tendency, while Apert[17] described the association of craniofacial synostosis with syndactyly. Some of the facial and cranial deformities are similar in Crouzon's and Apert's syndromes—increased height of the anterior face, widened and shallow orbits, maxillary hypoplasia, relative mandibular prognathism, and a nose that resembles a parrot's beak.

The most striking feature that differentiates patients with Apert's syndrome from those with Crouzon's syndrome is the complex syndactyly. The typical hand deformity in Apert's syndrome consists of bony fusion of the second, third, and fourth fingers, which often have a single common nail. There may be a similar deformity of the feet. There are a number of variations of Apert's syndrome. Additional features that may be more characteristic of Apert's syndrome than of Crouzon's syndrome include acne, submucous or overt cleft palate, high palatal arch, alveolar ridge hypertrophy, anterior open bite, absence of extraocular muscles (especially the superior rectus muscles), asymmetry in the orbital bone deformity, blepharoptosis, and overhanging of the upper frontal area with a transverse frontal skin furrow. Crouzon's syndrome is inherited in an autosomal dominant fashion; an estimated 25% of cases represent fresh mutations.[7,18-22] Apert's syndrome is less common than Crouzon's syndrome and has been estimated to occur in 1 in 160,000 live births.[23] Its hereditary pattern appears to be autosomal dominant with low penetrance and a much higher spontaneous mutation rate than Crouzon's syndrome. The majority of cases of Apert's syndrome occur sporadically.[24]

Ocular Abnormalities Associated with Crouzon's and Apert's Syndromes

Optic Atrophy (see above)

Exorbitism

The shallow orbits, not an excess of orbital contents, give the eyes of patients with craniofacial dysostosis a bulging appearance. In this context the term "exorbitism" has been preferred to "exophthalmos." It is remarkable how readily the globes may protrude between the eyelids with spastic closure of the eyelids behind the eyeball. Relatively minor manipulations around the globes, including scleral depression for detailed retinal examination, may precipitate luxation of the globe. As striking as this phenomenon is, generally the globe can be manually reposited without visual consequence (by anterior pressure on the conjunctiva on either side of the cornea and simultaneous retraction of the lids).

V-Pattern Strabismus in Crouzon's and Apert's Syndrome

Patients with Crouzon's or Apert's syndrome often have characteristic V-pattern strabismus (ie the eyes are further apart on upgaze and closer together on downgaze—hence the designation "V" pattern).[25-27] They may have exotropia in all fields of gaze, complete

esotropia in all fields of gaze, or exotropia in upward gaze and esotropia in downward gaze, but it is the V pattern that is so characteristic. With both eyes open, on versions the inferior oblique muscles appear to be markedly overactive. This observation is consistent with the usual V-pattern strabismus. However, if one eye is covered and the ductions of the other eye are studied, both the superior oblique muscles and the superior rectus muscles often appear to be weak. In many patients with craniofacial synostosis it has seemed reasonable to weaken the inferior oblique muscles with the usual techniques when a large V pattern is associated with marked overaction of the inferior oblique muscles and corresponding underaction of the superior oblique muscles. In this carefully selected circumstance, the V pattern and the apparent overaction of the inferior oblique muscles have been improved, but not necessarily eliminated, by surgical intervention.

It has been suggested that the V-pattern strabismus seen in craniofacial synostosis syndromes is not due to the usual imbalances of oblique muscles that ophthalmologists have become accustomed to seeing in association with infantile esotropia.[27-29] Of particular interest are the growing number of reports documenting the absence of superior rectus muscles in patients with Apert's syndrome.[30-33] Even more exceptionally, absent vertical rectus muscles *and* absent oblique muscles have been reported in a patient with Crouzon's syndrome[34] and in one with Apert's syndrome.[30] Clearly, the superior rectus muscle is not absent in most patients with craniofacial synostosis, but the fact that marked anomalies in the anatomy of the extraocular muscles may occur in these syndromes is worth keeping prominently in mind.

It has been noted that even when the extraocular muscles are present they may insert anomalously on the globe.[35] In an attempt to explain V pattern strabismus that cannot be fully explained by oblique muscle imbalance it has been suggested that the medial rectus muscles may insert higher than usual and that the lateral rectus muscles may insert lower than is normally expected.[36] It has even been suggested that in craniofacial dysostosis *all* the rectus muscles might insert anomalously: the medial higher than usual, the superior more temporal than usual, the lateral lower than usual, and the inferior more nasal than usual. To date, these questions have generated considerable discussion, but measurements and data are as yet insufficient to reach a conclusion on how frequently and in what settings extraocular muscles insert in anomalous positions. Increasing experience with direct observation at the time of surgical dissection will probably ultimately define how frequently abnormalities of the extraocular muscles occur in craniosynostosis patients. It should also be possible to predict the general region of the rectus muscle insertions on office examination by looking through the conjunctiva for the prominent ciliary vessels associated with the rectus muscles. This office evaluation can sometimes be facilitated by blanching the superficial vessels with a drop of phenylephrine. Computed tomography and magnetic resonance imaging may provide additional information.[27] These techniques could be expected to verify a major anomaly such as an absent muscle, but, to date at least, the position of the muscle behind the globe has not correlated with the position of its insertion on the globe itself.

It is usually recommended that surgical intervention for strabismus be deferred until after craniofacial reconstructive surgery is completed.[27,29] A change in ocular alignment with regard to strabismus is relatively infrequent in patients undergoing osteotomies for craniosynostosis, and some ophthalmologists have suggested that strabismus surgery in this setting need not necessarily await the outcome of skeletal repair.[25,26,37] In contrast there is more regularly a change in ocular alignment when craniofacial teams mobilize each orbit, as in the treatment of hypertelorism. We personally favor waiting until craniofacial procedures are completed before undertaking strabismus surgery for most craniofacial malformations, including the craniosynostoses. However, if the degree of craniofacial deformity is only moderate and the merits of the major interventions required for craniofacial reconstruction are dubious, strabismus surgery alone may, in some circumstances, provide substantial benefit for the patient. Regardless of the timing of strabismus surgery, it remains for the ophthalmologist an ongoing challenge to maintain an educational program that emphasizes the importance of early examination and the consistent application of well-known refractive and patching therapies for amblyopia.

References

1. Shillito J, Matson D. Craniosynostosis: a review of 519 surgical patients. *Pediatrics.* 1968;41:829-853.
2. Camfield PR, Camfield CS. Neurologic aspects of craniosynostosis. In: Cohen MM, ed. *Craniosynostosis.* New York, NY: Raven Press; 1986.
3. Winston K. Craniosynostosis. In: Wilkins RH, Rengachary SS, eds. *Neurosurgery.* vol 3. New York: McGraw-Hill, 1985;2173-2191.
4. Tessier P. Relationship of craniostenoses to craniofacial dysostoses and to faciostenosis. *Plast Reconstr Surg.* 1971;48:224-237.
5. Cohen MM. History, terminology, and classification of craniosynostosis. In: Cohen MM, ed. *Craniosynostosis: Diagnosis, Evaluation, and Management.* New York, NY: Raven Press; 1986.
6. Virchow HR. Ueber den Cretinismus, namentlich in Franken, and uber pathologische Schadelforamen. *Ver Phys Med Ges Wurzburg.* 1852;2:230-271.
7. Blodi FC. Developmental anomalies of the skull affecting the eye. *AMA Arch Ophthalmol.* 1957;57:593-610.
8. Howell SC. The Craniostenoses. *Am J Ophthalmol.* 1954;37:359-379.
9. Koziak PH Craniostenosis. *Am J Ophthalmol.* 1954;37:380-390.
10. Parks MM, Costenbader FD. Craniofacial dysostosis (Crouzon's disease). *Am J Ophthalmol.* 1950;33:77-82.
11. Seelenfreund M, Gartner S. Acrocephalosyndactyly (Apert's syndrome). *Arch Ophthalmol.* 1967;78:8-11.
12. Puliafito CA, et al. Optic atrophy and visual loss in craniometaphyseal dysplasia. *Am J Ophthalmol.* 1981;92(5):696-701.
13. Robb RM, Boger WP III. Vertical strabismus associated with plagiocephaly. *J Pediatr Ophthalmol Strab.* 1983;20(2):58-62.
14. Myrianthopoulos NC. Concepts, definitions and classification of congenital and developmental malformations of the central nervous system and related structures. In: Vinken PJ, Bruyn GW, eds. *Handbook of Clinical Neurology.* vol 30. Amsterdam; Elsevier North-Holland; 1977;1-13.
15. Cohen MM. Perspectives on craniosynostosis. In: Cohen MM, ed. *Craniosynostosis, Diagnosis, Evaluation, and Management.* New York, NY: Raven Press; 1986.
16. Crouzon MO. Dysostose cranio-faciale hereditaire. *Soc Med Dis Hosp Paris.* 1912;10:545-555.
17. Apert ME. De l'acrocephalosyndactylie. *Soc Med Dis Hosp Paris.* 1906;21:1310-1330.
18. Dodge HW, et al. Craniofacial dysostosis: Crouzon's disease. *Pediatrics.* 1959;23:98-106.
19. Lake MS, Kuppinger JC. Craniofacial dystosis (Crouzon's disease) report of three cases. *AMA Arch Ophthalmol.* 1950;44:37-46.
20. Schiller JG. Craniofacial dysostosis of Crouzon: a case report and pedigree with emphasis on heredity. *Pediatrics.* 1959;23:107-112.
21. Pinkerton OD, Pinkerton FJ. Hereditary craniofacial dysplasia. *Am J Ophthalmol.* 1952;35:500-506.
22. Flippen JH Jr. Cranio-facial dystosis of Crouzon: report of a case in which the malformation occurred in four generations. *Pediatrics.* 1950;5:90-96.
23. Blank CE. Apert's syndrome (a type of acrocephalosyndactyly); observations on a British series of thirty-nine cases. *Ann Hum Genet.* 1960;24:151-164.
24. Gorlin RJ, Pindborg JJ, McKusick VA. *Syndromes of the Head and Neck.* New York, NY: McGraw-Hill; 1964.
25. Greaves B, et al. Disorders of ocular motility in craniofacial dysostosis. *J R Soc Medicine.* 1979;72:21-24.
26. Walker J, Wybar K. Ocular motility problems in craniofacial dysostosis. In: Moore S, et al. eds. *Orthoptics, Past, Present, Future.* New York, NY: Stratton Intercontinental Medical Book Corp.; 1976.
27. Prazansky S, Miller MT. Ocular defects in craniofacial syndromes. In: Renie WA, Goldberg MF, eds. *Goldberg's Genetic and Metabolic Eye Disease.* Boston, Mass: Little, Brown; 1986;241-255.
28. Margolis S, et al. Structural alterations of extraocular muscle associated with Apert's syndrome. *Br J Ophthalmol.* 1977;61:683-689.
29. Miller MT. Ocular findings in craniosynostosis. In: Cohen

MM, ed. *Craniosynostosis.* New York, NY: Raven Press; 1986.

30. Diamond GR, et al. Variations in extraocular muscle number and structure in craniofacial dysostosis. *Am J Ophthalmol.* 1980;90(3):416-418.

31. Weinstock FJ, Hardesty HH. Absence of superior recti in craniofacial dysostosis. *Arch Ophthalmol.* 1965;74:152-153.

32. Cuttoone J, Drazis PT, Miller MT, et al. Absence of the superior rectus in Apert's syndrome. *J Pediatr Ophthalmol Strab.* 1979;16:349-354.

33. Carruthers J. Strabismus in craniofacial dysostosis. *Graefe's Arch Clin Exp Ophthalmol.* 1988;226:230-234.

34. Snir M, et al. An unusual extraocular muscle anomaly in a patient with Crouzon's disease. *Br J Ophthalmol.* 1982;66(4):253-257.

35. Caputo AR, Lingua RW. Aberrant muscular insertions in Crouzon's disease. *J Pediatr Ophthalmol Strab.* 1980; 17(4):239-241.

36. Diamond G. Surgical treatment of overelevation in adduction in patients with craniostenosis. Presented at annual meeting of the American Association for Pediatric Ophthalmology and Strabismus.

37. Diamond GR, et al. Ocular alignment after craniofacial reconstruction. *Am J Ophthalmol.* 1980;90(2):248-250.

Chapter 169

Crouzon's Syndrome

Craniofacial Dysostosis

DAVID S. FELDER and JOHN D. BULLOCK

Crouzon's syndrome (craniofacial dysostosis) is a rare congenital anomaly involving premature fusion of the cranial sutures, maxillary hypoplasia, and shallow orbits. It was first described by Crouzon[1] in 1912 and now belongs to a group of closely related craniofacial syndromes that includes Apert's syndrome, Carpenter's syndrome, Pfeiffer's syndrome, and Saethre-Chotzen syndrome. The absence of hand and foot anomalies helps to differentiate Crouzon's syndrome from the others, which have craniosynostosis as a common feature. Crouzon's syndrome generally is not associated with extracranial defects.

Premature fusion of the sagittal, coronal, and lambdoidal sutures may give rise to a brachycephalic-looking cranium (one with a decreased anteroposterior diameter and a wide lateral dimension). However, depending on the rate, order, and severity of the suture fusion, there may be a diverse spectrum of cranial deformities. It is generally agreed that there is no characteristic calvarium appearance in Crouzon's syndrome. Brachycephaly is most commonly observed, but trigonocephaly and scaphocephaly may be seen.[2] Although infrequent, there may be no cranial suture involvement in some patients with Crouzon's syndrome.

Common findings in Crouzon's syndrome include exophthalmos, hypertelorism, divergent strabismus, maxillary hypoplasia, a narrow, high-arched palate, and

a "parrot-beak" nose (Fig. 169-1). Maxillary hypoplasia may be associated with a relatively prognathic mandible, a short upper lip, and a drooping lower lip (Fig. 169-2).

Oral manifestations include cleft (hard or soft) palate, bifid uvula, crowding of the upper teeth, V-shaped maxillary dental arch, and severe malocclusions that cause the maxilla to lie within the mandibular arch. Other reported findings include a narrow nasopharynx, oropharyngeal obstruction, frequent upper respiratory infections, and conductive hearing loss.

Crouzon's syndrome follows an autosomal dominant mode of transmission, with a high degree of penetrance and variable expression. Many sporadic cases due to fresh mutations also have been reported. There is an equal incidence in males and females. Although Juberg and Chambers[3] described two patients they believed inherited Crouzon's syndrome in an autosomal recessive manner, their conclusions have been disputed.[4] Although the cause of Crouzon's syndrome is not known, Burdi et al.[5] suggest that the premature fusion of the cranial sutures may be a secondary pathogenic event. They postulated that the premature fusion of the sphenoethmoidal synchondrosis was the primary event, which leads to hypoplasia and retrodisplacement of the midface. Certainly, more investigation must be undertaken before any definite conclusions can be made.

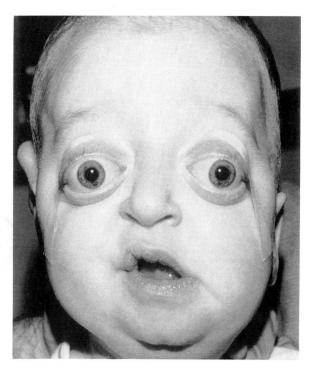

FIGURE 169-1 Characteristic facies of Crouzon's syndrome demonstrating exophthalmos, divergent strabismus, hypertelorism, and "parrot-beak" nose. (Courtesy of G. Frank Judisch, M.D.)

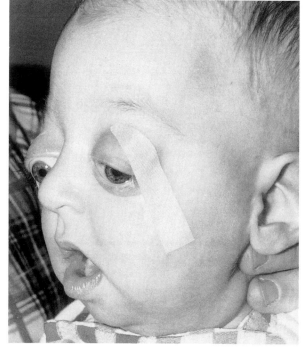

FIGURE 169-2 Side view of the Crouzon's syndrome patient in Figure 169-1 shows maxillary hypoplasia, mandibular prognathism, short upper lip, and drooping lower lip. (Courtesy of G. Frank Judisch, M.D.)

The incidence of mental deficiency is reported to vary from 10 to 13%. It may be that premature synostosis of the cranial sutures prevents normal brain maturation.

Ocular Manifestations

The ophthalmic findings associated with Crouzon's syndrome are principally considered to be secondary to bony malformations of the skull and orbits. Proptosis, a characteristic feature, is due to extremely shallow orbits and reduction of the anteroposterior axis. Hypertelorism is frequently observed and may arise from abnormal sphenobasilar synchondrosis.

Strabismus, usually exotropia, commonly accompanies Crouzon's syndrome. In one study, Nelson et al.[6] reported a 61% incidence of V patterns (where the eyes are more divergent in upgaze and closer together on downgaze) in patients with Crouzon's syndrome. Muscle imbalance in patients with craniosynostosis may be secondary to several underlying abnormalities. Morax felt that changes in the orbital architecture resulted in asymmetric muscle origins and insertions along with

mechanical alterations in the extraocular muscles.[7] Margolis et al. described several structural abnormalities found in an extraocular muscle in a patient with Apert's syndrome.[8] These observations suggest that strabismus in association with Crouzon's syndrome is a consequence of many factors.

Optic atrophy has been reported in at least 30% of patients and may be due to chronic papilledema secondary to increased intracranial pressure, traction on the optic nerves, or compression of the optic nerves by narrowed optic foramina. Other less frequent ocular abnormalities associated with Crouzon's syndrome are nystagmus, glaucoma, keratoconus, microcornea, cataract, ectopia lentis, aniridia, anisocoria, and medullated nerve fibers.[9]

References

1. Crouzon O. Dysostose cranio-faciale hereditaire. *Bull Mem Soc Med Hop.* 1912;33:545.
2. Singh M, Hadi F, Aram GN, et al. Craniosynostosis—Crouzon's disease and Apert syndrome. *Indian Pediatr.* 1983;20(8):608.

3. Juberg RC, Chambers SR. An autosomal recessive form of craniofacial dysostosis (the Crouzon syndrome). *J Med Genet.* 1973;10(1):89.

4. Cohen MM Jr. An etiologic and nosologic overview of craniosynostosis syndromes. *Birth Defects.* 1975;11(2):137.

5. Burdi AR, Kusnetz AB, Venes JL, et al. The natural history and pathogenesis of the cranial coronal ring articulations: implications in understanding the pathogenesis of the Crouzon craniostenotic defects. *Cleft Palate J.* 1986;23(1):28.

6. Nelson LB, Ingoglia S, Breinin GM. Sensorimotor disturbances in craniostenosis. *J Pediatr Ophthalmol Strab.* 1981;18(5):32.

7. Morax S. Oculo-motor disorders in craniofacial malformations. *J Maxillofac Surg.* 1984;12(1):1.

8. Margolis S, Pachter BR, Breinin GM. Structural alterations of extraocular muscle associated with Apert's syndrome. *Br J Ophthalmol.* 1977;61(11):683.

9. Wolter JR. Bilateral keratoconus in Crouzon's syndrome with unilateral acute hydrops. *J Pediatr Ophthalmol.* 1977;14(3):141.

Section B

Facial Malformation Syndromes

Chapter 170

Facial Hemiatrophy

Parry-Romberg Syndrome

MICHAEL E. MIGLIORI and ALLEN M. PUTTERMAN

The syndrome of hemifacial atrophy (Parry-Romberg syndrome) is characterized by insidious onset of slowly progressive atrophy of subcutaneous fat and soft tissues of one side of the face. The syndrome typically has its onset during the first 2 decades of life. The active phase of the disease usually lasts 2 to 10 years.[1] Although subcutaneous fat and connective tissue are most severely affected, skin, muscle, cartilage, and bone may also be involved.[2] In some cases the atrophy may extend to the ipsilateral side of the neck, upper extremity, and occasionally the entire side of the body. The syndrome may be accompanied by Jacksonian seizures, ipsilateral trigeminal neuralgia, alopecia, and enophthalmos. The condition is bilateral in 5% of cases.[1] Females are affected more often than males, in a ratio of 3:2. The left side of the face is more commonly involved.[2]

Most cases are sporadic, although a few familial cases have been reported.[3] The cause is unknown. The most popular hypothesis is that the syndrome is the result of a disorder of the trophic sympathetic nervous system.[1,2,4] A similar condition has been produced in rats by unilateral cervical sympathectomy.[5] Although many patients have reported suffering trauma prior to the onset of symptoms, it is difficult to identify trauma as a specific pathogenic factor. It is possible that trauma damages the peripheral sympathetic nervous system, interfering with its normal trophic function.[1] Wartenberg[6] postulated that the primary disturbance is in higher centers, resulting in the loss of the regulatory functions that control the peripheral nervous system. The frequent associations with heterochromia iridis, pupillary abnormalities, and pigmentary disturbances of the skin all support disturbances of the sympathetic nervous system as a causal mechanism in the pathogenesis of hemifacial atrophy.

A well-recognized manifestation of this disorder is the sharply demarcated line between normal and abnormal skin on the forehead called the *"coup de sabre"* deformity. This sign is also frequently seen in linear scleroderma of childhood. There is considerable clinical overlap between the syndromes of hemifacial atrophy and linear scleroderma. Both have their onset in the first 2 decades of life. Wartenberg[6] suggested that scleroderma *en coup de sabre* may be an arrested form of hemifacial atrophy. Many authors feel that they are two distinct but similar diseases. Antinuclear antibodies were found in one patient with hemifacial atrophy.[3] These antibodies are commonly found in patients with scleroderma. Histologically, the only difference between hemifacial atrophy and linear scleroderma is that the elastic tissue is more frequently preserved in hemifacial atrophy.[1,4]

The acquired, progressive nature of hemifacial atrophy distinguishes it from congenital facial hypoplasia and oculoauriculovertebral dysplasia (Goldenhar's syndrome).

Systemic Manifestations

The earliest changes are seen in the paramedian areas of the face. The left side of the face is affected more often than the right.[2] There is atrophy of subcutaneous fat, connective tissue, and muscle (Fig. 170-1). The overlying skin is not always affected. There may be atrophy of the upper lid, tongue, neck, trunk, and extremities.[2] If the onset occurs during the 1st decade of life, before the skull and facial bones are fully developed, there may be atrophy of the underlying cartilage and bone as well.[1] Frequently, the overlying skin is hyperpigmented, although hypopigmentation and vitiligo have been reported.[1-4] Alopecia may be evident near the midline on the affected side. There may be loss of the medial eyebrow and lashes. Poliosis has also been described in connection with this syndrome.[2]

Contralateral Jacksonian seizures, ipsilateral trigeminal neuralgia, facial paresthesias, facial paralysis, Horner's syndrome, and migraine headaches are the most common central nervous system manifestations.[1,2,4]

Ocular Manifestations

Ocular findings are present in 10 to 35% of cases.[1] Enophthalmos usually results from the atrophy of orbital fat on the affected side. Oculomotor palsies, both paralytic and restrictive, have been reported.[1] Fibrosis of the orbital content may account for the restrictive motility disturbances. Pupillary abnormalities include miosis, mydriasis, Adie's, Argyll Robertson, and oval pupils. Horner's syndrome, heterochromia iridis, and iris atrophy may be present.[1,2]

Ocular inflammation is frequently seen. Iritis, iridocyclitis, Fuch's heterochromic iridocyclitis, and papillitis

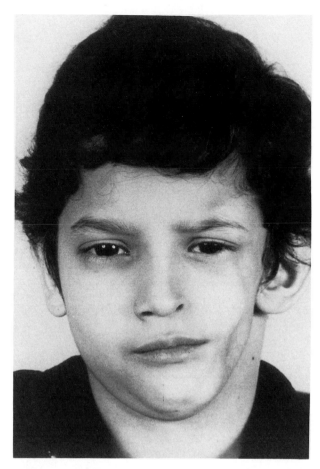

FIGURE 170-1 A 12-year-old boy with hemifacial atrophy. (From Miller et al.[1])

have all been observed.[1,2,4] Neuroparalytic keratitis is secondary to corneal hypesthesia and lagophthalmos. Posterior segment abnormalities include choroidal atrophy, pigmentary disturbances, and optic nerve hypoplasia.[1,7] Extraocular muscle palsies, optic atrophy, ptosis, miosis, and mydriasis have all been reported in the contralateral eye.[1]

Treatment is directed at augmenting the atrophic areas after the disease has stabilized. Liquid silicone injections, bone grafts, dermis fat grafts, and pedicle flaps have all been utilized with varying degrees of success.[4]

References

1. Miller MT, Sloane H, Goldberg MF, et al. Progressive hemifacial atrophy (Parry-Romberg disease). *J Pediatr Ophthalmol Strab.* 1987;24:27-36.
2. Gorlin RJ, Pindborg JJ, Cohen MM. Hemifacial atrophy. In: Gorlin RJ, Pindborg JJ, Cohen MM, eds. *Syndromes of the*

Head and Neck, 2nd ed. New York, NY: McGraw-Hill; 1976;341-344.

3. Lewkonia RM, Lowery RB. Progressive hemifacial atrophy (Parry-Romberg syndrome), report with review of genetics and nosology. *Am J Med Genet.* 1983;14:385-390.

4. Dedo DD. Hemifacial atrophy, a review of an unusual craniofacial deformity with a report of a case. *Arch Otolaryngol.* 1978;104:538-541.

5. Moss ML, Crikelair GF. Progressive facial hemiatrophy following cervical sympathectomy in the rat. *Arch Oral Biol.* 1960;1:254-258.

6. Wartenberg R. Progressive facial hemiatrophy. *Arch Neurol Psychiatr.* 1945;54:75-96.

7. vanDalen JTW. Hemifacial atrophy systemic and ophthalmological anomalies. *Fortschr Ophthalmol.* 1986;83:302-304.

Chapter 171

Mandibulofacial Dysostosis

Treacher Collins Syndrome

GARY DIAMOND, DONALD A. HOLLSTEN, and JAMES A. KATOWITZ

Mandibulofacial dysostosis was first described in the English literature by Thomson[1] in 1847, followed by Berry[2] in 1889 and Treacher Collins[3] in 1900. Since that time, over 300 cases of mandibulofacial dysostosis have been reported. It remained for Franceschetti and Zwahlen in 1944, however, to fully describe this condition as a distinct clinical entity. Franceschetti's[4] extensive article in 1948 described 65 cases; to this date it remains the most comprehensive discussion of the condition and its clinical manifestations.[4] As described by Franceschetti, the condition consists of abnormalities in the eydlids, hypoplastic facial bones, ear malformations, palatal and dental malformations, blind fistulas, abnormal hairline, and skeletal deformities, and it has an autosomal dominant inheritance pattern.

Etiology

Mandibulofacial dysostosis is inherited as an autosomal-dominant trait with high penetrance, marked variability in its expression, and an incidence of 1 in 10,000 live births. Hereditary transmission through four generations has been repeatedly described, with variable clinical presentations noted within families.[5] The underlying genetic defect has yet to be described. New mutations account for more than 50% of the cases, suggesting that an exogenous factor in the uterine environment may be a cause. Close scrutiny of family members is required to look for mild manifestations before a patient can be considered to represent a new mutation. A lethal effect of the gene is possible, since miscarriage and early postnatal death are common.[6]

The visible manifestations represent defective development of the parts of the face derived from visceral mesoderm, an inhibiting process that acts before or about the 7th week of fetal life when bony facial structures are developing. With delayed ossification of the temporal orbital wall, bony displacement occurs, and secondary hypoplasia and malposition of associated soft tissues follows. Developmental arrest at this stage can cause micrognathia, macrostomia, and downward displacement of the external ear as well. Although the cause of the developmental anomaly is unknown and could be any combination of inductive, neurotropic, or circulatory disturbances, McKenzie and Craig have described a defect in the development of the stapedial artery that, in their opinion, causes the syndrome.[7] Poswillo has also induced a phenocopy with hypervitaminosis A in an animal model that produces neural crest cell injury.[8]

Systemic Manifestations

Mandibulofacial dysostosis is associated with a variety of systemic manifestations, the most common and most striking of which involves the bones of the face. The

---

The opinions or assertions contained herein are the private views of the authors and are not to be construed as official or as reflecting the views of the Department of the Army or the Department of Defense.

fully expressed condition gives the patient a convex facial profile with a prominent nose and retrusive chin, which are caused by a variety of underlying bony abnormalities.[9]

Marsh et al.[9] examined the skulls of 14 patients in vivo with varying phenotypes of mandibulofacial dysostosis using three-dimensional computed tomographic (CT) techniques. Facial skeletal defects consistently present were (1) absence or severe hypoplasia of the zygomatic process of the temporal bone, (2) deformity, often with clefting, of the orbital rim, (3) deformity of the zygoma, (4) deformity of the mandible, and (5) deformity of the medial pterygoid plates and hypoplasia of the medial pterygoid muscles. There was much variability between patients as well as individual asymmetry. Inconsistent findings included absence of the external auditory meatus (43% of patients) and presence of a cleft palate with or without a cleft alveolus (36% of patients).[9]

The external ear is frequently involved with a crumpled, misplaced pinna. The external auditory canal is frequently absent. Middle ear dysplasia with fusion or absence of one or more of the ossicles is often found. Appropriate treatment is essential to allow for maximal intellectual development. Mental retardation has been noted in some patients but is unusual; thus it is essential to correct hearing deficits as early as possible to avoid intellectual impairment.[6]

Micrognathia is usually present and is seen with a variety of other oral manifestations, such as cleft palate, a high-arched palate, and malocclusion.[6] Extra ear tags and blind fistulas may be found anywhere between the angle of the mouth and the tragus.[6] These patients typically demonstrate atypical hair growth in the form of tongue-shaped processes of the hairline extending toward the cheeks.[4]

As in any wide-ranging syndrome complex, many occasional abnormalities have also been described. These include choanal atresia, absence of the parotid gland, congenital heart disease, cervical vertebral malformations, renal anomalies, cryptorchidism, hydrocephalus, dolichocephaly, small sella turcica, syndactyly, forearm and hand malformations, spine deformities, and in one case, associated neurofibromatosis.[6,4,10,11]

Early respiratory problems can also occur because of a narrow airway. This may require a temporary tracheostomy and presents a challenge to tracheal intubation for surgery.[12,13]

Ocular Manifestations

The ophthalmic manifestations of mandibulofacial dysostosis are related to the associated facial skeletal defects. The dysplasia of the bony orbit causes an

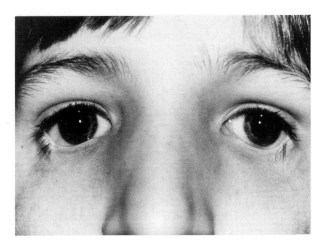

FIGURE 171-1 Patient with mandibulofacial dysostosis demonstrates pseudocoloboma of the left lower lid and true coloboma of the right lower lid.

antimongoloid slant of the soft tissues (lids) which is reflected in the clinical appearance. The inferotemporal orbital cleft is often associated with a pseudocoloboma rather than a true lid coloboma. The pseudocoloboma is so named because the marginal lid structures (tarsus, lashes, glands) are intact despite the aplasia and recession of the soft tissues into the orbital cleft. The pseudocoloboma gives the palpebral opening a triangular shape (Fig. 171-1).

True colobomas of the lower lid can also occur. They are manifested by absence of lashes, glands, and tarsus at the lid margin. An upper lid notch can occur also, but it is not as common as lower lid involvement.

Other ophthalmic findings that occur with much less frequency include absent lower lid cilia nasal to the coloboma, absent lower lid lacrimal puncta, iris coloboma, and microphthalmia. Strabismus and other ophthalmic defects are rarely seen in this syndrome.[4,6,10]

References

1. Thomson A. Notice of several cases of malformation of the external ear together with experiments on the state of hearing in such patients. *Monthly J Med Sci.* 1847;76:729.
2. Berry G. Note on a congenital defect (? coloboma) of the lower lid. *Ophthalmol Hosp Rep.* 1889;12:255.
3. Treacher Collins E. Case with symmetrical congenital notch in the outer part of each lower lid and defective development of the malar bone. *Trans Ophthalmol Soc UK.* 1900;20:190.
4. Franceschetti A, Klein D. The mandibulo-facial dysostosis: a new heredity syndrome. *Acta Ophthalmol (Copenh)* 1949;27:143-224.

5. Debusmann H. Familiare kombinierte Gesichtsmissbildung im Bereich des ersten Visceralbogens. *Arch Kinderheilk.* 1940;120:133.

6. Goodman RM, Gorlin RJ. Mandibulofacial dysostosis. In: *The Malformed Infant and Child: An Illustrated Guide.* New York, NY: Oxford University Press; 1983;272-273.

7. McKenzie E, Craig F. Stapedial artery defect causing mandibulofacial dysostosis. *Arch Dis Child.* 1955;30:391.

8. Poswillo D. The pathogenesis of the Treacher-Collins syndrome (mandibulofacial dysostosis). *Br J Oral Surg.* 1975;13:1-26.

9. Marsh JL, Celin SE, Vamier MW, et al. The skeletal anatomy of mandibulofacial dysostosis (Treacher Collins syndrome). *Plast Reconstr Surg.* 1986;78(4):460-468.

10. Duke-Elder S. Mandibulofacial dysostosis. In: Duke-Elder S, ed. *System of Ophthalmology; Congenital Deformities,* Vol. III. St. Louis, Mo: CV Mosby; 1963; 1013.

11. Kahana M, Romen M, Schewachi-Millet M, et al. Treacher-Collins syndrome and neurofibromatosis. *Int J Dermatol (USA).* 1986;25(2):115-116.

12. Roa NL, Moss KS. Treacher-Collins syndrome with sleep apnea: anesthetic considerations. *Anesthesiology.* 1984;60:71-73.

13. Shprintzen PJ, Berkman M. Pharyngeal hypoplasia in Treacher-Collins syndrome. *Arch Otolaryngol.* 1979;105:127.

Chapter 172

Oculodentodigital Dysplasia

A. MARTIN-CASALS

In 1957 Meyer-Schwickerath et al.[1] described two patients with a malformation syndrome involving the eyes, nose, teeth, and bones. He called the condition "dysplasia oculo-dento-digitalis." The characteristic features of oculodentodigital dysplasia (ODD) are: (1) striking facies dominated by a long, thin nose with a prominent bridge, hypoplastic alae nasi, and anteverted nostrils (Fig. 172-1); (2) microphthalmia and/or microcornea with other variably present ocular anomalies; (3) enamel dysplasia of teeth; (4) syndactyly, usually involving the fourth and fifth fingers; and (5) skeletal anomalies, the most common of which is a deficiency of the fifth finger middle phalanx. ODD is considered to be an autosomal dominant disorder. The following variably present ocular anomalies have been described: palpebral fissure hypoplasia, epicanthal folds, iris dysplasia, glaucoma, and strabismus.

A review of the literature was published in 1979 by Judisch et al.[2]

References

1. Meyer-Schwickerath G, Gruterich E, Weyers H. Microphthalmussyndrom. *Klin Monatsbl Augenheilkd.* 1957;131:18–30.

2. Judisch GF, Martin-Casals A, Hanson JW, et al. Oculodentodigital dysplasia. *Arch Ophthalmol.* 1979;97:878-884.

FIGURE 172-1 Typical facies of ODD in mother and son.

Chapter 173

Oculoauriculovertebral Dysplasia

Goldenhar-Gorlin Syndrome

FREDERICK M. WANG and ROSALIE B. GOLDBERG

In 1952, Goldenhar[1] described the associated triad of epibulbar dermoids, auricular appendages, and pretragal fistulas. Gorlin[2] emphasized that many patients with Goldenhar's triad have microtia plus mandibular and vertebral abnormalities that place them in the spectrum of oculoauriculovertebral dysplasias (OAV). Included in this constellation of OAV are hemifacial microsomia and the Goldenhar-Gorlin syndrome. A spectrum of severity is suggested by the many transitional forms; hemifacial microsomia is at one end of the spectrum and Goldenhar syndrome at the other.

Although the phenotype is usually quite characteristic (Figs. 173-1, 173-2), there are no agreed-upon minimal diagnostic criteria. The expression (severity) and range of anomalies vary widely. Ear malformations or preauricular tags are rarely absent and may, on occasion, be the only finding.

The syndrome is estimated to occur between 1 in 5600 and 1 in 26,500 births.[3] The entity is usually sporadic. Discordance is frequently reported in monozygotic twins. Familial cases are known, and some pedigrees demonstrate autosomal dominant inheritance, but others seem to be autosomal recessive and many are multifactorial. Expression may vary widely within families. Rarely, aberrant karyotypes have been associated with variants of the phenotype.[3] Recurrence risk counseling should be provided on an individual family basis.

Multiple causes have been reported; causation remains a matter of speculation. An animal model of hemifacial microsomia has been produced by a pharmacologic agent, triazine, inducing hematoma formation in the developing first and second branchial arch region.[4] The severity of the anomaly produced was related to the size of the hematoma. Thalidomide and retinoic acid have produced similar anomalies.[4,5] The OAV phenotype has been noted to occur in the offspring of diabetic mothers.[5]

Systemic Manifestations

Facial asymmetry is present in the majority of patients and is severe in 20%. Asymmetry results from malposition or displacement of soft tissue (eg, the pinna) and/or underlying facial skeletal anomalies. Bilateral involvement is common. The maxillary, temporal, and malar bones frequently are hypoplastic and flattened. Anteroinferior displacement of the temporomandibular joint occurs.

Abnormality of the external ear is the most suggestive finding. The spectrum of anomalies ranges from small preauricular tags of skin and cartilage to anotia. Tags, which may be multiple and bilateral, are located anywhere from the tragus to the angle of the mouth. Preauricular sinuses are frequently noted. Hypoplasia of the ossicles has been reported. Hearing loss, either sensorineural or conductive, is found in about half the patients.[6]

Oral manifestations are common. Agenesis of the mandibular ramus is frequently associated with lateral facial clefts. Hypoplasia of palate and tongue muscles, parotid gland, and salivary glands occur. Cleft lip or palate is found in up to 15%.[7] Velopharyngeal insufficiency may be noted, with resulting hypernasal speech.[8]

Malformations of the cervical spine and skull are noted in approximately one third of patients.[9] Most frequent are fusion of cervical vertebrae, platybasia, and occipitalization of the atlas. Hemivertebrae, hypoplastic

FIGURE 173-1 Goldenhar-Gorlin syndrome: Note right-sided limbal dermoid, grade II microtia, pretragal tag, repaired cleft lip, and hemifacial microsomia. The left side was less severely affected and exhibited a seventh nerve paresis.

FIGURE 173-2 Goldenhar-Gorlin syndrome: note right-sided limbal dermoid, right-sided inferotemporal lipodermoid, left iris coloboma, bilateral grade III microtia, and mandibular deficiency with resultant dental malocclusion.

vertebrae, Klippel-Feil anomaly, and other vertebral dysmorphologies are common and may lead to scoliosis. Anomalous ribs, club feet, and radial limb deformities have been reported.

Nervous system abnormalities include microcephaly, Arnold-Chiari malformation, hydrocephalus, aplasia of cranial nerve nuclei V and VII, and unilateral arhinencephaly.[10] Mental retardation has been noted in approximately 10% of all patients. Bony abnormalities near the facial canal may result in seventh nerve facial weakness.

Pulmonary, cardiac, renal, and gastrointestinal malformations are not uncommon.[3]

Ocular Manifestations

Choristomas (congenital overgrowth of normal tissues in abnormal locations) on the outer coats of the globe are one of the features of Goldenhar's triad. Because dermoids were diagnostic criteria for Goldenhar's syndrome, the older literature reported a very high incidence. In the total spectrum of the Goldenhar-Gorlin syndrome, they appear in about one third of patients.[11] The most common type straddles the limbus, usually inferotemporally, consisting primarily of collagenous connective tissue covered by epidermoid epithelium. Various dermal appendages (eg, hair and glandular tissue) may be found within these lesions. Subconjunctival choristomas consisting primarily of adipose tissue (lipodermoids) and usually located near the fornices are frequent. The tumor may extend posteriorly around the globe. A spectrum of location and histologic features exists between these two major types.[12] Bilaterality and multiple choristomas in one eye are not uncommon.

The lids may be affected with upper, or less frequently, lower lid colobomas, ptosis, and palpebral fissure narrowing. Orbital asymmetry due to facial asymmetry or microphthalmos may be noted.

Lacrimal system dysfunction has multiple causes.[11] Anatomic abnormalities include nasolacrimal duct obstruction with or without sinus and fistula formation, canalicular obstruction, ectopic puncta, and aberrant lacrimal drainage into the oropharynx. Seventh nerve weakness may affect lacrimal drainage.

Corneal sensation may be reduced by trigeminal nerve abnormalities. Hairs from epibulbar dermoids may rub on the cornea. Tear film disturbances from facial nerve weakness and epibulbar dermoids may further compromise the cornea.

Microphthalmos and anophthalmos occur. Mental retardation may be more frequent in patients with small eyes.

Involvement of the posterior segment of the eyes is not uncommon.[13] It may consist of optic nerve hypoplasia, tilted optic disc, tortuous retinal vessels, macular hypoplasia and heterotopia, peripapillary choroidal hyperpigmentation, and iris and retinal coloboma. These findings are frequently associated with microphthalmos.

Strabismus is reported in one fourth of patients. Esotropia and exotropia are most common. Duane's syndrome is associated with significant frequency.[14] The defect that produces the Goldenhar-Gorlin syndrome may disturb normal extraocular innervation and lead to aberrant innervation which may underly Duane's syndrome.

Vision is compromised with microphthalmos and optic nerve and retinal anomalies. Epibulbar dermoids may produce a significant irregular astigmatism or a more regular oblique astigmatism with resultant anisometropia and amblyopia.[11,12] Strabismic amblyopia is also found.

Excision of epibulbar dermoids should be reserved for significantly large ones and should be performed judiciously. Partial excision of limbal dermoids may not change the eye's appearance significantly and can lead to corneal thinning. Excision of lipodermoids should be performed conservatively so as to avoid violating surgical planes and possibly restricting ocular motility. Ptosis repair should also be conservative, considering the factors in this syndrome that may result in corneal compromise. Cryoepilation of hairs that rub on the cornea may be necessary.

References

1. Goldenhar M. Associations malformatives de l'oeil et de l'oreille en particularre le syndrome dermoide epibulbaire-appendices auriculaires-fistula auris congenita et ses relations avec la dysostose mandibulo-faciale. *J Genet Hum.* 1952;1:243-282.
2. Gorlin RJ, Jue KL, Jacobsen V, et al. Oculoauriculovertebral dysplasia. *J Pediatr.* 1963;63:991-999.
3. Rollnick BR, Goldberg RB. Oculoauriculovertebral dysplasia. In: Cohen MM, Gorlin RJ, eds. *Syndromes of the Head and Neck.* 3rd ed. Oxford: Oxford University Press; 1988.
4. Poswillo D. The pathogenesis of the first and second branchial arch syndrome. *Oral Surg.* 1973;35:302-329.
5. Grix A Jr. Malformations in infants of diabetic mothers. *Am J Med Genet.* 1982;13:131-137.
6. Budden SS, Robinson GC. Oculoauricular vertebral dysplasia. *Am J Dis Child.* 1973;125:431-433.
7. Rollnick BR, Kaye CI, Nagatoshi K, et al. Oculoauriculovertebral dysplasia and variants: phenotypic characteristics of 294 patients. *Am J Med Genet.* 1987;26:361-375.
8. Shprintzen RJ, Croft CB, Berkman MD, et al. Velopharyngeal insufficiency in the facio-auriculo-vertebral malformation complex. *Cleft Palate J.* 1980;17:132-137.
9. Figueroa AA, Friede H. Craniovertebral malformations in hemifacial microsomia. *J Craniofac Genet Dev Biol Suppl.* 1985;1:167-178.
10. Aleksic S, Budzilovich G, Greco MA, et al. Intracranial lipomas, hydrocephalus and other CNS anomalies in oculoauriculo-vertebral dysplasia (Goldenhar-Gorlin syndrome). *Childs Brain.* 1984;11:285-297.
11. Mansour AM, Wang FM, Henkind P, et al. Ocular findings in the facio-auriculovertebral sequence (Goldenhar-Gorlin syndrome). *Am J Ophthalmol.* 1985;100:555-559.
12. Baum J, Feingold M. Ocular aspects of Goldenhar's syndrome. *Am J Ophthalmol.* 1973;75:250-257.
13. Margolis S, Aleksic S, Charles N, et al. Retinal and optic nerve findings in Goldenhar-Gorlin syndrome. *Ophthalmology.* 1984;91:1327-1333.
14. Miller MT. Association of Duane retraction syndrome with craniofacial malformations. *J Craniofac Genet Dev Biol Suppl.* 1985;1:273-282.

Chapter 174

Oculomandibulodyscephaly

Hallermann-Streiff Syndrome

RONALD V. KEECH

Oculomandibulodyscephaly (OMD, also called Hallermann-Streiff syndrome, Hallermann-Streiff-Francois syndrome) is a congenital syndrome characterized primarily by facial malformation, dental anomalies, short stature, and congenital cataracts. Originally described in the late 19th century, it was not recognized as a specific syndrome until, in the ophthalmology literature, Hallermann[1] and Streiff[2] further defined the characteristics that separated it from the other mandibulofacial disorders. Comprehensive reviews of previously reported cases are given by Francois[3] and by Steele and Bass.[4]

The cause of OMD is unknown. The majority of reported cases have been sporadic. An occasional familial case has occurred, which implies the possibility of autosomal dominant or autosomal recessive transmission. Regardless of the initiating event, the characteristic features of OMD suggest an aberration in the embryogenesis of the surface ectoderm, the tissue layer responsible for the development of the skin, hair, tooth enamel, and lens. Neural crest cells, which are derived from ectoderm, are the primary source for the cranial and facial bones, portions of the teeth and eyes, and the branchial arches, all of which are involved in this disorder.

Systemic Manifestations

The most striking systemic feature is the characteristic facial dysmorphia. This is due primarily to a hypoplastic mandible and a thin tapered nose resulting in a bird- or parrotlike facies (Figs. 174-1, 174-2). The mandibular hypoplasia occasionally is severe enough to cause pain on mastication and to require surgery. Associated with mandibular hypoplasia is microstomia and a narrowed, high-arched palate.

The abnormal shape of the head adds to the dysmorphia. The skull is frequently brachycephalic with frontal bossing, but other variations have been reported, including scaphocephaly and microcephaly. Another common finding is delayed closure of the fontanels with widening of the longitudinal and lambdoidal sutures.

The unusual facial appearance is enhanced by changes in the skin and hair. The skin is clinically atrophic. It is often described as being dry and fine in appearance with multiple telangiectasias. On histologic examination the facial skin—as well as skin elsewhere on the body—demonstrates decreased cohesion of collagen fibers and fragmentation of elastic fibers. The hair is scanty over the entire body, especially on the scalp, eyelashes, and eyebrows. Hair that is present is fine textured and hypopigmented.

Dental anomalies occur in virtually all cases of OMD. Teeth may be present in the newborn or neonate. In older children and adults, teeth may be absent or malformed, and have premature caries. The dental abnormalities are usually severe and often require partial or total extraction and fitting of dentures.

Numerous other systemic features have been reported in association with OMD. A proportionate reduction in overall body size occurs in the majority of cases. Height is approximately 2.5 standard deviations below the normal mean. Mental retardation, a less common finding, occurs about 15% of the time in this syndrome. Other rarely associated anomalies have included syndactyly, winging of the scapula, scoliosis, spina bifida, and hypogenitalism.

Ocular Manifestations

The most common ocular condition found with OMD is congenital cataracts. They may be total or partial and have an unusual tendency to degenerate spontaneously with marked intraocular inflammation. Infants are reported to have shrunken degenerative cataractous lenses,[5] whereas specimens from adult eyes revealed marked reabsorption.[6] All patients with this syndrome have subnormal vision, and most have less than 20/200 visual acuity. The most common causes for poor vision are deprivational amblyopia from delayed surgery and

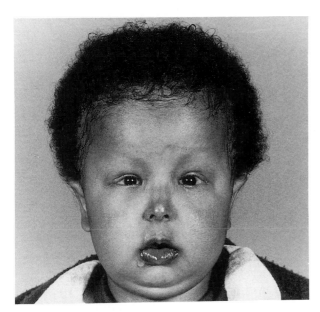

FIGURE 174-1 Typical facies of oculomandibulo-dyscephaly shows mandibular hypoplasia, thin tapered nose, and microphthalmos.

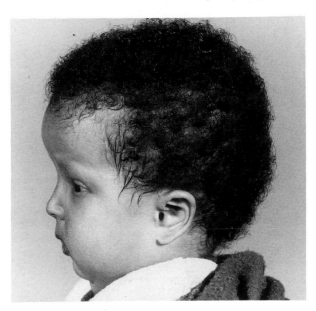

FIGURE 174-2 Side view demonstrates the beaklike nose, frontal bossing, and hypotrichosis.

complications of the degenerative lens changes. For these reasons, early cataract surgery should be considered for all patients with OMD.

Glaucoma associated with this syndrome has been discussed in several reports.[7,8] The mechanism of the glaucoma appears to be dual. Abnormal angles with anomalous iris insertions and mesodermal tissue filling the iridocorneal angle have been described.[9] The most common mechanism for glaucoma, however, is thought to be inflammatory, with peripheral anterior synechiae and posterior synechiae associated with lens reabsorption. Filtration surgery has been successful in some cases.

Many other ocular anomalies have been associated with OMD. Microphthalmia of varying degrees probably occurs in all cases. This has been suggested in the older literature with corneal measurements and has been confirmed more recently by ultrasonographic measurement of axial length. Other less common ocular anomalies have included blue scleras, antimongoloid eyelid fissures, strabismus, optic nerve colobomas or dysplasia, retinal folds, and chorioretinal scars.

The natural history of this syndrome is not completely understood. Deaths in infancy and childhood from respiratory or feeding difficulties have been reported. Most patients, however, live into adulthood and can have a normal life span.

Distinguishing OMD from other mandibulofacial and ectodermal dysplasias usually is not difficult.[10] Patients with this syndrome lack the chronic arthritis and pre-mature arteriosclerosis seen with progeria. Moreover, patients with progeria generally lack eye anomalies. The ear and eyelid anomalies that occur commonly with other first branchial arch anomalies such as Treacher Collins or Goldenhar's syndrome are absent in OMD. Oculodentodigital syndrome has a characteristic facial configuration similar to that of OMD, but it uniformly lacks the hypoplastic mandible and cataracts. In addition, oculodentodigital syndrome has associated syndactyly and other phalangeal aberrations.

In summary, OMD is a congenital syndrome of unknown cause that primarily involves anomalies of the face and eye. Its principal clinical disorder is decreased vision due to congenital cataracts and associated ocular conditions. While rare, OMD is especially important to the ophthalmologist, who can help establish the diagnosis and treat the ocular problems.

References

1. Hallermann W. Vogelgesicht and Cataracta congenita. *Klin Monatsbl Augenheilkd.* 1948;113:315.
2. Streiff EB. Dysmorphic mandibulo-faciale (tete d'oiseau) et alteration oculaires. *Ophthalmologica.* 1950;120:79.
3. Francois J. A new syndrome: dyscephalia with bird face and dental anomalies, nanism, hypotrichosis, cutaneous atrophy, microphthalmia and congenital cataract. *Arch Ophthalmol.* 1958;60:842.
4. Steele RW, Bass JW. Hallermann-Streiff syndrome. Clinical and prognostic considerations. *Am J Dis Child.* 1970;120: 462.

5. Donders PC. Hallermann-Streiff syndrome. *Doc Ophthalmol.* 1977;44:161.

6. Wolter J, Jones D. Spontaneous cataract absorption in Hallermann-Streiff syndrome. *Ophthalmologica.* 1965;150:401.

7. Hopkins DJ, Horan EC. Glaucoma in the Hallermann-Streiff syndrome. *Br J Ophthalmol.* 1970;54:416.

8. Falls HF, Schull WJ. Hallermann-Streiff syndrome. A dyscephaly with congenital cataracts and hypotrichosis. *Arch Ophthalmol.* 1960;63:409.

9. Aracena T, Sangueza P. Hallermann-Streiff-Francois syndrome. *J Pediatr Ophthalmol.* 1977;14:373.

10. Suzuki Y, Fujii T, Fukuyama Y. Hallermann-Streiff syndrome. *Develop Med Child Neurol.* 1970;12:496.

Chapter 175

Pierre Robin Sequence

SHAUNA McKUSKER and BURTON J. KUSHNER

In 1923, Pierre Robin, a French authority on diseases of the oral cavity, described a syndrome consisting of micrognathia, cleft palate, and respiratory difficulties.[1] Subsequent articles emphasized the respiratory and feeding problems of children with this set of findings, and glossoptosis was added to the oral cavity abnormalities in what was referred to as "Pierre Robin syndrome."[2,3]

Embryogenesis

The branchial arches appear around the 4th week of gestation and are usually fully differentiated by the 6th to the 7th week.[4] The first branchial arch divides into a maxillary portion and a mandibular portion, the latter forming the mandible, lower lip, and chin. The initiating event in the Pierre Robin syndrome is thought to be hypoplasia of the mandible occurring prior to 9 weeks' gestation.[5] Although the tongue is usually of normal size, the hypoplasia of the mandible results in the tongue being inserted posteriorly, producing glossoptosis. The combination of the small mandible and the abnormal tongue position prevents closure of the posterior palatal shelves, which normally must "grow over" the tongue to meet in the midline and form the palate.[3,5] Failure of these shelves to meet results in a cleft palate.

Because the three primary characteristics of Pierre Robin syndrome, micrognathia, glossoptosis, and cleft palate, result in sequence from the initial hypoplasia of the mandible, it has been suggested that the name be changed to "the Robin sequence."[5]

Because Pierre Robin syndrome is primarily a problem of the posterior oral cavity, cleft lip rarely occurs. Clefting is usually confined to the soft palate. In most cases the jaw ultimately develops to normal size and shape; however, in some cases it remains small.[3]

Systemic Manifestations

Infants with the Robin sequence typically present shortly after birth with difficulty swallowing, respiratory difficulty, and cyanosis. Although the Robin sequence most commonly occurs in isolation, numerous ocular and nonocular associations have been reported with it—low-set ears, anomalies of the feet, neurologic abnormalities, congenital heart defects and others (Fig. 175-1).[3,5,6]

Early therapy for the nonocular manifestations of the Pierre Robin sequence centers on control of the respiratory obstruction and feeding problems. The glossoptosis may produce a ball-valve type of respiratory obstruction of varying severity on inspiration, even to the point of causing exhaustion, failure to thrive, or death. In mild cases the posterior airway obstruction may be controlled by placing the infant in a slanted position with the head down. In more severe cases the tongue may be held forward with a retainer or by suturing it in place. Severely affected patients may require a tracheostomy, which may need to stay in place for many months. This can be a difficult procedure in newborns.[3] Later the cleft palate may be repaired, but it is recommended that this be deferred until the palatal shelves have stopped growing toward the midline.[5]

The Pierre Robin sequence may occur alone; however, it also can occur in conjunction with other syndromes of multiple congenital anomalies, such as trisomy 18. The causal relationship between the Robin sequence and these other anomalies is not known.

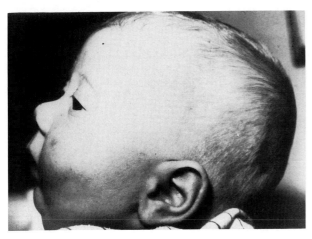

FIGURE 175-1 Typical appearance of a child with the Robin sequence demonstrating micrognathia. This child has the Robin sequence associated with Stickler's syndrome.

Most cases of isolated Robin sequence are sporadic, though it has been reported in association with both dominant and recessive inheritance patterns.[6,7] When the Robin sequence occurs as part of a genetic syndrome, its inheritance pattern is that of the specific associated genetic syndrome.

Ocular Associations

Numerous ocular abnormalities have been associated with the Robin sequence. They include congenital glaucoma, high myopia, vitreoretinal degeneration, retinal detachment, megalocornea, congenital cataract, microphthalmos, and esotropia.[7,8] The glaucoma associated with the Robin sequence has been described as being similar to primary congenital glaucoma, and therefore goniotomy is recommended as the first surgical approach.[8] This is important because a child with the Robin sequence who survives to 4 to 6 years of age may be expected subsequently to have a normal life. Because a child with the Robin sequence and congenital glaucoma may frequently need to undergo multiple general

anesthesias in the early newborn period, the added risk general anesthesia poses to these children is of particular concern. Because of the glossoptosis, respiratory obstruction, sometimes resulting in death, is a real danger as such children emerge from general anesthesia.

It has been estimated that one third of patients with the Robin sequence also have Stickler's syndrome (hereditary progressive arthro-ophthalmopathy).[9,10] Specifically it is likely that patients with high myopia, vitreoretinal degeneration, and retinal detachment associated with Robin sequence have Stickler's syndrome. Stickler's syndrome has an autosomal dominant inheritance pattern with a high degree of penetrance, unlike the Robin sequence alone, which is typically sporadic.

As with the systemic anomalies that are frequently associated with the Robin sequence, the causal relationship between the ocular anomalies and the Robin sequence is obscure.

References

1. Robin P. Backward lowering of the root of the tongue causing respiratory disturbances. *Bull Natl Acad Med.* 1934;89:37.
2. Robin P. Glossoptosis due to atresia and hypotrophy of the mandible. *Am J Dis Child.* 1934;48:541.
3. Randall P, Hamilton R. The Pierre Robin syndrome. In: Grabb W, Rosenstein S, Broch K, eds. *Cleft Lip and Palate. Surgical, Dental and Speech Aspects.* Boston, Mass: Little, Brown; 1971;559.
4. Arey LB. *Developmental Anatomy.* 5th ed. Philadelphia, Pa: WB Saunders; 1946.
5. Smith D. *Recognizable Patterns of Human Malformation.* 3rd ed. Philadelphia, Pa: WB Saunders; 1982;172.
6. Feingold M, Gellis S. Ocular abnormalities associated with first and second arch syndromes. *Surv Ophthalmol.* 1969;14(1):30-42.
7. Perkins T. Pierre Robin syndrome. *Trans Ophthalmol Soc UK.* 1970;90:179-180.
8. Crandall A. Developmental ocular abnormalities and glaucoma. *Intl Ophthalmol Clin.* 1984;24(1):73-86.
9. Blair NP, Albert DM, Liberfarb RM, et al. Hereditary progressive artho-ophthalmopathy of Stickler. *Am J Ophthalmol.* 1979;88:976.
10. Sargent RA, Ousterhout DK. Ocular manifestations of skeletal diseases. In: Harley R, ed. *Pediatric Ophthalmology.* Philadelphia, Pa: WB Saunders; 1983;1030.

Chapter 176

Waardenburg's Syndrome

J. W. DELLEMAN and M. J. van SCHOONEVELD

Since it was described in 1951 by Waardenburg[1] this autosomal dominant condition has intrigued many physicians. The principal features include outer displacement of the lower lacrimal orifices (dystopia canthorum), fusion of the eyebrows (synophrys), a broad and prominent nasal root, unilateral or bilateral deafness, heterochromia iridis, unilateral hypopigmentation of the fundus, leucism, and a white or discolored hair lock or premature graying (Figs. 176-1 to 176-5).[2,3] A distinction is made between type I (with dystopia canthorum) and type II (without dystopia canthorum). Differences in clinical features between these two types are shown in Table 176-1. The most important difference, also with regard to genetic counseling, is that the prevalence of deafness in patients with type II disease is about 50%.

In Waardenburg's syndrome we encounter in one condition: morphologic anomalies of the ocular adnexa and the midface, pigmentary disturbances of the eyes, skin, and skin derivatives, and congenital perceptive deafness.

The role of the neural crest in embryogenesis and in malformation syndromes has been emphasized in many

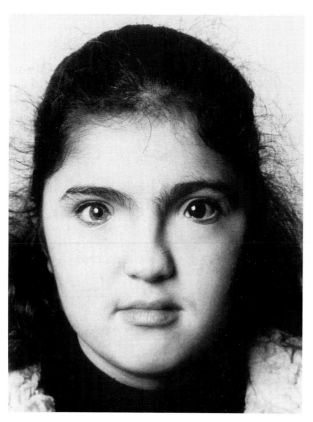

FIGURE 176-1 Frontal view of person with Waardenburg's syndrome: note synophrys, dystopia canthorum, and heterochromia iridis.

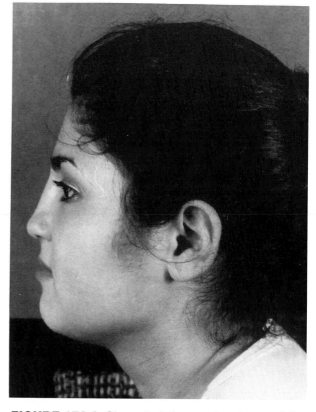

FIGURE 176-2 Characteristic prominent root of the nose is visible in profile view.

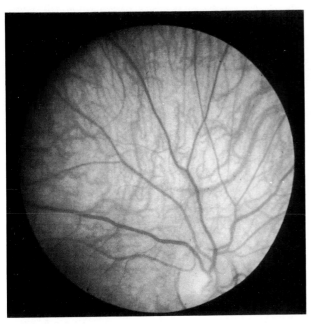

FIGURE 176-3 Right fundus of patient in Figure 176-1 shows hypopigmentation.

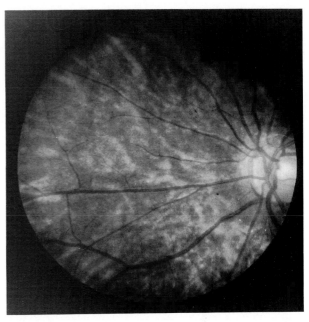

FIGURE 176-4 Left fundus of the same patient shows normal pigmentation.

publications during the last decade[4,5]; however, in 1959 Fisch[6] postulated a primary role for the neural crest in the pathogenesis of Waardenburg's syndrome. The most cephalad part of the neural crest is indeed involved in craniofacial development, and disturbances can lead to malformations such as facial cleft, an anomaly that also occurs in Waardenburg's syndrome.

The melanoblasts, progenitors of melanocytes, are derived from the neural crest. These melanocytes provide melanin pigment for the iris and the choroid. Therefore, a relationship between Waardenburg's syndrome and the neural crest seems plausible. Deafness, too, is considered a principal constituent of neurocristopathies by Liang et al.[7] Additionally, occasional pa-

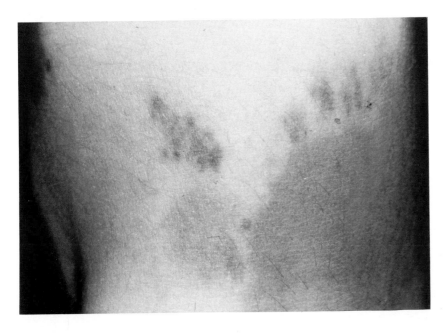

FIGURE 176-5 Leucism: circumscribed hypopigmentation of the skin.

TABLE 176-1 Percentage of Clinical Symptoms in 230 Patients with Type I and 112 Patients with Type II Waardenburg's Syndrome (without Probands)

	Type I (%)	Type II (%)
Dystopia canthorum	99	0
Prominent broad nose root	75	31
Synophrys	65	23
Iris heterochromia	21	25
Iris hypoplasia	9	21
White hair lock	28	38
Early graying	10	8
Leucism	7	16
Deafness		
Bilateral	18	41
Unilateral	8	4

tients with Waardenburg's syndrome have been described who have Hirschsprung's disease, which also points to a disturbance in the differentiation of the neural crest.[7]

Accepting the concept of Waardenburg's syndrome as a "neurocristopathy," we are left with the question,

Why is the symptom complex of Waardenburg's syndrome so typical and why is it possible to differentiate Waardenburg's syndrome from other neurocristopathies, as for instance, from neurofibromatosis? To answer these questions, future investigations, including experimental embryologic models, are required.

References

1. Waardenburg PJ. A new syndrome combining developmental anomalies of the eyelids, eyebrows and nose root with pigmentary defects of the iris and headhair and with congenital deafness. *Am J Hum Genet*. 1951;(3):195-255.
2. Hageman MJ, Delleman JW. Heterogeneity in Waardenburg syndrome. *Am J Hum Genet*. 1977;(29):468-485.
3. Arias S. Genetic heterogeneity in the Waardenburg syndrome. *Birth Defects: Original Article Series*, 1971;7(4):87-101.
4. Bolande RP. The neurocristopathies: a unifying concept of disease arising in neural crest maldevelopment. *Hum Pathol*. 1974;(5/4):409-429.
5. Beauchamps GR, Knepper PA. Role of the neural crest in anterior segment development and disease. *J Pediatr Ophthalmol Strab*. 1984;(21/6):209-214.
6. Fisch L. Deafness as part of an hereditary syndrome. *J Laryngol Otol*. 1959;(73/6):355-382.
7. Liang JC, Juarez CP, Goldberg MF. Bilateral bicolored irides with Hirschsprung's disease. *Arch Ophthalmol*. 1983;(101):69-73.

Section C

Generalized Skeletal Disorders

Chapter 177

Apert's Syndrome

Acrocephalosyndactyly

DAVID S. FELDER and JOHN D. BULLOCK

Apert's syndrome (acrocephalosyndactyly) is an uncommon disorder consisting of a skull deformity (most often oxycephaly or "tower skull") in association with symmetric syndactyly of the hands and feet. Although Apert[1] was the first to describe several cases of acrocephalosyndactyly in 1906, the original description was probably presented by Troquart in 1886.[2]

Systemic Manifestations

Oxycephaly is believed to result from premature fusion of the cranial sutures, particularly the coronal suture. The cranium grows mainly in the vertical direction, and growth is limited in its anteroposterior diameter. These distinctive growth patterns result in the characteristic egg-shaped appearance of the Apert's cranium, with high, steep frontal bones, underdeveloped occiput, and shallow orbits. The maxilla is hypoplastic and symmetrically retrognathic, causing a "dish-face" appearance.

The mandible is commonly normal but may appear prognathic relative to the recessed maxilla. Malocclusion is common secondary to crowding of the teeth in the hyoplastic maxilla. Also typical of patients with Apert's syndrome are a high-arched, narrow palate that may be cleft, low-set ears, a "parrot-beak" nose, and a bifid uvula (Fig. 177-1). Skull films may show premature closure of the coronal sutures and digital impressions (Fig. 177-2).

The second major abnormality in Apert's syndrome is a symmetric deformity of the hands and feet. The syndactyly may range from partial to complete fusion of all the fingers and toes (Fig. 177-3). In the majority of cases, the hands display cutaneous and osseous fusion of the second, third, and fourth fingers with varying degrees of synonychia (a single common abnormal nail). This abnormality has been termed "mitten-hand" deformity.[3] Temtamy and McKusick[4] have described and categorized five different types of acrocephalosyndac-

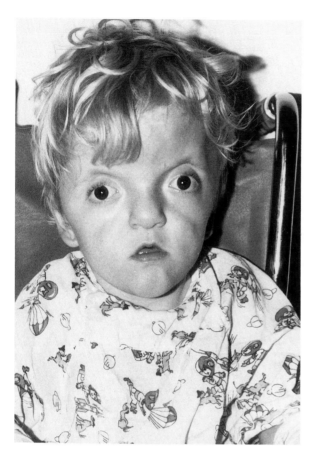

FIGURE 177-1 Typical facies of Apert's syndrome. Note antimongoloid slanting of the palpebral fissures, exophthalmos, hypertelorism, oxycephaly, and midfacial hypoplasia. (Courtesy of G. Frank Judisch, M.D.)

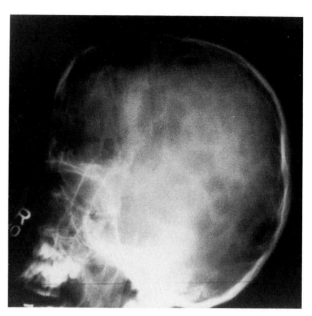

FIGURE 177-2 Skull x-ray shows characteristic changes of oxycephaly and prominent digital impressions. (Courtesy of Meinhard Robinow, M.D.)

tyly: (1) Apert's syndrome (as described); (2) Vogt's cephalodactyly (syndactyly with Crouzon's facies); (3) Saethre-Chotzen syndrome (asymmetric cranium with minimal syndactyly); (4) Waardenburg type (acrocephaly, asymmetric cranium, cleft palate, mild cutaneous syndactyly, and pointed nose); and (5) Pfeiffer's syndrome (acrocephaly, mild syndactyly, broad thumbs and great toes). Other skeletal abnormalities include aplasia and ankylosis of the shoulders, hips, elbows, and cervical spine.

Apert's syndrome may also be associated with several nonskeletal abnormalities, including polycystic kidneys, ectopic anus, pyloric stenosis, ventricular septal defects, and agenesis of the corpus callosum.[5,6] These features are, at best, inconsistent, and it may very well be that fetuses with more severe abnormalities of inter-

nal organs do not survive into the neonatal period. Many patients with Apert's syndrome appear to have normal intelligence, although varying degrees of mental impairment have been reported.[5] Since the majority of patients with severe mental deficiency had been discovered in institutions, that number may have been overestimated. It is possible that premature closure of the cranial sutures prevents the brain from developing normally.

The incidence of Apert's syndrome has been estimated to range from 1 in 160,000 to 1 in 200,000 births.[5] It has been reported to occur equally in males and females, and there is no definite racial predilection.[5] The life expectancy, although probably lower than average, may be within normal limits. Apert's syndrome is generally considered to be an autosomal dominant disorder with a high degree of penetrance. Studies by Blank[5] and Margolis[7] have suggested that most cases of Apert's syndrome are the result of sporadic mutations of an autosomal dominant gene. Blank[5] also found a statistically significant correlation between incidence and increased paternal age. The increase of sporadic dominant mutations noted in association with advancing paternal age, along with confirmed pedigrees of dominant inheritance, provide some evidence that most Apert's patients demonstrate a fresh mutation. There has been only one reported abnormal karyotype in a patient with Apert's syndrome, which has led most investigators to conclude

Ocular Manifestations

Ophthalmic complications associated with Apert's syndrome are thought of as being secondary to the abnormal structure of the bony orbits.[10] The orbits are shallow as a result of forward displacement of the greater wing of the sphenoid bone along with a steeply inclined orbital roof. This causes a decrease in the volume of the orbit, leading to proptosis and occasionally to exposure keratitis.

Loss of visual acuity is often a result of severe optic atrophy, which may be due to compression of the optic nerves by narrowed optic foramina, traction on the nerves due to downward displacement of the base of the brain, or to papilledema from increased intracranial pressure.

Hypertelorism with antimongoloid palpebral fissures is commonly present. Strabismus is a common feature, vertical muscle imbalance being more frequent than horizontal. When horizontal muscle imbalance is present, exotropia is more frequent than esotropia. The V pattern (eyes further apart on gaze up than gaze down) is the most typical extraocular muscle disturbance, with overaction of one or both inferior oblique muscles and weakness of one or both superior obliques. Morax[10] postulated that the gross disturbances of the size and position of the orbital structures result in asymmetric muscle origins and insertions in addition to mechanical restriction of ocular motility.

Although the abnormal bony orbits may account for many of the ocular motility problems frequently encountered, there are other factors that may also contribute. Weinstock and Hardesty[11] described absence of the superior rectus muscle in a patient with Apert's syndrome. A fibroadipose band was attached to the globe in the region of the usual insertion of the superior rectus muscle.

More recently, however, Margolis et al.[12] examined an inferior oblique muscle of a patient with Apert's syndrome. Light- and electron microscopy revealed alterations in the muscle fibers, myoneural junctions, and intramuscular nerves. Thus, ocular motility abnormalities may be multifactorial in origin.

Margolis et al. also described five cases of Apert's syndrome associated with albinism. The ocular findings consisted of iris transillumination defects, depigmented fundi, diffuse or absent foveal reflexes, and photophobia. None of these manifestations was severe, and visual acuity remained essentially normal in all of their patients.

Refractive errors may be noted, particularly hyperopia, which is caused by axial shortening of the globe.[13] Keratoconus also has been associated infrequently with Apert's syndrome.[14] Other ocular abnormalities associated with craniosynostosis with and without syndactyly

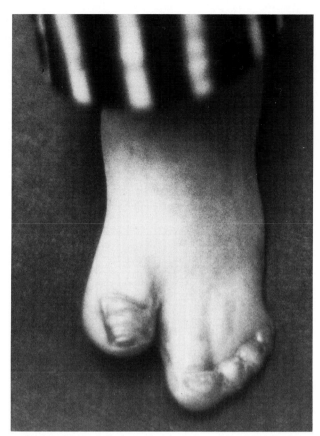

FIGURE 177-3 Left foot shows marked syndactyly and a central nail mass. (Courtesy of Meinhard Robinow, M.D.)

that there are no common chromosomal abnormalities.

Although the etiology of this syndrome is not known, several hypotheses have been formed. The leading theory is that the basic malformation occurs during the critical teratogenic period for differentiation of the skull and hands. This period appears to be the 29th to the 35th days of embryonic life. The evidence for this theory was derived from studies on thalidomide-induced phocomelia, in which the sensitive period for the drug action was shown to be between the 28th and 42nd days following conception.[8]

Stewart et al.[9] concluded from their observations and a review of the literature that the craniofacial abnormalities found in Apert's syndrome are the result of a malformation at the base of the skull. This malformation leads secondarily to premature fusion of the cranial sutures. The synostosis is thus thought of as a symptom and not the cause of the typical Apert's facies.[9]

are colobomas of the iris, nystagmus, ptosis, medullated nerve fibers, retinal detachment, megalocornea, congenital cataract, and subluxation of the lens.

References

1. Apert E. De l'acrocephalosyndactylie. *Bull Soc Med Paris.* 1906;23:1310.
2. Troquart R. Syndactylie et malformations diverses. *Mem Bull Soc Med Chir Bordeaux.* 1886;15;69.
3. Rubin MB, Pirozzi DJ, Heaton CL. Acrocephalosyndactyly. Report of a case, with review of the literature. *Am J Med.* 1972;53(1):127.
4. Temtamy S, McKusick VA. Synopsis of hand malformations with particular emphasis on genetic factors. *Birth Defects.* 1969;5:125.
5. Blank CE. Apert's syndrome (a type of acrocephalosyndactyly)—observations on a British series of thirty-nine cases. *Ann Hum Genet.* 1960;24:151.
6. Musallam SS, Poley JR, Riley HD Jr. Apert's syndrome (acrocephalosyndactyly). A description and a report on seven cases. *Clin Pediatr.* 1975;14(11):1054.
7. Margolis S, Siegel IM, Choy A, et al. Depigmentation of hair, skin, and eyes associated with the Apert syndrome. *Birth Defects.* 1978;14(6B):341.
8. Lenz W, Knapp K. Die Thalidomid-Embryopathie. *Dtsch Med Wochnschr.* 1962;87:1232.
9. Stewart RE, Dixon G, Cohen A. The pathogenesis of premature craniosynostosis in acrocephalosyndactyly (Apert's syndrome). *Plast Reconstr Surg.* 1977;59(5):699.
10. Morax S. Oculo-motor disorders in craniofacial malformations. *J Maxillofac Surg.* 1984;12(1):1.
11. Weinstock FJ, Hardesty HH. Absence of superior recti in craniofacial dysostosis. *Arch Ophthalmol.* 1965;74:152.
12. Margolis S, Pachter BR, Breinin GM. Structural alterations of extraocular muscle associated with Apert's syndrome. *Br J Ophthalmol.* 1977;61(11):683.
13. Krueger JL, Ide CH. Acrocephalosyndactyly (Apert's syndrome). *Ann Ophthalmol.* 1974;6(8):787.
14. Seelenfreund M, Gartner S. Acrocephalosyndactyly (Apert's syndrome). *Arch Ophthalmol.* 1967;78:8.

Chapter 178

Conradi's Syndrome

HELEN MINTZ-HITTNER and FRANK L. KRETZER

Conradi's syndrome is a heterogeneous cluster of diseases with an incidence of at least 1 in 500,000 births. The syndrome is characterized by congenital stippling of the epiphyses and the extraepiphyseal cartilages. Synonymns include chondrodystrophia calcificans congenita, congenital stippled epiphyses, and chondrodysplasia punctata. Radiography must be performed early to establish this diagnosis, since the stippling may disappear with time. There are at least three genetic subtypes. In 1971, Spranger, Opitz, and Bidder[1] differentiated a lethal, autosomal recessive, rhizomelic type from a mild, heterogeneous, autosomal dominant Conradi-Hunermann type. In 1979, Happle[2] character-ized an X-linked dominant subgroup from the mild type. The X-linked dominant type is highly variable owing to lyonization in females, and it is uniformly lethal in males. In addition to these genetic forms, ingestion of warfarin by a pregnant woman during the 6th to 8th weeks of gestation was recognized in 1976 to produce a phenotypically similar disorder.[3]

Systemic Manifestations

All three genetic types have characteristic facies consisting of short neck, frontal bossing, saddle nose, high-arched palate, and hypertelorism. Each genetic type has unique manifestations of the degree and symmetry of bone involvement, skin with and without hair and nail abnormalities, neuronal atrophy with and without mental retardation, and vascular anomalies.

The lethal, autosomal recessive rhizomelic type has severe, symmetric proximal shortening of the extremities with marked metaphyseal changes. Flocculent mate-

This research was supported by grants from the Retina Research Foundation, Research to Prevent Blindness, and the Retinitis Pigmentosa Foundation Fighting Blindness. The authors acknowledge the ultrastructural data base generated by Rekha S. Mehta and Evelyn S. Brown, the photographic expertise of Alexander Kogan and Gilma Miranda, and the literature search by Dorothy Carr.

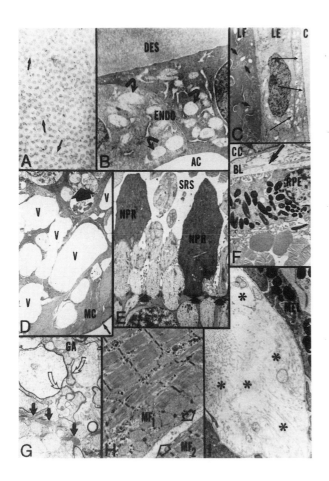

FIGURE 178-1 These nine transmission electron micrographs demonstrate the ocular manifestations of rhizomelic Conradi's syndrome. (**A**) The corneal stroma contains an abnormal, heterogeneous population of collagen fibrils (*arrows;* original magnification ×54,700). (**B**) Mitochondrial atrophy (*curved arrows*) occurs within the corneal endothelium (ENDO) between Descemet's layer (DES) and the anterior chamber (AC; original magnification ×12,200). (**C**) Lens fibers (LF) contain inclusion bodies (*short arrows*), which are presumably abnormal mucopolysaccharides. The lens epithelium (LE) has mitochondrial atrophy (*long arrows*) and secretes an intact capsule (C; original magnification ×4600). (**D**) A few remnant ganglion cells (*large black arrowhead*) remain within the peripheral retina. Extensive ganglion cell atrophy results in large vacuoles (V) between the Müller cell (MC) foot processes in the nerve fiber layer at the retina-vitreous interface (*small arrow;* original magnification ×4000). (**E**) Necrotic photoreceptors (NPR) in the peripheral retina extend from the outer limiting membrane (*arrows*) into the subretinal space (SRS; original magnification ×5300). (**F**) Bruch's layer (BL) between the choriocapillaris (CC) and retinal pigment epithelium (RPE) is fragmented (*large arrow;* original magnification ×5000). (**G**) The optic nerve has fragmented connective tissue septae (*arrows*) and some necrotic ganglion axons (GA) despite intact myelination (*curved arrows;* original magnification ×5800). (**H**) There is a scarcity of connective tissue (*open arrows*) between adjacent myofibers (MF1 and MF2) of the extraocular muscles (original magnification ×8600). (**I**) The iris stroma contains scarce, fragmented collagen fibrils (*asterisk*) between normal melanocytes (MEL; original magnification ×11,400).

rial, possibly abnormal acid mucopolysaccharides, is seen within the rough endoplasmic reticulum and produces distention of the chondrocytes in a matrix containing nonbanded, thin collagen fibrils. This leads to an insufficiently developed zone of ossification with a decreased number of osteoblasts and abnormal calcification.[4,5] The skin manifestation is congenital ichthyosis with hypertrophy of the horny layer. The adherence of keratinized epithelial cells has been shown in other tissues to be dependent on an extracellular mucopolysaccharide secreted by maturing keratinocytes.[6] Abnormal mucopolysaccharides could explain the ichthyosis. The hair and nails are unaffected. Neuronal atrophy with severe retardation is uniform. Vascular abnormalities include defective valves which may reflect an abnormal mucopolysaccharide and collagen matrix. Infants uniformly expire within the first few years of life.

The mild, X-linked dominant type has minimal asymmetric shortening of the femur, humerus, and other tubular bones, abnormalities of the vertebral column, and contractures of the joints.[2,7,8] The pathogenic mechanism has not been elucidated. Moderate or severe scoliosis results from asymmetric shortening of the legs. Additionally, hexadactyly occurs uniquely in this type. Skin, hair, and nail abnormalities, the hallmark of this type, can be used to establish the diagnosis even without radiographs. The skin abnormality consists of a distinctive congenital ichthyosis. Thick, yellow, adherent keratinized plaques distributed as linear whorls or blotchy areas over the entire body may be intensely erythematous. Histologically, there is hyperkeratosis which disappears in the first few weeks of life and is superseded by follicular atrophoderma. Patchy alopecia with irregularly twisted and lusterless hair, sparse eyebrows, and flattened nail plates with splitting of the nails into layers are present. All of these cutaneous manifestations are present to varying degrees because of lyonization. Neuronal manifestations, mental retardation, and vascular abnormalities are absent. Life expectancy is normal if heart and lung function are not impaired by severe scoliosis.

The mild, heterogeneous, autosomal-dominant Conradi-Hunermann type has predominantly epiphyseal (frequently asymmetric) calcifications and dysplastic changes. Probably, multiple pathogenic mechanisms are included in this type, and none has been elucidated by morphologic analyses. Aside from the characteristic facies, no pathognomonic skin, neural, or vascular changes unify these cases, and life expectancy is normal.

Ocular Manifestations

Ocular pathology has been reported only for the lethal, autosomal-recessive, rhizomelic type.[9-14] Clinically, the cataracts are congenital, dense, and symmetric.[10-12] They have been attributed to altered carbohydrate metabolism, which produces the ocular counterpart of the systemic mucopolysaccharide abnormality in the epiphyses. Furthermore, when the cataracts are removed, optic nerve atrophy is clinically apparent. The optic nerve atrophy is the ocular manifestation of the neuronal atrophy that accounts for severe mental retardation.[13-15]

Ocular pathophysiologic parameters for the rhizomelic type are shown in Figure 178-1 for an infant who was originally described by Kretzer, Hittner, and Mehta[9] in 1981. The abnormal collagen noted systemically is manifested ocularly by an abnormal heterogeneous population of collagen fibrils in the corneal stroma (Fig. 178-1,A), by the fragmented Bruch's layer (Fig. 178-1,F), by the fragmented connective tissue septae in the optic nerve (Fig. 178-1,G), by the scarcity of extracellular collagen between myofibers in extraocular muscles (Fig. 178-1,H), and by a paucity of collagen fibers in the iris stroma (Fig. 178-1,I). The neuronal abnormality noted systemically is manifested ocularly by peripheral atrophy of ganglion cells (Fig. 178-1,D), by necrotic photoreceptors in the peripheral retina (Fig. 178-1,E), and by necrotic axons in the optic nerve despite intact myelination (Fig. 178-1,G). Mitochondrial atrophy is present in the corneal endothelium (Fig. 178-1,B) and in the lens epithelium (Fig. 178-1,C). The corneal keratocytes contain bloated rough endoplasmic reticulum whose cisternal spaces are filled with flocculent material that is presumed to be abnormal mucopolysaccharide.[9] Similarly, the lens fibers possess inclusion bodies (Fig. 178-1,C), which presumably are abnormal mucopolysaccharides.

Clinically, cataracts are the only ocular finding for the mild, X-linked–dominant type. These cataracts are acquired, mild, and asymmetric. This asymmetry, which may even be considered a unilateral occurrence, is due to lyonization.[11] Since the lens is the only epithelial derivative in the eye, lens abnormalities will probably be found eventually to be related to the basic skin lesions. In contrast, no ocular manifestations, including cataracts, have been associated with the mild, autosomal-dominant type.

References

1. Spranger JW, Opitz JM, Bidder U. Heterogeneity of chondrodysplasia punctata. *Humangenetik.* 1971;11:190.
2. Happle R. X-linked dominant chondrodysplasia punctata. *Hum Genet.* 1979;53:65.
3. Pauli RM, Madden JD, Kranzler KJ, et al. Warfarin therapy initiated during pregnancy and phenotypic chondrodysplasia punctata. *J Pediatr.* 1976;88:506.
4. Zieger G, Conter C. Morphology of chondrodysplasia calcificans congenita. *Beitr Path Bd.* 1973;150:76.
5. Bosman C, Bonucci E, Gugliantini P, et al. Ultrastructural aspects of chondrodystrophia calcificans congenita (syndrome of Conradi-Hunermann). *Virchows Arch A.* 1977;373:23.
6. Squier CA, Rooney L. The permeability of keratinized and nonkeratinized oral epithelium to lanthanum in vivo. *J Ultrastruct Res.* 1976;54:286.
7. Manzke H, Christophers E, Wiedemann HR. Dominant sex-linked inherited chondrodysplasia punctata: a distinct type of chondrodysplasia punctata. *Clin Genet.* 1980;17:97.
8. Mueller RF, Crowle PM, Jones RAK, et al. X-linked dominant chondrodysplasia punctata: a case report and family studies. *Am J Med Genet.* 1985;20:137.
9. Kretzer FL, Hittner HM, Mehta R. Ocular manifestations of Conradi and Zellweger syndromes. *Metabol Pediatr Ophthalmol.* 1981;5:1.
10. Ryan H. Cataracts of dysplasia epiphysialis punctata. *Br J Ophthalmol.* 1970;54:197.
11. Happle R. Cataracts as a marker of genetic heterogeneity in chondrodysplasia punctata. *Clin Genet.* 1981;19:64.
12. Abedi S. Syndromes with congenital cataract (Conradi-Hunermann syndrome): a case report. *Ann Ophthalmol.* 1982;14:595.
13. Armaly MF. Ocular involvement in chondrodystrophia calcificans congenita punctata. *Arch Ophthalmol.* 1957;57:491.
14. Levine RE, Snyder AA, Sugarman GI. Ocular involvement in chondrodysplasia punctata. *Am J Ophthalmol.* 1974;77:851.
15. Billson FA, Hoyt CS. Optic nerve hypoplasia in chondrodysplasia punctata. *J Pediatr Ophthalmol.* 1977;14:144.

Chapter 179

Fibrous Dysplasia

DAVID SEVEL

Fibrous dysplasia (osteitis fibrosa disseminata, Albright's syndrome) is a disease of unknown cause characterized by thickened, sclerosed, and brittle bone.[1]

Systemic Manifestations

The condition, which commences in the 1st decade, involves both males and females and retrogresses slowly with exacerbations during puberty. Femoral and pelvic fractures are followed by deformities (Fig. 179-1). Fibrous dysplasia has a variety of clinical presentations. A single focus of monostotic fibrous dysplasia usually involves the skull and may have associated irregular brown, patchy skin pigmentation (as seen in neurofibromatosis) and endocrine abnormalities such as sexual precocity. McCune-Albright syndrome is characterized by polyostotic foci of fibrous dysplasia and features both dermal and endocrine abnormalities.

Clinically, the areas of fibrous dysplasia replace the medulla of the bone and contain areas of cartilage, calcification, hemorrhage, or cysts.[2] This involved bone is characteristically soft and fragile.

Histologic examination reveals congenital arrest in bone formation. Poorly formed bone fragments have irregular, feathery borders. These bone spicules show formative and resorptive features and are surrounded by fibrillar and loose fibrous tissue. No osteoblastic reaction is present (Fig. 179-2). The bone is weakly birefringent with polarized light. Sarcomatous change may occur in areas of fibrous dysplasia.[3-5]

Radiographically the bones are widened, with cortical thinning. Both the pelvis and the femur are usually included in the fibrous dysplastic reaction. In the skull, the facial bones and the frontal, sphenoidal, and temporal bones are involved.

The brown dermal pigmentation has a café-au-lait appearance similar to that in neurofibromatosis and is ipsilateral to the bone involvement.

The cause of the endocrine disturbance in the McCune Albright disease is unclear. The one theory suggests hypersecretion of one or more hypothalamic-releasing hormones,[6-14] while the other theory suggests that the involved end-organs are especially sensitive to trophic hormones or that they function autonomously.[15-17]

Ocular Manifestations

The skeletal disorders of the skull bones are the source of ophthalmic complications (Fig. 179-3).[18-24] The ocular complications occur more frequently with the monos-

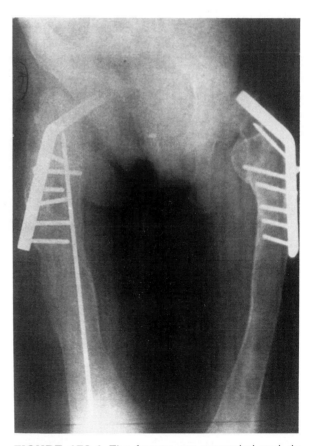

FIGURE 179-1 The femurs are expanded and demineralized. There is bilateral hip nailing with blade-plate prostheses, and a right femoral intramedullary rod is present.

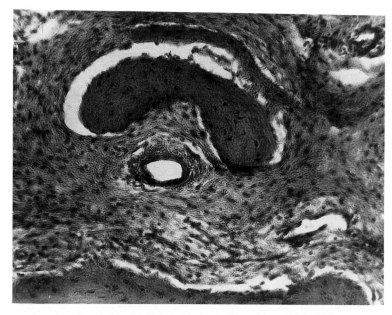

FIGURE 179-2 Poorly formed bone fragments are present surrounded by loose fibrous tissue.

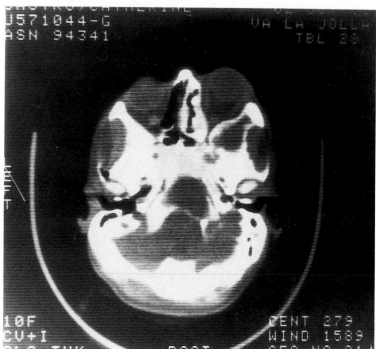

FIGURE 179-3 Computed tomography shows fibrous dysplastic tissue invading the right orbit and causing proptosis.

totic form of fibrous dysplasia and are rare with the polyostotic form. Characteristically, the proptosis that occurs is directed inferiorly and laterally (see Fig. 179-3).[22,23] Papilledema followed by optic atrophy may result. The optic nerve complications result from stretching of the ophthalmic artery with compromise of the

blood supply to the optic nerve. Constriction of the optic canal may occur if the greater wing of the sphenoid is involved by fibrous dysplasia. Finally, a tumor comprised of diseased bone may form in the orbit, resulting in an increase in the intraorbital pressure with pressure on the optic nerve or compromise of the blood supply of the optic nerve. Diplopia may result following proptosis or impingement on muscle by orbital bony exostosis. Facial asymmetry may be accompanied by headache, anosmia, and hearing defects.

Fibrous dysplasia is relentlessly progressive. Fractures of the femur and the pelvis usually require orthopedic surgical treatment.

Orbital tumors that cause compression should be excised but characteristically are difficult to shell out and are very hemorrhagic.[2] When compression of the optic nerve is due to fibrous dysplasia involving the sphenoid bone, an attempt should be made to decompress the optic canal.[18]

References

1. Jaffee HL. Fibrous dysplasia of bone. *Bull NY Acad Med.* 1946;22:588.
2. Graf CJ, Perrett GE. Spontaneous recurrent hemorrhage as an unusual complication of fibrous dysplasia of the skull. *J Neurosurg.* 1980;52:570.
3. Coley BL, Stewart FW. Bone sarcoma in polyostotic fibrous dysplasia. *Ann Surg.* 1945;121:872.
4. Perkinson NG, Higgenbotham NL. Osteogenic sarcoma arising in polyostotic fibrous dysplasia. *Cancer.* 1955;8:396.
5. Schwartz DT, Alpert M. The malignant transformation of fibrous dysplasia. *Am J Med Sci.* 1964;247:1.
6. Jakobiec FA, Tannenbaum M. The ultrastructure of orbital fibrosarcoma. *Am J Ophthalmol.* 1974;77:899.
7. Aarskog D, Treteroas E. McCune-Albright's syndrome following adrenalectomy for Cushing's syndrome in infancy. *J Pediatr.* 1968;73:89.
8. Danon M, Robboy SJ, Kim S. Cushing syndrome, sexual precocity and polyostotic fibrous dysplasia (Albright syndrome) in infancy. *J Pediatr.* 1975;87:917.
9. Scurry MT, Bicknell JM, Fajars SS. Polyostotic fibrous dysplasia and acromegaly. *Arch Intern Med.* 1964;114:40.
10. Joishy SK, Morrow LB. McCune-Albright syndrome associated with a functioning pituitary chromoprobe adenoma. *J Pediatr.* 1976;89:77.
11. Ehrig V, Wilson DR. Fibrous dysplasia of bone and primary hyperparathyroidism. *Ann Intern Med.* 1972;77:234.
12. Carr D, Mathie IK, Manor AR. Hypoprolactinemia in a patient with McCune-Albright syndrome. *Br J Obstet Gynecol.* 1979;86:330.
13. Hall R, Warrick C. Hypersecretion of hypothalamic releasing hormones: a possible exploration of the endocrine manifestations of polyostotic fibrous dysplasia (Albright's syndrome). *Lancet.* 1972;1:1313.
14. Giovanelli G, Bermasconi S, Barchin G. McCune-Albright in a male child: a clinical and endocrinologic enigma. *J Pediatr.* 1978;92:220.
15. Lightner ES, Perry R, Frasier SD. Growth hormone excess and sexual precocity in polyostotic fibrous dysplasia (McCune-Albright syndrome). Evidence of abnormal hypothalamic function. *J Pediatr.* 1975;87:922.
16. Danon M, Crawford JD. Peripheral endocrinopathy causing sexual precocity in Albright's syndrome. *Pediatr Res.* 1974;8:368.
17. DiGeorge AM. Albright syndrome: Is it coming of age? *J Pediatr.* 1975;87:1018-1020. Editorial.
18. Moore RT. Fibrous dysplasia of orbit. *Surv Ophthalmol.* 1969;13:321.
19. Sevel D, James HG, Burns R, et al. McCune-Albright syndrome (fibrous dysplasia) associated with an orbital tumor. *Ann Ophthalmol.* 1984;16:283.
20. Calderon M, Brady HP. Fibrous dysplasia of bone. *Am J Ophthalmol.* 1969;68:513.
21. Sassis JF, Rosenberg RN. Neurological complications of fibrous dysplasia of the skull. *Arch Neurol.* 1969;18:363.
22. Mortada A. Fibrous dysplasia of orbital bone. *Br J Ophthalmol.* 1961;45:737.
23. Schorder A. Fibrous dysplasia of bone with proptosis. *Am J Dis Child.* 1977;131:678.
24. Laikos GM, Walker CB, Carruth JAS. Ocular complications in craniofacial fibrous dysplasia. *Br J Ophthalmol.* 1979;63:611.

Chapter 180

Hypophosphatasia

STUART T. D. ROXBURGH

Hypophosphatasia is a rare heritable metabolic bone disease characterized by low serum and tissue alkaline phosphatase (ALP) activity and increased blood and urine levels of phosphoethanolamine and inorganic pyrophosphate. Skeletal ALP is believed to promote bone mineralization by hydrolyzing inorganic pyrophosphate, which acts as a crystal poison, inhibiting hydroxyapatite crystal formation.[1,2] In hypophosphatasia osteoblasts cannot incorporate calcium into otherwise normal bone matrix, so bone maturation is prevented.

Defective bone mineralization presents clinically as rickets in children and as osteomalacia in adults.[3] The disease was first described by Anspach, and the name was first applied by Rathbun.[4,5] Depending on the age at which the disease presents, hypophosphatasia is classified as infantile, childhood, or adult.[6]

Infantile

Infantile hypophosphatasia is manifest before 6 months of age and is inherited as an autosomal recessive trait.[7] Failure to thrive is associated with characteristic rickets-like changes and craniostenosis, which predisposes affected infants to recurrent pneumonia and increased intracranial pressure. Renal complications occur secondary to hypercalcemia and nephrocalcinosis, and the disease is often fatal.

Childhood

Childhood hypophosphatasia is recognized later in childhood and has a variable, more benign clinical course. The cardinal features are rachitic skeletal defects, premature loss of deciduous teeth, and delayed developmental milestones. Joint pains and swelling occur with typical valgus deformities of the knees and costochondral beading (Fig. 180-1). Craniostenosis with its attendant neurologic complications occurs, and renal damage is seen. Studies have suggested either an autosomal recessive or autosomal dominant inheritance pattern.[7]

Adult

Adult hypophosphatasia is the least severe form. The disorder usually begins with loss of adult teeth and is followed by recurrent fractures secondary to osteomalacia. Some persons do not develop clinically apparent skeletal disease.[8,9] Autosomal dominant inheritance with variable penetrance has been described.[7-9] The clinical severity of all three forms of hypophosphatasia is related to the magnitude of the underlying defect in bone mineralization, reflecting quantitatively the deficiency in skeletal ALP activity.[10] Patients with the disease show great heterogeneity in the clinical findings, and it has been suggested that more than one genetic defect may underlie the disorder.

There is evidence to suggest that adult and childhood hypophosphatasia represent the heterozygous state for ALP deficiency, which, when homozygous, results in the severe infantile form of the disease.[10]

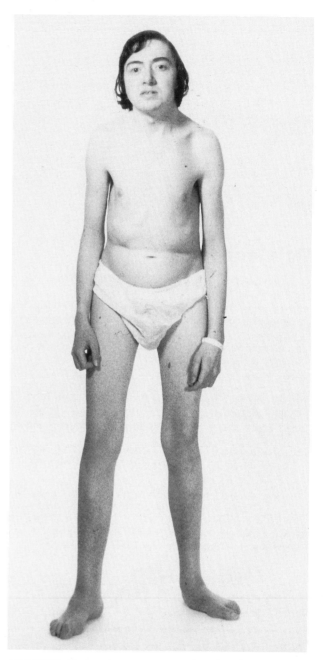

FIGURE 180-1 A case of childhood hypophosphatasia. (The patient is now 26 years old.) Note the rachitic features of costochondral beading, valgus deformities of the knees, and flat feet. This patient also had craniostenosis and retinitis pigmentosa.

Ocular Manifestations

Ocular manifestations are rare in the disease and have been described only in the infantile and childhood forms. Fraser[6] noted three instances of blue sclera and prominent eyes in a review of 35 cases. Bethune and Dent[11] described two sisters who survived to adulthood. One had cataracts and the other had pale optic discs. Lessel and Norton[12] described band keratopathy and conjunctival calcifications in an infant with hypophosphatasia; a case of blue sclera, malformed orbit, pathologic lid retraction, and craniostenosis in an infant was described by Brenner et al.[13] Atypical retinitis pigmentosa has been described in one case.[14] The eye signs may be summarized as follows:

1. Blue sclera
2. Complications of craniostenosis—proptosis, shallow orbits, papilledema, optic atrophy
3. Complications of hypercalcemia—conjunctival calcification and band keratopathy
4. Cataract
5. Retinitis pigmentosa

References

1. Anderson HC. Matrix vesicles of cartilage and bone. In: Bonne GH, ed. *Biochemistry and Physiology of Bone.* vol. IV. New York, NY: Academic Press; 1976;135-157.
2. Fleisch H, Russel RGG, Straumann F. Effect of pyrophosphate on hydroxyapatite and its implication in calcium homeostasis. *Nature.* 1966;212:901-903.
3. Rasmussen H. Hypophosphatasia. In: Stanbury JB, Wyngaarden JB, Frederickson DS, et al., eds. *The Metabolic Basis of Inherited Disease.* 5th ed. New York, NY: McGraw-Hill; 1983;1497.
4. Anspach WE, Clifton WM. Hyperthyroidism in children. *Am J Dis Child.* 1939;58:540-557.
5. Rathbun JC. Hypophosphatasia a new developmental anomaly. *Am J Dis Child.* 1948;75:822-831.
6. Fraser D. Hypophosphatasia. *Am J Med.* 1957;22:730-745.
7. McKusick VA. *Mendelian Inheritance in Man.* 5th ed. Baltimore, Md: The John Hopkins University Press; 1978;221-222, 551-552.
8. Whyte MP, Teitelbaum SL, Murphy WA, et al. Adult hypophosphatasia. Clinical, laboratory, and genetic investigation of a large kindred with review of the literature. *Medicine.* 1979;58:329-347.
9. Whyte MP, Murphy WA, Fallon MD. Adult hypophosphatasia with chondrocalcinosis and arthropathy. Variable penetrance of hypophosphatasia in a large Oklahoma kindred. *Am J Med.* 1982;72:631-641.
10. Fallon MD, Teitlebaum SL, Weinstein RS, et al. Hypophosphatasia: clinicopathologic comparison of infantile, childhood, and adult forms. *Medicine.* 1984;63:12-24.
11. Bethune JE, Dent CE. Hypophosphatasia in the adult. *Am J Med.* 1960;28:615-622.
12. Lessel S, Norton EWD. Band keratopathy and conjunctival calcification in hypophosphatasia. *Arch Ophthalmol.* 1964;71:497-499.
13. Brenner RL, Smith JL, Cleveland WW, et al. Eye signs in hypophosphatasia. *Arch Ophthalmol.* 1969;81:614-617.
14. Roxburgh STD. Atypical retinitis pigmentosa with hypophosphatasia. *Trans Ophthalmol Soc UK.* 1983;103:513-516.

Chapter 181

Marfan's Syndrome

PAUL DAVID REESE

Marfan's syndrome is a heritable disorder of connective tissue with characteristic skeletal, cardiovascular, and ocular manifestations. Dislocation of the lens (ectopia lentis) is a frequent finding, and patients suspected of having the syndrome are often referred to the ophthalmologist for help in establishing the diagnosis. Since the syndrome is inherited in an autosomal dominant fashion with a high degree of penetrance, other family members are frequently found to be affected.

The essential biochemical defect of Marfan's syndrome remains unknown, but a single primary defect in collagen is suspected. No diagnostic laboratory test exists.

Marfan's syndrome exemplifies how a single defect in a widely distributed substrate (presumably collagen) may produce a constellation of clinical signs in seemingly disparate anatomic systems. Close collaboration among clinicians of diverse disciplines is required for management of the broad spectrum of clinical problems posed by this syndrome.

Systemic Manifestations

The cardinal skeletal manifestation of the syndrome is the increased length of the extremities, particularly the fingers and toes, produced by excessively long tubular bones. The patients are taller for their age than other members of their family. There is an abnormally low ratio of upper to lower body segment and the arm span exceeds the height. Redundant ligaments, tendons, and joint capsules result in lax and hyperextensible joints, kyphoscoliosis, flat feet, and repeated dislocations of the hips, patellas, clavicles, and mandible. Chest deformities include sternal displacement outward (pectus carinatum) or inward (pectus excavatum), resulting from longitudinal overgrowth of the ribs.[1]

The life-threatening cardiovascular effects of Marfan's syndrome include aortic dilatation, dissecting aortic aneurysm, and mitral regurgitation. Bacterial endocarditis may develop on structurally altered cardiac valves, and vascular rupture may occur during participation in contact sports or isometric exercises. Profound aortic regurgitation may precede dilatation of the aorta seen on normal chest radiographs.[1,2] As expected, cardiac complications are the leading cause of death in these patients. In a life-table study of survivorship, half of all male patients were dead by age 41, half of women by age 49; cardiac causes accounted for 52 of the 56 deaths.[3]

What is the common factor of the skeletal and cardiovascular systems affected in Marfan's syndrome? Histopathologic evidence suggests that the basic defect of the aorta involves either the elastic or collagen fibers. Collagen appears the more likely candidate for the primary defect because no abnormality of elastic tissue has been found in the trachea, spinal ligaments, or intervertebral discs. The pathologic changes observed in elastic tissue of the aorta may be secondary to weakening of the aortic collagenous skeleton, since the interconnections between the elastic lamellae of the aorta are collagenous.[3]

A defect in collagen may also play a role in many of the skeletal abnormalities, such as lax joint capsules and weak ligaments. The abnormally long, thin extremities may reflect a derangement in periosteal collagen. Attached to the epiphyseal cartilage (the locus for longitudinal growth), periosteum appears to control the length of long bones; if a cuff of periosteum is removed from the circumference of a growing long bone, the length of the bone will ultimately be increased.[3] An inherent weakness in periosteal collagen has been invoked as a possible mechanism for the unbridled longitudinal growth that produces the characteristic Marfan's habitus.

Ocular Manifestations

Subluxation of the lens is the most common ocular abnormality, occurring in 50 to 80% of patients (Fig. 181-1).[4] It is almost invariably bilateral and symmetric. The dislocation may be complete, with the lens floatingfree within the vitreous cavity, or it may be so

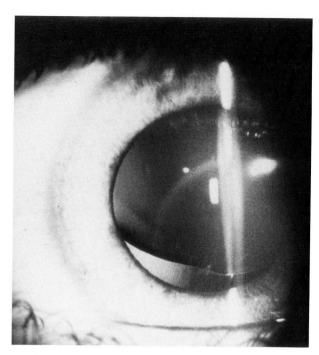

FIGURE 181-1 Superotemporal dislocation of the lens in Marfan's syndrome. Note notch in lens border inferonasally and parallel orientation of the zonular fibers.

subtle as to escape all but the most meticulous slit lamp examination. Iridodonesis (tremulousness of the iris) may occur from nonsupport of the overlying iris by the lens. The lens may opacify, or it may tilt forward to produce corneal touch and secondary decompensation of the corneal endothelium. Young children are particularly susceptible to the development of amblyopia, owing to optical changes engendered by lens subluxation; the refraction of these children must be carefully measured and they must be given their optimal phakic or (if necessary) aphakic correction.

The axial length of the globe is increased in Marfan's patients; hence, moderate to severe myopia is the most common refractive error. The corneal diameter may also be increased (megalocornea).[5,6] Retinal detachment is more common than in the normal population, and the incidence of detachment increases with axial length.[5] Glaucoma is not a frequent problem but when it does occur, further stretching of the globe may be followed by the development of retinal detachment.[5] Enlargement of the globe in Marfan's syndrome is presumed to result from stretching of the sclera, the principal component of which is collagen.

The pathophysiologic mechanism underlying ectopia lentis is obscure. The zonular fibers are collagenous, but unlike most collagen they are digested by chymotrypsin and not by collagenase. McKusick rightly observed that if we knew the common factor shared by the suspensory ligament of the lens and the tunica media of the aorta, the basic defect of Marfan's syndrome might finally be understood.[2]

References

1. Prockop DJ. Heritable disorders of connective tissue. In: Braunwald E, et al, eds. *Harrison's Principles of Internal Medicine*, 11th ed. New York, NY: McGraw-Hill;1987;1687-1688.
2. McKusick VA. Heritable disorders of connective tissue. 4th ed. St. Louis, Mo: CV Mosby; 1972;91-138.
3. Murdoch JL, Walker BA, Halpern BL, et al. Life expectancy and causes of death in the Marfan syndrome. *N Engl J Med.* 1972;286:804.
4. Nelson LB, Maumenee IH. Ectopia lentis. *Surv Ophthalmol.* 1982;27:143.
5. Maumenee IH. The eye in the Marfan syndrome. *Trans Am Ophthalmol Soc.* 1981;79:684.
6. Stephenson WV. Anterior megalophthalmos and arachnodactyly. *Am J Ophthalmol.* 1954;8:315.

Chapter 182

Osteogenesis Imperfecta

OI, Fragilitas Ossium, Maladie de Lobstein

MOURAD KHALIL

Osteogenesis imperfecta (OI) is a heritable disorder of connective tissue with clinical manifestations in the skeleton, ear, joints and ligaments, teeth, skin, and eyes. The three main signs of OI—blue scleras, deafness, and bone fractures—were described by Van der Hoeve in 1918. There is no known racial, ethnic, or sex predilection. The frequency of OI has been estimated to be more than 1 in every 20,000 births.[1]

Pathogenesis

OI is a phenotype with clinical and biochemical heterogenicity. The tissues that are abnormal are composed primarily of type I collagen. Recent advances in the molecular biology of collagen defined the abnormalities at the level of type I collagen gene, and individual variants result from a structural or regulatory abnormality of the alpha$_1$ or alpha$_2$ chains of type I collagen. The disorder in collagen metabolism results in the failure of type I collagen fibrils to mature to their normal diameters, possibly owing to the increased level of hydroxylysin. Electron microscopic studies of scleras of patients with OI revealed a reduction in collagen fiber diameters and a change in their typical cross-striation periodicity.[2] Histomorphometric profile of iliac bone in patients with OI is characterized by cortical and trabecular osteoporosis with decreased activity of individual osteoblasts.[3]

Systemic Manifestations

The terms osteogenesis imperfecta congenita and -tarda have been replaced by a classification proposed by Sillence[1] that recognizes four types:

OI Type I

The mildest and most common form is inherited as an autosomal dominant trait, and there may be two subtypes, one with and one without dental abnormalities. Bone fragility and nondeforming multiple fractures may be present at birth. The susceptibility to fractures decreases after puberty but returns later with pregnancy and menopause. Deafness occurs from otosclerosis. Loose joints and tendons result in flat feet, kyphosis, and recurrent joint dislocations. Hypoplasia of dentine and pulp causes small irregular yellowish blue teeth (dentinogenesis imperfecta). Blue scleras are almost a constant finding.

OI Type II

Type II may be inherited as a recessive trait or can occur as a sporadic new dominant mutation. The most severe form of the disease, it is uniformly lethal. Almost all bones break in utero, causing marked deformity of limbs and skull and beaded ribs. The scleras are dark blue.

Connective tissue abnormalities may be so severe that dismemberment may occur during birth.

Both types I and II can be diagnosed in utero by ultrasonography and radiography.

OI Type III

In some patients inheritance appears to be autosomal recessive, but it is undoubtedly genetically heterogeneous. At birth, the infants' weight and length are normal, the bones are better developed, but numerous fractures are often present. Children who survive infancy have very short stature and develop progressive bone deformities and severe kyphoscoliosis, which may lead to respiratory failure. More variable are the blue scleras, hearing difficulty, and dentinogenesis imperfecta. Joint laxity is moderate.

OI Type IV

Inheritance by recessive or dominant modes may occur in this least severe and most heterogeneous type of the disease. Blue scleras, lax joints, and hearing impairment are less common in this group of OI patients.

Associated systemic features: in addition to the four principal features; blue scleras, fractured bones, deafness, and dentinogenesis imperfecta, other tissues may also manifest symptoms of the disease. The skin is thin and translucent and appears prematurely aged. Torn skin at the corners of the mouth and groin lesions are present. In some patients skin and joint changes are indistinguishable from those of Ehler-Danlos syndrome. A few patients have cardiovascular manifestations such as aortic regurgitation, floppy mitral valves, and fragility of large and medium-sized blood vessels, leading to increased susceptibility to large subcutaneous hematomas or even intracranial hemorrhages.

Ocular Manifestations

Blue scleras are probably most frequent in OI type I. The color can vary from normal to a slightly bluish or slate color to bright blue. OI type II scleras are dark blue. The blueness is caused by the thinness and transparency of the collagen fibers of the sclera that allow visualization of the underlying uveal layer. The sclera is reduced in thickness by about 50%. These patients have significantly lower ocular rigidity measurements than control subjects matched by age, sex, and refractive error.[4] Blue scleras can be an inherited trait in some families without any evidence of increased bone fragility.

The central corneal thickness was found to be reduced in OI.[5] Associated keratoconus, megalocornea, and anterior embryotoxin were reported.

Other ocular manifestations sporadically reported with OI include congenital glaucoma, zonular cataract, dislocated lens, partial color blindness, choroidal sclerosis, and retinal and subhyaloid hemorrhages.[6] Optic neuropathy and optic nerve atrophy are mostly secondary to compression of the visual pathways by the deformities and fractures of the skull bones. In the literature, syndromic associations of OI and congenital blindness have been described in sporadic families. These are probably cases of the severe form of the disease in which skull deformities resulted in optic nerve compression and eventually in atrophy and blindness.[7]

References

1. Rowe DW. In: Wyngaarden JB, Smith L, eds. *Cecil Textbook of Medicine.* Philadelphia, Pa: WB Saunders; 1985;1151-1152.
2. Chan CC, Green WR, de la Cruz Zenaida C, et al. Ocular findings in osteogenesis imperfecta congenita. *Arch Ophthalmol.* 1982;100:1459-1463.
3. Jones CJ, Cummings C, Ball J, et al. Collagen defect of bone in osteogenesis imperfecta type I. An electron microscopic study. *Clin Orthop.* 1984;183:208-214.
4. Kaiser-Kupfer MI, McCain L, Shapiro JR, et al. Low ocular rigidity in patients with osteogenesis imperfecta. *Invest Ophthalmol Vis Sci.* 1981;20:807-809.
5. Pedersen U, Bramsen T. Central corneal thickness in osteogenesis imperfecta and otosclerosis. *ORL J Otorhinolaryngol Relat Spec.* 1984;46:38-41.
6. Khalil M. Subhyaloid hemorrhage in osteogenesis imperfecta tarda. *Can J Ophthalmol.* 1983;18:251-252.
7. Beighton P, Winship I, Behari D. The ocular form of osteogenesis imperfecta; a new autosomal recessive syndrome. *Clin Genet.* 1985;28:69-75.

Chapter 183

Osteopetrosis

Marble Bone Disease

JOSEPH H. CALHOUN

Osteopetrosis, or marble bone disease, is a generic term for several entities that have in common a defect in the osteoclastic function of bone resorption leading to bone sclerosis and fragility. This condition was first described in 1904 by a German radiologist, Heinrich Albers-Schonberg, in a 26-year-old man with generalized sclerosis of the skeleton and multiple fractures.

Osteopetrosis is characterized by failure of bone absorption. The basic defect is in the osteoclast, a multinucleated giant cell that resorbs bone and mineralized cartilage. The osteoblasts continue in their normal function but produce excessive bone without the balance of bone resorption and remodeling normally performed by the osteoclasts. The result is the production of generalized bone sclerosis that paradoxically is more brittle than normal, leading to excessive fractures.

The diagnosis is strongly suggested or confirmed by radiographic studies. Fractures of the long bones are common, healing is accompanied by abundant callus formation. Sclerosis is common, especially in flat bones. The vertebral bodies have a characteristic sandwich appearance because of the sclerosis at the upper and lower portions. All bones, especially of the extremities, may have a bone-within-a-bone appearance. There is often encroachment of the sclerotic bone into the foramina of the skull. The medullary cavity is often crowded or absent, with the development of pancytopenia as a secondary hematopoietic effect.

There are two inheritance patterns for this disease, a relatively mild autosomal dominant form and several autosomal recessive types, which vary from relatively mild to severe and lethal.

The mild, benign, adult, "tarda," or autosomal dominant, form is frequently asymptomatic. It is often discovered incidentally by x-ray examinations taken for other reasons. Bone pain is another common complaint, usually in the lumbar area. Fractures from mild trauma or no trauma are common, especially in the long bones. These heal rapidly, with prominent callus formation. Osteomyelitis occurs in about 10% of patients, especially in the mandible, typically as a result of or following the removal of carious teeth, another feature of this entity. Cranial nerve palsies occur in about one tenth of these patients secondary to bony encroachment on the foramina through which the nerves exit the skull. The third, fourth, and seventh cranial nerves are more commonly involved, but the optic nerve may be affected.

The first form of the autosomal recessive variety is the more common, variously estimated to occur 5 to 10 times in a million births. As in other varieties of osteopetrosis, the failure of bone resorption by the defective osteoclasts in the face of normal bone formation by the osteoblasts results in the deposition of excessive bone, mineralized osteoid, and cartilage. Encroachment on the marrow spaces leads to secondary but inadequate extramedullary hematopoiesis. As a result there is secondary enlargement of the liver and spleen. Prominent features are anemia, thrombocytopenia, and leukopenia with immature red and white blood cells in the peripheral blood.

There is a characteristic appearance to these patients owing to their frontal bossing and hypertelorism. Various signs of the thrombocytopenia are common—petechiae, ecchymosis, and bleeding from the mouth, nose, and gums. Many patients are pale from the anemia.

Most patients have some ocular manifestation, especially optic atrophy, with resultant nystagmus and strabismus. Exophthalmos is also common. Any cranial nerve may be involved; deafness, anosmia, and facial palsy are frequent. Hypotonia and psychomotor retardation are also frequent. Hydrocephalus may develop secondary to hemorrhage or to constriction around the venous drainage channels. A primary retinal degeneration has been reported to be the cause of blindness in some of these infants.[1]

Until recently treatment for this form of osteopetrosis has been only symptomatic and palliative. High doses of calcitriol and a low-calcium diet can increase osteoclast activity and bone resorption while promoting increased calcium excretion.[2] Definitive therapy is probably achieved only with bone marrow transplantation. However, suitable donors are needed.[3,4]

A second form of autosomal recessive osteopetrosis is associated with renal tubular acidosis, cerebral calcification, and deficiency of carbonic anhydrase II. In general, affected persons develop normally in early infancy. Failure to thrive may become apparent any time up to midchildhood. Anemia, if present, is mild, unlike that of the lethal form of autosomal recessive osteopetrosis. Skeletal abnormalities and cranial nerve dysfunction are variable, and generally milder than in the lethal form. The renal tubular acidosis has some characteristics of both type I and type II. Cerebral calcification involves primarily the basal ganglia and the periventricular white matter and is best revealed by computed tomographic scanning. Affected patients have no detectable levels of carbonic anhydrase II, and obligate heterozygotes have depressed levels. The renal tubular acidosis has been treated successfully for the short term with bicarbonate, but long-term follow-up is not available.[5,6] As expected, intravenous acetazolamide has no effect on intraocular pressure in these patients.[7]

There is a third (and mild) form of autosomal recessive osteopetrosis. In addition to the osteopetrosis demonstrated radiographically, there is short stature, macrocephaly, recurrent fractures, dental and mandibular abnormalities with frequent osteomyelitis of the mandible, and mild to moderate anemia with secondary extramedullary hematopoiesis. Typically, there is flattening of the midface with prominence of the zygomatic arch. There is disproportionate shortness of the limbs, altering the upper segment–lower segment ratio and reducing the arm span.[8]

References

1. Keith CG. Retinal atrophy in osteopetrosis. *Arch Ophthalmol.* 1968;79:234.
2. Key L, Carnes D, Cole S, et al. Treatment of congenital osteopetrosis with high-dose calcitriol. *N Engl J Med.* 1984;310:409.
3. Soreel M, Kapoor N, Kirkpatrick D, et al. Marrow transplantation for juvenile osteopetrosis. *Am J Med.* 1981;79:1280.
4. Coccia PF, Krivit W, Cervenka J, et al. Successful bone-marrow transplantation for infantile malignant osteopetrosis. *N Engl J Med.* 1980;302:701.
5. Sly WS, Whyte MP, Sundaram V, et al. Carbonic anhydrase II deficiency in 12 families with the autosomal recessive syndrome of osteopetrosis with renal tubular acidosis and cerebral calcification. *N Engl J Med.* 1985;313:139.
6. Ohlsson A, Cumming W, Paul A, et al. Carbonic anhydrase II deficiency syndrome: Recessive osteopetrosis with renal tubular acidosis and cerebral calcification. *Pediatrics.* 1986;77:371.
7. Krupin T, Sly WS, Whyte MP, et al. Failure of acetazolamide to decrease intraocular pressure in patients with carbonic anhydrase II deficiency. *Am J Ophthalmol.* 1985;99:396.
8. Kahler SG, Burns JA, Aylsworth AS. A mild autosomal recessive form of osteopetrosis. *Am J Med Genet.* 1984;17:451.

Chapter 184

Paget's Disease

Osteitis Deformans

STEPHEN J. BOGAN and GREGORY B. KROHEL

Paget's disease (osteitis deformans) is a chronic, progressive disorder of bone characterized by abnormal osteoclastic activity. Excessive bone resorption is followed by the production of abnormal new bone that is coarse, soft, and very vascular.[1,2] These pathologic changes produce structural alterations of the skeleton and adjacent tissues. The clinical manifestations depend on the extent and location of skeletal involvement, which is variable. Estimated incidence in the adult population ranges from 0.1 to 3%, and increases to 5 to 11% by the 9th decade of life.[3]

The cause of Paget's disease remains uncertain. James Paget felt that it was a chronic inflammatory process.[4] Intranuclear inclusions resembling those produced by paramyxoviruses have been found in the osteoclasts of pagetic bone.[5,6] This raises the possibility

Supported by an unrestricted grant from Research to Prevent Blindness, Inc.

of a slow virus–type infection. Other theories regarding causation include abnormal connective tissue biosynthesis, hormone dysfunction, autoimmune processes, vascular abnormalities, and neoplastic transformation.[6]

The pathophysiology of Paget's disease may be divided into three phases. In the osteolytic phase, there is excessive bone resorption and increased vascularity of the bones. This is followed in the mixed phase by chaotic deposition of abnormal lamellar bone (pagetic bone). Finally, the osteoclastic phase results in the formation of hard, dense, less vascular bone that exhibits a typical "mosaic" pattern of cement lines.[1,3] All of the phases may occur simultaneously in single or multiple foci of one or many bones.

Since most patients with Paget's disease remain asymptomatic, the disease is often discovered on radiographic examination for an unrelated problem. Radiographic findings vary from a predominantly lytic pattern to the characteristic cotton-wool appearance due to adjacent areas of lysis and sclerosis (Fig. 184-1). Other laboratory findings include elevated levels of serum alkaline phosphatase and urine hydroxyproline.

Systemic Manifestations

Clinical manifestations may be divided into two categories: local complications due to abnormal, deformed bone, and systemic complications due to metabolic or vascular derangements. Local complications are often characterized by pain, pathologic fractures, deformities, and neurologic symptoms that can result from the pressure of distorted bone on the brain, spinal cord, cranial nerves, and peripheral nerves. Involvement of the skull occurs in approximately 50% of patients. Patients may complain of changes in hat size or headache exacerbated by coughing or straining.[7,8] Hearing loss and prominence of superficial temporal arteries due to temporal bone deformities are characteristic. Cranial nerves may be compromised by narrowing of their respective foramina or by traction from bony displacement of intracranial contents. Trigeminal neuralgia, hemifacial spasm, and Paget's disease have been observed as a triad of findings in association with basilar impression which may result simply from the weight of the brain on the soft pagetic bone of the occiput and foramen magnum.[9] Brain stem dysfunction, cerebellar dysfunction, and hydrocephalus can also result from basilar compression. Spinal cord and nerve root compression occur most commonly in the thoracic and lumbar regions and can produce numbness and weakness of the extremities and, at later stages, flexor spasms and sphincter difficulties. Peripheral nerve entrapment in the carpal and tarsal tunnels has been reported.[6]

Systemic manifestations of Paget's disease may also

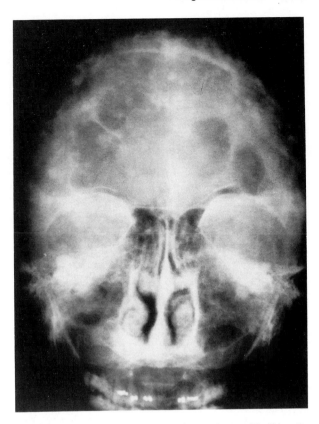

FIGURE 184-1 Skull x-ray of a patient with Paget's disease reveals typical increased patchy density of the calvarium and skull base.

be attributed to metabolic derangements and cardiovascular complications. Although serum concentrations of calcium and phosphorus are usually normal, urinary calcium excretion may be high. Renal calculi may develop, particularly in association with fractures or immobilization.[1]

Vascular abnormalities associated with Paget's disease are influenced by a variety of factors. Increased perfusion due to the vascular hyperplasia and ectasia of pagetic bone may lead to high-output congestive heart failure, especially if more than 30% of the skeleton is involved.[1] Heart block may result from intracardiac calcifications. Arterial calcifications of the Monckeberg type (medial involvement with preservation of a lumen) are common.[1] Finally, vascular ischemia of neural structures can occur as a combined consequence of compression of vessels by deformed bone, shunting of blood through hypervascular pagetic bone (pagetic "steal"), or premature atherosclerosis. The latter two phenomena may account for the observation of neurologic impairment in the absence of an adjacent demonstrable bone deformity.

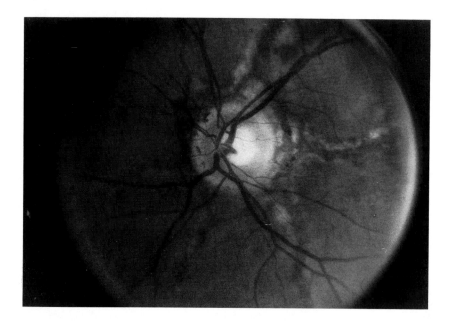

FIGURE 184-2 Angioid streaks associated with Paget's disease.

Ocular Manifestations

The ocular findings in Paget's disease, as with complications in other parts of the body, may be due to local bone deformities or to systemic vascular or metabolic disturbances. Distortions involving the orbital or periorbital bone may lead to exophthalmos, epiphora due to lacrimal duct obstruction, or extraocular muscle palsies due to pressure on cranial nerves III, IV, or VI at the superior orbital fissure.[10]

Angioid streaks are reported to occur in 8 to 15% of patients with Paget's disease and can lead to subretinal neovascularization (Fig. 184-2).[11,12] Histopathologically, the angioid streaks associated with Paget's disease represent linear breaks in a calcified and thickened Bruch's membrane.[12]

The incidence of optic neuropathy in Paget's disease is not known but it may be more common than we realize. Eretto et al.[13] found visual field defects in 9 of 22 patients with Paget's disease, yet only two showed evidence of optic canal abnormality on radiographic examination. This supports the concept that neurologic impairment, including optic atrophy, may result from factors other than direct mechanical compression by deformed bone.

Since most patients with Paget's disease remain asymptomatic, therapy is usually unnecessary. The agents currently used to treat symptom-producing cases include calcitonin, disodium phosphate etidronate, and mithramycin.[3] All three agents suppress the action of osteoclasts, though through different mechanisms. The resulting decrease in bone resorption is accompanied by a decrease in new bone formation. Calcitonin has also been shown to decrease the vascularity of pagetic bone and therefore may theoretically reduce ischemia produced by the pagetic "steal" syndrome.[14] Surgical debulking in cases of compressive neuropathies may help in some instances, although bleeding and recurrent compression are frequent complications. Ocular manifestations usually are not improved with medical therapy. The complications of exophthalmos, lacrimal duct obstruction, subretinal neovascularization, and diplopia may respond to surgical or laser treatment in selected cases.

References

1. Singer RF, Schiller AL, Pyle EB, et al. Paget's disease of bone. In: Aviloi LV, Krane SM, eds. *Metabolic Bone Disease.* New York, NY: Academic Press; 1977.
2. Jaffe HL. The classic Paget's disease of bone. *Clin Orthop.* 1977;127:4.
3. Krane SM. Paget's disease of bone. *Clin Orthop.* 1977;127:24.
4. Paget J. On a form of chronic inflammation of the bones (osteitis deformans). *Lancet.* 1876;18:714.
5. Harvey L. Viral etiology of Paget's disease of bone: a review. *J R Soc Med.* 1984;77:943.
6. Singer FR, Mills GM. The etiology of Paget's disease of bone. *Clin Orthop.* 1977;127:37.
7. Feldman RG, Culebras A, Wallach S, et al. Neurological aspects of Paget's disease. In: Vinken PJ, Bruyn GW, eds. *Handbook of Clinical Neurology.* vol. 15. Amsterdam: North Holland Publishing Co.; 1979; 361.

8. Schmidek HH. Neurologic and neurosurgical sequelae of Paget's disease of bone. *Clin Orthop.* 1977;127:70.

9. Friedman P, Sklaver N, Klawans HL. Neurologic manifestations of Paget's disease of the skull. *Dis Nerv Sys.* 1971;32:809.

10. Walsh FB, Hoyt WF. *Clinical Neuro-ophthalmology.* Baltimore, Md: Williams & Wilkins; 1964;1064-1068.

11. Clarkson JG, Altman RD. Angioid streaks. *Surv Ophthalmol.* 1982;26:235.

12. Gass JDM, Clarkson JG. Angioid streaks and disciform macular detachment in Paget's disease (osteitis deformans). *Am J Ophthalmol.* 1973;75:576.

13. Eretto PA, Krohel GB, Shihab ZM, et al. Optic neuropathy in Paget's disease. *Am J Ophthalmol.* 1984;97:505.

14. Chen J, Rhee R, Wallach S, et al. Neurologic disturbances in Paget's disease of bone: response to calcitonin. *Neurology.* 1979;29:48.

Chapter 185

Stickler's Syndrome

Hereditary Progressive Arthro-ophthalmopathy

SCOTT R. SNEED and THOMAS A. WEINGEIST

Stickler's syndrome is an autosomal dominant inherited disorder of connective tissue with variable penetrance characterized by orofacial abnormalities, ocular disease, and skeletal changes. Stickler[1] originally described an autosomal dominant "progressive arthro-ophthalmopathy" consisting of high myopia, retinal detachment, and bony articular disturbances. Subsequent authors have described hearing loss,[2,3] orofacial abnormalities,[3] mitral valve prolapse,[4] and cataracts and vitreoretinal degeneration[5,6] associated with Stickler's syndrome. The classic findings in Stickler's syndrome include midfacial flattening, generalized epiphyseal dysplasia, myopia, vitreoretinal degeneration, and retinal detachment. Some authors consider Wagner's vitreoretinal degeneration to be a variant of Stickler's syndrome; however, the relative lack of systemic abnormalities and the low incidence of myopia and retinal detachment in patients with Wagner's vitreoretinal degeneration have led others to regard Stickler's syndrome and Wagner's disease as separate entities.[5] The clinical spectrum of Stickler's syndrome is consistent with a defect in collagen metabolism, and there is evidence suggesting that the mutation that causes it may be closely linked with the structural gene for type II collagen (COL2A1).[7]

Systemic Manifestations

The systemic manifestations of Stickler's syndrome originally described by Stickler consisted of "premature degenerative changes in various joints; and in addition, abnormal epiphyseal development and slight hypermobility." Additional reports have stressed the common radiographic finding of generalized epiphyseal dysplasia, particularly in the tibia, distal femur, and distal radius. Radiographic changes may be present in patients with asymptomatic joint disease. Anterior wedging of the vertebral bodies is often detected radiographically. Joint laxity and hyperextensibility, and abnormal bony enlargement of the ankles, knees, and wrists may also occur. Most patients over the age of 10 years complain of joint problems, which range from morning stiffness to generalized discomfort or discomfort after exercise. The knees, ankles, wrists, hands, and hips tend to be most commonly affected.

Palate abnormalities are common and may include complete clefts of the secondary palate, submucous cleft, high-arched palate, and bifid uvula.[3]

Dental anomalies including maleruption, natal teeth, enamel hypoplasia, and noneruption have been described in Stickler's syndrome, but they seem to be uncommon. Weingeist et al.[5] found no evidence of dental abnormalities in their series of 16 patients, although dentures seemed to be needed at an earlier age than in the normal population.

The characteristic facies in some, but not all, patients with Stickler's syndrome consists of midfacial flattening with a flat bridge of the nose. Despite the abnormal clinical appearance of some patients, Weingeist et al.[5]

reported satisfactory facial bone development on cephalometric radiographic evaluation in their series. Another roentgencephalometric study of Stickler's patients revealed a "markedly shortened cranial base length, midfacial depth and height, maxillary depth, and mandibular depth but significantly larger total and lower facial height dimensions."[8] Maxillary and mandibular hypoplasia, epicanthus, and long philtrum have also been associated with Stickler's syndrome. Several authors have described the Pierre Robin anomaly (micrognathia, glossoptosis, and cleft palate) in patients with Stickler's syndrome.[9] Although many of these patients are of normal stature, Stickler's patients' habitus may range from Marfanoid to Weill-Marchesani-like.

Sensorineural or mixed hearing loss has been reported in these patients but appears to be uncommon.[3] The hearing loss tends to occur in the higher frequencies and may be due to abnormalities in the pigment epithelium of the ear or secondary to chronic otitis media, which may develop in children owing to eustachian tube dysfunction.[5]

One study documented a 45.6% incidence of mitral valve prolapse in 57 patients with Stickler's syndrome.[4] Only one was symptomatic, and 34.6% of those with mitral valve prolapse had a click-murmur syndrome. These authors recommend cardiac auscultation, electrocardiography, and cardiac echography in all patients with Stickler's syndrome in order to identify and counsel those patients with mitral valve prolapse.

Ocular Manifestations

The characteristic eye findings in Stickler's syndrome include high myopia, cataracts, vitreous liquefaction, chorioretinal pigmentary changes, and complicated retinal detachments.

A high degree of axial myopia is found in the majority of these patients. The myopia in Stickler's original pedigree ranged from −8.00 to −18.00 diopters. Weingeist et al.[5] found myopia ranging from −2.75 to −16.50 diopters and an average axial eye length of 28.1 mm in their series.

Cataracts are very common, often developing in more than 50% of these patients. Posterior subcapsular cataracts, progressive presenile nuclear sclerotic cataracts, and stationary peripheral cortical opacities have been described. Presenile cataracts develop in the 4th and 5th decades and may cause significant vision loss. Ectopia lentis has been reported in five cases; the lens was displaced superotemporally in two bilateral congenital cases, and in three cases it was displaced into the vitreous.[6]

Abnormal anterior chamber angles have been described by several authors. These changes vary from abnormal angle vessels, fine membranes over the trabecular meshwork, atrophic patches of the iris root with absent iris processes, and long, thick iris processes.

Marked vitreous liquefaction is present in most eyes. The vitreous may appear optically empty and may not be visible with the slit lamp. Vitreous membranes, which often insert into the peripheral retina, are found in many eyes.

Hypopigmentation of the fundus, straightening of the retinal vessels, and a myopic conus are more apparent with higher degrees of myopia; however, staphyloma, lacquer cracks, and choroidal neovascular membranes, which are often seen in high myopia, are rare in these patients. Peripheral vascular sheathing, attenuation, nonperfusion, and abnormal vascular leakage on fluorescein angiography have been described.[6] Lipid exudates may develop in areas of vascular leakage. Perivascular retinal pigmentary changes and choroidal hypopigmentation are common. Latticelike degeneration, with or without small retinal breaks, is present in many eyes.

Patients with Stickler's syndrome are at high risk of developing complicated retinal detachments at an early age (often in the first 2 decades of life). Retinal breaks are often multiple and posterior and may be associated with lattice degeneration. Giant retinal tears are not unusual in these patients. The surgical management of these retinal detachments is often difficult. Pars plana vitrectomy and scleral buckling are often necessary to successfully repair retinal detachments with multiple posterior breaks or giant retinal tears.

Electroretinography in some patients has been reported as normal; however, a decrease in the amplitude of the scotopic B wave has been described in patients with high myopia. These ERG changes appear to be related to the degree of myopia.[5]

References

1. Stickler GB, Belau PG, Farrell FJ, et al. Hereditary progressive arthro-ophthalmopathy. *Mayo Clin Proc.* 1965;40:433-455.
2. Stickler GB, Pugh DG. Hereditary progressive arthro-ophthalmopathy. II. Additional observations on vertebral abnormalities, a hearing defect, and a report of a similar case. *Mayo Clin Proc.* 1967;42:495-500.
3. Lucarini JW, Liberfarb RM, Eavey RD. Otolaryngological manifestations of the Stickler syndrome. *Int J Pediatr Otorhinolaryngol.* 1987;14:215-222.
4. Liberfarb RM, Goldblatt A. Prevalence of mitral-valve pro-

lapse in the Stickler syndrome. *Am J Med Genet.* 1986;24:387-392.

5. Weingeist TA, Hermsen V, Hanson JW, et al. Ocular and systemic manifestations of Stickler's syndrome: a preliminary report. *Birth Defects.* 1982;18:539-560.
6. Spallone A. Stickler's syndrome: a study of 12 families. *Br J Ophthalmol.* 1987;71:504-509.
7. Francomano CA, Liberfarb RM, Hirose T, et al. The Stickler syndrome: evidence for close linkage to the structural gene for type II collagen. *Genomics.* 1987;1:293-296.
8. Saksena SS, Bixler D, Yu P-L. Stickler syndrome: a cephalometric study of the face. *J Craniofac Genet Develop Biol.* 1983;3:19-28.
9. Turner G. The Stickler syndrome in a family with the Pierre Robin syndrome and severe myopia. *Austral Paediatr J.* 1974;10:103-108.

Chapter 186

Weill-Marchesani Syndrome

G. FRANK JUDISCH

Weill-Marchesani syndrome (WMS) is a rare heritable generalized connective tissue disorder that is presumed to be due to an unknown metabolic defect.[1] Brachymorphism, microspherophakia, lenticular myopia, ectopia lentis, shallow anterior chamber (Fig. 186-1), and secondary glaucoma are the salient features. Although brachymorphism, especially brachydactyly and microspherophakia are usually considered requisites for making the diagnosis of WMS, they are not pathognomonic.[2-10]

Systemic Manifestations

When the syndrome is fully expressed, the pyknic physique, with short stature (1.4 to 1.6 m for adult males, 5 to 10 cm less for females[4]), brachycephaly, and brachydactyly, is easily recognized. Limitation of joint mobility, especially in the hands and wrists, is not infrequently observed. It should be emphasized that there may be considerable variation in the expression of these physical findings, to the point where they may be very subtle and easy to miss. Occasionally, radiographic examination, particularly of the metacarpals, is necessary to substantiate the brachydactyly.[5]

WMS is usually an autosomal recessive condition.[11] Most of the reported pedigrees reflect a high incidence of consanguinity, affected persons belonging to one generation. The possibility of genetic heterogeneity has been suggested by a few reports of autosomal dominant transmission.[10,11] Obfuscating matters is the fact that simple ectopia lentis may be inherited as either an autosomal recessive or dominant disorder and both may occur in persons whose short stature is not pathologic.[1] After reevaluating one of the families reported as an example of WMS with autosomal dominant transmission, McKusick concluded that the pedigree represented simple autosomal dominant ectopia lentis in a family with generally short stature.[1]

It has also frequently been suggested that those who are heterozygous for WMS manifest only short stature. Short people do tend to marry short people. Pseudodominance resulting from homozygous-heterozygous marriages is another possible explanation for these occasional reports of autosomal dominant transmission in WMS.[7] Determination of the metabolic defect(s) in WMS should resolve the heritability question.

Ocular Manifestations

From a functional standpoint, the most debilitating sequelae of WMS are ocular. Most patients have sought attention by the end of the 1st decade of life because of poor vision, usually due to an often progressive lenticular myopia, which may range from −3.00 to −20.00 diopters. Microspherophakia is almost always apparent at the first examination, permitting visualization of the entire lens equator with wide pupillary dilation (Fig. 186-2). The unusual spherical shape of the lens is to be emphasized, as it is a major factor in the pathogenesis of glaucoma, which develops so commonly and is so potentially devastating. While the equatorial diameter is diminished by 2.0 to 2.25 mm,[6] the anteroposterior (sagittal) dimension is increased up to 1.0 mm or more. The elongated irregular zonules that are frequently

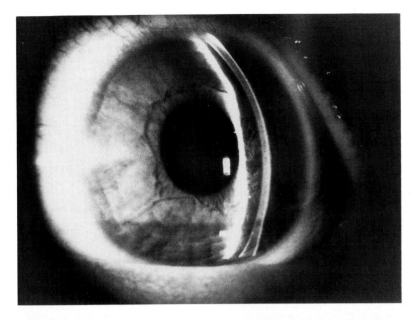

FIGURE 186-1 Slit lamp photograph of a 15-year-old girl with WMS and myopia of −15.00 diopters. The anterior chamber depth measured 1.7 mm, about 50% of normal for emmetropes of this age.

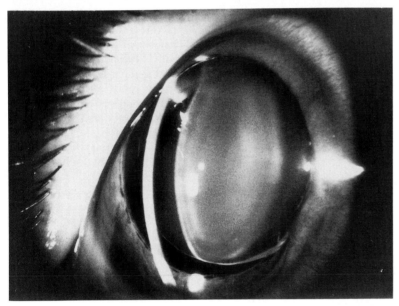

FIGURE 186-2 The same patient after pupillary dilation demonstrates the spherical shape of the lens and the visibility of the lens equator. The anteroposterior lens diameter measured 4.8 mm (normal is 3.7 ± 0.26 mm).

observed do not entirely explain the lens dysmorphia, as the lens volume has been found to be reduced by 25 to 40%.[6,8] It is suspected that the pathogenic metabolic defect affects both lenses and zonules. Laxity or rupture of the zonules probably accounts for the hypoaccommodation often reported. Vitreoretinal abnormalities are not a feature of WMS.

Some degree of ectopia lentis is usually apparent by the end of the 2nd decade. Spontaneous migration into the anterior chamber seldom occurs without pupillary

dilatation. However, lax zonules often permit slight forward movement. The dislocated lens may also rotate into the pupil, producing pupillary block.

Because of the propensity to develop glaucoma, vision loss in WMS occurs earlier and is more devastating than in other ectopia lentis syndromes.[2,3,5,6,8] The exact incidence of glaucoma is unknown, but the percentage is high, perhaps approaching 100% if untreated patients are followed long enough. The glaucoma is often acute and probably almost always phacogenic.

Although a few patients have been reported to have congenital angle anomalies, far more gonioscopic descriptions have failed to note such changes.[9]

Pupillary block glaucoma may occur with or without lens dislocation. In the former instance, the lens with its increased sagittal diameter and lax zonules moves forward, expanding the area of iris-lens apposition, which increases resistance to aqueous flow from the posterior to anterior chambers. This may result in pupillary block leading to iris bombe and angle-closure glaucoma. In patients with normally positioned lenses, miotics may precipitate or aggravate angle-closure glaucoma, resulting in so-called inverse glaucoma. In such patients, miotics contract the ciliary body, further loosening the zonules, while iris sphincter constriction draws the iris more tightly over the lens face, increasing iris-to-lens apposition and the resistance to aqueous flow from posterior to anterior chamber.

It is presumed that prolonged or repeated attacks of angle closure are responsible for the occasional patient's peripheral anterior synechiae or elevated intraocular pressure in spite of unremarkable chamber angles and widely dilated pupils. In the latter instance, it is suggested that previous angle-closure attacks produced microscopic damage to the trabecular meshwork resulting in an increase in outflow resistance and ultimately to elevated intraocular pressures. Once it is luxated, the lens may produce pupillary block by becoming wedged in the pupil or by migrating into the anterior chamber.

Determining the lens position and status is very important in the management of glaucoma in WMS. If the intraocular pressure fails to rise in response to miotics, it may be assumed that the lens is dislocated or that any remaining zonules are incapable of altering lens position. In attacks of acute angle-closure glaucoma, miotics should not be given unless the lens is known to be dislocated or essentially free of zonular support, to avoid inverse glaucoma. Hopefully, in cases where the lens-zonule status is unknown, hyperosmotics, carbonic anhydrase inhibitors, and timolol would relieve the attack and permit performance of a peripheral iridectomy to reestablish communication between the anterior and posterior chambers.[5] Today, laser iridotomy would

seem to be the procedure of choice. Surgical complications were quite common in the past following surgical peripheral iridectomy or cataract extraction using the intracapsular technique. After iridotomy, miotics may be used to constrict the pupil, preventing the lens from migrating into the pupil or the anterior chamber.

As lens dislocation and acute glaucoma are so common in WMS, it has been proposed that prophylactic laser iridotomy performed when the patient is young would prevent pupillary block glaucoma. The subsequent use of topical thymoxamine has been suggested as the miotic of choice to prevent migration of the lens into the anterior chamber.[5]

The need for careful lifelong ophthalmologic follow-up of WMS patients cannot be overemphasized.

References

1. McKusick VA. *Heritable Disorders of Connective Tissue.* 2nd ed. St Louis, Mo: CV Mosby; 282-291.
2. Wright KW, Chrousos GA. Weill-Marchesani syndrome with bilateral angle-closure glaucoma. *J Pediatr Ophthalmol Strab.* 1985;22:129.
3. Johnson GL, Bosanquet RC. Spherophakia in a Newfoundland family: 8 years' experience. *Can J Ophthalmol.* 1983;18:159.
4. Schmidt P, Bernth-Petersen P. Marchesani's syndrome. *Ophthalmologica Basel.* 1981;183:110.
5. Ritch R, Wand M. Treatment of the Weill-Marchesani syndrome glaucoma. *Ann Ophthalmol.* 1981;13:665.
6. Jensen AD, Cross HE. Ocular complications in the Weill-Marchesani syndrome. *Am J Ophthalmol.* 1974;77:261.
7. Gorlin RJ, L'Heureux PR, Shapiro I. Weill-Marchesani syndrome in two generations: genetic heterogeneity or pseudodominance? *J Pediatr Ophthalmol Strab.* 1974;11:139.
8. Willi M, Kut L, Cotlier E. Pupillary-block glaucoma in the Marchesani syndrome. *Arch Ophthalmol.* 1973;90:504.
9. Feiler-Ofry V, Stein R, Godel V. Marchesani's syndrome and chamber angle anomalies. *Am J Ophthalmol.* 1968; 65:862.
10. Rosenthal J, Kloepfer H. The spherophakia-brachymorphia syndrome. *Arch Ophthalmol.* 1956;55:28.
11. McKusick VA. *Mendelian Inheritance in Man.* 7th ed. Baltimore, Md: The Johns Hopkins University Press; 1986;1300-1301.

Section D

Miscellaneous Developmental Disorders

Chapter 187

Aarskog's Syndrome

KENNETH V. CAHILL, ANNA MARIE SOMMER, and JOHN A. BURNS

Aarskog's syndrome is the eponym for the facial-digital-genital syndrome described by Aarskog[1] in 1970. It affects the skeletal and connective tissue systems, producing a characteristic set of anomalies. The most apparent signs are hypertelorism, short stature, small broad hands and feet, protruding umbilicus, and variable presence of inguinal hernias, cryptorchidisim, and a scrotal skin fold that hoods the penis. Although mental retardation has been present in some family pedigrees, Aarskog's syndrome is generally associated with normal mentation. Life expectancy is also normal.

Inheritance

Initially, transmission of Aarskog's syndrome was thought to be X-linked recessive, because only males were involved in the first families described. Subsequent reports showed that females can also be affected, although they usually show only mild signs of Aarskog's syndrome[2]. Involvement of females suggests that the inheritance may be X-linked semidominant. Male-to-

male transmission documented in one family is evidence that sex-influenced autosomal dominant inheritance also occurs.[3] It is possible that more than one hereditary pattern can be involved in the transmission of Aarskog's syndrome, but autosomal dominant inheritance now appears to be the most common cause.

Systemic Manifestations

Nonocular facial anomalies are part of this syndrome. The frontal hairline frequently forms a widow's peak. A horizontal crease is often found below the lower lip. The superior helixes of the ears are typically thickened and have a flattened curvature.

Short stature is a constant skeletal anomaly in Aarskog's syndrome. No underlying metabolic defect has been identified. The hands and feet are typically broad with short digits and mild interdigital webbing. The short fingers may display a single ventral crease, and the palm may have a simian crease. Additional skeletal anomalies can occur, such as cervical vertebral deformities, pectus excavatum, cubitus valgus, metatarsus adductus, and internal tibial torsion. These may warrant orthopedic consultation, and in some cases surgical correction is necessary.[4]

Urogenital anomalies are frequently seen in males who exhibit this syndrome. A fold of scrotal skin may extend over the dorsum of the penis (shawl scrotum), forming a saddlebag deformity. Cryptorchidisim and infertility are occasionally present.

Protruding umbilicus is a constant finding in males and females.[5] Inguinal hernias are seen in half of the cases, regardless of sex.

Ocular Manifestations

The ophthalmologist is likely to see patients with this syndrome because of their facial involvement. Hypertelorism without evidence of epicanthal folds is a constant finding. The inner canthal distance measures 40 mm or more, and the interpupillary distance is similarly increased (Fig. 187-1). This is generally not a significant cosmetic or functional problem, so surgical correction is unlikely to be indicated. There may be mild facial asymmetry with unequal levels of the eyes. An antimongoloid fissure slant is sometimes present.

Congenital ptosis is seen in about 50% of affected persons. It is usually unilateral and shows the typical characteristics of levator muscle dystrophy seen in isolated cases of congenital ptosis. The degree of ptosis is mild to moderate. It can be managed using the same principles as those employed for the management of isolated congenital ptosis.

Since these patients generally have normal mentation and normal life expectancy, they are able to lead a normal life. Reconstructive surgery may be necessary to correct congenital ptosis, orthopedic anomalies, urogenital defects, protruding umbilicus, and inguinal hernias. Otherwise these patients are not prone to develop other medical problems. Genetic counseling of affected persons is recommended so that they understand the possibilities for passing on skeletal and connective tissue anomalies to their children.

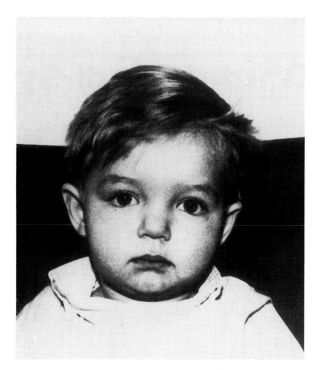

FIGURE 187-1 This boy with Aarskog's syndrome demonstrates hypertelorism, antimongoloid fissure slant, a horizontal crease below the lower lip, and thickened helixes of the ears with a flattened curvature.

References

1. Aarskog D. A familial syndrome of short stature associated with facial dysplasia and genital anomalies. *J Pediatr.* 1970;77:856.
2. Furukawa CT, Hall BD, Smith DW. The Aarskog syndrome. *J Pediatr.* 1983;81:117-1122.
3. Van De Vorren MJ, Neirmeijer MF, Hoogeboom AJM. The Aarskog syndrome in a large family, suggestive for autosomal dominant inheritance. *Clin Genet.* 1983;24:439-445.
4. Hurst DL. Metatarsus adductus in two brothers with Aarskog syndrome. *J Med Genet.* 1983;20:477.
5. Friedman JM. Umbilical dysmorphology. *Clin Genet.* 1985;28:343-347.

Chapter 188

Bardet-Biedl Syndrome

SAMUEL G. JACOBSON

Historical Perspective

The Bardet-Biedl syndrome, also known as the Laurence-Moon-Biedl syndrome and by other combinations of these eponyms, is a multisystem disorder consisting of retinal degeneration, obesity, anomalies of the digits, mental retardation, abnormalities of sexual development, and nephropathy. The syndrome is considered among the rarer forms of recessively inherited retinitis pigmentosa (RP). Although the condition became widely acknowledged only about 60 years ago, the original observations date back more than 120 years.

In 1864, Höring[1] described two siblings, a 5-year-old boy and his 9-year-old sister, with RP, polydactyly, and defective intellect. This lead Stör[2] in 1865 to report the coincidence of RP, polydactyly, mental retardation, and obesity in a 23-year-old woman. Laurence and Moon,[3] in 1866 described four siblings (one girl and three boys, aged 7 to 20 years) who had short stature, mental retardation, hypogenitalism, and paraparesis. Their ocular condition was characterized by early onset of nyctalopia and poor visual acuity, nystagmus, full visual fields, and fundus findings of both central and peripheral retinal degeneration. There was no mention of polydactyly in these otherwise well-described patients.

Between 1866 and 1920 a number of cases of retinal degeneration associated with polydactyly, mental retardation, obesity and hypogenitalism were reported (reviewed in reference 4). Bardet[5] in 1920 published a thesis that described an 11-year-old girl with evidence of the first four of these five features. In 1922, Biedl[6] reported the constellation of five findings in two sisters and was subsequently credited, along with Laurence and Moon, as a discoverer of the syndrome.[7]

There has been some debate as to the most appropriate name for the condition.[8] If Höring, Stör, and the many other early clinical observers who recognized the association of retinal degeneration with digital anomalies, obesity, and retardation are not to be credited, then it would seem appropriate to continue to use the term Bardet-Biedl syndrome, considering the more atypical nature of the condition originally described by Laurence and Moon (ie, spastic paraplegia and/or cerebellar findings, full visual fields even at age 20 years, and presumed absence of polydactyly).

Genetics

There is strong evidence from a number of studies that the Bardet-Biedl syndrome is inherited as an autosomal recessive trait.[4,9] Pleiotropism has been used to explain the multiple phenotypic effects in this syndrome.[10]

Systemic Manifestations

Among the more common nonocular manifestations of the syndrome are obesity, polydactyly, mental retardation, hypogenitalism, and renal abnormalities.[4,9] There is, however, a considerable degree of heterogeneity in the syndrome—and some overlap with other syndromes. Certain authors have attempted to subclassify patients by the number of typical and atypical manifestations.[9]

Although the truncal obesity and the abnormalities of sexual development have provoked much speculation about the pathogenesis of the syndrome (reviewed in reference 8), no convincing unifying endocrinologic mechanism has been demonstrated. The most frequent digital anomaly is postaxial polydactyly, but syndactyly and brachydactyly have been described.[4,9] The degree of mental retardation is also variable: some patients are severely retarded and others are likely to be only educationally deprived because of their early-onset visual disability. Defective renal function has been reported in certain series to occur in as many as 90% of patients with the Bardet-Biedl syndrome.[11] A urologic evaluation certainly seems warranted for all patients suspected of having the syndrome, since many of the renal abnormalities are detectable with urography and since renal failure is often the cause of death.[4]

Other syndromes that share features with the Bardet-Biedl syndrome include the Alström-Hallgren (retinal degeneration, obesity, diabetes mellitus, deafness) and Biemond II (polydactyly, iris coloboma, mental retardation, obesity, hypogenitalism) syndromes.[9]

Ocular Manifestations

Retinal degeneration, the principal ocular manifestation of the Bardet-Biedl syndrome, is relatively severe compared to that in many of the more typical forms of RP. In general, both night vision and central vision are affected early, and the visual field can already be very constricted by the 2nd or 3rd decade of life. The presentation of such severe retinopathy in a young person should suggest the diagnosis.[9]

A variety of fundus appearances have been described in patients with the Bardet-Biedl syndrome.[9] The amount and type of pigmentary disturbance in the peripheral retina is variable, as it can be in typical RP. Correlating with the serious and early central visual disturbances in these patients are a number of funduscopic changes in the macular region, notably bull's-eye type lesions and geographic atrophy of the retinal pigment epithelium and choriocapillaris.[12]

The application of electrophysiologic and psychophysical tests of visual function to these patients has led to some debate over the pattern of rod and cone dysfunction in the syndrome. The electroretinogram (ERG) signal is most often nondetectable. When it is recordable, there have been reports of rod>cone, cone>rod and rod=cone abnormalities.[13-16] Apart from the interest in subclassifying patients by retinal function, an abnormal ERG can be of diagnostic aid in a child with other manifestations of the syndrome who has no obvious fundus findings.

Psychophysical testing is restricted, of course, to patients with the intellectual capacity to perform such tests reliably. When dark adaptometry (at various rental loci) has been tested, it most often has shown severely impaired rod function.[12,13] Dark-adapted two-color perimetry, a technique that has been used to define subtypes of typical RP,[17] has been applied to patients with Bardet-Biedl syndrome. The different patterns of rod and cone dysfunction found in these patients suggest that there may be functional heterogeneity in the syndrome.[14]

Conclusions

Bardet-Biedl syndrome is definitely a clinically recognizable entity, but it is a condition marked by heterogeneity of systemic and retinal manifestations. There is already some evidence that there are different patterns of retinal degeneration within the syndrome. With the advent of sophisticated noninvasive retinal testing techniques, there may be an opportunity to define subtypes of the retinopathy in young cooperative patients, such as is being attempted in those with other forms of RP.[18] This might make the future task of molecular geneticists easier.

References

1. Höring Dr. Retinitis pigmentosa. *Klin Monatsbl Augenheilk.* 1864;2:233.
2. Stör Dr. Retinitis pigmentosa. *Klin Monatsbl Augenheilk.* 1865;3:23.
3. Laurence JZ, Moon RC. Four cases of retinitis pigmentosa occurring in the same family and accompanied by general imperfections of development. *Ophthalmic Rev.* 1866;2:32.
4. Bell J. The Laurence-Moon syndrome. In: *The Treasury of Human Inheritance*, V, III, Cambridge: Cambridge University Press; 1958;51.
5. Bardet G. Sur un syndrome d'obésité congenitale avec polydactylie et rétinite pigmentaire. These de Paris. 1920.
6. Biedl A. Ein Geschwisterpaar mit adiposo-genitaler Dystrophie. *Dtsch Med Wochenschr.* 1922;48:1630.
7. Solis-Cohen S, Weiss E. Dystrophia adiposogenitalis, with atypical retinitis pigmentosa and mental deficiency—the Laurence-Biedl syndrome. *Am J Med Sci.* 1925;169:489.
8. Stiggelbout W. *The (Laurence-Moon) Bardet Biedl Syndrome.* Van Gorcum & Comp, Assen Medical collection 207, 1969.
9. Klein D, Amman F. The syndrome of Laurence-Moon-Bardet-Biedl and allied diseases in Switzerland; clinical, genetic and epidemiological studies. *J Neurol Sci.* 1969;9:479.
10. Thompson JS, Thompson MW. *Genetics in Medicine.* 4th ed. Philadelphia, Pa: WB Saunders; 1986;68.
11. Churchill DN, McManamon P, Hurley RM. Renal disease—a sixth cardinal feature of the Laurence-Moon-Biedl syndrome. *Clin Nephrol.* 1981;16:151.
12. Campo RV, Aaberg TM. Ocular and systemic manifestations of the Bardet-Biedl syndrome. *Am J Ophthalmol.* 1982;94:750.
13. Katsumi O, Tanino T, Hirose T, et al. Laurence-Moon-Bardet-Biedl syndrome: electrophysiological and psychophysical findings. *Jpn J Ophthalmol.* 1985;29:282.
14. Jacobson SG, Borruat FX, Apáthy PP. Patterns of rod and cone dysfunction in the Bardet-Biedl syndrome. *Invest Ophthalmol Vis Sci Suppl.* 1986;27:310.
15. Berson EL, Gouras P, Gunkel RD. Progressive cone-rod degeneration. *Arch Ophthalmol.* 1968;80;68.
16. Rizzo JF, Berson EL, Lessel S. Retinal and neurologic findings in the Laurence-Moon-Bardet-Biedl phenotype. *Ophthalmology.* 1986;93:1452.
17. Jacobson SG, Voigt WJ, Parel JM, et al. Automated light- and dark-adapted perimetry for evaluating retinitis pigmentosa. *Ophthalmology.* 1986;93:1604.
18. Kemp CH, Jacobson SG, Faulkner DJ. Two types of visual dysfunction in autosomal dominant retinitis pigmentosa. *Invest Ophthalmol Vis Sci.* 1988;29:1235.

Chapter 189

Carpenter's Syndrome

ANDREW E. CHOY and A. GLEN RICO

Acrocephalopolysyndactyly type II, or Carpenter's syndrome, is a rare developmental disorder of undetermined cause.[1,2] The salient features of this autosomal recessive syndrome include acrocephaly, peculiar facies, obesity, hypogonadism, variable syndactyly of the hands, and preaxial polysyndactyly of the feet.[3] The latter characteristic distinguishes Carpenter's syndrome from other autosomal recessive craniosynostoses. The common denominator appears to be premature closure of sutures resulting in cranial, orbital, and skeletal deformities, though this does not explain the multiple systemic abnormalities that have been described in association with the syndrome.[4,5] No biochemical, chromosomal, or hormonal aberration has been uncovered.

Systemic Manifestations

Synostosis, or premature closure of the cranial sutures, results in myriad abnormalities. The coronal and lambdoidal sutures close first, followed by the sagittal sutures, producing acrocephaly. Increased intracranial pressure and seizures may ensue if they are not surgically corrected early. Unilateral involvement of the sutures may lead to marked cranial asymmetry. "Midface hypoplasia"—flattened nasal bridge, high-arched palate, and maxillary and dental abnormalities—is a common finding.

Mental retardation was once thought to be a consistent feature of Carpenter's syndrome. Normal intelligence is now a well-established finding.[6,7] The cranial deformities may not be responsible for mental retardation, since many patients who have early surgical correction have persistent mental dysfunction.[8]

Recurrent serous and purulent otitis media is associated with a short cranial base and abnormal middle ear anatomy. Low-set ears as well as conductive and neurosensory hearing loss are also encountered sporadically.

Atrial septal defect, ventricular septal defect, persistent patent ductus arteriosus, pulmonic stenosis, and tetralogy of Fallot comprise the congenital cardiac defects seen in approximately one third of patients with this syndrome.[4] Abdominal hernia or omphalocele and accessory spleen are incidental clinical findings.

Endocrine dysfunctions frequently include hypogonadism.[9] Isolated cases of empty sella syndrome[10] and precocious growth[11] have been noted. Short stature is the rule, and patients usually rank below the 25th percentile for height. It is not established whether obesity is endocrine mediated.

Preaxial polysyndactyly of the feet is a consistent and distinguishing characteristic in Carpenter's syndrome. Preaxial polysyndactyly may also be seen in Apert's syndrome.[12] However, Apert's syndrome is an autosomal dominant inherited disorder with acrocephaly but lacking obesity and hypogonadism.[3] Metatarsus varus and soft tissue syndactyly of the feet are also common in Carpenter's syndrome.

Abnormalities of the hand consist of variable soft tissue syndactyly, brachydactyly (short hands), brachymesophalangy (agenesis of the middle phalanx), and clinodactyly (abnormal bending). Various common abnormalities of the knee and hip include lateral displacement of the patella, genu valgum, flat acetabulum, and coxa valga. Scoliosis, absence of coccyx, and spina bifida occulta are frequent spine disorders.

Ocular Manifestations

The orbital anomalies in Carpenter's syndrome are the consequence of premature closure of the integrated orbital and cranial sutures. Shallow orbits, exophthalmos, and hypertelorism or hypotelorism are most notable. Resultant exposure keratopathy and lagophthalmos are common. Epicanthal folds and antimongoloid slant of the palpebral fissures are also frequently observed. Papilledema may be seen as a consequence of increased intracranial pressure. Optic atrophy can result from chronic papilledema, small optic foramina, or stretching of the optic nerve secondary to posterior displacement of the brain.[13] Esotropia, exotropia, and hypertropia are the extraocular muscle imbalances occasionally associated with this syndrome.

Earlier reports on "acrocephalosyndactylism" have documented other ocular abnormalities, notably microcornea, corneal leukoma, coloboma of the iris and choroid, congenital cataract, lens subluxation, nystagmus, and retinal detachment.[14,15] In essence, a wide variety of ocular abnormalities may be expected. Performing a complete orbital and ocular examination is a responsibility incumbent on the ophthalmologist.

A multidisciplinary approach to treatment cannot be overemphasized.[16] The primary step is correction of the cranial deformity with craniosynostectomy during infancy. Treatment of individual ocular abnormalities as well as orbital decompression may be necessary. Genetic counseling must be addressed.

References

1. McKusick VA. *Mendelian Inheritance in Man. Catalog of Autosomal Dominant, Autosomal Recessive and X-Linked Phenotypes.* 4th ed. Baltimore, Md: Johns Hopkins University Press;1975:334-335.
2. Carpenter G. Two sisters showing malformations of the skull and other congenital abnormalities. *Rep Soc Study Dis Child London.* 1901;1:110-118.
3. Temtamy SA. Carpenter's syndrome: Acrocephalopolysyndactyly. An autosomal recessive syndrome. *J Pediatr.* 1966;69:111-120.
4. Cohen MM. An etiologic and nosologic overview of craniosynostosis syndromes. *Birth Defects,* 1975;11(2):137-189.
5. Pfeiffer RA, Seemann KB, Tünte W. et al. Akrozephalo-Polysyndaktylie (Akrozephalosyndaktylie, Typ II McKusick) (Carpenter syndrome), Bericht über 4 Fälle und eine Beobachtung des Typs von Marshall-Smith. *Klin Padiatr.* 1977;189:120-130.
6. Robinson LK, James HE, Mubarak S et al. Carpenter syndrome: natural history and clinical spectrum. *Am J Med. Genet.* 1985;20:461-469.
7. Frias JL, Felman AH, Rosenbloom AL, et al. Normal intelligence in two children with Carpenter syndrome. *Am J Med Genet.* 1978;2:191-199.
8. Cohen MM. Craniosynostosis and syndromes with craniosynostosis: incidence, genetics, penetrance, variability and new syndrome updating. *Birth Defects.* 1979;15(5B):13-63.
9. Eaton AP, Sommer A, Kontras SB, et al. Carpenter syndrome—acrocephalopolysyndactyly type II. *Birth Defects.* 1974;10(9):249-260.
10. Verdy M, Dussault RG, Verrault R, et al. Carpenter's syndrome with empty sella and abnormal LRH and TRH response. *Acta Endocrinol (Copenh).* 1983;104(1):6-9.
11. White J, Boldt DB, David DJ, et al. Carpenter's syndrome with normal intelligence and precocious growth. *Acta Neurochir (Wien).* 1981;57(1–2):43-49.
12. Yonenobu K, Tada K, Tsuyugushi Y. Apert's syndrome—a report of five cases. *The Hand.* 1982;14(3):317-325.
13. Koziak PH. Craniostenosis-report of 22 cases. *Am J Ophthalmol.* 1954;37:380-390.
14. Sunderhaus E, Wolter JF. Acrocephalosyndactylism. *J Pediatr Ophthalmol.* 1968;5:118-120.
15. Howell SC. The Craniostenoses. *Am J Ophthalmol.* 1954;37:359-379.
16. Munro IR, Orbito-cranio-facial surgery: the team approach. *Plast Reconstr Surg.* 1975;55(2):170-176.

Chapter 190

Cohen's Syndrome

CHRISTINA RAITTA and REIJO NORIO

A person with Cohen's syndrome[1] usually appears to the doctor as a mentally retarded patient with a peculiar habitus; the visual handicap and highly typical ophthalmic features may totally escape notice. We have investigated 14 patients and have published reports of 6, together with a review of 25 cases reported by others.[2] Since then more than 10 new cases have been published. Cohen's syndrome may be more common than is believed. The autosomal recessive inheritance can be taken for granted. Cohen's syndrome clearly belongs to the so-called Finnish disease heritage, about 30 rare recessively inherited disorders that are overrepresented in Finland.

Systemic Manifestations

A typical patient is slightly or moderately mentally retarded. The milestones of psychomotor development are more or less delayed. Microcephaly is common. The ability to communicate varies from normal speech to a few invented words, the social adaptation level from that of special schools and sheltered workshops to institutions for the mentally retarded.

The best clue to a specific diagnosis is derived from the facial appearance, which is easy to spot in children but less easy in infants and adults (Fig. 190-1,A–C). Flame-shaped or wave-shaped lid openings and thick

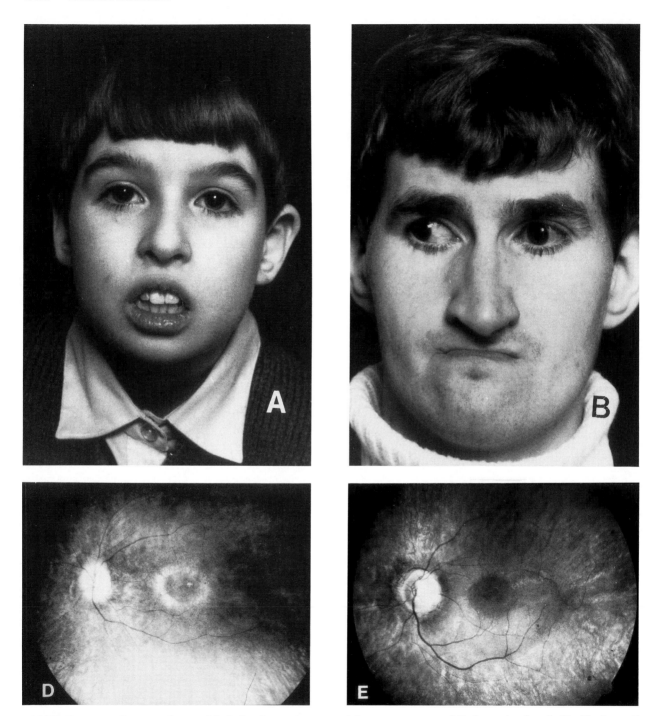

FIGURE 190-1 Three patients with Cohen's syndrome and the appearance of their ocular fundi: (**A, D**) at age 7 years; (**B, E**) at age 22 years; (**C, F**) at age 47 years.

hair, thick eyebrows, and long eyelashes give young patients a charming appearance. The root of the nose is prominent, the philtrum short, and the upper central incisors prominent. The palate arch is high. The slant of the eyes does not seem to be a significant characteristic.

Truncal obesity has probably been emphasized too much in the literature: it may be nonexistent but is usually moderate and hardly ever resembles that of Prader-Willi syndrome. Fingers and toes are slender; the space between the first and second toes is wide. Muscu-

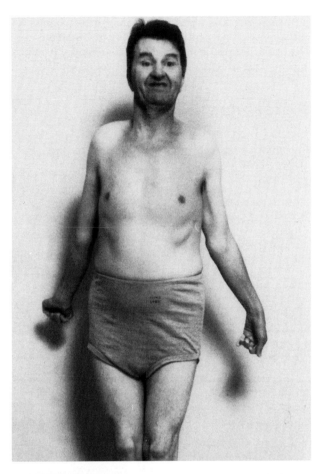

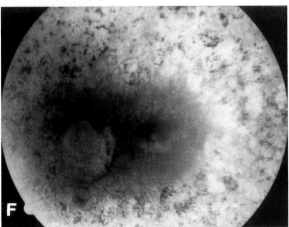

lar or ligamental hypotonia leads to pes planovalgus, genu valgum, or thoracic kyphosis. Tendon reflexes are, however, brisk rather than weak.

Puberty is often delayed. Sporadic endocrinologic studies have not produced logical results that could be generalized. Height is often less than 2 SD below normal. The beautiful face in infancy and childhood assumes a senile appearance in adults.

A peculiar finding is moderate leukopenia, especially granulocytopenia. It is not apparent in every count and does not harm the patients at all.

The patients are typically active, cooperative, sociable, and cheerful and, so, seldom need institutional care.

Ocular Manifestations

The ocular manifestations have been uniform and almost pathognomonic in all our 14 patients. They consist of symptoms and signs of myopia and tapetoretinal degeneration. All but one patient had moderate or severe myopia ranging from −2 to −11.5 diopters. One fourth of the patients had exotropia. Signs of slight or moderate, slowly progressive nightblindness and visual field restriction were found in nearly all of the mentally retarded patients when the parents or nursing staff were specifically questioned. In the majority, visual acuity was reduced; even after correction of refraction it ranged from nearly normal to finger counting at 0.5 m.

The most typical finding was retinochoroidal dystrophy with a bull's-eye macular lesion (Fig. 190-1,D–F). It started at the periphery and showed patches of varying sizes with choroidal atrophy and pigmentary dystrophy of the retina. Bone spiculelike pigment was absent in the youngest patients and later was present in varying degrees. The retinochoroidal dystrophy was most severe in the oldest patients, whereas it could not be detected ophthalmoscopically in young ones. On the other hand, the electroretinogram (ERG) was isoelectric in all patients. Thus an ERG should be included in the examination of all patients suspected of having Cohen's syndrome. Because the patients are very cooperative despite their mental retardation, in gentle hands the examinations can usually be managed without anesthesia.

Not even our oldest patients (46 years old) were blind; they were all able to move about in their own familiar surroundings.

Discussion

The diagnosis of Cohen's syndrome is easy and certain, at least in a typical Finnish case, especially if sufficient ophthalmic investigations, including ERG, are conducted. On the other hand, many of the symptoms and signs are common as such, which may easily lead to overdiagnosis.

Ocular manifestations are essential in the delineation of Cohen's syndrome. The problem of possible heterogeneity arises easily: are there at least two entities, one with and one without retinochoroidal dystrophy? Most case reports from outside Finland mention nonspecific or ill-documented ocular findings, such as mottled pigmentation of the fundus, or no findings at all. An exception is the 1986 report by Resnick et al.,[3] who, without knowing about our article, reported a perfect combination of "Finnish" ocular manifestations in their patient. Because taking a case history and making a proper ophthalmologic investigation of myopic, mentally retarded patients are difficult, the patient evaluations may have remained superficial or failed completely. At present the question of heterogeneity cannot be answered. For every person suspected of having Cohen's syndrome the importance of a thorough, careful, well-documented ophthalmologic investigation, including ERG and fundus photography, cannot be over-stressed. Also the already published cases should be reexamined. The common metabolic denominator for the ocular, central nervous, connective tissue, hematopoietic, and structural findings is as yet unknown.

References

1. Cohen MM Jr, Hall BD, Smith DW, et al. A new syndrome with hypotonia, obesity, mental deficiency, and facial, oral, ocular, and limb anomalies. *J Pediatr.* 1973;83:280-284.
2. Norio R, Raitta C, Lindahl E. Further delineation of the Cohen syndrome: report on chorioretinal dystrophy, leukopenia, and consanguinity. *Clin Genet.* 1984;25:1-14.
3. Resnick K, Zuckerman J, Cotlier E. Cohen syndrome with bull's eye macular lesion. *Ophthalmic Paediatr Genet.* 1986; 7:1-8.

Chapter 191

Prader-Labhart-Willi Syndrome

RICHARD ALAN LEWIS

The Prader-Labhart-Willi syndrome (PLWS) is a complex congenital syndrome characterized by infantile hypotonia with feeding problems in early infancy; hyperphagia and obesity developing between ages 1 and 5 years; hypogonadism; intellectual impairment and mental retardation; and relatively short stature with proportionately small hands and feet.[1,2]

PLWS was first described in 1956.[3] In their initial nine patients, the investigators recognized all the defining features that are still considered characteristic of this syndrome, although more than 600 cases have been described subsequently. It is the most common dysmorphic form of human obesity, having a reported prevalence of 1 in 10,000 live births.[2] Approximately 95% of the reported individuals are white and of European descent. The paucity of Blacks and Orientals may represent ascertainment and reporting bias rather than differential occurrence.

Most cases of PLWS occur sporadically within families, although recurrence in sibships and concordance in monozygotic twins have been recorded. In 1976, Hawkey and Smithies reported the first well-documented PLWS associated with a 15/15 chromosomal translocation.[4] Subsequently, Ledbetter et al., using high-resolution prometaphase banded karyo-types, demonstrated the deletion of chromosome 15q11.2-q13 in more than 50% of PLWS patients.[5,6] Other cytogenetic studies of PLWS patients have demonstrated, in descending order of frequency: normal chromosomes; translocation of 15p/proximal 15q; apparently balanced 15/15 translocation; marker chromosome 15 with proximal 15q trisomies and tetrasomies; and mosaic 15q11-q13 interstitial deletion. The reported incidence of chromosomal abnormalities in PLWS subjects varies from 60 to 100%, depending on the patient population, the clinical criteria for diagnosis, and the cytogenetic techniques.[7] The paternal chromosome 15 has been identified as the source of the defective chromosome.[8] Rearrangements of proximal chromosome 15q may produce variable gene expression as a result of position effects or submicroscopic deletions, even in patients with apparently normal chromosomes. On the other hand, an autosomal recessive mode of inheritance has been suggested to account for PLWS individuals who have normal chromosomes.

The recurrence risk for PLWS in siblings has been estimated empirically at 1.5 to 3%; however, all PLWS individuals should undergo prometaphase chromosomal analysis of at least 30 chromosomal spreads (Fig. 191-1). If chromosomal rearrangement is identified, pa-

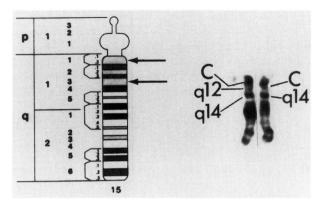

FIGURE 191-1 Prometaphase karyotype of chromosome 15q11-q13 deletion (*right*) compared to normal chromosome 15. Idiogram is included for reference. (Courtesy of David H. Ledbetter, Ph.D.)

rental chromosomes should be analyzed, in a search for a balanced translocation carrier, which would clearly alter the risk of recurrence in subsequent offspring.

Systemic Manifestations

Hypotonia

Reduced fetal activity is noted in 90% of pregnancies, and breech presentation occurs in 25 to 35% of vaginal births. This antenatal history appears to arise from hypotonia, which in the neonatal period complicates feeding with inadequate suck reflex.

Hypotonia is generalized and apparently superspinal or central in origin. Most patients experience feeding difficulty and failure to thrive. Birth weight generally is below the 50th percentile, and frequently below the 3rd percentile until 6 to 12 months of age.

Obesity

Obesity becomes apparent between age 6 months and 5 years (average 2 years). Fat is centrally distributed on the trunk, buttocks, and thighs, with relative sparing of the arms and legs and permanent sparing of the hands and feet. Hyperphagia appears to be the proximal cause of obesity; most patients behave as though they have no sense of satiety. Complications include Picwickian somnolence and hypoventilation with right-sided heart failure, and non–insulin dependent diabetes mellitus with onset between age 3 and 16 years.

Management of the obesity is predominantly dietary, frequently necessitating rigid environmental controls, including locking cupboards and refrigerators, educating immediate and extended family, and teaching behavior modification techniques that encourage self-

control, especially prior to age 6 years. Since the non–insulin dependent diabetes mellitus is associated with the obesity, it resolves with adequate weight loss.

Hypogonadism

Both small penis and cryptorchidism occur in virtually all male PLWS subjects. It is unclear whether these anomalies are the result of intrinsic testicular failure or of lack of central stimulation by gonadotropins. In females, the detection of small clitoris and hypoplastic labia are extremely subjective and are often overlooked; therefore the frequency of such variations is unknown. Sexual maturation in both males and females is delayed and incomplete; sparse axillary and pubic hair develops in adolescence. Menstruation occurs in approximately 40% of PLWS females. Exogenous testosterone supplementation for micropenis and gonadotropin for cryptorchidism in males have been suggested for the latter part of the 2nd decade of life.

Intelligence and Development

PLWS individuals are typically described as happy, affectionate, placid, and friendly infants who experience delayed psychomotor development. Crawling typically occurs between 16 and 17 months, walking alone at age 28 months, and speech is delayed until after 40 months of age. Some of the developmental delay has been ascribed to hypotonia, which is not a sufficient explanation for delay in language and social skills. The mean IQ for PLWS subjects is between 60 and 65, but approximately 12% have "normal" intellect, and 30% "borderline." It is estimated that 40% are mildly retarded and another 12 to 15%, moderately retarded.

Stature

Short stature is a cardinal feature of PLWS, reported in 90 to 95% of affected individuals in most series. Height is below the 50th percentile (corrected for midparental height) and frequently less than the 10th percentile for age. The average adult height is 61 inches for males and 59 inches for females. No specific deficiency of growth hormone or other recognizable cause has been established.

From midchildhood, some 90% of PLWS patients are characterized as having small hands and feet, which appearance is further emphasized by the relative sparing of hands and feet from the obesity. Fat distribution often appears to end at the ankle or wrist, giving rise to a "Rubensesque" appearance.

Hypopigmentation

In 1982, Hittner et al., described nine PLWS patients with cutaneous hypopigmentation and low hair bulb tyrosinase activity.[9] Subsequent studies suggest that approximately one half of PLWS subjects have both cutaneous and ocular hypopigmentation, including the presence of Type I or Type II skin, the lightest skin in the family by history, and iris transillumination defects, with or without nystagmus and reduced visual acuity. The presence of hypopigmentation was correlated with a small interstitial deletion of proximal chromosome 15q, although the same deletion was found in some persons who do not have hypopigmentation.[10]

These observations suggest the presence of a locus on proximal 15q that has a role in the dermal pigment system, although there is no evidence for a defect in neural crest migration.

Ocular Manifestations

Strabismus occurs in approximately 63% of PLWS individuals and is approximately equally distributed between hypopigmented and normally pigmented subjects.[10] Strabismus is the nonaccommodative, or "infantile," type.

Most PLWS patients have blue irides, and all reported Caucasian patients with chromosome deletion have blue irides; about three fourths of patients without visible deletions have blue irides and one fourth have hazel-green to brown ones. Iris transillumination defects and nystagmus, especially accentuated on lateral gaze, occur more frequently in patients with chromosome 15q deletion. Although their skin is hypopigmented, it does not fulfill the defining criteria for albinism.

Misrouting of fibers with excessive crossing in the optic chiasm from the temporal retina to the contralateral geniculate nucleus and then on to the visual cortex has been well documented.[11] Presumably, the strabismus results from unequal visual stimuli from each eye into the brain, but whether this represents a direct effect of hypomelanosis or a secondary defect from a timing error in ontogeny remains unresolved. Approximately one fourth of PLWS patients develop myopia and require refractive correction. Retinal pigment epithelial hypopigmentation and hypoplasia of the fovea occasionally occur; they are more apparent in patients with chromosome 15q deletion.

Differential Diagnosis

Because of the relatively mild dysmorphic features, the differential diagnosis of PLWS in infancy largely revolves around the diagnosis of hypotonia, including neuromuscular diseases, brain injuries, cerebral malformations, and chromosomal defects. In adolescents, the Laurence-Moon-Bardet-Biedl (LMBB) syndrome (see Chap. 188), an autosomal recessive disorder, shares with PLWS mental retardation, truncal obesity, and hypogonadism. Infantile hypotonia is not a component of the Bardet-Biedl syndrome, which, however, has retinitis pigmentosa, postaxial hexadactyly, and relative or absolute macrocephaly, none of which occurs in PLWS.

References

1. Cassidy SB. Prader-Willi syndrome. *Current Problems in Pediatrics.* 1984;14:1-55.
2. Holm VA. The diagnosis of Prader-Willi syndrome. In: Holm VA, Sulzbacher S, Pipes PI, eds. *The Prader-Willi Syndrome.* 1981.
3. Prader A, Labhart A, Willi H. Ein Syndrom vol Adipositas, Kleinwuchs, Kryptochismus und Oligophrenie nach myatonieartigem Zustand im Neugeborenenalter. *Schweiz Med Wochenschr.* 1956;86:1260-1261.
4. Hawkey CJ, Smithies A. The Prader-Willi syndrome with a 15/15 translocation. *J Med Genet.* 1976;13:152-163.
5. Ledbetter DH, Riccardi VM, Airhart SD, et al. Deletion of chromosome 15 as a cause of the Prader-Willi syndrome. *N Engl J Med.* 1981;304:325-329.
6. Ledbetter DH, Mascarello JI, Riccardi VM, et al. Chromosome 15 abnormalities and the Prader-Willi syndrome: a follow-up report of 40 cases. *Am J Human Genet.* 1982;34:278-285.
7. Labidi F, Cassidy SB. A blind prometaphase study of Prader Willi Syndrome: frequency and consistency in interpretation of deletion 15q. *Am J Human Genet.* 1986;39:452-460.
8. Butler MG, Palmer CG. Parental origin of chromosome 15 deletion in Prader-Willi syndrome. *Lancet.* 1983;1:1285-1286.
9. Hittner HM, King RA, Riccardi VM, et al. Oculocutaneous albinoidism as a manifestation of reduced neural crest derivatives in the Prader-Willi syndrome. *Am J Ophthalmol.* 1982;94:328-337.
10. Wiesner GL, Bendel CM, Olds DP, et al. Hypopigmentation in the Prader-Willi syndrome. *Am J Hum Genet.* 1987;40:431-442.
11. Creel DJ, Bendel CM, Wiesner GL, et al. Abnormalities of the central visual pathways in Prader-Willi syndrome associated with hypopigmentation. *N Engl J Med.* 1986;314:1606-1609.

Chapter 192

Progeroid Syndromes

Werner's, Hutchinson-Gilford, Cockayne's, Rothmund's Syndromes

GEORGE R. BEAUCHAMP

Werner's, Hutchinson-Gilford, Cockayne's, and Rothmund's syndromes display clinical features commonly observed in elderly persons. These and similar variants are known as progeroid syndromes.[1] Studies of cell biologic processes in these diseases should provide insight into mechanisms of premature senescence in certain cell systems and diseases. Among genetic progeroid syndromes of man, Martin[2] classifies syndromes as unimodal or segmental (Table 192-1); progeroid syndromes are most likely to be segmental (ie, where multiple aspects of senescent phenotype are involved). Their ocular manifestations are summarized in Table 192-2.

Werner's Syndrome

Werner's syndrome, a rare autosomal recessive condition of young adulthood, has been studied at the cellular level; laboratory investigations relate disorders of extracellular components and genetic alterations to clinical observations. Therefore, Werner's syndrome has become an important model for linking basic cell biologic inquiry with clinical disease.

The relationship of Werner's syndrome to aging is illustrative, and it remains a matter of some controversy. Some authors suggest the syndrome is an acceleration of the aging process; others are doubtful. Epstein summarizes these discussions and concludes, "There is no substantial evidence to support and considerable evidence to oppose equating the Werner's syndrome with aging,"[3] suggesting instead that the disease is a caricature of aging. Despite the controversy, Werner's syndrome is a disease worthy of study as a genetic and biologic model for degenerative diseases, many of which are associated with aging. However, "Forcing the Werner's syndrome into the mold of premature aging does nothing to enhance its value as a subject for study

and, . . . serves only to detract from . . . investigations."[3]

Clinical Presentation

Twelve principal clinical characteristics of Werner's syndrome were described by Thannhauser[4] and summarized by Salk[5]: "(1) shortness of stature in a characteristic habitus (thin extremities with stocky trunk); (2) premature graying of the hair; (3) premature baldness; (4) patches of apparently stiffened skin, particularly in the face and lower extremities ("scleropoikiloderma"); (5) trophic ulcers of the legs; (6) juvenile cataracts; (7) hypogonadism; (8) tendency to diabetes; (9) calcification of blood vessels; (10) osteoporosis; (11) metastatic calcification; and (12) tendency to occur in siblings."[5] Additional features noted subsequently by several authors include thin, high-pitched or hoarse voice; laryngeal abnormalities; increased incidence of neoplasia; flat feet; hyperreflexia; irregular dental development; early cessation of growth; excessive urinary excretion of hyaluronic acid; and premature death. Growth typically ceases at age 13 and stigmata of aging are noted particularly in the twenties. The average age of death is 47, the principal causes being malignancy and myocardial or cerebral vascular accidents.

The principal ocular finding of Werner's syndrome is early onset (twenties or thirties) posterior subcapsular cataracts. Other eye findings are reviewed by Bullock[6] and Petrohelos[7]: proptosis, keratoconjunctivitis, blue scleras, nystagmus, cloudy corneas, presbyopia and astigmatism, telangiectasia of the iris, senile macular degeneration, chorioretinitis, and retinitis pigmentosa. An increased incidence of postoperative complications leads several authors to advise caution in recommending and performing cataract surgery on patients with Werner's syndrome. Complications include delay or failure of wound healing, wound dehiscence, and cor-

TABLE 192-1 Selected Criteria of Genetic Progeroid Syndromes of Man*

Criteria	Unimodal Syndromes†	Segmental Syndromes‡
Increased susceptibility to one or more types of neoplasms of relevance to aging	Porokeratosis of Mibelli	Werner's
Premature graying or loss of hair	Prematurely white hair	Progeria
Increased deposition of lipofucsin pigments	Neuronal ceroid lipofuscinosis (Parry type)	? Cockayne's
Degenerative vascular disease	Amyloidosis, cerebral, arterial	Werner's
Cataracts	Cataract, nuclear total	Werner's
Regional fibrosis	Antitrypsin deficiency of plasma with chronic obstructive pulmonary disease	Werner's
Variations in amount and/or distribution of adipose tissue	Adiposis dolorosa	Progeria

* (Modified from Martin.[2])
† Unimodal: predominantly one aspect of senescent phenotype involved
‡ Segmental: multiple aspects of senescent phenotype involved

neal complications including decompensation with opacification.[8] A defect in wound healing associated with a biochemical abnormality in collagen and matrix metabolism has been hypothesized. While diabetes mellitus is a common accompaniment of the syndrome, diabetic retinopathy has not been a significant complication.

Genetic Aspects

The most compatible inheritance pattern for Werner's syndrome is autosomal recessive, which is suggested by several features: analysis of sibships using simple sib and truncate ascertainment loss methods, increased incidence of consanguinity rates in affected families, a

sex ratio of 1:1, and absence of birth order effect.[5] Further, heterozygote manifestations are notably absent, with the possible exceptions of increased rate of cancer deaths and early graying of hair. The homozygote frequency has been estimated to be between 1 and 22 cases per million population; in Japan, the frequency has been estimated to be between 3 and 45 cases per million population.

Chromosomal alterations in Werner's syndrome reveal a pattern of pseudodiploidy with multiple, variable, stable chromosome rearrangements that are clonal (variegated translocation mosaicism).[5] Although every chromosome has been observed to be involved, certain sites called "hot spots" have been observed to demonstrate an increased number of identifiable break points; chief among them are 1q12, 1q44, 5q12, and 6 cen. The previously observed relationship between site 1q12 and neoplasia in non-Werner's patients may be relevant.

The relationship between chromosomal aberrations and reduced growth potential may be related to chromosomal instability; losses of genetic material and changes in gene regulation due to position effects may result in balanced rearrangements, the effects of which may be reduced growth. However, "It is possible that the relationship between chromosome aberrations and reduced growth potential is not one of cause and effect, but rather that they are two distinct manifestations of a single underlying genetic defect."[5]

In summary, the genetic characteristics of Werner's syndrome are autosomal recessive inheritance pattern, chromosomal instability, and increased incidence of neoplasia. This pattern is not unique to Werner's syndrome; for example, it is also characteristic of ataxia telangiectasia and xeroderma pigmentosum.

Laboratory Studies

Werner's syndrome offers opportunity as a system for the investigation of a number of age-related pathologies. Clinical and laboratory features suggest impairment in the development and metabolism of connective tissue: (1) atrophy of connective tissue, musculature, and fat, (2) development of hyperkeratoses, (3) mesenchymal tumors such as fibroliposarcomas, osteogenic sarcomas, and meningiomas, and (4) excessive excretion of hyaluronic acid in urine.[9] Connective tissue is principally formed of collagen, proteoglycans, and glycoproteins. A number of micromorphologic changes suggest defective collagen synthesis: elevated collagen content of the skin, variable collagen fibril size, fraying of collagen fibrils, and incompletely processed collagen from cultured fibroblasts of Werner's patients.

Proteoglycans, important links between extracellular and intracellular compartments, have been implicated as well. Proteoglycans form a major portion of the extracel-

TABLE 192-2 Ocular Features of Progeroid Syndromes

	Werner's	Hutchinson-Gilford	Cockayne's	Rothmund's
Cataract	++	−	+	++
Defective wound healing	++	−	−	−
Retinopathy	+	−	++	−
Involutional macular degeneration	+	−	−	−
Chorioretinitis	+	−	−	−
Optic atrophy	−	−	+	−
Nystagmus	+	−	+	−
Corneal degeneration	+	−	+	+
Blue scleras	+	−	−	−
Keratoconjunctivitis	+	−	−	−
Proptosis	+	−	−	−
Loss of eyebrows and lashes	−	+	−	−
Iris telangiectasia	+	−	−	−
Iris hypoplasia	−	−	+	−
Pupillary abnormalities	−	−	+	−
Vitreous degeneration	−	−	+	−
Ametropia	+	−	+	−

Key: ++, core, prominent feature; +, reported; −, not reported.

lular matrix of cartilage and contribute to its elasticity. Reductions in chondroitin sulfate species through time are balanced by a rise in keratan sulfate; and the relative proportion of chondroitin sulfate 4 decreases as chondroitin sulfate 6 increases. Similarly, proportions of glycosaminoglycans (GAG; both hyaluronic acid and dermatan sulfate) steadily decrease. Heparan sulfate increases on cell surfaces with time, suggesting cellular aging may be regulated by cell surface heparan sulfate. Tissue hyaluronic acid (HA) content rises with age, and may represent a normal characteristic of aging.

Skin samples from Werner's syndrome patients reveal an absolute increase in hexosamines (perhaps reflective of connective tissue replacing subcutaneous fat), marked decrease in dermatan sulfate, and reduction in hyaluronic acid. Studies of urinary excretion of GAG reveal an increased proportion of hyaluronic acid secretion. Cultured skin fibroblasts show an increased accumulation of hyaluronic acid and sulfated GAG in intracellular, pericellular, and extracellular material. Such studies of the synthesis, characteristics, and structural modifications of proteoglycan molecules in pathologic states such as Werner's syndrome offer hope for reciprocal understanding of normal aging.

In summary, Werner's syndrome patients offer opportunities to study the biology of a number of pathologic conditions, some of which are characteristic of the

aging process. Cultured mesenchymal cells exhibit accelerated aging in vitro. Chromosomal mutations are frequent, and alterations of neuroendocrine function are common. This suggests an inappropriate gene action in the domain of development, perhaps a genetically defective pool of stem cells; mesenchymal cells would appear to be the principal target. Understanding these processes may be of major importance in understanding certain inborn errors (known as chromosomal instability syndromes), that pose increased risks for diverse neoplasms. The molecular lesion may represent a defect in enzyme function(s) more characteristic of autosomal recessive disease. The principal interest of these investigations to ophthalmologists will relate to the cell biology of cataract formation.

Hutchinson-Gilford Progeria Syndrome

The Hutchinson-Gilford syndrome (or progeria) shares little clinically with Werner's syndrome, yet common cytochemical defects suggest similar pathophysiologic mechanisms. Progeria is a rare genetic disease with an estimated incidence of approximately 1 in 8 million. While patients generally appear normal at birth, growth retardation presents at approximately 1 year of age.[10]

Clinical Presentation

Features seen in the first few years of life include balding, loss of eyebrows and eyelashes, pervasive loss of subcutaneous tissue, prominent veins over the scalp, and aged-looking skin. Extreme shortness of stature and thinness result in average height of 40 inches and weight no more than 25 or 30 pounds, a very low weight-to-height ratio. Additional features include a thin, high-pitched voice; absence of sexual maturation; and a typical facial appearance resembling a plucked bird, with prominent eyes and beaked nose. Facial disproportion is the result of a small jaw and large cranium. The combination gives the aged appearance. Bones show frequent resorption with fibrous replacement of the clavicles, resorption of the terminal finger bones, stiffening of finger joints, elbow and knee enlargement, and coxa valga, often accompanied by aseptic necrosis of the head of the femur and hip dislocation.[11]

Intelligence is normal to above average. The median age of death is 12 years, most often as a result of myocardial infarction or congestive heart failure. Atherosclerosis is widespread with interstitial fibrosis of the heart. The thymus gland is often markedly enlarged; however, tumors, cataracts, diabetes, and hyperlipidemia (all often associated with normal aging) are conspicuously absent.

Genetic Aspects

Progeria apparently results from a sporadic dominant mutation. The lack of consanguinity in family trees, incidence between 1 and 4 per million births, increased paternal age effect yielding a secondary age peak, and the low proportion of affected sibs argue against a recessive condition.[11]

Laboratory Studies

Laboratory research in progeria has yielded inconsistent results.[10-12] Abnormally increased thermolabile enzymes in progeria fibroblasts, altered immune function by a greatly reduced HLA cell surface molecule concentration, abnormalities of DNA repair capacity after x-ray damage, and decreased binding of insulin to nonspecific receptors have been reported; their significance is unclear. Tissue hyaluronic acid (HA) content and excretion are elevated in progeria (and in Werner's syndrome) and may represent a potentially unique marker. Concomitantly, as in Werner's syndrome, hyaluronic acid elevation is accompanied by a pronounced alteration of GAG production (ie, non–HA-containing GAGs decrease as a function of cell density). These findings suggest that initial synthesis is relatively unimpaired, but that a degradation pathway abnormality follows. The relationship of hyaluronic acid and GAG production is potentially important at several levels. First, total GAG production is greater in old than in young people, possibly owing to an abnormal level of synthetase. Both HA and GAG are considered important mediators in morphogenesis during embryogenesis. HA in particular is related to elaboration of the primary mesenchyme. In addition, there is "a striking correlation between hyaluronic synthesis and cell movement and proliferation"[11] related to HA degradation and differentiation. Finally, HA apparently acts as an antiangiogenesis factor during embryogenesis. It may be expected to play "an equally important role as an antiangiogenesis factor during maturation and aging."[11]

The accumulation of abnormal or modified proteins in fibroblasts and aging cells reflects abnormal accumulations of "defective enzymes." These may result from the diminished ability to respond to environmental stresses, although their relationship vis à vis cause and effect is not known.[5]

Progeria is a disease genetically most consistent with a sporadic dominant mutation with elevated levels of HA excretion (like Werner's syndrome). Cultured cells accumulate excessive HA and one may hypothesize, as did Brown et al.,[11] that failure of patients with progeria (and Werner's syndrome) may be the result of inadequate vasculogenesis related to excessive HA. The clinical features that support this hypothesis include the presence of scleroderma, decreased numbers of blood vessels, and increased rate of cardiovascular deaths.[13-16]

Cockayne's Syndrome

Cockayne's syndrome is clinically characterized by dwarfism, skeletal and neurologic abnormalities, cutaneous photosensitivity, and pigmentary degeneration of the retina.[17] The mechanism by which these changes occur is quite different from that proposed for progeria and Werner's syndrome. Cultured fibroblasts demonstrate hypersensitivity to lethal effects of ultraviolet radiation. Therefore, the inherited defect appears to be one of repair of DNA damaged by ultraviolet radiation. Certain other diseases, notably xeroderma pigmentosum, apparently share this defect.

Clinical Characteristics

Affected persons appear quite normal at birth. Severe cachectic dwarfism usually presents first between 6 and 12 months of age. The facies is characterized by a thin, prominent nose, sunken eyes, prognathism, and lack of subcutaneous fat. Fine hair and anhidrosis are commonly noted. Males appear sexually infantile, with undescended testes. Females have oligomenorrhea and incomplete development of secondary sexual characteristics, including small breasts. Acute sun sensitivity, manifested by photosensitivity dermatitis, is characterized by desquamation and scarring. Despite laboratory similarities to xeroderma pigmentosum, clinical features such as pigmentary and sunlight–induced neoplasms do not occur in Cockayne's syndrome. Musculoskeletal features include disproportionately increased hand and limb size contrasted with a small trunk. Radiographically, tarsal and carpal bones are enlarged; abnormalities of vertebral bodies, clavicles, and ribs may be present; and increased density of the skull bones is observed.

Microcephaly is apparent in the 2nd and 3rd years of life. Growth retardation antedates mental deterioration, accompanied by normal-pressure hydrocephalus. Other neurologic abnormalities such as sensorineural hearing loss, ataxia, choreoathetosis, spasticity, myoclonus, and gait disturbance have all been noted. Neuropathologically, cerebellar atrophy, demyelination, and pericapillary calcification are found in the basal ganglia, cortex, cerebrum, and cerebellum.[17]

A salt-and-pepper retinopathy (with waxy optic atrophy and narrowed retinal vessels) is perhaps the most consistent ocular feature of Cockayne's syndrome. Electroretinograms may be low to normal in amplitude. Optic atrophy, heterogeneous cataract, band keratopathy, nystagmus, pupillary unresponsiveness, hypermetropia, anhidrosis, irregular pupils, vitreous floaters, and hypoplastic irides have all been observed.[18,19]

Genetic Features

Cockayne's syndrome is a rare autosomal recessive disease in which genetic heterogeneity has been demonstrated.[17] Utilizing the rate of semiconservative DNA synthesis after ultraviolet radiation, three genetic complementation groups have been identified (types A, B, and C). The significance of these findings remains to be determined.

Laboratory Studies

Skin fibroblasts cultured from patients with Cockayne's syndrome and treated with ultraviolet radiation show lower survival than normal fibroblasts. Similar low survivability is also noted after treatment with agents that mimic ultraviolet–radiation. Sister chromatid exchanges increase abnormally after exposure to ultraviolet radiation as well. Host cell reactivation studies show decreased host cell reactivation, indicating DNA repair in the cells was defective. DNA synthesis is markedly delayed compared with the temporary depression in the rate of semiconservative DNA synthesis in normal cell lines.[18,20]

These observations suggest that in Cockayne's syndrome there is accumulated damage in neuronal DNA that cannot be repaired. Therefore, death of photoreceptor cells, myelin-producing cells, and fat cells results from their inability to repair DNA damaged by certain intracellular metabolites. Since these cells share the feature of high intracellular lipid content, DNA damage leading to cell death may result from metabolites involved in the synthesis or degradation of lipids. Clinically, the retinitis pigmentosa–like picture, demyelinization, and cachexia may be related to such laboratory defects.[21,22]

DNA is similarly defective in other syndromes, including xeroderma pigmentosum, ataxia telangiectasia, Fanconi's anemia, and Bloom's syndrome.[17,20-22] These conditions should be considered in a differential diagnosis of progeroid syndromes, as certain clinical features may mimic those discussed. In addition, they represent important biological models for understanding mechanisms of DNA repair in human cells.

Rothmund's Syndrome

Rothmund[23] first described the syndrome that bears his name in 1868. There are three principal features: dermatosis, cataract, and hypogenitalism.

Clinical Features

Rothmund's syndrome begins with the appearance shortly after birth of large, ill-defined areas of skin erythema. The process ultimately leads to atrophy and pigmentary changes with telangiectasia. The pigmentary changes are alternately hypopigmented and hyperpigmented, and they are aggravated (as in Cockayne's syndrome and xeroderma pigmentosum) by exposure to light. Most patients are dwarfs with dystrophies of nails and teeth, loss of hair, small hands and fingers, and hypogonadism. Cataracts usually appear between the 2nd and 6th year of life; they are bilateral and develop rapidly.[24,25] Francois[26] has described degenerative lesions of the cornea.

Heredity

The condition is considered a rare autosomal recessive syndrome with approximately 70% of patients being female.

Laboratory Studies

To date laboratory studies have not suggested a basic pathophysiologic mechanism.

Conclusion

It has been suggested that longevity is encoded by a limited number of genes, perhaps as few as 20 to 50.[11] The genetic basis of aging may therefore be studied by looking at specific genetic mutants that appear to affect the aging process. Several mutations appear to accelerate many but not all features of the aging process. The Werner and Hutchinson-Gilford syndromes (progeria of adulthood and progeria of childhood, respectively) are two genetic diseases with features strikingly suggestive of accelerated aging. Their basic mutations are not known; however, patients with both diseases appear to excrete an excessive amount of hyaluronic acid. The final common pathway for production of disease may be the result of inhibition of vascular development by the excessive hyaluronic acid. The mechanism for Cockayne's syndrome is apparently quite different; defective repair of DNA damaged by ultraviolet irradiation seems to account for observed clinical features.

References

1. Martin GM. Genetics and aging: the Werner syndrome as a segmental progeroid syndrome. *Adv Exp Med Biol.* 1985;190:161-170.
2. Martin GM. Syndromes of accelerated aging. *Natl Cancer Inst Monogr.* 1982;60:241-247.
3. Epstein CJ. Werner's syndrome and aging: a reappraisal. *Adv Exp Med Biol.* 1985;190:219-228.
4. Thannhauser SJ. Werner's syndrome (progeria of the adult) and Rothmund's syndrome: Two types of closely related heredofamilial atrophic dermatoses with juvenile cataracts and endocrine features: a critical study with five new cases. *Ann Intern Med.* 1945;23:559-626.
5. Salk D. Werner's syndrome: a review of recent research with an analysis of connective tissue metabolism, growth control of cultured cells, and chromosomal aberrations. *Hum Genet.* 1982;62:1-15.
6. Bullock JD. Werner syndrome. *Arch Ophthalmol.* 1973;90:53-56.
7. Petrohelos MA. Werner's syndrome. A survey of three cases, with review of the literature. *Am J Ophthalmol.* 1963;56:941-953.
8. Jonas JB, Ruprecht KW, Schmitz-Valckenberg P, et al. Ophthalmic surgical complications in Werner's syndrome: report on 18 eyes of nine patients. *Ophthalmic Surg.* 1987;18(10):760-764.
9. Bryant E, Salk D, Wight T. Proteoglycans in the Werner syndrome and aging: a review and perspective. *Adv Exp Med Biol.* 1985;190:553-565.
10. Brown WT, Kieras FJ, Houck GE, Jr, et al. A comparison of adult and childhood progerias: Werner syndrome and Hutchinson-Gilford progeria syndrome. *Adv Exp Med Biol.* 1985;190:229-244.
11. Brown WT, Zebrower M, Kieras FJ. Progeria, a model disease for the study of accelerated aging. *Basic Life Sci.* 1985;35:375-396.
12. Gracy RW, Chapman ML, Cini JK, et al. Molecular basis of the accumulation of abnormal proteins in progeria and aging fibroblasts. *Basic Life Sci.* 1985;35:427-442.
13. Ishii T. Progeria: autopsy report of one case, with a review of pathologic findings reported in the literature. *J Am Geriatr Soc.* 1976;24(5):193-202.
14. Shozawa T, Sageshima M, Okada E. Progeria with cardiac hypertrophy and review of 12 autopsy cases in the literature. *Acta Pathol Jpn.* 1984;34(4):797-811.
15. Ogihara T, Hata T, Tanaka K, et al. Hutchinson-Gilford progeria syndrome in a 45-year-old man. *Am J Med.* 1986;81:135-138.
16. Baker PB, Baba N, Boesel CP. Cardiovascular abnormalities in progeria. *Arch Pathol Lab Med.* 1981;105(7):384-386.
17. Otsuka F, Robbins JH. The Cockayne syndrome—an inherited multisystem disorder with cutaneous photosensitivity and defective repair of DNA. *Am J Dermatopathol.* 1985;7(4):387-392.
18. Levin PS, Green R, Victor DI, et al. Histopathology of the eye in Cockayne's syndrome. *Arch Ophthalmol.* 1983;101:1093-1097.
19. Lewis JM. Cockayne's syndrome. *Clin Pediatr.* 1987;26(3):156. Letter.
20. Schwaiger H, Hirsch-Kauffmann M, Schweiger M. DNA repair in human cells: in Cockayne syndrome cells rejoining of DNA strands is impaired. *Eur J Cell Biol.* 1986;41:352-355.
21. Cleaver JE. DNA repair and replication in xeroderma pigmentosum and related disorders. *Basic Life Sci.* 1986;39:425-438.
22. Schweiger M, Auer B, Burtscher HJ, et al. DNA repair in human cells: Biochemistry of the hereditary diseases Fanconi's anaemia and Cockayne syndrome. *Biol Chem Hoppe Seyler.* 1986;367:1185-1195.
23. Rothmund A. Ueber cataracten in Verbindung mit einer eigenthumlichen hautdegeneration. Graefes Arch Klin Exp Ophthalmol. 1868;14:159.
24. Spaeth G, Nelson LB, Beaudoin AR. Ocular teratology. In: Duane TD, Jaeger EA, eds. *Biomedical Foundations of Ophthalmology.* vol. 1. Philadelphia, Pa: Harper & Row; 1983;51.
25. Morris DA. Cataracts and systemic disease. In: Duane TD, Jaeger EA, eds. *Clinical Ophthalmology.* vol. 15. Philadelphia, Pa: JB Lippincott; 1986;5.
26. Francois J. *Heredity in Ophthalmology.* St. Louis, Mo: CV Mosby; 1961;641-643.

Chapter 193

Rubinstein-Taybi Syndrome

NORMAN N. K. KATZ, KENNETH N. ROSENBAUM, and
MICHELE R. FILLING-KATZ

In 1963, Rubinstein and Taybi[1,2] reported a similar constellation of findings in seven unrelated patients. The syndrome that now bears their names includes mental and motor retardation, broad thumbs and broad first toes, significant growth retardation, microcrania, characteristic facies, ocular anomalies, high-arched palate, and cryptorchidism in males.

The frequency of Rubinstein-Taybi syndrome in institutionalized persons over 5 years of age is said to be 1 in 300 to 500. The exact mode of inheritance has not been established. Although a chromosomal abnormality is possible, high-resolution cytogenetic studies have thus far not revealed any specific abnormality.[3] The history of consanguinity and incest in some cases and involvement of various members of families and of monozygotic twins suggest a genetic mechanism, perhaps autosomal recessive.[4,5]

Systemic Manifestations

The characteristic facies includes hypoplastic maxillae, narrow, high-arched palate, and beaked nose with the nasal septum extending below the alae nasi (Figs. 193-1 to 193-4). The auricles may be low set and malformed. Abnormal dentition is frequent. Low IQ, electroencephalographic abnormalities, and gait problems may be present. In addition to the typical broad and radially angulated thumbs and first toes, skeletal involvement includes vertebral, sternal, and pelvic anomalies. A large foramen magnum may be present. Unusual dermatoglyphics with excessive dermal ridge pattern have been found in the thenar and first interdigital areas of the palm. Genitourinary, abdominal visceral, and cardiac (pulmonary stenosis) abnormalities have been described. There is great variability in the clinical presentation.

These patients exhibit feeding problems during infancy. Allergies and respiratory and urinary tract infections are common. Keloids often form in surgical scars, and cardiac arrhythmias have been noted when succinylcholine is used during anesthesia.[6] Histologic examination of muscle tissue has revealed light- and electron microscopic abnormalities suggestive of denervation atrophy.[7]

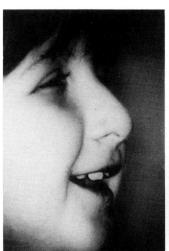
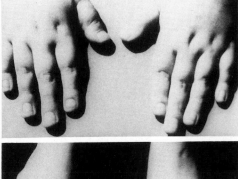

FIGURE 193-1 Composite photograph of a 7-year-old girl with Rubinstein-Taybi syndrome. The characteristic beaked nose and elongated nasal septum are seen, as are wide thumbs and wide first toes with radial angulation.

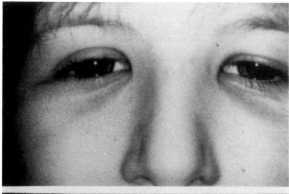

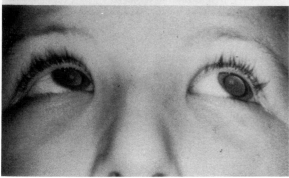

FIGURE 193-2 A composite photograph of the patient in Figure 193-1 demonstrates the various abnormalities in Rubinstein-Taybi syndrome: antimongoloid palpebral fissures, an overacting right inferior oblique muscle (exophoria was present), and abnormal dentition.

Ocular Manifestations

Ocular and adnexal involvement are common.[8] Antimongoloid slant of the palpebral fissures is characteristic. Epicanthal folds may occur. Congenital lacrimal excretory obstruction is common and was present in all three patients studied by the authors. This is probably related to structural abnormalities in the nasal and maxillary regions. Ptosis and strabismus (especially exotropia) may occur. Refractive errors are common. Micro-

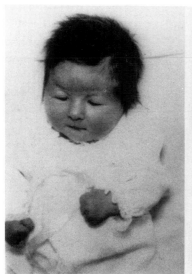

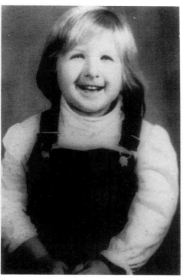

FIGURE 193-3 The patient in Figure 193-1 at birth and 3½ years later. The broad thumb is seen at birth. Characteristic facial changes have evolved during growth.

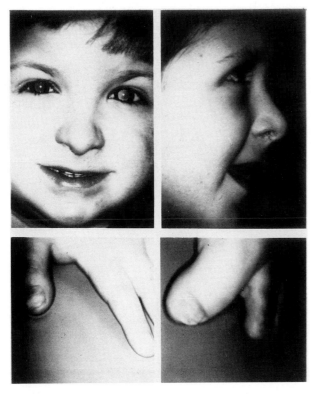

FIGURE 193-4 A 6½-year-old boy with Rubinstein-Taybi syndrome has broad thumbs and first toes, elongated nasal septum, and slight downward slant of the outer canthi.

ophthalmos, various colobomas, cataracts, and optic nerve atrophy have been described. Glaucoma is a rare complication.[9]

Diagnosis

The diagnosis is based on clinical findings. Isolated case reports describe confusion of the Rubinstein-Taybi syndrome with the congenital rubella syndrome, trisomy 13 syndrome, mandibulofacial dysostosis, and Pfeiffer's craniostenosis (another syndrome in which broad thumbs and broad first toes may be present).

Management

Management of these patients includes long-term multi-specialty, symptomatic and supportive therapy, especially physical therapy and special education. Corrective orthopedic surgery for severe skeletal problems, especially the broad, radially angulated thumbs (hitch-hiker's thumb), has been successful in restoring good function. Dental care and maxillofacial surgery are also extremely beneficial.

The majority of patients have good vision if their refractive errors are appropriately corrected. Lacrimal excretory obstruction, which is often refractory to conventional therapy, has become more manageable with the availability of silicone lacrimal intubation techniques. Other problems such as strabismus, ptosis, cataract, and glaucoma are managed by standard methods.

References

1. Rubinstein JH, Taybi H. Broad thumbs and toes and facial abnormalities. *Am J Dis Child*. 1963;105:588-608.
2. Rubinstein JH. The broad thumb syndrome—progress report 1968. *Birth Defects Original Article Series*. 1969;V(2):25-41.
3. Wulfsberg EA, Klisak IJ, Sparkes RS. High resolution chromosome banding in the Rubinstein-Taybi syndrome. *Clin Genet*. 1983;23:35-37.
4. Gillies DRN, Roussounis SH. Rubinstein-Taybi syndrome: further evidence of a genetic aetiology. *Develop Med Child Neurol*. 1985;27:751-755.
5. Baraitser M, Preece MA. The Rubinstein-Taybi syndrome: occurrence in two sets of identical twins. *Clin Genet*. 1983;23:318-320.
6. Stirt JA. Succinylcholine in Rubinstein-Taybi syndrome. *Anesthesiology*. 1982;57:429. Letter.
7. Der Kaloustian VM, Afifi AK, Sinno AA, et al. The Rubinstein-Taybi syndrome: clinical and muscle electron microscopic study. *Am J Dis Child*. 1972;124:897-902.
8. Roy FH, Summitt RL, Hiatt RL, et al. Ocular manifestations of the Rubinstein-Taybi syndrome: case report and review of the literature. *Arch Ophthalmol*. 1968;79:272-278.
9. Shihab ZM. Pediatric glaucoma in Rubinstein-Taybi syndrome. *Glaucoma*. 1984;3:288-290.

PART 19

Skin and Mucous Membrane Disorders

Section A

Benign Proliferative and Neoplastic Disorders

Chapter 194

Basal Cell Nevus Syndrome

JAN W. KRONISH and DAVID T. TSE

The basal cell nevus syndrome is an uncommon, autosomal dominant, multisystem disorder with high penetrance and variable expressivity. This disease complex, also called "Gorlin's syndrome," is characterized by the following major features: multiple nevoid basal cell carcinomas, odontogenic keratocysts of the jaw, congenital skeletal anomalies, ectopic calcification, and pits of the hands and feet. A diagnosis may be established by the presence of any two of these characteristics and is supported by a positive family history. Other associated manifestations have been described that involve almost every body system (Table 194-1).[1,2]

The inheritance pattern of basal cell nevus syndrome supports the theory that a defective gene is the underlying cause of the disorder; however, chromosome abnormalities have been found in only a small number of patients. A high spontaneous mutation rate may exist. Males and females are affected with equal frequency, and the condition is most commonly found in Caucasians.[3] Genetic counseling is an important aspect of patient management, as are frequent examinations to avoid potentially disfiguring and lethal complications.

Systemic Manifestations

Multiple basal cell carcinomas that appear early in life are one of the hallmarks of basal cell nevus syndrome. The cutaneous tumors usually present as pigmented or flesh-colored smooth, round papules between the 2nd and 3rd decade, although they may develop in the first few years of life. During childhood, such lesions remain quiescent until puberty, when they increase in number and demonstrate more rapid and invasive growth patterns. Their distribution is widespread, unlike the more common acquired basal cell carcinomas that typically develop in sun-exposed areas. The central facial region and trunk are most often affected; the scalp, neck, and extremities may also be involved. Only approximately 50% of affected patients exhibit basal cell carcinomas, which may number from a few to several hundred. All histologic types of basal cell carcinoma have been found, and their morphologic appearance is indistinguishable from those of patients who do not exhibit the syndrome. Osteoid formation and calcification may be present more frequently.[2,4]

TABLE 194-1 Basal Cell Nevus Syndrome: Common Clinical Manifestations

Skin
  Multiple basal cell carcinomas with calcification or bone or osteoid formation
  Pits of hands and feet
  Milia

Face and mouth
  Multiple jaw cysts
  Mild mandibular prognathism
  Broad nasal root
  Prominent supraorbital ridge

Skeletal anomalies
  Ribs: bifurcation, splaying, synostosis, partial agenesis, or rudimentary cervical ribs
  Bridging of sella turcica
  Vertebrae: scoliosis, cervical or upper thoracic fusion, spina bifida occulta
  Frontal and temporoparietal bossing
  Long bone cysts
  Brachymetacarpalism

Ophthalmic anomalies
  Hypertelorism
  Dystopia canthorum
  Strabismus
  Congenital blindness (coloboma of choroid and optic nerve, corneal opacities, cataracts, glaucoma)
  Myelinated nerve fibers
  Optic atrophy
  Nystagmus
  Meibomian cysts

Central nervous system anomalies
  Calcification of dura (falx, tentorium), petroclinoid ligament
  Medulloblastoma
  Mental retardation
  EEG changes

Endocrine anomalies
  Male hypogonadism
  Ovarian fibroma or cyst

Other findings
  Lymphomesenteric cysts

The potentially invasive and destructive nature of large numbers of lesions in multiple areas may lead to dysfunction of vital structures, disfigurement, and death. These issues must be addressed when planning a treatment protocol for the management of these patients. Curettage and electrocoagulation, cryosurgery, and topical chemotherapy can be used to treat small lesions. Mohs' microscopically controlled excision technique is recommended for large, invasive, or recurrent lesions. Two recent experimental modes of therapy include long-term administration of systemic isotretinoin,[5] and photoradiation therapy with hematoporphyrin derivative as a photosensitizer. It should be stressed that ionizing radiation is contraindicated because of the predisposition of such patients to malignant transformation.[6]

Jaw cysts are found in approximately 70% of patients and often appear in the first decade of life. Symptoms of pain, swelling, drainage, and displacement of teeth provide the basis of the presenting complaint in about half of the cases. The cysts arise in the mandible twice as frequently as in the maxilla, and are best imaged by panoramic radiography of the jaw. They are usually multiple and bilateral; they vary in size and can grow several centimeters, producing pathologic fractures. Histologically, these benign odontogenic keratocysts reveal a thin, corrugated, stratified squamous epithelium with varying degrees of keratinization surrounded by a thick fibrous capsule.[1,3] Surgical removal of the cysts is the treatment of choice. Radiation therapy is, again, contraindicated in view of reports of radiation-induced malignant sarcomatous degeneration.

Various developmental skeletal anomalies are common though nonspecific manifestations of the basal cell nevus syndrome. Rib anomalies, such as bifurcations, splaying, synostoses, and partial agenesis, are found in approximately half of the patients. Kyphoscoliosis constitutes the major associated spine deformity, followed in frequency by spina bifida, hemivertebrae, and incompletely segmented vertebrae. Skull anomalies include bridging of the sella turcica, broadening of the nasal root, and frontal and biparietal bossing. Brachymetacarpalism (shortened fourth metacarpals) and bone cysts may be detected by radiographic studies.[1,2]

Ectopic calcification is present in 80% of syndrome patients. Calcification of the dura, particularly of the falx cerebri, is distinctive in its lamellar appearance and is asymptomatic. Basal cell carcinomas, subcutaneous tissues, jaw cysts, lymphomesenteric cysts, ovarian and uterine cysts, and the sacrotuberous ligaments may also reveal calcification.

The development of pits of the palms and soles is a unique feature of basal cell nevus syndrome. They appear as shallow, erythematous depressions, 2 to 3 mm in diameter and consist of focal areas of defective keratinization. Histologically, the pits demonstrate partial or complete absence of the stratum corneum with underlying dilated small vessels in the upper dermis. Approximately 50% of patients are affected. Like the cutaneous tumors in this syndrome, the pits range in number from a few to several hundred, usually appear during the 2nd decade of life, and increase in frequency with age.[2]

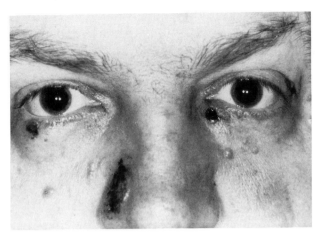

FIGURE 194-1 The typical facies of a patient with basal cell nevus syndrome demonstrates orbital hypertelorism, prominent supraorbital ridges, a broad nasal root, and multiple cutaneous basal cell carcinomas.

Ocular Manifestations

The most common and serious ophthalmic manifestation in basal cell nevus syndrome is basal cell carcinomas of the eyelids and periocular structures. Management of these lesions, with the goals of preserving normal lid function and preventing recurrence of disease, is often difficult, given the multiplicity of tumors and their aggressive growth pattern. Lid malpositions, blindness, facial nerve palsies, and death have been associated with recurrent and uncontrollable growth of periocular basal cell carcinomas. Southwick and Schwartz[7] reported on three patients who required unilateral or bilateral orbital exenterations, two of whom developed direct intracranial infiltration of tumor. Surgical excision of these lesions with microscopic monitoring of tumor margins, followed by meticulous reconstruction, is recommended for treatment of eyelid neoplasms in this syndrome.

Orbital hypertelorism and an associated lateral dis-placement of the medial canthi, referred to as dystopia canthorum, are found in the majority of patients. These anomalies, along with prominent supraorbital ridges, frontoparietal bossing, a broad nasal root, and mild mandibular prognathism, produce a characteristic facies (Fig. 194-1).

Most ocular anomalies associated with basal cell nevus syndrome are discovered early in life. Strabismus is frequently encountered and is usually an exodeviation or an esodeviation. Congenital blindness due to glaucoma, corneal opacities, cataracts, and choroidal and optic nerve colobomas have also been reported. Unilateral or bilateral myelinated nerve fibers have been found to be a feature of this syndrome. Other less common signs include meibomian cysts, optic atrophy, and various patterns of nystagmus.[2,8] Early recognition of these ocular findings allows for appropriate management in patients with this multisystem disease complex.

References

1. Gorlin RJ, Vickers RA, Kelln E, et al. The multiple basal-cell nevi syndrome. *Cancer.* 1965;18:89.
2. Gutierrez MM, Mora RC. Nevoid basal cell carcinoma syndrome: a review and case report of a patient with unilateral basal cell nevus syndrome. *J Am Acad Dermatol.* 1986;15:1023.
3. Olson RAJ, Stroncek GG, Scully JR, et al. Nevoid basal cell carcinoma syndrome: review of the literature and report of a case. *J Oral Surg.* 1981;39:308.
4. Clendenning WE, Block JB, Radde IC. Basal cell nevus syndrome. *Arch Dermatol.* 1964;90:38.
5. Peck GL, Gross EG, Butkus D. Chemoprevention of basal cell carcinoma with isotretinoin. *J Am Acad Dermatol.* 1982;6:815.
6. Tse DT, Kersten RC, Anderson RL. Hematoporphyrin derivative photoradiation therapy in managing nevoid basal-cell carcinoma syndrome. *Arch Ophthalmol.* 1984;102:990.
7. Southwick GJ, Schwartz RA. The basal cell nevus syndrome: disasters occurring among a series of 36 patients. *Cancer.* 1979;44:2294.
8. DeJong PTVM, Bistervels B, Cosgrove J, et al. Medullated nerve fibers: a sign of multiple basal cell nevi (Gorlin's) syndrome. *Arch Ophthalmol.* 1985;103:1833.

Chapter 195

Hemangiomas

BARRETT G. HAIK and BRYANT LUM

Capillary hemangiomas are common cutaneous tumors that usually occur as single lesions on the trunk or extremities and pose no more problem than a transient cosmetic defect. Unfortunately, when these tumors occur in the facial region, they may have profound ophthalmic and systemic consequences. They are not true neoplasms but hamartomatous proliferations of primitive vasoformative tissues. Ultrastructurally they consist of anastamosing vascular channels lined by a single layer of endothelial cells surrounded by pericytes.

The majority of orbital capillary hemangiomas present in the first few months of life, and one third are present at birth.[1] Pregnancy and birth history are typically normal, and there is no evidence of an associated genetic defect. The female-to-male ratio is approximately 3:2. The most frequent presentation is that of a subcutaneous tumor, which appears as a bluish purple, spongy mass underlying an apparently normal skin surface (Fig. 195-1). The second common presentation occurs when the lesion is superficial and appears on the skin (Fig. 195-2). It is called a "strawberry nevus," since the tumor is bright red and has an elevated, irregular, dimpled surface. The least frequent presentation (approximately 5% of patients) is a deep orbital lesion. Affected children typically present with proptosis alone; no discoloration and no visible or palpable lesion is evident clinically.

The tumors undergo a period of rapid growth that lasts from 3 to 12 months. It is followed by a period of stabilization, succeeded by an involuting phase that ordinarily begins by the first birthday. Involution is gradual, and in small lesions usually complete, leaving little or no cosmetic sequelae by 5 years of age. Unfortunately, some of the larger lesions, principally those with a combined subcutaneous component, may produce objectionable cosmetic sequelae and functional disabilities.

Systemic Manifestations

The ophthalmologist should be aware of the rare systemic complications that can arise in patients with large facial capillary hemangiomas. These complications may be related to infection, airway obstruction, or hematologic or cardiac disorders.

Large hemangiomas are subject to superficial ulcerations and may rarely become secondarily infected, resulting in sepsis. Hemangiomas extending into the neck may produce airway compromise from tracheal or pharyngeal obstruction or aspiration secondary to esophageal compression. A death has been reported following hemorrhage of a supraglottic hemangioma during intubation, so anesthesiologists should be forewarned to thoroughly examine patients with facial hemangiomas prior to inducing general anesthesia.[2]

Hemangioma with thrombocytopenia (Kasabach-Merritt syndrome) is a rare disorder characterized by a coagulopathy secondary to platelet or fibrinogen entrapment and consumption of other clotting factors.[3] Large ecchymoses are usually found in and around the hemangiomas, but they may occur anywhere, including the viscera. Life-threatening complications of this syndrome include sudden acute hemorrhage or hemorrhage into the lesion with resulting compression of nearby vital structures. Microangiopathic hemolytic anemia has also been noted in patients with extensive hemangiomas. Extensive hemangiomas may act hemodynamically as large arteriovenous shunts and may result in life-threatening high-output congestive heart failure.

Ocular Manifestations

Capillary hemangiomas are among the most frequently seen orbital tumors in the pediatric population. Visual disturbances are common, appearing in 50 to 75% of patients who have orbital or adnexal involvement.[4,5] Approximately 50% of patients become amblyopic secondary to stimulus deprivation, anisometropia, or strabismus. Partial or even complete occlusion of the visual axis is not uncommon when the tumor involves the eyelids. Almost all cases of occlusion for longer than 1 month will produce some degree of amblyopia. Anisometropia may be caused by astigmatism secondary to pressure from the tumor or relative myopia in the affected eye. Strabismus occurs in about one third of the patients and may be due to either direct involvement of

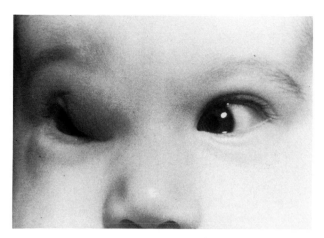

FIGURE 195-1 Subcutaneous hemangiomas of the eyelid.

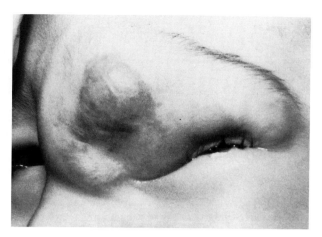

FIGURE 195-2 Superficial eyelid hemangiomas.

the muscles by tumor, by displacement of the muscle, or most commonly secondary to amblyopia.

Proptosis is found in 30 to 40% of cases. Although it is usually mild in degree, it can be severe enough to cause exposure keratitis. Other ocular complications resulting from the mass effect of the tumor and compression or displacement of orbital tissues include optic atrophy, ptosis, occlusion of the lacrimal system, and diplopia.

Treatment of many small periocular hemangiomas is not necessary. Spontaneous regression occurs, and although therapy may speed the regression of the tumor, it will not significantly affect the final cosmetic outcome. If however, a specific ocular or systemic complication exists that would be modified by more rapid tumor resolution, treatment should be implemented immediately. A number of techniques are effective for the treatment of periocular capillary hemangiomas, including radiation therapy, corticosteroid therapy, cryotherapy, diathermy, vascular occlusive techniques, and surgical excision. In our experience, the safest therapeutic approaches with the greatest benefit-to-risk ratio are local corticosteroid therapy and radiotherapy.[6,7]

References

1. Haik BG, Jakobiec FA, Ellsworth RM, et al. Capillary hemangiomas of the lids and orbit: an analysis of the clinical features and therapeutic results in 101 cases. *Ophthalmology.* 1979;86:760-789.
2. Yee RD, Hepler RS. Congenital hemangiomas of the skin with orbital and subglottic hemangiomas. *Am J Ophthalmol.* 1973;75:876.
3. Kasabach HH, Merritt KK. Capillary hemangioma with extensive purpura. *Am J Dis Child.* 1940;59:1063.
4. Robb RM. Refractive errors associated with hemangiomas of the eyelids and orbit in infancy. *Am J Ophthalmol.* 1977;83:52-58.
5. Stigmar G, Crawford JS, Ward CM, et al. Ophthalmic sequelae of infantile hemangiomas of the eyes and orbit. *Am J Ophthalmol.* 1978;85:806-813.
6. Kushner BJ. Intralesional corticosteroid injection for infantile adnexal hemangioma. *Am J Ophthalmol.* 1982;93:496-506.
7. Haik BG, Jones IS, Ellsworth RM. Vascular tumors of the orbit. In: Hornblass A., ed. *Ophthalmic and Orbital Plastic & Reconstructive Surgery.* Baltimore, Md: Williams & Wilkins; 1989.

Chapter 196

Juvenile Xanthogranuloma

W. L. BURT

Juvenile xanthogranuloma is a benign histiocytic inflammatory disorder affecting mainly infants and children, usually under 3 years of age. It affects races and sexes equally and involves mainly the skin and occasionally the eyes. A hereditary factor has not been identified. The disorder has previously been labeled "nevoxanthoendothelioma," but the lesion is neither nevoid nor endothelial. This label is a misnomer and should not be used.[1]

In addition, in the past, juvenile xanthogranuloma has been considered to be related to the histiocytosis-X group of histiocytic proliferative disorders; however, present information based on clinical, histocytologic, and histochemical grounds indicates that juvenile xanthogranuloma is distinct from these differentiated systemic histiocytoses.[2-5] Clinically, the latter have frequent systemic manifestations, rare ocular complications, an overall poor prognosis, and often a fatal outcome. On the other hand, in juvenile xanthogranuloma, systemic manifestations other than the skin lesions are rare, ocular involvement is relatively common, and the disease follows a benign course followed by spontaneous regression. Juvenile xanthogranuloma does not evolve into any of the histiocytosis-X conditions, and there are no transitional forms between the two disorders. Pathologically, histiocytosis-X is regarded as an abnormal proliferation of Langerhan's cells that belong to a subdivision of the mononuclear phagocytic system, or perhaps to a cell lineage independent of the mononuclear phagocytic system. The Langerhan's cells and the tumor cells of histiocytosis-X have in common the presence of Birbeck's granules on electron microscopy, and strong staining for S-100 protein immunohistochemically. The tumor cells of juvenile xanthogranuloma reveal no Birbeck's granules and no such staining. Juvenile xanthogranuloma, therefore, should no longer be included in the histiocytosis-X group of proliferative disorders.

Juvenile xanthogranuloma appears to be an idiopathic and essentially benign histiocytic inflammatory proliferation and aggregation of cells of the macrophagic or phagocytic (M) cells. Although it has been widely claimed, based on ultrastructural and cytochemical characteristics, that juvenile xanthogranuloma is a reactive histioxanthomatous response to some local tissue injury rather than a neoplastic growth, its true pathogenesis remains obscure.

Systemic Manifestations

Juvenile xanthogranuloma may be present at birth (30%) or may appear within the first year of life. (It rarely presents in adults.) The skin and the eye are the two most common sites of involvement. Rarely lesions have been noted in soft tissue, skeletal muscle, salivary glands, kidneys, testes, periosteum, bone, colon, ovaries, pericardium and myocardium, lungs, and central nervous system.[6] Both skin and visceral lesions undergo spontaneous regression and disappear within 3 to 6 years. The patient's general health is seldom impaired, and physical and mental development are normal. Metabolic disturbances have not been identified, skeletal x-ray surveys are negative, and results of laboratory examinations of blood cholesterol and other lipids are normal.

The vast majority (possibly 90% or more) of juvenile xanthogranuloma patients have self-limited skin lesions only. The skin lesions may be present at birth or may arise during the first 12 months of life and are rarely seen in adults. There may be as few as one or two lesions or as many as several hundred. The distribution is characteristic, involving the scalp, face, trunk and proximal extremities, mainly on the upper body. The lesions are somewhat variable. They may be elevated, round or oval, or they may be flat and barely palpable. They vary from 1 to 20 mm in diameter. They may increase in size and number during the first year or year and a half, and then begin to resolve spontaneously over the next 3 or 4 years. In their declining phases, the skin lesions first lose their mass, becoming wrinkled as a result, and often turning more brownish in the process. They shrink until a flat or even slightly depressed atrophic scar is formed, which may be hyperpigmented or hypopigmented.[7]

Histologically the cutaneous lesions show the usual configuration of xanthomas, with a nodular histiocytic proliferation in the subepidermal location, usually with frank foam cell formation. Eosinophils and other inflammatory cells may be admixed. The Touton giant cell, with its orderly circular arrangement of nuclei and foamy peripheral cytoplasm, is seen with relative frequency in cutaneous lesions. The observation of lipid vacuoles in smooth muscle cells of erector pili and in Schwann cells of nerves and in mast cells, supports the concept that the condition results from local tissue injury that evokes a histioxanthomatous response.[7]

The clinical story of the vast majority of juvenile xanthogranuloma patients with this self-limited cutaneous disease would be considered completely innocuous were it not for certain less than benign ocular complications described below and were it not for some very curious and more recently appreciated relationships to certain generalized disease processes.[8] A retrospective survey of patients with juvenile xanthogranuloma has revealed that many, perhaps a majority, have café-au-lait macules on the trunk, or have relatives who have them or other features of neurofibromatosis. In fact, patients have been described with juvenile xanthogranuloma (cutaneous and ocular) and café-au-lait pigmentation, who have subsequently developed neurofibromas. In addition, patients with juvenile chronic myeloid leukemia (JCML) a distinctive form of childhood leukemia that constitutes approximately 2% of all cases of childhood leukemia, may have xanthomas that have been interpreted as being juvenile xanthogranuloma.[9] Some of these JCML patients (perhaps up to 21%) also have multiple café-au-lait spots and a family history of some type of neurofibromatosis involvement, including fully developed neurofibromatosis. There is one case report of a patient with cutaneous neurofibromatosis and JCML. JCML patients are normolipemic, and the xanthomatous lesions lack evidence of malignancy, have many of the clinical manifestations of sporadic juvenile xanthogranuloma, and are histologically indistinguishable from those of juvenile xanthogranuloma. The skin lesions, however, are more likely to be multiple and to be confluent. Others have concluded there is an increased risk of leukemia in children with neurofibromatosis.[8,9]

It is evident that there is an intriguing and complex relationship between leukemia, xanthomas, and neurofibromatosis. It appears that normocholesterolemic skin xanthomas in infants and young children may have a sporadic origin and a benign course or in some cases may arise in a setting of peculiar constitutional disease related in some fashion to the relatively benign neurofibromatosis or to the more critically severe JCML. Juvenile xanthogranuloma has also been described in a rare association with Neimann-Pick disease and urticaria pigmentosa.

Ocular Manifestations

Ocular involvement is the most frequent extracutaneous manifestation of juvenile xanthogranuloma. The ocular lesions may precede, follow, or occur concurrently with the cutaneous lesions. The lesions are usually unilateral but may be bilateral. The anterior uvea (iris and ciliary body) is most typically involved, whereas the posterior uveal tract (choroid) and retina are rarely involved. Other rare ocular complications include leisons of the eyelids, the orbit, including orbital bone and rectus muscles of the globe, and epibulbar lesions of the cornea, conjunctiva, and sclera.[10]

The most noteworthy lesion involves the anterior uvea. The reason for the localization of this peculiar lesion in the iris is unknown. The iris lesion may occur as a localized solitary nodule or mass (Fig. 196-1), or it may diffusely infiltrate and thicken the iris. The lesion consists of an infiltrate of large, uniform histiocytes with small, regular nuclei and clear eosinophilic cytoplasm containing fat. Touton giant cells, eosinophils, and other inflammatory cells are present in varying numbers. The iris lesions may contain many engorged, thin-walled blood vessels that have a marked tendency to bleed spontaneously into the aqueous of the anterior chamber (hyphema).

The cutaneous lesions of juvenile xanthogranuloma tend to regress spontaneously, running a benign course. The iris lesions, however, may be anything but benign. The tendency of the lesion to hemorrhage spontaneously into the anterior chamber may result in secondary glaucoma and ultimately blindness or loss of the eye by enucleation.

Infants typically present to the ophthalmologist because of unilateral redness of an eye. A spontaneous hyphema may be apparent, or the cornea may be enlarged in diameter and cloudy (from corneal edema) due to glaucoma, or the eye may simply be red and inflamed from uveitis. On occasion an asymptomatic iris tumor or heterochromia iridis may be the presenting feature.[7]

Periorbital skin lesions are similar to cutaneous lesions elsewhere. The rare orbital lesions may present as unilateral proptosis owing to orbital infiltration. The epibulbar lesion may present as a yellowish conjunctival or episcleral mass that may extend across the limbus to involve the cornea.

Management

In the 74% of juvenile xanthogranuloma patients who present with cutaneous lesions only, no active intervention is indicated, although careful ophthalmic follow-up is indeed necessary. Ocular involvement limited to the eyelid or epibulbar tissue also often requires no specific therapy. When there are uveal lesions, however, prompt treatment is necessary to prevent serious ocular consequences. The treatment modalities available for the

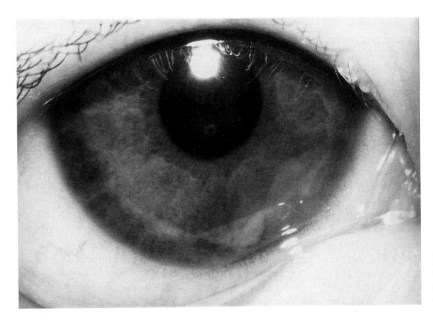

FIGURE 196-1 Iris mass extends from the 3 o'clock to the 6 o'clock position and projects into the anterior chamber inferiorly.

ocular lesions of juvenile xanthogranuloma include topical or systemic corticosteroid therapy, irradiation, and surgery. One must choose the treatment regimen best suited to the individual case.[7] When a more generalized disease process is apparent, therapy must of course be tailored to the specific disease process or the specific organ site involved.

References

1. Helwig EB. Histiocytic and fibrocytic disorders. In: Graham JH, Johnson WC, Helwig EB, eds. *Dermal Pathology*. New York, NY: Harper & Row; 1972;715.
2. Sanders TE. Intraocular juvenile xanthogranuloma. *Trans Am Ophthalmol Soc*. 1960;58:59-74.
3. Favara BE, McCarthy RC, Mierau GW. Histiocytosis X. *Human Pathol*. 1983;14:663-676.
4. Sonada T, Hashimoto H, Enjoji M. Juvenile xanthogranuloma. *Cancer*. 1985;56:2280-2286.
5. Gonzalez-Crussi F, Campbell RJ. Juvenile xanthogranuloma ultrastructural study. *Arch Pathol*. 1970;89:65-72.
6. Roper SS, Spraker MK. Cutaneous histiocytosis syndromes. *Paediatr Dermatol*. 1985;3:19-30.
7. Cadera W, Silver MM, Burt L. Juvenile xanthogranuloma. *Can J Ophthalmol*. 1983;18:169-174.
8. Crocker AC. The histiocytosis syndromes. In: Fitzpatrick TB, Freedberg IM, eds. *Dermatology in General Medicine*. New York, NY: McGraw-Hill; 1987;1937-1946.
9. Cooper PH, Frierson HF, Kayne AL, et al. Association of juvenile xanthogranuloma with juvenile myeloid leukemia. *Arch Dermatol*. 1984;120:371-375.
10. Zimmerman LE. Ocular lesions of juvenile xanthogranuloma. *Am J Ophthalmol*. 1965;60:1011-1035.

Chapter 197

Xeroderma Pigmentosum

JAMES WIENS and W. BRUCE JACKSON

Xeroderma pigmentosum (XP) is a rare, autosomal recessive condition, that has an estimated incidence of 1 in 250,000 persons. The disease is characterized by solar damage to the skin resulting in pigmentation changes and malignancies, especially in exposed areas.[1]

Ultraviolet (UV) light is known to cause damage to nuclear DNA by forming dimers between two adjacent pyrimidine molecules. Normally, this is repaired by a series of five enzymatic steps initiated by endonuclease.[1,13] In XP an inherited defect in this initial step results in a decreased rate of DNA repair.[1,13] This same repair mechanism operates to repair DNA damage induced by some carcinogenic chemicals such as BCNU(1,3-bis[2-chloroethyl]-1-nitrosourea). A separate cellular DNA repair mechanism is used for x ray–induced damage.[1]

Cell fusion studies using fibroblasts from different persons with XP have shown the genetic heterogeneity of XP. Fibroblasts from two persons can be fused, and the rate of DNA repair in the new cell can be studied. If the cell is then able to repair UV–induced DNA damage, the two persons are of different complementation groups. If there is no improvement in DNA repair, they are of the same complementation group.[1,4] While these studies have been done mainly on fibroblasts, this DNA repair defect is felt to be present in all nucleated cells.

To date, nine different complementation groups have been identified.[2] Each is characterized by its own rate of DNA repair. The delay in DNA repair occurs in conjunctival cells as well.[5]

To try to account for this genetic heterogeneity in a condition that involves only one enzyme, a co-recessive inheritance model has been proposed that hypothesizes that some genetic diseases are expressed clinically only if the individual is homozygous or hemizygous for more than one allele at different gene loci.[3]

Systemic Manifestations

XP primarily affects the skin, especially in sun-exposed areas. In early infancy patients are acutely sensitive to sun, presenting with a picture resembling sunburn despite minimal sun exposure. This sensitivity decreases with age. Freckles then develop in the first several years of life, and the skin later becomes dry and scaly. Eventually telangiectasia and areas of hypopigmentation develop. With time, actinic keratoses and verrucous papules form in severely sun-damaged areas.[1,4,6]

Histologically, dilation and inflammation of the superficial blood vessels are seen in the erythematous stage.[6] Pigmented areas show pigment granules in the basal layer of the epithelium.[6,15] Atrophic lesions show epidermal atrophy, primarily of the prickle cell layer, without pigment accumulation.[6,15] Hyperkeratinization and a minimal increase in dermal elastotic tissue also occur.[1,15] This lack of dermal elastotic change aids in separating XP lesions from those due to excessive sun exposure. In these latter cases there is a prominent dermal elastotic component.[1]

Squamous cell carcinomas, basal cell carcinomas, and melanomas may form in these sun-damaged areas of skin. Usually beginning in childhood, they are multiple,

recurrent, and may metastasize. Basal cell tumors are the most common of the three.[1]

Two thirds of patients die before 20 years of age of either metastatic disease or infection. There is greater susceptibility to infections that involve the skin and the respiratory or urinary tracts, resulting in sepsis and subsequent death.[12]

Ocular Manifestations

Ocular changes, occurring in 30% of patients, may involve the eyelids, conjunctiva, cornea, and occasionally the iris (Table 197-1).[7,8,11] Initial eye complaints of photophobia, lacrimation, and mild conjunctivitis develop in almost all patients at an early stage.[1,7,15] The conjunctivitis may be associated with a serous or mucopurulent discharge. Blepharospasm may also be present.[1,7,15]

Early evidence of ocular involvement is usually found in the lower lid. Skin changes include erythema, pigmentation, atrophy, and malignancies similar to those seen on the rest of the body. Atrophic skin changes develop, mainly in the lower lid, beginning at the lid margin, with a progression in some to total loss of the lower lid. Madarosis, as well as entropion or ectropion formation occur during this process.[6]

In addition, cutaneous epithelial invasion of the palpebral conjunctiva, forming an irregular mucocutaneous border, may be seen.[6] Epitheliomas of the lower lid border also occur.[6,7] With loss of the lower lid, symblepharons and ankyloblepharons form.[6-8]

Clinically the conjunctiva appears dry and hyperemic, with areas of pigment deposition and keratin formation. These appear mainly in the sun-exposed interpalpebral fissure area.[7]

Benign conjunctival tumors are seen in about 20% of patients. They include pseudopterygia, phlyctenular growths, and pingueculae. Histologically, inflammatory cells with subsequent epithelial hyperplasia are noted.[7]

Squamous cell carcinoma of the conjunctiva, seen in 13% of patients as reviewed by El-Hefnawi, may be both recurrent and multiple.[7-9] Early lesions usually begin at the limbal area and spread onto the cornea. Involvement of the cornea by primary squamous cell carcinoma is rare; when documented, it has been seen in areas where prior corneal ulceration has occurred.[6,7]

Melanosis of the conjunctiva may occur at the limbus, caruncle, or lid margin, occasionally becoming malignant.[7] A fungating granulomatous mass with associated areas of angiosarcoma, developing at the limbus, has also been described.[15]

Corneal opacities have been noted in about 40% of patients. They may be secondary to corneal exposure due to lower lid loss, dryness, or ulceration with sub-

TABLE 197-1 Ocular Manifestations of
Xeroderema Pigmentosum

Lids
  Madarosis
  Entropion, ectropion
  Lid loss
  Symblepharon, ankyloblepharon
  Squamous cell carcinoma, basal cell carcinoma,
    melanoma

Conjunctiva
  Xerosis
  Hyperemia
  Pigmentation
  Benign lesions
    Pseudopterygium
    Phlyctenules
    Pingueculae
  Malignant lesions
    Squamous cell carcinoma
    Basal cell carcinoma
    Melanoma
    Angiosarcoma

Cornea
  Exposure keratitis
  Xerosis
  Ulceration
  Band-shaped nodular dystrophy
  Opacification
  Tumors

Iris
  Iris atrophy
  Iritis

sequent extensive vascularization and opacification.[7] An associated iritis may lead to iris atrophy or synechiae formation.[1,7] Development of band-shaped nodular corneal dystropy has been reported in black patients.[14]

Neuro-ophthalmologic findings have not been noted, but neurologic abnormalities frequently occur. These were originally described by De Sanctis and Cacchione, who reported on a series of XP patients with microcephaly and progressive mental deficiency, hearing loss, choreoathetosis, ataxia, and quadriparesis. Most XP patients do not have the complete syndrome but only some components.[1] Recent studies indicate that neurologic defects may occur more frequently in certain complementation groups.[10] There have been no studies to indicate whether ocular findings are more likely to occur in certain complementation groups.

Treatment is limited, consisting of wearing sunscreen lotions to protect the skin from UV light. Regular follow-up of both skin and eyes is necessary to detect the earliest signs of malignancy. Once detected, they are usually removed with surgery or irradiation. The drug 5-fluorouracil has also been used, but while it is initially effective, the lesions become refractory with time.[4]

References

1. Robbins JH, Kraemer KH, Lutzner MA, et al. Xeroderma pigmentosum: an inherited disease with sun sensitivity, multiple cutaneous neoplasms, and abnormal DNA repair. *Ann Intern Med.* 1974;80:221-248.
2. Fischer E, Keijzer W, et al. A ninth complementation group in xeroderma pigmentosum, XP I. *Mutation Res.* 1985;145:217-225.
3. Lambert WC, Lambert MW. Co-recessive inheritance: a model for DNA repair, genetic disease and carcinogenesis. *Mutation Res.* 1985;145:227-234.
4. Carter DM, O'Keefe EJ. Xeroderma pigmentosum. In: Moschella ST, Hurley HT, eds. *Dermatology.* 2nd ed. vol. 2. Philadelphia, Pa: WB Saunders; 1985;1206-1208.
5. Newsome DA, Kraemer KH, Robbins JH. Repair of DNA in xeroderma pigmentosum conjunctiva. *Arch Ophthalmol.* 1975;93:660-662.
6. Reese AB, Wilber IE. The eye manifestations of xeroderma pigmentosum. *Am J Ophthalmol.* 1943;26:901-911.
7. El-Hefnawi H, Mortada A. Ocular manifestations of xeroderma pigmentosum. *Br J Dermatol.* 1965;77:261-276.
8. Giller H, Kaufmann WC. Ocular lesions in xeroderma pigmentosum. *Arch Ophthalmol.* 1959;62:130-133.
9. Gaasterland DE, Rodrigues MM, Moshell AN. Ocular involvement in xeroderma pigmentosum. *Ophthalmology.* 1982;89:980-986.
10. Cleaver JE, Zelle B, Hashem N, et al. Xeroderma pigmentosum patients from Egypt: II. preliminary correlations of epidemiology, clinical symptoms and molecular biology. *J Invest Dermatol.* 1981;77:96-101.
11. Stenson S. Ocular findings in xeroderma pigmentosum: a report of two cases. *Ann Ophthalmol.* 1982;14:580-585.
12. Rook A. Xeroderma pigmentosum. In: Rook A, Wilkinson DS, Ebling FJG, eds. *Textbook of Dermatology.* 3rd ed. Oxford: Blackwell Scientific Publications; 1979;124-127.
13. Cleaver JE. Defective repair replication of DNA in xeroderma pigmentosum. *Nature.* 1968;218:652-656.
14. Freedman J. Xeroderma pigmentosum and band-shaped nodular corneal dystrophy. *Br J Ophthalmol.* 1977;61:96-100.
15. Bellows RA, et al. Ocular manifestations of xeroderma pigmentosum in a black family. *Arch Ophthalmol.* 1974;92:113-117.

Section B

Disorders of Connective Tissue

Chapter 198

Ehlers-Danlos Syndrome

DUANE C. WHITAKER and JEFFREY A. NERAD

Ehlers-Danlos syndrome is a multisystem genetic disorder that may affect skin, eyes, vasculature, and joints. At least nine subtypes of Ehlers-Danlos syndrome are distinguishable on the basis of inheritance, clinical features, and ultrastructural and biochemical defects. Phenotypic characteristics seen in Ehlers-Danlos syndrome include hyperextensibility of skin and joints, blue scleras, blood vessel and visceral fragility, fragile skin with poor wound healing, cutaneous pseudotumors most prominent over elbows and knees, and subcutaneous calcifications.[1] Premature rupture of fetal membranes and concomitant complications at birth may be an early sign of the disease. Ehlers-Danlos syndrome is a disorder of collagen and connective tissue synthesis whose main eye abnormality is ocular fragility.[2] The most common types of Ehlers-Danlos syndrome are I, II, and III, all of which have an autosomal dominant inheritance pattern. These three are classified by severity of expression.

Systemic Manifestations and Classification

Ehlers-Danlos syndrome I (*gravis*) is associated with premature rupture of fetal membranes, hyperextensibility of joints and skin, pseudotumors, subcutaneous nodules, easy bruisability, poor wound healing, varicosities, and hernias.

Ehlers-Danlos syndrome II (*mitis*) has mild expression of the abnormalities described in type I, and in fact, may go undiagnosed.

In *Ehlers-Danlos syndrome III* (*benign hypermobile*), hypermobility is the predominant feature. Skin findings are variable.

Ehlers-Danlos syndrome IV (*ecchymotic*), also referred to as the arterial form, is inherited as an autosomal-recessive trait.[4] Rupture of large arteries, bowel, and viscera are the primary risks. The main clinical feature is hypermobility of joint digits. Inability to synthesize type III collagen is a known biochemical defect.

Manifestations of *Ehlers-Danlos syndrome V* (*X-linked*) are similar to those of type I (gravis), and lack of sufficient lysyl oxidase appears to be the defect.

In *Ehlers-Danlos syndrome VI* (*ocular-scoliotic*) ocular fragility, intraocular bleeding, and other eye findings (see below) are prominent. This form presents the most severe risk for blindness. Scoliosis as well as skin and joint hypermobility and poor wound healing are seen. The mode of inheritance is autosomal recessive, and deficiency of lysyl hydroxylase is evident in this form.

Patients with *Ehlers-Danlos syndrome VII* (*arthrochalasis multiplex congenita*) have short stature, joint laxity, and congenital dislocations. Cutaneous findings show skin laxity and bruisability. The inheritance is autosomal recessive and the abnormality appears to be a mutation in the pro-alpha$_2$ chain that blocks enzymatic conversion of procollagen to collagen.[2,3]

Ehlers-Danlos syndrome VIII (*periodontal type*) is transmitted as an autosomal dominant trait. Clinical features are severe periodontitis, moderate skin fragility, and joint laxity. Biochemical and ultrastructural defects are unknown.

Ehlers-Danlos syndrome IX is described as mental retardation type and fibronectin type by several authors.[4] Further subdivisions of this type have been proposed.[2] Ehlers-Danlos syndrome IX usually refers to features of autosomal recessive inheritance, skin and joint fragility, severe mental deficiency, and inguinal hernias and protuberant ears. A platelet aggregation abnormality is caused by dysfunction of plasma fibronectin.[4]

Light microscopy shows no abnormality in the majority of Ehlers-Danlos patients, either in skin thickness or quality or quantity of collagen or elastic fibers. The pseudotumors may show fibrosis, capillaries, and accumulations of foreign body giant cells.[4] The subcutaneous nodules usually show calcification and adipose tissue. Scanning electron microscopy has shown abnormalities in the caliber and organization of collagen fibers.[4]

Ocular Manifestations

A variety of ocular features of Ehlers-Danlos syndrome have been described. In the majority of cases they are minor—epicanthal folds, myopia, microcornea, blue scleras. In rare cases, serious complications leading to blindness may result from an increase in ocular fragility. Signs of ocular fragility in Ehlers-Danlos syndrome include keratoconus, lens dislocation, and retinal detachment.[1]

In a series of 100 patients with Ehlers-Danlos syndrome, Beighton[5] identified 27 with epicanthal folds, 7 with blue scleras, 7 with strabismus, and 8 with myopia.

Frequently noted features included redundant skin on the eyelids, telecanthus, and easily everted lids (Metenier's sign).[6] No patients in this series had serious eye problems, however. Later, Beighton[6] described two siblings with Ehlers-Danlos syndrome who became blind because of ocular rupture after minimal eye trauma.

Severe ocular fragility is part of the type VI Ehlers-Danlos syndrome. In addition to the familiar joint and skin disturbances these patients develop severe scoliosis and eye problems. Rupture of the sclera and retinal detachment from relatively minor trauma are frequent findings.[7]

In type VI, the ocular-scoliotic form of Ehlers-Danlos syndrome, the enzyme that catalyzes the hydroxylation of lysine in the collagen polypeptide chain, lysyl hydroxylase, is defective. This results in an abnormal collagen cross-linking[8] and abnormally soluble skin collagen. This enzyme defect was first found in two sisters whose features of Ehlers-Danlos syndrome included severe scoliosis, recurrent joint dislocations, stretchable skin, premature rupture of the fetal membranes, and floppiness early in life. One sister had one eye enucleated at 9 years of age after a car accident. McKusick et al.[9] reported on a second family with this enzyme defect. A brother and sister were affected; both were blind from ocular complications of their disease. He proposed the term "fragilitas oculi."

In 1976 Judisch[10] reported a case of ocular Ehlers-Danlos syndrome with normal lysyl hydroxylase activity. In 1977 Behrens-Baumann et al.[11] reported a similar case.

Generally Ehlers-Danlos syndrome is inherited as a dominant trait. Beighton[6] described a family in which the transmission of the syndrome was consistent with recessive inheritance of the trait. In the same publication, McKusick noted that type VI Ehlers-Danlos syndrome shares many features with fragilitas oculi and he expressed uncertainty as to whether fragilitas oculi and Ehlers-Danlos syndrome type VI were separate and distinct entities. Ocular fragility is now a well-recognized part of the Ehlers-Danlos syndrome type VI.

Weakened connective tissue is the cause of many of the other eye problems associated with Ehlers-Danlos syndrome. These less serious eye problems include blue scleras, myopia, lax lids and lid skin, keratoconus, keratoglobus, ectopia lentis, and angioid streaks. The blue sclera of Ehlers-Danlos syndrome is secondary to thinning of the sclera with underlying dark choroid showing through causing a bluish hue. Although no information is known regarding the myopia associated with Ehlers-Danlos syndrome, one may postulate that weakened scleras allow ocular enlargement or perhaps a staphyloma, causing myopia. Lax lids and lid skin are comparable to the soft, hyperextensible skin seen over the rest of the body in association with Ehlers-Danlos syndrome.

The corneas seem to be especially weak in Ehlers-Danlos syndrome patients, which explains the frequent occurrence of corneal rupture in the face of minor eye trauma. Keratoglobus, a bilateral general thinning and anterior protrusion of the cornea, is often associated with the blue scleras and hyperextensible joints of Ehlers-Danlos syndrome type VI. Corneal thickness is reduced, often to about one-third normal.[12] Spontaneous central breaks in Descemet's membrane may appear at any time, producing marked corneal edema (acute hydrops). Amblyopia may result from the refractive abnormalities associated with the steep corneal curvature. Minor eye trauma may cause corneal and scleral rupture. Protective spectacles are recommended.[13]

Similarly, keratoconus (localized corneal thinning) may result from the abnormal cross-linkage of collagen in Ehlers-Danlos syndrome.[14] Visual disturbances result from irregular astigmatism of the cornea. Spectacles and contact lenses may give satisfactory vision in early stages. Stromal scarring may occur in keratoconus. Corneal transplantation is generally successful in advanced cases of keratoconus.

Ectopia lentis, or subluxation of the crystalline lens, may occur in Ehlers-Danlos syndrome. It is a common manifestation of other connective tissue systemic disorders, including Marfan's syndrome, homocystinuria, Weill-Marchesani syndrome, hyperlysinemia, and sulfite oxidase deficiency.[15]

A variety of retinal abnormalities may be seen in Ehlers-Danlos syndrome. Bonnet[16] reported depigmentation in the pre-equatorial region of the fundus. Retinal detachment and retinitis proliferans were observed in a 19-year-old girl with Ehlers-Danlos syndrome.[17] The occurrence of angioid streaks in two persons with Ehlers-Danlos syndrome alone has been reported.[18] A defect in the elastic layer of Bruch's membrane of the retina is thought to be the cause of angioid streaks in pseudoxanthoma elasticum. It is not certain how the collagen abnormality in Ehlers-Danlos syndrome may cause angioid streaks. Since angioid streaks have been observed in only two patients with Ehlers-Danlos syndrome alone it is difficult to draw any firm conclusions as to the cause or relationship between the formation of angioid streaks and the collagen organizational abnormality in Ehlers-Danlos syndrome.

References

1. Pinnell SR, Murad S. Disorders of collagen. In: Stanbury JB, Wyngaarden JB, Frederickson DS, et al., eds. New York, NY: McGraw-Hill; 1983;1434-1440.
2. Renie WA, ed. *Goldberg's Genetic and Metabolic Eye Disease*. 2nd ed. Boston, Mass: Little, Brown; 1986;511-513.
3. Fitzpatrick TB, Eisen AZ, Wolff K, et al., eds. *Dermatology in General Medicine*. 2nd ed. New York, NY: McGraw-Hill; 1979;1149-1150.
4. Lever WF, Schaumberg-Lever G. *Histopathology of the Skin*. Philadelphia, Pa: JB Lippincott; 1983;78-79.
5. Beighton P. The Characteristics of the Ehlers-Danlos Syndrome. London: University of London. Thesis.
6. Beighton P. Serious ophthalmological complications in the Ehlers-Danlos syndrome. *Br J Ophthalmol*. 1970;54: 263-268.
7. McKusick VA. Multiple forms of the Ehlers-Danlos syndrome. *Arch Surg*. 1974;109:475-476.
8. Pinnell SR, Krane SM, Kenzora JE, et al. Heritable disorder with hydroxylysine-deficient collagen. Hydroxylysine-deficient collagen disease. *N Engl J Med*. 1972;286:1013-1020.
9. McKusick VA. *Heritable Disorders of Connective Tissue*. 4th ed. St. Louis, Mo: CV Mosby; 1972.
10. Judisch GF, Waziri M, Krachmer JH. Ocular Ehlers-Danlos syndrome with normal lysyl hydroxylase activity. *Arch Ophthalmol*. 1976;94:1489-1491.
11. Behrens-Baumann W, Gebauer H-J, Langenbeck U. Blane-Sklera-Syndrome and Keratoglobus (Oculaerer Typ des Ehlers-Danlos-Syndromes). *Graefes Arch Klin Exp Ophthalmol*. 1977;204:235-246.
12. Waring GA III, Rodriques MM. Congenital and neonatal corneal abnormalities. In: Duane TD, Jaeger EA, eds. *Biomedical Foundations of Ophthalmology*. vol 1. Philadelphia, Pa: Harper & Row; 1983;9:2-3.
13. Kenyon KR, Fogle JA, Grayson M. Dysgeneses, dystrophies, and degenerations of the cornea. In: Duane TD, Jaeger EA, eds. *Clinical Ophthalmology*. vol. 4. Philadelphia, Pa: Harper & Row, 1986;4.
14. Maumanee IH. The cornea in connective tissue disease. *Ophthalmology (Rochester)*. 1978;85:1014.
15. Renie WA, ed. *Goldberg's Genetic and Metabolic Eye Disease*. 2nd ed. Boston, Mass: Little, Brown; 1986;391.
16. Bonnet MP. Les manifestations oculaires de la maladie d'Ehlers-Danlos. *Bull Soc Ophthalmol France*. 1953;6:623-626.
17. Bossu A, Lambrecht. Manifestations oculaires du syndrome d'Ehlers-Danlos. *Ann Oculist*. 1954;187:227-236.
18. Green WR, Friedman-Kien A, Banfield WG. Angioid streaks in Ehlers-Danlos syndrome. *Arch Ophthalmol*. 1966;76:197-204.

Chapter 199

Pseudoxanthoma Elasticum

JEFFREY A. NERAD and DUANE C. WHITAKER

Pseudoxanthoma elasticum (PXE) is a systemic disease in which the primary defect appears to be production of abnormal elastic fibers with secondary calcification. Manifestations of the disease are seen in the skin, cardiovascular system, and eyes. The defects in skin elasticity characterized by redundant folds of soft, wrinkled, and lax skin are generally the most obvious feature of the disease, but the cutaneous features are the least damaging. Abnormalities in the elastic tissues of blood vessels and the eye may cause hemorrhage, peripheral vascular disease, and blindness.

In 200 case reports of PXE reviewed by Eddy and Farber,[1] organ involvement was found to be as follows: skin alone, 8%; skin and eye alone, 8%; skin, eye, and other organ systems, 45.5%; skin and other organs, 3%; eye and other organs, 8.5%; and eye alone, 5.5%. PXE generally begins by age 30, although it may be seen in very young and very old persons. Usually the incidence in women is twice that in men.[1] There is no racial predilection.

The exact cause of PXE remains unclear. The disease is classified as a "heritable disorder of connective tissue."[2] Disorders are transmitted in a simple mendelian manner and have generalized defects primarily involving one of the elements of connective tissue. Three principal components of connective tissue are (1) cells (fibroblasts); (2) fibrous elements, the two main components being collagen and elastin; and (3) the so-called ground substance, which like the fibrous elements is elaborated by the fibroblast.[3] In PXE the primary defect appears to involve the elastic fiber component of connective tissue, primarily in the skin, blood vessels, and eyes. Electron microscopic studies[4] and enzyme digestion studies using elastase and collagenase on tissue[5] from patients with PXE support the concept that PXE is caused by an alteration in the elastic fibers and not the collagen fibers of the connective tissue.[6,7]

Histopathologically the earliest change is calcification of the elastic fibers in the skin, blood vessels, and in Bruch's membrane beneath the retinal pigment epithelium. Fragmentation of the elastic fibers occurs in later stages, giving rise to the classic angioid streaks of the ocular fundus. Foreign body giant cells are sometimes found in these areas. Calcific changes of elastic fibers may be seen in other disease states, such as chronic inflammation, atherosclerosis, hyperphosphatemia, and Paget's disease of bone; however, the deposition is thought to be secondary to the degenerative process or to hypercalcemia causing metastatic calcification. In PXE the deposition of calcium in elastic fibers seems to be part of the primary disease process,[8] present in all lesions, occurring upon elastic fibers that are not altered in any other visible way.

Genetics

The classic disorder of PXE which involves skin, vessels, and eye is inherited as an autosomal recessive trait (type I); however, both recessive and dominant forms have been characterized. Pope's[9,10] Type II recessive form is very rare and is characterized by generalized skin changes with no blood vessel or ocular manifestations. Pope's Type I dominant form is characterized by classic *peau d'orange* skin changes, by severe vascular complications such as angina, claudication, and hypertension, and by prominent angioid streak formation. The Type II dominant form is four times more common than the other dominant form. It is characterized by focal skin changes that render the skin excessively stretchable and by myopia, high-arched palate, blue scleras, and loose joints.

All forms of PXE are uncommon. Pedigrees and case reports describe several hundred patients. The incidence is roughly estimated at about 1 in 160,000 people, suggesting that there are 1000 or 2000 patients in the United States.[11]

Systemic Manifestations

Cutaneous findings in PXE are very helpful in the recognition and diagnosis of this disorder. The characteristic lesions are found in skin folds, most prominently in the sides of the neck, the axillas, and the antecubital and inguinal areas. Lesions are yellowish and xanthomatous and consist of coalescing *peau d'orange* plaques. There may also be scattered yellow papules several millimeters in diameter. Other areas of skin folds, including genital, popliteal, and periumbilical regions, may be involved. On palpation the skin may be velvety and either thickened or atrophic. Mucosal lesions can been seen, particularly on the lower lip, rectum, and vagina. The skin in affected areas appears wrinkled and redundant.

In addition to these clinical findings, skin biopsies are also valuable in establishing the diagnosis. The characteristic histologic finding is fragmentation and clumping of elastin fibers in the middle and lower dermis. Because of the presence of calcium, these abnormal fibers are seen on hematoxylin and eosin and with calcium stains. Sometimes calcification is seen in the skin lesions on routine radiography.[12]

Abnormalities of the elastic tissues of the vascular system may lead to serious complications from hemorrhage and peripheral vascular disease. In a summary of the accumulated experience of the Mayo Clinic, 10 of 74 patients with PXE had experienced gastrointestinal bleeding. Six of these patients bled from the large bowel, three had ulcerative colitis, and three internal hemorrhoids.[13] Four bled from the upper gastrointestinal tract, one had atrophic gastritis, one had a duodenal ulcer, and in two patients no specific lesion was found.

Gastrointestinal hemorrhage in PXE has been reported as early as age 6½ years.[14] Gastrointestinal hemorrhage usually occurs early in the course of the disease, often at a time when ocular and cutaneous changes are minimal. Kaplan[15] reported the average age at hemorrhage to be 26 years. Urinary tract and uterine hemorrhage have also been reported.[16]

The peripheral vessels and heart are frequently affected by PXE. Absent or reduced pulses in extremities and radiographically demonstrable arterial calcification are found in a majority of PXE patients by age 30.[12] Arteries show fragmentation of the elastica of the media with surrounding reaction which has swollen the media and occluded the lumen of the artery. Characteristically, occlusion in the upper extremities is not accompanied by symptoms due to collateral vessel formation. Intermittent claudication and cramps of the lower extremities are common. Intermittent claudication has been seen as early as age 9. Hypertension often occurs due to renal artery narrowing.

A common cardiac manifestation of PXE is premature coronary atherosclerosis. Angina and myocardial infarctions have been reported in patients less than 20 years of age.[17] Valvular heart disease is also common, including mitral valve prolapse, which can be detected in over 70% of patients.[18] Restrictive cardiomyopathy attributed to PXE has been described in a patient.[19] Fibrous thickening of the endocardium and atrioventricular valves has been documented at autopsy.[20]

Neurologic complications of cerebrovascular disease include lacunar infarcts, aneurysms, progressive intellectual deterioration, and psychic and mental disturbances that may be due to cortical atrophy. Subarachnoid and intracerebral hemorrhage are frequent causes of death. Hypertension and alteration of the integrity of the cerebral vessels are the two basic pathophysiologic mechanisms responsible for the neurologic complications of PXE.[21]

Ocular Manifestations

Angioid streaks are the hallmark of PXE. In 1889 Doyne[22] described the jagged, irregular lines of angioid streaks in a patient who had received a "poke in the eye" and sustained choroidal hemorrhage in each eye. Plange[23] described a second case, and Knapp[24] coined the term "angioid streak," which reflected the consensus at that time that the streaks were vascular. Kofler[25] in 1916 correctly interpreted the location of the streaks to be at the level of Bruch's membrane.

Gronblad, an ophthalmologist, and Strandberg, a dermatologist, separately reported a patient they had studied together who exhibited both the skin changes of PXE and angioid streaks of the retina. These two manifestations of the disease subsequently have been referred to as "Gronblad-Strandberg syndrome."[26] PXE is complicated by the formation of angioid streaks in the fundus in approximately 85% of patients. It is complicated by subsequent loss of central vision in 70% of patients so affected.[8,13]

Angioid streaks are bilateral asymmetric, clinically manifest cracks in an abnormal Bruch's membrane at the choroidal-retinal interface. Characteristically, they form an incomplete ring around the optic disc and they radiate anteriorly toward the equator of the globe. The streaks taper as they extend forward. The color varies from reddish-brown to dark brown, depending on the color of the overlying retinal pigment epithelium (RPE). In some cases the defects may appear grayish white if fibrous tissue has grown into the cracks.

The same elastic fiber abnormality exhibited elsewhere in the body apparently is seen in the eye. Bruch's membrane lies between the retinal RPE and the choroid of the eye. It is composed of an elastic layer sandwiched between two collagen layers. Primary changes in Bruch's membrane in PXE patients develop in the elastic layer, where rupture occurs and calcium is deposited. The "cracks" of angioid streaks allegedly fall coincident with the lines of force of the intraocular and extraocular muscles, as proposed by Adelung.[27] Histopathologically there is calcification and basophilia of Bruch's membrane with breaks in it that correspond to the clinical location of the streaks.[28] In some cases fibrous or fibrovascular proliferation from the underlying choroid may be seen extending through the breaks into the subretinal space. Hyperplastic or degenerative changes may occur in the RPE. Fibrovascular ingrowth may be accompanied by serous and hemorrhagic detachment of the RPE and retina.

The natural course of angioid streaks is varied. They may remain stationary or there may be periodic progressions in length and number. The cracks may widen as the cracked elastic membrane shrinks. While visual symptoms don't occur unless the angioid streaks cross the macula, secondary complications often cause acuity to decrease. These include hemorrhage into the macula, macular scarring, choroidal atrophy and sclerosis, and progressive RPE atrophy. Patients with complications of angioid streaks usually present with signs and symptoms of macular dysfunction including metamorphopsia or loss of central vision.[28-30]

Fluorescein angiography delineates angioid streaks well. Early transmitted choroidal hyperfluorescence indicates retinal pigmentary alterations. Persistent fluorescence of the streaks is due to scleral fluorescence and possible staining of collagen tissue in Bruch's membrane. Fluorescein angiography may be useful in demonstrating subtle streaks that are not recognized funduscopically. Detailed fluorescein angiography studies have been described.[31,32]

Visual loss in PXE is primarily due to complications of angioid streaks, such as macular hemorrhage and scarring. Less significant fundus changes of PXE include diffuse mottling of the RPE or *peau d'orange* akin to the skin changes, and irregularly shaped punched-out lesions in the periphery.[28,33]

Angioid streaks may be seen in sickle cell anemia, Paget's disease of bone, and other disease states (familial hyperphosphatemia, metastatic calcification, idiopathic thrombocytopenic purpura, Ehlers-Danlos syndrome, and lead poisoning). PXE is the condition most commonly associated with angioid streaks. The presence of angioid streaks in the eye should prompt an evaluation of elastic tissues elsewhere in the body and strongly suggests a diagnosis of PXE.

References

1. Eddy DD, Farber EM. Pseudoxanthoma elasticum. Internal manifestations: a report of cases and a statistical review of the literature. *Arch Dermatol*. 1962;86:729-740.
2. McKusick VA. The cardiovascular aspects of Marfan syndrome. *Circulation*. 1955;11:321-342.
3. Maumenee IH. The cornea in connective tissue diseases. Symposium: the eye and inborn metabolic disorders. *Ophthalmology*. 1978;85:1014-1017.
4. Loria PR, Kennedy CB, Freeman JA, et al. PXE. *Arch Dermatol*. 1957;76:609.
5. Huang S, Steele H, Kumar G, et al. Ultrastructural changes of elastic fibers in pseudoxanthoma elasticum: a study of histogenesis. *Arch Pathol*. 1967;83:108.
6. Findlay GH. On elastase and the elastic dystrophies of the skin. *Br J Dermatol Symp*. 1954;66:16.
7. Rodnan GP, Fischer ER, Warren JE. Pseudoxanthoma elasticum: clinical findings and identification of the anatomic defect. *Clin Res*. 1958;6:236. Abstract.
8. Goodman RM, Smith EW, Paton D, et al. Pseudoxanthoma elasticum: a clinical study and histopathological study. *Medicine*. 1963;42:297-334.
9. Pope FM. Two types of autosomal recessive pseudoxanthoma elasticum. *Arch Dermatol*. 1974;110:209-212.
10. Pope FM. Autosomal dominant pseudoxanthoma elasticum. *J Med Genet*. 1974;11:152-157.
11. Engelman MW, Fliegelman MT. Pseudoxanthoma elasticum. *Cutis*. 1978;21(6):837-840.
12. Fitzpatrick TB, Eisen AZ, Wolff K, et al., eds. *Dermatology in General Medicine*. 2nd ed. New York, NY: McGraw-Hill; 1979;1144-1154.
13. Connor PJ, Juergens JL, Perry HO, et al. PXE and angioid streaks. A review of 106 cases. *Am J Med*. 1961;30:537.
14. McKusick VA. *Medical Genetics 1958-1960*. St Louis, Mo: CV Mosby; 1961;437.
15. Kaplan L, Hartman SW. Elastica disease: case of Gronblad-Strandberg syndrome with GI hemorrhage. *Arch Intern Med*. 1954;94:489.
16. Heaton JPW, Wilson JWL. PXE and its urological implications. *J Urol*. 1986;135(4):776-777.
17. Mendelsohn G, Bulkley BH, Hutchins GM. Cardiovascular manifestations of pseudoxanthoma elasticum. *Arch Pathol Lab Med*. 1978;102:298-302.
18. Lebwoh MG, Distefano D, Prioleau PG, et al. Pseudoxanthoma elasticum and mitral valve prolapse. *N Engl J Med*. 1982;307:228-231.
19. Navarro-Lopez F, Llorian A, Ferrer-Roca O, et al. Restrictive cardiomyopathy in PXE. *Chest*. 1980;70:113-115.
20. Huang S, Kumar G, Steele HD, et al. Cardiac involvement in PXE. Report of a case. *Am Heart J* 1967;74:680-686.
21. Iqbal A, Alter M, Lee SH. PXE: a review of neurological complications. *Ann Neurol*. 1978;4(1):18-20.
22. Doyne RW. Choroidal and retinal changes. The result of blows on the eyes. *Trans Ophthalmol Soc UK*. 1889;9:128.
23. Plange O. Uber Streifenförmige Pigmentbildung mit Secundären Veränderungen der Netzhaut infolge von Hämorrhagien. *Arch Augenheilkd*. 1891;23:78.
24. Knapp H. On the formation of dark angioid streaks as an unusual metamorphosis of retinal hemorrage. *Arch Ophthalmol (N.Y.)*. 1892;21:289.
25. Kofler A. Beitraege zur Kenntnis der Angioid streaks. *Arch Augenheilkd*. 1917;82:134-149.
26. Gronblad E. Angioid streaks—pseudoxanthoma elasticum. *Acta Ophthalmol (KbH)*. 1929;7:329.
27. Adelung JC. Zur genese der Angioid streaks. *Klin Monatsbl Augenheilkd*. 1951;119:241-250.
28. Clarkson JG, Altman RD. Angioid streaks. *Surv Ophthalmol*. 1982;26(5):235-246.
29. Hogan JF, Heaton CL. Angioid streaks and systemic disease. *Br J Ophthalmol*. 1973;89(4):411-416.
30. Percival SPB. Angioid streaks and elastorrhexis. *Br J Ophthalmol*. 1968;52:297-309.
31. Hull DS, Aaberg TM. Fluorescein study of a family with angioid streaks and pseudoxanthoma elasticum. *Br J Ophthalmol*. 1974;58:738-745.
32. Rosen E. Fundus in pseudoxanthoma elasticum. *Am J Ophthalmol*. 1968;66(2):236-244.
33. Shimizu K. Mottled fundus in association with PXE. *Jpn J Ophthalmol*. 1961;5:1-13.

Section C

Hyperkeratotic Disorders

Chapter 200

Ichthyosis

JONATHAN C. HORTON and NEIL G. DREIZEN

Ichthyosis (Greek *ichthys,* a fish) is a physical finding common to a heterogeneous group of systemic diseases characterized principally by abnormal scaling of the skin. In most types of ichthyosis, the mechanisms responsible for excessive scale formation are poorly understood. In the past, ichthyosis was classified entirely on the basis of clinical presentation, giving rise to a plethora of confusing, descriptive terms in the medical literature. Over the past 20 years, the old nomenclature has been replaced by a new classification system (Table 200-1) that recognizes four primary diseases: ichthyosis vulgaris, X-linked ichthyosis, lamellar ichthyosis, and epidermolytic hyperkeratosis. Each has a distinct phenotypic appearance and a defined mode of inheritance.[1,2]

Systemic Manifestations

Ichthyosis vulgaris is the most common disease, with a prevalence of at least 1 per 5300 persons.[3] Scales usually develop within a few years after birth. Mildly affected patients exhibit only fine white scales, accompanied by dryness and pruritus. In more severely affected persons, scales appear more prominent and may involve a large portion of the skin. The scales are usually concentrated over the trunk and extensor surfaces of the extremities, sparing the axillary, antecubital, and popliteal fossas. There is frequently associated eczema, hay fever, or asthma. The disorder is inherited in an autosomal dominant fashion; the basis of the genetic defect is unknown.

X-linked ichthyosis is estimated to have a prevalence of 1 per 6190 males[3]; female carriers are phenotypically normal. The disease is generally more severe than ichthyosis vulgaris. Scales appear at birth or within the first few months of life, covering the scalp, face, neck, trunk, and limbs, including flexor surfaces. The scales are large, thick, and imbued with a dark yellow color that imparts a dirty appearance to the skin. No association with atopy has been reported. The disease has been linked to a deficiency of steroid sulphatase, an enzyme encoded by a gene on the X chromosome.[4] It remains to be elucidated how deficiency of this enzyme leads to the skin manifestations of ichthyosis.

599

TABLE 200-1 Principal Forms of Ichthyosis

Disease	Inheritance	Onset	Prevalence	Scales	Distribution	Ocular Findings
Ichthyosis vulgaris	Autosomal dominant	Childhood	1 in 5300	Fine and white	Trunk and limbs; spares flexor areas	Scales on lashes and eyelids
X-linked ichthyosis	X-linked recessive	Birth to 6 months	1 in 6190 males	Large and yellow	Neck, trunk, and limbs; includes flexor areas	Scales on lashes and eyelids, pre-Descemet's stromal deposits
Lamellar ichthyosis	Autosomal recessive	Birth	1 in 300,000	Large and polygonal	Generalized	Scales on lashes and eyelids, ectropion, lagophthalmos, corneal and conjunctival exposure
Epidermolytic hyperkeratosis	Autosomal dominant	Birth	1 in 300,000	Coarse and verrucous May be associated with bullae	Generalized or localized	Scales on lashes and eyelids

Lamellar ichthyosis is an exceedingly rare autosomal recessive disorder with a prevalence of about 1 per 300,000 persons.[3] The clinical appearance is striking. Infants are born encased in a characteristic, shiny, collodionlike membrane, which desquamates over the first weeks of life. Eventually the body is completely covered by large, polygonal scales, raised at the edges and affixed centrally. Hyperkeratosis of the palms and soles is present, with thickening and scarring of the nails.

Epidermolytic hyperkeratosis is a rare autosomal dominant disease with an estimated prevalence of 1 per 300,000 persons.[5] It is characterized by erythema, hyperkeratosis, and scaling, usually accompanied by localized blister formation. The disease is usually evident at birth. Bullae between 3 and 30 mm in diameter appear spontaneously, often recurring in certain regions of the skin. When they rupture, moist raw subepidermis is exposed, which becomes secondarily infected by gram-positive cocci, giving the skin an unpleasant odor.

Ocular Manifestations

Ichthyosis vulgaris has no significant ocular findings, although scales may be noted occasionally on the lashes and eyelids.

In X-linked ichthyosis scales are frequently observed on the lashes and eyelids. The cornea exhibits fine, irregular, punctate opacities evenly distributed in the deep stroma, just anterior to Descemet's membrane. The deposits can be quite extensive, forming a solid layer that gives the stroma a frosted appearance; however, they are usually subtle and may escape detection on routine slit lamp examination. The histopathology of these corneal lesions has never been described; why they occur in X-linked ichthyosis is unknown. Visual acuity and corneal function are entirely unimpaired. The chief significance of the corneal opacities lies in their diagnostic utility. They have been reported to occur in all patients with X-linked ichthyosis and in most female carriers, whereas they are generally absent in patients with ichthyosis vulgaris.[6] Thus the ophthalmologist may be asked to perform a slit lamp examination to help distinguish between these two forms of ichthyosis. It should be noted that some investigators have observed corneal deposits in only 50% of patients with X-linked ichthyosis, so absence of corneal deposits does not exclude the diagnosis of X-linked ichthyosis.[7,8] In cases where doubt remains after careful review of the physical findings and family history, measurement of leukocyte steroid sulphatase activity may establish the diagnosis.

In lamellar ichthyosis the tight collodionlike membrane covering newborns often produces bilateral ectro-

FIGURE 200-1 Child with lamellar ichthyosis has extensive scaling of the eyelids and bilateral ectropion. (Courtesy of Elias I. Traboulsi, M.D.)

pion of the upper and lower eyelids. This should be managed conservatively with topical lubricants to avoid exposure, as the condition often resolves spontaneously when the membrane peels. Later, however, extensive lamellar ichthyosis involving the eyelids may create excessive tautness of the skin, resulting in cicatricial ectropion (Fig. 200-1). Some degree of ectropion is present in at least half of patients with lamellar ichthyosis.[6,7] Exposure often ensues, with hyperemia, thickening, and keratinization of the conjunctiva. Lagophthalmos may result in exposure keratitis, corneal ulceration, and corneal neovascularization. Treatment is aimed at softening the skin by application of hydrating emollients like petrolatum jelly. Salicylic acid may be useful as a keratolytic to remove scales. Tretinoin (retinoic acid) cream retards epidermal cell turnover, thereby reducing scale formation. If these measures prove inadequate, surgical ectropion repair may be necessary.

In epidermolytic hyperkeratosis scales may be found on the lashes and eyelids. No other important ocular findings are present.

Other Ichthyosiform Disorders

In addition to the four principal forms of ichthyosis described above, a diverse group of approximately 20 diseases, most encountered infrequently, may exhibit ichthyosis as a physical finding. Only those with significant ocular manifestations are mentioned here.

Refsum's syndrome is an autosomal recessive disorder associated with elevated serum phytanic acid levels. The cardinal findings are nerve deafness, cerebellar ataxia, polyneuropathy, and mild ichthyosis. Night-

blindness, progressive visual field constriction, and loss of visual acuity occur in all patients, owing to an atypical retinitis pigmentosa.[9] Lens opacities and nystagmus have also been reported in some cases.

Sjögren-Larsson syndrome is another autosomal recessive disorder. It is characterized by spastic diplegia, mental retardation, and ichthyosis. Of the 28 patients originally described by Sjögren and Larsson, 11 patients underwent ophthalmologic evaluation. Three patients exhibited degenerative changes of the macular retinal pigment epithelium accompanied by reduction in visual acuity.[10]

More recently a new syndrome consisting of keratitis, ichthyosis, and deafness (KID) has been recognized.[11] The cornea is prone to repeated episodes of keratitis, leading eventually to opacification and vascularization. The cause and inheritance pattern of this syndrome remain unknown.

Finally, ichthyosis may appear at any age as a manifestation of occult malignancy. Hodgkin's disease, lymphosarcoma, multiple myeloma, and carcinoma of lung or breast are the most common associations. Therefore, in patients with acquired ichthyosis the ophthalmologist should be alert to the possibility of orbital or intraocular metastases.

References

1. Wells RS, Kerr CB. Genetic classification of ichthyosis. *Arch Dermatol.* 1965;92:1-6.
2. Frost P, Van Scott EJ. Ichthyosiform dermatoses. *Arch Dermatol.* 1966;94:113-126.
3. Wells RS, Kerr CB. Clinical features of autosomal dominant and sex-linked ichthyosis in an English population. *Br Med J.* 1966;1:947-950.
4. Shapiro LJ, Weiss R, Webster D, et al. X-linked ichthyosis due to steroid-sulphatase deficiency. *Lancet* 1978;1:70-72.
5. Baden HP. Ichthyosiform dermatoses. In: Fitzpatrick TB, Eisen AZ, Wolff K, et al. eds. *Dermatology in General Medicine.* New York, NY: McGraw-Hill; 1979;252-263.
6. Sever RJ, Frost P, Weinstein G. Eye changes in ichthyosis. *JAMA.* 1968;206:2283-2286.
7. Jay B, Blach RK, Wells RS. Ocular manifestations of ichthyosis. *Br J Ophthalmol.* 1968;52:217-226.
8. Lykkesfeldt G, Hoyer H, Ibsen HH, et al. Steroid sulphatase deficiency disease. *Clin Genet.* 1985;28:231-237.
9. Refsum S. Heredopathia atactica polyneuritiformis phytanic acid storage disease (Refsum's disease) with particular reference to ophthalmological disturbances. *Metabol Ophthalmol.* 1977;1:73-79.
10. Sjogren T, Larsson T. A clinical and genetic study of oligophrenia in combination with congenital ichthyosis and spastic disorders. *Acta Psychiatr Neurol Scand.* 1957;32(Suppl 113):1-112.
11. Skinner BA, Greist MC, Norins AL. The keratitis, ichthyosis, and deafness (KID) syndrome. *Arch Dermatol.* 1981;117:285-289.

Chapter 201

Psoriasis

ALAN SUGAR

Psoriasis is a common skin disease that affects about 2% of people in the United States.[1] It occurs at any age, but begins most frequently in older childhood or young adulthood. Although the cause is unknown, there is a family history of psoriasis in about one third of cases, and a strong association with several HLA antigen types (B13, B17, B27, B37, CW6, D11, DR7).[2] Autosomal dominant and polygenic inheritance have been suggested.

Systemic Manifestations

Characteristic skin lesions are large, well-defined, multiple, pink to red patches covered by silver-white scales. The lesions occur most often on the scalp and extensor surfaces, such as the back and elbows. They often begin at the site of skin trauma (Koebner's phenomenon). As many as 50% of patients have finger and toenails affected with ridging, pits, and separation of nails from the nailbed. Several clinical patterns of psoriasis occur; the classic form is psoriasis vulgaris. Guttate psoriasis occurs in children, and is characterized by multiple papules; it often follows streptococcal pharyngitis. In pustular psoriasis plaques are surrounded by pustules. Pustules are frequently present on the palms and soles, but they may be generalized. Rarely an exfoliative erythroderma develops, with severe systemic complications.[1,2]

Histopathologic examination reveals acanthosis and hyperkeratosis. There are increased basal layer mitoses and white cell infiltrates in the epidermis. The rate of epithelial cell turnover is greatly increased, and cells in all layers are correspondingly immature when compared to normal skin.[3]

The most frequent nondermal condition associated with psoriasis is arthritis, occurring in 5 to 7% of patients.[2] Of these, 70% have asymmetric monoarticular or oligoarticular arthritis of digits. Fifteen percent have polyarthritis of small joints similar to rheumatoid arthritis. The remaining 15% have ankylosing spondylitis, "sausage digit" distal arthritis, or severely destructive osteoarthritis mutilans. These are all seronegative forms of arthritis. Crohn's disease, ulcerative colitis, and Reiter's syndrome are also associated with psoriasis, probably through their linkage to HLA B27.

Psoriasis therapy has been an area of active research. It is important ophthalmologically because of known and potential ocular toxicities of psoriasis treatment regimens. The standard therapy for psoriasis has been the use of either topical steroids or topical coal tar derivatives, although liberal use of bland emollient creams and ointments is useful in mild cases. Steroids are applied as ointments, with or without occlusive dressings. The fluorinated compounds frequently used are very potent, and significant systemic absorption may occur. Ocular hypertension and cataracts may occur from chronic application around the eyes. Topical coal tar and its derivatives are widely used. Although their mode of action is unknown, they may act partly by increasing sensitivity to ultraviolet (UV) light. Anthrolin, another topical antimitotic agent, is a potent ocular irritant.

The observation that UV light helps psoriasis led to the development of photochemotherapy, the combined use of photosensitizing agents and controlled exposure to long-wave UV light (UVA, 320 to 400 nm), known as PUVA therapy. Oral methoxypsoralen is given 2 hours prior to a 20-minute exposure to UVA irradiation. While it is effective in 90% of patients, PUVA has the potential for late development of actinic skin disease, including carcinomas.[4] Because psoralens bind to lens proteins there is the potential for cararact formation.[5-10] As psoralens have an effect for up to 24 hours, patients are cautioned to wear near UV–blocking glasses (UV400, Blak-Ray, NOIR, etc.) during that time. A stellate cortical cataract has been reported in a patient treated with PUVA for vitiligo, but a significant increase in cataract has not been found in prospective studies of psoriasis patients who wear protective glasses during PUVA therapy.[7,10] Transient visual field defects have been reported in three PUVA patients.[11] Punctate keratitis, conjunctival injection, and dry eyes have been reported to occur in almost 50%.[9]

The retinoids are vitamin A derivatives effective in treating various disorders of keratinization. Oral etretinate (Tegison), a synthetic retinoid, has recently been approved for treatment of psoriasis. Its use has been associated with lash and brow hair loss, lid erythema, conjunctival injection, and dryness. Reversible retinal function abnormalities have been documented with other synthetic retinoids.[12] The systemic antimetabolite methotrexate has been widely used for severe psoriasis. Besides significant systemic toxicity, this drug causes ocular irritation, redness, and tearing.[13] Recently, studies of cyclosporin A for treatment of psoriasis have been initiated.

Ocular Manifestations

Ocular involvement in psoriasis can take several forms. Psoriatic patches may involve the lids and can extend onto the conjunctiva. Chronic blepharitis and lash loss frequently occur. Nonspecific conjunctivitis with mild ocular irritation and tearing is common.[14] Rarely, cicatricial conjunctival changes can occur with symblepharon, trichiasis, and dry eyes.[15,16]

Psoriatic keratitis can occur in a mild punctate form, frequently in association with conjunctivitis. Peripheral corneal opacities in the superficial or deep stroma with thickening of the overlying epithelium are reported, as are scattered subepithelial infiltrates.[14,15] Peripheral ulcerative keratopathy with stromal melting occurs rarely.[17] These forms of keratitis may resemble rosacea keratitis, or they may be quite nonspecific, suggesting some uncertainty in their relation to psoriasis. The incidence of cataracts is not increased in psoriasis patients, other than as a potential toxic reaction to steroid or PUVA therapy.[18]

Uveitis is the ocular complication of psoriasis most likely to cause vision loss, although it is infrequent. Bilateral indolent anterior uveitis with heavy vitreous debris sensitive to relatively low doses of systemic steroids has been described.[19] Retinal involvement with edema or vasculitis can occur, as can hypopyon iridocyclitis.[20] There is debate as to the relationship of iritis to psoriatic arthritis. In some series all patients with iritis also had arthritis, while in others none did.[19-21]

References

1. Baker H, Wilkinson DS. *Textbook of Dermatology.* 3rd ed. Oxford: Blackwell Scientific Publications; 1979.

2. Anderson TF. Psoriasis. *Med Clin North Am.* 1982;66:769.

3. Mehregan AH. Pinkus' guide to dermatohistopathology. 4th ed. Norwalk, Conn: Appleton, Century, Crofts; 1986.

4. Bickers DR. Position paper—PUVA therapy. *J Am Acad Dermatol.* 1983;8:226.

5. Lerman S, Megaw J, Willis I. Potential ocular complications from PUVA therapy and their prevention. *J Invest Dermatol.* 1980;74:179.

6. Woo TY, Wong RC, Wong JM, et al. Lenticular psoralen photoproducts and cataracts of a PUVA-treated psoriatic patient. *Arch Dermatol.* 1985;121:1307.

7. Cyrlin MN, Pedvis-Leftick A, Sugar J. Cataract formation in association with ultraviolet photosensitivity. *Ann Ophthalmol.* 1980;12:786.

8. Glew WB, McKeever G, Roberts WP, et al. Photochemotherapy and the eye: photoprotective factors. *Trans Am Ophthalmol Soc.* 1980;78:243.

9. Backman HA. The effects of PUVA on the eye. *Am J Optom Physiol Optics.* 1982;59:86.

10. Stern RS, Parrish JA, FitzPatrick TB. Ocular findings in patients treated with PUVA. *J Invest Dermatol.* 1985;85:269.

11. Fenton DA, Wilkinson JD. Dose-related visual field defects in patients receiving PUVA therapy. *Lancet.* 1983;1:1106.

12. Kaiser-Kupfer MI, Peck GL, Caruso RC, et al. Abnormal retinal function associated with Fenretinide, a synthetic retinoid. *Arch Ophthalmol.* 1986;104:69.

13. Doroshow JH, Locker GY, Gaasterland DE, et al. Ocular irritation from high-dose methotrexate toxicity. *Cancer.* 1981;48:2158.

14. Stuart JA. Ocular psoriasis. *Am J Ophthalmol.* 1963;55:615.

15. Eustace P, Pierse D. Ocular psoriasis. *Br J Ophthalmol.* 1970;54:810.

16. Kaldeck R. Ocular psoriasis. *Arch Dermatol Syph.* 1953;68:44.

17. Boss JM, Peachey RDG, Easty DL, et al. Peripheral corneal melting syndrome in association with psoriasis: a report of two cases. *Br Med J.* 1981;282:609.

18. Wancher B, Vesterdal E. Syndermatotic cataract in patients with psoriasis. *Acta Dermatovener.* 1976;56:397.

19. Knox DL. Psoriasis and intraocular inflammation. *Trans Am Ophthalmol Soc.* 1979;77:210.

20. Hatchome N, Tagami H. Hypopyon-iridocyclitis as a complication of pustular psoriasis. *J Am Acad Dermatol.* 1985;13:828.

21. Catsarou-Catsari A, Katsambas A, et al. Ophthalmological manifestations in patients with psoriasis. *Acta Dermatol Venereol (Stockh).* 1984;64:557.

Section D

Miscellaneous Skin Disorders

Chapter 202

Acne Rosacea

GARTH STEVENS, Jr., and MICHAEL A. LEMP

Acne rosacea is a dermatologic disorder characterized by erythema and telangiectasias of the sun-exposed areas of the face and upper chest that often has associated ocular problems. The dermatologic lesions of rosacea are characterized by erythema, telangiectasias, papules, pustules, and sebaceous gland hypertrophy distributed in the flush areas over the face and chest. The skin lesions are remarkable for the absence of comedones, distinguishing it from acne vulgaris. Rosacea occurs most commonly in the age range of 20 to 60 years and is rare in children.

Ocular involvement occurs in more than 50% of patients.[1] Borrie[2] examined two populations of rosacea patients, one from an eye hospital and one from a dermatologic service. He found that in 20% of patients with rosacea the eyes were affected first, in 27% ocular and skin manifestations occurred simultaneously, and in 53% skin lesions antedated ocular involvement. Ocular symptoms most commonly begin with foreign body sensation, pain, and burning.[3] Ocular signs include recurrent or chronic blepharitis, meibomitis, styes, chalazia, papillary conjunctivitis, an evanescent nodular conjunctivitis, episcleritis, superficial punctate keratopathy, and, in severe cases, infiltrative and ulcerative keratitis. The severe keratitis in rosacea is described as a symptomless peripheral vascularization with more central subepithelial opacities. The infiltrates are confined to the inferior two thirds of the cornea; they may become ulcerated and, ultimately, vascularized.[4] More commonly, the keratitis is limited to a superficial punctate epitheliopathy associated with blepharitis, conjunctivitis, or tear deficiency. Lemp et al.[5] reported aqueous tear deficiencies in 36% of patients with rosacea who presented to an ophthalmologist.

The clinical findings of rosacea blepharitis are nonspecific and consist primarily of telangiectatic vessels along the lid margins and meibomian gland alterations. Meibomitis may be manifested by excessive secretions, plugging of ducts, and dilation of ducts. Chalazia are frequently associated with rosacea.[6] In the tear film, sudsing is common, and superficial punctate keratopathy may occur. Less often, episcleritis and iritis have been reported.

The diagnosis of rosacea is based on clinical findings

of typical skin changes, though in 20% of cases the ocular symptoms may precede the skin changes, making the diagnosis difficult.[2] The ocular changes are nonspecific, but when seen in combination with typical skin findings they allow a firm diagnosis of ocular rosacea. Jenkins et al.[3] noted that 8 of 29 patients with ocular rosacea (responsive to tetracycline) had minimal objective signs despite marked symptoms. Skin signs, when they do occur, may be subtle, and underdiagnosis is common.

Diagnostic testing, unfortunately, has not been useful. Tear pH has been examined by several authors.[7-9] Despite initial suggestions that rosacea tears were alkaline or acid, subsequent study indicates that tear pH is probably normal in cases of rosacea.[9] Schirmer's test has been useful in identifying patients with rosacea who also have tear deficiencies. Lemp et al.[5] have hypothesized that the high incidence of low Schirmer's test results in patients with rosacea may be a coincidental finding due to patient selection. Patients with these two common ocular problems combined may seek earlier ophthalmic intervention. Numerous investigators have attempted to relate infestation of the parasite *Demodex folliculorum* to acne rosacea. The presence or absence of the parasite, however, does not appear to be related to the disease process.[10] Browning and Proia[11] suggested a point system for use in the diagnosis of acne rosacea that gives significant emphasis to the skin findings and less emphasis to the nonspecific ocular findings.

Histopathology

Histopathologic findings in rosacea dermatitis demonstrate vascular dilatation of the small vessels with a perivascular infiltration of histiocytes, lymphocytes, and plasma cells. Sebaceous gland hypertrophy commonly occurs in rhinophyma but is not a prominent feature in other lesions of rosacea.[10]

Dermal changes include loss of integrity of the superficial dermal connective tissue with edema, disruption of collagen fibers, and frequently severe elastosis.[10]

Immunoglobulin and complement deposition at the dermoepidermal junction has been reported in skin biopsies and in conjunctiva from rosacea patients.[12-15] Similar deposition was noted by Jablonska et al.[12] in patients with idiopathic telangiectasias, and the authors hypothesized that the vascular changes led to nonspecific immune complex deposition.

Ocular pathologic findings include conjunctival and corneal infiltration with chronic inflammatory cells, including lymphocytes, epithelioid cells, plasma cells, and giant cells.[16] Nonspecific corneal scarring has also been noted.[11]

Pathogenesis

The pathogenesis of rosacea is unknown, but it is presumed to be a genetically determined anomalous vascular response that develops mostly in the 3rd to 6th decades of life. The hypothesis that vasodilatation or a flushing disorder is the basic pathogenesis of the disease is based on several findings. The first is that the disease appears to be more frequent and prevalent in northern climates, where light-skinned persons are more exposed to cold. In addition, association with a flushing disorder is noted in the carcinoid syndrome, where recurrent flushing over the face and upper chest has resulted in the full-blown rosacea syndrome.[1] Marks[10] reported a higher degree of solar elastosis in rosacea biopsy specimens. He hypothesized that the resulting altered connective tissue gives inadequate support for the small vessels, resulting in prolonged vasodilatation and secondary immune complex deposition. A variety of hypotheses relating to presumed underlying gastrointestinal and psychosomatic disorders are not supported by scientific evidence.[17]

The concept that rosacea is a sebaceous gland disease is supported by the observation in severe rosacea that rhinophyma may result from pronounced sebaceous gland hypertrophy; however, sebaceous gland hypertrophy is not a typical feature of early rosacea lesions.[10] Meibomitis is a frequent accompaniment of rosacea, and McCulley and Sciallis[18] have suggested a diffuse sebaceous gland abnormality to be the cause of the meibomitis. *Demodex folliculorum* has been considered a causative agent of rosacea, as it feeds on sebum, and some reports of treatment of *Demodex* infestation have noted that rosacea also improved.[11,19] However, in a review of 74 biopsies of rosacea papules, Marks[10] noted evidence of *D. folliculorum* in only 19% of specimens. Tissue-fixed immunoglobins have been reported in patients with chronic inflammation of rosacea, but other evidence of an autoimmune process is lacking.[14,15] A bacterial cause for the disease has been hypothesized, but no consistent findings of bacteria, either in the skin or the lid margins, have been demonstrated.

In summary, the pathogenesis of rosacea remains unknown; no single hypothesis can adequately explain both the vascular changes and the inflammatory reaction.

Oral tetracycline appears to be the most effective therapy for ocular or cutaneous rosacea. Jenkins and associates[3] reported successful therapy with tetracycline in 36 of 37 patients. Erythromycin and ampicillin have also been suggested for rosacea.[20]

The mode of action of tetracycline is unclear.[21] It decreases chemotaxis of neutrophils and this may have a salutary effect on the inflammatory component of the disease.[22] It is not clear whether tetracycline significantly alters the lipid component of secretions directly or whether there is an indirect effect secondary to alteration of the bacterial flora. Treatment with tetracycline involves the use of 250-mg capsules two to four times a day and has been reported to show marked improvement after 3 days' to 3 weeks' therapy. In addition, warm compresses and lid scrubs are frequently recommended. More recently, doxycycline 100-mg per day has been found to be as effective as tetracycline with fewer side effects.[23]

In summary, acne rosacea is a common dermatologic and ocular disorder of unknown origin that often causes ocular symptoms from blepharitis, conjunctivitis, or punctate keratitis. In more severe forms the facial lesions may be cosmetically significant and the ocular disease may threaten sight. Therapy with tetracycline is effective in most cases, and long-term low-doses of therapy is often useful in maintaining prolonged remissions.

References

1. Starr PAH, McDonald A. Oculocutaneous aspects of rosacea. *Proc R Soc Med.* 1969;62:9.
2. Borrie P. Rosacea with special reference in its ocular manifestations. *Br J Dermatol.* 1953;65:458.
3. Jenkins MA, Brown SI, Lempert SL, et al. Ocular rosacea. *Am J Ophthalmol.* 1979;88:618.
4. Goldsmith AJB. The ocular manifestations of rosacea. *Br J Dermatol.* 1953;65:448.
5. Lemp MA, Mahmood MA, Weiler HH. Association of rosacea and keratoconjunctivitis sicca. *Arch Ophthalmol.* 1984;102:556.
6. Lempert SL, Jenkins MS, Brown, SI. Chalazia and rosacea. *Arch Ophthalmol.* 1979;97:1652.
7. Abelson MB, Sadun AA, Udell IJ, et al. Alkaline tear pH in ocular rosacea. *Am J Ophthalmol.* 1980;90:866.
8. Jaros PA, Coles WH. Ocular surface pH in rosacea. *CLAO J.* 1983;9:333.
9. Browning DJ. Tear studies in ocular rosacea. *Am J Ophthalmol.* 1985;99:530.
10. Marks R, Harcourt-Webster JN. Histopathology of rosacea. *Arch Dermatol.* 1969;100:682.
11. Browning DJ, Proia AD. Ocular rosacea. *Surv Ophthalmol.* 1986;31:145.
12. Jablonska S, Chorzelski T, Maciejowska E. The scope and limitations of the immunofluorescence method in the diagnosis of lupus erythematosus. *Br J Dermatol.* 1970;83:242.
13. Manna V, Marks R, Holt P. Involvement of immune mechanisms in the pathogenesis of rosacea. *Br J Dermatol.* 1982;107:203.
14. Nunzi E, Rebora A, Hamerlinck F, et al. Immunopathological studies on rosacea. *Br J Dermatol.* 1980;103:543.
15. Brown SI, Shahinian L. Diagnosis and treatment of ocular rosacea. *Ophthalmology.* 1978;85:779.
16. Foster CS. Ocular surface manifestations of neurological and systemic disease. *Int Ophthalmol Clin.* 1979;69(2):229.
17. Wilkin JK. Rosacea. *Int J Dermatol.* 1983;22:343.
18. McCulley JP, Sciallis GF. Meibomian keratoconjunctivitis. *Am J Ophthalmol.* 1977;84:788.
19. Rufli T, Buchner SA. T-cell subsets in acne rosacea lesions and the possible role of *Demodex folliculorum. Dermatologica.* 1984;169:1.
20. Marks R, Wilkinson OS. Rosacea and perioral dermatitis. In: Rook A, Wilkinson DS, Ebling FJG, et al., eds. *Textbook of Dermatology.* Oxford: Blackwell Scientific Publications; 1986;1610.
21. Salamon SM. Tetracyclines in ophthalmology. *Surv Ophthalmol.* 1985;29:265.
22. Martin RR, Warr GA, Couch RB, et al. Effects of tetracycline on leukotaxis. *J Infect Dis.* 1974;129:110.
23. Frucht-Pery J, Chayet AS, Feldman ST, et al. The effect of doxycycline on ocular rosacea. *Am J Ophthalmol.* 1989;107:434.

Chapter 203

Atopic Dermatitis

PETER A. RAPOZA and JOHN W. CHANDLER

Atopic dermatitis is a chronic, pruritic, erythematous inflammation of the skin that occurs in conjunction with a variety of ocular manifestations. The disorder is often associated with increased serum levels of IgE and a family or personal history of allergic rhinitis or asthma. The prevalence of atopic dermatitis in the general population is nearly 5%. It typically develops before age 5 years (90% of cases), and most often (60%) by 1 year of age.

The pathogenesis of atopic dermatitis is largely unknown. Historically, the disorder has been termed "*neurodermatitis,*" suggesting its purported psychogenic origin; it is now believed that psychologic changes are caused by exacerbations of the disease. Although the factors necessary to produce atopic dermatitis are not known, the major hypotheses include abnormalities in the immune system and blockage of β-adrenergic receptors.[1,2]

Approximately 80% of patients with atopic dermatitis have elevated levels of serum IgE. The remainder have normal or decreased levels of that immunoglobulin. Most studies support a relationship between disease severity and disease activity. The level of IgE is most elevated when other intercurrent atopic disorders are present.

Abnormalities in cell-mediated immunity have also been recognized. No consistent relationship to specific HLA status has been demonstrated. Decreased numbers and diminished function of T lymphocytes usually parallel disease activity. Defects in the chemotactic activity of neutrophils and monocytes have also been demonstrated. These defects increase in proportion to advancing age.

Blockage of β-adrenergic receptors with subsequent decreased amounts of intracellular cyclic adenosine monophosphate has been noted in atopic dermatitis. This finding may explain several of the manifestations of atopic dermatitis, including exaggerated vasoconstriction on exposure to cold; white dermographism in response to stroking the skin with a blunt instrument; delayed blanch response and increased sweating in response to intradermal injection of acetylcholine; and lichenification of the skin as epidermal mitotic activity is inhibited by β-adrenergic stimulation.

The protean manifestations of atopic dermatitis have produced confusion in diagnosing the disease. Diagnostic criteria have been established to allow for consistent assessment and comparability of patients (Table 203-1).[1]

Systemic Manifestations

Nonocular involvement in atopic dermatitis is usually limited to the skin.[2,3] The primary symptom is intense itching. According to some sources, predilection to infections leads to an increased prevalence of pneumonitis and encephalitis.

Clinically, the disease has been subclassified into three chronologic periods. The infantile phase (birth to age 2 years) has as its initial lesion erythema of the cheeks, followed by the formation of exudative papules on the forehead and extensor surfaces of the extremities (Fig. 203-1). Examination of histopathologic sections reveals hyperkeratosis, parakeratosis, acanthosis, and intercellular and intracellular accumulations of fluid. The epidermis and dermis are infiltrated by lymphocytes, monocytes, and macrophages. There is a spontaneous remission rate of 40% by age 5 years.

Atopic dermatitis during the childhood phase (ages 2

TABLE 203-1 Diagnostic Criteria for Atopic Dermatitis[1]

A. Absolute features (all of the following must be present)
  1. Pruritus
  2. Typical morphology and distribution
    a. Flexural lichenification in adults
    b. Facial and extensor involvement in infancy
  3. Tendency toward chronic or chronically relapsing dermatitis

  plus

B. Two or more of the following features
  1. Personal or family history of atopic disease (asthma, allergic rhinitis, atopic dermatitis)
  2. Immediate skin test reactivity
  3. White dermographism and/or delayed blanch to cholinergic agents
  4. Anterior subcapsular cataracts

  or

C. Four or more of the following features
  1. Xerosis/ichthyosis/hyperlinea of the palms
  2. Pityriasis alba
  3. Keratosis pilaris
  4. Facial pallor and infraorbital darkening
  5. Dennie-Morgan infraorbital fold
  6. Elevated serum IgE
  7. Keratoconus
  8. Tendency toward nonspecific hand dermatitis
  9. Tendency toward repeated cutaneous infections

to 12 years) may be an initial presentation or may follow from the infantile stage. Xerosis of the skin is a relatively constant feature. Elevated papules or confluent plaques of lichenification are found, especially in the flexural areas of the extremities and around the eyes and mouth.

Initial presentation during the adolescent or adult phase (after 12 years of age) is rare, although the disease may persist from an earlier period. Lichenification of the flexural surfaces becomes even more pronounced. The face, neck, hands, and feet are especially involved (Fig. 203-2). Histopathologic examination reveals the presence of hyperkeratosis, dyskeratosis, and a greatly thickened epidermis. The epidermis has a slight lymphocytic infiltration, whereas the dermis has more moderate accumulations of lymphocytes, monocytes, macrophages, and mast cells.[4]

Intercurrent infection with herpes simplex virus or *Vaccinia* can result in Kaposi's varicelliform reaction, a disseminated vesicular eruption that usually occurs in areas that have previously exhibited cutaneous manifestations of atopic dermatitis. Exposure of atopic dermatitis patients to smallpox vaccine or herpes simplex virus should therefore be minimized. Bacterial infections of the skin are often superimposed on the lesions of atopic dermatitis.

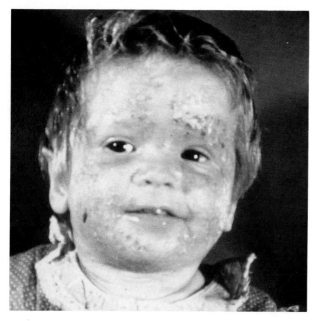

FIGURE 203-1 Exudative papules of infantile atopic dermatitis. (Courtesy of Derek Cripps, M.D.)

Ocular Manifestations

Ocular manifestations of atopic dermatitis are well recognized.[5] Loss of the lateral portions of the eyebrows may occur secondary to rubbing. Dennie-Morgan folds, prominent folds of the lower eyelids beginning at the inner canthi, are found in lids affected by chronic atopic dermatitis.[5,6]

Keratoconjunctivitis is usually bilateral and is associated with cutaneous disease activity.[7,8] There is diffuse papillary hypertrophy of hyperemic, edematous conjunctiva. Clear epithelial inclusion cysts and Trantas' dots are occasionally present. Symblepharon, entropion, and trichiasis can be found in rare advanced cases. Corneal involvement includes pannus, superficial epithelial keratitis, and superficial stromal haze. Marginal ulceration and corneal opacification and vascularization have been reported. The relationship between atopic dermatitis and keratoconus remains controversial.

Cataracts have been found in up to 25% of patients with atopic dermatitis. Most frequently described are anterior cortical shieldlike cataracts and posterior subcapsular cataracts. The severity of atopic dermatitis may relate to an increased prevalence of cataracts. Less common associations have been noted between atopic dermatitis and uveitis, retinal detachment, and corticosteroid-induced glaucoma with visual field loss (related to protracted application of topical steroids to the eyelids).

FIGURE 203-2 Lichenification of the face, neck, and hands in adolescent/adult atopic dermatitis. (Courtesy of Derek Cripps, M.D.)

Treatment

The goal of treatment of atopic dermatitis is to control pruritus and restore the skin to its normal condition. Avoidance of any identifiable inciting agents, proper hygiene, application of moisturizing agents, and use of topical steroids or antibiotics are usually sufficient. Topical chromolyn may reduce itching, inflammation, and lichenification of the skin. It also has been used successfully to control keratoconjunctivitis. Systemic steroids, systemic antibiotics, or PUVA treatments are sometimes necessary in refractory cases.

References

1. Hanflin JM, Lobitz WC Jr. Newer concepts of atopic dermatitis. *Arch Dermatol.* 1977;113:663-670.
2. Champion RH, Parish WE. Atopic dermatitis. In: Rook A,

Wilkinson DS, Ebling FJG, eds. *Textbook of Dermatology.* Oxford: Blackwell Scientific Publications; 1986;419-434.

3. Blaylock WK. Atopic dermatitis. In: Stone J, ed. *Dermatologic Immunology and Allergy.* St Louis, Mo: CV Mosby; 1985;315-322.

4. Mihm MC Jr, Soter NA, Dvorak HF, et al. The structure of normal skin and the morphology of atopic eczema. *J Invest Dermatol.* 1976;67:305-312.

5. Garrity JA, Liesegang TJ. Ocular complications of atopic dermatitis. *Can J Ophthalmol.* 1984;19:21-24.

6. Uehara M. Infraorbital fold in atopic dermatitis. *Arch Dermatol.* 1981;117:627-629.

7. Hogan MJ. Atopic keratoconjunctivitis. *Trans Am Ophthalmol Soc.* 1952;50:265-280.

8. Braude LS, Chandler JW. Atopic corneal disease. *Int Ophthalmol Clin.* 1984;24:145-156.

Chapter 204

Ectrodactyly, Ectodermal Dysplasia, and Cleft Lip and Palate

EEC Syndrome

EDWARD JAY GOLDMAN

Ectrodactyly

Ectrodactyly, split (cleft) hand and food or lobster-claw deformity, is the congenital absence of one or more digits of the central rays on the hand or foot (Fig. 204-1). A hereditary basis for this deformity can be found in 90% of cases, and the incidence is reported to be 1 in 90,000.[1] It is generally considered to be an autosomal dominant disorder but has rarely been reported with an autosomal recessive pattern. It occurs in two forms.

Type I is characterized by bilateral cleft feet with or without cleft hands. This form has 100% penetrance.
Type 2 is more variable. The feet are not always affected. This type shows incomplete penetrance and may skip generations.

The most common anomalies seen in association with ectrodactyly are ectodermal dysplasia, tear duct abnormalities, and cleft palate.[2] Patients with lobster-claw deformity of the hands usually have normal, well-coordinated manual skills and legible script. Ectrodactyly may also be associated with syndactyly (fusion of adjacent fingers or toes).

Ectodermal Dysplasia

Ectodermal dysplasia has four major clinical features: trichodysplasia (abnormal hair), abnormal dentition, onychodysplasia (abnormal nails), and dyshidrosis. Some authors have combined these features in their various permutations, and at least 32 different ectodermal dysplasias are currently described in the literature.[3] The incidence of ectodermal dysplasia occurring independently of other disorders is reported to be as high as 1 in 10,000 live births. The spectrum of associated diseases includes abnormalities of nonectodermal origin as well as lacrimal punctal atresia and cleft lip and palate.

Cleft Lip with Cleft Palate

The entity of cleft lip with cleft palate is separate and distinct from cleft palate alone (Fig. 204-1).[4] Their frequency varies greatly in the literature but has been estimated by the World Health Organization[5] to be approximately 1 in 1000 for cleft lip with cleft palate and 1 in 5000 for cleft palate alone. Most cases of cleft lip with

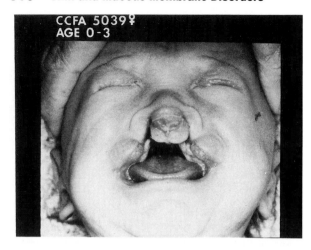

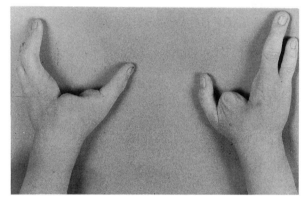

FIGURE 204-1 (**Top**) Cleft lip and palate. (**Bottom**) Ectrodactyly. (Courtesy of the Center for Craniofacial Anomalies, University of Illinois.)

cleft palate are felt to be sporadic. The association of facial clefting and multiple malformations has been emphasized in the literature, but probably not more than 14% of patients with facial clefts have such additional problems.

EEC Syndrome

In 1963 Walker and Clodius[1] reported a case of ectrodactyly with cleft lip and palate that was later noted to have features of ectodermal dysplasia.[2] The syndrome of ectrodactyly, ectodermal dysplasia, and cleft lip with cleft palate was formally described and named EEC syndrome by Rudiger, Haase, and Passarge[6] in 1970. Since that time numerous descriptions of the full syndrome have appeared in the literature.

If each abnormality is considered on an independent basis, it is difficult to postulate the syndrome occurring as frequently as it has been reported. Ectrodactyly

(1:90,000) X ectodermal dysplasia (1:10,000) X cleft lip and cleft palate (1:1,000) would be expected to occur approximately once in 10^{12} persons. The EEC syndrome generally occurs as an autosomal dominant disorder with incomplete penetrance.[7] Sporadic cases have been described, and a single case associated with a chromosomal abnormality has been reported.

The lacrimal sac and duct are formed during the overlapping of the lateral nasal and maxillary processes, trapping a fold of epithelium within. Later the fold canalizes and forms the lacrimal duct, while subsequent outpouching forms the canaliculi and punctae between the 6th and 9th weeks of gestation. During the 6th week, the hand and foot plates are forming and the digits are beginning to develop. The threshold period for closure of the palate occurs on approximately the 47th day. It is therefore likely that whatever is influencing these associated malformations is occurring during the 6th or 7th week of gestation.

The epithelium within the fold that later develops into the lacrimal drainage system may assume pathologic significance when the ectoderm is abnormal. It is known that lacrimal defects may be inherited independently in a dominant fashion, but EEC patients probably have this as part of their syndrome by involvement of related structures.

Ocular Manifestations

It has been observed that EEC patients have abnormally small numbers of meibomian gland orifices on the lid margin.[8] These glands normally develop from ectodermal cells as they invaginate in the lid margins. The oily secretion of these holocrine glands functions to stabilize the tear film. These oils increase the thickness and smooth the surface of the tear film, thereby improving its optical properties. The interaction between the meibomian lipids and the underlying mucin layer reduces the surface tension of tears and increases the ability of the film to spread. The major components of the meibomian secretion retard the evaporation of water from tears. The sebum coating the lids acts as a nonpolar barrier, discouraging overspill of liquid tears beyond the lid margin.[9] It is not surprising that derangement of such a delicate system has effects on lid hygiene, tear function, and vision.

A series of these patients were studied and detailed eye examinations were performed.[10] All patients seem to have severe photophobia. In nearly every case the lid margins are affected with blepharitis, which usually leads to partial corneal vascularization, opacification, and scarring. The tear film abnormalities are complicated by the frequent finding of lacrimal punctal atresia.

The poor drainage system spares the patient with dry

eyes some complications, but contributes to the problems of the patient with otherwise normal aqueous tear film production. Most patients with EEC do not have dry eyes on Schirmer's test. With no drainage outlet, the tear lake enlarges. In normal persons, the oily barrier secreted by the meibomian glands lining the lid margin discourages overspill of tears onto the lashes. In EEC patients, the tear film is stagnant, overflows onto the lashes, and exacerbates the blepharitis. The severe corneal problems these patients may have are probably secondary to the chronic blepharitis, but they could also be a primary manifestation of the ectodermal dysplasia.

An abnormal lacrimal drainage system may be associated with canaliculitis. The combined lacrimal and lid margin inflammatory and infectious processes that affect the conjunctival goblet cells cause further tear film instability. Meibomian gland involvement is seen in every patient with this disease; however, the inflammation may vary, and the secondary disruption of the cells producing the mucin and aqueous layer of the tear film is not constant. The variation in lid margin involvement leads to a spectrum of changes in tear production that corrrespond to the different degrees of severity of corneal pathology as described in the literature.

These patients are born with a number of serious malformations. The clefting can be managed well by plastic surgery, and the dental problems have been ameliorated with prostheses. Fortunately the limb disfigurement presents no functional handicap. Patients are generally in stable health by the 2nd decade and are then faced with the constant problem of blepharitis and tear film abnormalities. They must be monitored closely to avoid the potential ocular complications of this disease.

References

1. Walker JC, and Clodius L. The syndrome of cleft lip, cleft palate, and lobster claw deformities. *Plast Reconstr Surg.* 1963;32:627.
2. Preus M, Fraser FC. The lobster claw defect with ectodermal defects, cleft lip-palate, tear duct anomaly and renal anomalies. *Clin Genet.* 1973;4:369.
3. Freire-Maia N. Ectodermal dysplasia. *Hum Hered.* 1971;21:309.
4. Rosenmann A, Shapira T, Cohen MM. Ectrodactyly, ectodermal dysplasia and cleft palate (EEC syndrome). *Clin Genet.* 1976;9:347.
5. Warkany J. *Congenital Malformations.* Chicago, Il: Year Book Medical Publishers; 1971;628-647.
6. Rudiger RA, Haase W, Passarge E. Association of ectrodactyly, ectodermal dysplasia, and cleft lip-palate. *Am J Dis Child.* 1970;120:160.
7. Bixler D, Spivack J, Bennett J, et al. The ectrodactyly-ectodermal dysplasia-clefting (EEC) syndrome. *Clin Genet.* 1971;3:43.
8. Pries C, Mittelman D, Miller M, et al. The EEC syndrome. *Am J Dis Child.* 1974;127:840.
9. Holly F. Formation and stability of the tear film. The preocular tear film and the dry eye syndromes. *Intl Ophthalmol Clin.* 1971;13:73.
10. Goldman EJ, Miller MT, Mittelman D. Ectrodactyly ectodermal dysplasia and cleft lip and palate (EEC): ophthalmologic findings (unpublished data).

Chapter 205

Erythema Nodosum

S. LANCE FORSTOT and JOSEPH Z. FORSTOT

Erythema nodosum is a cutaneous inflammation usually associated with an underlying systemic disease.[1-4] Classically erythema nodosum is characterized by multiple, bilateral, erythematous, tender, nonulcerating, nonscarring nodules on the extensor surfaces of the lower legs. The lesions may also be found on the extensor surfaces of the forearms. They are not pruritic. They usually attain maximum size, between 0.5 and 2.0 cm in diameter, within 1 day. They last from 1 to 3 weeks, progressing from red to a darker purple hue, passing through color changes that include yellow, green, and blue. As noted, there is no ulceration and the lesions do not leave scars or pigmentary changes.

On histopathologic examination of early lesions of erythema nodosum, the inflammatory cellular infiltrate is mainly polymorphonuclear leukocytes mixed with lymphocytes. Later the lesions are composed predominantly of lymphocytes and macrophages. Perivascular inflammation may occur around small veins of the septa of the fat lobule with cells present in the walls of the veins. Inflammatory involvement of the arterioles is rare. Hemorrhage and a granulomatous reaction with

TABLE 205-1 Etiologic Associations with Erythema Nodosum

Infectious diseases
  Bacterial
    Streptococci
    Yersinia
    Tuberculosis
    Leprosy
  Chlamydial
    Psittacosis
    Lymphogranuloma venereum
  Fungal
    Histoplamosis
    Coccidioidomycosis
    North American blastomycosis
    Trichophytosis
  Other
    Cat scratch disease

Systemic diseases
  Sarcoidosis
  Behçet's syndrome
  Inflammatory bowel disease
    Ulcerative colitis
    Regional enteritis (Crohn's disease)

Drugs
  Sulfonamides
  Oral contraceptives
  Bromides
  Iodides

Idiopathic

giant cell formation may occur. Basically, erythema nodosum is a panniculitis involving the connective tissue septa of the fat lobules of the subcutaneous tissue.[5] The dermis is involved only to a mild degree.

Although idiopathic erythema nodosum may occur in an otherwise healthy person, the disease is usually associated with a systemic disease (either infectious or inflammatory) or with drug therapy (Table 205-1).

Systemic Manifestations

Erythema nodosum may appear before the symptoms of the underlying disease or concurrently with it. Systemic manifestations are only those of the specific associated disease and not of erythema nodosum itself. The immunopathogenesis of erythema nodosum is not clear. The role of cell-mediated and immune complex–mediated mechanisms has not been established.

Ocular Manifestations

Erythema nodosum has no specific or characteristic ocular findings; however, ocular manifestations are not uncommon findings in the associated systemic diseases (eg, sarcoidosis, Behçet's syndrome, tuberculosis, leprosy, histoplasmosis, cat scratch disease). The ocular findings are more specific for the underlying disease and are not associated with isolated erythema nodosum. Erythema nodosum occurs only in a minority of the cases of these underlying diseases, and it has no direct relationship to their ocular manifestations.

References

1. Blomgren SE. Erythema nodosum. *Semin Arthritis Rheum*.1974;4:1-24.
2. Cupps TR, Fauci AS. Miscellaneous syndromes. In: Smith LH, ed. *The Vasculitides*. Philadelphia, Pa: WB Saunders, 1981;147-149.
3. White JW. Erythema nodosum. *Derm Clin North Am*. 1985;3:119-127.
4. Bullock WE. The clinical significance of erythema nodosum. *Hosp Pract*. 1986;March:102E-102X.
5. Winkelmann RK, Forstrom L. New observations in the histopathology of erythema nodosum. *J Invest Dermatol*. 1975;65:441-446.

Chapter 206

Focal Dermal Hypoplasia

Goltz' Syndrome

G. FRANK JUDISCH

In 1962, Goltz et al.[1,2] reported on three girls with linear areas of cutaneous connective tissue deficiency, which they referred to as "focal dermal hypoplasia" (FDH). By 1981, approximately 100 cases of FDH, also known as Goltz' or Goltz-Gorlin syndrome, had been recognized. Although the skin lesions are an obligate feature, numerous other congenital anomalies of tissues of mesodermal and ectodermal origin are usually present.

The cause of FDH is not entirely clear. Most cases are isolated, and the great majority occur in females. Female-to-female transmission through four generations has been reported, as has father-to-daughter transmission. In most cases the mode of transmission is thought to be X-linked dominant and to be lethal for males.[3] Father-to-daughter transmission would not be incompatible with this mode of inheritance if the father were mosaic or if the mutation arose in a gamete or in a very early postzygotic stage.[4]

Systemic Manifestations

The pathognomonic feature of FDH is patchy but widespread dermal dysgenesis, which varies from mild hypoplasia to virtual aplasia.[5-11] The anomalous dermis is manifested in a variety of ways. Herniation of subcutaneous fat through the dermal defects may produce soft reddish yellow nodules. Streaks of telangiectasia and linear or reticular areas of hyper- or hypopigmentation are common. Angiofibromas commonly develop periorifically and on mucous membranes. If the dermis is totally absent, the slightest trauma causes ulceration, which tends to heal normally by scarring. Areas of total skin absence have been reported in a few patients. The fingernails and toenails are usually dystrophic or absent.

The second most frequently involved system is the skeleton. Bilateral syndactyly, especially between the third and fourth fingers, is a most common finding. Polydactyly, oligodactyly, clinodactyly, adactyly, and hemimelia are also common. Not infrequently spinal deformities and microcephaly have been observed. Of special note are striations in the metaphyseal regions of the long bones seen on radiographic examination, which are a characteristic feature of FDH.[12] It is presumed that the striations, which result from two populations of chondroblasts producing matrix with two different calcium levels, like the linear or patchy skin lesions, are a manifestation of Lyon's hypothesis. The random inactivation of an X chromosome results in two populations of somatic cells, one with the normal X chromosome remaining active, the other with the normal X chromosome becoming the Barr body.

A variety of other less common systemic abnormalities have been reported. The dental anomalies include hypodontia, microdontia, retarded eruption, and enamel fragility. Cleft lip and palate are occasionally seen. A few patients have had aural defects. Varying

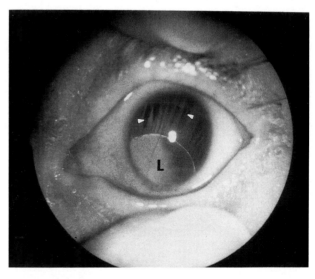

FIGURE 206-1 Left eye of a patient with FDH. There is severe iris hypoplasia, almost to the point of true aniridia. The lens (*L*) is displaced inferiorly. Numerous, elongated ciliary processes (*arrowheads*) are drawn inferiorly and are attached to the lens by active blood vessels. The fellow eye showed atypical coloboma involving the iris, retina, uvea, and optic nerve.

degrees of mental retardation and seizures are infrequent but serious manifestations.

Ocular Manifestations

It is estimated that about 40% of patients with FDH have ocular abnormalities; however, as so few of the cases have been reported in the ophthalmic literature this percentage may be suspect.[5,6,10] Photophobia is said to be the most common ocular symptom. A wide range of ocular abnormalities have been described, many of them reflecting developmental aberrations during early gestation. These include colobomas of the iris, retina, choroid, and optic nerve; microphthalmos; anophthalmos; aniridia; pupil irregularities; lens subluxations; retinal pigmentary aberrations; cloudy cornea and vitreous; strabismus; nystagmus; ptosis; blue scleras; and nasolacrimal duct obstruction (Fig. 206-1). No data are available on the number of patients who suffer significant vision impairment from these defects.

References

1. Goltz RW, Peterson WC, Gorlin RJ, et al. Focal dermal hypoplasia. *Arch Dermatol.* 1962;86:52.

2. Goltz RW, Henderson RR, Hitch JM, et al. Focal dermal hypoplasia syndrome. *Arch Dermatol.* 1970;101:1.

3. McKusick VA. *Mendelian Inheritance in Man.* 7th ed. Baltimore, Md: The Johns Hopkins University Press; 1986;1351-1352.

4. Burgdorf WHC, Dick GF, Soderberg MD, et al. Focal dermal hypoplasia in a father and daughter. *J Am Acad Dermatol.* 1981;4:273.

5. Warburg M. Focal dermal hypoplasia. *Acta Ophthalmol.* 1970;48:525.

6. Willetts GS. Focal dermal hypoplasia. *Br J Ophthalmol.* 1974;58:620.

7. Ruiz-Maldonado R, Carnevale A, Tamayo L, et al. Focal dermal hypoplasia. *Clin Genet.* 1974;6:36.

8. Toro-Sola MA, Kistenmacher ML, Punnett HH, et al. Focal dermal hypoplasia syndrome in a male. *Clin Genet.* 1975;7:325.

9. Gorlin RJ, Pindborg JJ, Cohen MM. Syndromes of the head and neck. 2nd ed. New York, NY: McGraw-Hill; 1976;310-314.

10. Thomas JV, Yoshizumi MO, Beyer CK, et al. Ocular manifestations of focal dermal hypoplasia syndrome. *Arch Ophthalmol.* 1976;5:1997.

11. Kegel MF. Dominant disorders with multiple organ involvement. *Dermatol Clin.* 1987;5:205.

12. Happle R, Lenz W. Striation of bones in focal dermal hypoplasia: manifestation of functional mosaicism. *Br J Dermatol.* 1977;96:133.

Chapter 207

Malignant Atrophic Papulosis

Degos' Syndrome

DAVID A. LEE and W. P. DANIEL SU

Malignant atrophic papulosis (Degos' syndrome) is a rare disorder that usually presents with typical skin lesions and may later progress to involve multiple organ systems, such as the gastrointestinal and central nervous systems. Köhlmeier[1] first described a patient with clinical features of this syndrome in 1941. Degos et al.[2] described a similar disorder and coined the term "atrophic papulosquamous dermatitis." This disease was characterized by papular elements with a tendency to atrophy and by an episode of violent intestinal bleeding that resulted in death. Later the term was changed to "malignant atrophic papulosis," to emphasize the fatal prognosis. It is characterized by an asymptomatic papular cutaneous eruption on the trunk and extremities that later evolves into lesions with atrophic porcelain-white centers surrounded by an erythematous telangiectatic rim varying from 2 to 10 mm in diameter. Degos' syndrome is generally regarded as a fatal disease; however, there appears to be a chronic, benign variant of the syndrome that neither involves multiple systems nor has a fatal outcome.[3]

Systemic Manifestations

Malignant atrophic papulosis may involve multiple organ systems of the body. It has been stated that males

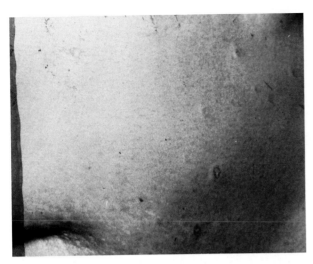

FIGURE 207-1 The porcelain-white depressed centers and erythematous borders of the skin lesions in malignant atrophic papulosis. (From Lee et al.[15])

may be more often affected than females and that males may have a worse prognosis.[4] The disease does not appear to have an age predilection; however, it is difficult to make generalizations on the sex and age distribution because of its rarity. The cutaneous eruption is the most characteristic manifestation, but the more severe lesions occur in the gastrointestinal tract and the central nervous system.

The skin lesions usually occur in episodes and are often the first sign of the disease. Isolated, noncontiguous lesions pass through progressive stages from small, pink-grey-yellow papules that later become umbilicated and centrally depressed. The lesions are not painful and are usually located on the trunk and upper body, sparing the head and peripheral extremities (Fig. 207-1). Although they may be delayed for several years, the gastrointestinal manifestations are generally the most pronounced and tend to occur shortly after the skin lesions develop.[5-7] Vague abdominal discomfort, abdominal distension, and alternating diarrhea and constipation are frequently the main features. Radiographic examination usually shows no abnormality, but at the time of surgical exploration, multiple "white infarcts" may be found throughout the small intestine with an intact mucosa.

The central or peripheral nervous system may be affected, usually after several bouts of the skin disease.[8,9] Multiple foci may be involved with vascular thrombosis, giving a wide array of neurologic signs, including meningoencephalitis, cranial nerve involvement, hemiparesis, and motor or sensory nerve abnormalities. Other organ systems that may be involved less

frequently are the cardiovascular, pulmonary, and genitourinary systems. Death can occur from gastrointestinal hemorrhage and perforation or from cerebral infarction and hemorrhage within a few years after onset of the disease.

The cause and pathogenesis of the disease are still unknown. The common vascular lesion seen in all of the involved organ systems is an obliterating arteriolitis or endovasculitis with secondary thrombosis and slow tissue necrosis.[8] Roenigk, Kay, and Farmer[10] suggested that intravascular coagulation could be a factor in the pathogenesis of the disease. Black, Nishioka, and Levene[11] suggested that a disturbance of endothelial function leads to impairment of the normal fibrinolytic activity, causing slow occlusion of arterioles. Using electron microscopy Howard and Nishida[12] found cytoplasmic inclusions within endothelial cells and suggested a possible viral cause. Results of various immunologic tests usually have been negative.[11] Soter et al.[13] suggested that the cutaneous lesions of malignant atrophic papulosis resulted from a lymphocyte-mediated necrotizing vasculitis that affects the entire cutaneous microvasculature. Vasculitis with lymphocytes as the predominant inflammatory cells is a prominent finding in patients with Degos' syndrome. Lymphocyte-mediated necrotizing vasculitis, which is a prominent cutaneous change, is probably analogous to the generalized vasculitis observed in patients with systemic lupus erythematosus in whom skin lesions similar to those of malignant atrophic papulosis have been observed.[14]

Differential diagnosis of the skin lesions seen in Degos' syndrome should include systemic lupus erythematosus. In that disease the typical skin lesions are atrophic, hyperkeratotic, erythematous plaques without porcelain-white centers.

Ocular Manifestations

Ocular involvement by malignant atrophic papulosis has been reported to affect eyelids, bulbar conjunctiva (Fig. 207-2), retina (Fig. 207-3), and choroid.[4,12,15,16] Also, ocular manifestations such as diplopia, visual field defects, ophthalmoplegia, ptosis, papilledema, and optic atrophy may occur secondary to involvement of the central nervous system. The ophthalmic manifestations are probably all secondary to the systemic vasculitis.

Treatment

What is the best treatment for this disease is unclear. Two of the patients reported by Su et al.[3] were treated with some benefit with a combination of dipyridamole (Persantin) and buffered aspirin (Ascriptin). Delaney and Black[17] reported beneficial effects from fibrinolytic

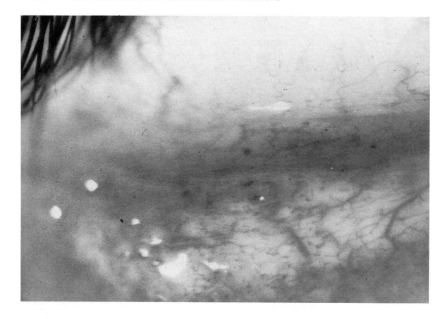

FIGURE 207-2 A telangiectatic conjunctival lesion seen in malignant atrophic papulosis. (From Lee et al.[15])

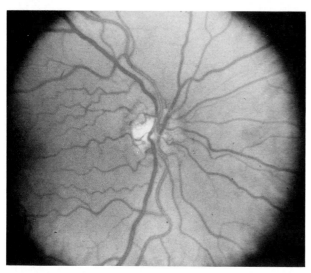

FIGURE 207-3 A fundus photograph showing an optic disc with vascular tortuosity and cilioretinal arteries anastomosing to retinal arteries at the optic disc with optic disc pallor inferotemporally. (From Lee et al.[15])

therapy (phenformin and ethylestrenol) in suppressing skin lesion formation in a patient. Systemic and topical steroids do *not* seem to help. Surgical intervention at first sign of gastrointestinal hemorrhage or perforation may allow resection of the affected intestine and prevent fatal peritonitis.

References

1. Köhlmeier W. Multiple Hautnekrosen bei Thrombangiitis obliterans. *Arch Dermatol Syphilol.* 1941;181:783-792.
2. Degos R, Delort J, Triot R. Dermatite papulo-squameuse atrophiante. *Bull Soc Fr Dermatol Syphiligr.* 1942;49:148-150.
3. Su WPD, Schroeter AL, Lee DA, et al. Clinical and histologic findings in Degos' syndrome (malignant atrophic papulosis). *Cutis.* 1985;35:131-138.
4. Henkind P, Clark WE II. Ocular pathology in malignant atrophic papulosis: Degos' disease. *Am J Ophthalmol.* 1968;65:164-169.
5. Degos R, Delort J, Tricot R. Papulose atrophiante maligne (syndrome cutanéo-intestinal mortel). *Bull Mem Soc Méd Hôp (Paris).* 1948;64:803-806.
6. Black MM, Jones EW. Malignant atrophic papulosis (Degos' syndrome). *Br J Dermatol.* 1971;85:290-292.
7. Strole WE Jr, Clark WH Jr, Isselbacher KJ. Progressive arterial occlusive disease (Köhlmeier-Degos): a frequently fatal cutaneosystemic disorder. *N Engl J Med.* 1967;276: 195-201.
8. Winkelmann RK, Howard FM Jr, Perry HO, et al. Malignant papulosis of skin and cerebrum; a syndrome of vascular thrombosis. *Arch Dermatol.* 1963;87:54-62.
9. Gever SG, Freeman RG, Knox JM. Degos' disease (papulosis atrophicans maligna): report of a case with degenerative disease of the central nervous system. *South Med J.* 1962;55:56-60.
10. Roenigk HH Jr, Kay, Farmer RG. Degos' disease (malignant papulosis); report of three cases with clues to etiology. *JAMA.* 1968;206:1508-1514.
11. Black MM, Nishioka K, Levene GM. The role of dermal blood vessels in the pathogenesis of malignant atrophic papulosis (Degos' disease); a study of two cases using enzyme histochemical, fibrinolytic, electron microscopi-

cal and immunological techniques. *Br J Dermatol.* 1973;88:213-219.

12. Howard RO, Nishida S. A case of Degos' disease with electron microscope findings. *Trans Am Acad Ophthalmol Otolaryngol.* 1969;73:1097-1112.

13. Soter NA, Murphy GF, Mihm MC Jr. Lymphocytes and necrosis of the cutaneous microvasculature in malignant atrophic papulosis: a refined light microscope study. *J Am Acad Dermatol.* 1982;7:620-630.

14. Dubin HV, Stawiski MA. Systemic lupus erythematosus resembling malignant atrophic papulosis. *Arch Intern Med.* 1974;134:321-323.

15. Lee DA, Su WPD, Liesegang TJ. Ophthalmic changes of Degos' disease (malignant atrophic papulosis). *Ophthalmology.* 1984;91:295-299.

16. Howard RO, Klaus SN, Savin RC, et al. Malignant atrophic papulosis (Degos' syndrome). *Arch Ophthalmol.* 1968;79:262.

17. Delaney TJ, Black MM. Effect of fibrinolytic treatment in malignant atrophic papulosis. *Br Med J.* 1975;3:415.

Chapter 208

Linear Nevus Sebaceus Syndrome

H. MICHAEL LAMBERT

The linear nevus sebaceus syndrome is a congenital, nonhereditary oculoneurocutaneous disorder belonging to a group of diseases generally called by ophthalmologists the phakomatoses, meaning mother spot, or birthmark. Also included in this genre of diseases are Sturge-Weber syndrome, tuberous sclerosis, neurofibromatosis, von Hippel-Lindau disease, and ataxia-telangiectasia.

The syndrome was first described in 1962 by Feuerstein and Mims,[2] who described two patients with a characteristic skin lesion (the linear nevus sebaceus of Jadassohn), seizures, and mental retardation. Since that time 20 well-documented cases have been reported, with all races and each sex represented.[2-18] The syndrome has been found to consist of the triad of midline facial linear nevus sebaceus of Jadassohn, neurologic abnormalities that may but do not necessarily include seizures and mental retardation, and ocular abnormalities (Table 208-1).

Systemic Manifestations

The hallmark of this syndrome is the characteristic facial lesion, the linear nevus sebaceus of Jadassohn. A smooth to verrucous pale yellow plaque is noted at birth on one side of the patient's face, extending only to the midline. It may involve any of the structures of the side of the head and face. Alopecia is usually present on the side of the lesion (Fig. 208-1). Histopathologic examination of the lesion reveals papillomatous hyperplasia and hyperkeratosis of the epidermis with small, poorly formed pilosebaceous units and multiple closely packed sebaceous glands. After puberty malignant transformation within the linear nevus is possible,[19] basal cell carcinoma being the most common secondary tumor. Treatment of the skin lesions consists of serial excisions of the nevus sebaceus and rotation of hair-bearing flaps for areas of alopecia.

TABLE 208-1 Reported Abnormalities in the Linear Nevus Sebaceus Syndrome

Cutaneous:	Neurologic:	Ocular:
Midline facial linear nevus of Jadassohn	Arachnoid cysts	Ptosis
	Cerebral hypoplasia	Strabismus
	Cerebellar hypoplasia	Amblyopia
	Unilateral cortical atrophy	Lid coloboma
	Unilateral extremity atrophy	Iris coloboma
	Hydrocephalus	Lens coloboma
	Seizures	Choroidal coloboma
	Mental retardation	Disc coloboma
		Choroidal osseous choristoma
		Subretinal neovascular membrane

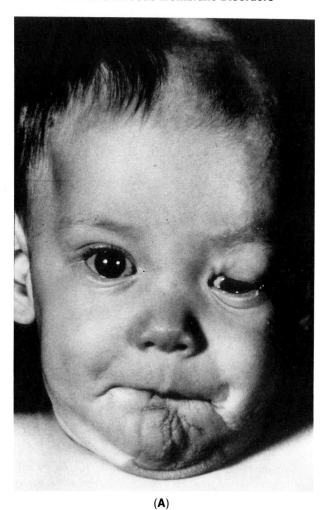

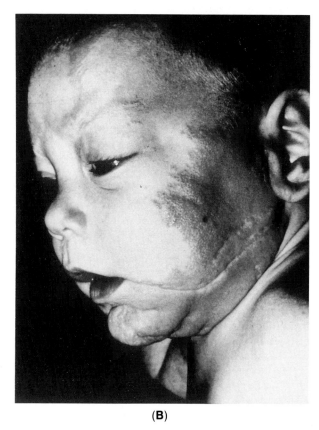

(A)   (B)

FIGURE 208-1 Involvement of the left side of the head and face by linear nevus sebaceus. The lesion extends to the midline and was smooth in the parietal region, becoming more verrucous in the area of the jaw. The left lid was notched and held in a narrowed position by linear nevus involvement. There was marked alopecia of the frontoparietal region. (From Lambert et al.[18])

Neurologic abnormalities are the rule: 16 of the 20 patients had electroencephalogram changes and seizures. Onset of seizures in all cases was before age 1, and they varied from apneic spells to jacksonian and grand mal seizures. Mental retardation was apparent in the same 16 patients and varied from mild to severe. CT abnormalities reported include cerebral and cerebellar hypoplasia and arachnoid cysts. Unilateral cortical atrophy, unilateral atrophy of the extremities, hydrocephalus, vitamin D–resistant rickets, and cardiac abnormalities have also been reported. Four of the 20 patients reported to date have normal to above normal intelligence and are seizure free.

Ocular Manifestations

The extraocular manifestations of the syndrome consist mainly of involvement of structures by the nevus sebaceus and choristomatous lesions around the eye. The nevus sebaceus may involve the lids, in which case ptosis may be present. Colobomas of the lid have been reported in several cases. The choristomatous lesions may appear as conjunctival lipodermoids (Fig. 208-2) and may actually surround and infiltrate the extraocular muscles (in one case a superior rectus muscle could not be found), with resulting strabismus and amblyopia. Histopathologic examination of one of these lesions

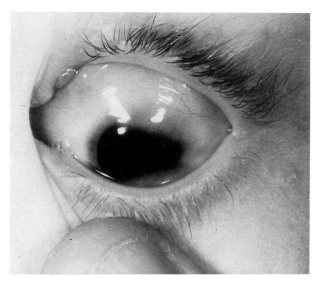

FIGURE 208-2 Conjunctival choristomatous lesions of left eye extend into cornea. (From Lambert et al.[18])

revealed a cartilaginous choristomatous mass ensheathed in a fibrous pseudoperichondrium with interspersed small, ectopic lacrimal glands. Angiomas of the orbit have also been reported.

Intraocular lesions reported with this syndrome include colobomas of the iris, choroid, and disc, and choroidal osseous choristomas (Fig. 208-3). In one case the osseous choristoma was associated with an overlying choroidal neovascular membrane.

References

1. Palena PV. Phakomatoses. In: Duane TD, Jaeger EA, eds. *Clinical Ophthalmology*. vol. 3. Philadelphia, Pa: Harper & Row; 1985;1.
2. Feuerstein RC, Mims LC. Linear nevus sebaceus with convulsions and mental retardation. *Am J Dis Child.* 1962;104:675-679.
3. Marks JG, Tomasovic JJ. Linear nevus sebaceus syndrome—a case report. *J Am Acad Dermatol.* 1980;2:31-32.
4. Marden PM, Venters HD. A new neurocutaneous syndrome. *Am J Dis Child.* 1966;112:79-81.
5. Monahan RH, Hill CW, Venters HD. Multiple choristomas, convulsions and mental retardation as a new neurocutaneous syndrome. *Am J Ophthalmol.* 1967;64:529-532.
6. Moynahan EJ, Wolff OH. A new neuro-cutaneous syndrome (skin, eye, brain) consisting of linear naevus, bilateral lipo-dermoid of the conjunctivae, cranial thickening, cerebral cortical atrophy and mental retardation. *Br J Dermatol.* 1967;79:651-652.
7. Soloman LM, Fretzin DF, Dewald RL. The epidermal nevus syndrome. *Arch Dermatol.* 1968;97:273-285.
8. Sugarman GI, Reed WB. Two unusual neurocutaneous

9. disorders with facial cutaneous signs. *Arch Neurol.* 1969;21:242-247.
9. Bianchine JW. The nevus sebaceus of Jadassohn. *Am J Dis Child.* 1970;120:223-228.
10. Herbst BA, Cohen ME. Linea nevus sebaceus. *Arch Neurol.* 1971;24:317-322.
11. Lansky LL, Funderburk S, Cuppage FE, et al. Linear sebaceous nevus syndrome (cases 1 and 2) AM. *Am J Dis Child.* 1972;123:587-590.
12. Holden KR, Dekaban AS. Neurological involvement in nevus unis lateris and nevus linearis sebaceus. *Neurology.* 1972;22:879-887.
13. Lovejoy FH, Boyle WE. Linear nevus sebaceus syndrome:

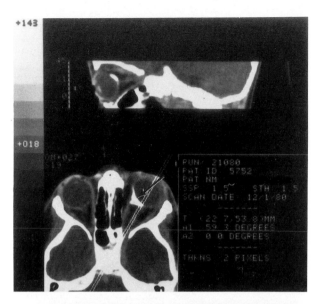

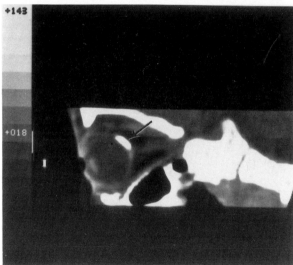

FIGURE 208-3 CT scan demonstrates bilateral choroidal osteomas.

report of two cases and a review of the literature. *Pediatrics.* 1973;52:382-387.

14. Moorjani R, Shaw DG. Feuerstein and Mims syndrome with resistant rickets. *Pediatr Radiol.* 1976;5:120-122.

15. Campbell WW, Buda FB, Sorensen G. Linear nevus sebaceus syndrome: neurological aspects documented by brain scans correlated with developmental history and radiographic studies. *Milit Med.* 1978;143:175-178.

16. Wilkes SR, Campbell RJ, Waller RR. Ocular malformation in association with ipsilateral facial nevus of Jadassohn. *Am J Ophthalmol.* 1981;92:344-352.

17. Shochot Y, Romano A, Barishak YR, et al. Eye findings in the linear sebaceous nevus syndrome: a possible clue to the pathogenesis. *J Craniofac Genet Dev Biol.* 1982;2:289-294.

18. Lambert HM, Sipperley JO, Shore JW, et al. Linear nevus sebaceus syndrome. *Ophthalmology.* 1987;94:278-282.

19. Jones EW, Heyl T. Naevus sebaceus. *Br J Dermatol.* 1970;82:99-117.

Section E

Pigmentary Disorders

Chapter 209

Chediak-Higashi Syndrome

ROBERT CASTLEBERRY and FREDERICK J. ELSAS

Chediak-Higashi syndrome (CHS) in man is an autosomal recessive disorder characterized by partial oculocutaneous albinism, nystagmus, photophobia, increased susceptibility to pyogenic infections, and the frequent occurrence of an accelerated phase as a terminal event.[1,2] The pathognomonic feature is the presence of giant granules in all granule-containing cells (Fig. 209-1).[2-6]

Beguez-Cesar[3] and Steinbrinck[4] initially described the abnormal leukocyte granules associated with CHS, but it was not until 1955 that the pathologic description of this familial leukocyte disorder by Chediak[5] and the clinical features of congenital gigantism of peroxidase granules reported by Higashi[6] were recognized as defining a single syndrome, to which their names were appended.

The membrane and cytoplasmic dysfunction in CHS is postulated to be a manifestation of aberrant microtubule organization and function.[7] Microtubules are associated with the motility as well as secretory and degranulation functions of cells. The mechanism by which abnormalities in microtubules occur in CHS is unknown, but experimental evidence suggests a role of cyclic nucleotides, which in the normal cell are associated with and may regulate microtubular assembly.[7] Specifically, elevated levels of cyclic 3',5'-guanosine monophosphate (cGMP) enhance microtubular polymerization and function whereas increased levels of cyclic adenosine monophosphate (cAMP) are inhibitory.[8] In CHS, levels of cAMP have been noted to be increased,[9] and improved CHS neutrophil function has been reported following exposure in vitro to cholinergic agents and ascorbic acid, both of which are known to increase levels of cGMP.[7,9-11] These observations support the hypothesis that abnormal control of microtubule assembly is likely the basis of CHS.

Systemic Manifestations

The pathophysiologic consequences of microtubule dysfunction are expressed in various cell lines. In the case of neutrophils, abnormalities of chemotaxis, phagocytosis, granulogenesis, and bacteriocidal activity secondary to delayed emptying of abnormal granules into phago-

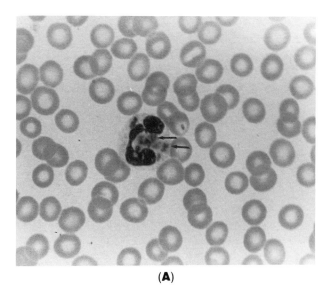

(A)

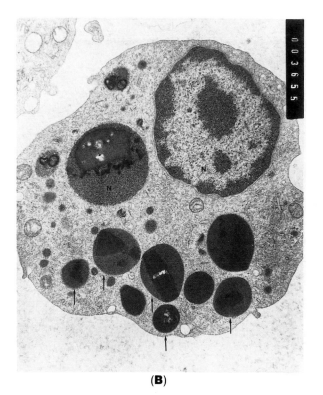

(B)

FIGURE 209-1 (A) Blood neutrophil from a patient with CHS shows large cytoplasmic granules (*arrows*). (Wright's stain, original magnification ×1700.) (B) This blood eosinophil from a child with CHS contains two nuclear profiles (N) and several giant cytoplasmic granules (*arrows*). (Original magnification ×10,000)

somes are well-described.[12-14] (The latter abnormality is a prominent factor in the increased prevalence of pyogenic infections.) Lymphocytes may demonstrate decreased natural killer cell function,[15] and platelets show abnormal aggregation which may lead to a hemorrhagic diathesis.[16,17] Abnormal distribution of melanosomes in the skin and retina leads to the apparent oculocutaneous albinism and photophobia. Similarly, the aberrant organization of neural connective tissue may be responsible for the neurologic features of CHS.[18]

The diagnosis of CHS is confirmed by the presence of light-colored skin, a silver sheen to the hair, photophobia, and the presence of giant granules in granule-containing peripheral blood and bone marrow leukocytes. Patients may also have a number of neurologic complaints, including sensory or motor neuropathies and ataxia. Infections occur principally in the skin, respiratory tract, and mucous membranes, and septicemia occurs in up to 10% of children, usually as a consequence of the accelerated phase of the disorder. Although numerous bacterial and fungal organisms have been cultured from these patients, *Staphylococcus aureus* is the most prevalent one. Even in the absence of demonstrable infection, patients frequently experience febrile episodes of undetermined cause.

Mild anemia has been observed, particularly during infections, but, in general, aplastic or hemolytic anemias are seen only during the accelerated phase. Studies of plasma coagulation proteins and platelet counts are normal. Platelet aggregation with collagen and thrombin is abnormal, but normal in the presence of adenosine diphosphate. For approximately 85% of patients the disease progresses to an accelerated phase characterized by lymphadenopathy, hepatosplenomegaly, high-spiking fevers, and pancytopenia. Histopathologically, the liver, spleen, and lymph nodes demonstrate lymphohistiocytic infiltration. Most children succumb to the disease during this phase because of infectious complications, including septicemia, hemorrhage, and progressive neurologic abnormalities. Few patients survive beyond the 2nd decade.

Ocular Manifestations

Ocular manifestations of CHS include marked photophobia and nystagmus, presumably caused by the aberrant distribution—and often diminished amount—of melanin in the retinal pigment epithelium and choroid.[2] Iris pigmentation is usually reduced and an albinotic reflex may be seen with transillumination. These findings are variable, however, and patients have been reported who have normal-looking irides, choroidal pigmentation, and no photophobia or nystagmus.[19,20] Papilledema and infiltration of the optic nerve with lymphocytes may occur in the accelerated stages of the disease.[21] The electroretinogram results and visual evoked potentials have been found to be abnormal.[20] The predisposition of these patients to pyogenic infections can lead to periorbital cellulitis, particularly if the lacrimal drainage system is compromised.[2] Typical giant intracytoplasmic lysosomal granules may be found in conjunctival fibroblasts.[20]

Treatment

The management of CHS is principally supportive care, with response to infectious and hemorrhagic manifestations as they occur. There is no evidence that the prophylactic use of antibiotics alters the incidence of either infectious or febrile episodes. Three patients have been treated with ascorbic acid in an attempt to reduce manifestations of the disease.[10,22] While the number of febrile episodes was attenuated in one case, other infectious and hemorrhagic events were not altered. Therefore, the role of ascorbic acid in the management of CHS remains to be determined.

References

1. Baehner RL. Disorders of granulocyte function. In: Miller DR, Pearson HA, Baehner RL, McMillan CW, eds. *Blood Diseases of Infancy and Childhood.* St Louis, Mo: CV Mosby; 1978.
2. Blume RS, Wolff SM. The Chediak-Higashi syndrome: studies in four patients and a review of the literature. *Medicine.* 1972;51:247.
3. Beguez-Cesar AB. Neutropenia cronica maligna familiar con granulaciones atipicas de los leucocitos. *Bol Soc Cubana Pediatr.* 1943;15:900.
4. Steinbrinck W. Uber eine neue Granulationsanomalie der Leukocyten. *Dtsch Arch Klin Med.* 1948;193:577.
5. Chediak M. Nouvelle anomalie leucocytaire de caractere constitutionnel et familial. *Rev Hematol.* 1952;7:362.
6. Higashi O. Congenital gigantism of peroxide granules. The first case ever reported of qualitative abnormality of peroxidase. *Tohoku J Exp Med.* 1954;59:315.
7. Oliver JM. Impaired microtubule function correctable by cyclic GMP and cholinergic agonists in the Chediak-Higashi syndrome. *Am J Pathol.* 1976;85:395.
8. Zurier RB, Weismann G, Hoffstein S, et al. Mechanisms of lysosomal enzyme release from human leukocytes. II. Effects of cAMP and cGMP, autonomic agonists and agents which affect microtubule function. *J Clin Invest.* 1974; 53:297.
9. Boxer LA, Watanabe AM, Rister M, et al. Correction of leukocyte function in Chediak-Higashi syndrome by ascorbate. *N Engl J Med.* 1976;295:1041.
10. Weening RS, Schoorel EP, Roos D, et al. Effect of ascorbate on abnormal neutrophil, platelet, and lymphocyte function in a patient with the Chediak-Higashi syndrome. *Blood.* 1981;57:856.
11. Boxer LA, Manfred R, Allen JM, et al. Improvement of Chediak-Higashi leukocyte function by cyclic guanosine monophosphate. *Blood.* 1977;49:9.
12. Clark RA, Kimball HR. Defective granulocyte chemotaxis in the Chediak-Higashi syndrome. *J Clin Invest.* 1971;50:2645.
13. Stossel TP, Root RK, Vaughan M. Phagocytosis in chronic granulomatous disease and the Chediak-Higashi syndrome. *N Engl J Med.* 1972;286:120.
14. Rausch PG, Pryzwansky KB, Spitznagel JK. Immunocytochemical identification of azurophilic and specific granule markers in the giant granules of Chediak-Higashi neutrophils. *N Engl J Med.* 1973;298:693.
15. Haliotis T, et al. Chediak-Higashi gene in humans. I. Impairment of natural-killer function. *J Exp Med.* 1980;151:1039.
16. Buchanan GR, Handin RI. Platelet function in the Chediak-Higashi syndrome. *Blood.* 1976;47:941.
17. Boxer GJ, Holmsen H, Robkin L, et al. Abnormal platelet function in Chediak-Higashi syndrome. *Br J Haematol.* 1977;35:521.
18. Sung JH, Meyers JP, Stadlan EM, et al. Neuropathological changes in Chediak-Higashi disease. *J Neuropathol Exp Neurol.* 1969;28:86.
19. Kinnear PE, Jay B, Witkop CJ. Albinism. *Surv Ophthalmol.* 1985;30:75.
20. BenEzra D, Mengistu F, Cividalli G, et al. Chediak-Higashi syndrome: ocular findings. *J Pediatr Ophthalmol Strab.* 1980;17:68.
21. Spencer WH, Hogan MJ. Ocular manifestations of Chediak-Higashi syndrome. *Am J Ophthalmol.* 1960;50:1197.
22. Gallin JI, Elin RJ, Hubert RT, et al. Efficacy of ascorbic acid in Chediak-Higashi syndrome: studies in humans and mice. *Blood.* 1979;53:226.

Chapter 210

Incontinentia Pigmenti

STEVEN I. ROSENFELD and MORTON E. SMITH

Incontinentia pigmenti is a rare inherited dermatologic disorder that affects females and presents in the first few weeks of life. The characteristic pigmentary changes of the skin are often associated with other mesodermal defects, including ocular, dental, skeletal, cardiac, and central nervous system abnormalities. The disease is transmitted as an X-linked dominant trait that is lethal to the majority of affected males. Thus, the disease is uncommon in living males, and female carriers tend to have a high rate of aborted pregnancies.[1]

The cause of the disease is postulated to be a generalized pigmentary disorder that primarily affects melanocytes.[2,3] On histologic examination the skin in the final stages shows an absence or a decreased amount of melanin in the basal layers of the epithelium and an increase in pigment in the upper dermis, findings that prompted Bloch to deduce that basal cells are "incontinent" of melanin and allowed pigment to drop down into the dermis.[4,5] Still undefined alterations in the retinal pigment epithelium (RPE) are felt to be the basis for most of the ocular abnormalities.

Systemic Manifestations

The skin lesions are seen at birth or within the first 2 weeks of life and can be divided into three stages.[6,7] The initial lesions consist of linearly arranged inflammatory bullas and papules along the extremities. Weeks to months later, dark, warty outgrowths occur along the extremities in the second stage. After the previous skin lesions have disappeared, the third stage consists of blotchy and whorl-like, flat, pigmented lesions on the trunk, along the nevus lines of Blaschko (Fig. 210-1). The pigmentation may fade or disappear over the course of years.[4,6,7]

The different stages can be found simultaneously.[5,7] Verrucous lesions may be absent in up to 30% of cases.[5,7] The inflammatory stages do not necessarily precede the typical pigmentation.[5] The cutaneous lesions of incontinentia pigmenti are self-limiting and require only symptomatic treatment during the bullous and verrucous phases of the disease.[1]

Dental involvement includes absent or delayed den-

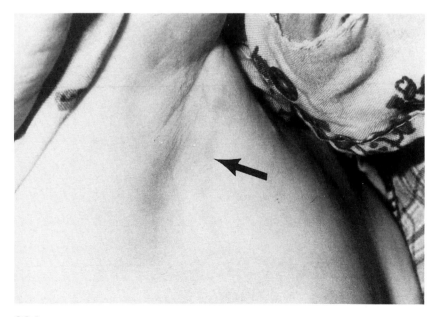

FIGURE 210-1 Pigmented lesions on the trunk.

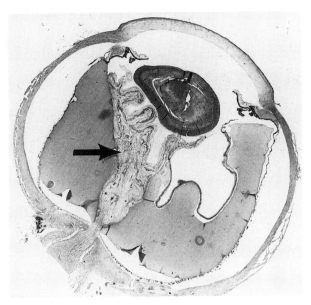

FIGURE 210-2 Low-power magnification of entire globe showing total retinal detachment (*arrow*).

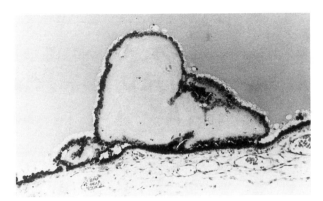

FIGURE 210-3 Nodular proliferation of retinal pigment epithelium (original magnification ×600).

tition, pegged teeth, impactions, and abnormalities of crown formation.[6,7] Central nervous system abnormalities occur in 30% of cases and include seizures, spastic disorders, mental retardation, paresis, and deformities of the calvaria (microcephaly and hydrocephalus).[2,6,7] Abnormalities of the nails and hair (alopecia) as well as congenital heart disease have been found in some patients.[6,7]

Ocular Manifestations

Ocular abnormalities are found in about 35% of cases, and it is these complications that present some of the more troublesome long-term problems.[7] Many affected persons have markedly reduced vision, and 7% become blind.[7] As part of this generalized pigmentary disorder, alterations in the RPE contribute to many of these ocular complications.

Strabismus, cataracts, and nystagmus are most commonly seen.[1,6,7] Other clinical findings include corneal opacities, uveitis, chorioretinitis, posterior synechiae, blue sclera, microphthalmia, myopia, retinal detachment, retrolental mass, pigmentary changes of the fundus, vascular anomalies of the fundus, pseudoglioma, optic atrophy, proliferative changes of the fundus, ablatio falciformis, and phthisis bulbi.[2,3,5,7,9]

The histopathology of the ocular findings is nonspecific. Retinal detachment (Fig. 210-2), intraocular hemorrhage, retinal dysplasia, and RPE proliferation are common.[8] The RPE proliferation is displayed in two forms: diffuse placoid RPE proliferation or nodular aggregates of pigment-filled macrophages containing melanin and lipofuscin with overlying RPE proliferation (Fig. 210-3).[2,3]

Since the RPE exerts an organizing influence on the developing retina, a primary alteration of the RPE could explain the retinal dysplasia or retinal detachment seen in the disease. These findings support the concept of a generalized pigmentary disorder affecting retinal and skin melanocytes.

References

1. Goldberg MF. The skin and the eye. In: Goldberg MF, ed. *Genetic and Metabolic Eye Disease.* Boston, Mass: Little, Brown; 1974;502.
2. Rosenfeld SI, Smith ME. Ocular findings in incontinentia pigmenti. *Ophthalmology.* 1985;92:543-546.
3. Mensheha-Manhart O, Rodrigues MM, Shields JA, et al. Retinal pigment epithelium in incontinentia pigmenti. *Am J Ophthalmol.* 1975;79:571-577.
4. Lever WF, Schaumburg-Lever G. *Histopathology of the Skin.* 6th ed. Philadelphia, Pa: JB Lippincott; 1983:83-84.
5. Berbich A, Dhermy P, Majbar M. Ocular findings in a case of incontinentia pigmenti (Bloch-Sulzberger Syndrome). *Ophthalmologica.* 1981;182:119-129.
6. Mausolf FA. Dermatology and the eye. In: Mausolf FA, ed. *The Eye and Systemic Disease.* St. Louis, Mo: CV Mosby; 1975;311-313.
7. Carney RG. Incontinentia pigmenti: a world statistical analysis. *Arch Dermatol.* 1976;112:535-542.
8. Yanoff M, Fine BS. *Ocular Pathology; A Text and Atlas.* 2nd ed. Philadelphia, Pa: Harper & Row; 1982;879.
9. Watzke RC, Steven TS, Carney RG, Jr. Retinal vascular changes in incontinentia pigmenti. *Arch Ophthalmol.* 1976;94:743-746.

Chapter 211

Oculodermal Melanocytosis
Nevus of Ota

IRA SNOW JONES

In 1930 Ota[1] described a lesion that he called "naevus fuscocaeruleus ophthalmomaxillaris." It consists of a unilateral bluish discoloration of the periorbital skin, including temple, forehead, malar area, and nose. The sclera, conjunctiva, iris, and choroid are involved with increased pigmentation, either completely or in a sectorial manner (Fig. 211-1). Although it is usually unilateral, in about 10% of cases it is bilateral. It is more common in Orientals and blacks. There is female preponderance. It may be noted at birth or may become obvious only later in life. It may be progressive. The descriptive term for nevus of Ota is "oculodermal melanocytosis."

The involved areas show elongated melanocytes with dendritic extensions that are located mostly in the upper third of the reticular dermis but may extend deeper.[2] The dopa reaction tends to be positive with poorly pigmented melanocytes and negative with strongly pigmented ones. Thickened or elevated areas may resemble the blue nevus on histologic examination. It is assumed that, unlike the Mongolian spot, the lesions represent a hamartomatous nevoid lesion.

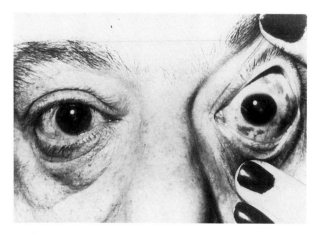

FIGURE 211-1 Skin, conjunctival, and scleral pigmentation in unilateral nevus of Ota.

Clinical Manifestations
Oral and Nasal Melanosis

Although not classically part of the nevus of Ota, oral and nasal melanosis appear to be extensions of the described picture.[3]

Nevus of Ito

Although it may appear alone, nevus of Ito may often be associated with ipsilateral or bilateral nevus of Ota. The location is deltoid, supraclavicular, and scapular. The clinical and histopathologic picture is similar to that of nevus of Ota.[4]

Mongolian Spot

Occasionally in Caucasians, and more frequently in black and Oriental infants, there is a flat bluish skin discoloration in the sacrococcygeal area, the Mongolian spot.[5] It is present at birth and usually disappears by age 4 years. In bilateral nevus of Ota the spots may persist.

Blue Nevus

The skin areas of nevus of Ota may rarely show infiltrates leading to discrete nodules from millimeters to centimeters in diameter. These have the clinical appearance of blue nevi, and the histopathology conforms to the expected picture.[6]

Melanoma

It is generally accepted that persons with nevus of Ota are at greater risk for melanoma.[7] These malignant transformations are most likely to occur in the uveal tract of the eye but may involve skin, orbit, or central nervous system. About 1.24% of patients with uveal melanoma have ocular or oculodermal melanocytosis. This is 30 times the expected association. Development of melanoma in nevus of Ota is most likely to occur in Caucasians, although rarely, it may appear in others.

Glaucoma

Most patients with nevus of Ota do not develop glaucoma. The association is nevertheless well-established under the term "benign melanocytic glaucoma." The mechanism is apparently invasion of the filtration angle by melanocytes.[8]

References

1. Ota M, Tanino H. Naevus fusco-caeruleus ophthalmomaxillaris. *Tokyo Med J.* 1939;63:1243.

2. Lever W, Lever GS. *Histopathology of the Skin.* 6th ed. Philadelphia, Pa: JB Lippincott; 1983.

3. Mishima Y, Mevorah B. Nevus Ota and Nevus Ito in American Negroes. *J Invest Dermatol.* 1961;36:133.

4. Hidano A, Kajima H, Endo Y. Bilateral nevus Ota associated with nevus Ito. *Arch Dermatol.* 1965;91:357.

5. Hidano, A, Kajima H, Ideda, S, et al. Natural history of nevus Ota. *Arch Dermatol.* 1967;95:187.

6. Kopf AW, Weidman WI. Nevus of Ota. *Arch Dermatol.* 1962;85:195.

7. Velazquez N, Jones IS. Ocular and oculodermal melanocytosis associated with uveal melanoma. *Ophthalmology.* 1983;90(12):1472.

8. Foulks GN, Shields MB. Glaucoma in oculodermal melanocytosis. *Am J Ophthalmol.* 1977;9:1299.

Chapter 212

Vitiligo

MARK J. WEINER, MICHAEL D. WAGONER, and DANIEL M. ALBERT

Vitiligo is characterized clinically by progressive, well-circumscribed, chalk-white macules on the skin. The typical lesion is millimeters to centimeters in diameter, round to oval in configuration, with fairly distinct margins (Fig. 212-1). The extent of involvement varies from a single macule to near total amelanosis. During the course of the disease, existing macules may progressively enlarge and new macules may develop. Typically, there are periods of slow progression or stability and occasional episodes of rapid progression. Characteristically involved areas include the regions around the eyes, nose, mouth, and rectum. In addition, extensor bony surfaces, anterior tibial areas, flexor surfaces of the wrists, axillas, and the low back are frequently involved. Traumatized areas may develop a macule corresponding to the area of injury. A Wood's lamp may be used to highlight vitiliginous patches in lightly pigmented zones. The prevalence of vitiligo is most frequently cited as being about 1%; there is no sexual preponderance. Vitiligo has been observed in all races. About one third of affected patients have a family history of the disease. The inheritance pattern is ill-defined, possibly multifactorial or autosomal dominant with incomplete penetrance.[1,2]

Associated Systemic Disorders

A correlation between vitiligo and a variety of autoimmune disorders has been noted for some time. It has been associated with hyperthyroidism, hypothyroidism, Graves' disease, and thyroiditis. Vitiligo may begin before, during, or after thyroid disease, and treatment of the thyroid disease does not appear to alter the course of the vitiligo. Additionally, diabetes mellitus, Addison's disease, pernicious anemia, alopecia areata, and multiglandular insufficiency have all been associated with an increased frequency of vitiligo. Vitiligolike leukoderma is an associated disorder seen in patients with malignant melanoma. Depigmentation occurs within, around, and at a distance from the primary melanoma and may have favorable prognostic significance for the patient's survival.[1,3]

Pathophysiology

Vitiligo is caused by destruction of the pigment cells (melanocytes). There is considerable evidence that this

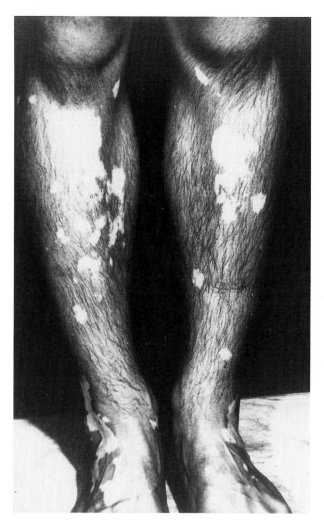

FIGURE 212-1 Typical skin lesion of vitiligo. (Courtesy of David Mosher, M.D.)

destruction is immune mediated. Although routine light microscopic studies typically reveal no abnormalities except the absence of melanocytes, lymphocytes are present at the borders of the vitiliginous macules in some cases. In addition, a number of associated autoimmune disorders (enumerated above) are seen frequently in patients with vitiligo. Antimelanocyte antibodies have been found in some patients with vitiligo, whereas antibodies to antigens of normal melanocytes were found in other affected patients. In patients with melanoma, antibodies also have been detected that were cytotoxic to both normal melanocytes and melanoma cells in culture.[1,3]

The actual mechanism of melanocyte destruction is probably more complex than a simple autoimmune cause since autoantibodies are not universally demon-

strable. The segmental distribution of the involvement, which often follows a dermatomal distribution, the sparing of paralyzed limbs from vitiliginous changes, and the neural control mechanisms of pigmentation in certain animal models have all been used to support the hypothesis that a neurochemical mediator destroys melanocytes or inhibits melanin production. Others have proposed that the normal mechanism for melanocyte destruction goes awry or that toxic metabolites in melanin synthesis cause melanocyte destruction.

Ocular Manifestations

Disorders of the eye in association with vitiligo have been reported for over a century. The most widely studied association of vitiligo with eye disease is the Vogt-Koyanagi-Harada syndrome. This disease occurs primarily in heavily pigmented young adults, typically Asians. Patients present with chronic, bilateral, exudative uveitis and choroiditis, often leading to retinal detachment. Areas of depigmentation and pigment clumping are often observed in the retina. Vogt-Koyanagi syndrome was classically described as a predominantly anterior uveitis, while Harada's syndrome patients manifested more marked choroiditis and uveitis. Recent literature has recognized these entities as part of a continuous spectrum of disease manifestation. Frequently associated signs include alopecia and poliosis, as well as vitiligo and dysacusis. Less frequently, meningeal symptoms with cerebrospinal fluid pleocytosis occur. Sympathetic ophthalmia has been noted to bear a close clinical and histopathologic relationship to the Vogt-Koyanagi-Harada syndrome, and instances of vitiligo in patients with sympathetic ophthalmia have been reported. However, sympathetic ophthalmia always follows ocular trauma and has no racial preference.[4,5]

A variety of less severe ocular disturbances involving both populations of pigmented cells (the uvea and retinal pigment epithelium) occur in association with cutaneous vitiligo (Fig. 212-2). There is a statistically increased prevalence of uveitis in a random population of patients with vitiligo as well as a statistically increased prevalence of chorioretinal scars. An increased number of areas of retinal pigment epithelium hypopigmentation or degeneration occurs in patients with vitiligo compared to a control population; these areas may be focally, sectorally, or geographically distributed and are frequently overlooked or neglected on routine examination. Vitiligolike leukoderma has also been associated with uveitis, heterochromia, halo nevi of the choroid, and hypopigmentation or atrophy of the retinal pigment epithelium or choroid, with an incidence of ocular involvement apparently greater than in patients with

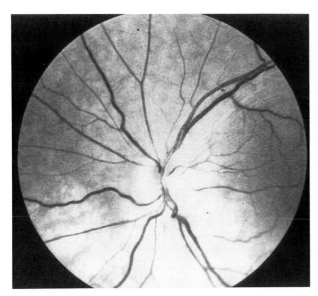

FIGURE 212-2 Hypopigmentation of the retinal pigment epithelium and the choroid in association with vitiligo.

cutaneous vitiligo. The pathophysiology of the ocular involvement is somewhat uncertain, since it is not clear that the mechanism that causes destruction of the skin melanocytes would simultaneously be responsible for the ocular pigmentary disturbances. The frequent association of ocular findings in patients with vitiligo and vitiligolike leukoderma makes ophthalmic examination of these patients advisable.[6,7]

References

1. Fitzpatrick TB, Eisen AZ, Wolff K, et al. eds. *Dermatology in General Medicine. Textbook and Atlas.* 3rd ed. New York, NY: McGraw-Hill; 1987;810-821.
2. Nordlund JJ, Lerner AB. Vitiligo. It is important. *Arch Dermatol.* 1982;118:5.
3. Chang MA, Fournier G, Koh HK, et al. Ocular abnormalities associated with cutaneous melanoma and vitiligolike leukoderma. *Graefe's Arch Clin Exp Ophthalmol.* 1986;224:529-535.
4. Albert DM, Nordlund JJ, Lerner AB. Ocular abnormalities occurring with vitiligo. *Ophthalmology.* 1979;86:1145-1158.
5. Perry HD, Font RL. Clinical and histopathologic observations in severe Vogt-Koyanagi-Harada syndrome. *Am J Opthalmol.* 1977;83:242-254.
6. Albert DM, Sober AJ, Fitzpatrick TB. Iritis in patients with cutaneous melanoma and vitiligo. *Arch Ophthalmol.* 1978;96:2081-2084.
7. Wagoner MD, Albert DM, Lerner AB, et al. New observations on vitiligo and ocular disease. *Am J Ophthalmol.* 1983;96:16-25.

Section F

Vesiculobullous Disorders

Chapter 213

Acrodermatitis Enteropathica

J. DOUGLAS CAMERON, CRAIG J. McCLAIN, and
DONALD J. DOUGHMAN

Acrodermatitis enteropathica is the congenital clinical syndrome caused by intestinal malabsorption of zinc and transmitted as an autosomal recessive defect.[1] The exact biochemical abnormality of the zinc absorption process has not been determined.[2] Zinc is an essential trace element that is necessary for RNA and DNA synthesis and for the function of a variety of zinc metalloenzymes (carbonic anhydrase, alkaline phosphatase, lactic dehydrogenase, carboxypeptidase, and alcohol dehydrogenase). The greatest concentration of zinc in the body is found in the eye, principally in the choroid.[3]

The disease usually presents in early infancy, often after weaning from breast milk. Untreated, the clinical course follows a fluctuating pattern of bullous pustular dermatitis of the extremities and about the body orifaces, chronic diarrhea associated with malabsorption and failure to thrive, central nervous system abnormalities, and impaired immune function with frequent infections.[4]

Many of the zinc-dependent processes of the body have been identified because of the clinical manifestations of this disease. Alteration of amino acid metabolism leads to impairment of skin wound healing. Impaired development of the central nervous system is expressed as psychomotor retardation. Thymic atrophy early in life leads to impairment of immune function and frequent infections.[4]

Acrodermatitis enteropathica is now treated with oral zinc supplementation, and the manifestations of the disease improve or disappear with therapy.[1] It remains important to recognize the clinical manifestations of this rare congenital disease so that treatment can be instituted promptly. It is also important to recognize that the manifestations of acquired zinc deficiency may arise in more commonly encountered diseases and circumstances, including alcoholism with or without liver disease, regional enteritis, sprue, short bowel syndrome, sickle cell anemia, certain types of cancer, pregnancy, and during total parenteral nutrition (TPN).[5-7]

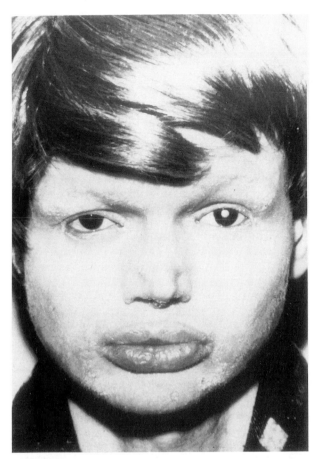

FIGURE 213-1 This 22-year-old man with acrodermatitis enteropathica has classic skin lesions of zinc deficiency around the eyes, nose, and mouth. He also has total alopecia and dense bilateral cateracts (note white pupil in his left eye).

Systemic Manifestations

The complications of zinc deficiency can affect nearly every organ and metabolic system in the body. The fetus appears to be supplied adequately by transplacental nutrition and through early infancy by ligands in breast milk that appear to enhance intestinal zinc absorption.

The initial manifestation of acrodermatitis enteropathica often involves the skin as a vesicular-pustular eruption of acral and periorificial regions of the body that is followed by acral pigmented patches (Fig. 213-1). Alopecia may also develop. Diarrhea is often a prominent feature; others include psychomotor retardation, apathy, dwarfism, and hypogonadism.[4] Other sensory symptoms include dysgeusia, hyposmia, and dysosmia.[8]

Ocular Manifestations

Ocular symptoms include photophobia, gaze aversion, blurred vision, abnormal dark adaptation, and decreased visual acuity.

The dermatitis frequently involves the lateral canthal area and lids as a vesicobullous eruption evolving into a psoriasiform reaction (Fig. 213-1). The cilia of the brow and lid margin may be lost following the onset of the dermatitis. Punctal stenosis may develop. Conjunctivitis frequently accompanies the dermatitis. Linear subepithelial corneal opacities, corneal epithelial thinning, anterior corneal scarring and vascularization, and prominent corneal nerves have been reported. Cataract formation and retinal pigment epithelial abnormalities have been observed.[9-13]

References

1. Moynahan EJ, Barnes PM. Zinc deficiency and a synthetic diet for lactose intolerance. *Lancet* 1973;1:676-677.
2. Lombeck I, Schnippering HG, Ritzl F, et al. Adsorption of zinc in acrodermatitis enteropathica. *Lancet* 1975;1:855.
3. Leopold IH. Zinc deficiency and visual impairment. *Am J Ophthalmol.* 1978;85:871-875.
4. Sunderman FW. Current status of zinc deficiency in the pathogenesis of neurological dermatological and musculoskeletal disorders. *Am Clin Lab Sci.* 1975;5:132-145.
5. Morrison SA, Russell RM, Carney EA, et al. Zinc deficiency: a cause of abnormal dark adaptation in cirrhotics. *Am J Clin Nutr.* 1978;31:276-281.
6. McClain CJ, Soutar C, Zieve L. Zinc deficiency: a complication of Crohn's disease. *Gastroenterology.* 1980;78:272-279.
7. McClain CJ. Trace metal abnormalities in adults during hyperalimentation. *J Parenteral and Enteral Nutr.* 1981;5:424-429.
8. Henkin RI, Schechter PJ, Hoye R, et al. Ideopathic hyogeusia with dysgeusia, hyposmia and dysosmia. A new syndrome. *JAMA.* 1971;217:434-440.
9. Wirsching L. Eye symptoms in acrodermatitis enteropathica. *Acta Ophthalmol.* 1962;40:567-574.
10. Matta CS, Felker GV, Ide CH. Eye manifestations in acrodermatitis enteropathica. *Arch Ophthalmol.* 1975;93:140-142.
11. Warshawsky RS, Hill CW, Doughman DJ, et al. Acrodermatitis enteropathica. Corneal involvement with histochemical and electron micrographic studies. *Arch Ophthalmol.* 1975;93:194-197.
12. Racz P, Kovacs B, Varga L, et al. Bilateral cataract in acrodermatitis enteropathica. *J Pediatr Ophthalmol Strab.* 1979;16:180-182.
13. Cameron JD, McClain CJ. Ocular histopathology of acrodermatitis enteropathica. *Br J Ophthalmol.* 1986;70:662-667.

Chapter 214

Cicatricial Pemphigoid

MELVIN I. ROAT and RICHARD A. THOFT

Cicatricial pemphigoid (CP) is a rare, chronic, scarring, bullous disease of mucous membranes and skin. CP affects women twice as often as men, usually presenting in the 5th or 6th decade.[1] It is believed to be an autoimmune disease and has been associated with HLA-B12 and an elevated serum IgA level in some patients.[1]

Pathology

The lesions of CP are believed to result from a type II hypersensitivity response, with binding of circulating antibodies against the lamina lucida of stratified squamous epithelial basement membrane. Complement fixation occurs and an inflammatory infiltrate consisting of polymorphonuclear neutrophils, acutely, or mononuclear cells, chronically, form in the subepithelial tissue. Complement, proteolytic enzymes, monokines, and lymphokines from effector cells produce disruption of the lamina lucida of the basement membrane, which leads to subepithelial bulla formation in oral mucosa and skin. The same products of inflammation cause epithelial hypermitosis, fibrocyte activation, and hyperproliferation. This results in abnormal epithelial differentiation, scarring, and fibrosis and shrinkage of conjunctiva, skin, oral, gastrointestinal, and genitourinary tract mucosa.

The CP antigen appears to be heterogeneous and localizes to the portion of the lamina lucida adjacent to the lamina densa of the basement membrane of stratified squamous epithelium.[2,3] Circulating antibodies to CP antigen are detected in 25% of patients.[2,4] When the CP antibody is present, there appears to be a correlation between the titer of circulating antibodies and clinical activity in CP.[5]

Linear deposits of immunoglobulin (most commonly IgG) and often complement are frequently detected at the epithelial basement membrane. This is found in 40% of skin, 80 to 97% of oral mucosa, and 50 to 84% of conjunctival biopsy specimens (Fig. 214-1).[1,2,6,7]

Histologic sections of cutaneous and oral lesions demonstrate dermal-epidermal separation of the basement membrane zone, subdermal cleft (eg, wall and roof of epithelial cells, base composed of lamina densa of basement membrane) and bulla formation.[4,6] Cutaneous, oral, and conjunctival lesions demonstrate a subepithelial inflammatory infiltrate of lymphocytes, plasma cells, mast cells, macrophages, and polymorphonuclear leukocytes.[1,4,6,7] Symblepharon and scar in CP consists of dense connective tissue.[1]

In active CP there is a decrease in conjunctival goblet cells and an increase in conjunctival epithelial cell mitosis.[1,8]

The same pathologic processes occur in all the body's epithelium. The variation of presentation at the different epithelial sites is the result of variation of structure of each epithelium (eg, different concentration of CP antigen, thickness of epithelium, density of hemidesmosomes, type of underlying connective tissue).

Clinical Features

Skin is involved in about 25% of patients with CP. There are two types of lesions: recurrent, nonscarring, generalized, tense bullous eruptions and localized erythematous plaques with recurrent bullae that scar.[1,4]

Oral mucosal involvement is present in up to 90% of patients with CP (Fig. 214-2).[1,4] Two types of oral lesions are found: a diffuse desquamative gingivitis and a vesiculobullous eruption. The oral bullae develop rapidly then rupture after a few days, resulting in large areas of epithelial denudation.[1,6]

The eyes are involved in approximately 90% of patients with CP (Fig. 214-3).[4] Ocular cicatricial pemphigoid is a bilateral asymmetric disease that usually begins as chronic conjunctivitis. With intermittent episodes of acute activity, subepithelial fibrosis of the conjunctiva often progresses to frank symblepharon, usually involving the inferior fornix first.[1] Corneal and conjunctival epithelial defects occur, usually preceding corneal neovascularization and scarring. Decreased vision is the result of corneal opacification. The scarring and contraction may cause entropion, trichiasis, lagophthalmos, abnormal blink, and exposure. As the lacrimal gland orifices are obliterated by scarring a severe, aqueous-deficient dry eye results. The scarring may progress to an end-stage eye with obliteration of the fornices, ankyloblepharon, corneal opacification, and a dry, keratinized surface.[1,7]

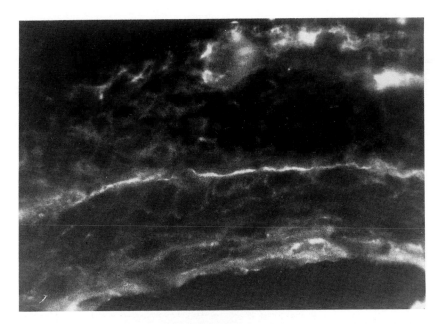

FIGURE 214-1 Cicatricial pemphigoid. Linear binding of IgG to the conjunctival basement membrane is demonstrated by direct immunofluorescence.

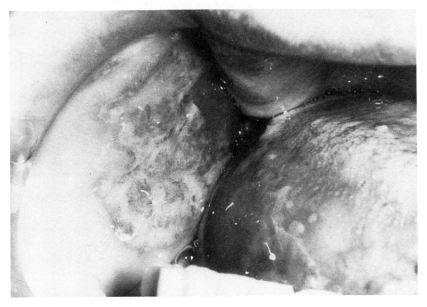

FIGURE 214-2 Cicatricial pemphigoid. Oral involvement with epithelial erosion. (Courtesy of Dr. Abell.)

Cicatricial involvement of the larynx or urogential mucosa is reported in 20% of patients; esophageal or anal involvement occurs occasionally. Cicatricial lesions may result in stricture with severe complications.[1,4,7]

CP and the more common blistering disease bullous pemphigoid (BP) are closely related. They share similar clinical, histologic, and immunopathic characteristics. As in CP, BP may involve skin and mucous membrane.

In contrast to CP, skin is always involved in BP. The cutaneous lesions are similar but in BP there is no associated scarring. Oral lesions, which are rare in BP, are also nonscarring. Ocular involvement in BP is uncommon and is usually a chronic conjunctivitis; however, cicatrizing conjunctivitis may occur with clinical, histologic, immunopathic, and prognostic features and a therapeutic response identical to that of CP. Basement membrane–bound immunoglobulins and circulating antibodies to lamina lucida are present more frequently, and the circulating antibodies present in higher titers in BP than in CP.[4,6] Highly sensitive indirect immunoelectron microscopy has detected BP

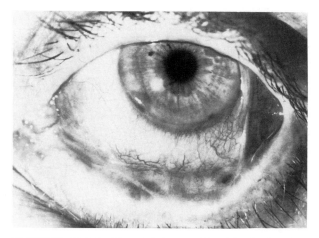

FIGURE 214-3 Cicatricial pemphigoid. Ocular involvement with symblepharon and loss of the inferior fornix.

antigen at the lamina lucida and basal cell hemidesmosomes.[2,3]

Treatment

The need for therapy depends on the activity of the disease, which may manifest itself by progression of symblepharon or corneal neovascularization, acute non-infectious conjunctivitis, or an increased conjunctival mitotic rate.

Systemic cytotoxic immunosuppression (SCI) has proven effective in decreasing inflammation and arresting progressive scarring. Monitored properly, SCI is a safe treatment with fewer serious irreversible side effects than long-term systemic steroids. Systemic steroids may be used as an adjunct to SCI.[7]

References

1. Mondino BJ, Brown SI. Ocular cicatricial pemphigoid. *Ophthalmology*. 1981;88:95.
2. Fine J, Neises GR, Katz SI. Immunofluorescence and immunoelectron microscopic studies in cicatricial pemphigoid. *J Invest Dermatol*. 1984;82:39.
3. Fine J. Epidermolysis bullosa: variability of expression of cicatricial pemphigoid, bullous pemphigoid, and epidermolysis bullosa acquisita antigens in clinically uninvolved skin. *J Invest Dermatol*. 1985;85:47.
4. Lever WF. Pemphigus and pemphigoid. *J Am Acad Dermatol*. 1979;1:2.
5. Franklin RM, Fitzmorris CT. Antibodies against conjunctival basement membrane zone. *Arch Ophthalmol*. 1983;101:1611.
6. Tyldesley WR. Oral pemphigoid. *Br J Oral Maxillofac Surg*. 1985;23:155.
7. Foster CS, Wilson LA, Ekins MB. Immunosuppressive therapy for progressive ocular cicatricial pemphigoid. *Ophthalmology*. 1982;89:340.
8. Thoft RA, Friend J, Kinoshita S, et al. Ocular cicatricial pemphigoid associated with hyperproliferation of the conjunctival epithelium. *Am J Ophthalmol*. 1984;98:37.

Chapter 215

Epidermolysis Bullosa

WILLIAM A. BOOTHE, BARTLY J. MONDINO, and PAUL B. DONZIS

Epidermolysis bullosa (EB) includes a broad spectrum of diseases that have in common the characteristic of extreme skin fragility and blister formation in response to insignificant trauma.[1] While some forms cause only minimal disability, others are extremely disabling and disfiguring. All but one are inherited and have their onset at birth or in early childhood. Generally the various types are classified according to whether they cause nondystrophic, atrophic, or dystrophic changes of the skin and mucous membranes.

The major types of EB are best classified on the basis of the anatomic location of blister formation.[2] The epidermolytic blisters of EB simplex or nondystrophic type occur intraepidermally, superficial to the basement membrane complex, and rarely cause scarring. In the atrophic or junctional type, the blisters form within the basement membrane complex in the lamina lucida; the intact epidermis forms the roof of the blister and the lamina densa becomes its floor. Although scarring is uncommon, skin atrophy often occurs in affected areas. The more severe dystrophic or dermolytic type manifests blister formation in the superficial dermis, immediately subjacent to the basement membrane complex, and results in scarring.[1] Within these categories are multiple variants based on clinical, genetic, histologic, and biochemical factors.

Except for EB acquisita (a dystrophic, acquired, autoimmune disease), the pattern of inheritance of these diseases is mendelian. The transmission pattern of non-dystrophic diseases is autosomal dominant, and that of junctional diseases is autosomal recessive. Two autosomal dominant groups and one autosomal recessive group of diseases are included in the dystrophic category of EB. The recessive dystrophic variants of EB are significantly more disfiguring than the others.

Systemic Manifestations

Complications are few in the EB simplex, or nondystrophic, group. Blistering increases when the weather becomes warmer. Infection of the affected areas of the skin is the most common complication. Occasionally the oral mucosa may develop blisters and erosions. Abnormalities of the teeth, neurologic and musculoskeletal systems, anemia, and growth retardation do not occur in this group. Although the pathogenesis of this group of EB is not known, skin fibroblasts from patients with generalized EB simplex have decreased amounts of gelatinase, an enzyme involved in collagen degradation.[3]

In junctional EB, there may be severe generalized blistering that heals without scarring or milia formation, unless secondary infection supervenes. The nails may become involved by paronychia (inflammation involving the tissue surrounding the fingernails) and dystrophy. Oral lesions may be present. Enamel defects and dysplastic teeth may occur. Pattern baldness and scalp atrophy are prominent. In the generalized form the prognosis is good, but the majority of patients with Herlitz's variant die before 2 years of age. Electron microscopy reveals reduced numbers of abnormal hemidesmosomes of the skin in these patients.[4]

In the dominant types of dystrophic EB, the skin blisters may be localized or generalized and hypertrophic or atrophic scars and milia may form. Dystrophic or absent nails are common but scalp involvement is unusual. Small flesh-colored papules (albopapuloid lesions) occur spontaneously on the trunk in Pasini's variety. Recurrent oral blisters and erosions occur with rare sequelae.

Recessive dystrophic EB often results in severe systemic and functional complications. One of the most disabling complications is the progressive formation of limb contractures due to repetitive scar deposition (Fig. 215-1). The digits of the hand can become completely encased in a keratinaceous shell, resulting in the so-called mitten deformity. Eventually, the digital bones can be resorbed if the deformity is not corrected surgically in a timely fashion. Severe mucous membrane involvement is common, causing dysplastic teeth, esophageal and anal strictures, and phimosis. Anemia is also common, owing to loss of blood through the skin, malnutrition, and the chronic disease state. Growth

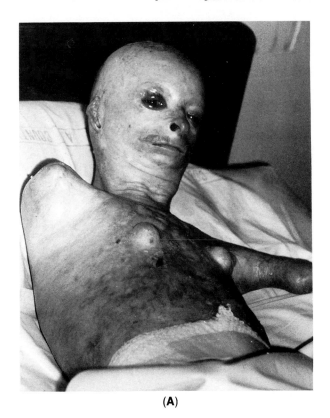

(A)

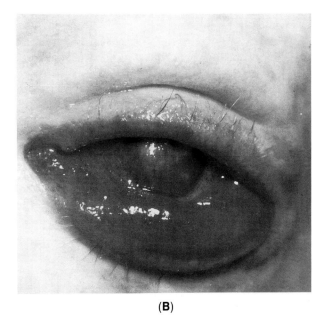

(B)

FIGURE 215-1 **(A)** A patient with severe dystrophic epidermolysis bullosa. **(B)** The same patient with cicatricial ectropion, exposure keratitis, and corneal scarring.

retardation can occur in severe cases. The incidence of skin cancer in areas of previous blistering and scarring is higher in all subtypes of dystrophic EB.

The mechanism involved in the pathogenesis of dystrophic EB may involve increased collagenase activity, which has been observed in organ cultures of blistered skin.[5] In addition, it has been shown by electron microscopy that anchoring fibrils are absent or diminished in blistered and unaffected skin in recessive dystrophic EB.

Ocular Manifestations

Ocular findings have been observed in each of the major types of EB. Ocular manifestations are rare in the nondystrophic types. Granek[6] discovered small cystic lesions at the depth of the epithelial basal cell layer in a ring distribution in the midperiphery of the cornea of both eyes of a teenaged boy and his mother. The son had symptoms, including pain in the right eye from cystic blebs that had migrated to the surface causing punctate defects that stained with fluorescein. Mild inferior stromal infiltrates were present in both eyes. The anatomic location of these blebs in the corneal epithelium corresponded to the location of blister formation in the epidermis. He also had refractory blepharoconjunctivitis. These corneal manifestations may be more prevalent than was previously realized, since often they may not cause symptoms. Severe ocular involvement has not been reported in this group of diseases.

Spontaneous recurrent corneal erosions from bleb formation with fine superficial stromal opacities were described in a patient with junctional EB by Hammerton et al.[7] Aurora et al.[8] described the following changes in another case of junctional EB: bullae of the eyelid skin, edematous cysts in the iris, variable focal changes in the retina and pigment epithelium, a focal breach in Bruch's membrane where the pigment epithelium was severely damaged, edema of the trabecular meshwork, ciliary body, lens, and optic nerve, complete detachment of the retina, and focal edema and congestion of the choroid. The basal cells of the cornea showed occasional hydropic changes, and a few endothelial cells showed vacuolar changes.

Though they are less common than oral mucosal lesions, ocular manifestations occur more often in recessive dystrophic EB than in other types. Involvement of the corneal epithelium by diffuse, spotty, and granular clouding has been histologically correlated with vacuolization in the basal cell layer in one case.[9] Conjunctival vesicles localized between the epithelium and the subjacent connective tissue with absence or diminution of elastic fibers were histologically documented in another case.[10] Other complications of dystrophic EB are blepharitis, cicatricial entropion, ectropion, trichiasis, pseudomembranes, symblepharon, corneal erosions and ulceration, vascularization, granulomatous infiltration, and perforation. Similar changes can occur in the sclera.[11] All these ocular complications appear to be either caused by or the result of the same sequence of recurrent vesicle formation and scarring that occurs elsewhere in the skin and mucous membranes (Fig. 215-1).

Treatment

Treatment of the mild types of the disease is supportive. In the severe types, treatment of the underlying disease with drugs such as vitamin E, systemic steroids, and phenytoin may be successful in some cases.[3]

References

1. Fine JD. Epidermolysis bullosa: clinical aspects, pathology, and recent advances in research. *Int J Dermatol.* 1986;25: 43-57.
2. Gedde-Dahl T, Anton-Lamprecht I. Epidermolysis bullosa. In: Emory AEH, Rimoin DL, eds. *Principles and Practice in Medical Genetics.* New York, NY: Churchill Livingstone; 1981.
3. Cooper TW, Bauer EA, Briggaman RA. The mechanobullous diseases (epidermolysis bullosa). In: Fitzpatrick TB, Eisen AZ, Wolf K, et al., eds. *Dermatology in General Medicine.* 3rd ed. New York, NY: McGraw-Hill; 1987:610-626.
4. Hashimoto I, et al. Ultrastructural studies in epidermolysis bullosa hereditaria. IV. Recessive dystrophic type with junctional blistering. *Arch Dermatol Res.* 1976;257:17.
5. Stricklin GP, et al. Human skin collagenase in recessive dystrophic epidermolysis bullosa. *J Clin Invest.* 1982;69:1373.
6. Granek H, Baden HP. Corneal involvement in epidermolysis bullosa simplex. *Arch Ophthalmol.* 1980;98:469-472.
7. Hammerton ME, et al. A case of junctional epidermolysis bullosa (Herlitz-Pearson) with corneal bullae. *Aust J Ophthalmol.* 1984;12:45-48.
8. Aurora AL, Madhavan M, Rao S. Ocular changes in epidermolysis bullosa letalis. *Am J Ophthalmol.* 1975;79:464-470.
9. Forgacs, J, Franceschetti A. Histologic aspect of corneal changes due to hereditary, metabolic, and cutaneous affections. *Am J Ophthalmol.* 1959;47:191-202.
10. Cohen M, Sulzberger MB. Essential shrinkage of conjunctiva in a case of probable epidermolysis bullosa dystrophica. *Arch Ophthalmol.* 1935;13:374-390.
11. Duke-Elder S. *System of Ophthalmology.* 5th ed. vol 8. London: Henry Kimpton; 1965;524.

Chapter 216

Erythema Multiforme

C. STEPHEN FOSTER

Erythema multiforme is a complex immunologic syndrome that in its mildest form is characterized by typical "target" type skin lesions and that sometimes has associated erythematous macules and papules, wheals, bullae, and erosions of mucous membranes.[1] In its most severe form, known as Stevens-Johnson syndrome, it is a life-threatening disorder that extensively affects mucous membranes, including the conjunctiva, and is accompanied by fever and constitutional signs and symptoms secondary to involvement of the kidneys, heart, gastrointestinal tract, and central nervous system.[2]

The prevalence is approximately 1% among patients attending dermatology clinics; milder forms of erythema multiforme have no sexual predilection, but more males than females have Stevens-Johnson syndrome. The disorder affects children and young adults primarily.

The immune complex vasculitis and/or the lymphocyte-dependent effector pathways responsible for the lesions may be precipitated by microbial, neoplastic, or pharmacologic agents. Seasonal variation has been noted, most cases occurring in the first half of each calendar year, possibly associated with an increased incidence of viral respiratory illness during that period. The disorder is generally acute and self-limited.

The earliest histologic change observed in skin lesions of patients with erythema multiforme is perivasculitis of the superficial dermal vessels, which are surrounded by mononuclear cells; leukocytoclastic vasculitis with fibrinoid necrosis of vessel walls does not occur.[3] Immunoglobulin M (IgM) and complement component 3 (C3) are found in the walls of skin vessels in the earliest phases of lesion evolution.[4] A possible explanation for the prominent absence of neutrophils in the vasculitic process has been proposed by Wuepper et al., who suggested that the immune complex deposits peculiar to Stevens-Johnson syndrome may bear C3d receptors and hence attract monocytes and lymphocytes that have C3d receptor sites, in contrast to neutrophils that have receptors for C3b.[5]

Helper/inducer T lymphocytes and Langerhans cells form the vast majority of cells in the inflammatory infiltrate in the dermis of affected skin; cytotoxic/suppressor cells predominate in the epidermis.[6] Such lymphocytes typically are not found in the epithelium of affected conjunctiva, but a conjunctival immune complex vasculitis does exist when conjunctiva is affected in Stevens-Johnson syndrome. In the substantia propria of the conjunctiva the lymphocyte infiltrate is almost exclusively helper/inducer T cells, regulatory suppressor T cells being almost completely absent.[7]

Approximately 15 to 79% of the cases of Stevens-Johnson syndrome affecting the skin occur 1 to 3 weeks after episodes of recurrent herpes simplex labialis or genitalis. Other notable provocateurs include *Mycoplasma pneumoniae*, particularly in very young children, and sulfonamides or other pharmacologic preparations. In truth, it is frequently difficult to know with certainty whether or not a pharmacologic agent has been the initiating stimulus, since most agents are typically given in response to the patient developing a microbial process, such as an upper respiratory infection; the microbial agent responsible for the infection itself could have been the initiator of the erythema multiforme.

The cutaneous manifestations of erythema multiforme may be preceded by fever, malaise, and sore throat. The geographic areas most typically affected by the target skin lesions are the abdomen, extensor surfaces of the forearms and legs, and dorsa of the hands and feet. The eruption may be intensely pruritic, skin lesions developing in crops that subside within 2 to 6 weeks after onset of the disease. Mucosal lesions are present in 20 to 45% of patients with erythema multiforme; conjunctiva, oral mucosa, and mucosa of the lips are the sites most often affected. Nose, genitals, and anus are infrequently involved, but involvement of those areas may lead to stenosis of the urethra, vagina, or rectum. In 50% of patients with Stevens-Johnson syndrome the mucosal surface of the trachea is involved. Tracheal constriction and aspiration may result in asphyxiation, but renal and central nervous system damage account for most of the deaths in fatal cases of Stevens-Johnson syndrome. In spite of the extensive use of high-dose corticosteroid therapy in the treatment of this disorder, the mortality rate is reported to range between 5 and 20%.[8] A recurrent variety of the disease also exists, herpes simplex virus being the most commonly implicated antigen in patients with recurrent erythema multiforme. Ten to 15% of patients may experience recurrences. It is now clear that in rare instances recurrent immune-mediated conjunctival inflammation may also occur in patients with a history of erythema multiforme.[9]

Therapy for this disease is generally supportive. There is little evidence to suggest that high-dose corticosteroid therapy affects the course of the disease. Erythromycin or another appropriate antibiotic is typically used in cases in which *M. pneumoniae* is implicated. Systemic antihistamines may be palliative for the severe itching. Pharmacologic agents strongly suspected as initiators of the process are typically discontinued.

Ocular Manifestations

Although Stevens-Johnson syndrome is generally an acute and self-limited disorder, it can produce devastating long-term ocular consequences because of conjunctival scarring and damage to adnexal structures with resultant tear insufficiency, conjunctival epithelial squamous metaplasia, trichiasis, distichiasis, lagophthalmos, epithelial defects, corneal ulceration, and corneal neovascularization. Therapy for these problems associated with the scarring produced by the acute exanthem include tear replacement, punctal occlusion, ocular lubricants, cryoablation of aberrant lashes, and surgical correction of entropion or lagophthalmos. Though little evidence supports the hypothetical efficacy of topical retinoids in this condition, the use of such agents is not unreasonable, given the pronounced squamous metaplasia typically seen in the conjunctiva of these patients. If recurrent herpes simplex labialis or genitalis is suspected as a stimulus for recurrent inflammation in the eyes, long-term suppressive doses of oral acyclovir may be advisable. Recurrent, immunologically driven inflammatory activity in the conjunctiva that is proven by biopsy to represent recurrent immune complex vasculitis not associated with a microbe stimulus may respond to long-term immunosuppressive chemotherapy.

References

1. Soter NA, Freedberg IM, Erythema multiforme and toxic epidermal necrolysis. In: Samter M, Talmage DW, Rose B, et al., eds. *Immunological Diseases.* Boston, Mass: Little, Brown; 1978;984-992.
2. Stevens AM, Johnson FC. A new eruptive fever associated with stomatitis and ophthalmia: report of two cases in children. *Am J Dis Child.* 1922;24:526.
3. Champion RH. Disorders affecting small blood vessels. Erythema and telangectasia. In: Rook A, Wilkinson DS, Ebling FJG, eds. *Textbook of Dermatology,* 3rd ed. London: Blackwell Scientific Publications; 1972;885.
4. Kazmierowski JA, Wuepper KD. Erythema multiforme: immune complex vasculitis of the superficial cutaneous microvasculature. *J Invest Dermatol.* 1978;71;366-369.
5. Wuepper KD, Watson PA, Kazmierowski JA. Immune complexes in erythema multiforme and the Stevens-Johnson syndrome. *J Invest Dermatol.* 1980;74:368-371.
6. Margolis RJ, Tonneson MG, Harrist TJ, et al. Lymphocyte subsets and Langerhans cells/indeterminant cells in erythema multiforme. *J Invest Dermatol.* 1983;81:403-406.
7. Bahn AK, Fujikawa LS, Foster CS. T cell subsets and Langerhans cells in normal and diseased conjunctiva. *Am J Ophthalmol.* 1982;94:205-212.
8. Claxton RC. Review of 31 cases of Stevens-Johnson syndrome. *Med J Aust.* 1963;1:963.
9. Foster CS, Fong LP, Asar D, et al. Episodic conjunctival inflammation after Stevens-Johnson syndrome. *Ophthalmology.* 1988;95:453-462.

Chapter 217

Hydroa Vacciniforme

CLEVELAND KIRKLAND, Jr., and JEFFREY DAY LANIER

Systemic Manifestations

Hydroa vacciniforme is a severe, recurrent, self-limiting, vesicular skin eruption related to sun exposure. It occurs in childhood, first manifesting itself before 4 years of age and usually disappearing soon after puberty. The disease has a 2 to 1 predilection for males. Skin lesions are limited to sun-exposed areas, mainly face, ears, neck, hands, and arms.

Symptoms antedate the skin eruption. Usually there is mild stinging or burning in the exposed skin sites 1 to 2 hours following sun exposure. Photophobia and tearing are also common complaints.

The skin lesions appear first as tense erythematous papules that develop into discrete vesicles filled with clear, then cloudy, fluid. Umbilication and necrosis of the vesicles occur, and finally they evolve into hypopigmented depressed scars similar in appearance to those of vaccinia. Evolution of the skin lesions occurs for several days following the initial sun exposure. Eruptions are most common in the spring and summer months. Histologically, focal intraepidermal vesicle formation is noted. Epidermal keratocytic necrosis and reticular degeneration progress to pandermal necrosis with a dermal and perivascular lymphohistiocytic infiltration.[1] The vessels themselves are normal. Tissue porphyrins are not present, and immunofluorescence studies reveal no evidence of complement or immunoglobulins.[2]

The cause of the condition, other than sun exposure, is unknown. There is no established inheritance pattern. Patients have no related systemic illnesses. An identical skin eruption can be seen in patients with congenital porphyria or erythropoietic protoporphyria, but porphyrin metabolism in patients with hydroa vacciniforme is normal.[3,4] Another similar group of photodermatoses, polymorphous light eruption and hydroa aestivale, have comparable skin lesions but heal without scarring.

Great confusion has been evident in the early literature concerning hydroa vacciniforme, as its relationship to the porphyrias was poorly understood. The generalized condition was first described by Bazin[5] in 1862. McCrae[6] established hydroa vacciniforme as a separate entity, distinct from similar skin eruptions where there was abnormal porphyrin metabolism. A major review of the world literature revealed that all but seven of the patients reported through 1923 had abnormal urine porphyrins.[7] In 1961, hydroalike skin eruptions were described in patients with abnormal protoporphyrin metabolism whose urine porphyrin studies were negative, while erythrocyte and fecal porphyrin studies were markedly abnormal; this is due to the water insolubility of the protoporphyrins.[4] This discovery casts doubt on the diagnosis in the other seven cases from the early literature, in which fecal or erythrocyte porphyrin screening was not done. Recent reviews of the literature reveal several cases of hydroa vacciniforme in which porphyrin metabolism has proved normal.

Ocular Manifesations

Though they are unusual, ocular complications may accompany hydroa vacciniforme. Tearing and photophobia are often present with no evident ocular abnormality. The conjunctiva may show evidence of limbal chemosis or vesiculation. Severe scarring is probably less common than was previously thought. Scleral involvement is not known to occur in true hydroa vacciniforme. Corneal involvement, which ranges from vesicle or ulcer formation to band-shaped keratopathy was noted in three cases.[6,8,9] One case report noted bilateral deep corneal neovascularization with deep infiltration and ulceration leading to dense stromal opacities and loss of vision.[10] Involvement of the uveal tract or retina has not been reported. Cicatrization of the lids due to skin involvement can lead to secondary changes in the conjunctiva or cornea.[9]

Treatment of this disorder chiefly centers around using sunscreen agents and protective clothing during exposure to sun. Oral beta carotene has also been used with some success. In difficult cases, hydroxychloroquine has been shown to have some value. The disease tends to be self-limited and usually abates after puberty.

References

1. Bickers DR. Hydroa vacciniforme. *Arch Dermatol.* 1978;114:1193-1196.
2. Goldgeier MH. Hydroa vacciniforme: diagnosis and management. *Arch Dermatol.* 1982;118:588-591.
3. Sevel D, Burger D. Ocular involvement in cutaneous porphyria. *Arch Ophthalmol.* 1971;85:580-585.
4. Magnus I. Erythpoietic protoporphyria: a new porphria syndrome with solar urticaria due to protoporphyrinaemia. *Lancet.* 1961;2:448-451.
5. Bazin E. Lecons theoretiques et cliniques sur les affectations generiques que le peau. Paris: Delahay; 1862;1:132.
6. McGrae JD Jr, Perry HO. Hydroa vacciniforme. *Arch Dermatol.* 1963;87:618-625.
7. Senear FE, Fink HW. Hydroa vacciniforme sen aestivale. *Arch Dermatol Syphilol.* 1923;7:145-162.
8. Bertail MA. Photodermatose bulleuse a evolution pseudocheloidienne avec keratite en bande horizontale: hydroa vacciniforme? *Ann Dermatol Venerol (Paris).* 1982;109:743-744.
9. Stokes WH. Ocular manifestations in hydroa vacciniforme. *Arch Ophthalmol.* 1940;23:1131-1145.
10. Crews SJ. Hydroa vacciniforme affecting the eye. *Br J Ophthalmol.* 1959;43:629-633.

Chapter 218

Pemphigus

MELVIN I. ROAT and RICHARD A. THOFT

Pemphigus vulgaris (PV) is a chronic bullous skin and mucous membrane disease that affects men and women equally, usually in the 5th and 6th decades of life.[1] In the past, PV was reported almost exclusively in Jews; now it has been documented in members of all ethnic groups.[1] PV is believed to be an autoimmune disease and there is a strong association with the HLA-A10 and HLA–DRw4 phenotypes.[1,2] It may be associated with increased risk for other autoimmune diseases and for malignancy.[1,3]

Pathology

The bullae of PV are believed to result from the binding of circulating antibodies to the glycocalyx (intercellular cement substance) on or near the cell membrane of stratified squamous epithelium. Antibody binding induces the secretion or activation of a protease, causing dissolution of intercellular attachments, loss of adherence of desmosomes, acantholysis, and subsequent blister formation.[1,3,4] The intraepithelial nature of the disease does not lead to activation of subepithelial fibrocytes with subsequent scarring. In contrast to most autoimmune diseases, studies indicate that antibody binding alone, without complement or inflammatory cells, is sufficient to cause tissue injury in PV.[1,3] Anhalt et al.[6] demonstrated conclusively that it is the circulating pemphigus antibody that causes clinical disease. Passive transfer of human PV antibody by intraperitoneal injection in neonatal mice resulted in clinical, histologic, and immunologic disease similar to human PV.

The identity of the PV antigen remains unknown. It has been partially characterized and is produced by human epidermal cells in tissue culture.[3]

Circulating antibodies to stratified squamous epithelium intercellular substance (PV antibody) have been detected in 80 to 95% of patients.[1,4] Most studies have found a positive correlation between circulating PV antibody titer and clinical activity.[1,3,4] Antibody titer has been observed to fall or to become undetectable with successful therapy.[1,3,4]

Tissue-bound PV antibody (Fig. 218-1) and often complement components are detected in skin biopsies of 90 to 95% of patients with PV. The immunoglobulin is localized to the intercellular substance of the epithelium of skin, conjunctiva, and oral mucous membrane.[1,3-7]

The characteristic histology of cutaneous, conjunctival, and oral lesions is acantholysis, intraepithelial clefts (walls, roof, and base of epithelial cells), and blisters. Free-floating cells (Tzanck cells) may be seen in the blister fluid.[1,3-6]

Clinical Features

PV may affect skin and mucous membranes, including those of the conjunctiva, vagina, anus, urethra, mouth, and nose.[1-9] The bullous lesions of a stratified squamous epithelium heal slowly and without scarring.[4,7,8] Cutaneous PV (Fig. 218-2) is characterized by thin flaccid bullae on normal-looking skin. The fluid in the blister is clear at first but may become hemorrhagic or seropurulent later. Bullae may rupture spontaneously or with slight pressure or rubbing (Nikolsky's sign). The painful ruptured bullas form raw oozing erosions that bleed easily and show little tendency to heal.[1,4,6] Pemphigus foliaceus, a variant of pemphigus, differs from PV by the presence of tissue-bound antibody exclusively in the uppermost intercellular layers of skin, with shorter-lived bullae, more severe ulceration, and exfoliation of skin.[4]

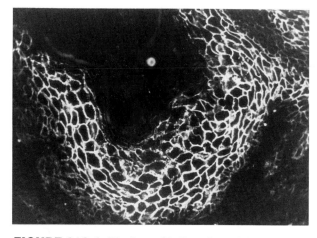

FIGURE 218-1 Binding of IgG to the epithelial intercellular substance in a skin biopsy specimen from a patient with pemphigus vulgaris is demonstrated by direct immunofluorescence. (Courtesy of Dr. E. Abell.)

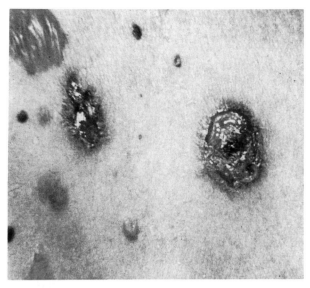

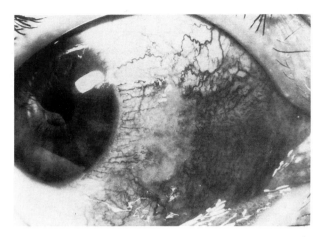

FIGURE 218-3 Conjunctival involvement in pemphigus vulgaris with hyperemia and edema.

FIGURE 218-2 Erosions and flaccid bullae of pemphigus vulgaris on normal-looking skin.

Oral mucosal involvement, which is present in 50% of patients with PV, usually antedates cutaneous manifestations by many months.[5,9] The short-lived bullae quickly rupture to form painful, raw erosions. Pharyngeal involvement may result in hoarseness and difficulty with swallowing.[4,9]

Conjunctival involvement occurs in pemphigus foliaceus and pemphigus vulgaris.[1,4,7,8] The bullae are transient, usually producing a purulent or pseudomembranous conjunctivitis (Fig. 218-3).[7,8] The conjunctiva and other stratified squamous epithelia do not scar in this disease. Active cutaneous eyelid involvement, however, may lead to exposure or trichiasis due to distortion of the normal eyelid architecture by the crusting lesions.

Therapy

Before the use of antibiotics and anti-inflammatory drugs, PV was universally fatal: 50% of patients died in the first year of the disease from fluid and electrolytic imbalance, malnutrition, or sepsis.[1,4] Now, with the use of high-dose systemic steroid therapy, the mortality rate is significantly reduced (8-10% per 5 years).[1] Most of the deaths are the result of complications of long-term steroid use. Current recommendations for therapy include the use of cytotoxic immunosuppression to allow steroids to be used at lower doses for shorter periods. Therapy should be adjusted based on clinical disease and serum PV antibody titer.[1]

References

1. Patel HP, Anhalt GJ, Dias LA. Bullous pemphigoid and pemphigus vulgaris. *Ann Allerg.* 1983;50:144.
2. Parks MS, Terasaki PI, Ahmed AR, et al. HLA-DRw4 in 91% of Jewish pemphigus vulgaris patients. *Lancet.* 1979;2:441.
3. Sams WM Jr, Gammon WR. Mechanism of lesion production in pemphigus and pemphigoid. *J Am Acad Dermatol.* 1982;6:431.
4. Flowers FP, Sherertz EF. Immunological disorders of the skin and mucous membranes. *Med Clin North Am.* 1985; 9:657.
5. Cohen L. Ulcerative lesions of the oral cavity. *Int J Dermatol.* 1980;19:362.
6. Anhalt GJ, Labib RS, Voorhees JJ, et al. Induction of pemphigus in neonatal mice by passive transfer of IgG from patients with the disease. *N Engl J Med.* 1982;306:1189.
7. Bean SF, Holubar K, Gillet RB. Pemphigus involving the eyes. *Arch Dermatol.* 1975;11:1484.
8. Michel B, Thomas CI, Levine M, et al. Cicatricial pemphigoid and its relationship to ocular pemphigus and essential shrinkage of the conjunctiva. *Ann Ophthalmol.* 1975;7:11.
9. Laskaris G, Sklavounou A, Stratigos J. Bullous pemphigoid, cicatricial pemphigoid and pemphigus vulgaris. *Oral Surg.* 1982;54:656.

Chapter 219

Toxic Epidermal Necrolysis

Lyell's Syndrome

CRAIG M. MORGAN

Toxic epidermal necrolysis (TEN), also known as Lyell's syndrome, is one of the most devastating generalized skin disorders.[1,2] The disease has a high mortality rate (25 to 50%), and survivors often develop significant sequelae that may include severe scarring, contractures, blindness, and esophageal strictures.[3]

The cause of TEN remains unknown, although many factors have been incriminated.[4] Drugs are considered to be the responsible agent most frequently, and almost any class of drug is capable of provoking TEN. Other possible causal factors include systemic infections, lymphoma, and vaccinations. Most patients with TEN receive multiple forms of therapy for complex medical problems; thus in some instances, TEN probably results from several coexisting factors.[2]

The pathogenesis of TEN seems to be an underlying immune mechanism, but its exact nature is unclear.[4] TEN has been reported during acute graft-versus-host disease following bone marrow transplantation. Additionally, lymphopenia resulting from a reduced number of inducer-helper T lymphocytes frequently occurs during the early phase of the disease.[5] It is possible, therefore, that promotion of cytotoxic cells directed against autologous epidermis by an abnormal immunologic balance could be the underlying cause of TEN.

Systemic Manifestations

TEN is characterized by a severe skin reaction that is confined to the epidermis. Initially, there is an inflammatory reaction, which is followed by necrosis of all the epidermal layers.[5-7] Later, subepidermal bulla formation develops, leading to full-thickness epidermal separation, a process that can occur over a major portion of the body.[3-5] Since the entire thickness of the epidermis is damaged, the morbidity and mortality rates of the disease are similar to those seen with extensive second-degree burns (Figs. 219-1, 219-2).[2]

Patients often have prodromal symptoms such as fever and malaise.[8] The cutaneous manifestations develop acutely and are usually associated with a high fever and leukocytosis. Initially a morbilliform rash appears, predominantly on the face and extremities, which rapidly becomes confluent, leading to diffuse erythema.[4] Subsequently, clear bullae appear and become confluent to form large flaccid bullae that rupture easily, allowing the epidermis to come off in large sheets.[3] The mucous membranes are severely involved as well, and the lips and oral, genital, and anal mucous membranes show diffuse erythema, vesiculation, and widespread erosions.[2] The fingernails and toenails may slough. Additionally, there may be severe involvement of the respiratory and gastrointestinal tracts. Healing generally takes place over a 2- to 4-week period, although relapses may occur. Virtually all deaths are

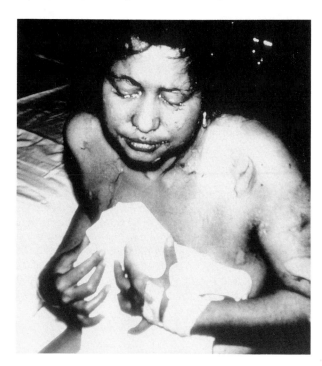

FIGURE 219-1 Note the scalded appearance of the skin of a patient with TEN. (From Ostler et al.[8])

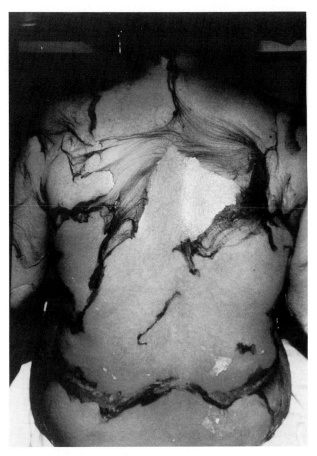

FIGURE 219-2 Sloughing skin in a TEN patient. (From Ostler et al.[8])

secondary to overwhelming sepsis, but gastrointestinal hemorrhage, renal failure, pulmonary edema, and fluid and electrolyte imbalance are additional major complications that may be fatal.[2,3]

Treatment is initially supportive; supplementary specific treatment is employed as indicated for secondary complications.[2] The use of porcine skin xenografts early in the course of the disease has been shown to reduce both mortality rate and long-term morbidity.[3]

Ocular Manifestations

Ocular complications are frequently the most serious sequelae of TEN.[2] The flaccid bullae can develop on the eyelids, and the eyebrows and cilia may slide off with the epidermis of the eyelids.[4] With healing, entropion or ectropion can develop.

As with the other mucous membranes, the conjunctiva is affected concomitantly with the epidermis, or later.[8] The reaction may be only mild conjunctivitis, but occasionally severe injection of the conjunctiva with a purulent discharge may develop and a pseudomembrane may form.[6] The subsequent scarring can result in trichiasis and symblepharon formation.

Keratoconjunctivitis sicca is an additional prominent ocular feature of TEN. It results from scarring and occlusion of the orifices of the ducts of the lacrimal gland and of the accessory lacrimal secretors.[6,8] Punctate staining of the cornea followed by loss of the corneal epithelium, pannus formation, opacification and scarring, corneal ulcers, or spontaneous perforation of the globe can occur secondarily, with resultant blindness.

Treatment is directed at replacing the tear loss and lubricating the eye. Long-term administration may be required. Synechiae may be prevented by early lysis with a glass rod. Secondary infections should be treated appropriately.

References

1. Lyell A. Toxic epidermal necrolysis: an eruption resembling scalding of the skin. *Br J Dermatol.* 1956;68:355.
2. Snyder, RA, Elias PM. Toxic epidermal necrolysis and staphylococcal scalded skin syndrome. *Derm Clin North Am.* 1983;1:235.
3. Heimbach DM, Engrav LH, Marvin JA, et al. Toxic epidermal necrolysis, a step forward in treatment. *JAMA.* 1987;257:2171.
4. Fritsch PO, Elias PM. Toxic epidermal necrolysis. In: Fitzpatrick TB, Eisen AZ, Wolff K, et al. eds. *Dermatology in General Medicine.* 3rd ed. New York, NY: McGraw-Hill; 1987;563-567.
5. Roujeau JC, Moritz S, Guillaume JC, et al. Lymphopenia and abnormal balance of T-lymphocyte subpopulation in toxic epidermal necrolysis. *Arch Dermatol Res.* 1985;277:24.
6. Bennett TO, Sugar J, Sahgal S. Ocular manifestations of toxic epidermal necrolysis associated with allopurinol use. *Arch Ophthalmol.* 1977;95:1362.
7. Roujeau JC, Koso M, Andre C, et al. Sjögren-like syndrome after drug-induced toxic epidermal necrolysis. *Lancet.* 1985; 2:609.
8. Ostler HB, Conant MA, Groundwater J. Lyell's disease, the Stevens-Johnson syndrome, and exfoliative dermatitis. *Trans Am Acad Ophthalmol Otolaryngol.* 1970;74:1254.

PART 20

Vascular Disorders

Chapter 220

Aortic Arch Syndrome

JAMES P. BOLLING, RICHARD GOLDBERG, and GARY BROWN

Aortic arch syndrome is the result of chronic occlusive disease of one or more of the major branches of the aortic arch—the innominate artery, the left common carotid artery, the left subclavian artery (Fig. 220-1)—which can be caused by atherosclerosis, connective tissue disease, or syphilitic aortitis.[1]

The ocular and cerebral effects of aortic arch diseases depend on the location of the obstructions and the pattern of collateral flow that subsequently develops.[2] For example, the blood supply to the occipital cortex comes from the carotid arteries when severe vertebrobasilar artery disease is present. Similarly when the left subclavian artery or the innominate artery is occluded at its origin, retrograde flow of blood through the vertebral artery on the affected side can supply the upper extremity. In such a pattern of circulation, use of the upper extremity can produce symptoms of cerebral ischemia (subclavian steal syndrome). Selective arteriography is critical in the evaluation of patients with the aortic arch syndrome. Of 100 consecutive digital subtraction angiograms ordered for suspected carotid artery disease by the retinal vascular service at Wills Eye Hospital, eight showed aortic arch involvement (unpublished data).

Systemic Manifestations

The signs and symptoms of the aortic arch syndrome are caused by marked impairment of blood flow to the arms and head. Symptoms of carotid occlusive disease, such as monocular blindness or hemiparesis, as well as symptoms of vertebrobasilar insufficiency can be seen with the aortic arch syndrome. Claudication and paresthesias in the upper extremity are commonly encountered. Complete occlusion of the left subclavian artery or the innominate artery can result in carotid or vertebrobasilar symptoms that are precipitated by the use of one arm. It is important to check radial pulses and blood pressure in both arms in any case of suspected carotid artery disease.

Takayasu's arteritis is an inflammatory disease of large arteries. The branches of the aorta are frequently involved and can cause aortic arch syndrome. Frequently the sedimentation rate is markedly elevated; the histopathology of this disease may resemble that of temporal arteritis. In contrast to temporal arteritis, Takayasu's arteritis usually affects children and young women and is most common in Japan. The disease

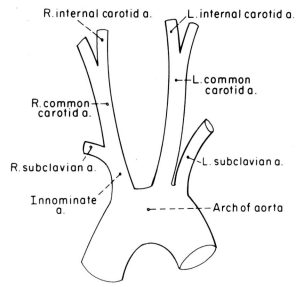

FIGURE 220-1 This line drawing shows the most common order of the major branches of the aorta. Any vessel may be involved at its origin in the aortic arch syndrome.

typically has an acute and a chronic phase. The acute phase is characterized by signs and symptoms of an acute inflammatory process. Patients have fever, malaise, weakness, myalgias, and night sweats. Takayasu's disease is thought to be the result of an autoimmune mechanism. There is an increased frequency of human leukocyte antigens B5, A10, and Bw52 in patients with Takayasu's disease.

Ocular Manifestations

Of 42 patients with aortic arch syndrome seen over a 10-year period at the Mayo Clinic, five had scintillating scotomas and three had classic amaurosis fugax (unpublished data). Figure 220-2 shows an arteriogram of a patient who developed scintillating scotomas when using the left arm.

The aortic arch syndrome can produce chronic ocular ischemia. The typical signs of anterior segment ischemia such as corneal edema, aqueous flare, cataract, neovascularization of the iris, glaucoma, and hypotony are seen and are most common when severe bilateral carotid involvement is present.[3] The most common signs of posterior segment involvement are low retinal artery pressure and arterial narrowing. Venous dilatation, microaneurysm formation, cotton-wool spots, cobblestone degeneration, and retinal hemorrhages are also seen.[4] The findings of ocular ischemia do not occur in all patients with aortic arch syndrome, or even in all patients with carotid occlusive disease. It has been suggested that in addition to decreased carotid perfusion a cerebral steal of collateral flow is necessary to produce these findings.[5]

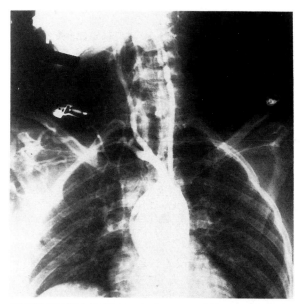
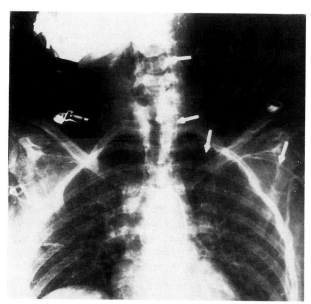

FIGURE 220-2 This arteriogram of the aortic arch demonstrates the subclavian steal syndrome. The view on the left shows complete occlusion of the left subclavian artery at its origin. The view on the right shows retrograde filling of the left vertebral artery supplying the axillary artery and left arm (*arrow*). The patient developed scintillating scotomas with exercise of the left arm.

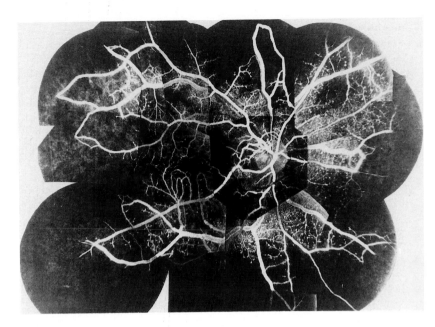

FIGURE 220-3 Fluorescein angiogram of the right eye shows that arteriovenous shunts have formed through dilatation of capillaries below the macula, through communication at arteriovenous crossings of major retinal vessels, and by formation of vascular loops around the optic disc. Numerous microaneurysms are present from vessels lying on the arterial side. (From Shimizu.[6])

Takayasu's disease may present with scleritis or iritis in the acute phase. In the chronic phase the fundus findings can be quite marked and have been extensively studied with fluorescein angiography.[6] Schimizu has divided the fundus findings in Takayasu's disease into three stages. The third stage has a unique feature in the development of wreathlike arteriovenous anastomoses in front of and around the optic nerve (Fig. 220-3). Additional findings are nonspecific changes of ocular ischemia, such as arteriolar narrowing, microaneurysms, nerve fiber layer infarcts, retinal hemorrhages, venous dilatation, and retinal nonperfusion.

References

1. Ross RS, McKusick VA. Aortic arch syndromes: diminished or absent pulses in arteries arising from the arch of the aorta. *Arch Intern Med.* 1953;92:701-740.
2. Tour RL, Hoyt WF. The syndrome of the aortic arch. *Am J Ophthalmol.* 1959;47:35-48.
3. Knox DL. Ischemic ocular inflammation. *Am J Ophthalmol.* 1965;60:995-1002.
4. Kahn M, Green WR, Knox DL, et al. Ocular features of carotid occlusive disease. *Retina.* 1986;6:239-252.
5. Kahn M, Knox DL, Green WR. Clinicopathologic studies of a case of aortic arch syndrome. *Retina.* 1986;6:228-233.
6. Shimizu K. Fluorescein microangiography of the ocular fundus. Baltimore, Md: Williams & Wilkins; 1973;45-52.

Chapter 221

Arteriosclerosis

SARKIS SOUKIASIAN and MOSHE LAHAV

Arteriosclerosis is a general term that refers to hardening and thickening of the arterial wall. It is highly prevalent and constitutes a major cause of morbidity and death in the United States and in Western societies.[1]

The terminology associated with arteriosclerosis has often been incorrectly interchanged. The disorder is composed of at least three different processes, each of which affects vessels of a different size and has different pathologic characteristics. *Atherosclerosis* affects the larger arteries and involves the intimal layer. *Medial sclerosis* affects the medium-sized arteries by causing calcification of the medial muscular layer. *Arteriolosclerosis* affects the smaller arteries and the arterioles and involves both intimal and medial layers.

The arterial vessel wall is comprised of three layers. The tunica intima, the innermost layer, consists of endothelium and the internal elastic lamina. The tunica media, the middle layer, is composed of circularly arranged smooth muscle surrounded by the external elastic lamina. Finally, the coating layer, the tunica adventitia, is a mixture of collagen and elastic fibers, with occasional smooth muscle cells and fibroblasts. In larger vessels, it contains the vasa vasorum and nerves. The relative contribution of each layer to the arterial wall varies according to the size of the vessel.

Although there is some overlap, the size and function of the arteries are intimately related. The large vessels have a thick intimal layer composed primarily of elastin. Their ability to stretch and passively contract helps to maintain an even blood pressure in the arterial system between contractions of the heart. The walls of medium-sized vessels consist mostly of smooth muscle cells. They regulate the flow of blood to different organs of the body, according to their needs, via reflex phenomena. The smaller vessels, the arterioles, which measure approximately 100 μm, have a smaller lumen-to-vessel wall ratio. They act as a valve between the high systemic pressure and the low pressure of the capillary bed, and their tone also regulates the degree of pressure within the arterial system.[2]

Pathophysiology

Atherosclerosis affects the larger arteries, such as the abdominal aorta and its branches, the coronary arteries, the carotid vessels, and the cerebral vasculature. It primarily involves the intimal layer, with patchy nodular lesions. Pathologically, they are classified as: (1) fatty streaks, accumulations of lipid-filled smooth muscle cells, macrophages, and fibrous tissue seen as early as the first decade of life; (2) fibrous plaques, firm, elevated, dome shaped lesions consisting of a central core of lipid and necrotic cell debris covered by a fibrous muscular cap; and (3) the complicated lesion, a calcified fibrous plaque containing various degrees of necrosis, thrombosis, and ulceration.[3] Progression of this last lesion may lead to weakness and rupture of the intima with subsequent formations of aneurysms, intramural hemorrhage, and arterial emboli from dislodged plaques.[3]

There are a few theories of atherogenesis. The endothelial injury theory implicates a reaction of the endothelial cell lining to injury by chemicals (in hypercholesterolemia and homocystinuria), mechanical stresses (hypertension) and immune-mediated processes.[2] The monoclonal hypothesis suggests uncontrolled proliferation of individual smooth muscle cells. The lysosomal theory implicates altered cellular enzyme function in the production of these lesions.[2] Chemotactic and growth factors released from endothelium, smooth muscle cells, platelets, and monocytes also play a role in atherosclerosis.[4]

Medial sclerosis, also termed focal calcific arteriosclerosis or Mönckeberg's sclerosis, affects medium- and small-sized muscular arteries and has a predilection for the lower extremities. Histologically, there is degeneration of the smooth muscle cells, and there is radiographic evidence of calcification of the media. Although the lesion does not cause luminal narrowing, the common association with atherosclerosis in the lower extremities may cause arterial occlusion. These changes are commonly seen in elderly persons but may be more severe and accelerated in diabetes patients. The pathogenesis may be related to sympathetic denervation of the smooth muscle cells.[2]

Arteriolosclerosis affects both the intima and the media of small arteries and arterioles, especially in the spleen, pancreas, adrenal, kidney, and retina. Pathologically, one may see varying degrees of endothelial hyperplasia, intimal and subintimal hyalinization, medial hypertrophy, and fibrosis in a patchy distribution.[2,5] There are two major types of arteriolosclerosis, proliferative and degenerative.

Proliferative arteriolosclerosis is present when hyperplastic changes are seen. These include thickening of the arteriole wall, muscle hyperplasia, increase in fibrous and elastic components, as well as concentric lamellar intimal thickening of the arterioles (onion skin degeneration).[6] This type is usually observed with acute and severe elevations of blood pressure. The degenerative type, which is the most common form of arteriolosclerosis, is called hyaline degeneration. There is deposition of extracellular lipohyaline material and collagenization in the intima and media.[5-7] This type is more commonly present in the elderly, and some have termed it involutional sclerosis.[7] It is accelerated and pronounced in patients with diabetes mellitus and hypertension. This is demonstrated in the kidney as arteriolar nephrosclerosis.[3] An acute and severe degenerative form, fibrinoid necrosis, is seen only with extreme blood pressure elevation. Histologic examination reveals thickening and loss of structural detail in the vessel wall with deposition of fibrinoid and eosinophilic staining amorphous material.

Systemic Manifestations

Atherosclerosis plays a major role in the pathogenesis of cardiovascular diseases, which are a leading cause of death in the United States.[1] Risk factors include aging, male sex, genetic predisposition, cigarette smoking, hypertension, obesity, hyperlipidemia, hyperglycemia, and low blood levels of high-density lipoprotein (HDL).[2,3]

Ischemic heart disease (coronary artery disease) correlates best with atherosclerosis but not arteriolosclerosis. The atherosclerotic lesion in the coronary arteries, with or without thrombosis, compromises the vascular supply to the myocardium, leading to myocardial ischemia, which may cause fatal arrhythmias or myocardial infarction. The clinical symptoms include angina, syncopal episodes, or palpitations and evidence of heart failure.

Cerebral vascular disease (stroke) is another manifestation of atherosclerosis. The carotid and vertebrobasilar arteries are most frequently involved. Decreasing perfusion due to luminal narrowing or emboli from degenerative plaques may cause cerebral ischemia or infarction. There is a spectrum of clinical manifestations from transient ischemic attacks (TIAs) with return of total function to irreversible neurologic deficits (strokes) and death. Transient or permanent dysarthria, diplopia, hemiparesis, hemiplegia, or severe cognitive dysfunction can be seen as part of the clinical presentation. Carotid disease may be detected by noninvasive techniques such as carotid auscultation, palpation, and ultrasound and Doppler studies. Invasive methods such as carotid angiography can be used to make a definitive diagnosis.

Decreased perfusion and emboli may affect the peripheral vasculature, especially in the lower extremities. Clinically, it manifests as intermittent claudication, peripheral nonhealing ulcers, and gangrene. If the genital circulation is involved impotence may be present (Leriche's syndrome). Atherosclerotic renal artery disease can lead to severe and refractory hypertension.

Medial sclerosis is rare before the age of 50 and is seen with equal frequency in both sexes. It is more often identified in the lower extremities (pelvic and femoral arteries) as radiographically opaque, regular, concentric calcifications. The palpable vessels, such as the radial arteries, can be felt as "rigid tubes."[2] Since the lumen is not narrowed, circulation is not compromised unless atherosclerosis is superimposed.

Arteriolosclerosis manifests in the smaller vessels of the spleen, pancreas, and kidney. Arteriolar nephrosclerosis is probably the most clinically evident.[3] Up to 70% of persons over age 60 years without hypertension have some degree of arteriolosclerosis.[3] There is an intimate association between arteriolar nephrosclerosis and both hypertension and the normal aging process. It is therefore difficult to differentiate the normal arteriolar changes of aging from those due to hypertension. These vascular changes can affect renal plasma flow and glomerular filtration, leading to decreased functional reserve, making the elderly more sensitive to volume depletion and the toxic effects of drugs (eg, digoxin).[2] Effects on other organ systems are usually subclinical.

Ocular Manifestations

Only atherosclerosis and arteriolosclerosis affect the eye. Atheromatous plaques occur in the ophthalmic and central retinal arteries, up to the level of the lamina cribrosa, but are infrequently visible. Suggestive clinical findings include narrowing of the retinal arteriolar tree, increased number of visible branches at the disc margin, and occlusion of the central retinal artery or vein.[8,9] Rarely have the plaques actually been seen within the first branch of the central retinal artery.[10] Emboli from atheromatous plaques elsewhere, especially in the carotid artery, are responsible for a significant portion of central retinal artery occlusions.[9,11] Small yellow refractile particles composed of cholesterol (Hollenhorst's plaques) can be seen in the branch retinal arterioles. They usually originate from the carotid artery, frequently without evidence of retinal infarction. Small, gray platelet-fibrin plaques, from cardiac or carotid origin, can occasionally cause branch retinal artery occlusions resulting in transient or permanent field defects. White calcific retinal emboli usually originate from a cardiac source. Recurrent noninfected emboli may lead to retinal vasculitis.[12] Severe occlusive atherosclerotic carotid disease may cause ocular ischemia, with midperipheral retinal hemorrhage, hypotony, anterior chamber cells and "flare," iris neovascularization, cataracts, and corneal edema.[13] Ocular evidence of embolization or ischemia should encourage carotid and cardiac valve evaluation.

Neuro-ophthalmologic findings can be manifestations of cerebral and carotid vascular disease. Transient cerebral and brain stem ischemia, as well as ischemic strokes, may be manifested as amaurosis fugax, visual field defects, cranial nerve palsies, gaze disturbances, diplopia, nystagmus, and neuroparalytic keratopathy.

Arteriolosclerosis directly affects the retinal vessels, which are approximately 100 μm in diameter and are by definition arterioles. There is no correlation with atherosclerosis elsewhere. The retina is unique in that it is the only tissue in which arterioles can be examined under direct visualization in vivo. Involutional sclerosis is the change seen with aging. However, it can be associated with and accelerated by hypertension, and separating the two processes is difficult.[7] The earliest signs of arteriolosclerosis are changes in the arteriolar light reflex due to altered density of the vessel wall and changes in its refractive index.[5] The less bright, more diffuse arteriolar reflex may progress in the presence of hypertension to changes described as "copper wiring." "Silver wiring" occurs only in severe cases of longstanding hypertension, where the blood column is obscured by the pathologic process in the vessel wall.[5] The vessels become narrower and straighter at branching points owing to flow reduction and vessel wall replacement changes. At the crossing of the arteriole and venule, where a common adventitia is shared, hiding, nicking, and deflection of the venule are seen. Banking of the venous blood column with evidence of obstruction to venous flow may result in retinal branch vein occlusion. The crossing changes are more prominent with hypertension and are due to alteration in the arteriole and venule walls as well as to perivascular glial cell proliferation.[5] Finally, intimal damage by emboli or the degenerative arteriolosclerotic process may induce retinal macroaneurysm formation, with secondary retinal hemorrhages, exudate, and edema.

In conclusion, the direct and indirect effects of arteriosclerosis on the visual system are significant. No direct treatment of arteriolosclerosis is available, yet modification of accelerating factors may be important in preventing its complications. More important is correct identification of the ocular manifestations associated with atherosclerosis, which can be potentially sight- and lifesaving with timely medical and surgical intervention.

References

1. National Center for Health Statistics. *Vital Statistics Report. Final Mortality Statistics.* 1982.
2. Boerman EL. Atherosclerosis and other forms of arteriosclerosis. In: Stanbury J., et al., eds. *Harrison's Principles of Internal Medicine.* 11th ed. New York, NY: McGraw-Hill; 1987;1014-1024.
3. McGill HC. Persistent problems in the pathogenesis of atherosclerosis. *Atherosclerosis.* 1984;4:443.
4. Ross R. The pathogenesis of atherosclerosis—an update. *N Engl J Med.* 1986;314:488-500.
5. Arteriosclerosis. In: Spencer W, ed. *Ophthalmic Pathology.* 3rd ed. Philadelphia, Pa: WB Saunders; 1985;77-79,1034-1045.
6. Ashton N. The eye in malignant hypertension. *Tr Am Acad Ophthalmol Otolaryngol.* 1972;76:17-39.
7. Leishman R. The eye in general vascular disease. Hypertension and arteriosclerosis. *Br J Ophthalmol.* 1957;41:641.
8. Green WR et al. Central retinal vein occlusion: a prospective histopathologic study of 29 eyes in 28 cases. *Retina.* 1981;1:27-55.
9. Appen R, Wray S, Cogan DG. Central retinal artery occlusion. *Am J Ophthalmol.* 1975;79:374-381.
10. Brownstein S, et al. Atheromatous plaques of the retinal blood vessels. *Arch Ophthalmol.* 1973;90:49-52.
11. Arruga J, Sanders MD. Ophthalmologic findings in 70 patients with evidence of retinal embolism. *Ophthalmology.* 1982;89:1333-1347.
12. Patrinely JR, Green WR, Randolph ME. Retinal phlebitis with chorioretinal emboli. *Am J Ophthalmol.* 1982;94:49-57.
13. Young LHY, Appen RE. Ischemic oculopathy: a manifestation of carotid artery disease. *Arch Neurol.* 1981;38:358-361.

Chapter 222

Carcinoid Syndrome

THOMAS C. BURTON

Carcinoid is the anglicized form of the German *karzinoide,* a term proposed by Orberndorfer in 1907 to denote a tumor of intestinal origin with a malignant histologic appearance but no metastatic potential. Carcinoid cells were found to contain cytoplasmic granules capable of reducing silver salts, establishing a link with Kultschitzky cells in the crypts of small intestine villi. The derivation from the enterochromaffin system was suggestive of endocrine dysfunction, but several decades passed before 5-hydroxytryptamine (serotonin) was discovered in carcinoid tumor tissue.[1] Slow to develop, also, was the realization that carcinoid tumors have locally invasive and lethal metastatic properties.

Carcinoid tumors have a histologic appearance similar to that of endocrine tumors, with solid nests and cords, lack of pleomorphism, and no mitoses, and they may resemble amelanotic melanomas.[2] Electron microscopic studies reveal electron-dense cytoplasmic neurosecretory granules.

Early classifications were dependent on the tumor position along the primitive endoderm (ie, foregut [bronchus, stomach, biliary passages, and pancreatic ducts], midgut [lower duodenum, jejunum, ileum, appendix, and cecum], and hindgut [colon, rectum]). Many carcinoids share cytochemical and ultrastructural features of islet cell tumors and medullary thyroid carcinomas, leading to a theory of neural crest origin. Known as apudomas, these endocrine tumors secrete active polypeptides.[3]

Approximately 80% of carcinoids originate in the intestine (half in the appendix) and 15%, in the bronchi. Most are physiologically silent and are discovered incidentally during surgical procedures and postmortem examinations. The tumors that arise in the foregut and midgut are more likely to be pharmacologically active, secreting serotonin and several other amine and polypeptide substances, such as catecholamines, histamine, growth hormone–like factor, kallikreins, and bradykinins. Serotonin is responsible for increased intestinal motility and paradoxical changes of blood pressure. Kallikreins are proteolytic enzymes with acute inflammatory and late fibrotic effects. Bradykinins cause vasodilation, increased capillary permeability, and bronchoconstriction.[4] The most useful laboratory value for detecting a functioning carcinoid tumor is an increased level of urinary 5-hydroxyindoleacetic acid, which is metabolized from serotonin.

Systemic Manifestations

Mean age of onset at initial diagnosis is in the 5th or 6th decade, although several cases have been reported in children. A person may have symptoms referable to the tumor for many months or years before it is diagnosed. Intestinal lesions produce abdominal pain, diarrhea, rectal bleeding, obstruction, and infarction.[5] Bronchial lesions produce dyspnea, chronic cough, hemoptysis, asthma-like episodes, and pneumonia.[1]

The carcinoid syndrome was described by Thorson in 1954. Always produced by metastatic lesions, usually with massive hepatic involvement and frequently accompanied by increased urinary 5-hydroxyindoleacetic acid, this symptom complex includes cutaneous flushing, diarrhea, bronchoconstriction, edema of extremities, and right-sided heart failure from endocardial fibrosis.[6,7]

Five-year survival rates, depending on site of origin, range from 99% (in the appendix) to 33% (in the colon). Deeply invasive tumors have a mortality rate exceeding 50%. A relatively small number of patients (10 to 20%) develop the carcinoid syndrome, but all of them succumb to the disease, usually within 3 years.

The principal therapy is surgical removal of tumor tissue. Some carcinoids are responsive to radiation. A number of chemotherapeutic agents, such as tryptophan, corticosteroids, serotonin blockers, adrenergic blockers, and cytotoxic drugs, are usually palliative. Exaggerated symptoms, even fatal carcinoid crisis, may be provoked by tumor cell lysis from cytotoxic agents. Excessive surgical manipulation of tumor tissue can liberate increased amounts of serotonin, resulting in vasomotor instability, especially acute hypertension and tachycardia, in the presence of anesthetic agents.[4]

Ocular Manifestations

Regular features of the carcinoid flush are injection of bulbar conjunctiva, periorbital edema, and lacrimation, probably due to the peripheral vasodilation and increased capillary permeability induced by bradykinins. Impaired vascular tone may be reflected in changes of the ocular fundus, including a cyanotic reflex, arteriolar narrowing, venous dilatation, and intravascular sludging. The vascular changes are accompanied by decreased systemic blood pressure and ophthalmic artery pressure.[8]

Most of the ocular manifestations are secondary to metastatic lesions of the choroid (usually bronchial in origin) and the orbit (usually from the ileum) and are indistinguishable from other types of metastatic tumors. Signs and symptoms of orbital lesions include decreased visual acuity, pain, diplopia, periorbital edema, exophthalmos, ptosis, impaired motility, and a palpable mass. Choroidal metastases are usually pale yellow or yellowish gray in appearance, producing decreased acuity and visual field defects by exudative retinal detachment.[9] Only 27 metastatic carcinoids with ocular signs have been described; 15 were orbital, 10 choroidal, and 1 each occurred in the iris and the optic nerve. Although acromegaly with increased size of the pituitary fossa has been reported, there were no visual field defects.

Effective treatment is limited. Local surgical excision is rarely possible. Because long-term survival is expected for many patients with carcinoid tumors, there are advocates of enucleation and exenteration; however some tumors respond to irradiation, which should be attempted first.[10,11]

References

1. Hajdu SI, Winawer SJ, Myers WP. Carcinoid tumors. A study of 204 cases. *Am J Clin Pathol.* 1974;61:521-528.
2. Toker C. Observations on the ultrastructure of a bronchial adenoma (carcinoid type). *Cancer.* 1966;19:1943-1948.
3. Pearse AGE, Polak JM, Heath CM. Polypeptide hormone production by "carcinoid" apudomas and their relevant cytochemistry. *Virchows Arch [B].* 1974;16:95-109.
4. Mason RA, Steane PA. Carcinoid syndrome: its relevance to the anesthetist. *Anesthesia.* 1976;31:228-242.
5. Moertel CG, Sauer WG, Docherty MG, et al. Life history of the carcinoid tumor of the small intestine. *Cancer.* 1961;14:901-912.
6. Thorson AH. Studies on carcinoid disease. *Acta Med Scand.* 1958;161(Suppl. 334):1-122.
7. Davis Z, Moertel CG, McIlrath DC. The malignant carcinoid syndrome. *Surg Gynecol Obstet.* 1973;137:637-644.
8. Wong VG, Melmon KL. Ophthalmic manifestations of the carcinoid flush. *N Engl J Med.* 1967;277:406-409.
9. Riddle PJ, Font RL, Zimmerman LE. Carcinoid tumors of the eye and orbit: a clinicopathologic study of 15 cases, with histochemical and electron microscopic observations. *Hum Pathol.* 1982;13:459-469.
10. Gaitan-Gaitan A, Rider WD, Bush RS. Carcinoid tumor-cure by irradiation. *Int J Radiat Oncol Biol Phys.* 1975;1:9-13.
11. Gragoudas ES, Carroll JM. Multiple choroidal metastasis from bronchial carcinoid treated with photocoagulation and proton beam irradiation. *Am J Ophthalmol.* 1979;87:299-304.

Chapter 223

Carotid Artery Insufficiency

SOHAN SINGH HAYREH

There is a large body of clinical evidence that emboli or hypoperfusion, secondary to carotid artery disease, are significant causes of ocular and cerebral ischemic lesions. It is, therefore, very important that the ocular lesions produced by carotid insufficiency be properly diagnosed and managed by ophthalmologists.

Pathophysiologic Mechanisms

The lesions in the carotid artery produce cerebral and ocular manifestations by one of two mechanisms: *Embolization* is produced by atheromatous or thrombotic debris discharged into the carotid arteries. *Reduction of blood flow* results from occlusion or hemodynamically significant stenosis of the internal carotid artery. In the intraocular vascular bed, blood flow depends on (1) the perfusion pressure, which is equal to the mean blood pressure (ie, diastolic blood pressure plus one third of the difference between the systolic and diastolic blood pressures) minus the intraocular pressure, and (2) peripheral vascular resistance (blood flow equals perfusion pressure divided by peripheral vascular resistance). Occlusion or severe stenosis of the internal carotid artery produces a drop in the mean blood pressure in the intraocular arteries and thus reduces the blood flow, resulting in chronic ocular ischemia. Similarly in the cerebral circulation, with fall of mean blood pressure in the cerebral arteries, distal circulatory failure occurs in areas of brain located farthest from the site of stenosis or occlusion, and this results in the development of stagnation thrombosis at these distant sites, producing infarction.[1] Of the two mechanisms, evidence suggests that embolism is the more common.

The most common lesions in the carotid artery are atherosclerotic, and it is well known that most significant atherosclerotic lesions are usually situated at the bifurcation of the common carotid artery and/or at the origin of the internal carotid artery; but they can develop anywhere in the extracranial as well as intracranial parts of the internal carotid artery. Plaque formation is an important feature of atherosclerotic disease. The plaques may be fibrous or fibromuscular and may encroach on the lumen. Calcification may occur within the plaque. Plaques may ulcerate on the luminal surface, exposing necrotic debris to the lumen of the vessel, and the content of the plaques (eg, cholesterol crystals, cholesterol esters, cellular debris) may embolize. Platelet fibrin aggregation may also occur on the surface of the complex plaques and can result in thrombus formation. Hemorrhage into the plaque may cause narrowing of the lumen of the vessel or rupture of the hemorrhagic material into the lumen; intraplaque hemorrhage is the most important morphologic characteristic associated with cerebral or ocular symptoms. Smooth atheromas may also produce emboli. Transient attacks, either ocular or cerebral, are less common in patients with occlusion of the internal carotid artery than in those with stenosis; furthermore, these attacks may abate when a stenotic artery becomes occluded. Occlusion due to pure atherosclerosis is uncommon, and it usually results from a superimposed thrombus.[2] Stenosis of the ophthalmic artery may be seen in addition to or independent of the carotid artery disease and may be responsible for the ocular manifestation.

In addition to the atherosclerotic lesions, nonatherosclerotic lesions (eg, kinking of the internal carotid artery, fibromuscular dysplasia, spontaneous cervicocephalic dissection of the internal carotid artery, aneurysms of the carotid artery, radiation arteritis of the carotid arteries, and Takayasu's arteritis) can also produce ocular complications through thromboembolic and hemodynamic mechanisms.

Systemic Manifestations

The systemic manifestations of carotid artery disease are neurologic. Since atherosclerosis is the major cause of strokes, these patients usually also have other manifestations of atherosclerosis such as cardiac and peripheral vascular occlusive disease.

Neurologic Manifestations

Neurologic manifestations essentially fall into three categories:

Hemispheric Transient Ischemic Attacks or Transient Cerebral Ischemic Attacks (TIAs) TIAs are temporary focal neurologic deficits, presumably related to ischemia, that last less than 24 hours.[3] It is well recognized that TIAs are a warning of stroke, because 50 to 75% of patients with stroke have had prior TIAs. The general manifestations are weakness and/or numbness of parts or all of the contralateral side of the body, with or without speech disturbance. The most common constellation of symptoms includes motor and sensory dysfunction of the contralateral limbs, followed by pure motor and pure sensory dysfunction, and, last, by isolated dysphasia.[4] The parts of the body that suffer most consistently are the contralateral hand and the distal arm. This may be due to either embolism or hypoperfusion. Typically TIAs last less than 15 minutes. There is a strong association between the TIAs and carotid artery disease (significant carotid occlusive disease in 30 to 50%).

Strokes Cerebral infarction is due either to embolism or distal circulatory flow failure (see above). The former is more common and accounts for about two thirds of strokes with internal carotid occlusion; the latter accounts for the remaining one third.[5] Many patients with embolic stroke experience no obvious TIAs prior to their stroke and sustain a moderate to severe clinical deficit, whereas those with distal flow failure experience more frequent TIAs before their stroke and have milder stroke than those with embolism.[5] In stroke due to carotid artery disease, common symptoms include weakness, paralysis, numbness, tingling and clumsiness in one or more fingers or hand, wrist, or arm and leg. There may be transient impaired ocular motility. Disturbances in higher functions, such as speech, writing, and behavior, have been reported.

"Silent" Cerebral Infarcts and Cerebral Atrophy With the advent of computed tomography (CT) and magnetic resonance imaging (MRI) came increasing evidence of "silent" cerebral infarction and cerebral atrophy, lesions that are detectable only by these two techniques. The "silent" infarcts may be due either to subclinical cerebral damage, or to infarcts in the silent areas, or because some TIAs occur during sleep and may go unnoticed. Nicolaides et al.[6] found "silent" cerebral infarction in 43% of patients with amaurosis fugax and cerebral atrophy in 25%; in patients with hemispheric TIAs the rates were 48 and 22%, respectively. These studies indicate that patients with asymptomatic internal carotid artery stenosis, amaurosis fugax, and TIAs have a high incidence of small, "silent" cerebral infarcts and cerebral atrophy, and there is an increased incidence of these with increasing carotid stenosis. These authors also showed that there is an appreciable incidence of these lesions in patients with so-called normal carotids (17% in those with amaurosis fugax and 31% in those with TIAs). This strongly indicates that showers of small emboli occur that, because of their size, do not produce symptoms unless they enter the retinal circulation but do produce small, clinically silent cerebral infarcts and atrophy.[6] Thus, contrary to the prevalent impression, amaurosis fugax and TIAs are not benign disorders.

Other Associated Systemic Manifestations of Atherosclerosis

These are important in atherosclerotic carotid artery disease since it is well established that a majority of these patients suffer from myocardial ischemia and/or intermittent claudication and that the majority of them eventually die from myocardial infarction and *not* cerebrovascular accident. Carotid artery occlusive disease and its associated ocular and neurologic manifestations are markers for increased risk of death from myocardial infarction.[7]

Ocular Manifestations

The ocular manifestations may be due either to embolism or to chronic ocular ischemia.

Ocular Embolic Disorders

Carotid artery disease is the major cause of various types of embolic disorders in the eye, which may vary from being totally asymptomatic to transient monocular blindness (amaurosis fugax), central or branch retinal artery occlusion, or anterior ischemic optic neuropathy. For embolism, the presence of stenosis or occlusion of the carotid arteries is not at all essential, because plaques in the carotid arteries without any significant stenosis can produce emboli.

Retinal Emboli

CHOLESTEROL EMBOLI.[8,9] The so-called Hollenhorst plaques belong to this category and are the most common emboli seen in the eye (reported to constitute 87% of all retinal emboli).[9] They are usually refractile, glistening, bright yellow, orange, or copper-colored, irregular, globular or rectangular, and they contain cholesterol. In patients with atherosclerotic lesions of the carotid arteries the intimal plaques may ulcerate, and cholesterol-containing material from the ulcer may be swept off as emboli into the circulation. These emboli may or may not produce an obstruction, and the degree of obstruction depends on the angle at which the flat crystalline structure is impacted. They have a tendency to move distally in the course of a few days, and usually disappear entirely over a 3-month period. A highly characteristic feature of cholesterol emboli is the localized sheathing reaction that develops in retinal arteries at the point of impaction, which may persist for months or years. These emboli are associated with atheromatous lesions of the carotid artery and much less frequently with atherosclerotic disease elsewhere.

PLATELET-FIBRIN OR THROMBOTIC EMBOLI. Composed of fibrin and platelets, these emboli are dull gray-white and readily migrate through retinal arterioles, so that they may produce only fleeting visual symptoms. They are usually seen with intracardiac thrombosis and may also come from carotid thrombosis.

CALCIFIC EMBOLI. Calcific emboli are typically dull white, chalky, nonrefractile, solid, and ovoid. They tend to block the retinal arteriole completely. They constitute only 4% of retinal emboli.[9] They usually originate from calcific cardiac valves but may come from the common carotid artery. Arruga and Sanders[9] reported that only patients with cholesterol emboli complained of amaurosis fugax, whereas all patients with calcific or stationary platelet-fibrin emboli experienced permanent visual loss.

Amaurosis Fugax
Amaurosis fugax[7] is a well-recognized and important manifestation of carotid artery disease. It is almost always monocular. It is due to transient ischemia or vascular insufficiency of the retina or optic nerve. Patients describe diminished or absent vision in one eye that progresses for a few seconds and lasts from seconds to a few minutes, followed by complete recovery of vision. The visual obscuration may progressively involve the entire visual field, often starting at the upper field so that the patient complains of a "blind pulled over the eye from above," and less frequently at other places in the visual field. Some patients may describe it as patchy or sectoral visual loss. Frequently the recurrent attacks follow a similar pattern. The majority are due to emboli originating from the common carotid artery and its branches (internal or external carotid arteries or the occluded stump of either artery). Carotid angiographic studies of patients with amaurosis fugax, particularly those over 50 years of age, revealed that atheromatous disease of the ipsilateral carotid artery is present in 75% of cases.[10] Amaurosis fugax is rarely accompanied by neurologic deficits.

Central Retinal Artery Occlusion
Sudden, painless visual loss is the major symptom, occasionally preceded by a history of amaurosis fugax. Our studies[11] revealed that if the retinal circulation is not restored within 100 minutes, retinal damage is permanent. Thus, this is a medical emergency, usually associated with devastating visual loss unless the eye has a major cilioretinal artery supplying the macular region. Central retinal artery occlusion is one of the easiest diagnoses to make. During the acute phase, within a few minutes the retina becomes opaque in the macular region (producing the classic cherry-red spot in the fovea) and in the posterior part of the eye, while the peripheral retina looks almost normal. However, after 6 to 8 weeks the entire retina looks normal, but there is optic atrophy. The classic "box carring" of the blood in the retinal vessels may or may not be present, depending on whether the occlusion was permanent or temporary. Central retinal artery occlusion associated with visual loss is almost invariably complete, and the concept of "partial" occlusion in such cases has no scientific basis. An embolus at the optic disc within the central retinal artery may rarely be seen. In at least two thirds of patients with central retinal artery occlusion, carotid artery disease is responsible for the occlusion.[12]

Branch Retinal Artery Occlusion
This may sometimes be preceded by transient monocular blindness when the embolus first obstructs the central retinal artery and soon migrates into one of the branch retinal arteries. An embolus may often be seen at the site of occlusion, which is usually a bifurcation of the retinal arteries. The site and extent of visual loss depend on the occluded branch retinal artery.

Anterior Ischemic Optic Neuropathy
Embolism of one of the posterior or short posterior ciliary arteries may produce anterior ischemic optic neuropathy, with or without choroidal infarction, depending on the size of the artery involved and the availability of collateral circulation. During the acute phase, the optic disc is swollen and there is an associated optic disc-related visual field defect; in about 2 months the involved part of the disc becomes atrophic.

Chronic Ocular Ischemia

The least recognized manifestation of carotid artery disease, chronic ocular ischemia is due to carotid insufficiency.[12,13] The ischemia is due to a fall in perfusion pressure in the intraocular vascular bed, secondary to occlusion or severe stenosis of the internal carotid artery (see above).

Manifestations

ANTERIOR SEGMENT ISCHEMIA. Anterior segment ischemia is the most common and important manifestation of chronic ocular ischemia. It results in the development of neovascularization of the iris and angle of the anterior chamber, frequently resulting in neovascular glaucoma, which is usually the presenting feature of chronic ocular ischemia.

POSTERIOR SEGMENT ISCHEMIA. As discussed above, the blood flow in the intraocular vascular beds depends on the perfusion pressure (see above). With the fall of mean blood pressure in the retinal, choroidal, and/or optic nerve head arteries and the rise of intraocular pressure due to neovascular glaucoma, the perfusion pressure in the retinal, choroidal, and optic nerve head vascular bed could fall to a level at which the tissues have inadequate circulation. A transient fall in perfusion pressure produces amaurosis fugax, whereas a permanent, marked fall can result in central retinal artery occlusion, anterior ischemic optic neuropathy, and/or choroidal infarction. Since the blood flow in the central retinal vein also depends on the perfusion pressure in the retinal arteries and the intraocular pressure, retinal vein occlusion may also develop; this (ie, venous stasis retinopathy[13]) has erroneously been attributed to simple retinal hypoxia. There may also be retinal and/or optic disc neovascularization.

Thus a patient with chronic ocular ischemia may present with (1) iris, angle, retinal, and/or optic disc neovascularization, (2) neovascular glaucoma, (3) central retinal artery occlusion, (4) anterior ischemic optic neuropathy, (5) choroidal infarction, and/or (6) central retinal vein occlusion. If chronic ocular ischemia is not recognized as the primary underlying factor in all these changes, a mistaken diagnosis may be made and the condition may not be managed properly. These ocular manifestations may be the earliest clinical signs of serious carotid artery disease, requiring urgent attention to prevent cerebral ischemic lesions, myocardial infarction, or other seriously disabling complications.

One of the manifestations of posterior segment ischemia secondary to carotid insufficiency that is stressed in the literature is the presence of so-called venous stasis retinopathy. Our studies on central retinal vein occlusion and carotid artery disease indicate that it is in fact a nonischemic central retinal vein occlusion produced by the mechanism mentioned above. The controversy on the subject is discussed in detail elsewhere.[13]

ORBITAL PAIN. An occasional patient with carotid occlusive disease may complain of moderate to severe pain in the eye, orbit, temple, and upper part of the face on the affected side because of ischemia.

References

1. Romanul FCA, Abramowicz A. Changes in brain and pial vessels in arterial border zones. *Arch Neurol*. 1964;11:40-56.
2. Fisher M. Occlusion of the carotid arteries. Further experiences. *Arch Neurol Psychiatry*. 1954;72:187-204.
3. Genton E, Barnett HJM, Fields WS, et al. XIV Cerebral ischemia: the role of thrombosis and of antithrombotic therapy. Joint Committee for Stroke Resources. *Stroke*. 1977;8:147-175.
4. Pessin MS, Duncan GW, Mohr JP, et al. Clinical and angiographic features of carotid transient ischemic attacks. *N Engl J Med*. 1977;296:358-362.
5. Pessin MS, Hinton RC, Davis KR, et al. Mechanisms of acute carotid stroke. *Ann Neurol*. 1979;6:245-252.
6. Nicolaides AN, Papadakis K, Grigg M, et al. Amaurosis fugax: data from CT scans—the significance of silent cerebral infarction and atrophy. In: Bernstein EF, ed. *Amaurosis Fugax*. Heidelberg: Springer-Verlag; 1988;200-226.
7. Amaurosis Fugax Study Group. Amaurosis fugax (transient monocular blindness): a consensus statement. In: Bernstein EF, ed. *Amaurosis Fugax*. Heidelberg: Springer-Verlag; 1988;286-301.
8. Russel RWR. The source of retinal emboli. *Lancet*. 1968;2:789-792.
9. Arruga J, Sanders MD. Ophthalmologic findings in 70 patients with evidence of retinal embolism. *Ophthalmology*. 1982;89:1336-1347.
10. Harrison MJG. Angiography in amaurosis fugax. In: Bernstein EF, ed. *Amaurosis Fugax*. Heidelberg; Springer-Verlag; 1988;227-235.
11. Hayreh SS, Kolder HE, Weingeist TA. Central retinal artery occlusion and retinal tolerance time. *Ophthalmology*. 1980;87:75-78.
12. Hayreh SS, Podhajsky P. Ocular neovascularization with retinal vascular occlusion. II. Occurrence in central and branch retinal artery occlusion. *Arch Ophthalmol*. 1982;100:1585-1596.
13. Hayreh SS. Chronic ocular ischemic syndrome in internal carotid artery occlusive disease: controversy on "venous stasis retinopathy." In: Bernstein EF, ed. *Amaurosis Fugax*. Heidelberg; Springer-Verlag; 1988;135-158.

Chapter 224

Carotid Cavernous Fistula

F. JANE DURCAN

A carotid cavernous fistula (CCF) occurs when there is communication between the carotid artery and the cavernous sinus. Subsequently, shunting of arterialized blood into the venous system produces a characteristic syndrome of pulsating exophthalmos, orbital bruit, and epibulbar congestion. Several classification systems have been proposed for CCFs. They can be grouped pathogenically (traumatic versus spontaneous), hemodynamically (high-flow versus low-flow), or anatomically (direct versus dural).[1] The anatomic classification is perhaps the most useful to the physician. Knowledge of the anatomy is crucial to understanding the pathogenesis, hemodynamics, and clinical manifestations of CCF.

The cavernous sinus is a plexus of veins surrounding the internal carotid siphon contained in the triangular space between the dura and the periosteum of the parasellar area. The oculomotor and trochlear nerves, and the ophthalmic and maxillary divisions of the trigeminal nerve run in the lateral wall of the cavernous sinus.[2] The abducens nerve passes within the sinus lateral to the carotid artery, and sympathetic fibers surround the artery.[2] Posteriorly, the sinus envelopes Meckel's cave, which contains the gasserian ganglion. The cavernous sinus receives venous drainage from the superior and inferior ophthalmic veins, central retinal vein, cerebral veins, and the middle meningeal vein.[2] It communicates with the contralateral sinus by intercavernous sinuses and drains into the superior and inferior petrosal sinuses, the basilar plexus, and the pterygoid plexus (Fig. 224-1).

Branches of the internal carotid artery (ICA) within the sinus are the meningohypophyseal trunk and the artery of the inferior cavernous sinus.[2,3] These thin-walled arteries, in conjunction and anastomosis with meningeal branches of the external carotid artery, supply blood to the dura.[3] Should one of them rupture as they pass through the venous channels of the cavernous sinus, the rupture would produce a dural cavernous fistula (DCF).[4] Flow from the ICA into the cavernous sinus produces a direct CCF.

CCFs may be traumatic in origin, or they can occur spontaneously. Seventy-five percent are the result of trauma.[5] They tend to occur most commonly in young men with basal skull fractures or deep penetrating orbital trauma caused by a bullet, knife, or knitting needle, for example.[5] These injuries traumatize the ICA and produce a direct, high-flow fistula with the cavernous sinus and prominent clinical symptoms. Iatrogenic CCFs have been seen after sphenoidectomy, transsphenoidal hypophysectomy, Fogarty catheter thromboendarterectomy and percutaneous retrogasserian procedures.

Twenty-five percent of CCFs are spontaneous and tend to occur in middle-aged to older women.[1,3,6] They are theorized to result from shunting of blood through congenital malformations or from the rupture of a previously weakened vessel.[3] Spontaneous fistulas often involve the small, thin-walled dural vessels of the internal or external carotid arteries and produce low-flow shunts with mild symptoms. Spontaneous rupture of an intracavernous carotid aneurysm can produce a direct, high-flow fistula with marked clinical changes.[3]

Vascular weakening is commonly due to degenerative change, usually atherosclerosis.[1,3] Trauma and infections such as syphilis have also been implicated. Connective tissue diseases (eg, fibromuscular dysplasia, Ehlers-Danlos syndrome, pseudoxanthoma elasticum, osteogenesis imperfecta) are another risk factor.[1] A strong association between CCF and pregnancy exists in young women.[5]

Communication between the internal carotid artery and the cavernous sinus produces a low-resistance shunt. Arterial pressure is reduced in the supracavernous carotid artery, leading to ischemic complications in the central nervous system and the eye. In addition, venous tributaries of the cavernous sinus are exposed to arterial pressure; as a result, blood flow is slowed—and sometimes reversed—in these veins. CCFs often drain via the ophthalmic veins because of their relatively low resistance.[1,6] The increase in venous pressure can produce congestion in the orbit, brain, and cavernous sinus. The degree of ischemia and venous congestion that occur is related to the volume of flow through the fistula.[1]

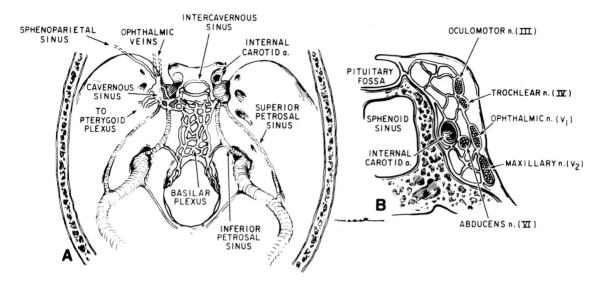

FIGURE 224-1 **(A)** Venous sinuses in the base of the skull. The right cavernous sinus is open to show the internal carotid artery. **(B)** Coronal section through the caudal cavernous sinus illustrates vascular and neural relationships.

Systemic Manifestations

The systemic manifestations of CCF are restricted to the head and neck. Headache, which presents early in the course, is often frontal or temporal.[3,4] Orbital and periorbital pain are common.[4,5,7] A subjective or objective bruit, which stops on carotid compression, is heard in up to 99% of direct CCFs. Facial pain or numbness is caused by vascular compromise to the gasserian ganglion or compression of fifth nerve branches by dilated veins.[7] Transient seventh nerve palsy has also been reported. Severe epistaxis can occur from engorged nasal mucosa or erosion of the cavernous sinus into the sphenoid sinus.[5] Enlarged venous loops can erode through the base of the skull and present as a nasopharyngeal mass. Three percent of patients get intracerebral hemorrhages and present with focal seizures, unilateral hemispheric dysfunction, or death. Other hemorrhagic complications include rupture of the cavernous sinus and subarachnoid hemorrhage.[8] Death can occur from progressive intracranial arterial insufficiency. High-output cardiac failure is an extremely rare complication.[5]

Ocular Manifestations

Ophthalmic manifestations of CCF can be ipsilateral or contralateral to the involved cavernous sinus, or they may be bilateral.[1,5] The external manifestations include proptosis, pulsation of the globe, orbital bruit, episcleral and conjunctival vascular engorgement, and chemosis.[3,5] The chemosis may be so severe that the lids become everted and the conjunctiva drys out, bleeds, or becomes infected.[5] Lid edema and ptosis occur, and dilated veins may be visible in the lid (Fig. 224-2).[5]

Motility abnormalities are common and are caused by engorgement and ischemia of the extraocular muscles in the orbit and by compression or ischemia of the cranial nerves in the cavernous sinus.[5,7] Paresis of the sixth nerve is most common[6]; however complete third nerve palsy with ptosis and pupillary signs, and fourth nerve palsies may also occur.[1,5] Ocular ischemia or Horner's syndrome can alter pupillary reactions and lid position.

Glaucoma usually is attributed to increased episcleral venous pressure.[1,4] Neovascular glaucoma develops in cases of chronic ocular ischemia. Increased intraocular pressure has also been reported secondary to angle closure from pupillary block.

Anterior segment changes from venous engorgement include conjunctival and episcleral injection, chemosis, and iris vessel engorgement. Those resulting from ischemia are corneal edema, hypotony, anterior chamber cell and flare, cataract, and rubeosis iridis.[4,5] Posterior segment abnormalities include optic disc edema, congested tortuous retinal veins, and retinal and vitreous hemorrhages.[4,5] Other less common complications include choroidal detachment, serous and exudative retinal detachment, retinal neovascularization, central retinal artery occlusion, and central retinal vein occlusion.

Optic neuropathy often results from damage to the optic nerve at the time of injury in traumatic CCF. A secondary neuropathy is caused by infarction from acute ischemia, progressive atrophy from glaucoma or chronic

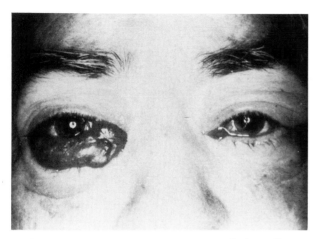

FIGURE 224-2 Proptosis and chemosis in a direct, high-flow, right carotid cavernous fistula following trauma. (Courtesy of N. Schatz, M.D.)

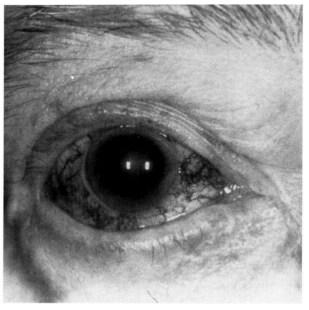

FIGURE 224-3 An 82-year-old woman with mild proptosis and epibulbar congestion from a spontaneous, low-flow, dural cavernous fistula. (Courtesy of Department of Ophthalmology, University of Iowa.)

hypoxia, or optic nerve compression by distention of the cavernous sinus.

The manifestations of a DCF are usually less severe than those of a direct CCF. Ocular pulsation is almost always absent; the bruit is absent in up to 70% of patients and it may be transient.[4,7] Patients often present with mild headache or orbital pain, transient diplopia, and dilated episcleral vessels.[3] This can result in misdiagnosis of conjunctivitis, iritis, or Graves' disease (Fig. 224-3).[1,4]

The major complication in CCF is visual loss. This is generally reported to occur in 25 to 50% of patients, but rates as high as 89% have been reported.[8] This number reflects the relatively high number of direct high-flow traumatic fistulas. Direct fistulas rarely close spontaneously. DCFs have a much lower complication rate; however, severe visual loss does occur and patients should be followed very closely.[4,6] A significant number of spontaneous fistulas eventually close without intervention.[1,3,4,6,7] A temporal association between spontaneous closure and performance of angiography has been noted by some authors.[1,3,4,7]

Indications for treatment include neurologic deterioration, visual loss, obtrusive diplopia, intolerable bruit, and malignant proptosis. Selective angiography, using subtraction and magnification techniques, is often necessary to distinguish a direct from a dural shunt, and to locate the feeding vessels and anastomosis in a DCF patient prior to treatment.[3,6] Carotid artery ligation can increase vascular compromise in an already ischemic globe and result in further deterioration of vision. Recently, therapy has been directed toward actual closure

of the fistula by embolization or balloon catheterization. Enucleation in a patient with CCF should be undertaken with caution, owing to the risk of severe hemorrhage.

References

1. Zimmerman RD, Russell EJ. Angiography in the evaluation of visual disturbances. *Int Ophthalmol Clin.* 1986;26:187-213.
2. Harris FS, Rhoton AL Jr. Anatomy of the cavernous sinus. *J Neurosurg.* 1976;45:169-180.
3. Slusher MM, Lennington BR, Weaver RG, et al. Ophthalmic findings in dural arteriovenous shunts. *Ophthalmology.* 1979;86:720-731.
4. Phelps CD, Thompson HS, Ossoinig KC. The diagnosis and prognosis of atypical carotid-cavernous fistula (red-eyed shunt syndrome). *Am J Ophthalmol.* 1982;93:423-436.
5. Hamby WB. *Carotid-Cavernous Fistula.* Springfield, Il: Charles C Thomas; 1966.
6. Newton TH, Hoyt WF. Dural arteriovenous shunts in the region of the cavernous sinus. *Neuroradiology.* 1970;1:71-81.
7. Nukui H, Shibasaki T, Kaneko M, et al. Long-term observation in cases with spontaneous carotid-cavernous fistulas. *Surg Neurol.* 1984;21:543-552.
8. Turner DM, Vangilder JC, Mojtahedi S, et al. Spontaneous intracerebral hematoma in carotid-cavernous fistula. *J Neurosurg.* 1983;59:680-686.

Chapter 225

Hereditary Hemorrhagic Telangiectasia

Rendu-Osler-Weber Disease

ELISE TORCZYNSKI

Hereditary hemorrhagic telangiectasia (HHT) is an uncommon inherited disorder of blood vessels manifested by bleeding from telangiectasias of the skin, mucous membranes, and viscera. Rendu (1896), Osler (1901),[1] and Weber (1907)[2] described the principal features of the disease, delineating it as an entity that was different from hemophilia and other bleeding diatheses. Hanes (1909)[3] named it hereditary hemorrhagic telangiectasia.

Different types of vascular lesions occur.[1] The first are pinpoint purple to red dots that may easily be overlooked and are frequently numerous. Flat spider nevi with a central dot from which branches radiate are the most common. Discrete macules, red to violet in color, up to 3 mm in size, and flat or slightly raised, lying beneath the epidermal or mucosal epithelium are also common. The nodules may arise from the center of a spider nevus and may be confluent (Fig. 225-1). Nodular lesions look like a split pea on the skin or mucosal surface. The lesions are easily traumatized and bleed readily. They blanch with pressure and do not pulsate.[4] Arteriovenous fistulas are common in the pulmonary circulation.[4] The vascular malformations are found on the face, neck, chest, trunk, extremities (especially on the fingers and under the nails), the mucosal surface of the lips, tongue, nasal septum, nasopharynx, buccal surface, gums, larynx, gastrointestinal tract including the liver, urinary tract, spleen, respiratory tract, adrenal gland, brain, spinal cord, meninges, bones, and only occasionally in the conjunctiva and retina.[1-7] Patients may develop hernias, varicose veins, and hemorrhoids.[8]

Systemic Manifestations

Clinically, epistaxis is the first manifestation of hereditary hemorrhagic telangiectasia in more than 50% of patients, often beginning at or before puberty; 78% of patients eventually develop epistaxis.[7] Daily or weekly nosebleeds are common. Mild trauma from coughing, sneezing, or blowing the nose may initiate bleeding, or it may start spontaneously. Nosebleeds may be only a mild nuisance that stops with finger- or tamponade pressure, or they can be life threatening, requiring many transfusions. Patients who are repeatedly treated with cauterization develop perforated nasal septa.[5] Gastrointestinal bleeding, while it is the first symptom in one fourth of the HHT patients, starts late, often in the 4th or 5th decade.[7] Melena, hemoptysis, or hematuria may be the first sign of the disease. Bleeding from skin lesions is unusual and rarely troublesome.[1,5] Anemia occurs secondary to blood loss. The skin lesions are less apparent in anemic persons. Patients with pulmonary arteriovenous fistulas may present with polycythemia, clubbing of the fingers, and cyanosis. HHT should be considered when any patient experiences unexplained bleeding from any part of the body.[5] About 10% of patients with HHT do not have bleeding episodes. Important symptoms such as those associated with brain abscesses found in patients with pulmonary arteriovenous fistulas may be more important than the symptoms associated with telangiectasias themselves. In 4% of patients the disease is lethal, often in the 5th decade.[5] Most patients enjoy a normal life span but suffer varying degrees of discomfort and morbidity from the vascular lesions.[5,7]

The vascular lesions usually are not evident at birth. In 30% of those afflicted the telangiectasias appear in the first decade, and about 5% of the initial lesions appear every decade thereafter through age 70.[7] Some skin lesions wax and wane over the years whereas others remain relatively stable over periods as long as 20 years.[1] After surgical excision or cauterizations new lesions may appear immediately adjacent to the treated area.

The disease is hereditary, occurring as an autosomal dominant trait with complete penetrance.[2] Males and females are affected in equal numbers, and both sexes

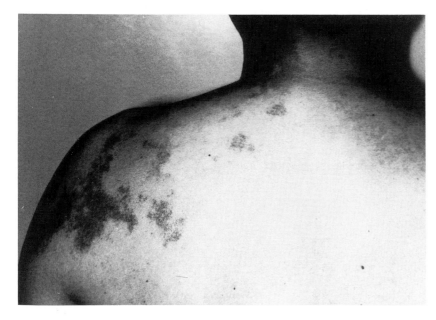

FIGURE 225-1 Skin lesions of confluent spider nevi on the shoulder and neck. (Courtesy of Robert A. Hardy, M.D.)

transmit the disease to their offspring. The homozygous state is lethal. The disease has been recorded through six generations in afflicted families.[4] So-called skip generations are reported, but they probably represent very mild involvement.[4] For 10 to 20% of patients the family history for HHT is negative. The disease is seen less frequently in blacks.[8]

Laboratory data may indicate microcytic and hypochromic anemia in patients who have frequent and serious bleeding episodes. Often a normal hemogram is reported.[5] Bleeding and coagulation tests are normal. Histologically the angiomatoid lesions consist of tortuous dilated endothelium-lined spaces that lack muscular walls and elastic fibrils. The telangiectasias lie just beneath the surface epithelium and may break through the epithelium.[5] The antinuclear antibodies (ANAs) and anticentromere antibodies (ACAs) are negative in HHT, whereas the latter is positive in the CREST syndrome (calcinosis, Raynaud's phenomenon, esophageal hypomotility sclerodactyly, telangiectasia), which may simulate HHT.[6]

The diagnosis may be delayed as long as 20 years.[7] The differential diagnosis includes other vascular malformations involving the skin and the mucous membranes, von Willebrand's disease, hereditary familial purpura simplex, lead poisoning, aortic arch syndrome, pulseless disease, and the CREST variant of scleroderma.[7]

Ocular Manifestations

Ocular changes are not common in the disease but have been reported in a number of instances.[2,9-11] The skin of the lids and the conjunctiva, both tarsal and bulbar, may develop the angiomatoid nodules or the spiderlike vascular malformations and the flat telangiectasias (Fig. 225-2). Bloody tears and subconjunctival hemorrhages may occasionally occur as a result of bleeding from the conjunctival telangiectasias[10,11] and may be extensive. Wolper and Laibson[12] described filamentary keratitis and bloody tears in a patient with HHT. Small petechial hemorrhages develop adjacent to the conjunctival vessels. Biopsy specimens of the conjunctival vessels showed tortuous abnormally dilated vessels in the epithelium.[12] Conjunctival telangiectasias have been reported in 9% of patients afflicted.[7,12] The nodular telangiectasias in the conjunctiva measure 1.0 to 3.0 mm.[9] Conjunctival lesions are more common than retinal ones.[9]

Retinal lesions have included telangiectatic tufts and patches adjacent to the retinal arteries or veins, tortuosity of the retinal veins with a twisted, cordlike appearance, varices of the retinal veins, and nodular lesions.[9,13,14] Occasionally neovascularization occurs. Vitreous hemorrhage may result.[9] Perivenous hemorrhages in the fundus, sheathing of the vessels, leakage from abnormal vessels, and staining on fluorescein angiography have been reported. In areas with flat tufts of newly formed vessels, the distal venular segment may be dilated to twice the size of the preangiomatous vein.[13] The changes thought to be specific in the fundus include tortuosity and segmental dilatations of the retinal veins, retinal hemorrhages, often perivenular, neovascularization of the retina and the optic disc (Fig. 225-3), telangiectasia of the retinal vessels, and vitreous hemorrhage.[14] Davis and Smith,[14] reporting in 1972, accepted seven cases from the world literature as having retinal

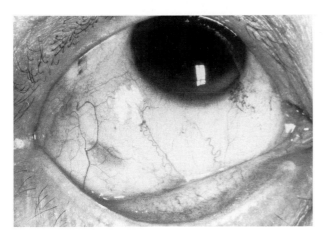

FIGURE 225-2 Conjunctival telangiectasia at limbus of the patient shown in Figure 225-1. Fine pinpoint angiomas are visible inferonasally and inferotemporally, beneath the tangled vessels. (Courtesy of Robert A. Hardy, M.D.)

FIGURE 225-3 Dilated vessels in the superficial retinal layers extend onto the retinal surface. (Courtesy F.C. Blodi.)

involvement, an incidence of less than 1 in 100 cases of HHT.

Visual problems from cerebral involvement have been reported. Visual field defects, and disturbances of pupillary response and ocular movement may result from intracranial hemorrhage.[15] Other central nervous system visual problems arise from brain abscesses, which develop secondary to the pulmonary arterial fistulas. The symptoms are not specific and include papilledema, visual field defects, ophthalmoplegia, ptosis, and anisocoria.

The differential diagnosis of the retinal lesions includes diabetic retinopathy, hypertensive retinopathy, Coats' disease, Eale's disease, perivasculitis, and the phakomatoses.[7] Ataxia-telangiectasia (Louis-Bar) is associated with conjunctival telangiectasias.[12]

No treatment is necessary for the conjunctival lesions unless excessive and repeated bleeding occurs, but to date this has not been reported. If vision threatening bleeding occurs from the posterior segment, cryotherapy, photocoagulation, and vitrectomy may be used.

References

1. Osler W. On family form of recurring epistaxis associated with multiple telangiectases of skin and mucous membranes. *Bull John Hopkins Hosp.* 1901;12:333-337.
2. Weber FP. Multiple hereditary developmental angiomata (telangiectases) of the skin and mucous membranes associated with recurring haemorrhages. *Lancet.* 1907;2:160-162.
3. Goldstein HI. Goldstein's heredofamilial angiomatosis with recurring familial hemorrhages (Rendu Osler Weber's disease). *Arch Intern Med.* 1931;48:836-865.
4. Hodgson CH, Burchell HB, Good CA, et al. Hereditary hemorrhagic telangiectasia and pulmonary arteriovenous fistula: survey of a large family. *N Engl J Med.* 1959;261:626-636.
5. Faculty of the Mayo Graduate School of Medicine: Hereditary hemorrhagic telangiectasia: Rendu-Osler-Weber syndrome. *Minnesota Med.* 1967;50:233-237.
6. Fritzler MJ, Arlette JP, Behm AR, et al. Hereditary hemorrhagic telangiectasia versus CREST syndrome: can serology aid diagnosis? *J Am Acad Dermatol.* 1984;10:192-196.
7. Reilly PJ, Nostrant TT. Clinical manifestations of hereditary hemorrhagic telangiectasia. *Am J Gastroenterol.* 1984;79:363-367.
8. Smith JL, Lineback ML. Hereditary hemorrhagic telangiectasia: Nine cases in one negro family with special reference to hepatic lesions. *Am J Med.* 1954;17:41-49.
9. Landau J, Nelken E, Davis E. Hereditary haemorrhagic telangiectasia with retinal and conjunctival lesions. *Lancet.* 1956;271:230-231.
10. Garner LL, Grossmann EE. Hereditary hemorrhagic telangiectasis: With beta irradiation of a conjunctival lesion. *Am J Ophthalmol.* 1956;41:672-679.
11. Miles NE. Hereditary hemorrhagic telangiectasia. *Am J Ophthalmol.* 1952;35:543-546.
12. Wolper J, Laibson PR. Hereditary hemorrhagic telangiectasis (Rendu-Osler-Weber disease) with filamentary keratitis. *Arch Ophthalmol.* 1969;81:272-277.
13. Forker EL, Bean WB. Retinal arteriovenous aneurysms in hereditary hemorrhagic telangiectasia. *Arch Intern Med.* 1963;111:778-783.
14. Davis DG, Smith JL. Retinal involvement in hereditary hemorrhagic telangiectasia. *Arch Ophthalmol.* 1971;85:618-623.
15. Press OW, Ramsey PG. Central nervous system infections associated with hereditary hemorrhagic telangiectasia. *Am J Med.* 1984;77:86-92.

Chapter 226

Hypertension

SOHAN SINGH HAYREH

Arterial hypertension is a very important public health problem in developed countries, where it is responsible for much premature death and disability. Of the Caucasian population in the Framingham study, about 20% of subjects' blood pressure (BP) was higher than 160/95 mm Hg, and 45% were over 140/90 mm Hg. There is evidence that the higher the BP, the worse the prognosis. Hypertension is nearly twice as prevalent in blacks as Caucasians in the United States, and morbidity is higher in blacks.

What do we mean by the terms "normal BP" and "arterial hypertension"? Pickering[1] concluded that BP is a continuous variable. There is evidence that in the western population BP rises each decade after 20 years, and there is also evidence that the rise of BP with age is a pathologic phenomenon and that all deviations above the mean BP of the younger age range should be regarded as hypertension.[2] It is reasonable to consider a BP higher than 140/90 mm Hg (especially the diastolic pressure) abnormal. This, combined with other risk factors for developing hypertension and for developing cardiovascular morbidity and mortality, in the long run determines the course of the disease and which persons are at risk and require treatment for hypertension.

Classification

Arterial hypertension is classified etiologically as essential (primary or idiopathic) or secondary. The cause of BP elevation in the essential type of hypertension is unknown, but the fundamental mechanism is an increase in peripheral vascular resistance. Secondary arterial hypertension may be due to renal, endocrine, neurogenic, mechanical, exogenous, and miscellaneous other causes, in addition to which it is seen with toxemia of pregnancy. Although in some forms of secondary hypertension the underlying mechanism is known, the term "secondary" does not necessarily mean that the cause or pathogenesis is always known. Clinically, arterial hypertension may be classified according to severity—mild, moderate, severe, or malignant. Hypertension has also been classified as borderline, labile, sustained, malignant, or accelerated, with purely arbitrary definitions for each of these types.

Pathophysiology

This is still not fully understood in spite of extensive research on the subject, as is evident from an exhaustive review by Genest et al.[3] Page[4] proposed a mosaic theory for the mechanism of hypertension, according to which many factors are involved. BP regulation is an extremely complex process. In simple terms, BP depends on the blood flow and vascular resistance to that flow in various organs, and an increase in either would increase BP. Blood flow and/or vascular resistance may be influenced by a variety of humoral, neural, or other mechanisms, resulting in various types of hypertension. The raised arterial BP can set in motion a succession of responses and adjustments that may cause the hypertension to continue (regardless of its cause), change its characteristics, and to a large extent determine its course and complications.[5] Raised arterial BP is a hemodynamic variable that to a great extent determines the evolution and treatment of hypertension. What follows is an extremely abbreviated account of the pathophysiology of hypertension, based on the excellent review by Genest et al.[3]

Essential Hypertension

Essential hypertension is considered to be a multifactorial disease secondary to interaction of genetic and environmental factors. There is compelling evidence that sodium is involved with the mechanisms of essential hypertension. The increase in the sodium level in the body may result from both excessive salt intake and inability of the kidneys to eliminate sodium. The latter tendency may be inherited. It is postulated that sodium retention causes an increase in concentration of a circulating sodium transport inhibitor (also known as natriuretic hormone) which is most probably secreted by the hypothalamus; this hormone increases urinary sodium excretion. Increase of intracellular sodium augments intracellular calcium. The important role of calcium as a primary regulator in vascular muscle is well-recognized, and the rise in cytoplasmic calcium is considered to be the primary trigger for contraction. The rise in calcium causes increased tone of the arteriolar smooth muscle, resulting in increased vascular resistance and hypertension. The vasodilatory and hypotensive effects of calcium channel blockers are due to the block in influx of calcium into smooth muscle cells. Potassium also plays a key role in the complex regulation of BP and has an important effect on the secretion of aldosterone and the release of renin, as well as a direct vasodilatory effect.

Renin, a primary cardiovascular regulator, is synthesized and stored in the granular juxtaglomerular cells of renal arterioles. Renin release is regulated by extracellular fluid volume (expansion suppresses), sodium (increase suppresses), potassium (increase suppresses), chloride (increase suppresses), and calcium (increase ? stimulates), angiotensin II (increase suppresses), β-adrenergic nerves (increased activity stimulates), α-adrenergic nerves (increased activity ? suppresses), and dopaminergic nerves (increased activity ? suppresses). Thus any one or more of these can influence renin release. Renin acts on angiotensinogen to liberate angiotensin I. Angiotensin-converting enzyme changes angiotensin I to angiotensin II, which metabolizes to angiotensin III. Angiotensin II is the most powerful vasoconstrictor known, and it also potentiates the vasoconstrictor activity of norepinephrine and leads to stimulation of aldosterone secretion. Angiotensin III is a less potent pressor agent than angiotensin II. Renin is present not only in the kidney but also in a number of extrarenal tissues, such as brain, adrenal gland, and blood vessel walls. From this it is evident that the renin-angiotensin system is distributed widely in the body and probably has broader functions than it was previously thought to have. Locally generated angiotensin may be involved in cardiovascular control as a tissue hormone or as a neuromodulator or neurotransmitter. Angiotensin II is important for both systemic and local vascular homeostasis. The exact role of angiotensin II in the production of essential hypertension is still not known. Another enzyme, called tonin, acts on angiotensinogen and angiotensin I to form angiotensin II directly.

Aldosterone plays an important role in the regulation of BP. It is the major regulator of extracellular fluid volume (through a direct effect on renal tubular transport of sodium, with the water following passively) and a major determinant of potassium metabolism. Aldosterone secretion is controlled mainly by adrenocorticotropic hormone (ACTH, which stimulates), potassium (potent stimulus), and the renin-angiotensin system (important regulator, angiotensin II probably most important). The renin-angiotensin-aldosterone volume-regulation mechanism works in the following way: Angiotensin II → aldosterone release → renal Na^+ retention → increase in circulating blood volume → increased renal perfusion pressure → renin release by action on juxtaglomerular cells → increased formation of angiotensin II, which also increases BP. Excessive secretion, excretion, and/or elevation of circulating levels of at least 10 different adrenocortical hormones, other than aldosterone, has been observed in human hypertension and experimental hypertension in animals.

There is evidence that prostaglandins participate in BP regulation and that they may be causal factors in the production of hypertension.

The nervous system influences BP in a number of ways—including the production of natriuretic hormone (most probably by hypothalamus) and vasopressin (by posterior pituitary), by autonomic control of the cardiovascular system, and through vasomotor centers (arterial baroreceptors, low-pressure receptors, chemoreceptors, and somatic afferents which transmit stimuli to the vasomotor centers). The vasomotor centers are influenced or modulated by higher integrative centers located in the hypothalamus, the limbic system, and the cortex. Hypertension may be mediated by increased neurogenic vasomotor tone. There is evidence that the suprachiasmatic nucleus tends to keep BP down.

Recent studies of essential hypertension have revealed that cell membrane abnormalities occur widely and that they play a key role in the pathogenesis of hypertension. This manifests itself in defective membrane control over intracellular sodium and calcium concentration.

Available evidence indicates that there is a powerful genetic influence on the BP that includes genetically transmitted membrane changes described above, heritable factors for renin and aldosterone secretion, and renal excretion of sodium and potassium. Genetic effects on BP are polygenic. Sodium chloride is a common environmental factor that interacts with genotype.

Malignant Arterial Hypertension

This has a fulminating course and is accompanied by severe vascular complications in many organs of the body. In 1914 Volhard and Fahr[6] stated that in malignant arterial hypertension there is severely elevated stationary BP, markedly impaired renal function, arterial lesions (mainly in the kidney) consisting of fibrinoid necrosis in arterioles and intimal proliferation in small arteries, and usually death from uremia. Keith, Wagener, and Barker[7] in 1939 described the presence of hypertensive retinopathy and optic disc edema as important findings in these cases. All the clinical manifestations of malignant arterial hypertension originate from lesions in arteries and arterioles. By some ill-understood mechanism or mechanisms, the levels of circulating endogenous vasoconstrictor agents (eg, angiotensin II, norepinephrine, vasopressin) increase. Kincaid-Smith[8] put forward the following schema for the various vascular changes seen in malignant arterial hypertension: Increase of vasoactive agents (angiotensin, vasopressin, norepinephrine) → "sausage string" effect in arteries → turbulence and endothelial separation and damage → platelet deposition on endothelium → release of thromboxane, serotonin, histamine → microangiopathic hemolytic anemia and intravascular coagulation → further platelet and fibrin deposition → mitogenic and migration factors released by platelet aggregation → myointimal proliferation and organization of thrombi within vessels → renal ischemia → increase in vasoactive agents. Thus, a vicious circle is set up in malignant hypertension. The renin-angiotensin-aldosterone system plays a very important role in the pathogenesis of malignant hypertension.

Renovascular Hypertension

One of the most frequent types of secondary hypertension, renovascular hypertension is due to increased liberation of renin by the kidney in response to renal artery stenosis. The role of the renin-angiotensin-aldosterone system in the pathogenesis of hypertension is discussed above.

Systemic Manifestations

The majority of patients with arterial hypertension are asymptomatic, and raised BP is usually discovered during a routine medical examination. The symptoms depend on the type of hypertension. Mild or moderate essential hypertension usually has no symptoms. Headache has frequently been mentioned as a symptom of hypertension, but available evidence indicates that when present it is usually associated with severe hypertension, coming on usually early in the morning, commonly occipital, and is relieved almost immediately with lowering of BP. Bleeding from the nose and/or kidney may be seen in some hypertensives. The patient may complain of palpitations, dizziness, lightheadedness, ringing in the ears, weakness, syncope, breathlessness, or angina pectoris. Malignant hypertension may present as a medical emergency, with oliguria, severe headache, vomiting, transient paralysis, lethargy, seizures, visual disturbances, and coma.

Patients with hypertension die unduly early, most from cardiac causes and others from cerebral or renal causes. Increased peripheral vascular resistance puts a marked strain on the left ventricle of the heart, causing hypertrophy and later dilatation, resulting in heart failure. Because of myocardial hypertrophy and/or associated coronary artery disease, myocardial ischemia (ie, angina pectoris or infarction) may be seen later on in these patients. Thus the cardiac causes of death usually include congestive heart failure or myocardial infarction. Other cardiovascular complications of hypertension include dissecting or saccular aneurysm of the aorta and peripheral vascular disease. Arterial hypertension produces arteriosclerosis of the renal arterioles, which interferes with glomerular and tubular functions; these changes finally result in proteinuria, microscopic hematuria, and renal failure. Neurologic complications of arterial hypertension are important; they include hypertensive encephalopathy (due to decompensation of cerebral vascular autoregulation), cerebral hemorrhage (from raised BP and development of cerebral berry aneurysms or microaneurysms of the Charcot-Bouchard type), and cerebral infarction (from thrombosis or embolism). Many patients also show evidence of microangiopathic hemolytic anemia.

Ocular Manifestations

These depend on the type of hypertension. There are usually no visual symptoms, although some patients (mainly those with malignant hypertension) may complain of blurred vision, blind spots, or even partial or complete visual loss. "Hypertensive retinopathy" is generally the only ocular manifestation mentioned in most discussions of arterial hypertension. No specific fundus sign is characteristic of essential hypertension, and the changes tend to be similar to those seen in arteriosclerosis and aging. General and focal narrowing of the retinal arteries has been described in the literature as the sign most specific of hypertension. There is no evidence that arteriovenous crossing changes, frequently attributed to arterial hypertension, are really specific to this disease.

The classification of fundus changes in hypertension advocated by Keith, Wagener and Barker[7] is widely used by physicians, but I[9] feel that it has been made obsolete by the great advances in our knowledge of the pathophysiology of arterial hypertension and its ocular manifestations; moreover, with the availability of highly effective antihypertensive therapy, it is of no clinical usefulness. The same is true of the several subsequent modifications of it. Similarly antiquated terms, such as "copper wiring" and "silver wiring," are so subjective and inexact as to be of little help in understanding the pathophysiology of retinal arterial changes or the clinical management of arterial hypertension.

In contrast to essential hypertension, in malignant arterial hypertension the fundus changes constitute an important manifestation of the disease. A colossal amount of literature on the fundus changes in malignant arterial hypertension has accumulated since Liebreich[10] first described them under the designation of "albuminuric retinitis" in 1859. Ever since, hypertensive retinopathy has been widely considered to be the sole clinical entity that represents all the fundus changes. Our detailed studies on malignant renovascular arterial hypertension have clearly shown that fundus changes in this condition fall into three distinct categories: hypertensive retinopathy, hypertensive choroidopathy, and hypertensive optic neuropathy.[11-22]

Hypertensive Retinopathy

Classically "hypertensive retinopathy"[17-22] is described as including the following fundus lesions: focal or generalized narrowing of retinal arteries[19], cotton-wool spots,[20] white deposits (so-called hard exudates[21]), macular star,[22] macular and/or retinal edema,[22] retinal hemorrhages,[17] and optic disc edema. Unfortunately none of these, individually or collectively, are specific for malignant hypertension; they can be seen in many other ocular and systemic conditions. In our studies, during the acute phase of malignant hypertension we could find no evidence of retinal arterial narrowing[19]: although on ophthalmoscopy narrowing of the retinal arteries was seen in some eyes, fluorescein fundus angiography failed to reveal any such narrowing.[19] We found that this discrepancy between the ophthalmoscopic and angiographic findings was due to retinal edema obscuring the retinal arteries on ophthalmoscopy. Rarely, central retinal artery occlusion may develop. In contrast to that, in the chronic phase the arterial changes were commonly seen, due to secondary permanent structural changes in the arteries.

Our studies revealed that focal intraretinal periarteriolar transudate (FIPT) was the only specific lesion of acute malignant hypertension.[14] FIPTs were one of the earliest lesions in hypertensive retinopathy. On ophthalmoscopy, they are usually pinpoint- to pinhead-sized, round or oval, dull white, and situated in deeper layers of the retina and beside the major retinal arteries and their main branches. On fluorescein fundus angiography, FIPTs show multiple punctate foci of fluorescein leakage from dilated precapillary retinal arterioles, and there is no focal retinal capillary obliteration. They usually last for 2 to 3 weeks, and on resolution leave no ophthalmoscopic, angiographic, or microvascular abnormality. In the past they have been confused with cotton-wool spots,[20] though the two have fundamentally very different pathogeneses, location, shape and size, color, fluorescein angiographic patterns, natural histories, and resolution patterns.[14,20]

Hypertensive Choroidopathy

The choroidal vascular bed shows impaired circulation and extensive occlusive and ischemic changes on fluorescein fundus angiography and histopathology.[11,16] In hypertensive choroidopathy retinal pigment epithelial lesions and serous retinal detachment are the classic ophthalmoscopic lesions.[16] The retinal pigment epithelial lesions consist of initial acute focal lesions (due to focal retinal pigment epithelial infarction) and degenerative lesions, which develop later and are progressive, maximally involving the macular and peripheral regions of the fundus. The so-called Elschnig's spots and Siegrist's streaks are retinal pigment epithelial lesions. The retinal detachment develops most commonly in the posterior pole and infrequently involves the peripheral retina. Available evidence indicates that hypertensive choroidopathy is due to choroidal ischemia and that hypertensive choroidopathy and hypertensive retinopathy are two independent and unrelated manifestations of malignant hypertension, because of the fundamentally different properties of the two vascular beds and their responses to the accelerated hypertension.

Hypertensive Optic Neuropathy

Optic disc edema has been described as an essential manifestation of malignant hypertension and as being of great clinical importance in the evaluation of malignant hypertension.[7] Optic disc edema is the initial manifestation of hypertensive optic neuropathy.[12,15] On followup, mild to marked pallor of the optic disc is noted in a number of these patients. All the available clinical and pathologic findings in our studies indicated that hypertensive optic neuropathy represents a form of anterior ischemic optic neuropathy, and that pathogenetically hypertensive optic neuropathy is a distinct entity and not simply a part of hypertensive retinopathy.[12,15] It is important to remember that for patients with malignant hypertension and hypertensive optic neuropathy it is dangerous to reduce BP precipitously, as that can produce sudden, complete and permanent blindness.

References

1. Pickering GW. High blood pressure. 2nd ed. London: Churchill Livingstone; 1968.
2. Peart WS. General review of hypertension. In: Genest J, Kuchel O, Hamet P, et al. *Hypertension: Physiopathology and Treatment*. 2nd ed. New York, NY: McGraw-Hill; 1983;3-14.
3. Genest J, Kuchel O, Hamet P, et al. Physiopathology of experimental and human hypertension. In: Genest J, Kuchel O, Hamet P, eds. *Hypertension: Physiopathology and Treatment*. 2nd ed. New York, NY: McGraw-Hill; 1983;1-675.
4. Page IH. The mosaic theory of arterial hypertension—its interpretation. *Perspect Biol Med*. 1967;10:325-333.
5. Tarazi RC. Hemodynamics, salt, and water. In: Genest J, Kuchel O, Hamet P, eds. *Hypertension: Physiopathology and Treatment*. New York, NY: McGraw-Hill; 1983;15-42.

6. Volhard F, Fahr KT. Die brightsche Nierenkranheit. Berlin: Springer-Verlag; 1914.

7. Keith NM, Wagener HP, Barker NW. Some different types of essential hypertension. Their course and prognosis. *Am J Med Sci.* 1939;197:332-342.

8. Kincaid-Smith P. Malignant hypertension: mechanisms and management. *Pharmacol Therap.* 1980;9:245-269.

9. Hayreh SS. Classification of hypertensive fundus changes and their order of appearance. *Ophthalmologica.* 1989;198:247-260.

10. Liebreich R. Ophthalmoskopischer Befund bei Morbus Brightii. *Graefes Arch Clin Exp Ophthalmol.* 1859;5(2):265-268.

11. Kishi S, Tso MOM, Hayreh SS. Fundus lesions in malignant hypertension—I. Pathologic study of experimental hypertensive choroidopathy. *Arch Ophthalmol.* 1985;103:1189-1197.

12. Kishi S, Tso MOM, Hayreh SS. Fundus lesions in malignant hypertension—II. Pathologic study of experimental hypertensive optic neuropathy. *Arch Ophthalmol.* 1985;103:1198-1206.

13. Hayreh SS, Servais GE, Virdi PS, et al. Fundus lesions in malignant hypertension—III. Arterial blood pressure, biochemical, and fundus changes. *Ophthalmology* 1986;93:45-59.

14. Hayreh SS, Servais GE, Virdi PS. Fundus lesions in malignant hypertension—IV. Focal intraretinal periarteriolar transudates. *Ophthalmology.* 1986;93:60-73.

15. Hayreh SS, Servais GE, Virdi PS. Fundus lesions in malignant hypertension—V. Hypertensive optic neuropathy. *Ophthalmology.* 1986;93:74-87.

16. Hayreh SS, Servais GE, Virdi PS. Fundus lesions in malignant hypertension—VI. Hypertensive choroidopathy. *Ophthalmology.* 1986;93:1383-1400.

17. Hayreh SS, Servais GE. Retinal hemorrhages in malignant arterial hypertension. *Int Ophthalmol.* 1988;12:137-145,197.

18. Hayreh SS. Hypertensive retinopathy. Introduction. *Ophthalmologica.* 1989;198:173-177.

19. Hayreh SS, Servais GE, Virdi PS. Retinal arteriolar changes in malignant arterial hypertension. *Ophthalmologica.* 1989;198:178-196.

20. Hayreh SS, Servais GE, Virdi PS. Cotton-wool spots (inner retinal ischemic spots) in malignant arterial hypertension. *Ophthalmologica.* 1989;198:197-215.

21. Hayreh SS, Servais GE, Virdi PS. Retinal lipid deposits in malignant arterial hypertension. *Ophthalmologica.* 1989;198:216-229.

22. Hayreh SS, Servais GE, Virdi PS. Macular lesions in malignant arterial hypertension. *Ophthalmologica.* 1989;198:230-246.

Chapter 227

Lymphedema

HENRY D. PERRY, ALFRED J. COSSARI, and ERIC D. DONNENFELD

Hereditary lymphedema with onset at or near birth is called Milroy's disease; that of later onset is called Meige's disease.[1] Although there is often an association between congenital lymphedema and Turner's syndrome, the lymphedema in these patients usually disappears by the first year of life. Persistence beyond this age is rare.[2,3]

Systemic and Ocular Manifestations

Marked edema of the extremities is an indication of the presence of Milroy's disease, Meige's disease, or a combination, depending on its initial age of onset. The occurrence of lymphedema at a young age should suggest the possible association of Turner's syndrome.[4,5] The lower extremities are involved more often than the upper, and involvement may be unilateral or bilateral.

There is usually painless chronic edema, which may, at times, involve part of the trunk as well as the extremities.

Izakovic[6] conducted a study of 26 children with congenital lymphedema, which revealed that 18 out of 22 girls had the 45× karyotype. Henriksen[7] reported two cases of Turner's syndrome associated with congenital lymphedema in newborn babies. Conjunctival lymphedema is rare, especially at an early age.[8] Klein and Doret[9] reported a case of congenital lymphedema with the development of chemosis at age 50 years.[9] Tabbara and Baghdassarian[10] reported on a 7-year-old female with congenital lymphedema of the conjunctiva and extremities. This girl also had moderate amblyopia of the right eye and esotropia; however, they described their patient as otherwise normal. A request for biopsy was refused by the patient's family.[10]

Perry and Cossari[11] reported on a 3½-year-old female with Turner's syndrome and Nonne-Milroy-Meige dis-

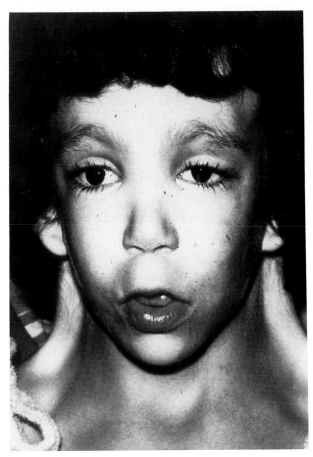

FIGURE 227-1 Full-face external photograph showing nasal conjunctival cysts and prominent webbed neck.

ease (Fig. 227-1). Her history was positive for bilateral inguinal hernias, atonic bladder, congenital hip dysplasia, and congenital lymphedema. She had diffuse conjunctival edema with bullae present nasally in the right eye and temporally in the left eye. Further examination revealed lymphedema of the left hand and right foot. The diagnosis of Nonne-Milroy-Meige disease was made from the association between the conjunctival chemosis and lymphedema of the left hand and right foot. Exotropia and bilateral ptosis were also present.

A conjunctival biopsy was taken at the time of strabismus surgery, affording a unique opportunity to evaluate the histopathologic conjunctival changes in Nonne-Milroy-Meige disease in association with Turner's syndrome. Pertinent findings were limited to the substantia propria, where there were multifocal areas of dilated vascular channels with delicate septa lined on both sides by noncontiguous, flattened endothelial cells. These changes are similar to those found in the lower extremities in Milroy's disease. Most of the spaces contained acellular fibrinous debris; only occasional red blood cells were noted. The surrounding stromal collagen fibers were thickened and proved to be the most striking differentiating feature from normal conjunctiva. This thickened stroma resembled that seen in conjunctival dermoids.

There have been two reports of persistent chemosis in this disorder, an ophthalmic finding not previously noted.[10,11] The conjunctival change was stable and unchanging in appearance for the seven years that both patients were followed.

References

1. Wheeler ES, Chan V, Wassman R, et al. A familial lymphedema praecox: Meige disease. *Plast Reconstr Surg.* 1981;67:362-364.
2. Gordon R, Eileen R, O'Neill M. Turner's infantile phenotype. *Br Med J.* 1969;i:483-485.
3. Turner HH. A syndrome of infantilism, congenital webbed neck and cubitus valgus. *Endocrinology.* 1938;23:566-574.
4. Chrousos GA, Ross JL, Chrousos G, et al. Ocular findings in Turner's syndrome: a prospective study. *Ophthalmology (Rochester).* 1984;91:926-928.
5. Nelson WE, ed. *Textbook of Pediatrics.* Philadelphia, Pa: WB Saunders; 1979;461.
6. Izakovic V. Congenital lymphedema and monosomy X. *Bratisl Lek Listy.* 1979;72:530-534.
7. Henriksen HM. Turner's syndrome associated with lymphedema, diagnosed in the newborns. *Z Geburtshilfe Perinatol.* 1980;184:313-315.
8. Duke-Elder S. *System of Ophthalmology.* London: Kimpton; 1963;3:909.
9. Klein D, Doret M. Chronic hereditary lymphedema. *Modern Problems in Ophthalmology.* 1957;1:576.
10. Tabbara KF, Baghdassarian SA. Chronic hereditary lymphedema of the legs with congenital conjunctival lymphedema. *Am J Ophthalmol.* 1972;73:531-532.
11. Perry HD, Cossari AJ. Chronic lymphangiectasis in Turner's syndrome. *Br J Ophthalmol.* 1986;70:396-399.

Chapter 228

Shy-Drager Syndrome

VERINDER S. NIRANKARI

The Shy-Drager syndrome, or multiple system atrophy (MSA), was first described in 1960.[1] It is a rare neurologic disorder characterized not only by autonomic dysfunction but also by involvement of the central motor tracts, including the corticobulbar, corticospinal, extrapyramidal, and cerebellar systems.[2]

Clinical Features
Systemic Manifestations

The most striking feature of the syndrome is the development of severe orthostatic hypotension that appears in the 4th to 6th decade of life. Males are affected about twice as commonly as females. Other features include impotence, which may precede other evidence of autonomic dysfunction by more than a decade; bladder involvement manifested by urinary frequency, urgency, and increased residual urine; severe constipation that develops gradually over several years; and occasional fecal incontinence and diarrhea. Orthostatic hypotension can be extremely disabling to the patient. Low blood pressure causes weakness, reduced exercise tolerance, and blurring of vision. Syncope results when the magnitude and rate of the blood pressure drop exceeds the limits in which cerebral perfusion can be preserved. Respiratory abnormalities and sleep apnea often develop. Laryngeal stridor occurs, caused by vocal fold abductor paresis. This may cause severe respiratory distress necessitating tracheostomy. A syndrome resembling parkinsonism is commonly present.

Two clinical subtypes of MSA can be distinguished: striatonigral degeneration (SND), which causes bradykinesia, loss of associated movements, and rigidity disproportionate to resting tremor, and olivopontocerebellar atrophy (OPCA) with truncal and limb ataxia, intention tremor, and slurred speech. Some patients have features common to both SND and OPCA. Exaggerated tendon reflexes and Babinski's responses are often observed. There is no evidence of sensory impairment or dementia in this disease, and intellectual function is preserved until the later stages of the illness.

Long-term follow-up suggests a gradual evolution of symptoms and signs and slowly progressive deterioration. Disability may result from any one of the major manifestations of the illness.[3] Patients gradually succumb 7 to 10 years after the onset of neurologic symptoms. Cardiac arrhythmias and sleep apnea contribute to their premature demise.

Ocular Manifestations

Ocular findings of the Shy-Drager syndrome were first described by Shy and Drager in 1960.[1] They included anisocoria and Horner's syndrome. Other authors have also described ocular signs, including iris atrophy, pupillary and convergence abnormalities, keratitis sicca, and nystagmus.[4-6] One study reviewed the eye findings in five patients with the Shy-Drager syndrome, including examination of palpebral fissure size, pupillary reactions, and extraocular movements.[7] Parasympathetic testing included use of dilute 0.0625% pilocarpine to test pupillary reactions, Schirmer's test to assess the precorneal tear film, and testing of corneal sensation with the esthesiometer of Cochet and Bonet. Sympathetic testing included testing the ciliospinal reflex and pupillary drug testing using 4% cocaine and 1% hydroxyamphetamine. The details of the testing have been described before.[8-11] Results of testing in these five patients are shown in Tables 228-1 and 228-2.

Ophthalmologic examination revealed generalized sympathetic and parasympathetic dysfunction. Three patients showed alternating Horner's syndrome. Three of the five patients had decreased tearing, and two had decreased corneal sensation. A combination of infrequent blinking, decreased tearing, and corneal hypesthesia can predispose to the development of keratoconjunctivitis sicca.[4,7]

Pathophysiology

Shy and Drager[1] demonstrated the diffuse nature of the degenerative process involving the autonomic nervous system, the corticobulbar and corticospinal tracts, some of the basal ganglia, and parts of the cerebellar system.

There is neuron loss or gliosis in a number of central

TABLE 228-1 Ocular Sympathetic Function Tests in the Shy-Drager Syndrome

Patient	Ciliospinal Reflex	Cocaine	Hydroxyamphetamine
1	Absent	No response OU	Not done
2*	Sluggish	Dilation OD, no response OS	Dilation OD, no response OS
3*	Sluggish	No response OU	No response OU
4	Absent	Dilation OU	Dilation OU
5*	Normal	No response OU	Dilation OU

* Showed clinical evidence of alternating Horner's syndrome.

TABLE 228-2 Ocular Parasympathetic Function Tests in the Shy-Drager Syndrome

Patient	0.0625% Pilocarpine	Schirmer's (mm/5 min)	Corneal Sensation
1*	Miosis OU	6 OU	Normal OU
2*	Miosis OD, no response OS	4 OD, 1 OS	Depressed OU
3*	Miosis OU	9 OU	Normal OU
4	No response OU	20 OU	Normal OU
5	No response OU	6 OU	Depressed OU

* Showed profuse tearing with complete wetting of Schirmer's strips after subcuntaneous administration of bethanechol chloride.

nervous system areas, including substantia nigra, locus ceruleus, caudate, putamen, cerebellar cortex, pontine nucleus, inferior olives, dorsal vagal nucleus, anterior and lateral horns of the spinal cord, and pyramidal tract.[12] Loss of neurons in the sacral cord may destroy voluntary sphincter control.[13]

Patients with the Shy-Drager syndrome have normal resting levels of plasma norepinephrine, normal pressor responses to an indirectly acting sympathomimetic amine, normal tissue levels of norepinephrine, and normal sensitivity to exogenously administered norepinephrine; however, they have low plasma levels of dopamine β-hydroxylase and produce an insufficient increase in plasma norepinephrine during postural changes or exertion, suggesting an inability to activate an otherwise intact peripheral sympathetic nervous system.[14] Further evidence of intact peripheral sympathetics is provided by the fact that there is an exaggerated response to tyramine infusion.[2]

Treatment
Systemic

Simple measures, such as increased salt and fluid intake, compressive garments, and sleeping in reverse Trendel-

enburg's position may be helpful in mild cases of orthostatic hypotension. The multitude of symptoms that result from autonomic dysfunction justify the use of a polypharmacy treatment approach.[2] Orthostatic hypotension may be treated with a number of medications, including fluorocortisone, indomethacin, propranolol, and clonidine. Supine hypertension, which Shy-Drager patients develop, probably limits the usefulness of these drugs. Further development and testing of a sympathetic neural prosthesis may help to resolve this therapeutic dilemma. Anticholinergic drugs may improve the extrapyramidal features without affecting blood pressure.

Ocular

Decreased tearing, especially associated with corneal hypesthesia and infrequent blinking, may be treated by artificial tear replacement therapy. Subcutaneous or oral administration of bethanechol, a muscarinic receptor agonist, improved lacrimation in three patients with decreased tearing, and improvement in cholinergic functions in other organs was observed.[7] This drug may be useful if topical medications are not, though long-term studies of this treatment regimen need to be assessed.

References

1. Shy GM, Drager GA. A neurological syndrome associated with orthostatic hypotension: a clinico-pathologic study. *Arch Neurol.* 1960;2:511.

2. Polinsky RJ. Multiple system atrophy: clinical aspects, pathophysiology, and treatment. *Neurol Clin North Am.* 1982;2(3):487-498.

3. Schatz IJ. Orthostatic hypotension: I. Functional and neurogenic causes. *Arch Intern Med.* 1984;144(4):773-777.

4. Rosen J, Brown SI. New ocular signs in the Shy-Drager syndrome. *Am J Ophthalmol.* 1974;78:1032-1033.

5. Khurana RK, Nelson E, Azzarelle B, et al. Shy-Drager syndrome: diagnosis and treatment of cholinergic dysfunction. *Neurology.* 1980;30:805-809.

6. Khurana R. Clinical assessment of the autonomic nervous system. In: Altzman PL, Katz DD, eds. *Human Health and Disease.* Bethesda, Md: FASEB; 1972; 277-282.

7. Nirankari VS, Khurana RK, Lakhanpal V. Ocular manifestations of the Shy-Drager syndrome. *Ann Ophthalmol.* 1982;14(7):635-638.

8. Purcell JJ, Krachmer JM, Thompson HS. Corneal sensation in Adies syndrome. *Am J Ophthalmol.* 1977;84:496-500.

9. Bourgon P, Pilley SFJ, Thompson HS. Cholinergic supersensitivity of the iris sphincter in Adies tonic pupil. *Am J Ophthalmol.* 1978;85:373-377.

10. Thompson HS, Mensher JH. Adrenergic mydriasis in Horner's syndrome. *Am J Ophthalmol.* 1971;72:472-480.

11. Grimson BS, Thompson HS. Drug testing in Horner's syndrome. In: Glaser JS, Smith JL, eds. *Neuro-ophthalmology.* vol 8. St. Louis, Mo: CV Mosby; 1975;265-270.

12. Oppenhiemer D. Neuropathology of progressive autonomic failure. In: Banniester R, ed. *Autonomic Failure: A Textbook of Clinical Disorders of the Autonomic Nervous System.* New York, NY: Oxford University Press; 1983.

13. Sung JH, Mastro AR, Segal E. Pathology of Shy-Drager syndromes. *J Neuropathol Exp Neurol.* 1979;38:353-368.

14. Ziegler MG, Lake CR, Kopin IJ. The sympathetic nervous system defect in primary orthostatic hypotension. *N Engl J Med.* 1977;296:293-297.

PART 21

Vitamin and Nutritional Disorders

Chapter 229

Hypovitaminoses and Hypervitaminoses

PAUL G. STEINKULLER

Deficiencies of vitamins A, B_1 (thiamine), B_2 (riboflavin), B_3 (niacin), B_6 (pyridoxine), B_{12}, C, and D may cause ocular signs and symptoms by producing changes in the cornea, retina, or optic nerve. Of these deficiency states, hypovitaminosis A is the most important as a cause of human ocular disease. Except in the context of severe malabsorption disorders or dietary disease of psychogenic origin, most vitamin deficiency states occur in a setting of general malnutrition in the Third World or under conditions of prolonged civil or political unrest.[1-4] Ophthalmologists who are involved or who anticipate being involved in medical or surgical care or preventive activities in developing countries must be keenly aware of the ocular manifestations of vitamin deficiency states and of their systemic implications.

Vitamin A

Vitamin A is a fat-soluble substance found in animal source foodstuffs such as eggs, meat, liver, and milk. It is ingested, usually in the form of its ester, retinyl palmitate, and is absorbed from the small intestine after hydrolyzation into the alcohol form, retinol. Humans can also metabolize the precursors of vitamin A, the carotenes. These provitamin A carotenoids are found in yellow fruits, green leafy vegetables, and in red palm oil.[5] The carotenoids are reduced to retinaldehyde, hydrolyzed into retinol in the mucosa of the small intestine, and absorbed. After absorption, retinol travels in the ester form to the liver, where 95% of the total vitamin A in the body is stored in the form of retinyl palmitate. Although the carotenoids are found in foods that are generally available in developing countries, they are much less biologically active than ingested retinyl palmitate, and humans must ingest up to six times as much, by weight, to obtain the same metabolic effect.[6,7] One gram of retinol equivalent (RE) equals 3.30 international units of vitamin A. (The recommended daily allowance is noted in Table 229-1.[8])

Vitamin A is involved in at least three metabolic

TABLE 229-1 Recommended Daily Allowances

Group	Age	Vitamin A (μg RE)	Vitamin B₁ (mg)	Vitamin B₃ (mg)	Vitamin C (mg)	Vitamin D (μg)
Infants	0–0.5	420	0.3	6	35	10
	0.5–1	400	0.5	8	35	10
Children	1–3	400	0.7	9	45	10
	4–6	500	0.9	11	45	10
	7–10	700	1.2	16	45	10
Adult males	11–14	1000	1.4	18	50	10
	15–18	1000	1.4	18	60	10
	19–22	1000	1.5	19	60	7.5
	23–50	1000	1.4	18	60	5
	51+	1000	1.2	16	60	5
Adult females	11–14	800	1.1	15	50	10
	15–18	800	1.1	14	60	10
	19–22	800	1.1	14	60	7.5
	23–50	800	1.0	13	60	5
	51+	800	1.0	13	60	5
Pregnant women		+200	+0.4	+2	+20	+5
Lactating women		+400	+0.5	+5	+40	+5

Food and Nutritional Board of the National Research Council.[8]

processes in the eye. First, it participates in rod outer segment turnover and in the phagocytosis of outer segment material by the retinal pigment epithelium. Second, in the retina the two forms of vitamin A, retinol or vitamin A₁, and 3-dehydroretinol or vitamin A₂, combine with two forms of protein opsins to yield the four main types of photosensitive pigments. These pigments are involved in the initiation of the neural impulse from the photoreceptors of the retina to the occipital cortex.[6] Third, it participates in glycoprotein and RNA synthesis and is needed for proper maintenance of the conjunctival mucosa and the corneal stroma.[1]

In 1976 the World Health Organization (WHO) published a classification scheme for xerophthalmia, which was modified in 1982 and has since become universally accepted as the clinical basis for categorizing ocular manifestations of vitamin A deficiency (Table 229-2).[9] Nightblindness (hemeralopia, nyctalopia) is the earliest and most common manifestation of vitamin A deficiency in humans, and it usually occurs without any other signs of the disease.[10] If the deficiency state progresses, the conjunctiva becomes overtly dry (conjunctival xerosis, X1A) and Bitot's spots (X1B) may appear on the bulbar conjunctiva in the temporal or nasal quadrants. These are plaques of keratinized debris commonly containing saprophytic bacteria (xerosis bacilli) overlying areas of xerotic conjunctiva.[11] The first sign of corneal involvement is inferior superficial punctate keratopathy (SPK) which then progresses superiorly and becomes con-

fluent as the entire epithelial surface becomes dry (X2). Loss of stromal integrity usually appears first as a small, sharply demarcated, noninfected-looking ulcer (X3A), which may progress to involve more than one third of the corneal surface (X3B). Once corneal involvement (X2, X3A, X3B) occurs, the rate of significant vision loss is at least 50% if no treatment is given.[12,13] Sudden severe stromal decompensation with perforation—keratomalacia—is a blinding condition that most commonly occurs in a setting of chronic severe malnutrition exacerbated by an acute febrile illness, usually measles, in a child under age 2 years. Among children who receive no medical care for their systemic nutritional disease, the overall mortality rate in cases of keratomalacia is between 50 and 90%; most often the cause of death is concurrent pneumonia or diarrhea and dehydration.[11] Thus, keratomalacia represents a significant marker of mortality as well as a visual calamity.

If the child survives the acute illness associated with keratomalacia, attention can be focused on the ocular disorder. Occasionally a small ulcer (X3A) will heal without major anterior segment disruption, and the patient retains some vision around a corneal scar (XS). More commonly the corneal perforation is sealed by iris, and a flat anterior chamber results, with ensuing secondary glaucoma and corneal staphyloma. If the condition is bilateral, as is commonly the case, the result is blindness.

Xerophthalmic fundus (XF) is an uncommon finding in hypovitaminosis A, consisting of peripheral

TABLE 229-2 World Health Organization Classification Scheme
for Xerophthalmia

Classification Code	Clinical Description	Prevalence Levels Among Preschool Children Indicating Significant Public Health Problems (%)
XN	Nightblindness	> 1
X1A	Conjunctival xerosis	
X1B	Bitot's spots	> 0.5
X2	Corneal xerosis	
X3A	Corneal ulceration or keratomalacia involving less than 1/3 of the corneal surface	> 0.01
X3B	Corneal ulceration or keratomalacia involving 1/3 or more of the corneal surface	
XS	Corneal scar	> 0.05
XF	Xerophthalmic fundus	
Biochemical Criterion	Plasma vitamin A 0.35 μmol/L (10 μg/dl) or less	> 5

(From World Health Organization.[9])

yellowish-white dots, perhaps representing focal defects in the retinal pigment epithelium.[14,15]

Recently Sommer et al.[12] described a relationship between mild vitamin A deficiency and increased mortality rate in young children in Indonesia, highlighting the role of vitamin A in maintaining the integrity of respiratory and intestinal epithelium. Children with vitamin A deficiency are more susceptible than normal ones to respiratory and intestinal infections.[13]

Treatment of the ocular complications of xerophthalmia is often dramatic for the XN, X1A, and X1B groups, and usually is completely unsatisfactory for the X3A, X3B, and XS groups. Reconstructive corneal surgery is difficult in children in industrialized countries and almost hopeless in the socioeconomic situations in which keratomalacia exists (ie, the Third World). Donor material is usually unavailable, as are adequate operating microscopes, surgical supplies, and, for that matter, corneal surgeons. Added to this mournful litany is an almost universal and complete lack of patient compliance, all of which create a virtually insurmountable obstacle to vision rehabilitation. From an ocular as well as systemic standpoint, *prevention* of this situation is infinitely preferable to treatment. There are three possibilities for prevention of vitamin A deficiency: periodic dosing during childhood; fortification of routinely consumed foodstuffs; and increasing the intake of normal dietary sources of vitamin A. There are advantages and disadvantages to each, and local political, economic, and social factors dictate which, if any, is feasible and appropriate in a given geographic area. Periodic dosing requires not only the constant and perennial availability of the drug but also a permanent and unfailing system of distribution to the sector of society which most needs it: the rural poor, always the hardest to reach and the most difficult to motivate. Fortification of normally consumed foodstuffs is often impractical: "purchased food" and "rural poor" are usually mutually exclusive terms. Increasing the intake of normal dietary sources of vitamin A requires a broad-based two-front war: social change through education and agricultural reform. Realistically, both processes require generation-long evolutions.

A therapeutic and prophylactic dosage schedule (Table 229-3) has been established and promulgated by WHO[16] that takes into account age, xerophthalmia classification, general nutritional status, and concurrent disease.[16]

Excessive vitamin A intake in humans most commonly causes intermittent bone and joint pain, fatigue, insomnia, hair loss, anorexia, weight loss, hepatosplenomegaly, dryness and fissuring of the lips and other epithelial effects, anemia, and headache. Pseudotumor cerebri has occurred with strabismus and diplopia. No deaths have been reported, despite documented doses as high as 100,000 units daily for 5 years or 1 million units daily for 25 days. The smallest dose of vitamin A reported to produce symptoms in adults is 50,000 units daily for 18 months.[4,17]

TABLE 229-3 World Health Organization
Recommended Vitamin A Prophylaxis Schedule

Clinical Population	Dose and Frequency
Pregnant or lactating women	20,000 IU/week or 5,000 IU/day
Newborns	50,000 IU at birth
Children under age 1	100,000 IU every 4 to 6 months
Children over age 1, adults	200,000 IU every 4 to 6 months

(From World Heath Organization.[9,16])

Vitamin B₁ (Thiamine)

Vitamin B_1 (thiamine) is a water-soluble, heat-stable substance consisting of a pyrimidine ring joined to a thiazole by a methylene bridge. It is present in almost all animal and plant tissues, especially in the outer layers of the cereal grains, and is absorbed from the diet by both active and passive transport. The daily allowances are noted in Table 229-1. Thiamine functions mainly as a coenzyme in the cleavage of carbon-carbon bonds and is also active in the decarboxylation of α-keto acids and keto analogs of leucine, isoleucine, and α-ketogluturate, and in the transketolase reaction in the pentose phosphate pathway.[7] Vitamin B_1 deficiency may occur in any of several settings: general starvation, as in prisoner-of-war camps; in malabsorption syndromes; in chronic alcoholism; in high-carbohydrate diets, especially during pregnancy when metabolic requirements are increased; and as a result of a high intake or tea or coffee, via the formation of inactive thiamine disulphide.[5,18]

The systemic disease resulting from vitamin B deficiency is beriberi (Singhalese, "I cannot," ie, the patient is unable to do anything),[1] of which several forms may occur.[19] In *wet* beriberi the cardiovascular system is affected, involving peripheral vasodilatation, biventricular myocardial failure, and sodium and water retention.[7] *Dry* beriberi, the neurologic form, similarly may involve three separate entities: Wernicke's encephalopathy, peripheral neuropathy, and Korsakoff's syndrome. Mixed varieties may occur. Four thiamine-responsive metabolic disorders have been described: anemia, a branched-chain ketoaciduria, a pyruvicacidemia, and a form of subacute necrotizing encephalomyelopathy (Leigh's disease). Of these four, only subacute necrotizing encephalomyelopathy has been reported in any number, and the disease appears to have an autosomal recessive inheritance pattern.[20,21]

The ocular effects of thiamine deficiency include corneal epithelial changes, ophthalmoplegia, nystagmus, and most significantly, optic atrophy. Although thiamine deficiency may be specifically involved in so-called alcohol amblyopia, its role remains unclear.[1,22]

Certainly some alcoholic patients have appeared to respond positively to vitamin B_1 therapy. In subacute necrotizing encephalomyelopathy, or Leigh's disease, dietary intake and stores of thiamine are typically normal but are underutilized, and megadoses of vitamin B_1 have been recommended.[23,24]

Whatever the specific underlying cause, optic atrophy is the major ocular side effect of thiamine deficiency. Treatment should be at least 50 mg per day until clinical resolution ceases.[1] No ocular manifestations of hypervitaminosis B_1 have been reported, although transient systemic hypersensitivity reactions have been observed.[5,10]

The relationship of thiamine deficiency to the optic atrophy of starvation as seen in prisoner-of-war camps ("camp eye") is unclear, owing to the limited clinical and laboratory capabilities that have been available to evaluate such patients in times of military and political upheaval and to the broad and nonspecific nature of the dietary deficiencies to which such patients have been subjected. The most one could say *definitively* is that optic atrophy occurs occasionally in adults under starvation conditions and that some early cases apparently have been at least partially reversed by general dietary improvement plus specific thiamine supplementation.[25-27]

Vitamin B₂ (Riboflavin)

Vitamin B_2 (riboflavin) is an isoalloxazine with a side chain of ribitol. It is present in milk and in many plant and animal foodstuffs. Riboflavin participates in a number of oxidative-reduction reactions and is important in the structural makeup of certain enzymes, such as succinate dehydrogenase and monoamine oxidase.[1,7,10] Experimental animals receiving a riboflavin-deficient diet may develop cataracts, an effect not observed in humans.[25,28] Riboflavin was originally grouped with a number of other factors in a single group as the vitamin B_2 group, or vitamin G.[29]

Riboflavin deficiency in humans is manifested by angular stomatitis, cheilosis, hyperemia and edema of the oral and pharyngeal mucous membranes, glossitis, normocytic anemia, and seborrheic dermatitis.

Ocular manifestations of riboflavin deficiency include peripheral corneal vascularization and angular blepharoconjunctivitis with sensations of burning and itching. These findings are often slight, vague, and inconclusive. Hypervitaminosis B_2 produces no ocular effects.[1,10]

Vitamin B₃ (Niacin)

Vitamin B_3 (niacin, nicotinic acid, pyridine carboxylic acid) is a heat-stable, water-soluble substance present in most plant and animal tissues, especially meat and fish. It is a component of nicotinamide-adenine dinucleotide

(NAD) and nicotinamide-adenine dinucleotide phosphate (NADP), which act as coenzymes in a number of oxidation-reduction reactions. Since humans can form niacin from the essential amino acid tryptophan, it is not a true vitamin.[5]

Pellagra (Italian for rough skin) is the human disease that results from dietary deficiency of niacin or of tryptophan.[19] Optic neuropathy has been reported in association with pellagra, but it is uncommon in this disease of the "3 Ds" (*d*iarrhea, *d*ementia, and *d*ermatitis), and in fact the optic neuropathy may be due to other B-complex deficiencies.[25,30,31] Rabbits fed a niacin- and tryptophan-deficient diet have been shown to develop retinal ganglion cell degeneration, corneal vascularization, limbal pigmentation, and separation of the basal cell layer of the conjunctiva.[32] To date these findings have not been documented in humans. Reversible atypical cystoid macular edema has been reported in patients who take high doses of nicotinic acid in an effort to lower serum lipids.[33]

Vitamin B$_6$ (Pyridoxine)

Vitamin B$_6$ (pyridoxine) is a water-soluble, heat-stable substance consisting of a group of related compounds, pyridoxamine, pyridoxol, and pyridoxal. It is widely distributed in most foods, including vegetables, meat, liver, and grains.[5,10] It is stored in muscle as phosphorylase and is involved in the metabolism of various amino acids.

Owing to its wide distribution and availability, naturally occurring pure pyridoxine deficiency has not been described in humans, although it may be induced artificially, as by utilization of certain specifically deficient infant formulas or by administration of pyridoxine antagonists. The latter group includes drugs such as isoniazid, pheniprazine, apresoline, cycloserine, and penicillamine.[1,34] Signs and symptoms of the deficiency state include EEG abnormalities, seizures, cheilosis, glossitis, seborrheic dermatitis, nausea, vomiting, weakness, and dizziness.[7] Optic neuritis and angular blepharoconjunctivitis may result from pyridoxine deficiency, although the association has not been established.[35] Several hereditary metabolic disorders are at least partially responsive to pyridoxine at 5 to 50 times the normal daily requirement, although none is associated with a measurable pyridoxine deficiency; in this group is homocystinuria.[1,36] Hypervitaminosis B$_6$ does not cause human disease.

Vitamin B$_{12}$

Vitamin B$_{12}$ is a porphyrinlike structure with a central atom of cobalt. Vitamin B$_{12}$ in food is liberated by peptic enzymes and by hydrochloric acid in gastric juice. It is then bound to proteins and absorbed from the small intestine.[37] Vitamin B$_{12}$ is important for normal production of red blood cells in the bone marrow and is also necessary for the detoxification of cyanide. Optic nerve dysfunction may accompany pernicious anemia, but only rarely.[38] Because tobacco smoke contains cyanide and because vitamin B$_{12}$ is required to detoxify this compound, it has been postulated that accumulation of cyanide may be responsible for the optic neuropathy occasionally associated with pernicious anemia[39-42]; however, cyanide is not a cumulative poison, and ingestion of sublethal amounts of cyanide only rarely leads to optic nerve damage.[1,43]

A peculiar form of nutritional amblyopia is associated with the ataxic neuropathy of the Nigerian and Jamaican amblyopias.[44,45] The Nigerian diet contains a high proportion of cassava, which is rich in cyanide and poor in vitamin B$_{12}$. The exact mechanisms of the optic nerve disease in pernicious anemia, tropical amblyopia and in so-called tobacco amblyopia are still unknown.[46-50] Vitamin B$_{12}$ deficiency may occur owing to poor dietary intake of the vitamin or to poor intestinal absorption, as in achlorhydria. Other causes of vitamin B$_{12}$ deficiency include intestinal parasites, surgical resection of the small intestine, and fad diets.[41,51] Hypervitaminosis B$_{12}$ has not been implicated in any ocular or neurologic disease.[10] Vitamin B$_{12}$ is discussed in greater detail in Chapter 230.

Vitamin C (Ascorbic Acid)

Vitamin C, 3-keto-1-glucofuranolactone, is a water-soluble compound found in various fruits and vegetables. It is involved in numerous enzymatic reactions, including the formation of norepinephrine, the reduction of the ferric ion, the conversion of proline to hydroxyproline, and the synthesis of chondroitin sulfate.[52] It is a structural component of the monopolysaccharides and is present in human tears.[5,53] Dietary deficiency of vitamin C results in the clinical condition of scurvy, a disorder that involves the mesenchymal tissues, including growing bones, teeth, and blood vessels. Numerous superficial and deep hemorrhages occur, in skin, joints, gums, the conjunctiva, and occasionally the orbits.[7,54] The patients are weak, and demonstrate dyspnea and aching in bones and joints. The gums bleed, and the teeth are loose and may fall out. Numerous petechiae are present, and anemia is common. Ocular involvement includes hemorrhages in the skin of the eyelids, in the conjunctiva, the anterior chamber, and the retina.[1] Proptosis may occur in infantile scurvy, secondary to subperiosteal hemorrhage within the orbits.[55] Large doses of vitamin C can lower intraocular pressure by osmotic means, but other drugs work better and have fewer side effects.[56] There are no ocular or neurologic effects of excessive vitamin C ingestion.[1]

Vitamin D

Vitamin D, actually a group of related sterols needed for calcium and phosphorus metabolism, is found in dairy products and fish in liver and viscera.[5] Vitamin D deficiency in adults produces osteomalacia and in children produces rickets.[7] While rickets is uncommon in industrialized nations, it is still seen with distressing regularity in developing countries.

It was formerly thought that zonular cataracts were caused by vitamin D deficiency, but this is now felt to be not strictly true. Although zonular cataracts are occasionally seen in rickets, they are apparently due to low serum calcium and not to vitamin D deficiency per se.[10,57] Other reports of proptosis in rickets patients probably represent scurvy-induced exophthalmos in clinical settings of combined deficiencies of vitamins C and D.[1] Scleral, conjunctival, and corneal calcifications (band keratopathy) have resulted from hypervitaminosis D.[58]

References

1. Hoyt CS. Vitamin metabolism and therapy in ophthalmology. *Surv Ophthalmol.* 1979;24(3):177.
2. Foster A, Sommer A. Childhood blindness from corneal ulceration in Africa: causes, prevention, and treatment. *Bull WHO.* 1986;64(5):619.
3. Tielsch JM, West KP, Katz J, et al. Prevalence and severity of xerophthalmia in southern Malawi. *Am J Epidemiol.* 1986;124(4):561.
4. Stimson WH. Vitamin A intoxication in adults. *N Engl J Med.* 1961;265(8):369.
5. Marks J. *The Vitamins: Their Role in Medical Practice.* Lancaster: MTP Press; 1985.
6. Steinkuller PG. Nutritional blindness inAfrica. *Soc Sci Med.* 1983;17(22):1715.
7. Wilson JD. Vitamin deficiency and excess. In: *Harrison's Principles of Internal Medicine.* 11th ed. New York: McGraw-Hill; 1987;410.
8. Food and Nutritional Board of the National Research Council. *Recommended Daily Allowances.* 9th ed. Washington, DC: National Academy of Science; 1980.
9. World Health Organization. Report of a Joint WHO/ UNICEF/USAID/HKI/IVACG Meeting, Control of Vitamin A Deficiency. WHO, Geneva 1982, *Tech Ser Rep 672.*
10. Bietti G. Ocular manifestations of vitamin deficiencies and disordered vitamin metabolism. *Metab Ophthalmol.* 1977;1:8.
11. Tielsch JM, Sommer A. The epidemiology of vitamin A deficiency and xerophthalmia. *Ann Rev Nutr.* 1984;4:183.
12. Sommer A, Hussaini G, Tarwotjo I, et al. Increased mortality in children with mild vitamin A deficiency. *Lancet.* 1983;2:583-585.
13. Sommer A, Katz J, Tarwotjo I. Increased risk of respiratory disease and diarrhea in children with pre-existing mild vitamin A deficiency. *Am J Clin Nutr.* 1984;40:1090.
14. Marmor MF. Fundus albipunctatus: a clinical study of the fundus lesions, the physiologic deficit, and vitamin A metabolism. *Doc Ophthalmologica.* 1977;43(2):277.
15. Carr RE, Margolis S, Siegel IM. Fluorescein angiography and vitamin A and oxalate levels in fundus albipunctatus. *Am J Ophthalmol.* 1976;82(4):549.
16. World Health Organization. *Field Guide to the Detection and Control of Xerophthalmia.* 2nd ed. Geneva: World Health Organization; 1982.
17. Morrice G, Havener WH, Kapetansky F. Vitamin A intoxication as a cause of pseudotumor cerebri. *JAMA.* 1960;173(16):1802.
18. Hoyt CS III. Low-carbohydrate diet optic neuropathy. *Med J Aust.* 1977;1:65.
19. *Dorland's Illustrated Medical Dictionary.* 26th ed. Philadelphia, Pa: WB Saunders; 1985.
20. Rosenberg LE. Vitamin-responsive inherited metabolic disorders. *Adv Hum Genet.* 1976;6:1.
21. Scriver CR. Vitamin-responsive errors of metabolism. *Metabolism.* 1973;22:1319.
22. Dowling JE. Nutritional and inherited blindness in the rat. *Exp Eye Res.* 1964;3:348.
23. Pincus JH, et al. Thiamine derivatives in subacute necrotizing encephalomyelopathy. *Pediatrics.* 1973;51:716.
24. Pincus JH. Subacute necrotizing encephalomyelopathy (Leigh's disease). A consideration of clinical features and etiology. *Dev Med Child Neurol.* 1972;14:87.
25. Venkataswamy G. Ocular manifestations of vitamin B complex deficiency. *Br J Ophthalmol.* 1967;51:749.
26. Bloom SM, Merz EH, Taylor WW. Nutritional amblyopia in American prisoners of war liberated from the Japanese. *Am J Ophthalmol.* 1946;29:1248.
27. Dekking HM. Tropical nutritional amblyopia ("camp eyes"). *Ophthalmologica.* 1947;113(2):65.
28. Miller ER, Johnson RL, Hoefer JA, et al. The riboflavin requirement of the baby pig. *J Nutr.* 1954;52:405.
29. Day PL, Langston WC, O'Brien LS. Cataract and other ocular changes in vitamin G deficiency. *Am J Ophthalmol.* 1931;14:1005.
30. Fine M, Lachman GS. Retrobulbar neuritis in pellagra. *Am J Ophthalmol.* 1937;20:708.
31. Irinoda R, Yamada S. Ocular manifestations in niacin deficiency and its reciprocity to riboflavin deficiency. *J Vitaminol.* 1956;2:83.
32. Yamada S. The ocular manifestations of niacin deficiency. *Acta Soc Ophthalmol Jpn.* 1954;58:1136.
33. Gass JD. Nicotinic acid maculopathy. *Am J Ophthalmol.* 1973;76(4):500.
34. Jones OW. Toxic amblyopia caused by pheniprazine hydrochloride. *Arch Ophthalmol.* 1961;66:55.
35. Irinoda K, Mikami H. Angular blepharoconjunctivitis and pyridoxine (vitamin B_6) deficiency. *AMA Arch Ophthalmol.* 1958;60:303.
36. Rosenberg LE. Vitamin-responsive inherited metabolic disorders. *Adv Hum Genet.* 1976;6:1.
37. Heaton JM. Vitamin B_{12} and the eye. *Proc Nutr Soc.* 1960;19:100.
38. Stambolian D, Behrens M. Optic neuropathy associated with vitamin B_{12} deficiency. *Am J Ophthalmol.* 1977;83(4):465.

39. Foulds WS, Chisholm IA, Bronte-Stewart J, et al. Vitamin B$_{12}$ absorption in tobacco amblyopia. *Br J Ophthalmol.* 1969;53:393.

40. Victor M, Dreyfus PM. Tobacco-alcohol amblyopia. *Arch Ophthalmol.* 1965;74:649.

41. Watson-Williams EJ, Bottomley AC, Ainley RG, et al. Absorption of vitamin B$_{12}$ in tobacco amblyopia. *Br J Ophthalmol.* 1969;53:549.

42. Foulds WS, Chisholm IA, Pettigrew AR. The toxic optic neuropathies. *Br J Ophthalmol.* 1974;58:386.

43. Lessell S. Experimental cyanide optic neuropathy. *Arch Ophthalmol.* 1971;86:194.

44. Osuntokun BO, Osuntokun O. Tropical amblyopia in Nigerians. *Am J Ophthalmol.* 1971;72(4):708.

45. MacKenzie AD, Phillips CI. West Indian amblyopia. *Brain.* 1963;19:249.

46. Agamanolis DP, Chester EM, Victor M, et al. Neuropathology of experimental vitamin B$_{12}$ deficiency in monkeys. *Neurology.* 1976;26:905.

47. Hamilton HE, Ellis PP, Sheets RF. Visual impairment due to optic neuropathy in pernicious anemia: report of a case and review of the literature. *Blood.* 1959;4:378.

48. Dreyfus PM. Blood transketolase levels in tobacco-alcohol amblyopia. *Arch Ophthalmol.* 1965;74:617.

49. Carroll FD. The etiology and treatment of tobacco-alcohol amblyopia. Part I. *Am J Ophthalmol.* 1944;27:713.

50. Carroll FD. The etiology and treatment of tobacco-alcohol amblyopia. Part II. *Am J Ophthalmol.* 1944;27:847.

51. Rothstein TB, Shapiro MW, Sacks JG, et al. Dyschromatopsia with hepatic cirrhosis: relation to serum B$_{12}$ and folic acid. *Am J Ophthalmol.* 1973;75(5):889.

52. Know WE, Goswami MND. Ascorbic acid in man and animals. *Adv Clin Chem.* 1961;4:121.

53. Paterson CA, O'Rourke MC. Vitamin C levels in human tears. *Arch Ophthalmol.* 1987;105:376.

54. Sabatine PL, Rosen H, Geever EF, et al. Scurvy, ascorbic acid concentration, and collagen formation in the guinea pig eye. *Arch Ophthalmol.* 1961;4:56-61.

55. Dunnington JH. Exophthalmos in infantile scurvy. *Arch Ophthalmol.* 1931;6:731.

56. Virno M, Bucci MG, Pecori-Giraldi J, et al. Oral treatment of glaucoma with vitamin C. *Eye Ear Nose Throat Monthly.* 1976;46:1502.

57. Von Bahr G. Experimentelle Untersuchungen uber den Schichstar und seinen Zusammenhang mit Rachitis und Tetanie. *Acta Ophthalmol.* 1936;36:205.

58. Gifford ES, Maguire EF. Band keratopathy in vitamin D intoxication. *Arch Ophthalmol.* 1954;52:106.

Chapter 230

Vitamin B$_{12}$ Disorders

RICHARD M. ROBB

Vitamin B$_{12}$ (cobalamin) deficiency in man occurs in several forms, two of which are described here because of their ophthalmic manifestations. One is caused by an acquired intestinal absorption defect, the other by an inherited disorder of intracellular metabolism. The absorption deficit, primary pernicious anemia, is most commonly present in adults over 30 years and is accompanied by gastric achlorhydria and absence of glycoprotein intrinsic factor that is normally secreted by the gastric mucosa and is necessary for absorption of vitamin B$_{12}$ from the intestine.[1] Serum levels of the vitamin are low despite adequate dietary intake. Megaloblastic anemia and neurologic signs and symptoms are prominent and progressive. The inherited metabolic disorder involves the two known metabolically active forms of vitamin B$_{12}$, methylcobalamin and adenosylcobalamin. These two forms act as coenzymes in intracellular metabolism of methylmalonic acid and homocysteine. In this disorder intestinal absorption and serum levels of vitamin B$_{12}$ are normal but there is a block in the formation of the active intracellular enzymes, resulting in early growth retardation and developmental delay associated with a megaloblastic anemia.[2] For clarity these two forms of vitamin B$_{12}$ deficiency are considered separately.

Pernicious Anemia
Systemic Manifestations

The anemia is macrocytic and the bone marrow shows megaloblastic changes that are believed to be due to impaired DNA synthesis.[1] Patients may experience weakness, numbness, and tingling in the extremities.[3] The tongue may be red and sore. Degeneration of the posterior columns of the spinal cord may cause loss of reflexes and impaired position and vibration sense while involvement of the lateral spinal columns results in spastic paraplegia of varying severity. Mental changes range from mood disturbances to overt psychoses. These neurologic signs and symptoms are due to patchy degeneration of the white mater of the brain and spinal cord,[4] the result of undefined metabolic derangements due to the vitamin deficiency.

Ocular Manifestations

The retina may show scattered flame-shaped hemorrhages, often with white centers, but they are a nonspecific finding in any profound anemia. More serious is the uncommon but well-documented gradual loss of central vision in some patients.[5] Optic atrophy may not be recognizable at the onset of symptoms, but it develops in time. Central or cecocentral scotomas are found on perimetry. Occasionally the visual symptoms may antedate the anemia and other neurologic abnormalities. Treatment is with parenteral injection of vitamin B_{12}, and improvement can be anticipated if the diagnosis is made before axonal loss is severe.

Inherited Defect in Vitamin B_{12} Metabolism
Systemic Manifestations

There may be several separate genetic defects that can cause the absence of both methylcobalamin and adenosylcobalamin in tissues, but the common clinical manifestations of these enzyme defects are megaloblastic anemia with normal serum vitamin B_{12} levels, poor growth and development in infancy, and mental retardation.[2,6] Laboratory examination reveals methylmalonic aciduria and homocystinuria, reflecting abnormally elevated blood levels of these substrates of the missing cobalamin coenzymes. To date fewer than one dozen patients with this disorder have been recognized.[2] Several died in infancy. A few survived into the second decade, perhaps because their enzyme derangement was less severe. Some clinical improvement has resulted from treatment with intramuscular injections of hydroxycobalamin, but too few patients have been treated at an early age to allow investigators to assess the full potential of this form of therapy.

Ocular Manifestations

Two kinds of ocular abnormalities have been identified in the small number of patients known to have this disorder of cobalamin metabolism. One is epileptiform blinking of the eyelids and simultaneous upward deviation of the eyes for periods of a few seconds.[7] These movements have been associated with abnormal electroencephalographic discharges. They have been recognized so far in three patients. When the epileptiform movements are not present ocular motility has seemed normal. The second abnormality, found in one, or possibly two, patients, is progressive pigmentary degeneration of the retina. In the better-documented case the retinal changes were more prominent in the posterior pole and were associated with an increasingly abnormal electroretinogram.[8] The second patient was said to have poor vision and "salt and pepper" retinopathy, but fundus photographs and electroretinography were not obtained prior to her death.[9] Further clinical observations are necessary to be certain that these ophthalmic findings are characteristic of the metabolic disorder described. Diagnosis at birth or shortly thereafter will be necessary to determine whether early treatment with metabolically active vitamin can improve the otherwise dismal outlook for these patients.

References

1. Babior BM, Bunn HF. Megaloblastic anemias. In: Braunwald E, Isselbacher KJ, Petersdorf RG, et al., eds. *Harrison's Principles of Internal Medicine.* 11th ed. New York, NY: McGraw-Hill; 1987;1498-1504.
2. Rosenberg LE. Disorders of propionate and methylmalonate metabolism. In: Stanbury JB, Wyngaarden JB, Frederickson DS, et al., eds. *The Metabolic Basis of Inherited Disease.* 5th ed. New York, NY: McGraw-Hill; 1983;474-497.
3. Walsh FB, Hoyt WF. *Clinical Neuro-ophthalmology.* 3rd ed. vol II. Baltimore, Md: Williams and Wilkins; 1969;1235-1238.
4. Adams RD, Kubik CS. Subacute degeneration of the brain in pernicious anemia. *N Engl J Med.* 1944;231:1-9.
5. Hamilton HE, Ellis PP, Sheets RF. Visual impairment due to optic-neuropathy in pernicious anemia. *Blood.* 1959;14: 378-385.
6. Levy HL, Mudd SH, Shulman JD, et al. A derangement in B_{12} metabolism associated with homocystinuria, cystathioninemia, hypomethioninemia and methylmalonic aciduria. *Am J Med.* 1970;48:390-397.
7. Cogan DG, Shulman JD, Porter RJ, et al. Epileptiform ocular movements with methylmalonic aciduria and homocystinuria. *Am J Ophthalmol.* 1980;90:251-253.
8. Robb RM, Dowton SB, Fulton AB, et al. Retinal degeneration in vitamin B_{12} disorder associated with methylmalonic aciduria and sulfur amino acid abnormalities. *Am J Ophthalmol.* 1984;97:691-695.
9. Carmel R, Bedros AA, Mace JW, et al. Congenital methylmalonic aciduria-homocystinuria with megaloblastic anemia: observations on response to hydroxycobalamin and on the effect of homocysteine and methionine on the deoxyuridine suppression test. *Blood.* 1980;55:570-579.

Index

Note: Page numbers followed by *t* or *f* indicate tables or figures, respectively.

ISBN 0-397-50722-4

90000